The Bottlenose Dolphin

PHOTO: Bottlenose dolphins chasing fish off Socorro Island, Revilla Gigedo Islands, Mexico (Jeff Jacobsen, EARTHVIEWS).

The Bottlenose Dolphin

Edited by

Stephen Leatherwood
Oceans Unlimited
San Diego, California, USA

Randall R. Reeves
Okapi Wildlife Associates
Hudson, Québec, Canada

Academic Press, Inc.
Harcourt Brace Jovanovich, Publishers
San Diego New York Berkeley Boston London Sydney Tokyo Toronto

This book is printed on acid-free paper. (∞)

Academic Press, Inc.
San Diego, California 92101

United Kingdom Edition published by
Academic Press Limited
24–28 Oval Road, London NW1 7DX

Library of Congress Cataloging-in-Publication Data

The Bottlenose dolphin / edited by Stephen
Leatherwood and Randall R. Reeves
p. cm.
Includes index.
ISBN 0-12-440280-1 (alk. paper)
1. Atlantic bottlenose dolphin. I. Leatherwood, Stephen.
II. Reeves, Randall R.
QL737.C432B67 1989
599.5'3--dc19 88-35093
CIP

Printed in the United States of America
90 91 92 93 9 8 7 6 5 4 3 2 1

This volume is dedicated to our friends and colleagues David K. Caldwell and Melba C. Caldwell. By their hard work, integrity, and insight, the Caldwells have inspired many of us in our studies of bottlenose dolphins and other cetaceans.

Contents

Part VIII

Research Techniques

Part IX

Management

Preface

Bottlenose dolphins, *Tursiops truncatus,* are among the most familiar cetaceans. Because of their exposure in captivity in marine parks and dolphinaria, movies, and television programs, as well as in the wild along tropical and warm temperate beaches around the world, they are, to most schoolchildren and landlubbers, the quintessential dolphins. They are also well known to field and laboratory scientists. Studies of wild bottlenose dolphins have been encouraged by their relative accessibility. In many lagoons, gulfs, and embayments and along discrete stretches of exposed outer coast, populations of bottlenose dolphins can be observed and studied throughout much of the year. Since they are relatively easy to obtain and thrive in captivity, where they may survive at least as long as in the wild and reproduce successfully, bottlenose dolphins have been used in a great variety of studies. Work with these dolphins in captivity has provided insight about sensory mechanisms, communication systems, energetics, reproduction, anatomy, and many other aspects of cetacean biology generally.

This book is intended to serve several purposes. First, we felt it would be useful to publish a series of papers on the most widely studied genus in the family Delphinidae, to demonstrate at once how much and how little is known about these animals. In that sense, we wanted the book to be a benchmark for researchers who, by seeing the variety of topics and approaches presented under one cover, might get a better sense of how to direct their own further work with this and other delphinid species. For example, many of us have begun to ask how important or relevant additional area population inventories of bottlenose dolphins would be. To date, local or regional population studies, using what have become standard procedures for identifying individuals with photographs, have been conducted in at least the United States, Mexico, Argentina, Australia, southern Africa, and Portugal. We know from fishery catches, strandings, and sightings surveys that bottlenose dolphins are present in virtually all tropical and temperate marine waters. Local concentrations of bottlenose dolphins have provided, and will continue to provide, attractive opportunities for students seeking advanced degrees. The increasing creativity shown in approaches to basic population studies encourages newcomers to go beyond what once passed for discovery.

Another reason for initiating this project was that we were aware of some good work that seemed to be taking too long to reach print. By providing an outlet for their contributions, we wanted to stimulate colleagues to write their results for formal presentation and scrutiny. Too much interesting and useful information was languishing in files and contract reports, where access was limited to insiders only.

The first section, "Evolution and Fossil Record," consists of a single chapter. Lawrence G. Barnes outlines and evaluates the fossil record of bottlenose dolphins and offers several new interpretations. The genus is represented by fossils dating to the early Pliocene epoch, about 4 million years ago. Since the earliest fossils referable to *Tursiops* are from North America, the genus probably did not originate in the Mediterranean, as has been postulated. The bottlenose dolphin is a primitive member of the subfamily Delphininae, sharing some anatomical characters with the more primitive stenine delphinids and the extinct Kentriodontidae, the probable Miocene ancestors of the Delphinidae.

In the second section, "Anatomy and Physiology," there are three chapters. Sentiel Rommel has used the excellent collection of skulls and skeletons in the United States National Museum to describe and illustrate the osteology of the bottlenose dolphin. Some skull and skeletal features have been defined ambiguously in the literature. The author has developed more precise definitions and a standard terminology for the different parts of the dolphin skull and skeleton.

Problems of identifying and describing muscle groups in other mammals are compounded in cetaceans, whose body plans have been uniquely modified for efficient aquatic locomotion. Also using specimens in the United States National Museum, D. Ann Pabst has brought some much-needed order to this branch of cetology. In addition to defining and illustrating the various muscle groups, she has, for the first time, treated the major connective tissues as discrete structures. Her comprehensive treatment of the bottlenose dolphin's axial morphology is intended as a template for similar reviews of other cetaceans. Preliminary comparisons of bottlenose dolphins with other odontocetes suggest that the general axial body plan is very conservative.

As a veterinarian and scientist who has treated and studied bottlenose dolphins for more than 25 years, Sam H. Ridgway is well qualified to review what is known about their central nervous system. Besides critically summarizing the literature on dolphin sensory systems and brain anatomy, he offers comparisons with other mammalian groups, including humans, and suggests ideas and approaches for further research.

Part III, "Systematics and Taxonomy," consists of two chapters. First, Graham J. B. Ross and Victor G. Cockcroft describe bottlenose dolphins in Australian waters by analyzing data mainly from museum specimens and captive live animals, supplemented by new data from populations off southern Africa. An important conclusion is that Ross's previous recognition of *Tursiops aduncus* as a species separate from *T. truncatus* is retracted. *T. aduncus* is a junior synonym of *T. truncatus*. The considerable variability in such features as extent of ventral spotting, body and skull length, and snout length and breadth nevertheless emphasizes the clinal differences between bottlenose dolphin populations living in different habitats, e.g., inshore versus offshore, high versus low latitudes. The approach taken in this study may provide a useful model for understanding the inshore–offshore differences in bottlenose dolphin populations in other parts of the world. Feeding rates for a captive animal suggest that bottlenose dolphins live in conditions close to or below the lower limit of their thermoneutral zone for most of the year, and consequently increase their blubber mass during colder periods.

Sandra L. Hersh and Deborah A. Duffield report that offshore bottlenose dolphins in the Northwest Atlantic have two electrophoretically distinguishable hemoglobins, whereas coastal bottlenose dolphins from the same regions have only one hemoglobin. The offshore dolphins are larger and have proportionately smaller flippers than the nearshore animals, and the two forms can be distinguished further by certain skull features. The authors suggest that habitat may explain these biochemical and morphological differences. The offshore ecotype lives in cool, deep water; the coastal ecotype, in shallow, warm water.

There are three chapters on "Life History and Biology" (Part IV). Very little is known about the biology of small odontocetes in the eastern South Pacific, despite the fact that large numbers have been taken in fisheries off Peru and Chile. Koen Van Waerebeek and his colleagues began a program in the mid-1980s to document the distribution, natural history, and exploitation of small cetaceans in this region. They point out that bottlenose dolphins occur in coastal waters from at least southern Ecuador to Concepción, Chile, at 37°S. Preliminary evidence indicates the presence of two forms, coastal and offshore, differing in cranial morphology, diet, and parasite fauna. Coastal animals also have significantly wider teeth than offshore animals. It is estimated that hundreds of bottlenose dolphins, in addition to thousands of other dolphins, are killed annually by the artisanal fisheries in central Peru.

Investigators in Florida have been collecting information on stranded cetaceans systematically since the mid-1970s. Using a sample of 170 beached bottlenose dolphins collected from the Indian and Banana River complex, January 1978 through December 1983, Sandra L. Hersh and associates have analyzed mortality patterns in the local dolphin population. They estimate an annual mortality rate of about 7–9%. Mortality is seasonally uniform in most years, with noticeable increases only during exceptionally cold winters. Newborn dolphins have a somewhat higher mortality rate than other age classes. The authors emphasize that this kind of study can be done only in situations where carcass salvage effort is intensive and consistent throughout the year, and when information is available on the size and age/sex composition of the living population.

Many of us were stunned when in 1987 and 1988 approximately 750 bottlenose dolphins washed ashore on the east coast of the United States, most having died from an insidious cause. This sad series of events was officially said to have been caused by poisoning from red-tide toxins in combination with bacterial and viral infections. Whatever its actual cause, the die-off brought clearly into focus the need for reliable information on the basic biology, diseases, parasites, and feeding habits of the bottlenose dolphins.

Since the early 1970s, James G. Mead and Charles W. Potter of the United States National Museum have been studying the biology of bottlenose dolphins stranded between Massachusetts and South Carolina. They report their basic findings from examination of 248 carcasses, most representing the coastal form. The dolphins had fed mainly on three fish species (croakers, *Micropogon undulatus,* spot, *Leiostomus xanthurus,* and sea trout, *Cynoscion* sp.) and were often infested with the trematode *Braunina cordiformis.* In this area, bottlenose dolphins are born at a mean length of 117 cm, reach an asymptotic length of 250 cm at an age of 12 years, and become sexually mature at 234 cm, 11 years, and a body weight of 150 kg (females); or 160 cm, 11 years, and a single testis weight of about 100 g (males). Skull maturity was found to be expressed at a mean age of 3.46 years in both sexes.

The largest section of the book, Part V, covers topics within the broad subject areas of "Behavior and Ecology." The study of dolphin sound production has been one of the most fruitful and provocative aspects of cetological research. Hearing is, without a doubt, these animals' most important sense. In addition to the well-known echolocation clicks and burst pulsed sounds often described as squawks, yelps, or barks, bottlenose dolphins produce frequency-modulated, narrow-band sounds called whistles. In 1965, David K. and Melba C. Caldwell first noted that captive animals produce individually distinctive, stereotyped whistles. In their chapter, the Caldwells, with co-author Peter Tyack, present and analyze a large sample of whistle recordings made from 126 different dolphins to develop and test their signature-whistle hypothesis. Signature whistles have proven difficult to define quantitatively, since their stereotypy involves the contour configuration of loops (repetitive elements) as well as more easily measured features. One function of signature whistles presumably is to allow members of a social group to broadcast their identity to their fellows. However, our appreciation of the social function of specific whistle contours produced by particular individuals must await the development and application of techniques for identifying which dolphin in an interacting group produces each whistle. The number of loops in a given signature whistle can vary, depending on such factors as the behavioral context and the dolphin's age.

Detailed local studies have significantly advanced our understanding of the behavior and ecology of coastal bottlenose dolphins. Michael D. Scott and associates present a summary of the methods and findings of their long-term research program in Sarasota Bay on Florida's Gulf Coast, uninterrupted since 1970. Although this chapter offers little that is new, we felt that the study's significance, both for the development of research methods and for the insights obtained about bottlenose dolphin societies, merited the inclusion of this brief review. The Sarasota program has become the prototype for area population studies of bottlenose dolphins. Individual dolphins have been followed over several generations, and changes in behavior and association patterns have been documented as calves mature into adults. The study has illuminated pervasive differences between the reproductive strategies of male and female dolphins.

Susan H. Shane worked in 1985 and 1986 in an area off western Florida south of Sarasota. By instantaneous sampling of focal groups, she was able to identify seven distinct types of feeding, each associated with specific ecological conditions. Diurnal and seasonal activity patterns differed from those in other areas where bottlenose dolphins have been studied. Dolphin behavior was found to be affected by a complex web of interacting environmental features rather than by any one feature.

Lisa T. Ballance studied a community of 155 identifiable bottlenose dolphins in Kino Bay, northeastern Gulf of California, Mexico, during 1984. She presents her findings concerning residence patterns, group stability, and surfacing associations. The dolphins exhibited different degrees of site fidelity and three different patterns of group membership. Individuals in some groups remained together over several months; some individuals were seen only once in the presence of a group with frequently seen individuals; and some individuals were often present but not consistently associated with the same individuals. After considering differences and similarities between the dolphins in Kino Bay and those elsewhere, Ballance speculates that the degree of site fidelity in this species may be highest in closed habitats, such as bays and lagoons, and lowest in open habitats, such as exposed coasts. Synchrony in surfacing patterns was short-lived and did not always involve long-term associates.

In the next chapter, Peter J. Corkeron compares the behavior of bottlenose dolphins and Indo-Pacific humpbacked dolphins, *Sousa chinensis,* in Moreton Bay, Australia, based on his observation of 334 and 50 photographically identified individuals of the two species, respectively, over nearly three years. The two species occurred in mixed groups, found in association with trawlers and dominated by bottlenose dolphins. Evidence suggests that the size of food patches created by trawling and possibly the threat of predation significantly affect the size of dolphin groups.

Three chapters offer new information on feeding habits of bottlenose dolphins. From examination of stomachs of 127 dolphins which died after entangling in shark nets off southern Natal, South Africa, Victor G. Cockcroft and Graham J. B. Ross document feeding on 72 species of fishes, elasmobranchs, and cephalopods. They found some seasonal variation in proportions of the various species, increases in prey length with predator length, lower feeding rates in calves than in other animals, and differences in feeding by age/sex class. The last finding is important in understanding home range and social structure.

Nélio B. Barros and Daniel K. Odell analyzed the stomach contents of 76 bottlenose dolphins stranded in the southeastern United States. They found otoliths from 43 fish species, including nine prey taxa not previously reported for bottlenose dolphins in this area, and cephalopod beaks representing three families. Observed geographic variation in diet appears consistent with regional differences in relative availability of prey. Inshore dolphins feed mainly on bottom-dwelling fish; offshore dolphins, on cephalopods. Most fish species consumed are known to be conspicuous sound producers, and the authors suggest that bottlenose dolphins use passive listening to detect prey.

In many areas, bottlenose dolphins feed in association with fishing operations. Peter J. Corkeron and colleagues took advantage of such an association in Moreton Bay, Australia, to evaluate food preferences. They noted how the dolphins following a trawl net selected prey. At the completion of trawls, they offered selected portions of the catch to nearby dolphins. The authors discovered that social factors which vary among groups affect the degree of choice available to any given individual, thereby affecting the composition of its diet. Benefits of feeding around trawlers, e.g., easy access to food, are at least partly offset by increased risk of capture in nets and increased threat of shark predation.

Unsolicited approaches to humans made by wild dolphins are emotionally charged events, recounted enthusiastically in mythology and factual literature. Christina Lockyer reviews the phenomenon of "sociability" in wild dolphins, treating incidents by area, species, and type of interaction. We asked her to prepare this chapter because bottlenose dolphins do affiliate with humans in various parts of their range, and this aspect of their behavior has been widely publicized (and occasionally exaggerated or distorted) in popular books and articles. Lockyer's first-hand experience with what she calls "sociable" wild bottlenose dolphins and her reputation as a scientist make her well qualified to comment on what this extraordinary behavior might mean.

With increasing scientific study of wild dolphins, progressively greater care is being taken to replace subjective accounts with quantitative descriptions of behavior. Richard C. Connor and Rachel A. Smolker provide a quantitative description of a female dolphin's behavior toward her deceased calf.

Resightings of recognizable individuals have been at the core of many recent studies of bottlenose dolphins. Among other things, they provide insight into long-term site fidelity and changes in habitat use. In their chapter, Bernd Würsig and Graham Harris report that ten individually identified dolphins in Golfo San José, Argentina, were associated with the same companions in 1984 and 1986 as they had been in 1974 through 1976. The natural marks permitting identification had persisted for up to 12 years. However, the dolphins had shifted their pattern of movement away from the Gulf and occurred there only sporadically from 1978 to 1986. Eight of the ten recognizable animals were presumed to be females, judging by their proximity to calves born in the previous summer or fall.

Part VI, "Distribution, Movements, and Abundance," contains four chapters. Particularly in view of the recent mass die-off of bottlenose dolphins along the east coast of the United States mentioned above, the extensive surveys of cetaceans on the continental shelf between Cape Hatteras, North Carolina, and the northern Gulf of Maine conducted between 1978 and 1983 by the University of Rhode Island have provided a valuable baseline of information on dolphin populations. Robert D. Kenney summarizes what was learned in this program about the bottlenose dolphin's distribution and population size. He estimates that in the early 1980s the population of bottlenose dolphins off the northeastern United States (Cape Hatteras north) numbered about 10,000–12,000. Only about 3–4% of this total is considered to belong to the migratory nearshore stock which was fished intensively during the 1880s. As Kenney points out, either the nearshore stock is much smaller now than it was in the late nineteenth century, the current population has been significantly underestimated, the surveys did not extend far enough south to census the majority of dolphins in the coastal stock, and/or the nineteenth century fishery also exploited the offshore stock. The offshore stock, centered along the shelf break, has a wider distribution and apparently tolerates a greater range of water temperatures than the inshore stock.

Michael D. Scott and Susan J. Chivers have used the extensive data collected during tuna-fishing and research expeditions to the eastern tropical Pacific to examine hypotheses about how and why bottlenose dolphin herd structure might vary in coastal and pelagic habitats. Small groups (median about 10 animals) are the norm in both kinds of habitat, and herd size does not increase with distance offshore. The greater range of herd sizes (from single individuals to aggregations of several thousand) distinguishes pelagic from coastal bottlenose dolphins. There is also a tendency for dolphins farther offshore to associate more frequently with other cetaceans, although no more than 30% of the sightings involved such associations even in the areas farthest from shore.

Based on analyses of sightings (1970–1983) and data from photoidentification surveys (1981–1983), Larry J. Hansen estimates that the southern California coastal population of bottlenose dolphins numbers about 240 animals. They normally range from Ensenada, Mexico, to Seal Beach, California, but occur much farther north in some years. About 17% of identified animals exhibited some site fidelity in San Diego County, where they seemed to prefer a particular stretch of coast.

Randall S. Wells and several co-authors report sightings in the Monterey Bay area, central California, of bottlenose dolphins previously photoidentified only in southern California. They discuss similar patterns of association in the two areas as evidence of long-term social bonds and speculate on environmental causes of these long-distance movements.

Part VII contains four chapters on "Husbandry and Captive Breeding." The future of captive dolphin programs may well depend on successful breeding. J. Pete Schroeder presents two chapters, the second co-authored with Karl V. Keller, describing the U.S. Navy's efforts to define reproductive cycles of male and female bottlenose dolphins and then to use that knowledge to enhance captive breeding programs by monitoring hormones, inducing ovulation, and artificially inseminating females in estrus.

In the next chapter, Victor G. Cockcroft and Graham J. B. Ross describe the physical and behavioral development of a bottlenose dolphin from its birth in captivity to two years of age. The calf generally decreased his milk and energy intake with age. He began taking solid food at an age of 321 days but continued sucking facultatively until well past his first year and probably beyond his second year. He apparently received a progressively decreasing energetic contribution from his mother's milk. Energy intake increased substantially immediately after weaning.

Successful maintenance of dolphins in captivity depends on good health care programs. Edward D. Asper and colleagues present normal values and ranges for 31 hematology and serum chemistry clinical tests of blood from captive and wild bottlenose dolphins. Age-related differences exist in some parameters, sex-related differences in others. However, in general, values for captive dolphins do not differ significantly from those for wild dolphins except insofar as the latter have a greater antigenic challenge and different diets and exercise regimes.

The next section (Part VIII) has seven chapters describing and evaluating recently developed "Research Techniques." Various kinds of tags, including natural marks, freeze-brands, and radiotags, have been useful in studies of dolphins. Michael D. Scott and associates review the history of tagging and marking studies, pointing out strengths and weaknesses of the various approaches. Daniel K. Odell and Edward D. Asper provide details of their study in the Indian and Banana rivers, Florida, in which natural marks and freeze-brands were used to study the distribution, movements, and discreteness of the dolphin population inhabiting this sea-flushed lagoon network. The fact that methods of tagging used required capturing and handling dolphins, some more than once, permitted the authors to conduct biological sampling, and they report some results of that sampling.

Recognizing that the particular research methods employed may be as significant as actual behavior and population characteristics of the dolphins in affecting the outcome of a behavior study, Susan H. Shane compares the approaches used in and results from her own studies in Texas and western Florida. The *ad libitum* sampling used in Texas, and indeed in most studies of bottlenose dolphin behavior, may be biased because it tends to overemphasize the more dramatic social behavior while underemphasizing the less spectacular behavior. The focal group sampling (following Altmann's classic 1974 paper) used in Florida more accurately represents the full range of the animals' behavior. Even allowing for the differences in research methods, however, the dolphins in Texas spent less time feeding than their counterparts in Florida. The Texas dolphins used two of seven identified feeding modes predominantly, while the Florida dolphins used each of the seven modes with approximately equal frequency. The fewer recognizable animals in Texas were identified up to 156 times each, while the larger number of recognizable animals in Florida were seen no more than 19 times each. Flexibility in adapting to different regional conditions is cited as a principal factor in the widespread distribution of bottlenose dolphins in coastal environments, including those with high densities of human activity.

Harold W. Goforth, Jr. describes an experimental method for studying various physiological and metabolic responses to exercise by captive dolphins. A bottlenose dolphin was conditioned to place its rostrum against a load cell (force transducer), swim in place for 3-minute and 8-second periods, and submit to blood sampling and muscle biopsies. The 3-minute trials were used to measure tailbeat force during sustained swimming; the 8-second trials, to measure maximum tailbeat force. The dolphin's maximum thrust on a single tailbeat was 4730 newtons. Its maximum burst effort was 3263 newtons per tailbeat; its maximum sustained swimming level, 1324 newtons per tailbeat. Maximum burst speeds and sustained swimming speeds of wild dolphins, about 21 and 15 knots, respectively, are summarized and discussed. This study demonstrates some of the scientific values of working with trained captive animals in controlled situations.

Increasingly, layering patterns in teeth are being used to estimate ages of dolphins. Such studies depend on knowing the rates at which layers (called Growth Layer Groups, or GLGs) are deposited and what factors influence those rates. Aleta A. Hohn reviews the history of attempts to define GLGs, summarizes her own work on known-age, free-ranging bottlenose dolphins, and points out some pitfalls of reading and interpreting layers in teeth. She also offers counsel on how best to conduct studies and report findings so as to increase their usefulness to future investigators. In the next chapter, Albert C. Myrick, Jr., and Lanny H. Cornell report the results of their age-estimation work with captive dolphins. They establish that one dentinal GLG is deposited per year in the teeth of bottlenose dolphins, and they identify principal factors affecting the rate of dentine deposition.

Deborah A. Duffield and Jan Chamberlin-Lea report that each of 66 bottlenose dolphins they sampled had its own karyotypic signature. They conclude that variability in R-banding of karyotypes and profiles of hemoglobins are useful in studies of populations and paternities.

The final section, Part IX, consists of only one chapter on "Management." The live-capture fishery for bottlenose dolphins in the southeastern United States is managed by the National Marine Fisheries Service under the Marine Mammal Protection Act of 1972. Gerald P. Scott describes the agency's goals and strategies. Annual quotas of 2% or less of the estimated population are set for regional management units, thought to correspond roughly with local stock boundaries. Readers should be advised that this chapter was current only through early 1987. The subsequent die-off of dolphins in the eastern United States has changed the atmosphere in which the management program operates. We expect the current period of re-evaluation, influenced by the sobering recognition that many animals in the coastal bottlenose dolphin population north of central Florida died within a period of less than a year, to result in an even more conservative approach to management of the live-capture fishery.

Inevitably, there are subjects which should have been addressed in a book like this one but were not. We had initially hoped for somewhat broader regional coverage. Although some areas not represented in the book, such as coastal southern Africa, have been treated in previous publications, there are still large parts of the bottlenose dolphin's range for which little has been published. We had hoped to include additional chapters directed at resolving the problems of *Tursiops* systematics. As noted by the Subcommittee on Small Cetaceans of the International Whaling Commission's Scientific Committee in 1974, there appear to be a number of geographical races of this species, defined on the basis of body and tooth size as well as different distributions relative to sea temperature and depth. The names *Tursiops aduncus, T. gillii, T. nuuanu,* and *T. gephyreus* are still used in certain areas as though they represent valid species, so their status begs to be clarified. We also would like to have included more review chapters, covering such topics as echolocation, sensory systems, language training, pathology and sensitivity to pollutants, exploitation in fisheries and conservation problems, habitat considerations, and world distribution. Some of these are adequately covered elsewhere, but others, such as the last four, are much needed.

In view of widening concern about the often unforeseen, potentially devastating impacts of pollution and environmental degradation on dolphins and their prey, the conventional approaches to management and conservation of bottlenose dolphins and other species will need to be reevaluated in the coming years. We trust this book will give that process a boost.

We are particularly grateful to two groups of colleagues for helping to create this book. First and most obviously, the authors have done the research, analyses, and writing. We asked many of them to rethink, reanalyze, and rewrite portions of their chapters after review, and we greatly appreciate the professional manner and positive spirit in which they have done so. The contributions by the reviewers are less obvious but crucial to ensuring a high standard of scholarship. We followed a peer review process similar to that used for scientific journals, sending all manuscripts to at least two anonymous reviewers. We, as editors, took into account the reviewers' comments in deciding to accept or reject a manuscript, and we sent these and our own comments to authors with requests for revision.

The reviewers were invariably generous and constructive in sharing their opinions, insights, and, in some instances, relevant new data. It is with sincere thanks that we list here (in alphabetical order) those who reviewed manuscripts submitted for this book: Frank T. Awbrey, Lawrence G. Barnes, Kurt W. Benirschke, Peter B. Best, Michael A. Bigg, Daryl J. Boness, Jeffrey M. Breiwick, Michael J. Bryden, Robert L. Brownell, Jr., John Calambokidis, Christopher W. Clark, Victor G. Cockcroft, Richard C. Connor, Lanny H. Cornell, James C. Cubbage, Randall W. Davis, Andrew E. Dizon, Robert W. Elsner, John K. B. Ford, R. Ewan Fordyce, David E. Gaskin, Jonathan Gordon, Nicholas Graham, Jack S. Grove, John D. Hall, Philip S. Hammond, Larry J. Hansen, John Harwood, Sandra L. Hersh, Russ Hoelzel, Wayne Hoggard, Aleta Hohn, Jay Hyman, A. Blair Irvine, Ronald J. Jameson, Robert E. Jones, Brian E. Joseph, Steven K. Katona, Dennis Kelly, Margaret Klinowska, Scott D. Kraus, David M. Lavigne, Richard M. Laws, Alfredo A. Lichter, Christina Lockyer, Lloyd F. Lowry, Helene Marsh, William Medway, Albert C. Myrick, Jr., Mary K. Nerini, Daniel K. Odell, Seiji Ohsumi, Thomas J. O'Shea, Josephine E. Pemberton, William F. Perrin, Wayne

Perryman, August Pivorunas, Charles W. Potter, Karen Pryor, Katherine Ralls, Clayton E. Ray, John E. Reynolds, Dale W. Rice, Sam H. Ridgway, Keith Rittmaster, Graham J. B. Ross, David J. Rugh, Victor B. Scheffer, William E. Schevill, Gary D. Schnell, Michael D. Scott, David E. Sergeant, Ingrid K. Shallenberger, Susan H. Shane, Gregory P. Silber, Donald B. Siniff, Tim D. Smith, David J. St. Aubin, Brent S. Stewart, Robert E. A. Stewart, Timothy Strickler, John C. Sweeney, C. Richard Taylor, Peter Tyack, R. V. Walker, William A. Walker, Marc A. Webber, Randall S. Wells, Bradley N. White, Terrie M. Williams, Forrest G. Wood, Bernd Würsig, and Pamela K. Yochem.

In addition to the authors and reviewers, we owe a special debt of gratitude to several other individuals. Lanny H. Cornell, Sea World of California, provided funding to initiate the project and marshaled the Sea World staff to complete their contributions. Kerry Pinchbeck, our production editor at Academic Press, cheerfully and professionally shepherded the unwieldy manuscript through the production process. Copy editor Connie Parks improved the book markedly by her thorough attention to fine detail. Kathy Kangas was a valuable assistant throughout the project. She watched over the files and kept our voluminous correspondence with authors, reviewers, and each other from getting bogged down. Randi Olsen kept the Reeves family together while we worked on the book, and we appreciate her dedication to that unheralded but much-valued endeavor.

San Diego, California
6 April 1989

PHOTO: Lateral and dorsal views of skull of *Tursiops cortesii* (from Sacco, 1893: plate 2, Figures 10 and 11; see Barnes, Chapter 1, this volume, for full citation) (courtesy Kellogg Library, Department of Paleobiology, National Museum of Natural History, Smithsonian Institution).

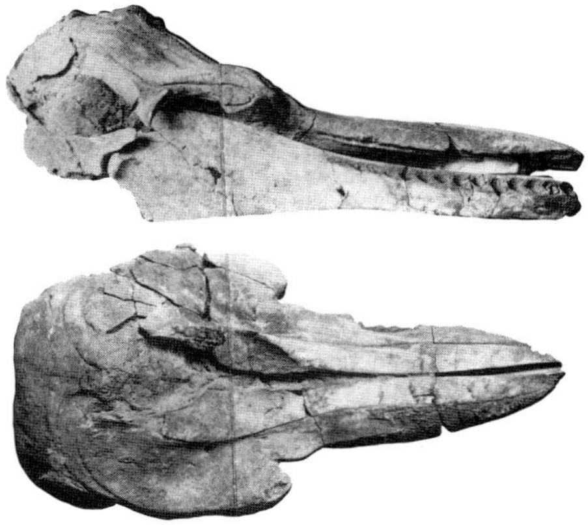

Part I

Evolution and Fossil Record

The Fossil Record and Evolutionary Relationships of the Genus *Tursiops* 1

Lawrence G. Barnes
Section of Vertebrate Paleontology
Natural History Museum of Los Angeles County
Los Angeles, California 90007

INTRODUCTION

Species in the genus *Tursiops* are generalized delphinids, judging by their morphology, evolution, and behavior. They share anatomical features with certain primitive fossil dolphins. Unlike the fossil records of most of the Recent (=extant) genera in the family Delphinidae, that of the genus *Tursiops* is moderately good and extends back several million years in geologic time. Much of the research on fossils of *Tursiops* dates from the 1800s, so this chapter provides an opportunity for a much-needed review and summary. All of the named fossil species of *Tursiops* are listed and reviewed in this chapter. Various other fossils attributed to the genus are summarized as well.

Referral of fossils to any extant delphinid genus must be evaluated with caution because modern names often have been misapplied to fossil specimens (e.g., the many reputed fossil species of the genus *Delphinus*). The fossil record of *Tursiops* is nonetheless good enough to help elucidate parts of the evolutionary history of the family Delphinidae, the subfamily Delphininae, and the genus *Tursiops* itself.

The currently accepted hypothesis is that the ancestors of all Cetacea were terrestrial mammals that adapted to a totally aquatic existence. This is consistent with the fact that modern cetaceans have features, even if they are only vestiges, of the anatomical and physiological qualities of terrestrial mammals. Studies of fossil cetaceans can document changes and adaptations in skeletal anatomy that provide clues to ancestry and the course of evolution of the various lineages.

Comparative osteology is presently the only widely available method of comparing fossil and extant species. By analogy with the living animals, we can make inferences about the soft anatomy, physiology, and behavior of the fossil species from their bones. The important features in paleontological analyses include overall body size, stage of maturity, shapes of individual bones (with emphasis on the skull), positions of foramina that mark the courses of nerves and blood vessels through bones (especially those of the cranium), distribution of air sinuses extending from the middle ear into other parts of the skull, shape and number of teeth, numbers of fused neck vertebrae, numbers of vertebrae and ribs, length and numbers of digits in the pectoral limb, and shape, size, and location of muscle scars and articular surfaces on bones. Characters are judged to be primitive or derived according to when they appear in the fossil record, their degree of complexity, and their occurrence in other organisms ("outgroups") not belonging to the group in question. Sound comparative osteological work is a prerequisite for the vertebrate paleontological studies that can relate the history of *Tursiops* to the geologic time scale, to the evolutionary history of Cetacea, and to the principles of classification and biogeography.

Some evolutionary patterns are discernible from the fossil record of *Tursiops*, and inferences can be made regarding the genus's area of origin, the effects of climate on the evolution and distribution of its species through time, and the morphological characters useful in analyzing relationships among different taxa. The fossil record is not yet sufficient to answer all questions about the origin, evolution, and classification of bottlenose dolphins, but a review at this time can at least indicate areas for future research. The purpose of this chapter, therefore, is to review the fossil record of bottlenose dolphins, to put these animals into evolutionary and taxonomic perspective, and to determine needs for future work.

PHYLOGENETIC POSITION OF *TURSIOPS*

Overview of Cetacean Evolution

Cetaceans were perhaps the earliest mammals to adapt to a marine existence, and the order is the most diverse of aquatic mammals. Whales appear in the fossil record about 50 million years ago (mya), in Early or Middle Eocene time. The earliest appearance in the fossil record of any species referable to the genus *Tursiops* is relatively recent, at about 5 mya. To place *Tursiops* in evolutionary perspective and to indicate the reasons why certain osteological features are used in determining relationships and stages of evolution, I summarize briefly the evolutionary history of cetaceans. This summary concentrates on those groups that are either directly on or close to the lineage leading to *Tursiops*. The taxa mentioned are outlined in the section on classification (see p. 9, 11).

Information derived from modern mammals about blood composition, chromosomes, uterine morphology, insulin, fetal blood sugar, and tooth-enamel microstructure points to closest relationships between cetaceans and ungulates, especially the Artiodactyla. Flower (1883) was one of the first to propose this relationship, based on comparative gross morphology. Van Valen (1966) and Szalay (1969a,b) argued on paleontological grounds that archaeocetes evolved from terrestrial ungulates known as mesonychid condylarths. The condylarths, in the broad sense, are also regarded as the ultimate ancestors of all later ungulates. This leads to the inference that cetaceans and artiodactyls share a common ancestor, as yet unknown or unrecognized in the fossil record, among the condylarths. This inference is supported by the morphological and physiological similarities that have been noted for many years. Of course, this does not imply that cetaceans evolved from the artiodactyls. Barnes and Mitchell (1978), Evans (1987), Gaskin (1982), Gingerich *et al.* (1983), and Rice (1984) have discussed origins of cetaceans (see also references cited therein).

All living whales, dolphins, and porpoises share certain basic anatomical and physiological characteristics with other mammals. Living cetaceans have remnant vestigial features (e.g., facial hairs, olfactory nerves, and hind-limb bones) that are clearly inherited from their terrestrial ancestors. Living cetaceans have become obligate aquatic animals, with a fusiform body shape, blubber, pectoral flippers, a short neck, horizontal caudal flukes supported by fibro-cartilage, and a dorsal fin (a derived character which is secondarily lost in some groups). All of these characters, except the dorsal fin, are convergent with such different groups of mammals as the sirenians and pinnipeds, and if considered individually they are not necessarily diagnostic of cetaceans. All cetaceans, fossil and living, share the following unique suite of derived osteological characters (also called apomorphies):

1. Anterior palatine foramina lost
2. Falcate processes of the basioccipital large
3. Peribullary and pterygoid air sinuses present as diverticula from the middle-ear air sinus
4. Tympanic bulla involuted and inflated
5. Supraorbital process of the frontal large and tabular
6. Hypoglossal foramen in the basioccipital located either at the apex of or inside the jugular notch
7. Mandibular foramen large
8. Scapula with reduced supraspinatus fossa and acromion and coracoid processes that are parallel and directed anteriorly

Skeletal evolution in the Cetacea is also characterized by telescoping of the cranial bones, isolation of the ear bones from the other cranial bones by air sacs or fat bodies, shortening and/or fusion of cervical vertebrae, addition of thoracic, lumbar, and caudal vertebrae, and hyperphalangy. The highest point on the skull is usually behind the nares, and this is called the cranial vertex.

Telescoping of the cetcean skull (see Miller, 1923) is a process that involves posterior movement of the narial openings, linked with the posterior extension of the bones comprising the rostrum (the elongate maxillae, premaxillae and vomer, and the mesethmoid cartilage, which lies in a mesorostral gutter), ultimately reaching the top of the braincase around the cranial vertex. In most groups of cetaceans, the occipital shield also projects forward a certain amount. A major step in the evolution of both the odontocetes and the mysticetes was the progressive movement of the maxillae over and/or under the front of the supraorbital processes of the frontals as the nares moved posteriorly on the rostrum. The orientation and location of the mammalian infraorbital foramen were thereby changed from its primitive position on the cheek in front of the zygomatic arch, and became expressed as multiple maxillary foramina on the dorsal surface of the face and on the posterior part of the rostrum and, in the Odontoceti, as premaxillary foramina as well.

For general reviews of cetacean evolution, the reader may wish to consult Winge (1921), Miller (1923), Kellogg (1928, 1938), Slijper (1936, 1979), Rice (1967, 1984), Fordyce (1980), Gaskin (1982), Barnes (1984b, in press a,b), Barnes *et al.* (1985), and Evans (1987).

In most traditional classifications (e.g., Simpson, 1945; Hall and Kelson, 1959; Romer, 1966; Gaskin, 1982; Evans, 1987) the order Cetacea is divided into suborders: The primitive Eocene and Oligocene Archaeoceti, the Oligocene (~36 mya) to Recent Odontoceti (toothed whales), and the Oligocene to Recent Mysticeti (baleen whales). Archaeocetes are early whales that retained many primitive structures. Mysticetes and odontocetes evolved special structures, many of which are correlated with the method of feeding.

The Protocetidae are the oldest and most primitive of the archaeocetes, known from Middle Eocene fossils from Pakistan, India, Egypt, Nigeria, and the southeastern United States (see Barnes and Mitchell, 1978; Gingerich *et al.*, 1983). Van Valen (1968) suggested that the later odontocete and/or mysticete suborders were derived directly from the early protocetids rather than from any of the later archaeocetes, as was suggested by Barnes and Mitchell (1978) and Fordyce (1980). The protocetids were relatively small-bodied animals (for whales), less than about 3 m long, having primitive characters such as nontelescoped skulls, nostrils on the anterior part of the snout, and the normal mammalian dental formula, with two roots on P1-2/ and three roots on P3/-M3/. At least one species of protocetid is known which retained facets on its sacral vertebrae for articulation with large innominate bones, suggesting that it probably had external, if not functional, hind limbs. It might even have been amphibious! In all of the later cetaceans, the bones of the pelvic girdle are reduced and the facets on the sacral vertebrae are lost (derived characters).

The more highly evolved family Basilosauridae includes medium-sized, generalized archaeocetes in the subfamily Dorudontinae and the giant archaeocetes of the subfamily Basilosaurinae. Apparently, the later Eocene basilosaurids dispersed throughout most of the world's oceans, because their fossils have been found in Africa, Great Britain, Australia, Antarctica, and the southeastern United States. Compared with the more primitive Protocetidae, the skulls were somewhat more telescoped, and the molars and premolars had achieved a more derived state, with only two roots and with accessory denticles on the anterior and posterior edges of the crowns. They had also lost the M3/ (derived character). *Basilosaurus cetoides*, a huge Late Eocene (~37 mya) archaeocete, retained an innominate bone and a femur, but these were reduced in size. Generalized dorudontines have been suggested as the possible ancestors of all mysticetes and odontocetes (see Barnes and Mitchell, 1978; Fordyce, 1980). This conflicts with Van Valen's (1968) idea that later cetaceans instead evolved from the Protocetidae, but it is important to note the widely supported theory that all of the later cetaceans must have evolved from some group of archaeocetes (see Fordyce, 1980; Barnes, 1984b).

The two derived suborders, Odontoceti and Mysticeti, are considered a monophyletic unit (an evolutionary clade) because they share the following derived characters not found in archaeocetes:

1. Multiple maxillary foramina derived from the infraorbital foramen
2. Advanced stages of cranial telescoping
3. Vomer extended backward and exposed on the basicranium to cover the basisphenoid/basioccipital suture
4. Zygomatic process of the squamosal contacting the postorbital process of the frontal or connected to it by a ligament
5. Monophyodonty
6. Elbow joint nonrotational with anteroposterior position of radius and ulna
7. Olecranon fossa of the humerus lost
8. Hyperphalangy

Some published diagnoses of the two living cetacean suborders were based solely on the characters of living animals, and, unfortunately, these excluded the numerous fossil representatives. For example, the suborder Odontoceti has been diagnosed as those cetaceans having asymmetrical skulls, and the mysticetes as those having baleen instead of teeth (e.g., Rice, 1967), but this would place in limbo those (numerous) fossil odontocetes which had symmetrical skulls as well as the early fossil mysticetes which still had teeth. Comprehensive diagnoses for these groups must include both the modern and fossil taxa and be based on osteology.

Odontocetes always have teeth present, the rostrum has thick lateral edges, the maxillae do not project beneath the orbits as infraorbital plates, and dorsally do not interlock with the frontals, but spread outward over the expanded supraorbital processes. The facial region is occupied by a fatty organ, the melon, and the internal nares are two bony openings (derived feature) while the external nares form a single opening (the blowhole). The posterior mandibular foramen is greatly enlarged and filled with adipose tissue, and its lateral wall (the pan-bone) is thin. All hyoid bones are separate.

Mysticetes have baleen and vestigial teeth that are lost before birth (except in the primitive Aetiocetidae, which had functional teeth); the rostrum has thin lateral edges. The maxillae have infraorbital plates and laterally projecting antorbital processes; they interlock with the frontals and do not spread over the supraorbital process. There is no melon, and the internal naris is a single bony opening (primitive) while the external nares are a pair of openings. The posterior end of the mandible is not greatly excavated (primitive), and there is no articulated or ankylosed mandibular symphysis but instead a ligamental attachment marked by a lingual groove (derived characters). Mysticetes never developed the echolocation capabilities of odontocetes and do not have cranial asymmetry.

Mysticetes and odontocetes were widespread and diverse in the Oligocene and afterward. Some primitive members of each of these groups are both chronologic and morphologic intermediates between archaeocetes and more highly evolved cetaceans. By the the start of the Miocene (~25 mya), all the major modern adaptive types of cetaceans had appeared, but the earliest known fossil records of the family Delphinidae (which includes *Tursiops*) and some of the other extant families are considerably more recent, dating from the Late Miocene (~11 mya) (see Barnes, 1976, 1984b; Barnes *et al.*, 1985).

Evolutionary Position of *Tursiops* among Toothed Whales

In this section I place *Tursiops* in context within the suborder Odontoceti. First, some functional and morphological attributes of odontocetes are discussed, then the genealogy of *Tursiops* is traced, as deduced from the fossil record, from primitive odontocetes of the Oligocene to the first *Tursiops*-like animals in the Late Miocene.

Odontocetes have retained the apparently primitive feeding mode employed by archaeocetes (and many other marine predators) of selecting and capturing individual prey. This contrasts with the bulk-feeding methods used by mysticetes. Teeth are obviously useful to odontocetes for holding slippery animals. From the primitive heterodont mammalian tooth formula of the archaeocetes, the trend in the derived odontocetes has been toward simplification of the teeth by development of single roots and conical crowns (homodonty) and increase in their numbers (polydonty). Some highly derived odontocetes have secondarily lost or reduced parts of their dentition or have developed specialized teeth with enamel rugosities or large teeth in the form of tusks. Some modern odontocetes that have secondarily lost teeth have acquired cornified structures on the palate in their stead. The species of *Tursiops* have tooth counts in the middle range for odontocetes, and upper and lower teeth are fully functional.

In general, progressive evolutionary trends in the odontocetes are toward further homodonty, polydonty (sometimes secondary reduction), cranial asymmetry, expansion of the middle-ear air-sinus system into various parts of the skull (Fraser and Purves, 1960), telescoping with an emphasis on movement of rostral bones over the face toward the occiput, shortening and/or fusion of cervical vertebrae, and increase in number of vertebrae. These characters evolved to varying degrees at different times in separate lineages.

Odontocetes have evolved sound-making abilities, which aid them in catching prey, communicating, and echolocating in dark waters. Sound, in the form of clicks, is produced during the movement of recycled air within complex diverticula, sacs, and valves of the nasal passages (see Lawrence and Schevill, 1956; Schenkken, 1973) and by friction/stiction of the nasal plugs (Evans and Maderson, 1973). The current working hypothesis is that this sound is projected into the water after passing through a fatty melon on the face which acts as an acoustic lens. Most of the sacs and associated structures are in the flesh of the facial region and, therefore, cannot be detected by examining only a skull. However, the positions of two of the largest of these, the paired premaxillary sacs, can be detected on a fossil or modern odontocete skull by the presence of a distinctive flat surface (premaxillary sac fossa) next to the naris at the back end of each premaxilla. The former existence of the premaxillary sacs can be inferred in even the earliest odontocetes of the Oligocene (~30 mya), indicating that either these animals actively echolocated or they were at least preadapted to do so. These premaxillary sacs are primitively symmetrical in most fossil odontocetes but assymmetrical in most living odontocetes. *Tursiops* has the typical delphinid, derived condition of assymmetrical sacs and premaxillae.

The external acoustic meatus, the primary path of sound conduction to the inner ear in land mammals, is closed in cetaceans, and hearing is by a different method. Sound waves in the water (whether from another source or produced by the odontocete and reflected back off another object) are received by the head of the animal and transmitted to the ear region by a thin bony area in the posterior part of the lower jaw. The ear bones are acoustically isolated from bone and muscle by fat bodies and air sacs, and this allows directional hearing (see Norris, 1968, for a summary).

The field of sound emitted by some living odontocetes has been shown to be asymmetrical. Norris *et al.* (1971) have also demonstrated asymmetrical movements of soft tissues in and around the nasal passages of dolphins during sound production, and Mead (1975) linked these with the specialized sounds made during echolocation. Mead also postulated that in the course of odontocete evolution the development of assymmetrical soft tissues preceded asymmetry of the bones. The majority of early fossil odontocetes did not have cranial asymmetry, and the majority of later fossil and living ones do. Even in the franciscana, *Pontoporia blainvillei*, one of the few living odontocetes with cranial symmetry, the nasal sacs are asymmetrical (Schenkken, 1973). In most odontocetes with asymmetry, the bones of the top of the skull around the nares are always offset to the left side. In one group of delphinoids, the pithanodelphine kentriodontids, however, the cranial vertex is displaced to the right side (Barnes, 1985b). At least six odontocete lineages, and possibly as many as ten (see Barnes *et al.*, 1985, their Fig. 1), must have acquired cranial asymmetry independently, either because they appear to have evolved from different ancestors which had symmetrical skulls or because the bones on the top of the skull have been modified in different ways.

The evolutionary lineage leading from the primitive odontocetes to the Delphinidae, including *Tursiops*, is moderately well understood from the fossil record. The suborder Odontoceti includes five major evolutionary groupings, usually recognized as the superfamilies Squalodontoidea (fossils only, totally extinct), Platanistoidea (river dolphins, etc.), Delphinoidea (belugas, true dolphins, porpoises, etc.), Ziphioidea (beaked whales), and Physeteroidea (sperm whales, etc.) (see Barnes *et al.*, 1985).

The extinct squalodontoid family Agorophiidae is the most primitive recognized family in the suborder Odontoceti and includes animals that are in many ways morphologically (and also chronologically) intermediate between archaeocetes and the family Squalodontidae. All known agorophiids are Oligocene in age (Whitmore and Sanders, 1976), not Late Eocene as sometimes reported in the literature. These animals are not known by many postcranial bones, and the few known skulls demonstrate that they had only a moderate degree of telescoping (the nares were still well anterior to the orbits) and that the cheek teeth had multiple roots and accessory denticles on the crowns.

The extinct family Squalodontidae is believed to have evolved directly from agorophiids. Simpson (1945) was one of the first to imply this relationship when he classified Agorophiidae and Squalodontidae in the same superfamily, Squalodontoidea. Rothausen (1968) further developed this idea of relationships and showed a phylogeny of squalodontoids with various grades of evolution. Squalodontids are probably the group from which most later odontocetes, principally "dolphins" in the broad sense, have their ultimate ancestry. Many species of squalodontids have been named, and skulls and skeletons are frequently represented in museum collections and in the literature. Most squalodontids were relatively large animals with bodies 3 m or more in length. Their crania were almost fully telescoped, with the nares located on top of the head, between the orbits, and with the intertemporal constriction much reduced. The dentition was polydont, but still heterodont with long, pointed (caniniform) anterior teeth and wide, multiple-rooted cheek teeth with many denticles.

The superfamily Delphinoidea represents a major evolutionary radiation of "dolphins" that started in the Late Oligocene and includes the majority of living cetacean species. This group apparently evolved directly from squalodontids. The currently recognized families in the Delphinoidea are the extinct Miocene Kentriodontidae and Albireonidae, and the extant Delphinidae (true dolphins), Phocoenidae (porpoises), and Monodontidae (belugas and narwhals) (Barnes *et al.*, 1985). All three of the living delphinoid families have fossil records that may be traced back to Late Miocene time (~11–12 mya), and they are characterized by having asymmetrical cranial vertices. In each group this asymmetry affects the bones of the vertex in different ways and was possibly acquired independently in each family.

Most kentriodontids were small dolphins approximately 2 m or less in length. They had skulls with short- to medium-length rostra, and most had symmetrical cranial vertices, the primitive odontocete condition. Their teeth were polydont and homodont, although some primitive species had small denticles on the posterior teeth. The family was relatively diverse in the Middle and Late Miocene, represented by fossil species in both the Atlantic and Pacific oceans, but there are no published records of any species that survived more recently than about 10 mya (Barnes, 1978).

The family Delphinidae might have evolved directly from Middle Miocene kentriodontids. In fact, such primitive living delphinids as species of *Steno*, *Sotalia*, and *Sousa* share numerous osteological similarities with kentriodontids. The crania of the latter differ mostly by not being asymmetrical. Delphinidae is the most diverse living cetacean family and includes what are commonly known as dolphins, killer whales, pilot whales, etc. Several extinct species have been named, primarily in extant genera, and most are *Tursiops* from Pliocene deposits in Europe. The fossil record of the family extends back in time no more than approximately 11–12 mya. Supposed earlier records of the family (see Kellogg, 1928; Simpson, 1945; Mchedlidze, 1976) are either erroneous or based on fossils that are simply indeterminable or do not have the diagnostic characters of the family (Barnes, 1978, 1985a). Especially in the first part of this century, many small fossil odontocetes (most notable exceptions being sperm whales and beaked whales) were assigned to the Delphinidae (Simpson, 1945, p. 216). Delphinids (*sensu stricto*) are, rather surprisingly, not very abundant in the fossil record; however, since they are for the most part pelagic, fossil collecting in Pliocene formations of deep-water origin may yield more fossils.

Living delphinids have varying diets. Some are stenophagous, feeding exclusively on either fish or cephalopods, while others are generalists, feeding on a variety of animals, including crustaceans. The more primitive of the living delphinids (e.g., Steninae, some Delphininae) are generalists and have rostra of intermediate length and width. This latter group includes *Tursiops*. Rostra of generalized delphinids are most like those of the fossil kentriodontids, and this is one line of evidence suggesting relatively close relationships. The diverse living delphinids have a variety of skull, dental, and body adaptations that reflect their varying diets as well as their methods of locomotion. Broad-headed delphinids such as pilot whales (*Globicephala*) and Risso's dolphins (*Grampus*) eat cephalopods; narrow-headed ones such as common dolphins (*Delphinus*) and spinner dolphins (some *Stenella*) usually eat fish; and animals with intermediate-width skulls such as bottlenose dolphins, striped dolphins (some *Stenella*), and white-sided dolphins (*Lagenorhynchus*) usually are more generalized feeders. Among the delphinids, some rather extreme anatomical adaptations have evolved. Killer whales (*Orcinus*) have a relatively robust body with a high dorsal fin, while the fast-swimming right whale dolphins (*Lissodelphis* spp.) have long, slender bodies with no dorsal fin. False killer whales (*Pseudorca*) have only a few large teeth, common dolphins have many small teeth, but Risso's dolphins have totally lost the upper teeth (a feature independently evolved in some other families). In comparison with the various living delphinoids and the fossil kentriodontids, the bottlenose dolphins are generalized in their morphology and diet.

Classification

The following abridged classification is intended only to provide a systematic context for *Tursiops* within the Cetacea. Thus, several groups of cetaceans peripheral to the lineage leading to *Tursiops* have been omitted, but the major groups mentioned in the text are represented, including all of those that are probably involved in the direct ancestry of *Tursiops*. The classification is derived from that of Barnes *et al.* (1985) and indicates some synonyms that are occasionally encountered in the literature. For those groups which are not used here at the same rank as they were originally proposed, the name of the author who proposed the revised rank follows that of the original author, which is in parentheses. All taxa marked with a + are wholly extinct and therefore known only as fossils.

Order Cetacea Brisson, 1762
+Suborder Archaeoceti Flower, 1883 (archaeocetes)
  +Family Protocetidae Stromer, 1908
  +Family Basilosauridae Cope, 1868
    +Subfamily Dorudontinae (Miller, 1923) Slijper, 1936
    +Subfamily Basilosaurinae (Cope, 1868) Barnes and Mitchell, 1978
Suborder Mysticeti Flower, 1864 (baleen whales)
Suborder Odontoceti Flower, 1867 (toothed whales)
  +Superfamily Squalodontoidea (Brandt, 1872) Simpson, 1945
  Superfamily Delphinoidea (Gray, 1821) Flower, 1864 (incl. Monodontoidea Fraser and Purves, 1960)
    +Family Albireonidae Barnes, 1984a
    +Family Kentriodontidae (Slijper, 1936) Barnes, 1978
    Family Monodontidae Gray, 1821
    Family Delphinidae Gray, 1821 (incl. Holodontidae Brandt, 1873; + Hemisyntrachelidae Slijper, 1936; Grampidae Nishiwaki, 1964)
      Subfamily Steninae (Fraser and Purves, 1960) Mead, 1975 (incl. Sotaliinae Kasuya, 1973)
      Subfamily Delphininae (Gray, 1821) Flower, 1867
        Tursiops Gervais, 1855 (incl. *Hemisyntrachelus* Brandt, 1873)
        Grampus Gray, 1828
        Peponocephala Nishiwaki and Norris, 1966
        Lagenorhynchus Gray, 1846
        Lagenodelphis Fraser, 1956
        Stenella Gray, 1866
        Delphinus Linnaeus, 1758
      Subfamily Lissodelphinae Fraser and Purves, 1960
      Subfamily Cephalorhynchinae Fraser and Purves, 1960
      Subfamily Globicephalinae (Gray, 1866) Gill, 1872 [incl. Orcininae (Brandt, 1873) Slijper, 1936; Globidelphinidae Nishiwaki, 1963]

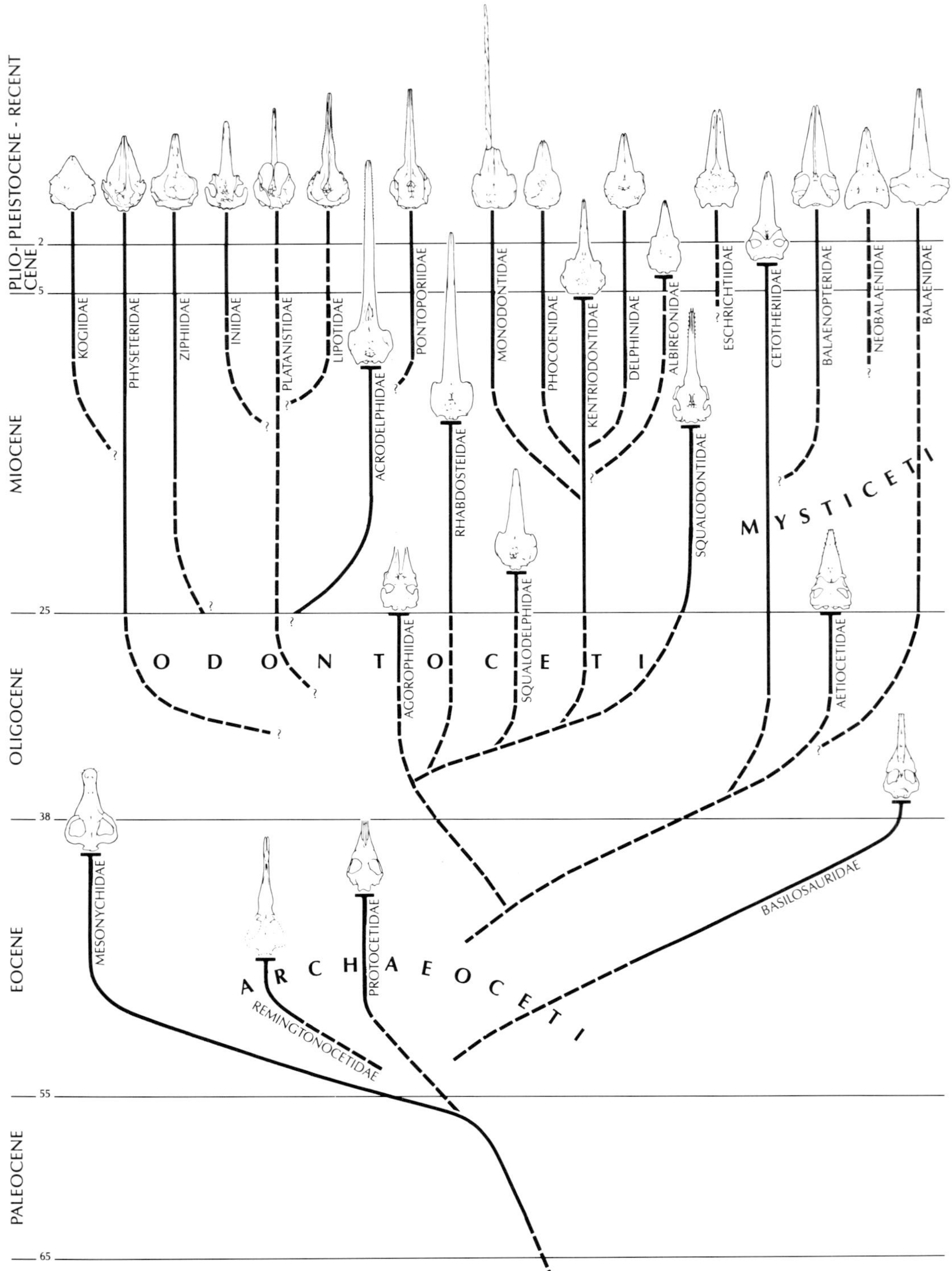

Figure 1 Phylogeny of Cetacea including all currently recognized fossil and living families. Solid lines indicate the known fossil record. Except for Delphinidae, Rhabdosteidae, Squalodelphidae, and Mesonychidae, the skulls used as examples belong to the type genus of the family. All skulls are in dorsal view and are not to scale. (Modified from Barnes *et al.*, 1985, their Fig. 1.)

Family Phocoenidae (Gray, 1825) Bravard, 1885
Superfamily Platanistoidea (Gray, 1863) Simpson, 1945
Superfamily Ziphioidea (Gray, 1865) Fraser and Purves, 1960
Superfamily Physeteroidea (Gray, 1821) Gill, 1872

EVOLUTIONARY HISTORY OF *TURSIOPS*

Synonymy of the Genus *Tursiops*

Genus *Tursiops* Gervais, 1855

Delphinus (part) Fabricius, 1780

Tursio Gray, 1843:xxiii, 105 (type species, *Delphinus truncatus* Montagu, 1821:75, plate 3, preoccupied by *Tursio* Fleming, 1822:211 = *Physeter* Linnaeus, 1758; *non Tursio* Wagler, 1830:30)

Tursiops Gervais, 1855:323 (new name for *Tursio* Gray, 1843); Trouessart, 1898:1029; Miller and Kellogg, 1955:658; Hall and Kelson, 1959:821; Hershkovitz, 1966:47

Tursio (part) Gray, 1866

Gudamu Gray, 1868:6 [subgenus of *Clymenia*, type species, *Clymenia gudamu* [sic] = *Delphinus (Steno) gadamu* Owen, 1866 = *Tursiops truncatus aduncus* Ehrenberg, 1832, by monotypy]

Hemisyntrachelus Brandt, 1873:vi, 240 (subgenus of *Delphinapterus*, type species, *Delphinus cortesii* = *Tursiops cortesii* Fischer, 1829; to include *Delphinus brocchii*); Slijper, 1936:553 (genus of the family Hemi-*syntrachelidae)*

Type species: *Tursiops tursio* of authors, by tautonymy (=*Delphinus truncatus* Montagu, 1821)

The *Hemisyntrachelus* Controversy

Brandt (1873, p. 240) proposed the subgenus *Hemisyntrachelus*, having *Delphinus cortesii* as its type species and also including the species *Delphinus brocchii*. At the time, neither species had yet been allied with the extant genus *Tursiops*, so Brandt's proposal was a relatively radical shift from previous authors, most of whom had assigned these two species to the genus *Delphinus*. Even more unusual was Brandt's assignment of *Hemisyntrachelus* to *Delphinapterus*, the genus of modern belugas or white whales (*D. leucas*).

Brandt's classification of these two species was not accepted by subsequent authors such as Van Beneden and Gervais (1880), Capellini (1882), Sacco (1891, 1893), del Prato (1898), Trouessart (1898), and Simonelli (1911), who put one or both of these fossil species in *Tursiops*. *Hemisyntrachelus* was resurrected, however, by Slijper (1936), who elevated it to full generic status and disassociated it from both the Delphinidae and the Monodontidae (which includes *Delphinapterus*) by naming a new family, the Hemisyntrachelidae. Slijper, however, did not assign to the Hemisyntrachelidae the fossil species *Tursiops osennae*, which has a high tooth count and cranial morphology close to the living *T. truncatus*. He did transfer *Tursiops capellinii* to *Hemisyntrachelus* and added the fossil Miocene odontocete *Lophocetus calvertensis* (Harlan) to the family.

Simpson (1945, pp. 102–103) listed the Hemisyntrachelidae with the included genera *Hemisyntrachelus* and *Lophocetus*, but the family name fell out of usage with subsequent authors. Barnes (1978) transferred *Lophocetus* to another extinct delphinoid family, the Kentriodontidae. This served to reduce the Hemisyntrachelidae to monotypy, and Barnes (1978, p. 4), noting that the species of *Hemisyntrachelus* have always been considered as having affinities with the Delphinidae, pointed out the need for a restudy of the group. Neither Hemisyntrachelidae nor *Hemisyntrachelus* was recognized by Pilleri (1979, 1980, 1985; Pilleri and Pilleri, 1982). Barnes *et al.* (1985) synonymized Hemisyntrachelidae with the Delphinidae.

Review of Nominal Fossil Species of *Tursiops*

Following are discussions of all of the nominal species (and subspecies) of fossil *Tursiops*. These are simply listed in alphabetical order, regardless of chronology or taxonomic priority. My new observations and conclusions are based on the published record because I was unable to examine the original specimens personally prior to this writing.

***Tursiops cortesii* var. *astensis* Sacco, 1891**

Tursiops Cortesii (Desm.) var. *astensis* Sacco, 1891:703, 704, 710; Sacco, 1893

Tursiops cortesii var. *astensis* Sacco. Pilleri, 1985

Tursiops astensis Sacco. Trouessart, 1898:1029 (synonym of *T. cortesii*)
Tursiops cortesi (var. *astensis* Sacco, 1891). Pilleri, 1980:41

The subspecies (variety) *Tursiops cortesii* var. *astensis* was described by Sacco (1891) based on a partial cranium found near Valdondona (not at Cortandone as listed in Pilleri, 1985, his Table 2), in the Astigiana in the Piedmont region, Italy. Sacco (1891) concluded that it came from the yellow sand of Valldunga, which represents the lower beds of the Astian. These beds are early Late Pliocene in age (~3–3.5 mya).

Although Sacco (1891) proposed the variety *astensis* (today we would probably use the term subspecies), he speculated that it could be perhaps a distinct species (Sacco, 1891, p. 704). The holotype cranium of *T. cortesii astensis* was differentiated by Sacco from the holotype of *T. cortesii* based on several principal characters, mostly relating to different degrees of development of the cranial crests and fossae. All of these, as Sacco admitted, could reflect merely age, sex, or individual variability. Three of the characters, however, seem to be taxonomically significant. These are differently shaped occipital condyles and nasal bones and the presence of two fossae (possibly for air sinuses) on the ventral surface of the sphenoid. These latter characters indicate that Sacco's taxon probably should be retained as valid, and I recognize it at the species level. It has 14 teeth on each side of each jaw.

Trouessart (1898, p. 1029) listed *T. astensis* as a synonym of *T. cortesii*. The species was mentioned by both Slijper (1936) and Pilleri (1985) and, as listed by the latter, is known only from the original specimen, stored at the Institute of Geology of the University of Turin. Pilleri (1985, his Fig. 4) phylogenetically showed the taxon as derived from typical *T. cortesii* and ancestral to *T. cortesii capellinii*, but he presented no morphological evidence to support these proposed relationships.

Tursiops brocchii (Balsamo-Crivelli, 1842)

Delphinus Brocchii Balsamo-Crivelli, 1842:132; 1843:629. Capellini, 1863a:264, 269; 1863b:86; Barnes, 1978:4
D.(elphinus) Brocchi. Capellini, 1863a:267
Delphinapterus (*Hemisyntrachelus*) *Brochii* [sic] Crivelli. Brandt, 1873:241
non Tursiops Brocchii (part). Van Beneden and Gervais, 1880:588–589, plate LIX, Fig. 5 (holotype of *T. cortesii*)
Tursiops Cortesii (Desm.) . . . variety *Brocchii* Bals. Sacco, 1891:710
Tursiops Brocchii Crivelli. Trouessart, 1898:1029
Tursiops Brocchii. Sacco, 1893; del Prato, 1898
Delphinapterus Brocchii Bals. del Prato, 1898:2
T.(ursiops) Brocchii (Bals. Criv.). Simonelli, 1911:251
Tursiops brochii [sic] Crivelli. Slijper, 1936:552
Hemisyntrachelus brochii [sic] Crivelli. Slijper, 1936:553
Delphinus brochii [sic] Balsamo-Crivelli, 1842. Barnes, 1978:4
Tursiops brocchii (Balsamo-Crivelli, 1842). Pilleri and Pilleri, 1982:298–299, plate VIII; Pilleri, 1985:15
Tursiops cortesii brocchii. Pilleri, 1985:15 (Table 2)

Balsamo-Crivelli (1842) proposed the new species, *Delphinus Brocchii*, for a mandible and teeth found by Cortesi at Montezago in 1804 (see also Capellini, 1882, p. 571). This was the holotype specimen and is apparently omitted in Pilleri's (1985) Table 2. It was, however, mentioned by Pilleri and Pilleri (1982, p. 299) as having been destroyed when the Milan Museum was bombed during World War II.

The illustration presented by Pilleri and Pilleri (1982, their Fig. 4) shows the outlines of two skulls. The caption explains that one is *Tursiops brocchii*(?) and the other *T. cortesii* (holotype). Aside from the fact that the authors do not explain which outline is which specimen, neither can in fact be *T. brocchii* because a skull has never been identified in print as belonging to that taxon. On comparison, I have concluded that the skull represented by the heavy lines in their illustration is the same skull from San Lorenzo in Collina referred to *T. cortesii* by Simonelli (1911), while the lighter outlines represent the holotype of *T. cortesii*.

A second specimen in the Giovanni Capellini Museum, Bologna University Geological Institute, from San Lorenzo in Collina, Orciano Pisano, Tuscany, Italy, was referred to *T. brocchii* by Capellini (1863a,b). This fossil consists of a partial rostrum, mandible, teeth, tympanic bulla, cervical and other vertebrae, and rib fragments. Slijper (1936, p. 552) gave the age of this specimen as Early Pliocene, and Pilleri and Pilleri (1982, their Figs. 3, 4, plate 8) reillustrated it and gave some measurements.

Capellini contrasted his referred specimen with *T. cortesii* by its having 16 teeth in each dentary in contrast to 14 in *T. cortesii*. The dentary has a very short symphysis (see Capellini, 1863a, his plate 2, Fig. 2), and a higher coronoid crest than that of *Tursiops cortesii*. Judging by the variability in the tooth counts of individuals of Recent *T. truncatus* (range 19–26; see Hall and Kelson, 1959, p. 821; Nishiwaki, 1963, p. 102; Leatherwood *et al.*, 1982, pp. 176–177; Baker, 1983, p. 107), *T. cortesii* and *T. brocchii* are, in this feature, within the expected range of variation for one species. Unless other characters can be shown to separate these two fossil species, the two might be considered as one species. Pilleri (1985) discussed *Tursiops brocchii*, but listed it in his Table 2 as a subspecies (=variety) of *T. cortesii*. Pilleri (1985, p. 15) also stated, ". . . *T. brocchii* is difficult to distinguish from *T. cortesii*. . . ." Certainly, since subspecies are virtually impossible to recognize in the fossil record, even with large samples, Pilleri's recognition of *T. cortesii brocchii* is at this time not acceptable. I do agree with Pilleri and Pilleri (1982) that the taxonomic distinction of *T. brocchii* is questionable, especially now that the holotype has been destroyed. The solution, however, is not to recognize it at the subspecies level.

***Tursiops capellinii* del Prato, 1898**

Il delfino pliocenico di Camerano Casasco (Astigiana). Sacco, 1893
Tursiops Capellinii Sacco. del Prato, 1898:1, 5, plate 1
T. Capellinii. del Prato, 1898:12, 14
Tursiops Capellinii. Simonelli, 1911:251
T. Capellinii Sacco. Simonelli, 1911:251
Tursiops capellinii Sacco. Slijper, 1936:552; Barnes, 1978:4
Hemisyntrachelus capellinii Sacco. Slijper, 1936:553
Tursiops cortesii ("var. *capellinii* Sacco, 1893"). Pilleri and Pilleri, 1982:plate 6
Tursiops cortesii capellinii (Sacco, 1893). Pilleri, 1985:16, his Table 2 (p. 15)

Surprisingly, *Tursiops capellinii,* attributed by virtually all subsequent authors to Sacco (1893), was not actually named by Sacco. At no place in his text, or even in his illustration captions, did he use the binomial, *Tursiops capellinii,* or refer to the species as *capellinii.* Only on p. 14, the last page of the paper, did he write, ". . . mi permetto di imporgli il nome dell' illustre Prof. Capellini, a cui la Cetologia italiana gia deve tante scoperte e da cui essa attende una Monografia generale." The first author actually to apply the Latin binomial to the fossil described by Sacco apparently was del Prato (1898), and it is because of this that he should be credited with authorship of the taxon.

This species was named subsequent to Brandt's (1873) assignment of various earlier-named fossil species to *Hemisyntrachelus.* By the time of Sacco's (1893) writing, the genus *Tursiops* had become firmly accepted for both fossil and living taxa (cf. Van Beneden and Gervais, 1880), and Sacco (1891) had already described the fossil *Tursiops cortesii* var. *astensis.* Slijper (1936) complicated the taxonomy, however, by assigning *T. capellinii* to Brandt's genus Hemisyntrachelus and to the new family Hemisyntrachelidae. Except for Pilleri's work (Pilleri and Pilleri, 1982; Pilleri, 1985), authors have recognized the taxon at the species level. Pilleri (1985, p. 16) wrote in conjunction with his use of the trinomial, *T. cortesii capellinii,* that the "subspecies level remains uncertain."

The holotype cranium and mandible are from Valle dello Stramonte, tributary of the Chiavenna River, Casa Ghia, Piacenza region, Italy, and are stated to be middle Pliocene in age. This would make it about 3.5 million years old. Pilleri (1985, his Table 2) indicated that the (holotype?) specimen should be at the Geological Institute of the University of Parma but is apparently lost. The specimen, as illustrated by Sacco (1893), is excellent, and clearly shows many important features including sutures, the number of alveoli, and the morphology of the narial region. The tooth count is 16 on each side of the palate and 15 in each dentary, 2 and 1 more, respectively, than in *T. cortesii.*

Another specimen was described by del Prato (1898) from the Camerano Cascasco area of Riero, Val Camerano, Piedmont region, Italy. He did not illustrate it fully, but it included the cranium, vertebrae, ribs, sternebrae, and limb bones.

***Tursiops cortesii* (Fischer, 1829)**

Not named. *"Delphinus Phocaena."* Cortesi, 1819:48
Dauphin de (M.) Cortesi. Cuvier, 1824:312; Desmoulins, 1824:360
D.(elphinus) Cortesii Fischer, 1829:512–513
Delphinus Cortesi oder *platyrhynchus.* Keferstein, 1834:203
Delphinus Cortesi Cuvier. Balsamo-Crivelli, 1842:129, 1843:629
Phocaena Cortesii. Laurillard, 1844:634
D.(elphinus) Cortesii. Capellini, 1863a:264
D.(elphinus) Cortesi. Capellini, 1863a:266–267
Delphinapterus (Hemisyntrachelus) Cortesii. Brandt, 1873:240
Tursiops Brocchii (part). Van Beneden and Gervais, 1880:588–589, plate LIX, Fig. 5 (holotype of *T. cortesii*)
Tursiops Cortesii (Desm.). Sacco, 1891
Tursiops Cortesii. Sacco, 1893
Tursiops Cortesii Desmoulins, 1824. Capellini, 1882; Sacco, 1891, 1893; del Prato, 1898; Trouessart, 1898; Simonelli, 1911; Slijper, 1936:552 (as *Tursiops cortesii* Desm.)
Hemisyntrachelus cortesii Desm. Slijper, 1936:553
Tursiops cortesi Keferstein, 1834. Pilleri, 1979, 1980
Tursiops cortesii Keferstein, 1834. Pilleri and Pilleri, 1982:294–296, Fig. 1, plates 1–6
Tursiops cortesii (Fischer, 1829). Kellogg, 1928:69; Barnes, 1978:4; Pilleri, 1985

Tursiops cortesii is the first fossil representative of the genus *Tursiops* to have been recognized. Capellini (1882) provided a history of the taxon (see also Pilleri, 1979), and some of the synonymy above is extracted from his paper. The holotype skeleton was found in 1793 (not 1783 as stated by Pilleri, 1979, p. 71) by Giuseppe Cortesi, who later described the fossil in 1806 and 1819. It was not named, however, until Fischer later (1829) called it *"D. cortesii."* The holotype skeleton was from Colle della Torrazza, near the town of Chiavenna, Piacenza, Italy, and was, unfortunately, destroyed at the Museo Civico of Milan by a bomb in World War II (Pilleri, 1979, 1985, p. 14). Luckily, plaster-cast replicas were long ago distributed to various other European institutions (Sacco, 1891). Pilleri reported that one cast of the holotype cranium had been located in the Paleontological Institute of the University of Parma and that casts of the skull and atlas vertebra are at the University of Turin Paleontological Institute. Pilleri and Pilleri (1982, p. 296) reported casts of the left dentary and cervical vertebrae in the Giovanni Capellini Museum of Paleontology in the Bologna University Geological Institute.

The age of the holotype is commonly cited as being Early Pliocene and from the Plaisancian Stage (Cuvier, 1824; Kellogg, 1928, p. 69; Slijper, 1936, p. 552). However, the Plaisancian is correlative with the Astian Stage (Berggren and Van Couvering, 1974, p. 53), which would make the specimen Late Pliocene in age (between approximately 1.75 and 3.5 mya).

The holotype skull (see Cortesi, 1819; Cuvier, 1824, plate 23, Figs. 1–2; Kellogg, 1928, p. 69; Pilleri, 1979, Figs. 3–6, plates 1–3) had 14 teeth on each side of the rostrum and in each dentary, and the atlas and axis vertebrae were fused. The illustration that Pilleri (1979, his Fig. 2) indicates to be Cuvier's (1824) representation of *Tursiops cortesii* actually includes specimens of various other fossil and Recent odontocetes. Only Figs. 1–2 of Cuvier's plate portray *T. cortesii.*

This is the most commonly reported fossil species of *Tursiops* from Italy, with more than ten specimens recorded in the literature (Capellini, 1882; Pilleri and Pilleri, 1982; Pilleri, 1985, his Table 2). Earlier referrals of the species to the genus *Delphinus* reflect the then-broader definition of that genus, not a radically different interpretation of the relationships of the fossil.

Capellini (1882, not 1881 as given by Pilleri and Pilleri, 1982, p. 296) referred to *Tursiops cortesii* a second specimen, consisting of the facial part of the cranium, rostrum, and mandible. This specimen (listed as number 2 in Pilleri and Pilleri, 1982, p. 296) was found at Mombercelli, Astigiano, Piedmont region and, although supposedly in the Museum at Turin, was not located by Pilleri and Pilleri. Capellini's identification seems to have been correct because the cranial, dental, and mandibular features in his (1882) plate (reproduced by Pilleri and Pilleri, 1982, Fig. 1) are similar to those of the holotype of *T. cortesii.*

Unfortunately, as noted by Pilleri and Pilleri (1982, p. 296), most of the remaining specimens attributed to, or attributable to, *T. cortesii* are either incomplete, lack cranial parts that are directly comparable with the two published skulls, or have imprecise collection data. For example, their identification of the limb bones, their items 4 and 5, as *T. cortesii* is based on Sacco's illustrations of similar bones of *T. capellinii.* Also, unfortunately, both published skulls have been destroyed and/or are lost, and no neotype has been designated.

Tursiops miocaenus Portis, 1885

Tursiops n. sp. (miocaenus) Portis, 1885:344, Fig. 106
Tursiops miocaenus Portis, 1885:344
Tursiops miocaenus Portis, 1886. Trouessart, 1898:1029
Tursiops miocaenus. Dal Piaz, 1977:49; Pilleri, 1985:19, Fig. 4, 5a,b

Although Portis (1885, p. 344) referred to *Tursiops miocaenus* as ". . . a species which still remains to be found," he did propose the name, and he described and illustrated specimens. These specimens, therefore, are the type material (=syntypes), namely, six isolated teeth on which Pilleri (1985, p. 19) commented. Pilleri also repeated Portis' description and reillustrated two teeth from Portis' work. The type material is stated to be Middle Miocene in age (which would be ~12–15 mya) and from Rosignano in the Piedmont region of Italy.

Although Portis favorably compared them "without hesitation" to those of *Tursiops cortesii,* such teeth are usually considered nondiagnostic in the fossil record, and the taxon is realistically a *nomen dubium.* Unfortunately, Pilleri (1985) uncritically accepted the previously proposed relationships, showed the taxon in a phylogeny, and even proposed that *T. miocaenus* was an ancestor of later species of *Tursiops.*

Tursiops osennae Simonelli, 1911

Tursiops osennae Simonelli, 1911:259
Tursiops osennae Simonelli, 1911; Kellogg, 1928:69; Slijper, 1936:553; Pilleri and Pilleri, 1982:296–298, plates 7–8; Pilleri, 1985

Tursiops osennae was described by Simonelli (1911) on the basis of a partial cranium, mandible, teeth, tympanic bulla, periotic, and fused atlas and axis vertebrae found in glauconitic clay at Il Palazzo Estate, San Quirico d'Orcia, in the Province of Siena, Tuscany, Italy. This species has met with little taxonomic controversy. Simonelli's specimen, the holotype, is in the Giovanni Capellini Museum, Bologna University Geological Institute, and was reillustrated by Pilleri and Pilleri (1982, their Figs. 2, 3, plates 7–8), who also provided some measurements. Either the measurements (Pilleri and Pilleri, 1982, their Table 2) or the scale lines on their illustrations are incorrect, because they are not compatible. The specimen does, however, appear to be larger than living *T. truncatus.*

The specimen is stated to be from the Sicilian Marine Stage, which was once considered to be Late Pliocene in age but is now considered to be middle or early Late Pleistocene (~0.5–0.8 mya; Berggren and Van Couvering, 1974). Kellogg (1928, p. 69; also cited by Slijper, 1936, p. 553) considered *T. osennae* to be a "true" *Tursiops.* It has not been referred to the genus *Hemisyntrachelus,* in contrast to most other nominal fossil species of *Tursiops.*

Tursiops osennae has 21 teeth on each side of the rostrum, which is within the range of 19–29 for the tooth count of Recent *Tursiops truncatus* (Hall and Kelson, 1959, p. 821; Nishiwaki, 1963, p. 102; Leatherwood *et al.*, 1982, pp. 176–177; Baker, 1983, p. 107). It has a much higher tooth count than the other Italian fossil species of *Tursiops*. The holotype of *T. osennae* resembles Recent *T. truncatus* because of a similar transverse expansion of the premaxillae anterior to the nares, similarities in the shape of the lateral edges of the facial region, the shape of the rostrum (see Pilleri and Pilleri, 1982, their Fig. 2), the short mandibular symphysis, and the shape and degree of fusion of the atlas and axis vertebrae. It differs by having deeper and narrower antorbital notches, a periotic with a shorter, blunter anterior process and a smaller, more laterally deflected posterior process, more slender teeth, and shorter transverse processes on the atlas vertebrae. Interestingly, in each of these characters, *T. osennae* is more derived than Recent *T. truncatus*. Unfortunately, the holotype cranium lacks the basicranium and the facial bones around the nares, rendering some critical morphology unobservable in the species.

Pilleri (1985) referred to the species a newly discovered specimen found along the Marecchia River, which is now in the Verona Natural History Museum. In reference to the age, he wrote, "Chronostratigraphically speaking, we are at the beginning of the lower Pliocene–middle Pliocene." He asserted but, aside from a speculation that the tooth count of the new fossil would have been similar to that of the holotype, did not demonstrate that the two specimens belong to the same species. This is unfortunate, because Pilleri's new specimen is apparently significantly older geologically than the holotype, and this should have dictated caution in making such an identification.

The morphology of the new specimen described by Pilleri from Marecchia River is unclear because of the incompleteness both of the fossil itself and of the preparation, and because the illustrations and descriptions do not elucidate its anatomy. It does appear to represent a *Tursiops*-like delphinid, and it has teeth like other fossil and Recent *Tursiops*. Like the holotype of *T. osennae*, it apparently is a larger individual than other species of *Tursiops*. For example, Pilleri's (1985, plates 1–2) illustrations indicate the bones to be approximately 50% larger than those of a Recent *T. truncatus*. The humerus and ulna of the fossil both appear to have longer, more slender shafts than do the corresponding bones of *T. truncatus*.

Following the earlier suggestion by Simonelli (1911), Pilleri (1985) considered *T. osennae* to be ancestral to Recent *T. truncatus* (including *T. aduncus*), and showed *T. osennae* on a phylogeny (his Fig. 4) in a position intermediate between the extant taxa and fossil *T. cortesii*. Unfortunately, except for the unconvincing references to tooth counts, Pilleri offered no other morphological evidence to support either his phylogeny or a postulated worldwide biogeographical history (his Fig. 6) of the genus. As noted above, my comparisons with the holotype of *T. osennae* indicate that it may be more derived than *T. truncatus*; this would rule out any possibility of an ancestral–descendant relationship between the two.

***Tursiops cortesii pedemontana* Sacco, 1891**

Tursiops Cortesii (Desm.) . . . variety *pedemontana* Sacc. Sacco, 1891:708–710

The taxon *Tursiops cortesii pedemontana* has not been cited subsequently, but it was proposed by Sacco (1891) for specimens from Cortandone identified as *Tursiops cortesii* by Portis (1885, p. 96, 97 of reprints, plate ix, Figs. 103–105). Sacco used the phrases (as translated): "var. *pedemontana* of *T. cortesii*" and "*Tursiops cortesii* (Desm.) presents until now three varieties, namely: *Brocchii* Bals., *astensis* Sacc., and *pedemontana* Sacc." Whether or not he was justified in recognizing such a taxon, it is one which must at least be acknowledged. In the classification I synonymize it with *Tursiops cortesii*.

Other Fossil Records of *Tursiops*

There are pre-Pleistocene fossils of *Tursiops* which have been reported yet are not named. Barnes (1976, p. 334, his Table 6) reported the occurrence of a dolphin species having affinities with *Tursiops* from California. The fossil is of Late Pliocene age, correlative with the Blancan North American Land Mammal Age, approximately 2–4 mya, and was recovered from rocks stratigraphically high in the Capistrano Formation in Orange County.

A slightly older occurrence of a species of *Tursiops* is known from the North Atlantic realm. Whitmore (in press) reports relatively complete cranial material from Lee Creek Mine near Aurora, North Carolina. The fossils are derived from the Yorktown Formation, which is removed as overburden in the mine, and is of latest Miocene age, about 4–7 mya, and correlative with the Hemphillian North American Land Mammal Age.

Fossil Records of *Tursiops truncatus*

There have been several discoveries of fossil *Tursiops* of Pleistocene age (<1.75 mya) throughout the world. These reflect the "Ice Age" distribution of the modern species. Bones attributed to *Tursiops truncatus*, apparently of Pleistocene age, have been recovered from the bottom of the North Sea (Kortenbout van der Sluijs, 1983). Other specimens from the Atlantic realm, and thus within the present range of the species, have been found on Edisto Island, South Carolina (Roth and Laerm, 1980). These are definitely Pleistocene in age, and were found with fossils of other Late Pleistocene vertebrates (~10,000–50,000 years old).

A partial skeleton of a dolphin, identified as "*Tursiops* sp., near *T. truncatus* (Montagu), but probably not identical with that species," was reported by Blake (1939) from Maryland. The specimen consisted of associated ribs and vertebrae, and undoubtedly represents the remnants of what was formerly a complete skeleton. The fossil was found in a well-studied Pleistocene marine bed that contains abundant mollusk fossils, at Wailes Bluff, St. Marys County. The bed was called Bed 1 by Mansfield (1928), who concluded that the associated fauna of littoral marine mollusks included some brackish-water species and was indicative of quiet, shallow water protected from the sea, with a muddy bottom, and water temperatures the same as at the locality today or slightly warmer. This postulated environment is similar to that which inshore populations of *Tursiops truncatus* are known to frequent today, in coastal bays along the Atlantic coastal plain.

From the Pacific realm there is an excellently preserved skull of *Tursiops* described by Tsao (1978) from Kiangsu Province. The specimen was apparently derived from an uplifted coastal Pleistocene deposit. Tsao identified the specimen as *Tursiops* sp., and discussed both *Tursiops truncatus* and *T. catalonia* as modern species. Since the latter is a junior synonym, that comparison is a moot point. The fossil skull is morphologically within the range of *Tursiops truncatus,* and I reidentify it more precisely as *Tursiops* sp., cf. *T. truncatus.*

DISCUSSION AND CONCLUSIONS

Problems with Interpreting the Fossil Record

Research on fossil organisms can include the study of paleobiogeography, but, unfortunately, some of the taxonomic conclusions of past paleontological researchers have been biased by the land mass on which the fossils were found. This is no less true with fossil cetacean research, and systematic works have in some cases been written without full consideration of the ancient ocean basin and paleoenvironment in which the animals once lived. Many living cetacean species are widely distributed, and some are even cosmopolitan. It is reasonable, therefore, to expect fossil species to have had similar distributions, and the practice of naming separate fossil taxa simply because they are separated geographically should be questioned.

Similarly, time differences should not necessarily bias taxomomic decisions. A species of fossil cetacean, as with those in other groups of mammals, might be expected to have existed for hundreds of thousands or even a million years but, usually, not for several millions of years. A species may evolve into another, or may become extinct, and it is this random turnover of species that provides the record of a sequence of fossil animals through the rocks, the data used in biostratigraphy. For this reason, the identification by Pilleri (1985) of an Early to middle Pliocene fossil dolphin as *Tursiops osennae,* a species previously known only from the Pleistocene holotype, may be questionable. Specimens of *T. cortesii* have been reported from rocks of both Early and Late Pliocene age, and even this time range is more than might be expected for one species. Larger samples and detailed studies would be necessary to confirm the identifications.

Comparisons among many of the named fossil species of *Tursiops* can be hindered by the lack of directly comparable skeletal parts. If different parts of fossil skeletons are designated as holotypes, then problems might arise in the systematic treatment of the group. Sample sizes of the named fossil species of *Tursiops* are simply not statistically significant, and in fact some species are known by only a single occurrence. An ideal fossil record would consist of stratigraphically superimposed samples of several successive species from rocks at different locations throughout the world, but the fossil record of *Tursiops* does not approach this ideal. Instead, it is limited mainly to Italy, and is spotty geographically, stratigraphically, and chronologically.

Pilleri (1985) reviewed the fossil occurrences of *Tursiops* in Italy, the source of all named fossil species. His review is of limited use, however, because new taxonomic combinations are proposed without justifications, little morphology is discussed, and the phylogeny is not based on objective morphologic characters. Further, his Table 2, which lists various taxa and occurrences, does not indicate type specimens, authorship of previously published specimens, or geologic ages.

Like some previous authors, Pilleri has recognized subspecies of *Tursiops*. Subspecies are not a useful taxonomic unit when interpreting the fossil record of *Tursiops*, because it is unjustifiable to recognize them with so scanty a record, especially when subspecies are unrecognizable among most living dolphin species, even using large samples. Accordingly, the following summary recommends against use of subspecies (or varieties).

Valid Fossil Species of *Tursiops*

For a number of reasons, some of the nominal fossil species of *Tursiops* have uncertain or invalid taxonomic status. *Tursiops miocaenus* must be designated as a *nomen dubium* because the type specimens, isolated teeth, are nondiagnostic.

Tursiops osennae is a valid species, and is demonstrably different from Recent *T. truncatus*. It has 21 relatively slender teeth on each side of the jaw, more than in any other fossil species. Compared with *T. truncatus*, its mandibular symphysis is shorter, its antorbital notch is deeper and narrower, its periotic has a blunt anterior process and a smaller and more laterally deflected posterior process, and the transverse processes of the atlas are shorter. These are the types of differences that are commonly used to differentiate other fossil and Recent dolphin species.

Tursiops capellinii is a valid, distinct species. It has more teeth than *T. cortesii*, is from a geological horizon that is chronologically older than that which yielded the holotype of *T. cortesii*, and has been traditionally recognized by past writers. It is not known by enough material (i.e., enough anatomical characters) to allow placing it in the cladogram (see Appendix).

Tursiops cortesii is valid by its priority and is more primitive than either *T. osennae* or *T. capellinii*. Recognizing its nominal subspecies, *T. cortesii pedemontana*, lacks utility because the reputedly distinguishing characters are either not diagnostic or are subject to individual, sexual, or ontogenetic variation in related Recent taxa. Likewise, it solves no problems to recognize *Tursiops brocchii* either as a separate species or as a subspecies of *T. cortesii*. First of all, its tooth count is only one or two more than for the holotype of *T. cortesii*, clearly within the expected range of variability for one species. Second, the holotype of *T. brocchii*, being a mandible, is not comparable with the type material of *T. cortesii astensis*, and is not demonstrably different from the mandibles of *T. capellinii* and *T. cortesii*.

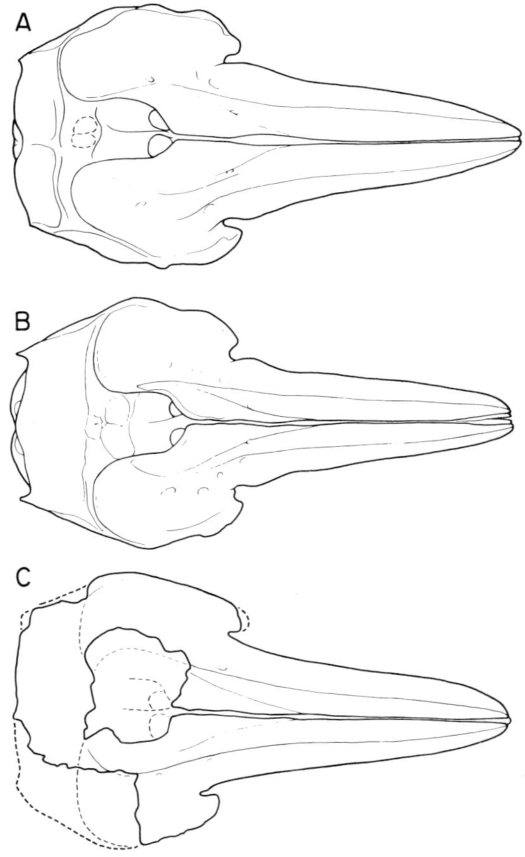

Figure 2 Crania of species of fossil and Recent *Tursiops* in dorsal view. A, *Tursiops cortesii* (Fischer, 1829) after Van Beneden and Gervais, 1880, pl. 34, and Pilleri and Pilleri, 1982, pl. 5; B, *Tursiops truncatus* (Montagu, 1821) after True, 1889, pl. 10, Fig. 2; C, *Tursiops osennae* (Simonelli, 1911), after Simonelli, 1911, Fig. 1, and Pilleri and Pilleri, 1982, pl. 7. All reproduced at the same braincase length.

Though contradicting most authors' acceptance of the species, I conclude that it is a synonym of *T. cortesii.*

Tursiops cortesii astensis differs from the holotype of *T. cortesii* by several characters and, provisionally, should be recognized at the species level and called *Tursiops astensis.*

Table 1 summarizes the current status of fossil *Tursiops* species. Only *Tursiops cortesii, T. astensis, T. capellinii,* and *T. osennae* are valid fossil species of *Tursiops.* These conclusions are, of course, based on a literature review. A detailed reassessment of the group, using all available specimens, is still desirable. It is hoped that this chapter will stimulate such a study. The following outline classification places the fossil species of *Tursiops* in systematic context and summarizes their geographic and geologic ranges.

Order Cetacea Brisson, 1762

- Suborder Odontoceti Flower, 1867
  - Superfamily Delphinoidea (Gray, 1821) Flower, 1864
    - Family Delphinidae Gray, 1821
      - Subfamily Delphininae (Gray, 1821) Flower, 1867
        - *Tursiops* Gervais, 1855
          - *T. cortesii* (Fischer, 1829) [incl. *T. cortesii pedemontana* Sacco, 1891; and *T. brocchii* (Balsamo-Crivelli, 1842)] Late Pliocene (~1.75–3.5 mya), possibly also Early Pliocene (as old as 5 mya for *T. brocchii*), Italy
          - *T. astensis* Sacco, 1891, early Late Pliocene (~3–3.5 mya), Italy
          - *T. capellinii* del Prato, 1898, Middle Pliocene (~3.5 mya), Italy
          - *T. truncatus* (Montagu, 1821) Pleistocene to Recent, temperate and warm oceans of the world
          - *T. osennae* Simonelli, 1911, Middle Pleistocene or early Late Pleistocene (~0.5–0.8 mya), Italy

Table 1
List of All Named Fossil Species and Subspecies of *Tursiops* Gervais, 1855, and Their Status as Determined in This Chapter[a]

Original Species or Subspecies Name	Status
Tursiops cortesii var. *astensis* Sacco, 1891	Valid, *T. astensis*
Delphinus brocchii Balsamo-Crivelli, 1842	Synonym of *T. cortesii*
Tursiops capellinii del Prato, 1898	Valid
Delphinus cortesii Fischer, 1829	Valid, *T. cortesii*
Tursiops miocaenus Portis, 1885	*Nomen dubium*
Tursiops osennae Simonelli, 1911	Valid
Tursiops cortesii pedemontana Sacco, 1891	Synonym of *T. cortesii*

[a] Taxa are listed as originally proposed by first authors.

Paleobiogeography and Paleoclimatology

Pilleri (1985) proposed a Mediterranean center of origin of *Tursiops.* While some of the older records of the taxon, and certainly the majority of them, are from this realm, it is unwise to postulate that this is the geographic center of origin. The geologically oldest occurrence of the genus, in fact, is from Lee Creek Mine in North Carolina. Other Pliocene and Pleistocene fossil records, from both the Pacific and Atlantic basins, suggest that the genus *Tursiops* probably has been as widespread throughout much of past time as it is now. The earlier Kentriodontidae, from which the delphinids probably arose, are now known also to have been equally widespread (Barnes, 1978). It is therefore unsafe to postulate a precise site of origin for *Tursiops* or even for the Delphinidae, in light of the available fossil evidence.

Living bottlenose dolphins have a temperate and tropical distribution in the world's oceans, and all of the fossil records of the genus *Tursiops* fall within the range of the modern animals. The genus is not among those cetacean groups described by Davies (1963) as having antitropical distributions (i.e., those with northern- and southern-hemisphere pairs of genera, species, or infraspecific populations). Unlike antitropical cetaceans, therefore, populations of bottlenose dolphins do not seem to have undergone separation at the equatorial latitudes, and presumably would not have been much influenced by Pleistocene glacial–interglacial fluctuations, which in certain other species appear to have influenced speciation via climatically induced allopatry. Any fossil species of *Tursiops* might therefore be suspected of having been nearly cosmopolitan, and to have had a much wider distribution than indicated by the presently available fossil record.

Evolutionary Origin and Relationships

There is no conclusive fossil evidence bearing directly on the origin of *Tursiops*. The fossil and extant species of the genus are clearly generalized Delphininae. Among delphinine genera, species of *Tursiops* are perhaps morphologically most similar to members of the Steninae, especially the genus *Steno*. In turn, *Steno* has some osteological features (especially rostral and postcranial proportions, facial configuration, tooth count, long mandibular symphysis) which it shares with many of the generalized fossil species of the Kentriodontidae. *Steno* does have, as one might expect, some derived characters (rugose enamel on teeth, enlarged antorbital air sinus, etc.) that separate it from *Tursiops*, and in fact from most Kentriodontidae.

Logic would place the origin of the Delphininae at some time in the Late Miocene, about 11 mya. References in the literature to earlier occurrences of delphinids are in error. For example, contrary to Pilleri's ideas (1985), the European fossil genera *Oligodelphis* and *Protodelphinus* are not delphinids, and are not relevant to a discussion of the ancestry of *Tursiops*. *Oligodelphis* (see Mchedlidze, 1976) is probably a primitive kentriodontid (see Barnes, 1985a), and *Protodelphinus* is known only by a lower jaw (see Dal Piaz, 1977), which is hardly enough material to assess its relationships with confidence.

The anatomical similarities cited above suggest that *Tursiops* evolved from some ancestral group of extinct fossil Delphininae which had smooth tooth enamel, a short mandibular symphysis, and an unenlarged antorbital sinus. This ancestral group in turn might have been related to some member of the subfamily Steninae, which might have evolved from Kentriodontidae.

Using sets of derived character states, it is possible to relate the fossil and living species of *Tursiops* in a scheme that also includes the other supra-specific groups of delphinoids, and the cladogram accompanying the Appendix shows these relationships.

ACKNOWLEDGMENTS

R. Ewan Fordyce and Clayton E. Ray both provided very helpful reviews of a draft of the manuscript. I also thank Frank C. Whitmore for comments on the manuscript and for providing me with copies of some relatively obscure Italian literature and English translations thereof. I thank Susan E. Barnes, Sharmayne M. Padilla, and Patricia Madrigal for help in typing and editing the manuscript. Pieter A. Folkens produced all of the artwork for this chapter.

APPENDIX

Cladogram showing relationships between *Tursiops* and other cetacean groups. Time is not implied in the diagram. The numbered dichotomies represent suites of derived characters that are understood to join all animals beyond that point on the cladogram. The polarities (primitive versus derived) of the characters are based on their presence or absence in primitive taxa.

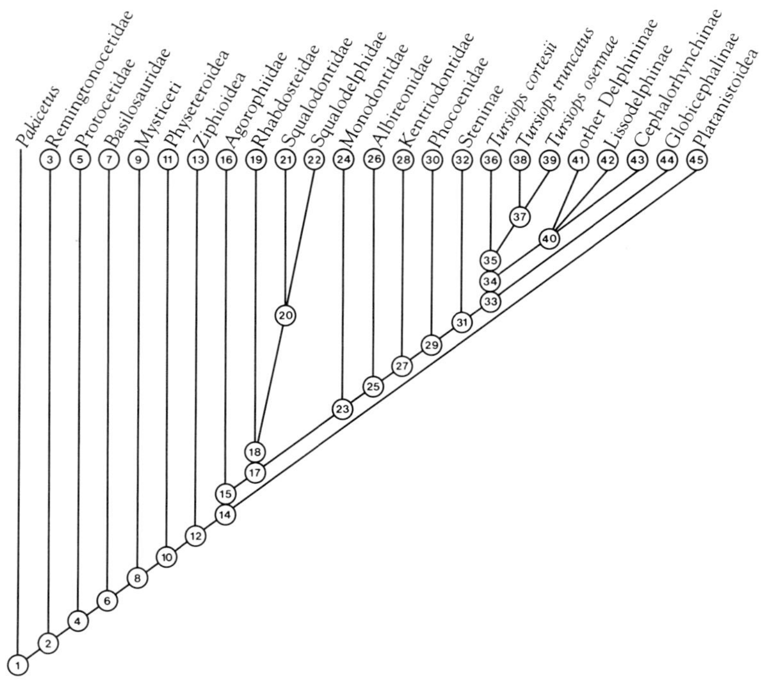

1. Facial region elongated, interorbital region narrow and elongated, sagittal and nuchal crests high, tympanic bulla involuted and inflated, crowns of cheek teeth transversely compressed (*Pakicetus* and all other Cetacea).
2. Anterior palatine foramina lost, basioccipital crests enlarged, peribullary and pterygoid air sinuses present as diverticula from the middle-ear air sinus, falciform process of squamosal present lateral to anterior process of periotic, periotic and bulla separated on their medial sides from basioccipital, bulla separated posteriorly from exoccipital, supraorbital process of the frontal large, horizontal, and tabular, hypoglossal foramen in the basioccipital located either at the apex of or inside the jugular notch, mandibular foramen enlarged, scapula with the supraspinatus fossa reduced and with the acromion and coracoid processes parallel and directed anteriorly, and postzygapophyses on thoracic and lumbar vertebrae reduced (true Archaeoceti, Mysticeti, and Odontoceti).
3. Rostrum elongated, cheek teeth elongated anteroposteriorly, spaces form between cheek teeth, orbit greatly reduced in size, palate exceptionally convex, pseudo-anterior sinus forms, pterygoid sinus enlarged (convergence), basioccipital crest directed laterally (family Remingtonocetidae).
4. Basioccipital crest thickened and aligned anteroposteriorly, pelvic limbs reduced (first demonstrable occurrence of character).
5. Rostrum elongated (family Protocetidae).
6. Nares and anterior aperture of infraorbital foramina retracted posteriorly, incisors caniniform (first demonstrable appearance of character), posterior premolars and molars acquire accessory denticles on anterior and posterior margins, innominate loses contact with transverse processes of sacral vertebrae.
7. M^3 lost (family Basilosauridae).
8. Cranium telescoped, multiple maxillary foramina derived from the infraorbital foramen, vomer exposed on the basicranium and covering the basisphenoid/basioccipital suture, zygomatic process of the squamosal contacting the postorbital process of the frontal or connected to it by a ligament,

monophyodonty, elbow joint nonrotational with anteroposterior position of radius and ulna and formation of discrete flattened articular facets on distal end of humerus (one for radius, two for ulna), olecranon fossa of the humerus lost, hyperphalangy (Mysticeti and Odontoceti).

9. Antorbital process of maxilla present, lateral margins of maxillae thin, maxillae extend posteriorly beneath supraorbital processes of frontals, temporal crest moved to dorsal surface of supraorbital process, mandibular symphysis loose, with only ligamental attachment, marked by anteroposterior groove, atlas with single transverse process (suborder Mysticeti).
10. Lateral margins of rostrum thick, maxillae extend posteriorly over supraorbital processes of frontals, antorbital notch opens anteriorly, premaxillary foramen present (as an aperature of the infraorbital foramen complex), lacrymal and jugal fused, middle sinus present, dentition monophyodont and polyodont (suborder Odontoceti).
11. Facial region of cranium skewed asymmetrically to left side, left naris greatly enlarged, right naris reduced, nasal bones reduced or lost, supracranial (spermaceti) basin formed, basin surrounded laterally and posteriorly by crests formed of maxillae, frontals, and occipital, temporal fossa reduced, elliptical in shape, and opening posterodorsally, zygomatic process of jugal shortened by anterior retraction and losing contact with zygomatic process of squamosal, dentition homodont (single-rooted with conical crowns), internal acoustic meatus of periotic circular in shape, accessory ossicle of tympanic bulla fused to anterior process of periotic, posterior process of tympanic bulla greatly enlarged and fissured, cervical vertebrae 2–7 fused, atlas vertebra with single transverse process, trochlear surface on humerus for articulation with ulna reduced from two discrete facets to one (superfamily Physeteroidea).
12. Posterior ends of maxillae broad and curved, nasal bones short, thick, and forming apex of cranium (cranial vertex), premaxillary sac fossae form on each premaxilla anterolateral to each naris, premaxillary foramen confluent with anteromedial, posteromedial, and posterolateral sulci, jugal becomes very slender (rodlike), posterior (mastoid) process of periotic compact, dense, and shortened.
13. Facial region of cranium skewed asymmetrically to left side, cranial vertex elevated, dentition primitively homodont, secondarily heterodont in more derived taxa, posterior teeth with conical crowns, crowns of anterior mandibular teeth expanded anteroposteriorly and dorsoventrally, atlas vertebra with one transverse process, atlas and axis vertebrae fused (superfamily Ziphioidea).
14. Anterior sinus and posterior sinuses develop, mastoid process of tympanic bulla loses contact with squamosal.
15. Pterygoid sinus enclosed by bony lamina except for opening at posterolateral side, anterior sinus extends external to maxilla and pterygoid, bones comprising lateral lamina of pterygoid sinus fossa reduced through atrophy, basioccipital crest with flattened ventromedial surface, diverging posteriorly, paroccipital process wide and projecting laterally nearly as far as zygomatic process of squamosal, zygomatic process of squamosal elongate, arched uniformly anterodorsally, with rounded dorsal surface, prominent posterolateral corner, and large, ventrally projecting postglenoid process.
16. Distal ends of rostrum and mandible curved ventrally (family Agorophiidae).
17. Parietals excluded from surface exposure on cranial vertex, frontals reduced and touching nuchal crest, maxillae expanded posteriorly to reach nuchal crest, and separated from occipital shield by only a narrow exposure of frontals, maxillary foramina (part of infraorbital foramen system) moved to dorsal position over supraorbital process of frontal.
18. Anterior process of periotic elongate, nearly cylindrical, with elongate groove for accessory ossicle of bulla on ventral side, pterygoid hamulus with a solid, posteriorly directed termination.
19. Rostrum greatly elongated, cranial hiatus incipiently formed by narrow fissure linking middle and posterior lacerate foramina, dentition homodont (single-rooted), extreme polyodonty, teeth small (family Rhabdosteidae).

20. Coracoid process of scapula reduced.
21. Teeth enlarged (family Squalodontidae).
22. Cranial vertex skewed asymmetrically to left side, dentition homodont (single-rooted) (family Squalodelphidae).
23. Dorsal exposure of lacriminal bones limited, olfactory fontanelles closed on posterior wall of nares, pterygoid sinus invades the palatine, cranial hiatus formed dorsal to periotic (petrosal) by complete merger of middle and posterior lacerate foramina, basioccipital crest becomes thin and excavated by peribullary air sinus, tympanic bulla to squamosal contact lost, dentition homodont (single-rooted with conical crowns, although sometimes with accessory cusps), atlas vertebra loses lower transverse process, anterior and posterior processes of periotic much shortened, deltoid crest of humerus reduced or lost (superfamily Delphinoidea).
24. Cranium expanded transversely, rostrum shortened, premaxilla transversely expanded distally between maxillae, facial surface anterior to cranial vertex slightly convex, maxilla extensively exposed on dorsal surface of cranium medial to premaxilla around posterior part of mesorostral gutter and anterior margins of nares, cranial vertex formed into a narrow, elevated prominence, tympanic bulla with anterior styliform process ventral to eustachian tube, teeth with elongate roots and crowns, reduced in numbers, olecranon process of ulna reduced and ultimately lost in derived taxa, trochlear surface on distal end of humerus for articulation with ulna reduced from one surface to two (family Monodontidae).
25. Zygomatic process of squamosal reduced, posterior part of palate and pterygoid hamulus region transversely compressed, cervical vertebrae shortened anteroposteriorly.
26. Cranial vertex elevated, premaxillary eminences anterior to nares low, narrow, and smoothly rounded, areas of premaxillary sac fossae, posterior ends of premaxillae, and nasals flattened, not raised above surrounding maxillary surfaces, nares large, circular in outline and passing nearly vertically through cranium, periotic compact, globose, thoracic and lumbar vertebrae with anteroposteriorly compressed centra and long spinous and transverse processes, humerus, radius, and ulna short, flattened, and greatly expanded anteroposteriorly, radius extremely wide distally, olacranon process of ulna reduced, trochlear facet for ulna on distal end of humerus reduced to one facet from two (family Albireonidae).
27. Posterior air sinus enlarged, antorbital and postorbital lobes of pterygoid air sinus develop.
28. Nasals rounded and thick, mandibular symphysis elongate (family Kentriodontidae).
29. Cranial vertex skewed assymmetrically to left side, humerus, radius, and ulna shortened, head of humerus enlarged and spherical, distal end of humerus expanded transversely.
30. Posterior ends of premaxillae retracted anteriorly, premaxillary eminences form anterior to each narial opening, branch of preorbital lobe of air sinus system extends from orbit posterodorsally into a recess between the frontal bone and the facial surface of the maxilla, tooth crowns expanded apically (family Phocoenidae).
31. Posterior end of left premaxilla always significantly narrower and shorter than right and retracted anteriorly away from anterolateral corner of left nasal, mesethmoid spreads dorsally on posterior walls of nares to overlap anterior surfaces of nasals, nasals moved posteriorly to crowd frontals from exposure on cranial vertex, pterygoid sinuses closely approximated medially at posterior end of palate by medial compression of palatines (family Delphinidae).
32. Antorbital lobe of air sinus system enlarged and invades region between lacrymal and frontal dorsal to orbit, manibular symphysis lengthened (subfamily Steninae).
33. Mandibular symphysis shortened, crowns of teeth smooth.
34. Total number of vertebrae increased to more than 60.
35. Cranium broadened, width across postorbital processes significantly wider than across preorbital processes, total number of vertebrae between 61 and 66 (genus *Tursiops*).
36. Dentition comprised of 14 teeth on each side of mandible and rostrum (*Tursiops cortesii*).
37. Dentition comprised of more than 19 teeth on each side of mandible and rostrum.

38. Dentition comprised of 19–26 teeth on each side of mandible and rostrum, humerus, radius, and ulna shortened and broadened (*Tursiops truncatus*).
39. Dentition comprised of 21 teeth on each side of mandible and rostrum, mandibular symphysis shorter, antorbital notch deeper and narrower, anterior process of periotic shorter and more blunt, posterior process of periotic smaller and more laterally deflected, teeth of lesser diameter, transverse process of atlas shorter (*Tursiops osennae*).
40. Total number of vertebrae increased to more than 68.
41. Dentition comprised of more than 22, and usually more than 30, teeth on each side of rostrum and mandible (other Delphininae).
42. Total numbers of vertebrae increased to between 88 and 90, vertebrae greatly compressed anteroposteriorly, dentition comprised of between 40 and 43 upper teeth on each side and 42 and 46 lower teeth on each side, dorsal fin lost (genus *Lissodelphis*).
43. Acquisition of some neotenous cranial characters, occipital shield convex, rostrum shortened and slightly broadened, right premaxilla retracted from anterolateral corner of right nasal, premaxillary sac fossae and other parts of premaxillae around nares elevated (subfamily Cephalorhynchinae).
44. Cranium greatly expanded transversely with concomitant widening of component bones, teeth enlarged but reduced in numbers to 13 or less on each side of rostrum and mandible (subfamily Globicephalinae).
45. Maxillary foramina moved posteriorly over supraorbital process, maxillary crest forms over orbit, lacrimal with enlarged dorsal exposure near antorbital notch, postnarial region of facial part of cranium expanded with extensive dorsal exposure of maxilla, expansion of lamina of bone lateral to pterygoid sinus (accomplished in varying amounts by pterygoid, palatine, and maxilla), anterior sinus extends into posterior part of rostrum with laminae of maxilla and/or pterygoid lateral to it, rostrum, mandible, and symphysis elongate (superfamily Platanistoidea).

REFERENCES

Baker, A. N. (1983). "Whales and dolphins of New Zealand and Australia. An Identification Guide." Victoria University Press, Wellington, New Zealand. 133 pp.

Balsamo-Crivelli, G. (1842). Memoria per servire all' illustrazione dei grandi mammiferi fossili esistenti nell' I. R. Gabinetto di Santa Teresa in Milano e cenno sovra due mammiferi fossili trovati nella lignite di Leffe nella provincia di Bergamo. *Giorn. R. Ist. Lombardo di Sci. e Let. Arti e Biblio. Italiana* **3,** 297–319.

Barnes, L. G. (1976). Outline of eastern North Pacific fossil cetacean assemblages. *Syst. Zool.* **25**(4), 321–343.

Barnes, L. G. (1978). A review of *Lophocetus* and *Liolithax* and their relationships to the delphinoid family Kentriodontidae (Cetacea: Odontoceti). *Nat. Hist. Mus. Los Angeles County Sci. Bull.* **28,** 1–35.

Barnes, L. G. (1984a). Fossil odontocetes (Mammalia: Cetacea) from the Almejas Formation, Isla Cedros, Mexico. *PaleoBios, Mus. Paleontol. Univ. California, Berkeley* **42,** 1–46.

Barnes, L. G. (1984b). Whales, dolphins and porpoises: Origin and evolution of the Cetacea. *In* "Mammals. Notes for a short course organized by P. D. Gingerich and C. E. Badgley" (T. W. Broadhead, ed.), pp. 139–154. University of Tennessee, Department of Geological Sciences, Studies in Geology, Vol. 8, pp. i–iv, 1–234.

Barnes, L. G. (1985a). Review: General features of the paleobiological evolution of Cetacea. Guram Andreevich Mchedlidze. *Mar. Mamm. Sci.* **1**(1), 90–93.

Barnes, L. G. (1985b). The Late Miocene dolphin *Pithanodelphis* Abel, 1905 (Cetacea: Kentriodontidae) from California. *Contrib. Sci., Nat. Hist. Mus. Los Angeles County* **367,** 1–27.

Barnes, L. G. (in press a). Cetacea. *In* "Textbook of Vertebrate Paleontology" (K. Thomson, ed.), in press. University of Chicago Press, Chicago and London.

Barnes, L. G. (in press b). Cetacea. *In* "The Tertiary Mammals of North America" (C. Janis, L. Jacobs, and K. Scott, eds.), in press. University of Cambridge Press, London.

Barnes, L. G., and Mitchell, E. D. (1978). Cetacea. *In* "Evolution of African Mammals" (V. J. Maglio and H. B. S. Cooke, eds.), pp. 582–602. Harvard University Press, Cambridge, Massachusetts, and London.

Barnes, L. G., Domning, D. P., and Ray, C. E. (1985). Status of studies on fossil marine mammals. *Mar. Mamm. Sci.* **1**(1), 15–53.

Berggren, W. A., and Van Couvering, J. A. (1974). The Late Neogene. Biostratigraphy, geochronology and paleoclimatology of the last 15 million years in marine and continental sequences. *Paleogeog., Paleoclimatol., Paleoecol.* **16,** i–xi, 1–216.

Blake, S. F. (1939). A Pleistocene porpoise (*Tursiops* sp.) from Maryland. *Proc. Biol. Soc. Washington* **52,** 99.

Brandt, J. F. (1873). Untersuchungen uber die Fossilen und Subfossilen Cetaceen Europa's. *Mem. Acad. Imp. Sci. St.-Petersbourg, Ser. 7* **20**(1), i–viii, 1–372, plates 1–34.

Capellini, G. (1863a). Sui delfini fossili del Bolognese. *Mem. dell'Accad. delle Sci. dell'Ist. di Bologna, Ser. 2* **3,** 245–273, plates 1–3. (Separately reprinted in 1864 as "Sui delfini fossili del Bolognese," with separate pagination.)

Capellini, G. (1863b). (Sopra un Delfino fossile scoperto a S. Lorenzo in Collina.) *Rend. Accad. Sci. Bologna, 1862–1863,* pp. 85–87.

Capellini, G. (1882). Del *Tursiops Cortesii* e del delfino fossile di Mombercelli nell'Astigiano. *Mem. Accad. Sci. Bologna, Ser. 4* **3,** 569–578, plate (unnumbered). [Abstr., *Rend. Accad. Sci. Bologna, 1881–1882,* pp. 88–89.]

Cortesi, G. (1804–1806). Sulle ossa fossili di grandi animali terrestri e marini scopertesi su colli Piacentini. *Nuova Scelta Opuscoli interessanti sulle Scienze e sulle Arti* **1,** 289–305, plates ix–xi; **2,** 379–391, plate xii.

Cortesi, G. (1819). Saggi geologici degli Stati di Parma e Piacenza. "Maino, Piacenza." x + 165 pp., 7 plates.

Cuvier, G. (1824). "Recherches sur les Ossemens Fossiles, ou l'on Retablit les Caracteres de Plusieurs Animaux dont les Revolutions du Globe on Detruit les Especes," 2nd Ed., 5 volumes; atlas, 2 volumes. Paris.

Cuvier, G. (1829). "Le Regne Animal Distribue d'Apres son Organization," 2nd Ed., 5 volumes (1829–1830).

Dal Piaz, G. (1977). Gli Odontoceti del Miocene Bellunese. Parti Settima. *Protodelphinus capellinii. Mem. Ist. Geol. R. Univ. Padova, Allegato* **4,** 41–52, plate 1.

Davies, J. L. (1963). The antitropical factor in cetacean speciation. *Evolution* **17**(1), 107–116.

del Prato, A. [1898 (for 1897)]. Il *Tursiops capellinii* Sacco del Pliocene, Piacentino. *Palaeontol. Italica* **3,** 1–14, plate 1.

Evans, P. G. H. (1987). "The Natural History of Whales and Dolphins." Facts On File Publications, New York and Oxford. xvi + 343 pp.

Evans, W. E., and Maderson, P. F. A. (1973). Mechanisms of sound production in delphinid cetaceans: A review and some anatomical considerations. *Am. Zool.* **13,** 1205–1213.

Fischer, J. B. (1829). "Synopsis Mammalium." Sumtibus J. G. Cottae, Stuttgart.

Flower, W. H. (1883). On whales, present and past and their probable origin. *Proc. Zool. Soc. London 1883,* pp. 466–513.

Fordyce, R. E. (1980). Whale evolution and Oligocene southern ocean environments. *Paleogeog., Paleoclimatol., Paleoecol.* **31,** 319–336.

Fraser, F. C., and Purves, P. E. (1960). Hearing in cetaceans. Evolution of the accessory air sacs and the structure and function of the outer and middle ear in Recent cetaceans. *Bull. British Mus. (Nat. Hist.), Zool.* **7**(1), 1–140, frontispiece, plates 1–53.

Gaskin, D. E. (1982). "Ecology of Whales and Dolphins." Heinemann Educational Books, Ltd., London.

Gingerich, P. D., Wells, N. A., Russell, D. E., and Shah, S. M. I. (1983). Origin of whales in epicontinental remnant seas: New evidence from the Early Eocene of Pakistan. *Science* **220,** 403–406.

Hall, E. R., and Kelson, K. R. (1959). "The Mammals of North America," Vol. II. Ronald Press, New York.

Keferstein, W. (1834). Die Naturgeschichte des Erdkorpers in ihren ersten Grundzugen. *Zweiter Theil. Die Geologie und Palaontologie, Leipzig,* iv + 896 pp.

Kellogg, A. R. (1928). The history of whales—Their adaptation to life in the water. *Quart. Rev. Biol.* **3,** 29–76, 174–208.

Kellogg, A. R. (1938). Adaptation of structure to function in whales. *Carnegie Inst. Washington Publ.,* No. 501, pp. 649–682.

Kortenbout van der Sluijs, G. (1983). De resten van zoogdieren uit de Noordzee. [The mammalian remains from the North Sea.] *Grondboor Hamer, 1983,* No. 1, pp. 4–7. [In Dutch, English summary, text figures.]

Lawrence, B., and Schevill, W. E. (1956). The functional anatomy of the delphinid nose. *Bull. Mus. Comp. Zool. Harvard College* **114**(4), 103–151.

Leatherwood, S., Caldwell, D. K., and Winn, H. E. (1976). Whales, dolphins, and porpoises of the western North Atlantic. A guide to their identification. *Natl. Oceanic Atmospheric Administration Technical Rep., Natl. Mar. Fisheries Service, Circular* **396,** i–iv, 1–176.

Leatherwood, S., Reeves, R. R., Perrin, W. F., and Evans, W. E. (1982). Whales, dolphins, and porpoises of the eastern North Pacific and adjacent Arctic waters. A guide to their identification. *Natl. Oceanic Atmospheric Administration Technical Rep., Natl. Mar. Fisheries Service, Circular* **444,** i–v, 1–245.

Mansfield, W. C. (1928). Notes on Pleistocene faunas from Maryland and Virginia and Pliocene and Pleistocene faunas from North Carolina. *U.S. Geol. Surv. Prof. Paper* **150,** 129–140, plates 24–25.

Mchedlidze, G. A. (1976). "Osnovnyye Cherty Paleobiologicheskoy Istorii Kitoobraznykh." [Basic features of the paleobiological history of cetaceans.] Akademia Nauk Gruzinskoi S.S.R., Institut Paleobiologii, "Metsniereba" Press, Tbilisi, Georgian S.S.R. [In Russian, English summary.]

Mead, J. G. (1975). Anatomy of the external nasal passages and facial complex in the Delphinidae (Mammalia: Cetacea). *Smithsonian Contrib. Zool.* **207,** i–iv, 1–72.

Miller, G. S., Jr. (1923). The telescoping of the cetacean skull. *Smithsonian Misc. Coll.* **76**(5), 1–71.

Nishiwaki, M. (1963). Taxonomial consideration on genera of Delphinidae. *Sci. Rep. Whales Res. Inst., Tokyo* **17,** 93–103.

Norris, K. S. (1968). The evolution of acoustic mechanisms in odontocete cetaceans. *In* "Evolution and Environment: A Symposium Presented on the Occasion of the 100th Anniversary of the Foundation of Peabody Museum of Natural History at Yale University" (E. T. Drake, ed.), pp. 297–324. Yale University Press, New Haven, Connecticut.

Norris, K. S., Dormer, K. J., Pegg, J., and Liese, G. J. (1971). The mechanism of sound production and air recycling in porpoises: a preliminary report. *In* "Proceedings of the Eighth Annual Conference on Biological Sonar and Diving Mammals," pp. 1–11. Stanford Research Institute, Menlo Park, California.

Pilleri, G. (1979). Taxonomic status of "Cortesi's dolphin" (*Tursiops cortesii* Keferstein, 1834) from the lower Pliocene of Northern Italy. *Investigations on Cetacea* **10,** 71–83.

Pilleri, G. (1980). The fossil odontocetes (Cetacea) in the Museum of Paleontology of the University of Turin. *Investigations on Cetacea* **11,** 39–53, plates 1–20.

Pilleri, G. (1985). Second record of *Tursiops osennae* (Cetacea: Delphinidae) in a Pliocene horizon of the Romagna Appennines, central Italy, and the phylogeny of *Tursliops. Investigations on Cetacea* **17,** 11–30.

Pilleri, G., and Pilleri, O. (1982). Catalogue of the fossil odontocetes (Cetacea) in the Bologna Giovanni Capellini Museum of Paleontology with description of a new species of *Hoplocetus* (Physeteridae). *Mem. di Scienze Geologiche gia Mem. degli Istituti di Geologia e Mineralogia dell'Universita di Padova* **35,** 293–317, plates 1–18.

Portis, A. (1885). Catalogo descrittive dei Talassoterii rinvenuti nei terreni terziari del Piemonte e della Liguria. *Mem. R. Accad. Sci. di Torino* (*Ser.* 2) **37,** 247–365.

Rice, D. W. (1967). Cetaceans. *In* "Recent Mammals of the World" (S. Anderson and J. K. Jones, Jr., eds.), pp. 291–324. Ronald Press, New York.

Rice, D. W. (1984). Cetaceans. *In* "Orders and Families of Recent Mammals of the World" (S. Anderson and J. K. Jones, Jr., eds.), pp. 447–490. Wiley, New York.

Romer, A. S. (1966). "Vertebrate Paleontology," 3rd Ed. University of Chicago Press, Chicago and London.

Roth, J. A., and Laerm, J. (1980). A Late Pleistocene vertebrate assemblage from Edisto Island, South Carolina. *Brimleyana* **3,** 1–29.

Rothausen, K. (1968). Die Systematische Stellung der europaischen Squalodontidae (Odontoceti, Mamm.). *Palaontoiogische Zeitschrift* **42,** 83–104, plates 11–12.

Sacco, F. (1891). Sopra un cranio di *Tursiops Cortesii* (Desm.) var. *astensis* (Sacc.) dell'Astigiana. *Att. R. Accad. Sci. Torino* **26,** 703–712, plate 13.

Sacco, F. (1893). Il delfino policenico di Camerano Casasco (Astigiana). *Mem. Soc. Italiana Sci., Naples* (*Ser.* 3) **9**(5), 1–14, 2 plates.

Schenkken, E. J. (1973). On the comparative anatomy and function of the nasal tract in odontocetes (Mammalia, Cetacea). *Bijdragen tot de Dierkunde* **43**(2), 127–159.

Simonelli, V. (1911). Avanzi di "*Tursiops*" del Pliocene Senese. *Mem. R. Accad. Sci. Ist. de Bologna, Cl. Sci. Fisiche* (*Ser.* 6) **8,** 249–260, 1 plate.

Simpson, G. G. (1945). The principles of classification and a classification of mammals. *Bull. Am. Mus. Nat. Hist.* **85,** i–xvi, 1–350.

Slijper, E. J. (1936). Die Cetaceen. Vergleichend-anatomisch und systematisch. *Capita Zool.* **7**(1–2), i–xv, 1–590.

Slijper, E. J. (1979). "Whales" (transl. A. J. Pomerans), 2nd Ed. Cornell University Press, Ithaca, New York.

Szalay, F. S. (1969a). The Hapalodectinae and a phylogeny of the Mesonychidae (Mammalia, Condylarthra). *Am. Mus. Novitates,* No. 2361, pp. 1–26.

Szalay, F. S. (1969b). Origin and evolution of function of the mesonychid condylarth feeding mechanism. *Evolution* **23,** 703–720.

Trouessart, E. L. (1898). "Catalogus Mammalium tam Viventium quam Fossilium," Vol. II, pp. 665–1264. R. Friedlander and Sohn, Berlin.

True, F. W. (1889). Contributions to the natural history of the cetaceans, a review of the family Delphinidae. *Bull. United States Nat. Mus.* **36,** 1–191, pls. 1–47.

Tsao, Ke-Ching (1978). The discovery of *Tursiops* fossil from Sarchow, Kiangsu. [Also cited as: "A discovery of a fossil *Tursiops* in Sha-chou County, Kiangsu Province."] *Vertebrata Palasiatica* **16**(4), 264–266, plate 1. [In Chinese, sketch map plate, 11 references.]

Van Beneden, P. J., and Gervais, P. (1880). "Osteographie des Cetaces Vivants et Fossiles, Comprenant la Description et l'Iconographie du Squelette et du Systeme Detaire de ces Animaux; Ainsi que des Documents relatifs a leur Histoire Naturelle." A. Bertrand, Paris. viii + 634 pp., atlas.

Van Valen, L. (1966). Deltatheridia, a new order of mammals. *Bull. Am. Mus. Nat. Hist.* **132**(1), 1–126.

Van Valen, L. (1968). Monophyly or diphyly in the origin of whales. *Evolution* **22**(1), 37–41.
Whitmore, F. C. (1989). The Cetacea of Lee Creek Mine. "Smithsonian Contributions to Paleobiology, Remington Kellogg Memorial Volume" (C. E. Ray, ed.), Vol. 3.
Whitmore, F. C., and Sanders, A. E. (1976). Review of the Oligocene Cetacea. *Syst. Zool.* **25**(4), 304–320.
Winge, H. (1921). A review of the interrelationships of the Cetacea (transl. G. S. Miller, Jr.). *Smithsonian Misc. Coll.* **72**(8), 1–97.

PHOTO: Stranded specimens have been an important source of material for biological studies. Here a carcass of a stranded bottlenose dolphin is weighed before being transported to the Smithsonian Institution in Washington, D.C. (courtesy James G. Mead).

Part II

Anatomy and Physiology

Osteology of the Bottlenose Dolphin **2**

Sentiel Rommel
3927 Norway Drive
Bar Harbor, Maine 04609

INTRODUCTION

This chapter describes some features of the skeleton of the bottlenose dolphin, *Tursiops truncatus.* The terminology I use generally follows that of Flower (1867, 1885). I have described the osteological features by examining specimens in the collection of the United States National Museum (USNM) (Table 1). Additional specimens were examined to verify controversial or questionable features. Most of the specimens were dolphins that stranded along the coasts of North Carolina and Virginia (see Mead and Potter, Chapter 9, this volume).

In the figures, specific bone names are capitalized; other features are in lower case. In the text, specific bone names are in boldface type (e.g., **orbitosphenoid**); commonly used alternate names are in parentheses. Distinct features of the skull that are self-explanatory (e.g., postorbital process of the **frontal**) are not described in the text unless clarification is required. Cranial nerves and digits of the manus are referred to by Roman numerals.

SKULL

The discussion below relates primarily to the skull of *Tursiops* (Figs. 1–5). General features shared with odontocetes and other cetaceans are so indicated.

Telescoping

The cetacean skull is telescoped (elongated antorbitally and compressed postorbitally) by the extreme remodeling of the "typical" mammalian

Table 1
Postcranial Vertebral Position of Skeletal Features for Ten Specimens in the U.S. National Museum Collection[a]

1	USNM ID number	22304	396165	500863	504326	504618	504726	504766	504906	550109	550375
2	Total length, cm	—	303	261	203	280	298	309	288	220	310
3	Skeleton weight, kg	9.5	13.2	7.5	3.9	6.3	11.4	9.1	9.1	5	16.7
4	Sex	F	M	M	F	F	M	M	M	F	M
5	Teeth: Upper left	26	20	21+	24	22	25	18+	23	24	24
6	Upper right	26	21	23	24	22	24	23	22	24	25
7	Lower left	24	18	21	24	21	24	21	22	23	22
8	Lower right	24	18	22	24	21	24	19+	20	23	18+
9	Premaxillary teeth	1	0	1E	3	0	1	1	—	3	2
10	Vertebrae, total number	64+	65	63+	60+	62+	62+	65	63+	60	65+
11	Cervical fusions	1–2	1–2	1–2	1–2	1–2	1–2, 4–5	1–2	1–3	1–2	1–2[b]
	Thoracic vertebrae										
12	Distinct synovial joint	19	19	19	18	19	18	19	19	18	19
13	Last enlarged T-Proc[c] tip	21	19E	21	19	20	21	21	20	19/18	21
14	First metapophysis	10	11	11	10	10	12	11	11	10	10
	Postthoracic vertebrae										
15	Perforation in T-Proc	47	44	47/46	43	46	47	47	46/45	41	46
16	Vertebrarterial foramen	49	45/46	49	44/45	45/46	47	47/48	48/47	43/42	48/47
17	Last neurapophysis	54	54	55E	51	53	54	54	55	49	55
18	Last T-Proc	49	48	50	46	48	48	49	49	45	49
19	Fluke, transition vertebra	55	55	56	52	54	55	55	56	51	56
20	Chevron facets: Start	39	36	39E	35	37E	37E	38	38	35	37E
21	Stop	57	60	58	53	56	56	57	58	54	57
22	Chevrons: Total	18+	19+	14+	17+	17+	21+	17+	18+	21	23+
23	Start	—	0	—	—	—	—	38	—	35	—
24	Stop	—	0	—	—	—	—	—	—	56	—
	Transitional vertebrae										
25	Diaphragmatic vertebra	13	10E	12	13	11E	10	13	10	13	11E
26	Metapophyses: Stop	22	21	21	28	21	23	20	24	29	28
27	Start	37	35E	38	34	36	38	39	38E	35	37
28	First anticlinal vertebra	29	32	32	27	28	29	33	31	29	30
29	Second anticlinal vertebra	48	45	50	40	45	45	48	47	39	47
30	Longest T-Proc	24	23	22	23	23	22	23	23	23	22
31	Shortest T-Proc	8, 12	8, 11	8, 11	8, 11	10	11	8, 10	12	8, 11	8, 11
	Ribs										
32	Double-headed vertebral	5	5	5	5	5	5	5	6	4	5
33	Single-headed vertebral	9	9	8+	7+	8+	9	7	7+	8	7+
34	Sternal	9	8+	8	8	8	9	8	8+	7	8
35	Sternabrae: Total	4	3	4	3	3	4	3	4	3	4
36	Fused	1–3	1–3	1–3	0	1–3	1–3	1–3	1–3	0	1–3
37	Fused basihyal–thyrohyals	No	Yes	Yes	No	Yes	Yes	—	Yes	Yes	Yes

[a] E represents an equivocal observation, a dash, no observation. A slash (/) means the feature on the right is on a different vertebra from the feature on the left side of the specimen.

[b] #550375 has cervical vertebrae 4–5 and 6–7 ankylosed from an arthritic condition.

cranial arrangement (Miller, 1923, p.2). The process of telescoping has obscured relative positions and changed the size and shape of many bones. Romer (1966, p. 297) described the cetacean skull, "Osteologically there is no top to the skull; it is all front and back." The narial passages have a vertical orientation. This orientation eliminates the role of the **nasals** as roofing bones of the nasal passages. Cetacean **nasals** are small vestiges at the apex of the dorsal skull; they remain just posterodorsal to the external narial opening. The palate, formed by the **premaxillae**, **maxillae**, **vomer**, **palatines**, and **pterygoids**, is the roof of the mouth but is not the floor of the nasal passages.

Cranial Asymmetry

The typical odontocete skull is bilaterally asymmetrical. Odontocetes are the only mammals either living or extinct in which cranial asymmetry is the normal condition (Howell, 1930, p. 115). This is primarily a superficial phenomenon on the dorsal aspect of the skull. It is a contralateral difference in size and shape of the following bones: the **ethmoids**, the lateral margins of the **vomer**, the ventromedial margins of the **frontals**, the ascending processes of the **maxillae** and **premaxillae**, the **nasals**, and the **interparietal**. There is also a lateral bend (to the animal's right) in the caudal portion of the mesorostral groove (median groove) and in the anterior aspect of the **presphenoid**. Asymmetry is also seen in the arrangement of infraorbital (antorbital or preorbital) foramina of the **premaxillae** and the **maxillae**.

The tooth rows, **sphenoids** (except the **presphenoid**), **pterygoids**, **palatines**, and **jugals** appear to be contralaterally symmetrical, as do the foramina for the cranial nerves, the occipital condyles, mandibular fossae, paroccipital processes, and basioccipital crests. The ventral, contralateral variations that are observed are no more extensive than those found in other mammals.

The ankylosed sutures which lie roughly along the nuchal (supraoccipital) and temporal crests are quite irregular and asymmetrical, even though the crests themselves are fairly regular. The sutures almost meander; this may be an artifact of the way the thin margins of adjacent bones overlap to form squamous joints (Hildebrand, 1982, p. 405; Jollie, 1973, p. 41). This contrasts with the loose, open sutures between the **frontals**, the ascending processes of the **maxillae**, and the **lacrimals**.

Rostrum

The rostral tip is relatively pointed in young specimens of *T. truncatus* (Fig. 1); older specimens have blunter, more robust tips. The **premaxillae**

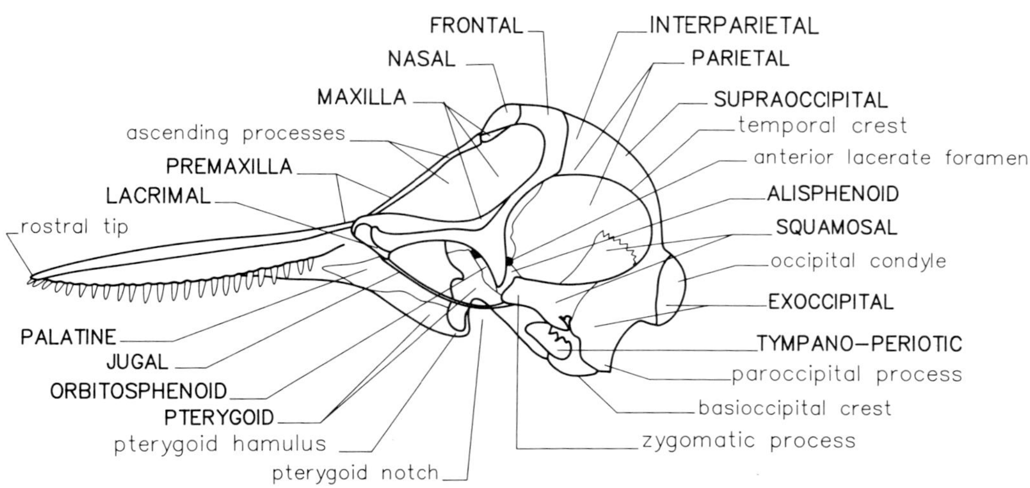

Figure 1 Left lateral view of the skull of the bottlenose dolphin. In this and subsequent figures, bone names are in boldface upper case; other features are in lower case.

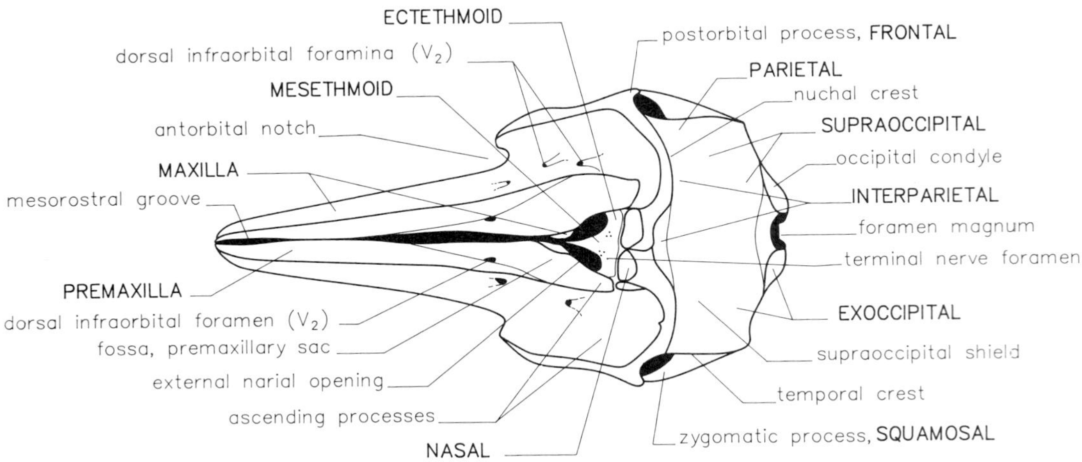

Figure 2 Dorsal view of the skull of the bottlenose dolphin.

extend from the rostral tip to the base of the **nasals** at the dorsal apex of the skull. They are bordered laterally by the **maxillae** and medially by the mesorostral groove, **vomer**, and small lengths of **maxillae** near the narial opening. Each **premaxilla** may bear up to three teeth (line 9, Table 1).

The **maxillae** extend from near the anterior tip of the **premaxillae** posterior to the temporal crest and the lateral margins of the nuchal crest. The ascending processes spread laterally over most of the **frontals**. Each **maxilla** bears 20–25 teeth (lines 5 and 6, Table 1), the roots of which may extend into the **premaxilla**. There are two shallow fossae on the dorsal surfaces of the **premaxillae** for the premaxillary sacs (Mead, 1975, p. 59); the right fossa is substantially larger than the left (Fig. 2).

In most other mammals the infraorbital (preorbital or antorbital) foramina (n.V) are single, horizontal openings in the **maxillae**; in odontocetes these foramina are vertical (*supra*orbital) complexes in the **maxillae** and **premaxillae** (Figs. 2 and 3).

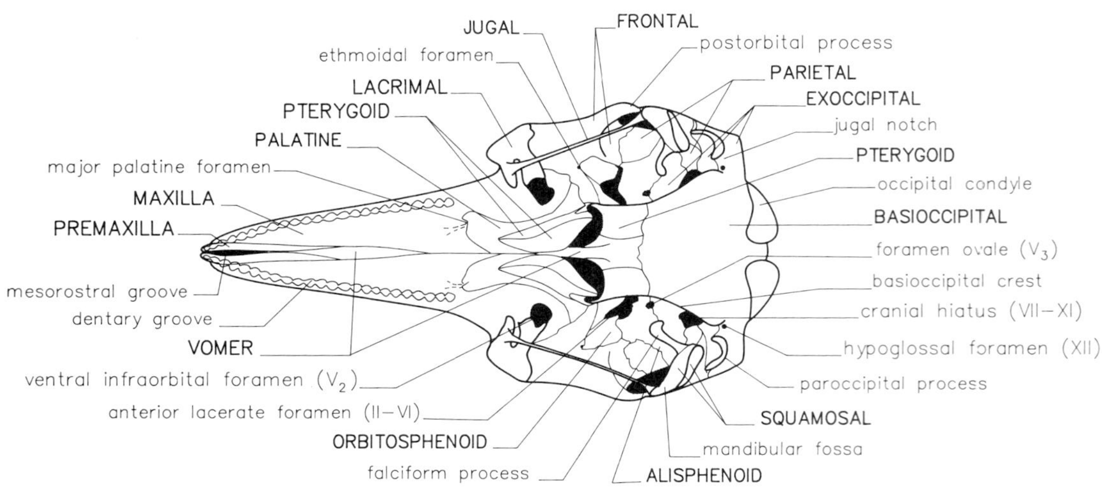

Figure 3 Ventral view of the skull of the bottlenose dolphin.

The mesorostral cartilage (mesethmoid cartilage, Flower, 1885, p. 214) is the nonbony extension of the **presphenoid** (note that Schulte and Smith, 1917, p. 384, considered this to be the **mesethmoid**) that fills the mesorostral groove; it is bordered by the **vomer**, **maxillae**, and **premaxillae**. USNM # 504288, a very young specimen, has a **presphenoid** that fits loosely in the mesorostral groove of the **vomer**; there is no indication of a separate **mesethmoid** in this specimen. Neonates have secondary ossifications on the **vomer** that appear to be the precursors to the **ethmoids**.

Nasal Passages and Nasal Bones

The dorsal circumferences of the nasal passages (anterior, dorsal, or external narial openings) are lined by the **ethmoids**, **frontals**, **premaxillae**, and **maxillae**. The ventral circumferences (posterior, ventral, or internal narial openings) are bounded by the **vomer**, **pterygoids**, and small portions of the **palatines**. The **presphenoid**, **ethmoids**, and **vomer** form the bony septum between the nasal passages.

The **nasals** are very asymmetrical within individuals and show considerable variation in size and shape between individuals. They are located in distinct fossae surrounded by variable amounts of **frontals** posteriorly and **ethmoids** laterally and anteroventrally. The **nasals** are loose in young specimens but become ankylosed to the surrounding bones in older individuals.

The presence of a separate **mesethmoid** and unique **ectethmoids** is debatable. Flower (1885, p. 209) considered the **mesethmoid** to be present. In other mammals the **mesethmoid** is a secondary ossification and appears late in development (Jollie, 1973, p.40). Kellogg (1928, p. 202) argued well for the presence of a **mesethmoid** and **ectethmoids**; however, some of the more recent (since Kellogg's tenure at the USNM) specimens in the USNM collection could be used to argue equally convincingly that these ossifications may be medial growths of the **vomer** and **frontals**. The **ethmoids** help close off the embryonic fontanelle between the **frontals** and above the lateral wings of the **vomer**. Some neonatal specimens in the USNM offer evidence in support of the **mesethmoid** and **ectethmoids** as separate ossifications. These ossifications first appear as secondary growths on the **vomer** and then later as secondary ossifications on the **frontals**; they do not ankylose to the **presphenoid** until much later.

On either side of the **presphenoid**, the lateral wings of the **vomer** form part of the anteroventral braincase (Figs. 4 and 5). Schulte and Smith (1917, p. 382) pointed out that in *Kogia simus* the **maxillae** and the right **premaxilla** form a large part of the anterior wall of the braincase; this is also the arrangement in *K. breviceps*. The **vomer**, **maxillae**, and **premaxillae** are normally not found in the walls of braincases of other mammals, and these exceptions may be unique to some of the odontocetes.

Orbit, Tympanic Bulla, and Periotic Region

The orbit is bounded dorsally by the **frontal** and **lacrimal**, caudally by the postorbital process of the **squamosal**, medially by the **maxilla**, **palatine**, **pterygoid**, **alisphenoid**, and **orbitosphenoid**, and ventrolaterally by the **jugal**.

The **jugal** (**malar**, **zygomatic**) extends from the **lacrimal** (with no apparent suture, even in neonates) and loosely articulates with the zygomatic process of the **squamosal**; in some older specimens it may be firmly ankylosed to the **squamosal**. The zygomatic arch consists entirely of the **jugal**; it is very slender and is often broken during preparation. The **lacrimal** varies considerably (within and between individuals) in the extent to which it covers the adjacent **frontal** and **maxilla**.

The posterior processes of the **tympanic** and **periotic** (Flower, 1885, p. 214; Kasuya, 1973, p. 2; Kellogg, 1928, p. 203; Oelschlager, 1986a, p. 158; Yamada, 1953, pp. 19, 22) extend together as a tonguelike process that attaches to the "mastoid process" of the **squamosal** via a fibrous joint (Kellogg, 1928, p. 206). In some specimens this joint has a thin synostosis or bony connection.

Fraser and Purves (1960, pp. 77–80) described three "mastoid processes" on three separate bones: the mastoid process of the **periotic**, the mastoid process of the **tympanic**, and the mastoid process of the **squamosal**. It would be less confusing to use posterior process in place of mastoid process for the **tympanic** and **periotic**.

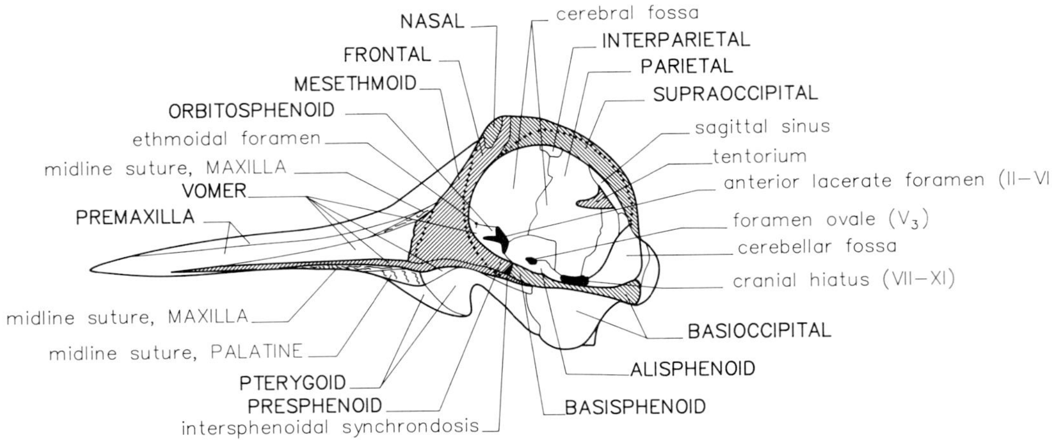

Figure 4 Left lateral view of the midsagitally sectioned skull of the bottlenose dolphin. The outlines of the dorsal braincase and the narial passages are dotted. Some of the sutures between adjacent bones are obliterated with age. Hatching distinguishes the cut surfaces of each bone or bone group.

Oelschlager (1986b, p. 361) described (in *Lagenorhynchus acutus)* a distinct suture between the **squamosal** and its mastoid process (represented schematically by Fraser and Purves, 1960, p.78). I have been unable to verify this unequivocally with either immature or adult specimens of *T. truncatus* in the collection of the USNM. I have difficulty with some of the sutures described by Oelschlager; they could be partially enclosed grooves for the sinuses or vessels of the adjacent tissues. Provisionally, I would suggest that the region which is termed **mastoid** be only that which has direct (synostotic only?) contact of the posterior process of the **periotic** with the **squamosal.**

The sigmoid process of the **tympanic** is aligned with the deep, ventrolateral groove (tympano-squamosal recess) between the **squamosal** and **exoccipital**. This groove parallels the proximal portion of the external auditory meatus.

The distal tip of the falciform process of the

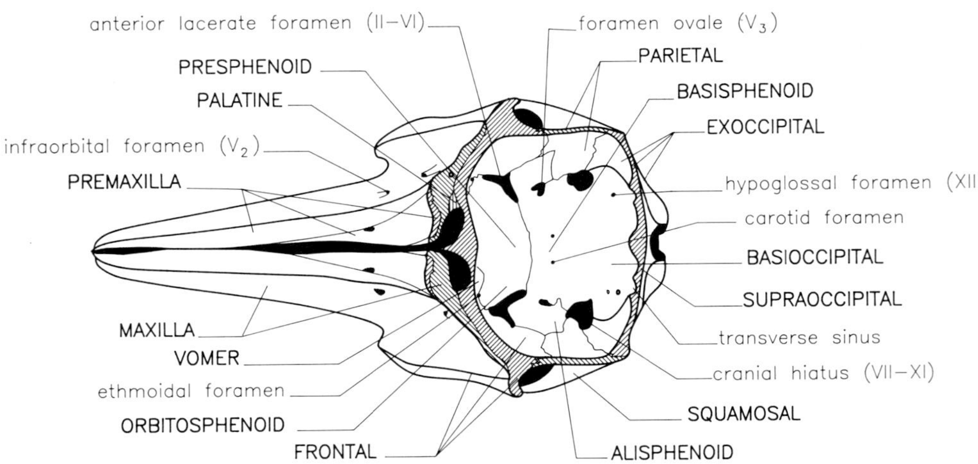

Figure 5 Dorsal view of the horizontally sectioned skull of the bottlenose dolphin. Some of the sutures between adjacent bones are obliterated with age. Hatching distinguishes the cut surfaces of each bone or bone group.

squamosal points caudally and inward toward the anterior margin of the tympanic bulla; in the intact head the tip of the falciform process is attached to the anterolateral margin of the bulla. In neonates the **periotic** lies snugly against the falciform process of the **squamosal**. The falciform process supports the soft tissue of the posterolateral wall of the pterygoid sinus.

Occipital Region

The cranial vault is roofed by the **frontals, parietals, interparietal,** and **supraoccipital** (Fig. 4). I use **interparietal** instead of **postparietal** because the **parietals** are placed laterally to this bone in the bottlenose dolphin; in many other mammals **postparietal** is more appropriate (Jollie, 1973, p. 40) because the **parietals** abut on the dorsal midline in front of the **postparietal**. Other bones of the cranial vault include the following: the **vomer, frontals,** and **ethmoids** anteriorly; the **orbitosphenoids, alisphenoids,** and **exoccipitals** ventrolaterally; the **presphenoid, basisphenoid,** and **basioccipital** ventromedially (Figs. 4 and 5). The **ethmoids** develop late; they fill in the large frontal fontanelle (Kellogg, 1928, p. 201; Eales, 1950, p. 21) dorsal to the **presphenoid** and **vomer** and medial to the **frontals**.

The occiput has three pairs of curved, crestlike regions projecting from its ventral surface. The medial basioccipital crests extend from the pterygoid notch to the bullae, forming the medial margins of the pterygoid fossae. More lateral and caudal are the crests of the paroccipital (exoccipital) processes that border the posterior and peribullar sinuses; they form the region of hyoid attachment. The most lateral are the crestlike mandibular (glenoid) fossae in which the condyles of the **dentaries** articulate.

Anterior and medial to these crests are the **pterygoids,** which serve to anchor the large pterygoid muscles (Seagars, 1982, p. 51). They are also the anterior extremities of the large pterygoid sinuses. Each pterygoid begins near the caudal extremity of the maxillary tooth row. The lateral and medial lamellae of the **pterygoid** are quite variable in size and shape. The posterior margin of the **pterygoid** forms the deep pterygoid notch which borders the posterior narial opening.

The **palatines** also vary in size and shape. On rare occasions each of the **palatine** bones is made up of two distinct and separate ossifications (USNM # 504218, # 504527, and # 504311). The lateral lamellae of the **palatines** may extend well into the anteromedial aspects of the orbits. The caudal-most extension of the **vomer** covers the suture between the **presphenoid** and **basisphenoid** (intersphenoidal synchrondosis; Fig. 4).

There is substantial variation in the extent and orientation of the portion of the **squamosal** that overlaps the **parietal** (in or near the temporal fossa). The zygomatic process of the **squamosal** is robust; it supports the mandibular fossa and the posterior extremity of the **jugal**.

Foramina of the Skull

In most specimens there is an unnamed, relatively large fenestration in the very thin wall between the orbit and the nasal passage. It is situated at the junction of the **pterygoid** and **palatine** along the margin of the **vomer** (Fig. 1). It perforates the posterolateral aspect of the bony nasal passage approximately midway between the external and internal narial openings. Boenninghaus (1904, p. 243) illustrated a similar opening in *Phocoena* that he described at the junction of the **pterygoid, palatine,** and **presphenoid** (this region of the **presphenoid** is easily confused with the **vomer** or **frontal;** my examinations of *Phocoena* specimens place this foramen in a position similar to that in *Tursiops*). At the anterior margins of the **palatines** are the major palatine foramina (Fig. 3) (Evans and deLahunta, 1980, p. 231; posterior palatine canal, Kernan, 1918, p. 367).

There is a small arterial foramen that enters the braincase at the common confluence of the sutures between the **frontal, orbitosphenoid,** and **vomer** (Figs. 4 and 5). The exit from the skull (Fig. 3) is dorsal to the optic groove; the location is similar to that of the ethmoidal foramen of the dog (Evans and deLahunta, 1980, p. 238; labeled canalis n. optici by Hosokawa and Kamiya, 1965, p. 128).

In very young individuals there is a large frontal fontanelle in the anterior wall of the braincase; the posterior braincase wall is perforated by the foramen magnum. There are three openings on each side of the ventral skull for the other cranial nerves: the large anterior lacerate foramen and cranial hiatus, plus the small hypoglossal (condylar) foramen. As the animal ages the two larger openings are subdivided into two or more foramina each.

The relatively large, irregular fissures for the cranial nerves are prominent, but extremely variable, features of the skull. There are 2–11 small nasal or cribriform foramina for the terminal nerve (Ridgway *et al.*, 1987, p. 208), some of which remain patent after the frontal fontanelle is closed off by the **ethmoids** and **frontals** (Fig. 2). These may be analogous to the foramina for the nasal nerves described by Schulte and Smith (1917, p. 378). This description contradicts Kellogg (1928, pp. 201–202), who implied that all the foramina are closed off with the formation of the caudal walls of the nasal passages. These foramina are located near the posterior aspects of the external narial openings; some of them are continuous with the more ventral openings on the anterior aspect of the anterior wall of the braincase. Telescoping has oriented them dorsoventrally; they are most easily seen by viewing a light directed down the dorsal narial openings while looking into the braincase through the foramen magnum of well-cleaned specimens.

The optic nerve (n. II) exits the braincase (accompanied by the ophthalmic artery and ophthalmic rete; McFarland *et al.*, 1979, pp. 3, 10) through the optic foramen and continues anteriorly along a distinct groove in the **orbitosphenoid** (and occasionally the **frontal**) bone. This optic groove, which forms early in development, may serve to locate other foramina; in the intact head its margins anchor the eye stalk. Nerves III (oculomotor), IV (trochlear), and VI (abducens) exit the braincase (caudal to the optic nerve) through the anterior lacerate foramen; only a vestige of n. V_1 (ophthalmic branch of the trigeminal) remains (Langworthy, 1931, p. 385). The optic foramen and the anterior lacerate foramen are often confluent, particularly in younger specimens. The foramen rotundum (located on the anterior aspect of the **alisphenoid**) for the maxillary branch (n. V_2) of the trigeminal may also be confluent with the anterior lacerate foramen. The foramen ovale for the mandibular branch (n. V_3) of the trigeminal develops from a notch in the posterior margin of the **alisphenoid**; it is often a well-formed opening with heavy, rounded margins in older individuals. It is confluent with the cranial hiatus in some specimens.

The infraorbital branch (n. infraorbitalis trigemini; Huber, 1934, p. 120) of the trigeminal nerve (a subdivision of the mandibular branch, n. V_2) reenters the skull via the (ventral) infraorbital foramen; it branches within the **maxilla** and reexits the skull through the (dorsal) infraorbital foramina of the **premaxilla** and the **maxilla**. The infraorbital nerve is accompanied by an arterial complex that branches from the internal maxillary artery and parallels the nerve on the dorsal aspect of the rostrum. The large (ventral) infraorbital foramen (really a *supra*orbital complex of foramina) is located anteromedial in the orbit and is bordered by the **maxilla** and (occasionally) the **lacrimal**; this complex is confluent with the maxillary sinus (Hosokawa and Kamiya, 1965, p. 107).

There is an additional dorsal exit from this infraorbital complex. The channels in the bones of disarticulated specimens and the angiograms of Boulay and Verity (1973, pp. 140–141) indicate that this is an arterial foramen, though it is probably also for a branch of the infraorbital nerve. The ventral extremity of the channel leading to this foramen is often continuous with the channel leading to the major palatine foramen (posterior palatine channel; Kernan, 1918, p. 367) at the anterior margin of the **palatine** where it joins the **maxilla** (Fig. 3). This foramen is found in or on the dorsal margin of the **palatine** (occasionally the **maxilla**) where it joins the **frontal–ethmoid** and the **maxilla** on the lateral aspect of the internal bony nares.

The cranial hiatus is located at the base of the skull lateral to the basioccipital crest, medial to the mandibular fossa, and anterior to the paroccipital process. It is an irregular opening for the auditory nerve (n. VIII) and the internal portion of the facial nerve (n. VII) which exit the braincase via the fundus of the internal auditory meatus of the **periotic**. Younger individuals tend to have a very large, open fissure in this region so that the dorsomedial portion of the **periotic** supports part of the floor of the braincase. The bony processes from the surrounding **basioccipital**, **exoccipital**, **alisphenoid**, and **parietal** may substantially reduce this opening in older individuals; this reduction in the size of the hiatus may result in 3–6 distinct openings for the nerves and blood vessels.

The facial nerve (n. VII) exits the **periotic** along

the facial canal at the caudal extremity of the **periotic** near the junction of the **periotic** and the **tympanic**. This is dorsal to the site of the **tympanohyal** described by Oelschlager (1986a, p. 159). The facial nerve passes ventral and lateral to the external auditory meatus and progresses toward the rostrum medial to the eye and lateral to the mandible. It parallels the anterior portion of the zygomatic arch. It is ventromedial to the ventrolateral aspect of the eyeball; anteriorly it loops up and around the supraorbital bones in the antorbital notch to provide sensory innervation to the face.

The glossopharyngeal (n. IX), vagus (n. X), and accessory (n.XI) exit the skull along the posterior margin of the cranial hiatus. These three nerves may exit the braincase through a distinct posterior lacerate foramen in specimens in which the cranial hiatus is closed in to form two (or more) separate openings.

The hypoglossal foramina (condylar foramina) for the hypoglossal nerves (n. XII) are at or near the apex of the deep notch (jugal notch or paroccipital notch) between the basioccipital crest and the paroccipital process (Fig. 3). There may be one or two foramina on the medial surface of the braincase (Fig. 4) but only a single exit in each jugal notch.

The internal carotid foramen is located on the floor of the braincase, lateral to the pituitary fossa (Fig. 5). The external carotid foramen is not visible in any of the views given; it is located just caudal to the **basioccipital–pterygoid** suture, lateral to the basioccipital crest, and medial to the foramen ovale.

There is some disparity between the labels in Figs. 3–5 and those for the same bones and foramina in Fraser and Purves (1960, p. 25 and plates 43 and 44), Hosokawa and Kamiya (1965, p. 128), and McFarland *et al.* (1979, p. 73). My opinion is based on my own dissections of the head and skull, the descriptions of the cranial nerves given by Breathnach (1960, p. 192) and Langworthy (1931, p. 384), and photographs of the brain in McFarland *et al.* (1979, p. 65) and Morgane *et al.* (1980, pp. 29, 65). I have traced the cranial nerves (except the terminal nerve and the vestige of n. V_1) from the brain to where they exit the skull. I was unable to trace n. VI beyond its confluence with n. V near the anterior lacerate foramen. I have dissected nn. V_2, V_3, and VII along most of their lengths to determine the relationships described above; in general my dissections of these nerves agree with Fig. 14 of Hosokawa and Kamiya (1965, p. 132).

Intersphenoidal Synchrondosis

The space between the **presphenoid** and the **basisphenoid** located at the ventromedial aspect of the skull (Fig. 4) remains partially filled with cartilage, even in old specimens. This region is covered on its ventromedial aspect with a thin layer of the **vomer** which is often fenestrated to expose the synchrondosis; laterally it is covered by thin lamellae of the **pterygoids**.

In young individuals there is a loose suture between the thin layer of **vomer** and the adjacent **basioccipital–presphenoid**; in older specimens the **vomer** becomes ankylosed to the adjacent **pterygoids** and/or the **basioccipital–presphenoid**. This correlates with ankylosis of the vertebral epiphyses in the midthorax of the specimens listed in Table 1 and may be useful for age estimation (see Mead and Potter, Chapter 9, this volume).

Mandible

The **dentary** bones of the mandible are essentially hollow beams with no vertical processes. There is a distinct coronoid ridge and a slightly rugose coronoid process to which the temporal muscle attaches.

The tooth-bearing portion of each dentary is hollow and filled with a fatty material similar to that of the melon. The non-tooth-bearing portion of the **dentary** (the vertical "ramus" in other mammals) has very thin walls that form the fat-filled "pan bone" (a common term derived from whalers, but not a unique ossification center).

The **dentary** is open medially, posterior to the tooth row (mandibular foramen, inferior dental canal, or mandibular hiatus) through which enter the mandibular nerve and vessels. Each **dentary** has 18–24 teeth (lines 7 and 8, Table 1). The tip of the articulated mandible protrudes beyond the tip of the rostrum. Note that when the mouth is closed the coronoid crest and coronoid process are both medial and *above* the midpoint of the orbit and the eyeball. In some terrestrial mammals the coronoid process fills much of the temporal fossa; in odontocetes the coronoid process is located in the orbit.

There are 2–6 mental foramina that transmit the mandibular branch of the trigeminal (n. V_3) plus the mandibular arteries and veins. These foramina are irregularly positioned in a row along the anterolateral aspect of the mandible beneath the tooth row.

There is a relatively short (for odontocetes) mandibular symphysis (10–15% of the **dentary** length). In some other odontocetes (e.g., *Mesoplodon* spp.) the size of this symphysis is used as a taxonomic feature.

The imaginary lines along the distal tips of the tooth rows are curved in the dorsoventral direction, a bit more so than the dental canals because the longest teeth are in the middle of the tooth row. Each **dentary**, when viewed dorsoventrally, is straight from tip to condyle, except for the protuberance of the condyle itself. The tooth rows of both jaws are also quite straight in that plane.

Dentition

The teeth are simple and peglike with single, open roots. They have a smooth enamel surface on their crowns; they are all homodont and nondeciduous. It is a mammalian characteristic to have a heterodont dentition (Reynolds, 1897, p. 7; Hyman, 1942, p. 181), so the homodont condition of odontocetes is exceptional. It probably is an adaptation to their type of feeding: prey are grasped and bolted without mastication. In older individuals many of the teeth are worn down; some teeth exhibit distinct wear patterns, caused by the abrasion of teeth against each other. W. A. McLellan (pers. commun.) pointed out to me that Montagu's original description (Montagu, 1821, p. 78) of the species *T. truncatus* refers to the truncated or worn-down teeth.

Teeth are loosely socketed in irregularly shaped alveoli. It may be better to use the term alveolar trough or dentary groove (Kellogg, 1928, p. 60)—dental canal was used by Howell (1927, p. 9); in young animals, groove or trough is more appropriate because the alveoli are poorly defined. Older individuals have more distinct alveoli. There may be up to three teeth in each of the premaxillae. There is neither a mandibular nor a maxillary diastema, though there is often a small gap at the anterior end of the mandibular symphysis. The open roots are gradually filled by layers of dentine as the animal ages (see Hohn, Chapter 33, this volume); distinct layers of cementum also form on the circumference of the roots.

Hyoid Apparatus

Flower (1885, pp. 211–217) used the hyoid names that are conventional today. He justified them on the basis of the synovial capsules found in the cartilage between the dorsal hyoids and the ventral hyoids. My own dissections of this region show synovial joints at both ends of the cartilaginous rod between the stylohyal and the basihyal.

The dorsal hyoids are the **stylohyals**. They articulate, by cartilage, to the paroccipital processes (but see Oelschlager, 1986a, p. 159). The ventral hyoids are the **basihyal** (hyoid body) plus the **thyrohyals**. In older individuals the ventral hyoids become ankylosed (line 37, Table 1). The caudal horns (cornua) of the thyrohyals articulate with the thyroid cartilages of the laryngeal complex.

Articulation of Skull with Spine

The occipital condyles of the **exoccipitals** are robust, hemispherical projections. They are covered with articular cartilage and fit into the large but shallow fossa of the atlas. This single hemispherical joint allows the skull three degrees (axes) of mechanical freedom; in most other mammals the same degrees of freedom require both the occipital–atlas and the atlas–axis joints. Loss of the atlas–axis joint with cervical fusion requires the spherical surface to accommodate all the degrees of freedom normally associated with both the occipital–atlas joint and the axis–atlas joint. There are large transverse processes and a large neural spine on the atlas–axis complex; in most other mammals there is almost no neural spine on the **atlas** and almost no transverse processes on the **axis**.

POSTCRANIAL SKELETON

Axial Skeleton

Skeletal features in each section of the spine are indicated by Arabic numerals to denote postcranial vertebra number or by position within a region (e.g., number 12 is postcranial vertebra 12 and is the fifth thoracic or **T5**). The definitions of lumbar, sacral, and caudal are equivocal in Cetacea because many of the vertebral landmarks, so distinct in the spines of other mammals, are variable or lacking (Hepburn and Waterson, 1901, p. 680). De Smet (1977, p. 64) suggested changing the terminology, but many of his arguments are not convincing. Vertebral formulae for twelve specimens are given in Table 2.

Table 2
Vertebral Formulae for the Bottlenose Dolphin[a]

USNM#	Cervical	Thoracic	Lumbar	Caudal	Total
22304	7	14	19	24+	64+
396165	7	12E	18	28	65
500863	7	14	19E	23+	63+
504326	7	12E	17	27+	60+
504618	7	13	18	24+	62+
504726	7	14	17E	24+	62+
504766	7	14	18	26	65
504906	7	13E	19	24+	63+
550109[b]	7	12	17	25	60
550375	7	14	17	27+	65+
Flower[c]	7	13	16	27	65
Slijper[d]	7	13	16	26	62

[a] USNM numbers refer to specimens in Table 1. These values are indicative, not definitive; definition of these concepts for any species should be determined from dissection plus radiographs of the anterior-most chevrons. A plus sign after an entry implies that there may be one additional vertebra; E, equivocal observation.
[b] For #550109, numbers of lumbars and caudals are based on radiographs of specimen, not data sheet from dissection.
[c] Flower (1885).
[d] Slijper's (1946) Table 3. His value for number of caudals should be changed by one to accommodate the difference in definition of first caudal.

Vertebral Processes

The mammalian vertebra has a body, several processes, and articular facets (Fig. 6). The body is the centrum (a diaphysis) plus terminal end plates or epiphyses. In the bottlenose dolphin, as in most other mammals, the vertebral bodies vary in length and diameter in different regions of the spine (Fig. 7). A unique feature of the Cetacea is the superficial location of the neurocentral sutures (between the base of the neural arch and the vertebral centrum); in all other mammals this suture lies within the centrum (Flower, 1885, p. 61). This may be observed only in very young cetaceans (fetuses or neonates) that have disarticulated (or very recently ankylosed) pedicles or laminae.

TRANSVERSE PROCESSES

All the lateral processes could be termed transverse; however, transverse process is usually reserved for the most prominent paired lateral process of each vertebra (Howell, 1927, p. 13). When two transverse processes are present on the same vertebra, the superior one is called the dorsal lamina (this term is easily confused with the lamina of the neural arch) and the ventral one is called the ventral lamina (this is contrasted in **C4** and **T11**, Fig. 6). The dorsal lamina is a diapophysis and is sometimes associated with the articulation of the tuberculum of a rib (**T2**, Fig. 6). The ventral lamina is a parapophysis and is sometimes associated with the articulation of the capitulum of a rib. In some works diapophysis refers *only* to the articular facet (costal facet) for the tubercle of the rib and parapophysis refers *only* to the articular facets (on the anterior and posterior centra and epiphyses) for the capitulum of the rib.

In the bottlenose dolphin the most prominent transverse processes appear to be the diapophyses. The positions of the transverse process roots or bases vary in different regions of the spine. Their positions shift from a relatively distal location on the neural arches of the neck and anterior thorax to a more proximal location on the centrum in the posterior regions of the spine.

The orientations of the transverse processes change in the region of the dorsal fin. In the thorax they are horizontal, their roots emerge dorsal to the centra, and their tips are behind the centra.

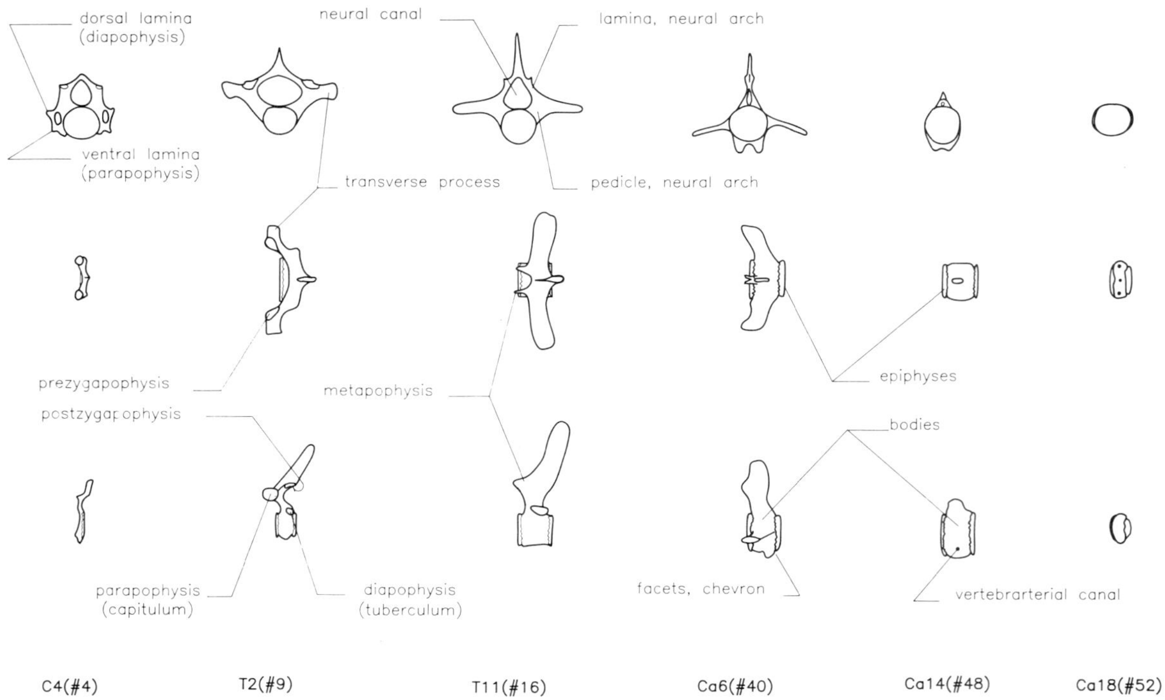

Figure 6 Anterior, dorsal, and left lateral views of selected vertebrae of the bottlenose dolphin. Vertebral positions are indicated at the bottom with the postcranial vertebra number in parentheses. Vertebrae are positioned with the bottoms of the neural canals in line and/or the center line of the anterior view aligned with the surface of the anterior epiphysis.

More caudally in the spine, the roots emerge on the centra (eliminating the pedicles), some of the processes have a negative dihedral, and some of the tips are actually in front of the vertebral centra. The transverse processes also vary in length; the shortest are in the neck and anterior thorax (line 31, Table 1), and the longest are under the dorsal fin (line 30, Table 1).

METAPOPHYSES

Oblique processes that tend to be smaller and subtend different angles than the more prominent transverse processes are termed metapophyses. Metapophyses tend to point forward and outward and may be rather robust depending on the forces acting on them (**T11**, Fig. 6). They often have

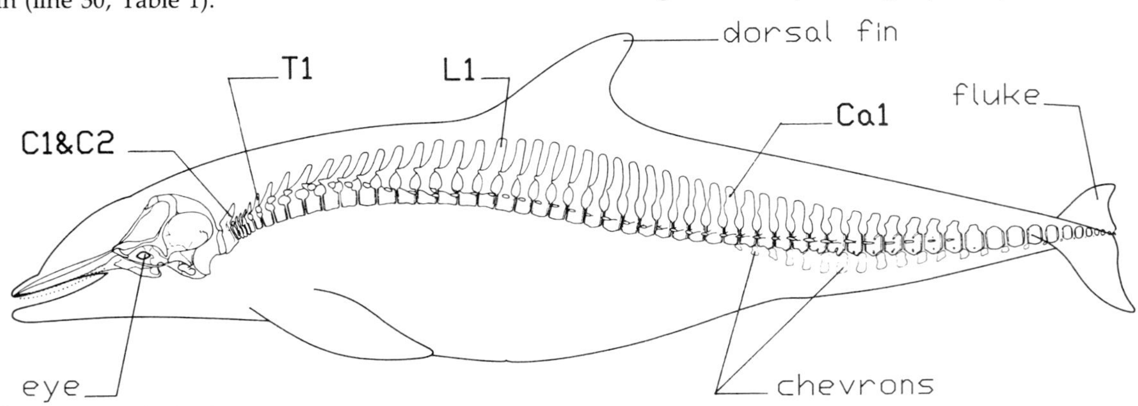

Figure 7 Vertebral column and skull in relation to the outline of the animal. Posture and outline are representative of a live bottlenose dolphin approximately 220 cm in length.

prezygapophyses on their dorsal or medial surfaces.

The first distinct metapophyses are found on the transverse processes of **T3–T5** (line 14, Table 1). They shift medially and dorsally so that they are on the distal portion of the neural arch in the lumbar region and on the proximal neural spine in the caudal region. The metapophyses, as paired anteriorly directed projections, are noticeably reduced starting at about the level of the anterior insertion of the dorsal fin. They disappear (line 26, Table 1) at or before the posterior insertion of the dorsal fin, and they reappear (line 27, Table 1) at the level of the first chevrons and the pelvic vestige. This reflects some of the changes in the dorsal musculature in this region (see Pabst, Chapter 3, this volume).

ZYGAPOPHYSES

The zygapophyses are often referred to as oblique processes, but technically they are articular surfaces (facets) rather than processes (Flower, 1885, p. 21). The "zygapophysial surfaces" have synovial joints and extend only to **T6–T7**; the rest of the metapophyses are only attachment points for tendons. The anterior or prezygapophyses face dorsally in the anterior spine and medially in the more caudal spine. The posterior or postzygapophyses point ventrally in the anterior regions and laterally in the more caudal regions.

There is usually a single transitional vertebra that has relatively horizontal prezygapophyses and oblique or relatively vertical postzygapophyses; it is the diaphragmatic vertebra (Slijper, 1961, p. 79) since it is found near the dorsal attachment of the diaphragm in some species. The diaphragmatic vertebra in the ten specimens examined was located at **T3–T6** (line 25, Table 1). Slijper (1946, his Tables 3 and 5) gave its position in *T. truncatus* as **T5**. The definition should probably include synovial joints between adjacent zygapophyses, at least in the odontocetes; this may be a more accurate definition than the orientation of articular facets.

The prezygapophyses of the cervical vertebrae are located on the dorsal aspects of the neural arch laminae or transverse processes (**C4**, Fig. 6). They shift to the medial surfaces of the metapophyses around **T5** and disappear about **T7** (**T11**, Fig. 6). The only distinct postzygapophyses are in the neck and anterior thorax (**T2**, Fig. 6). Posterior to this, there are small, modified regions on the posterior aspects of the neural spines that may accommodate the movement of the tendons attached to the corresponding metapophyses.

Neurapophyses

The neurapophyses are the dorsal projections of the vertebra; they include the neural arches, neural spines plus the associated accessory processes, and articular facets. The neural arch, which surrounds the neural canal, is formed by two (relatively vertical) pedicles which project dorsally from the centrum and which support the transverse processes. The pedicles (when present; **T2** and **T11**, Fig. 6) attach to the neural spine via the (relatively horizontal) dorsal laminae (not to be confused with the laminae of the transverse processes). In regions where the transverse processes arise from the body of the vertebra the entire neural arch is made up of the two (relatively vertical) laminae (**Ca6**, Fig. 6).

The pedicles (or laminae) are notched in front and behind, with the posterior notch usually deeper. These notches border the intervertebral foramina that allow the spinal nerves to exit the spinal cord. The first spinal nerve (suboccipital nerve; Cunningham, 1877, p. 214) produces a deep notch in the anterior face of the atlas. When a bony bridge arches over this and forms a distinct *vertebral* foramen, it is sometimes mistaken for an *intervertebral* foramen and an additional vertebra may be counted. The neural spines (spinous processes) are the midsagittal processes dorsal to the neural arch. Occasionally the tip of a neural spine is bifid. The shape and texture of the tips of the neural spines change with age; those of older individuals tend to be less smoothly rounded and more roughly splayed with sharper corners. The longest neural spines are located at or near the dorsal fin. The neural spines have different orientations in different sections of the vertebral column (**T2** and **Ca6**, Fig. 6). Transitional neural spines are anticlinal, that is, they have an intermediate inclination (Gambaryan, 1974, p. 121). There are two anticlinal neural spines: the first near the posterior insertion of the dorsal fin (line 28, Table

1), the second near the longest chevron in the caudal region (line 29, Table 1; Fig. 7). The neural spines carry metapophyses from the anterior thorax posteriorly. In contrast to young specimens, older specimens have neural spines that are inclined more strongly toward the dorsal fin, and the metapophyses are more robust.

Vertebral Epiphyses

The vertebral epiphyses are bony plates on the anterior and posterior faces of the centra of each vertebra. They ankylose to the centra as the animal matures. The midthoracics are chronologically the last to ankylose. As in other mammalian backbones, the fusions progress posteriorly from the atlas and anteriorly from the tip of the tail. These epiphyseal sutures close circumferentially in a rather irregular manner. The anterior and posterior epiphysial sutures disappear from the same vertebra at slightly different ages. In the thorax the anterior epiphysis fuses one or two vertebrae in front of the posterior epiphysis. In the posterior spine the posterior epiphysis may be fused as many as six vertebrae in front of the anterior-most anterior epiphysis.

The intervertebral faces of adjacent epiphyses are parallel in a neutrally postured animal (Fig. 7); relaxed or static spinal curvature is achieved by variation of the angles between the ends of each vertebral centrum. The range of dynamic postures, achieved with muscle activity, is determined by intervertebral size and shape and the mechanical limits of the vertebral processes and articular facets (Slijper, 1961, p. 86).

The lengths of the intervertebral disks are quite different in each region of the spine. They are short to moderate in the neck (see radiographs in McFarland *et al.*, 1979, pp. 42–43), and relatively shorter in the anterior thorax where there are double-headed ribs. The *relative* intervertebral disk size then increases gradually to where, in the flukes, the disks are longer than their adjacent vertebrae.

Vertebral epiphyses vary in thickness in different portions of the spine as do the intervertebral disks. There is also a difference in thickness between epiphyses of the same vertebra, the caudal epiphysis often being substantially thicker than the cranial epiphysis. Epiphyses which are completely disarticulated demonstrate that each epiphyseal plate has a particular pattern which accommodates a unique fit of each epiphysis to its own diaphysis.

Axial skeletal maturity is determined by the closure of all the vertebral epiphyseal sutures; this is the termination of longitudinal growth of the individual. Age and appendicular skeletal maturity may be more conveniently determined by the sequences of epiphyseal fusions of proximal bones in the flippers (see Mead and Potter, Chapter 9, this volume).

Cervical Vertebrae

The cervical vertebrae are distinguished from the thoracics by not having ribs that attach to the sternum. Occasionally, there are short but distinct, movable cervical ribs on the cervical vertebrae. The bottlenose dolphin has seven cervical vertebrae (Table 2). The first two cervicals, the **atlas** (**C1**) and **axis** (**C2**), are completely ankylosed to each other. The blunt vestige of the odontoid process projects from the open center of the **atlas**. There are often distinct sutures that indicate ankylosis of the two vertebrae. Occasionally **C3** is ankylosed to the first two; other vertebrae may also be serially fused, but this is not common in the bottlenose dolphin (line 11, Table 1). Older specimens may exhibit secondary (pathological) serial ankyloses (Table 1, #504726 and #550375).

The bases of the transverse processes of the cervical vertebrae (except the **atlas** and **axis**) may be perforated by transverse foramina (transverse canals or vertebrarterial canals) for the arteries and nerves of the neck. This produces two lamellae or roots (a diapophysis and a parapophysis; Fig. 6) of the transverse processes which are variable in size and shape; they are rather thin anteriorly and more robust posteriorly.

Thoracic Vertebrae

The first thoracic (**T1**) is that vertebra whose transverse processes (diapophyses) articulate with

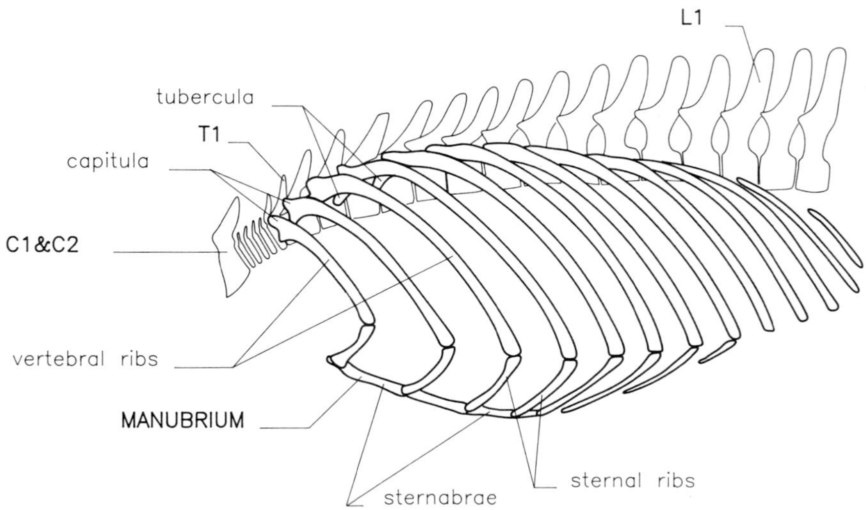

Figure 8 Thoracic region of the axial skeleton of the bottlenose dolphin.

the tubercula of the first ribs (Fig. 8). These ribs articulate, via the first sternal ribs, with the **manubrium**. The capitulum of the first rib articulates with the parapophysis of either the sixth or seventh cervical.

The Cetacea are unique among mammals in having the single-headed ribs attach to the transverse processes instead of the centra. This means that the capitula rather than the tubercula are reduced as the ribs change from double- to single-headed.

The bottlenose dolphin has 12–14 thoracic vertebrae (Table 2). The ventral aspects of the thoracic vertebral centra are concave, in some cases substantially so. The tips of the thoracic transverse processes are modified to accommodate synovial joints (costal facets) with the tubercula of the ribs (line 12, Table 1). The distal tips of the last two to three thoracic transverse processes do not have substantial modifications (e.g., enlarged tips; line 13, Table 1) that help indicate the presence of ribs. The posterior-most ribs are distantly and ligamentously attached to the tips of the last thoracic transverse processes. Therefore, the definition of the posterior limit of the thoracic region, in the absence of the ribs, may be equivocal.

Lumbar Vertebrae

The first lumbar vertebra (**L1**) is the first vertebra posterior to those bearing vertebral ribs. There are 16–18 lumbar vertebrae (Table 2). The lumbar region is often erroneously labeled because the terminal ribs are missing or the location of the first chevron is in error. The lumbar vertebrae start at about the level of the insertion of the dorsal fin. The metapophyses are reduced in the dorsal fin region of the lumbar spine (lines 26 and 27, Table 1).

The ventral aspects of the lumbar vertebral centra develop convex surfaces in contrast to the concave centra of the thoracics. There are no true keels or hypapophyses as are found in *Kogia breviceps* and *K. simus.*

Sacral Region

There are no sacral vertebrae per se in Cetacea. Cunningham (1877, p. 225) showed a distinct sacral complex of nerves for the "porpoise," but there are no readily found bony landmarks. Slijper (1936, pp. 260, 262; 1961, p. 86) offered similar information for the harbor porpoise (*Phocoena phocoena*) and the common dolphin (*Delphinus delphis*). Hepburn and Waterson (1901, p. 681) offered evidence of a pelvic cavity in the region of the five vertebrae immediately anterior to the first caudal, and they labeled these vertebrae sacral.

Caudal Vertebrae

I define the first caudal vertebra (**Ca1**) as the first vertebra with a chevron immediately *posterior* to its caudal epiphysis (Fig. 7); this contrasts with some other authors. Slijper (1961, p. 86) and De Smet (1977, p. 73) considered the first chevron to be anterior to the first caudal vertebra (**Ca1**). I follow the definitions used by Flower (1867, p. 344; 1885, p. 75) and Omura (1957, p. 7) because there is more vertebral modification to accommodate the chevron on the posterior aspect of the vertebral body (**Ca6**, Fig. 6).

There are 25–27 caudal vertebrae, 7–10 of which occur within the flukes (lines 10, 19, and 20, Table 1). The epiphyses of the caudal vertebrae may be modified along with the centra to accommodate the chevron facets. Start and stop locations of distinct vertebral *facets* (bifurcated inferior carina) for chevron attachment are given in Table 1, lines 20 and 21; lines 23 and 24 give the start and stop locations of *chevrons* that were determined with radiographs of the intact region. The chevrons continue into the flukes, even though the articular facets disappear. There may be as many as or more than 21 chevron bones (line 22, Table 1). The range of values in line 20, Table 1 clearly indicates variation in the location of the first chevron; radiographic evidence from a large number of specimens is required prior to definitive description of the lumbar–caudal transition.

Flukes

The caudal-most vertebrae continue into the flukes, ending just anterior to the fluke notch. The vertebrae anterior to the insertion of the flukes are laterally compressed (**Ca14**, Fig. 6); those in the flukes are dorsoventrally compressed (**Ca19**, Fig. 6). The transition occurs abruptly (usually a single vertebra) but varies between individuals (line 19, Table 1). Notice the relatively large intervertebral spacing in this region (Fig. 7). Coincidentally, the intervertebral disks in the neck and flukes are roughly the same length as their respective vertebrae.

Chevrons

The chevron bones are ventral *intervertebral* ossifications found in the caudal region of many vertebrates. By definition they are associated with the vertebra anterior to them. They articulate via the paired vertebral facets of the preceding vertebra. The anterior-most chevrons are positioned nearer to their antecedent vertebrae. Further posterior in the tail the chevrons are nearer to their subsequent vertebrae. The paired elements of the first one or two and the last two or three chevrons are unfused bilaterally. The number and position of these small, terminal elements are best determined with radiographs.

At the sixth or seventh chevron (**Ca6–Ca7**) the distance from the ventral tip of the longest chevron to the transverse process is about the same as the distance from the metapophysis to the middle of the same transverse process. Caudal to this region the chevrons and neural spines appear to diminish in size at the same rate.

The caudal vertebrae grow around the vertebral arteries to produce distinct features in the bones of the tail. The transverse processes (line 15, Table 1) and the ventrolateral pedicles (line 16, Table 1) are pierced by vertebrarterial foramina for these blood vessels. These occurrences are variable among individuals and often between sides of the same individual (**Ca14**, Fig. 6).

Ribs

Bony ribs that attach to the vertebral column are vertebral ribs; ribs that attach to the sternum are sternal ribs; floating ribs are vertebral ribs that do not touch their respective transverse processes (see Fig. 8). There are 12–14 vertebral ribs in the bottlenose dolphin (lines 32 and 33, Table 1). The vertebral ribs attach to the vertebral column with synovial joints in the anterior thorax and with simple fibrous connections in the caudal thorax (floating ribs). The anterior-most 4–5 vertebral ribs are double-headed (line 32, Table 1); they each have a proximal capitulum and a distal tuberculum.

The first vertebral rib differs from the others in being short and stout, with a very robust sternal rib which is unique because of its articulation with the **manubrium.** The capitula of the first rib pair articulate ventral and somewhat *posterior* to their tubercula to accommodate the location of the diapophyses of the first thoracic vertebra. The posterior-most 7–9 ribs lose their capitula and have single heads attached by the tubercula (line 33, Table 1). The last rib pair is positioned far down on the lateral body curve, several centimeters from the articulation of the penultimate pair; they are floating ribs. A few mammals like the koala (*Phascolarctos cinereus*) and rhinoceros (*Rhinoceros unicornis*) appear to have the terminal rib attached only to the transverse process, but these are isolated and ambiguous cases. In the caudal thorax of the crocodilians both the capitulum and tuberculum attach to the transverse processes.

In the bottlenose dolphin the last three ribs in the series have reduced tubercula; the last two have reduced necks. The terminal rib is significantly shorter than the penultimate rib; this feature can be used to determine if the terminal ribs are present in a disarticulated specimen.

Sternal ribs attach between the distal ends of the vertebral ribs and the sternum; there are 7–9 bony sternal ribs (line 34, Table 1). It should be noted that Slijper (1936, p. 306) illustrated a schematic thorax of the bottlenose dolphin with 10 sternal ribs.

Each double-headed rib has its articular facets positioned so that the entire rib can hinge about a line connecting the two attachment points (parapophyses on the centra and diapophyses on the transverse processes). In fact, individual double-headed ribs, with most of the muscles removed but the articular ligaments intact, can be rotated about this axis for more than 45°. This articulation allows the entire rib cage to flex and accommodate the extreme changes in volume associated with deep diving (Ridgway *et al.*, 1969, p. 1651; Ridgway, 1971, p. 133).

Sternum

Sternebrae are the bony, serial segments of the sternum. There are three or four sternebrae including the presternum (line 35, Table 1). The anterior-most sternal elements serially ankylose as the animal ages (line 36, Table 1). Sternebrae occasionally retain a midline fenestration that serves as a reminiscence of the paired embryonic origin of the sternal elements. The first sternebra (**presternum** or **manubrium**) has the first sternal rib articulating at its anterolateral margin, the next two or three sternal ribs articulate between adjacent sternebrae, and the rest of the sternal ribs attach via fibrous connective tissue at the posterior end of the sternum. There is no caudal projection of the sternum (**xiphisternum**).

Pectoral Girdle

Scapula

The fan-shaped **scapula** of the bottlenose dolphin (Fig. 9) is typical of odontocetes. The coracoid is elongate and robust; it extends anteriorly from the distal extremity of the medial surface of the **scapula** roughly parallel to the long body axis. The acromion is substantially larger than and parallel to the coracoid. It also is robust and flat. Its surface is raised lateral to that of the coracoid and is an extension of the lateral surface of the **scapula.**

There is almost no supraspinous fossa; it is merely a groove on the anterior edge of the blade. The "spine" of the **scapula** is at the anterior extremity of the scapular blade; Klima *et al.* (1980, p. 293) interpreted the **scapula** of *T. truncatus* as having no spine. The interosseous membrane between the coracoid and the anterior margin of the blade effectively produces a functional supraspinous fossa for the supraspinatus muscle. The infraspinous region is nearly the entire blade of the **scapula.**

The **scapula** is positioned against the ribs and costal muscles, in a location similar to that in terrestrial mammals. The glenoid fossa and the humeral head are roughly in line with the presternum, like those of many terrestrial mammals (Evans and deLahunta, 1980, p. 4; Nickel *et al.*,

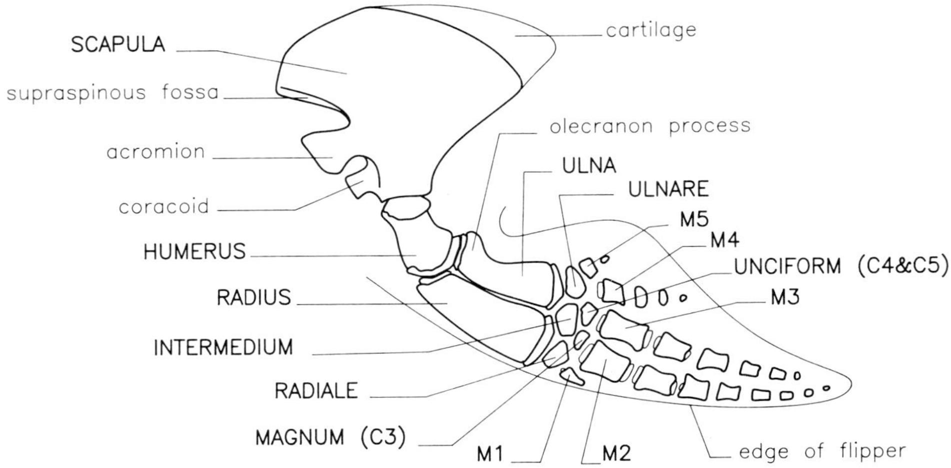

Figure 9 Forelimb of the bottlenose dolphin.

1977, pp. 24–28). The very short neck of cetaceans makes it appear to be positioned directly behind the skull.

Proximal Forelimb

The **humerus** is relatively short and very robust (Fig. 9). The head (almost entirely epiphysis) of the **humerus** is oriented dorsomedially. There is no apparent relative motion between the **ulna**, **radius**, and **humerus** since the flipper bones are embedded in a cartilaginous matrix and surrounded by a heavy fibrous sheath.

The humeral shaft is amedullary (Felts, 1966, p. 259) and almost as wide as it is long. The flattened surfaces of the distal humeral epiphysis abut the proximal epiphyses (see Mead and Potter, Chapter 9, this volume) of the relatively large **radius** and small **ulna**; there is no hinge joint as found in the same position in other mammals. The proximal processes of the **humerus**, such as the greater tubercle and lesser tubercle, are less distinct than the discrete processes of other mammals; Klima *et al.* (1980, p. 294) referred to these as a common tubercle.

There is a distinct olecranon process on the proximal **ulna**. The **radius** is located at the leading edge of the flipper; it is substantially larger than the **ulna**. The **ulna** is located along the trailing edge of the flipper. The margin between the **radius** and **ulna** is almost straight, whereas the anterior surface of the **radius** is convex and the posterior surface of the **ulna** is concave.

Manus

The cetaceans, unique among the mammals, exhibit hyperphalangy, that is, they have more than the standard mammalian number of phalanges (Howell, 1930, p. 259). In the bottlenose dolphin the **pollex** usually has a single phalanx (**M1**), but on occasion there are two. The longest digit is usually the second. The phalangeal formulae (which include the metacarpals—some cetologists exclude metacarpals in these formulae; I follow Flower, 1885, and Eales, 1950) for several USNM specimens are given in Table 3. Pisiform (sesamoid) bones are occasionally found in the region between the fifth digit and the **ulna**.

There are usually five carpals (Table 3). In the proximal row there are three: the **radiale** (**scaphoid** or **naviculare**), **intermedium** (**lunar** or **semilunare**), and **ulnare** (**cuneiform** or **triquetrum**). In some older specimens there may be secondary ossifications on the **radiale** or **ulnare** (Table 3). The distal or second carpal row has two bones: the **magnum** (**capitatum, C3**) and the **unciform** (**hamatum**). The **unciform** is considered to be the fusion

Table 3
Phalangeal Formulae for the Bottlenose Dolphin[a]

Specimen	Number of phalanges per digit					Number of carpals
	I	II	III	IV	V	
USNM #484992	1	8	6	4	2	6[b]
USNM #504109	1	8	6	4	1	6[c]
USNM #504123	1	9	7	4	2	5
USNM #504273	2	9	7	4	3	5
USNM #504315	1	8	6	4	2	5
USNM #504326	1	8	6	4	2	5[d]
USNM #504348	1	7	6	3	1	4
USNM #504501	1	8	6	4	2	5
USNM #540121	1	8	6	4	2	5
USNM #550855	1	8	6	3	2	5
USNM #550858	1	8	6	3	1	5
vBen and Gerv	2	8	7	3	2	5
Kükenthal	2 (0)	12 (8)	10 (6)	5 (2)	2	5
Felts	1 (2)	7 (6)	5 (4)	2 (3)	1 (2)	5

[a] The numbers of phalanges in each digit include metacarpals. The formulae for the specimens in the USNM collection are from radiographs only. The other formulae are from figures in the publications cited. If the number of phalanges differs from the number of ossification centers, the number of actual ossifications is given in parentheses (i.e., Kükenthal and Felts). Nonoriginal data are from van Beneden and Gervais (1868–1879, Fig. 14), Kükenthal (1889, p. 300), and Felts (1966, p. 260).
[b] USNM #484992 has six distinct carpal ossification centers, two of which (ulnare and radiale) also have secondary ossifications; the extra carpale is ankylosed to the proximal epiphysis of M1.
[c] USNM #504109 has the "extra" carpal adjacent to the carpale(s).
[d] USNM #504326 has secondary ossifications on the ulnare(s).

of the fourth (**C4**) and fifth (**C5**) carpals. There is some ambiguity with distal carpals **C1** and **C2**—one of them fuses with **M1** early in development, the other is absent (Eales, 1954, p. 209).

There may be epiphyses associated with most of the proximal phalanges of the first four digits. Digits II (**index**) and III (**medius**) commonly have epiphyses on both ends of the proximal phalanges (including the metacarpals); the number of phalanges (in these two digits) that have epiphyses increases with age. Digits I (**pollex**) and IV (**annularis**) have proximal epiphyses on their metacarpals (**M1** and **M4**) in some specimens; I have found no epiphyses on digit V (**minimus**). The names I use for the carpals follow Flower (1885, p. 302); this opinion is based on the review of the early (and rather controversial) literature by Eales (1954, p. 202) in which she criticized some of the interpretations of Kükenthal (1889) and others.

Pelvic Girdle

The pelvic vestiges are all that remain of the innominate bones. Homologies with specific bones of the pelvis have not been established. The small pelvic vestiges may be larger in males than in females. These small bones are embedded in the pelvic muscles and connective tissues and are often lost in preparation. They are positioned dorsal to the anus; the anterior tip is in line with the first chevron bones.

ACKNOWLEDGMENTS

Thanks to J. G. Mead, C. W. Potter, and W. A. McLellan for specimens, facilities, and review of the manuscript. A. Pabst offered many helpful suggestions on the manuscript. C. Bell provided photographic materials used to prepare the figures and criticized early versions of the drawings. Thanks to Jim and Becky Mead for providing board and room while I worked at the U.S. National Museum.

Figures 1–6, 8, and 9 were first drawn by hand, then digitized and edited with AUTOCAD on IBM-compatible, desk-top microcomputers; they were plotted on an HP 7580B pen plotter. Figure 7 is a photoreduction of a pen and ink drawing that was labeled with AUTOCAD. H. Hyman provided technical assistance in customizing the AUTOCAD software. Plotter time was donated by Brecher-Hyman Associates. I thank the Newman Foundation for providing funding for some of the equipment used in this study.

Meristic and morphometric data (Table 1) were collected with the help of several former students: K. Anderson, L. Baraff, L. Burton, P. Girton, A. Handley, M. Hare, C. Kinton, J. Long, W. McLellan, R. Marshal, E. Roos, J. Rothal, S. Swann, S. Wendt, and S. Whitman.

REFERENCES

van Beneden, P. J., and Gervais, F. L. P. (1868–1879). "Osteographie des Cetaces Vivants et Fossiles, Comprenant, la Description et l'Iconographie, du Squelette et du Systeme Dentaire, de ces Animaux," Par MM, Atlas, plates XXIV (Figs. 1–9) and XXXV (Figs. 1–14). Librarie de la Societe de Geographie, Paris.

Boenninghaus, G. (1904). Das Ohr des Zahnwales, zugleich ein Beitrag zur Theorie der Schalleitung. *Zool. Jahrb.* **19,** 13–360.

Boulay, G. H. du, and Verity, P. M. (1973). "The Cranial Arteries of Mammals." Heinemann Medical Books, Ltd., London, ix + 299 pp.

Breathnach, A. S. (1960). The cetacean central nervous system. *Biol. Rev.* **35,** 187–230.

Cunningham, D. J. (1877). The spinal nervous system of the porpoise and dolphin. *J. Anat. Physiol. (London)* **11,** 209–228, 1 plate.

De Smet, W. M. A. (1977). The regions of the cetacean vertebral column. *In* "Functional Anatomy of Marine Mammals" (R. J. Harrison, ed.), Vol. 3, pp. 58–80. Academic Press, New York. 428 pp.

Eales, N. B. (1950). The skull of the foetal narwhal *Monodon monoceros*. *Philos. Trans. R. Soc. London Ser. B,* **235,** 1–33.

Eales, N. B. (1954). The manus of the narwhal, *Monodon monoceros* L. *Proc. Zool. Soc. London* **124**(2), 201–211.

Evans, H. E., and deLahunta, A. (1980). "Miller's Guide to the Dissection of the Dog." Saunders, Philadelphia, Pennsylvania, 318 pp.

Felts, W. J. L. (1966). Some functional and structural characteristics of cetacean flippers and flukes. *In* "Whales, Dolphins, and Porpoises" (K. S. Norris, ed.), pp. 252–276. University of California Press, Los Angeles. 789 pp.

Flower, W. H. (1867). On the osteology of the cachalot or sperm-whale (*Physeter macrocephalus*). *Trans. Zool. Soc. London* **6,** 309–372, plates 55–61.

Flower, W. H. (1885). "An Introduction to the Osteology of the Mammalia" (revised with the assistance of H. Gadow), 3rd Ed. Macmillan and Co., London [reprinted by A. Asher & Co., Amsterdam (1966)]. 344 pp.

Fraser, F. C., and Purves, P. E. (1960). Hearing in cetaceans: Evolution of the accessory air sacs and the structure and function of the outer and middle ear in Recent cetaceans. *Bull. Br. Mus. (Nat. Hist.) Zool.* **7**(1), 1–140, 53 plates.

Gambaryan, P. P. (1974). "How Mammals Run: Anatomical Adaptations" (transl. from the Russian by H. Hardin). Wiley, New York. 367 pp.

Hepburn, D., and Waterson, D. (1901). The pelvic cavity of the porpoise (*Phocoena communis*) as a guide to the determination of a sacral region in Cetacea. *Rep. 71st Med. Br. Assoc. Adv. Sci.,* pp. 680–681.

Hildebrand, M. (1982). "Analysis of Vertebrate Structures," 2nd Ed. Wiley, New York. 654 pp.

Hosokawa, H., and Kamiya, T. (1965). Sections of the dolphin's head (*Stenella caeruleoalba*). *Sci. Rep. Whales Res. Inst., Tokyo* **19,** 105–133.

Howell, A. B. (1927). Contribution to the anatomy of the Chinese finless porpoise *Neomeris phocaenoides*. *Proc.* U.S. Natl. Mus. **70,** 1–43, plate 1.

Howell, A. B. (1930). "Aquatic Mammals." Charles C Thomas, Baltimore, Maryland. 338 pp.

Huber, E. (1934). Anatomical notes on Pinnipedia and Cetacea. *Carnegie Inst. Wash. Publ.,* No. 447, pp. 105–136.

Hyman, L. H. (1942). "Comparative Vertebrate Anatomy," 2nd Ed. University of Chicago Press, Chicago, Illinois. 544 pp.

Jollie, M. (1973). "Chordate Morphology." Krieger, Huntington, New York, 478 pp.

Kasuya, T. (1973). Systematic consideration of Recent toothed whales based on the morphology of tympano-periotic bone. *Sci. Rep. Whales Res. Inst., Tokyo* **25,** 1–103.

Kellogg, R. (1928). The history of whales–Their adaptation to life in the water. *Quart. Rev. Biol.* **3**(1), 29–76 and **3**(2), 174–208.

Kernan, J. D. (1918). The skull of *Ziphius cavirostris*. *Bull. Am. Mus. Nat. Hist.* **38,** 349–394, plates XX–XXXII.

Klima, M., Oelschlager, H. A., and Wunsch, D. (1980). Morphology of the pectoral girdle in the Amazon dolphin *Inia geoffrensis* with special reference to the shoulder joint and movements of the flippers. *Sonderdeuck aus Zeitschrift fur Saugetierkunde* **45**(5), 288–309.

Kükenthal, W. (1889). "Vergleichend Anatomiche und Entwickelungsgeschichtliche Untersuchunger an Walthieren. Kapitel II; Die Hand der Cetaceen." Gustav Fischer, Jena.

Langworthy, O. R. (1931). Central nervous system of the porpoise, *Tursiops truncatus*. *J. Mammal.* **12,** 381–389.

McFarland, W. L., Jacobs, M. S., and Morgane, P. J. (1979). Blood supply to the brain of the dolphin, *Tursiops truncatus,* with comparative observation on special aspects of the cerebrovascular supply of other vertebrates. *Neurosci. Behav. Rev.* **3**(1), 1–93.

Mead, J. G. (1975). Anatomy of the external nasal passages and facial complex in the Delphinidae (Mammalia: Cetacea). *Smithsonian Contrib. Zool.* **207,** 1–72.

Miller, G. S. (1923). The telescoping of the cetacean skull. *Smithsonian Misc. Coll.* **76**(5), 1–62, 8 plates.

Montagu, G. (1821). Description of a species of *Delphinus,* which appears to be new. *Mem. Wernerian Nat. Hist. Soc.* **III,** 75–82, plate III.

Morgane, P. J., Jacobs, M. S., and McFarland, W. L. (1980). The anatomy of the brain of the bottlenose dolphin (*Tursiops truncatus*): Surface configurations of the telencephalon of the bottlenose dolphin with comparative anatomical observations in four other cetacean species. *Brain Res. Bull.* **5**(3), 1–107.

Nickel, R., Schummer, A., Seiferle, E., Frewein, J., and Wille, K. H. (1977). "Lehrbuch der Anatomie der Haustiere; Band I, Benegunsapparat." Paul Parey, Berlin. 539 pp.

Oelschlager, H. A. (1986a). Tympanohyal bone in the toothed whales and the formation of the tympanoperiotic complex (Mammalia: Cetacea). *J. Morphol.* **188,** 157–165.

Oelschlager, H. A. (1986b). Comparative morphology and evolution of the otic region in toothed whales (Cetacea, Mammalia). *Am. J. Anat.* **177,** 353–368.

Omura, H. (1957). Osteological study of the little piked whale from the coast of Japan. *Sci. Rep. Whales Res. Inst., Tokyo* **12,** 1–21, plates 1–8.

Reynolds, S. H. (1897). "The Vertebrate Skeleton." Cambridge University Press, London. 559 pp.

Ridgway, S. H. (1971). Buoyancy regulation in deep diving whales. *Nature (London)* **232,** 133–134.

Ridgway, S. H., Scronce, B. L., and Kanwisher, J. (1969). Respiration and deep diving in the bottlenose porpoise. *Science* **166,** 1651–1654.

Ridgway, S. H., Demski, L. S., Bullock, T. H., and Schwanzel-Fukuda, M. (1987). The terminal nerve in odontocete cetaceans. *In* "The Terminal Nerve (Nervus Terminalis); Structure, Function, and Evolution" (L. S. Demski and M. Schwanzel-Fukuda, eds.). *Ann. N.Y. Acad. Sci.* **519,** 201–212.

Romer, A. S. (1966). "Vertebrate Paleontology," 3rd ed. University of Chicago Press, Chicago. pp. 297–301.

Schulte, H. von W., and Smith, de F. M. (1917). The skull of *Kogia breviceps* Blainv. *Bull. Mus. Comp. Zool.* **37,** 361–404.

Schulte, H. von W., and Smith, de F. M. (1918). The external characters, skeletal muscles, and peripheral nerves of *Kogia breviceps* (Blainville). *Bull. Am. Mus. Nat. Hist.* **38,** 7–72.

Seagars, D. J. (1982). Jaw structure and functional mechanics of six delphinids (Cetacea: Odontoceti). M.S. thesis, San Diego State University, San Diego, California.

Slijper, E. J. (1936). "Die Cetacean: Vergleichend-anatomisch und Systematisch." [reprinted by A. Asher & Co., Amsterdam (1973).] 590 pp.

Slijper, E. J. (1946). Comparative biologic–anatomical investigations on the vertebral column and spinal musculature of mammals. *Verh. K. Ned. Akad. Wet. Afd. Natuurk. Sect. 2* **42,** No. 5.

Slijper, E. J. (1961). Locomotion and locomotory organs in whales and dolphins (Cetacea). *Symp. Zool. Soc. London* **5,** 77–94.

Yamada, M. (1953). Contribution to the anatomy of the organ of hearing of whales. *Sci. Rep. Whales. Res. Inst., Tokyo* **8,** 1–79.

Axial Muscles and Connective Tissues of the Bottlenose Dolphin

3

D. Ann Pabst
Department of Zoology
Duke University
Durham, North Carolina 27706

I do not think that anyone who has not dissected the spinal muscles of a cetacean can quite appreciate the impracticability of homologizing exactly all of the elements.

A. B. HOWELL (1930)

INTRODUCTION

The Question

What are the identities of the axial muscles of cetaceans? This has proved to be a difficult question to answer—a fact that can be appreciated by any student of cetacean morphology who, upon reviewing the literature, discovers there are almost as many interpretations of axial muscle identities as there are published anatomies (see Table 1). The confusion is understandable: morphology is an interpretive science. When an anatomist describes a structure, he or she is not stating a truth, but rather presenting an often simplified interpretation of a complex morphology.

Cetacean axial muscles are particularly open to the vagaries of interpretation for a number of reasons. First, the axial muscles of mammals, in general, are not well understood; their metameric morphology and habit of interdigitating with other muscles complicate the identification of distinct muscle groups. Second, although there exist international standards of nomenclature for the axial muscles of tetrapod mammals (see below), these are not universally followed, due mostly to the identification problems noted above. Third, cetacean axial muscles are highly modified from those of tetrapod mammals.

There is, in fact, one more obstacle to the interpretation of the existing published accounts of the anatomy of cetacean axial muscles. Whereas the anatomy of a large number of species (many of them quite rare) has been described (Table 1), few investigators have studied more than one species (but see comments on Slijper, 1936, below). It is thus difficult to determine whether the differing anatomical descriptions of cetacean axial muscles are due to real species differences or are rather due to interpretational differences among authors.

The cetacean axial musculature is thus especially difficult to interpret, for both biological and operational reasons. These roadblocks have hampered the development of a unified, synthetic view of cetacean axial morphology.

The Solution

A comprehensive view of cetacean axial morphology should have two components. First, muscles should be described and identified according to stated criteria for homology (but see Roth, 1984, for comments on limitations of operational criteria for homology). We already know that this is an especially difficult task when dealing with cetacean axial muscles. Second, previous anatomies should be reviewed and synonymous muscles reidentified (i.e., standardized) with this nomenclature.

Two recently published cetacean anatomies (Smith *et al.*, 1976; Strickler, 1980) have offered a more unified view of the axial morphology by reviewing some earlier works in the context of their own studies. While these reviews are valuable, it is unfortunate that Smith *et al.* (1976) and Strickler (1980) standardized the earlier anatomies using different nomenclatures. Ironically, these authors thus perpetuated some of the identification problems noted above.

Smith *et al.* (1976) suggested that the confusion in the literature was due mostly to *de novo* nomenclature and suggested that the solution was to follow the standards of *Nomina Anatomica* (NA) (1966) and *Nomina Anatomica Veterinaria* (NAV) (1968). While the NA and NAV establish standard muscle names, they do not establish standard criteria for homology. Smith *et al.* (1976) made decisions about homology, but they unfortunately did not describe the criteria used.

Strickler (1980) followed the nomenclature of Slijper (1936) (based on Eisler, 1912). Slijper's (1936) *Die Cetaceen* is indisputably one of the most thorough, detailed studies of cetacean anatomy ever done. His work was comparative (he described and illustrated the axial musculature of four species of odontocetes and noted differences among them), and his nomenclature was based on homologies established by innervation pattern and muscle position. Unfortunately, Slijper's nomenclature is very complex and not always based on internationally accepted standards. Strickler recognized these problems and offered a simplified version of Slijper's nomenclature.

I follow the nomenclature of Strickler (1980). It is based on stated criteria for homology. Although these criteria are certainly not indisputable, they are consistent with my current understanding of muscle homologies. Strickler's nomenclature also follows NA and NAV for those muscles with obvious tetrapod homologs but, like Slijper's, uses unique names when muscle organizations are novel to cetaceans (e.g., m. extensor caudae lateralis).

Using Strickler's nomenclature, I describe the axial morphology of the bottlenose dolphin, *Tursiops truncatus*. It does seem ironic that the axial muscle anatomy of many rare species of cetaceans has been described, while that of *T. truncatus*, the quintessential dolphin, has never been thoroughly discussed. It has, however, been diagrammed by Sokolov and Rodionov (1978) and mentioned by Slijper (1936) and Schulte and Smith (1918). The histochemical properties of the axial muscles are reported by Bello *et al.* (1985). Rommel's superbly illustrated osteology of the bottlenose dolphin (Chapter 2, this volume) provides a detailed description of the skeletal structures I discuss.

A synoptic table, comparing the work of thirteen anatomists who have investigated the axial morphology of odontocetes, follows the descriptive anatomy presented here (see Table 1). I have included authors reviewed by Slijper (1936, pp. 234–235) because my interpretations often differ from those he presented.

The Approach

I am a functional morphologist, with biases about how to approach anatomy. When I look at the backs of dolphins, I see an outstanding mammalian locomotor design (*sensu* Lauder, 1982). Dolphins swim using robust muscles that develop contractile force and complicated systems of tendons that deliver that force to the skeleton. If the descriptive anatomy is to be the basis for a functional analysis of this locomotor system, it should include all the system's components: muscles, nerves, connective tissues, and bones.

The anatomy of the bottlenose dolphin that I present is therefore different from all previous anatomies in one important way: I describe all major connective tissues as discrete structures. By doing so, I do not mean to imply that these connective tissues are autonomous from their muscles. I wish simply to point out their general pattern of organization, as is always done for muscles. I also introduce the subdermal connective tissue sheath (SDS), a collagenous membrane that surrounds the axial muscles.

Because these connective tissues have never been described as discrete structures, I cannot consistently follow established nomenclature. When tendon names do not exist, I have named these structures as simply as possible using relative positional names (e.g., medial tendon, posterior connective tissue sheath).

The axial muscles are innervated by the spinal nerves. I refer the reader to Slijper (1936) for identification of the branches of the spinal nerves that supply the muscles.

DESCRIPTIVE ANATOMY

Axial muscles are long muscle bodies composed of many shorter muscle fibers. For the most part, the fibers do not arise off a single skeletal element, but off many skeletal elements along the length of the muscle. Likewise, muscles do not insert via a single tendon, but most often insert via a series of long tendons that run throughout the length of the muscle. I therefore define the *origin* of a muscle as the structure(s) on which the anterior ends of its muscle fibers insert. The *insertion* of a muscle is the structure(s) on which the posterior ends of its muscle fibers insert. The *run* of a muscle (or connective tissue) fiber is its path from its origin toward its insertion.

The connective tissue structures I describe are organized into multilayered sheaths and tendon systems. The sheaths surround, and the tendons are derived from, the axial muscles. The *origin* of a tendon system is the muscle from which it emerges. The *insertion* of a tendon system is the skeletal element(s) to which the tendon fibers attach. As with muscles, the origin of a tendon is anterior to its insertion.

Perhaps the most vital part of an anatomy is the illustrations. I illustrate most muscle structures and connective tissue structures separately, as they appear in a forward-directed, oblique left-lateral view of the dolphin. These structures are, of course, not really separated physically or functionally. To illustrate their positions relative to one another I use whole-body cross sections. (Appendix 1 contains the abbreviations and conventions used; Appendix 2 lists the specimens; Appendix 3 explains the procedure used to generate figures.)

Dorsal Connective Tissues

Connective Tissue Sheaths

The subdermal connective tissue sheath (SDS), a structure formed by many interwoven layers of collagen fibers, surrounds the axial muscles of cetaceans (Wainwright *et al.*, 1985; Pabst, 1987). The connective tissue fibers that form the SDS come from many sources: tendons of abdominal and axial muscles, ligamentous fibers, and connective tissue fibers from the blubber. This structure is piecewise homologous to the fascia thoracolumbalis. By this I mean that some of the connective tissues forming the SDS and the fascia thoracolumbalis are from homologous sources (e.g., abdominal muscles) but that the SDS is also composed of connective tissues from additional sources (e.g., blubber).

The SDS is a crossed, helically wound, fiber array, that is, the fibers within each layer are wrapped around the body at oblique angles to the long axis of the animal (Fig. 1). Helically wound fiber arrays that are wrapped around cylinders (or cylindrically shaped animals, like cetaceans) have unique mechanical properties (Wainwright *et al.*, 1976). The mechanical property germane to this descriptive anatomy is that a helically wound fiber array which is under biaxial tension is stiff. The SDS is under biaxial tension in *T. truncatus* (D. A. Pabst, unpublished data). The stiffened SDS appears to be a skeletal element in dolphins; it acts as a site of origin for some muscle fibers and as a site of insertion for other muscle fibers.

The SDS also provides connective tissue layers that separate muscle groups. The SDS, as indicated above, is formed by alternate layers of

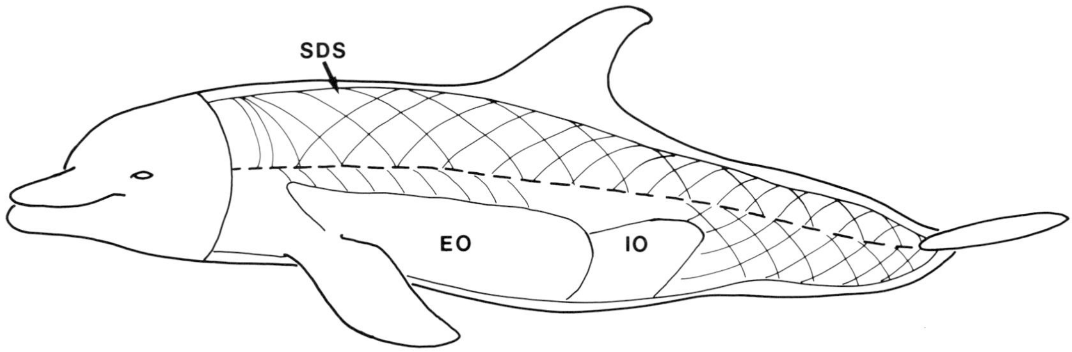

Figure 1 Subdermal connective tissue sheath (SDS) of a bottlenose dolphin (blubber and m. cutaneus trunci removed). Approximate connective tissue fiber angles are indicated. See Appendix 1 for abbreviations.

collagen fibers that run in right- and left-handed helices. Two of these layers (sheaths) leave the peripheral position of the SDS and dive medially through the dorsal muscles to insert onto the vertebral column as the lateral and posterior sheaths (Fig. 2A,B). The lateral sheath is a vertically oriented septum that runs virtually the entire length of the dolphin and separates the m. longissimus from the m. (ilio)costalis and the m. intertransversarius caudae dorsalis (Fig. 4C,D). The posterior sheath is a horizontally oriented septum that separates the m. extensor caudae lateralis from the m. extensor caudae medialis from Ca1 posteriorly (Figs. 2A and 4D). (The position of the posterior sheath varies slightly between individual bottlenose dolphins; although it usually begins at Ca1, it can begin as far posterior as Ca3 to Ca4.)

The SDS and lateral sheath are skeletal structures and are referred to as sites of origin and/or insertion for the axial muscles. (See Ventral Connective Tissues for more on the SDS.)

Dorsal Tendon Systems

There are three major dorsal tendon systems: the superficial tendon, the deep tendon, and the tendons of the extensor caudae lateralis. Each of these tendon systems originates within the mass of the dorsal musculature and inserts posteriorly onto the vertebral column.

SUPERFICIAL TENDON

Origin Dorsolateral border of (posterior region of) m. semispinalis. Medial belly of m. longissimus.

Insertion Neural spines of thoracic vertebrae (posterior to T7) and all lumbar vertebrae. Transverse processes of posterior lumbar vertebrae and all caudal vertebrae anterior to insertion of flukes (Ca18 for specimens I have examined; see Rommel, Chapter 2, this volume, for variability in the position of fluke insertion).

Relations Dorsal to m. multifidus. Ventral to SDS. Within body of m. longissimus.

Description The superficial tendon (Figs. 2C and 4B–D) is composed of many long, flat tendon fibers. All tendon fibers originate within the dorsal muscle mass; the most anterior fibers originate within the m. semispinalis, and the posterior fibers originate within the m. longissimus (as the tendinous insertional fibers of these muscles). The tendon fibers run dorsoposteriorly and medially at a low angle to the long axis of the body, eventually emerge from their muscles, and run over the dorsal surface of the m. multifidus toward the dorsal midline. These tendon fibers are organized into a curved sheet (aponeurosis). This curved sheet incompletely divides the m. longissimus into medial and lateral parts and covers the m. multifidus (Fig. 4C). The most anterior of these connective tissue fibers appears on the dorsal surface of the m. multifidus at T8.

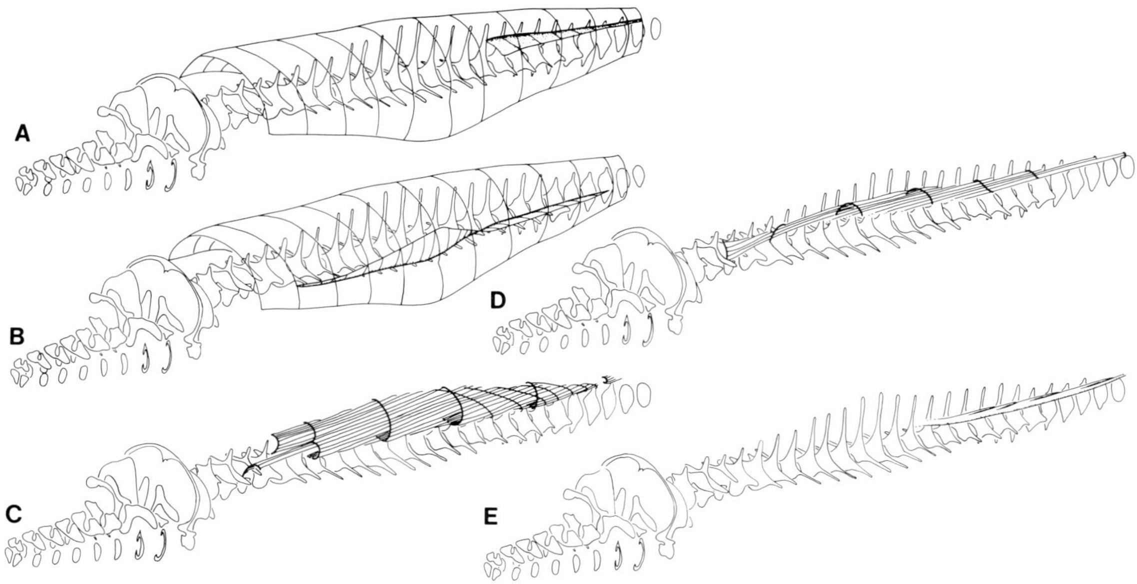

Figure 2 Dorsal connective tissues. (In these orientations you are standing above the dolphin and looking down at its dorsal and left-lateral surfaces.) (A) Posterior sheath. The posterior and lateral sheaths are seen through a "transparent" subdermal connective tissue sheath (SDS). Only the contours of the SDS and the sheaths are indicated, not the orientation of connective tissue fibers. (B) Lateral sheath. (C) Superficial tendon. The connective tissue fibers that form the superficial and deep tendons appear as thin, longitudinal lines. The medium-thickness lines indicate connective tissue fibers that change their direction and turn laterally to run toward the transverse processes. The thick, circumferentially oriented lines indicate the cross-sectional shape of the tendon sheaths. (D) Deep tendon. (E) Tendons of the extensor caudae lateralis, which are cylindrically shaped tendons. Only five of the seven tendons are shown; the posterior two separate just beyond the boundary of the figure.

All the tendon fibers run toward the dorsal midline; as they approach the dorsal midline they do one of three things. In the thoracic and lumbar regions, the tendons (1) insert onto the neural spines or neural septum or (2) cross the dorsal midline, enter the SDS, and continue toward the transverse processes on the other side of the dolphin. Beginning in the posterior lumbar region, many of the tendon fibers (3) change their direction just lateral to the middorsal line. These tendon fibers make a lateral turn, enter the SDS, and run (at oblique angles to the long axis of the dolphin) through the SDS toward the ipsilateral transverse processes. The tendon fibers insert either (a) onto the transverse processes of the posterior lumbar vertebrae or (b), in the caudal peduncle, onto the neural spines of the caudal vertebrae by way of the posterior connective tissue sheath.

DEEP TENDON

Origin Ventrolateral aspect of the m. multifidus. Medial aspect of (posterior region of) m. semispinalis.

Insertion Metapophyses of T1 through all vertebrae anterior to fluke insertion.

Relations Dorsal to m. longissimus. Ventral to m. multifidus.

Description The deep tendon (Figs. 2D and 4B–D) is composed of many long, flat tendon fibers similar to the superficial tendon. These tendon fibers originate on the ventrolateral aspect of the m. multifidus and the medial aspect of the m. semispinalis (as the insertional fibers of these muscles). The tendons run ventroposteriorly and medially to insert onto the metapophyses of all the vertebrae from T1 to Ca18. These tendon fibers are organized into a curved sheet that forms the floor of the m. multifidus and separates the m. multifidus from the m. longissimus.

TENDONS OF THE M. EXTENSOR CAUDAE LATERALIS

Origin Belly of m. longissimus in posterior lumbar vertebrae as most ventromedial of the superficial tendon fibers.

Insertion Dorsal surfaces of bodies of Ca16 to Ca24.

Relations Ventral to posterior sheath. Within body of m. extensor caudae lateralis.

Description The tendons of the m. extensor caudae lateralis (ECL) (Figs. 2E and 4D) originate in the posterior lumbar region of the m. longissimus as the most ventromedial fibers of the superficial tendon. Instead of running dorsoposteriorly and medially, as described above, these tendons run only posteriorly and coalesce to form seven distinct, cylindrically shaped tendons (Fig. 2E). (Thus, the morphology of these tendons is unlike the curved-sheet morphology of the superficial and deep tendons.) In the caudal peduncle these tendons are separated from the rest of the superficial tendon by the posterior sheath (Fig. 4D).

The seven tendons emerge from the dorsal surface of the ECL. The most anterior tendon to emerge is the most anterior tendon to insert, that is, tendon 1 runs to Ca16, tendon 2 runs to Ca18, etc. The tendons split at their posterior end so that one tendon may insert onto two to three adjacent vertebrae.

Dorsal Division Axial Muscles (Musculi Dorsi)

M. SEMISPINALIS

Origin On the skull: across the dorsal surfaces of the supraoccipital and interparietal regions. Anterior tendon. Connective tissue septa within m. semispinalis.

Insertion Anterior-most superficial tendon fibers. Anterior-most deep tendon fibers. Connective tissue septa within m. semispinalis.

Relations Dorsal to m. rectus capitis. Ventral to superficial tendon (posterior to T5/7) and SDS. Medial to m. longissimus. Lateral to m. multifidus.

Description Slijper (1936) defined the m. semispinalis (Figs. 3A and 4A,B) as running from the occiput to T8, and thus defined a posterior termination to this muscle in the thoracic region. Strickler (1980) defined the m. semispinalis as continuing posteriorly with the m. longissimus through the caudal vertebrae. I follow Slijper's interpretation and define the termination of this muscle in the midthoracic vertebrae.

The m. semispinalis originates as a laterally broad muscle off the entire dorsal border of the exoccipital bones, just posterior to the nuchal ridge, as muscle fibers or via a strong, flattened tendon that runs on the ventral surface of the muscle. Posterior to the occiput, it deepens ventrally and contacts the dorsal surface of ribs 1–3. Posterior to this the ventral portions of the m. longissimus and m. multifidus broaden and displace the m. semispinalis off the ribs. As the m. semispinalis continues posteriorly it (1) narrows, as the m. multifidus and m. longissimus continue to broaden (it is "squeezed out" by the muscles surrounding it), and (2) fuses its borders with the m. multifidus and the m. longissimus. (Fusion is the crossing over of muscle fibers from one muscle body to another.) Fusion begins along the shared ventral borders of the muscles and progresses dorsally. By T4/5, the dorsal edges of the m. semispinalis and m. longissimus are fused; by T5/6, the dorsal borders of the m. semispinalis and m. multifidus are fused. The m. semispinalis, identifiable as an autonomous muscle, terminates at T9/11.

The muscle fibers of the m. semispinalis run longitudinally. Muscle fibers originate off the skull or the anterior tendon. Muscle fibers also originate off connective tissue septa that traverse the m. semispinalis at oblique angles to the long axis of the muscle. Muscle fibers insert onto either connective tissue septa or onto the superficial or deep tendons.

The morphology of the m. semispinalis suggests that its action is to change the attitude of the skull relative to the vertebral column (skull extension and lateral flexion). The muscle also helps tense the anterior superficial and deep tendon fibers.

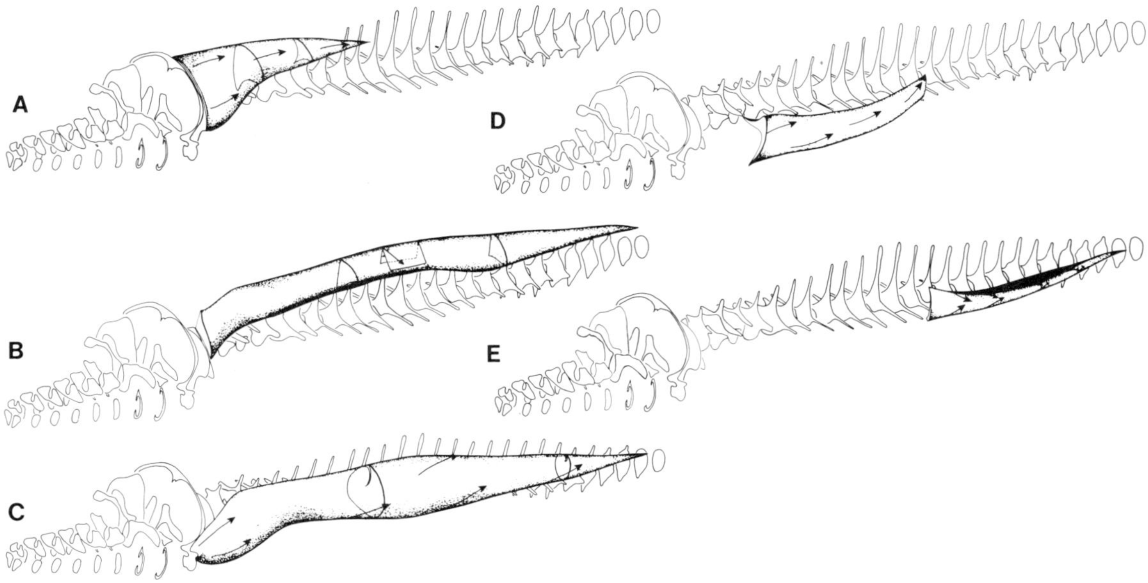

Figure 3 Dorsal muscles. (In these orientations you are standing above the dolphin and looking down at its dorsal and left-lateral surfaces.) Arrows indicate approximate orientations of muscle fibers. Cross-sectional shape is indicated at various points along the muscles, In B, D, and E the anterior extent of the muscle either was not "captured" in a cross section or was difficult to identify unequivocally. Thus, these muscles seem to appear with a well-defined cross section. (A) M. semispinalis. (B) M. multifidus. (C) M. longissimus. (D) M. (ilio)costalis. (E) M. intertransversarius caudae dorsalis. © D. A. Pabst, reproduced by permission of the author.

M. MULTIFIDUS AND M. EXTENSOR CAUDAE MEDIALIS (ECM)

Origin Posterior edge of neural spine of atlas–axis complex. Lateral aspect of neural spines (dorsal to metapophyses) from C3 to Ca1. Deep surface of superficial tendon (that covers m. multifidus). Deep surface of SDS just lateral to neural spines from midlumbar vertebrae to Ca1.
ECM: Lateral aspect of neural spines (dorsal to metapophyses) from Ca1 to Ca10/12. Lateral surface of neural septum from Ca10/12 to Ca15/17. Deep surface of superficial tendon (that covers ECM). Deep surface of SDS just lateral to middorsal line from Ca1 to Ca15/17.

Insertion Deep tendon.
ECM: Deep tendon.

Relations Dorsal to m. longissimus from T7 to Ca1. Dorsal to posterior sheath from Ca1 to Ca15/17. Ventral to superficial tendon (posterior to T7). Medial to m. semispinalis from Ca1/2 to T9/11. Medial to m. longissimus (posterior to T11).

Description Slijper (1936) defined and named the caudal continuation of the m. multifidus as the m. extensor caudae medialis (ECM) (Figs. 3B and 4A–D). A clear distinction between these two muscles is not obvious in the bottlenose dolphin. Following convention (Slijper, 1936; Strickler, 1980), I define the beginning of the ECM at the level where the posterior sheath dives medially (Ca1).

The m. multifidus arises off the posterior edge of the neural arch of the atlas–axis complex and runs, as the ECM, farther posteriorly than any other dorsal muscle (muscle fibers run to Ca18/19). Its ventral border is defined by the deep tendon, which inserts onto the metapophyses of all vertebrae posterior to the cervical vertebrae. The metapophyses are placed low on the neural arch in the anterior thoracic vertebrae and move progressively dorsad onto the neural spine of posterior vertebrae. Concomitantly, the floor of the m. multifidus begins low and wide across the thoracic vertebrae. As the m. multifidus runs poster-

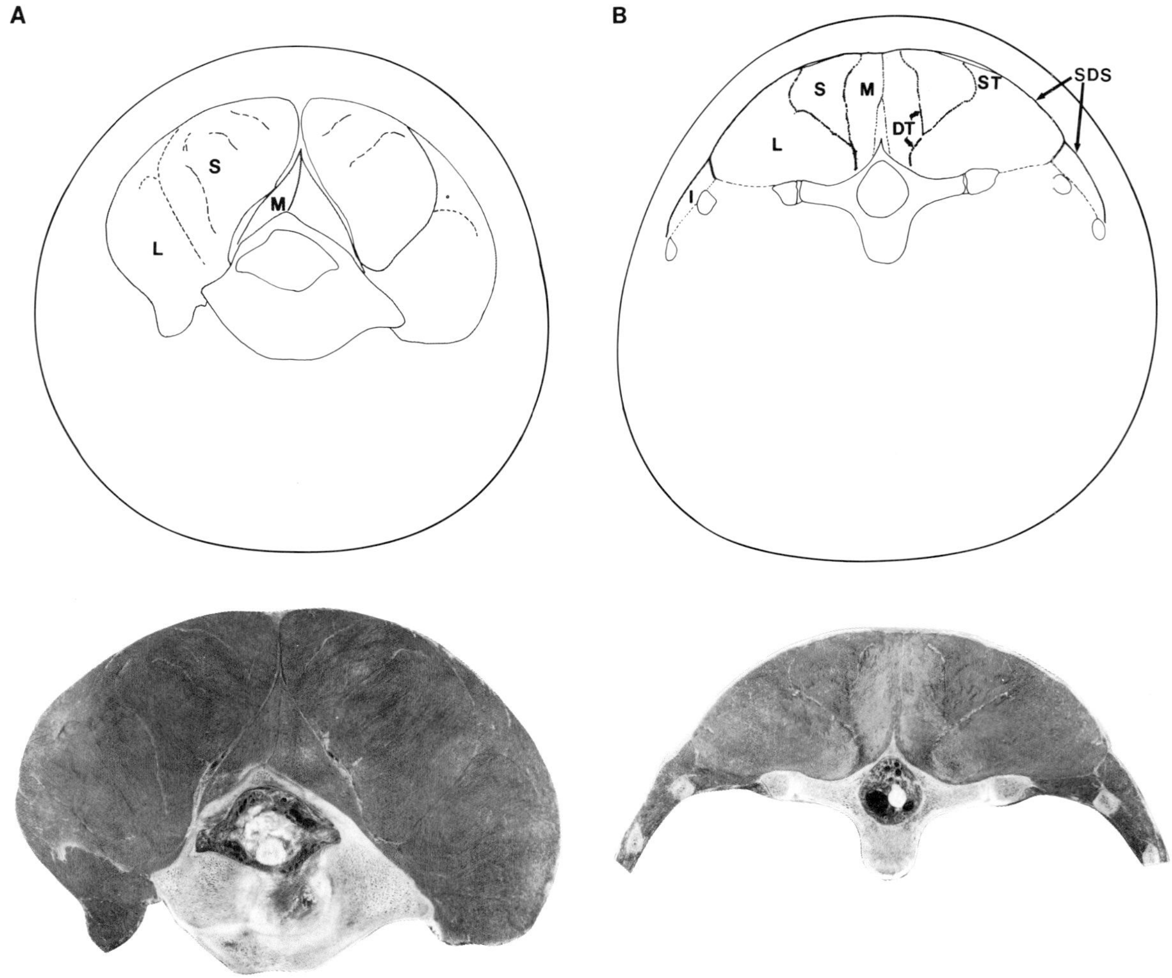

Figure 4 Whole body cross sections. Scale (bars = 5 cm) is constant in diagrams of cross sections, but note differences in scale between photos. Muscles are labeled on the left half of the cross section, connective tissue structures on the right. Thin, dashed lines that are not labeled indicate connective tissue septa. (A) Section through cervical region. (B) Section through anterior thoracic region. (C) Section through midlumbar region. (D) Section through anterior caudal region. © D. A. Pabst, reproduced by permission of the author.

iorly, the floor of the muscle moves dorsad on the neural spines and the muscle broadens laterally so that, in cross section, the changes in shape that the m. multifidus undergoes are reminiscent of those of a tulip bud blooming (Fig. 4A–C).

The muscle fibers of the m. multifidus are short relative to the fibers of the m. longissimus. The m. multifidus fibers arise off (1) the lateral aspect of the neural spines, (2) the deep surface of the superficial tendon, and (3) the deep surface of the SDS just lateral to the neural spines. The muscle fibers insert onto the deep tendon.

The morphology of the m. multifidus suggests that its action is to extend the caudal peduncle by applying force to the neural spines of the vertebrae anterior to the insertion of the flukes. This force is transmitted by way of the deep tendon. The m. multifidus also tenses the SDS just lateral to the neural spines.

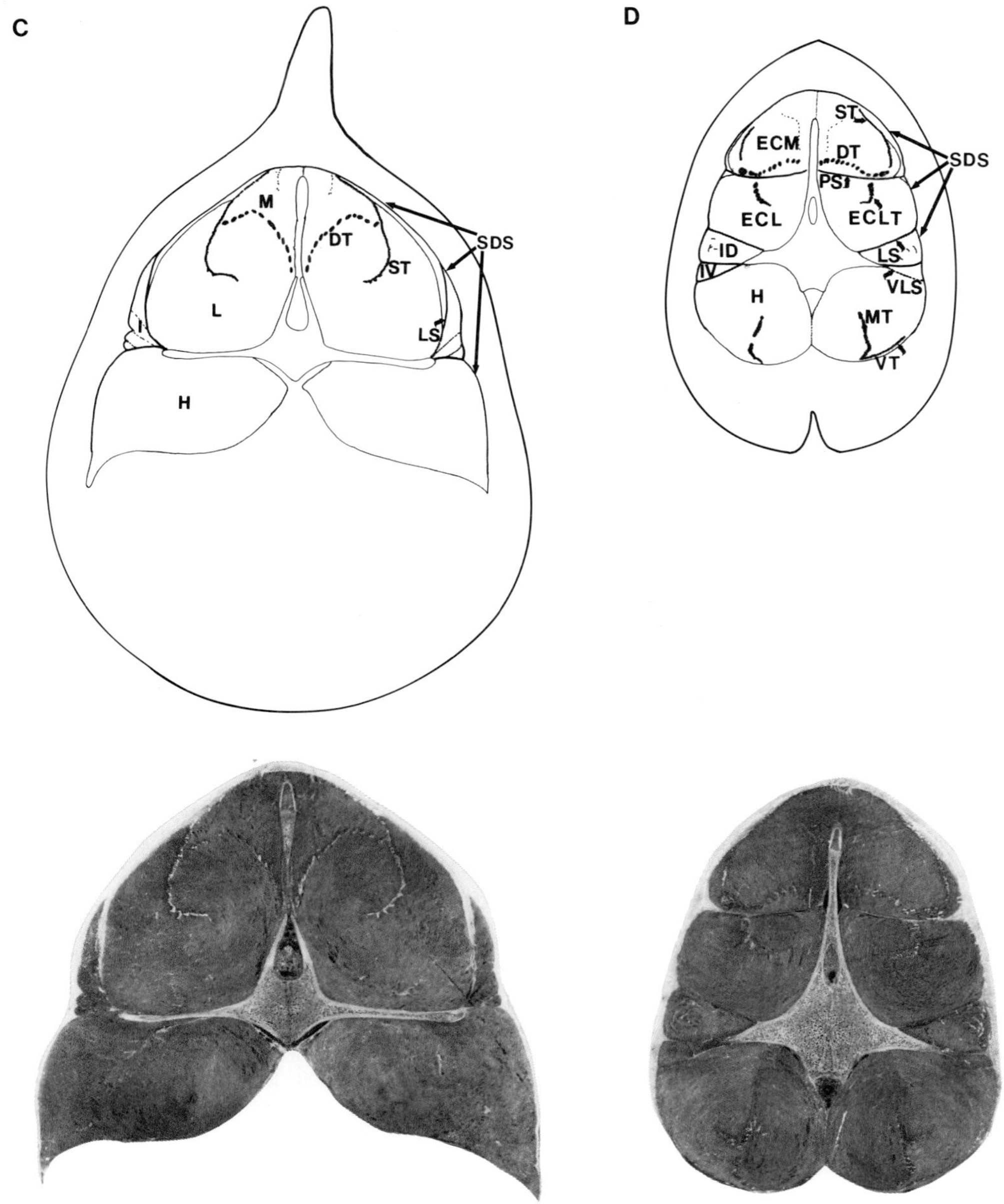

Figure 4 (continued)

M. LONGISSIMUS AND M. EXTENSOR CAUDAE LATERALIS (ECL)

Origin Capitis: On the skull: posterior edge of zygomatic process of squamosal and extreme lateral margin of exoccipital bones.

Cervicis: Lateral aspect of neural arch of atlas–axis complex and lateral aspect of C3 thru C7.

Dorsi: Transverse processes of all thoracic vertebrae and anterior lumbar vertebrae. Proximal aspect of dorsal surface of ribs. Neural arches of posterior thoracic vertebrae and anterior lumbar

vertebrae. Deep surface of lateral sheath (near insertion of sheath onto transverse processes) from T1 to posterior lumbar vertebrae. Ventral surface of deep tendon.
ECL: Complete shelf of transverse processes of posterior lumbar vertebrae and caudal vertebrae (thru Ca13). Neural arches of posterior lumbar vertebrae and caudal vertebrae (thru Ca15/17). Deep surface of lateral sheath from posterior lumbar vertebrae to Ca15.

Insertion Capitis: Anterior SDS. Shared insertion with cervicis.
Cervicis: Transverse processes of anterior thoracic vertebrae. Superficial tendon.
Dorsi: Superficial tendon.
ECL: Tendons of ECL.

Relations Ventral to m. multifidus from midthoracic vertebrae to Ca1. ECL ventral to posterior sheath from Ca1 to Ca15/17. Capitis medial to m. splenius. Cervicis and dorsi medial to m. (ilio)costalis from T1 to L6/8 (midregion of dorsal fin). Dorsi medial to m. intertransversarius caudae dorsalis posterior to L8. Capitis lateral to m. obliquus capitis superior. Dorsi lateral to (posterior) m. semispinalis and m. multifidus.

Description Slijper (1936) defined and named the caudal extension of the m. longissimus as the m. extensor caudae lateralis (ECL) (Figs. 3C and 4A–D). (See comments above for ECM.)

The m. longissimus originates on the skull as a laterally compressed but dorsoventrally elongated muscle. As it runs posteriorly, it broadens out laterally to occupy the entire shelf of the transverse processes (from their distal tips to the neural arches). It is a very robust muscle.

The relatively long muscle fibers of the m. longissimus arise off the transverse processes, neural arch, lateral sheath, and deep tendon. These muscle fibers converge posteriorly to insert onto the superficial tendon. The posterior-most muscle fibers form the ECL and insert onto the tendons of the ECL.

The morphology of the m. longissimus suggests that its action is to assist the m. semispinalis in changing the attitude of the skull relative to the vertebral column. It also assists the m. multifidus in extending the caudal peduncle by transmitting its contractile force (via the superficial tendon) to the vertebrae anterior to the fluke insertion. The morphology of the ECL suggests that its action is to extend, or control, the angle of the flukes by way of its seven terminal tendons.

M. (ILIO)COSTALIS

Origin Lateral aspect of m. longissimus capitis. Lateral edge of transverse process of the atlas–axis complex. As slips off dorsolateral aspect of all ribs and off posterior border of terminal floating rib. Superficial surface of lateral sheath anterior to midregion of dorsal fin.

Insertion SDS anterior to midregion of dorsal fin.

Relations Medial to SDS (dorsal to transverse processes). Medial to m. cutaneus trunci (commonly called panniculus) (ventral to transverse processes). Lateral to lateral sheath (dorsal to transverse processes). Lateral to m. intercostales externi (ventral to transverse processes).

Description The name m. iliocostalis reflects the evolutionary history of this muscle, but it is not literally correct in cetaceans, since this muscle does not terminate on the ilium [which has not been conclusively identified in odontocetes (see Rommel, Chapter 2, this volume)], or even on the vestigial pelvis. I will indicate this by putting the nondescriptive "ilio-" in parentheses.

The anterior head of the m. (ilio)costalis (Figs. 3D and 4A–C) is bifurcate; it originates as thin slips off the transverse process of the atlas–axis complex and the superficial aspect of the m. longissimus cervicis. These two fingers of muscle run through the body of the m. longissimus cervicis, to join the posterior portion of the m. (ilio)costalis. The m. (ilio)costalis and m. longissimus cervicis thus appear to share some muscle fibers anterior to T1; posterior to T1, the lateral sheath separates these two muscles. Therefore, the above-described muscle slips are difficult to identify in cross section (Fig. 3D). The anterior extend of the m. (ilio)costalis is shown where the muscle can first be unequivocally identified.

The costal portion of the m. (ilio)costalis originates off the dorsolateral aspect of the ribs as laterally flattened muscle slips. The muscle fibers run dorsoposteriorly and terminate via tendon fibers that sweep dorsally and join the SDS. The tendons of the costal portion of the m. (ilio)costalis are very long. The most terminal of these tendons runs through the SDS into the peripheral connective tissue of the dorsal fin.

Dorsal to the midlateral line, other m. (ilio)costalis muscle fibers originate off the superficial surface of the lateral sheath. These muscle fibers are enclosed in a connective tissue tunnel formed by the SDS and lateral sheath (Fig. 4C). The muscle fibers run posterolaterally and insert onto the deep surface of the SDS either directly or via short tendons. This muscle is composed of rather discrete muscle fiber bundles with numerous short tendons throughout.

The morphology of the m. (ilio)costalis suggests that its action is to draw the rib cage posteriorly. It might also act to compress the vertebral ribs.

M. INTERTRANSVERSARIUS CAUDAE DORSALIS

Origin Superficial surface of lateral sheath posterior to midregion of dorsal fin. Dorsal surface of transverse processes (lateral to lateral sheath) from midlumbar vertebrae to Ca13/14. Lateral aspect of vertebral body from Ca13/14 to Ca15/16.

Insertion SDS directly lateral to m. intertransversarius caudae dorsalis. Anterior edges of transverse processes. Lateral aspect of vertebral bodies from Ca15/16 to Ca19.

Relations Medial to SDS. Lateral to lateral sheath.

Description The m. intertransversarius caudae dorsalis (Figs. 3E and 4D) originates at the level of the termination of the m. (ilio)costalis in the midregion of the dorsal fin. It begins as a dorsally elongated, laterally flattened muscle. As it runs posteriorly, it broadens laterally to occupy a larger surface area of the transverse processes. Like the m. (ilio)costalis, it runs through a connective tissue tunnel formed by the lateral sheath and the SDS (Fig. 4D).

Its muscle fibers originate off the superficial surface of the lateral sheath and the dorsal aspect of the transverse processes, and they insert via short tendons that run through the body of the muscle. The tendons insert either onto the deep surface of the SDS (just lateral to the muscle) or onto posterior transverse processes. In the posterior caudal vertebrae, the tendons insert onto the bodies of the vertebrae. The muscle fibers, like those of m. (ilio)costalis, are relatively short, generally crossing one to two intervertebral joints.

The morphology of the m. intertransversarius caudae dorsalis suggests its action is to aid the m. multifidus and m. longissimus in extending the caudal peduncle. If it contracts unilaterally, its action would be to draw the caudal peduncle laterally.

Ventral Connective Tissues

Ventral Connective Tisssue Sheaths

The ventral axial muscles (m. hypaxialis lumborum and m. intertransversarius ventralis) are also wrapped by the SDS, but not as completely as the dorsal muscles. Anterior to the anus (where the m. hypaxialis lumborum lies within the abdominal cavity), the SDS covers only the lateral surface of the m. hypaxialis lumborum. Posterior to the anus, the SDS completely surrounds the ventral axial muscles (Fig. 1).

The SDS provides a connective tissue layer, the ventrolateral sheath, that dives through the ventral muscles to insert onto the vertebral column (Figs. 4D and 5A). This horizontally oriented septum separates the m. intertransversarius caudae ventralis from the m. hypaxialis lumborum; it runs from Ca3 to Ca15/17.

The surface of the m. hypaxialis lumborum that lies within the abdominal cavity is covered by the abdominal sheath, a connective tissue sheath that is autonomous from the SDS (unlike the posterior, lateral, and ventrolateral sheaths). The abdominal sheath is formed by two layers of connective tissue fibers (Fig. 5B). The fibers in one layer are oriented at oblique angles to the long axis of the dolphin and to the fibers in the other layer (which also run at oblique angles to the body long axis). The abdonial sheath originates within the m. hypaxialis lumborum and inserts onto the ventral aspects of the vertebral bodies. It forms a connective tissue girdle over the ventromedial surface of the m. hypaxialis lumborum.

Ventral Tendons

There are two major ventral tendon systems: the ventral tendon and the medial tendon. Each of

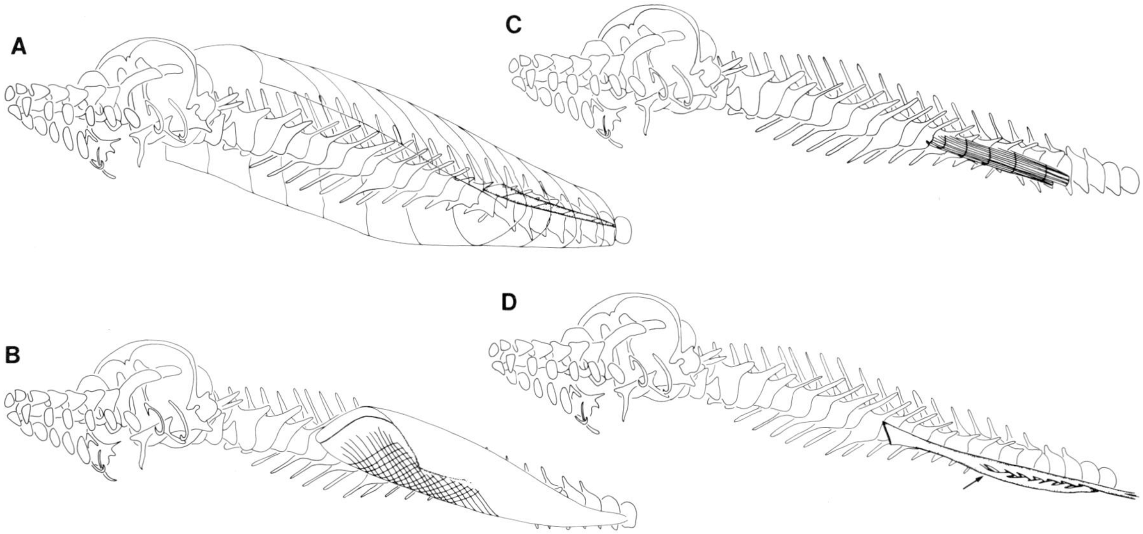

Figure 5 Ventral connective tissues. (In these orientations, you are standing beneath the dolphin and looking up at its ventral and left-lateral surfaces.) (A) Ventrolateral sheath. The ventrolateral sheath is seen through a "transparent" subdermal connective tissue sheath (SDS). Only the contours of the SDS and the ventrolateral sheath are indicated, not the orientation of connective tissue fibers. (B) Abdominal sheath. The m. hypaxialis lumborum is outlined in this view (see Fig. 6B). The thin, obliquely oriented lines indicate the connective tissue fibers that form the abdominal sheath. (C) Ventral tendon. The thin, longitudinal lines indicate the connective tissue fibers that form the tendon. The thick, circumferentially oriented lines indicate the cross-sectional shape of the tendon sheath. (D) Medial tendon. Anteriorly, the tendons form a laterally compressed two-dimensional sheet; they coalesce posteriorly to form cylindrically shaped tendons. The first tendon to separate from the coalesced tendons (arrow) runs ventrally to the other tendons; it fans out medially and dorsally to form a tunnel through which the more posteriorly placed tendons travel.

these tendons originates within the m. hypaxialis lumborum and inserts posteriorly onto the vertebral column.

VENTRAL TENDON

Origin Ventral belly of m. hypaxialis lumborum.

Insertion Ventral aspects of vertebral bodies from midlumbar vertebrae to Ca9. Chevrons from Ca1 to Ca9. Deep surface of SDS from Ca5/6 to Ca9.

Relations Within body of m. hypaxialis lumborum.

Description The ventral tendon (Figs. 4C, D and 5C) is composed of many long, flat tendon fibers that arise in the anterior lumbar region of the m. hypaxialis lumborum. These fibers run medially (at a low angle to the long axis of the body) and insert onto the lumbar and caudal regions of the vertebral column, and further posteriorly, onto the deep surface of the SDS. The tendon fibers are organized into a curved sheet that cuts through the muscle horizontally and incompletely divides the m. hypaxialis lumborum into dorsal and ventral parts.

MEDIAL TENDON

Origin Dorsal belly of m. hypaxialis lumborum.

Insertion Ventral aspect of vertebral bodies and chevrons of Ca4/5 to Ca26. Deep surface of ventral SDS between Ca5/6 and Ca26.

Relations Within body of m. hypaxialis lumborum. Medial to SDS from Ca9/10 posteriorly.

Description The medial tendon (Figs. 4C, D and 5D) is composed of many long, flat tendon fibers

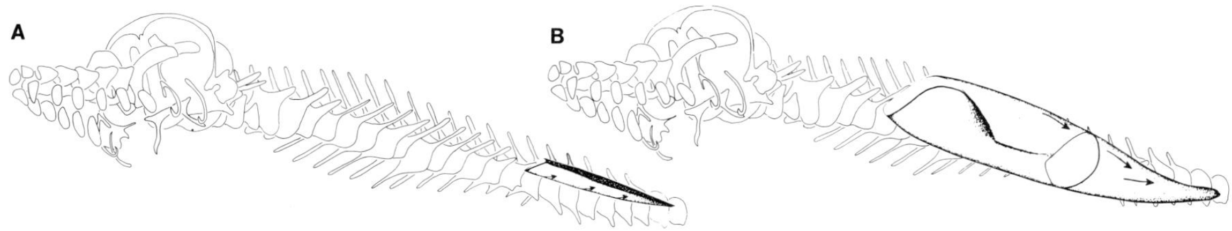

Figure 6 Ventral muscles. (In these orientations, you are standing beneath the dolphin and looking up at its ventral and left-lateral surfaces.) Arrows indicate approximate orientations of muscle fibers. Cross-sectional shape is indicated at various points along the muscles. In A and B the anterior extent of the muscles was not "captured" in a cross section. Thus, these muscles seem to appear with a well-defined cross section. (A) M. intertransversarius caudae dorsalis. (B) M. hypaxialis lumborum. © D. A. Pabst, reproduced by permission of the author.

that are of larger diameter than the fibers of the ventral tendon. These fibers originate deep within the dorsal portion of the m. hypaxialis lumborum (discrete tendons first appear in the midlumbar region). Anteriorly, these tendons are arranged in a vertical sheet that incompletely divides the m. hypaxialis lumborum into medial and lateral parts.

As the tendon fibers run posteriorly they coalesce to form cylindrically shaped tendons. The most superficial (ventrally placed) six to seven of these thickened tendons flatten posteriorly to insert broadly across the chevron of Ca4 to Ca8 (shared insertion with the ventral tendon). The deeper (more dorsally placed) six to seven tendons run posterior to Ca8 and split at their insertions to insert onto two to three adjacent vertebrae.

The deep six to seven tendons have an interesting pattern of insertion. The more superficial tendons flatten out and insert broadly across the chevrons and bodies of the caudal vertebrae and onto the deep surface of the ventral SDS. In so doing, they from a continuous connective tissue tunnel (retinaculum) through which the deeper tendons travel to their terminal insertions. The deepest of the medial tendon fibers run into the flukes to Ca26.

Ventral Division Axial Muscles (Musculi Coccygei and Musculi Membri Inferioris)

M. HYPAXIALIS LUMBORUM

Origin Ventral aspect of T9/11 to Ca18 (including chevrons). Medial surface of rib(s) 10/11 to terminal rib. Deep surface of overlying SDS.

Insertion Abdominal sheath. Ventral tendon. Medial tendon.

Relations Dorsal to m. rectus abdominus and m. levator ani. Ventral to m. intertransversarius ventralis and ventrolateral sheath. Medial to SDS.

Description The m. hypaxialis lumborum (Figs. 4C, D and 6B) is a continuous muscle mass. Although different regions of the muscle insert via different tendon systems, it is difficult in the bottlenose dolphin to identify, and thus name, discrete parts, as Strickler (1980) has done for the m. hypaxalis lumborum (note spelling difference) of *Pontoporia blainvillei*. [The m. hypaxialis lumborum of *Delphinus delphis* and *Phocoenoides dalli*, though, are more easily separated into discrete parts (D.A. Pabst, unpublished data).] Therefore, I discuss general regions of muscle but do not name them as separate entities.

The m. hypaxialis lumborum originates off the deep surface of the ribs and thoracic vertebrae as a dorsoventrally flattened muscle. It deepens posteriorly, using the deep surface of the SDS as an added site of origin. The robust muscle tapers to its termination at Ca16/17.

The muscle fibers of the m. hypaxialis lumborum run from their origins at low angles to the long axis of the body and insert via three connective tissue systems. In the ventrolateral region of the m. hypaxialis lumborum, the muscle fibers insert via the abdominal sheath. The muscle fibers in the ventromedial region of the m. hypaxialis lumborum insert via the ventral tendon. The muscle fibers dorsal to the ventral tendon insert via the medial tendon.

The morphology of the m. hypaxialis lumborum suggests that it has at least three actions. (1) The ventrolateral muscle fibers tense the abdominal sheath. The stiffened abdominal sheath might act to resist bulging of the m. hypaxialis when the muscle contracts (i.e., as a girdle). (2) The ventromedial muscle fibers use the ventral tendon to flex the anterior caudal peduncle. (3) The medial muscle fibers use the medial tendon to flex the posterior caudal peduncle and to flex, or control, the angle of the flukes.

M. INTERTRANSVERSARIUS CAUDAE VENTRALIS

Origin Ventral aspect of transverse processes (lateral to ventrolateral sheath) of Ca3 to Ca15/16. Dorsal surface of ventrolateral sheath from Ca3 to Ca15/16.

Insertion Deep surface of SDS directly lateral to m. intertransversarius caudae ventralis. Lateral aspect of vertebral bodies from Ca15/16 to Ca18.

Relations Dorsal to m. hypaxialis lumborum and ventrolateral sheath. Medial to SDS.

Description The m. intertransversarius caudae ventralis (Figs. 4D and 6A) originates off the ventral aspect of the transverse processes of Ca3 and the dorsal surface of the ventrolateral sheath as a dorsoventrally flattened muscle, which expands as it runs posteriorly. The muscle is similar to the m. intertransversarius caudae dorsalis in its construction, that is, it is composed of short muscle bundles that insert via short tendons onto the deep surface of the SDS or onto the lateral aspect of posterior caudal vertebrae.

The morphology of the m. intertransversarius caudae ventralis suggests that its action is to aid the m. hypaxialis lumborum in flexion of the caudal peduncle. If it contracts unilaterally, its action would be to draw the caudal peduncle laterally.

DISCUSSION

I stated that a comprehensive view of cetacean axial morphology would (1) identify and describe the muscle and connective tissue structures (in this case, of the bottlenose dolphin) and (2) review and synonymize previously published axial anatomies. Table 1 addresses the second requirement. I could determine muscle synonymies for each of the 14 species, representing (conservatively) five families of odontocetes, presented in Table 1. I have dissected the axial muscles of 7 of the species listed in Table 1, and although there are (often subtle) species-specific features of the axial muscles, the general axial body plan is very conservative. As Strickler (1980) pointed out, the axial anatomy in many species of cetaceans is "remarkably similar" (a sentiment echoed by Berzin, 1972, Murie, 1871, and Parry, 1949).

I would argue that the axial morphology of all species of odontocetes can be adequately described using the nomenclature presented here. The value of adopting such a nomenclature is at least threefold. First, an established nomenclature offers a foundation from which one can challenge the criteria for homology used by Slijper (1936), Strickler (1980), and herein. As stated at the start, these decisions on homology are disputable. Second, the recognition of the basic conservative structure of the axial msucles of odontocetes offers a foundation from which to investigate interesting species-specific differences in muscle arrangement. Third, a unified view of the axial morphology of odontocetes offers a common ground for the discussion of the locomotor function of these muscles and connective tissues (Slijper, 1946; Smith *et al.*, 1976; Strickler, 1980). The identification of interactive muscle and connective tissue structures is the first step in understanding the design of this unique mammalian locomotor system.

ACKNOWLEDGMENTS

I would like to thank James Mead, Charley Potter, and William McLellan of the Marine Mammal Program, Smithsonian Institution, for the generous use of their laboratory facilities and access to most of the specimens used in this study. I would also like to thank Sea World of Orlando for making a specimen available to me. Stephen A. Wainwright offered constant support and helpful criticism. Kathleen Smith generously offered use of her time, computer, and the three-dimensional reconstruction program used for the figures. Steven Vogel, S. A. Rommel, August Pivorunas, and an anonymous reviewer made helpful suggestions on early drafts of the

Table 1
Axial Muscle Synonymies of Some Odontocetes

Authors, Species	Muscle						
	Semispinalis	Multifidus and ECM	Longissimus and ECL	(Ilio)costalis	Intertransversarius Caudae Dorsalis	Hypaxialis	Intertransversarius Caudae Ventralis
Murie (1870, 1871, 1873), *L. albirostris, G. griseus, G. melas*	Splenius (may include complexus)	Spinalis dorsi; levator caudae internus	Transversalis cervicis; longissimus dorsi; levator caudae externus; deep semispinalis fibers	Sacrolumbalis; cervicis ascendens (sans cranial extension)	Supracaudal [1870: transversarius superior]	Sacro-coccygeus [1870: infracaudalis]	Infracaudal [1870: transversarius inferior]
Purves and Pilleri (1978), *P. crassidens*	Semispinalis	Spinalis dorsalis	Longissimus capitis; longissimus dorsi	Cervicalis ascendens; anterior lumbocostalis	Supracaudal	Quadratus lumborum; psoas major and minor; sacrococcygeus	Infracaudal
Pilleri *et al.* (1976), *D. delphis, P. indi (= minor)*	Semispinalis capitis	Spinalis dorsi	Longissimus capitis; longissimus dorsi	Iliocostalis	Supracaudal	Quadratus lumborum; psoas major and minor; pubo- and sacrococcygeus	Infracaudal
Agarkov and Lukhanin (1970), *D. delphis*	—	Spinalis	Longissimus	Iliocostalis	Intertransversarius superior	Hypaxialis	Intertransversarius inferior
Slijper (1936), *P. phocoena, D. delphis*	Cranial extension of multifidus; semispinalis	Multifidus; longissimus dorsal to deep tendon	Longissimus capitis; longissimus cervicis; longissimus dorsi ECL	Iliocostalis cervicis and dorsi; intertransversarius dorsalis from midlumbars to anterior caudals	Iliocostalis lumborum; intertransver-sarius caudae dorsalis	Hypaxialis lumborum (flexor caudae medialis and lateralis)	Intertransversarius caudae ventralis
Sokolov and Rodionov (1978), delphinid spp.	Semispinalis capitis	Spinalis et semispinalis; ECM = longissimus pars externus	Longissimus and longissimus pars internus	Iliocostalis pars dorsalis (sans cranial extension)	Iliocostalis pars lumborum and intertransver-sarii superiores	Hypaxialis pars lateralis; hypaxialis pars medialis	Intertransversarii inferiores
Parry (1949), *P. phocoena*	Epiaxial muscle	Epiaxial muscle	Epiaxial muscle	Lateral muscle	Lateral muscle	Hypoaxial muscle	Lateral muscle
Howell (1927, 1930), *N. phocaenoides, M. monoceros*	Semispinalis capitis; [1927: includes biventer cervicis and complexus] [1930: prefers longissimus capitis]	Longissimus dorsi	[1927: longissimus capitis et cervicis; iliocostalis lumborum] [1930: iliocostalis capitis and lumborum]	[1927: anterior intertransversarius superior] [1930: intertransversarius thoracic]	[1927: posterior intertransver-sarius superior] [1930: posterior intertransver-sarii]	Hypaxialis	Intertranversarius inferior
von Schulte and Smith (1918), *K. breviceps*	Semispinalis capitis	Multifidus; spinalis dorsi; longissimus (from midthoracics caudally)	Anterior extension of longissimus; iliocostalis thoracis and lumborum	Anterior transversarius superior	Posterior transversarius superior	Hypaxial muscle	Transversarius inferior
Berzin (1972), *P. catodon*	Splenius	Spinalis dorsi	Longissimus; longissimus dorsi	Iliocostalis (sans cranial extension)	Intertransversarii	Iliolumbar muscle (with many subdivisions)	Intertransversarii
Zhou and Qian (1981), *L. vexillifer* (figure only)	Semispinalis; spinalis capitis	Spinalis	Longissimus dorsi	Iliocostalis	Supracaudalis	Hypaxialis	Infracaudalis
Strickler (1980), *P. blainvillei*	Anterior semispinalis	Multifidus; posterior semispinalis fibers	Longissimus capitis, cervicis and dorsi; ECL; most posterior semispinalis fibers	Iliocostalis	Intertransversarius caudae dorsalis	Hypaxalis lumborum (pars dorso-lateralis, pars ventro-medialis)	Intertransversarius caudae ventralis

manuscript. William McLellan helped transport and process many specimens, and almost single-handedly cross-sectioned two dolphins. This study was funded in part by a Smithsonian Graduate Student Fellowship, a Sigma Xi Grant-in-Aid of Research, a Lincoln Ellsworth Fund grant, and a Training Grant in Morphology from the Cocos Foundation. All figures are copyrighted by the author.

REFERENCES

Agarkov, G. B., and Lukhanin, V. Ya. (1970). Propulsive tail musculature of the Atlantic common dolphin. *Bionika* **4,** 77–81.

Agarkov, G. B., and Sych, V. F. (1974). Morphofunctional analysis of the epaxial musculature of the propulsive complex of the porpoise. *Bionika* **8,** 69–76.

Bello, M. A., Roy, R. R., Martin, T. P., Goforth, H. W., Jr., and Edgerton, V. R. (1985). Axial musculature in the dolphin (*Tursiops truncatus*): Some architectural and histochemical characteristics. *Mar. Mamm. Sci.* **1**(4), 324–336.

Berzin, A. A. (1972). "The Sperm Whale (*Kashalot*)" (A. V. Yablokov, ed.). [transl. from Russian by E. Hoz and Z. Blake. Israel Program for Scientific Translations, Jerusalem.]

Eisler, P. (1912). "Die Muskelin des Stammes," pp. 384–385. Gustav Fischer, Jena.

Howell, A. B. (1927). Contribution to the anatomy of the Chinese finless porpoise, *Neomeris phocaenoides*. *Proc. U.S. Nat. Mus.* **70**(13), 1–43.

Howell, A. B. (1930). Myology of the narwhal (*Monodon monoceros*). *Am. J. Anat.* **46**(2), 187–215.

Lauder, G. V. (1982). Historical biology and the problem of design. *J. Theor. Biol.* **97,** 57–67.

Murie, J. (1870). Notes on the white-beaked bottlenose, *Lagenorhynchus albirostris*, Gray. *J. Linn. Soc., Zool.* **11,** 141–152.

Murie, J. (1871). On Risso's grampus, *G. rissoanus* (Desm.). *J. Anat. Physiol.* **5,** 118–138.

Murie, J. (1873). On the organization of the caaing whale, *Globicephalus melas*. *Trans Zool. Soc. London* **8**(4), 235–301.

Nomina Anatomica. (1966). Excerpta Medica Foundation, New York. [As cited in Smith *et al.* (1976). I used *Nomina Anatomica* (1983), 5th Ed. Williams & Wilkins, Baltimore, Maryland.]

Nomina Anatomica Veterinaria. (1968). New York State Veterinary College, Ithaca, New York. [as cited in Smith *et al.* (1976). I used *Nomina Anatomica Veterinaria* (1972), 2nd Ed. International Committee on Veterinary Anatomical Nomenclature. New York State Veterinary College, Ithaca, New York.]

Pabst, D. A. (1987). A membranous force transmission system in dolphins. *Am. Zool.* **27**(4), 86A.

Parry, D. A. (1949). The anatomical basis of swimming in whales. *Proc. Zool. Soc. London* **119**(I), 49–60.

Pilleri, G., Gihr, C., Kraus, C., Purves, P. E., and Zbinden, K. (1976). Comparative study of the skin and general myology of *Platanista indi* and *Delphinus delphis* in relation to hydrodynamics and behavior. *Investigations on Cetacea* **6,** 90–127.

Purves, P. E., and Pilleri, G. (1978). The functional anatomy and general biology of *Pseudorca crassidens* (Owen) with a review of the hydrodynamics and acoustics in Cetacea. *Investigations on Cetacea* **9,** 67–227.

Roth, V. L. (1984). On homology. *Biol. J. Linn. Soc.* **22,** 13–29.

Schulte, H. von W., and Smith, de F.M. (1918). The external characters, skeletal muscles and peripheral nerves of *Kogia breviceps* (Blainville). *Bull. Am. Mus. Nat. Hist.* **38**(2), 7–72.

Slijper, E. J. (1936). Die Cetaceen, vergleichend-anatomisch und systematisch. *Capita. Zool. Bd VI and VII* **7,** 1–600.

Slijper, E. J. (1946). Comparative biologic–anatomical investigations on the vertebral column and spinal musculature of mammals. *Verh. Akad. Ned. Wet., Amst., Sect. 2* **42,** 1–128.

Smith, G. J. D., Browne, K. W., and Gaskin, D. E. (1976). Functional myology of the harbor porpoise, *Phocoena phocoena* (L.). *Can. J. Zool.* **54,** 716–729.

Sokolov, V. Ye., and Rodionov, V. A. (1978). Weight characteristics of muscles of Black Sea dolphins. *Zool. Z.* **57**(2), 272–279.

Strickler, T. L. (1980). The axial musculature of *Pontoporia blainvillei*, with comments on the organization of this system and its effect on fluke-stroke dynamics in the Cetacea. *Am. J. Anat.* **157,** 49–59.

Wainwright, S. A., Biggs, W. D., Currey, J. D., and Gosline, J. M. (1976). "Mechanical Design in Organisms." Princeton University Press, Princeton, New Jersey.

Wainwright, S. A., Pabst, D. A., and Brodie, P. F. (1985). Form and possible function of the collagen layer underlying cetacean blubber. *Am. Zool.* **25**(4), 146A.

Zhou, K., and Qian, W. (1981). The myology of the baiji, *Lipotes vexillifer* (Odontoceti: Lipotidae) I: Skin muscles, forelimb muscles, trunk and tail muscles. *Acta Theriol. Sinica* **1**(2), 117–126.

APPENDIX 1

CONVENTIONS AND ABBREVIATIONS

C,	Cervical
T,	Thoracic
L,	Lumbar
Ca,	Caudal
Ca3/4,	A slash (/) indicates that there is some individual variation in the placement of the feature.
Ca3 to Ca4,	Feature is continuous between these two landmarks.
S,	M. semispinalis
M,	M. multifidus
ECM,	M. extensor caudae medialis
L,	M. longissimus
ECL,	M. extensor caudae lateralis
I,	M. (ilio)costalis
ID,	M. intertransversarius caudae dorsalis
IV,	M. intertransversarius caudae ventralis
H,	M. hypaxialis lumborum
EO,	M. obliquus externus abdominis
IO,	M. obliquus internis abdominis
SDS,	Subdermal connective tissue sheath
PS,	Posterior connective tissue sheath
LS,	Lateral connective tissue sheath
ST,	Superficial tendon
DT,	Deep tendon
ECLT,	Tendons of the m. extensor caudae lateralis
VLS,	Ventrolateral connective tissue sheath
VT,	Ventral tendon
MT,	Medial tendon

APPENDIX 2

TURSIOPS TRUNCATUS SPECIMENS EXAMINED

Lateral dissections were done on all specimens except where noted. WAM 140 was fixed in buffered formalin; all other specimens were dissected fresh.

Identification number	Total length (cm)	Sex
USNM 550772	270.0	F
USNM 550827	168.5	F
USNM 550832 (cross sectioned)	101.7	M
USNM 550835	106.0	F
USNM 550852	247.5	M
No number (cross sectioned)	170.9	F
CWP 248	228	F
NZPX-3080	201.5	M
RAB 073	172.3	M
WAM 140	109.0	M
WAM 162	250+	F
WAM 164	240	M
WAM 198	273	M
WAM 282	157.5	M
WAM 183	100+	M

APPENDIX 3

EXPLANATION OF FIGURES

The "no number" specimen (female, 170.9 cm) was cross-sectioned with a bow saw into sections 3–5 cm thick. These sections were photographed, computer digitized, and transformed into three-dimensional reconstructions (PC3D, by Jandel Scientific). The images were rendered in pen and ink to "flesh out" the muscles relative to the skull and skeleton (which are shown as series of sections). Fluke vertebrae are not included, and not all cervical, thoracic, lumbar, and caudal vertebrae are represented in the figures (as there was often more than one vertebra per cross section). The irregular cross-sectional shapes of the vertebrae are artifacts of the placement of the saw cut; chevrons are included in the outline of the caudal vertebrae.

The Central Nervous System of the Bottlenose Dolphin

4

Sam H. Ridgway
Naval Ocean Systems Center
San Diego, California 92132

INTRODUCTION

Tursiops is among the most ecologically diverse cetacean genera, with different populations occupying offshore and inshore niches. Inshore groups sometimes spend hours, days, or even weeks in freshwater rivers or lakes connected to the sea while offshore groups may be seen hundreds of miles at sea. The genus *Tursiops* also has a wide distribution: different individuals from various populations might be seen over a north–south range covering about 80% of the distance from the Arctic and Antarctic circles to the equator during the course of a year.

Dolphin brains probably have evolved in response to their ecological niches (Wood and Evans, 1980; Ridgway and Wood, 1988). Since bottlenose dolphins are diverse in distribution, ecology, and adult body size, a comparative study of brain size among adult dolphins of different bottlenose dolphin populations around the world might shed light on the relationship between brain size, body size, and ecology. I have not completed such a study, but in this chapter I review published information on the brain of the genus *Tursiops* and provide some comparative data with other genera and among a few different bottlenose dolphin populations. In addition, I intend this chapter to guide readers toward the relevant brain literature and introduce them to some of the problems posed by the large, convoluted brain of the bottlenose dolphin (also see Ridgway, 1986a). In this brief discussion, I have not mentioned the thalamus (see Kruger, 1959), the ventricles (McFarland *et al.*, 1969), the fissure pattern of the convexity hemispheres (Morgane *et al.*, 1980), many nuclei such as the red nucleus, the basal ganglia, and other structures. These features are described in some of the literature I have cited.

BRAIN SIZE

In keeping with their diverse body sizes and habitats, cetaceans have variably sized brains, ranging from 200 g in adult franciscanas, *Pontoporia blainvillei,* and susus, *Platanista gangetica,* to a reported high of 9200 g for the largest odontocete, the sperm whale (Kojima, 1951). Although 16 sperm whale brains averaged just over 7800 g and brains of mysticetes can weigh 2600–7800 g, certain species of the family Delphinidae, a group of small- to medium-sized odontocetes which includes the bottlenose dolphin, have the highest brain size to body size ratio among cetaceans. Only in this family have animals with a body size within two or three multiples of human body size attained similar brain weights, for example, *Tursiops* (Fig. 1) about 1500 g, *Steno* 1400 g, and *Lagenorhynchus* 1300 g.

The adult male bottlenose dolphin of 172 kg

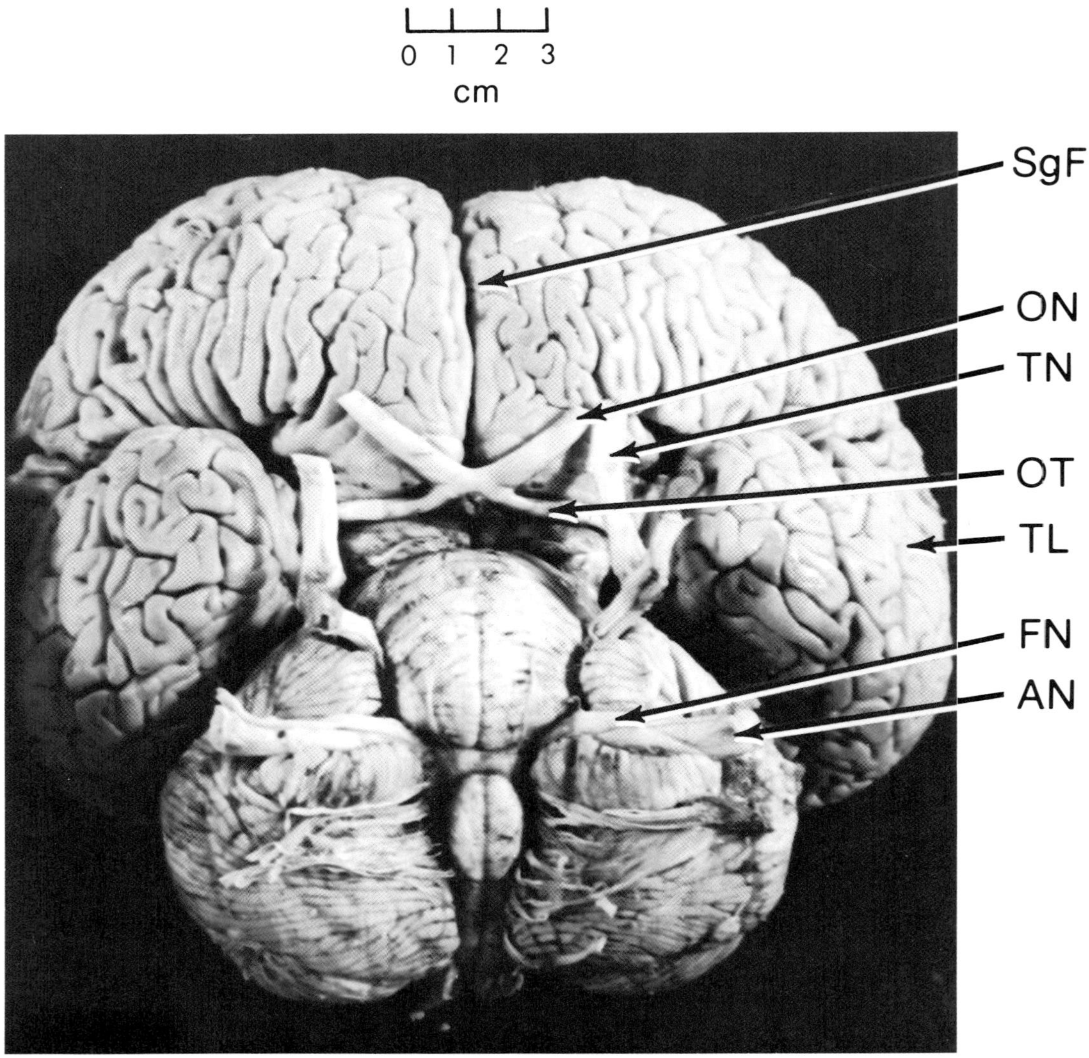

Figure 1 Ventral surface of the brain of an adult bottlenose dolphin. SgF, Sagittal fissure; ON, optic nerve; TN, trigeminal nerve; OT, optic tract; TL, temporal lobe; FN facial nerve; AN, auditory nerve.

body weight listed in Table 1 had a brain of 1629 g with an encephalization quotient almost four and one-half times larger than an average mammal of the same size (cf. Jerison, 1973; Wood and Evans, 1980). The bottlenose dolphin's closest rival in brain size, other than a few related delphinids and humans, is the gorilla, *Gorilla gorilla*. A male gorilla of 172 kg had a brain of 570 g (Jerison, 1973, p. 393).

While some authors (e.g., Lilly, 1962, 1964) have viewed the large dolphin brain as an indication of high intelligence and possibly even of language capability, others such as Tomilin (1968) have regarded the large dolphin brain simply as an adaptation to a specialized aquatic life. Questions about dolphin intelligence and language remain largely unanswered, and there is no general agreement as to what adaptations to aquatic life could be responsible for a large brain. Sirenians, mammals also totally adapted to an aquatic existence, have relatively small brains.

Some data on brain and body sizes of adult dolphins (Sergeant *et al.*, 1973, list maturity criteria) from the coastal western North Atlantic and Gulf of Mexico, North Pacific, and eastern North Atlantic or Mediterranean Sea are given in Table 1.

BRAINS OF CAPTIVE DOLPHINS

Pilleri (1983) suggested that dolphins' brains shrink as much as 30% when the animals are maintained in captivity. This hypothesis is not based on dolphin data but instead mainly on comparisons of domestic and wild species of land mammals. In reviewing some of Pilleri's sources, I found that the early land mammal studies mentioned in his table had been questioned by Hemmer (1976), who concluded that these studies often were based on unfounded comparisons between wild and domestic groups; for example, domestic animals were compared with species that were not closely related in phylogeny. Moreover, a study by Fram and Stephan (1976) on the tree shrew, *Tupaia glis*, provides evidence which contradicts Pilleri's suggestion that captivity causes a reduction in brain size. Fram and Stephan dissected and weighed the brains of a group of recently trapped wild tree shrews. When they compared these brains to the brains of captive-reared offspring whose parents came from the same population as the wild group, Fram and Stephan found that the brains of the captive-reared tree shrews were significantly larger.

Although Pilleri's hypothesis about captive dolphins appears questionable, it is true that poor nutrition at critical stages of early brain development can result in retardation of brain growth. For example, animals removed from the wild before weaning might be vulnerable to such retardation if their nutritional requirements are not met during the first weeks or months of captivity. In addition, mammalian brain size depends on age. An individual's brain reaches full size just before or soon after sexual maturity and then slowly shrinks. In humans 70–80 years old the brain is more than 100 g smaller, on average, than in 19–20 year olds (Spector, 1956). We know something about the growth of dolphins' brains to maturity (Ridgway, 1986a), but we know nothing of their shrinkage in old age. The question of whether a dolphin's brain shrinks with age can only be answered if careful records are kept on an adequate sample of specimens whose brains are examined and weighed at death.

Because Pilleri's (1983) statement on dolphin brain atrophy has been repeated by others (e.g., Meith, 1985), it prompted me to analyze postmortem records to test whether or not bottlenose dolphins' brains became relatively smaller with increasing numbers of years in captivity. Figure 2 is a regression of body length to brain weight ratio against years in captivity for 18 adult Atlantic bottlenose dolphins. Note that there is no significant difference in this ratio between newly captured animals and those that died after having been in captivity for 5, 10, or more years.

SHAPE AND GROWTH OF THE DOLPHIN BRAIN

In the embryonic stage, the dolphin brain, like those of other mammals, is somewhat oblong in shape (longer than wide) (Kamiya and Pirlot, 1974). By the midfetal stage the odontocete brain appears globular, and by the late fetal stage the brain is wider than long, as it is in the adult (Figs. 1

Table 1
Brain and Body Size of Adult Bottlenose Dolphins[a]

Category		Sex	BdL (cm)	BdW (kg)	BrnW (kg)	Source
E. North Atlantic and Mediterranean			300	240	2.020	Pilleri and Gihr (1970)
			285	200	1.930	Pilleri and Gihr (1970)
			290	220	2.230	Pilleri and Gihr (1970)
			290	220	2.240	Pilleri and Gihr (1970)
			340	250	2.010	Pilleri and Gihr (1970)
		F	296	278	1.886	Weber (1897)
	Mean		300	235	2.053	
	S.D.		18	25	0.137	
E. North Pacific		M	291	262	1.901	Ridgway (unpublished)
		F	283	255	1.672	Ridgway (unpublished)
		M	261	210	1.555	Ridgway (unpublished)
		M	285	268	1.691	Ridgway and Brownson (1979)
	Mean		280	249	1.705	
	S.D.		11	23	0.125	
W. North Atlantic coastal		F	240	140	1.588	Kruger (1959)
		F	259	156	1.685	Kruger (1959)
		F	256	154	1.707	Kruger (1959)
		F	240	177	1.558	Ridgway and Brownson (1979)
		F	251	151	1.673	Ridgway and Brownson (1979)
		F	244	164	1.446	Ridgway and Brownson (1979)
		M	257	186	1.849	Ridgway and Brownson (1979)
		M	244	170	1.408	Ridgway and Brownson (1979)
		M	252	152	1.625	Ridgway and Brownson (1979)
		M	240	170	1.524	Ridgway and Brownson (1979)
		F	245	172	1.629	Ridgway and Brownson (1979)
		F	235	120	1.296	Ridgway *et al.* (1966)
		F	244	160	1.359	Ridgway *et al.* (1987)
		M	238	165	1.378	Ridgway *et al.* (1987)
		F	227	140	1.112	Ridgway *et al.* (1987)
		F	240	145	1.487	Ridgway *et al.* (1987)
		F	252	197	1.609	Ridgway *et al.* (1987)
		F	242	225	1.410	Ridgway *et al.* (1987)
		M	263	198	1.724	Ridgway (unpublished)
	Mean		246	165	1.530	
	S.D.		8.9	24	0.172	

[a] Data are given for sex, body length (BdL), body weight (BdW), and brain weight (BrnW) of adult bottlenose dolphins from three different geographic areas: eastern North Atlantic and Mediterranean, eastern North Pacific, and western North Atlantic Coastal (including Gulf of Mexico). Specimens from the literature were judged to be mature if they exceeded the mean adult length criteria of Sergeant *et al.* (1973). Some animals that did not exceed the mean adult length are included because they were determined to be sexually mature by other means. For example, the 227 cm female had two offspring with 10 years intervening between the two. The average bottlenose dolphin female would not be sexually mature at 227 cm.

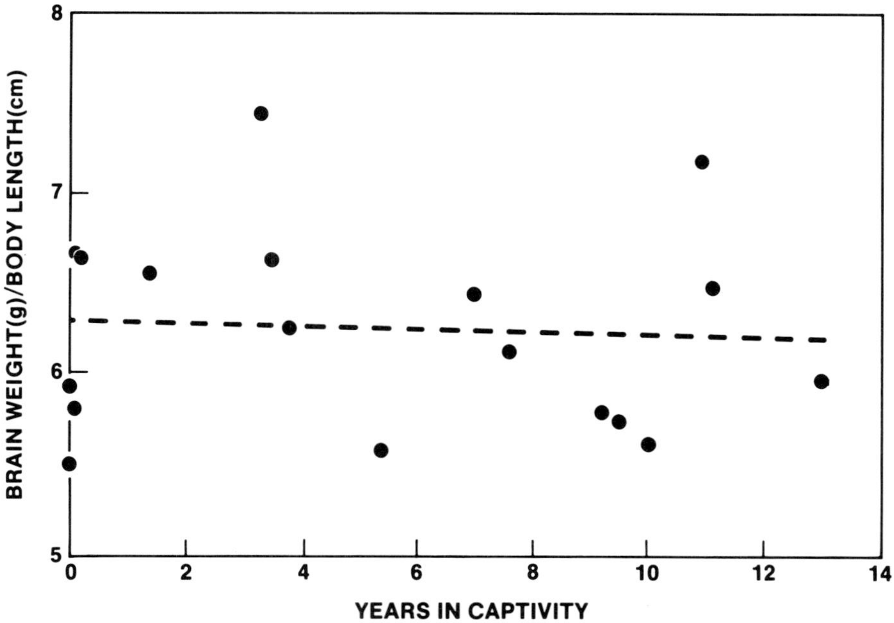

Figure 2 Each point represents the ratio of brain weight to body length for a single bottlenose dolphin brain. The dashed line is a linear regression which shows no significant difference in the brain weight to body length ratio over periods up to 13 years in captivity.

and 3). In postnatal mysticete brains, length and width are more similar. A dolphin brain viewed from the medial aspect of a sagittal section shows that the height is also great compared with the length. An S-shaped curving of the neural axis leaves the axis of the forebrain almost perpendicular to that of the midbrain, giving the impression that the forebrain rotated ventrally during development.

The human brain grows from just under 25% of its adult weight at birth to full development in about 17 or 18 years. J. T. Bonner states, "In man there seems to be an extraordinary prolongation of youth, and this permits the brain to continue its expansion. The period of dependency on parents is increased so that the period when learning can occur becomes relatively long" (Bonner, 1980). Kesarev (1971) has pointed out that human cortex develops mainly in the postnatal years, but that some lower primates are born with cortical formation about 80% of the adult complement. Kesarev states, "The earliest to appear and the latest to conclude their development in man are those formations of the neocortex which may be considered phylogenetically as the most recent and are associated with the most complex forms of cerebral activity, present only in man, the integrative and analytical functions."

Delphinids are also born with brains at an advanced stage of development. A long gestation period, about 12 months in bottlenose dolphins, is needed so the neonate can swim and engage in activities requiring a more mature brain. In eight Atlantic coastal bottlenose dolphins, neonatal brain weight averaged 42.5% of mean adult weight (Ridgway and Brownson, 1984) (Fig. 3). By weaning age at 18 months, bottlenose dolphin brains are over 80% of mean adult weight, a stage not reached by human brains until 3 or 4 years. Male bottlenose dolphins do not reach sexual maturity

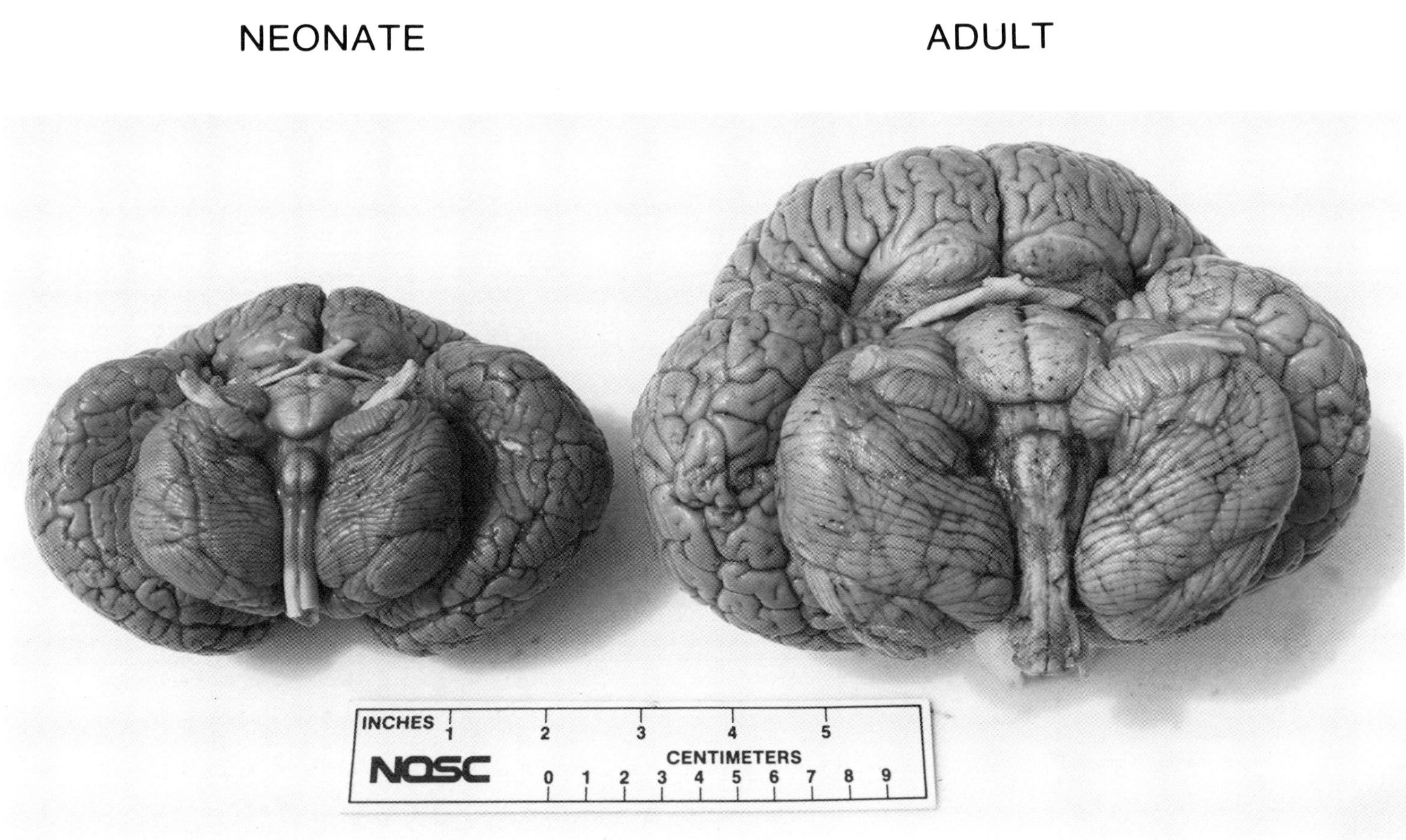

Figure 3 The small brain (662 g) on the left is from an 18-day old male bottlenose dolphin. The larger brain (1411 g) on the right is from a female of the same species that was an adult or nearing adulthood.

until about 13 years of age, while females mature somewhat earlier (Sergeant *et al.*, 1973; also see Schroeder, Chapter 25, this volume). From the sparse data available on animals of known age, I estimate that Atlantic coastal bottlenose dolphins reach full brain development in 9 or 10 years, about half the total growth period required for human brains but considerably longer than for brains of most higher mammals.

THE CEREBRAL HEMISPHERES

The greatest expansion in the nervous system in the course of evolution has involved the cerebral hemispheres. The hemispheres of cetaceans are large and enormously convoluted. Although it is often stated that the human brain is the most convoluted of all brains (e.g., Teyler, 1977), odontocetes exceed humans, and all other animals, in convolutedness or fissurization of the cerebral cortex; the cetacean pattern of gyri and sulci is more complicated than in the human cerebral cortex. A much larger proportion of the dolphin's cortical surface is buried beneath the exterior of each cerebral hemisphere. The main sulci that arch along the dorsal and dorsolateral cerebral surface are very deep and incised with microgyri. Toward the core from the sylvian cleft lies an extremely large insula consisting of radial gyri arising fanlike from the transverse insular gyrus, and covered completely by the frontal, parietal, and temporal opercula. In a series of bottlenose dolphins, mean cortical surface area was 3745 cm^2 compared with 2275 cm^2 in a similar number of human brains (Elias and Schwartz, 1969; also see Haug, 1969, 1970; Ridgway and Brownson, 1984); however, the dolphin's cortex was 1.3–1.8 mm in thickness, considerably thinner than the human cortex which is about 3.0 mm thick. The total volume of bottlenose dolphin cortex was about 80% of that found in humans.

In cetaceans, the most primitive of neocortical regions (Kesarev, 1969), the limbic lobe, is extensive, although the hippocampal formation is much smaller than that of man. In the bottlenose dolphin's limbic lobe, the total number of cells is greater than in the limbic lobe formations of primates, including man (Morgane *et al.*, 1982). Nevertheless, the limbic lobe of the dolphin brain is relatively small compared with the remainder of the neocortex. In the common dolphin, *Delphinus delphis*, Filimonoff (1965) found that periarchicortex constituted about 0.4% of total neocortical surface area as compared with 1.3% in man, 2.7% in dog, and 17.6% in hedgehog, *Erinaceus*. Although these percentages are quite different in dolphin and hedgehog, some authors have drawn parallels in brain structure between these very remotely related mammals based on microstructure and lamination characteristics of the cerebral cortex (Kesarev and Malofeyeva, 1969; Morgane *et al.*, 1986; Glezer *et al.*, 1988).

Following principles of brain organization established by Filimonoff (1949, 1965), Kesarev and Malofeyeva (1969) have suggested that the ancestors of dolphins left land for the sea 70–90 million years ago, before land mammals developed many modern features of brain structure. These authors have pointed out various similarities in brain microstructure and organization between dolphins and primitive insectivores such as the hedgehog. Morgane and associates (1986), who have done even more extensive histological studies, agreed with Kesarev (1969) and suggested that the cetacean cortex as a whole does not appear to have reached the latest stage of cortical evolution, retaining all the conservative characters seen in primitive terrestrial and aerial forms such as hedgehogs and bats. Cortical layer IV (granular layer) is absent or at least difficult to identify. Areas of cortical surface identified as primary sensory cortex by physiological mapping studies are agranular and not distinguished on histological grounds as primary sensory cortex. Glezer *et al.* (1988) elaborated these ideas into an initial brain hypothesis, which holds that the dolphin brain is the result of repeated elaboration of a primitive "initial brain," contrasted with more complex development seen in the brains of land mammals, especially primates.

Proponents of the initial brain hypothesis have generally placed the move of ancestral cetaceans from land to sea much earlier than the fossil record indicates (see van Valen, 1966; Gingerich *et al.*, 1983), and very little independent evidence supports the proposed cetacean–insectivore connection (Ridgway and Wood, 1988). We suggested that brains of cetaceans have evolved in response to their various ecological niches and that the ancestors of cetaceans entered the sea to take advantage of its rich food resources after they had developed large bodies and a predatory lifestyle on land, in marked contrast to the small insectivores. Ridgway and Wood (1988) proposed that to the extent that ontogeny recapitulates phylogeny (Gould, 1977), developmental studies on the cetacean brain would seem to be a rewarding approach toward understanding dolphin brain phylogeny. Garey and Leuba (1986) reported finding a thin layer IV in the visual cortex of two young bottlenose dolphins, one 3 years of age and the other 18 days old. Layer IV was absent in brains of animals age 12, 22, 26, and 33 years. If these findings hold up, they suggest that granular layer IV is a regressing characteristic of cetacean brain evolution and would not seem compatible with the initial brain hypothesis.

As mammalian brains increase in size, the neuron density decreases. According to at least three studies (Hawkins and Olszewski, 1957; Tower, 1954; Morgane *et al.*, 1982), the ratio of glial cells per nerve cell is correlated directly with brain size. In mice, the glial/neuron ratio is around 0.2, in humans approximately 1.8, in bottlenose dolphins about 2.1, and in fin whales, *Balaenoptera physalus*, 4.5. Thus, there is at least a 20-fold decrease in neuron density from mouse to whale.

THE CORPUS CALLOSUM

The corpus callosum (CC) is a large bundle of nerve fibers that connects the two cerebral hemispheres. These fibers apparently are important communication links between the two hemispheres and are necessary for transmission of learned behavior from one hemisphere to the other (Myers, 1956).

Although the average adult bottlenose dolphin has a brain weighing more than that of the average adult human, the CC of the dolphin brain is considerably smaller. Nieto *et al.* (1976) measured the cross-sectional surface area of the CC in a smaller brained dolphin *Stenella graffmani* (*Stenella attenuata*, the pantropical spotted dolphin, Perrin *et al.*, 1987). The dolphin brain of 832 g was compared with a human brain of 1085 g. The cross-sectional surface area of the human CC was 991 mm^2 compared to 180 mm^2 for the *S. graffmani* CC. The CC of the larger bottlenose dolphin brain probably has more cross-sectional area; however, even if brains of identical size were compared, I estimate that the bottlenose dolphin CC has no more than about one-quarter the cross-sectional area of the human CC.

The finding of a smaller cross-sectional area of the dolphin CC leads me to conclude that the dolphin CC is made up of fewer total fibers or that the average fiber diameter is much smaller. This in turn implies that, in comparison with the human brain, communication between hemispheres is at a reduced overall level owing to fewer fibers or is slower because of smaller fiber diameters. Either condition would result in greater hemispheric independence and probably a lesser degree of hemispheric coordination in the dolphin brain.

DOLPHIN BRAIN METABOLISM

Hockett (1978) states, "Brains are metabolically expensive and don't get bigger (phylogenetically) unless in some fashion they are more than paying for their upkeep." This statement, supported by measures of brain metabolism in many vertebrates, gives rise to these questions: Might the cetacean brain represent a departure from this general trend? Could the dolphin brain be large, but metabolically less active than the brains of other mammals?

Many authors have found that dolphins have a higher metabolic rate than average terrestrial mammals of similar size (Irving *et al.*, 1941; Pierce, 1970; Ridgway and Patton, 1971; Hampton *et al.*, 1971). Opposing evidence has been presented by Karandeeva *et al.* (1973) and Lavigne *et al.* (1986), who contend that bottlenose dolphins do not have high rates of metabolism. I am skeptical of the results published by Karandeeva and colleagues because I question their method: they employed a mask and rubber bag placed over the blowhole to collect expired air from Black Sea bottlenose dolphins, later measuring the volume and oxygen content with a gasometer. Karandeeva *et al.* (1973) reported mean respiratory volumes ranging from about 18 to 23 ml/kg, values less than half those reported by Irving *et al.* (1941) and Ridgway *et al.* (1969) for Atlantic bottlenose dolphins and by Kooyman and Cornell (1981) for Pacific bottlenose dolphins. Their measures of oxygen consumptions are therefore equivalently reduced. Based on my own experience in trying to collect bottlenose dolphin lung volume with a mask and rubber bag, I suspect the difference in reported oxygen consumption values between Black Sea and Atlantic bottlenose dolphins may be methodological. Lavigne *et al.* (1986) gave more credence to the Karandeeva data for bottlenose dolphin metabolic rates than to the measurements reported by others (cited at the beginning of this paragraph). In addition, I doubt the findings of Lavigne *et al.* because they employed allometric methods, procedures which I believe may be questionable [see reviews by McNab (1988) and Smith (1984)], with respect to scaling the rate of metabolism and applying such scaling to individual species.

In addition to oxygen consumption data, there is other evidence for a high rate of metabolism. The high-protein diet (dolphins consume almost no carbohydrates) contributes considerably to a high metabolic rate. Bottlenose dolphins have large thyroid glands (Harrison, 1969) and a high level of circulating thyroid hormones (Ridgway and Patton, 1971). Long-term food consumption of captive animals correlates well with observed levels of oxygen consumption. The bottlenose dolphin studied by Ridgway and Patton (1971) required over 6000 k cal/day to maintain body weight. Even higher levels of daily caloric consumption have been reported by other authors (Shapunov, 1973; Van Dyke and Ridgway, 1977; Cates and Schroeder, 1986).

Bottlenose dolphins are diving mammals with the demonstrated capability to reach depths as great as 547 m (see Ridgway, 1986b, for a review of diving depth studies) and to remain submerged for more than 7 minutes. Alveolar gas tensions after long dives by Atlantic coastal bottlenose dolphins (Ridgway *et al.*, 1969) suggest that the brain may be capable of short periods of anaerobic metabolism, a capability not found in adult land mammals that have been studied. Since this observation in bottlenose dolphins, anaerobic metabolism by the seal brain has been demonstrated during the later stages of a maximal dive (Kerem *et al.*, 1971; Simon *et al.*, 1974).

I was unable to confirm anaerobic metabolism during hypoxic periods of up to 6 minutes in anesthetized dolphins. Arteriovenous differences in oxygen and glucose across the dolphin brain were as great as or greater than those in the brains of anesthetized humans. Based on my own data and those of Soviet researchers, I conclude that there is no basis for assuming dolphin brains have a lower overall rate of metabolism than do human brains. The great fissurization of the dolphin brain creates a large cortical surface area; the large area may be essential to provide sufficient surface contact with the abundant meningeal blood supply (Wilson, 1933; Ries and Langworthy, 1937), supporting the cortex's need for oxygen. The view that brain size and fissurization in odontocetes are related to diving and the need to sustain brain function during prolonged hypoxia is, however, weakened by the observation that seals (Scholander, 1940) and sirenians (Scholander and Irving, 1941), two groups of animals that also dive for long periods, have smaller, less convoluted brains and a thicker cortex.

THE VISUAL SYSTEM

Bottlenose dolphins have well-developed optic nerves and large eyes, and they appear to have good vision. Rods and cones are present in the retina. However, there is no distinct fovea centralis, the retina is thick, and a layer of giant ganglion cells (cell bodies up to 150 μm in diameter) appears to serve most of the central retina. These giant ganglion cells support giant dendrites and myelinated optic nerve fibers that range up to 8 or 9 μm in diameter (Dawson *et al.*, 1982).

In most mammals, each eye projects the majority of its nerve fibers to the opposite (contralateral) cerebral hemisphere; however, considerable numbers of fibers project to the cerebral hemisphere on the same side (ipsilateral side). The dolphin may be an exception to the general rule that crossed and uncrossed fibers arise from each eye. Each of the dolphin's eyes probably projects only to the contralateral hemisphere. This conclusion is somewhat tenuous because it results mainly from two anatomical studies of one animal each. First, Hatschek (1903, 1960) studied the brain of a captured dolphin that had lost one eye. Using a Weigert-stained series of brain stem sections, he observed total decussation of the optic nerve across the optic chiasm. From this information, he concluded that dolphins lacked binocular vision. Second, Jacobs *et al.* (1975) described a silver degeneration study carried out on a male bottlenose dolphin that had been enucleated and then vitally perfused after surviving for 14 days. With some reservations concerning the long survival time and the 19% discrepancy between numbers of fibers in the optic nerve and optic tract of their specimen, the researchers concluded that the dolphin exhibited an apparently total decussation of optic nerve fibers, a condition atypical of mammals but common in other vertebrates.

Some physiological evidence also supports complete crossing of optic nerve fibers. Supin *et al.* (1978) mapped the sensory projection zones on the dorsal and dorsolateral surface of dolphin cortex. They found that large evoked potential responses were obtained from visual cortex when light flashes were presented to the contralateral eye, but no such responses were obtained when equivalent flashes were presented to the ipsilateral eye.

Although certainly incomplete, the anatomical and physiological evidence suggests that dolphins do not have binocular vision. Langworthy (1932), like Hatschek (1903, cited in Breathnach, 1960), doubted that bottlenose dolphins had stereoscopic vision, stating that "the development of the visual cortex compares well with that described by Mott for other mammals having panoramic vision." It appears to me that bottlenose dolphins have some degree of visual overlap ventrally and rostrad as well as dorsally and slightly caudad. Despite these small areas of overlap in the visual field, the eyes appear to move independently. McCormick (1969) observed that one eye might look forward and dorsally, while the other might look rearward and ventrally. Dawson *et al.* (1981) reported a physiologic investigation which confirmed that dolphin eye movements are not conjugate. Power spectral density analyses of dolphin eye movements showed maximal power around 0.1 Hz, lower than the human counterpart (Dawson *et al.*, 1981).

The dolphin corneal reflex appears well developed, and the pupil reacts to light. The retractor bulbi is large, and the other extrinsic ocular muscles are present and well developed. The oculomotor nerves are small compared with the fifth, seventh, and eighth cranial nerves but are by no means rudimentary. The presence of an Edinger-Westphal nucleus is questionable. Description of the oculomotor nuclei has been complicated by the presence of a large encapsulated group of cells just dorsal to the third nucleus, the nucleus ellipticus (Breathnach, 1960).

In other mammals, except elephants and possibly seals, the mesencephalic elliptic nucleus apparently has no homolog. The nucleus appears to represent an extensive bilateral elaboration of ventral and ventrolateral periaqueductal gray matter. In at least two studies (see Breathnach, 1960) of odontocetes, the nucleus was described as consisting of a larger dorsal part and a smaller ventromedial portion. One report suggested that the medial portion of the elliptic nucleus contributed fibers to the third nucleus and that the dorsal part possibly corresponded with the nucleus of Darkschewitsch of other mammals. Fibers from the elliptic nucleus have been traced in a caudal direction in a large medial tegmental tract as far as the rostral levels of the inferior olivary complex.

The functional significance and anatomical connections of the nucleus ellipticus, an unusual structure of the cetacean brain, remain obscure. A parallel has been drawn between the complex nasal sac structure of dolphins and the large trunk of the elephant, suggesting that in each the nucleus ellipticus may be involved in innervating nasal musculature (Breathnach, 1960). However, the nucleus ellipticus appears to have no clear connections to the facial complex; its association with the mesencephalic reticular formation, a matrix of tracts and specific nuclei, perhaps makes identification of definite connections more difficult.

The superior colliculus forms an obvious prominence in all cetacean species. In many, such as bottlenose dolphins, the superior colliculus is much smaller than the inferior (acoustic) colliculus, a reversal of the size relationship seen in humans and other land mammals. The lateral geniculate is readily identifiable, but in bottlenose dolphins and common dolphins at least, there is reported to be no true lamination such as there is in mammals with binocular connections (Kruger, 1959, 1966). Mapping studies by Soviet investigators have located visual cortex not at the occipital pole, but in a superior and medial position high in parietal cortex (Supin *et al.*, 1978).

DOLPHIN SLEEP

Shurley *et al.* (1969; see also Serafetinides *et al.*, 1972) were the first to find evidence supporting Lilly's (1964) claim that cetaceans could sleep with half the brain awake. More extensive research by Soviet investigators found marked asymmetries in electroencephalograms (EEG) from right and left hemispheres of bottlenose dolphins (Mukhametov *et al.*, 1977, 1987; Mukhametov, 1984). Mukhametov (1984) recognized three stages in the dolphin EEG: stage 1, desynchronization; stage 2, intermediate synchronization including sleep spindles and theta and delta waves; and stage 3, maximal synchronization, when delta waves of maximal amplitude occupied not less than 50% of each scoring interval. Stages 1 and 2 occurred bilaterally or unilaterally. Stage 3 occurred in only one hemisphere at a time. Wakefulness or bilateral EEG desynchronization (EEG stage 1 in both hemispheres) occupied 50–60% of the recording time. Stage 2, intermediate synchronization, was sometimes recorded bilaterally, and at such times dolphins displayed EEG patterns typical of terrestrial mammals. Mukhametov *et al.* (1987) report, "The whole sequence of respiratory movements of dolphins may be observed during bilateral intermediate EEG synchronization without arousal." Bilateral intermediate EEG synchronization (stage 2 in both hemispheres) occupied only a small percentage of the recording time. Unihemispheric slow-wave sleep (stage 3 in only one hemisphere), the main type of sleep observed in the dolphin brain, occupied 30–40% of recording time. It was stated (Mukhametov, 1984),

> We have never recorded bilateral delta waves in the dolphin brain during natural sleep. The unilateral sleep episodes can last more than 2 h. . . . Simultaneous recordings from the parietal, occipital, and frontal fields of the hemispheres confirm that a hemisphere is always synchronized or desynchronized as a unit. Records from thalamic nuclei demonstrate that the thalamus can also generate slow wave activity unilaterally and concurrently with the ipsilateral neocortex. Thus, unilateral slow wave sleep is not only a cortical, but also a subcortical phenomenon.

In the most recent paper on bottlenose dolphins by the group (Mukhametov *et al.*, 1987), they quantified sleep studies of several days' duration. Based on EEG signs, the dolphins spent an average of 33.4% of each day asleep.

Dreams certainly can be regarded as mental events, and dreaming in humans has been firmly connected to physiological events described as rapid eye movement sleep (REM). REM (also called paradoxical or dream sleep) has been reported in many mammals, but in the dolphin there is definitely room for controversy about its presence. Mukhametov (1984) states flatly that he has never observed REM in many days and nights of recording dolphin EEG, involving 30 experimental subjects over a period of 8 years: "*Tursiops truncatus* and *Phocoena phocoena* are with the echidna among the three species in which no paradoxical sleep has been revealed. It is noteworthy that the echidna is a secondary terrestrial mammal and its ancestors were aquatic mammals" (Mukhametov, 1984; the basis for the assertion that the echidna is a secondary terrestrial mammal is not given). Six minutes of REM was observed in a single night's recording of a pilot whale, *Globicephala macrorhynchus* (Shurley *et al.*, 1969). Flanigan (1974, 1975) observed bottlenose dolphin male nocturnal erections during apparent sleep, a sign of REM in man and other mammals.

BLOOD SUPPLY TO THE BRAIN

In most mammals the internal carotid artery is the major supplier of blood to the brain. This is not the case in dolphins. Their brains are supplied through the massive thoracicospinal retia (Viamonte *et al.*, 1968). In its course from heart to brain, blood passes through the retia. This arrangement is thought to have a pressure-damping function for the cerebral circulation (Nagel *et al.*, 1968). Neither the internal carotid artery nor vertebral arteries supply blood to the dolphin brain, and the circle of Willis is absent (McFarland *et al.*, 1979). The entire cerebral blood supply comes through a thoracicospinal rete, supplied by intercostal and posterior thoracic arteries.

Kovalzon and Mukhametov (1982) have shown that cerebral temperature varies during unihemispheric slow-wave sleep. Temperature drops several tenths of a degree centigrade in the hemisphere showing slow waves but remains constant in the opposite hemisphere. The variation in temperature during unihemispheric slow-wave sleep may result from changes in metabolism in the cerebral hemisphere or from changes in the circulation to the hemisphere. Independent circulation may be essential for the development of the marked asymmetries in brain-wave patterns and in brain temperature that have been observed.

BRAIN ASYMMETRY

In recent years a good deal of attention has been focused on asymmetry between the two hemispheres of the human brain, and various right-brain, left-brain theories have been exaggerated to the point of scientific cultism. Still, there are firm bases for both anatomical and physiological differences between the hemispheres that are manifest in behavioral differences. Between 70 and 80% of human brains have a larger planum temporale in the left hemisphere than in the right (Geschwind and Levitsky, 1968; Witelson, 1977). Studies of normal and brain-damaged people have demonstrated that the two cerebral hemispheres are specialized for different mental abilities. The body of literature on human laterality is extremely large. I mention the findings only in a brief generalization. In most right-handed adults, language material is processed in the left hemisphere, and material concerned with spatial relationships is more dependent on right-hemisphere processing (Neville and Hillyard, 1982; Levy, 1982). As Hillyard and Bloom (1982) point out, "Some authors have proposed that cerebral laterality occurs in direct proportion to the evolution of language, of handedness, of executive control, or the existence of a conscious self."

Human cortical specializations do not mature until at least puberty (Neville and Hillyard, 1982). Most of the dolphins used in behavioral experiments have been immature animals, thus we might not expect a full display of lateral specialization. In addition, I am aware of no attempt to demonstrate any lateral specializations in behavioral experiments and know of only one study on anatomical differences between the hemispheres. There are, however, at least two observations that may suggest laterality preferences: (1) the marked proclivity in confinement to swim in a counterclockwise direction and (2) the tendency to favor one eye over the other. Studies of cortical surface area have revealed that the right hemisphere is slightly larger than the left in bottlenose dolphins (Ridgway and Brownson, 1979, 1984).

Counterclockwise Swimming

Townsend (1914) observed that three newly captured bottlenose dolphins circled counterclockwise and two went clockwise. Kellogg and Kohler (1952) observed six animals at Marineland of Florida "almost always swimming in a clockwise direction against the current." Lilly (1962) observed that two dolphins, one female and one male, in his laboratory always circled to the left (counterclockwise). Reports from three different oceanaria in Florida revealed that newly introduced bottlenose dolphins and spotted dolphins, *Stenella plagiodon*, swam counterclockwise around the pool regardless of direction of water flow (Caldwell *et al.*, 1965). At one oceanarium, all newly introduced bottlenose dolphins swam counterclockwise between 1954 and the winter of 1964, but in the winter of 1964 six of seven animals circled clockwise (Caldwell *et al.*, 1965). In observations of dolphins swimming in concrete tanks at Point Mugu, California, and in wooden tanks in San Diego, California, I noticed that all newly introduced dolphins (27 bottlenose dolphins and 8 Pacific white-sided dolphins, *Lagenorhynchus obliquidens*) recently captured in the wild swam in a counterclockwise direction. Newly captured dolphins at Point Mugu swam in counterclockwise circles even when they were apparently asleep (Ridgway, 1972). After the animals had been in captivity for a time, the direction of swimming became less stereotyped; however, the counterclockwise direction usually predominated, especially during after-hours periods when the animals swam in a leisurely manner neither chasing, playing with other dolphins, nor attending to their human trainers. In observations of seven bottlenose dolphins in three different tanks at Point Mugu, I found that switching the direction of water current in the pool, after the dolphins' swimming direction had been well established, did not change the direction of swimming.

Soon after capture the bottlenose dolphins swam almost continuously, even while apparently asleep; however, as time passed most of the dolphins ceased this continuous swimming and settled into a surface hanging behavior (see McCormick, 1969; Ridgway, 1972; Flanigan, 1974) during periods of rest or sleep. I have observed six females, adapted to such surface hanging behavior, that became pregnant and delivered live offspring; in each case, the surface hanging behavior ceased immediately after parturition, and mother and calf swam continuously in a counterclockwise circle. Only occasionally did the animals reverse direction. Females that had relatively straight dorsal fins developed a distinct leftward tilt in the fin as did their offspring. The majority of bottlenose dolphins at Sea World in San Diego had left-tilted dorsal fins and swam counterclockwise (L. H. Cornell, pers. commun., 1984).

According to Balonov *et al.* (1981; also see Harnad and Doty, 1977) the direction of rotational tendencies indicates greater activity of the opposite hemisphere. When dolphins swim in a counterclockwise direction, the rotational tendency is toward the left and away from the right hemisphere, presumably the one with greater activity. Those dolphins that circle while asleep should have alternation of activity between the hemispheres according to the findings of Soviet investigators (Mukhametov *et al.*, 1977; Mukhametov, 1984); yet they do not change the direction of circling. These findings seem to confound the idea that the dolphins turn in the opposite direction from the hemisphere with greater activity.

Eye Preference

In 1964, Adolph Frohn (then of the Miami Seaquarium and an early dolphin trainer who trained at the first oceanarium, Marine Studios of Florida) told me, referring to bottlenose dolphins, "A porpoise has a looking eye and it is more often the right one." I have not made a thorough study of eye preference, but in observing 27 recently captured bottlenose dolphins, I noted that 19 approached and examined my outstretched hand with the right side and eye facing me. Right eye preference is by no means universal, and the animals often appear to attend the trainer with both eyes. In the early stages of training, the dolphin often stations in front of the trainer; since the majority of trainers are right-handed, the dolphin might be expected to develop a bias toward the left eye.

Right eye preference seems consistent with the counterclockwise swimming that we observe in the majority of captive dolphins. This direction of swimming places the right eye toward the enclosure wall and toward any human outside the tank who might provide food, other reward, or stimulus. Since the optic chiasma is completely crossed, right eye preference may suggest a leading role for the left hemisphere in vision, whereas the counterclockwise turning might suggest a leading role for the right hemisphere in that movement. The case for counterclockwise turning is certainly much stronger than that for right eye preference.

With captive bottlenose dolphins that I have observed, the neonate, especially in the hours soon after birth, usually swam on the mother's left, the side away from the enclosure walls, as she swam counterclockwise. This may be a manifestation of the mother's desire to keep the neonate away from the tank wall and other outside influences such as people. In the wild, priorities might be different. Maxwell (1960) observed wild bottlenose dolphins with calves off the coast of northern Scotland, noting that the calves swam to the right of the mothers.

THE TERMINAL NERVE, OLFACTORY APPARATUS, AND CRANIAL ASYMMETRY

Kellogg (1928) suggested that during the aquatic evolution of the odontocetes, the olfactory nerves were effectively shut off from the narial passages and did not survive the rearrangement of the skull and migration of the nasal opening to the top of the head. Neurologists have questioned Kellogg's idea (e.g., Edinger, 1955). At this time, we are still lacking an entirely satisfactory explanation for the absence of the peripheral olfactory apparatus. In early fetal stages, the olfactory bulb, nerve, and tracts are present. As the fetus develops, the structures degenerate, and they are completely absent from mature odontocete brains. Although the more peripheral olfactory components of the cetacean nervous system are absent or rudimentary, other brain structures traditionally thought to be involved in olfaction are present in the mature dolphin. The olfactory lobes and septal areas are large; the hippocampus and subiculum are small. The well-developed entorhinal cortex is more anteriorly located in the temporal lobe, and the presubiculum is more posterior than in other mammals; in general, the rhinencephalon displays the same basic structural arrangement as in primates or carnivores (Breathnach, 1960; Jacobs *et al.*, 1971). The seemingly paradoxical presence of a "nosebrain" without peripheral connections or sensory olfactory endings is one of the more intriguing features of the cetacean central nervous system. Despite the complete absence of olfactory nerves in adult odontocetes, the tiny terminal nerves that have been observed to accompany the olfactory nerves in many vertebrate species are present in fetal and adult dolphins (Figs. 4 and 5).

Toothed whales, but not baleen whales, generally have an asymmetrical cranium (Ness, 1967; Rommel, Chapter 2, this volume). The nasal opening is shifted to the left of the cranial midline to varying degrees, and several skull and head structures on the right are larger than those on the left. There is no known reversal of the family-specific asymmetries on the part of any individual odontocete. A few early authors reported asymmetries of the brain (see review by Breathnach, 1960). These observations often have been discounted because of the possibility that such large brains might be liable to deformation during long periods of fixation. However, Ridgway and Brownson (1984), using freshly fixed material, found significant differences in the surface area of cerebral cortex between left and right hemispheres in bottlenose dolphins. The right hemisphere had a significantly greater surface area.

The asymmetry of the odontocete skull is apparently the result of the enlargement of dorsal elements on the right side. In bottlenose dolphins (and most other odontocetes), the blowhole is shifted to the left of the midline; the nasal plug, nasal-plug muscle, premaxillary sac, premaxillary fossa, and nasofrontal sacs are all much larger on the right. Wood (1964) proposed that the cranial asymmetry related to the development of a sound-producing organ and represented a modification of the skull having to do with echolocation (see also Wood and Evans, 1980).

Sleptsov (1939) proposed another interesting explanation for the asymmetry of the skull and brain of odontocetes based on his studies of common dolphin, white whale, *Delphinapterus leucas*, and harbor porpoise, *Phocoena phocoena*, embryos. Sleptsov found, in the dolphin embryo, that the left olfactory nerve and lobe degenerate more rapidly than those on the right, resulting in the cerebral and cranial asymmetries.

Johnson (1914) reported that the terminal nerves or nervus terminalis (nT) were present in the harbor porpoise. From a single gross specimen he illustrated seven strands running in the pia on the ventromedial surface of the frontal cortex of one cerebral hemisphere. This brain had been removed from the cranium by someone else, and Johnson could only describe the strands to the point on the frontal brain surface where they had been cut off (see Fig. 4). Sinclair (1951a,b) first described the dural part of nT, the large stellate ganglia (see Fig. 5). Having only a skull of a bottlenose dolphin from which the brain had been removed, Sinclair described the ganglia as being large and asymmetrical, and he illustrated the left ganglion as being 3 mm wide. [Sinclair called the species he studied a "porpoise." At least one recent report (Oelschläger and Buhl, 1985) has mistakenly inferred that Sinclair's specimen was from a harbor porpoise. In Europe, the term porpoise is used almost exclusively for the harbor porpoise and its close relatives; however, this usage of cetacean vernacular names does not hold in the United States. Harbor porpoises are absent from the Texas coast where Sinclair obtained his specimen, but bottlenose dolphins are common there and are usually called porpoise locally. Furthermore, Sinclair's notes in the Blocker History of Medicine Collections, Moody Library, University of Texas Medical Branch, Galveston, indicate that his porpoise specimens were bottlenose dolphins.] Sinclair was apparently unaware of Johnson's work on the harbor porpoise, since he did not cite the earlier observation. Yoshikawa and Nakamura

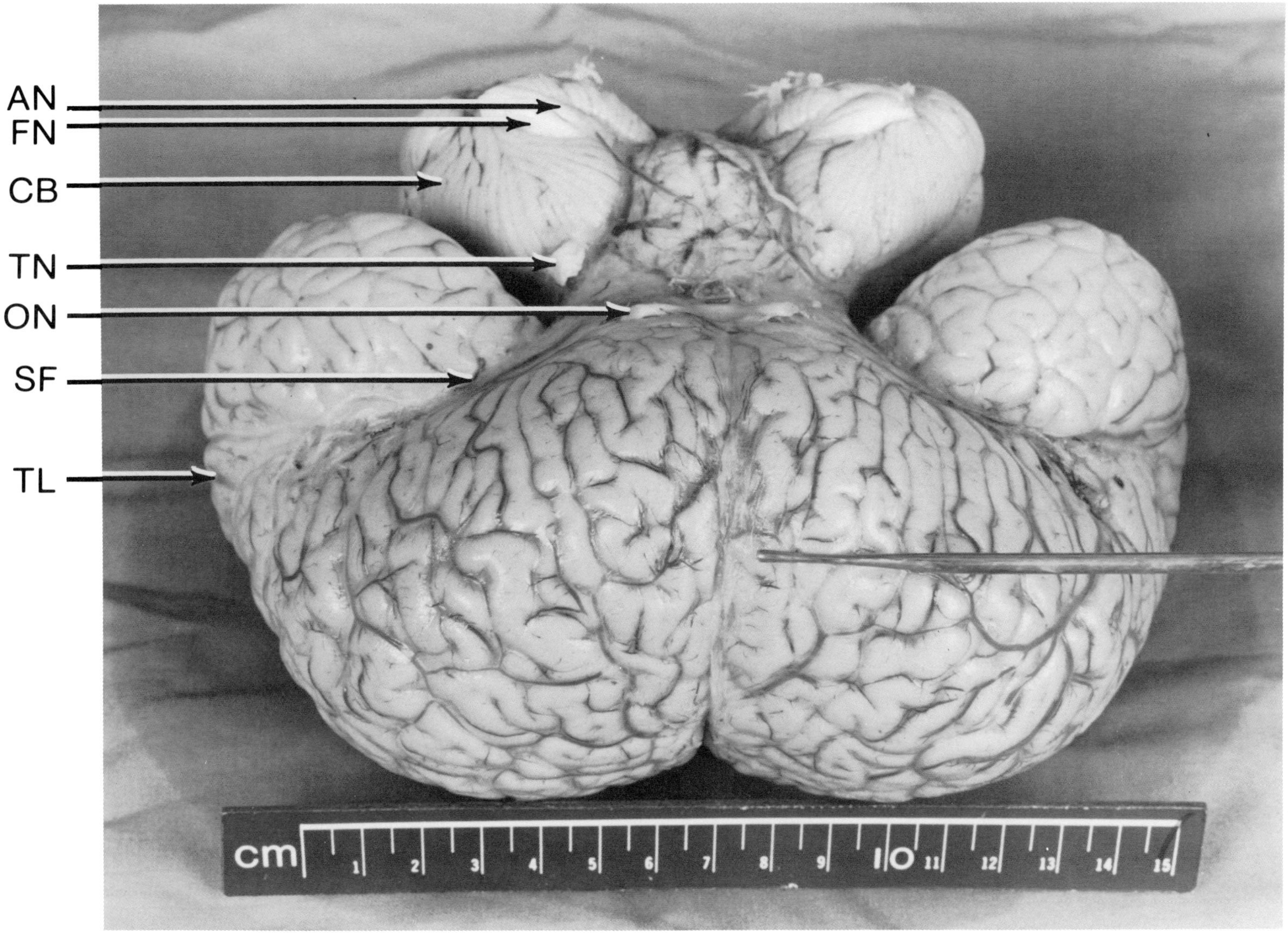

Figure 4 Anterior aspect of a bottlenose dolphin brain (ventral surface up). The probe points to a strand of the right terminal nerve. TL, Temporal lobe; SF, sylvan fissure; ON, optic nerve; TN, trigeminal nerve; CB, cerebellum; FN, facial nerve; AN, auditory nerve.

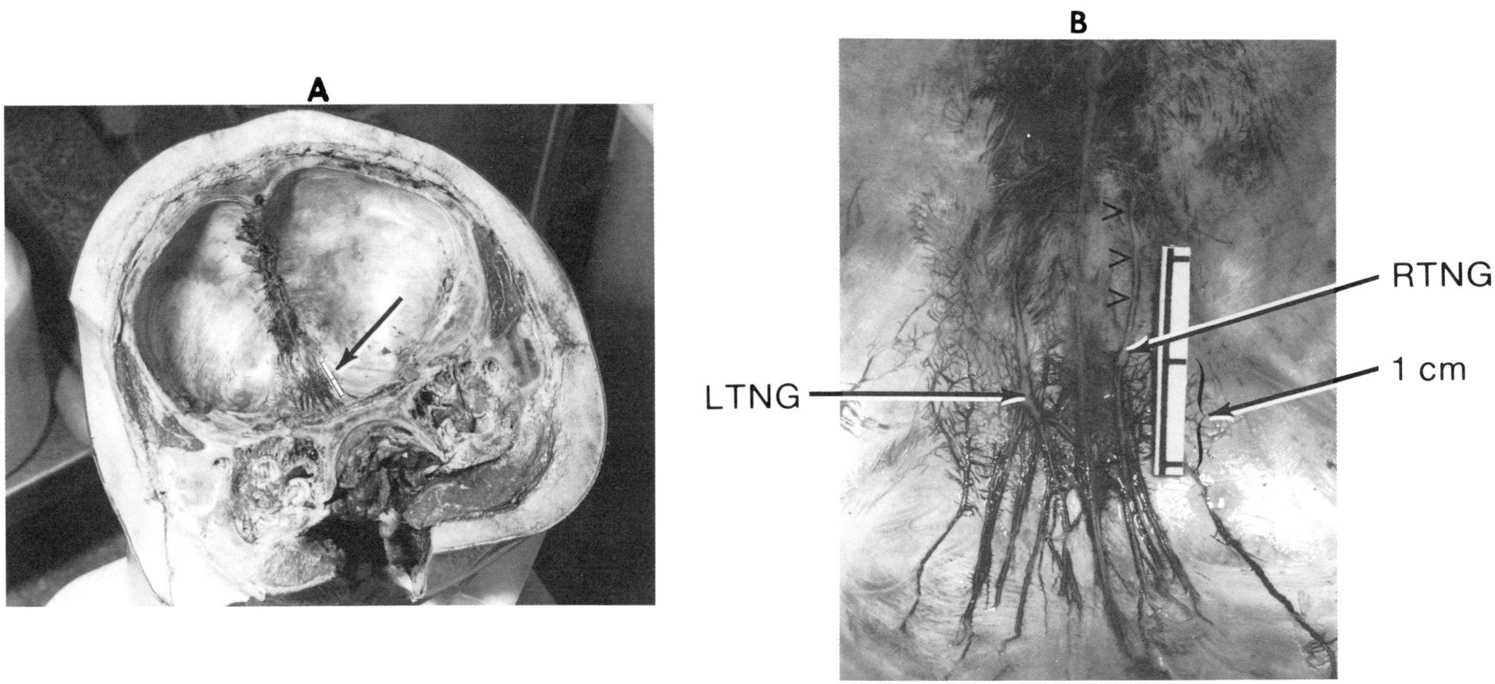

Figure 5 (A) Section through the cranial vault of a bottlenose dolphin. The arrow points to the scale positioned just lateral to the right terminal nerve ganglion. (B) Closeup of A showing the left (LTNG) and right (RTNG) terminal nerve ganglia accompanied by numerous blood vessels of the vascular plexus that courses through the area.

(1952) found in the bottlenose dolphin what Johnson (1914) had found in the harbor porpoise; they traced nT strands about 7.5 cm, concluding that the nervus terminalis very likely goes into the brain just anterior to the optic fasciculus.

Sinclair (1966) studied the development and subsequent atrophy of the olfactory complex in the delphinid genus *Stenella*. In a 27-mm embryo, he traced the nT into the olfactory bulb. He found that later in development the olfactory bulb atrophies, leaving the nT alone coursing through the cribriform area. Buhl and Oelschläger (1986) identified the nT system in embryos of several species of odontocetes. In this study and in an earlier series on the olfactory system of the embryonic harbor porpoise (Oelschläger and Buhl, 1985), they found no evidence of a vomeronasal system. Thus, it appears that of the trio of nerve complexes that course through the cribriform area in most mammals—olfactory, vomeronasal and terminal—only the terminal nerve system survives in odontocetes.

The nT ganglia of the bottlenose dolphin are large (Fig. 5) and contain several thousand cells each. There are at least two large myelinated neuron types, round cells 30–50 μm in diameter and elongate fusiform cells about 15 μm across (Ridgway *et al.*, 1987). Many of the fusiform cells contain luteinizing hormone-releasing hormone (LHRH), which has been observed in terminal nerves of various species and has led some observers to suggest that the nT influences sexual reproduction (Demski and Schwanzel-Fukuda, 1987). The nT ganglia of toothed whales are the largest nT structures so far identified and to me suggest an intriguing evolutionary question: Why did the terminal nerve survive the migration of the nares to the top of the head when the olfactory nerves did not?

THE GUSTATORY SENSE

Because dolphins are regarded as anosmic, because the animals usually swallow fish and other food whole without mastication, and because, until recently, taste buds had not been found in the tongue of several dolphin species, some cetologists have doubted that the animals possess the sense of taste. A presumed gustatory nucleus of the thalamus and the elaboration of the seventh and ninth cranial nerves suggested to neuroanatomists, however, that the gustatory sense is well developed (cf. Kruger, 1959).

Histological studies have produced descriptions of taste buds within five to eight pits that form a V-shaped row on the posterior dorsum of the tongue (Donaldson, 1977; Yamasaki *et al.*, 1978). In one study of *Stenella coeruleoalba*, taste buds were found in the tongue pits of young dolphins but not in those of adults (Yamasaki *et al.*, 1978); in another study, a nerve supply to the buds could not be demonstrated (Donaldson, 1977). There still seems room for doubt that cetaceans have taste buds such as those present in man and most other mammals. Nonetheless, there appears to be psychophysical evidence for chemoreception by sensors in the tongue or mucous membranes of the mouth since trained bottlenose dolphins reportedly can detect chemicals dissolved in seawater (Nachtigall and Hall, 1984). More study is needed on the epithelium and sensory endings of the tongue and oropharynx and especially on the V-shaped row of pits found in the tongue of dolphins.

In most mammals, the facial nerve, which is well developed in cetaceans, attends taste to the anterior two-thirds of the tongue and carries secretomotor fibers to salivary glands. Pores of glands present within the pits, and numerous gland orifices clearly visible on the dolphin tongue and oropharynx, secrete a viscous mucoid material which does not readily dissolve in water. The presence of salivary glands has been doubted by several authors (see Breathnach, 1960), although it is possible that some of the small glands in the tongue or pharyngeal area are salivary. For animals with a diet almost completely devoid of carbohydrates, and which swallow their food whole, the absence of salivary secretion would not seem a great handicap. In dozens of dissections, I have not found salivary glands in the bottlenose dolphin; moreover, I have observed that a triangular structure along the dorsal margin of the mastohumeralis muscle, identified as a parotid salivary gland in at least one earlier report on the pilot whale (Murie, 1874), has the histological structure of a lymph node.

THE SENSES OF TOUCH, TEMPERATURE, AND PAIN

The second largest cranial nerve, the trigeminal nerve, which is exceeded in size only by the auditory nerve, has a prominent ganglion. The sensory trigeminal complex is better represented than the general body somatosensory representation. The tactile thalamic region is reduced compared with that of many other mammals. The arcuate division is more developed than the external division, reflecting a greater representation of the face in animals that have relatively large heads.

Encapsulated nerve endings, especially numerous about the head and snout, around the blowhole, and encircling the anus and genital slit, are found in the superficial portion of the dermis (Palmer and Weddell, 1964; Harrison and Thurley, 1974). Fetal bottlenose dolphins have several hairs on either side of the snout or upper lip, but these fall out around the time of birth leaving only whisker pits barely visible in adult animals. On histological section, these pits can be seen to contain the remnants of the hair and a good nerve supply.

Several studies have been done on tactile sensivity of small odontocetes (Lende and Welker, 1972; Kolchin and Bel'kovich, 1973; Ridgway and Carder, 1989). Evoked potentials to stimuli such as vibrating, tapping, stroking, or dripping water on the skin have been recorded from the contralateral somatosensory cortex of the bottlenose dolphin (Lende and Welker, 1972). The areas of greatest skin sensitivity were on the head: the upper and lower lip near the commissure, around the eyes, and around the blowhole. Stimuli to the body trunk and tail produced very minimal evoked potentials. Soviet investigators used the galvanic skin response (GSR) to stimuli produced by a 0.3 mm weighted wire to make a partial map of body skin sensitivity in the common dolphin (Kolchin and Bel'kovich, 1973). Of the body portions studied, separate circular areas of about 5 cm diameter around the blowhole and eyes were most sensitive (10 mg/mm^2) while the snout, lower jaw, and melon were found to be somewhat less sensitive (10–20 mg/mm^2); the skin along the back both anterior and posterior to the dorsal fin was even less sensitive (20–40 mg/mm^2). The authors considered their values for the threshold of sensitivity to touch around the dolphin eyes and blowhole to be similar to those of a human in the most sensitive skin areas, namely, the tactile surfaces of the fingers, the skin of the eyelids, and the lips (Kolchin and Bel'kovich, 1973). Physiological mapping studies have located the somatosensory cortex of bottlenose dolphins in an area posterolateral to and bordering the motor cortex and anterior to and bordering the visual and auditory cortex (Supin *et al.*, 1978).

No systematic studies have been done on pain perception or sensitivity to temperature; however, I have made rough observations during the veterinary care of bottlenose dolphins and have compared these with similar observations of domestic animals such as pigs or horses with which I have also had some experience. I have found that bottlenose dolphins are sensitive to pain to roughly the same extent as are pigs or horses and must be anesthetized for surgical procedures.

THE FACIAL NERVE

The main facial nucleus lateral to the superior olive is large and conspicuous (Breathnach, 1960). The dolphin facial nerve is well developed and apparently supplies the extensive musculature of the blowhole, nares, and nasal-sac system. Langworthy (1932) states, "The facial nucleus is large and the cells are subdivided into groups; each group apparently supplies a single muscle of the facial group." Since sound production is so important to animals that use echolocation and rely on sound for underwater communication, such development is not surprising.

THE SPINAL CORD

Because the spinal cord carries fibers to and from most of the body, most of the regulatory and body monitoring functions of the brain should be represented in the cord. Most reptiles and fish have more spinal cord than brain, while mammals have brains that are much heavier than the spinal cord. Worden (1951) proposed the ratio of brain weight to spinal cord weight as an index of encephalization that might be correlated with intelligence across species. The brain to spinal cord ratio is higher in the bottlenose dolphin (Fig. 6) than in any other species measured thus far except for *Homo sapiens*. The ratio is about 50 : 1 in humans, 40 : 1 in bottlenose dolphins, 34 : 1 in Pacific white-sided dolphins, 26 : 1 in Dall porpoises, *Phocoenoides dalli*, 8 : 1 in the macaque monkey, *Macacus*, and about 4 : 1 in cats, *Felis domesticus* (Ridgway *et al.*, 1966). [Wirz (1950) presented an index of cerebralization relating different parts of the brain that ranked odontocetes along with humans at the highest level]. Of course, whether or not the ratio of brain weight to spinal cord weight can be correlated with intelligence across species must be evaluated by independent behavioral and physiological means.

The spinal cord of one bottlenose dolphin ended at about the third lumbar vertebra (Ridgway, 1968). The spinal cord is nearly cylindrical throughout. The ventral columns of white matter are strongly developed. Clarke's columns are fused into a single median nucleus, at least in the caudal part. The pyramidal tract is very small. The cortical projection into the cord is rather small; however, there seems no doubt that corticospinal fibers are present and form a crossed projection. Muscle spindles have been identified in dorsal epaxial muscle; the morphology of the spindles and neuromuscular junctions appeared similar to those of other mammals (Tulsi, 1975).

THE PINEAL, MAMMILLARY BODIES, AND PITUITARY

In numerous dissections of odontocete brains, I have never found the pineal body, which has been considered absent by most observers of the odontocete brain (Breathnach, 1960). There is one report of a small pineal (displacing about 1 cm^3) from a mysticete, the humpback whale, *Megaptera novaeangliae* (Gersch, 1938). The organ is also present in embryonic stages of the blue whale, *Balaenoptera musculus*. Gross examination of the bottlenose dolphin brain does not reveal mammillary bodies, but mammillary nuclei are recognized histologically. The pituitary is well developed. The adenohypophysis and neurohypophysis are completely separated by a dural septum.

THE CEREBELLUM

The absolute size of the cerebellum of the largest mysticetes, without free digits and presumably without fine movements, is huge, approximately

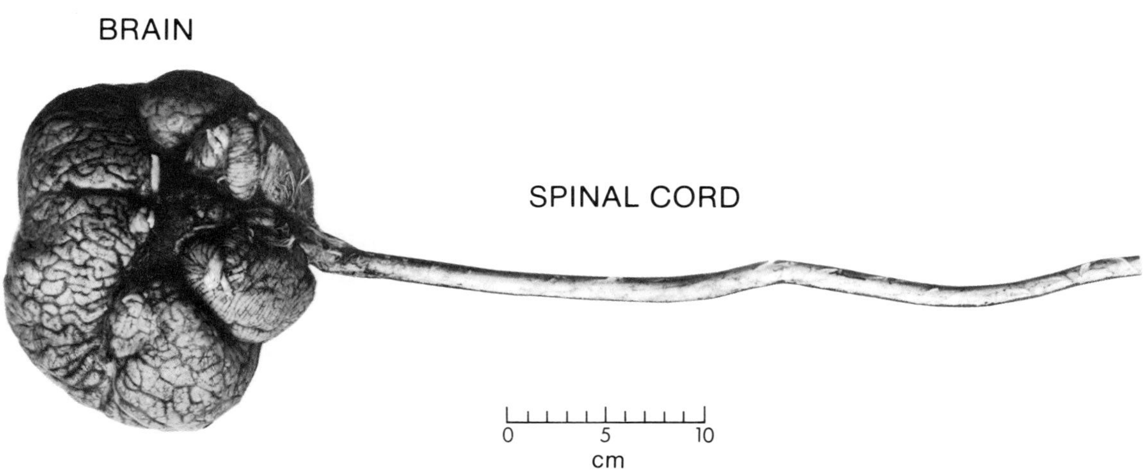

Figure 6 Bottlenose dolphin brain and spinal cord which was dissected out as a unit. The cord ended at about the third lumbar vertebra. (From dissections by Robert F. Green and the author.)

1500 g. In these species, the cerebellum may be more than 20% of total brain weight; in the bottlenose dolphin, the cerebellum is more typically about 15% of total brain weight compared with about 11% in humans. In the embryonic stage, the dolphin cerebrum and cerebellum are about equal in width. According to paleontologists, the early forebears of modern cetaceans were the first giant mammals adapted to life-long swimming. Fossil studies have shown that these ancient cetaceans already had a very large cerebellum that was much wider than the cerebrum (Edinger, 1955). In modern odontocetes, the cerebrum has so enlarged as to cover the anterior portion of the cerebellum.

In comparison with the anatomy of the cerebellum of terrestrial mammals, the lobus simplex and paramedian lobule of the cetacean are much larger (Jansen and Jansen, 1969). The paraflocculus and nucleus interpositus, in these comparative terms, are enormous. The hemispheral parts of the anterior lobe, the flocculonodular lobe, the lateral nucleus, and the ansiform lobule are relatively small. In the same terms, the medial nucleus is moderately developed.

Lange (1975) has studied cell numbers and cell densities in the cerebellar cortex. He made comparisons between humans and various mammals, attaching particular significance to the ratio of Purkinje cells to granule cells. Lange found ratios of 1 : 2991 in human, 1 : 1898 in rhesus monkey, *Macaca mulatta,* 1 : 1812 in bottlenose dolphin, 1 : 1790 in pilot whale, 1 : 1508 in cat, and 1 : 609 in hedgehog.

THE AUDITORY SYSTEM

Auditory structures are generally larger in odontocete brains than in mysticete brains (Breathnach, 1960). Hypertrophy of the auditory system may be the primary reason for the dolphin's large brain. The medial geniculate is about 7 times, the inferior colliculus (Fig. 7) is 12 times, and the nucleus of the lateral lemniscus of the dolphin is over 250 times as large as the equivalent structures in humans (see Bullock and Gurevich, 1979). The ventral cochlear nucleus and some other brain stem nuclei also appear massive when compared with the human equivalents. The cetacean auditory nerve has several times as many fibers as the eighth nerve of man. Studies of fiber spectra have revealed larger myelinated fibers in odontocetes than in mysticete whales which in turn have larger eighth nerve fibers than humans. Fiber diameters have been measured using light microscopy. Plots of fiber diameters have shown that the largest numbers of fibers in the eighth nerve of the sperm whale, *Physeter catodon,* are aboout 9 μm in diameter, in the bottlenose dolphin about 7 μm, and in the fin whale about 5 μm (Jacobs and Jensen, 1964). Auditory tracts reaching the cerebral cortex are extensive. Some observers of the dolphin brain suggest that the cerebral cortex may have reached its great development on the basis of acoustic input (Langworthy, 1932; Wood and Evans, 1980).

Mapping the Auditory Cortex

Soviet investigators located extensive auditory projection areas on the dorsal surface of each hemisphere about 1.5–3.0 cm lateral to the sagittal fissure between the hemispheres (Supin *et al.*, 1978). Thus, compared with the brains of most land mammals, a difference appears in the dolphin's brain which indicates an apparent shifting of the auditory area from the temporal to the parietal lobe and the dorsum of the hemisphere. Data from evoked potential studies suggest the presence of both primary and secondary auditory cortex, whereas histological investigation does not. On physiological grounds, bats also exhibit complex organization of the auditory cortex (Suga, 1984). Yet, like the dolphin, bats reveal to the histologist a more primitive level of cortical development than, for example, carnivores or primates (Morgane *et al.*, 1986; Glezer *et al.*, 1988). The extent of auditory cortex in dolphins may be greater than that indicated in the Soviet mapping experiments. A tonotopic map of the cochlear projections on the dolphin cortex has not been done; nor have complex stimuli of different delays or rise times or other acoustic characteristics been used in attempts to find areas of cortex specialized for specific information resembling those found in bat cortex. Therefore, it is entirely possible that further auditory projection areas will be found in temporal cortex, which, in the dolphin, is less accessible than the dorsal area that has been mapped.

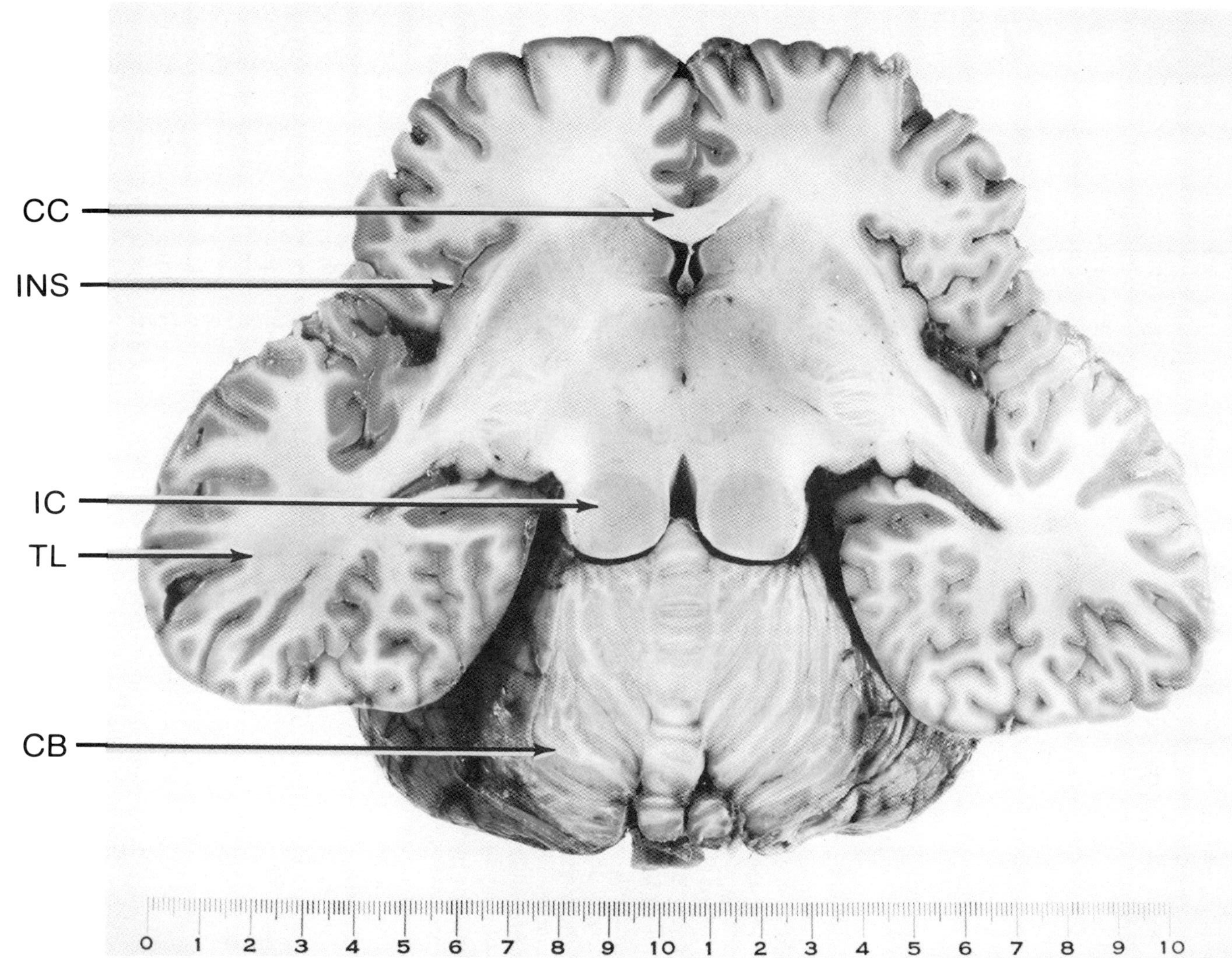

Figure 7 Horizontal section through the brain (1901 g) of a Pacific bottlenose dolphin. CC, Corpus callosum; INS, insula; IC, inferior colliculus; TL, temporal lobe; CB, cerebellum. (Section by Robert F. Green.)

Hearing and the Ear

The mammalian ear evolved for hearing in air. In assuming the aquatic mode, cetaceans have undergone modification of this aerial system, and hearing apparently has become their most highly developed sense. When the distant ancestors of cetaceans entered the water, they almost certainly could hear high-frequency sounds by bone conduction. Human divers can hear high frequencies underwater, but, because of the upper-frequency limit of human hearing, they have no pitch discrimination in the ultrasonic region above about 20 kHz. The evolution of the cetacean ear apparently has been a process of enhancing those characteristics which allow for greater high-frequency hearing sensitivity and complex auditory processing. Audiograms developed by both behavioral and electrophysiological techniques have shown that bottlenose dolphins respond to frequencies as high as 150 kHz, with greatest sensitivity between about 40 and 100 kHz (Johnson, 1966; McCormick *et al.*, 1970).

The pinna is lacking, and the external auditory meatus apparently has become a vestige (Norris, 1968; Bullock *et al.*, 1968). The ossicular chain is stiff, and the malleus does not attach directly to the tympanic membrane (McCormick *et al.*, 1970). The basilar membrane of the cochlea is relatively long (35–40 mm) and very narrow near the basal end (25 μm). The number of inner hair cells in the dolphin cochlea is about the same as, and the number of outer hair cells only slightly greater than, those in the human cochlea, but the number of ganglion cells is much greater. The ratio of ganglion cells to hair cells (inner and outer) is roughly 5:1 compared with 2:1 in humans (Wever *et al.*, 1971).

Echolocation

Research on dolphin echolocation done since the late 1950s has confirmed an earlier suspicion. Men who collected the first permanent captive colony of bottlenose dolphins at Marine Studios in Florida in the late 1930s and 1940s suspected that these animals possessed some sort of sonar or echolocation ability (McBride, 1956). Subsequent studies (Kellogg, 1958; Norris *et al.*, 1961; Evans and Powell, 1967) have shown that blindfolded bottlenose dolphins can make extremely fine discriminations over underwater distances up to at least 100 m (Au, 1980) by employing trains of high-frequency clicks emitted from the nasal system and directed forward in a fairly narrow beam (Au and Moore, 1984; Au *et al.*, 1986).

Acoustic Processing in the Brain

Electrophysiological experiments with dolphins have shown that temporal resolution of successive sounds is extremely rapid and that very small changes in stimulus frequency altered evoked potential (EP) amplitude and waveform (Bullock, *et al.*, 1968). The typical midbrain EP of the bottlenose dolphin appears specialized for ultrasonic, ultrabrief, fast-rising, closely spaced sounds like the echolocation clicks. However, at several cerebral locations (mainly in the posterior lateral temporal cortex) long latency, long duration, slowly recovering EPs have been evoked by lower frequencies, with either fast or slowly rising acoustic envelopes (Bullock and Ridgway, 1972). This suggests dual analysis systems, one specialized for the ultrasonic click and the other for lower-frequency sounds, such as the whistles and squeals produced by many of the smaller odontocetes (see Caldwell *et al.*, Chapter 10, this volume).

The auditory brain stem response (ABR) can be recorded from electrodes placed on the skin surface or placed subcutaneously over the dolphin cranium. The click-evoked ABR of the bottlenose dolphin consists of 7 waves within 10 mseconds, numbered by the positive peaks at the vertex (Fig. 8). This ABR corresponds well, in respect to number and order of the peaks, with ABRs in other mammals. The dolphin ABR waves are consistently large, sometimes reaching an amplitude of 10 μV. The waves decrease in amplitude and increase in latency as click stimuli are attenuated. This feature makes it possible to employ the ABR to test dolphin hearing (Ridgway *et al.*, 1981; Popov *et al.*, 1984). The latency–intensity function is rather flat, compared with those of other species. Brain stem transmission time (BTT) is considerably faster in the bottlenose dolphin (with a mean brain weight around 1500 g) than in humans or domestic cats and is similar to that of the rat

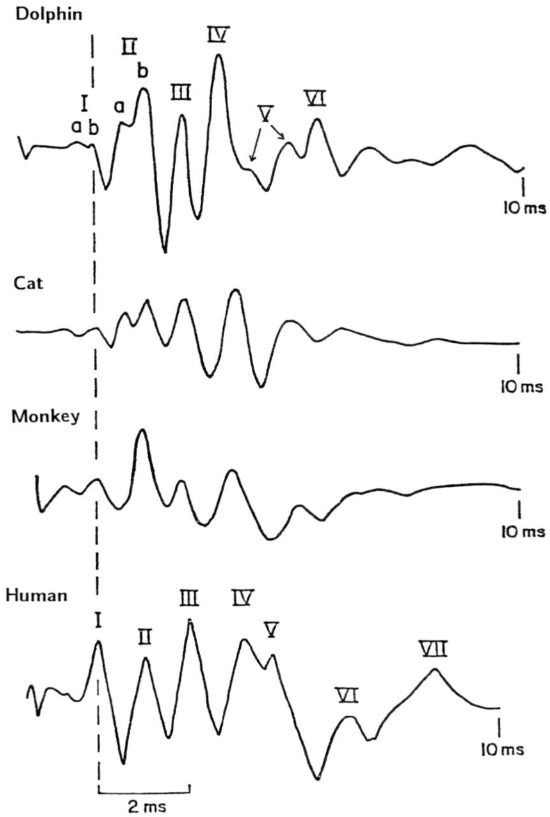

Figure 8 Comparison of a typical dolphin auditory brain stem response (ABR) with ABRs from cat, monkey, and human. All responses represent a 10-msecond time base which is slightly different for each species. The vertical broken line at the left passes through presumably similar components, the auditory nerve deflection, Wave I. (From Ridgway *et al.*, 1981.)

(Ridgway *et al.*, 1981). Despite a much longer nerve pathway, BTT in the dolphin is equal to or faster than that in much smaller-brained species. It has been suggested that, owing to larger auditory myelinated-fiber diameters, the axonal conduction velocity is higher by just enough to compensate for a longer path in the dolphin brain stem (Ridgway *et al.*, 1981).

An interesting problem that might be investigated with the ABR concerns the increase in fiber diameter with increasing brain size. The largest delphinid, the killer whale, *Orcinus orca*, has a brain about four times larger than that of the bottlenose dolphin (6000 versus 1500 g) and perhaps ten times larger than those of some smaller delphinids. BTT in this large brain may be kept short by increased fiber diameters in the auditory pathway.

The above findings suggest that the dolphin brain is specialized for rapid processing of auditory stimuli (see also Bullock *et al.*, 1968; Bullock and Ridgway, 1972). If given enough time, the human auditory system seems to perform as well on some echolocation tasks (Fish *et al.*, 1976). Indeed, when pulses similar to dolphin echolocation pulses were projected at targets by instrumented divers and the received echoes stretched 128 times (tantamount to a slowed-down tape recording and therefore reduced equivalently in frequency), human divers performed with as few errors as bottlenose dolphins in distinguishing metal targets of copper, brass, or aluminum and geometrical aluminum shapes covered with neoprene rubber (Fish *et al.*, 1976).

Studies with static metal targets may give only limited detail of the dolphin's echolocation processing ability. For example, the unnatural targets could well be a factor. However, if the findings of Fish *et al.* (1976) are a fair comparison of echolocation ability and based on a sonar discrimination task that is difficult for dolphins (and I have some doubt that they are), we must conclude that the major accomplishment in the sonar processing component of the dolphin auditory system is the ability to process sound rapidly.

Mackay (1967) has suggested that dolphins may employ phase information in sonar detection. Johnson (1967) has elaborated on the advantages of phase detection for echolocation by dolphins (Evans and Powell, 1967; Au, 1980). Wever *et al.* (1971) suggested that "the representation of detailed high frequency sounds, and especially their time and phase relations, may well be the basis for the dolphin's remarkable facility in echolocation." Since no proof existed, however, that the dolphin auditory system is sensitive to phase at the high frequencies used in echolocation target detection, we attempted to test the dolphin capability for phase detection using the ABR (Ridgway *et al.*, 1981). Reversing the phase of single sine waves delivered to the hydrophone revealed a small but consistent difference in response between initial compression clicks versus initial rarefaction clicks, suggesting that the dolphin auditory system might be capable of detecting phase even at high frequencies.

Suga's (1984) research has revealed an intricate organization of auditory cortex in bats, and he has suggested not only that equally complex organization may be present in other auditory systems of animals but also that human auditory cortex may show similar arrangements for processing the complex sounds of speech. In Suga's bats, complex sounds are processed by neurons tuned to combinations of information-bearing elements or parameters in the sounds. For example, areas of cortex are tuned to particular echo delays and particular echo amplitudes. Neurons tuned to specific information-bearing parameters (IBPs) or combinations of IBPs are clustered in areas of the cerebral cortex and systematically arranged along axes or in coordinate systems for representation of sounds that are biologically important to the species (Suga, 1984).

It seems likely that equally complex sound processing takes place in the dolphin cortex. If so, the mapping experiments conducted to date (Supin *et al.*, 1978), though they showed certain differences of response in three areas of auditory cortex, may not have been of sufficient acoustic detail to reveal systematically arranged axes of specific delay-sensitive or amplitude-sensitive neurons and thus do not reveal the full extent of cortical organization. Possibly, surface segments that were not responsive in the mapping experiments (Supin *et al.*, 1978) contain neurons tuned to more specific acoustic parameters than those tested.

Detailed organization of auditory cortex described for bats may be equally or even more complex in dolphins. If so, the dolphin cortex must map an extensive surrounding area of the environment with specific acoustic parameters. According to Suga (1984), mustached bats (*Pteronotus parnellii*) map an area of only about 3 m, and target distances of 50–140 cm are best represented; on the other hand, dolphins, with an echolocation range of 100 m or more, probably map a large area of their surroundings for specific acoustic parameters. Since sound travels over four times as fast in water, neurons performing equal echo delays of X mseconds will represent Y m in a bat and $4Y$ m in a dolphin. Specific neurons forming such axes may take up considerable space in the cortex and could be a major reason for the great expansion of neocortex in the dolphin. In addition, I have mentioned various findings, for example, the large diameter of auditory nerve fibers, the short latency and the relatively flat latency–intensity function of the dolphin ABR waves, and the rapid temporal resolution of successive sounds, that lend support to my view that much of the great hypertrophy of the dolphin auditory system—and perhaps of the entire cerebrum—results from the animal's need for great precision and speed in processing sound.

REFERENCES

Au, W. L. (1980). Echolocation signals of the Atlantic bottlenose dolphin (*Tursiops truncatus) in open waters. In* "Animal Sonar Systems" (R. G. Busnel and J. F. Fish, eds.), pp. 251–282. Plenum, New York. 1135 pp.

Au, W. W. L., and Moore, P. W. B. (1984). Receiving beam patterns and directivity indices of the Atlantic bottlenosed dolphin (*Tursiops truncatus*). *J. Acoust. Soc. Am.* **75**, 255–262.

Au, W. W. L., Moore, P. W. B., and Pawloski, D. (1986). Echolocation transmitting beam of the Atlantic bottlenose dolphin. *J. Acoust. Soc. Am.* **80**(2), 688–694.

Balonov, L., Deglin, V., Kaufman, D., and Nikolaenko, N. (1981). Functional asymmetry of the animal brain. *J. Evolut. Biochem. Physiol.* **17,** 163–170. [In Russian, English transl. (1982). Plenum, New York.]

Bonner, J. T. (1980). "The Evolution of Culture in Animals." Princeton University Press, Princeton, New Jersey. 216 pp.

Breathnach, A. S. (1960). The cetacean central nervous system. *Biol. Rev.* **35,** 187–230.

Buhl, E. H., and Oelschläger, H. A. (1986). Ontogenetic development of the nervus terminalis in toothed whales. *Anat. Embryol.* **173,** 285–294

Bullock, T. H., and Gurevich, V. (1979). Soviet literature on the nervous system and psychobiology of Cetacea. *Int. Rev. Neurobiol.* **21,** 48–127.

Bullock, T. H., and Ridgway, S. H. (1972). Evoked potentials in the central auditory system of alert porpoises to their own and artificial sounds. *J. Neurobiol.* **3,** 79–99.

Bullock, T. H., Grinnell, A. D., Ikezono, E., Kameda, K., Katsuki, K., Nomoto, M., Sato, O., Suga, N., and Yanagisawa, K. (1968). Electrophysiological studies of central auditory mechanisms in cetaceans. *Zeitschr. Vergleich. Physiol.* **59,** 117–156.

Caldwell, M. C., Caldwell, D. K., and Siebenaler, J. B.

(1965). Observations on captive and wild Atlantic bottlenosed dolphins, *Tursiops truncatus*, in the Northeastern Gulf of Mexico. *Los Angeles County Mus. Contrib. Sci.* **91,** 1–10.

Cates, M. B., and Schroeder, J. P. (1986). Nutrition of acclimated vs newly captured *Tursiops truncatus*. *Aquatic Mamm.* **12**(1), 17–20.

Dawson, W. W., Carder, D. A., Ridgway, S. H., and Schmeisser, E. T. (1981). Synchrony of dolphin eye movements and their power density spectra. *Comp. Biochem. Physiol.* **68A,** 443–449.

Dawson, W. W., Hawthorne, M., Jenkins, R., and Goldston, R. (1982). Giant neural systems of the inner retina and optic nerve of small whales. *J. Comp. Neurol.* **205,** 1–7.

Demski, L. S., and Schwanzel-Fukuda, M., eds. (1987). The terminal nerve (nervus terminalis) structure, function and evolution. *Ann. N.Y. Acad. Sci.* **519,** 1–469.

Donaldson, B. J. (1977). The tongue of the bottlenosed dolphin (*Tursiops truncatus*). *In* "Functional Anatomy of Marine Mammals" (R. J. Harrison, ed.), Vol. 3, pp. 175–197. Academic Press, London.

Edinger, T. (1955). Hearing and smell in cetacean history. *Mschr. Psychiat. Neurol.* **129,** 37–58.

Elias, H., and Schwartz, D. (1969). Surface areas of the cerebral cortex of mammals determined by stereological methods. *Science* **166,** 111–113.

Evans, W. E., and Powell, B. A. (1967). Discrimination of different metallic plates by an echolocating delphinid. *In* "Animal Sonar Systems" (R. G. Busnel, ed.), pp. 363–383. Laboratoire De Physiology Acoustique, INRA–CNRZ, Jouy-en-Josas, France. 1233 pp.

Filimonoff, I. N. (1949). "Stravnitel'naya Anatimya kory Bol'shoge Mozga Miekopitayushchikh: Paleokorteks, Arkhikorteks, i Mechutochnaya kora." Moscow Academy of Medical Sciences, Moscow. 262 pp.

Filimonoff, I. N. (1965). On the so-called rhinencephalon in the dolphin. *J. Hirnforsch.* **8,** 1–23.

Fish, J. F., Johnson, C. S., and Ljungblad, D. K. (1976). Sonar target discrimination by instrumented human divers. *J. Acoust. Soc. Am.* **59,** 602–606.

Flanigan, W. F. (1974). Nocturnal behavior of captive small cetaceans. I. The bottlenosed porpoise *Tursiops truncatus*. *Sleep Res.* **3,** 84.

Flanigan, W. F. (1975). More nocturnal observations of captive, small cetaceans. I. The killer whale, *Orcinus orca*. *Sleep Res.* **4,** 139.

Fram, H., and Stephan, H. (1976). Vergleichende volumenmessungen an hiren von wild- und gefangenschaftstieren des spitzhörnchens *Tupaia*. *J. Hirnforsch.* **17,** 449–462.

Garey, L. J., and Leuba, G. (1986). A quantitative study of the neuronal and glial numerical density in the visual cortex of the bottlenosed dolphin: Evidence for a specialized subarea and changes with age. *J. Comp. Neurol.* **247,** 491–496.

Gersch, I. (1938). Note on the pineal gland of the humpback whale. *J. Mammal.* **19,** 477–480.

Geschwind, N., and Levitsky, W. (1968). Human brain: Left–right asymmetries in temporal speech region. *Science* **161,** 186–187.

Gingerich, P. D., Wells, N. A., and Russell, D. E. (1983). Origin of whales in epicontinental remnant seas: New evidence from the early Eocene of Pakistan. *Science* **220,** 403–406.

Glezer I. I., Jacobs, M. S., and Morgane, P. J. (1988). Implications of the "initial brain" concept for brain evolution in Cetacea. *Behav. Brain Sci.* **11,** 75–116.

Gould, S. J. (1977). "Ontogeny and Phylogeny." Harvard University Press, Cambridge, Massachusetts.

Hampton, I. F. G., Whittow, G. C., Szekercezes, J. and Rutherford, S. (1971). Heat transfer and body temperature in the Atlantic bottlenosed dolphin, *Tursiops truncatus*. *Int. J. Biometeorol.* **15,** 247–253.

Harnad, S., and Doty, R. (1977). Introduction. *In* "Lateralization of the Nervous System" (S. Harnad, R. Doty, L. Goldstein, J. Jaynes, and G. Krauthamer, eds.), p. 17. Academic Press, New York.

Harrison, R. J. (1969). Endocrine organs: Hypophysis, thyroid, and adrenal. *In* "The Biology of Marine Mammals' (H. T. Anderson, ed.), pp. 349–389. Academic Press, New York.

Harrison, R. J., and Thurley, K. W. (1974). Structure of the epidermis in *Tursiops, Delphinus,* and *Phocoena*. "Functional Anatomy of Marine Mammals" (R.J. Harrison, ed.), Vol. 2, pp. 45–71. Academic Press, London.

Hatschek, R. (1903). Sehnervenatrophie bei einem Delphin. *Arb. neurol. Inst. Univ. Wien*, 10, 223–229.

Haug, H. (1969). Vergleichende, quantitative untersuchungen an den gehiren des menschen, des elefanten und einiger zahnwale. *Medizinsche Monatschrift* **23,** 201–205.

Haug, H. (1970). Qualitative und quantitative untersuchungen an den gehirnen des menschen, der delphinoideae und des elefanten. *Ergebnisse der Anatomie und Entwicklungsgeschichte* **43,** 1–70.

Hawkins, A., and Olszewski, J. (1957). Glia/nerve cell index for cortex of the whale. *Science* **126,** 76–77.

Hemmer, H. (1976). Man's strategy in domestication—A synthesis of new research trends. *Experientia* **32,** 663-666.

Hillyard, S. A., and Bloom, F. E. (1982). Brain functions and mental processes. *In* "Animal Mind–Human Mind" (D. R. Griffin, ed.), pp. 13–32. Springer-Verlag, Berlin. 427 pp.

Hockett, C. F. (1978). In search of Jove's brow. *Am. Speech* **53,** 243–313.

Irving, L., Scholander, P. F., and Grinnell, S. W. (1941).

The respiration of the porpoise, *Tursiops truncatus. J. Cell. Comp. Physiol.* **17,** 145–168.

Jacobs, M. S., and Jensen, A. V. (1964). Gross aspects of the brain and fiber analysis of cranial nerves in the great whale. *J. Comp. Neurol.* **123,** 55–72.

Jacobs, M. S., Morgane, P. J., and McFarland, W. L. (1971). The anatomy of the brain of the bottlenosed dolphin *Tursiops truncatus.* Rhinic lobe (rhinencephalon). I. The paleocortex. *J. Comp. Neurol.* **141,** 205–272.

Jacobs, M. S., Morgane, P. J., and McFarland, W. L. (1975). Degeneration of visual pathways in the bottlenosed dolphin. *Brain Res.* **88,** 346–352.

Jansen, Jan, and Jansen, J. K. S. (1969). The nervous system of Cetacea. *In* "Biology of Marine Mammals" (H. T. Andersen, ed.), pp. 175–252. Academic Press, New York.

Jerison, H. J. (1973). "Evolution of the Brain and Intelligence." Academic Press, New York.

Johnson, C. S. (1966). Auditory thresholds of the bottlenosed dolphin *Tursiops truncatus* Montagu. *U.S. Naval Ordinance Test Station NOTSTP,* No. 4178. 25 pp.

Johnson, C. S. (1967). Discussion *In* "Animal Sonar Systems" (R. G. Bushel, ed.), pp. 384–398. Laboratoire de Physiol. Acoustique, INRA–CNRZ, Jouy-en-Josas, France. 1233 pp.

Johnson, J. B. (1914). The nervus terminalis in man and animals. *Anat. Rec.* **8,** 185–198.

Kamiya, T., and Pirlot, P. (1974). Brain morphogenesis in *Stenella coeruleoalba. Sci. Rep. Whales Res. Inst., Tokyo* **26,** 245–253.

Karandeeva, O. G., Matisheva, S. K. and Shapunov, V. M. (1973). Features of external respiration in the delphinidae. *In* "Morphology and Ecology of Marine Mammals. Seals, Dolphins, Porpoises" (K. K. Chapskii and V. E. Sokolov, eds.), pp. 196–206. Wiley, New York. 232 pp.

Kellogg, R. (1928). The history of whales—Their adaptation to life in the water. *Quart Rev. Biol.***3,** 29–76, 174–208.

Kellogg, W. N. (1958). Echo ranging in the porpoise. *Science* **128,** 982–988.

Kellogg, W. N., and Kohler, R. (1952). Reactions of the porpoise to ultrasonic frequencies. *Science* **116,** 250–252.

Kerem, D., Elsner, R., and Wright, J. (1971). Anaerobic metabolism in the brain of the harbor seal during the late stages of a maximum dive. *Fed. Proc., Fed. Am. Soc. Exp. Biol.* **30,** 484.

Kesarev, V. S. (1969). Structural organization of the limbic cortex in dolphins. *Arkh. Anat. Gistol. Embriol.* **56,** 28–35.

Kesarev, V. (1971). The inferior brain of the dolphin. *Soviet Sci. Rev.* **2,** 52–58.

Kesarev, V. S., and Malofeyeva, L. I. (1969). Structural organization of the dolphin motor cortex. *Arkh. Anat. Gistol. Embriol.* **56,** 48–55 (Neuroscience Transls. No. 12, Federation of American Societies for Experimental Biology).

Kojima, T. (1951). On the brain of the sperm whale (*Physeter catodon* L.) *Sci. Rep. Whales Res. Inst., Tokyo* **6,** 49–72.

Kolchin, S., and Bel'kovich, V. (1973). Tactile sensitivity in *Delphinus delphis. Zoologicheskiy Zhurnal* **52,** 620–622.

Kooyman, G. L., and Cornell, L. H. (1981). Flow properties of expiration and inspiration in a trained bottlenosed porpoise. *Physiol. Zool.* **54;** 55–61.

Kovalzon, V. M., and Mukhametov, L. M. (1982). Temperature variations in the brain corresponding to unihemispheric slow wave sleep in dolphins. *J. Evolut. Biochem. Physiol.* **18,** 307–309. [English transl. (1983) Plenum, New York.]

Kruger, L. (1959). The thalmus of the dolphin *Tursiops truncatus* and comparison with other mammals. *J. Comp. Neurol.* **3,** 133-194.

Kruger, L. (1966). Specialized features of the cetacean brain. *In* "Whales, Dolphins, and Porpoises"(K.S. Norris, ed.), pp. 232–254. University of California Press, Berkeley, California. 789 pp.

Lange, W. (1975). Cell number and cell density in the cerebellar cortex of man and some other mammals. *Cell. Tiss. Res.* **157,** 115–124.

Langworthy, O. R. (1932). A description of the central nervous system of the porpoise *(Tursiops truncatus) J. Comp. Neurol.* **54,** 437–499.

Lavigne, D. M., Innes, S., Worthy, G. A. J., Kovacs, K. M., Schmitz, O. J., and Hickie, J. P. (1986). Metabolic rates of seals and whales. *Can J. Zool.* **64,** 279–284.

Lende, R. A., and Welker, W. I. (1972). An unusual sensory area in the cerebral neocortex of the bottlenose dolphin, *Tursiops truncatus. Brain Res.* **45,** 555–560.

Levy J. (1982). Mental processes in the nonverbal hemisphere. *In* "Animal Mind–Human Mind" (D. R. Griffin, ed.), pp. 57–73. Springer-Verlag, Berlin.

Lilly, J. C. (1962). *In* "Interhemispheric Relations and Cerebral Dominance" (V. Mouncastle, ed.), pp. 112–114. Johns Hopkins Press, Baltimore, Maryland.

Lilly, J. C. (1964). Animals in aquatic environments: Adaptation of mammals to the oceans. *In* "Handbook of Physiology—Environment" (D. B. Dill, ed.), pp. 741–747. American Physiological Society, Washington, D. C.

McBride, A. F. (1956). Evidence for echolocation by cetaceans. *Deep-Sea Res.* **3,** 153–154.

McCormick, J. G. (1969). Relationship of sleep, respiratin, and anesthesia in the porpoise: A preliminary report. *Proc. Nat. Acad. Sci. U.S.A.* **62,** 697–703.

McCormick, J. G., Wever, E. G., Palin, J., and Ridgway, S. H. (1970). Sound conduction in the dolphin ear. *J. Acoust. Soc. Am.* **48,** 1418–1428.

McFarland, W. L., Morgane, P. J., and Jacobs, M. S.

(1969). Ventricular system of the brain of the dolphin, *Tursiops truncatus,* with comparative anatomical observations and relations to brain specializations. *J. Comp. Neurol.* **135,** 275–368.

McFarland, W. L., Jacobs, M. S. and Morgane, P. J. (1979). Blood supply to the brain of the dolphin, *Tursiops truncatus,* with comparative observations on special aspects of the cerebrovascular supply of other vertebrates. *Neurosci. Biobehav. Rev.* Suppl. 1. 93 pp.

Mackay, R. S. (1967). Experiments to conduct in order to obtain comparative results. *In* "Animal Sonar Systems" Laboratoire De Physiology Acoustique, INRA–CNRZ, Jouy-en-Jocas, France. 1233 pp.

McNab, B. K. (1988). Complications inherent in scaling the basal rate of metabolism in mammals. *Quart. Rev. Biol.* **63,** 25–54.

Maxwell, G. (1960). "Ring of Bright Water." Longmans, London.

Meith, N. (1985). "Marine Mammals." United Nations Environment Programme, Geneva.

Morgane, P. J., Jacobs, M. S., and McFarland, W. L. (1980). The anatomy of the brain of the bottlenose dolphin *(Tursiops truncatus).* Surface configurations of telencephalon of the bottlenose dolphin with comparative anatomical observations in four other cetacean species. *Brain Res. Bull.* **5**(3), 1–108.

Morgane, P. J., McFarland, W. L., and Jacobs, M. S. (1982). The limbic lobe of the dolphin brain: A quantitative cytoarchitectonic study. *J. Hirnfosch.* **23,** 465–552.

Morgane, P. J., Jacobs, M. S., and Galaburda, A. (1986). Evolutionary morphology of the dolphin brain. *In-* "Dolphin Cognition and Behavior: A Comparative Approach" (R. Schusterman, J. Thomas, and F. Wood, eds.), pp. 5–29. Erlbaum, Hillsdale, New Jersey. 393 pp.

Mukhametov, L. M. (1984). Sleep in marine mammals. *Exp. Brain Res. Suppl.* **8,** 227–238.

Mukhametov, L. M., Supin, A. Y., and Polyakova, I. G. (1977). Interhemispheric asymmetry of the electroencephalographic sleep patterns in dolphins. *Brain Res.* **134,** 581-584.

Mukhametov, L. M., Oleksenko, A. I., and Polyakova, I. G. (1987). Duration of EEG stages in dolphin cerebral hemispheres. *Dokl. Akad. Nauk* **294,** 748–751. (In Russian.)

Murie, J. (1874). On the organization of the caaing whale, *Globiocephalus melas. Trans. Zool. Soc. London* **8,** 235–301.

Myers, R. E. (1956). Function of corpus callosum in interocular transfer. *Brain* **79,** 358–363.

Nachtigall, P. E., and Hall, R. W. (1984). Taste reception in the bottlenosed dolphin. *Acta Zool. Fennica* **172,** 147–148.

Nagel, E. L., Morgane, P. J., McFarland, W. L., and Galliano, A. E. (1968). Rete mirabile of dolphin: Its pressure-damping effect on cerebral circulation. *Science* **161,** 898-900.

Ness, A. R. (1967). A measure of asymmetry of the skulls of odontocete whales. *J. Zool. (London)* **153,** 209-221.

Neville, H. J., and Hillyard, S. A. (1982). Neurophysiological approaches, state of the art report. *In* "Animal Mind–Human Mind" (D. R. Griffin, ed.), pp. 333–353. Springer-Verlag, Berlin. 427 pp.

Nieto, A., Nieto, D., and Pacheco, P. (1976). Possible phylogenetical significance of the corpus callosum with special reference to the dolphin brain (*Stenella graffmani*). *Acta Anat.* **94,** 397–402.

Norris, K. S. (1968). The evolution of acoustic mechanisms in odontocete cetaceans, *In* "Evolution and Environment" (E. T. Drake, ed.) pp. 293–323. Yale University Press, New Haven, Connecticut.

Norris, K. S., Prescott, J. H., Asa-Dorian, P. V., and Perkins, P. (1961). An experimental demonstration of echolocation behavior in the porpoise *Tursiops truncatus* (Montagu). *Biol. Bull.* **1202,** 163–176.

Oelschläger, H. A., and Buhl, E. H. (1985). Development and rudimentation of the peripheral olfactory system in the harbor porpoise *Phocoena phocoena. J. Morphol.* **184,** 351–360.

Palmer, E., and Weddell, G. (1964). The relationship between structure, innervation and function of the skin of the bottlenosed dolphin (*Tursiops truncatus*). *Proc. Zool. Soc. London* **143,** 553–568.

Perrin, W. F., Mitchell, E. D., Mead, J. G., Caldwell, D. K., Caldwell, M. C., Van Bree, P. J. H., and Dawbin, W. H. (1987). Revision of the spotted dolphins, *Stenella* spp. Marine Mammal Science 3(2), 99–170.

Pierce, R. W. (1970). Design and operation of a metabolic chamber for marine mammals. Ph. D. dissertation. University of California, Berkeley.

Pilleri, G. (1983). Cetaceans in captivity. *In* "Investigations on Cetacea" (G. Pilleri, ed.), pp. 221–249. Brain Anatomy Institute, Berne, Switzerland. 258 pp.

Pilleri, G., and Gihr, M. (1970). The central nervous system of the mysticete and odontocete whales. *In* "Investigations on Cetacea" (G. Pilleri, ed.), Vol. II, pp. 89–129.

Popov, V. V., Ladygina, T. F., and Supin, A. Ya. (1984). Evoked potentials of the auditory cortex of *Phocoena phocoena. J. Comp. Physiol.* **158,** 705–711.

Ridgway, S. H. (1968). The bottlenosed dolphin in biomedical research. *In* "Methods of Animal Experimentation"(W. I. Gay, ed.), Vol. III, pp. 387–446. Academic Press, New York. 469 pp.

Ridgway, S. H. (1972). Homeostasis in the aquatic environment. *In* "Mammals of the Sea: Biology and Medicine" (S. H. Ridgway, ed.), pp. 590–747. C. C. Thomas, Springfield, Illinois. 812 pp.

Ridgway, S. H. (1986a). Physiological observations on the dolphin brain. *In* "Dolphin Cognition and Behavior: A Comparative Approach" (R. Schusterman, J. Thomas, and F. Wood, eds.), pp. 31–59. Erlbaum, Hillsdale, New Jersey. 393 pp.

Ridgway, S. H. (1986b). Diving by cetaceans. *In* "Diving in Animals and Man" (A.O. Brubakk, J. W. Kanwisher, and G. Sundness, eds.), pp. 33–62. Tapir, Trondheim, Norway. 308 pp.

Ridgway, S. H., and Brownson, R. H. (1979). Brain size and symmetry in three dolphin genera. *Anat. Rec.* **193,** 664.

Ridgway, S. H., and Brownson, R. H., (1984). Relative brain sizes and cortical surface areas of odontocetes. *Acta Zool. Fennica* **172,** 149–152.

Ridgway, S. H., and Carder, D. A. (1989). Tactile sensitivity, somatosensory responses, skin vibrations, and the skin surface ridges of the bottlenosed dolphin, *Tursiops truncatus*. *In* "Cetacean Sensory Abilities: Field and Laboratory Evidence" (J. Thomas and R. Kastelein, eds.). Plenum, New York (in press).

Ridgway, S. H., and Patton, G. S. (1971). Dolphin thyroid: Some anatomical and physiological findings. *Zeitschr. Vergleich. Physiol.* **71,** 129–141.

Ridgway, S. H., and Wood, F. G. (1988). Dolphin brain evolution. *Behav. Brain Sci.* **11,** 99.

Ridgway, S. H., Flanigan, N. J., and McCormick, J. G. (1966). Brain–spinal cord ratios in porpoises: Possible correlations with intelligence and ecology. *Psychonomic Sci.* **6,** 491–492.

Ridgway, S. H., Scronce, B. L., and Kanwisher, J. (1969). Respiration and deep diving in a bottlenose porpoise. *Science* **166,** 1651–1654.

Ridgway, S. H., Bullock, T. H., Carder, D. A., Seeley, R. L., Woods, D., and Galambos, R. (1981). Auditory brainstem response in dolphins. *Proc. Acad. Sci. U.S.A.* **78,** 943–1947.

Ridgway, S. H., Demski, L. S., Bullock, T. H., and Schwanzel-Fukuda, M. (1987). The terminal nerve in odontocete cetaceans. *Ann. N.Y. Acad. Sci.* **519,** 201–212.

Ries, F. A., and Langworthy, O. R. (1937). A study of the surface structure of the brain of the whale (*Balaenoptera physalus* and *Physeter catadon*). *J. Comp. Neurol.* **68,** 1–47.

Scholander, P. F. (1940). Experimental investigations on the respiratory function in diving mammals and birds. *Hvalradets Skrifter* **22,** 1–131.

Scholander, P. F., and Irving, L. (1941). Experimental investigations on the respiration and diving of the Florida manatee. *J. Cell. Comp. Physiol.* **17,** 169–191

Serafetinides, E., Shurley, J., and Brooks, R. (1972). Electroencephalogram of the pilot whale, *Globicephala scammoni*, in wakefulness and sleep: Lateralization aspects. *Int. J. Psychobiol.* **2,** 129–135.

Sergeant, D. E., Caldwell, D. K., and Caldwell, M. C. (1973). Age, growth and maturity of bottlenose dolphins *(Tursiops truncatus)*. *J. Fisheries Res. Board Can.* **30,** 1009-1011.

Shapunov, V. M. (1973). Food requirements and energy balance in the Black Sea bottlenose dolphin *Tursiops truncatus* Ponticus Barabasch. *In* "Morphology and Ecology of Marine Mammals. Seals, Dolphins, Porpoises" (K. K. Chapskii and V. E. Sokolov, eds.), pp. 207–212. Wiley, New York. 232 pp.

Shurley, J., Serafetinides, E., Brooks, R., Elsner, R., and Kenney, D. (1969). Sleep in cetaceans. I. The pilot whale, *Globicephala scammoni*. *Psychophysiology* **6,** 230.

Simon, L. M., Robin, E. D., Elsner, R., Van Kessel, A., and Theodore, J. (1974). A biochemical basis for differences in maximal diving time in aquatic mammals. *Comp. Biochem. Physiol.* **47B,** 209–215.

Sinclair, J. G. (1951a). The terminal olfactory complex in the porpoise. *Texas J. Sci.* **3,** 251.

Sinclair, J. G. (1951b). A ganglionic plexus in the human falx cerebri. *Texas Rep. Biol. Med.* **9,** 348–352.

Sinclair, J. G. (1966). The olfactory complex of dolphin embryos. *Texas Rep. Biol. Med.* **24,** 426–431.

Sleptsov, M. M. (1939). On the problem of the asymmetry of the skull in odontoceti. *Zoologicheskiy Zhurnal* **18,** 367–386.

Smith, R. J. (1984). Allometric scaling in comparative biology: Problems of concept and method. *Am. J. Physiol.* **246,** R152–R160.

Spector, W. S. (1956). "Handbook of Biological Data,"p. 162. Saunders, Philadelphia, Pennsylvania.

Suga, N. (1984). Neural mechanisms of complex-sound processing for echolocation. *Trends NeuroSci.* **7,** 20–27.

Supin, A. Y., Mukhametov, L. M., Ladygina, T. F., Popov, V. V., Mass, A. M., and Poliakova, E. G. (1978). Electrophysiological Study of the Dolphin Brain. Nauka, Moscow. (In Russian.)

Teyler, T. J. (1977). An introduction to the neurosciences. *In* "The Human Brain" (M. C. Wittrock, ed.), p. 50 Prentice Hall, Englewood Cliffs, New Jersey.

Tomilin, A. G. (1968). Factors promoting powerful development of the brain in odontoceti. *Tr. Veses. Skh. Inst. Zaochn. Obraz.* **31,** 191–200.

Tower, D. (1954). Structural and functional organization of the mammalian cerebral cortex: The correlation of neurone density with brain size. *J. Comp. Neurol.* **101,** 19–52.

Townsend, C. H. (1914). The porpoise in captivity. *Zoologica* **1,** 289–299.

Tulsi, R. S. (1975). Observations of the structure of the dorsal muscle in the bottlenosed dolphin *(Tursiops truncatus)*. *J. Anat.* **119,** 39–48.

Van Dyke, D., and Ridgway, S. H. (1977). Diets for marine mammals. *In* "Handbook Series in Nutrition and Food" (M. Rechcigl, ed.) pp. 595–598. CRC Press, Cleveland, Ohio. 645 pp.

Van Valen, L. (1966). Deltratheridia, a new order of mammals. *Bull. Am. Mus. Nat. Hist.* **132,** 1–126.

Viamonte, M., Morgane, P. J., Galliano, R. E., Nagal, E. L., and McFarland, W. L. (1968). Angiography in the living dolphin and observations on blood supply to the brain. *Am. J. Physiol.* **214,** 1225–1249.

Weber, M. (1897). Vorstudien über das hirngewicht der śugethiere. *Festschr. 70 Geb. C. Gegenbaur, III Bd.*, pp. 105–123. W. Engelmann, Leipzig.

Wever, E. G., McCormick, J. G., Palin, J., and Ridgway, S. H. (1971). The cochlea of the dolphin *Tursiops truncatus:* Hair cells and ganglion cells. *Pro. Nat. Acad. Sci. U.S.A.* **68,** 2908–2912.

Wilson, R. B. (1933). The anatomy of the brain of the whale *(Balaenoptera sulfurea). J. Comp. Neurol.* **101,** 19–51.

Wirz, K. (1950). Studien über die cerebralization: Zur quantitativ en bestimmung der rangordnung beisäugetieren. *Acta Anat.* **9,** 134–196.

Witelson, S. F. (1977). Anatomic asymmetry in the temporal lobes: Its documentation, phylogenesis, and relationship to functional asymmetry. *In* "Evolution and Lateralization of the Human Brain" (S. J. Dimond, and D. A. Blizard, eds.), pp. 328–354. *Ann. N. Y. Acad. Sci.* **299.**

Wood, F. G., Jr. (1964). General discussion. *In* "Marine Bio-Acoustics" (W. N. Tavolga, ed.), pp. 395–396. Pergamon Press, New York.

Wood, F. G., and Evans, W. E. (1980). Adaptiveness and ecology of echolocation in toothed whales. *In* "Animal Sonar Systems" (R. Busnell and J. Fish, eds.), pp. 381–426. Plenum, New York. 1135 pp.

Worden, C. J. (1951). Animal intelligence. *Sci. Am.* **185,** 64.

Yamasaki, F., Komatsu, S., and Kamiya, T. (1978). Taste buds in the pits at the posterior dorsum of the tongue of *Stenella coeruleoalba. Sci. Rep. Whales Res. Inst.*, Tokyo **30,** 285–290.

Yoshikawa, T., and Nakamura Y. (1952). On the problems of the olfactory bulb in the dolphin. *Acta Anat. Nippon.* **27**(3), 9 (abstr.).

PHOTO: In many parts of the world there are coastal and offshore forms of bottlenose dolphins, but for the present, at least, only the one species, *Tursiops truncatus*, is considered valid (Port Elizabeth Oceanarium, Republic of South Africa; courtesy *Die Oosterlig*).

Part III

Systematics and Taxonomy

Comments on Australian Bottlenose Dolphins and the Taxonomic Status of *Tursiops aduncus* (Ehrenberg, 1832)

5

Graham J. B. Ross
Victor G. Cockcroft
Port Elizabeth Museum
Humewood 6013, South Africa

INTRODUCTION

Bottlenose dolphins, *Tursiops* spp., have been intensively studied in a wide variety of disciplines, both in captivity and in their natural environment. They are undoubtedly the best-known of the delphinids. The taxonomy of the genus, however, is poorly understood, and opinions vary on the status of nominal species. Most authors have accepted that there is a single species, *Tursiops truncatus* Montagu, 1821 (e.g., Tomilin, 1957; Mitchell, 1975; Rice, 1977). Some have suggested that there are two species, *T. gilli* Dall, 1873, and *T. truncatus*, the latter comprised of two subspecies, *T. t. truncatus* and *T. t. aduncus* (Rice and Scheffer, 1968; Hershkovitz, 1966). Van Bree (1966, pers. commun.) held that these subspecies warrant full specific status, while Pilleri and Gihr (1972a,b, 1973–1974) concluded that *T. truncatus, T. gilli, T. aduncus* (Ehrenberg, 1832), and *T. gephyreus* Lahille, 1908, are all valid species.

For the most part, the taxonomic uncertainty stems from the lack of series of material, particularly from more remote areas, suitable for systematic study. A few detailed regional investigations of variation in local *Tursiops* populations have been undertaken. These studies have identified a number of distinct morphological forms, apparently distributed allopatrically for the most part. Thus, in southern African waters, the distinctness of two forms in size and morphological features led Ross (1977) to accord *T. truncatus* and *T. aduncus* full specific status. The smaller *T. aduncus* inhabits the coastal waters along the southeast coast. *Tursiops truncatus* is found in inshore waters (and possibly offshore) on the west coast but occurs only offshore on the south and southeast coasts. W. A. Walker (1981) distinguished three forms of *Tur-*

siops in the eastern North Pacific, two of which correspond to the descriptions of *T. nuuanu* (Andrews, 1911) and *T. gillii*. The latter is a large coastal form distributed along the coasts of southern California and Mexico. It differs from the two closely related offshore forms which are separable on skull length and size at sexual maturity. The separations, however, are modal rather than absolute.

Additional regional studies will be necessary before the taxonomic significance of the various forms can be fully understood. There is also a need for research on the effects of environmental parameters on the morphology of dolphins, as factors in the development and maintenance of such forms. Mitchell (1975) commented that forms of *Tursiops* vary primarily "in body size . . . distributed differentially relative to sea temperature and depth." Other authors have hypothesized that variation in body size and form in *Stenella attenuata*, *S. frontalis* and *Cephalorhynchus commersonnii* reflect the influence of environmental factors, especially sea temperature and availability of food (Perrin, 1975; Perrin *et al.*, 1987; Robineau, 1986; Schnell *et al.*, 1986). Coastal populations of *Tursiops* appear to be the most suitable candidates for testing these hypotheses. Ideally, studies should involve contiguous populations that are resident or semiresident along an extensive stretch of coast on which gradients in environmental parameters such as sea temperature are considerable but gradual.

This chapter describes bottlenose dolphins occurring in Australian waters, based primarily on museum specimens and captive live animals, supplemented with new data gathered from *Tursiops* populations elsewhere. Morphological variation is related to mean surface water temperature, the only environmental parameter for which data are readily available.

Published comments on the specific status of Australian *Tursiops* are few and brief. Gray (1862) described *T. catalania* based on two dolphins collected off Cape Melville and Cape Flattery, northeast Queensland. This species was synonymized subsequently with *T. aduncus* (Ross, 1977). Heinsohn (1978) referred dolphins from the same area to *T. truncatus*, as did Lear and Bryden (1980) and Paterson (1986), following Watson (1981), for material from southern Queensland. Scott and Lord (1919) identified animals from Tasmania as belonging to a southern form of *T. tursio* (=*T. truncatus*), distinguishing them from *T. catalania*, the only species then admitted to the list of Australian cetaceans (Beddard, 1900). Iredale and Troughton (1934) simply renamed *T. tursio* as a new species, *T. maugeanus*, but provided no justification for doing so. Dolphins along the South Australian coast have been referred to *T. truncatus* by Wood Jones (1925), Aitken (1971), and Harrison and Fanning (1973–1974).

METHODS AND MATERIALS

Between April and June 1983, G.J.B.R. examined 103 specimens of *Tursiops* from Australian waters in 11 museums or institutions (abbreviations in parentheses): Western Australian Museum, Perth (WAM); Australian National University, Canberra (ANU); CSIRO Division of Wildlife and Rangelands Research, Canberra (CM); South Australian Museum, Adelaide (SAM); National Museum of Victoria, Melbourne (NMV); Tasmanian Museum and Art Gallery, Hobart (TM); Queen Victoria Museum, Launceston (QVM); Australian Museum, Sydney (AM); Queensland Museum, Brisbane (QM); Anatomy Department, University of Queensland, St. Lucia (UQ); and Department of Zoology, James Cook University, Townsville (JCU). Single skulls held by the National Parks and Wildlife Service of South Australia (SANPWS) and Tasmania (TNPWS) and the Department of Veterinary Science, University of New South Wales, Cambden (UNSW), and a skull from Western Australia in the British Museum, London (BM), were also examined. Measurements and data on the cotypes of *T. catalania* were taken from True (1889).

Localities and catalog numbers for this material are as follows (* indicates specimens grouped as subadults/adults for analysis purposes, as described below):

Queensland: Townsville (JCU temporary no. 1*, JCU temporary no. 3*, QM J4155, AM S1778); Magnetic Island (JCU MM72*, JCU MM33); Percy Island (JCU temporary no. 2); near Bundaberg (QM J7015*); Moreton Bay (QM

JM1230*, UQ C19*, QM J14.2647); Stradbroke Island (QM J6421); Burleigh Heads (QM J3849, UQ C6*)

New South Wales: Pt. Plomer (UNSW no no. *); near Crowdy Head (AM M10852); Port Stephens (AM 279*); Terigal (AM no no.); Newport Beach (AM M12402); Port Jackson (AM B10527*); Botany Bay (AM S1555*); Wollongong (CM 10215); Nowra (AM M10853*); Termeil (ANU 1858); Kioloa Beach (ANU no no.); NSW coast (AM 278*, Am 280*)

Victoria: Thalia Point (NMV C11271); Lakes Entrance (NMV C24966); Bass River (NMV C24960*); Stingaree Bay (NMV C25071*); Port Phillip Bay (NMV FWD12950*); Werribee (NMV C25706*); Lorne (NMV C23490, NMV C24987, NMV C7799); Danger Point (NMV C24989*); Killarney Bay (NMV C24990)

Tasmania and Bass Strait: Ralph's Bay (TM A1289*); Maria Island (TM A1288); Lisdillon (TM A198*); Eaglehawkneck (TNPWS); NW Tasmania (QVM 1361*); Bass Strait (QVM 1972.1.35*); Flinders Island (QVM 1974.1.19*, QVM 1362*); Esk River (QVM 1946.8, QVM 1365*); Launceston (QVM 1360*)

South Australia: Port Stanvac (SAM M8385*, SAM M10639, SAM M10638*); Sellicks Bay (SAM M2666); Wood's Well (SAM M5902*); Nalpa (SAM M1075*); Destee Bay, Kangeroo Island (SAM M10101*); Port Lincoln (SAM M5078*); Goolwa (SAM M6038*); Port Pearce (SAM M3268); Memory Cove (SAM temporary no. 5, temporary no. 12*, temporary no. 10, temporary no. 6*, temporary no. 11*); Kangeroo Island (SANPWS); Meningie (SAM M1078*); Cowell (SAM M2130); Henley Bay (SAM M4819); Parson's Beach (SAM M10098*); West Island (SAM M7479)

Western Australia: Madura, Nullabor (NMV C21940); Esperance (WAM M4794*); Albany district (WAM Albany "A", WAM Albany "B"*); Albany (WAM M7871*); Parry Inlet (WAM M15246); Cape Leeuwin (WAM M1228); Yallingup (WAM M7028); Rockingham (WAM no no.*); Cape Peron (WAM M7499); Garden Island (WAM M7584*); Perth (WAM M1192*); Swanbourne (WAM M6395); Swan River (WAM prep. no. 6888); Trigg Island (WAM JH2); Perth district (WAM temporary "A"*, WAM no no.); Rottnest Island (WAM M5723*); Yanchep (WAM prep. no. 6743) MDNM; Jurien Bay (WAM M15249*); Boranup (WAM M7870); Houtman Abrolhas Islands (WAM M7044*, AM M2097); Tamala Station, Shark Bay (WAM M11940*); Nanga Bay, Shark Bay (WAM M11941*); Peron Peninsula (WAM M6384*); Shark Bay (WAM prep. no. 2959); 130 km S of Carnarvon (WAM M6786); Carnarvon (BM 1906.10.5, WAM M6845); Barrow Island (WAM M11387*); Meerup Beach (WAM M2677*); Elwood (NMV C24944*)

Though most specimens consisted of skeletal material only, other data occasionally available were sex, date of death, locality, external appearance and coloration (photographs or notes), external measurements, body mass, reproductive state, and stomach contents (after Norris, 1961).

Alveolar counts and 35 skull measurements were taken to the nearest millimeter, following Perrin (1975, his Table 2, Nos. 1–40, excluding 18 and 23, his Fig. 2). Two additional measurements were made: length of mandibular symphysis and width of cancellous bone along the lateral edge of the premaxillae at the midlength of the rostrum (after Ross, 1977). If less than about 20 mm of the rostral tip was broken off, the amount was estimated and added. Corrections were made to rostral measurements for rostra which had splayed during preparation (Ross, 1977). Rudimentary teeth and their alveoli were included in analyses of tooth counts and in measurements of tooth row length, respectively. Four nonmetric cranial features described by Ross (1977) were classified as follows:

1. In lateral view, the width and apex of the pterygoid notch was narrow, acute; narrow, rounded; broad, acute; or broad, rounded.
2. The size of the subtriangular squamosal bone appearing in the wall of the posttemporal fossa was graded in five categories from small to large. Its "tilt" (in degrees) was estimated as the angle subtended between a line through its apex and midpoint of its base and a line perpendicular to the long axis of the skull.
3. The width and margins, respectively, of the posterior part of the vomer were graded in five categories each, from narrow to broad and parallel to flaring posteriorly.
4. In anterior view, arching of the premaxillae at midlength of the rostrum was graded in five categories from weak to strong.

Several criteria were used to group specimens into five age classes (calf, juvenile, subadult, subadult/adult, adult), in order to distinguish fully grown animals for skull morphometric comparisons. Of necessity, these groupings were approximate, but probably corresponded to unweaned calves, juveniles less than 5 years old, animals 5 years or older but actively growing, large and apparently fully grown animals for which data on sexual maturity state were not available, and sexually mature animals. In order of decreasing value, the criteria were as follows: fusion of epiphyses of thoracic vertebrae, known reproductive status, known or minimum age of captive animals, the degree to which the tooth pulp cavity was obliterated (open, closing, pinhole, closed), the degree to which the maxillary plates were fused to the cranium (free, loose, fused and firm), the state of tooth wear (unworn, tooth tips only, extensively worn), and the state of the mandibular alveoli (deep and often with thin septa, shallow and filled basally). Subsequent comparison indicated that the subadult/adult and adult groups were very similar morphometrically, and they were treated as a single group for analysis purposes.

To detect potential taxonomic entities in the sample of skulls, principal component analyses of skull measurements were undertaken, using Statgraphics (1986, Statistical Graphics Corporation, Inc.). In order not to reduce this small sample further, the analysis was constrained considerably by selection for variables with fewest missing values. Thus, all mandibular measurements were omitted, as many specimens were represented by crania only.

Total and regional counts of vertebrae were made following the recommendations of de Smet (1977). However, identification of intermediary thoracolumbar vertebrae could not be made with certainty, and these have been grouped with the X (lumbar) vertebrae.

Measurements and meristics are available on request and provision of a formatted $5\frac{1}{4}$–inch disk as ASCII files or in hard copy from the authors. With the authors' permission, copies of the data may be obtained also from the Department of Mammalogy, Western Australian Museum, Perth, or the Department of Vertebrate Zoology, U.S. National Museum, Washington, D.C.

All photographs and measurements available in April 1983 of bottlenose dolphins caught in the gillnet fishery operating in the Timor and Arafura Seas were made available to the authors by the Australian National Parks and Wildlife Service, Canberra, providing valuable data on size and appearance of bottlenose dolphins in areas not represented by museum material. Captive animals in Atlantis Marine Park, South Australian Marineland, King Neptune Park, Pet Porpoise Pool, Seaworld, Surfer's Paradise, and African Lion Safari Park provided data on external appearance for sections of the coasts of Western and South Australia, northern New South Wales, and southern Queensland.

To facilitate comparisons between material from different regions, the Australian coastline was divided into 26 areas approximately along lines of latitude or longitude at 5° intervals (Fig. 1, Table 1). Additional sampling areas were designated within Shark Bay (area 22) and at the eastern limit of area 16. The division between the areas to the north (area 20) and east (area 19) of Cape Leeuwin does not follow lines of latitude or longitude. Mean annual sea surface temperatures were estimated for coastal waters in most of these areas from data in Allen and Ferguson Wood (1950), Bennett and Pope (1953), Endean *et al.* (1956), Womersley and Edmonds (1958), Wyrtki (1971), Rochford (1975), O'Connor (1979), T. Walker (1981), Rainier (1984), Kimmerer *et al.* (1985), Nunes and Lennon (1986), and Pearce (1986). Minimum temperatures only were estimated for areas 1–4, 25, and 26, owing to insufficient data (Pearce, 1986).

Data were collated on daily feeding rate of a 236-cm adult female *T. aduncus* (sensu Ross, 1977) in relation to water temperature, for the period March 1977 to September 1985. This dolphin had been maintained for 24 years in the Port Elizabeth Oceanarium, South Africa. Throughout the study period, she was held in a large pool containing 4.5×10^6 liters chlorinated seawater along with two to four female or immature male dolphins. She participated freely and consistently in twice-daily performances of uniform intensity, suggesting that activity levels remained comparable during the experiment. Intervals exceeding 45 days between weighing the dolphin were excluded from the analysis, as were any intervals in which she was sufficiently ill to have required medication, or showed inappetence. Water temperatures were taken daily between 0800 and 1000 hours. Components of the diet were weighed before and after each feeding session. Energy values for each prey species were determined using bomb calorimetry of freeze-dried samples collected at intervals from the food (Table 2). Three-dimensional

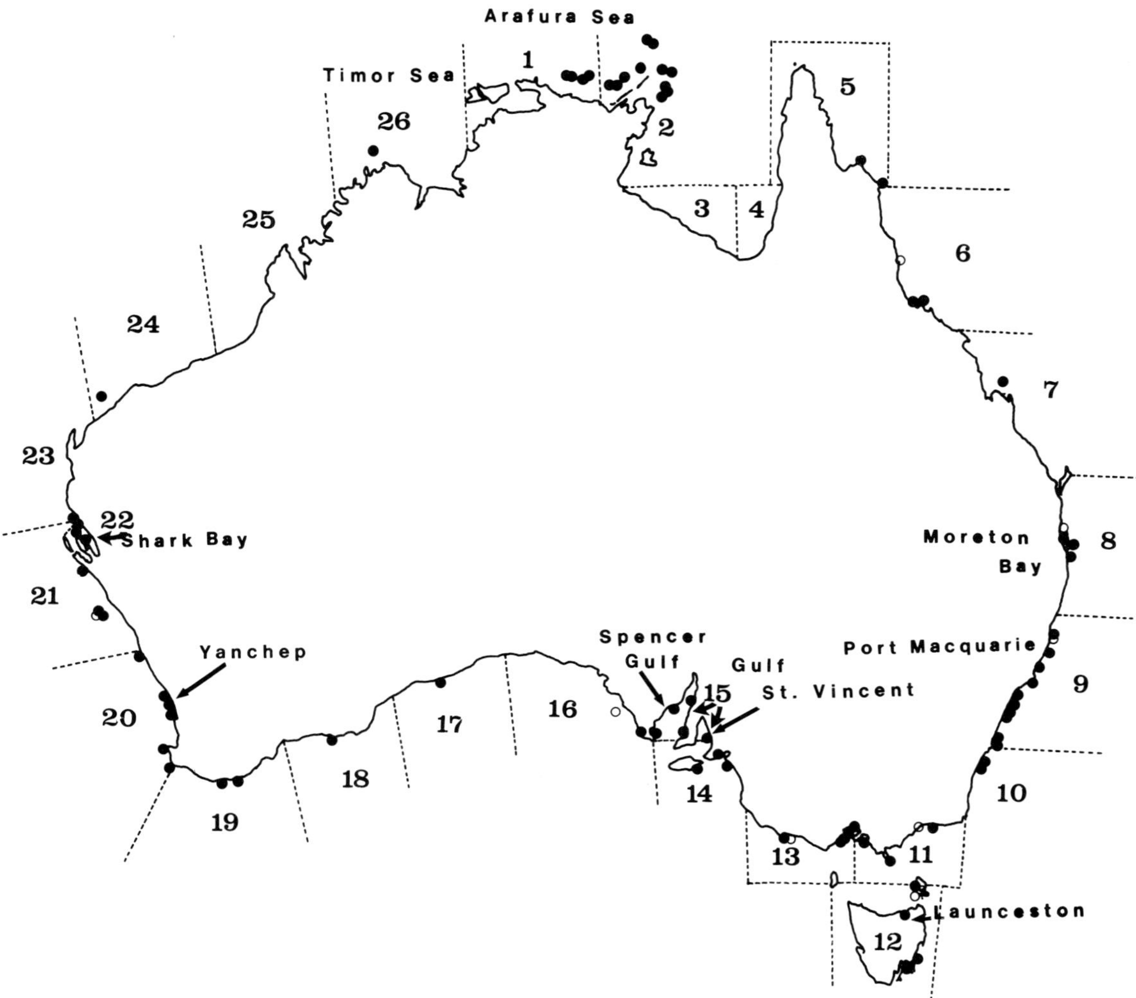

Figure 1 Distribution of bottlenose dolphin specimens: capture localities for live animals (●) and photographic and published records (○) on the Australian coast. Dashed lines indicate boundaries of the 26 regions described in the text.

plots of the results were drawn with minimal smoothing from mean values weighted in favor of the nearest neighbors.

Data on color pattern, body and blubber mass, and reproductive state were collected from bottlenose dolphins captured in shark nets protecting bathing beaches along the Natal coast, South Africa.

RESULTS

Distribution

The distribution of bottlenose dolphins along the Australian coast shown in Fig. 1 is based on stranding or capture localities of museum specimens, animals captured for oceanaria or during

Table 1
Number of *Tursiops* Specimens Examined and Estimated Mean Annual Surface Water Temperature for Each 5° Square

Locality code (Fig. 1)	NW corner of 5° square		Number of specimens examined		Mean surface water temperature (°C)
	Latitude (°S)	Longitude (°E)	Total	Adult	
1	10	130	0	0	28.0[a]
2	10	135	0	0	28.0[a]
3	15	135	0	0	28.0[a]
4	15	140	0	0	27.0[a]
5	10	141	1	1	27.0
6	15	145	7	4	26.5
7	20	147	2	1	23.5
8	25	150	6	3	22.5
9	30	150	9	5	19.0
10	35	150	2	0	18.5
11	35	145	3	1	17.0
12	39	143	11	8	14.5
13	35	140	8	4	15.0
14	35	135	7	5	17.0
15	30	135	13	8	18.5
16	30	130	1	0	17.0
17	30	125	1	0	17.0
18	30	120	1	1	18.0
19	34	115	5	2	19.5
20	30	115	14	7	20.5
21	25	110	2	1	21.5
22	26	110	5	3	23.0
23	20	110	2	0	24.5
24	20	115	1	1	26.5
25	15	120	0	0	28.0[a]
26	10	125	0	0	28.0[a]

[a] Minimum estimate.

fishing operations, photographs in museum files, and some published records (True, 1889; Ellis, 1987; Australian National Parks and Wildlife Service, 1986, 1987). Not unexpectedly, most of the data points are concentrated around major cities, where material comes to the notice of the museums and interested biologists more readily. However, given the low human population densities over most of this vast coastline, there are a sufficient number of records for other areas to suggest that this genus is distributed continuously around the continent. In support of this, cetacean sightings made during aerial surveys for dugongs, *Dugong dugon*, in the inshore waters of Queensland between Hinchinbrook Island (18° 10′ S) and Moreton Bay (27° 30′ S) from 1974 to 1979 indicated that bottlenose dolphins occurred throughout this area and were the most commonly seen cetacean, particularly in the southern part of the survey area (Heinsohn, 1979).

Bottlenose dolphins occur offshore well beyond the continental shelf in other parts of the world (Scott and Chivers, Chapter 22, this volume). However, information on the offshore distribution of bottlenose dolphins around Australia is very limited, other than the general comment that all

Table 2
Energy Values for Species of Food Fish and Squid and Quantities of Each Fed to a Captive Bottlenose Dolphin

Species	Energy values (MJ/kg)		Number of		Total mass (kg)
	Mean	Range	Samples	Fish	
Trachurus trachurus	5.63	4.42–6.68	11	28	6261
Merluccius capensis (cutlets)	4.08	3.91–4.20	4	8	5927
Sardinops ocellatus (whole)	7.66	5.84–10.55	9	27	5847
Sardinops ocellatus (cutlets)	6.10	5.68–6.81	9	17	—
Pagellus bellotti	5.29	5.04–5.64	6	11	452
Loligo reynaudii	4.55[a]	—	—	—	437
Cheilodonichthys capensis	5.60[b]	—	—	—	187
Emmelichthys nitidus	7.51	—	1	8	89
Scomber japonicus	5.60[b]	—	—	—	8

[a] After Heath and Randall (1986).
[b] Generalized value for fish, after Batchelor and Ross (1984).

sightings and specimens captured or taken offshore were over the continental shelf. Most of these were close to shore, but their significance in distributional terms is uncertain without surveys to determine inshore–offshore distributional patterns. Observations from two areas are of particular interest.

At Monkey Mia, Shark Bay, Western Australia (Fig. 1), a group of bottlenose dolphins has become habituated progressively to humans since the 1950s. Some dolphins in this group regularly come into water less than 1 m deep, where they make contact with humans and accept dead fish. Resightings of individually recognizable members of this group have shown that this area forms part of the group's range throughout most of the year, implying residency (Gawain, 1981; Conner and Smolker, 1985; Smolker, 1985, pers. commun., 10 March 1986).

At the southern border of Queensland, Moreton, North Stradbroke, and South Stradbroke islands form a chain enclosing and separating Moreton Bay (Fig. 1) and the interconnected Broadwater to the south from the open sea, except for passages between the islands. Bottlenose dolphins occur in Moreton Bay throughout the year, with a slight increase in numbers in winter (Lear and Bryden, 1980), and repeated sightings of individually recognizable animals in the bay suggest that groups are resident in this area (P. Corkeron, pers. commun. and Ch. 14 and 17, this volume). Staff of Seaworld at Southport, with over two decades of experience in dolphin capture, report two forms of bottlenose dolphin in this area, which may be distinguished by external features described below. The inshore form is found in the Broadwater, while the offshore animals have been captured from the Queensland border to Pt. Lookout, North Stradbroke Island, and out to sea to a maximum distance of 16 km over water up to 100 m deep (R. Clapp, pers. commun.).

External Appearance

Australian bottlenose dolphins resemble those from other regions in their basic color pattern (Leatherwood and Reeves, 1983), the form of the dark gray dorsal cape and appendages, the paler gray dorsal field overlay extending onto the flanks, the off-white ventral region, and the spinal blaze (terminology of Mitchell, 1970, and Perrin, 1972). Variation in the darkness of the dorsal cape appears to be uncorrelated with latitude or a particular region. The eye-to-flipper stripe is generally poorly defined. However, it was strongly demarcated on its anterior edge in at least one of the bottlenose dolphins mass-stranded at Memory Cove, South Australia.

Most regional variation noted in this study occurred in the absence or extent of spotting on the ventral surface, from unspotted in the south to spotted forms in the north on the east and west coasts. This feature is described in some detail because of its potential value in distinguishing different populations and, by analogy with development of spotting in coastal bottlenose dolphins in Natal waters (Ross, 1984), as a probable indicator of the onset of sexual maturity and adulthood.

None of the adult dolphins captured in the Gulf region of South Australia and observed in Marineland of South Australia by Harrison and Fanning (1973–1974) and G.J.B.R., or those stranded at Memory Cove, showed spots on the ventral surface. The description of two adult animals stranded near Launceston, Tasmania, strongly suggests that no ventral spots were present, despite indications that some darkening of the skin had occurred (Scott and Lord, 1919).

Information on ventral spotting is lacking for most of southwest Australia. Spots were absent in all seven dolphins captured within 30 km each side and 5 km offshore of Yanchep, Western Australia, and held in Atlantis Marine Park. When examined within 2 years of capture (April 1983) these animals were approximately 210–230 cm long, but may not have been fully adult. At the time of writing (July 1988), spotting was still lacking in this group of dolphins (N. Gales, pers. commun.); two of these were pregnant for the first time, reflecting either their attainment of sexual maturity or the sexual competence of males in the group. No ventral spots were seen on four to seven dolphins clearly visible in a series of photographs taken underwater off Houtman Abrolhos Island (P. Baker, 17 May 1975, WAM files). The group of dolphins at Monkey Mia, Shark Bay, do

Figure 2 Distribution of ventral spotting on an adult male bottlenose dolphin, 2.14 m total length, captured in the gill net fishery, Arafura Sea (11° 39′ S, 137° 09′ E), 4 December 1981. (Photograph courtesy of ANPWS, Canberra.)

show ventral spotting (Smolker, 1985, pers. commun., 10 March 1986). Smolker's preliminary studies suggested that spots began to develop at about 9–10 years of age and could be a useful indicator of the onset of sexual maturity. Animals that were moderately or intensely spotted were always fully grown.

Adult bottlenose dolphins captured in gill nets in the Arafura Sea off northern Australia (between 9° 38′ and 11° 42′ S, 134° 08′ and 137° 22′ E) are extensively spotted ventrally. The elongate dark gray spots are distributed from the tip of the lower jaw, where they are fewest, between the flippers to the anal region and adjacent flanks, where they are most intense (Figs. 2 and 3). The capture together of unspotted juveniles and larger spotted animals in the same net suggests that spotting develops with age. This observation is supported by data from 17 measured animals, of which 5 lacked ventral spots [three males, two females; total length (TL) 123–156.5 cm], 9 were recorded as spotted (three males, one female; TL 188–207 cm), and 3 males were intensely spotted (TL 214–218 cm).

Dolphins in oceanaria provided most information on ventral spotting for the east coast of Australia. One male and two female dolphins, estimated at about 8, at least 10, and 15 years old, respectively, were examined at the King Neptune Park, Port Maquarie, New South Wales (31° 30′ S). They were caught within 30 km of Port Maquarie close to shore, or in one instance in a nearby river

Figure 3 Distribution of ventral spotting on an adult male bottlenose dolphin, 2.15 m total length, captured in the gill net fishery, Arafura Sea (11° 42′ S, 137° 14′ E), 4 December 1981. (Photograph courtesy of ANPWS, Canberra.)

(R. Smallman, pers. commun.). None showed any form of ventral spotting. Another adult female, "Sally," captured in 1973 with a male calf at Pt. Plomer near Port Maquarie, and held at the African Lion Safari Park, Warragamba, showed numerous fine spots ventrally posterior to the flippers and flanks adjacent to the anal region. The skull of her calf, which died aged 6 years, was included in the measured sample (UNSW no no.). A second mother and calf, captured at Nambucca Heads (~30° 40′ S) in November 1971 (G. Pickering, pers. commun.), were examined at the Porpoise Pet Pool, Coffs Harbour. The mother was spotted over the ventral surface and flanks, and had been so when captured. In 1983, her male calf, 160 cm long and still suckling at capture, showed small or moderate-sized elongate spots ventrally, especially in the anal region.

Nine animals caught in southern Queensland waters were examined at Seaworld, Southport. Their ages (in 1983) are based on estimates of minimum age at capture plus years in captivity. Five of these were "inshore" animals, caught in the Broadwater (R. Clapp, pers. commun.). Three adult males, more than 15–20 years old, were spotted ventrally with dark gray, elongate blotches. One sexually mature female (9+ years old) also had ventral spotting, and one female caught as an immature animal in 1975 lacked ventral spotting. None of four adult dolphins caught in the "offshore" area described above (one male, two females, and one animal not sexed, aged 16+, 19+, 20+, and 10+ years, respectively) showed any ventral spotting.

The flanks, ventral surface, and flippers of the cotypes of *T. catalania* from northern Queensland were covered with longitudinally elongate blotches of dark gray pigment (True, 1889). The larger of the two dolphins was pregnant and therefore sexually mature, and the anterior teeth of both were well worn, suggesting adulthood.

A reevaluation of the conclusions of Ross (1984) concerning the development of spotting in *Tursiops* from Natal is appropriate here, in view of its value to behavioral and other studies when no other information on maturity state is available. Data on the extent of spotting on 73 dolphins for which lengths and sexual maturity state were available are summarized in Table 3. Sexual maturity of males was based on body length (>235 cm) and a combined testes mass exceeding 500 g (Ross, 1984), and that for females on the presence of one or more corpora albicantia in the ovaries or active lactation. The correlation between nonspotting and spotting with immature and mature animals, respectively, is almost exact, confirming the development of ventral spotting at about the time of sexual maturation. The testes mass of one of the three spotted males indicated immaturity (54 g, TL 242 cm). The other two were the only males with testes in the development phase (158 g, TL 221 cm; 230 g, TL 233 cm), supporting the conclusion that

Table 3
Extent of Ventral Spotting in 73 Bottlenose Dolphins from the Coast of Natal, South Africa

Sex	Extent of spotting	Immature dolphins		Mature dolphins	
		Number	Length (cm)	Number	Length (cm)
Males	Absent	10	125–224	—	—
	Light	2	233–242	2	236–250
	Moderate	1	221	2	237–245
	Intense	—	—	20	233–257
Females	Absent	13	130–228	1	242
	Present	—	—	3	230–236
	Light	—	—	6	229–243
	Moderate	—	—	3	230–240
	Intense	—	—	10	232–249

spotting develops during sexual maturation. However, the single mature female lacking spotting was pregnant with her first calf, implying some caution in the use of spotting as an indicator of sexual maturity. This dolphin was longer (TL 244 cm) than the estimated length at sexual maturity (210–230 cm), and therefore possibly atypical in her sexual development or her spotting.

These data are insufficient to make more than tentative conclusions concerning changes in spotting with age beyond sexual maturation. Means (and ranges) of testes mass (g) in the sample for lightly, moderately, and intensely spotted mature males were 1130 (910–1320), 1185 (1170–1200), and 1386 (730–2315), respectively, suggesting that spotting intensifies with increasing testes mass.

Two captive females from Natal in Seaworld, Durban, provide additional evidence that spotting intensifies with age and that spotting may be useful for individual recognition. Both animals were spotted when captured in June 1979. Their state of sexual maturity at capture is uncertain; the earliest dates of impregnation in captivity for each dolphin were about February 1982 and October 1983, respectively. Comparison of photographs showed that (a) spots present in June 1980 were unchanged in size or position in May 1985 and February 1988, (b) spots that were pale gray in 1980 had become dark gray by 1985, and (c) a number of new spots had appeared with time. Both animals were classified as lightly spotted in 1980, moderately spotted in 1985, and moderately/intensely spotted in 1988.

Total length appears to be the most variable feature of external form (Table 4). Dolphins from northern Australia (ANPWS data) were classified as adults on the basis of presence of ventral spotting or extreme tooth wear. Other measurements of value in distinguishing between the two southern African forms of *Tursiops* (length of snout, length and width of flipper; Ross, 1977) are included in Table 4. Several points of interest are indicated when comparing adult length with locality (Fig. 4). There appears to be an increase in body length with increasing latitude, the largest animals occurring in Tasmania, southern Victoria, and the open coast of South Australia. However, dolphins from Spencer Gulf, South Australia (area 15), are distinctly smaller than those in adjacent areas. There is a marked difference in length between dolphins from the "inshore" region (adult males, 210 and 236 cm TL) and the "offshore" region (immature female, 223 cm TL, captured 7 km offshore of Burleigh Heads, water depth ~100 m; adult female, 289.5 m) of southern Queensland (area 8).

There is a significant correlation between total length and the estimated mean water temperature (Fig. 5), though the Spencer Gulf animals fall well below a linear regression line drawn through the points:

$$Y = 376.5 - 6.27X \qquad (r^2 = 0.59)$$

where Y is the total length (cm) and X the water temperature (°C). As water temperatures increase considerably toward the heads of Spencer Gulf and Gulf St. Vincent, and are very variable seasonally (Nunes and Lennon, 1986), it is quite likely that the estimated water temperature for area 15 is not representative of the Port Broughton region where those animals were captured (Harrison and Fanning, 1973–1974). This point is considered further in the Discussion section.

The appearance of the snout also varies in different areas, becoming proportionately shorter and broader in higher latitudes. Dolphins from northern Australia (Fig. 2), Shark Bay (Gawain, 1981), off Yanchep, Western Australia (Fig. 6A), the "offshore" (Fig. 6B) and "inshore" (Fig. 6C) regions of southern Queensland, and northern New South Wales (Fig. 6D,E) have long, attenuate snouts. R. Clapp (pers. commun.) noted that the snout of the "offshore" form in southern Queensland was distinctly longer than that of the "inshore" form, as shown in Fig. 6B,C, but larger samples are required to verify this. On the south coast, the snouts of animals from Albany, Western Australia (Fig. 7), and the open coast of South Australia (Fig. 8; Ellis, 1987) were proportionately shorter and broader at the base. This stubby appearance is greatest in the adult male (TM 1065) from Launceston, Tasmania (Scott and Lord, 1919, their plate 23), for which the snout length was estimated at 3% of body length. The snouts of animals from the Gulf region of South Australia are elongate (4.7–5.4% TL) but have a broad base resembling that of other south-coast animals (Fig. 9; Harrison and Fanning, 1973–1974).

There are too few data to comment on regional variation in flipper length. Comparison of flipper

Table 4
Total Length and Selected External Measurements (cm) for Adult Australian Bottlenose Dolphins

Catalog no. or source	Area code (Fig. 1)	Total length	Tip of snout to		Flipper	
			Eye	Apex of melon	Anterior length	Width
BVKP 20/7/82[a]	1	193	31			
BZAP 17/01/83[a]	1	197				
BVQX 8/10/81[a]	2	178				
BVXW 19/2/82[a]	2	188				
BVBL 4/12/81[a]	2	198				
BVQX 8/10/81[a]	2	199				
BVQX 8/10/81[a]	2	202				
BVAB 19/2/82[a]	2	203	27			
BYNC 19/11/82[a]	2	206				
BVAA 10/2/82[a]	2	207				
BVXW 4/12/81[a]	2	214				
BVBL 4/12/81[a]	2	215				
BVAA 12/12/81[a]	2	218				
BVAB 8/10/81[a]	2	233				
BM (NH) 1391a	5	229	30.5			
BM (NH) 1391b	6	206				
Demon[b]	8	210		10		
UQ C6	8	223	36.5	12.2	38.5	13.4
Freckles[b]	8	236		11		
Splash[b]	8	290				
Sally[c]	9	232				
TM 1360	12	275				
QVM 1974/1/19	12	315				
TM 1365	12	326				
SAM M10101	14	285			33	
AD5[d]	15	201	30	9.5		14
AD2[d]	15	201	30	9.5		13
AD1[d]	15	202	30	9.5		14
AD7[d]	15	223	32	12		19
AD6[d]	15	223	32	11		16.5
SAM M10098	15	224	34.5		41	18
SAM temp. 3	16	272	40		43	16
SAM temp. 11	16	277	38		41	15
SAM temp. 2	16	282	37		40	16
SAM temp. 1	16	285	40		44	17
SAM temp. 4	16	285	38		42	16
SAM temp. 9	16	285	36		42	15
SAM temp. 12	16	285	37		42	15
SAM temp. 7	16	290	39		44	16
SAM temp. 6	16	310	39		42	17
SAM temp. 8	16	310	36		46	16
Connor[e]	22	229				

[a] Data on gill net catches, from ANPWS.
[b] Held at African Lion Safari Park, Warragamba, New South Wales; from Seaworld.
[c] Held at Seaworld, Southport, Queensland.
[d] Data from Harrison and Fanning (1973–1974).
[e] R. C. Connor (pers. commun., 28 August 1985).

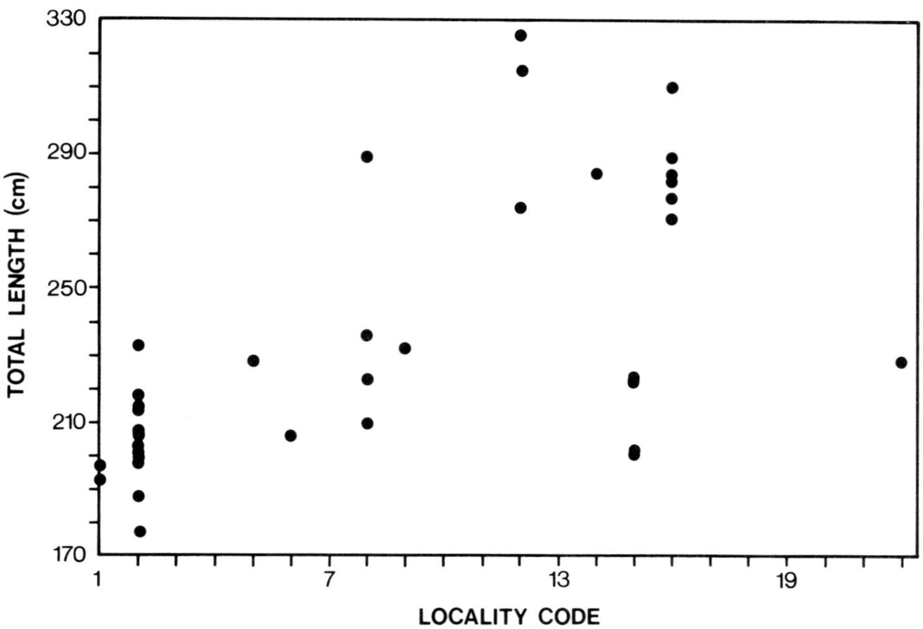

Figure 4 Variation in total length of adult bottlenose dolphins (Table 4) in relation to locality.

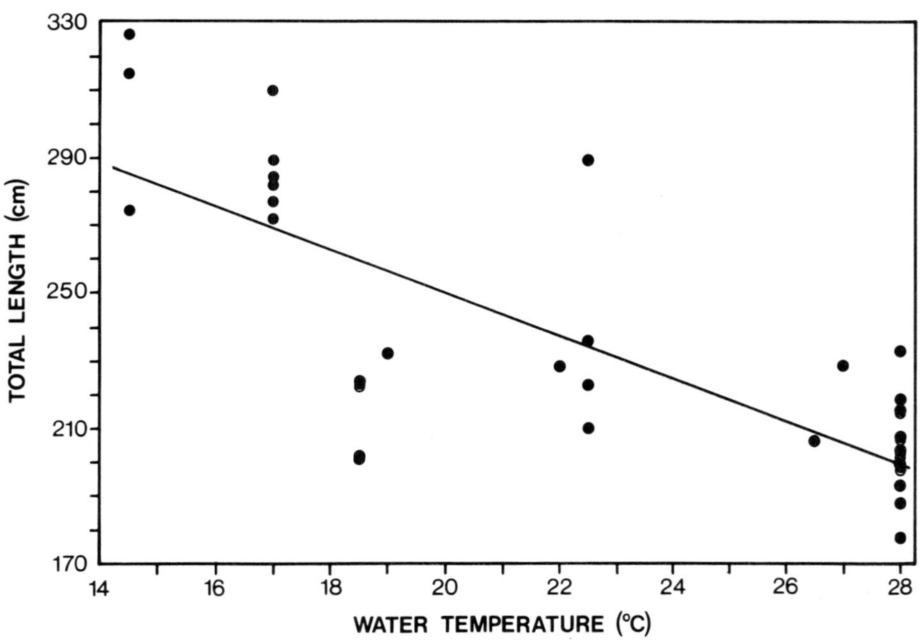

Figure 5 Relationship between total length of adult bottlenose dolphins and estimated mean annual surface temperature. See text for the equation of the regression line through the points.

Figure 6 Variation in head shape, ventral color pattern, and capture localities of adult bottlenose dolphins: (A) ventral spotting absent, off Yanchep, Western Australia; (B) dorsal aspect of snout, "offshore" region of southern Queensland; (C) dorsal aspect of snout, "inshore" region of southern Queensland; (D) ventral spotting present, off Nambucca Heads, New South Wales; (E) ventral spotting absent, off Port Macquarie, New South Wales. (Photographs by G. J. B. Ross.)

Figure 7 Adult bottlenose dolphin stranded at Albany, Western Australia, April 1983, illustrating the short, broad-based snout. (Photograph courtesy V. Milne.)

width with total length indicates a close relationship for all lengths, with the notable exception of the three adult dolphins from Spencer Gulf, which had very broad flippers.

Skull Features

Variation in condylobasal length (CBL) was marked in the sample of 58 skulls classified as adult, changing with locality (Fig. 10) in a similar way to total length. In general, smaller and larger skulls were associated with low and high latitudes, respectively. Animals from the Gulf region of South Australia (area 15) were notable exceptions, in having small skulls.

In the principal components analysis, the initial 33 variables were examined in a series of analyses, in which those with smaller component weightings were eliminated, until the final analysis was based on standardized data for five variables: CBL, width of the rostrum 60 mm anterior to its base, the greatest width of the premaxillae, the least

Figure 8 Adult bottlenose dolphin, one of 55 animals mass-stranded at Memory Cove, South Australia, 2 May 1977, illustrating the short, broad-based snout. (Photograph by P. Aitken, courtesy South Australian Museum.)

Figure 9 Adult bottlenose dolphin caught off Adelaide within Gulf St. Vincent, South Australia, illustrating the short snout. (Photograph by G. J. B. Ross.)

width between the posttemporal fossae, and the major diameter of the temporal fossa. The five principal components described by this analysis accounted for 64.6, 21.7, 8.3, 3.3, and 2.1% of the variance, respectively. Their component weights are shown in Table 5.

Principal component 1 (PC1) was most strongly influenced by variation in size, and individual scores and CBL were highly correlated ($Y = 0.0398X - 19.1$, where X is the CBL in millimeters and Y the PC1 score; $r^2 = 0.81$; $n = 51$). A plot of PC1 against PC2, however, showed no evidence of groupings indicative of different taxa (Fig. 11), though some grouping of specimens from different localities was apparent, reflecting similarity in size. Comparisons between the other principal components produced similar results. The importance of size as the main source of variation in the sample is particularly interesting in view of the close relationship between PC1 and estimated mean water temperature (Fig. 12), which indicates clearly a decrease in skull size with increasing water temperature.

Proportional skull measurements (as % CBL) for the present sample were compared with those for groups assigned to *T. aduncus* and *T. truncatus* from southern Africa (Ross, 1977). For all measurements the ranges of the Australian material were equivalent to the combined ranges of both southern African forms, and the data indicated a general pattern of allometry in skull proportions.

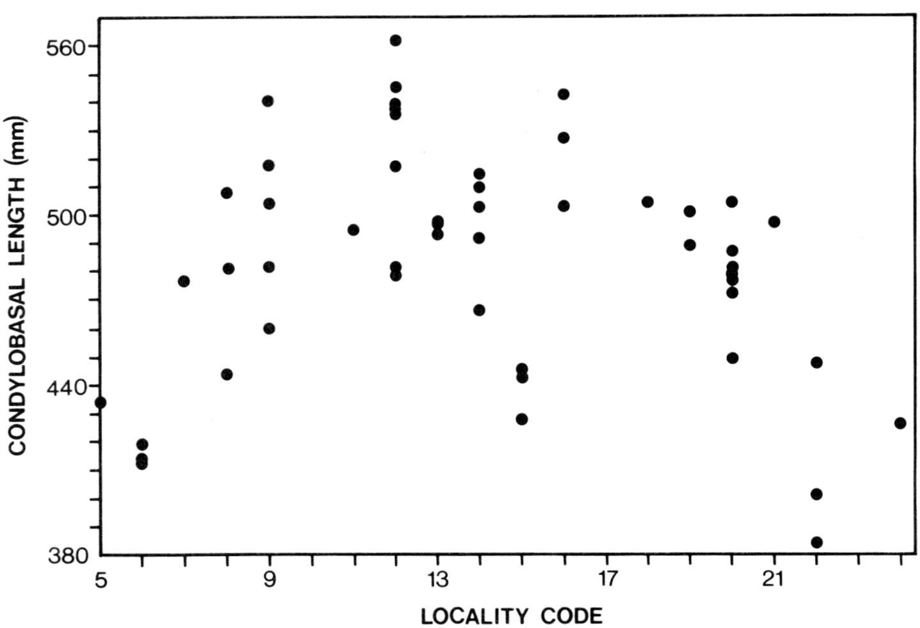

Figure 10 Variation in condylobasal length of Australian bottlenose dolphins in relation to locality.

Table 5
Component Weights and Eigenvalues of Principal Components for Five Skull Variables

Variable	Component weights 1	2	3	4	5
Condylobasal length	0.504	0.156	−0.373	0.762	0.048
Rostral width, 60 mm from base	0.523	0.007	−0.305	−0.456	−0.653
Greatest width of premaxillae	0.530	−0.093	−0.060	−0.407	0.736
Width between posttemporal fossae	0.391	−0.507	0.719	0.207	−0.172
Major diameter of temporal fossa	0.196	0.843	0.497	−0.057	−0.026

These included measurements found to differ significantly between the two southern African forms (Ross, 1977), illustrated by the following examples.

In Australian animals, the proportional width of the skull across the parietals declines with increasing CBL (Fig. 13). The ranges of the southern African material overlap each end of this cline (*T. aduncus* 33.7–40.4%, CBL < 500 mm; *T. truncatus* 33.1–37%, CBL > 500 mm). The proportional height of the posttemporal fossa also declined as CBL increased (Fig. 14), covering the ranges of

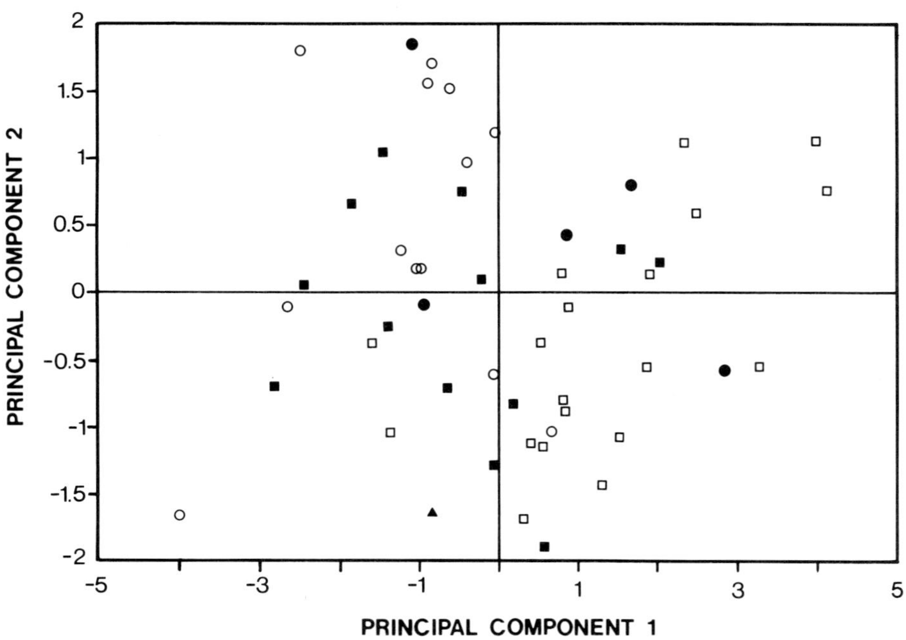

Figure 11 Relationship between scores for principal components 1 and 2 for five skull variables in 51 adult bottlenose dolphins. Symbols and locality codes for specimens from different regions are as follows: codes 5–10 ■; codes 11–14 □; code 15, ▲; codes 16–19, ●; codes 20–24, ○.

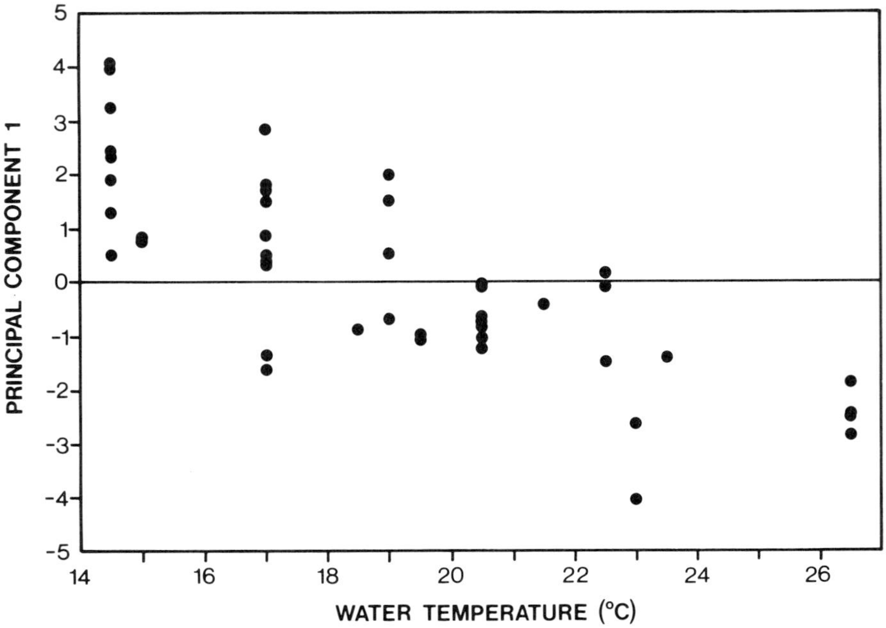

Figure 12 Relationship between principal component 1 and estimated mean water temperature for adult bottlenose dolphins. The regression line was drawn from the equation $Y = 7.6 - 0.393X$ ($r^2 = 0.61$, $n = 47$) fitted by the least-squares method.

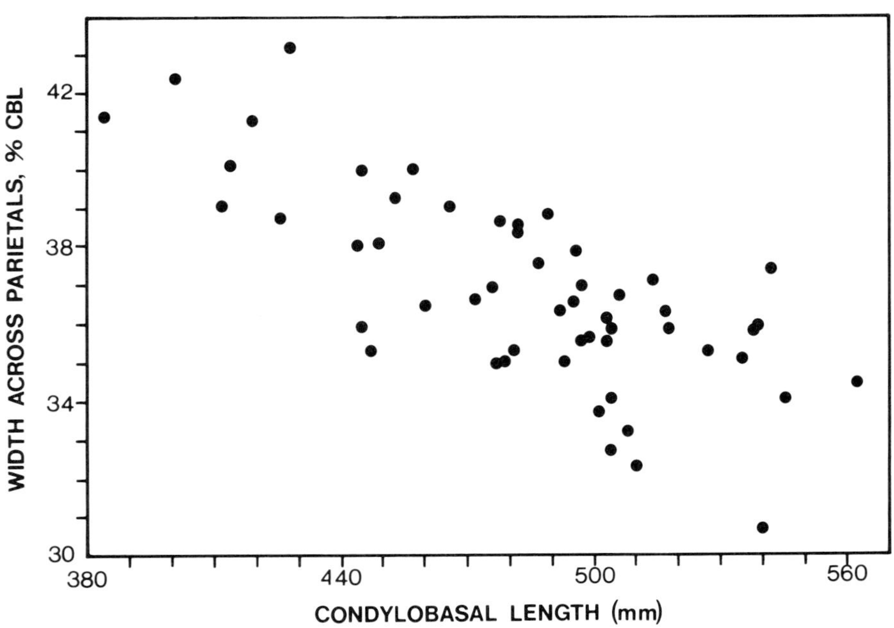

Figure 13 Variation in the proportional width across the parietal bones with increasing skull size in Australian bottlenose dolphins.

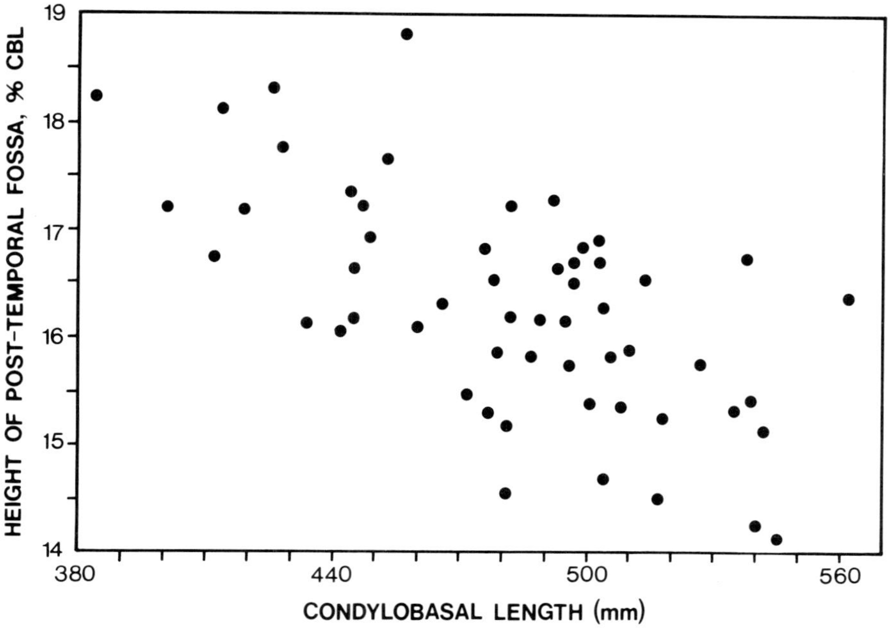

Figure 14 Variation in the proportional height of the posttemporal fossa with increasing skull size in Australian bottlenose dolphins.

southern African *T. aduncus* (14.3–18.8% CBL) and *T. truncatus* (14.4–16.6% CBL). The length of the antorbital process of the jugal increased proportionately with increasing CBL (Fig. 15), closely matching the ranges of southern African *T. aduncus* (8.2–11% CBL) and *T. truncatus* (10.1–12.6% CBL).

The tendency for numbers of upper and lower teeth to decrease slightly with increasing skull length in Australian animals appeared to be of no taxonomic significance. Maximum and minimum numbers of mandibular teeth in adult specimens (20–27) ranged most widely in animals of 450–500 mm CBL, and they were very similar to the ranges recorded for southern African *T. aduncus* (23–29) and *T. truncatus* (21–24). Though tooth counts in the specimens examined appear to be slightly higher in dolphins from lower latitudes, counts of erupted teeth only for 33 dolphins captured in gill nets in the Arafura Sea show that these range as widely (maxillary teeth, 21–29; mandibular teeth, 19–28) as those elsewhere in Australia, and illustrating very well the variation within a localized population. The addition of rudimentary teeth would probably have increased these counts by 1 or 2, making them comparable to southern African *T. aduncus*.

Clinal changes were apparent also in the cranial features proposed by Ross (1977) as potentially useful for distinguishing between the southern African forms. The posterior part of the vomer was progressively broader and more flared as CBL increased, and the premaxillae on the dorsal surface of the rostrum were less strongly arched in the larger skulls.

The size of the squamosal bone in the posttemporal fossa was equally variable, regardless of the skull length. The angle of tilt of this bone, however, showed a marked increase with increasing CBL over the whole sample (Fig. 16). Animals from the west and east coasts appeared to differ in the degree of tilt, which was limited to less than 20° in west coast animals regardless of CBL (Fig. 17). This difference is of particular interest and warrants further investigation as a genetic marker. Almost identical patterns were shown by the width of the cancellous bone on the lateral margin of the premaxilla in relation to CBL and locality.

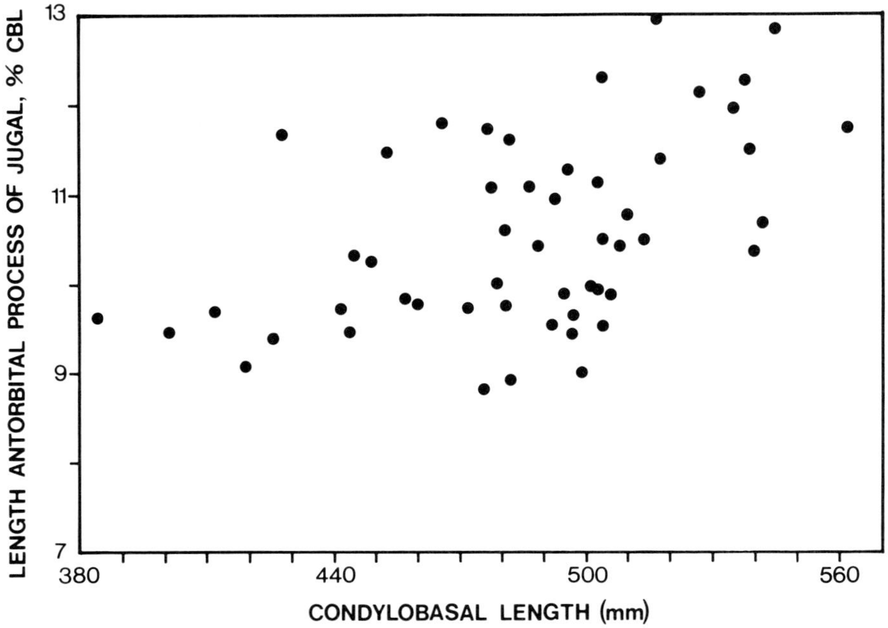

Figure 15 Variation in the proportional length of the antorbital process of the jugal bone with increasing skull size in Australian bottlenose dolphins.

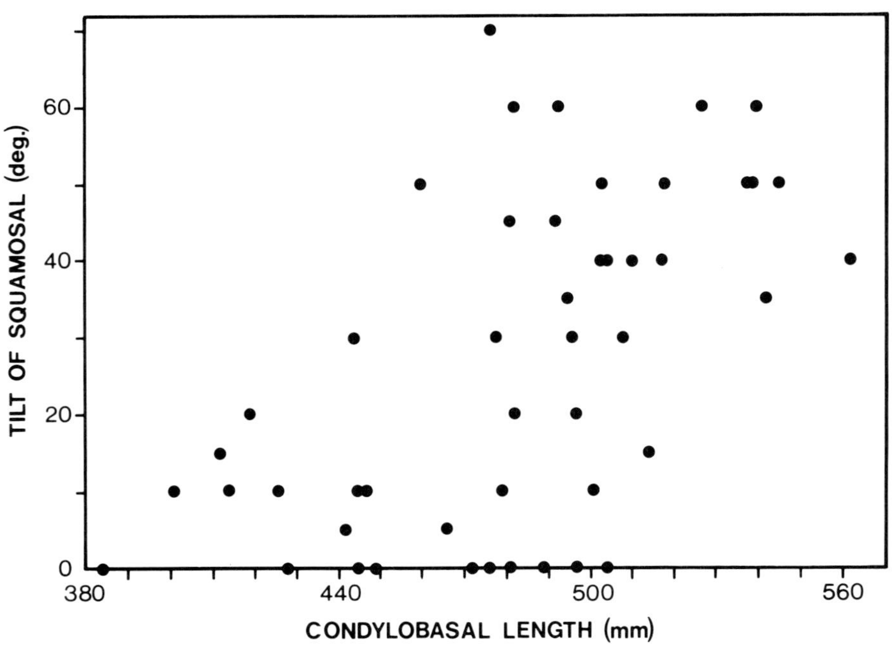

Figure 16 Relationship between tilt of the squamosal bone in the posttemporal fossa and condylobasal length in Australian bottlenose dolphins.

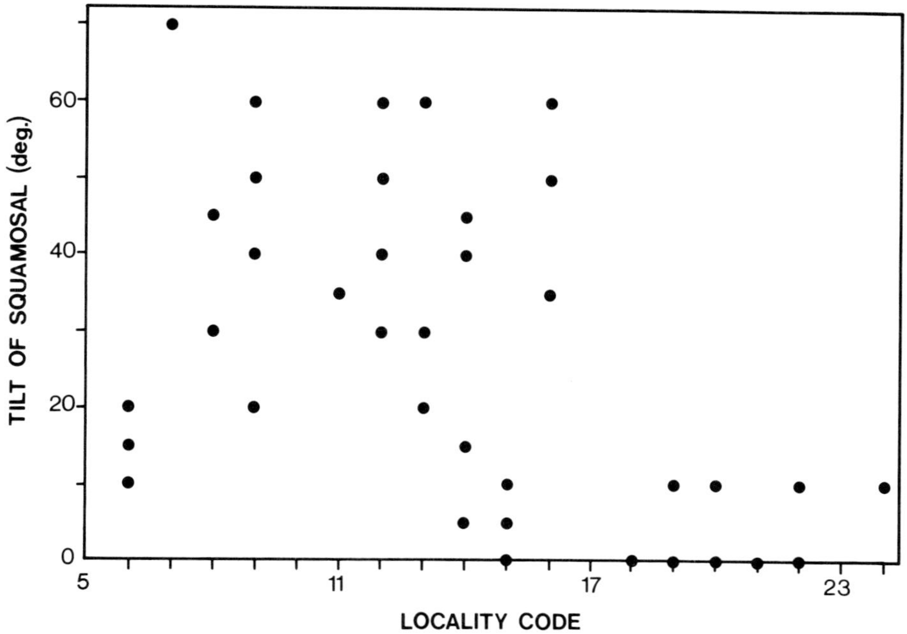

Figure 17 Relationship between tilt of the squamosal bone in the posttemporal fossa and locality in Australian bottlenose dolphins.

Broader cancellous bone was strongly associated with increasing tilt of the squamosal bone, indicating a high level of co-occurrence.

Postcranial Skeleton

Useful vertebral material was available from 10 specimens only, for which data are given in Table 6. Variation in the numbers of *X* and *Y* vertebrae may reflect different ages and subsequent interpretation in these specimens. Though the data are few, there is a suggestion of two groups in the material which have the first vertical perforating foramen in vertebra 40–41 or 43–44 and a total of 59–60 or 63–67 vertebrae, respectively. The relationship between body size and vertebral count is also speculative, though there is some indication that animals with larger skulls have greater numbers of vertebrae. Groupings with similar numbers of vertebrae were found to distinguish different forms of *Tursiops* off southern Africa (Ross, 1977). Additional postcranial material for Australian *Tursiops* may have considerable potential as a means of establishing whether different forms do exist along the coast.

Response to Water Temperature

Two data sets provide insight into the responses of bottlenose dolphins to changes in water temperature. The effect of changes in body mass and water temperature on energy intake, in terms of feeding rate, was examined in a captive dolphin over an 8½-year period. Mean monthly water temperature ranged from 15°C in August to 25°C in February. The relationship for mean feeding rate (kg/day) as a function of mean body mass and mean water temperature is shown in Fig. 18 for 51 periods of 45 days or less. The most striking feature was the flattened region of the plot above a mean body mass of some 155 kg and water temperatures of more than 20°C, which appeared to represent the thermoneutral zone for this dolphin. Below these values, feeding rate rose rapidly from about 6 kg/day to a maximum of 8 kg/day, particularly as temperature decreased. A similar plot emerged for

Table 6
Data on the Vertebral Column of 10 Australian Bottlenose Dolphins

Catalog no.	Locality code	CBL (mm)	Number of vertebrae						
			Cerv.	Thor.	*X*	*Y*	*Z*	Total	*A*[a]
QM JM1230	8	444[b]	7	12	14	17	9	59	42[c]
AM 279	9	504[b]	7	12	18	17	9 + 1–2	64–65	—
S1555	9	482[b]	7	12	17	18	9 +1?	63–64	44
AM 278	8–10	457[b]	7	12	13	17	11	60	40
ANU 1858	10	379	7	11	16	16	9	59	40
QVM 1360	12	538[b]	7	12	17	20	11	67	43
NMV C25071	13	496[b]	7	12	11	19	10 + 1?	59–60	—
NMV C23490	13	484	7	13	15	19	9	63	44
SAM M10638	15	445[b]	7	12	16	16	1 + ?	—	40
SAM M5078	15	442[b]	7	12	16	16	9	60	41

[a] Number of the first vertebral centrum with a vertical perforating foramen.
[b] Classified as adult for analysis.
[c] Two preceding vertebrae have very small foramina.

the equivalent relationship for energy intake per day (Fig. 19), in which the thermoneutral zone was somewhat narrower and the increase in energy intake more rapid below 155 kg body mass and 20°C, from about 34 MJ/day to a maximum of 44 MJ/day. The lowest feeding rates in terms of mass and energy occurred at approximately 157 kg body mass and about 23°C, and these may be the optimum values for this particular individual, though her mean body mass over the entire period was 153 kg. Evidence that the observed decrease in feeding rate with increasing water temperature was not an artifact of a captive feeding regimen is given in Fig. 20, which shows the proportion of the daily ration offered and accepted by this dolphin in relation to decreasing or increasing water temperature. The lower proportion of food accepted as water temperature increased, that is,

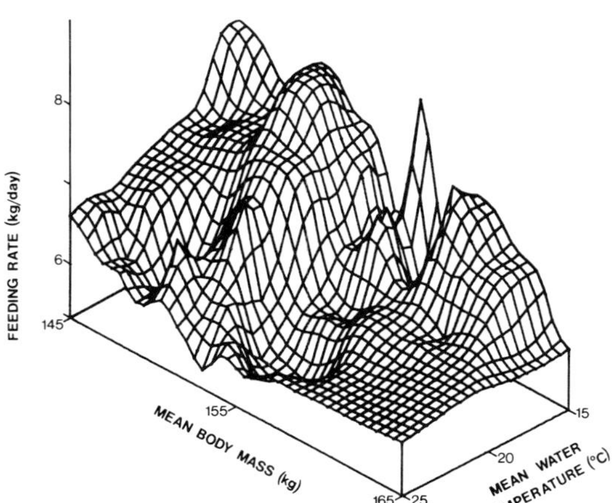

Figure 18 Relationship between feeding rate, body mass, and mean monthly water temperature in a captive bottlenose dolphin from Algoa Bay, South Africa.

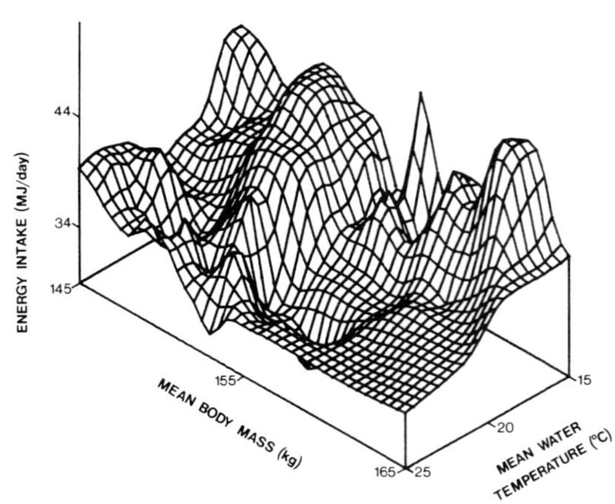

Figure 19 Relationship between energy intake, body mass, and mean monthly water temperature in a captive bottlenose dolphin from Algoa Bay, South Africa.

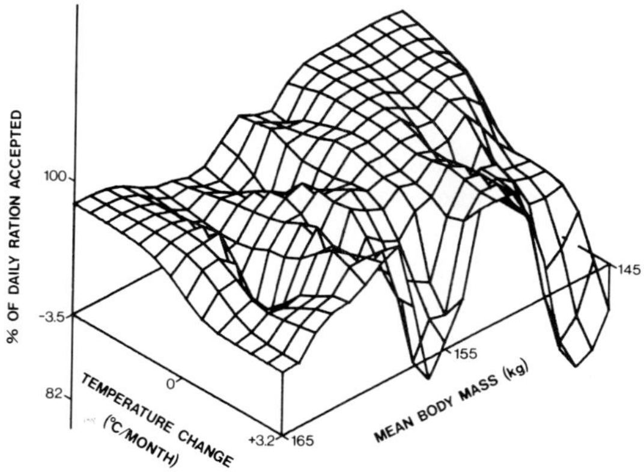

Figure 20 Proportion of daily ration accepted by a captive bottlenose dolphin in relation to body mass and water temperature.

as summer approached, indicated an innate response by the dolphin for reducing energy intake, rather than the possible influence of her trainer.

It is interesting to note that the mean annual sea surface temperature for Algoa Bay, where this dolphin was caught, is 18°C, with monthly means ranging from 15° to 21°C (Ross, 1984). Such conditions suggest that these dolphins live in conditions close to or below the lower limit of their thermoneutral zone for most of the year.

Decreased heat loss, reducing the need for increased food intake, is most readily accomplished by increased insulation. Data from 58 bottlenose dolphins caught in the Natal shark nets indicated that those living in tropical waters conserve heat by this means, thus extending their range of suitable environmental temperature. Immature animals below 2.0 m long and adults, excluding lactating females, showed a marked increase in the proportional mass of blubber during the winter months, amounting to some 3% of body mass or about 5 kg. This increase was inversely related to mean monthly surface water temperature (Fig. 21). Interestingly, most lactating females did not show a similar increase, perhaps reflecting the greatly increased nutritional load faced by these females during lactation (Cockcroft and Ross, Chapter 15, this volume). The mean

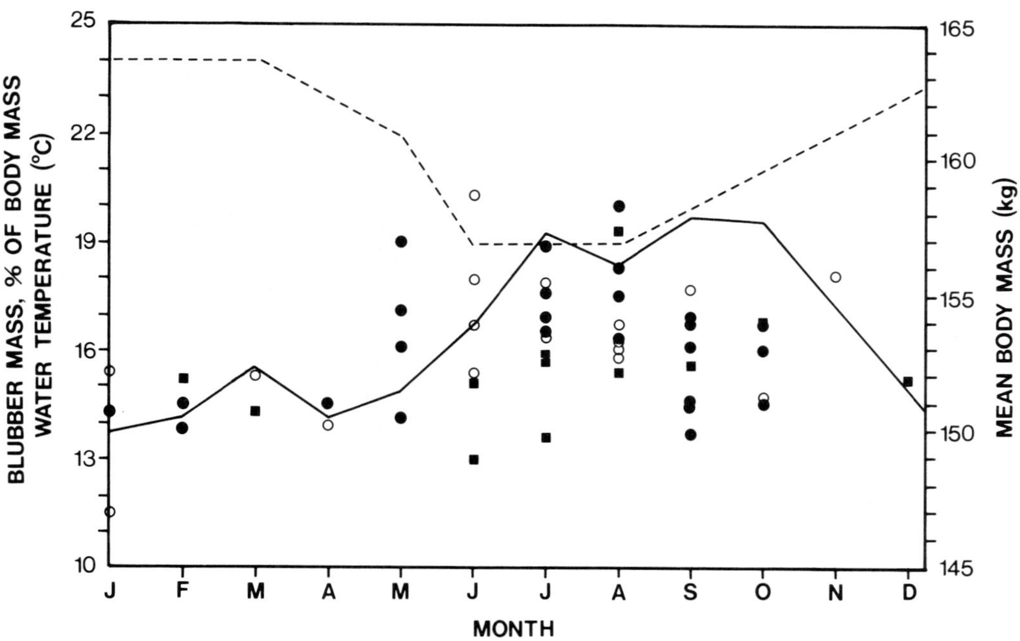

Figure 21 Seasonal change in the proportion of blubber mass (% body mass) in bottlenose dolphins from Natal coastal waters: immature animals, ○; adults, excluding lactating females, ●; lactating females, ■. Mean monthly water temperature (°C) at Durban is indicated by the dashed line. The solid line shows seasonal change in the mean monthly body mass (kg) of a captive female dolphin (see text).

body mass of the captive female dolphin showed a similar increase of approximately 8 kg more in winter than in summer (Fig. 21). It is very probable that some or all of this increase was due to an increase in her blubber mass.

DISCUSSION

At the individual and local population level, the evidence that bottlenose dolphins in tropical/subtropical waters live close to the lower limit of their thermoneutral zone implies certain requirements in coping with seasonal changes in water temperature, if they are not to incur an excessive energy deficit through heat loss. Where seasonal changes range over a few degrees centigrade, an increase in body insulation in the form of blubber, probably coupled with a moderate increase in feeding rate, appear to be adequate for populations of bottlenose dolphins along the Natal coast. In regions where seasonal temperature changes are more extreme, such as in the higher-latitude regions of warm, western boundary currents, seasonal movements between areas with equitable conditions may also be necessary. The seasonal migration of bottlenose dolphins past Cape Hatteras (Mead, 1975) is probably a response to such temperature changes.

Migration would not seem to be an appropriate option for most coastal populations of bottlenose dolphins along the Australian coast, where seasonal differences in coastal water temperatures are small (3–7°C) and longshore changes are gradual (Rochford, 1975; Pearce, 1986). One area where seasonal migration may occur is the northern part of Spencer Gulf, South Australia, where temperatures may fluctuate by up to 12°C. (Nunes and Lennon, 1986), though there is presently no evidence for such movements. The alternative option of residency is supported by the limited information available. In such circumstances each population would have assumed an optimum body size for the local temperature regime, with the resultant formation of clines in body size and other dimensions on a continental scale, in accordance with Bergman's rule.

The presence of such clines from north to south on both sides of the Australian continent is of considerable interest taxonomically, particularly as there was no morphological differentiation of taxonomic entities at the species level within the sample. Comparison of skull measurements, tooth counts, and other skull features with data for the two most distinct forms of *Tursiops* off southern Africa showed that the data ranges for the latter occupied opposite ends of the Australian clines. These southern African forms have been treated previously as two species, *T. truncatus* and *T. aduncus* (Ross, 1977, 1984). In the present perspective, however, there is little doubt that they should be treated as a single species, *T. truncatus*, and that Australian bottlenose dolphins should be assigned to this species also.

There may be grounds for considering the *T. aduncus* form as a subspecies of *T. truncatus*, to include populations characterized by adults with a spotted ventral surface. Spotting appears to be restricted to the Indian Ocean and adjacent tropical western Pacific Ocean, though some spots may develop on the ventral surface of old females in the Atlantic Ocean (Leatherwood *et al.*, 1976). Dolphins in the eastern tropical Pacific are not spotted ventrally (Leatherwood *et al.*, 1982). Spotted adult bottlenose dolphins have been recorded from the following locations: southeast coast of southern Africa (Ross, 1984); the Zanzibar Channel, off Malindi, Kenya, and Djibouti (Robineau and Rose, 1984); Mammelle Island, Seychelles (Leatherwood and Ross, unpublished data); the Red Sea (Ruppell, 1842—type of *T. absulam*); Trevandrum, southern India (Lydekker, 1904), and Visagapatam, eastern India (Owen, 1866); off Sri Lanka (A. Alling, pers. commun., 12 May 1982; Leatherwood, 1986); Java Sea (Tas'an and Leatherwood, 1984; Tas'an, 1985); the coast of Taiwan (Hung-chia Yang, 1976; R. Abel, pers. commun.); and the Australian coast. It is uncertain if bottlenose dolphins examined by Pilleri and Gihr (1973–1974) in the Gulf of Siam were spotted ventrally. These authors did not mention spotting specifically, and their single photograph of an adult was taken at an awkward angle.

Such a distribution is most readily explained through geographical isolation of Indian Ocean bottlenose dolphins during the Holocene series of glacial maxima. At such times, the Subtropical

Convergence shifted northward about 2° and 5° latitude, respectively, on the western and eastern boundaries of the Indian Ocean. Though the Agulhas Current would have been weakened, the thermal barrier presented at the interface of the Agulhas and Benguela currents would have been maintained (Prell *et al.*, 1980). On the Australian coast, part of the West Wind Drift would have been deflected northward, forming a significant eastern boundary current, cooling and effectively blocking access to the southern coastline for bottlenose dolphins. During these periods, water temperatures at lower latitudes were probably less than 2°C cooler than at present (Prell *et al.*, 1980). At the same time, lowered sea levels would have closed several of the access routes between the Indian and Pacific Oceans north of Australia, reducing genetic interchange. In such circumstances, the Indian Ocean would have been an isolated basin in which development of spotted forms could have occurred. During interglacial periods, such barriers would have been removed, reestablishing contact with Pacific populations via the southern and northern Australian coasts.

Given the influence of environmental factors on morphological features, traditional morphological techniques are limited in their ability to investigate such suggestions satisfactorily. Other techniques, such as comparisons of mitochondrial DNA, may provide more satisfactory answers. The present results also raise several other queries, indicating that factors other than water temperature need to be examined. For example, surface water temperature does not account for the differences in body size of the inshore and offshore forms in southern Queensland. Interestingly, an analogous situation occurs off Natal, where offshore bottlenose dolphins along the edge of, or in, the Agulhas Current are considerably larger (>300 cm long) than the coastal animals (<260 cm) (Ross, 1984). Surface temperatures in the Agulhas Current range seasonally from 22° to 27°C, and are usually 1–2°C higher than those along the coast (Schumann, 1987). However, temperatures in the current decline rapidly with depth to below 15°C at 200 m or less (Pearce, 1977). Very limited stomach-content data suggest that these dolphins feed on mesopelagic prey or demersally along the shelf edge (Ross, 1984), thus experiencing water temperatures similar to the southern African west coast, where bottlenose dolphins are comparable in size. The possibility that body size in dolphins is more closely correlated with colder temperatures experienced by them needs further investigation.

The need to resolve such queries for cetacean taxonomy as a whole underlines the value of coastal bottlenose dolphins as a model on which the taxonomy of more oceanic dolphins can be interpreted, and one which should be exploited in the field situation using modern techniques currently available. In the Australian context, determination of the relationships between the "inshore" and "offshore" populations of southern Queensland, and those within and outside the Gulf region of South Australia, would be valuable contributions.

ACKNOWLEDGMENTS

We thank the directors and curators of Australian institutions and oceanaria listed in the text, for ready access to specimens, live animals, and other important data. In particular we thank the following, for the willingness with which they assisted: R. Abel, L. Agostino, P. Andrews, J. Bannister, M. Bryden, J. Calaby, R. Clapp, H. Cogger, P. Corkeron, W. Dawbin, T. Dennis, J. Dixon, S. van Dyk, N. Gales, L. Gibson, H. Goodall, R. Green, M. Harwood, G. Heinsohn, L. Huxley, R. Hyne, D. Kitchener, C. Kemper, J. Ling, Q. MacNamara, H. Marsh, V. Milne, R. Molnar, D. Needham, G. Pickering, G. Shaughnessy, P. Shaughnessy, R. Smallman, R. Tedman, C. Tidemann, E. Turner, H. Wapstra, and R. Warneke. G. J. B. R. is especially grateful to J. Bannister for the invitation to visit Australia, to A. Pearce and B. Phillips for essential sea-temperature data, and to all those who assisted him in his travels. We wish to thank also the curatorial staff of the Port Elizabeth Oceanarium for collection of feeding-rate data, and the director of Natal Sharks Board for access to dolphin material. D. Venter and the Departments of Computer Science at University of Port Elizabeth and Rhodes University provided invaluable assistance with computation.

Travel funds for G. J. B. R., provided by the Cape Provincial Administration, the University Research Division of the South African Council for Scientific and Industrial Research, the Port Elizabeth Museum Board of Trustees, and the Australian Museum, Sydney, are gratefully acknowledged.

REFERENCES

Aitken, P. F. (1971). Whales from the coast of South Australia. *Trans. R. Soc. South Aust.* **95,** 95–103.

Allen, F. E., and Ferguson Wood, E. J. (1950). Investigations of underwater fouling. II. The biology of fouling in Australia: results of a year's research. *Aust. J. Mar. Freshwater Res.* **1**(1), 92–109.

Australian National Parks & Wildlife Service. (1986). Australia, progress report on cetacean research, June 1984 to May 1985. *Rep. Int. Whal. Commn* **36,** 143–149.

Australian National Parks & Wildlife Service. (1987). Australia, progress report on cetacean research, June 1985 to May 1986. *Rep. Int. Whal. Commn* **37,** 159–166.

Batchelor, A. L., and Ross, G. J. B. (1984). The diet and implications of dietary change of Cape Gannets on Bird Island, Algoa Bay, *Ostrich* **55,** 45–53.

Beddard, F. E. (1900). "A Book of Whales." Putnam, New York. 320 pp.

Bennett, I., and Pope, E. C. (1953). Intertidal zonation of the exposed rocky shores of Victoria, together with a rearrangement of the biogeographical provinces of temperate Australian shores. *Aust. J. Mar. Freshwater Res.* **4,** 105–159.

van Bree, P. J. H. (1966). On a skull of *Tursiops aduncus* (Ehrenberg, 1833) (Cetacea, Delphinidae) found at Mossel Bay, South Africa in 1904. *Ann. Natal Mus.* **18**(2), 425–427.

Connor, R. C., and Smolker, R. S. (1985). Habituated dolphins *Tursiops* sp.) in Western Australia. *J. Mammal.* **66**(2), 398–400.

Ellis, R. (1987). Australia's southern seas. *Nat. Geographic* **171**(3), 286–319.

Endean, R., Kenny, R., and Stephenson, W. (1956). The ecology and distribution of intertidal organisms on the rocky shores of the Queensland mainland. *Aust. J. Mar. Freshwater Res.* **7**(1), 88–146.

Gawain, E. (1981). "The Dolphins' Gift." Boobook Publications, Sydney, Australia. 212 pp.

Gray, J. E. (1862). Notice of a new species of dolphin *(Delphinus catalania),* discovered in north Australia by Mr. John Macgillivray. *Proc. Zool. Soc. London 1862,* pp. 143–145.

Harrison, R., and Fanning, J. C. (1973–1974). Anatomical observations on the South Australian bottlenose dolphin *(Tursiops truncatus). Invest. Cetacea* **5,** 203-217.

Heath, R. G. M., and Randall, R. M. (1985). Growth of jackass penguin chicks *(Spheniscus demersus)* hand reared on different diets. *J. Zool., London* **205,** 91–105.

Heinsohn, G. E. (1978). Marine mammals of the northern Great Barrier Reef region. Workshop on the northern sector of the Great Barrier Reef, Great Barrier Reef Marine Park Authority Workshop Series No. 1, pp. 315–335.

Heinsohn, G. E. (1979). Biology of small cetaceans in North Queensland waters. Unpubl. Grant Report on Great Barrier Reef Marine Park Authority, December, 1979. 24 pp.

Hershkovitz, P. (1966). Catalog of living whales. *Bull U.S. Nat. Mus.* **246,** 1-259.

Hung-chia Yang (1976). Studies on the whales, porpoises and dolphins of Taiwan. *Annu. Rep. Sci., Taiwan Mus. Taipei* **19,** 131–178.

Iredale, T., and Troughton, E. le G. (1934). Checklist of mammals recorded from Australia. *Memoirs Aust. Mus.* **6,** 1–122.

Kimmerer, W. J., McKinnon, A. D., Atkinson, M. J., and Kessel, J. A. (1985). Spatial distribution of plankton in Shark Bay, Western Australia, *Aust. J. Mar. Freshwater Res.* **36,** 421–432.

Lear, R. J., and Bryden, M. M. (1980). A study of bottlenose dolphin *Tursiops truncatus* in eastern Australian waters. *Occas. Papers Aust. Nat. Parks Wildlife Serv.* **4,** 1–25.

Leatherwood, S., and Reeves, R. R. (1983). "The Sierra Club Handbook of Whales and Dolphins." Sierra Club Books, San Francisco, California. xiv +302 pp.

Leatherwood, S. (1986). Whales, dolphins and porpoises of the Indian Ocean Sanctuary. A catalogue of available information. *Hubbs Marine Research Center Tech. Rep.,* 87–197. San Diego, California.

Leatherwood, S., Caldwell, D. K., and Winn, H. E. (1976). Whales, dolphins and porpoises of the western North Atlantic. A guide to their identification. NOAA Technical Report NMFS CIRC-396. iv + 176 pp.

Leatherwood, S., Reeves, R. R., Perrin, W. F., and Evans, W. E. (1982). Whales, dolphins, and porpoises of the eastern north Pacific and adjacent Arctic waters. A guide to their identification. NOAA Technical Report NMFS CIRC. 444. v + 245 pp.

Lydekker, R. (1904). On two dolphins from Madras. *J. Bombay Nat. Hist. Soc.* **15,** 408–414.

Mead, J. G. (1975). A preliminary report on the former net fisheries for *Tursiops truncatus* in the western North Atlantic. *J. Fish Res. Board Can.* **32,** 1155–1162.

Mitchell, E. D. (1970). Pigmentation pattern evolution in delphinid cetaceans: An essay in adaptive coloration. *Can. J. Zool.* **48,** 717–740.

Mitchell, E. D., ed. (1975). Report on the meeting on smaller cetaceans, Montreal, April 1–11, 1974. *J. Fish. Res. Bd., Can.* **32**(7), 889-983.

Norris, K. S., ed. (1961). Standardized methods for measuring and recording data on the smaller cetaceans. *J. Mammal.* **42**(4), 471–476.

Nunes, R. A., and Lennon, G. W. (1986). Physical property distribution and seasonal trends in Spencer Gulf, South Australia: An inverse estuary. *Aust. J. Mar. Freshwater Res.* **37,** 39–53.

O'Connor, C. (1979). Reproductive periodicity of a *Penaeus esculentus* population near low islets, Queensland, Australia. *Aquaculture* **16,** 153–162.

Owen, R. (1866). On some Indian Cetacea collected by Walter Elliott. *Esq. Trans. Zool. Soc. London* **6,** 17–47.

Paterson, R. A. (1986). A list of specimens of the order Cetacea in the Queensland Museum. *Mem. Qd. Mus.* **22**(2), 309–311.

Pearce, A. F. (1977). Some features of the upper 500 m of the Agulhas Current *J. Mar. Res.* **35,** 731–753.

Pearce, A. (1986). Sea temperatures off Western Australia. *Fins* **19**(2), 6–9.

Perrin, W. F. (1972). Colour patterns of spinner porpoises (*Stenella* cf. *S. longirostris*) of the eastern Pacific and Hawaii, with comments on delphinid pigmentation. *Fishery Bull.* **70**(3), 983–1003.

Perrin, W. F. (1975). Variation of spotted and spinner porpoise (genus *Stenella*) in the eastern Pacific and Hawaii. *Bull. Scripps Inst. Oceanogr.* **21,** 1–206.

Perrin, W. F., Mitchell, E. D., Mead, J. G., Caldwell, D. K., Caldwell, M. C., van Bree, P. J. H., and Dawbin, W. H. (1987). Revision of the spotted dolphins, *Stenella* spp. *Mar. Mamm. Sci.* **3**(2), 99–170.

Pilleri, G., and Gihr, M. (1972a). Contribution to the knowledge of the cetaceans of Pakistan with particular reference to the genera *Neomeris, Sousa, Delphinus,* and *Tursiops* and description of a new porpoise *Neomeris Lasiaeorientalis*). *Invest. Cetacea* **4,** 108–162.

Pilleri, G. and Gihr, M. (1972b). On the record and taxonomy of *Tursiops gephyreus* Lahille, 1908, off the Playa Coronilla, Uruguay. *Invest. Cetacea* **4,** 173–181.

Pilleri, G., and Gihr, M. (1973–1974). Contributions to the knowledge of the cetaceans of southwest and monsoon Asia (Persian Gulf, Indus Delta, Malabar, Andaman Sea and Gulf of Siam). *Invest. Cetacea* **5,** 95–149.

Prell, W. A., Hutson, W. H., Williams, D. F., Be, A. W. H., Geitzenauer, K., and Molfino, B. (1980). Surface circulation of the Indian Ocean during the last glacial maximum, approximately 18,000 yr. B.P. *Quartenary Research*, **14,** 309–336.

Rainier, S. F. (1984). Temporal changes in a demersal fish and cephalopod commmunity of an unexploited coastal area in northern Australia. *Aust. J. Mar. Freshwater Res.* **35,** 747–768.

Rice, D. W. (1977). A list of the marine mammals of the world (3rd ed.). *NOAA Technical Report* **NMFS SSRF-711.**

Rice, D. W., and Scheffer, W. B. (1968). A list of the marine mammals of the world. U. S. Fish Wildlife Service, Special Scientific Report Fisheries, No. 579. 16 pp.

Robineau, D. (1986). Valeur adaptative des caracteres morphologiques distintifs (taille et pigmentation) d'une population isolee d'un dauphin subantarctique, *Cephalorhynchus commersonii* (Lacepede, 1804). *Mammalia* **50**(3), 357–368.

Robineau, D., and Rose, J.-M. (1984). Les cetaces de Djibouti. Bilan des connaissances actuelles sur la faune cetologique de la mer Rouge et du golfe d'Aden. *Bull. Mus. Nat. Hist. Nat., Paris* **6,** 219–249.

Rochford, D. (1975). Oceanography and its role in the management of aquatic ecosystems. *Proc. Ecol. Soc. Aust.* **8,** 67–83.

Ross, G. J. B. (1977). The taxonomy of bottlenosed dolphins *Tursiops* species in South African waters, with notes on their biology. *Ann. Cape Prov. Mus. (Nat. Hist.)* **11**(9), 135-194.

Ross, G. J. B. (1984). The smaller cetaceans of the southeast coast of southern Africa. *Ann. Cape Prov. Mus (Nat. Hist.)* **15**(2), 173-410.

Ruppell. (1842). Beschreibung mehrere neuer Saugetiere, in der zoologischen Sammlung der Senkenbergischen naturforschenden Gesellschaft befindlich. *Mus. Senkenbergianum* **3,** 140–144.

Schnell, G. D., Douglas, M. E., and Hough, D. J. (1986). Geographic patterns of variation in offshore spotted dolphins *(Stenella attenuata)* of the eastern tropical Pacific Ocean. *Mar. Mamm. Sci.* **2,** 186–213.

Schumann, E. H. (1987). The coastal ocean off the east coast of South Africa. *Trans. R. Soc. S. Afr.* **46,** 213–229.

Scott, H. H., and Lord, C. E. (1919). Studies of Tasmanian Cetacea. Part III. *Tursiops tursio. Southern Form. Pap. Proc. R. Soc. Tasmania, 1919,* pp. 96–109.

de Smet, W. M. A. (1977). The regions of the cetacean vertebral column. *In* "Functional Anatomy of Marine Mammals" (R. Harrison, ed.), Vol. 3. pp. 59–80. Academic Press, London.

Smolker, R. (1985). Association patterns of wild Indian Ocean bottlenose dolphins. Sixth Biennial Conference on the Biology of Marine Mammals, Vancouver, Nov. 1985.

Tas'an. (1985). *Tursiops aduncus* (Ehrenberg, 1833) from Java Sea. Symposium on Endangered Marine Animals and Marine Parks, Cochin, India, 12–16 January 1985, 13 pp.

Tas'an and Leatherwood, S. (1984). Cetaceans live-captured for Jaya Ancol Oceanarium, Djakarta, 1974–1982. *Rep. Int. Whaling Comm.* **34,** 485–489.

Tomilin, A. G. (1957). "Mammals of the U. S. S. R. and adjacent countries. Vol. IX. Cetacea." Israel Program for Scientific Translations, Jerusalem, 1967. 717 pp.

True, F. W. (1889). Contributions to the natural history of the cetaceans, a review of the Delphinidae. *Bull. U.S. Nat. Mus.* **36,** 1–191.

Walker, T. (1981). Annual temperature cycle in

Cleveland Bay, Great Barrier Reef Province. *Aust. J. Mar. Freshwater Res.* **32,** 987–991.

Walker, W. A. (1981). Geographical variation in morphology and biology of bottlenose dolphins (*Tursiops*) in the eastern North Pacific. NMFS, Southwest Fisheries Center Administrative Report, No. LJ-81-03C. 17 pp. + appendices.

Watson, L. (1981). "Sea Guide to Whales of the World." Hutchinson and Co., London.

Womersley, H. B. S., and Edmonds, S. J. (1958). A general account of the intertidal ecology of South Australian coasts. *Aust. J. Mar. Freshwater Res.* **9,** 217–260.

Wood Jones, F. (1925). "The Mammals of South Australia." Part III. Government Printer, Adelaide.

Wyrtki, K. (1971). "Oceanographic Atlas of the International Indian Ocean Expedition." National Science Foundation, Washington, D.C. 531 pp.

Distinction between Northwest Atlantic Offshore and Coastal Bottlenose Dolphins Based on Hemoglobin Profile and Morphometry

6

Sandra L. Hersh
Division of Biology and Living Resources
Rosenstiel School of Marine and Atomospheric Science
University of Miami
Miami, Florida 33149
Present address: National Marine Fisheries Service, Southeast Fisheries Center, 75 Virginia Beach Drive, Miami, Florida 33149

Deborah A. Duffield
Department of Biology
Portland State University
Portland, Oregon 97207

INTRODUCTION

Bottlenose dolphins (genus *Tursiops*) from waters of the southeastern United States are usually ascribed to the species *T. truncatus* (Hershkovitz, 1966; Marcuzzi and Pilleri, 1971; Nishiwaki, 1972; Pilleri and Gihr, 1972; Mitchell, 1975; Brownell *et al.*, 1982). However, the systematics of *Tursiops* are poorly defined, and there is need for a revision of the genus (Hershkovitz, 1966; Leatherwood and Reeves, 1982).

The presence in these waters of both a large offshore form and a small coastal form has been suggested by several authors (Mitchell, 1975; Leatherwood and Reeves, 1982; Perrin, 1984). Offshore and coastal forms of *Tursiops* can be differentiated by hemoglobin concentration (Hb), packed cell volume (PCV), red blood cell (RBC) count, and type of hemoglobin present (Duffield *et al.*, 1983; Duffield, 1986). The offshore form or ecotype exhibits higher values than the coastal ecotype for all three hematological parameters. The coastal dolphins of southeastern U.S. waters have a single hemoglobin, while offshore dolphins have two electrophoretically distinguishable hemoglobins. Several bottlenose dolphins that have stranded along the Florida coast in areas with free access to open, deep water have been identified as offshore animals based on their hemoglobin and hematological profiles (Rodriquez *et al.*, 1984; Hersh *et al.*, 1985; Duffield, 1986).

As blood samples for the evaluation of the hemoglobin and hematological parameters are not always available from beached carcasses, and because determination of the ecotype of these dolphin specimens is important for population studies, another means of distinguishing the two forms is needed. Morphometric differences, if found to exist, would be useful in this determination. This chapter documents differences in skull and body measurements exhibited by coastal and offshore bottlenose dolphins from the waters of eastern Florida.

Methods and Materials

Postmortem blood samples were taken from fresh, intact carcasses of bottlenose dolphins found beached (stranded) along the east coast of Florida. Samples were refrigerated and shipped on ice to Portland State University (D. A. D.) for hemoglobin electrophoresis. If a sample was not examined immediately, the red blood cells were stabilized by the addition of 40% buffered glycerol (19.5 g potassium citrate, 3.6 g potassium phosphate dibasic, 2.82 g potassium phosphate monobasic, 400 ml glycerol, 600 ml water) and frozen. Hemolysates were prepared by dilution of an aliquot of postmortem blood with three parts water. Cellulose acetate electrophoresis (Helena Laboratories, Beaumont, TX 77704) was done at pH 7.6, using a Tris–EDTA–boric acid buffer (10.2 g Trisma base, 0.6 g EDTA, 3.2 g boric acid, 1 liter water), and run at 300 V for 10 minutes. The cellulose acetate plates were stained with Ponceau S (Helena Laboratories).

Four adult bottlenose dolphins from Florida were identified as having a two-hemoglobin electrophoretic profile. These animals had all stranded in areas with free access to open water: one at Canaveral Beach, one at Sebastian Inlet, one at Virginia Key (Miami), and one at Long Key in the Florida Keys. Skull and body morphometrics from these animals were compared to those of bottlenose dolphins that had stranded within the Indian/Banana River (IBR) complex on the east coast of Florida (Fig. 1). Five of the IBR animals used for this comparison were known to have single-hemoglobin electrophoretic profiles. In addition, of the over 100 bottlenose dolphins from within this estuarine complex which have been evaluated by hemoglobin electrophoresis, all exhibited a single hemoglobin, typical of coastal bottlenose dolphins. We therefore feel confident that all animals within the IBR complex are of the coastal ecotype.

A series of 28 skull measurements (based on Perrin, 1975; Appendix 1) and 20 body measurements (based on Norris, 1961; Appendix 2) were recorded for the dolphins. All animals were 10 years of age or older as determined by examination of tooth growth layer groups (GLGs), following the methodology of Hohn (1980). They were therefore assumed to be physically mature. *t*-Tests were used to compare body measurements and arcsin-transformed proportional body measurements of the two groups. Skull measurements and proportional skull measurements were not compared statistically since these were available from only two of the offshore animals. Since there is only a minor degree of sexual dimorphism in bottlenose dolphins from these areas (Hersh, 1987), males and females were not treated separately.

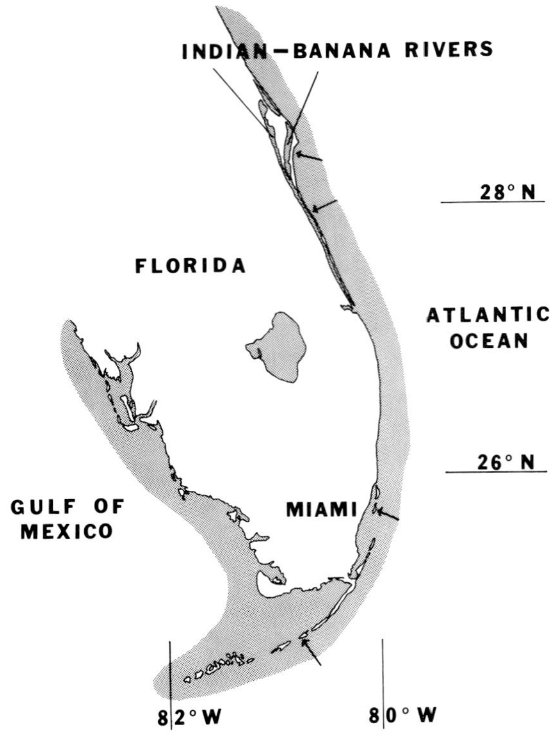

Figure 1 Map of Florida showing the Indian/Banana River complex on the east coast and locations where offshore bottlenose dolphins have stranded (arrows).

Because blood samples for biochemical analysis are only sporadically available from beached bottlenose dolphins, it is likely that some offshore animals which strand along the coast go unrecognized. To examine this possibility, cluster analyses (using the SPSSX program CLUSTER), based on measurements found to differ in the comparison of offshore and coastal dolphins, were used to identify animals that appeared morphologically similar to the two-hemoglobin animals. Squared euclidean distance was chosen as the measure and average linkage between groups as the clustering method.

RESULTS

Offshore bottlenose dolphins from the Northwest Atlantic have two hemoglobins that can be distinguished by electrophoresis. The electrophoretically fast hemoglobin comprises approximately 30% of the total, while the electrophoretically slow hemoglobin comprises about 70% of the total (Fig. 2). Coastal Northwest Atlantic and Gulf of Mexico bottlenose dolphins typically have only one hemoglobin, electrophoretically matching the fast hemoglobin of the offshore animals.

The results of *t*-tests applied to body measurements and comparisons of skull measurements from offshore and coastal IBR animals indicate distinct morphological differences between the two groups. Differences are seen in 10 of 32 skull measurements (Table 1) and in 6 of 27 proportional skull measurements, expressed as a percentage of condylobasal length (Table 2). Four of the 18 body measurements were found to be significantly ($p < .05$) larger in offshore animals (Table 3). Eight of 17 arcsin-transformed proportional body measurements, expressed as percentage of total body length, differed significantly in coastal and offshore animals (Table 4). In general, offshore dolphins appear longer, with a proportionately shorter snout and proportionately smaller flippers. Their skulls appear wider, with wider rostrums, than those of coastal specimens.

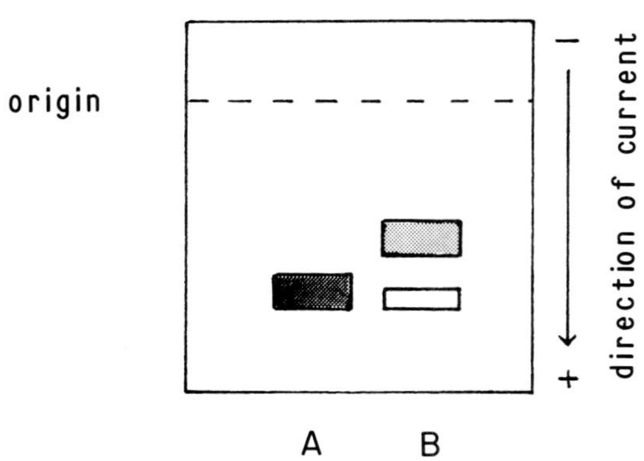

Figure 2 Electrophoretic analysis (cellulose acetate, pH 7.6) of hemoglobins in coastal and offshore bottlenose dolphins from Florida waters. (A) Coastal hemoglobin profile. (B) Offshore hemoglobin profile.

Cluster analysis (based on selected proportional skull and body measurements) of all stranded *Tursiops* from the study area for which morphological data were available indicated that several dolphins for which no blood data were available were morphologically similar to dolphins with offshore hemoglobin profiles. Four animals (A, B, C, and D) were singled out in the cluster based on body measurements (Fig. 3). All four stranded in areas with free access to open, deep water, and two of the four (A and B) stranded alive. Dolphin A also bore scars from the pelagic cookie-cutter shark, *Isistius* sp. Two dolphins (B and E) were singled out by the cluster based on skull measurements (Fig. 4). Both stranded in areas with open-water access, and one (B) stranded alive. Two of the four dolphins singled out in the first cluster based on body measurements (A and C) also appear close to the two-hemoglobin animals, which clustered together in the analysis based on skull measurements. Another animal (F) that appears similar to the offshore dolphins also stranded alive.

DISCUSSION

Hematological differences between offshore and coastal bottlenose dolphins have been observed previously (Duffield *et al.*, 1983). The higher hemoglobin concentrations, hematocrits, and red blood cell counts of offshore animals suggest an adaptation for greater oxygen-carrying capacity to facilitate longer and possibly deeper routine dives. Such hematological differences have also been seen between cetacean species that differ in the

Table 1
Comparison of Skull Measurements from Indian/Banana River (IBR) Bottlenose Dolphins and Dolphins Identified as Offshore Animals (OFF) Based on Hemoglobin Profile[a]

Measurement	*N* (IBR)	*N* (OFF)	*x* (IBR)	*x* (OFF)	S.D. (IBR)
COBALEN	9	2	45.20	50.05	1.86
LENROS	9	2	24.91	27.45	1.23
WIDROSB*	9	2	11.02	13.05	0.70
WIDROSA*	9	2	8.81	10.15	0.34
WIDROSM	9	2	7.50	8.15	0.29
WIDPREM	9	2	3.87	4.30	0.32
ROSEXNA	9	2	29.89	32.60	1.44
ROSINNA	9	2	29.74	32.65	1.38
GPREWID*	9	2	20.01	23.35	0.82
GPOSWID*	9	2	23.42	26.55	1.04
LSUPWID*	9	2	19.80	22.60	0.87
GEXNAWD	9	2	5.51	5.65	0.27
GZYWID*	9	2	23.39	26.55	1.02
GPRMWID	9	2	8.71	8.90	0.35
GPARWID	8	2	18.04	19.25	0.51
GLLPF	9	2	10.29	10.85	0.42
GWLPF	9	2	8.10	8.40	0.40
LENLOR	8	2	6.26	6.30	0.25
LENLAP*	9	2	4.97	5.85	0.37
GWIDIN*	9	2	6.60	8.55	0.35
GLLPTY*	8	2	6.61	7.95	0.45
GLLBUL	7	2	3.73	3.65	0.13
GLLPER	7	2	3.13	3.40	0.11
LENULT	9	2	21.67	23.45	1.20
NUMUL	9	2	24.6	23.0	1.1
NUMUR	10	2	24.5	23.0	1.1
NUMLL	10	2	24.4	22.5	1.1
NUMLR	10	2	24.0	23.0	1.3
LENLLT	9	2	22.00	22.90	1.14
GLENLR	9	2	38.00	42.05	1.69
GHTLR	9	2	8.52	9.15	0.40
LENLMF*	9	2	12.41	13.85	0.54

[a] *N* is the sample size, *x* the mean (in cm), and S.D. the standard deviation. An asterisk next to the abbreviation indicates mutually exclusive ranges of the offshore animals and IBR animals. Abbreviations used for skull measurements are defined in Appendix 1. All animals were 10 years or older.

nature of their habitat (pelagic versus coastal; Ridgway and Johnson, 1966)

The morphological differences between coastal and offshore bottlenose dolphins may also be interpreted with respect to habitat differences. The smaller size of coastal animals may be an adaptation to shallow environments where maneuverability is important. This may also explain the proportionately larger size of the flipper in coastal animals. Ross (1984) described similar differences in the habitat and morphometry of two forms of bottlenose dolphin from the southeastern coast of Africa. In this region, the larger form (identified by Ross as *T. truncatus*) is found in water depths of 200 m or more while the smaller form (*T. aduncus*, according to Ross) is common closer to shore (see

Table 2
Comparisons of Proportional Skull Measurements (% of Condylobasal Length) from Indian/Banana River (IBR) Bottlenose Dolphins and Dolphins Identified as Offshore Animals (OFF) Based on Hemoglobin Profile[a]

Measurement	*N* (IBR)	*N* (OFF)	*x*P (IBR)	*x*P (OFF)
COBALEN	9	2	—	—
LENROS	9	2	0.551	0.548
WIDROSB	9	2	0.244	0.260
WIDROSA	9	2	0.195	0.203
WIDROSM	9	2	0.166	0.163
WIDPREM	9	2	0.086	0.086
ROSEXNA	9	2	0.661	0.651
ROSINNA	9	2	0.658	0.652
GPREWID	9	2	0.444	0.467
GPOSWID	9	2	0.519	0.531
LSUPWID	9	2	0.438	0.452
GEXNAWD	9	2	0.122	0.113
GZYWID	9	2	0.518	0.531
GPRMWID*	9	2	0.193	0.178
GPARWID	8	2	0.398	0.385
GLLPF*	9	2	0.229	0.217
GWLPF	9	2	0.180	0.169
LENLOR*	8	2	0.140	0.126
LENLAP	9	2	0.110	0.117
GWIDIN*	9	2	0.146	0.170
GLLPTY*	8	2	0.148	0.159
GLLBUL	7	2	0.082	0.073
GLLPER	7	2	0.069	0.068
LENULT	9	2	0.480	0.469
NUMUL	9	2	—	—
NUMUR	10	2	—	—
NUMLL	10	2	—	—
NUMLR	10	2	—	—
LENLLT*	9	2	0.486	0.457
GLENLR	9	2	0.840	0.840
GHTLR	9	2	0.189	0.183
LENLMF	9	2	0.275	0.277

[a] *N* is the sample size and *x*P the proportional mean. An asterisk next to the abbreviation indicates mutually exclusive ranges of the offshore animals and IBR animals. Abbreviations used for skull measurements are defined in Appendix 1. All animals were 10 years or older.

Ross and Cockcroft, Chapter 5, this volume). The smaller body size of coastal dolphins may be primarily an adaptation to warm-temperate and tropical waters, whereas smaller size may be thermally disadvantageous in cooler waters. This may explain why relatively large bottlenose dolphins are found close to shore along the western coasts of North America and Africa where cold water is present near shore. Large size may also be an advantage against the larger predators often found in pelagic environments.

Differences in skull morphology, similar to those noted in this study, are found in bottlenose dolphins off southern Africa. The more coastal *T. aduncus* has a narrower cranium and a more slender, tapering rostrum than the larger *T. truncatus* (Ross, 1977; also see Ross and Cockcroft, Chapter 5, this volume).

Differences in skull dimensions of coastal and offshore bottlenose dolphins may be related to food habits. Differences in the diet of coastal and offshore bottlenose dolphins have been noted by several authors (Ross, 1977; Walker, 1981; Barros, 1987; Barros and Odell, Chapter 16, this volume; Van Waerebeek *et al.*, Chapter 7, this volume). A relationship between observed differences in food habits and morphological variation has been suggested for several delphinid species (Perrin, 1984). Walker (1981) stated that the cranial characteristics which best distinguish coastal and offshore Pacific *Tursiops* are those related to feeding (e.g., tooth width, mandibular condyle size, and shape of rostrum). A similar observation has been made for Pacific common dolphins, *Delphinus delphis* (Banks and Brownell, 1969). Differences in skull characteristics associated with feeding have also been found between the offshore and inshore forms of spotted dolphins, *Stenella attenuata,* and spinner dolphins, *Stenella longirostris* (Perrin, 1975; Douglas *et al.*, 1984).

These morphological observations appear to imply population differentiation based on resource exploitation. Offshore dolphins seem to be exploiting food sources found in deep, cool waters and show appropriate structural divergence from coastal dolphins in the cranial characteristics associated with feeding.

Other differences between coastal and offshore bottlenose dolphins were noted during this study. Usually coastal bottlenose dolphins wash ashore dead; offshore bottlenose dolphins commonly strand alive, probably because those that die far offshore never reach shore. The offshore bottlenose dolphins found stranded in the study area also appeared to be darker in color than the coastal dolphins seen in the IBR complex. This difference does not appear to be an artifact of postmortem

Table 3
Results of *t*-Tests Performed on Body Measurements from Indian/Banana River (IBR) Bottlenose Dolphins and Dolphins Identified as Offshore Animals (OFF) Based on Hemoglobin Profile[a]

Measurement	*N* (IBR)	*N* (OFF)	x (IBR)	x (OFF)	S.D. (IBR)	S.D. (OFF)
TOTLEN*	9	4	247.7	289.3	7.0	13.5
SNEYE	9	4	31.2	31.0	1.4	3.4
SNMEL	9	4	10.00	9.80	1.12	1.71
GAPE	8	4	25.9	28.3	0.8	2.9
SNEAR	7	4	37.4	39.0	2.0	2.5
EYEEAR	6	4	8.08	8.00	0.74	1.08
EYEMO	7	4	6.93	7.13	0.89	0.63
EYEBLO*	7	4	21.4	23.8	1.0	0.5
SNBLO	7	4	34.1	33.3	1.2	2.4
SNFLP	8	4	50.9	50.8	2.6	2.2
SNTDF*	6	4	146.2	168.5	5.5	6.9
SNUMB*	7	4	113.1	125.8	3.2	6.3
PROJLJ	7	4	1.59	1.05	0.5	0.7
FLIPAN	9	4	41.1	43.5	2.9	2.5
FLIPPO	9	4	30.7	28.8	4.1	3.1
FLIPWID	9	4	16.6	15.6	0.9	1.4
HTDOR	9	4	23.7	26.5	2.5	2.9
FLUWID	9	3	63.3	65.7	5.2	1.5

[a] *N* is the sample size, x the mean (in cm), and S.D. the standard deviation. An asterisk next to the abbreviation represents a significant difference ($p < .05$) in the means. Abbreviations used for body measurements are defined in Appendix 2. All animals were 10 years or older.

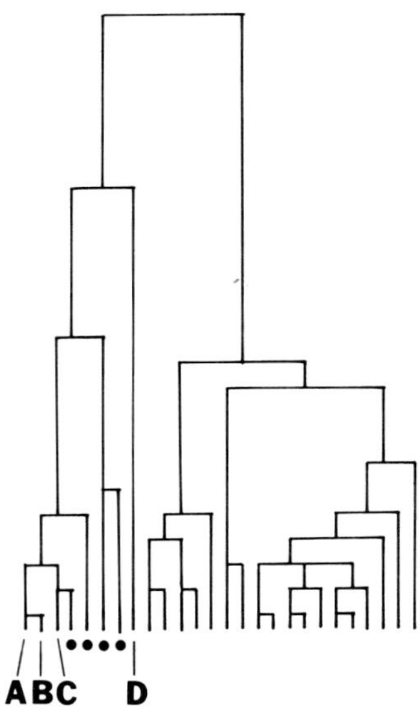

Figure 3 Cluster dendrogram of bottlenose dolphins based on proportional body measurements (SNEYE, SNEAR, SNBLO, SNFLP, and FLIPWID), using squared euclidean distance as the measure and average linkage between groups as the method. Abbreviations used for body measurements are defined in Appendix 2. Offshore animals are represented by circles.

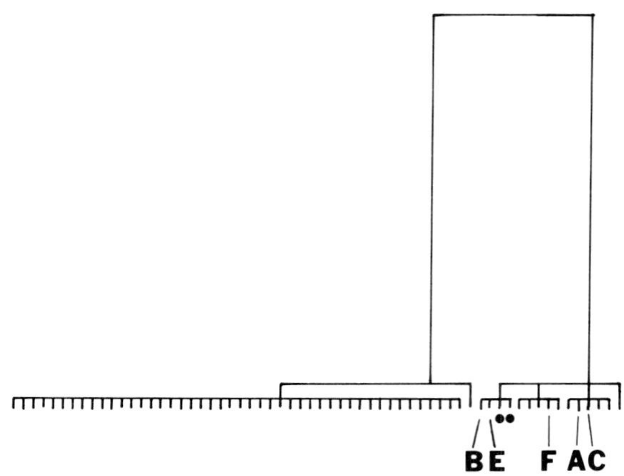

Figure 4 Cluster dendrogram of bottlenose dolphins based on proportional skull measurements (GPRMWID, GLLPF, LENLOR, GWIDIN, GLLPTY, and LENLLT), using squared euclidean distance as the measure and average linkage between groups as the method. Abbreviations used for skull measurements are defined in Appendix 1. Offshore animals are represented by circles.

Table 4
Results of *t*-Tests Performed on Arcsin-Transformed Proportional Body Measurements (% of Total Length) from Indian/Banana River (IBR) Bottlenose Dolphins and Dolphins Identified as Offshore Animals (OFF) Based on Hemoglobin Profile[a]

Measurement	N (IBR)	N (OFF)	xP (IBR)	xP (OFF)
TOTLEN	9	4	—	—
SNEYE*	9	4	0.126	0.107
SNMEL*	9	4	0.041	0.003
GAPE	8	4	0.104	0.100
SNEAR*	7	4	0.153	0.132
EYEEAR	6	4	0.031	0.027
EYEMO	7	4	0.028	0.025
EYEBLO	7	4	0.085	0.083
SNBLO*	7	4	0.139	0.115
SNFLP*	8	4	0.205	0.175
SNTDF	6	4	0.594	0.583
SNUMB	7	4	0.456	0.432
PROJLJ	7	4	0.008	0.001
FLIPAN*	9	4	0.162	0.150
FLIPPO*	9	4	0.124	0.100
FLIPWID*	9	4	0.068	0.055
HTDOR	9	4	0.096	0.091
FLUWID	9	3	0.256	0.231

[a] *N* is the sample size and *x*P the proportional mean. An asterisk next to the abbreviation represents a significant difference ($p < .05$) in the means. Abbreviations used for body measurements are defined in Appendix 2. All animals were 10 years or older.

darkening because the offshore animals collected during this study appeared dark gray while still alive or freshly dead. Offshore animals may also bear scars caused by bites of the pelagic cookie-cutter shark (S. L. Hersh, personal observation).

In 1983, a male bottlenose dolphin stranded live in the Florida Keys. This animal was rehabilitated and maintained for over 4 years by the Dolphin Research Center, Marathon Shores, Florida. It had the external features of an offshore dolphin: robust in overall body form with a short snout, a long body, and small flippers; dark gray in color. He had a two-hemoglobin electrophoretic profile and higher hemoglobin concentration, hematocrit, and red blood cell numbers than coastal Atlantic bottlenose dolphins (Rodriquez *et al.*, 1984; Duffield, 1986).

In conclusion, we believe that the suite of characteristics discussed in this chapter (hemoglobin type, hemoglobin concentration, packed cell volume, red blood cell count, overall body size, and certain skull and body proportions) reflects fundamental adaptations to the animal's environment. Bottlenose dolphins living in warm, shallow waters exhibit a relatively small body size and proportionately large flippers, perhaps in response to demands for increased maneuverability and heat dissipation. Their hematological profile, similar to those of many terrestrial mammals living at low altitude, suggests that diving and feeding in shallow-water environments may not put a high demand on the hematological parameters of oxygen storage and delivery.

Dolphins living in cool, deep waters appear to have a larger body size and proportionately smaller flippers than shallow-water animals, again perhaps reflecting different thermal and maneuverability requirements. Differences in skull characteristics may be related to differences in food habits. The increased hematological parameters of these offshore animals seem an appropriate adaptation in a habitat where deeper food sources may encourage deeper and/or longer dives and create greater oxygen demand.

We suggest that in other areas where morphological differences such as the ones described here are seen, corresponding differences in hematology are likely to be found once appropriate samples are analyzed. We suggest further that the "offshore" ecotype is not necessarily restricted to offshore waters but may be found relatively close to shore wherever cool, deep water is found. Similarly the "coastal" ecotype may be absent in near-shore areas where water temperatures are cold and/or shallow-water bottom fish are not abundant. It may be more appropriate to refer to these two ecotypes as cool, deep-water animals and warm, shallow-water animals.

SUMMARY

Offshore bottlenose dolphins (genus *Tursiops*) from the Northwest Atlantic have two electrophoretically distinguishable hemoglobins, whereas coastal bottlenose dolphins from this region (including the Gulf of Mexico) have only a single hemoglobin. This difference was used to distinguish offshore bottlenose dolphins found stranded along the east coast of Florida. Comparison of skull and body measurements from coastal and offshore dolphins indicates that offshore dolphins are larger and exhibit proportionately smaller flippers than coastal animals. In addition, their skulls are wider and exhibit several different proportions than those of coastal dolphins. Cluster analyses of morphometric data from stranded animals indicate that offshore bottlenose dolphins may occur more commonly along the coast than previously realized. Hematological, hemoglobin, and morphological differences appear to form a suite of characteristics that can be used to distinguish these two forms, more appropriately characterized as cool, deep-water and shallow, warm-water ecotypes.

ACKNOWLEDGMENTS

This study was made possible by the generous contribu tion of data by Dan Odell of the University of Miami Florida, and Ed Asper of Sea World Inc., Orlando Florida. We also extend our gratitude to the many private citizens, veterinarians, Marine Patrol officers Department of Natural Resources personnel, parl rangers, and employees of state and local agencie whose cooperation with the Southeastern U.S. Marin Mammal and Sea Turtle Stranding and Salvage Networl provided information on stranded animals. This study was supported in part by the Portland State Universit Biomedical Research Fund and Sea World, Inc.

APPENDIX 1

Tursiops skull measurements (after Perrin, 1975) used for morphometric comparisons and their abbrevi ations. (A) Dorsal view of skull. (B) Ventral view. (C) Lateral view of the left side of skull. (D) Internal latera view of the left ramus of the mandible. Measurements 22, 23, and 25–28 are not pictured.

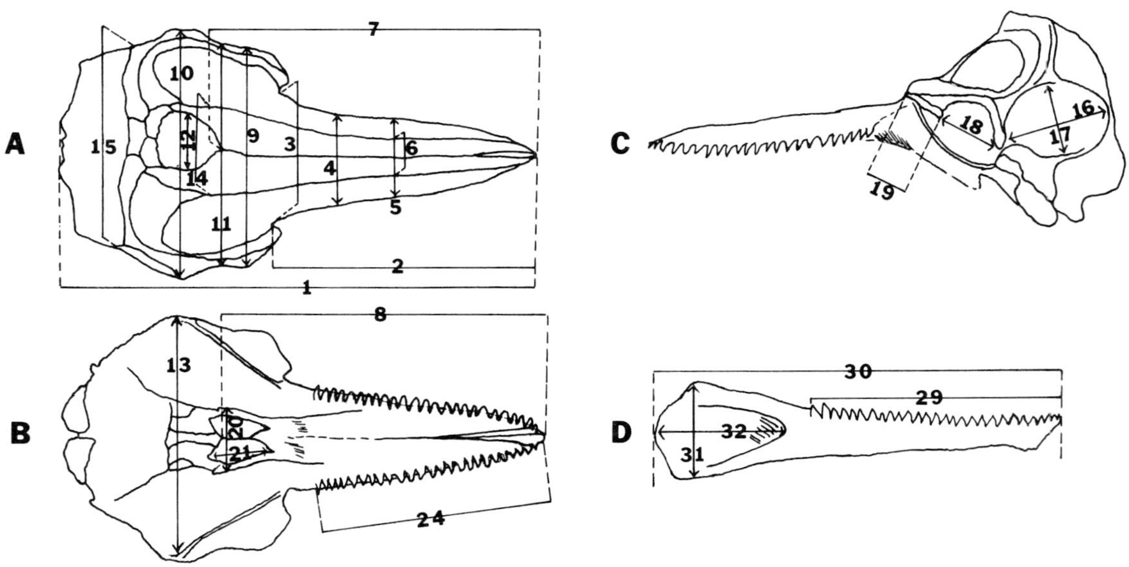

Measurement	Abbreviation
1. Condylobasal length, from tip of rostrum to hindmost margin of occipital condyles	COBALEN
2. Length of rostrum, from tip to line across hindmost limits of antorbital notches	LENROS
3. Width of rostrum at base, along line across hindmost limits of antorbital notches	WIDROSB
4. Width of rostrum at 60 mm anterior to line across hindmost limits of antorbital notches	WIDROSA
5. Width of rostrum at midlength	WIDROSM
6. Width of premaxillaries at midlength of rostrum	WIDPREM
7. Distance from tip of rostrum to external nares	ROSEXNA
8. Distance from tip of rostrum to internal nares	ROSINNA
9. Greatest preorbital width	GPREWID
10. Greatest postorbital width	GPOSWID
11. Least supraorbital width	LSUPWID
12. Greatest width of external nares	GEXNAWD
13. Greatest width across zygomatic processes of squamosal	GZYWID
14. Greatest width of premaxillaries	GPRMWID
15. Greatest parietal width, within posttemporal fossae	GPARWID
16. Greatest length of left posttemporal fossa	GLLPF
17. Greatest width of left posttemporal fossa, at right angles to measurement No. 16	GWLPF
18. Length of left orbit, from apex of preorbital process of frontal to apex of postorbital process	LENLOR
19. Length of antorbital process of left lacrimal	LENLAP
20. Greatest width of internal nares	GWIDIN
21. Greatest length of left pterygoid	GLLPTY
22. Greatest length of bulla of left tympanoperiotic	GLLBUL
23. Greatest length of periotic of left tympanoperiotic	GLLPER
24. Length of upper left toothrow, from hindmost margin of hindmost alveolus to tip of rostrum	LENULT
25. Number of teeth, upper left toothrow	NUMUL
26. Number of teeth, upper right toothrow	NUMUR
27. Number of teeth, lower left toothrow	NUMLL
28. Number of teeth, lower right toothrow	NUMLR
29. Length of lower left toothrow, from hindmost margin of hindmost alveolus to tip of mandible	LENLLT
30. Greatest length of left ramus	GLENLR
31. Greatest height of left ramus at right angles to measurement No. 30	GHTLR
32. Length of left mandibular fossa, measured to mesial rim of internal surface of condyle	LENLMF

APPENDIX 2

Tursiops body measurements (after Norris, 1961) used for morphometric comparisons and their abbreviations.

Measurement	Abbreviation
1. Total length, snout to notch	TOTLEN
2. Snout to center of eye	SNEYE
3. Snout to apex of melon	SNMEL
4. Length of gape	GAPE
5. Snout to ear	SNEAR
6. Center of eye to ear	EYEEAR
7. Center of eye to angle of mouth	EYEMO
8. Eye to center of blowhole (not pictured)	EYEBLO
9. Snout to center of blowhole	SNBLO
10. Snout to flipper	SNFLP
11. Snout to tip of dorsal fin	SNTDF
12. Snout to center of umbilicus	SNUMB
13. Projection of lower jaw	PROJLJ
14. Flipper length (anterior)	FLIPAN
15. Flipper length (posterior)	FLIPPO
16. Maximum width of flipper	FLIPWID
17. Height of dorsal fin	HTDOR
18. Width of flukes	FLUWID

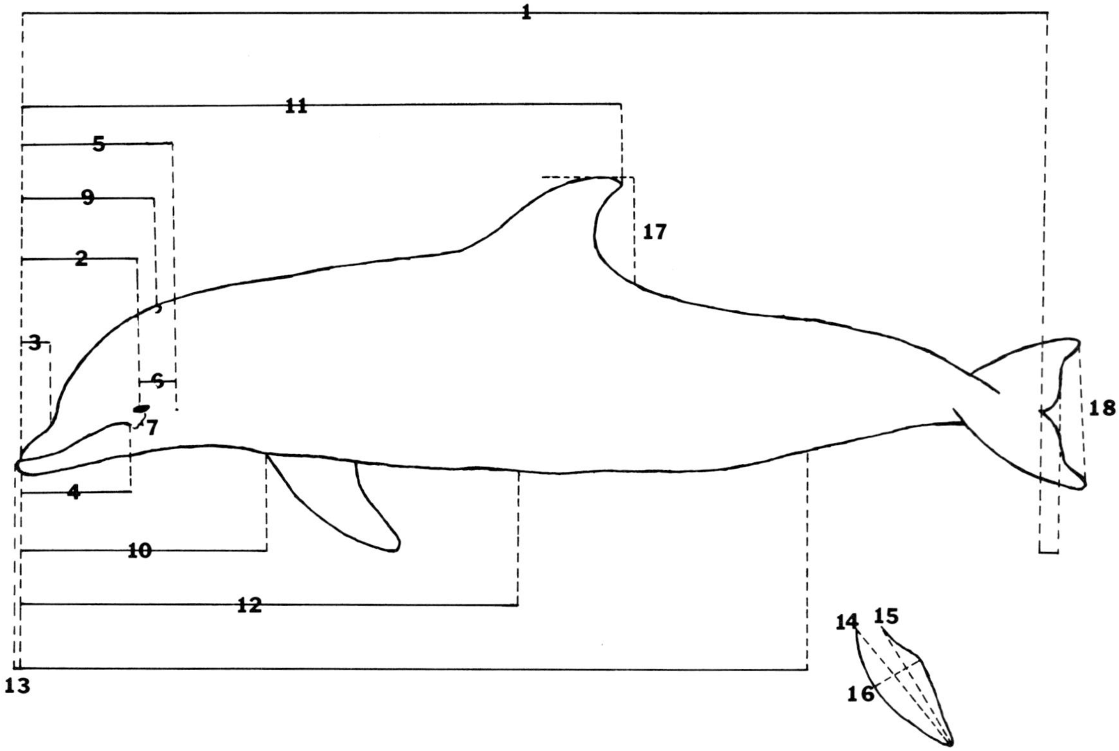

REFERENCES

Banks, R. C., and R. L., Brownell (1969). Taxonomy of the common dolphins of the eastern Pacific Ocean. *J. Mammal.* **50**(2), 262–271.

Barros, N. B. (1987). Food habits of bottlenose dolphins (*Tursiops truncatus*) in the southeastern United States with special reference to Florida waters. M.S. thesis, University of Miami, Coral Gables, Florida. 139 pp.

Brownell, R. L., Jr., Kasuya, T., Mead, J. G., Nishiwaki, M., Perrin, W. F., Rice, D. W., Rossolimo, O. L., and Wang, S. (1982). Family Delphinidae. *In* "Mammal Species of the World." (J. H. Honacki, K. E. Kinman, and J. W. Koeppl, eds.), pp. 291–297. Allen Press, Inc., Lawrence, Kansas.

Douglas, M. E., Schnell, G. D., and Hough, D. J. (1984). Differentiation between inshore and offshore spotted dolphins in the eastern tropical Pacific Ocean. *J. Mammal.* **65**(3), 375–387.

Duffield, D. A. (1986). Investigation of genetic variability in stocks of the bottlenose dolphin (*Tursiops truncatus*). Final report to NMFS/SEFC, Contract No. NA83-GA-C-00036.

Duffield, D. A., Ridgway, S. H., and Cornell, L. H. (1983). Hematology distinguishes coastal and offshore forms of dolphins (*Tursiops*). *Can. J. Zool.* **61,** 930–933.

Hersh, S. L. (1987). Characterization and differentiation of bottlenose dolphin populations (genus *Tursiops*) in the southeastern U.S. based on mortality patterns and morphometrics. Ph.D. dissertation. University of Miami, Coral Gables, Florida. 213 pp.

Hersh, S. L., Duffield, D. A., Asper, E. D., Odell, D. K., Rodriquez, J., and White, J. (1985). Morphological and biochemical distinction of offshore and coastal Atlantic bottlenose dolphins. *Sixth Biennial Conf. Biol. Mar. Mamm.*, Vancouver, British Columbia, Nov. 22–26, Abstr.

Hershkovitz, P. (1966). Catalog of living whales. *Bull. U.S. Nat. Mus.* **246,** 1–259.

Hohn, A. A. (1980). Age determination and age related factors in the teeth of coastal North Atlantic bottlenose dolphins, *Tursiops truncatus* (Cetacea, Odontoceti). M.S. thesis, University of Maryland, Baltimore. 74 pp.

Leatherwood, S., and Reeves, R. (1982). Bottlenose

dolphin, *Tursiops truncatus*, and other toothed cetaceans. *In* "Wild Mammals of North America" (J. A. Chapman and G. A. Feldhamer, eds.), pp. 369–414. Johns Hopkins University Press, Baltimore, Maryland. 1147 pp.

Marcuzzi, G., and Pilleri, G. (1971). On the zoogeography of Cetacea. *Invest. Cetacea* **3,** 101–170.

Mitchell, E., ed. (1975). Review of biology and fisheries for smaller cetaceans. *J. Fish. Res. Board Can.* **32**(7), 889–983.

Nishiwaki, M. (1972). General biology. *In* "Mammals of the Sea" (S. H. Ridgway, ed.), pp. 3–204. Charles C. Thomas, Springfield, Illinois. 812 pp.

Norris, K. S. (1961). Standardized methods for measuring and recording data on the smaller cetaceans. *J. Mammal.* **42**(2), 471–476.

Perrin, W. F. (1975). Variation of spotted and spinner porpoise (genus *Stenella*) in the eastern Pacific and Hawaii. *Bull. Scripps Inst. Oceanog.* **21,** 1–206.

Perrin, W. F. (1984). Patterns of geographical variation in small cetaceans. *Acta Zool. Fennica* **172,** 137–140.

Pilleri, G., and Gihr, M. (1972). Contributions to the knowledge of cetaceans of Pakistan with particular reference to the genera *Neomeris, Sousa, Delphinus,* and *Tursiops* and description of a new Chinese porpoise (*Neomeris asiaeorientalis*). *Invest. Cetacea* **4,** 107–162.

Ridgway, S. H., and Johnson, D. G. (1966). Blood oxygen and ecology of porpoises of three genera. *Science* **151,** 456–458.

Rodriquez, J., Rodriquez, M., Garrigan, C., Asper, E. D., Duffield, D. A., and Hersh, S. L. (1984). Different forms of *Tursiops* in the Atlantic. *15th Ann. Meet. Int. Assoc. Aq. Animal Med.,* Tampa, Florida, April 30–May 2, Abstr.

Ross, G. J. B. (1977). The taxonomy of bottlenosed dolphins. *Tursiops* species, in South African waters with notes on their biology. *Ann. Cape Prov. Mus. (Nat. Hist.)* **11,** 135–194.

Ross, G. J. B. (1984). The smaller cetaceans of the south east coast of southern Africa. *Ann. Cape Prov. Mus (Nat. Hist.)* **15**(2), 173–410.

Walker, W. A. (1981). Geographical variation in morphology and biology of bottlenose dolphins (*Tursiops*) in the eastern North Pacific. Final contract report to NMFS, Contract No. 03-7-208-35238. Administrative Report No. LJ-81-03C. 54 pp.

PHOTO: Bottlenose dolphins, such as these calves taken in a drift gillnet, are netted or harpooned in temperate and tropical waters around the world (Pitipana, Negombo, Sri Lanka, August 1985; Stephen Leatherwood).

Part IV

Life History and Biology

Preliminary Observations of Bottlenose Dolphins from the Pacific Coast of South America

7

Koen Van Waerebeek
Rijksuniversiteit Gent
Laboratorium voor Morphologie en Systematiek
Museum voor Dierkunde
Ledeganckstraat 35
9000 Gent, Belgium, and
Grupo Cetaceos-ECCO
Vanderghen 560-2A
Lima 27, Peru

Julio C. Reyes
Universidad Ricardo Palma
Lima, Peru, and
Grupo Cetaceos-ECCO
Vanderghen 560-2A
Lima 27, Peru

Andrew J. Read
Department of Zoology
University of Guelph
Guelph, Ontario, Canada N1G 2W1

Jeff S. McKinnon
Department of Zoology
University of Guelph
Guelph, Ontario, Canada N1G 2W1
Present address: The Biological Laboratories, Harvard University, Cambridge, Massachusetts 02138

INTRODUCTION

In this chapter we present some preliminary observations on the distribution, natural history, and exploitation of bottlenose dolphins from the Pacific coast of South America. The first records of bottlenose dolphins from Peru and Ecuador were, respectively, a fresh specimen from Talara (4° 34′ S) and a skull from Indefatigable Island, Galapagos, secured by the Leon Mandel Galapagos expedition in 1941 (Hershkovitz, 1963). Hershkovitz (1963) provisionally identified the specimens as *Tursiops nesarnack catalania* Gray 1862, only to reassign them later (Hershkovitz, 1966) to *T. truncatus aduncus* (Ehrenberg, 1832). In this paper, we assume the existence of a single cosmopolitan species of bottlenose dolphin, *T. truncatus* (Montagu, 1821), with a series of geographical forms, as proposed by Tomilin (1957, p. 562).

The "eastern tropical Pacific offshore population" of bottlenose dolphins is believed to extend as far south as the waters off Colombia and Ecuador, including the Galapagos Islands (Walker, 1981). Leveque (1963) described the species as abundant and probably resident in the Galapagos archipelago. Offshore sightings of bottlenose dolphins were recorded by U.S. National Marine Fisheries Service observers in the eastern tropical Pacific as far south as 15° S (Walker, 1981; Scott and Chivers, Chapter 22, this volume). Sightings in offshore waters near Peru have also been reported by Donovan (1984). From Chile, bottlenose dolphins are known from the offshore islands of San Ambrosio and San Felix (Gilmore, 1971; Aguayo, 1975), the Juan Fernandez Islands (Aguayo, 1975), the Salas and Gomez Islands (Cardenas *et al.*, 1986), and waters off central Chile (Clarke *et al.*, 1978).

Less information is available regarding the distribution of bottlenose dolphins in coastal waters.

Grimwood (1969) reported that bottlenose dolphins were frequently observed swimming in the surf along the Peruvian coast and noted that the species appeared to be common from Mancora (3° 45′ S) south to the Chilean border. Further south, the species has been reported from the Gulf of Arauco, south of Concepcion (Oliver, 1946).

The information we present here was collected as part of a study of the exploitation and biology of small cetaceans along the Pacific coast of South America, with emphasis on Peruvian waters, conducted in 1985–1987 (Read *et al.*, 1985, 1988). Information provided by fishermen in Peru suggested that bottlenose dolphins were captured either very close to shore or in tropical offshore waters beyond the coastal upwelling zone. This prompted us to adopt a working hypothesis of distinct coastal and offshore forms (*sensu* Perrin, 1984) in the area, as proposed for southern African (Ross, 1977), eastern North Pacific (Walker, 1981), and western North Atlantic waters (Hersh and Duffield, Chapter 6, this volume).

FISHERY INTERACTIONS AND DISTRIBUTION

Interviews with fishermen and port authorities in Puerto Bolivar, Ecuador (3°17′ S), during February 1986 indicated that bottlenose dolphins occur frequently in the mangrove areas around the port. The dolphins are occasionally trapped in nylon shrimp nets or small-mesh gill nets, but they are not landed because of strict enforcement of conservation measures and the lack of demand for dolphin meat. We found no skeletal material around the port to contradict the information obtained during interviews.

In Peru, dolphin meat is used for human consumption, and bottlenose dolphins and other small cetaceans are captured in drift and demersal gill nets, in purse seines, and occasionally with hand-thrown harpoons. Interviews with fishermen in northern Peru (see Fig. 1), using photographs of several common cetacean species, suggested bottlenose dolphins are present from Puerto Pizarro (3° 30′ S) to Pacasmayo (7° 24′ S). In addition, we found osteological remains of bottlenose dolphins on the beaches of Sechura (5° 34′ S) and Pimentel (6° 50′ S). During visits to the large industrial port of Chimbote (9° 5′ S) in March 1985 and July 1986, we observed numerous bottlenose dolphins being landed at the fish market, with larger numbers of common dolphins, *Delphinus delphis*. The animals were landed by pelagic purse seiners fishing for sardine, *Sardinops sagax*, and anchoveta, *Engraulis ringens*.

We retrieved two fresh heads of bottlenose dolphins from the wharf at Huacho (11° 7′ S), in central Peru, on 15 July 1986. Despite documenting a large directed fishery for dusky dolphins, *Lagenorhynchus obscurus*, at Ancon (11° 47′ S) during the winter of 1985, we did not observe bottlenose dolphins in the catch. We made no attempt to monitor the catch of small cetaceans at the major port of Callao (12° 5′ S).

We examined 57 bottlenose dolphins at the Pucusana (12° 29′ S) fish market between January 1985 and February 1987. All but 2 of these animals were landed during the summer months (December–March), when the coastal upwelling zone is fairly narrow (Zuta *et al.*, 1983). The majority of dolphins were caught at the edge of the upwelling zone or in oceanic waters by drift nets set for sharks, bonito, *Sarda chiliensis*, and dorado, *Coryphaena hippurus*. In addition, bottlenose dolphins were captured in coastal waters by a variety of methods. Of the total, 39 were captured in gill nets, 11 were harpooned, 2 were found stranded and transported to the market, 1 was captured in a shore seine, and 4 were taken by unknown means.

The Cerro Azul (13° 2′ S) fish market was monitored between March and August 1986, and we observed two bottlenose dolphins captured in nets. The species is captured occasionally in Tambo de Mora (13° 29′ S), where we recovered several weathered skulls from the beach and from a small refuse dump near the wharf. Bottlenose dolphins may be taken more frequently in San Andres (13° 51′ S), where we have collected fresh severed heads near the market on several occasions.

Observed catches, corrected for number of observation days (211) and stratified by month, yield an estimate of 33.7 (S.D. 0.75) bottlenose dolphins landed at Pucusana in 1986. Similar calculations for Cerro Azul, although with fewer observation days (134), provide an estimate of 3.0 (S.D. 1.4) bottlenose dolphins captured in 1986.

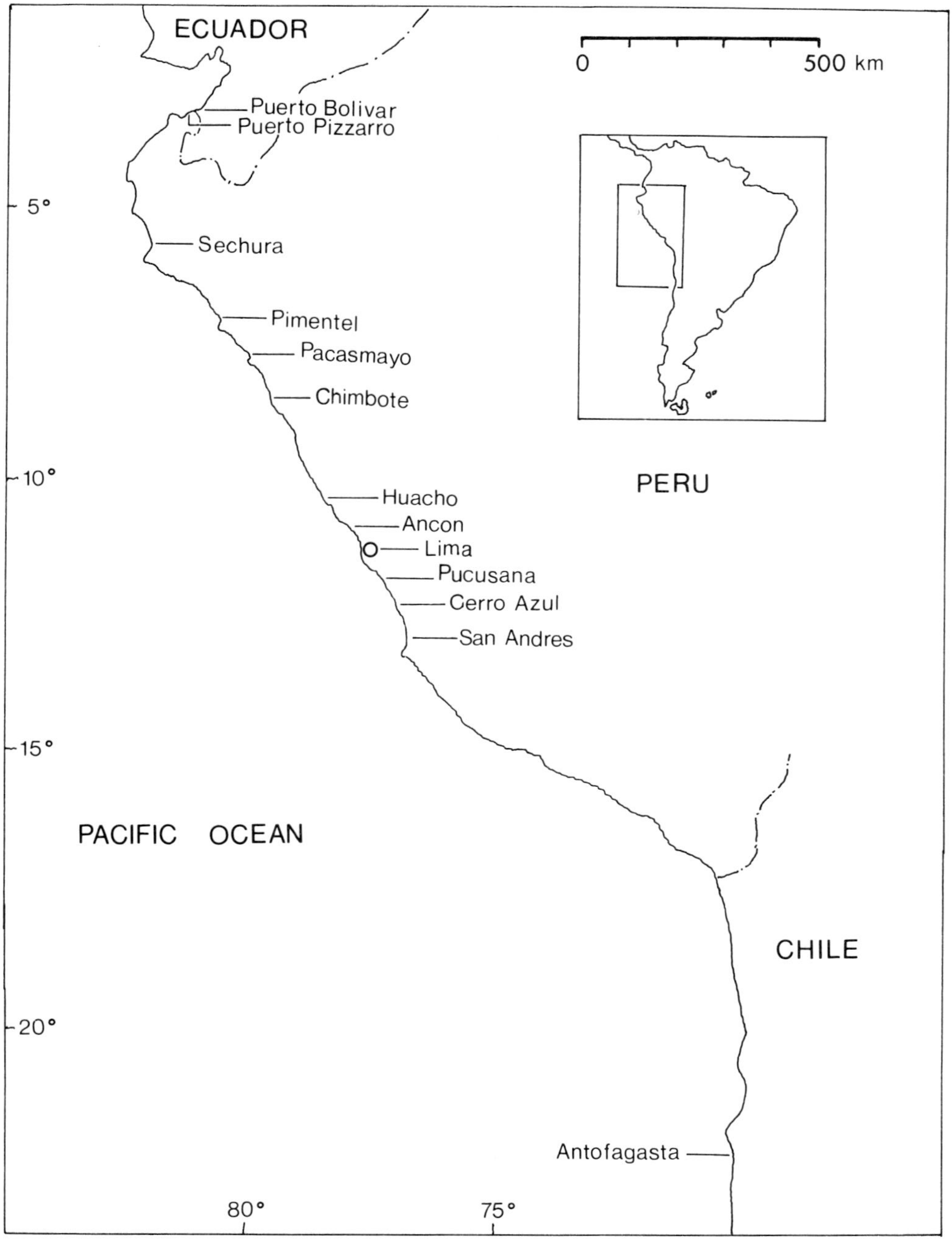

Figure 1 Map of the study area, with place-names mentioned in the text. The inset shows the area in relation to the remainder of South America.

The catch of bottlenose dolphins at Pucusana and Cerro Azul comprised only a small proportion of the total number of small odontocetes captured at both ports in 1986 (4.4 and 0.52%, respectively) (Read *et al.*, 1988). Unfortunately, we have insufficient data from other areas to estimate a total catch for the entire coast, but given that the catch of small cetaceans at several ports is much larger than at Pucusana and Cerro Azul the total is likely in the hundreds.

In northern Chile, bottlenose dolphins seem to be exploited only sporadically. A small incidental catch is made by the industrial purse seine fishery for sardine and anchoveta operating out of Mejillones (23° S) (Guerra *et al.*, 1987). On 11 June, a female bottlenose dolphin was harpooned near Punta Coloso, Antofagasta (23° 43′ S). This appears to be the first record of a directed catch and inshore sighting of the species from Chilean waters (Guerra *et al.*, 1987). No information is available regarding the incidental catch of bottlenose dolphins in other regions of Chile.

While working in central Peru, we recorded 40 sightings of bottlenose dolphins from shore and during coastal boat surveys. Dolphins were observed near Ancon, Punta Negra (12° 21′ S), Pucusana, Cerro Azul, and Paracas (13° 56′ S). Dolphins were observed exclusively within a kilometer from shore and usually just outside the surf zone. We did not survey waters outside the coastal upwelling zone. Mean group size was 6.5 (S.E. 1.0), including two large groups (20–50 animals) associated with feeding flocks of Peruvian boobies, *Sula variegata* and brown pelicans, *Pelecanus occidentalis.*

CRANIAL CHARACTERS

From a total of 50 available skulls, only 19 specimens were physically mature and from known capture locations (i.e., coastal or offshore). The distal fusion of maxillaries and premaxillaries was used as the criterion of physical maturity (Perrin *et al.*, 1979).

Thirty-five cranial measurements and meristics were recorded, following Schnell *et al.* (1985) except for three omissions (Nos. 17, 18, and 23) and two additions (Nos. 37 and 38) (Table 1). Measurements were taken with vernier calipers to the nearest 0.5 mm. Tooth widths were measured perpendicular to the ramus axis, 2 mm below the gum line, on two to four teeth with lowest wear from the middle of the mandibular tooth row (Walker, 1981). Each width was measured twice independently to the nearest 0.05 mm and then averaged. The sample sizes of tooth widths and tooth counts were increased by including material from subadult specimens (three skulls of the coastal form and seven sets of tooth widths from offshore animals).

The cranial measurements of offshore (n = 15) and coastal (n = 4) specimens are presented in Table 1. No obvious sexual dimorphism was present in the offshore specimens, and the sample sizes are too small to test statistically. Both sexes are tabulated together; all four coastal skulls are from males.

Walker (1981) found that tooth width was the best single criterion for differentiating coastal and offshore forms of bottlenose dolphins in the eastern North Pacific. In Peru, coastal animals have significantly wider teeth than offshore animals (t-test with unequal variances, p = .01), as illustrated in Fig. 2.

The samples of other cranial characters are too small to permit adequate statistical comparison, but some qualitative differences were noted between the two forms. Coastal animals feature a shorter antorbital process and narrower palatine bones, with no overlap observed between the two forms. The coastal skulls have fragile, thinly walled pterygoids; the pterygoid notch is wide with a rounded, obtuse apex. The coastal form also features a broad separation, quadrangular in shape, between the occipital condyles at the level of the basioccipital. The skulls of the offshore form have thickly walled pterygoid bones and a narrow pterygoid notch with a subacute apex, and the separation of condyles is narrow and triangular in shape. Representative inshore and offshore skulls are presented in Fig. 3.

Other skull characters, such as posterior width of the vomer, shape of squamosal, shape of premaxillary convexity, relative size of mandibular condyles, and tapering of rostrum, that were used by Ross (1977) and Walker (1981) to discriminate between forms of bottlenose dolphins in other areas were found to be extremely variable in Peruvian specimens.

Table 1
Cranial Measurements (in mm) and Meristics of Adult Bottlenose Dolphins from Peru[a]

Measurement	Offshore			Coastal		
	$\overline{X}$	n	Range	$\overline{X}$	n	Range
1. Condylobasal length	521.1	15	494–542	519.5	4	507–542
2. Length of rostrum (from base)	291.8	15	278–311	282.5	4	274–297
3. Length of rostrum (from pterygoid)	350.9	15	332–370	333.2	4	310–351
4. Width of rostrum (at base)	144.3	15	132–158	135.2	4	129–145
5. Width of rostrum (at ¼ length)	107.1	15	98–114	101.8	4	98–105
6. Width of rostrum (at ½ length)	87.5	15	82–95	88.5	4	86–91
7. Width of premaxilla (at ½ length)	51.0	15	45–59	52.0	4	50–53
8. Width of rostrum (at ¾ length)	64.5	15	62–74	68.5	4	64–71
9. Preorbital width	246.1	15	231–264	233.2	4	229–244
10. Postorbital width	273.1	15	257–288	268.8	4	260–281
11. Skull width (at zygomatic process)	276.3	15	263–294	273.8	4	261–290
12. Skull width (at parietals)	191.3	15	181–207	197.8	4	195–202
13. Height of braincase	152.4	14	137–170	158.2	4	149–178
14. Length of braincase	165.1	14	151–182	172.5	4	166–180
15. Maximum width of premaxilla	101.0	15	88–110	105.2	4	101–110
16. Width of external nares	61.3	15	58–67	61.0	4	57–65
19. Length of temporal fossa	117.7	15	107–126	121.2	4	115–126
20. Width of temporal fossa	82.5	15	73–96	82.0	4	78–91
21. Orbital length	71.6	15	65–78	73.0	4	70–76
22. Length of antorbital process	62.9	15	58–67	53.0	4	49–57
24. Width of internal nares	80.9	15	72–89	71.8	4	69–73
25. Length of left tympanic cavity	76.2	15	72–89	77.8	4	73–83
26. Length of right tympanic cavity	79.2	15	71–88	77.8	4	74–82
27. Width at pterygoid sutures	90.8	15	81–98	84.2	4	80–90
28. Length of upper toothrow	246.9	15	233–264	243.5	4	236–255
29. No. of teeth (upper left toothrow)	22.6	15	19–25	21.7	7	20–23
30. No. of teeth (upper right)	22.3	15	19–25	21.3	7	20–23
31. No. of teeth (lower left)	21.1	15	18–24	20.0	7	19–21
32. No. of teeth (lower right)	21.2	15	18–24	19.6	7	19–21
33. Length of lower toothrow	240.7	15	219–265	233.5	4	227–250
34. Height of ramus	96.6	15	92–101	98.0	4	94–100
35. Tooth width	7.4	22	6.55–8.55	9.2	7	8.55–9.6
36. Length of ramus	447.4	15	423–465	435.5	4	421–459
37. Mandibular condyle width	39.0	15	35.6–42.0	38.1	4	37.0–43.3
38. Maximum width of palatine	62.7	13	55.2–69.3	52.6	4	50.3–54.4

[a] Samples for tooth widths and number of teeth include subadults.

EXTERNAL MORPHOMETRICS AND COLORATION

Although the sample sizes are small, there appears to be sexual dimorphism in total length from both coastal and offshore waters. The largest offshore male measured 304.5 cm (n = 22), while the largest offshore female was 289.0 cm (n = 11). The largest coastal male was 308.0 cm in length (n = 5) and the largest coastal female 289.0 cm (n = 7). However, the largest specimen of *T. truncatus* from the Southeast Pacific is a 315.0-cm female (AMM-017, Instituto de Investigaciones Oceanologicas, Universidad de Antofagasta) from Punta Coloso, Chile (Guerra *et al.*, 1987).

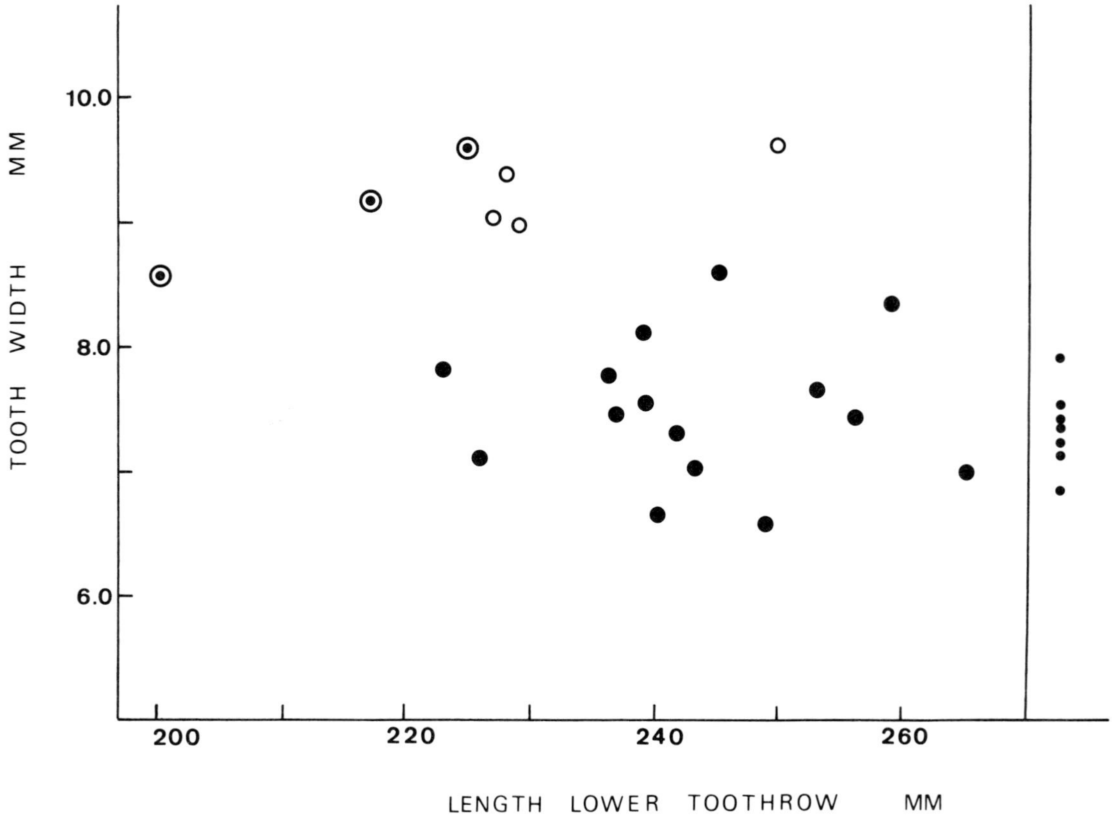

Figure 2 Comparative tooth widths of bottlenose dolphins from central Peruvian waters: adult inshore (○), subadult inshore (⊙), adult offshore (●), and subadult offshore (•). No values are available for the length of lower toothrows of subadult offshore specimens.

Photographs and field notes from 22 offshore and 8 coastal bottlenose dolphins from Pucusana and Cerro Azul were available for examination of pigmentation patterns. Males and females of both forms exhibited eye-to-apex and apex-to-blowhole bridle stripes and flipper stripes of varying intensity (terminology of Mitchell, 1970). In three offshore animals, two males (KVW-558 and JCR-901) and a female (KVW-520), a posterior vestige of an eye-to-anus stripe was observed. Two offshore specimens displayed yellowish hues, one on the thoracic area (KVW-519) and the other below its flipper stripe (KVW-558).

Coastal animals appeared to be lightly colored with a clearly distinguished dark cape system and light cape overlay (Perrin, 1972), resulting in a tripartite pattern. With few exceptions, offshore specimens were much darker, except for the ventral surface, with an obscured cape pattern. However, the observed difference in skin darkness should be interpreted with caution, since longer periods of exposure to the sun during offshore fishing trips might have resulted in a greater degree of postmortem darkening among offshore animals.

Figure 3 Dorsal, ventral, left lateral, and occipital views of typical adult skulls of coastal (JCR-670, A–D) and offshore (JCR-900, E–H) bottlenose dolphins from the central Peruvian coast. For both specimens the condylobasal length is 542 mm.

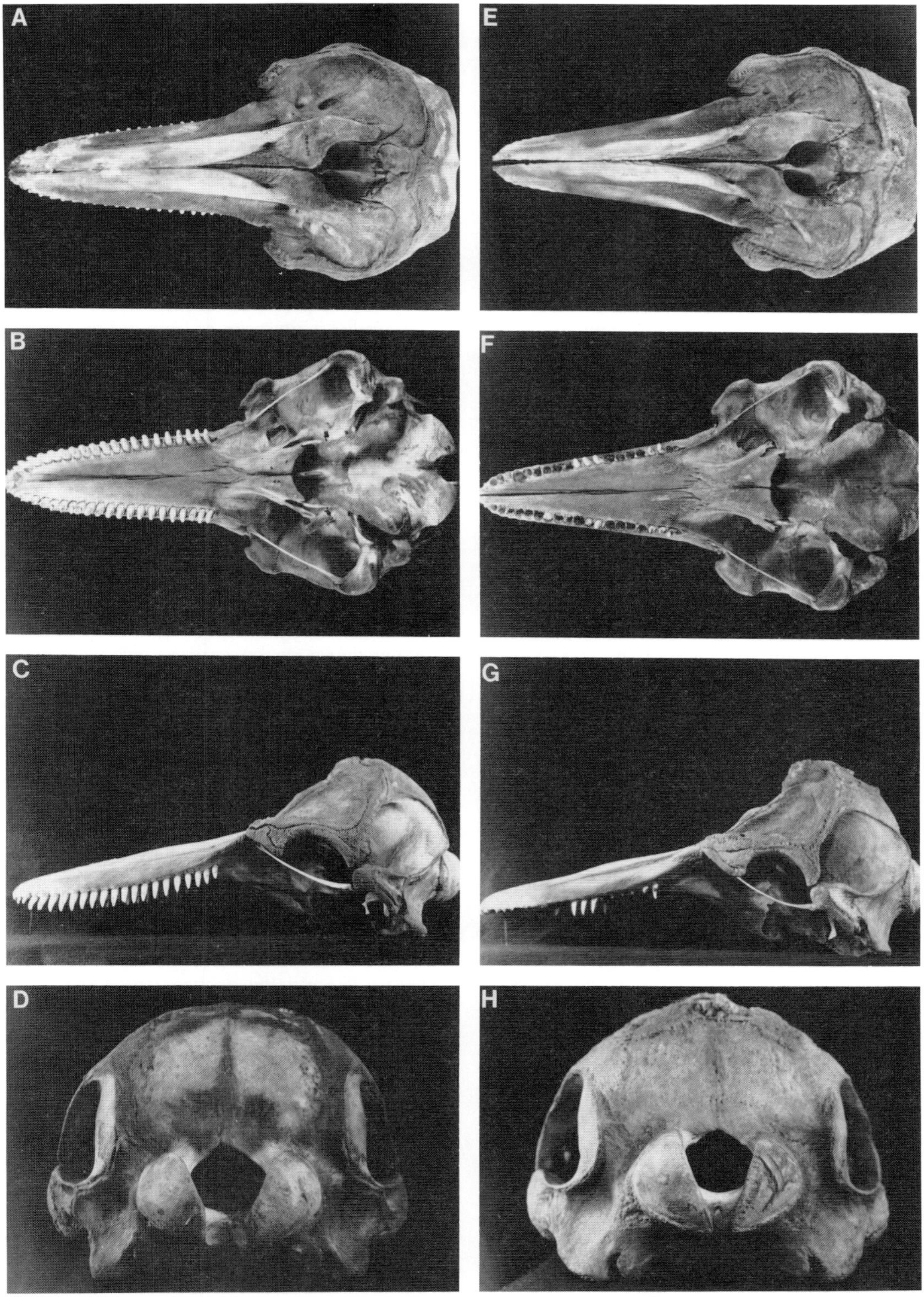
A
B
C
D
E
F
G
H

REPRODUCTION

Histological examination of gonads and age estimation procedures have not yet been performed, so information on reproduction presented here is restricted to that obtained in gross examination of female ovaries and reproductive tracts. Females were considered sexually mature if they had ovulated at least once, based on the presence of at least one corpus luteum or corpus albicans in the ovaries (Perrin and Reilly, 1984). Data are available from nine offshore and five coastal females.

The largest immature offshore female was 236 cm long, while the smallest mature female from offshore waters measured 275 cm. Two offshore females landed in February were pregnant; one 280-cm animal (KVW-293) bore a 23.3-cm (238 g) fetus and a 282-cm dolphin (KVW-190) had a 97-cm (10kg) near-term fetus.

The largest immature coastal female was 258.5 cm in length while the smallest mature coastal female measured 284 cm. A single coastal female landed in March (JSM-209) was pregnant with a 23-mm (7g) fetus. Another inshore female (AJR-019) was lactating and accompanied by a calf of unknown size when harpooned in February (J. S. McKinnon, pers. observation).

Seven calves were examined in Pucusana between December 1985 and February 1986. The specimens were judged to be young of the year on the basis of incomplete eruption of teeth, presence of milk in the forestomach, and total body length. The smallest calf was a 124.0-cm neonate with a fresh umbilicus; the largest, a 165.0-cm animal with teeth erupted in the lower jaw only. Three of the animals were from offshore waters and four from unknown capture locations.

FEEDING

A complete account of the feeding habits of Peruvian bottlenose dolphins is currently in preparation. Only an overview of the stomach contents of coastal and offshore animals is given here.

Stomachs were dissected from dolphins landed at Pucusana during the austral summer and early autumn of 1985, 1986, and 1987, and from a single animal landed at Huacho in the winter of 1986. Contents of the forestomachs were run through a series of sieves in a straining–decanting procedure from which squid beaks and otoliths were retained. Beaks of the squid *Loligo gahi* were identified through reference to a small comparative sample provided by F. Cardoso (Universidad Nacional Mayor de San Marcos, Lima), and with the assistance of S. Candela (University of Miami, Florida). Otoliths were identified through reference to a comparative collection and with the aid of P. Majluf (University of Cambridge, England).

Samples were obtained from the stomachs of 7 coastal and 20 offshore dolphins. Two additional offshore dolphins and a single coastal animal had empty forestomachs, and the stomach of another offshore animal contained only milk. Seven samples were obtained from animals for which no capture location (coastal or offshore) was available. One empty stomach and one containing only milk were also obtained from two individuals from unknown capture locations.

The most common prey of coastal bottlenose dolphins, in terms of both total prey items and frequency of occurrence, were anchoveta, *Engraulis ringens*, patagonian squid, *Loligo gahi*, a drum, *Cynoscion analis*, and Pacific sardine, *Sardinops sagax*. Other prey items included hake, *Merluccius gayi*, a toadfish, *Aphos porosus*, a second species of anchovy, *Anchoa* sp., mullet, *Mugil* sp., and three other sciaenids: *Sciaena deliciosa*, *Stellifer minor*, and *Menticirrhus ophicephalus*.

The stomachs of offshore animals contained a great variety of otoliths and squid beaks, only some of which we were able to identify successfully. Fish species present included anchoveta, hake, Pacific sardine, jack mackerel, *Trachurus symmetricus*, and the deepsea smelt, *Leuroglossus urotramus*. In addition, a pair of mullet otoliths was present in a single sample. Otoliths from one of the most common prey have not yet been identified to species but appear to be from the family Myctophidae (P. Majluf, pers. communication). Several other small otoliths are very common in the samples, and are probably from mesopelagic fishes such as myctophids and bathylagids. It should be noted that the mesopelagic fishes may have been prey of the squids and other fishes consumed by the dolphins (Perrin *et al.*, 1973;

Walker, 1981). Several types of large squid beaks occur in these samples and are currently being identified by S. Candela.

Three samples from unknown locations contained prey items similar to those found in the stomachs of known coastal dolphins. A jack mackerel otolith and two crustacean chelipeds, not present in the coastal samples, were also recovered. Three samples from unknown locations contained prey similar to those found in the stomachs of offshore dolphins. The sample from Huacho contained only sardine and could not be classified as coastal or offshore.

These data indicate that both coastal and offshore bottlenose dolphins consumed Pacific sardine and anchoveta, the common pelagic fish species present throughout the Peruvian upwelling system. Both forms also consumed hake, which is normally demersal but at times is semipelagic (Mejia and Jordan, 1979). The diets of coastal and offshore dolphins differed principally in that the inshore demersal fish species, such as the sciaenids and toadfish (Koepcke, 1966; Arntz, 1986), were present in the stomachs of coastal dolphins but not in the stomachs of offshore animals. In addition, the offshore dolphins consumed mesopelagic fishes and squids not found in the stomachs of coastal animals. This dichotomous pattern was also present in samples obtained from unknown capture locations. The presence of mullet, normally an inshore species (Koepcke, 1966), in the stomach of an offshore dolphin is surprising. The dolphin may have been foraging in inshore waters, or the mullet may have strayed offshore. In the Gulf of Mexico, mullet are sometimes found in offshore waters, where they reportedly spawn (Arnold and Thompson, 1958).

Our results are consistent with the descriptions of the diets of coastal and offshore bottlenose dolphins from other areas. Data presented by Ross (1977, 1984) from the southeastern coast of Africa indicate that inshore and offshore bottlenose dolphins have fairly discrete diets in that region. Walker (1981) observed that coastal bottlenose dolphins in the North Pacific fed principally on inshore benthic and demersal species, while dolphins from the eastern tropical Pacific consumed a variety of epipelagic fishes, offshore squids, and (perhaps secondarily introduced) mesopelagic fishes.

PARASITES AND PATHOLOGY

During necropsies of 27 offshore and 10 coastal bottlenose dolphins in Pucusana, several organs and tissues were routinely checked for the presence of parasites. In addition, 23 skulls from known capture locations were examined for the presence of bone lesions associated with infections of the nematode *Crassicauda* sp. (Perrin and Powers, 1980). Pathological observations were recorded when abnormal tissues were encountered.

The parasitological examinations yielded four trematode, three nematode, and three cestode species (Table 2). Identifications were based on Price (1932), Baer (1954), Yamaguti (1958), Skrjabin (1961), and Arnold (1973). One of the trematode species, *Pholeter gastrophylus,* was present in the main and pyloric stomachs and duodenal ampullae in 10 of 23 offshore specimens but absent from 8 coastal animals. The cestodes *Phyllobothrium delphini* and *Monorygma grimaldi* were present in the majority of offshore specimens but were not encountered in the few coastal specimens examined (Table 2). Bone erosion caused by parasites of the genus *Crassicauda* was found in 11 of 16 offshore specimens, but not in any of the 7 coastal specimens examined. Walker (1981) found that *Crassicauda* sp., *Phyllobothrium delphini,* and *Monorygma* sp. were useful in discriminating between offshore and coastal populations of bottlenose dolphins in the eastern North Pacific.

The ectocommensal barnacle *Xenobalanus globicipitis* was found on a single offshore animal out of 25 checked. No barnacles were found on 9 coastal animals examined. Two specimens carrying barnacles were among 10 dolphins examined from unknown capture locations.

Evidence of bone pathology was discovered in a few specimens, including two dolphins with the first cervical vertebrae fused to the occipital bone. One of these specimens also exhibited bone erosion in the form of perforations in the mandibular rami. These anomalies have previously been reported from bottlenose dolphins by van Bree and

Table 2
Parasites Found in Bottlenose Dolphins from Peruvian Waters

Parasite	Infection site	Inshore form			Offshore form		
		N_e[a]	N_i[b]	%[c]	N_e	N_i	%
Trematoda							
Pholeter gastrophylus	Main and pyloric stomachs	8	0	0.0	23	10	43.5
Synthesium tursionis	Pyloric stomach	6	1	16.7	18	0	0.0
Braunina cordiformis	Main and pyloric stomachs	8	1	12.5	22	3	13.6
Nasitrema globicephalae	Cranial sinuses and auditory complex	5	4	80.0	17	16	94.1
Nematoda							
Anisakis typica	Stomachs	8	3	37.5	21	13	61.9
Stenurus ovatus	Lungs	4	3	75.0	7	7	100.0
Crassicauda sp.	Cranial sinuses and blubber–muscle interface	7	0	0.0	16	11	68.8
Cestoda							
Tetrabothrius forsteri	Pyloric stomach and intestine	6	1	16.7	17	12	70.6
Phyllobothrium delphini	Blubber	3	0	0.0	10	7	70.0
Monorygma grimaldi	Blubber–muscle interface, bladder, and testicles	1	0	0.0	5	4	80.0

[a] N_e, Number examined.
[b] N_i, Number infected.
[c] %, Percentage occurrence.

Duguy (1970) and De Smet (1977). Other pathological observations included the following: a large vesicle in the left lung of one specimen, calcified nematode remains in the kidney renculi of another, and small cysts of unknown etiology in the liver and pancreas of one and two specimens, respectively.

CONCLUSIONS

Evidence presented here confirms Grimwood's (1969) observation that *T. truncatus* is common along the coast of Peru. In addition, the species' range likely forms part of a continuous distribution complex along the Pacific South American coast, from at least southern Ecuador to Concepcion, Chile (37° S).

In Peru, bottlenose dolphins have been sighted both offshore (Donovan, 1984) and close to the coast, often just outside the surf zone. Fishery data indicate that in central Peru bottlenose dolphins are captured by fishermen in coastal waters as well as in oceanic waters outside the upwelling zone, suggesting the existence of separate inshore and offshore forms. Preliminary evidence indicates that the two forms differ in tooth width, diet, and parasite fauna.

Peruvian coastal dolphins have significantly wider teeth than offshore animals, a result also obtained by Walker (1981) for bottlenose dolphins in the eastern North Pacific. Although both forms consume some widely distributed prey species, such as anchoveta, there are clear differences in their diets. Coastal bottlenose dolphins feed on a variety of inshore demersal fish species, while offshore animals prey on mesopelagic fish and squid found at the edge of or outside the upwelling zone. Several parasite species recorded only from offshore specimens may be useful in differentiating the two forms of the species in this region. These observations of differences in diet and parasite load are consistent with those reported by Walker (1981) and Mead and Potter (Chapter 9, this volume).

Bottlenose dolphins are captured off central Peru primarily by gill nets, purse seines, and harpoons. Because the meat is used for human consumption, there is no clear distinction between directed and incidental catch. The majority of dolphins appear to be taken from offshore waters. At the present time it is impossible to estimate the total catch in Peru, although it likely numbers in the hundreds each year.

ACKNOWLEDGMENTS

We wish to thank the following: B. A. Luscombe, who helped us with logistics throughout the duration of this study; Cabo Segundo V. H. Yepez Mercado of the Pucusana Port Authority, who provided us with fishery data; S. Candela, F. Cardoso, and P. Majluf for assistance with prey identification; Neti Vuckovic, and the families Farnworth, Reyes Robles, Yong, Velasquez, and Valle-Riestra for their kind and long-standing hospitality. We also wish to acknowledge the fishing communities of Pucusana and Cerro Azul for their cooperation and patience. Ricardo Huldisch V. typed an early version of the manuscript. David Gaskin and two anonymous reviewers made useful comments on drafts of the chapter.

This study was funded by the IUCN/UNEP Burmeister's Porpoise Project through D. E. Gaskin, Department of Zoology, University of Guelph; Natural Sciences and Engineering Research Council of Canada Operating Grant A-5683 through D. E. Gaskin; and 1985 and 1986 grants to K. V. W. by the "Leopold III-Fonds voor Natuuronderzoek en Natuurbehoud" (Brussels) through A. Capart and J. P. Gosse, Koninklijk Belgisch Instituut voor Natuurwetenschappen.

REFERENCES

Aguayo, A. (1975). Progress report on small cetacean research in Chile. *J. Fish. Res. Board Can.* **32,** 1123–1143.

Arnold, E. L., Jr., and Thompson, J. R. (1958). Offshore spawning of the striped mullet, *Mugil cephalus,* in the Gulf of Mexico. *Copeia 1958,* pp. 130–132.

Arnold, P. W. (1973). The lungworms (Metastrongyloidea, Pseudaliidae) of the harbour porpoise (*Phocoena phocoena* L.). M.S. thesis, University of Guelph, Guelph, Ontario. 91 pp.

Arntz, W. E. (1986). The two faces of E1 Nino 1982–83. *Sonderdruck aus Bd.* **31,** 1–46.

Baer, J. G. (1954). Revision taxinomique et etude biologique des cestodes de la famille des Tetrabothriidae, parasites d'oiseaux de haute mer et de mammiferes marins. *Mem. Univ. Neuchatel.,* pp. 1–122.

Bree, P. J. H. van, and Duguy, R. (1970). Sur quelques aberrations pathologiques chez les petits cetaces. *Der Zoologische Garten. (NF)* **39,** 11–15.

Cardenas, J. C., Stutzin, M. E., Oporto, J. A., Cabello, C., and Torres, D. (1986). "Manual de Identificacion de los Cetaceos Chilenos." Proyecto WH-455 "Cetaceos—Chile." CODEFF, Santiago. 102 pp.

Clarke, R., Aguayo, A., and Basulto del Campo, S. (1978). Whale observation and whale marking off the coast of Chile in 1964. *Sci. Rep. Whales Res. Inst., Tokyo* **30,** 117–177.

De Smet, W. M. A. (1977). The fate of old bottlenose dolphins, *Tursiops truncatus,* in nature as revealed by the condition of their skeletons. *Aquatic Mamm.* **5,** 78–86.

Donovan, G. P. (1984). Small cetaceans seen during the IWC/IDCR research cruise in the eastern tropical Pacific, and in particular off Peru. *Rep. int. Whal. Commn* **34,** 561–567.

Gilmore, R. M. (1971). Observations on marine mammals and birds off the coast of southern and central Chile, early winter 1970. *Antarct. J. U.S.* **6,** 10–11.

Grimwood, I. R. (1969). Notes on the distribution and status of some Peruvian mammals 1968. *N.Y. Zool. Soc. Spec. Publ.,* No. 21, pp. 1–86.

Guerra, C., Van Waerebeek, K., Portflitt, G. E., and Luna, G. S. (1987). Presencia de cetaceos frente a la segunda region de Chile. *Estudios Oceanologicos, Univ. Antofagasta, Chile* **6,** 87–96.

Hershkovitz, P. (1963). Notes on South American dolphins of the genera *Inia, Sotalia* and *Tursiops. J. Mammal.* **44,** 98–103.

Hershkovitz, P. (1966). Catalog of living whales. *Bull. U.S. Natl. Mus.* **246,** 1–259.

Koepcke, H. W. (1966). "Peces Comunes de la Costa Peruana." Serie de Divulgacion Cientifica No. 6, Servicio de Pesqueria, Ministerio de Agricultura, Lima, Peru.

Leveque, R. (1963). Notes sur quatre cetaces de l'Ocean Pacifique (Ecuador et Galapagos). *Mammalia* **27,** 608–609, plate 15.

Mejia, J., and Jordan, R. (1979). "La Situacion Actual del Stock de Merluza a Septiembre de 1978." Informe No. 57, Instituto del Mar del Peru, Callao, Peru.

Mitchell, E. D. (1970). Pigmentation pattern evolution in delphinid cetaceans: An essay in adaptive coloration. *Can. J. Zool.* **48,** 717–740.

Oliver, S. C. (1946). Catalogo de los mamiferos de la provincia de Concepcion. Orden Cetacea. *Bd. Soc. Biol. (Concepcion)* **21,** 79–83.

Perrin, W. F. (1972). Color patterns of spinner porpoises (*Stenella* cf. *S. longirostris*) of the eastern Pacific, with comments on delphinid pigmentation. *Fishery Bull.* **70,** 983–1003.

Perrin, W. F. (1984). Patterns of geographical variation in small cetaceans. *Acta Zool. Fennica* **172,** 137–140.

Perrin, W. F., and Powers, J. E. (1980). Role of a nematode in natural mortality of spotted dolphins. *J. Wildl. Man.* **44,** 960–963.

Perrin, W. F., and Reilly, S. B. (1984). Reproductive parameters of dolphins and small whales of the family Delphinidae. *Rep. int. Whal. Commn* (Special Issue 6), 97–133.

Perrin, W. F., Warner, R. R., Fiscus, C. H., and Holts, D. B. (1973). Stomach contents of porpoise, *Stenella* spp., and yellowfin tuna, *Thunnus albacares,* in mixed species aggregations. *Fishery Bull.* **71,** 1077–1092.

Perrin, W. F., Sloan, P. A., and Henderson, J. R. (1979). Taxonomic status of the "southwestern stocks" of spinner dolphin *S. longirostris* and spotted dolphin *S. attenuata*. *Rep. int. Whal. Commn* **29,** 175–184.

Price, E. W. (1932). The trematode parasites of marine mammals. *Proc. U.S. Natl. Mus.,* pp. 1–68.

Read, A. J., Van Waerebeek, K., Reyes, J. C., McKinnon, J. S., and Gaskin, D. E. (1985). Preliminary observations on the catch of small cetaceans in Peruvian coastal waters. *Sixth Biennial Conf. Biol. Mar. Mamm.,* Nov. 22–26, Vancouver, Canada, Abstr.

Read, A. J., Van Waerebeek, K., Reyes, J. C., McKinnon, J. S., and Lehman, L. C. (1988). The exploitation of small cetaceans in coastal Peru. *Biol. Conservation,* 53–70.

Ross, G. J. B. (1977). The taxonomy of bottlenose dolphins, *Tursiops* species, in South African waters. *Ann. Cape Prov. Mus. (Nat. Hist.)* **11,** 135–194.

Ross, G. J. B. (1984). The smaller cetaceans of the south east coast of southern Africa. *Ann Cape Prov. Mus. (Nat. Hist)* **15,** 173–410.

Schnell, G. D., Douglas, M. E., and Hough, D. J. (1985). Sexual dimorphism in spotted dolphins (*Stenella attenuata*) in the eastern tropical Pacific Ocean. *Mar. Mamm. Sci.* **1,** 1–14.

Skrjabin, K. I. (1961). "Key to Parasitic Nematodes. Vol. III. Strongylata." Israel Program for Scientific Translation, Jerusalem, 759 pp.

Tomilin, A. G. (1957). "Mammals of the USSR and Adjacent Countries. Vol. IX. Cetacea." [Transl. from Russian.] Israel Program for Scientific Translation, Jerusalem 1967. 717 pp.

Walker, W. A. (1981). Geographical variation in morphology and biology of bottlenose dolphins (*Tursiops*) in the eastern North Pacific. *NOAA, NMFS, SWFC Admin. Rep.* LJ-81-03C. 54 pp.

Yamaguti, S. (1958). *Systema helminthum.* "The Digenetic Trematodes of Vertebrates," Vol. I. Interscience Publishers, New York. 979 pp.

Zuta, S., Tsukayama, I., and Villanueva, R. (1983). El ambiente marino y las fluctuaciones de las principales poblaciones pelagicas de la costa peruana. "Proceedings of the Expert Consultation to Examine Changes in Abundance and Species Composition of Neritic Fish Resources, San Jose, Costa Rica, 18–29 April 1983" (G. D. Sharp and J. Csirke, eds.), pp. 179–254. FAO Fisheries Report No. 291, Vol. 2.

Bottlenose Dolphin Mortality Patterns in the Indian/Banana River System of Florida

8

Sandra L. Hersh
Division of Biology and Living Resources
Rosenstiel School of Marine and Atmospheric Science
University of Miami
Miami, Florida 33149
Present address: National Marine Fisheries Service,
Southeast Fisheries Center, 75 Virginia Beach Drive,
Miami, Florida 33149

Daniel K. Odell
Division of Biology and Living Resources
Rosenstiel School of Marine and Atmospheric Science
University of Miami
Miami, Florida 33149, and
Sea World of Florida
Orlando, Florida 32821

Edward D. Asper
Sea World of Florida
Orlando, Florida 32821

INTRODUCTION

The bottlenose dolphin, *Tursiops truncatus,* is an abundant and highly visible inhabitant of the Indian/Banana River (IBR) complex on the east coast of Florida (Leatherwood, 1979). Because this area is a major source of dolphins for research and public display, information on population parameters is needed for effective management. One source of such information is beached carcasses. In areas where a high proportion of beached dolphins is reported and examined, analysis of yearly, seasonal, and age- and sex-specific mortality patterns is possible. Results may then be evaluated in light of information on population size and structure from other sources (e.g., aerial surveys) to estimate mortality rates.

The study area (Fig. 1) is bounded on the north by Ponce de Leon Inlet (29° 4′ 30″ N) and on the south by St. Lucie Inlet (27° 10′ 0″ N). It covers approximately 225 km of coastline and includes both the Indian and Banana rivers as well as Mosquito Lagoon. The area is not truly riverine but rather an estuarine system protected from the open ocean by barrier islands on the east side and subject to seasonal freshwater runoff.

The protected waters of the IBR complex are open to the sea at four locations: Ponce de Leon, Sebastian, Fort Pierce, and St. Lucie inlets. A fifth opening, the Cape Canaveral locks, has not been known to permit passage of dolphins. The study area is relatively shallow (usually <3 m except in dredged channels) with numerous low spoil islands and sand bars. Water temperature varies from about 8° to 33°C and salinity from 25 to 33 ppt (D. K. Odell and E. D. Asper, unpublished data).

Sea World of Florida has been salvaging cetacean carcasses in the IBR area since 1974, in cooperation with various federal, state, county, and private agencies. Since most of the waterfront

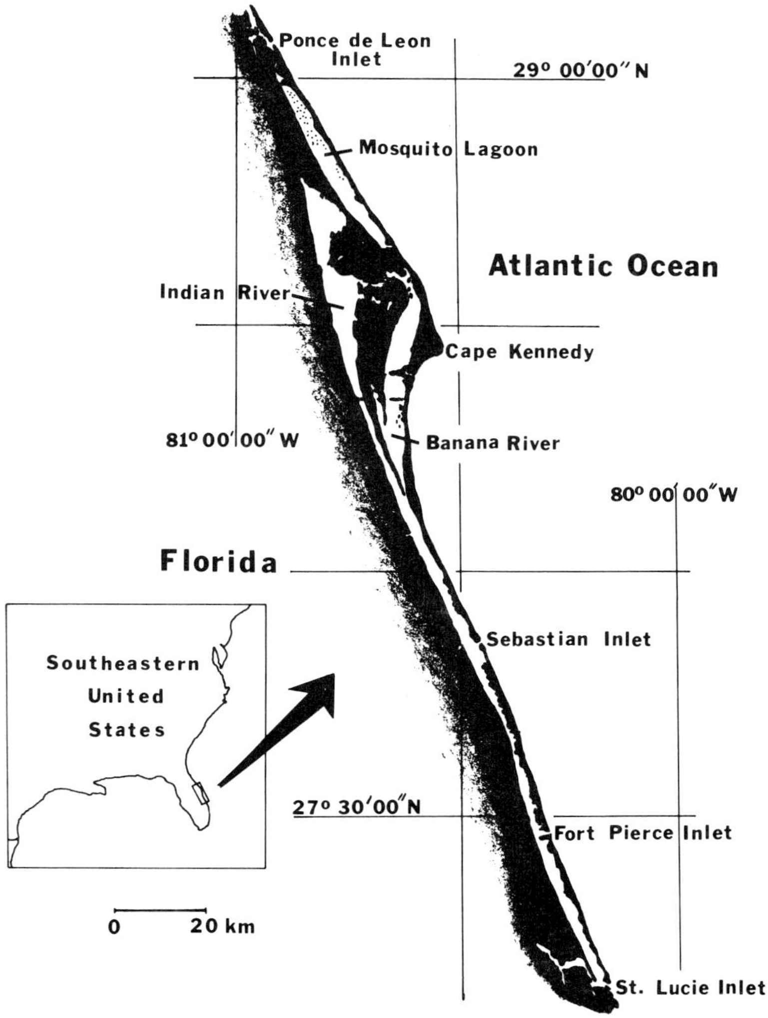

Figure 1 The Indian/Banana River study area, on the east coast of Florida.

along the IBR is within range of easy observation by either shoreside residents or local boaters, we feel confident that nearly every stranding event during the study period was reported and nearly every carcass examined.

METHODS

Data on sex, length, age, and date of discovery were collected for 170 beached bottlenose dolphin carcasses from the IBR complex from January 1976 through December 1983. Carcasses found on the ocean side of the barrier islands were not included in the sample because of evidence indicating that dolphins in the IBR complex may comprise a separate population from those in the Atlantic Ocean (Duffield, 1982).

Sex was determined by the position of the genital slit as well as internal examination of the gonads. Age was estimated from the number of growth layer groups (GLGs) in the dentine. Hui (1980) concluded that these layers were deposited annually (also see Hohn, Chapter 33, this volume). Teeth were sectioned, etched, and prepared for reading following the methodology of Hohn (1980). GLGs were counted in etched half-teeth using a dissecting microscope at a power magnification of 12. Each tooth was read three times by the same person (S. L. H.), and the mean count was used in statistical analysis.

Age categories were assigned using the following definitions. All specimens showing little or no dentine deposition beyond the neonatal line, which is deposited at birth, were classified as newborns or perinates. Those animals showing more dentine deposition than a newborn but less than 2 complete GLGs were considered yearlings (<2 years old). The third category consisted of animals with 2–5 GLGs, and the fourth of animals with more than 5 and up to 10 GLGs. The last category consisted of dolphins with more than 10 GLGs. These categories were intended to represent different life stages. Young bottlenose dolphins may be weaned at about 12–18 months, hence the division between yearlings and older animals. Females may become sexually mature as early as 6–7 years while males mature at about 10 or 12, hence the divisions among the remaining categories. These assumptions are based on Harrison *et al.* (1972), Sergeant *et al.* (1973), Odell and Asper (1982), and Schneyer (1983). Mead and Potter (Chapter 9, this volume) found females attain sexual maturity at 7–14 GLGs, with a mean of 10.64.

In some older animals, exact age could not be determined owing to occlusion of the pulp cavity. In these cases a minimum age estimate was used to place the animal in an age category. In a small number of cases, no tooth was available for sectioning, but data on the size, weight, or condition of the gonads, presence of a fetus or milk, and total length of the specimen were. In such cases the specimens were assigned to one of the five age categories by using an age–length curve derived from bottlenose dolphins stranded in the southeastern United States (Hersh, 1987).

An alternative scheme for categorizing specimens was based on sexual maturity. The first two divisions, perinate and yearling, were defined as in the first scheme. All remaining animals, however, were classified either as sexually immature juveniles or sexually mature adults. If no information from gonadal examination was available, females were assumed to be sexually mature at age 6, and males at ages greater than 10. Both of these schemes, the first based strictly on age and the second on sexual maturity, were used in the analysis of mortality patterns.

Stranding dates were grouped by quarters roughly equated to seasons (January, February, and March = winter; April, May, and June = spring; July, August, and September = summer; October, November, and December = fall) and by year. Associations between year and season of stranding, age or sexual maturity category, and sex were tested using chi-square and log-likelihood techniques. Multidimensional contingency tables were analyzed using log-linear modeling, a technique also used by O'Shea *et al.* (1985) in the analysis of manatee, *Trichechus manatus*, mortality patterns in Florida. These methods are nonparametric, independent of sample size, and assume random sampling. SPSS programs FREQUENCIES, CROSSTABS, LOGLINEAR, and HILOGLINEAR were employed.

RESULTS AND DISCUSSION

The sample for determining annual mortality rates of the IBR bottlenose dolphin population consisted of 156 beached carcasses found from January 1976 through December 1983. Specimens found in Mosquito Lagoon were not included in the analysis because available estimates of population size apply only to the Indian and Banana Rivers (Leatherwood, 1979, 1982; Leatherwood and Show, 1980; Thompson, 1981). Data from Mosquito lagoon were included, however, in analyses of age- and sex-specific mortality patterns and seasonal patterns of beaching. Several beachings that occurred from 1973 to 1976 and from 1983 to 1985 were also considered in the latter analyses. There was no reason to suspect that the inclusion of these data would produce any bias in sex ratios, age category distributions, or seasonal beaching frequencies.

Annual Mortality Rate for the Population

The mean number of beached dolphins recovered annually from the IBR complex during the designated 8-year period was 19.5 (Fig. 2). Thompson (1981) reported three monthly estimates of bottlenose dolphin abundance (and their associated

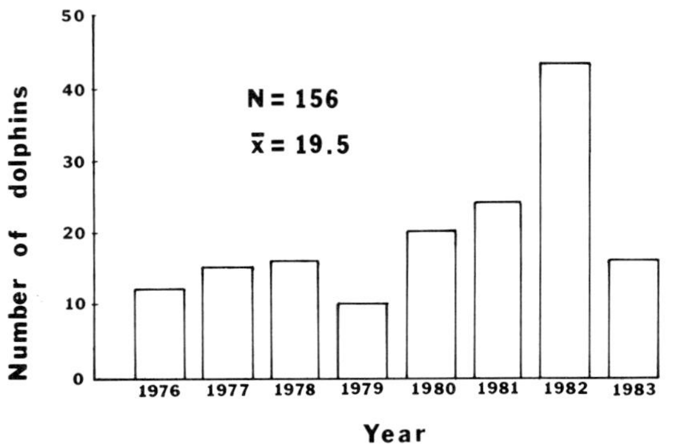

Figure 2 Numbers of beached bottlenose dolphins recorded in the Indian/Banana River area (excluding Mosquito Lagoon) from 1976 through 1983.

95% confidence intervals) for the IBR study area: 206 ± 170 (May 1980), 435 ± 172 (August 1980), and 202 ± 106 (November 1980). Using the mean of these three estimates (281) as the total population size and the average number of beached dolphins per year (19.5), one obtains a minimum annual mortality rate of 6.9%. However, since the confidence intervals of the population estimates are large, mortality estimates based on these figures could theoretically range from 3.2 to 54.2%.

Scott (Chapter 36, this volume) suggested that the August 1980 estimate included transient animals, and that the size of the resident stock was better represented by an average of the other four available estimates (Table 1). If this mean (211) is used as the total population size, the minimum annual mortality rate is 9.2%. It must be kept in mind, however, that the total number of beached animals includes all carcasses found in the area, whether they were transient (assuming Scott is right and there are transient animals in this area) or resident dolphins. No corrections were made for animals below the water surface during the aerial surveys, but the IBR area is relatively shallow so that any downward bias in population size estimates caused by failure to count submerged animals may be considered minor.

Table 1
Estimates of Abundance for Bottlenose Dolphins in the Indian/Banana River Area, Florida[a]

Date	Density (dolphins/km^2)	Abundance (95% CI)
November 1979	0.274	222 (±34)
January 1980	0.264	214 (±42)
May 1980	0.255	206 (±170)
August 1980	0.539	435 (±172)
November 1980	0.251	202 (±106)

[a] After Scott (Chapter 36, this volume).

Yearly Variations in Mortality

Before proceeding to a discussion of the differences in mortality rates among sex and age groups, as well as seasonal beaching frequencies, it is appropriate first to examine yearly variations in the data and to determine any significant interactions between year of beaching and the three other factors under investigation. A comparison of the yearly totals of beached carcasses to an equiprobability model shows that both 1979 and 1982 deviated from the model ($p < .001$). Neither the significantly lower mortality that occurred in 1979 nor the significantly higher mortality in 1982, as compared to other years of the study, appears to be correlated with air temperature, as these years had neither the coldest nor the warmest temperatures recorded during the 8-year study (Table 2).

A chi-square contingency table analysis of sex and year of beaching showed no significant association ($p > .10$). This suggests that the factors affecting yearly variations in mortality act equally on both sexes, assuming a sex ratio of unity in the population. A similar analysis of age category and year also showed no significant association ($p > .25$), implying that factors affecting yearly variations also act on age category in an unbiased manner.

There was, however, a significant association between year and season, indicating that seasonal influences on mortality vary from year to year. In fact, in the year with the highest mortality rate (1982) almost 50% of the beachings occurred in the spring quarter (April, May, and June). In the year with the lowest mortality (1979), beachings were evenly distributed throughout the year (Fig. 3). It must be noted here that the artificially constructed seasonal quarters used in these analyses may not correspond to actual seasons as measured by

Table 2
Lowest and Highest Air Temperatures (°C) and Month of Occurrence for Jacksonville and Miami, Florida, from 1976 to 1983[a]

	Low			High		
Year	J'ville	Miami	Month	J'ville	Miami	Month
1976	9.3	18.2	Jan	27.7	28.4	July
1977	6.7	16.2	Jan	28.2	28.7	July
1978	8.6	17.3	Feb	27.7	28.1	July
1979	8.8	18.3	Jan	27.8	28.4	July
1980	10.7	17.8	Feb	28.7	28.2	July/Aug
1981	8.1	15.4	Jan	29.1	29.4	July
1982	12.2	19.9	Jan	28.1	29.1	July
1983	9.4	19.6	Jan	28.1	29.4	July

[a] Information derived from *Monthly Climatic Data for the World*, NOAA.

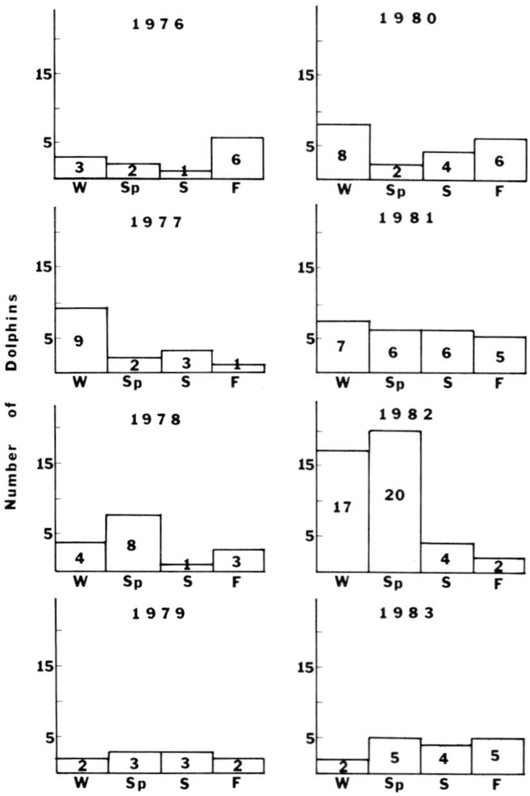

Figure 3 Seasonal distribution of beached bottlenose dolphins in the Indian/Banana River area (including Mosquito Lagoon) from 1976 through 1983, by year. Winter (W) = January, February, and March; spring (Sp) = April, May, and June; summer (S) = July, August, and September; fall (F) = October, November, and December.

environmental parameters. However, the winter quarter contains both the coldest air and water temperatures of the study period and the summer quarter contains both the warmest air and water temperatures of the study period (D. K. Odell and E. D. Asper, unpublished data).

A log-likelihood heterogeneity analysis performed on the eight separate seasonal distributions of the sample indicated only 1977 ($p < .05$) and 1982 ($p < .001$) to have been significantly different from a uniform distribution. The coolest winter air temperatures of the study period occurred in January 1977. Along the east coast of Florida temperatures dropped to 6.7°C in Jacksonville and 16.2°C in Miami. During January, February, and March of 1977, 60% of the year's beachings took place. In 1982, 74% of the beachings occurred in March, April, and May. The cause of this mortality was apparently localized in time, yet the peak of the beachings did not occur during the coldest months. When 1977 and 1982 are deleted from the 8-year sample, the overall seasonal distribution of the remaining six years of data shows no deviation from a uniform distribution ($p > .75$).

It would appear, then, that under normal circumstances mortality is relatively constant year-round. Temperatures in the study area are usually mild throughout the year. Bottlenose dolphins at higher latitudes are thought to migrate to warmer areas during the coldest months (Shane *et al.*, 1986). Whether such migrations are a response to cooler temperatures, changes in prey distributions, or other factors is not known.

Differences in Mortality by Sex and Age

Of 158 bottlenose dolphin carcasses recovered from the IBR system, 85 were males and 73 were females, indicating a sex ratio of 1.2 males per female. Comparison to an equiprobability model showed this ratio to be not significantly different from unity ($p = .73$). This implies that mortality rates are the same in males and females, assuming a one-to-one sex ratio in the population. In a sample of 157 bottlenose dolphins taken in a net fishery at Cape Hatteras, North Carolina, there were 82 males and 75 females (True, 1890), which tends to support this assumption for the population which migrates along the Atlantic coast. Though some local segregation by sex or age may occur within coastal bottlenose dolphin populations (True, 1890; Wells, 1978, 1986), it appears that, when considering an entire population, numbers of males and females are about equal.

The frequency distributions of carcasses by age and sexual maturity category are shown in Figs. 4 and 5. Of 161 known-age carcasses, 18 (11.2%) were considered perinates. Leatherwood (1979) and Asper and Odell (1980) both reported 8.1% calves of the year for the IBR population based on aerial surveys and mark–resighting studies, respectively. Odell and Asper (1982) reported 7.1% calves for this area based on mark–resightings.

If mortality rates were the same among all age categories, one would expect to find similar percentages of calves among beached animals and in population surveys. However, calves comprised a larger percentage of the beached animals. This appears to indicate that mortality in newborns is higher than expected given an equiprobability assumption. Given that some of the animals identified in either aerial surveys or mark–resighting studies as calves would have been classified, according to the scheme used in this study, as yearlings, the difference between the proportion of carcasses and the proportion of live sightings which are calves is probably greater still. This assumes the proportion of live calves in the population has not been underestimated.

If reporting of carcasses were biased by age group, one would expect perinates to be most often overlooked because of their smaller size, resulting in a downward bias in the estimate of perinate mortality. High mortality rates among young age groups conform to the typical U-shaped mortality curves seen in many other mammalian populations (Ralls *et al.*, 1980).

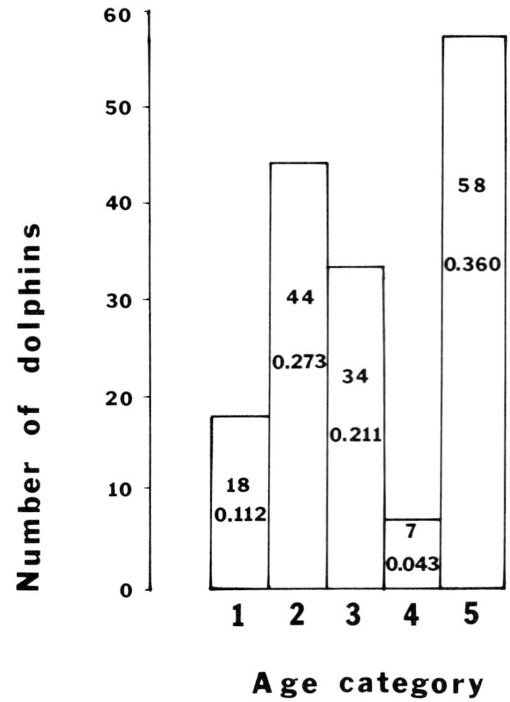

Figure 4 Age category distribution of 161 beached bottlenose dolphins from the Indian/Banana River area (including Mosquito Lagoon) from 1976 through 1983. Categories are as follows: 1, perinates; 2, yearlings (<2 years or GLGs); 3, 2- to 5-year-olds (2–5 GLGs); 4, 5- to 10-year-olds (5–10 GLGs); 5, animals older than 10 years (>10 GLGs). Numbers within the bars are the actual number of animals (top) and the proportion of the sample which falls in each category (bottom).

Yearlings comprised 27.3% of the beachings and 2- to 5-year-old animals, 21.1%. Five- to 10-year-olds comprised 4.3% while animals older than 10 comprised 36% of the total number of beached carcasses. Without knowledge of the actual age class distribution of the population, these figures cannot be interpreted.

Seasonal Variations in Mortality

The seasonal distribution of 170 beached bottlenose dolphins from the study area is shown in

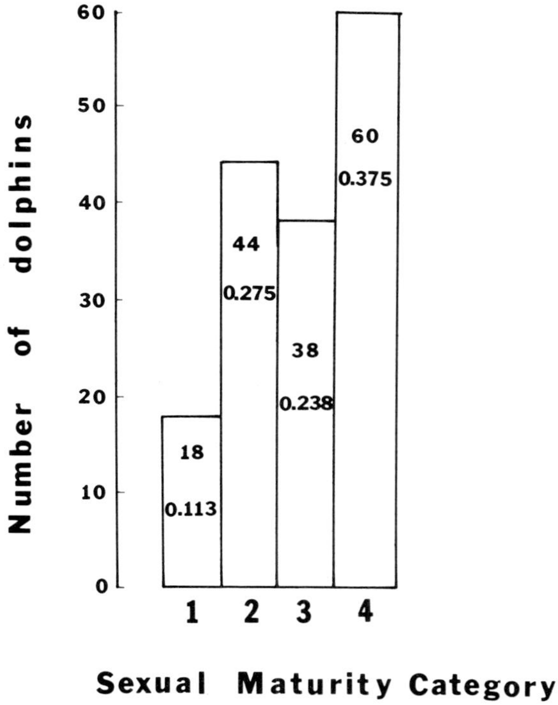

Figure 5 Sexual maturity category distribution of 160 beached bottlenose dolphins from the Indian/Banana River area (including Mosquito Lagoon) from 1976 through 1983. Sexual maturity categories are as follows: 1, perinates; 2, yearlings (<2 years or GLGs); 3, subadults (sexually immature); 4, adults (sexually mature). Numbers within the bars are explained in the legend to Fig. 4.

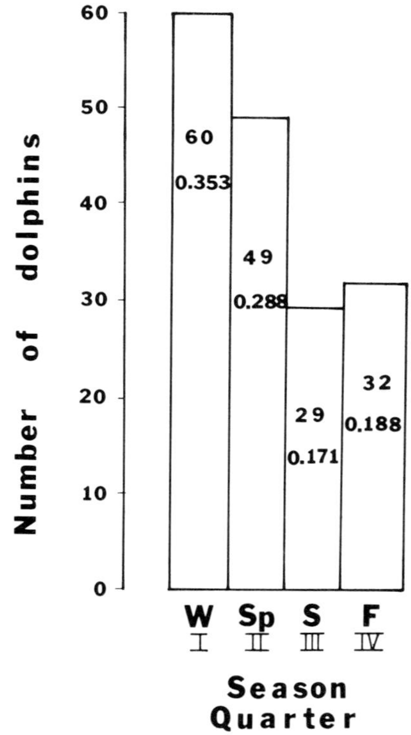

Figure 6 Quarterly (seasonal) distribution of 170 beached bottlenose dolphins from the Indian/Banana River area (including Mosquito Lagoon) from 1976 through 1983. I/W = January, February, and March; II/Sp = April, May, and June; III/S = July, August, and September; IV/F = October, November, and December. Numbers within the bars are explained in the legend to Fig. 4.

Fig. 6. The highest number of beachings (60) occurred in the winter quarter and the lowest (29) in the summer quarter. A comparison to an equiprobability model showed a significant deviation ($p = .002$) of the observed frequencies. Although the seasonal distribution of all 8 years of data proved to be different from an equiprobability distribution, a heterogeneity analysis pinpointed the years 1982 and 1977 as being the source of the deviation. Possible reasons for these patterns are discussed above. The remaining 6 years showed no significant deviation from a uniform distribution (including 1979, the year with the lowest mortality). This indicates that the high mortality seen in 1977 and 1982 were the cause of the significant deviation of the 8-year seasonal mortality pattern from an equiprobability model.

Interactions among Age, Sex, and Season

Hierarchical log-linear analysis indicated that the most parsimonious model which adequately describes the frequency distributions of the data based on age category, sex, and season involves no third-order interactions between the variables ($p = .12$). However, at least some second-order terms are necessary ($p = .01$) to describe the interaction of sex and age category and season and age category. Removal of these effects from the model causes a significant ($p < .05$) deviation from the observed frequencies. This model, including only two of the three possible interaction terms, shows no significant deviation ($p = .58$) from the actual frequency distributions.

Analysis of a chi-square contingency table indicated significant associations between sex and age category ($p = .02$). Only 23.7% of 17 perinate carcasses were males. This would appear to indicate a greater hardiness of male perinates, assuming the sex ratio at birth is 1 to 1.

The opposite pattern is observed for the yearling category, in which, of 44 beached animals, 68.2% were males. At this point in their lives, male bottlenose dolphins may become more vulnerable to various causes of mortality than females, thus accounting for the shift. Another possible explanation lies in the criteria for classifying some of these sample animals. Since the age of some animals was inferred from an age–length curve, it is possible that differences in the growth rates of newborn male and female bottlenose dolphins are affecting the results. Although Hersh (1987) found little or no sexual dimorphism in the lengths of IBR dolphins over the age of 2, it is possible that the mean length at birth differs for male and female bottlenose dolphins or that the sexes exhibit different growth rates at an early age. Both 2- to 5-year-olds and animals over 10 appear evenly distributed between the sexes (56.3/43.8% and 47.2/52.8% males to females, respectively). The sex ratio of 5- to 10-year-olds appears uneven (66.7/33.3% males to females). However, this could be an artifact of the very small sample size of this age category ($n = 6$).

Chi-square contingency table analysis of season and age category resulted in no significant association ($p = .056$). However, log-linear analysis implied a significant relationship between the two variables. No perinates were found in the fall, but 50% of all perinates were found in the winter. This may indicate that no births occur during October, November, and December, that all perinates survived during this period, which is highly unlikely, or that the carcasses simply were not found.

Assuming a calving season evenly distributed over winter, spring, and summer, the high number of perinate carcasses observed in the winter quarter may indicate higher perinate mortality during this season. Alternatively, if calving increases during the winter (probably in March), then the mortality rate during this season may be no different than at other times.

Of all yearling deaths, 40.9% occurred in the spring. However, the greatest number of deaths for all remaining age categories (35.3% of the 2- to 5-year-olds, 71.4% of the 5- to 10-year-olds, and 39.7% of animals older than 10) occurred in winter. Since all but two yearly seasonal distributions showed no deviation from a uniform distribution, these high seasonal frequencies are most likely a result of the winter 1977 and spring 1982 mortality peaks.

When sexual maturity category was used in place of age category, log-linear analysis indicated a model analogous to the one chosen in the previous analysis as that which best describes the data. The third-order effects were considered negligible ($p = .15$), as was the second-order effect of sex and season ($p = .11$). A model containing only the main effects of the three parameters plus second-order effects of sexual maturity category and sex, and sexual maturity category and season, is sufficient to describe the frequency distributions of the data ($p = .38$). Chi-square analyses using sexual maturity category in place of age category indicated significant associations between sex and sexual maturity category ($p < .01$) but not between season and sexual maturity category ($p = .09$).

The use of sexual maturity category in place of age category effectively produces a bias. Although the results of the statistical analyses using sexual maturity category were analogous to those using age category, the number of male subadults is increased relative to female subadults as males are believed to become sexually mature at a later age. For instance, in a chi-square contingency table analysis of sex and sexual maturity category, males comprised 61.1% of the subadult category while females comprised 38.9%. Of the sexually mature adults, 54.5% were females while 45.5% were males. However, because females spend fewer years as subadults than do males, they would be expected to show up in proportionately fewer beachings of that category. Therefore, the discrepancies for these life stages may not be due to differences in mortality rates between the sexes but rather to the differences in age at sexual maturity. However, more data are needed on the relative mortality rates of 5- to 10-year-old animals (when females are becoming mature but males are still sexually immature) to determine if this is so.

SUMMARY

1. The mean annual number of beached carcasses in the IBR area (excluding Mosquito Lagoon) from 1976 through 1983 was 19.5. This implies annual mortality rates of 0.069 to 0.092 (~7–9%), depending on the population estimate used.
2. The highest overall mortality occurred in 1982, the lowest in 1979. These extremes in the samples do not appear to be correlated with temperature.
3. Mortality appears to be seasonally uniform except under such unusual circumstances as extremely cold winter temperature.
4. Neither sex nor age category was significantly associated with year of beaching, suggesting that whatever factors affect yearly variations in mortality act equally on both sexes and all age categories.
5. The sex ratio of recovered carcasses was not significantly different from unity. Assuming a one-to-one sex ratio in the population, this suggests that sex ratios of beached carcasses may accurately reflect those in the population and that mortality rates are similar for the two sexes.
6. Perinates comprised 11.2% of all beached animals. Comparison of this figure to the estimated proportion of live calves in the population suggests a relatively high perinate mortality rate.
7. Males comprised only 23.7% of perinate carcasses, indicating a difference in perinate mortality between the sexes or a faster rate of growth in newborn males than in newborn females and thus the males' premature classification as yearlings, according to our criteria.
8. Males comprised 68.2% of the yearling carcasses. Two- to 5-year-olds and animals older than 10 were evenly distributed between the sexes.
9. Half of all stranded animals were found in the winter. For yearlings, peak mortality occurred in the spring.
10. Under certain conditions the examination of beached carcasses may provide information on overall as well as age- and sex-specific mortality rates. The limitations of using information collected in this manner to make inferences about the population under study are manyfold. Primarily, carcass salvage effort must be intensive and consistent, with no seasonal fluctuations or other sampling bias. Also, information on population size and age/sex composition must be available from other sources.

ACKNOWLEDGMENTS

We thank the officers of the Florida Marine Patrol District 5, numerous Sea World of Florida employees (especially D. Beusse, J. D. Lynch, R. Stevens, S. Searles, G. Young, R. Friday, P. Sassic, G. Case, J. Marsh, and J. Provancha), the staffs of Canaveral National Seashore and Merritt Island National Wildlife Refuge, local law enforcement agencies, and the many unknown individuals who reported beached dolphins. Preliminary findings based on these data were presented at a NMFS stock assessment workshop in Miami, Florida (contributing document SAW/84/MMT/20). This study was funded, in part, by NMFS Contracts NA79-GA-C-00027, NA80-GA-C-00063, NA80-GA-C-00032, and NA81-GA-C-00026 and Cooperative Institute for Marine and Atmospheric Science Contract NA80-RAH-00001. It is based on a doctoral dissertation by Hersh (1987). This is Sea World of Florida Technical Report 8715-F.

REFERENCES

Asper, E. D., and Odell, D. K. (1980). Bottlenose dolphin local herd monitoring: capture, marking, collection of biological data, and follow-up observations of marked animals. *Final Rep. Natl. Mar. Fish. Serv.*, Contract No. NA79-GA-C-00027. 163 pp.

Duffield, D. A. (1982). *Tursiops truncatus* genetics studies: Indian River 1980–1981. *Final Rep. Natl. Mar. Fish. Serv.*, Contract No. NA80-GA-C-00063, pp. 313–325.

Harrison, R. J., Brownell, R. L., and Boice, R. C. (1972). Reproduction and gonadal appearances in some odontocetes. *In* "Functional Anatomy of Marine Mammals" (R. J. Harrison, ed.), Vol. I, pp. 361–429. Academic Press, London. 451 pp.

Hersh, S. L. (1987). Characterization and differentiation of bottlenose dolphin populations (genus *Tursiops*) in the southeastern U.S. based on mortality patterns and morphometrics. Ph.D. dissertation, University of Miami, Coral Gables, Florida. 213 pp.

Hohn, A. A. (1980). Age determination and age related factors in the teeth of coastal North Atlantic bottlenose dolphins, *Tursiops truncatus* (Cetacea, Odontoceti). M.S. thesis, University of Maryland, Baltimore, Maryland. 74 pp.

Hui, C. A. (1980). Variability of dentin deposits in *Tursiops truncatus*. *Can. J. Fish. Aquat. Sci.* **37,** 712–716.

Leatherwood, S. (1979). Aerial survey of the bottlenosed dolphin, *Tursiops truncatus,* and the West Indian manatee, *Trichechus manatus,* in the Indian and Banana Rivers, Florida. *Fish. Bull.* **77,** 47–89.

Leatherwood, S. (1982). Size of bottlenose dolphin population(s) in the Indian River, Florida. *Rep. int. Whal. Commn* **32,** 567–568.

Leatherwood, S., and Show, I. T. (1980). Development of systematic procedures for estimating sizes of populations of bottlenose dolphins and estimates of sizes of populations of bottlenose dolphins in three geographic areas with incidental observations on the densities of West Indian manatees and marine turtles. *Final Contract Rep. Natl. Mar. Fish. Serv.,* Contract No. NA79-GA-C-0038. 98 pp.

Odell, D. K., and Asper, E. D. (1982). Live capture, marking and resighting of bottlenose dolphins, *Tursiops truncatus. Final Rep. Natl. Mar. Fish. Serv.,* Contract No. NA80-GA-C-00063. 325 pp.

O'Shea, T. J., Beck, C. A., Bonde, R. K., Kochman, H. I., and Odell, D. K. (1985). An analysis of manatee mortality patterns in Florida, 1976–1981. *J. Wildl. Manage.* **49**(1), 1–11.

Ralls, K., Brownell, R. L., and Ballau, J. (1980). Differential mortality by sex and age in mammals, with specific reference to the sperm whale. *Rep. int. Whal. Commn* (Special Issue 2), 233–243.

Schneyer, A. L. (1983). Development and validation of radioimmunoassays for pituitary gonadotropic hormones and their application to reproductive endocrinology of the bottlenosed dolphin. Ph.D. dissertation, University of Miami, Coral Gables, Florida. 83 pp.

Sergeant, D. E., Caldwell, D. K., and Caldwell, M. C. (1973). Age, growth, and maturity of bottlenosed dolphins (*Tursiops truncatus*) from northeast Florida. *J. Fish. Res. Board Can.* **30,** 1009–1011.

Shane, S. H., Wells, R. S., and Würsig, B. (1986). Ecology, behavior and social organization of the bottlenose dolphin: A review. *Mar. Mamm. Sci.* **2**(1), 34–63.

Thompson, N. B. (1981). "Estimation of Abundance of *Tursiops truncatus* in the Indian–Banana River Complex, May, August, and November, 1980." Data Analysis Division, Natl. Mar. Fish. Serv., Southeast Fisheries Center. 30 pp.

True, F. W. (1890). Observations on the life history of the bottlenose porpoise. *Proc. U.S. Natl. Mus.* **13,** 197–203.

Wells, R. S. (1978). Home range characteristics and group composition of Atlantic bottlenosed dolphins (*Tursiops truncatus*) on the west coast of Florida. M.S. thesis, University of Florida, Gainesville, Florida. 91 pp.

Wells, R. S. (1986). Structural aspects of dolphin societies. Ph.D. dissertation, University of California, Santa Cruz, California. 234 pp.

Natural History of Bottlenose Dolphins Along the Central Atlantic Coast of the United States

9

James G. Mead
Charles W. Potter
Division of Mammals
Smithsonian Institution
Washington, D.C. 20560

INTRODUCTION

Although bottlenose dolphins, *Tursiops* spp., are distributed worldwide in tropical and temperate waters and are often displayed and studied in captivity, surprisingly little is known about their natural history in the wild. Since 1972 we have studied the coastal population of bottlenose dolphins that frequents the Cape Hatteras, North Carolina, area, principally by examining stranded specimens. It was from this area that Townsend (1914) removed five animals alive for display at the New York Aquarium, and a beach seine fishery for bottlenose dolphins operated in the location until 1928 (Mead, 1975). Mitchell (1975, p. 85) estimated the population available to that fishery in the late 1880s as 17,000. Kenney (Chapter 21, this volume) estimated the current bottlenose dolphin population (including coastal and offshore dolphins from Cape Hatteras north) off the northeast coast of the United States as 10,000–12,000.

The National Museum of Natural History [also known as the United States National Museum (USNM)], Smithsonian Institution, initiated the Marine Mammal Program (MMP) in 1972. We began by developing a stranding network to notify us of stranded cetaceans on the mid-Atlantic coast (South Carolina to Massachusetts). The network has functioned well and is the source of most of the data reported in this chapter. In 1975 the museum started the Scientific Event Alert Network (SEAN), which facilitated the investigation of cetacean strandings and aided in collection of data about pinnipeds, sea turtles, sea otters, and manatees as well. Unfortunately the biological side of SEAN was terminated in August 1982, and there was no further systematic reporting of data on strandings until September 1983 when SEAN was replaced by the Marine Mammal Events Program (MMEP). The MMP also collects data on both living and dead specimens of bottlenose dolphins from periodic aerial beach surveys and from incidental observations of live bottlenose dolphins from the shore.

TAXONOMY AND SYSTEMATICS

The genus *Tursiops* is polymorphic; at least 20 nominal species have been described. All systematic work done to date has involved relatively small samples and restricted geographic areas. Most regional studies have demonstrated the presence of at least two discrete forms that are morphologically separable as adults (Norris and Prescott, 1961; Walker, 1981; Ross, 1977; Van Waerebeek *et al.*, Chapter 7, this volume). On the Pacific coast of North America these forms have been regarded either as separate species, with the coastal form called *Tursiops gilli* and the offshore form *T. nuuanu* (Norris and Prescott, 1961), or as different "ecotypes" of the same species (Walker, 1981). Ross (1977) regarded the two forms he identified off southern Africa as different species, *T. truncatus* (offshore) and *T. aduncus* (coastal) (but see Ross and Cockcroft, Chapter 5, this volume). Hersh and Duffield (Chapter 6, this volume) have defined morphological and biochemical differences between the coastal and offshore populations along the southeastern coast of the United States.

The bottlenose dolphins along the Atlantic coast of North America have always been referred to *T. truncatus*. Much of our knowledge of the bottlenose dolphin comes from studies of captive specimens taken from the coastal populations off Florida and along the Gulf of Mexico in the United States. The type specimen of *T. truncatus* (Montagu, 1821) is from the coastal population of *Tursiops* in northern Europe. The Northeast Atlantic bottlenose dolphin is a large animal with a modal body length of about 300 cm and a maximum length of about 400 cm (based on animals stranded on British coasts) (Fraser, 1974, p. 48). In our sample (N = 248) of the population from the Northwest Atlantic, the model length is 250–260 cm and the maximum length is 309 cm.

There is not enough information to compare quantitatively bottlenose dolphins from the Atlantic coast of North America with those from the type locality of *T. truncatus*. However, the differences observed in length–frequency between the two groups raise a series of questions: Are there two or more species of bottlenose dolphin in the North Atlantic? Are there two species or ecotypes off the Atlantic coast of North America? Is the North American form identical to *T. truncatus?*

There are reasons to believe that coastal and offshore populations are present along the Atlantic coast of North America. In summer, bottlenose dolphins can be seen from the beach with the naked eye from Florida to New Jersey. In winter, groups can be seen from Florida to Cape Hatteras, but usually not north of Hatteras. About 25 km offshore to the continental shelf edge, more groups of bottlenose dolphins are found. The geographic and temporal distribution of these offshore groups is discussed by Kenny (Chapter 21, this volume).

We hypothesize that the coastal population moves south from Cape Hatteras in winter, in response to the incursion of the cold Labrador Current, but that the offshore population remains in the domain of the Gulf Stream and is little affected by seasonal environmental changes. This offshore population, then, is likely the main source of the few winter strandings north of Cape Hatteras and the source of virtually all strandings from Cape Cod north.

Different populations of animals may be distinguishable by differences in their parasite burdens, with the parasite species serving as natural tags (McKenzie, 1983; Walker *et al.*, 1984; Van Waerebeek *et al.*, Chapter 7, this volume). For example, cetaceans commonly carry cysts of phyllobothriid cestodes, *Phyllobothrium* and *Monorhygma*, embedded in their tissues (Dailey and Brownell, 1972). The cysts can mature only if they are eaten by another animal (Williams, 1968), so once they are embedded in cetacean tissue, they remain there until the death of their host. Similarly, the brauninid trematode, *Braunina cordiformis*, encysts on the walls of dolphin stomachs. Both the phyllobothriid and brauninid parasites are ingested with food. Therefore, if the different populations of cetaceans eat different foods, they may well have different parasite burdens.

The results of a comparison of parasite burdens are shown in Table 1. We dissected five female bottlenose dolphins, 238–289 cm long, captured at sea in the North Atlantic in 1967 and 1969 on cruises led by Dr. Edward Mitchell and compared them with the sample of stranded bottlenose dolphins from South Carolina to New Jersey examined by the MMP from 1972 to 1986. From this evidence we suggest that the coastal form of the bottlenose dolphin usually carries *Braunina* but not *Phyllobothrium* or *Monorhygma* and that the offshore form usually carries *Monorhygma* and/or *Phyllobothrium* but not *Braunina*.

DISTRIBUTION

Coastal

In the western North Atlantic, the coastal form of the bottlenose dolphin is distributed from Florida to New Jersey. Bottlenose dolphins that strand north of the New York Bight are considered strays, probably from the offshore population. Strandings have occurred at Salisbury, New Brunswick (USNM 300190; Sergeant and Fisher, 1957, p. 93), and Milford, Nova Scotia (Sergeant *et al.*, 1970, p. 1912).

In Maine, bottlenose dolphins have stranded at Sullivan Harbor (Mairs and Scattergood, 1958, p. 78) and have been sighted at Harpswell (Packard, 1958, p. 9). The USNM has a skull (USNM 13293) from Portland; there are no data on the manner of collection.

Strandings are more frequent in Massachusetts. Little and Andrews (1982, p. 31) reported three strandings on Nantucket between 1964 and 1970. In the MMP files there is a drawing made by Dr. Eugenie Clark of a 231-cm male that stranded at Sagamore Highlands, Cape Cod Bay, 6 December 1947 (see Waters and Rivard, 1962, p. 118). There are records of three specimens in the Museum of Comparative Zoology (MCZ) (Harvard): a mounted specimen (MCZ 6724) from Yarmouth; the skeleton (MCZ 48565) of a stranded animal from Penzance Point, Buzzards Bay; and an unnumbered skull, probably from Massachusetts, received from the Boston Society of Natural History. MMEP received reports of two Massachusetts strandings, a 279-cm male at Eastham on 20 November 1984 and an approximately 350-cm animal at Truro in April 1985.

The USNM has a partial skeleton (USNM 49557) collected at Fort Adams, Rhode Island, 13 December 1899. Cronan and Brooks (1962, p. 79) stated that a bottlenose dolphin was "observed in Newport at an unknown date." MMEP received a report of a 265-cm female stranded at Warren, Rhode Island, 16 August 1983. There are no records from Connecticut.

The collections of the USNM (three specimens), the New York State Museum in Albany (four specimens; Connor, 1971, p. 39), the MCZ (one specimen), and the American Museum of Natural History (AMNH, New York) (one specimen) contain specimens of bottlenose dolphins from New York. SEAN received reports of two strandings in New York, a 206-cm male on Breezy Point (Gateway National Recreation Area), 1 February

Table 1
Incidence of Strigeid Trematode, *Braunina cordiformis*, and Cestode Cysticerci, *Monorhygma* and *Phyllobothrium*, in Tissues of North Atlantic Bottlenose Dolphins

	Coastal		Offshore	
Parasite	Present % (*N*)	Absent % (*N*)	Present % (*N*)	Absent % (*N*)
Braunina	84 (72)	16 (14)	0 (0)	100 (5)
Monorhygma	8 (4)	92 (48)	20 (1)	80 (4)
Phyllobothrium	6 (6)	94 (87)	60 (3)	40 (2)

1979, and a 262-cm female on Fire Island National Seashore, 3 April 1982. MMEP received a report of a 310-cm male that stranded on Fishers Island on 14 April 1984. Strandings in states south of New York are too numerous to mention individually (Table 2).

Recent (1979–1981) marine mammal surveys [Cetacean and Turtle Assessment Program (CETAP)] revealed that the coastal population of bottlenose dolphins occurs from Cape Lookout, North Carolina (the southern margin of the study area), to Cape Henlopen, Delaware (Winn, 1982, p. 186; Kenney, Chapter 21, this volume). Coastal bottlenose dolphins were once abundant around Cape May, New Jersey, and formed the basis for a fishery there (Mead, 1975).

Bottlenose dolphins are common in the mouth of Chesapeake Bay (Blaylock, 1984; J. G. Mead and C. W. Potter, unpublished data and personal observations) but rare in the inner reaches of the bay. We are aware of a total of 64 stranding records from the bay, including possible records from as far inland as Alexandria, Virginia [*The Evening Star* (Washington), 11 July 1883; Paradiso, 1969, p. 178], and Havre de Grace, Maryland (Paradiso, 1969, p. 178; Potter, 1984). There are five specimens from the Chesapeake Bay shores of Maryland in the USNM and one from Annapolis in the Academy of Natural Sciences in Philadelphia (ANSP) (No. 12903).

The bottlenose dolphin appears to be much less common in Delaware Bay, where it is represented by only four records: one specimen each at ANSP, AMNH, and USNM. A specimen from Bridgeton, New Jersey, was described by Ulmer (1980, p. 4), who also noted that a number of the museum specimens were from as far inland as the Delaware River.

Table 2
Total Records of Bottlenose Dolphins by State[a]

	Published records	Museum records	SEAN 1975–1982 records	MMEP 1983–1987 records
Maine	2	1	0	0
New Hampshire	0	0	0	0
Massachusetts	4	3	0	2
Rhode Island	2	1	0	0
Connecticut	0	0	0	0
New York	2	9	2	1
New Jersey	12	15	6	4
Delaware	0	2	1	1
Maryland	1	13	2	2
Virginia	8	24	47	58
District of Columbia	1	0	0	0
North Carolina	1	111	114	88
South Carolina	10	11	25	9
Georgia	26	3	28	24
Florida	44	107	216	165
Alabama	1	0	9	5
Mississippi	0	5	48	35
Louisiana	1	1	2	10
Texas	0	3	77	241

[a] The published records may include sightings. Published records have no corresponding museum records, but museum records may have been published; SEAN and MMEP records are not represented by either museum or published reports. A record represents one event (sighting or stranding); it may involve more than one animal.

Offshore

In addition to delineating range and seasonality of the coastal population, the CETAP data indicated an offshore population distributed along the 2000-m isobath from Cape Lookout to Georges Bank (Kenney, Chapter 21, this volume).

Seasonal

Our experience with the coastal population, including not only strandings but also shore-based and aerial sightings, indicates that bottlenose dolphins remain south of Cape Hatteras in the winter, then occupy the area north of the cape, along the coasts of Virginia, Maryland, and Delaware, in the spring and summer. We have too few fall observations to draw any conclusion. Generally, our own observations agree with those of Kenney (Chapter 21, this volume) and Blaylock (1984).

The offshore population follows a similar pattern of seasonal distribution. It reaches its maximum geographic extent in summer. Sightings north of Cape Hatteras decline in the winter (Winn, 1982, p. 180).

FOOD HABITS

True (1885) found common gurnard (*Prionotus carolinus*) in the stomachs of two adult females that he examined at Cape May, New Jersey. Townsend (1914) was told by fishermen that the Cape Hatteras animals fed primarily on squeteague or weakfish (*Cynoscion* sp.). Hoese (1971) observed bottlenose dolphins feeding on the shore of a tidal creek in Louisiana and determined independently that the most common fish species in the area were the silver perch (*Bairdiella chrysura*) and spot (*Leiostomus xanthurus*). Caldwell and Caldwell (1972) reported stargazers (*Astroscopus y-graecum*), sea catfish [*Arius* (*Galeichthys*) *felis*], croakers and drums (Sciaenidae), herring (Clupeidae), needlefish (Belonidae), eels, and squid as having been found in the stomachs of bottlenose dolphins. Odell (1975) reported that bottlenose dolphins steal king mackerel (*Scomberomorus cavalla*) from lines of commercial fishermen.

In his study of food habits of bottlenose dolphins in Texas, Gunter (1942) included anecdotal comments from fishermen as well as data from 34 stomachs that he examined, all from animals taken in a small-scale harpoon fishery. Fishermen claimed to have seen bottlenose dolphins take from their lines the following: Spanish mackerel (*Scomberomorus maculatus*), kingfish (*S. regalis*), tarpon [*Megalops* (*Tarpon*) *atlanticus*], sailfish (*Istiophorus americanus*), hammerhead shark (*Sphyrna zygaena*), speckled trout (*Cynoscion nebulosus*), roballo (*Centropomus undecimalis*), spotted eagle-ray [*Aetobatus* (*Stoasodon*) *narinari*], mullet (*Mugil* sp.), sea catfish (*Aris* [*Galeichthys*] *felis*), sheepshead (*Archosargus probatocephalus*), and flounder (*Paralichthys* sp.). Gunter (1942) identified the following species from intact fish found in 29 stomachs of adults (number in parentheses is the total number of individuals of that species that were found): *Mugil cephalus* (423), *Dorosoma cepedianum* (40), *Leiostomus xanthurus* (12), *Micropogon undulatus* (12), *Cynoscion arenarius* (13), *Sphoeroides marmoratus* (3), *Archosargus probatocephalus* (2), *Strongylura marina* (1), *Pogonias cromis* (1), *Cynoscion nebulosus* (1), and *Paralichthys lethostigmus* (1).

Although Kemp (1949) gave no details on the source of the Texas animals in his study, they were probably from the same fishery as those studied by Gunter (1942). Ten stomachs contained the following (number in parentheses is the number of animals that had remains in their stomachs): unidentified shrimp (1); unidentified crab (1); *Loligo* sp. (1); unidentified fish (1); mullet (*Mugil cephalus*) (4); flounder (4); ribbonfish (*Trichiurus lepturus*) (2); pinfish (*Lagodon rhomboides*) (1); unidentified parrotfish (Scaridae sp.) (1); lizard fish (*Synodus foetens*) (1); smooth puffer (*Lagocephalus laevigatus*) (1); and sea catfish (*Aris* [*Galeichthys*] *felis*) (1). For more data on the diet of bottlenose dolphins in the southeastern United States, see Barros and Odell (Chapter 16, this volume).

Our food-habit sample consists of 172 animals stranded on the coasts of North Carolina, Virginia, and Maryland during the period 1972–1980. These were known not to be infected with *Phyllobothrium* or *Monorhygma;* some were infected with *Braunina.* Sixty of the animals were recovered in the late winter and spring during 1976–78, when we made intensive surveys for stranded animals on the Outer Banks. During those years at least one of us was on the Outer Banks during the peak stranding season of December–April. Leatherwood *et al.* (1978) included nine of these animals in a brief report on food habits of bottlenose dolphins from the northeastern United States.

We examined the stomach contents of only one animal which had stranded well inside Chesapeake Bay. Five animals were collected just inside the mouth of the bay, at Fishermans Island on the north and from Cape Henry to Ocean View on the south. Only one of the animals was found on the Pamlico Sound side of the Outer Banks (GWO 008).

The few data that exist on the age or length at weaning suggest that weaning is gradual, with calves continuing to nurse facultatively for up to 1.5 years (McBride and Kritzler, 1951, p. 262; Caldwell and Caldwell, 1972, p. 35; Leatherwood *et al.*, 1978). The shortest animals that we examined which had remains of solid food in their stomachs were 144 and 155 cm long. We consider weaning to begin when calves are about 150 cm long.

Of the 172 animals examined, 90 were females and 82 were males; 117 were longer than 150 cm. Of these 117 specimens, 64 had identifiable food remains in their stomachs, 31 had empty stomachs, and 22 were not checked for food remains. It should be noted that a few of the empty stomachs were examined under adverse circumstances and may have had traces of remains that were missed.

Whenever possible stomachs were excised intact (the esophagus and duodenum were tied off), and brought back to be opened up in a laboratory situation. The contents were rinsed into white enamel pans. After removing large and dark-colored items, the residue was rinsed into a dark brown glass tray to sort out the otoliths. On the rare occasions when a stomach had to be opened on the beach, we tried to collect all of its contents; however, smaller trace elements may have been overlooked and not collected.

Food items were found in all three chambers of the stomach. All relatively fresh remains and most of the otoliths came from the forestomach. The main stomach often contained a few otoliths and fish bones. The pyloric stomach was usually empty but occasionally contained a few otoliths or squid beaks.

The frequency of occurrence of the four principal prey species is given in Table 3. Larger dolphins might be expected to take larger individuals of a particular prey species. We regressed mean modal estimated standard lengths of fish on total lengths of dolphins and obtained the following correlation coefficients: *Cynoscion* sp., r = 0.145; *Micropogon undulatus*, r = 0.239; *Leiostomus xanthurus*, r = 0.130. Thus, larger dolphins do not necessarily prey on larger fish.

The sample consisted entirely of stranded animals. There is reason to believe that most of these were sick or injured and that their stomach contents may not reflect the food habits of normal, healthy individuals. We separated the data for stomach contents of six animals that were captured incidentally in gill nets or haul seines and presumably were not debilitated or sick immediately before death:

USNM 504121, 249-cm female, taken in haul seine, 19 September 1974
- *Micropogon undulatus:* otoliths from 48 fish

USNM 504123, 232-cm female, taken in haul seine with 504121
- *Micropogon undulatus:* otoliths from 75 fish
- *Leiostomus xanthurus:* otoliths from 18 fish
- Unidentified hermit crabs: remains of about 11 individuals

USNM 504540, 206-cm female, taken in haul seine, 5 February 1977
- *Cynoscion* sp.: otoliths from 2 fish, estimated standard lengths (ESL) 5, 22 cm
- *Leiostomus xanthurus:* otoliths from 87 fish, ESL 4 cm; otoliths from 15 fish, ESL 9 cm

USNM 504551, 216-cm male, tangled in gill net, 11 February 1977
- *Cynoscion* sp.: otoliths from 13 fish, ESL 5–17 cm
- *Leiostomus xanthurus:* otoliths from 16 fish, ESL 4–10 cm
- *Bairdiella chrysura:* otoliths from 1 fish, ESL 21 cm

JGM 388, 242-cm female, tangled in gill net, 13 May 1978
- *Cynoscion* sp.: 2 partly digested fish, 25 and 60 cm standard length (SL); two pairs of otoliths from this species, ESL 9, 16 cm
- *Pomatomus saltatrix:* 2 fish identified by mandibles, ESL 20, 27 cm

Table 3
Fish in Bottlenose Dolphin Stomachs[a]

Prey species	Occurrence, abs./rel.	Number		Estimated modal length (cm)			Linear regression of greatest otolith length (mm) on standard body length (cm)			
		Mean	S.D.	Mean	S.D.	Range	SL =	range	r	n
Cynoscion cf. *C. regalis*	33/49	8.0	8.1	20.2	7.6	3–60	$2.31x - 8.9$	25–60	0.99	7
Micropogon undulatus	23/34	14.2	19.8	19.9	5.6	12–28	$1.87x + 2.4$	21–34	0.92	6
Leiostomus xanthurus	25/37	17.4	24.8	10.8	6.0	2–22	$2.85x - 3.7$	6.5–25	0.99	11
Bairdiella chrysura	10/15	3.1	3.3	—	—	—	—	—	—	—

[a] Occurrence is the number of stomachs in which a prey species occurred/percentage of that prey species in the total number of stomachs that contained identifiable food remains; number is the number of individual fish contained in stomachs; estimated modal length is of prey individuals; linear regression of greatest otolith length on standard length of the prey species is qualified by the correlation coefficient (r) and number of individuals measured (n). The formula given was used to compute the estimated lengths based on measurements of otoliths and was based on fresh otoliths taken from intact fresh fish.

Loligo sp.: 2 lower, 1 upper beaks of 2 squid, estimated mantle length 10, 15 cm

JGM 389, 189-cm female, tangled in gill net with JGM 388

Cynoscion sp.: 1 partly digested fish, ESL 12 cm; 1 pair of otoliths, ESL 8 cm

Leiostomus xanthurus: 14 intact fish, 20 partly digested fish, SL 9–12 cm; 20 otoliths, ESL 10–14 cm

Urophycis sp.: 6 pairs of otoliths, otolith length (OL) 2.4–3.7 mm

Anchoa sp.: 7 pairs of otoliths, OL 3.4–4.1 mm

Unidentified clupeids: 2 pairs of otoliths, OL 1.9, 2.7 mm

Unidentified ophidiids: 8 otoliths, partly digested, OL 2.5 mm

Unidentified shrimp: remains of two individuals, estimated lengths 5–7 cm

The species composition of prey from these specimens does not seem to differ from that of the stranded animals.

Though they show a consistent preference for sciaenids, scombrids, and mugilids, bottlenose dolphins seem to be rather catholic in their taste for "occasional items" (Table 4). Even the seemingly unpalatable stargazer (Uranoscopidae), toadfish (Batrachoididae), scorpionfish (Scorpaenidae), and small sharks have been found in the stomachs of bottlenose dolphins, including the stomachs of captured as well as stranded animals. It is possible that some of the more unlikely small prey animals, for example, silversides (Atherinidae), killifishes (Cyprinodontidae), and minnows (Cyprinidae), were introduced secondarily by way of stomachs of the predatory fish eaten by the dolphins.

We had been surprised by the lack of remains of mullet (*Mugil*) and bluefish (*Pomatomus*) in the

Table 4
Prey Genera of Atlantic Bottlenose Dolphins[a]

Prey	Locality				
	U.S. Atlantic[b]	Texas Atlantic	U.S. Atlantic	Africa Atlantic	U.S.S.R. Black Sea
Vertebrates					
Osteichthyes					
Ammodytidae					
Ammodytes	(1)				
Anguilidae					
Anguila	1				
Ariidae					
Arius		‡[c,d]	‡[e]		
Atherinidae					
Menidia	1				
Batrachoididae					
Opsanus	1				
Belonidae					
Unidentified			‡[e]		
Strongylura		†[c]			
Bothidae					
Bothus					‡[f,*g]
Paralichthys	1	†[c]			
Unidentified	1				
Carangidae					
cf. *Trachurus*	1				
Unidentified	1				
Centropomidae					
Centropomus		‡[c]			

Table 4 *(continued)*

Prey	Locality: U.S. Atlantic[b]	Texas Atlantic	U.S. Atlantic	Africa Atlantic	U.S.S.R. Black Sea
Clupeidae					
Brevoortia	1 (1)				
Dorosoma		*[c]			
Unidentified	5		‡[e]	‡[h]	‡[i]
Congridae					
Conger	2 (1)				
Cyprinidae					
Abramis					‡[f]
Cyprinodontidae					
Fundulus	(1)				
Elopidae					
Megalops		‡[c]			
Engraulidae					
Anchoa	5			‡[h]	
Engraulis					‡[f],*[g]
Gadidae					
Gadus					‡[f],*[g]
Laemonema				‡[h]	
Urophycis	10 (2)				
Gempylidae					
Gempylus				‡[h]	
Gerreidae					
Eucinostomus	1				
Haemulidae					
Haemulon	1				
Istiophoridae					
Istiophorus		‡[c]			
Labridae					
Unidentified	(1)				
Macrouridae					
Coelorhynchus				‡[h]	
Hymenocephalus				‡[h]	
Macruroplus				‡[h]	
Malacocephalus				‡[h]	
Merlucciidae					
Merluccius	(2)				
Mugilidae					
Mugil	1	*[c],‡[d]	‡[j,k]	*[h]	‡[f,g,i]
Mullidae					
Mullus					‡[f,g]
Ophichthidae					
Unidentified				‡[h]	
Ophidiidae					
Ophidion	3				
Unidentified	2				
Percidae					
Lucioperca					‡[f]
Pomatomidae					
Pomatomus	4			*[l]	‡[m]

(continued)

Table 4 *(continued)*

	Locality				
Prey	U.S. Atlantic[b]	Texas Atlantic	U.S. Atlantic	Africa Atlantic	U.S.S.R. Black Sea
Sciaenidae					
Bairdiella	11		?[n]		
Cynoscion	33	*[c]	*[o]		
Leiostomus	24	*[c]	?[n]		
Menticirrhus	1				
Micropogon	24	*[c]			
Pogonias		†[c]			
Umbrina					‡[g]
Unidentified			‡[e]		
Scombridae					
Euthynnus				‡[h]	
Sarda					‡[f]
Scomberomorus		‡[c]	‡[p]		
Unidentified			‡[q]		
Scorpaenidae					
Scorpaena					‡[f,g]
Serranidae					
Centropristes	1				
Sparidae					
Archosarqus		†[c]			
Lagodon		‡[d]			
Stromateidae					
Poronotus	2				
Synodontidae					
Synodus		‡[d]			
Tetraodontidae					
Lagocephalus		‡[d]			
Sphoeroides		†[c]			
Trichiuridae					
Trichiurus		‡[d]			
Triglidae					
Prionotus	1 (1)		‡[r]		
Uranoscopidae					
Astroscopus			‡[e]		
"Eels"			‡[e]		
Chondrichthyes					
Myliobatidae					
Aetobatus		‡[c]			
Rajidae					
Raja					‡[f,g]
Sphyrnidae					
Sphyrna		‡[c]			
"Shark"	2				
Invertebrates					
Cephalopoda					
Loliqo	20	‡[d]			
Squid			‡[e]	‡[h]	

Table 4 (*continued*)

Prey	Locality				
	U.S. Atlantic[b]	Texas Atlantic	U.S. Atlantic	Africa Atlantic	U.S.S.R. Black Sea
Crustacea					
Decapoda					
Crangon					‡[g]
"Shrimp"	3		*[s]	‡[h]	
Isopoda					
Idothea					‡[g]

[a] Key to symbols: *, abundant; †, trace; ‡, no data on relative frequency; ?, questionable. The numbers are the number of stomachs in which the species occurred in the sample reported in this chapter; the numbers in parentheses are from animals (USNM 504881, 504887) which we had reason to believe were from the offshore population. Data sources are indicated in the footnotes below.
[b] This chapter.
[c] Gunter (1942).
[d] Kemp (1949).
[e] Caldwell and Caldwell (1972).
[f] Tsalkin (1940).
[g] Kleinenberg (1938).
[h] Cadenat (1959).
[i] Malm (1933).
[j] Harris (1938).
[k] Mowbray (1952).
[l] Tayler and Saayman (1972).
[m] Morozov (1970).
[n] Hoese (1971).
[o] Townsend (1914).
[p] Odell (1975).
[q] Hoglund (1965).
[r] True (1885).
[s] Gunter (1951).

dolphin stomachs that we examined. Both species are common seasonally in the coastal waters of the study area and are reported as major components of the diet of bottlenose dolphins in other areas (Table 4). We found some fairly fresh bluefish in the stomach of USNM 504501 and attempted to extract the otoliths from their skulls. The otoliths were found to be substantially dissolved. We then reexamined the bones saved from the other bottlenose dolphin stomachs and searched for the highly distinctive mandibles of bluefish. Such mandibles were found in three other specimens' stomachs, none of which contained bluefish otoliths. The only trace of mullet considered of one otolith in the stomach of a juvenile male (USNM 504551).

There were two animals which we believed to have been from the offshore population, namely, USNM 504881 and 504887. They stranded in the winter and early spring (January and March) in New York and New Jersey, respectively. Their stomach contents were distinctly different from those of the rest of our sample. They included deeper-water families (Congridae, Gadidae, Merlucciidae) or families not usually represented (Ammodytidae, Labridae) in the coastal sample. These prey items are enclosed in parentheses in Table 4.

We had access to the stomach contents of four bottlenose dolphins taken offshore in the late 1960s on cruises led by Dr. Edward Mitchell. The stomach contents consisted of deep-water fish from the families Diretmidae, Gadidae, Macrouridae, Melamphaidae, Myctophiidae, Sternoptychidae, and Stromateidae and squid statoliths.

PARASITES AND COMMENSALS

We examined 240 dolphins from the study area for evidence of parasites. If no parasites were found in badly deteriorated specimens, we regarded the findings as suspect and scored the specimens as "not examined." If we found evidence of parasites, even in badly decomposed specimens, we noted the presence and species. For concise descriptions of the species mentioned, see Dailey and Brownell (1972).

Blubber

We routinely examined incisions that we made in the blubber for the presence of larval cestode cysts, *Phyllobothrium,* which are extremely common in some cetaceans. We found them present in 6% (6 of 94) of the specimens examined, typically as cysts about 1 cm in diameter located in the middle of the blubber layer and concentrated in the ventral posterior half of the animal. This parasite appears to be more common in pelagic animals that habitually feed on squid (see Table 1). It was used as a biological tag to differentiate coastal and offshore animals.

Digestive System

In cursory examinations of the digestive system, we found no gross indication of parasites in the mouth or esophagus, only occasional loose nematodes in the forestomach, and small ulcerations in the forestomach that may have been parasite related. The intestines were rarely examined. When they were, we found no parasites.

The only parasite that we observed consistently in the digestive system was the strigeid trematode, *Braunina cordiformis,* which was present in 83% (69 of 86) of the specimens examined. This parasite has been described previously as occurring in bottlenose dolphins (Schryver *et al.*, 1967). It occurs as globose individuals about 5 mm in diameter, attached by a slender stalk of host tissue to the wall of the stomach. The presence of the parasite was not correlated with age or length. Trematodes of this species were commonly present in the pyloric stomach, where they sometimes occurred in dense concentrations [e.g., 8 *Braunina* per square centimeter in the pyloric stomach of one specimen, USNM 504538, male, 237 cm, 5 GLGs (see below)]. The total area of the pyloric stomach was estimated to be 400 cm^2, giving a total estimated *Braunina* count of 3,200 individuals. *Braunina* were also observed in the main stomach of USNM 504538. This animal appeared to have been chronically ill. The high density of *Braunina* observed in its stomach was probably the result of a lowered resistance on the part of the host. Usually, infestations were less extensive, with 100 or fewer individuals. Frequently the walls of the pyloric stomach were thickened with what seemed to be focal masses of scar tissue. Since *Braunina* was the only parasite found commonly in the pyloric stomach, we assume that these were scars from previous infestations.

Trematodes (cf. *Campula*) were often found in the main ducts of the pancreas and liver. They were not correlated with liver damage but were correlated with damage to the pancreas. Fifty-five percent (20 of 47) of bottlenose dolphins examined had what appeared to be chronic pancreatic damage ranging from mild to extreme fibrosis. This condition did not appear to be correlated with the individual's age.

Other Abdominal Viscera

We found larval cysts of the cestode *Monorhygma* frequently in other cetaceans from the study area, but they were very rare in bottlenose dolphins. These parasites usually occur as rounded cysts (~2–3 cm in diameter) just beneath the peritoneum (Dailey and Brownell, 1972, p. 569). Only 4 of 52 (8%) animals that were checked had *Monorhygma* cysts.

Pterygoid and Accessory Air Sinuses

The trematode *Nasitrema* (Neiland *et al.*, 1970) was common in the pterygoid sinus system. Of 22 animals examined for this parasite, 11 (50%) were infected. The center of infestation appeared to be in the peribullary sinus. Commonly the infestations were light, with just a few (3–5) individuals present. This is a rather common parasite in cetaceans and has been implicated with brain damage in *Delphinus* (Cowan *et al.*, 1986, p. 339). None of the bottlenose dolphins that we examined had infestation levels approaching those found by Cowan *et al.* (1986). Brains were not examined.

Nematodes of the genus *Crassicauda* were found in 6 (27%) of 22 individuals examined. Dailey and Perrin (1973, p. 459) and Perrin and Powers (1980) reported *Crassicauda* in the pterygoid sinus system of *Stenella*. Geraci *et al.* (1978) reported the same genus in the mammary glands and associated muscle of *Lagenorhynchus acutus*. In 3 of our infected specimens *Crassicauda* were found in the mammaries only. In 2 individuals the parasites were found in the pterygoid air sinus. The location of the parasites in the sixth animal was not recorded. Although Perrin and Powers (1980) and Geraci *et al.* (1978) found indications of parasite-related mortality in their studies, none of the bottlenose dolphins we examined showed any signs of being adversely affected by *Crassicauda*.

The nematode *Stenurus* was even less common. It was present in only 1 (5%) of 21 animals examined. In that individual (USNM 504608, 266-cm female), the infestation was very heavy and associated with tissue erosion and ulceration. *Stenurus* is a fairly common parasite in the air sinuses of small cetaceans (Baylis, 1932).

Barnacles

Evidence of the presence of pseudo-stalked barnacles, *Xenobalanus globicipitis*, was frequently found on the appendages, as had been reported previously in this population (True, 1891). Of 46 animals examined, 30 (65%) bore either intact barnacles or their bases. These barnacles were present on the dorsal fin, flukes, and flippers. It was not unusual to find four or five individuals on each of the flippers, the same number on the dorsal fin, and about ten on the flukes. They nearly always occurred on the trailing edges of the appendages.

There is a possibility that a few of the barnacles that were identified as *Xenobalanus* were, in fact, the related barnacle, *Platylepas hexastylos*. A 228-cm male (USNM 504418) that stranded on Pea Island, NC, had two barnacles of this species on its flukes. No cases of the other common odontocete barnacle, *Conchoderma*, were encountered.

Parasites Not Found

No parasitic copepods, *Penella*, or cyamid amphipods were found on stranded bottlenose dolphins.

PREDATION

There was no evidence that any of the specimens we examined had been killed by a predator. However, seven specimens bore wounds or scars from what we interpreted to be previous encounters with sharks. In six of the seven animals, the wounds were fresh, but it was not possible to determine whether they had been inflicted while the animal was still alive. In the seventh animal there were healed scars around the genital area. We were not able even to guess at the species of shark involved, except for one case. In that animal (USNM 504273), we found teeth in the fresh wound which were identified by Robert Purdy (Dept. Paleobiology, USNM) as being from a shortfin mako (*Isurus oxyrinchus*).

Two of the seven scarred or wounded dolphins were newborns. If inflicted while the animals were alive, the wounds almost certainly would have been fatal. One of the neonates had been bitten twice by a large shark, causing a wound about 20 cm wide. A shark that size easily could have eaten the whole carcass.

Most of the shark wounds were on the ventral surface of the caudal peduncle, in the anal and genital areas. These are the same areas where concentrations of the larval cestode cysts of the genus *Phyllobothrium* are found in bottlenose dolphins and other cetaceans. The known definitive hosts of *Phyllobothrium* are sharks (Williams, 1968). Regardless of whether the shark "attack" is made on a live or a dead cetacean, the chances of larval cestodes making a successful transfer to the definitive host are greater if the cysts are in an area favored by the sharks.

AGE AND GROWTH

Neonatal Length

Animals classed as neonates (stranded animals with flaccid dorsal fins) ranged from 106 to 132 cm in total length (Table 5). We chose to use the condition of the dorsal fin (folded and flaccid versus erect) rather than the presence of fetal folds as indicating a newborn because fetal folds last up to several months (Caldwell and Caldwell, 1972, p. 33). We estimate the average length at birth as 117 cm, which is the mean of 13 animals that had the dorsal fin folded. The mean total weight of 9 animals with lengths between 111 and 132 cm was 20.4 kg, and the weight of the one animal that was exactly 117 cm was 21.2 kg. The longest fetus recorded was 113 cm. Data on newborn animals are presented in Table 5.

Age Determination

Ages of animals were estimated by reading and scoring teeth with definable growth layer groups (GLGs) (Perrin and Myrick, 1980) on longitudinal thin sections of undecalcified teeth (Sergeant *et al.*, 1973). Teeth in which the layers deposited late in life were not readable were assigned a value followed by a plus sign (e.g., 11+), indicating that

Table 5
Data on Newborn Bottlenose Dolphins

Total length (cm)	Weight (kg)	Date	Number	Dorsal Fin	Fetal Folds
92	NE[a]	04 18 78	JGM 375	Folded	NE[b]
105	NE	03 29 78	JGM 360	NE	NE
106	NE	05 20 83	550311	NE	NE
106	NE	04 10 77	JGM 329	Folded	NE
108	NE	03 06 78	JGM 351	NE	NE
110	11.5	04 17 78	JGM 374	Erect	Present
111	16	03 17 77	504598	Folded	NE
112	NE	04 23 78	JGM 378	Folded	NE
112	NE	04 17 78	JGM 373	Folded	Present
112	NE	03 23 77	JGM 299	Folded	NE
113	14.7	04 04 77	JGM 325F	Fetus	NE
114	18.5	03 19 77	504576	Erect	Present
115	NE	04 28 78	JGM 379	Folded	NE
115	17.5	03 25 77	JGM 300	Folded	NE
115	22	12 13 76	504527	NE	NE
116	NE	05 15 76	504398	NE	Probable
116	19.5	03 17 77	504599	Folded	NE
117	21.2	04 10 77	JGM 330	Folded	NE
117	NE	03 17 77	504597	Erect	NE
118	NE	05 15 76	504401	Folded	NE
118	NE	03 08 77	504582	Erect	NE
120	NE	03 17 77	JGM 290	NE	NE
120	NE	06 04 77	AL	NE	NE
126	NE	03 05 77	504560	Folded	NE
127	27.1	03 06 77	504573	Folded	NE
132	26.0	03 05 77	504562	Folded	NE

[a] NE, Not examined.
[b] This animal had traces of the umbilicus attached and may represent an aborted near-term fetus.

they had at least that many GLGs. Hohn's (1980) sample was a subsample of the one considered in the present study. We assumed (after Sergeant *et al.*, 1973; Perrin and Myrick, 1980) that bottlenose dolphins deposit 1 GLG per year (see Hohn, Chapter 33, this volume; Myrick and Cornell, Chapter 34, this volume). Detailed age data are presented in the Appendix.

Age–Length Relationship

No noticeable difference in length at age was evident between males and females (Fig. 1). Growth reached an asymptote at about 250 cm and 12 GLGs, which corresponds to Hohn's (1980, p. 45) finding that the mean length of animals with more than 4 GLGs was 250 cm. Sergeant *et al.* (1973) found a difference in the shapes of the age–growth plots, with males reaching an asymptotic length of 270 cm and females 250 cm. We did not find this to be true in our sample. Our age–length plot does not differ appreciably from that of Hohn (1980, p. 45). The oldest animal in our sample had more than 27 GLGs.

A comparison of our age–length plot (Fig. 1) with those from the Gulf of Mexico (Fig. 2, based on data from Solangi and Dukes, 1983, their Table 1), Japan (Fig. 3, data on the early stages of growth of captive animals from Kasuya *et al.*, 1986), and southern Africa (Fig. 4, data from Ross, 1977, 1984) gives the following results. The growth curve of the Gulf of Mexico animals (Fig. 2) appears to reach an asymptote at about 230 cm, when they are the same age as our 230-cm animals. The growth curve based on the Japanese sample (Fig. 3) appears to reach an asymptote at 280–290 cm. The average length at 4 GLGs is about 265 cm, much greater than in our sample at 4 GLGs (220 cm). The curve for the southern African population that Ross (1977, 1984) identified as *Tursiops aduncus* (Fig. 4) is more similar to the curve for our sample. It reaches an asymptote at about 240 cm. The two aged specimens from the population that Ross identified as *T. truncatus* fall above our curve, and more comfortably fit the curve of Kasuya *et al.* (1986).

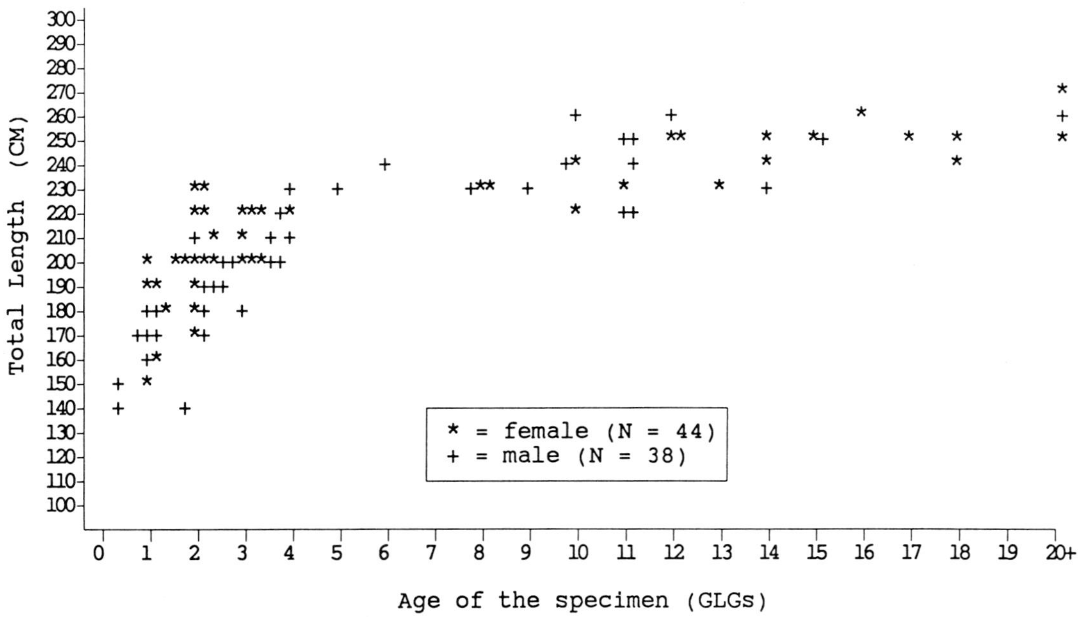

Figure 1 Age–length curve of our sample.

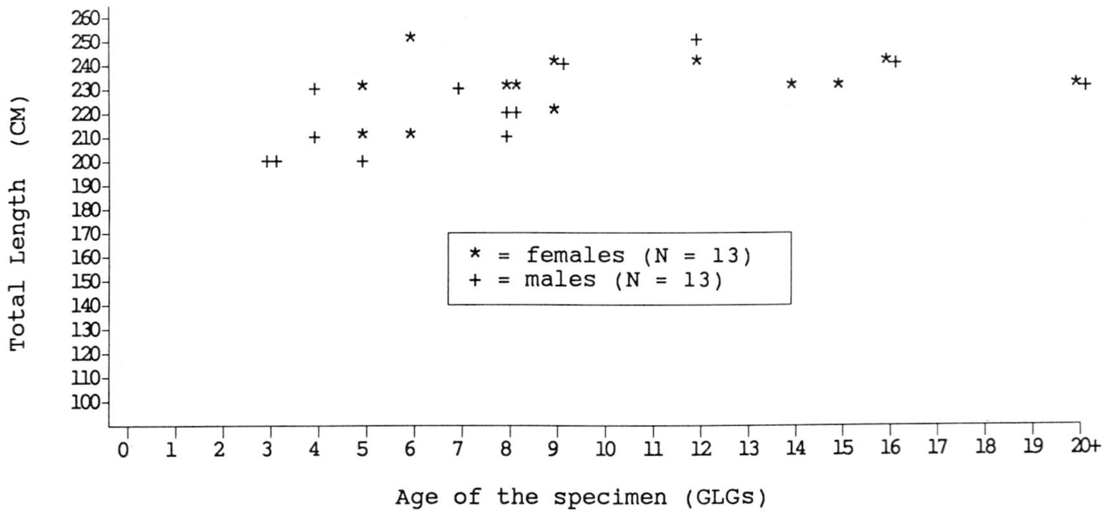

Figure 2 Age–length curve from Solangi and Dukes (1983, their Table 1).

Weight

Total weight is a parameter that is seldom available for stranded animals. We had 58 animals for which a total body weight was available (Fig. 5, Appendix). The lack of mature males is conspicuous. Some of these weights probably are too low because of chronic illness. We consider a line drawn along the upper left-hand margin of the scatter (i.e., taking the greatest weight for a given length) to represent the weight–length relationships of normal, healthy bottlenose dolphins.

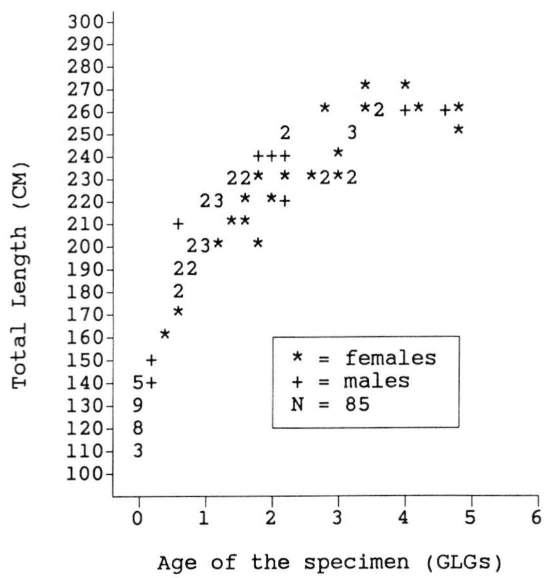

Figure 3 Age–length curve from Kasuya *et al.* (1986).

Physical Maturity

The age or size at which individuals stop growing and become physically mature is of general interest to biologists, and it is of special interest to systematists. In selecting specimens for species comparisons one must be certain that the apparent differences are not related to the ages of the individuals being compared. The degree of physical maturity of an individual cetacean can be determined by the degree of fusion of the vertebral epiphyses. If all epiphyses are fused to the centra, growth of the individual has ceased and the animal can be considered physically mature.

At the time of necropsy, we sectioned the face of a midthoracic vertebra, as it is generally believed that the thoracic vertebrae are the last ones to fuse. For individuals in which the vertebra was either obviously unfused or fused to the extent that the epiphyseal line had disappeared, scoring was immediate. Where there was any doubt,

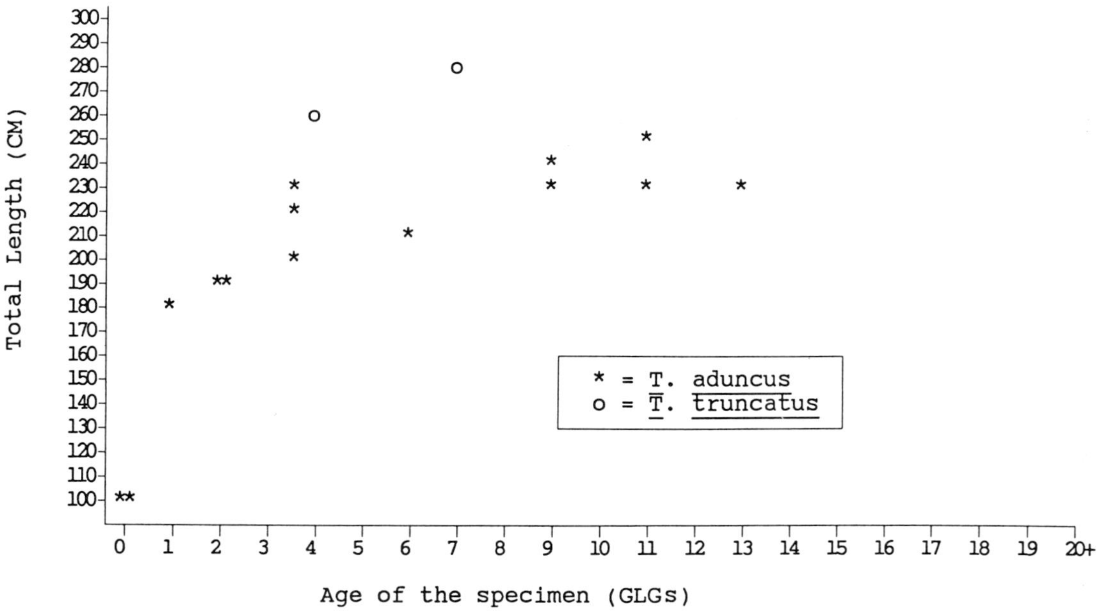

Figure 4 Age–length curve from Ross (1977, 1984; *Tursiops aduncus*).

decalcified histological sections were prepared from samples of the epiphyseal suture. Animals were scored as follows: 1, epiphyses open; 2, epiphyses closed, but epiphyseal line still visible; 3, epiphyses fused, epiphyseal line invisible. Class 3 animals were considered mature. Histological sections were usually prepared for animals receiving a score of 2; any in which there was any site where the bone of the centrum was in contact with the bone of the epiphysis, without any cartilage in between, also were considered mature.

Mean length and age at physical maturity (Cooke–Hohn method) were 245 cm (variance 0.123) and 13.09 GLGs (variance 2.506), respectively. The shortest mature animal was 202 cm; the longest immature animal, 264 cm. The youngest mature animal had 10 GLGs; the oldest immature animal, 15 GLGs.

Maturity also can be assessed by examining the condition of the premaxilla–maxilla suture on the distal portion of the rostrum (Dailey and Perrin, 1973, p. 465; Walker, 1981, p. 5). Dailey and Perrin correlated fusion of these bones with sexual maturity in *Stenella graffmani* (=*S. attenuata*). We found that care was necessary in judging the fusion of the maxilla and premaxilla because in some specimens the groove between them was held open by small nutrient vessels lying in the groove, even after the premaxilla and maxilla had fused. Such specimens were examined closely and classified as mature if "bridges" of bone were found connecting the premaxilla and maxilla. Special care was also necessary in the few older specimens in which osteoarthritic deposits masked the fusion of the bony elements.

We recorded fusion of the rostral elements as a portion (in centimeters) of the suture that was fused. Results were as follows:

Premaxillae and maxillae	Sexually immature	Sexually mature	Total sample	Percent misclassified
Unfused	37	1	53	2.6
Fused	18	32	75	36

Fusion of the rostral elements of the skull proved to be a fairly reliable means of classifying immature animals, but it misclassified, as mature, 36% of the immature animals. The mean age at skull maturity was 3.46 GLGs (variance 1.142), and the range of ages over which skulls became mature was 1–6 GLGs.

We do not mean to imply that this technique should not continue to be used as one criterion for

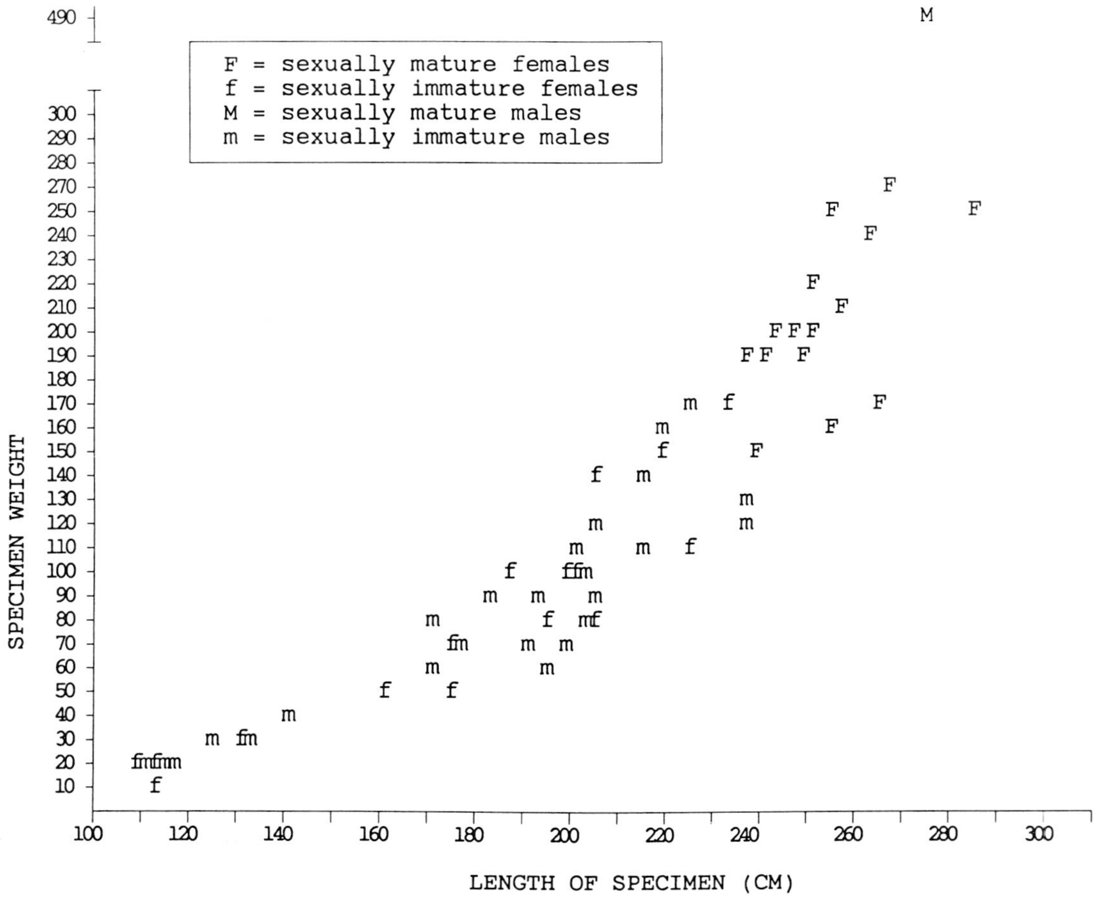

Figure 5 Total weight plotted against total length. Note that the weight axis is discontinuous. It happened that the only mature male we had (USNM 550446) was a very long (very heavy) animal which had stranded on Cape Cod; we believe it to be a member of the offshore population.

deciding whether to include specimens in a systematic series. Lacking other biological data on skeletal maturity, it is one way of eliminating young animals from the sample.

We also assessed physical maturity by examining X-rays of the flippers, following Hui (1977) and Ogden *et al.* (1981), and scoring the flippers according to the grading method of the latter authors. They used the appearance and fusion of epiphyseal centers of ossification on the distal end of the radius and ulna as follows: stage 0, no secondary ossification center present; stage 1, epiphyseal ossification apparent but less than 50% of the width of adjacent bone; stage 2, epiphyseal ossification center between 50 and 100% of the width of adjacent bone; stage 3, distance between epiphysis and bone beginning to diminish; stage 4, epiphysis beginning to fuse to bone; stage 5, epiphysis completely fused to the bone but epiphyseal suture still evident; stage 6, all traces of epiphyseal suture disappeared.

In many cases, we could discern distinct intermediate states, particularly in grades of 3 or higher. Accordingly, we graded such intermediate stages as 3.5 (between grades 3 and 4), 4.5, etc. These scores are presented in the Appendix and

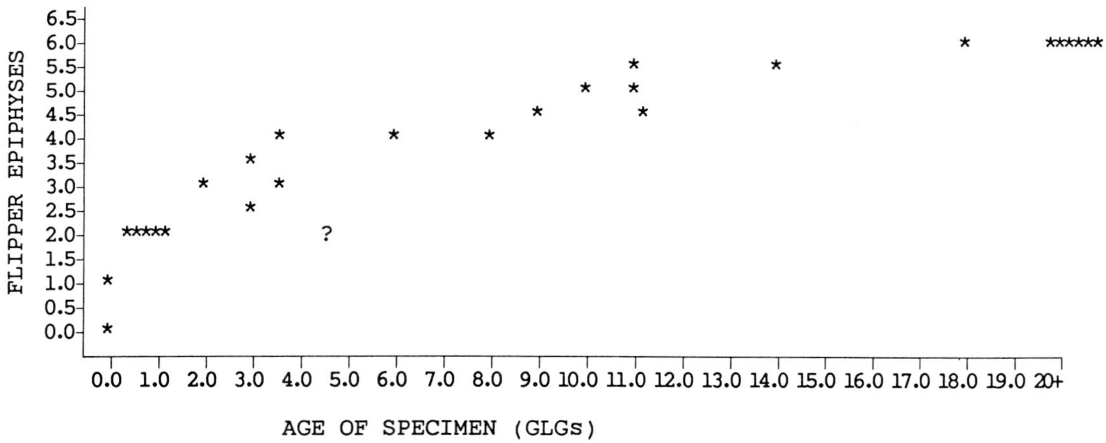

Figure 6 Status of flipper epiphyses versus age (GLGs).

are illustrated in Figs. 6 and 7. This method is of considerable interest because it is relatively noninvasive and can easily be used on living animals.

REPRODUCTION

The reproductive status of nine of the animals in our sample was briefly described by Leatherwood *et al.* (1978). Kasuya *et al.* (1986) provided sufficient data to allow some comparisons of aspects of life history of bottlenose dolphins off Japan and those off the east coast of the United States (Table 6). The Japanese animals are, on the average, 12% longer than those in our sample.

Sexual Maturity

Females

The youngest mature female had 8 GLGs. Using the Cooke–Hohn method (Hohn, 1985), the mean age at sexual maturity was 10.64 GLGs, with a

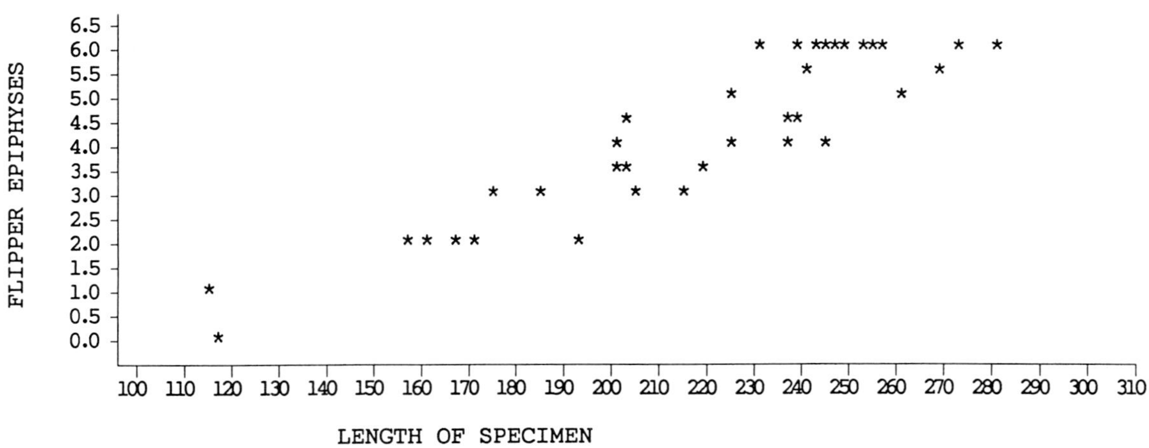

Figure 7 Status of flipper epiphyses versus total length.

Table 6
Comparison of Some Life History Parameters of a Japanese Sample of Bottlenose Dolphins to our Sample

	Japanese sample[a]	Our sample	Ratio
Mean length at birth	128	117	1.09
Mean length at sexual maturity (females)[b]	267	232	1.15
Modal length (both sexes)	290	250	1.16
Asymptotic length	280	250	1.12
Maximum length (both sexes)	328	306	1.07
		Mean	1.12

[a] Kasuya *et al.* (1986).
[b] Mean length at sexual maturity is the length at which the number of immature animals greater than that length is equal to the number of mature animals less than that length.

binomial variance of 0.53. These values were for a sample of 20 animals that fell into indeterminate age classes (classes in which there were both immature and mature animals). Our mean is not far below the stimate of Sergeant *et al.* (1973) of "about 12 years." Mean weight at attainment of sexual maturity in females was 150 kg (sample: 12 immature, 13 mature).

The shortest sexually mature female was 207 cm. The longest immature female was 251 cm. Mean length at sexual maturity (Cooke–Hohn, modified for length) was 233.5 cm, with a binomial variance of 0.08.

Males

Males were judged as sexually mature if they had evidence of spermatozoa in the testes, using Be-rg's stain (Berg, 1963). There was a dearth of adult males in this sample (Table 7), and only six showed a positive reaction with Berg's stain. The youngest of these (USNM 550820) had more than 8 GLGs (tooth was marginally readable); it was also the shortest (252 cm) of the six, and its gonads were not weighed. However, its gonad dimensions were consistent with an estimated single gonad weight of more than 500 g.

Table 7
Distribution of Gonad Weight and Dimensions in Males in our Sample

Length (cm)	Gonad Weight (g)	Gonad Dimensions (L × W × D, cm)	Age (GLGs)	Sexual Maturity[a]
108	1.8		0	I
114		4.6 1.0 1.0	0	I
117		3.2 1.1 0.9	0	I
127		5.6 1.0 1.0	0	I
152		5.7 1.2 1.1		I
159	3.5	8.0 1.5 1.0	0.5	I
172	8.1	7.2 1.5 1.1		I
172	4.4	6.6 1.2 0.8	1	I

Table 7 *(continued)*

Length (cm)	Gonad Weight (g)	Gonad Dimensions (L × W × D, cm)	Age (GLGs)	Sexual Maturity[a]
177	20.0	8.4 2.2 1.3	1	I
179	10.4	7.5 1.7. 1.3		I
184	10.3	7.5 1.7 1.7	1	I
185		7.5 1.3 1.1	2	I
188		7.7 1.6		I
191		7.0 2.0	2.5	I
194		12.0 1.0 1.0		I
198	15.0	8.0 2.0 1.7		I
201		10.5 1.5 1.3	3.5	I
202	13.0	8.5 1.5 1.2	2.25	I
202	17.6	7.5 2.2 1.7		I
203		9.0 2.5 1.5		I
205		6.0 2.2 1.5		I
206		8.8 2.2 1.9	3	I
206	7.8	6.8		I
206		8.7 2.0 1.6		I
210	15.0	12.0 2.0	3.5	I
215	31.0	13.0 2.6 1.9		I
216	2	10.0 2.3 1.9	4	I
216		10.7 2.0 1.3		I
222		8.0 1.7 1.0		I
226	29.5	10.3 2.3		I
228	36.3	12.2 2.7 1.7	11	I
232		9.5 2.4 1.7	4	I
237	13.0	9.0 1.5 1.2	5	I
237		9.5 3.1 2.0	4	I
239	60.0	12.0 3.0 3.0	9	I
241	55.0	19.0 2.5 1.7	11.25	I
249		12.0	6	I
249		14.6 4.8 2.8		I
250	68.0	13.5 3.5 1.8	11	?
252		24.8 9.9 8.5	8+	M
254		10.0 3.5 3.0		I
256	180.0	15.1 6.8 4.3	15	M
258	73.0	14.0 4.5 2.0	11	I
264	26.4	11.7 2.4 1.6	12	I
267		19.0 5.0	21	M
268		24.5	20+	M
268		28.0 8.0 5.0		M
272		30.0 9.0		M
274		26.0		M
277		20.3		?
279		26.0 8.0 6.0	24	M
279	230.0	18.6 6.1 2.4	24	M
284	52.2	13.5 3.9 1.6	16+	M
287	324.0	17.5 7.5 4.0		M

[a] Sexual maturity was judged by either positive results on sections stained with Berg's stain or gross size. I, Immature; M, mature; ?, unable to classify.

There are too few males in our sample to draw any firm conclusions, but some generalizations are possible. The oldest sexually immature males had about 11 GLGs, the largest immature testis weight was 73 g, and the longest immature testis was 19 cm. The youngest mature male had over 8 GLGs, the smallest mature testis was 230 g, and the shortest mature testis was 18.6 cm. This is discounting one specimen (USNM 504879), a 284-cm individual that had 16+ GLGs but whose testis was only 13.5 cm long and weighed 52.2 g. We believe there was something anomalous about this specimen. Our figures agree with the data of Sergeant *et al.* (1973), which show males maturing at 10–15 years. Juvenile males have testis weights of about 10 g. The largest measured testis weight in this sample was 324 g. An estimated weight of the largest measured testis (based on comparison with the dimensions of 180- and 324-g testes) was 460–690 g. Sergeant *et al.* (1973) gave 700 g as the maximum weight for a single testis.

Calving Seasonality

In a sample of stranded neonates (animals <140 cm long) there was a pronounced mode in the month of March. Of 32 such animals, 14 stranded in March, 6 in April, 8 in May, 2 in June, 1 in December, and 1 in January. We found no particular pattern in length of individuals as a function of time of stranding (i.e., one does not find that the larger "neonates" are stranded later in the year than the smaller).

This agrees with previous statements about breeding seasonality (e.g., Mead, 1975, p. 1159), which indicate that the calving season is prolonged, with a peak in the spring. We found no evidence of a second peak in the fall, as suggested by Hohn (1980, pp. 49, 57).

SUMMARY

Food Habits

The data presented above indicate that bottlenose dolphins in our study area feed primarily on three species of sciaenid fish: weakfish or sea trout, croaker, and spot. Gunter (1942) reported these three species as ranking third to fifth in importance in his sample of stomach contents from animals from the Gulf of Mexico; mullet (*Mugil cephalus*) and gizzard shad (*Dorosoma cepedianum*) were most and second-most abundant, respectively. Mullet has often been reported as a food of bottlenose dolphins [Harris, 1938; Malm, 1933 (cited in Kleinenberg, 1956, p. 190); Mowbray, 1952, p. 202]. Mullet were common off the North Carolina coast, and gizzard shad were taken infrequently by haul seiners in the years of our intensive beach survey, but neither species was found in the bottlenose dolphin stomachs we examined. Twenty of the individuals had remains of squid (*Loligo*) in their stomachs. Two of these had muscle fragments of squid, indicating that the dolphins had eaten the squid directly, that is the squid were not residues of the stomach contents of fish.

Parasites

The coastal population of bottlenose dolphins is characterized by the presence of *Braunina* in the stomach and the absence of *Phyllobothrium* and *Monorhygma* in other tissues. In a small sample of offshore dolphins, no *Braunina* were present, and *Monorhygma* and *Phyllobothrium* were present in a relatively high percentage of the specimens. We have used this difference to separate the coastal from the offshore population.

Age and Growth

The mean length at birth for the studied population is 117 cm. Growth continues until a length of 250 cm and an age of approximately 12 GLGs. This asymptotic length is comparable to that for bottlenose dolphins off southern Africa (~240 cm), slightly longer than that for bottlenose dolphins in the Gulf of Mexico (~230 cm), and slightly shorter than that for bottlenose dolphins off Japan (280–290 cm).

Physical maturity (judged by fusion of the vertebral epiphyses) is reached at a mean length of 245 cm and a mean age of 13.09 GLGs. Skull maturity, as evidenced by distal fusion of the maxilla and premaxilla, occurs over a range of 1–6 GLGs, with a mean age of 3.46 GLGs. The state of flipper epiphyses, determined from flipper radiographs (Ogden *et al.*, 1981), appeared to correlate very well with both age and length. This technique has the advantage of being relatively noninvasive, so that it can be used on live animals.

Reproduction

Odell (1975, p. 1056) summarized the data available for reproduction of the Florida population. He stated that females reach sexual maturity at 5–12 years of age and a length of 220–235 cm, males at 10–13 years and 245–260 cm. This agrees with our findings that females attain sexual maturity at 7–14 GLGs and 207–251 cm, with a mean age at sexual maturity of 10.64 GLGs. We do not have enough data to make meaningful comparisons for males. Although calving occurred in many months, we found a peak in the spring but no indication of a secondary peak in the fall.

Figure 8 shows the relationship between length at sexual maturity, physical maturity, and skull maturity in our sample. Skull maturity is clearly expressed at a shorter length than sexual or physical maturity. Length at physical maturity in males appears to be much greater than in females, although caution must be used because the sample of physically mature males was extremely small.

Age at sexual maturity in females (Fig. 9) has a wide range (7–13 GLGs) but a very narrow confidence interval about the mean of 10.64. As would be expected, the mean age at physical maturity is higher than the age at sexual maturity (13.09 versus 10.64 GLGs). Surprising was the relatively low mean age at skull maturity of 3.46 GLGs.

ACKNOWLEDGMENTS

Most of the material described in this chapter was collected by the authors or the following Smithsonian technicians: Jane Ailes, Arthur Bridgida, Paul Finnegan, Linda Gordon, Matthew Hare, Aleta Hohn, William McLellan, Joan McDonald, Carrie Olin, Randi Olsen, Ann Pabst, Barry Peers, and James Critchfield. We are also indebted to many people outside the Smithsonian for reporting strandings and assisting salvage efforts. We thank particularly Patrick Whaling, Peter Laurie, Guy Oliver, Robert Blaylock, and the staff of the Marine Mammal Stranding Center in Atlantic City, New Jersey, for contributing specimens to the project. Aleta Hohn wrote a master's thesis on the determination of age for part of this sample of bottlenose dolphins. We are indebted to her for determining the ages of most of the animals considered in this chapter and for providing

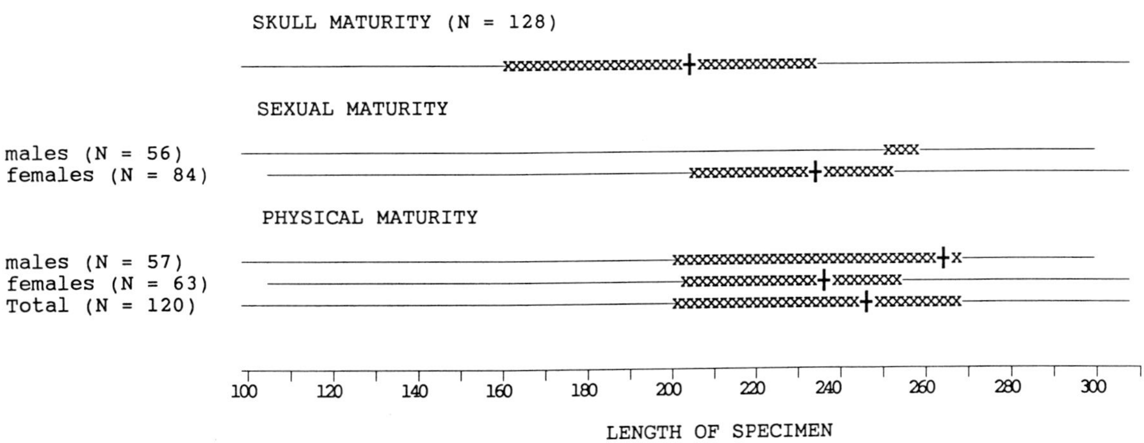

Figure 8 Graphic relationship of lengths at sexual maturity, physical maturity, and skull maturity in our sample. ——, Total range of sample; xxx, range of lengths between the smallest mature specimen and the largest immature specimen; **I**, mean length.

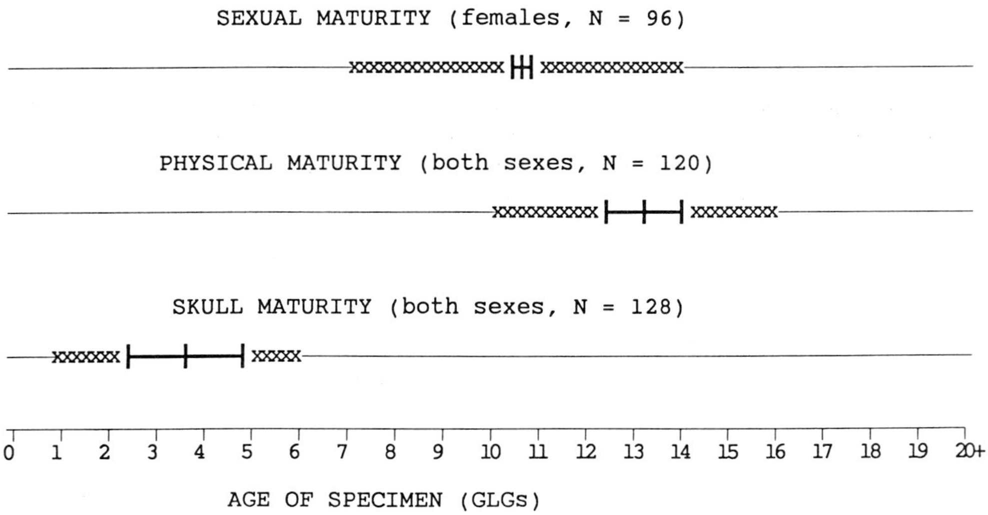

Figure 9 Graphic relationship of ages at sexual maturity, physical maturity, and skull maturity in our sample. ——, Total range of sample; xxx, range of ages between the youngest mature specimen and the oldest immature specimen; ——, 95% confidence interval; **I**, mean age.

fruitful discussions on delphinid life history in general. We are also indebted to the late John Fitch, who identified most of the otoliths in the stomach contents. We particularly thank the staff of the Cape Hatteras National Seashore; without their help this project would have taken many more years to complete.

APPENDIX
DATA ON ANIMALS USED IN THIS STUDY

Explanation of headings and abbreviations: Serial, USNM catalog number (NN, no number); Col. No., collector's number (NN, no number); Skull mat., skull maturity (length of fusion of premaxilla to maxilla in cm); Epip., thoracic vertebral epiphysis scoring [1, open; 2, closed (epiphysis in bony contact with vertebra); 3, fused (epiphyseal line absent)]; Sex. mat., sexual maturity (I, immature; M, mature); Age, number of tooth growth layer groups (GLGs) (age for neonates: 0, flukes and dorsal fin flaccid, fetal folds present, muscle light in color; 0+, either flukes or dorsal fin turgid, fetal folds absent, muscle dark; 0−, fetus); Pathol., pathology [A, absent; Q, questionable; P, probable; Y, present; Ph, *Phyllobothrium;* Mo, *Monorhygma;* Br, *Braunina;* Pa, pancreatic condition (f, fibrotic; n, normal); Na, *Nasitrema;* Cr, *Crassicauda* (m, present in mammaries; s, present in sinus); Xe, *Xenobalanus* (mention of a given pathogen means that an attempt was made to check for it)]; Stom. cont., stomach contents (C, collected; E, empty; P, present but not collected); Flip. epip., flipper epiphysis scoring (see p. 182).—, Data not available, either unknown or not examined.

Serial	Col. No.	Date (mo dy yr)	Sex	Length (cm)	Weight (kg)	Skull mat	Epip.	Gonad weight (g)	Gonad dim. (L×W×D, cm)	Sex. mat.	Age (GLGs)	Pathol.	Stom. cont.	Flip. epip.
484529	JGM 007	12 16 72	F	243	—	6	3	—	—	—	16+	—	—	—
484530	JGM 003	11 11 72	M	194	88	—	—	—	12.0 1.0 1.0	—	—	P: Br; Y: Xe	E	—
484992	JGM 112	12 8 73	M	239	120	8	1	60.0	12.0 3.0 3.0	I	9	A: Na, St; P: Br; Y: Xe	E	4.5
484993	JGM 113	12 4 73	F	256	—	—	—	—	—	—	—	Y: Xe	—	—
500857	CWP 74-1	1 22 74	M	189	—	0	—	—	—	—	3	Y: Na	—	—
504095	JGM 117	6 25 74	F	235	—	9	—	—	—	—	17+	—	—	—
504096	JGM 119	6 30 74	F	256	163	—	1	4.1	3.1 1.9 1.0	M	—	Q: Pa; P: Mo; Y: Br, Cr(m)	C	—
504118	NN	4 18 74	F	154	—	0	—	—	—	—	—	—	—	—
504121	BAP 003	9 19 74	F	249	—	2	2	—	—	—	24+	Y: Xe	P	6
504122	BAP 004	9 19 74	M	191	—	0	1	—	7.0 2.0	—	2.5	Q: Na, St; Y: Xe	P	3
504123	BAP 005	9 19 74	F	232	—	2	3	—	—	M	27+	Q: Na, St; Y: Xe	P	6
504133	BAP 010	11 15 74	F	250	—	—	2	—	2.5	—	—	A: Xe; Y: Br	E	—
504135	BAP 012	12 15 74	F	257	—	9	3	—	—	M	17	Y: Xe	P	—
504218	JGM 134	2 26 75	M	177	70	0	1	20.0	8.4 2.2 1.3	I	1	Q: Pa	—	2
504254	CWP 041	5 13 75	F	205	—	0	1	—	—	—	2	—	E	—
504255	CWP 042	5 13 75	M	185	—	2	1	—	—	—	1	—	E	—
504273	CWP 050	6 6 75	F	276	—	2	3	3.7	3.0. 2.1 0.7	M	24+	A: Xe; Y: Ph, Mo	P	6
504290	CWP 049	6 14 75	M	174	—	0	—	—	—	—	2.3	—	—	—
504291	JGM 145	6 27 75	M	210	—	0	1	15.0	12.0 2.0	I	3.5	A: Ph; Y: Xe	C	—
504295	NN	7 28 75	M	241	—	4	1	55.0	19.0 2.5 1.7	I	11.25	A: Ph, Mo, Br; Y: Pa(f), Na, Xe	E	4.5
504310	PMF 071	9 16 75	F	255	—	5	—	10.0	7.4 3.6 0.7	M	15+	Q: Na, St, Cr	—	6
504311	NN	8 18 75	M	159	—	0	1	3.5	8.0 1.5 1.0	I	0.5	A: Ph, Na, St, Xe	E	2
504313	NN	10 3 75	M	203	—	5	—	—	—	N	3	P: Br; Y:Xe	E	3.5
504315	CWP 051	11 11 75	F	243	—	3	—	—	—	—	18	A: Ph; Q: Na, St; Y: Xe	—	6
504325	JEA 456	1 16 76	F	241	—	5	—	17.2	—	M	14	Q: Na, St, Cr; Y: Pa(f)	E	5.5
504326	LKG 024	1 14 76	F	203	—	0	2	1.2	3.3 1.1 0.5	I	2	A: Ph; Q: Na, St, Cr; Y: Br, Xe	E	4.5
504348	LKG 030	—	F	233	172	3	—	—	—	—	13	—	—	—
504398	JGM 167	5 15 76	F	116	—	0	—	—	—	—	0	A: Xe	—	1
504399	JGM 168	5 15 76	F	177	55	0	—	—	—	—	2	A: Ph, Mo; Y: Br, Xe	P	3
504400	JGM 169	5 15 76	F	196	80	0	1	1.9	3.5 1.3 0.5	I	1	A: Ph, Mo, Na, St; Y: Br, Pa(s), Xe	E	2
504401	JGM 170	5 15 76	F	118	—	0	—	—	—	I	0	A: Xe	—	0
504402	JGM 171	—	M	197	63	0	—	—	—	—	2.25	—	P	—
504403	JGM 172	5 13 76	F	162	52	3	1	—	—	—	1.25	A: Ph, Na, St; Y: Br, Xe	C	2
504405	JGM 175	5 28 76	F	176	73	0	—	1.2	3.5 1.0 0.5	I	—	A: Ph, Xe; Y: Br	C	—
504418	LKG 031	6 13 76	M	228	—	2	—	36.3	12.2 2.7 1.7	I	11	A: St; Y: Na, Xe	C	5
504474	CWP 058	9 8 76	M	199	—	—	1	—	—	—	—	Y: Xe	C	—
504482	CWP 059	9 24 76	F	245	—	—	3	—	—	M	—	A: Ph, Mo, Cr; Y: Xe	E	—
504500	CWP 061	10 15 76	M	201	72	5	1	—	10.5 1.5 1.3	I	3.5	Y: Xe	E	4
504501	JGM 204	10 21 76	F	256	230	2	2	—	6.4 2.1 0.6	M	22	Y: Pa(f), Na, Xe	—	6
504527	JGM 223	12 13 76	F	115	22	0	1	—	—	I	—	A: Ph, Na, St, Xe	—	—
504528	JGM 224	12 29 76	F	193	—	0	1	—	4.5 1.4 0.5	I	2	A: Ph, Mo, Xe	E	—
504531	JGM 226	12 7 76	F	194	—	1	—	—	—	I	—	—	—	—
504532	JGM 227	11 21 76	F	252	—	3	—	—	—	M	12	—	—	—
504534	JGM 228	1 19 77	F	222	—	1	1	—	—	I	4	Y: Br, Pa(f), Xe	C	—
504535	JGM 229	1 20 77	M	210	—	0	—	—	—	I	2	—	E	—
504536	JGM 230	1 21 77	M	184	93	0	1	10.3	7.5 1.7 1.7	I	1	A: Ph, Na, St; Y: Br, Xe	C	—
504538	JGM 231	1 30 77	M	237	129	0	1	13.0	9.0 1.5 1.2	I	5	A: Ph; Y: Br, Pa(f), Xe	C	—
504539	JGM 232	12 — 76	F	183	—	0	1	—	—	I	1.5	Y: Br	C	—
504540	JGM 233	2 5 77	F	206	141	1	1	—	—	I	3	A: Ph, Y: Br, Na	C	—
504541	JGM 234	1 — 77	F	250	—	4	1	—	—	M	—	Y: Br, Pa(f)	P	6
504549	JGM 242	2 27 77	F	205	—	0	—	—	—	—	2	—	E	—
504550	JGM 243	2 27 77	F	201	103	0	1	—	—	I	2	A: Ph, St, Cr, Xe; Y: Br, Pa(n), Na	—	—
504551	JGM 244	2 11 77	M	216	145	1	1	20.0	10.0 2.3 1.9	I	4	A: Ph, St, Cr; Y: Br, Pa(f), Na, Xe	—	—
504552	JGM 245	2 28 77	F	217	—	1	—	—	—	—	—	A: Ph; Y: Br	E	—
504553	JGM 246	2 28 77	M	162	—	0	—	—	—	I	1	A: Ph; Y: Br	E	—
504559	JGM 252	3 1 77	F	252	220	1	3	—	—	M	14	A: Ph, St, Cr; Y: Pa(n), Na, Xe	C	—
504560	JGM 253	3 5 77	F	126	—	0	1	—	3.5 0.9 0.5	I	0	A: Ph	E	—
504561	JGM 254	3 6 77	M	206	118	0	1	—	8.8 2.2 1.9	I	3	A: Ph, Mo; Y: Br, Xe	C	—

(Continued)

Serial	Col. No.	Date (mo dy yr)	Sex	Length (cm)	Weight (kg)	Skull mat	Epip.	Gonad weight (g)	Gonad dim. (L×W×D, cm)	Sex. mat.	Age (GLGs)	Pathol.	Stom. cont.	Flip. epip.
504562	JGM 255	3 5 77	F	132	26	0	1	—	3.5 1.0 0.5	I	0	A: Ph, Xe	E	—
504565	JGM 258	3 5 77	F	192	—	0	—	—	—	—	1	Y: Br, Pa(f)	C	—
504567	JGM 260	3 5 77	M	174	—	0	—	—	—	I	1.0	—	E	—
504573	JGM 266	3 6 77	M	127	27	0	1	—	5.6 1.0 1.0	I	0	A: Xe	E	—
504576	JGM 294	3 19 77	M	114	19	0	—	—	4.6 1.0 1.0	I	0+	A: Ph, Xe	E	—
504578	JGM 298	3 19 77	M	146	—	0	1	—	—	I	0.5	Y: Pa(n)	E	—
504582	JGM 270	3 8 77	F	118	—	0	1	—	2.9 0.9 0.4	I	0+	A: Ph, Xe	E	—
504583	JGM 271	3 10 77	F	185	—	0	1	—	—	I	2	A: Ph; Y: Br, Pa(f)	E	—
504590	JGM 281	3 14 77	F	155	—	0	—	—	—	I	1	Y: Br, Pa(n)	C	—
504591	JGM 282	3 14 77	M	185	—	3	1	—	7.5 1.3 1.1	I	2	Y: Br, Pa(f)	P	—
504592	JGM 284	3 13 77	F	205	—	2	1	—	—	I	3	Y: Br, Pa(f)	P	—
504597	JGM 289	3 17 77	M	117	—	0	—	—	3.2 1.1 0.9	I	0+	—	E	—
504598	JGM 291	3 17 77	F	111	16	0	—	—	—	I	0	A: Xe	—	—
504599	JGM 292	3 17 77	M	116	20	0	—	—	—	I	0	A: Xe	—	—
504601	JGM 303	3 27 77	F	250	190	—	3	—	5.2 1.7 1.4	M	18	A: Ph, St, Xe; Y: Br, Pa(n), Na, Cr(s)	E	—
504607	JGM 325	4 4 77	F	251	198	—	2	—	5.4 2.4 1.3	M	12	A: Ph, St; Y: Br, Pa(f), Na, Cr(m), Xe	E	—
504608	JGM 326	4 4 77	F	266	170	3	3	—	6.5 2.2 1.2	M	16	A: Ph, Mo, Na, Cr; Y: Br, St, Xe	C	—
504609	JGM 328	4 10 77	F	274	—	5	3	—	6.0 4.5 1.0	M	25	Y: Pa(n)	E	—
504726	NN	10 26 77	M	298	—	5	2	—	—	—	12	—	C	—
504739	JGM 348	2 8 78	M	250	—	3	1	68.0	13.5 3.5 1.8	—	11	A: Ph, Mo, St, Cr; Y: Br, Pa(f), Na, Xe	C	—
504753	CWP 072	4 16 78	M	220	157	0	—	—	—	—	—	—	—	—
504754	JGM 357	3 24 78	M	258	—	3	1	73.0	14.0 4.5 2.0	I	11	—	C	—
504755	JGM 358	3 25 78	F	268	270	5	3	11.0	5.5 2.7 0.8	M	—	A: Ph, Mo, Br, Na, St; Y: Pa(n), Cr(m), Xe	C	—
504857	NN	5 14 78	F	306	—	—	3	—	—	—	—	Y: Ph	—	—
504870	JGM 406	1 8 79	M	264	—	—	1	26.4	11.7 2.4 1.6	—	12	A: Ph, Mo, St; Y: Br, Pa(f), Na, Cr, Xe	C	—
504879	JGM 411	1 27 79	M	284	—	5	2	52.2	13.5 3.9 1.6	M	16+	A: Mo, St, Cr; Y: Ph, Br, Pa(n), Na, Xe	C	6
504881	JGM 413	2 1 79	M	206	87	0	1	7.8	6.8	I	—	A: Ph, Mo, Xe; Y: Pa(n)	C	3
504884	GWO 001	2 22 79	M	287	—	—	2	324.0	17.5 7.5 4.0	—	—	A: Ph; Y: Xe	C	—
504887	JGM 416	3 8 79	F	286	—	6	3	—	5.1 3.2 1.7	M	—	A: Ph, Na, St, Cr; Y: Mo, Pa(n), Xe	P	—
504906	NN	5 10 79	M	288	—	3	—	—	—	—	—	—	P	—
504961	GWO 007	5 22 79	M	272	—	3	—	—	—	M	17+	—	—	—
550095	RGD 001	9 10 81	M	277	—	6	—	—	20.3	—	—	A: Mo; Y: Xe	—	—
550109	RGD 002	12 4 81	F	220	155	5	—	—	6.4 2.2	I	3	A: Ph, Mo; Q: Na, St, Cr; Y: Br, Xe	C	3.5
550130	CWP 128	5 15 82	M	268	—	3	1	—	24.5	—	10	Y: Xe	—	—
550176	CWP 129	6 1 82	F	221	—	0	—	—	—	I	2	A: Ph, Mo, Cr; Y: Br, Pa(f), Xe	C	—
550177	CWP 132	8 19 82	M	261	—	3	1	—	—	—	10	A: Ph, Mo, Cr, Xe; Y: Br	E	5
550198	WAW 780	—	M	263	—	3	—	—	—	—	—	—	—	—
550209	RGD 009	4 11 82	F	232	—	5	—	—	—	I	2	A: Xe; Q: Ph, Mo; Y: Cr	C	—
550213	CWP 127	5 6 82	F	257	—	4	3	—	—	M	—	A: Ph, Mo, Cr; Y: Br, Xe	—	—
550225	LKG 143	4 24 82	F	258	210	7	3	—	—	M	—	A: Ph, Mo, Br, Cr, Xe	E	—
550226	CWP 125	5 4 82	M	188	—	5	—	—	7.7 1.6	I	—	—	—	—
550309	CWP 213	5 17 83	M	249	—	2	1	—	12.0	—	6	—	—	4
550311	CWP 215	5 20 83	M	106	—	0	1	—	—	I	—	—	—	—
550313	CWP 217	6 6 83	F	246	—	—	—	—	—	M	—	Y: Xe	—	6
550314	CWP 216	6 7 83	M	238	—	2	—	—	—	—	8	—	—	—
550315	CWP 218	6 9 83	F	234	—	2	—	—	—	M	11	—	—	—
550354	JGM 429	8 3 83	F	202	99	0	1	2.0	4.3 1.5 0.5	I	—	Y: Br, Pa(f), Xe	—	3.5
550363	JMM 001	1 15 84	F	240	146	4	3	6.2	5.0 2.4 0.7	M	20+	Y: Br	C	6
550364	DAP 006	4 29 84	F	258	—	4	3	—	5.0 2.0 1.2	M	20+	A: Ph, Mo, Xe; Y: Br	C	6
550365	JMM 003	2 14 84	F	168	—	0	—	—	—	I	1	—	C	2
550372	DAP 007	5 1 84	M	227	—	8	—	—	—	—	—	—	E	—
550381	MPH 001	9 5 84	M	237	—	3	—	—	9.5 3.1 2.0	I	4	A: Ph; Y: Br, Pa(n)	C	—
550382	WAM 001	9 5 84	F	263	—	3	3	—	—	—	—	A: Ph, Br; Y: Pa(n)	C	—
550386	WAM 003	9 9 84	M	226	172	4	1	29.5	10.3 2.3	I	—	—	—	4
550387	WAM 004	9 9 84	M	267	—	7	—	—	19.0 5.0	M	21	A: Ph	C	—

Serial	Col. No.	Date (mo dy yr)	Sex	Length (cm)	Weight (kg)	Skull mat	Epip.	Gonad weight (g)	Gonad dim. (L×W×D, cm)	Sex. mat.	Age (GLGs)	Pathol.	Stom. cont.	Flip. epip.
550388	MPH 002	9 10 84	M	272	—	8	3	—	30.0 9.0	M	—	A: Ph; Y: Pa(n)	C	—
550394	JGM 437	9 16 84	F	203	—	0	—	—	—	I	—	—	—	—
550395	JGM 438	9 18 84	M	256	—	8	2	180.0	15.1 6.8 4.3	—	15	A: Br; Y: Xe	P	—
550398	MPH 003	10 17 84	M	171	—	0	—	—	5.0 3.0	I	—	A: Ph; Y: Br, Xe	E	—
550400	MPH 004	10 24 84	M	203	—	0	—	—	9.0 2.5 1.5	—	—	A: Ph; Y: Br, Xe	C	—
550401	WAM 008	11 1 84	M	268	—	4	3	—	28.0 8.0 5.0	—	—	A: Ph, Mo, Cr	C	—
550402	MPH 005	—	M	179	—	0	1	10.4	7.5 1.7 1.3	I	—	A: Ph, Mo; Y: Br, Xe	E	—
550403	WAM 009	11 18 84	M	202	—	0	2	—	—	—	2	A: Ph, Mo; Y: Xe	C	—
550406	WAM 011	12 13 84	M	257	—	6	—	—	—	—	10+	—	E	—
550422	WAM 014	3 13 85	M	279	—	7	2	—	26.0 8.0 6.0	M	24	—	C	—
550428	WAM 017	3 29 85	M	249	—	—	1	—	14.6 4.8 2.8	—	—	Y: Xe	C	—
550430	WAM 018	4 5 85	F	255	—	3	3	—	—	M	—	—	P	—
550432	MPH 009	4 11 85	F	239	190	2	1	13.1	6.5 3.4 0.9	M	8	A: Ph, Mo	C	4
550433	WAM 019	4 16 85	F	251	—	3	—	—	—	M	20+	Y: Mo	C	—
550434	MPH 010	4 18 85	M	165	—	—	—	—	—	—	—	Y: Br	—	—
550436	WAM 023	4 28 85	F	243	200	7	1	54.2	3.7 1.7 1.5	M	10	A: Ph, Mo; Y: Xe	C	—
550439	WAM 025	5 10 85	F	216	—	0	1	—	—	M	2.5	A: Ph, Mo; Y: Br	C	—
550440	WAM 028	5 17 85	F	264	240	3	3	35.0	2.2 1.2 0.5	M	—	A: Ph, Mo, Na, St,Cr; Y: Br, Xe	E	—
550442	WAM 029	6 3 85	F	231	—	—	—	—	—	M	—	—	—	—
550445	WAM 030	6 10 85	F	221	—	2	—	—	4.1 1.3 0.8	I	2	A: Ph, Mo; Y: Br, Xe	C	—
550446	MH84-339	11 20 84	M	279	494	7	2	230	18.6 6.1 2.4	M	24	Y: Ph	P	—
550454	WAM 032	7 — 85	M	222	—	0	—	—	8.0 1.7 1.0	—	—	Y: Br	C	—
550455	WAM 033	7 17 85	M	232	—	6	—	—	9.5 2.4 1.7	—	4	Y: Br	C	—
550481	MPH 016	9 20 85	M	205	—	—	1	—	6.0 2.2 1.5	I	—	A: Ph, Mo; Y: Xe	C	—
550493	MPH 023	11 16 85	F	259	—	—	—	—	—	—	—	—	—	—
550736	WAM 075	1 — 86	F	207	—	7	—	—	4.4 1.9 0.9	I	3	—	C	—
550747	JGM 441	2 7 86	F	270	—	9	—	—	—	M	—	—	—	—
550753	MPH 026	2 — 86	F	244	—	2	—	—	—	—	—	—	—	—
550756	WAM 079	2 20 86	F	251	—	—	3	—	4.5 2.1 1.5	M	—	Y: Xe	C	—
550772	WAM 080	2 28 86	F	270	—	2	2	—	4.1 2.1 1.2	M	11	A: Mo, St; Q: Cr; Y: Ph, Br, Pa(f), Na, Xe	C	5.5
550816	WAM 085	4 18 86	F	263	—	4	—	—	3.9 2.6 1.3	M	—	A: Ph, Mo, Br; Q: Na, St; Y: Xe	C	—
550817	WAM 086	4 — 86	M	206	—	0	—	—	8.7 2.0 1.6	I	—	Y: Br	C	—
550820	WAM 089	5 2 86	M	252	—	4	—	—	24.8 9.9 8.5	M	8+	—	C	—
550826	RAB 073	4 7 86	M	172	78	—	—	8.1	7.2 1.5 1.1	I	—	A: Xe; Y: Br	—	—
550834	RAB 079	6 9 86	M	202	—	—	1	17.6	7.5 2.2 1.7	I	—	Y: Br	C	—
550835	RAB 080	6 14 86	F	106	—	—	—	0.3	2.8 1.2 0.5	I	0	—	—	—
550838	WAM 096	8 30 86	F	196	—	0	1	—	—	—	—	A: Ph, Mo; Y: Br	E	—
550832	WAM 024	5 9 85	M	102	—	—	—	—	—	I	—	—	—	—
550851	WAM 097	10 8 86	F	166	—	—	1	—	—	—	—	A: Ph, Mo, Br, Xe	E	—
NN	AL	6 4 77	M	120	—	—	—	—	—	—	0+	—	—	—
NN	CWO 075	8 26 78	F	206	84	—	1	—	—	I	2.2	A: Ph, Mo; Y: Pa(f),Xe	C	—
NN	CWP 076	8 25 78	F	227	109	—	1	—	—	I	3	A: Ph, Mo; Y: Pa(f), Xe	C	—
NN	CWP 077	9 14 78	F	214	—	—	1	—	—	I	3	A: Ph, Mo	C	—
NN	CWP 079	9 14 78	F	255	—	—	1	—	—	—	—	—	C	—
NN	CWP 080	10 5 78	F	251	—	—	1	—	—	I	15	Y: Pa(f)	C	—
NN	CWP 081	4 19 79	M	274	—	—	2	—	26.0	M	—	A: Br	E	—
NN	CWP 083	3 — 79	F	265	—	—	3	—	—	M	—	—	C	—
NN	CWP 084	3 27 79	M	223	—	—	1	—	—	—	4	—	—	—
NN	CWP 086	3 25 79	M	113	—	—	—	—	—	—	—	—	—	—
NN	CWP 087	3 27 79	M	110	—	—	—	—	—	—	0	—	—	—
NN	CWP 088	5 15 79	M	240	—	—	—	—	—	—	10	—	—	—
NN	CWP 089	5 15 79	M	230	—	—	1	—	—	—	14	—	—	—
NN	CWP 090	5 16 79	—	252	—	—	—	—	—	—	—	—	—	—
NN	CWP 091	5 16 79	M	200	—	—	—	—	—	—	—	—	—	—
NN	CWP 094	5 21 79	F	200	—	—	—	—	—	I	1	A: Xe; Y: Pa(n)	C	—
NN	CWP 095	5 21 79	F	223	—	—	—	—	—	I	3	Y: Pa(n)	—	—
NN	CWP 100	1 8 80	M	226	—	—	—	—	—	—	11	—	—	—
NN	CWP 112	5 12 81	F	189	—	—	—	—	—	—	—	—	—	—
NN	CWP 116	5 30 81	M	202	109	—	1	—	—	I	—	A: Ph, Mo, Na, St, Cr; Y: Br, Xe	E	—

(*Continued*)

Serial	Col. No.	Date (mo dy yr)	Sex	Length (cm)	Weight (kg)	Skull mat	Epip.	Gonad weight (g)	Gonad dim. (L×W×D, cm)	Sex. mat.	Age (GLGs)	Pathol.	Stom. cont.	Flip. epip.
NN	CWP 122	11 10 81	M	219	—	—	1	—	—	—	—	Q: Xe; Y: Ph, Pa(f)	C	—
NN	DAP 002	3 22 83	F	227	—	—	2	—	—	M	10	A: Ph, Mo; Y: Pa(f), Xe	C	—
NN	DAP 008	5 20 85	F	113	—	—	—	—	—	I	—	—	—	—
NN	DAP 010	3 9 86	F	194	—	—	—	—	5.7 1.9 0.6	I	—	Y: Br, Xe	C	—
NN	DAP 009	5 20 84	M	103	—	—	—	—	—	I	—	—	—	—
NN	GWO 012	6 23 79	F	239	—	—	—	—	—	—	—	—	C	—
NN	JGM 290	3 17 77	F	120	—	—	—	—	—	I	—	—	—	—
NN	JGM 297	3 23 77	M	169	—	—	1	—	—	I	—	Y: Pa(f), Xe	C	—
NN	JGM 299	3 23 77	F	112	—	—	—	—	—	I	—	—	—	—
NN	JGM 300	3 25 77	M	115	18	—	1	—	—	I	0	A: Ph	—	—
NN	JGM 325F	4 1 77	F	113	15	—	—	—	—	—	0−	—	—	—
NN	JGM 328F	4 10 77	M	49	1	—	—	—	—	I	—	—	—	—
NN	JGM 329	4 10 77	M	106	—	—	—	—	—	I	0	—	—	—
NN	JGM 330	4 10 77	M	117	21	—	—	—	—	I	0	—	—	—
NN	JGM 338	1 18 78	F	239	—	—	1	3.0	4.2 1.3 0.8	I	8	A: Ph, Xe; Y: Br, Pa(f)	E	—
NN	JGM 342	1 24 78	F	240	—	—	—	3.5	—	I	—	Y: Br	C	—
NN	JGM 345	1 30 78	M	134	33	—	—	—	—	—	—	—	—	—
NN	JGM 347	2 4 78	M	204	83	—	—	—	—	—	—	Y: Xe	—	—
NN	JGM 350	2 25 78	F	190	—	—	—	1.9	—	I	—	A: Ph; Y: Br	E	—
NN	JGM 351	3 6 78	M	108	—	—	—	1.8	—	I	0+	A: Ph, Br, Xe	E	—
NN	JGM 359	3 29 78	M	192	73	—	—	—	—	—	2	Y: Xe	—	—
NN	JGM 384	5 7 78	M	202	105	—	1	13.0	8.5 1.5 1.2	I	2.25	A: Ph, Mo, Xe; Y: Br, Pa(n)	E	—
NN	JGM 388	5 12 78	F	242	188	—	3	—	—	M	16+	A: Ph, Mo; Y: Pa(n), Xe	C	—
NN	JGM 389	5 12 78	F	189	97	—	1	—	—	I	—	A: Ph, Mo; Y: Pa(n), Xe	C	—
NN	JGM 396	12 11 78	M	172	59	—	—	4.4	6.6 1.2 0.8	I	1	A: Xe; Y: Br, Pa(f)	E	—
NN	JGM 401	12 12 78	M	143	43	—	—	—	—	—	1.75	A: Br; Y: Xe	—	—
NN	JGM 404	1 2 79	M	203	—	—	—	—	—	—	3	Y: Br, Pa(n)	C	—
NN	JGM 420	10 — 80	F	200	—	—	1	—	—	I	—	A: Ph, Mo; Y: Br, Pa(f)	C	—
NN	JGM 422	3 6 81	M	216	—	—	—	—	—	—	—	A: Xe; Y: Br	C	—
NN	JGM 431	4 18 84	F	200	—	—	—	—	—	I	—	Y: Br, Pa(f), Xe	C	—
NN	JGM 443	2 8 86	M	254	—	—	—	—	10.0 3.5 3.0	—	—	—	C	—
NN	LKG 030F	4 29 76	M	94	—	—	—	—	—	I	—	—	—	—
NN	LKG 144	5 26 82	M	216	112	—	1	—	10.7 2.0 1.3	—	—	A: Ph, Mo, Cr; Q: St; Y: Br, Xe	E	3
NN	MPH 007	3 13 85	M	170	—	—	—	—	—	—	—	A: Ph; Y; Br, Xe	C	—
NN	MPH 011	4 18 85	—	193	—	—	—	—	—	—	—	—	—	—
NN	MPH 012	4 21 85	F	112	—	—	—	—	—	—	—	A: Ph, Br	E	—
NN	MPH 015	5 6 85	M	186	—	—	—	—	—	—	—	—	—	—
NN	MPH 018	9 23 85	M	153	—	—	1	—	—	—	—	—	—	—
NN	RAB 018	7 27 80	M	207	—	—	1	—	—	—	—	—	—	—
NN	RAB 021	8 23 80	—	180	—	—	—	—	—	—	—	—	—	—
NN	RAB 022	8 27 80	F	210	—	—	—	—	—		—	—	—	—
NN	RAB 023	9 2 80	F	181	—	—	1	—	—	—	—	A: Ph, Xe	—	—
NN	RAB 026	9 23 80	F	287	254	—	3	—	—	—	—	A: Ph, Br; Y: Xe	P	—
NN	RAB 027	10 2 80	F	253	—	—	—	—	—	—	—	—	—	—
NN	RAB 028	5 8 80	M	197	—	—	1	—	—	—	—	A: Ph; Y: Xe	C	—
NN	RAB 032	5 12 81	F	230	—	—	1	—	—	—	—	—	E	—
NN	RAB 033	5 12 81	M	203	—	—	—	—	—	—	—	A: Ph; Y; Xe	P	—
NN	RAB 036	6 8 81	—	185	—	—	—	—	—	—	—	—	—	—
NN	RAB 037	6 20 81	F	205	—	—	—	—	—	I	—	A: Ph, Mo; Y: Br, Xe	P	—
NN	RAB 038	7 22 81	M	233	—	—	—	—	—	—	—	A: Ph, Mo; Y: Xe	—	—
NN	RAB 041	9 11 81	M	189	—	—	—	—	—	—	—	—	—	—
NN	RAB 042	10 7 81	F	144	—	—	—	—	—	—	—	A: Xe	C	—
NN	RAB 047	5 7 82	F	204	—	—	2	—	—	I	—	A: Ph; Y; Xe	—	—
NN	RAB 050	7 20 82	—	170	—	—	—	—	—	—	—	—	—	—
NN	RAB 051	7 29 82	M	224	—	—	1	—	—	—	—	A: Xe	—	—
NN	RAB 052	10 20 82	M	197	—	—	—	—	—	—	—	A: Ph, Mo, Xe	—	—
NN	RAB 054	12 20 83	F	263	—	—	—	—	—	M	—	A: Xe	—	—
NN	RAB 056	5 17 84	M	216	—	—	2	—	—	—	—	A: Ph, Mo, Br, Xe	P	—
NN	RAB 057	5 31 84	M	127	—	—	—	—	—	—	—	—	—	—
NN	RAB 058	6 22 84	M	270	—	—	3	—	—	—	—	A: Ph, Mo; Y: Br	P	—
NN	RAB 059	6 24 84	—	200	—	—	—	—	—	—	—	—	—	—
NN	RAB 060	8 5 84	M	208	—	—	—	—	—	—	—	A: Xe; Y: Br	E	—
NN	RAB 061	8 5 84	F	209	—	—	—	—	—	—	—	A: Ph, Mo; Y: Br, Xe	E	—

Serial	Col. No.	Date (mo dy yr)	Sex	Length (cm)	Weight (kg)	Skull mat	Epip.	Gonad weight (g)	Gonad dim. (L×W×D, cm)	Sex. mat.	Age (GLGs)	Pathol.	Stom. cont.	Flip. epip.
NN	RAB 074	5 7 86	M	215	—	—	1	31.0	13.0 2.6 1.9	I	—	A: Ph; Y: Br, Xe	P	—
NN	RAB 075	5 9 86	M	198	—	—	—	15.0	8.0 2.0 1.7	I	—	Y: Xe	—	—
NN	RAB 076	5 11 86	M	245	—	—	1	—	—	—	—	—	—	—
NN	RAB 077	5 21 86	F	213	—	—	—	—	—	I	—	—	—	—
NN	WAM 005	11 14 84	M	152	—	—	—	—	5.7 1.2 1.1	I	—	A: Ph, Mo, Br, Xe	E	—
NN	WAM 020	4 16 85	F	99	—	—	—	—	—	—	—	—	—	—
NN	WAM 021	4 18 85	F	238	—	—	—	—	—	—	—	—	—	—
NN	WAM 022	4 20 85	F	124	—	—	—	—	—	—	—	—	—	—
NN	WAM 023F	4 28 85	M	107	—	—	—	—	—	I	0	—	—	—
NN	WAM 027	5 17 85	F	108	—	—	—	—	—	—	—	—	—	—
NN	WAM 029F	6 3 85	M	101	—	—	—	—	—	I	0–	—	—	—
NN	WAM 092	5 5 86	M	106	—	—	1	—	—	I	—	—	—	—

REFERENCES

Berg, J. W. (1963). Differential staining of spermatozoa in sections of testis. *Am. J. Clin. Pathol.* **23,** 513–515.

Baylis, H. A. (1932). A list of worms parasitic in Cetacea. *Discovery Rep.* **6,** 393–418.

Blaylock, R. A. (1984). The distribution and abundance of the bottlenose dolphin, *Tursiops truncatus,* in Virginia. M.S. thesis presented to the School of Marine Science, College of William and Mary. x + 56 pp.

Cadenat, J. (1959). Notes sur les delphinides Ouest-Africains. VI. Les gros dauphin gris (*Tursiops truncatus*), est il capable de faire des plongées profondes? *Bull. Inst. Fr. d'Afrique Noire, Ser. A* **21**(3), 1137–1143.

Caldwell, D. K., and Caldwell, M. C. (1972). "The World of the Bottlenosed Dolphin." Lippincott, 157 pp.

Connor, P. F. (1971). The mammals of Long Island, New York. *N. Y. State Mus., Bull.* No. 416. v + 78 pp.

Cowan, D. F., Walker, W. A., and Brownell, R. L., Jr. (1986). Pathology of small cetaceans stranded along southern California beaches. *In* "Research on Dolphins" (M. M. Bryden and R. J. Harrison, eds.), pp. 323–367. Clarendon Press, Oxford. xii + 478 pp.

Cronan, J. M., and Brooks, A. (1962). The Mammals of Rhode Island." Rhode Island Department of Agriculture and Conservation, Division of Fish and Game, Wildlife Pamphlet No. 6. ix + 133 pp.

Dailey, M. D., and Brownell, R. L., Jr. (1972). A checklist of marine mammal parasites. *In* "Mammals of the Sea, Biology and Medicine" (S. H. Ridgway, ed.), pp. 528–589. Thomas, Springfield, Illinois. xiii + 812 pp.

Dailey, M. D., and Perrin, W. F. (1973). Helminth parasites of porpoises of the genus *Stenella* in the eastern tropical Pacific, with descriptions of two new species: *Mastigonema stenellae* gen. et sp. n. (Nematoda: Spiruroidea) and *Zalophotrema pacificum* sp. n. (Trematoda: Digenea). *Fishery Bull.* **71**(2), 455–471.

Fraser, F. C. (1974). Report on Cetacea stranded on the British coasts from 1948 to 1966. Br. Mus. (Nat. Hist.), No. 14. iii + 65 pp., 9 maps.

Geraci, J. G., Dailey, M. D., and St. Aubin, D. J. (1978). Parasitic mastitis in the Atlantic white-sided dolphin, *Lagenorhynchus acutus,* as a probable factor in herd productivity. *J. Fish. Res. Board Can.* **35,** 1350–1355.

Gunter, G. (1942). Contributions to the natural history of the bottlenose dolphin, *Tursiops truncatus* (Montagu), on the Texas coast, with particular reference to food habits. *J. Mammal.* **23**(3), 267–276.

Gunter, G. (1951). Consumption of shrimp by the bottlenosed dolphin. *J. Mammal.* **32**(4), 465–466.

Harris, J. C. (1938). Porpoises feeding. *Life* **5**(12), 67.

Hoese, H. D. (1971). Dolphin feeding out of water in a salt marsh. *J. Mammal.* **52**(1), 222–223.

Hoglund, R. (1965). Porpoises feeding. *Underwater Naturalist* **3**(2), 3, 50–51.

Hohn, A. A. (1980). Age determination and age related factors in the teeth of western North Atlantic bottlenose dolphins. *Sci. Rep. Whales Res. Inst. Tokyo* **32,** 39–66.

Hohn, A. A. (1985). Comparison of methods to estimate the average age at sexual maturity in dolphins. *Sixth Biennial Conference Biol. Mar. Mamm.,* Abstr.

Hui, C. A. (1977). Growth and physical indices of maturity in the common dolphin, *Delphinus delphis. In* "Breeding Dolphins; Present Status, Suggestions for the Future" (S. H. Ridgway and K. Benirschke, eds.), pp. 231–260. Marine Mammal Commission, Report No. MMC-76-07 (NTIS PB-273 673). iii + 308.

Kasuya, T., Tobayama, T., Saiga, T., and Kataoka, T. (1986). Perinatal growth of delphinoids: Information from aquarium reared bottlenose dolphins and finless porpoises. *Sci. Rep. Whales Res. Inst., Tokyo* **37,** 85–97.

Kemp, R. J. (1949). Report on stomach analysis from June 1, 1949–August 31, 1949. *Annu. Rep. Mar. Lab. Texas Fish, Game Oyster Comm.,* Fiscal Year 1948–1949, pp. 111, 112, 126, 127.

Kleinenberg, S. E. (1938). Nekotorye dannye o pitanii *Tursiops tursio* Fabr. b Chernom more. *Bull. Soc. Nat. Moscou* **47,** 406–413.

Kleinenberg, S. E. (1956). Mammals of the Black Sea and the Sea of Azov. *In* "Results of the Joint Biological–Commercial Dolphin Whaling Studies." USSR Academy of Sciences Press, Moscow. 287 pp. [Original in Russian, transl. for Fisheries and Marine Service, Canada, Translation Series No. 4319, 1978.]

Leatherwood, S., Deerman, M. W., and Potter, C. W. (1978). Food and reproductive status of nine *Tursiops truncatus* from the northeastern United States coast. *Cetology,* No. 28. 6 pp.

Little, E. A., and Andrews, J. C. (1982). Drift whales at Nantucket: The kindness of Moshup. *Man in the Northeast* **23,** 17–38.

McBride, A. F., and Kritzler, H. (1951). Observations on pregnancy, parturition, and post-natal behavior in the bottlenose dolphin. *J. Mammal.* **32**(3), 251–266.

McKenzie, K. (1983). Parasites as biological tags in fish population studies. *Appl. Biol.* **7,** 251–331.

Mairs, D. F., and Scattergood, L. W. (1958). Recent Maine records of the bottlenose porpoise and the beluga. *Maine Field Naturalist* **14**(4), 78–80.

Malm, E. N. (1933). Delfini chernova morya. *Priroda* **1933**(2), 31–39.

Mead, J. G. (1975). Preliminary report on the former net fisheries for *Tursiops truncatus* in the western North Atlantic. *J. Fish. Res. Board Can.* **32**(7), 1155–1162.

Mitchell, E. D. (1975). Porpoise, dolphin and small whale fisheries of the world, status and problems. *Int. Union Conservation Nature Nat. Resources, Monograph* No. 3. 129 pp.

Montagu, G. (1821). Description of a species of *Delphinus,* which appears to be new. *Mem. Wernerian Nat. Hist. Soc., Edinburgh* **3,** 75–82, 1 plate.

Morozov, D. A. (1970). [Dolphins hunting.] *Rybnoe Khoziastvo* [*Fish. Ind., Moscow*] **46**(5), 16–17.

Mowbray, L. L. (1952). Certain citizens of the warm sea. *In* "The Book of Fishes," pp. 127–161. National Geographic Society, Washington, D.C. 243 pp.

Neiland, K. A., Rice, D. W., and Holden, B. L. (1970). Helminths of marine mammals, I. The genus *Nasitrema,* air sinus flukes of delphinid Cetacea. *J. Parasitol.* **56**(2), 305–316.

Norris, K. S., and Prescott, J. H. (1961). Observations on Pacific cetaceans of California and Mexican waters. *Univ. Calif. Publ. Zool.* **63**(4), 291–402, plates 27–41.

Odell, D. K. (1975). Status of aspects of the life history of the bottlenose dolphin, *Tursiops truncatus,* in Florida. *J. Fish. Res. Board Can.* **32**(7), 1055–1058.

Ogden, J. A., Conlogue, G. J., Rhodin, M. H. S., and Rhodin, A. G. J. (1981). Roentgenographic indicators of skeletal maturity in marine mammals. *Skeletal Radiol.* **7,** 119–123.

Packard, C. (1958). December, 1957, Mammal observations. *Maine Field Naturalist* **3**(1), 9.

Paradiso, J. L. (1969). Mammals of Maryland. *North Am. Fauna,* No. 66. 193 pp.

Perrin, W. F., and Myrick, A. C., eds. (1980). Age determination of toothed whales and sirenians. *Rep int. Whal. Commn, Special* Issue 3. viii + 229 pp.

Perrin, W. F., and Powers, J. E. (1980). Role of a nematode in natural mortality of spotted dolphins. *J. Wildl. Manage.* **44**(4), 960–963.

Potter, C. W. (1984). The marine mammals of Maryland. *In* "Threatened and Endangered Plants and Animals of Maryland" (A. W. Norden, D. G. Forester, and G. H. Fenwick, eds.), pp. 442–453. Maryland Natural Heritage Program, Baltimore, Maryland, Special Publication 84-1. iv + 475 pp.

Ross, G. J. B. (1977). The taxonomy of bottlenosed dolphins *Tursiops* species in South African waters, with notes on their biology. *Ann. Cape Prov. Mus. (Nat. Hist.)* **11**(9), 135–194.

Ross, G. J. B. (1984). The smaller cetaceans of the southeast coast of southern Africa. *Ann. Cape Prov. Mus. (Nat. Hist.)* **15**(2), 173–410.

Schryver, H. F., Medway, W. M., and Williams, J. F. (1967). The stomach fluke *Braunina cordiformis* in the Atlantic bottlenose dolphin. *J. Am. Vet. Med. Assoc.* **151**(7), 884–886.

Sergeant, D. E., and Fisher, H. D. (1957). The smaller Cetacea of eastern Canadian waters. *J. Fish. Res. Board Can.* **14,** 83–115.

Sergeant, D. E., Mansfield, A. W., and Beck, B. (1970). Inshore records of Cetacea for eastern Canada, 1949–68. *J. Fish. Res. Board Can.* **27,** 1903–1915.

Sergeant, D. E., Caldwell, D. K., and Caldwell, M. C. (1973). Age, growth and maturity of bottlenosed dolphin (*Tursiops truncatus*) from northeast Florida. *J. Fish. Res. Board Can.* **30,** 1009–1011.

Solangi, M. A., and Dukes, G. E. (1983). Atlantic bottlenose dolphin, *Tursiops truncatus,* herd studies in the Mississippi Sound, U.S.A.: Capture, freeze marking and biological sampling. *Fin. Rep. Natl. Mar. Fish. Serv.* Contract No. NA82-GA-C-00023. xi + 178 pp.

Tayler, C. K., and Saayman, G. S. (1972). The social organization and behavior of dolphins (*Tursiops aduncus*) and baboons (*Papio ursinus*): Some comparisons and assessments. *Ann. Cape Prov. Mus.* **9**(2), 11–49.

Townsend, C. H. (1914). The porpoise in captivity. *Zoologica* **1**(16), 289–299, 9 plates.

Tsalkin, V. I. Nekotorye nablyudeniya nad biologiei delfinov Azobskovo i Chernovo morei [Some observations of the biology of the Asoff and Black Sea dolphins]. *Bull. Soc. Nat. Moscou* **49** (1), 61–70.

True, F. W. (1885). The bottle-nose dolphin, *Tursiops tursio,* as seen at Cape May, New Jersey. *Science* **5**(116), 338–339.

True, F. W. (1891). Observations on the life history of the bottlenose porpoise. *Proc. U.S. Natl. Mus.* **13,** 197–203.

Ulmer, F. A. (1980). New Jersey's dolphins and porpoises. *N. J. Audubon Soc., Occas. Pap.* No. 137. 11 pp.

Walker, W. A. (1981). Geographical variation in morphology and biology of bottlenose dolphins (*Tursiops*) in the eastern North Pacific. Natl. Mar. Fish. Ser., Southwest Fish. Center, Administrative Report No. LJ-81-03C. 19 pp. + 6 tables.

Walker, W. A., Hochberg, F. G., and Hacker, E. S. (1984). The potential use of the parasites *Crassicauda* (Nematoda) and *Nasitrema* (Platyhelminthes) as biological tags and their role in the natural mortality of common dolphins, *Delphinus delphis,* in the eastern North Pacific. Nat. Mar. Fish. Ser., Southwest Fish. Center, Administrative Report No. LJ-84-08C. 15 pp., 7 figures, appendix.

Waters, J. H., and Rivard, C. J. (1962). "Terrestrial and Marine Mammals of Massachusetts and Other New England States." Standard-Moder Printing Co. Inc., Brockton, Massachusetts. 151 pp., 33 figs.

Williams, H. H. (1968). The taxonomy, ecology and host-specificity of some Phyllobothriidae (Cestoda: Tetraphyllidae), A critical revision of *Phyllobothrium* Beneden, 1849 and comments on some allied genera. *R. Soc. London, Philos. Trans. Ser. B* **253,** 231–301, plates 13–17.

Winn, H. E. (1982). A characterization of marine mammals and turtles in the Mid- and North Atlantic areas of the U.S. outer continental shelf. Final report of the Cetacean and Turtle Assessment Program, University of Rhode Island. Prepared for the Bureau of Land Management, U.S. Department of the Interior, Contract No. AA551-CT8-48. vii + 450 pp., appendices.

PHOTO: The sounds of a wild bottlenose dolphin calf are recorded while its mother, temporarily restrained in a stretcher, is examined by researchers (Sarasota Bay, Florida, April 1976; Michael D. Scott).

Part V

Behavior and Ecology

Review of the Signature-Whistle Hypothesis for the Atlantic Bottlenose Dolphin

10

Melba C. Caldwell
David K. Caldwell
Marineland of Florida
St. Augustine, Florida 32084

Peter L. Tyack
Woods Hole Oceanographic Institution
Woods Hole, Massachusetts 02543

INTRODUCTION

Bottlenose dolphins (*Tursiops truncatus*) have several potential channels for communication between conspecifics (e.g., see D. K. Caldwell and M. C. Caldwell, 1972a, 1977; Herman and Tavolga, 1980). These dolphins have highly developed senses of sight, hearing, and touch; different displays are detected using each of these senses. Most research on dolphin communication, including our own, has concerned the acoustic signaling system. The sounds produced by dolphins fall into three categories. Two are broad-band: the click sounds used in echolocation (see Nachtigall, 1980) and the burst pulsed sounds often described as squawks, yelps, barks, etc. (Caldwell and Caldwell, 1967). The third category, the whistle, is a frequency-modulated narrow-band sound which appealed to us because of its relative ease of analysis and categorization.

Since 1965, M. Caldwell and D. Caldwell have studied the whistles of 126 bottlenose dolphins of both sexes and a wide range of ages. For most dolphins, we have more than 100 whistles drawn from several recording sessions. Our total sample includes 22,278 whistles (Table 1). The results, which have not been published elsewhere, are summarized in this chapter; we clarify and elaborate the "signature-whistle hypothesis." This hypothesis arose from our observation, made in 1965, that each dolphin in a captive group tended to produce whistles which were individually distinctive and stereotyped in certain acoustic features. Since the individually distinctive features of these whistles were so striking to us, we called them signature whistles. While some acoustic features of an individual's whistles change as a function of behavioral context, as discussed below, others do not. We have hypothesized that the individually distinctive attributes of signature

Table 1
Data on Sex, Age Class, Size, Capture Location, and Whistles from the 126 Bottlenose Dolphins in This Study[a]

Animal number	Sex	Age class	Age at start (yr)	Size at start (cm)	Total no. of whistles	% Signature	% Aberrant	Duration (seconds) Average	Minimum	Maximum	Average no. loops	Maximum no. loops	Span of recordings (months)	No. of recording sessions
1	F	1	0	—	836	100	0	0.4	0.2	0.6	1	1	11.5	9
2	F	1	—	106	79	100	0	0.5	0.1	0.8	1	1	0	1
3	F	1	—	156	324	99	1	0.5	0.1	0.9	1	1	0	4
4	F	1	—	166	276	100	0	0.7	0.1	1.4	1	1	1	4
5	F	1	—	168	1453	100	0	1.3	0.06	3.1	1.3	5	0	5
6	F	2	—	170	30	100	0	1.1	0.5	1.9	1.9	3	1.4	3
7	F	2	—	178	35	100	0	0.5	0.1	1	1	1	0	1
8	F	2	—	180	100	97	3	0.7	0.4	1	3.4	5	0	1
9	F	2	—	185	219	100	0	0.7	0.3	1.1	1	1	0	1
10	F	2	—	185	220	95	5	1	0.1	1.8	1.8	5	0	10
11	F	2	—	192	102	97	3	0.3	0.1	0.6	1	1	0	4
12	F	2	—	197	129	95	5	0.9	0.5	1.2	1.8	4	25	7
13	F	2	—	200	119	95	5	0.9	0.1	1.5	3.1	5	0	1
14	F	2	—	202	100	100	0	0.8	0.4	1.5	1.1	2	0	2
15	F	2	—	205	89	98	2	0.7	0.3	0.9	1	1	0	1
16	F	2	—	208	100	90	10	0.8	0.1	1.3	1	2	7	10
17	F	2	—	—	139	100	0	1.4	0.6	2.2	1.8	3	0	2
18	F	3	—	211	140	100	0	1.3	0.1	2.9	2.9	5	1	3
19	F	3	—	212	100	100	0	1.1	0.8	1.5	2.5	5.5	1	4
20	F	3	—	213	90	93	7	0.9	0.3	3.8	2.3	8	0	1
21	F	3	—	214	41	100	0	0.7	0.1	2.1	1.5	6	0	4
22	F	3	—	215	100	97	3	0.2	0.2	1	1	1	0	2
23	F	3	—	216	105	93	7	1	0.2	1.3	1	1	1	5
24	F	3	—	217	86	100	0	1	0.7	2	2.6	5	0	1
25	F	3	—	218	590	100	0	0.8	0.2	1	1	1	0	3
26	F	3	—	218	100	98	2	0.9	0.1	2	1.5	4.5	0	1
27	F	3	—	222	31	100	0	1	0.1	2.4	2.2	7.5	0	1
28	F	3	—	225	23	100	0	1.2	0.6	2.9	—	—	4	3
29	F	3	—	226	102	95	5	1.6	0.3	4.6	3.9	12	0	1
30	F	3	—	226	52	71	29	1.2	0.3	2	3.5	6	9.5	3
31	F	3	—	227	45	100	0	1	0.4	2.2	3.4	8.5	0	2
32	F	3	—	229	2	—	—	—	—	—	—	—	0	1
33	F	3	—	229	106	100	0	1.6	1	2.4	3.2	5	0	1
34	F	3	—	229	13	100	0	1.2	0.6	1.8	2.3	3	0	1
35	F	3	—	230	100	100	0	1.1	0.4	2.5	2.5	4	3.3	3
36	F	3	—	231	63	98	2	0.8	0.1	1.6	3.1	5	0	2
37	F	3	—	231	11	100	0	1.1	0.5	1.6	2	3	0	1
38	F	3	—	231	76	97	3	1.8	0.5	3.1	3	7	0	1
39	F	3	—	234	17	—	—	0.9	0.1	2	2.9	4	0	1
40	F	3	—	231	125	100	0	1	0.2	1.6	3.2	8	23	8
41	F	3	—	234	120	100	0	1	0.4	2.3	1.9	4	20.5	7
42	F	3	—	234	100	100	0	0.9	0.5	1.5	3.1	5	5.7	5
43	F	4	—	234	124	100	0	1.8	0.8	4.2	5.6	13.5	14.5	5
44	F	4	—	235	141	96	4	0.4	0.1	1	1.8	4	17.4	9
45	F	4	—	238	129	99	1	1.2	0.1	2.8	2.5	6	0	2
46	F	4	—	239	100	98	2	0.9	0.5	1.2	1.5	3	6.7	5
47	F	4	—	239	100	98	2	0.7	0.1	1.1	2.1	5	0	2
48	F	4	—	235	24	83	17	0.9	0.3	1.6	1.5	2	0	1
49	F	4	—	235	5	—	—	—	—	—	—	—	0	2
50	F	4	—	235	87	100	0	1.5	0.1	2	4.1	7	3	7
51	F	4	—	235	77	100	0	1.7	1	2.9	4.4	7.5	11.5	4
52	F	4	—	240	132	99	1	1.3	0.3	2.5	2.2	4	15.5	7
53	F	4	—	241	74	100	0	0.4	0.1	1.1	1	1	1	4
54	F	4	—	243	100	93	7	0.8	0.6	1.9	1.9	3.5	0	2
55	F	4	—	244	115	99	1	0.9	0.2	1.2	2	4	25.7	8
56	F	4	—	244	100	98	2	0.9	0.3	1.3	1.4	3	7.2	4
57	F	4	—	246	19	89	11	0.8	0.3	1.1	1.1	2	0	1
58	F	4	—	248	100	100	0	0.7	0.1	1.3	2.2	4	3.5	2
59	F	4	—	250	92	87	13	1	0.3	2	1.6	5	9.7	3
60	F	4	—	251	2	—	—	—	—	—	—	—	0	1
61	F	4	—	231	100	88	12	1.7	0.1	5.4	3.4	8	7.5	3
62	F	—	—	—	490	84	16	0.6	0.1	1.4	1.1	2.5	0	5
63	M	1	0	109.8	688	5	95	0.9	0.1	2.7	1	4	1.5	9
65	M	1	0	100	245	0	100	0.6	0.4	0.9	1	1	14	6
66	M	1	0	106	159	82	18	0.8	0.1	1.3	1	1	2.5	3
67	M	1	0	152	1051	99	1	1	0.1	1.1	1	1	0	5
68	M	1	0	154.9	1059	96	4	1	0.1	2.3	2.7	12	37	13
69	M	1	0	—	355	100	0	0.4	0.1	0.5	1	1	2	5
70	M	1	0	158	945	99	1	0.6	0.4	0.9	1	1	0	4

Capture location	Recording site	Conditions of Recording									
		Water	Air	Alone	With others	Vet.	Stranded	Being moved	Feeding	Playing	Normal
Born MLF[b]	MLF	Y 6	Y 3	Y 2	Y	Y 2	N	N	N	N	Y 2
Born MLF	MLF	Y 1	N	—	—	N	N	N	N	N	Y 1
Destin, FL	Gulfarium	Y 3	Y 1	—	Y 2	N	Y 1	Y 1	N	N	Y 2
Steinhatchee, FL	MLF	N	Y 4	—	—	Y 4	N	N	N	N	—
Steinhatchee, FL	MLF	Y 5	N	—	—	N	N	N	N	Y 1	Y 3
Port St. Joe, FL	MLF	Y 2	Y 1	—	—	N	N	Y 1	N	N	Y 1
Steinhatchee, FL	MLF	N	Y 1	—	—	Y 1	N	N	N	N	—
Fernandina Beach, FL	MLP	N	Y 1	—	—	Y 1	Y 1	N	N	N	—
Horseshoe Beach, FL	Steinhatchee, FL	N	Y 1	—	—	—	—	—	Y 1	N	—
St. Augustine, FL	MLF	Y 7	Y 3	—	—	Y 2	N	Y 3	N	N	Y 4
St. Augustine, FL	MLF	Y 3	Y 1	—	—	Y 1	N	N	N	N	Y 1
St. Augustine, FL	MLF	Y 3	Y 4	—	—	Y 1	Y 1	Y 2	N	N	Y 2
Near Homasassa Springs, FL	MLF	N	Y 1	—	—	N	N	Y 1	N	N	—
Fernandina Beach, FL	MLF	Y 1	Y 1	—	—	N	N	Y 1	N	N	—
Steinhatchee, FL	Steinhatchee, FL	N	Y 1	—	—	N	N	Y 1	N	N	—
New Smyrna Beach, FL	MLF	Y 8	Y 2	—	—	N	N	Y 2	N	N	Y 8
Tampa Bay, FL	Niagara Falls Aq.	Y 2	N	—	—	N	N	N	N	N	Y 1
Crescent Beach, FL	MLF	N	Y 3	—	—	N	N	Y 3	N	N	—
Vero Beach, FL	MLF	Y 1	Y 3	—	—	Y 1	Y 1	Y 1	N	N	—
Crescent Beach, FL	MLF	N	Y 1	—	—	N	N	N	Y (forced)	N	—
Vero Beach, FL	MLF	Y 2	Y 2	—	—	Y 1	N	Y 1	N	N	—
Port St. Joe, FL	MLF	N	Y 2	—	—	Y 1	N	Y 1	N	N	—
New Smyrna Beach, FL	MLF	Y 2	Y 3	—	—	N	N	Y 1	N	N	Y 1
Crescent Beach, FL	MLF	N	Y 1	—	—	Y 1	N	N	N	N	—
Vero Beach, FL	MLF	Y 1	Y 2	—	—	Y 1	N	Y 1	N	N	—
Port St. Joe, FL	MLF	N	Y 1	—	—	Y 1	N	N	N	N	—
Steinhatchee, FL	MLF	N	Y 1	—	—	N	N	Y 1	N	N	—
Not known	MLP[b]	Y 3	N	—	Y 1	N	N	N	N	Y 1	Y 1
Port St. Joe, FL	MLF	N	Y 1	—	—	N	N	Y 1	N	N	—
Not known	MLP	Y 2	Y 1	—	—	Y 1	N	N	N	N	Y 2
St. Augustine, FL	MLF	N	Y 2	—	—	Y 2	N	N	N	N	—
Not known	MLF	N	Y 1	—	—	N	N	Y 1	N	N	—
Destin, FL	Gulfarium	Y 1	N	N	Y 1	N	N	N	N	N	—
Crescent Beach, FL	MLF	Y 1	N	—	—	N	N	N	N	N	—
Matanzas Inlet, FL	MLF	N	Y 3	—	—	N	N	Y 3	N	N	—
Not known	MLP	Y 1	Y 1	—	—	N	N	Y 1	N	N	—
St. Augustine, FL	MLF	N	Y 1	—	—	N	N	Y 1	N	N	—
Steinhatchee, FL	Steinhatchee, FL	N	Y 1	—	—	N	N	Y 1	N	N	—
Steinhatchee, FL	Steinhatchee, FL	N	Y 1	—	—	N	N	Y 1	N	N	—
Yankeetown, FL	MLF	Y 2	Y 6	—	—	Y 5	N	Y 1	N	N	Y 1
Crescent Beach, FL	MLF	Y 2	Y 5	—	—	Y 3	N	Y 1	N	N	—
St. Augustine, FL	MLF	N	Y 5	—	—	Y 4	N	Y 1	N	N	—
Not known	MLP	Y 4	Y 1	N	Y 4	Y 1	N	N	N	N	—
Steinhatchee, FL	MLF	N	Y 9	—	—	Y 9	N	N	N	N	—
Vero Beach, FL	MLF	Y 2	N	—	—	N	N	N	N	N	—
Born MLF	MLF	Y 1	Y 4	—	—	Y 4	N	N	N	N	—
Fernandina Beach, FL	MLP	Y 1	Y 1	—	—	Y 1	N	N	N	N	—
Not known	Aquatarium	N	Y 1	—	—	N	N	Y 1	N	N	—
Fernandina Beach, FL	MLP	Y 1	Y 1	—	—	N	Y 1	N	N	N	—
Not known	MLP	Y 7	N	N	Y 5	N	N	N	N	N	—
Not known	MLP	Y 4	N	—	Y 2	N	N	N	N	N	—
Matanzas Inlet, FL	MLF	Y 3	Y 4	—	—	Y 2	N	Y 2	N	N	Y 1
St. Augustine, FL	MLF	N	Y 4	—	—	Y 4	N	N	Y (forced)	N	—
Port Orange, FL	MLF	N	Y 2	—	—	N	N	N	N	N	—
St. Augustine, FL	MLF	Y 1	Y 7	—	—	Y 7	N	N	N	N	—
St. Augustine, FL	MLF	Y 1	Y 3	—	—	Y 1	N	Y 2	N	N	—
Crescent Beach, FL	MLF	N	Y 1	—	—	N	N	N	N	N	—
St. Augustine, FL	MLF	Y 2	N	—	—	N	N	N	N	N	Y 1
New Smyrna Beach, FL	MLF	N	Y 3	—	—	Y 3	N	N	N	N	—
St. Augustine, FL	MLF	N	Y 1	—	—	—	N	Y 1	N	N	—
St. Augustine, FL	MLF	Y 2	Y 1	—	—	Y 1	N	Y 1	Y 2	N	Y 2
Not known	Pt. Mugu, CA	Y 5	N	—	—	N	N	N	Y 1	N	Y 2
MLF, Captive born	MLF	Y 8	Y 4	Y 1	Y 2	N	Y 3	Y 1	N	N	Y 4
MLP, Captive born	MLP	Y 5	Y 1	—	Y 3	Y 1	N	N	N	N	Y 2
Aquatarium, Captive born	Aquatarium	Y 2	Y 1	—	—	N	N	Y 1	N	N	Y 1
Destin, FL	Gulfarium	Y 4	Y 1	—	Y 1	N	Y 1	N	N	N	—
Suwannee River, FL	MLF	Y 7	Y 6	—	Y	Y 3	N	Y 5	Y 1	Y 1	Y 3
Fernandina Beach, FL	MLP	Y 3	Y 2	—	—	Y 1	Y 1	Y 1	N	N	Y 1
Destin, FL	Gulfarium	Y 4	N	Y 1	—	N	N	N	Y 1	N	Y 1

(*continued*)

Table 1 (*continued*)

Animal number	Sex	Age class	Age at start (yr)	Size at start (cm)	Total no. of whistles	% Signature	% Aberrant	Duration (seconds) Average	Minimum	Maximum	Average no. loops	Maximum no. loops	Span of recordings (months)	No. of recording sessions
71	M	1	0	164	100	100	0	1.1	0.2	1.9	1	1	0	5
72	M	2	1	180	96	100	0	1.1	1	1.9	1.6	4	8.7	4
73	M	2	1	180	53	94	6	1.1	0.5	2.5	1.8	4	0	3
74	M	2	1	180	14	100	0	1.7	1.6	1.9	2.4	4	0	1
75	M	2	1	180	300	99	1	0.9	0.4	2	3	7	0	3
76	M	2	1	183	1568	99	1	0.8	0.1	4.5	1.4	5.5	2.2	10
77	M	2	1	183	89	100	0	0.7	0.3	1.7	3.8	5.5	2.7	5
78	M	2	1	183	100	100	0	0.7	0.1	1.1	1	1	0	1
79	M	2	1	186	254	86	14	1.6	0.6	2.4	3	9	14.5	6
80	M	2	1	190	1013	98	2	0.9	0.1	2.1	1.3	6	2.5	11
81	M	2	1	190	122	100	0	1	0.9	1.1	1	2	0	3
82	M	2	1	191	111	96	4	0.9	0.1	1.8	1.5	3	4.3	4
83	M	2	1	192	302	99	1	0.9	0.2	1.4	3.6	4.5	0	2
84	M	2	1	193	32	100	0	0.6	0.1	1.3	1	2	0	1
85	M	2	1	194	144	96	4	0.6	0.1	1.6	2.1	3	4.2	3
86	M	2	1	197	100	100	0	1.4	0.9	2.1	3.2	4	0	1
87	M	2	1	198	110	97	3	0.9	0.4	2.1	2.1	6	8.3	6
88	M	2	1	201	129	80	20	0.8	0.1	1.6	2.1	3	0	2
89	M	2	1	201	111	64	36	0.8	0.8	0.9	1.3	3	3.7	5
90	M	2	1	202	126	100	0	1	0.5	1.4	2.5	5	5.5	7
91	M	2	1	202	82	100	0	0.8	0.4	1.2	1	1	0	1
92	M	2	1	203	100	100	0	0.7	0.6	0.8	1	1	6.5	5
93	M	2	1	203	61	93	7	1.4	1.1	1.5	1.5	1.5	7	3
94	M	2	1	204	344	95	5	0.6	0.1	1	1.5	1.5	1.5	3
95	M	2	1	205	84	100	0	0.6	0.5	0.8	1	1	1.5	3
96	M	2	1	206	1074	96	4	1.7	0.2	3.9	3.6	7	9	8
97	M	3	—	—	130	97	3	1	0.1	2.5	1.7	4	0	2
98	M	3	—	210	35	100	0	1.2	0.1	1.9	2.5	4	0	1
99	M	3	—	213	100	100	0	0.9	0.2	1.9	2.1	4.5	0	2
100	M	3	—	213	120	99	1	1.2	0.1	1.9	2.4	5	0	3
101	M	3	—	213	271	97	3	1.2	0.3	1.7	3.3	5	9	8
102	M	3	—	215	153	85	15	1.1	0.1	3.2	2.6	7	14.5	6
103	M	3	—	215	129	89	11	0.8	0.2	1.5	1.8	4	0	1
104	M	3	—	219	110	100	0	0.8	0.3	1.3	2.0	6	7.2	6
105	M	3	—	220	66	100	0	1.5	0.4	2.2	2.6	4	0	4
106	M	3	—	222	30	100	0	0.8	0.8	1.1	1	1	0	2
107	M	3	—	225	94	98	2	1.1	0.1	2.4	2.2	6	9.3	4
108	M	3	—	226	100	100	0	1	0.5	2.2	2.3	5	0	1
109	M	3	—	226	126	95	5	0.6	0.1	1.6	1.1	4	1	2
110	M	3	—	226	105	100	0	1	0.6	1.4	3.2	4	0	1
111	M	3	—	230	13	77	23	1.4	0.4	2.4	1.9	3	0	1
112	M	3	—	231	49	90	10	0.5	0.1	1.2	1.5	3	9.2	6
113	M	3	—	—	5	100	0	0.7	0.4	1.3	2.5	5	0	1
114	M	3	—	232	86	93	7	0.9	0.1	2.1	1.6	4.5	0	2
115	M	3	—	241	2	—	—	—	0.5	0.5	1	1.5	3	2
116	M	3	—	241	110	96	4	1.1	0.1	1.5	1	1	6	6
117	M	3	—	242	18	83	17	1	0.1	2	3	5	0	1
118	M	3	—	243	34	80	20	0.7	0.3	0.9	4.4	7.5	9.5	4
119	M	4	—	248	58	100	0	0.8	0.4	1.8	2	5.5	4.3	4
120	M	4	—	250	114	99	1	1	0.6	3	2.8	9	12.3	10
121	M	4	—	257	131	100	0	0.9	0.5	2.3	1.9	5	9	4
122	M	4	—	265	11	91	9	0.6	0.3	1	1.6	4	1.5	3
123	M	4	—	267	26	100	0	1.2	0.6	2	2.8	4	0	1
124	M	4	—	268	53	94	6	1.1	1	1.8	1.2	5	6.5	5
125	M	4	—	247	60	50	50	0.9	0.1	2.6	2	8	0	1
126	M	—	—	—	19	100	0	0.8	0.7	0.8	1	1	0	1

[a] Total number of whistles refers to whistles coded as stereotyped or aberrant; other acoustic features may have smaller sample sizes. The span of recordings rounds off the number of months; for example, if the dolphin was recorded for less than 1 month, the span is indicated as 0 months. The rightmost 10 columns indicate the contexts in which the whistles were recorded, judged from notes accompanying the recording: Water, whistles recorded underwater; Air, whistles recorded in air. Some recording notes indicated whether the dolphin was alone or with other dolphins, or whether the context was considered "normal." It was more consistently noted whether a dolphin was undergoing veterinary (Vet.) inspection, stranded out of the water, being moved, feeding, or playing. If there were no recording sessions in a context for an individual dolphin, "N" appears as that entry. If the context was noted for any recording sessions, "Y" and the number of sessions are listed, when the latter could be determined. If neither the context nor its absence was noted for any of the recording sessions, the entry is marked with a dash (—).

[b] MLF, Marineland of Florida, St. Augustine; MLP, Marineland of the Pacific, Palos Verdes, California.

Capture location	Recording site	Conditions of Recording									
		Water	Air	Alone	With others	Vet.	Stranded	Being moved	Feeding	Playing	Normal
Steinhatchee, FL	MLF	Y 4	Y 1	—	—	N	N	Y 1	N	N	Y 3
St. Augustine, FL	MLF	Y 1	Y 3	—	—	Y 3	N	N	N	N	Y 1
Tampa Bay, FL	Niagara Falls Aq.	Y 3	N	—	Y 2	N	N	N	Y 1	N	Y 1
Unknown	Pt. Mugu, CA	Y 1	N	—	—	N	N	N	N	N	—
Fernandina Beach, FL	MLP	Y 1	Y 2	—	—	Y 2	Y 1	Y 1	N	N	—
Port St. Joe, FL	MLF	Y 9	Y 1	—	—	Y 1	N	N	Y 3	Y 1	Y 5
Port St. Joe, FL	MLF	Y 1	Y 4	—	—	Y 3	N	Y 1	N	N	—
Steinhatchee, FL	MLF	Y 1	N	—	—	N	N	N	N	N	Y 1
St. Augustine, FL	MLF	Y 2	Y 4	—	—	Y 4	N	N	N	N	Y 1
Port St. Joe, FL	MLF	Y 10	Y 1	—	—	N	Y 1	Y 1	Y 2	Y 3	Y 5
St. Augustine, FL	MLF	Y 1	Y 2	—	—	Y 2	N	N	N	N	Y 1
Port St. Joe, FL	MLF	Y 3	Y 1	—	Y 1	N	N	Y 1	N	N	Y 3
Destin, FL	Gulfarium	Y 2	N	—	—	N	N	N	N	N	—
Port St. Joe, FL	MLF	N	Y 1	—	—	N	N	Y 1	N	N	—
Port St. Joe, FL	MLF	Y 1	Y 2	—	—	N	N	Y 2	N	N	—
Steinhatchee, FL	MLF	N	Y 1	—	—	N	N	Y 1	N	N	—
Suwannee River, FL	MLF	Y 1	Y 5	—	—	Y 5	N	—	N	N	Y 1
Port St. Joe, FL	MLF	Y 2	N	—	—	N	N	N	N	N	Y 2
Steinhatchee, FL	MLF	N	Y 5	—	—	Y 3	N	N	N	N	—
Steinhatchee, FL	MLF	Y 1	Y 6	—	—	Y 6	N	N	N	N	Y 1
Port St. Joe, FL	MLF	N	Y 1	—	N	N	N	Y 1	N	N	—
New Smyrna Beach, FL	MLF	Y 1	Y 4	—	—	Y 1	N	Y 3	N	N	—
New Smyrna Beach, FL	MLF	Y 1	Y 2	—	—	Y 1	N	Y 1	N	N	—
Port St. Joe, FL	MLF	Y 2	Y 1	—	—	N	N	N	N	N	Y 2
New Smyrna Beach, FL	MLF	Y 1	Y 2	—	—	Y 1	N	Y 1	N	N	—
Cedar Key, FL	MLF	Y 6	Y 2	—	Y 1	Y 2	N	N	N	N	Y 3
Suwannee River, FL	MLF	Y 2	N	—	—	N	N	N	Y 2	N	Y 2
New Smyrna Beach, FL	MLF	N	Y 1	—	—	N	N	Y 1	N	N	—
New Smyrna Beach, FL	MLF	Y 1	Y 1	—	—	Y 1	N	Y 1	N	N	—
Port St. Joe, FL	MLF	Y 2	Y 1	—	—	N	N	Y 1	N	N	—
Cedar Key, FL	MLF	Y 3	Y 5	—	Y 1	Y 4	N	Y 1	N	N	Y 3
Steinhatchee, FL	MLF	Y 1	Y 5	—	—	Y 1	Y 1	N	N	N	—
Fernindina Beach, FL	MLF	Y 1	Y 1	—	—	N	N	Y 1	N	N	—
Steinhatchee, FL	MLF	N	Y 6	—	—	Y 6	N	N	N	N	—
Not known	MLP	Y 4	N	—	Y 2	N	N	N	N	N	Y 1
Destin, FL	Gulfarium	Y 2	N	—	Y 1	N	N	N	N	N	—
St. Augustine, FL	MLF	N	Y 4	—	—	Y 4	N	N	N	N	—
Matanzas Inlet, FL	MLF	Y 1	N	—	—	N	N	N	N	N	—
Crescent Beach, FL	MLF	N	Y 2	—	—	N	N	Y 2	N	N	—
Port St. Joe, FL	MLF	N	Y 1	—	—	N	N	Y 1	N	N	—
Matanzas Inlet, FL	MLF	N	Y 1	—	—	N	N	Y 1	N	N	—
St. Augustine, FL	MLF	N	Y 6	—	—	Y 6	N	N	Y (forced)	N	—
Not known	Aquatarium	N	Y 1	—	—	N	N	Y 1	N	N	—
Port St. Joe, FL	MLF	N	Y 2	—	—	Y 2	N	N	N	N	—
Marineland, FL	MLF	N	Y 2	—	—	Y 1	N	Y 1	N	N	—
New Smyrna Beach, FL	MLF	Y 3	Y 3	—	Y 1	Y 2	N	Y 1	N	N	—
Matanzas Inlet, FL	MLF	N	Y 1	—	—	N	N	Y 1	N	N	—
Port Orange, FL	MLF	Y 1	Y 3	—	—	Y 2	N	Y 1	N	N	Y 1
St. Augustine, FL	MLF	N	Y 4	—	—	Y 3	N	Y 1	N	N	—
Suwannee River, FL	MLF	Y 5	Y 5	—	—	Y 4	N	Y 2	N	N	Y 3
Crescent Beach, FL	MLF	N	Y 4	—	—	Y 4	N	N	N	N	—
Northeastern Florida	MLF	N	Y 3	—	—	Y 2	N	N	Y (forced)	N	—
Port Orange, FL	MLF	Y 1	N	—	Y 1	N	N	N	N	N	—
St. Augustine, FL	MLF	N	Y 5	—	—	Y 5	N	N	N	N	—
Tampa Bay, FL	Aquatarium	Y 1	N	—	—	N	N	N	N	N	Y 1
Not known	Pt. Mugu, CA	Y 1	N	—	—	N	N	N	N	N	Y 1

whistles function to broadcast the identity of the whistler. The more variable acoustic features of the same whistles may communicate other information and serve other functions.

Our primary objectives have been to define the acoustic structure of signature whistles and to determine for each individual the percentage of whistles which were stereotyped signature whistles. We also assessed variation in whistles as a function of age and sex and of the different conditions under which recordings were made. Considerable effort was devoted to obtaining whistles emitted under widely varying conditions and between lengthy time intervals. The whistles of several of these dolphins have been recorded over a period of more than a decade.

We have analyzed several, but by no means all, of the acoustic properties of each individual dolphin's whistles. The acoustic features measured include the following:

Percentage of stereotyped whistles
Percentage of aberrant whistles
Average and maximum numbers of loops
Average and maximum durations of loops
Minimum, average, and maximum whistle durations
Lowest and highest frequency components
Total frequency modulation

Most of the data were averaged for each individual, so the individual dolphin was the unit of most later analyses. An average of 178 whistles per individual, from one to eleven recording sessions, was analyzed for percent of stereotypy (see Table 1). However, there was a large range (5–1453) in sample sizes of whistles from different individuals for percent stereotypy. The mature males were the most difficult group from which to obtain an adequate sample because they whistled less frequently than females and young males. Most other acoustic features had lower sample sizes than percent stereotypy. Duration and number of loops are averages, usually but not always based on a minimum of 25 whistles from each individual. Frequencies are based on measurements from one or more spectrograms.

Review of the Acoustic Structure of Dolphin Whistles

Whistles of most dolphins exhibit a considerable rise and fall in frequency. Kellogg *et al.* (1953) first described the whistles of *T. truncatus* as ranging from 7 to 15 kHz, or slightly over an octave. This early paper noted the likelihood that whistles produced by different individuals vary considerably. Later descriptions of frequency modulation in whistles included approximate frequency ranges of 4–15 kHz (Dreher, 1961), 5–16 kHz (Dreher and Evans, 1964), 4–20 kHz (Evans and Prescott, 1962), 5–15 kHz (Herman and Tavolga, 1980), and 4–18 kHz (Lilly and Miller, 1961a).

Whistle durations have been cited as follows: approximately 0.5 seconds (Kellogg *et al.*, 1953; Kellogg, 1961); 0.25–3 seconds, mean 0.5 seconds (Dreher, 1961); 0.1–0.4 seconds, mean 0.25 seconds, but under some conditions 2–3 seconds (Lilly and Miller, 1961a); 0.1–3.6 seconds (Evans and Prescott, 1962).

Loudness

Whistles, even those from the same individual, vary widely in loudness. Tyack (1985) noted variation in loudness of underwater whistles of at least 15 dB, from 125 to 140 dB re 1 μPa at 1 m. Lilly (1963) reported similar variability in the loudness of in-air whistles. He noted that whistles produced in air with the blowhole open may be either loud or faint but that all of the piercingly loud ones were made with the blowhole open. Our observations confirm that in-air whistles emitted when the blowhole is closed are usually faint, while whistles are much louder in air when the blowhole is open.

Loops

In most bottlenose dolphins past infancy, each individual tends to produce a specific contour, or rising and falling pattern of frequency modulation. This contour is repeated a varying number of times in each whistle (see Fig. 1; also Caldwell *et al.*, 1973). In early studies, the repetitive elements of whistles were variously termed "warbles" (Dreher and Evans, 1964; Caldwell and Caldwell, 1965), "sinusoidal frequency oscillations" (Busnel and Dziedzic, 1966), and "humps" (Dreher, 1966). We call these repetitive elements "loops." Some authors have called each repeated loop a separate whistle (Tavolga, 1968, entry 94) or "squeal" (Schevill and Watkins, 1962, their Fig. 13). This is a perceptual problem (Caldwell *et al.*, 1973) as well as a conceptual one. For some dolphins there is often no detectable sound between loops; however, the intervals between loops are relatively constant. Rather than repeating a constant section of whistle, the dolphin repeats both a section of whistle and an interval of silence. For example, Schevill and Watkins (1962, their Fig. 13) illustrated what we would call a single multilooped whistle, even though the spectrogram indicates silent periods between repeats. In the Tavolga (1968) figure of what may well be the same animal (Schevill was credited with the recording, and the whistle contour looks similar), a signal is detectable between cycles on the sound spectrogram, but it is faint. Lang and Smith (1965) discussed this problem in presenting their results and resolved it by dotting in lines as they perceived them by ear.

When one works in depth with whistles of individual animals, the question of whether loops separated by intervals of silence are separate whistles almost always ceases to be a problem. The silent intervals between loops have much more consistent timing than intervals between whistles. Furthermore, variations are sometimes present in the first and last loops of the whistle. We call these introductory and terminal loops. When present, they also help discriminate between one multiloop whistle and several whistles of fewer loops. Many whistles consist of only a brief segment of the loop. Caldwell and Caldwell (1970) followed Dreher (1966) in calling these brief sounds "chirps," as they fit Griffin's (1959, p. 87) definition of audible sounds that sweep through a wide frequency range in a fraction of a second. Sounds intermediate between brief chirps and a whistle with a complete loop are called "partial whistles." We have not made a clear-cut quantitative distinction between chirps and partial whistles.

Amplitude Modulations

Amplitude modulations within individual whistles were discussed by Lilly and Miller (1961a), who noted that in most cases the amplitude quickly reaches a peak, then falls during the rest of the emission. This tends to be true for whistles of only a single loop. Our measurements suggest that the lower frequencies usually contain greater energy than the upper frequency portions. On continuing repetitive multilooped whistles, this same pattern is repeated with the greater energy repetitively displayed in the lower frequencies. More rapid amplitude modulation may appear, especially in sections of rapid frequency modulation. As Tyack (1989) noted, very rapid amplitude modulations may appear as sidebands on spectrograms of whistles, and they may be perceived as introducing an unusual timbre to the whistle. Specific patterns of amplitude modulation, or the lack of them, are characteristic features of the signature whistle of each individual.

Interpretations of Variation in Whistle Contours

Consistent differences in the duration and frequency ranges of whistles from two captive dolphins were noted by Lilly and Miller (1961b). Most dolphins have individually characteristic frequency ranges as well as rates of frequency modulation. Both of these parameters may shift slightly under some conditions or in some individuals more drastically, particularly when the animal is either under extreme duress or when first arousing from a quiescent period resembling sleep. In general, however, the frequency range and rate of frequency modulation remain relatively stable for each individual.

While there is general agreement concerning the acoustic structure of whistles, there is less agreement concerning the interpretation of variation in whistles. Schevill and Lawrence (1956) noted that the "whistle-like squeal" of bottlenose dolphins is not used in echolocation, but is produced in contexts indicative of social communication. This interpretation of dolphin whistles is either clearly stated or implied by virtually everyone who has worked with the sounds of this species. Although whistles are generally accepted as part of the communication system, the nature of the information communicated remains controversial.

Previous work on social functions of dolphin whistles indicates that whistles are often used to establish vocal or physical contact between individuals. For example, dolphins whistle when separated from parents, offspring, or other group members (McBride and Hebb, 1948; McBride and Kritzler, 1951; Wood, 1954). Dolphins have been reported to respond to whistles either by whistling themselves (Lang and Smith, 1965; D. K. Caldwell and M. C. Caldwell, 1968; Gish, 1979) or by approaching the whistler. When separated, female dolphins and their young calves exchange whistles until reunited.

Distress Whistles

Lilly (1961) may have been the first to hypothesize about the functions of specific dolphin whistles when he wrote of a distress call and an attention call. His approach of assigning discrete contours to specific behavioral situations continued to be explored by him (Lilly, 1963), Dreher (1961, 1966), and Dreher and Evans (1964). Association of specific whistles with specific objects was postulated for *Tursiops aduncus* by Tietz and Tayler (1964). Observations on three dolphins formed the origin of Lilly's postulated "distress call":

1. Lilly (1961, p. 55) stated that when a dolphin with apparent brain damage from anoxia was put into a pool with two other animals it emitted a very short, high-pitched whistle, in two parts, a rise and a fall in frequency.
2. The next occasion (Lilly, 1961, p. 84) apparently involved a dolphin that whistled when placed in a sling, when turned in the sling, and when the brain was being stimulated. Lilly wrote (p. 86): "The only vocalization that we were able to obtain by stimulating the negative system was the distress call, stereotyped, repetitive, and at times piercing." Again, on p. 87 for the same animal, Lilly wrote: "For every stimulus the dolphin emitted a distress call. She also emitted similar signals when her shoulder was cut accidentally and was held in an uncomfortable position."
3. Again, to quote from Lilly (1961, p. 90), an injured animal was placed into a tank with two animals and emitted "the distress call, the usual crescendo–decrescendo [*sic:* rising and falling in frequency] pair of whistles." It was this episode (p. 91) that "gave rise to our suspicions that these animals have a very complex language and know how to use it for descriptive and predictive purposes."

Lilly (1963) later described the fundamental frequency of the distress call as starting at 3–5 kHz rising to 8–20 kHz and falling back to 3–5 kHz. This broad description fits most, but not all, bottlenose dolphin whistles of one loop. Lilly (1963) also wrote, "The call itself is similar to other whistles in the 'vocal exchange' group of sounds." However, he added, "This call is one of many unique short whistle patterns and is not to be confused with general 'alarm whistling'. . . ." In summary, Lilly accurately described a widely generalized one-looped whistle of the species, but he assigned to it a specificity and uniqueness of context which was not tested. The evidence for both a context-specific distress call and a complex repertoire of context-specific whistles is weak.

Busnel and Dziedzic (1968) analyzed whistles from individuals of several dolphin species under situations of distress. They recorded 549 whistles from a harpooned male *T. truncatus,* and all were similar in structure. However, the whistles emitted by the harpooned male were so different from the distress whistle described by Lilly that in attempting to integrate those whistles into Lilly's (1963) concept of a context-specific "distress whistle," they were forced to postulate two different types of distress whistle. The writers postulated that the "distress call" published by Lilly is positively phonotaxic, possibly eliciting cooperation and aid by others of the school, whereas the whistle they recorded was negatively phonotaxic.

Signature Whistles

Caldwell and Caldwell (1965) presented a very different interpretation of dolphin whistles based on an analysis of differences in the whistles recorded from individual dolphins within a captive group. They found that each dolphin tended to produce an individually distinctive whistle contour. This contour might be repeated a variable number of times without a break, some sections of it might be deleted, or the contour might vary in duration, but the contour configuration remained distinctive in spite of these specific forms of variability (see Fig. 1). The striking distinctiveness of each individual dolphin's whistle contour led M. C. Caldwell and D. K. Caldwell (1968) to call these whistles "signature whistles" and to hypothesize that one of the functions of the whistle was to broadcast the individual identity of the whistler to other members of its social group.

The assumption behind the work on distress whistles was that dolphins share a large repertoire of stereotyped whistle contours, each of which is produced in a particular behavioral context. An alternate interpretation of these results is that each dolphin was producing its signature whistle in these stressful circumstances. In the 1970s and 1980s a somewhat sterile debate emerged between these two approaches. The original signature-whistle hypothesis did not propose that signature information was the *only* information broadcast by whistles. Yet a succession of researchers have addressed this issue by attempting to strike down a straw-man version of the signature-whistle hypothesis, a straw man which states that there is *no* variation in *any* acoustic feature of whistles as a function of behavioral context. It is difficult to understand where these straw men came from since most of them were constructed after Caldwell *et al.* (1970) had already integrated the signature-whistle hypothesis with observations of variation in several acoustic parameters of whistles with behavioral context.

For example, in the context of distress, we noted that signature whistles of some distressed dolphins tended to quaver, have abrupt discontinuities of contour similar to when a voice "breaks," or have sections with less frequency modulation than is typical of the individual. Furthermore, our qualitative observations have led us to wonder whether bottlenose dolphins might emit more, louder, and faster whistles when stressed than under normal conditions (Caldwell and Caldwell, 1965). The reader must bear in mind, however, that these variations are relatively independent of between-individual differences in whistle contour. The differences associated with distress do not interfere with our identification of the animal's signature whistle, and they are not easily interpretable unless one knows the signature whistle of the dolphin. A quantitative comparison of whistles produced in different contexts is presented in the Results section below. This comparison indicates that different individuals do not always show similar patterns of variation in whistles in different contexts. For example, some individuals may tend to produce whistles with more loops when in stressful circumstances, while other dolphins may produce whistles with fewer loops.

Mimicry

Several researchers have demonstrated that bottlenose dolphins can imitate electronically generated whistlelike sounds (Evans, 1967; M. C. Caldwell and D. K. Caldwell, 1972; Richards *et al.*, 1984). The mimicry work of Penner (cited in Evans, 1967) was probably based on work initiated by Batteau and Markey (1968) and included food reward for correct responses (Wood, 1973, p. 111). Richards *et al.* (1984) and Richards (1986) also shaped dolphin sounds by food reward. All of these authors noted spontaneous, unreinforced mimicry of whistlelike sounds. Caldwell and Caldwell (1979) also noted precise mimicry in an infant housed with its mother.

M. C. Caldwell and D. K. Caldwell (1972) deliberately eliminated any possible external reward for whistle mimicry. We recorded from a trailer where we could not be seen, usually after leaving the animal alone to listen to the computer-generated sounds. While we recorded, we did not know what whistles were emitted, and we could not reward the dolphin inadvertently in any way. The animal mimicked the tones to which it was exposed. Hence, it appears that mimicry is completely self-rewarding to the animal, although it can probably be shaped more rapidly by external rewards.

Captive dolphins spontaneously produce precise imitations of the signature whistles of poolmates. Burdin *et al.* (1975) and Gish (1979) noted occasional mimicry of the whistles produced by one adult dolphin in acoustic contact with another dolphin via a hydrophone–transducer link. Tyack (1986a) found rates of mimicry of 20–30% for stereotyped signature whistles in two captive bottlenose dolphins that were in actual contact. The two animals studied by Tyack (1986a) had been housed together for years without contact with other bottlenose dolphins, and they had a much higher rate of spontaneous mimicry than the animals studied by Burdin *et al.* and Gish. Social contact may be a factor in this higher rate of mimicry, for in the work of Burdin and Gish the dolphins could only interact acoustically.

METHODS

Age Estimation

The following age classes were assigned to dolphins by standard length measurements from the tip of the upper jaw to the deepest part of the fluke notch:

Age class	Length
I Infant	Less than 170 cm
II Juvenile	Between 170 and 210 cm
III Subadult	Between 210 and 235 cm for females; between 210 and 247 cm for males
IV Adult	Greater than 235 cm or known breeding for females; greater than 247 cm for males

Length is not a completely reliable indicator of age in bottlenose dolphins, but, as discussed by Caldwell and Caldwell (1979), animals classed as infants using these size ranges are estimated to range from birth to 1 year of age, juveniles from 1 to about 6 years of age, subadults from 6 to 12 (females) or 13 (males) years of age. Adult females were estimated to be 12 or more years old, while adult males were estimated to be 13 or more years of age. A dolphin was also scored as adult if it was known to have bred.

Equipment

Underwater whistles from dolphins usually were recorded using an Atlantic Research Corporation, Model LC-57, hydrophone with a special preamplifier designed and built by William W. Sutherland, then of the Lockheed–California Company. Tape recorders, used interchangeably, were all Uhers (Models 4000 Report-S, 4000 Report-L, 4000 Report-IC, or 4400 Report-Stereo). On some occasions a hydrophone and preamplifier designed and built by Thomas E. Poulter of the Stanford Research Institute were used in conjunction with the same Uher recorders. Those recordings which were made in air used the same series of Uher recorders with compatible microphones. All tape recordings were made at a tape speed of 19 cm (7.5 inches) per second.

Sound spectrograms (sonagrams) were prepared on a Kay SonaGraph, Model 662A, Sound Spectrograph Analyzer. When the recorded tape speed is reduced by half, the effective frequency response of the analyzer is increased to 24 kHz. Visual displays made on a Listening, Incorporated, MSA-1 Spectrum Analyzer were used as an intermediate step between the analysis by ear at one-eighth recorded tape speed and sonagrams made with the Kay machine. The accuracy of frequency measurements taken from Kay sonagrams was checked frequently on a Listening, Inc., spectrum analyzer. All sound spectrograms were prepared with an effective filter bandwidth of 600 Hz.

Statistical Methods

There is enormous variation in the number of whistles analyzed per individual in this study, and different whistle measures may also have different sample sizes for each individual (Table 1). This leads to statistical problems. If the data are analyzed by individual, should one weigh an average of 1500 whistles from one dolphin the same as an average based on only 5 whistles from another dolphin? The range of duration or frequency modulation is likely to be higher for a population sampled 1500 times than one sampled 5 times. The variance of the distributions of these variables is likely to differ greatly for individuals represented by such different sample sizes.

There are different problems if the data are lumped by age or sex class for chi-square contingency table analysis without averaging whistles from each individual. Treating each whistle rather than each individual as the fundamental unit of analysis assumes that each whistle is an independent event. Table 2 shows that whistles may occur in nonrandom sequences (chi-square = 287.2, $df = 63$, $p < 0.005$). This means that each whistle within a sequence may not be independent of those which preceded it. Counting all whistles analyzed as independent events may thus inflate the sample size if several whistles were analyzed within a short sequence. Several factors may reduce the magnitude of this potential bias. Most of the whistle data used in this chapter comes from several long recording sessions for each individual. Where results of these tests showed significant differences, the level of significance was high enough to be relatively insensitive to small reductions in sample size.

The data were originally tallied for each individual. For some measures, such as duration, it is impossible to reconstruct the data except as averaged by individual. For other measures, such as percentage of signature whistles, it has been possible to reconstruct original tallies. Where possible, the data have been analyzed both treating each individual and treating each whistle as the basic unit of analysis (e.g., Fig. 3). This allows one to test the sensitivity of the analysis to the assumptions inherent in either approach.

RESULTS

Analysis of Whistle Parameters

Table 1 indicates whistle parameters averaged for each individual as well as the number and contexts of recording sessions for each individual. Our total sample contains 22,278 whistles recorded from 126 animals. We analyzed an average of 178 whistles per individual for percent stereotypy. These dolphins covered a wide range of ages, which have been sorted into four age classes as described in the Methods section. The sample includes 62 females, 5 of which were infants, 12 juveniles, 25 subadults, 18 adults, and 2 of undetermined age. There were 64 males, of which 8 were infants, 25 juveniles, 22 subadults, 7 adults, and 2 of undetermined age. All dolphins with known capture locations were caught in waters off Florida. Nine dolphins of unknown capture location were studied in facilities in southern California, but the four dolphins from California facilities whose capture location was known were also from Florida. Seven of the dolphins studied were born in captivity.

Table 2
Transition Matrix for Sequences of Whistles[a]

Initial whistle	Following Whistle No. of loops								Total
	Ab	<1	1–<2	2–<3	3–<4	4–<5	5–<6	6–6+	1209
Ab	5	5	5	5	8	3	0	0	31
<1	7	22	10	14	12	8	4	1	78
1–<2	3	9	16	32	25	14	4	2	105
2–<3	5	14	33	93	70	31	12	8	266
3–<4	4	13	18	82	107	68	17	6	315
4–<5	7	9	15	29	64	55	27	11	217
5–<6	1	4	5	11	17	22	20	15	95
6–6+	1	3	2	7	9	10	12	5	49
Total	33	79	104	273	312	211	96	48	1156

[a] Subject dolphin is MLF 229, in isolation. Ab, Aberrant.

Stereotypy

The signature whistles of one dolphin may vary in duration, frequency, number of loops, etc. while maintaining a highly distinctive loop contour pattern. Figure 1 illustrates several signature whistles of two different animals. These indicate the variability of signature whistles according to our definition (Caldwell *et al.*, 1973). These stereotyped whistles may stop at any point; for example, there may be $2\frac{1}{2}$ or $3\frac{3}{4}$ or any such fractional number of loops. These abbreviated whistles may not have a terminal loop. Space limitations do not permit including abbreviated whistles but several were included in an earlier report (Caldwell *et al.*, 1970). Similar abbreviated loops have been shown for the Atlantic spotted dolphin, *Stenella attenuata* (Caldwell *et al.*, 1973). The major point is that each animal tends to produce a stereotyped loop contour (but see section on Mimicry below).

This pattern of contour configuration is difficult to quantify, although it is relatively easy for a trained worker to categorize whistles by visual inspection of sonagrams or by careful and repeated listening. We determined the contour configuration typical of each dolphin and then judged which whistles in each sample were not clearly stereotyped examples of this dolphin's signature whistle.

We did not measure the reliability of observer ratings of percent stereotypy from the same sample of whistles. Several observers often did score stereotypy for different whistle samples of an individual, and these usually were in close agreement. The largest discrepancy occurred for two observers who scored dolphin MLF 232. As discussed in the section on Ontogeny, this dolphin was isolated for almost 8 years, was exposed to a variety of man-made whistlelike sounds, and ultimately produced his signature whistle only rarely. This was unlike any other dolphin in the study. Either the percentage of stereotyped whistles was different for the two whistle samples analyzed by the two observers, or one observer had stricter criteria for judging a whistle to be the stereotyped signature whistle. As in other forms of taxonomy, perhaps one observer was a "splitter" and the other a "lumper." In order to test this worst-case discrepancy, a third observer was asked to judge the percentage of stereotyped whistles from early recording sessions of MLF 232. The results are as follows:

Observer	Number of whistles analyzed	% Signature
1	100	100
2	600	91
3	1059	96

This ±5% variation in scores is the worst case of disagreement we encountered. If different observers scored the same whistle sample, we would thus expect variation of only a few percentage points.

Of the 120 individuals analyzed for percentage of signature whistles, two showed a second stereotyped whistle. We have designated it a secondary whistle because it is emitted less frequently. M. C. Caldwell and D. K. Caldwell (1968) presented another example of secondary whistles in the common dolphin, *Delphinus delphis*. Most dolphins do not produce secondary whistles but do produce a varying percentage of whistles that do not resemble their typical signature whistles. These are variable. We cannot identify these whistles, which we term "aberrants," to individuals.

The whistles (n = 22,250) of 120 animals of all ages were analyzed for the presence of stereotypy (Fig. 2). We aimed to analyze a minimum of 100 whistles per individual for stereotypy, but the samples from 45 individuals contained fewer than this number (see Table 1). The overall mean for stereotypy averaged by individual is 94%. As Fig. 2 indicates, most individuals averaged 100% stereotyped whistles, and only 3 ranged below 51%, all of which were males. The percentage of stereotyped whistles was 97% for the 58 females and 92% for the 62 males. This difference was not statistically significant (Kruskal–Wallis test statistic = 2.13, p = .14), nor was there a significant difference in percentage of stereotyped whistles between age classes (Kruskal–Wallis test statistic = 1.44, p = .70).

Caldwell and Caldwell (1979) indicated that none of the calves they studied had stereotyped signature whistles at birth, but most had developed them by 4–6 months of age. Two of the eight male infants in the sample had 0 and 5% stereotyped whistles. These scores are outliers far re-

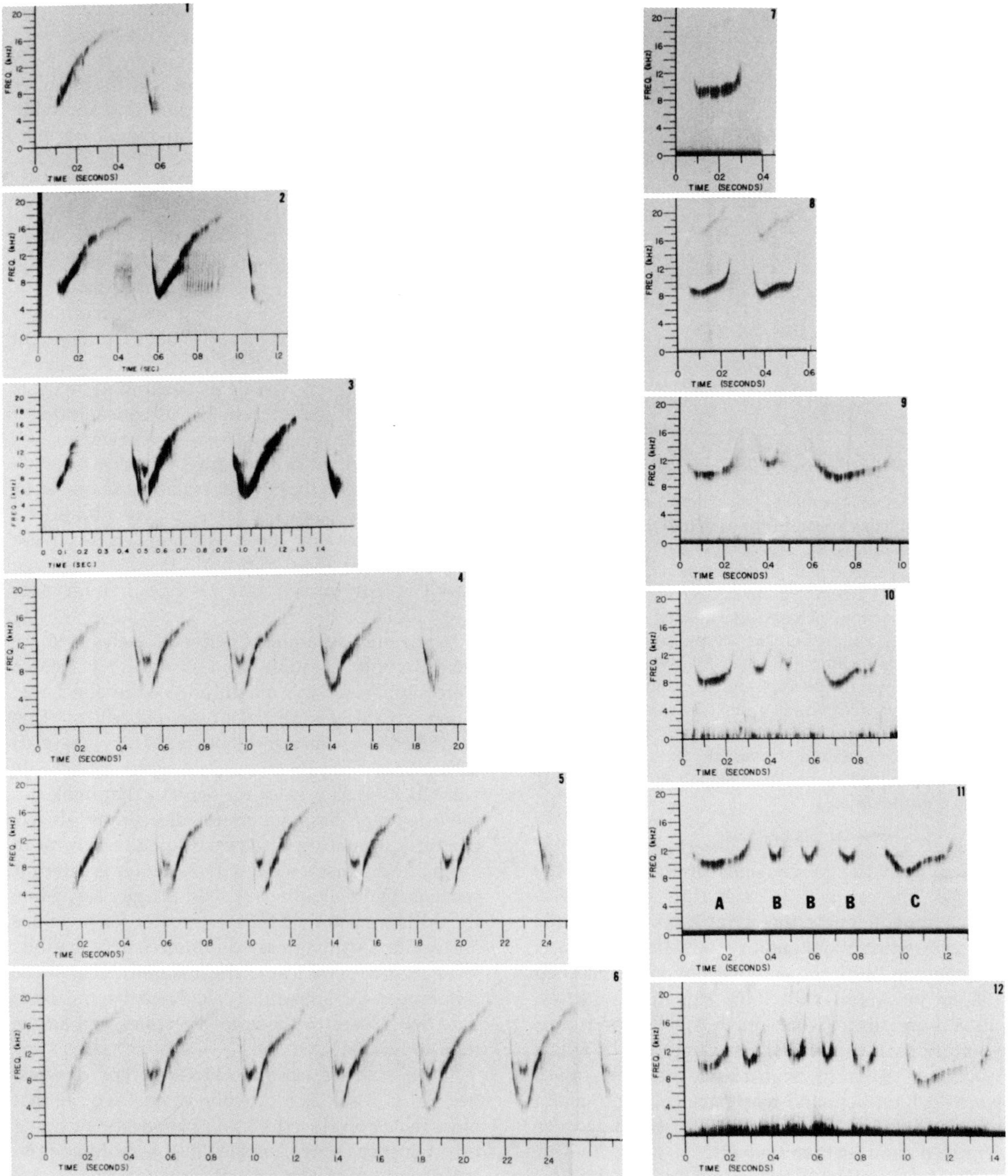

Figure 1 Comparison of signature whistles of two different Atlantic bottlenose dolphins. (Left) One to six loops (1–6) of a juvenile male (MLF 229). (Right) One to six loops (7–12) of a different juvenile male (MLF 232). Whistles of most bottlenose dolphins are as repetitive as those illustrated at left. A few have an introductory or terminal loop structure (11, A and C) different from the central ones (11, B,B,B).

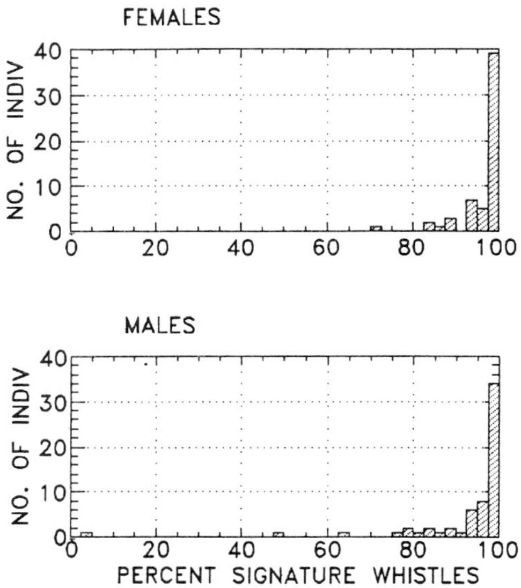

Figure 2 Percentage of whistles that are stereotyped signature whistles averaged for each individual dolphin. Of the total sample of individuals, 43% (51/120) produced only stereotyped signature whistles under all conditions (this group includes all age classes and both sexes). Seventy-four percent (88 animals) of the sample exhibited 95% or greater stereotypy, again under all conditions and including all age classes and both sexes. Eighteen percent of the total sample exhibited less than 90% stereotypy. Of these 21 animals, 4 were infants with undeveloped or developing stereotypy. Therefore, 14% of the total population examined, exclusive of those infants, produced stereotyped signature whistles only 50–90% of the time.

moved from the percentages of the rest of the sample. The calf with 0% signature whistles was 100 cm long, a length indicating a newborn, while the calf with 5% signature whistles was only 110 cm long. With the exception of one female calf 106 cm long (and with 100% signature whistles) these two calves were much smaller and thus younger than the rest of the calves, all of which were over 150 cm in length. Since these two calves were such outliers and may have been too young to have developed a signature whistle, they were removed from further analysis.

To test the effect of sex or age class on the type of whistle emitted by a dolphin, we counted the number of stereotyped versus aberrant whistles from each of the eight age–sex classes. This analysis differs from that of the previous paragraphs in that we do not weigh the percentages from each individual equally. The contribution from each individual is directly proportional to the number of whistles recorded from it. Contingency table analysis showed a significant dependence of percentage of stereotyped signature whistles on sex and age class (chi-square = 221.8, $df = 7$, $p << .001$).

Figure 3 indicates the percentage of signature whistles for each age and sex class, with data averaged by individual in Fig. 3A or pooled for each age–sex class in Fig. 3B. These percentages are relatively insensitive to whether the individual or each whistle is used as the unit of analysis. Since treating each whistle as the unit of analysis simplifies statistical analysis, let us concentrate on Fig. 3B. If we examine differences between sexes in the percentage of signature whistles for each age class, we see that the largest differences occur for age classes 1 and 3, infants and subadults. The sex differences for these two age classes were highly statistically significant (chi-square for age class 1 = 66.4, for age class 3 = 27.8, in both cases $p << .001$).

While the differences between males and females in both the infant and subadult age classes were significant, the magnitude of the effect was much greater for infants than subadults. This magnitude is measured by the odds ratio (Agresti, 1984). The odds in favor of a signature whistle were 21.7 times greater for females than males in the infant age class, where females almost always produced signature whistles. The odds in favor of a signature whistle were only 2.6 times greater for females than males in the subadult age class, while the odds were only 1.2 times greater for juveniles and adults. Although sex differences for subadults were highly significant, the magnitude of these differences for subadults calculated by the odds ratio was closer to those for juveniles and adults than to infants.

If we examine variation in stereotypy of whistles for each sex as a function of age, we see that older males produced a lower proportion of signature whistles, while females had a more complicated pattern of age-related changes. Female infants in this sample had an extraordinarily high proportion of signature whistles, showing only 3 aberrants out of 2968 whistles analyzed; juveniles

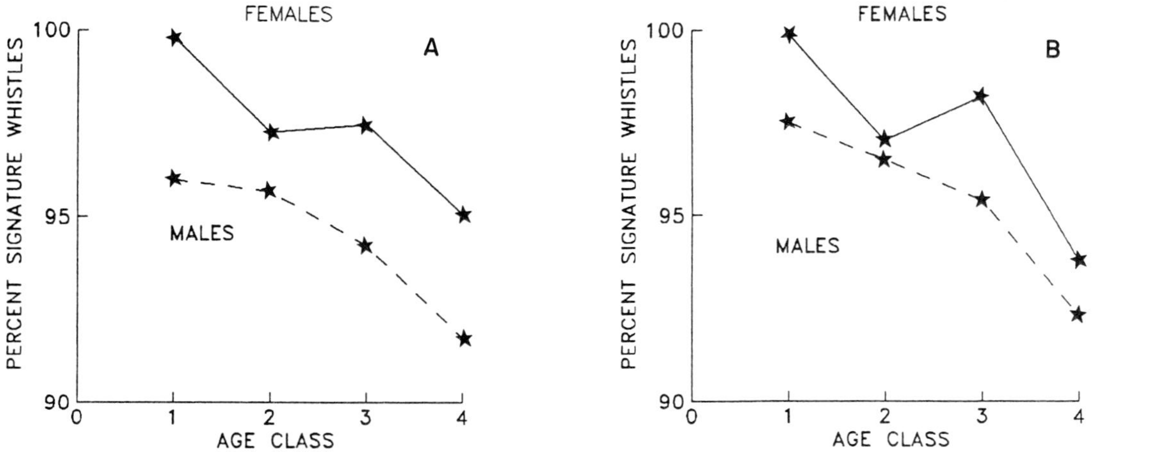

Figure 3 Overall percentages of stereotyped signature whistles separated by age class and sex as derived from cumulative data of 22,278 analyzed whistles. (A) Data analyzed by treating the individual dolphin as the unit of analysis. The percentages of stereotyped whistles were averaged for each individual, and then averages of these averages were made for each age–sex class. The dashed line indicates males (N = 62 individuals), the solid line females (N = 58). (B) The same data analyzed with each whistle treated as the unit of analysis. The numbers of stereotyped and aberrant whistles for each dolphin within each age–sex class were tallied to generate the overall percentage of stereotyped whistles for each age–sex class.

had 97% signature whistles, but subadults had 98% signature whistles, an increased percentage of stereotypy relative to juveniles. As in the males, adult females had the lowest percentage of signature whistles among the different age classes.

Loops

The sound loops of most individuals require about 0.3–0.5 seconds for completion, but some individuals repeat the loops more quickly or slowly. Some may cycle as rapidly as 0.2 seconds per loop, while others may average over a full second per cycle (Fig. 4). Even for the same individual, loops vary to some degree in duration (Fig. 5). The data on loops have been averaged by individual, and it is not possible to reconstruct the original tallies. The following analysis, therefore, must treat each individual as the unit of analysis, even though there may be differences in the number of whistles analyzed for each individual.

The individuals in this sample varied considerably in both average and maximum number of loops in their whistles (Figs. 6 and 7). The maximum number of loops is 13.5. No differences between sexes were noted in either the average number of loops (Kruskal–Wallis test statistic = 0.74, p = .39) or the maximum number of loops (Kruskal–Wallis test statistic = 0.09, p = .77). There was a strong difference in both parameters as a function of age class. A Kruskal–Wallis analysis of maximum loop number by age class yields a test statistic of 19.62, with a significance level of .0002. For average loop number, Kruskal–Wallis analysis shows that differences between age classes are significant to p = .00003. Only 3 of the 13 infants in our sample ever repeated more than one loop in a whistle, while all but 2 of the 23 adults emitted multiple-loop whistles. Even past infancy, these variables are dependent on age class (actually measured as animal size) to some extent as the following tabulation indicates:

		Means averaged by individual	
Age class	Sample size	Average loop number	Maximum loop number
1	13	1.2	2.4
2	37	1.9	3.5
3	44	2.3	4.7
4	23	2.3	5.2

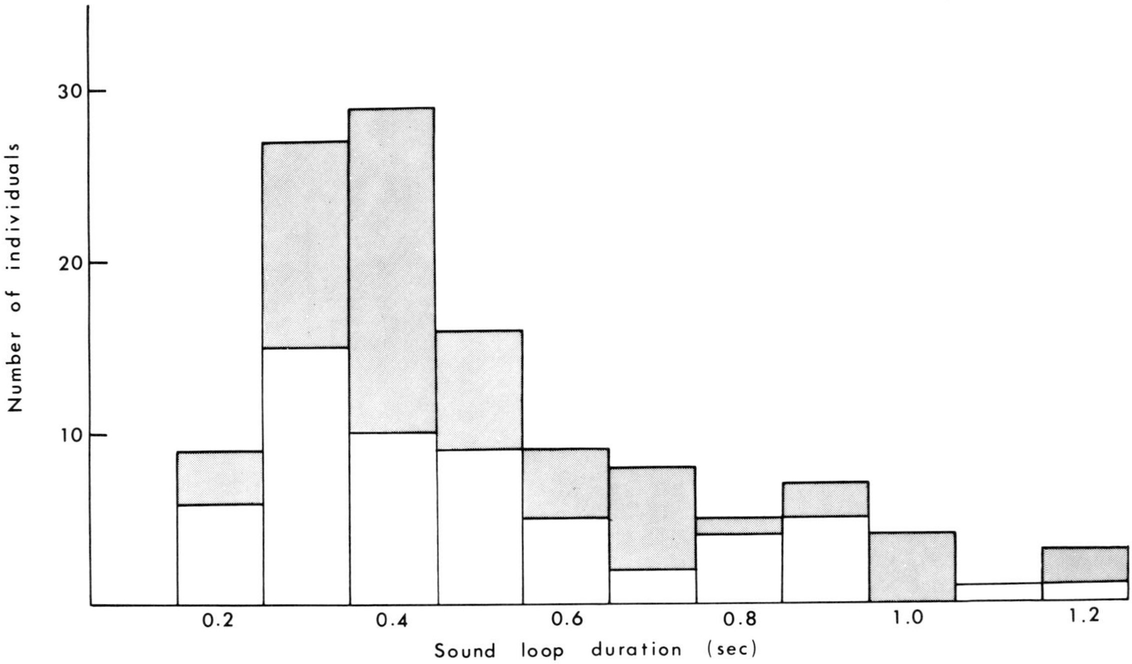

Figure 4 Average durations of the sound loops in whistles of Atlantic bottlenose dolphins. Durations of sound loops were averaged from a sample of whistles from each of the 118 dolphins used to make up this figure. Open bars indicate females ($N = 58$); shaded areas indicate males ($N = 60$). Total whistle durations are used for those individuals that did not produce multiloop whistles. Infants with no consistently patterned sound loops are omitted. No major differences between sexes are obvious in this parameter.

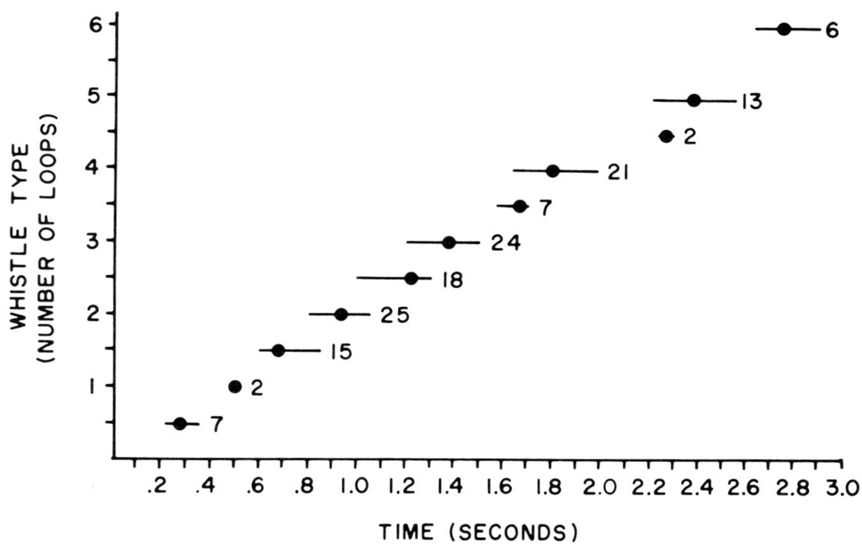

Figure 5 Average and extremes of durations of whistles of various numbers of sound loops, based on 140 whistles of a juvenile male (MLF 229). Figures to right of each line indicate numbers of whistles analyzed in each category.

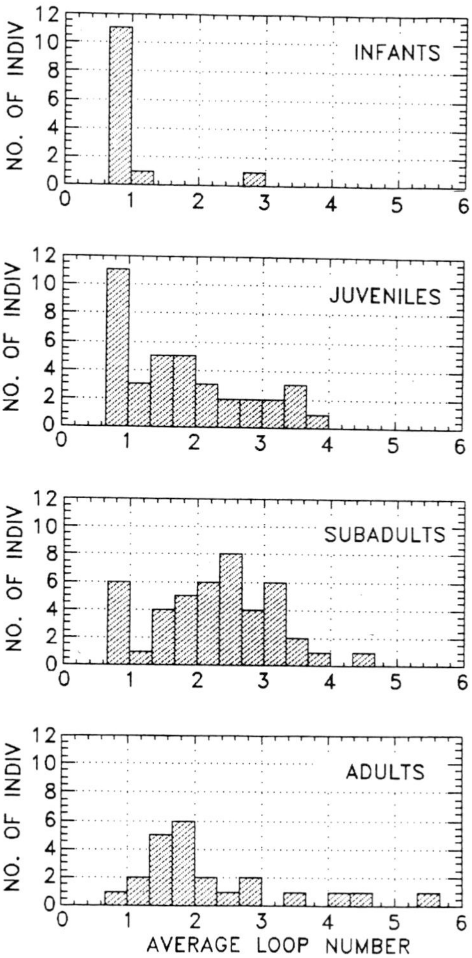

Figure 6 Differences between Atlantic bottlenose dolphins in average number of sound loops per whistle, averaged for each individual dolphin. The data are separated by age class. Sexes did not differ significantly in this parameter.

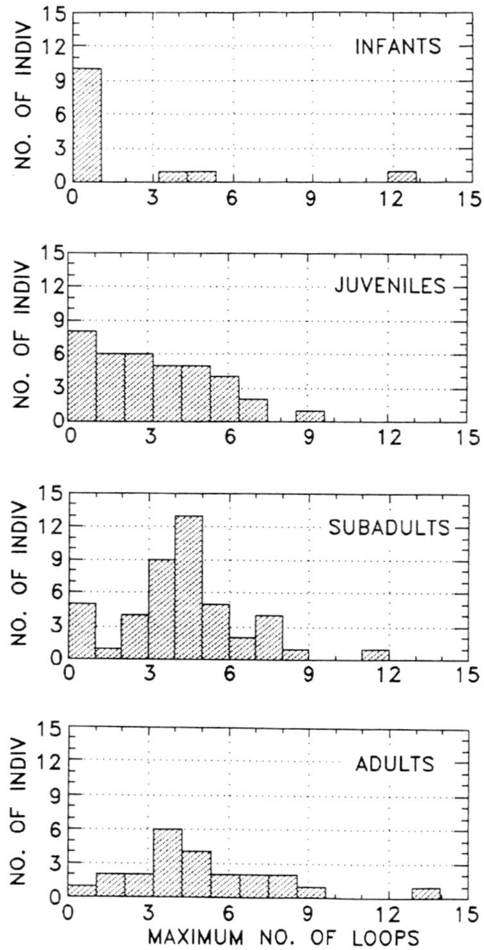

Figure 7 Differences between Atlantic bottlenose dolphins in maximum numbers of sound loops per whistle, averaged for each individual dolphin. The data are separated by age class. Sexes did not differ significantly in this parameter.

Frequencies

The total frequency sweep averaged by individual is plotted by age class in Fig. 8 for each individual's signature whistle. The mode for this parameter among the entire sample was 7 kHz. The lowest frequency component in the whistles ranged, according to individual, from 1 to 9 kHz (Fig. 9), with a mode of 5 kHz. The highest frequency component for each individual ranged from 8 to over 24 kHz (Fig. 10), with a modal value of about 13–15 kHz. No major differences between sexes were noted in any of the above parameters, either in lowest frequency (Kruskal–Wallis test statistic = 0.02, p = .89), highest frequency (Kruskal–Wallis test statistic = 0.36, p = .55), or degree of frequency modulations (Kruskal–Wallis test statistic = 0.68, p = .41). Looking at age-specific variations, there was a slight increase in lowest frequency between infant and juvenile as well as a decrease in lowest frequency between the older three age classes, but the latter was not

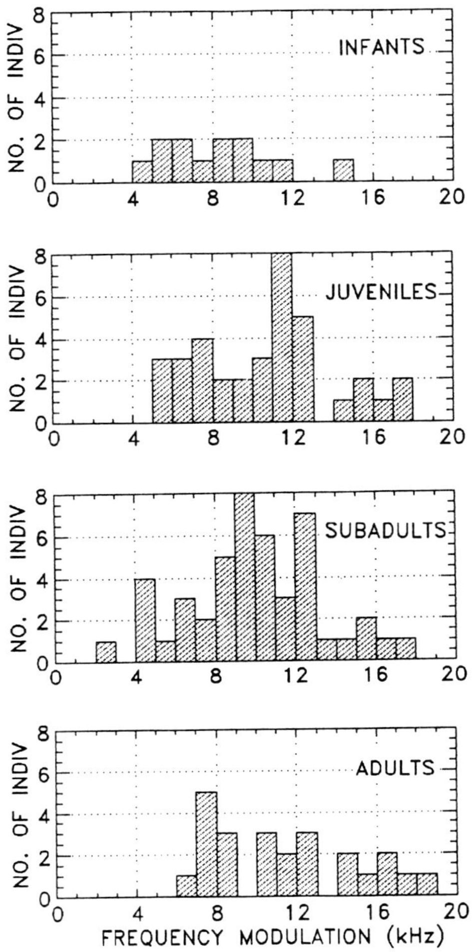

Figure 8 Extent of frequency modulation in whistles of Atlantic bottlenose dolphins. The frequency modulation of a sample of whistles was averaged for each of 122 individual dolphins. The distributions are separated by age class.

significant (Kruskal–Wallis test statistic = 3.34, p = .34). Highest frequency shows an increase between the infant and older age classes, but this was not statistically significant (Kruskal–Wallis test statistic = 6.38, p = .09). Frequency modulation was significantly age-dependent in this species, increasing with age (Kruskal–Wallis test statistic = 8.08, p = .04). This parameter combines the increase in highest frequency between infants and older dolphins and the decrease in lowest frequency between the older three age classes.

		Means averaged by individual		
Age class	**Sample size**	**Lowest Freq.**	**Highest freq.**	**Freq. mod.**
1	13	4.4	13.0	8.6
2	36	4.6	15.7	11.1
3	46	4.2	14.6	10.3
4	24	3.8	15.7	11.8

It is unusual for a dolphin's signature whistle to have a flat contour with little frequency modulation (Fig. 11). Many dolphins, however, emit whistles with characteristic plateaus of non-frequency-modulated sections (Fig. 12). This may occur in any part of the loop, but the location of the plateau is largely consistent in an individual. The whistles of some individuals achieve the highest frequency of the sound loop almost immediately. It is less common for whistles to reach the highest frequency near the end of the loop. Extremes of the average time taken for individuals to reach the highest frequency within sound loops varied from less than 0.1 to 0.9 second. These differences, in conjunction with non-frequency-modulated sections of both the rising and falling frequencies, are especially salient to us when we listen to whistles. This is particularly true at reduced tape speed and assists in identifying various individuals from hydrophone recordings of groups, once the whistle pattern has been established for each individual.

Durations

Durations are primarily, but not exclusively, a function of the number of loops per whistle (Fig. 5). Some animals emit loops faster than others, and many animals repeat sound loops slightly faster than their own individual norms when they are highly excited (Wood *et al.*, 1970). Durations of whistles averaged by individual are plotted by age class in Fig. 13. The mean of average whistle duration found in this sample, 0.96 second, was somewhat greater than many of the averages cited above from earlier papers, possibly reflecting the greater numbers of large adult animals in our sample versus the preference of research and oceanaria personnel for working with younger animals. Young animals in this sample generally

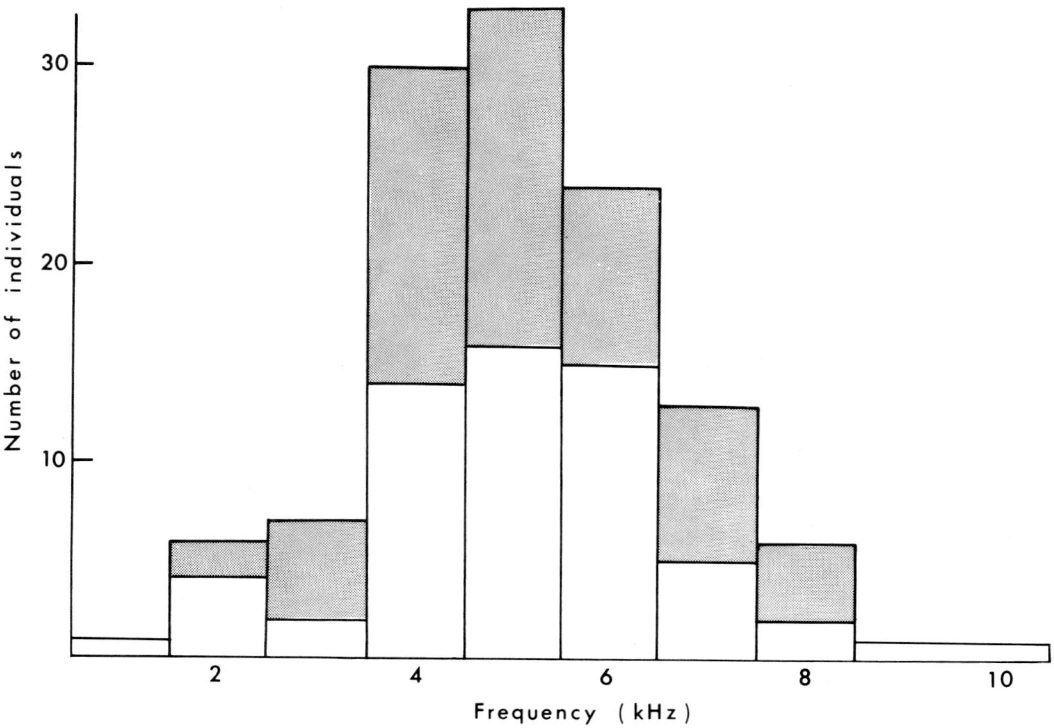

Figure 9 Lowest frequency component in whistles of Atlantic bottlenose dolphins. The lowest frequency from a sample of whistles was selected for each of 122 individual dolphins. Open bars indicate females (N = 60); shaded areas indicate males (N = 62). No major differences between sexes are evident in this parameter. Infants that were developing stereotyped whistles are included in the histograms.

emitted whistles of shorter duration as the following table indicates:

Age class	Sample size	Duration (seconds)		
		Average	**Minimum**	**Maximum**
1	13	0.8	0.2	1.4
2	37	0.9	0.4	1.6
3	45	1.0	0.3	2.0
4	23	1.0	0.4	2.1

Variation in the number of loops and thus whistle duration may communicate information other than individual identity, as discussed below. The minimum duration of whistles may be governed by the minimum time required to discriminate the signature whistles of different dolphins. One dolphin trained to discriminate between whistles of two conspecifics required approximately 0.5 second of signal duration in order to discriminate the whistles at a 100% correct level (Fig. 14). The average whistle durations for each age class were all longer than this interval required for reliable whistle discrimination.

None of the duration measures showed statistically significant differences by sex, but all three showed a significant increase in duration as a function of age class:

	Sex		Age class	
	K–W stat.	***p***	**K–W stat.**	***p***
Average duration	0.10	.75	9.83	.02
Minimum duration	0.71	.40	9.38	.02
Maximum duration	0.12	.73	11.10	.01

Maximum durations calculated by individual are plotted by age class (Fig. 15).

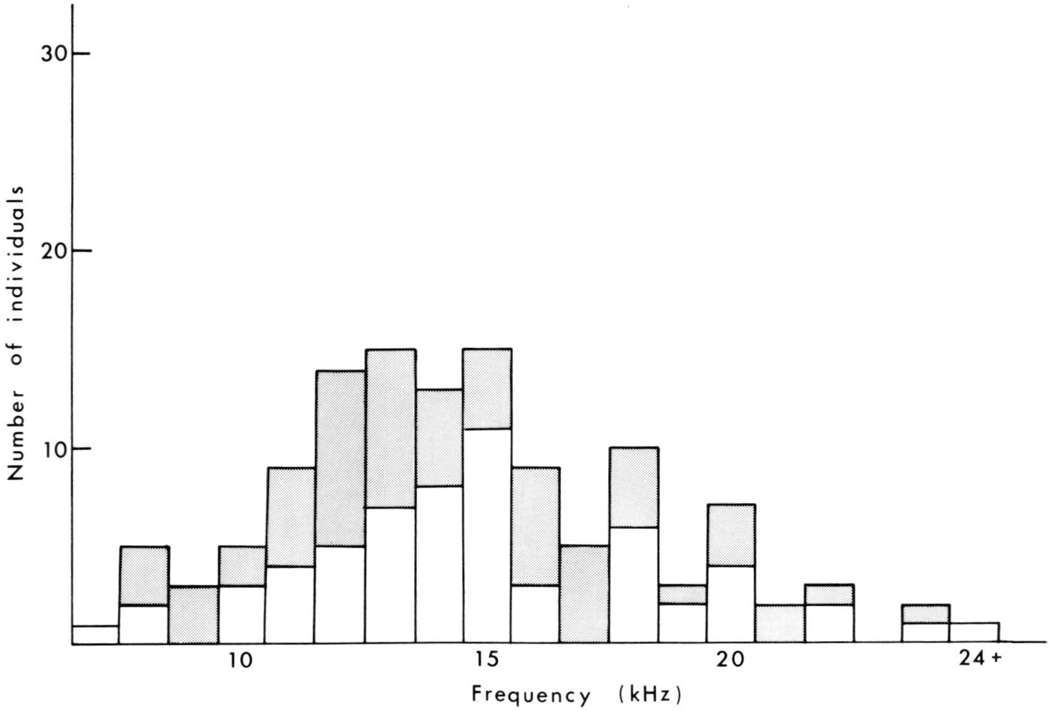

Figure 10 Highest frequency component in whistles of Atlantic bottlenose dolphins. The highest frequency from a sample of whistles was selected for each of 122 individual dolphins. Open bars indicate females ($N = 60$); shaded areas indicate males ($N = 62$). No major differences between sexes are evident in this parameter. Infants that were developing stereotyped whistles are included in the histograms.

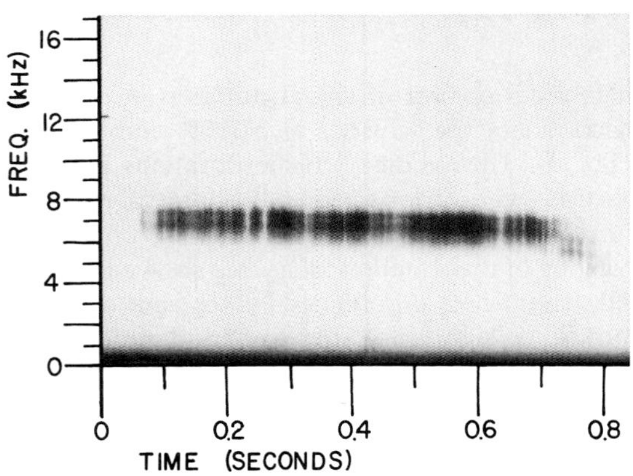

Figure 11 Signature whistle of a 218-cm female Atlantic bottlenose dolphin (MLF 290) recorded underwater while water in the tank was being lowered (effective filter bandwidth 600 Hz). Whistles of this animal ($N = 590$), recorded over a 4-day period under variable conditions, were categorized as 100% signature whistles. None were of more than one loop, which is unusual for animals of this size.

Contexts in Which Whistles Were Recorded

Although we have been said to have drawn most of our data on stereotypy from in-air recordings or from animals stranded or nearly stranded during health treatment or when their tanks were being cleaned (Herman and Tavolga, 1980), this is not at all the case. We have, indeed, examined many whistles of animals under these conditions, as this occasionally has been the only time that the animal was isolated and we could be certain it was the vocalizing individual. Additionally, these conditions were ones during which we felt different types of whistles would be most likely to appear. However, the vast majority of whistles in the sample were hydrophone recordings of animals usually, but by no means always, isolated. Whistles emitted under these different conditions are compared in the next section.

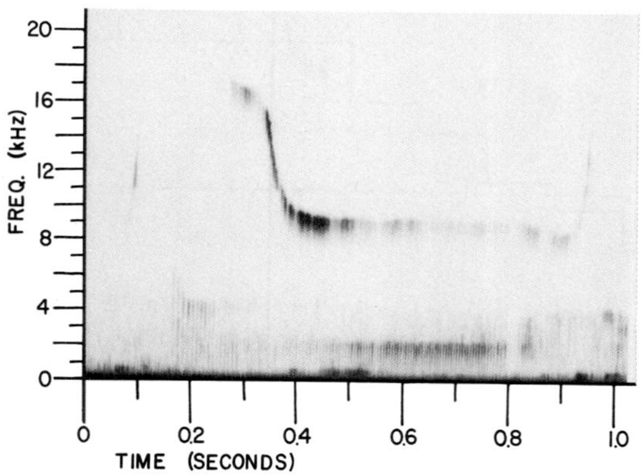

Figure 12 Signature whistle of one loop of an adult female ("Mable," at the Aquatarium, St. Petersburg Beach, Florida) Atlantic bottlenose dolphin recorded in air on 4 December 1971 (effective filter bandwidth 600 Hz). Whistles from only this one occasion were analyzed, when Mable was being moved with her 4-month-old infant. She emitted 24 whistles, 83% of which were categorized as signature whistles. Procedures such as stranding, moving, and, particularly separation from an infant may result in a higher percentage of aberrant whistles in some dolphins. Of the four aberrants recorded in the above session, three were variable aberrant chirps (of 0.2-, 0.2-, and 0.3-second durations). The likelihood of confusing such sounds with normal dolphin whistles is remote.

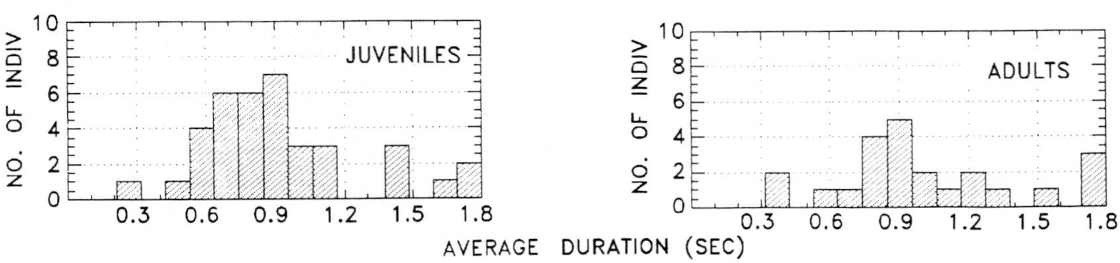

Figure 13 Average whistle durations for Atlantic bottlenose dolphins. Durations from a sample of whistles were averaged for each of 122 individual dolphins. The distributions are separated by age class. Sexes did not differ significantly in this parameter.

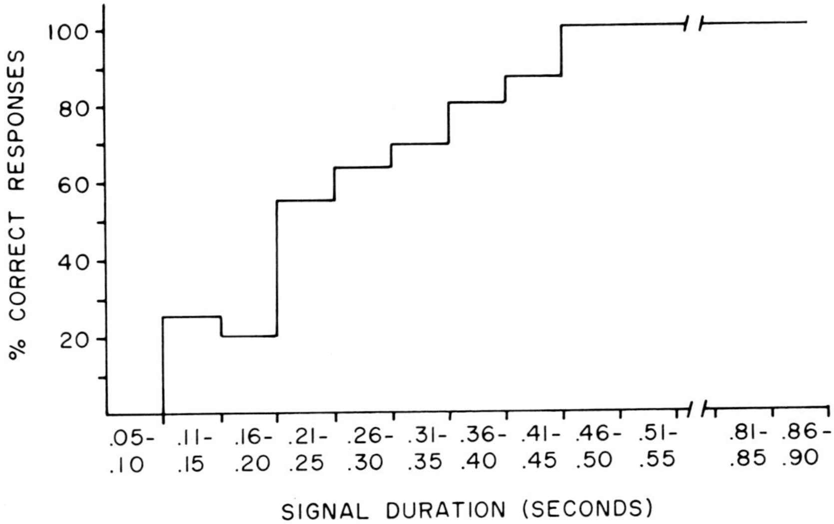

Figure 14 Relationship of signal duration to percentage of correct responses of an Atlantic bottlenose dolphin (MLF 232) in discrimination tests between whistles of two conspecifics. (Modified after Caldwell *et al.*, 1971a. their Fig. 9.)

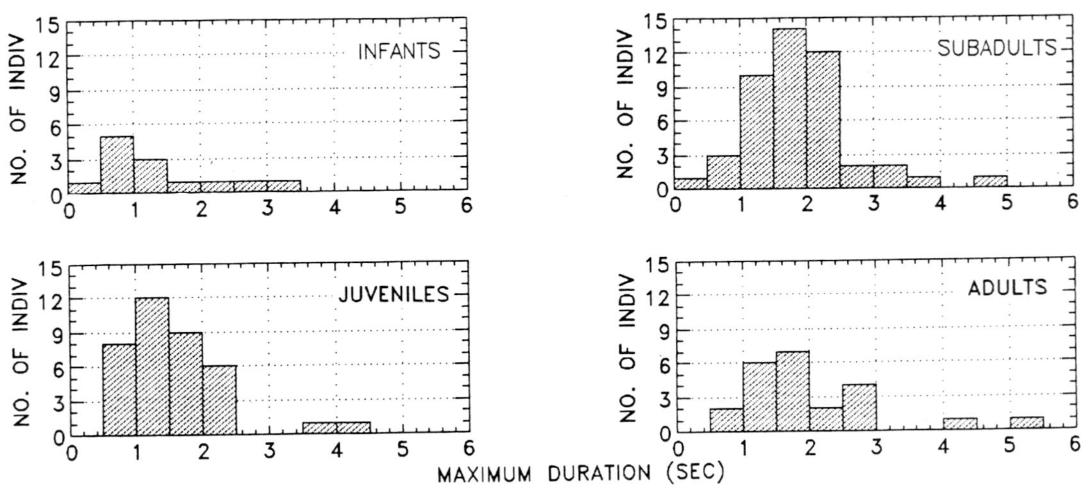

Figure 15 Maximum whistle durations for Atlantic bottlenose dolphins. The maximum duration from a sample of whistles was selected for each of 122 individual dolphins. The distributions are separated by age class. Sexes did not differ significantly in this parameter.

When dolphins were together in a tank, the individual who produced a particular whistle could occasionally be identified using several methods. If a dolphin is close to a tankside observer and whistles near the water surface, the observer occasionally can identify the whistling dolphin. Air may escape from the blowhole during whistling, particularly when the animal is excited and vocalizing loudly (D. K. Caldwell and M. C. Caldwell, 1972b). This event has enabled investigators at times to determine which dolphin is whistling, even when several dolphins are underwater in the same tank. This technique can be used as a last resort to identify whistles of individuals which cannot be isolated, but the expenditure of effort required to obtain sufficient data to make certain that a particular animal is emitting a particular whistle type can be very great. Also these data are never as trustworthy as those gathered from isolated animals. We were occasionally able to test the reliability of the data when we could change the composition of small groups of dolphins and study which whistles were present or absent as different individuals were added to or removed from a group (but see section on Mimicry for potential versus implications of this test). The development of techniques to identify whistles reliably from dolphins interacting underwater such as that described in Tyack (1985) are crucial for future research.

Whistle Variation Across Conditions

Variations exist in certain acoustic features of the signature whistle of each individual bottlenose dolphin. It has appeared to us that these characteristics might carry information on emotional state or behavioral context and might well extend across the species, that is, that all bottlenose dolphins might emit more, louder, and faster whistles when stressed (Caldwell and Caldwell, 1965). We examined variation in several acoustic features of signature whistles from several dolphins as a function of behavioral context in order to test whether these features vary systematically for different individuals with behavioral condition.

Four males of approximately the same size and locality of capture were selected for study. Recordings were made from April 1968 to July 1972 under circumstances selected to maximize variability in whistles: isolated when first captured and placed in a tank; isolated but habituated to captivity and undisturbed; stranded in air while subjected to venipuncture; and isolated and habituated to captivity and receiving human attention which appeared positively reinforcing (feeding and play). A fifth condition was included later to examine the potential effects of the presence of a conspecific. All four subjects had attained a multilooped whistle (see Figs. 1, 16, and 17). Table 3 lists the number of whistles that were aberrant and, for stereotyped whistles, the number with different numbers of loops.

To simplify statistical analysis, variation in the number of loops was not examined. Rather, we investigated variation in the percentages of aberrant whistles versus signature whistles (obtained by pooling all stereotyped whistles). Differences in the percentage of signature whistles were analyzed by contingency table analysis. The test for joint independence of the frequency of signature whistles as a function of individual and condition yields a chi-square statistic of 162.2 with 15 de-

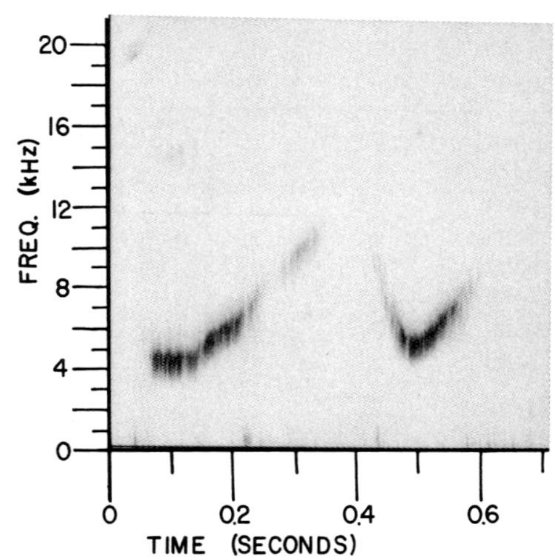

Figure 16 Signature whistle of about 1.5 loops by a 183-cm male Atlantic bottlenose dolphin (MLF 348) recorded in water under normal conditions on 18 May 1972 (effective filter bandwidth 600 Hz). Of the 1568 whistles recorded from him over a 2-month period under various conditions (see text), 99% were categorized as signature whistles.

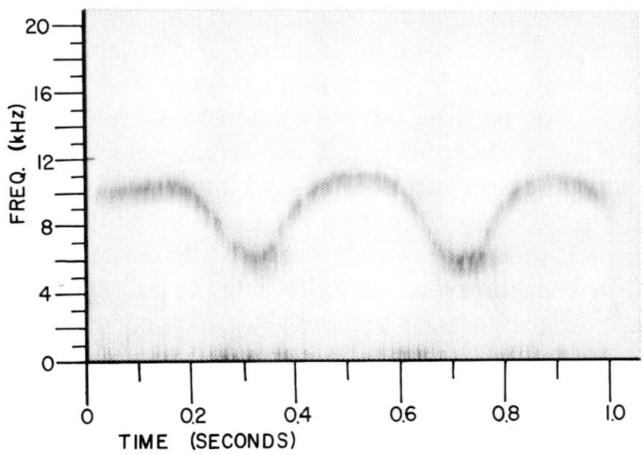

Figure 17 Signature whistle of about 2.5 loops by a 190-cm male Atlantic bottlenose dolphin (MLF 349) recorded in water under normal conditions on 18 May 1972 (effective filter bandwidth 600 Hz). Of the 101 whistles recorded from this individual over a 2.5-month period under various conditions, 98% were categorized as signature whistles (see text).

grees of freedom. The probability that the percentage of signature whistles is independent of individual and condition is vanishingly small ($p << .001$).

Tests of the partial independence of the percentage of signature whistles and condition for each individual yield the following results:

Individual	**Chi-square**	***p***
229	89.2	<<0.0001
232	15.2	0.0016
348	15.8	0.0012
349	8.2	0.0425

Each individual showed significant variation in percentage of signature whistles as a function of condition. If we investigate the differences between the four individuals in percentage of signature whistles for each behavioral condition, we obtain the following:

Table 3
Variation in Numbers of Aberrant and Stereotyped Whistles from Four Male Bottlenose Dolphins under Various Conditions

			Stereotyped whistles (numbers of loops)							
Condition[a]	Indiv.	Aberrant	<1	1–2	2–3	3–4	4–5	5–6	6+	Total
Isolated–naive	229	3	9	25	98	86	17	1	0	239
Isolated–naive	232	8	13	4	1	20	21	18	45	130
Isolated–naive	348	0	51	95	55	1	1	0	0	203
Isolated–naive	349	2	6	108	42	5	0	0	0	163
Isolated–habit	229	7	34	82	158	192	195	130	73	871
Isolated–habit	232	8	33	33	21	9	6	1	0	111
Isolated–habit	348	3	149	365	177	11	5	3	2	715
Isolated–habit	349	4	44	259	113	12	0	0	0	432
Stranding	229	19	44	32	24	17	3	1	2	142
Stranding	232	5	50	57	100	89	60	30	32	423
Stranding	348	1	1	19	26	7	0	0	0	54
Stranding	349	1	9	103	37	5	2	1	0	158
Feed/play	229	5	33	25	53	43	83	18	8	268
Feed/play	232	21	99	99	57	57	22	17	22	394
Feed/play	348	14	227	272	69	11	1	1	0	595
Feed/play	349	9	126	103	19	2	0	0	0	259
Total		110	928	1681	1050	567	416	221	184	5157

[a] Conditions: Isolated–naive, isolated when first captured and placed in a tank; Isolated–habit, isolated but habituated to captivity and undisturbed; Stranded, stranded in air while subjected to venipuncture; Feed/play, isolated and habituated to captivity and receiving human attention that appeared positively reinforcing.

Condition	Chi-square	p
Isolated, naive	18.7	0.0003
Habituated	46.9	<<0.0001
Stranded in air	113.5	<<0.0001
Feed/play	8.4	0.0388

While each dolphin showed a significant variation in percentage of signature whistles as a function of condition, the four dolphins showed significantly different patterns for each condition.

Each of the four dolphins appears to have developed its own distinctive pattern of varying the percentages of stereotyped and aberrant whistles in different behavioral conditions (Fig. 18). The highest percentage of aberrant whistles occurred when 229 was stranded, when 232 was isolated and habituated to the tank, and when 348 and 349 were feeding or playing. Dolphin 348 did not produce any aberrant whistles under the naive condition, while 232 produced 6% aberrant whistles in this condition.

To evaluate whether the different dolphins would vary other acoustic features of whistles in distinctive ways across behavioral condition, we measured seven variables in addition to percent aberrants. The eight acoustic variables measured for all whistles examined (n = 5216) were percent stereotypy (as previously defined), total duration, total duration minus the chirps and aberrants,

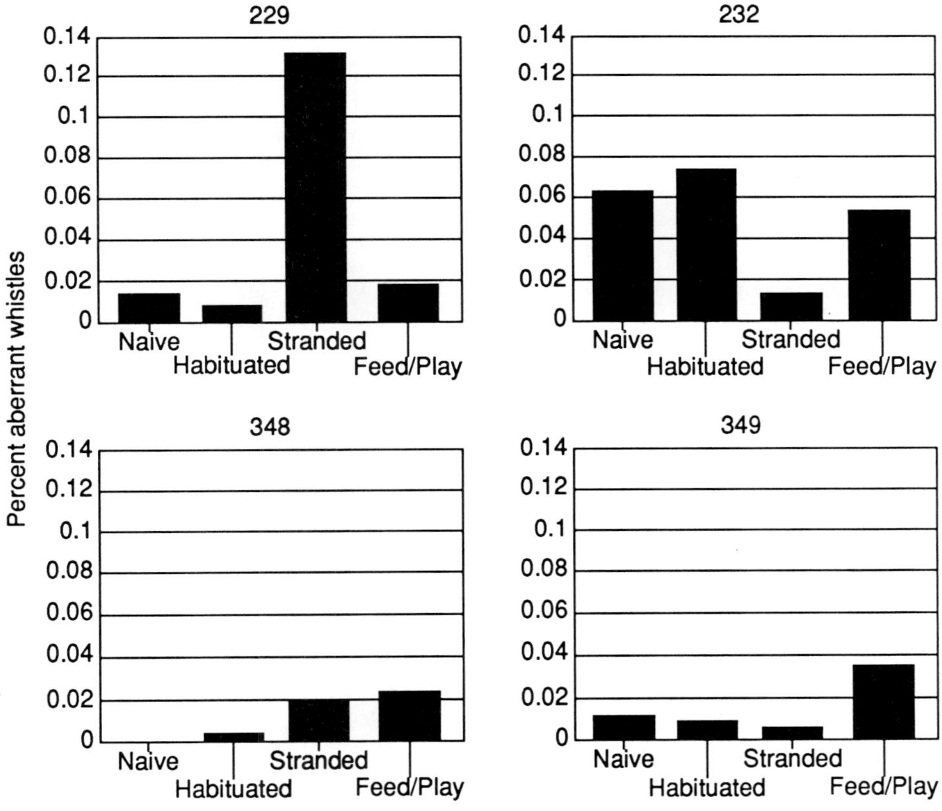

Figure 18 Percentage of aberrant whistles for each of four captive Atlantic bottlenose dolphins in four different behavioral conditions. A dolphin was considered naive when it had been recently caught and isolated in a tank. A dolphin was considered habituated when it had habituated to captivity and was isolated in a tank. Dolphins were also recorded stranded in air for veterinary attention. Recordings were used for the feed/play condition if the dolphin was being fed by or playing with humans. Dolphin 348 produced no aberrant whistles under the naive conditions.

duration of sound loops, number of sound loops in each whistle, number of whistles per minute, percent whistles with less than one loop, and percent aberrants.

Averages across the first four conditions were established in each parameter for each individual. The variation in whistles between conditions was compared for each individual against its own average. In only four of the 32 subsequent subsamples did the whistles of all four animals show the same pattern:

1. The number of whistles per minute was higher than average when the animals were first isolated in an unfamiliar environment
2. The percentage of whistles containing less than one loop (chirps and partials) was less than average when the animals were first isolated
3. The number of whistles of less than one loop (chirps and partials) was higher than average when the animals were being positively reinforced
4. The average number of loops (exclusive of chirps and partials) was reduced when the animals were being positively reinforced

The other variables did not show consistent changes across condition for different animals, and differences between sessions with the same animal under the same conditions were too variable for generalizations to be made.

The addition of a second animal into the tanks of recently captured and isolated animals resulted in a decreased rate of whistling in the two cases studied. There was also more variability in the number of loops per whistle in both animals, indicating some social effect on this parameter (Table 4). Note that this skewing is the same pattern seen when a dolphin is receiving moderate stimulation from a human. Dolphins are more likely to whistle when separated from others or when exposed to novel stimuli. When a dolphin is first isolated there is an increase in the number and duration of whistles. After the addition of a second animal the rate of whistling usually de-

Table 4
Differences in Numbers and Loop Structure of Whistles of Two Naive Atlantic Bottlenose Dolphins in Isolation and After Being Placed with One Other Dolphin of the Same Species

	Animal number 1 (MLF 349)		Animal number 2 (MLF 354)	
	Isolation	Post isolation	Isolation	Post isolation
Number (each based on five different recording sessions)	660	134	933	557
Whistles per minute	4.4	0.8	5.6	3.3
Number of loops (% of total)				
Aberrant	0.7	1.5[a]	0.1	0.7[a]
<1	12.1	30.6	27.3	37.7
1 to <2	49.8	41.8	55.9	39.5
2 to <3	33.1	20.8	13.6	12.4
3 to <4	3.4	2.2	2.1	6.8
4 to <5	0.1	0.7	0.5	1.4
5 to <6	0.4	1.4	0.3	1.1
6 or more	0.0	0.7	0.0	0.4

[a] Aberrant whistles from tanks containing two animals frequently represent only the listener's best judgment of the emitting animal.

creases. If one of these two animals is then moved to a new tank in isolation, it again tends to increase its rate of whistling. Conversely, the animal left behind in a familiar environment does not. For such animals isolated in a familiar environment, exposure to the whistles of another animal may increase the rate of whistling (Lang and Smith, 1965).

Effects of Stress and Arousal on Whistle Production

Having previously recorded naive dolphins when the water in their tanks was being drained, we were aware that the whistles appeared to be shifting progressively in several parameters, including number, duration, and intensity. As the water level is lowered, the silence in a tank containing naive dolphins is nearly complete. Not only do the dolphins cease to vocalize but their swimming and breathing become virtually noiseless. Newly captive dolphins appear to be very sensitive to changes in tank water level. That individual bottlenose dolphins are acutely aware of any change of water level in their tank was noted by Townsend (1914), in one of the first reports on dolphins in captivity. Townsend pointed out that "the entirely noiseless opening of a distant valve to lower the water, is apparent to them and may stop their play temporarily." They show slowly increasing signs of stress as the water is lowered, until the animals are semistranded. When the water level is raised again, there appears to be a slow reduction of stress. We report now on the effects of these changes in stress on the numbers of whistles emitted.

Whistles were recorded during several water lowerings in a tank containing five recently captured dolphins of assorted sizes and both sexes (Figs. 19–21). Tests were initiated in the morning, prior to feeding. Unless otherwise noted, the animals were undisturbed during these sessions except when the water level was read at 5-minute intervals from a stick gauge on the inside of the tank. The animals were first habituated to the equipment (hydrophone, cable, and gauge) for about an hour to establish a pretest baseline. Figures 19–21 indicate an immediate inhibition of vocalizations after the drain opened and the water level started to fall. It appears that the rate of whistling started to rebound when the water level had fallen below approximately 1m (~3 feet). The rate of whistling increased to original levels or higher after refilling commenced, but note that the rate of whistling had begun to increase before the drain was closed and water turned on. The animals appear rapidly to have learned that when the equipment associated with this experiment was put in the tank, the level was about to be lowered, as is indicated by the sharp rise in whistles prior to the opening of the drain in Figs. 20 and 21.

We have eliminated from these figures any test runs in which wind velocities suddenly increased (when numbers of whistles tended to increase sharply) or when such events occurred as unscheduled activity in an adjacent area. In spite of our efforts to keep conditions constant over the several hours necessary for a session, we were at times unsuccessful in preventing feeding. We include Fig. 22 chiefly to corroborate Powell's (1966) findings on the pronounced effects of feeding on the vocalizations of these animals. The effect of feeding was not so evident in Fig. 21 because the water level was still too low for the animals to eat.

There was some variation in whistles that is not evident in these whistle counts. For example, in Fig. 21, the animals were in a quiescent state resembling sleep during the first 15 minutes of the recording session. Although they continued to circle the tank quietly, there was no other overt activity. The whistles during the next 15 minutes were infrequent and brief, the frequency modulations of most were limited, and many were not recognizable to individual. We classified most as aberrants. These phased rather rapidly during the next few minutes into the normal whistles of the five individuals. Whistling of animals that are being euthanized frequently show similar but reverse trends as they are being put to sleep. The description of these whistles is also reminiscent of the aberrants emitted by a long-term isolated male with which we worked. In all of these recording conditions the animals could be described as being in a low state of arousal.

We interpret the variation in rates of whistling during changes of water level as effects of alarm or stress. Stressful stimuli may either inhibit or stimulate whistle production. In the initial stages of

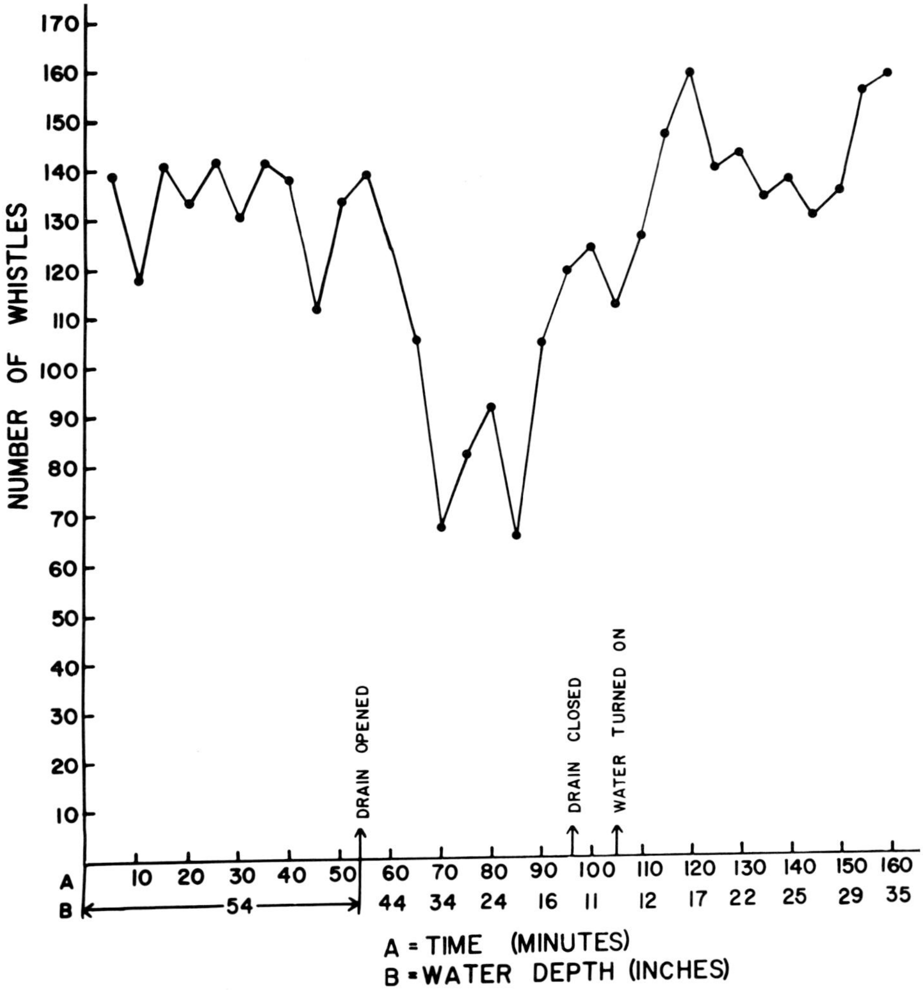

Figure 19 Number of whistles per progressive 5-minute interval emitted by five Atlantic bottlenose dolphins prior to, during, and following stranding in a tank [first session (after one incomplete pilot session on 12 November 1968), recorded on 26 November 1968].

dropping the water level, whistle production falls, but as the water drops below 1m (3 feet), whistle production rebounds. Initially the dolphins appeared to sensitize to dropping the water level. There was a potentiation of the decrease in whistling as the water was lowered between subsequent sessions. Captive animals, however, eventually become habituated to such stranding procedures. Their tanks are cleaned routinely, and these patterns eventually disappear.

Last, any ongoing vocalization may be sharply terminated by the sudden introduction of a novel stimulus. This caused us a considerable loss of time and film when we were trying to get X-ray photography of the nasal sacs of dolphins in the different phonatory modes (Hollien *et al.*, 1976). For an interminable number of exposures, regardless of whether the animal was whistling, squawking, or clicking, it would abruptly stop phonating when the X-ray timer began clicking. The dolphin sometimes promptly resumed vocalizing when the exposure was complete. Figure 23 illustrates the

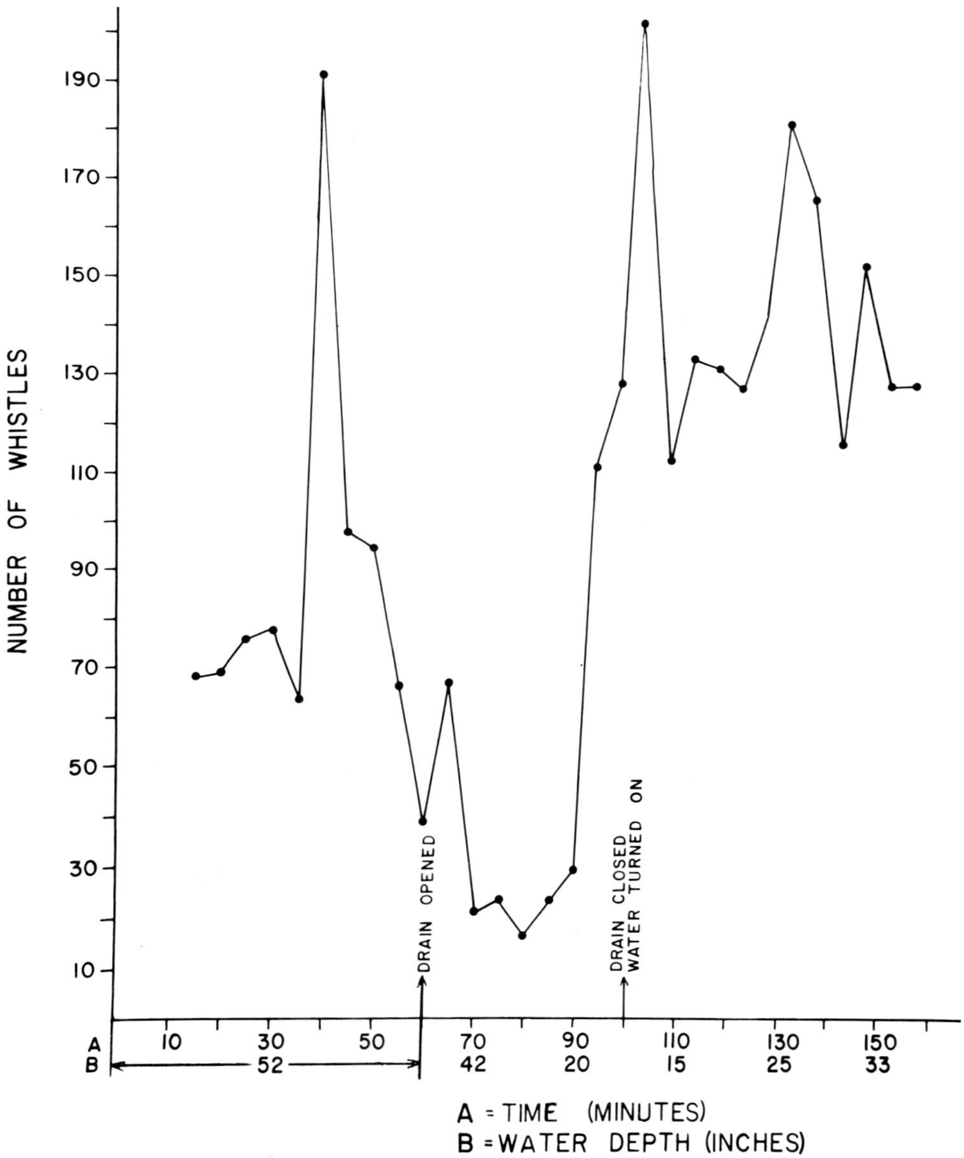

Figure 20 Number of whistles per progressive 5-minute interval emitted by the five bottlenose dolphins as in Fig. 19 prior to, during, and following stranding in a tank (second session, recorded on 27 November 1968).

rapidity with which these events occurred. The animal was clicking steadily during the initial time period of 0–0.5 second. The dolphin clicks extend to about 2 kHz. The onset and offset of the timer is indicated by the darker lines extending to about 4.5 kHz. Following the onset of the timer, the animal stopped clicking within 0.4 second and resumed about 0.3 second following cessation of the timer clicks. A similar cessation in click production in sperm whales, *Physeter catodon,* after exposure to the sounds of a pinger was noted by Watkins and Schevill (1975).

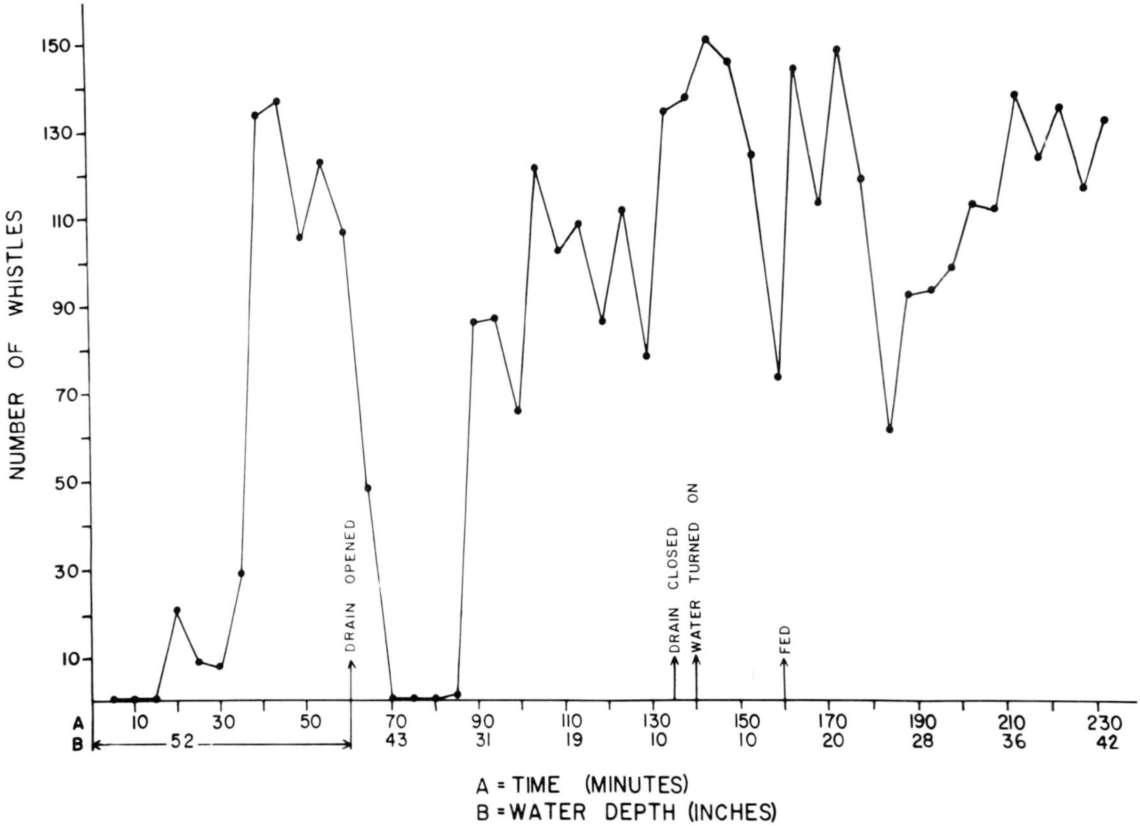

Figure 21 Number of whistles per progressive 5-minute interval emitted by the five bottlenose dolphins as in Fig. 19 prior to, during, and following stranding in a tank (fourth session, recorded on 3 January 1969).

Patterning of Successive Whistles

Table 2 presents a transition matrix indicating variation in the number of loops of a whistle depending on the number of loops of the whistle which preceded it for series of whistles recorded from an isolated dolphin. These transitions are strongly dependent on the initial whistle type (chi-square = 287.2, $df = 63$, $p < 0.005$). The two strongest patterns were for whistles with the same number of loops to repeat and for this animal to shift toward its own preferred number of three to four loops. For example, the most common transition both from aberrants and whistles of four to five loops (initial vocalizations) was to whistles of three to four loops (following vocalizations). There is no indication in our data of more complex stereotyped patterns in long series of whistles.

Ontogeny

We have studied the development of signature whistles in 14 infant bottlenose dolphins (Caldwell and Caldwell, 1979). Briefly, all animals studied had a whistle at birth, but it was tremulous and not classified as stereotyped. The earliest age at which an infant developed a stereotyped whistle was between 45 and 68 days. Only one animal in the sample had not developed a stereotyped whistle by the end of the first year. These results were obtained largely from longitudinal studies. As has been shown above, there are significant changes in whistles as a function of age class. These consist of a general decrease in percent stereotypy, an increase in number of sound loops and therefore longer durations, and greater frequency modulations with increasing age.

Two mature females and one juvenile male

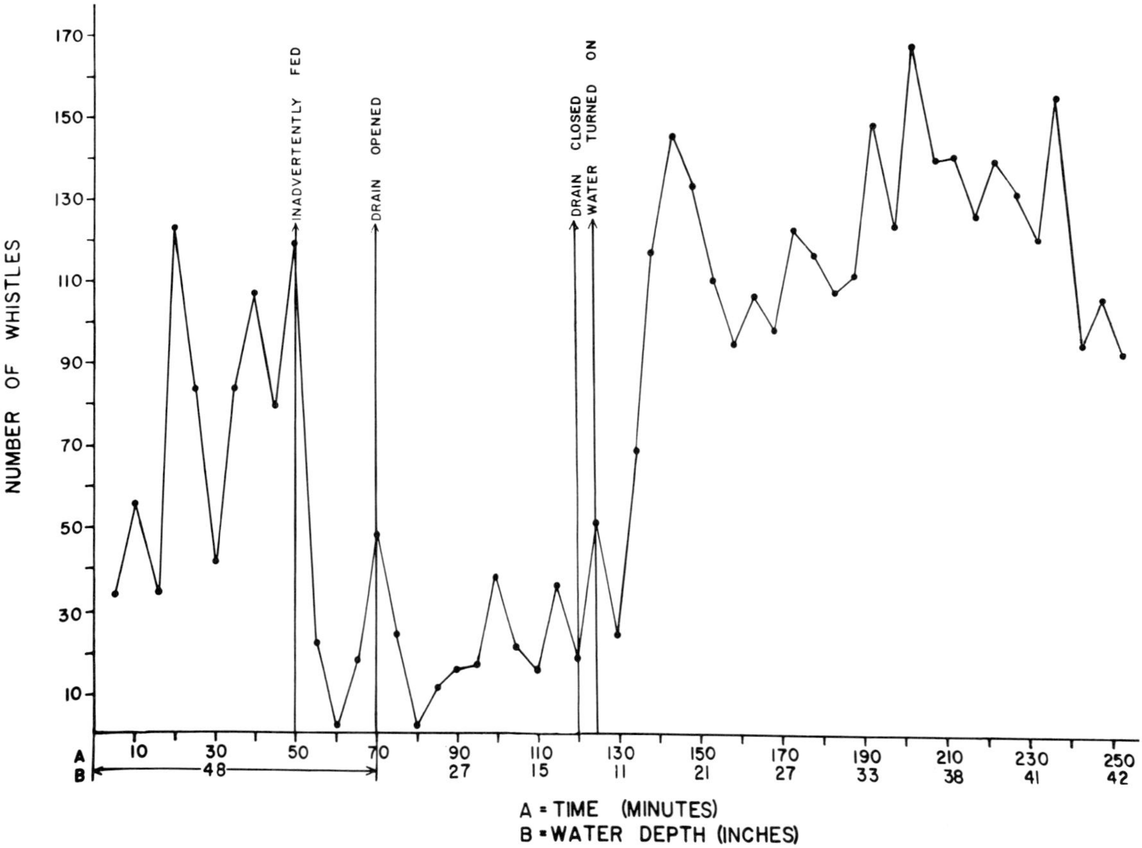

Figure 22 Number of whistles per progressive 5-minute interval emitted by the five bottlenose dolphins as in Fig. 19 prior to, during, and following stranding in a tank (fifth session, recorded on 10 January 1969).

provide long-term data concerning the permanence of signature whistles. Animal MLF 110 was 235 cm long when captured from the wild and first recorded on 4 April 1968. Her signature whistle had not changed when she was last recorded in December 1985 and therefore remained constant for a period approaching 18 years. Animal MLF 216, captured in the wild, was first recorded as a mature adult (245 cm long) in June 1968. Last recorded in July 1981, she had retained her signature whistle for 13 years. The third animal, MLF 331, was a juvenile male 204 cm long when first recorded in January 1972. His signature whistle was unusual in that it was a constant 1½ loops. In February 1980 this animal still emitted the same 1½-looped signature whistle, although we expected a multilooped whistle to have developed over this 8-year period.

All three of these dolphins were in acoustic contact with other dolphins, the first in a stadium pool and the other two in community tanks where they were in constant physical and acoustic contact with several other animals. These results suggest that signature whistles, once established, remain constant for dolphins in captive groups. Each animal in this study maintained a constant signature whistle, despite hearing different whistles from other animals following capture and as dolphins were moved in and out of their tank.

Results were very different for a juvenile male that was isolated for 7 years. MLF 232 was captured on 1 May 1968. He was 185 cm long at that time and was estimated to be 1–2 years of age. He was isolated in a tank on 6 February 1969, and from 15 April 1976 he was trained and exposed to a variety of recorded and synthesized dolphin whistles and to electronically produced pure tones. He virtually lost the signature whistle that he had

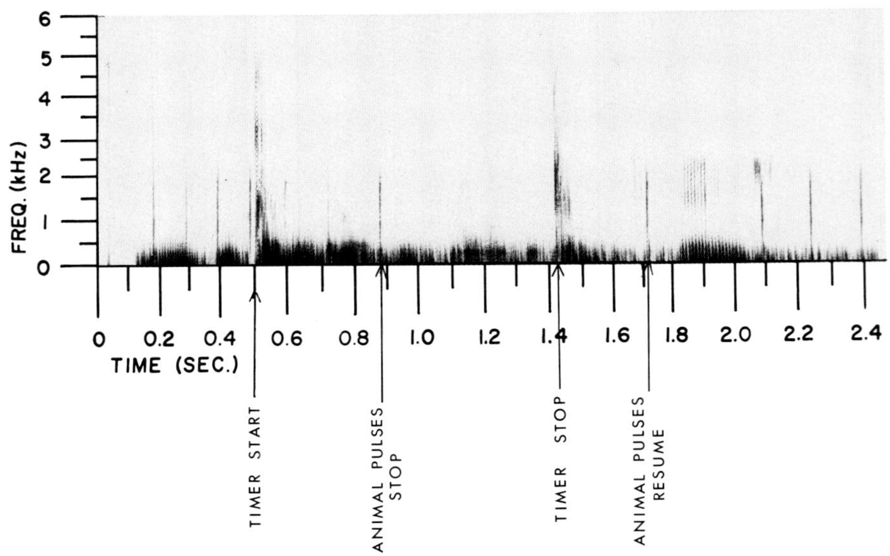

Figure 23 Inhibition of ongoing vocalizations by a late-infant male Atlantic bottlenose dolphin (MLF 232) through the sudden introduction of an external sound stimulus (X-ray timer) and subsequent release of some kind of vocalization following the cessation of the external sound (effective filter bandwidth 300 Hz).

produced prior to exposure to our playback stimuli. Instead he produced an almost unlimited number of other whistle sounds. Some of these were obviously imitations of the last tones to which he had been exposed. Others were multilooped whistles that were uncharacteristic of his signature whistle. These multilooped whistles also did not obviously match the signature whistles of other bottlenose dolphins. This observation alerts us to potential differences between the whistles of long-term captive experimental animals subjected to many sounds versus those in a relatively normal acoustic environment. This kind of abnormal behavior may help explain the controversy that has occasionally arisen over our concept of the signature whistle.

DISCUSSION

Stereotypy of Signature Whistles

We can sort spectrograms of signature whistles reliably, and dolphins are easily able to discriminate signature whistles of different individuals. The stereotypy of signature whistles has not been defined quantitatively, however, nor have indices for comparing the similarity of different whistles been developed. Development of quantitative indices of similarity will be important for studies of mimicry and the effect of early acoustic environment on the ontogeny of signature whistles. This will not be simple, for the stereotypy of signature whistles involves the contour configuration of the loop rather than the more simply measured features of duration, minimum and maximum frequency, number of maxima and minima in the contour, etc.

Variation in the number of loops in a dolphin's whistle may lead to major changes in the duration and number of maxima and minima in the contour with only minor changes in loop configuration. Tyack (1986a) showed examples of variation in duration of 30% for one-loop whistles of one dolphin. Even over this range, the contour configuration remains similar. Experiments where dolphins were trained to imitate whistlelike manmade sounds (R. Penner, cited in Evans, 1967; Richards *et al.*, 1984) show that bottlenose dolphins frequently will transpose whistles to a different frequency range than the models while maintaining other aspects of contour configuration. Thus measurements of absolute duration or frequency may be misleading unless accompanied by evaluation of contour configuration. The only way to validate our estimates of contour similarity is to compare them with tests in which dolphins themselves are trained to evaluate similarity of whistles.

Variation in many acoustic parameters of the whistle besides contour configuration may convey information other than signature. Several examples relating to whistle duration are obvious. It is likely that a listening dolphin would find a multiloop whistle more attention-demanding than a shorter vocalization. We also have found that a dolphin is better able to localize longer pure tones of 5 seconds than those of 1 second (Caldwell, *et al.*, 1971b). Again, long multiloop whistles would give a continuing sound source on which to localize an underwater moving target such as another dolphin.

Function of the Signature Whistle

Individual-specific social relationships are an important feature of dolphin social behavior (Tyack, 1986b). For example, bottlenose dolphin calves in the wild may remain closely associated with their mother for periods of up to 8 years (Wells *et al.*, 1987). The composition of bottlenose dolphin groups is characterized by a high degree of fluidity coupled with more stable subgroups (Shane *et al.*, 1986). The stability of these subgroups within more casual larger groups argues for the importance of social bonds between particular individuals. Individuals within these groups frequently spread out beyond sighting distance of one another. How do these individuals stay together when out of sight? The utility of being able to broadcast the location of an individual within these large groups should be obvious.

The signature-whistle hypothesis proposes that each member of a social group can broadcast its identity and location to other members of the group. Bottlenose dolphins tested on pure tones as well as clicks were able to localize such sounds very well (Caldwell *et al.*, 1971b; Renaud and Popper, 1975). Experiments have shown that a dolphin is capable of distinguishing between signature whistles of at least eight different individuals if allowed to hear only one complete whistle (Caldwell *et al.*, 1972). This same test animal had been tested earlier to determine the duration of the test whistles necessary for him to discriminate between the whistles of two different animals. The subject dolphin required only 0.45 second of these signature whistles to which he had been heavily exposed to be 100% correct in his identification (Fig. 14, and see Caldwell *et al.*, 1971a). This figure of about one-half second is approximately that of one full sound loop of the signature whistles of most, but not all, bottlenose dolphins (Fig. 4) and was the approximate duration of the loops used in the test whistles.

Evidence that a dolphin can distinguish between signature whistles does not demonstrate that dolphins learn to associate each signature whistle with the appropriate individual. Richards *et al.* (1984) have shown that dolphins have the cognitive skills required to associate arbitrary whistlelike sounds with arbitrary objects such as balls and pipes. However, no data presented to date demonstrate whether dolphins associate signature whistles with the appropriate individual. Nor is it known whether wild dolphins learn only the signature whistles of a few close associates or those of tens or hundreds of individuals within their community.

The signature-whistle hypothesis proposes a specific social function for dolphin whistles. It cannot be tested fully until we can investigate the social contexts in which one dolphin produces a particular whistle contour and how other group members respond. These investigations have been severely limited by our inability to identify which dolphin within an interacting group produces each whistle. As techniques are developed to solve this problem, and as we integrate studies of dolphin communication more fully with studies of social behavior, the social functions of specific whistle contours produced by particular individuals will be understood more completely.

Investigators must be sensitive to the potential importance of learning in this system of communication. Vocal learning must mediate the precise imitation of whistles and whistlelike sounds uncovered in dolphin mimicry research. If the signature-whistle hypothesis is correct, dolphins must also learn to make the association between each individual and its signature whistle. The role of learning in the ontogeny of signature whistles is not clear. If a major function of whistles is to broadcast individual identity, then one might expect dolphins to develop a signature whistle that is particularly distinctive with respect to its natal group. On the other hand, one of the young captive calves studied by Caldwell and Caldwell (1979) developed a whistle very similar to that of its mother. This would argue that some young dolphins may produce signature whistles particularly like those of their mother.

Variation in Whistles as a Function of Context

The data presented here on the variation in whistles across condition for four captive bottlenose dolphins indicate that different dolphins may vary their signature whistles in different ways as they change behavioral context. For example, dolphin 229 tended to produce whistles of 3–5 loops when isolated after habituation but produced abbreviated whistles of less than 1 loop during stranding (Table 3). Dolphin 232 showed the opposite tendency, with whistles of 0–2 loops most common when isolated after habituation versus 2–4 loops when stranded for veterinary attention (Table 3). While these patterns may be predictable for a given dolphin, they are not consistent for different dolphins. For another dolphin to learn what a particular pattern of whistling predicts, it may need to learn how the individual it is hearing tends to vary its whistles in different contexts.

Since the differences between individual dolphins in the number of loops per whistle as a function of behavioral context were not always consistent, it is not obvious why dolphins vary the number of loops. Dolphins tend to increase the variability in number of loops per whistle when they are with other dolphins versus in isolation (Table 4). This suggests that they produce more stereotyped whistles when establishing contact with another dolphin. Once contact is established, the higher variability in number of loops may reflect increased communication about information other than individual identity. However, it remains to be seen whether dolphins learn how their associates' whistles vary across contexts.

Nor is it obvious what limits the number of loops per whistle. Many young dolphins and some adults never produced whistles of more than one loop, while several other dolphins produced whistles with more than 10 loops. On some of these extremely long multiple-loop whistles, there is a suggestion of loss of energy on the last loops. Younger animals also tended to produce whistles with fewer loops. These observations suggest that some factor which scales with size may limit the number of loops a dolphin can produce.

The early papers of Lilly and Dreher on bottlenose-dolphin whistles assumed that each dolphin produces a specific set of contours and that all bottlenose dolphins use each contour in a particular context. While this may be typical of many systems of animal communication, it does not fit the data for bottlenose-dolphin whistles. Each of these dolphins appears to have its own individualized whistle pattern, with a whistle repertoire primarily based on variations of its signature whistle and including imitations of the signature whistles of other dolphins. Different dolphins may also vary their signature whistles in distinctively different ways as a function of behavioral context. Data on signature whistles and whistle mimicry point to a system of communication highly dependent on learning and to a social system based on individual-specific social relationships.

ACKNOWLEDGMENTS

Financial support for M. C. and D. C. for various phases of this study came primarily from the Office of Naval Research (Contracts N00014-67-C-0358 and N00014-70-C-0178), the National Science Foundation (Grant GB-1189), the National Institute of Mental Health (Grant MH-07509), the National Institute of Neurological Diseases and Stroke (Grant NS-09694), Marineland of Florida, and Biological Systems, Inc. Funding for P.T. for the analysis of changes in whistles as a function of age and sex came from the National Institute of Neurological and Communicative Disorders and Stroke (Grant 5 R29 NS25290). Andrew Solow of WHOI helped with statistical analysis of whistles, and performed the contingency table and chi-square analyses. We especially wish to thank Mary Ann Daher, Kurt Fristrup, Harry Hollien, Peter Marler, Kenneth S. Norris, Cheri Recchia, and William Watkins for their valuable advice and insight in reviewing earlier drafts of the manuscript. This is contribution no. 7156 from the Woods Hole Oceanographic Institution.

REFERENCES

Agresti, A. (1984). "Analysis of Ordinal Categorical Data." Wiley (Interscience), New York. 287 pp.

Batteau, D. W., and Markey, P. R. (1968). Man/dolphin communication. Final Report (15 December 1966–13 December 1967) prepared for U. S. Naval Ordnance Test Station, China Lake, California, under Contract Number N00123-67-C-1103 to Listening, Inc., Arlington, Massachusetts. 84 pp. (processed).

Burdin, V. I., Reznik, A. M., Skornyakov, V. M., and Chupakov, A. G. (1975). Communication signals of the Black Sea bottlenose dolphin. *Sov. Phys. Acoust.* **20,** 314–318.

Busnel, R.-G., and Dziedzic, A. (1966). Acoustic signals of the pilot whale *Globicephala melaena* and of the porpoises *Delphinus delphis* and *Phocoena phocoena*. *In* "Whales, Dolphins, and Porpoises" (K. S. Norris, ed.), pp. 607–646. Univ. California Press, Berkeley. 789 pp.

Busnel, R.-G., and Dziedzic, A. (1968). Étude des signaux acoustiques associés a des situations de détresse chez certains cétacés odontocètes. *Ann. Inst. Océan.* **46**(2), 109–144.

Caldwell, D. K., and Caldwell, M. C. (1968). The dolphin observed. *Nat. Hist.* **77**(8), 58–65.

Caldwell, D. K., and Caldwell, M. C. (1970). Etiology of the chirp sounds emitted by the Atlantic bottlenosed dophin: A controversial issue. *Underwater Naturalist* **6**(3), 6–8, 43.

Caldwell, D. K., and Caldwell, M. C. (1972a). Senses and communication. *In* "Mammals of the Sea; Biology and Medicine" (S. H. Ridgway, ed.), pp. 466–502. Charles C. Thomas, Springfield, Illinois. 812 pp.

Caldwell, D. K., and Caldwell, M. C. (1972b). "The World of the Bottlenosed Dolphin." J. B. Lippincott Co., Philadelphia. 157 pp.

Caldwell, D. K., and Caldwell, M. C. (1977). Cetaceans. *In* "How Animals Communicate" (T. A. Sebeok, ed.), pp. 794–808. Indiana Univ. Press, Bloomington. 1128 pp.

Caldwell, M. C., and Caldwell, D. K. (1965). Individualized whistle contours in bottlenosed dolphins (*Tursiops truncatus*). *Nature (London)* **207,** 434–435.

Caldwell, M. C., and Caldwell, D. K. (1967). Intraspecific transfer of information via the pulsed sound in captive odontocete cetaceans. *In* "Les Systèmes Sonars Animaux, Biologie et Bionique" (R.-G. Busnel, ed.), Vol. 2, pp. 879–936. Laboratoire de Physiologie Acoustique, Jouy-en-Josas, France.

Caldwell, M. C., and Caldwell, D. K. (1968). Vocalization of naive captive dolphins in small groups. *Science* **159,** 1121–1123.

Caldwell, M. C., and Caldwell, D. K. (1972). Vocal mimicry in the whistle mode by an Atlantic bottlenosed dolphin. *Cetology* **9,** 1–8.

Caldwell, M. C., and Caldwell, D. K. (1979). The whistle of the Atlantic bottlenosed dolphin (*Tursiops truncatus*)—Ontogeny. *In* "Behavior of Marine Animals" (H. E. Winn and B. L. Olla, eds.), Vol. 3, pp. 369–401. Plenum, New York. 438 pp.

Caldwell, M. C., Caldwell, D. K., and Turner, R. H. (1970). Statistical analysis of the signature whistle of an Atlantic bottlenosed dolphin with correlations between vocal changes and level of arousal. *Los Angeles County Mus. Nat. Hist. Found., Tech. Rept.,* No. 8, LACMNHF/MRL ONR Contract No. N00014-67-C-0358. 26 pp. (processed).

Caldwell, M. C., Caldwell, D. K., and Hall, N. R. (1971a). An experimental demonstration of the ability of an Atlantic bottlenosed dolphin to discriminate between playbacks of recorded whistles of conspecifics. *Stanford Res. Inst. Proc. Seventh Ann. Conf. Biol. Sonar Diving Mamm., 1970,* Menlo Park, Calif., pp. 141–158.

Caldwell, M. C., Hall, N. R., Caldwell, D. K., and Hall, H. I. (1971b). A preliminary investigation of the ability of an Atlantic bottlenosed dolphin to localize underwater sound sources. *Marineland Res. Lab. Tech. Rept.,* No. 6, prepared under ONR Contract No. N00014-70-0178. 6 pp. (processed).

Caldwell, M. C., Hall, N. R., and Caldwell, D. K. (1972). Ability of an Atlantic bottlenosed dolphin to discriminate between, and respond differentially to, whistles of eight conspecifics. *Proc. Eighth Annu. Conf. Biological Sonar Diving Mamm. 1971,* Fremont, Calif., pp. 57–65.

Caldwell, M. C., Caldwell, D. K., and Miller, J. F. (1973). Statistical evidence for individual signature whistles in the spotted dolphin, *Stenella plagiodon*. *Cetology* **16,** 1–21.

Dreher, J. J. (1961). Linguistic considerations of porpoise sounds. *J. Acoust. Soc. Am.* **33**(12), 1799–1800.

Dreher, J. J. (1966). Cetacean communication: Small-group experiment. *In* "Whales, Dolphins, and Porpoises" (K. S. Norris, ed.), pp. 529–541. Univ. California Press, Berkeley. 789 pp.

Dreher, J. J., and Evans, W. E. (1964). Cetacean communication. *In* "Marine Bio-acoustics," (W. N. Tavolga, ed.), Vol. 1, pp. 373–393. Pergamon, New York. 413 pp.

Evans, W. E. (1967). Vocalization among marine mammals. *In* "Marine Bio-acoustics" (W. N. Tavolga, ed.), Vol. 2, pp. 159–186. Pergamon, New York. 353 pp.

Evans, W. E., and Prescott, J. H. (1962). Observations of the sound production capabilities of the bottlenosed porpoise: A study of whistles and clicks. *Zoologica* **47**(3), 121–128, plates 1–4.

Gish, S. L. (1979). A quantitative description of two-way acoustic communication between captive Atlantic bottlenosed dolphins (*Tursiops truncatus*). Ph.D. dissertation, Univ. of California, Santa Cruz. 133 pp.

Griffin, D. R. (1959). "Echoes of Bats and Men." Doubleday and Co. (Anchor Books), Garden City, New York. 156 pp.

Herman, L. M., and Tavolga, W. N. (1980). The communication systems of cetaceans. *In* "Cetacean Behavior: Mechanisms and Functions" (L. M. Herman, ed.), pp. 149–209. Wiley (Interscience) New York. 463 pp.

Hollien, H., Hollien, P., Caldwell, D. K., and Caldwell, M. C. (1976). Sound production by the Atlantic bottlenosed dolphin *Tursiops truncatus*. *Cetology* **26,** 1–8.

Kellogg, W. N. (1961). "Porpoises and Sonar." Univ. of Chicago Press, Chicago. xiv + 177 pp.

Kellogg, W. N., Kohler, R., and Morris, H. N. (1953). Porpoise sounds as sonar signals. *Science* **117,** 239–243.

Lang, T. G., and Smith, H. A. P. (1965). Communication between dolphins in separate tanks by way of an electronic acoustic link. *Science* **150,** 1839–1844.

Lilly, J. C. (1961). "Man and Dolphin." Doubleday and Co., Garden City, New York. 312 pp.

Lilly, J. C. (1963). Distress call of the bottlenose dolphin: stimuli and evoked behavioral responses. *Science* **139,** 116–118.

Lilly J. C., and Miller, A. M. (1961a). Sounds emitted by the bottlenose dolphin. *Science* **133,** 1689–1693.

Lilly, J. C., and Miller, A. M. (1961b). Vocal exchanges between dolphins. *Science* **134,** 1873–1876.

McBride, A. F., and Hebb, D. O. (1948). Behavior of the captive bottle-nose dolphin, *Tursiops truncatus*. *J. Comp. Physiol. Psychol.* **41**(2), 111–123.

McBride, A. F., and Kritzler, H. (1951). Observations on pregnancy, parturition, and post-natal behavior in the bottlenose dolphin. *J. Mammal.* **32**(3), 251–266.

Nachtigall, P. E. (1980). Bibliography of echolocation papers on aquatic mammals published between 1966 and 1978. *In* "Animal Sonar Systems" (R.-G. Busnel and J. F. Fish, eds.), pp. 1029–1069. Plenum, New York. 1135 pp.

Powell, B. A. (1966). Periodicity of vocal activity of captive Atlantic bottlenose dolphins: *Tursiops truncatus*. *Bull. S. Calif. Acad. Sci.* **65**(4), 237–244.

Renaud, D. L., and Popper, A. N. (1975). Sound localization by the bottlenose porpoise *Tursiops truncatus*. *J Exp. Biol.* **63,** 569–585.

Richards, D. G. (1986). Dolphin vocal mimicry and vocal object labeling. *In* "Dolphin Cognition and Behavior: A Comparative Approach" (R. J. Schusterman, J. A. Thomas, and F. G. Wood, eds.), pp. 273–288. Lawrence Erlbaum Associates, Hillsdale, New Jersey. 393 pp.

Richards, D. G., Wolz, J. P., and Herman, L. M. (1984). Vocal mimicry of computer-generated sounds and vocal labeling of objects by a bottlenosed dolphin, *Tursiops truncatus*. *J. Comp. Psychol.* **98**(1), 10–28.

Schevill, W. E. (1964). Underwater sounds of cetaceans. *In* "Marine Bio-acoustics" (W. N. Tavolga, ed.), Vol. 1, pp. 307–316. Pergamon, New York. 413 pp.

Schevill, W. E., and Lawrence, B. (1956). Food-finding by a captive porpoise (*Tursiops truncatus*). *Breviora* **53,** 1–15.

Schevill, W. E., and Watkins, W. A. (1962). "Whale and Porpoise Voices." Woods Hole Oceanogr. Inst., Woods Hole, Massachusetts. Phonograph record and 24-page booklet.

Shane, S. H., Wells, R. S., and Wuersig, B. (1986). Ecology, behavior and social organization of the bottlenose dolphin: A review. *Mar. Mamm. Sci.* **2,** 34–63.

Tavolga, W. N. (1968). Marine animal data atlas. *Naval Training Device Center, Tech. Rep.*, NAVTRADEVCEN 1212-2: i–x, 1–239.

Tietz, R. M., and Tayler, C. K. (1964). Dolphin talk. *Sci. S. Africa* **1**(10), 385–390.

Townsend, C. H. (1914). The porpoise in captivity. *Zoologica* **1,** 289–299.

Tyack, P. L. (1985). An optical telemetry device to identify which dolphin produces a sound. *J. Acoust. Soc. Am.* **78,** 1892–1895.

Tyack, P. L. (1986a). Whistle repertoires of two bottlenosed dolphins, *Tursiops truncatus:* mimicry of signature whistles? *Behav. Ecol. Sociobiol.* **18,** 251–257.

Tyack, P. L. (1986b). Population biology, social behavior, and communication in whales and dolphins. *Trends Ecol. Evolut.* **1,** 144–150.

Tyack, P. L. (1989). Use of a telemetry device to identify which dolphin produces a sound: When bottlenosed dolphins are interacting, they mimic each other's signature whistles. *In* "Dolphin Societies: methods of study" (K. Pryor and K. S. Norris, eds.), Univ. of California Press, Berkeley.

Watkins, W. A., and Schevill, W. E. (1975). Sperm whales (*Physeter catodon*) react to pingers. *Deep-Sea Res.* **22,** 123–129.

Wells, R. S., Scott, M. D., and Irvine, A. B. (1987). The social structure of free-ranging bottlenose dolphins. *Curr. Mammal.* **1,** 247–305.

Wood, F. G., Jr. (1954). Underwater sound production and concurrent behavior of captive porpoises, *Tursiops truncatus* and *Stenella plagiodon*. *Bull. Mar. Sci. Gulf Caribbean* **3**(2), 120–133.

Wood, F. G., Jr. (1973). "Marine Mammals and Man." Luce, Washington, D.C. 264 pp.

Wood, F. G., Jr., Caldwell, D. K., and Caldwell, M. C. (1970). Behavioral interactions between porpoises and sharks. *In* "Investigations on Cetacea" (G. Pilleri, ed.), Vol. 2, 264–279.

A Long-Term Study of Bottlenose Dolphins on the West Coast of Florida

11

Michael D. Scott
Inter-American Tropical Tuna Commission
c/o Scripps Institution of Oceanography
La Jolla, California 92038

Randall S. Wells
Long Marine Laboratory
University of California at Santa Cruz
Santa Cruz, California 95060, and
Dolphin Biology Research Associates
Sarasota, Florida 34242

A. Blair Irvine
Dolphin Biology Research Associates
Sarasota, Florida 34242

INTRODUCTION

Researchers have been able to overcome many of the obstacles to studying dolphins. Radio and theodolite tracking, photoidentification of naturally marked individuals, tagging, and detailed observations of captive dolphins have given us more than just glimpses into the lives of bottlenose dolphins, *Tursiops truncatus*. However, while dealing with such day-to-day problems as salt-air corrosion of equipment and temperamental boat engines, researchers have had to continue grappling with a fundamental dilemma: How can one study the social structure of an animal that lives for several decades—essentially the time span of a researcher's working career? Deciphering the patterns of social relationships of a long-lived animal requires a long-term approach. Such a long-term study ideally involves the monitoring of recognizable individuals throughout their lifetimes, tracking their reproductive success, and investigating how that success is influenced by their feeding strategies and social roles and by the pressures of predation.

Since 1970, we have studied bottlenose dolphins near Sarasota, Florida. When the project began, it was designed to provide information on the home range of a few free-ranging individuals. At the time, most of the information available on dolphin biology had come from captive animals, and the only data on home ranges consisted of a few sporadic resightings of one naturally marked bottlenose dolphin in Florida and one in Georgia (Caldwell, 1955; Essapian, 1962; Caldwell and Golley, 1965).

Initially, we captured, tagged, and released 12 dolphins in the Sarasota Bay area. Resightings and observations of these animals' behavior provided some initial insights into their distribution and movements (Irvine and Wells, 1972). During 1975

–1976, we carried out a more ambitious tagging effort, primarily to answer questions relating to management (Irvine *et al.*, 1981). We wanted to know whether bottlenose dolphins were permanent residents of particular areas or were more transient. Such information was expected to help managers evaluate the likely impact of removing small numbers of dolphins from particular areas for public display or research. We freeze-branded and tagged 47 dolphins, and we radiotagged 10 of these for detailed monitoring of their movements.

Among the tagged animals were 11 of the 12 dolphins tagged in 1970–1971, indicating that at least some of the dolphins were long-term residents of this area. Regular surveys during 1975–1976 allowed us to observe the dolphins' daily activities, and much was learned about movements and social relationships within the community. The numerous resightings and occasional recaptures of the animals also allowed us to assess the utility and longevity of several tagging and marking methods.

Regular surveys were resumed in 1980 to observe how the marked animals had fared over time. We identified many of the previously tagged animals from freeze-brands or nicks and scars on their dorsal fins caused by tag shedding. We also increased our photographic effort to identify naturally marked dolphins. In addition to monitoring continued residency and movements of the dolphins, we have maintained regular observations of marked females as their calves have been born, raised, and become independent. From observations of these mother–calf pairs, we can estimate reproductive parameters such as the mean calving interval, and we can study the changes in social relationships that occur as an individual matures from a calf into an adult.

In 1984, we initiated another capture and tagging study to increase the number of identifiable dolphins. Calves of known females and naturally marked dolphins that had been previously identified were of particular interest because, with information on the sex, age, and other parameters of these individuals, we could better interpret their previous patterns of sightings and associations. Annual recaptures during 1984–1987 also allowed us to measure the growth rates of individuals, to estimate their ages from tooth samples, to evaluate their body condition from weights and ultrasonic measurements of blubber thickness, and to determine genetic relationships and the levels of reproductive hormones from blood samples. We have captured 60 individual dolphins, including 20 originally tagged in 1975–1976 and 5 tagged in 1970–1971. Fourteen additional dolphins were tagged in neighboring areas.

One important feature of long-term studies is that they build on themselves. As the original questions are answered, new questions elbow their way forward. Thus, our objectives and methods have changed over time. We began with basic descriptive questions: Can home ranges be defined for coastal bottlenose dolphins? If so, how extensive are these ranges? What constitutes a population? With answers to these questions presumably in hand, we could ask broader questions: What social units make up the population, and how do the ecological strategies of these social units differ? How do reproductive parameters, such as the calving interval and age of sexual maturity, vary among individual females? What interactions occur between dolphins in neighboring areas?

In this chapter, we summarize briefly the results of our studies and the directions in which current studies are headed. The long time span of these studies has allowed us to address a wide range of biological problems. Details of the results can be found in a series of published works. Initial tagging results and conclusions were reported by Irvine and Wells (1972). The second round of tagging resulted in an evaluation of tagging techniques (Irvine *et al.*, 1982; Scott *et al.*, Chapter 29, this volume), a more detailed study of movements and behavior (Irvine *et al.*, 1981), and an examination of social associations and home-range patterns (Wells, 1978; Wells *et al.*, 1980). More recent studies have focused on social structure (Wells, 1986; Wells *et al.*, 1987), population structure (Wells, 1986), and habitat use (Wells, in press). We also have initiated cooperative studies with other researchers on age estimation (Hohn *et al.*, in press), acoustics (Graycar, 1976; Sayigh *et al.*, in press), seasonal body condition (with Dan Costa and Graham Worthy, University of California at Santa Cruz, and Andy Read, University of Guelph), reproductive hormone levels (with Vicky Kirby, University of California at Santa Cruz), and genetic relationships (Duffield *et al.*, 1985; Duffield and Wells, 1986).

Summary of Results

Study Area and Dolphin Home Ranges

Wells (1986) considered the dolphins residing in the Sarasota Bay area to be a "community," based on their long-term residency in the area, their low number of associations with dolphins in neighboring areas, and their close genetic relationships. The home range of the Sarasota community encompasses a 40-km stretch of coastline that includes a system of bays, protected by a series of barrier islands, and the waters of the Gulf of Mexico up to about 1 km offshore of the islands (Fig. 1). The area is well suited for studying dolphins, for typically the waters are calm and the climate mild. The shallow depth (<1.5 m) in many places allows for relatively safe captures. Large numbers of small powerboats frequent the area, so the dolphins are acclimated to the presence of boats, making them easier to approach and observe. We have not noted any anomalous behavior of the dolphins caused by the cautious approach of observation boats (i.e., boats that have not been associated with the capture process). The relatively small number of about 100 dolphins also makes the community favorable for study. The dolphins are abundant enough to be sighted frequently, yet few enough to allow a large proportion of the community to be readily identified.

Most marked individuals are found to the south of Tampa Bay and north of the Big Pass area. This area encompasses several shallow seagrass meadows where fish are abundant and where dolphins are often observed feeding. Narrow, deep passes, through which sometimes-swift tidal currents flow, connect the bays to the Gulf of Mexico. The dolphins use these passes for traveling to and from the Gulf and for feeding on fish that move through the same passes. The area is traversed by dredged waterways which the dolphins also use for traveling.

Proportionately greater numbers of young calves are sighted in what we have termed "nursery areas." Palma Sola Bay, a nearly enclosed, protected bay, is one such area (Fig. 1). The bay is most frequented in the spring and summer, when many of the calves are born. We speculate that the bay may provide sufficient food in its extensive seagrass meadows for mothers to meet the energetic costs of lactation, and, at the same time, it affords protection for the calves from predation by sharks owing to the shallowness of the bay and its distance from deeper and more exposed waters.

Some marked dolphins have maintained fidelity to the Sarasota Bay area over the 17 years. Most individuals utilize particular core areas within the home range. Originally, we concluded that different age–sex classes were segregated spatially within the community's home range (i.e., the composite of individual home ranges). With additional years of data, however, we have modified our ideas. Marked females with calves initially seemed to range the most widely throughout the community's home range. We now consider, however, that these animals comprise several relatively discrete bands of females, with each band occupying different and relatively limited core areas that broadly overlap. We have identified four female bands to date: (1) the Anna Maria band includes 7 known females that frequently patrol the waters off the northern tip of Anna Maria Island and in Anna Maria Sound, (2) the Palma Sola band includes 14 known females that are found most often in Palma Sola Bay, (3) the Sarasota band includes 3 known females that travel throughout the area, although they are often sighted near Big Pass and New Pass to the south, and (4) the Manatee River band which includes 6 known females that are found predominantly in the northern portion of the area, but, unlike other female bands, frequent the Manatee River and Terra Ceia Bay as well.

The members of these female bands share a common home range and display a relatively high degree of association with one another. This does not mean, however, that band members are always found together. Membership of these bands is inferred from the day-to-day sightings of more ephemeral aggregations, termed "schools." Schools are defined as those aggregations of dolphins sighted within a 100-m radius of one another and engaged in similar activities (Wells *et al.*, 1987).

Adult males travel more widely throughout the area. They do not associate with any particular female band, but instead their ranges encompass the core areas of several female bands. Males are rarely seen in the deeper waters of Sarasota Bay; they are typically found in the northern portion of the study area, which contains more extensive seagrass meadows.

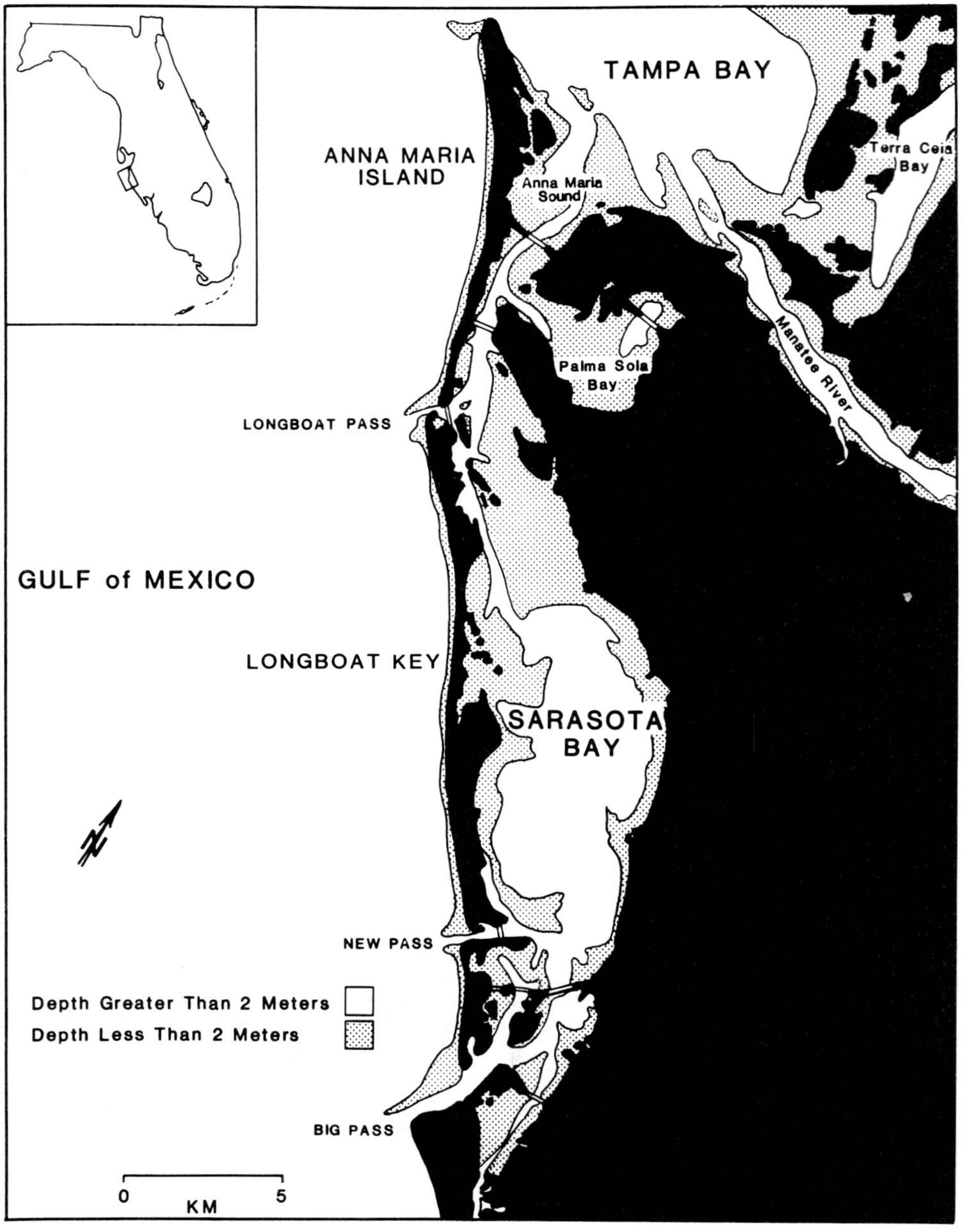

Figure 1 The Sarasota Bay area. The shallow areas less than 2 m deep that are inshore of the barrier islands approximately correspond to areas of seagrass meadows.

In 1975–1976, we found that the ranges of the five largest of the seven known subadult males were concentrated primarily along the edges of Sarasota Bay in the southern half of the community's home range. Three of these males, now of adult size, still have similar ranges. Five males that are currently of subadult size also have been sighted more often in the southern half of the area than other age–sex classes, although the pattern of spatial segregation seen during 1975–1976 is not so apparent. In general, subadult males tend to have smaller ranges than adult males and tend to frequent the edges of the community home range more often than do other classes.

The greatest differences in individual home ranges are seen between males and females. Females appear to occupy relatively limited home ranges centered around productive seagrass meadows. Males range more widely, with the adults traveling from one female band to the next. This pattern demonstrates the different strategies of the two sexes (cf. Trivers, 1972; Clutton-Brock *et al.*, 1982). The reproductive success of the females likely depends primarily on their procuring enough food to support the heavy investment of energy involved in raising and protecting their calves, while the success of the males likely depends mainly on their gaining access to the females.

Community Size and Discreteness

The community contains about 100 dolphins, and this number does not appear to have changed between 1976 and 1983. In 1976, the community size was estimated to be 102 dolphins based on 38 tagged dolphins (Lincoln Index, 95% confidence interval 90–117; Irvine *et al.*, 1981). In 1983, the estimate was 98 dolphins based on the sightings of 74 marked individuals (Schnabel estimator, 95% confidence interval 89–108; Wells, 1986).

We observed no evidence of mixing with dolphins from adjacent areas in 1975–1976. Since 1980, however, with increased surveys into the Gulf of Mexico and Tampa Bay and subsequent identification of over 350 naturally marked dolphins from these areas, we have found that mixing occurs more often than previously thought. We have observed that 17% of the schools that contained marked individuals from the Sarasota community also contained naturally marked dolphins from the adjacent Gulf or Tampa Bay communities. These mixed schools were found usually along the periphery of the Sarasota community's home range. We have seen such schools more frequently during the late summer and fall when the Sarasota animals tend to use the peripheral areas more often. The Sarasota males were more likely to be in such schools than were the Sarasota females.

The rates of immigration and emigration appear to be low for the Sarasota community. Wells (1986) estimated that 3% was the maximum annual rate for both immigration and emigration, but the actual rates are almost certainly less. Nonetheless, genetic heterogeneity is relatively high within the Sarasota community, suggesting that the community has not been a closed reproductive unit. This may have been due to the behavior of the adult males of the Sarasota and adjacent communities who appear to range widely enough to consort with females of other communities, yet do not appear to emigrate permanently from their own communities often.

Daily Activities

We have studied the dolphins' behavior using three methods. Radiotracking allows us to follow, usually from a distance, the daily movements of a few individuals over periods of days to weeks. Our regular surveys give us numerous opportunities to observe group activities for up to several hours at a time. Focal-animal behavioral observations give us detailed observations of individuals for periods of a half-hour or longer.

The dolphins appear to be active both night and day. They typically travel slowly in small, tightly knit groups along the channels, but often fan out as they begin to forage across the seagrass meadows. The dolphins feed in larger groups, apparently in a more coordinated fashion, when they are in the deeper and more open waters of the Gulf of Mexico. Their movements may be correlated with the movements of striped mullet, *Mugil cephalus,* their principal prey. Commercial fishermen report that both dolphins and mullet appear to come onto the shallow seagrass meadows with the flooding tide, entering deeper waters as the tide ebbs. We have seen the dolphins feed in various ways, but we have been impressed with how they use their entire bodies to corral and catch fish. "Pinwheeling," first described by Leatherwood (1975), consists of a dolphin's rapidly spinning around one or more fish, as if it were using its body to encircle and cut off the fish's escape. Wells (in press) described a behavior dubbed "fish-whacking," in which the dolphin makes a tight circle around the fish, lashes its flukes through the middle of the circle, and launches a fish through the air ahead of the dolphin (also see Shane, Chapter 12, this volume, who refers to this behavior as "fish kicking"). The dolphin then charges toward the spot where the fish reenters the water.

Radiotagging work in Sarasota Bay did not show significant daily variation in diving patterns, in contrast to most other telemetry studies of small cetaceans. Diving patterns in other cetaceans have been correlated with activities such as traveling, exploratory diving, and feeding (*Delphinus delphis:* Evans, 1971; *Phocoena phocoena:* Gaskin *et al.*, 1975; *Stenella attenuata:* Leatherwood and Ljungblad, 1979; Scott and Wussow, 1983), but data from our radiotagged bottlenose dolphins did not show such patterns. This is not surprising, however, since the shallow waters of Sarasota Bay make long dives unnecessary.

If and when dolphins rest during the day remain open questions. All we have to offer is anecdotal evidence. During some of the night radiotracks, the dolphins remained in one location for several hours, sometimes staying continuously at the surface. Their activities were unknown, but, unlike their typical movement patterns, their triangulated positions changed little. We have also sometimes seen the dolphins milling slowly in shallow areas when they did not appear to be engaged in feeding or social behavior. Similar observations of Hawaiian spinner dolphins, *Stenella longirostris,* have been interpreted as resting behavior (Norris and Dohl, 1980). We observed one daytime episode which hinted that the bottlenose dolphins had entered a state of rest or at least inattentiveness to their surroundings. We were recording the sounds of slow-moving dolphins as they began to fan out across a seagrass meadow. Few sounds in the 1–15 kHz range were heard as they approached our boat, which was drifting with the engine shut down. Two dolphins, a mother and calf, then approached the boat on a collision course until they were 2 m away. At that point, they startled and suddenly veered away, either in response to the closeness of the boat, or to movement or a noise by the boat crew. This behavior was unusual in that the dolphins in this area normally appear to be aware of all boats, whether drifting or not. These animals appeared to be traveling on "automatic pilot."

Seasonal Distribution

The distribution of dolphins within their home range changes seasonally. During late fall through early spring, the dolphins often frequent the passes between the barrier islands and the nearby Gulf waters. During the warmer months, the dolphins are generally found inside the bays. Palma Sola Bay is visited particularly in the spring and early summer, while other seagrass meadows are frequented more often during the late summer.

We have suggested three factors that may influence the distributional shifts: seasonal changes in prey distribution, predation pressure, and reproductive requirements (Wells *et al.*, 1980; Irvine *et al.*, 1981; Wells, in press). Mullet migrate during the fall from the bays into the Gulf to spawn, and the dolphins probably shift their distribution in response to the movements of their principal prey. In the early spring, the dolphins begin to spend most of their time inside the bays, again probably correlated with the migration of mullet. The appearance of bull sharks, *Carcharhinus leucus,* in the inshore waters during the summer may also prompt the dolphins to avoid the Gulf in favor of the more protected waters of the bays. Mothers with newborn calves, in particular, may seek out sheltered areas such as Palma Sola Bay during the spring when the calves are the most vulnerable (see below).

Mating and Reproduction

We still lack a detailed picture of the mating system of the bottlenose dolphins in our area. The roving nature of the adult males from one female band to the next and the brevity of male–female associations suggest a promiscuous mating system in which the males search for reproductively receptive females. The relatively large size of the testes is also an indication of a promiscuous mating system (Kenagy and Trombulak, 1986). The heavily scarred bodies of the older males, however, indicate that battles with other dolphins, characteristic of polygynous systems, are not uncommon. Male–male competition, then, may take the form of both dispersal to find receptive females and fighting to gain access to them.

Calving is broadly seasonal, peaking in the spring and early summer, but with a secondary peak occurring the late summer to early fall. The calves typically remain with their mothers for 3–6 years, although some associations last for longer. In one case, a calf remained with the mother for over 10 years (Wells *et al.*, 1987). We recently discovered that the young male was undersized and anemic, which may explain the long association. During the 10 years, the female was not observed with another calf. When she finally did produce another calf, the first calf was no longer with her but was associating with other subadults.

The opportunity to observe the growth of known-age individuals allows us to determine the age at which they become sexually mature. In 1976, one female was first tagged as a calf at a length of 198 cm, approximately the length of a 2-year-old, based on the age–length curves of Hohn (1980). This female gave birth to her first calf in 1982, indicating that she had become sexually mature at about the age of 7 (assuming a gestation period of 1 year). This age is probably less than the average age of sexual maturity because, in a study such as this, we initially obtain data only on "early bloomers" and not on animals that mature at greater ages. With time, however, we can monitor other known females as they grow and mature, and thus obtain a more representative picture.

Social Structure

The mean school size for the Sarasota community is 7.0 dolphins. This is typical of coastal bottlenose dolphins studied in other areas (Leatherwood and Reeves, 1982). School size tends to increase with water depth, with the largest schools occurring in the Gulf of Mexico or in the passes. Mean school size also tends to increase until midday, and then decreases in the late afternoon.

The mean size of schools containing different age or sex classes also varies. Females, particularly those accompanied by calves, tend to occur in larger schools than do males. As male calves become older, however, they are generally found in smaller and smaller schools. This decrease in school size continues until they become socially mature enough to associate with the female bands as potentially breeding adults.

The ratio of males to females appears to decrease with age, apparently owing to a higher mortality rate for males than females. Eight marked males are known or strongly suspected to have died, compared with only one female, since 1975. The sex ratio of all marked animals captured in 1975–1976 was 24 males : 23 females, but the ratio of this sample of dolphins declined with time. After 8 years, the sex ratio of the remaining animals had decreased to 11 males : 20 females. The higher loss of males from the population also has been noted for other dolphin species (see Perrin and Reilly, 1984).

Although the composition of schools may fluctuate over the course of a day, many associations are relatively stable and long-term. The mother–calf bond is a close one, and the relationship may persist for many years. Membership of the female bands is also relatively stable, while subadult and adult males may form lasting associations among themselves as well. We have observed little association between subadult males and adult males, which, combined with the spatial difference in home ranges between the two classes, supports the notion of segregation of subadult males.

Patterns of long-term associations may be understood in terms of the ontogenetic development of social behavior. In the female bands, mothers with calves of similar ages tend to form stable associations with one another. As a consequence, their calves have several years during which they may form stable associations among themselves. When the males become independent from their mothers and leave the female bands, they congregate in bachelor groups that appear to be based on the relationships established while they were calves. These relationships appear to be maintained even after the males attain sexual maturity, and some adult males have been sighted together over periods of several years. Subadult females may join these bachelor groups as well, but it appears that, on reaching sexual maturity, they return to the female band in which they were born.

This pattern has implications for the genetic relationships that exist among close associates. If daughters return to their mother's female band, the members of the band would tend to be closely related. If males that grew up together tend to form lasting bonds after they become independent, they also would likely be genetically related because of the potentially close genetic relationship of their mothers. If such genetic relationships among close associates do exist, instances of coordinated feeding, "babysitting" by "auntie" dolphins, or the aiding of distressed animals reported for this species may have evolved by kin selection (Hamilton, 1964). If close associates are likely to share the same genes, conditions will be favorable for the evolution of altruism or cooperative behavior.

Age Determination Methods

In 1984, we began collecting teeth from captured dolphins whenever possible (see Perrin and Myrick, 1980, and Hohn, Chapter 33, this volume, for discussions of estimating age from teeth). These samples have proved useful in two ways. First, we have used the teeth to test the validity of age-estimation techniques. Previously, the calibration of the layering pattern in the teeth had been based on a few known-age captive animals, but the collection of teeth from dolphins whose ages were known, because they were first sighted shortly after their birth to marked females, allowed the first calibration from free-ranging dolphins (Hohn *et al.*, in press; see Hohn, Chapter 33, this volume). Independent estimates of the ages of known-age dolphins made from teeth subsequently have proved to be accurate (Hohn *et al.*, in press; Hohn, Chapter 33, this volume).

Once the reliability of the age estimates has been established, the tooth samples provide an increasingly detailed picture of the age structure of the community. Growth rates, age-specific reproduction and mortality, and longevity can be examined once the ages of individuals are known. The oldest female we examined was estimated to be 46 years old in 1987, while the oldest male was estimated to be 34 years old.

CONCLUSIONS

We believe that the adult female bands comprise the stable core of the Sarasota community. The members of the bands appear to be linked by genetic ties and by long-term associations stretching over years or even decades. These associations appear to be promoted by seasonal reproduction, so that females with similar reproductive cycles tend to associate with one another most closely. The result of this synchrony is a cohort of calves that may continue to associate closely even after they become independent from their mothers.

The adult males, on the other hand, appear to be the vectors for genetic exchange. Males range more widely than females, and they sometimes may associate with members of other communities. As has been shown by Clutton-Brock *et al.* (1982) in red deer, *Cervus elaphus*, differences between the sexes in their breeding strategies have profound effects on virtually all aspects of their life histories. Differences between sexes in movement patterns, associations, mortality, and longevity all appear to be tied to their basic differences in reproductive strategies.

One of the lessons that we learned during this study is that conclusions based on short-term data tend to be simplistic and transitory. Collecting data for only 2 or 3 years is unlikely to give a complete picture of a complex society of long-lived animals, and we have been forced to revise our ideas with disconcerting regularity. We started with tentative ideas about the residency and movement patterns of the community. Observations of more tagged animals indicated a community that was discrete from neighboring communities and segregated by age and sex class. We now recognize that the community is less discrete than we thought, and we appreciate better the role of the social subunits in the community and the interplay of different male and female strategies.

We also learned the value of simultaneously pursuing multiple lines of investigation. One way that we have been able to broaden the scope of our study has been by conscripting biologists with different areas of expertise. As noted by Bartholomew (1985), the explosive growth of the scientific literature has meant that biologists can keep abreast of only a tiny fraction of the information available, but by employing a team approach, the disadvantages of this tendency toward parochialism can be diminished. The study has become more and more of a corporate affair, uniting biologists interested in behavior, life history, genetics, acoustics, reproduction, and population biology.

At the outset of the project, we were concerned mostly with the fates of individual dolphins—their movements, behaviors, and associations with other individuals. Now, we have expanded our interest to questions about higher levels of organization: What roles do population subunits, such as the female bands, have in the population as a whole? How do different populations interact? How is the community of dolphins interwoven into its greater community of predators, prey, and noncombatants? In addition, we are exploring physiological questions, such as those concerning reproductive cycles and seasonal body condition. A long-term study provides an opportunity to study organisms in reference to both higher and lower integrative levels, a process that Bartholomew (1985) advocated for promoting both an analytical approach for dealing with mechanism and a synthetic approach for appreciating biological significance.

ACKNOWLEDGMENTS

We would like to acknowledge the troupe of itinerant biologists who have helped us over the years with their efforts in the field and in the lab: D. Black, J. Chamberlin-Lea, S. Chivers, D. Duffield, A. Hohn, S. Kruse, A. Read, J. Sweeney, and P. Tyack. Larry Fulford has shown remarkable skill in capturing dolphins for us, and even more remarkable patience in dealing with large numbers of certifiable zanies. Invaluable help has been provided by an army of volunteers that has included R. Arden, L. Mayall, K. Miller, P. Page, E. Patterson, R. Spaulding, J. Zaias, and over 150 volunteers from Earthwatch, New College, University of Florida, and University of South Florida. The following have provided logistical support with constant and enduring cheerfulness: Cannon's Marina, Dick Wagner Realty, F. Wells, J. Wells, and F. Worl. The project has been supported over the years by the Mote Marine Laboratory, the Marine Mammal Commission, the National Marine Fisheries Service, the Inter-American Tropical Tuna Commission, the University of California at Santa Cruz, Earthwatch, the Denver Wildlife Research Center, Gainsville Laboratory, and Woods Hole Oceanographic Institution. The manuscript was reviewed by W. Bayliff, M. Hall, A. Hohn, and J. Joseph.

REFERENCES

Bartholomew, G. A. (1985). The role of natural history in contemporary biology. *Biosci.* **36**(5), 324–329.

Caldwell, D. K. (1955). Evidence of home range in the bottlenosed dolphin. *J. Mammal.* **36**(2), 304–305.

Caldwell, D. K., and Golley, F. B. (1965). Marine mammals from the coast of Georgia to Cape Hatteras. *J. Elisha Mitchell Sci. Soc.* **81**(1), 24–32.

Clutton-Brock, T. H., Guiness, F. E., and Albon, S. D. (1982). "Red Deer: Behavior and Ecology of Two Sexes." Univ. of Chicago Press, Chicago. 378 pp.

Duffield, D. A., and Wells, R. S. (1986). Population structure of bottlenose dolphins: Genetic studies of bottlenose dolphins along the central west coast of Florida, *Fin. Rep. Natl. Mar. Fish. Serv., Southeast Fish. Cen.* Contract No. 45-WCNF-5-00366. 10 pp.

Duffield, D. A., Chamberlin-Lea, J., Wells, R. S., Scott, M. D., and Irvine, A. B. (1985). Inheritance of an extra chromosome in a resident female band of bottlenose dolphins in Sarasota, Florida. *Proc. Sixth Biennial Conf. Biol. Mar. Mamm., Nov. 1985,* Vancouver, B.C., abstr.

Essapian, F. S. (1962). An albino bottlenosed dolphin, *Tursiops truncatus,* captured in the U.S. *Norsk Hvalfangst-tidende* **9,** 341–344.

Evans, W. E. (1971). Orientation behavior of delphinids: Radio telemetric studies. *Ann. N.Y. Acad. Sci.* **188,** 142–160.

Gaskin, D. E., Smith, G. J. D., and Watson, A. P. (1975). Preliminary study of movements of harbor porpoises (*Phocoena phocoena*) in the Bay of Fundy using radiotelemetry. *Can. J. Zool.* **53**(10), 1466–1471.

Graycar, P. (1976). Whistle dialects of the Atlantic bottlenosed dolphin, *Tursiops truncatus*. Ph.D. dissertation, Univ. of Florida, Gainesville. 90 pp.

Hamilton, W. D. (1964). The evolution of social behavior. *J. Theor. Biol.* **7**(1), 1–52.

Hohn, A. A. (1980). Age determination and age related factors in the teeth of western North Atlantic bottlenose dolphins. *Sci. Rep. Whales Res. Inst., Tokyo* **32,** 39–66.

Hohn, A. A., Scott, M. D., Wells, R. S., Sweeney, J. C., and Irvine, A. B. (In press). Growth layers in teeth from known-age, free-ranging bottlenose dolphins. *Mar. Mammal Sci.*

Irvine, B., and Wells, R. S. (1972). Results of attempts to tag Atlantic bottlenosed dolphins (*Tursiops truncatus*). *Cetology* **13,** 1–5.

Irvine, A. B., Scott, M. D., Wells, R. S., and Kaufmann, J. H. (1981). Movements and activities of the Atlantic bottlenose dolphin, *Tursiops truncatus,* near Sarasota, Florida. *Fish. Bull.* **79**(4), 671–688.

Irvine, A. B., Wells, R. S., and Scott, M. D. (1982). An evaluation of techniques for tagging small odontocete cetaceans. *Fish. Bull.* **80**(1), 135–143.

Kenagy, G. J., and Trombulak S. C. (1986). Size and function of mammalian testes in relation to body size. *J. Mammal.* **67**(1), 1–22.

Leatherwood, S. (1975). Some observations of feeding behavior of bottlenosed dolphins (*Tursiops trunctus*) in the northern Gulf of Mexico and *Tursiops* cf. *T. gilli* off southern California, Baja California, and Nayarit, Mexico. *Mar. Fish. Rev.* **37**(9), 10–16.

Leatherwood, S., and Ljungblad, D. K. (1979). Nighttime swimming and diving behavior of a radio-tagged spotted dolphin, *Stenella attenuata. Cetology* **34,** 1–6.

Leatherwood, S., and Reeves, R. R. (1982). Bottlenose dolphin (*Tursiops truncatus*) and other toothed cetaceans. *In* "Wild Mammals of North America: Biology, Management, Economics" (J. A. Chapman and G. A. Feldhamer, eds.), pp. 369–414. Johns Hopkins Univ. Press, Baltimore, Maryland. 1147 pp.

Norris K. S., and Dohl, T. P. (1980). The structure and functions of cetacean schools. *In* "Cetacean Behavior: Mechanisms and Functions" (L. M. Herman, ed.), pp. 211–261. *Wiley* (*Interscience*) New York. 463 pp.

Perrin, W. F., and Myrick, A. C., eds. (1980). Age determination of toothed whales and sirenians. *Rep. Int. Whal. Commn.,* Special Issue 3. 229 pp.

Perrin, W. F., and Reilly, S. B. (1984). Reproductive parameters of dolphins and small whales of the family Delphinidae. "Reproduction in Whales, Dolphins and Porpoises" (W. F. Perrin, R. L. Brownell, Jr., and D. P. DeMaster, eds.), pp. 97–133. *Rep. int. Whal. Commn.,* Special Issue 6. 495 pp.

Sayigh, L. S., Tyack, P. T., Wells, R. S., and Scott, M.D. (In press). Signature whistles of free-ranging bottlenose dolphins: Stability and mother–offspring comparisons. *Beh. Ecol. and Sociobiol.*

Scott, M. D., and Wussow, P. C. (1983). Movements of a Hawaiian spotted dolphin. *Proc. Fourth Int. Wildl. Biotelemetry Conf. Halifax,* Nova Scotia, pp. 353–364.

Trivers, R. L. (1972). Parental investment and sexual selection. *In* "Sexual Selection and the Descent of Man 1871–1971" (B. Campbell, ed.), pp. 136–179. Aldine, Chicago.

Wells, R. S. (1978). Home range characteristics and group composition of Atlantic bottlenosed dolphins, *Tursiops truncatus,* on the west coast of Florida. M.S. thesis, University of Florida, Gainesville. 110 pp.

Wells, R. S. (1986). Structural aspects of dolphin societies. Ph.D. dissertation, Univ. of California at Santa Cruz, Santa Cruz. 234 pp.

Wells, R. S. (In press). The marine mammals of Sarasota Bay: Present status and patterns of habitat use. *Proc. Sarasota Bay Area Scientific Inform. Symp. 24–26 April 1987,* Sarasota, Florida, in press.

Wells, R. S., Irvine, A. B., and Scott, M. D. (1980). The social ecology of inshore odontocetes. *In* "Cetacean Behavior: Mechanisms and Functions" (L. M. Herman, ed.), pp. 263–317. Wiley (Interscience), New York. 463 pp.

Wells, R. S., Scott, M. D., and Irvine, A. B. (1987). The social structure of free-ranging bottlenose dolphins. *In* "Current Mammalogy" (H. Genoways, ed.), Vol. 1, pp. 247–305. Plenum, New York. 519 pp.

Behavior and Ecology of the Bottlenose Dolphin at Sanibel Island, Florida

12

Susan H. Shane
West Coast Whale Research Foundation
250 Cottini Way
Santa Cruz, California 95060

INTRODUCTION

The behavior of bottlenose dolphins, *Tursiops truncatus*, is closely tied to local ecology, and behavior patterns change according to ecological factors such as prey distribution and tidal regime. Although no single comprehensive study of bottlenose dolphin behavior and ecology has been conducted, many studies have contributed insights into the relationship between behavior and ecology in this species.

Feeding is the activity most often associated with particular temporal and ecological conditions. Dolphins in South Africa (Tayler and Saayman, 1972; Saayman *et al.*, 1973) and Texas (Shane, 1977; Shane *et al.*, 1986) fed most in the early morning and late afternoon, possibly in response to the diurnal patterns of prey. A significant increase in the frequency of feeding in Texas in fall and winter compared with other seasons was associated with major migrations of fish (Shane, 1977; Gruber, 1981). Dolphins usually fed cooperatively offshore in deep, open waters in Argentina (Würsig and Würsig, 1979), South Africa (Saayman *et al.*, 1973), and the Gulf of Mexico off Florida (Irvine *et al.*, 1981; Wells *et al.*, 1980), whereas they fed individually or in small groups in shallow, nearshore waters in these same areas. Shane (1977) found feeding concentrated in 15-m-deep holes in Texas. In Georgia and South Carolina bottlenose dolphins moved into narrow tidal creeks during low tide and pursued fish onto exposed mud banks, often beaching themselves temporarily in the process (Hoese, 1971; Rigley, 1983). A variation on mud-bank feeding occurred in Portugal where dolphins threw mullet onto a beach and then stranded themselves to retrieve the fish (dos Santos and Lacerda, 1987).

Tidal flow has a complicated and contradictory association with bottlenose dolphin movements, depending on the geographical location. In Texas (Shane, 1977), San Diego Bay, California (Norris and Prescott, 1961), and Brittany (Hussenot, 1980) dolphins moved against the tidal flow whereas they moved with it at Sarasota, Florida (Irvine *et al.*, 1981). In Argentina dolphins were found in increasingly shallow water as the tide ebbed but moved variably in response to rising tides (Würsig and Würsig, 1979).

Bottlenose dolphins are particularly adept at taking advantage of human activities in their environment. Dolphins follow shrimp fishing boats to get food in many places (Gunter, 1942, 1954; Norris and Prescott, 1961; Caldwell and Caldwell, 1972; Hogan, 1975; Leatherwood, 1975; Shane, 1977; Gruber, 1981; Goodwin, 1985; Corkeron *et al.*, Chapter 17, this volume). In Hawaii dolphins have learned to steal fish from fishermen's lines (Schlais, 1984). Mullet fishermen in Mauritania and Australian aborigines in Queensland call dolphins by beating the water with sticks or spears and then let the dolphins pursue fish in the vicinity of their nets; some fish are captured by the dolphins and others are trapped in the nets (Busnel, 1973; Baker, 1983). A small group of bottlenose dolphins in Western Australia has been coming to shore to be fed and petted by people on the beach for over 20 years (Connor and Smolker, 1985).

This brief summary shows that dolphin behavior can be influenced by time of day, season, water depth, bottom topography, tidal flow, and human activities. Dolphins' responses to these ecological variables are somewhat unpredictable and can differ depending on the habitat in which the animals are studied.

I spent 1 year (April 1985–May 1986) studying the behavior and ecology of bottlenose dolphins at Sanibel Island, Florida (Fig. 1). I had three objectives for this study: (1) to quantify dolphin time budgets and the association between dolphin behavior and various physical and biological variables in their habitat; (2) to test two different methods for collecting data on dolphin behavior; and (3) to compare the behavior and ecology of dolphins at Sanibel with that of dolphins in Texas (Shane, 1977, 1980) and in other parts of the world. In this chapter I describe bottlenose dolphin behavior and ecology at Sanibel and assess two methods for collecting behavioral data. The comparison of dolphin behavior at the two sites is given elsewhere (Shane, Chapter 31, this volume).

MATERIALS AND METHODS

Study Area

Sanibel is an 18-km-long island with a northwest–southeast orientation lying off the southwestern coast of Florida (Fig. 1). My study area included the Gulf of Mexico up to 1.6 km offshore as well as inland waters between Sanibel and Ft. Myers Beach (adjacent to Matanzas Pass), Punta Rassa, and Pine Island. The waters around Captiva Island, just north of Sanibel, were also part of the study area but were covered less regularly than the Sanibel area. A total of approximately 140 km^2 of water, divided into 99 unequal sectors, was covered during this study.

The Gulf shore is a shallow, sandy area reaching a maximum of 6 m in depth 1.6 km offshore. The undeveloped bay shores are vegetated by red (*Rhizophora mangle*) and black (*Avicennia germinans*) mangroves, and there are abundant seagrass beds throughout the undredged sections of the bay. The Intracoastal Waterway (ICW) is the only major dredged channel in the study area, and its depth varies between about 3 and 6 m. The remainder of the bay area is 0.5–2.5 m deep with rare spots reaching 7 m in depth. There is a 5-km-long causeway intersecting the bay and leading from Punta Rassa to Sanibel.

Diurnal and semidiurnal tides, always strongly influenced by wind conditions, occur in the Sanibel area (Clark, 1976). The mean tidal range of 0.55 m can increase to 4.6 m during tropical storms or hurricanes (Clark, 1976).

Weather conditions throughout this year-long study were very good. Cloud cover varied from completely clear to one-quarter of the sky covered during 47% of the observations. Seventy-nine percent of all records were collected under Beaufort wind force conditions of 0–2.

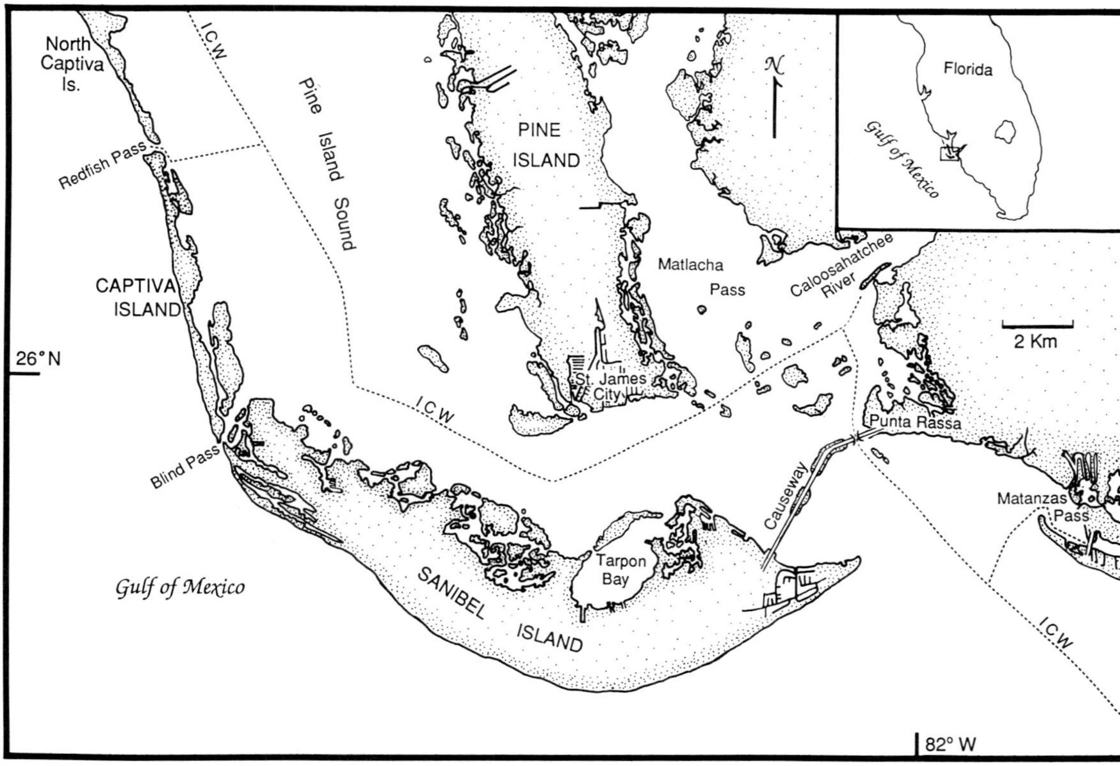

Figure 1 Bottlenose dolphin study area around Sanibel Island, Florida.

Collection of Data on Behavior

Almost all dolphin observations were made from a 5-m Boston Whaler equipped with a 65-hp Suzuki outboard; a few observations were made from land. My primary objective was to record the behavior of focal groups of dolphins (Altmann, 1974) throughout the daylight hours. Observed behavior was classified according to the definitions given in the Appendix. Pod size and composition (number of adults, juveniles, and calves), speed of travel relative to my boat's idle speed of 3 knots (5.5 km/hour), direction of movement, and pod geometry (shape) were noted at each sighting or when any variable changed. A pod was defined as any group of dolphins observed in apparent association, moving in the same direction and often, but not always, engaged in the same activity. I use the term "pod" in a general sense to designate a group of cetaceans just as one refers to a "school" of fish, a "gaggle" of geese, or a "pack" of dogs. Calves were defined as animals two-thirds or less the length of an adult they consistently swam beside and slightly behind. Large calves were those estimated to be 1.5–2 m long; small calves were estimated to be about 1.2 m long; newborns (<1 month old) had visible fetal folds, were lead gray in color, and usually head-slapped when breathing (McBride and Kritzler, 1951). Juveniles were approximately 2 m long and swam independently. I photographed and, thereby, identified individual dolphins with distinctive marks on their dorsal fins and backs (Würsig and Würsig, 1977) throughout the study.

I used three methods for collecting behavioral data (Altmann, 1974). (1) I conducted instantaneous sampling of focal-group activity every 3 minutes [instantaneous data (I.D.); 4086 records covering 204.25 hours]. The instantaneous sampling varied from that described by Altmann (1974) in two respects. Dolphins that fed or socialized at least once as they traveled during the 3 minutes preceding the sampling time were recorded in the travel/feed or social travel category. Also, if dolphins were submerged at the sampling time the last observed activity was recorded. (2) I recorded the duration of each activity performed by a focal group during a given observation period [duration data (D.D.); 2002 records covering 261.25 hours; duration records ranged from 0, meaning an activity was observed for less than a minute, to 144 minutes]. (3) I recorded all occurrences of surface behaviors in a focal group during randomly selected 15-minute sampling periods (surface data; 101 sessions covering 25.25 hours and including 1219 surface behaviors). I also recorded dolphin respirations to look for associations between activity, dive type, and respiration rate (respiration data; 45 sessions covering 11.6 hours, averaging 15.4 minutes/session). Respiration data were collected only on individually identifiable dolphins.

The two methods of collecting data on activities (I.D. and D.D.) produced different results. There was no obvious pattern to the disparity between the two data sets; sometimes they agreed and at other times they differed radically but inconsistently. I believe, however, that the I.D. best reflect the dolphins' behavior. The I.D. contained only sessions lasting 15 minutes ($N = 6$ instantaneous records) or longer; all shorter observation sessions were discarded. In contrast, only 296 of 2002 D.D. records lasted at least 15 minutes.

A possible problem with using short (<15 minutes) D.D. records is that the dolphins' behavior may be altered initially by the presence of the observer's boat, and the first minutes of observation may not accurately reflect natural behavior. Therefore, I converted those D.D. records lasting 15 or more minutes to 3-minute instantaneous samples and analyzed the proportion of time dolphins devoted to different activities, considering only the five most frequently recorded activities. I then compared those results with the appropriate I.D. results (Table 1). Although the two data sets proved to be significantly different from each other statistically ($\chi^2 = 52.95$, $df = 4$, $p < 0.001$), they did not appear to be significantly different from a biological viewpoint.

Some natural variation between the I.D. and D.D. sets can be expected. The important point is that the two data sets follow identical patterns in terms of the relative frequency of each activity. Therefore, I decided to combine the I.D. records (193.3 hours of data on five activities) with the 15+-minute D.D. records (116.5 hours of data on five activities) which I converted to frequencies and to use this combined data set ($N = 6196$) to describe dolphin activities at Sanibel. For some analyses I used only the I.D. because certain variables were never recorded in the D.D. Because the difference in proportion of records devoted to each of the five activities in the I.D. and D.D. varied between 1 and 6% and because I considered that amount of variation biologically insignificant, I considered only differences exceeding 6% to be biologically significant in the analyses of dolphin behavior according to temporal and environmental factors.

I recorded the two-dimensional geometry of dolphin pods at each sighting, as follows: "tight" refers to dolphins less than one body length apart, "loose" designates a separation of greater than one and less than five body lengths, and "widely

Table 1
Frequency and Percent Occurrence of Five Dolphin Activities at Sanibel Island, Florida[a]

	I.D.		D.D.	
	Frequency	%	Frequency	%
Travel	1765	46	1189	51
Feed	643	17	420	18
Socialize	261	7	98	4
Travel/feed	809	21	358	15
Social travel	388	10	265	11
N	3866		2330	

[a]D.D. represents duration data that were converted to frequency data. I.D. represents instantaneous data that were collected in frequency form. Only observation sessions lasting 15 minutes or longer were included in both data sets. The combined data sets cover 310 hours of direct observation of dolphins involved in the five activities considered here (the total data sets cover 326 hours).

dispersed" refers to pod members spread greater than five body lengths apart. Geometry was termed "mixed" when pod members were irregularly spaced.

I considered four time periods (morning, 0700–0959 hours; midday, 1000–1259 hours; afternoon, 1300–1559 hours; and evening, 1600–2059 hours) and four seasons (summer, June–August; fall, September–November; winter, December–February; and spring, March–May) in my analysis of the influence of temporal factors on activity.

Collection of Data on the Environment

Records of dolphin focal-group activity were accompanied by records of the following temporal and environmental variables: date, time (to the minute for I.D. and D.D., to the second for surface data and respiration data), location (sector number), water depth, distance offshore, birds or sport-fishing boats fishing within 100 m of the dolphins, my boat or any other boat within 10 m of the dolphins, fish seen visually or on the fish finder/fathometer (a 200-kHz Unimetrics Sea Hawk 20/20 chart recorder), cloud cover, Beaufort wind force, wind direction, water temperature (using the thermometer built into the chart recorder), tidal state, and current direction and speed. Surface current speed was judged subjectively as none, slow, medium, or fast by observing the flow of water past a stationary object such as a piling or crab-pot buoy.

I collected indirect data on food availability in two ways. First, I used my fish finder to test for the presence of fish in the vicinity of dolphins. Second, I used recorded environmental variables as secondary indicators of fish location, since fish distribution changes according to season, water temperature, and tidal state.

Time Spent in the Field

From April 1985 to May 1986, I spent a total of 881 hours in the field searching for dolphins and a total of 450 hours actually observing them (51% contact rate). Twenty-six percent of the search time was spent in the Gulf of Mexico and 74% in the bay. The overall contact rate in the Gulf (56%, monthly range 12–71%) was higher than in the bay (49%, monthly range 37–60%) ($\chi^2 = 1.98$, $df = 1$, not significant). During the study I alternated between remaining with a dolphin pod as long as possible and covering a large part of my study area (staying with pods only long enough to determine their activity).

RESULTS AND DISCUSSION

Respiration Rates

Bottlenose dolphins at Sanibel typically surface in four distinctly different ways prior to diving: in the "regular dive" only the blowhole, part of the back, and the dorsal fin are exposed; the "racing dive" resembles a regular dive performed at high speed with more of the dolphin's back exposed and white water forming as the animal surfaces; in the "tail-stock dive" the dolphin arches its back and exposes its peduncle but not its flukes as it submerges after breathing; and in the "flukes-up dive" the dolphin arches its back and exposes its flukes as it dives. Mean dive times and standard deviations (seconds) are tabulated below:

Dive type	N	Mean	S.D.
Regular	1223	21	17.2
Racing	36	24	13.8
Tail-stock	205	37	25.8
Flukes-up	24	49	29.1

Mean dive time varied significantly by dive type ($F = 55.36$, $df = 3$, $p < .001$). Specifically, flukes-up dives and tail-stock dives each varied significantly from regular dives and racing dives, but the latter two did not vary significantly from each other (Turkey's $q = 4.41$, $\alpha = 0.01$).

Mean dive time ranged from 20 seconds for dolphins that were social/traveling to 25 seconds for dolphins that were feeding, but the differences in dive time according to the dolphins' activity were not statistically significant ($F = 0.71$, $df = 3$; $N = 1575$). By averaging all dives recorded for a particular activity in this analysis, any dive patterns peculiar to that activity (e.g., one long dive followed by three short dives) would have been obscured. Würsig (1978) found a mean dive time of 21.8 seconds for bottlenose dolphins in Argentina where the animals were observed in water less than 10 m deep 92% of the time (Würsig and Würsig, 1979). At Sanibel 92% of the I.D. sightings occurred in 4.5 m or less.

The type of dive a dolphin made at Sanibel usually depended on the dolphin's activity. While regular dives were the dominant dive type for all activities (N = 1251 regular dives out of 1528 total dives), travel/feeding pods used racing dives most often (82% of 411 dives recorded during travel/feeding). Feeding dolphins most often made flukes-up dives (75% of 394 feeding dives), but tail-stock dives also made up a fairly high proportion (22%) of feeding dives. Since tail-stock and flukes-up dives are both strongly associated with feeding, I suggest that by arching its body as it dives, a dolphin is able to descend at a steeper slope and reach the bottom rapidly. It can then spend the maximum amount of dive time searching for benthic or epibenthic prey. Shane (1977) described the same four types of dives for dolphins in Texas. Tyack's (1976) "ragged slicing" for bottlenose dolphins in Argentina was equivalent to my racing dive, and his "humping" matched my tail-stock dive.

Because it was easier to record respirations on lone dolphins than on dolphins in groups, most records of dive times (77%) were recorded on solitary individuals, and the mean pod size for the respiration data was 1.7 (range = 1–15, N = 1630). Single animals were recorded much less frequently in the other data sets: 15% in the I.D. (N = 4086), 18% in the D.D. (N = 2002), and 12% in the surface data (N = 1219). Thus, the frequency of occurrence of lone dolphins is a result of sampling bias in the respiration data but is probably representative of dolphin grouping behavior in the other data sets.

Descriptive Observations of Feeding

Feeding Behavior

I interpreted the following kinds of behavior as feeding:

1. Dolphins, alone or in loose or widely dispersed geometry, commonly dove repeatedly in varying directions in one location, often making tail-stock dives or flukes-up dives. Occasionally, mud was stirred up, suggesting bottom feeding. Fish were never seen during these "subsurface feeding" bouts. I saw subsurface feeding several times at a current cline where Gulf and bay waters met during a tidal change.
2. Dolphins sometimes faced against a strong tidal current and remained in one place except when chasing and catching fish, often at the surface. This "against-current feeding" behavior was especially common under Causeway bridges where my fish finder indicated fish to be unusually abundant.
3. Dolphins frequently swam in circles just beneath the surface in two ways. First, a dolphin would swim rapidly in a circle on its side with its body bent forward, much like a cat chasing its tail. I observed individual fish fleeing just a few centimeters ahead of the dolphin's rostrum during such "horizontal circle feeding." Second, a dolphin would position itself almost vertically in the water column with its head up. The dolphin then would swing its head or, rarely, its whole body repeatedly in a 360° arc as one or more small fish fled in a circle at the tip of the dolphin's rostrum just beneath the surface. I observed dolphins pursuing only fish less than about 10 cm long using this "vertical circle feeding." On one such occasion, my hydrophone picked up continuous "rusty hinge" or "creaky door" echolocation clicks.
4. A dolphin would swim along the edge of an oyster bar, sand bar, submerged shoal, channel, or mangrove shoreline apparently hunting for food; my chart recordings indicated that fish aggregated on such sloping areas. Prey usually were caught by "edge-feeding" dolphins in the midst of a sudden acceleration and splash (a "feeding rush").
5. The feeding rush seen in edge feeding was seen in other situations as well. Typically, a dolphin would increase its speed suddenly (often with its fin constantly exposed) for 10–20 m heading toward a shoreline. Just before reaching the shore, the dolphin would lean on its side and spin in a circle or make a

hairpin turn as it caught its prey (called "pinwheeling" by Leatherwood, 1975). Although it usually was impossible to see the actual prey capture because the splashing water obscured it, I occasionally did see dolphins capturing fish at the end of a feeding rush. Feeding rushes were particularly common along the Gulf beach in spring and fall as well as in a cove off Punta Rassa.

6. "Fish kicking" (mentioned by Wells *et al.*, 1987) was the most spectacular type of feeding seen at Sanibel. A dolphin would use its tail flukes and/or tail stock to kick a fish near the water's surface high into the air; it appeared that both sideways and upward thrusts of the tail were used. Fish were kicked approximately 1.2–9.2 m high ($\bar{x}$ = 4.2 m) and 0–9.2 m forward ($\bar{x}$ = 4.8 m) (N = 11). Distances were estimated in feet in the field and later converted to meters. The fish that were kicked were often visibly wounded and bleeding. They were usually about 20 cm long and could often be identified as striped mullet, *Mugil cephalus*, or spotted seatrout, *Cynoscion nebulosus*. Fish kicking was usually done by a lone dolphin, which normally swam to where the fish landed and presumably consumed it. In one instance another dolphin appeared to eat a fish kicked by its podmate. Dos Santos and Lacerda (1987) saw dolphins in Portugal lofting medusae, *Catostylus tagi*, into the air with their flukes.
7. There was scant evidence for cooperative feeding at Sanibel. On a few occasions dolphins were side-by-side or single file while edge feeding, circle feeding, making feeding rushes, fish kicking, and subsurface feeding, and cooperation was possible. On 12 August 1985 an adult dolphin made repeated feeding rushes for 1.75 hours off Punta Rassa while another adult with a calf swam nearby. Although the dolphin with the calf also made occasional feeding rushes, the first adult often joined the adult and calf immediately after the apparent fish capture; it is possible that the feeding animal was sharing food with the other two. Similarly, on three occasions a single dolphin caught a large (~1 m long) fish and carried it in its mouth for up to 30 minutes; during that time its podmates clustered around it, again, possibly to share the food. Connor and Smolker (1985) described similar possible food sharing in bottlenose dolphins in Western Australia. Brower and Curtsinger (1979) reported seeing rough-toothed dolphins, *Steno bredanensis*, share large mahi-mahis, *Coryphaena hippurus*..
8. "Surface tail slaps," in which a dolphin barely cleared the surface with its flukes before bringing them down, creating a downward suction, were associated with feeding behavior. They usually were followed by feeding circles and feeding rushes.
9. On several occasions a dolphin lay at the surface and bounced up and down or moved its body with its head down as if manipulating something. Although I never observed fish during these instances, the dolphin may have been feeding.

The ways in which bottlenose dolphins fed at Sanibel appeared to depend on the location of the prey in the water column (e.g., benthic versus near the surface), the presence of physical barriers against which fish could be trapped, the size of the fish, the depth of the water, and other unknown factors. The seven types of feeding described in 1–6 were seen with approximately equal frequency during the study, while the latter three (7–9) were seen rarely.

Food

I often observed the fish that dolphins were pursuing or capturing. They ranged in estimated size from 2.5-cm-long unidentified fish to snook, *Centropomus undecimalis*, up to 1 m long ($\bar{x}$ = 22 cm, N = 54). Species identified visually in the field included needlefish, *Strongylura notata*, striped mullet, catfish, either *Arius felis* (sea or hardhead catfish) or *Bagre marinus* (gafftopsail catfish), spotted seatrout, snook, balao, *Hemiramphus balao*, and ladyfish, *Elops saurus*. Among the species tentatively identified were jack crevalle, *Caranx hippos*, Florida pompano, *Trachinotus carolinus*, pinfish, *Lagodon rhomboides*, lizardfish, *Synodus foetens* or *Trachinocephalus myops*, mackerel, *Scomberomorus* spp., and southern flounder, *Paralichthys lethostigma*. These species include bottom dwellers, surface dwellers, and midwater inhabitants (Clark, 1976). By analyzing stomach contents of bottlenose dolphins stranded in southwest Florida, Barros and Odell (Chapter 16, this volume) found that sand seatrout, *Cynoscion arenarius*, mullet, *Mugil* sp., and silver perch, *Bairdiella chrysoura* were the most common prey species.

After one feeding splash I saw a blowfish, *Sphoeroides nephelus*, lying inflated on the surface for 30–60 seconds before deflating and swimming away. Because the dolphin never returned to the fish, I presume that bottlenose dolphins do not feed on blowfish, which are usually poisonous (Allyn, 1982).

Prey Distribution

Many fishes change their distribution and movement patterns seasonally. Striped (black) mullet, considered the bottlenose dolphin's primary prey in the Gulf of Mexico (Gunter, 1942; Irvine *et al.*, 1981), spawn offshore during October–February (Collins 1985). Snook become sluggish at 15.6°C and die at 10°C (Stewart, 1985); these large, vigorous carnivores are, thus, probably most vulnerable to capture by bottlenose dolphins during the winter. Atlantic croakers, *Micropogonias undulatus*, also move offshore during winter (Lassuy 1983a). Some fishes move into deeper water in response to extreme temperatures (Harris *et al.*, 1983; Lassuy, 1983b). For instance, spotted seatrout move off shallow grass flats into deeper waters within estuaries to avoid extremely cold or extremely warm temperatures (Harris *et al.*, 1983). Sea catfish and gafftopsail catfish also move into deep channels when water temperatures drop below about 15°C (Muncy and Wingo, 1983).

Although I found statistically significant relationships between the mean depth at which dolphins fed or travel/fed and season ($F = 3.80$, $df = 3$, $p < .01$; $N = 325$) and four temperature ranges ($F = 6.17$, $df = 3$, $p < .0004$; $N = 325$), the season–depth model ($R^2 = 0.03$) and the temperature–depth model ($R^2 = 0.05$) were weak. Therefore, I could detect no biologically significant change in the depth of water in which dolphins fed according to season or water temperature (see Fig. 6 for monthly Sanibel water temperatures).

The availability of prey to bottlenose dolphins changes diurnally. Mullet feed during high tide in bays and estuaries (Harris *et al.*, 1983). Sea catfish concentrate in constricted areas where tidal currents are rapid (Muncy and Wingo, 1983).

Substrate type and vegetational cover also affect fish distribution. Spotted seatrout prefer seagrass beds as habitat (Lassuy, 1983b), and snook are most strongly associated with mangrove areas (Seaman and Collins, 1983). Striped mullet inhabit a wide variety of areas (Collins, 1985), while Florida pompano are found primarily over open sand and mud bottoms (Gilbert, 1986).

Since learning plays an important role in bottlenose dolphin behavior (Defran and Pryor, 1980) and there is an extended period of mother–calf association (Wells *et al.*, 1987), it is likely that bottlenose dolphin calves learn where and when to find specific types of prey and how to capture them by observing (and mimicking?) their mothers and other adults. Würsig (1986) contended that bottlenose dolphins and other odontocetes (e.g., killer whales, *Orcinus orca*) use feeding methods that must be based on the animals' learning where, when, and how to find prey.

The bottlenose dolphins inhabiting the Sanibel study area live in an environment which is subdivided into patches (e.g., dredged channels, seagrass flats, mangrove-lined shores, open beaches), each of which has its own faunal community that varies temporally and according to a variety of biological and physical influences. As predicted by Roughgarden (1972), animals faced with such a variable and unpredictable habitat should be selected to show behavioral flexibility. As I have shown, bottlenose dolphins conform to this prediction, as least as far as their feeding habits are concerned.

Descriptive Observations of Social Behavior

Greetings

I repeatedly observed that when two separate pods met, all of the dolphins dove and remained submerged for longer than the average dive time of 20–25 seconds. I quantified the dive time during greetings ($N = 5$) and found the dolphins remaining submerged for an average of 43 (±21.2) seconds. On some occasions when groups met, the dolphins swam quickly around each other just beneath the surface and occasionally tail-slapped or coughed.

Roughhousing

On five occasions I saw adult dolphins vigorously tumbling with and tossing small or newborn calves. In each case two or more adults focused their attention on the calf and roughhoused with it. At times, they used their rostrums and flukes to toss the calf into the air. The calf never appeared to try to escape this rough treatment. I have seen similar behavior in a Florida oceanarium and off the Texas coast. On one occasion dos Santos and Lacerda (1987) saw an adult dolphin, using its head and rostrum, throw a calf into the air; they interpreted this as disciplinary behavior.

Alloparental Care

I observed calves swimming with adults other than their mothers for extended periods (>1 hour) (also see Wells *et al.*, 1987). Typically, the "babysitter" played with the calf while its mother fed, often at a distance of a few hundred meters. I identified the mother as the adult who eventually departed alone with the calf and/or was a photographically identified individual seen with a calf on several days. These observations add a note of caution regarding assumptions about which dolphin might be the mother of a particular calf and make lengthy, repeated observations imperative for such determinations.

Dolphin Speed

Dolphins traveled at regular (5.5 km/hour) speed or slower 98% of the time that speed was recorded. Only 2% of 3107 records of dolphin speed were fast (i.e., faster than my boat's idle speed of 5.5 km/hour). Dolphins in Argentina averaged 6.1 km/hour (Würsig and Würsig, 1979) which is similar to the "regular" speed for Sanibel dolphins. Lockyer and Morris (1987) reported a much higher average speed for a juvenile male dolphin off Wales: 10–15 km/hour.

Activity Patterns: Temporal Factors

Activities varied significantly according to time of day (Fig. 2; $\chi^2 = 225.09$, $df = 12$, $p < .0001$; $N = 6378$), but they varied in a manner somewhat different from that observed elsewhere. Feeding and travel/feeding were both most common before 1300 hours, although feeding also increased in the evening following an afternoon lull. Socializing was most common in the evening, but social traveling was least common then. Tayler and Saayman (1972), Saayman *et al.* (1973), and Shane (1977) found morning and evening feeding peaks and a midday peak in "mating" (synonymous with socializing as defined here) despite the very different environments in which their studies were conducted. Würsig and Würsig (1979) noted an increase in aerial behavior associated with feeding and socializing in the afternoon in Argentina. Thus, bottlenose dolphins may budget

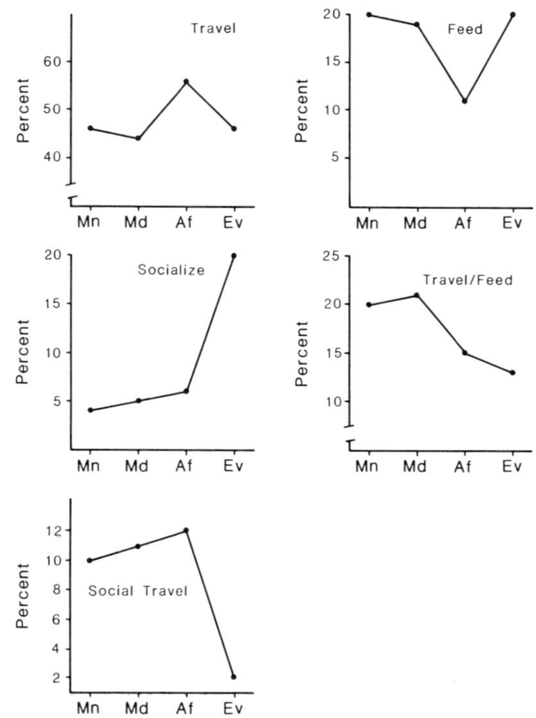

Figure 2 Percent occurrence of five bottlenose dolphin activities at Sanibel during four periods. Mn, Morning (0700–0959); Md, midday (1000–1259); Af, afternoon (1300–1559); Ev, evening (1600–2059).

their time differently in different habitats. Presumably, the diurnal movement patterns of their prey strongly affect the way dolphins apportion their time each day. Similarly, common dolphins, *Delphinus delphis*, off southern California dove to feed on deep scattering layer (DSL) organisms throughout the night, when the DSL was closest to the surface (Evans, 1971).

Dolphin activity patterns might be expected to vary seasonally owing to changes in prey abundance and distribution and because of changing energy requirements in the dolphins themselves (e.g., arising from reproduction or water temperature changes). At Sanibel activity patterns varied significantly by season (Fig. 3; $\chi^2 = 221.97$, $df = 12$, $p < .0001$; $N = 6378$), but the strength of association was weak (Cramer's $V = 0.11$). There was no distinct peak in socializing (which presumably included mating although copulation was never observed), but the fall–winter peak in social traveling could have been related to breeding. Captive and wild bottlenose dolphins have been reported to breed essentially year-round (McBride and Kritzler, 1951; Tavolga and Essapian, 1957; Layne, 1965; Harrison and Ridgway, 1971; Ridgway, 1972; Hohn, 1980; Irvine *et al.*, 1981; Wells *et al.*, 1987). The fall–winter peak in travel/feeding at Sanibel suggests that dolphins may feed more frequently during the cold season, perhaps to produce and maintain an adequate blubber layer. However, the lack of a distinct fall–winter peak in feeding casts doubt on this hypothesis. At two sites in Texas, traveling frequency peaked in winter or spring; feeding was twice as common in fall and winter as in spring and summer; and the frequency of "mating" (socializing) peaked in spring or summer (Shane, 1977; Gruber, 1981).

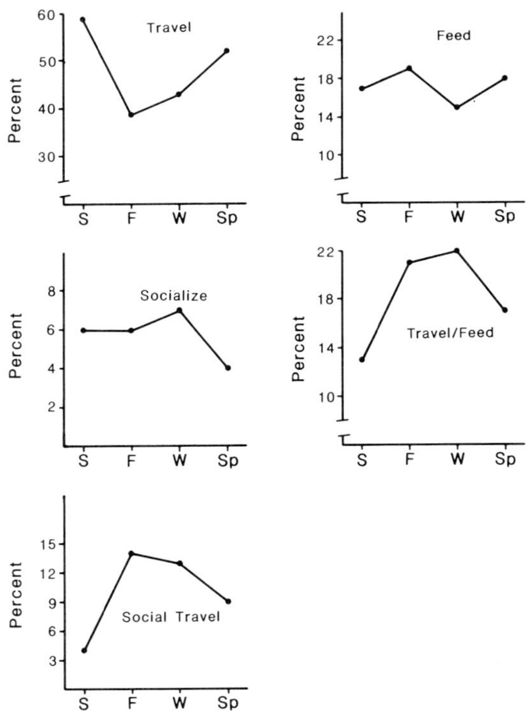

Figure 3 Percent occurrence of five bottlenose dolphin activities at Sanibel according to season. S, Summer (June–August); F, fall (September–November); W, winter (December–February); Sp, spring (March–May).

Activity Patterns: Group Characteristics

Group size in bottlenose dolphins depends on habitat and activity (Shane *et al.*, 1986). At Sanibel mean pod size varied significantly according to activity ($F = 461.97$, $df = 4$, $p < .0001$, $R^2 = 0.23$; $N = 6378$): social traveling pods were largest and travel/feeding pods were smallest (Fig. 4). All mean pod sizes were significantly different from each other (Tukey's $q = 4.61$, $\alpha = 0.01$). At Sanibel, as at Port Aransas, Texas (Shane, 1977), socializing pods were larger than those involved in other activities. In contrast, R. S. Wells, M. D. Scott, and A. B. Irvine (unpublished results, cited in Shane *et al.*, 1986) found socializing pods to be smaller than others. As discussed in the Respiration Rate section, solitary dolphins were relatively rare (12–18% of all sightings).

Pod geometry is not random but has functional significance which varies depending on the dolphins' activity. When traveling, dolphins usually (41% of all traveling pods for which geometry was recorded; $N = 1518$) formed a tight line-abreast group, allowing individuals to be in close contact yet cover a broad swath of the habitat. Feeding groups were either loose ovals (24%), widely dispersed (27%), or in mixed geometry (24%), indicating that individual rather than cooperative feeding was the norm at Sanibel ($N = 432$). Dolphins that were socializing formed tight ovals

(76%, N = 233), as did those that were social traveling (52%, N = 356). Travel/feeding dolphins were seen in all types of pod geometry (N = 583).

Activity Patterns: Physical Environmental Factors

Temperature

Dolphins entered human-made canal systems on Sanibel and Pine Island (see Fig. 1) only during December–March (N = 7). Canal temperatures ranged from 15.6 to 25.6°C during the winter (N = 12) and averaged 1.4°C (S.D. = 0.79) above bay temperatures. The dolphins' movement into the canals always involved feeding and, based on information from local residents, mullet and snook were probably the dolphins' primary prey in the canals. These fish, like many other species, seek out warmer, deeper water for refuge during the coldest time of year.

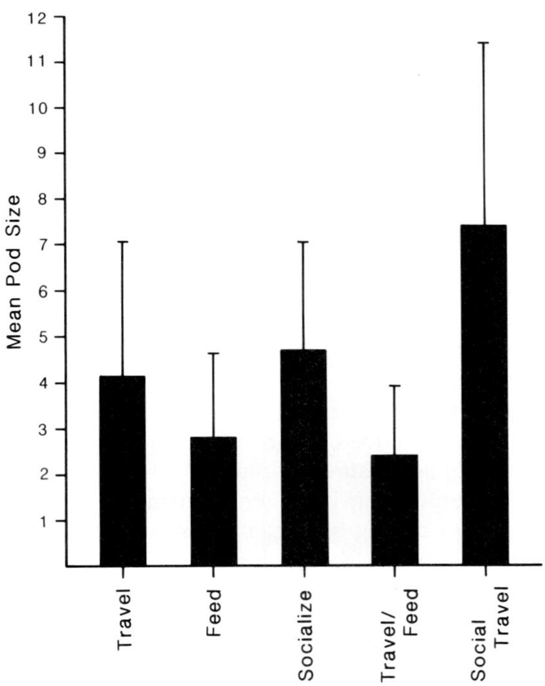

Figure 4 Mean pod size (±S.D.) for bottlenose dolphins engaged in five different activities at Sanibel.

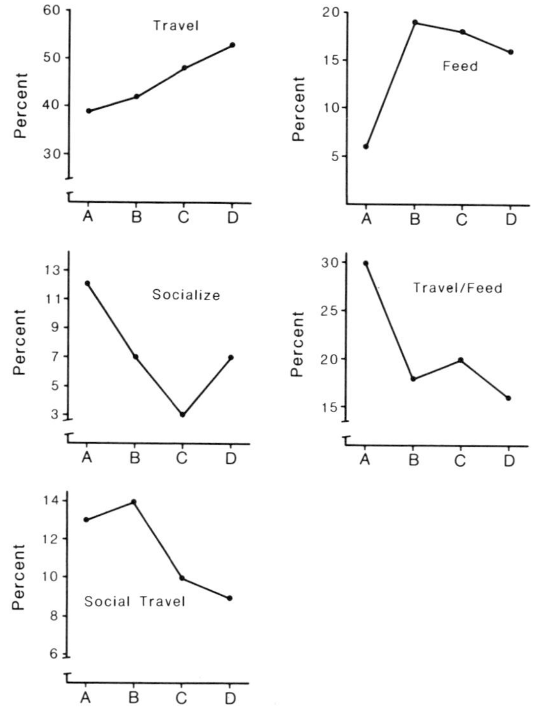

Figure 5 Percent occurrence of five bottlenose dolphin activities at Sanibel according to ranges of water temperatures. A, 50°–59°F (10°–15°C); B, 60°–69°F (15.6°–20.6°C); C, 70°–79°F (21.1°–26.1°C); D, 80°–89°F (26.7°–31.7°C).

Contrary to what would be expected if dolphin energy requirements increased during cold weather, feeding activity decreased as water temperature declined (Fig. 5; χ^2 = 163.15, df = 12, $p <$ .001; N = 6294). However, travel/feeding fit the expectation by declining as temperatures increased. Water temperature was not a very potent factor in explaining variation in activity patterns (Cramer's V = 0.09), perhaps because there were not any dramatic changes in water temperature throughout the year at Sanibel (Fig. 6).

Depth

Although water depth in the Sanibel study area is generally shallow, dolphins appeared to concentrate particular activities significantly more in par-

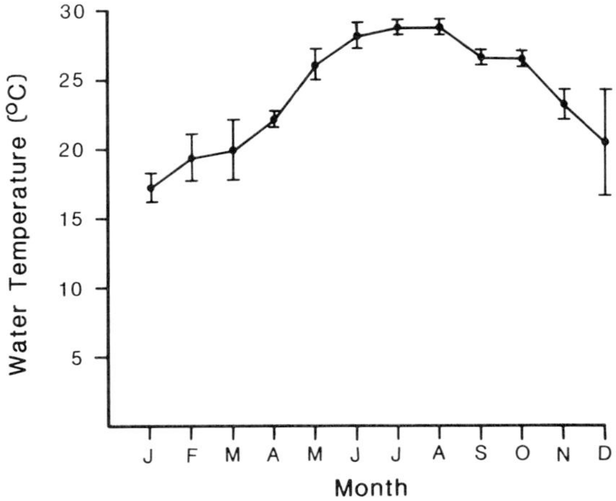

Figure 6 Mean water temperature (±S.D., in °C) by month during a 1-year (1985–1986) study of bottlenose dolphin behavioral ecology at Sanibel. The value for May is an average of temperatures in May 1985 and May 1986.

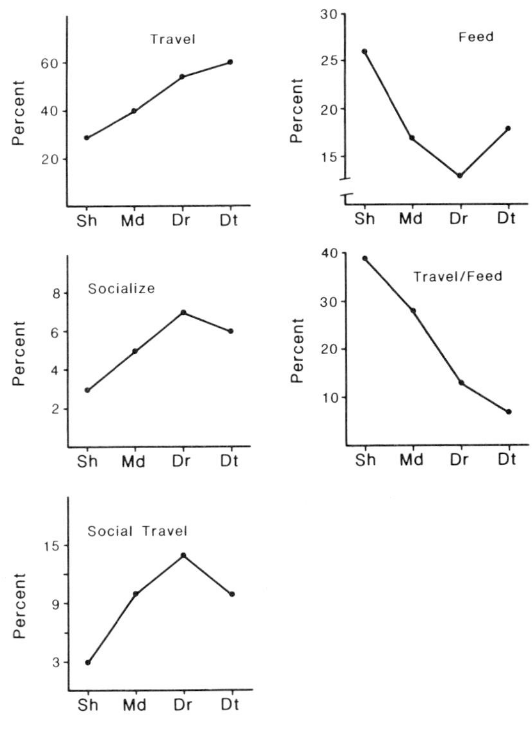

Figure 7 Percent occurrence of five bottlenose dolphin activities at Sanibel according to depth ranges. Sh, Shallow, 1–3 feet (0.3–0.9 m); Md, medium depth, 4–8 feet (1.2–2.4 m); Dr, deeper, 9–13 feet (2.7–4.0 m); Dt, deepest, 14–25 feet (4.3–7.6 m).

ticular depth ranges (Fig. 7; $\chi^2 = 484.90$, $df = 12$, $p < .001$; $N = 5969$). Traveling, socializing, and social traveling generally increased as depth increased. Most travel/feeding and feeding occurred in the shallowest depth range. The association between depth and activity was relatively strong (Cramer's $V = 0.17$).

Distance Offshore

Mangrove-lined shores provide shelter and food for many fishes, so dolphins might be expected to feed close to such shores. Indeed, feeding at Sanibel was concentrated close to shore, whereas socializing and traveling occurred farther offshore (Fig. 8; $\chi^2 = 438.09$, $df = 16$, $p < .001$, Cramer's $V = 0.17$; $N = 3799$). In Argentina, feeding and socializing usually took place in deeper, offshore waters, although individual feeding (similar to that seen at Sanibel) occurred close to shore (Würsig and Würsig, 1979).

Tidal Currents

Tidal currents can affect bottlenose dolphin behavior directly by assisting or impeding travel and indirectly by influencing prey movements. Dolphins moved significantly more against (64% of 854 pods) the tidal current than with (36%) it at Sanibel ($\chi^2 = 66.33$, $df = 1$, $p < .001$). The Sanibel study adds to the conflicting evidence regarding dolphin movements in relation to current flow. Dolphins moved predominantly against the current at Port Aransas, Texas (Shane, 1977, 1980), Port O'Connor, Texas (Gruber, 1981), and San Diego Bay, California (Norris and Prescott, 1961); however, at Sarasota, Florida, most dolphins moved with the current (Irvine *et al.*, 1981). The behavior described by Shane (1977) and Shane *et al.* (1986) as "resting" (dolphins maintaining one position facing against a strong current) was most likely not resting but against-current feeding described here for Sanibel dolphins.

The association between dolphin movements in relation to currents and activity (excluding socializing because of small sample size) was significant (I.D., $\chi^2 = 67.64$, $df = 3$, $p < .001$; $N = 1074$). All feeding dolphins moved against the current, and most traveling (64%), travel/feeding (70%), and social/traveling (59%) dolphins did so, as well.

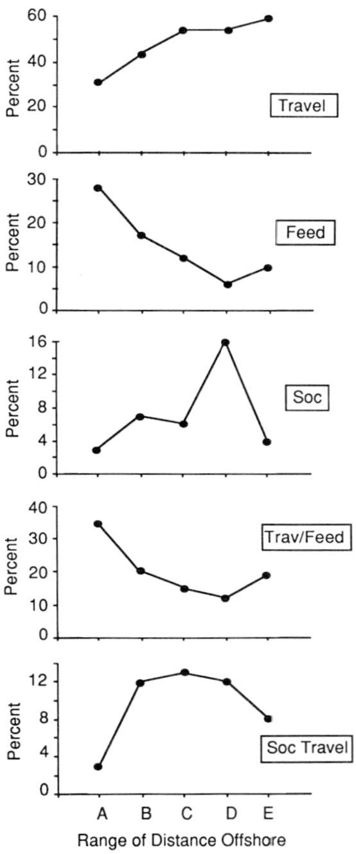

Figure 8 Percent occurrence of five bottlenose dolphin activities at Sanibel according to distance offshore. A, less than 100 m; B, 100 m to less than 0.25 mile (0.4 km); C, 0.25 to less than 0.5 mile (0.4–0.8 km); D, 0.5 to less than 1 mile (0.8–1.6 km); E, 1 mile (1.6 km) or farther. Data are from the I.D. set only.

Location

Since movement patterns and behavior of dolphins in Texas indicated that they recognized a boundary between bay and Gulf waters (Shane, 1977, 1980; Gruber, 1981), I wonder whether the behavior of dolphins in Florida varied in the bay as compared with the Gulf. I found that dolphins spent far more time traveling and less time feeding in the Gulf than in the bay (Fig. 9; I.D., $\chi^2 = 147.12$, $df = 8$, $p < .001$; $N = 3866$). These results suggest that the bay provided the most abundant and accessible food source and that the Gulf was used mostly as a transit area. Possibly, the diversity of habitats in the bay resulted in higher fish densities there than in the ecologically less diverse Gulf. Socializing and social/traveling occurred with approximately equal frequencies in the bay and Gulf.

The Sanibel study area was separable into 18 subareas, each of which could be described ecologically (Table 2). Dolphins presented with such a variety of habitat patches might be expected to concentrate particular activities in certain patches (Table 3). Therefore, I tested the null hypothesis that location had no effect on dolphin activity; this hypothesis was rejected (I.D., $\chi^2 = 824.97$, $df = 68$, $p < .001$, Cramer's $V = 0.23$; $N = 3866$).

The disproportionate occurrence of certain activities in certain locations at Sanibel was probably due to a mixture of factors. Feeding was concentrated in areas with many sport fishing boats (which presumably are found where there are fish) and strong currents or in areas with abundant seagrasses and mangroves. Strangely, socializing occurred most in the ICW, which had very heavy boat traffic; why dolphins would socialize in an area where they were subject to boat noise and the threat of being hit by boats is not known. In Texas, dolphins socialized most in areas free from heavy boat traffic (Shane, 1977). In the Gulf of California, dolphins within 5.5 km of the mouth of an estuary spent most of their time feeding while those more distant from estuaries spent most of their time traveling (Ballance, 1987). Other cetacean species are known to focus particular activities in specific types of habitat. For example, killer whales in the northeastern Pacific fed in areas of high bottom

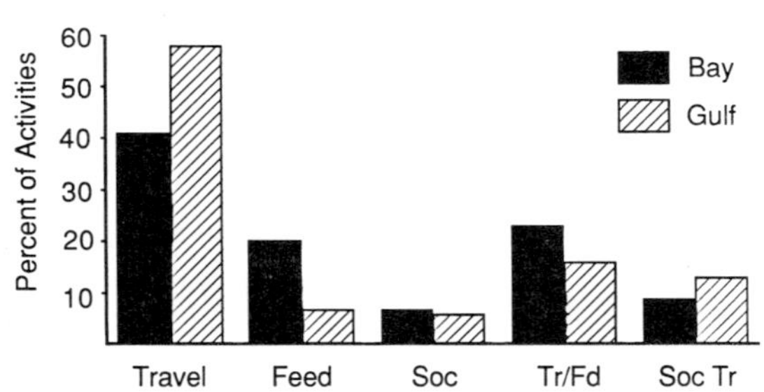

Figure 9 Percent occurrence of five bottlenose dolphin activities in the bay and in the Gulf of Mexico around Sanibel during 1985–1986. Data are from the I.D. set only.

Table 2
Habitat Descriptions for 18 Subareas of the Sanibel Study Area[a]

Area	Boat traffic	Fishing	Currents	Depth (m)	Seagrasses	Shoreline		
						Mangrove	Beach	Seawall
Sanibel Pass	+	+	+	<1–7	−	0	−	−
Causeway	+	+	+	<1–6	−	0	−	−
Drawbridge	+	+	+	<1–7	−	0	−	+
ICW	+	−	*	3–7	−	*	0	0
Sanibel Bay	*	+	*	<1–7	+	+	0	*
Small Islands	−	*	*	<1–3	+	+	0	0
Tarpon Bay Channel	*	+	+	<1–3	*	*	−	0
Ft. Myers Beach	−	−	*	<1–5	−	0	+	0
Matanzas Pass	+	*	+	<1–4	*	*	−	+
Matlacha Pass	−	−	−	<1–4	+	+	0	−
Pine Island Bay	*	+	−	<1–5	+	+	0	*
Mainland	−	*	−	<1–4	+	+	*	0
Tarpon Bay	−	*	−	<1–2	+	+	0	0
Blind Pass	−	*	+	<1–2	+	+	0	−
Captiva Bay	*	−	−	<1–4	+	+	0	−
Captiva Gulf	−	−	−	1–8	0	0	+	0
Undeveloped Sanibel Gulf	−	−	−	<1–7	0	0	+	0
Developed Sanibel Gulf	*	*	−	<1–7	0	0	+	0

[a] Symbols represent the relative attributes of each area: 0, none; −, small amount; +, medium amount; *, abundant (or strong for currents). Boat traffic, fishing, and currents were assessed subjectively during the study. Depth was taken from navigational charts, while seagrass coverage and shoreline type were determined from maps compiled from aerial surveys by Harris *et al.* (1983). Note: Not all subareas are labeled on Figure 1.

Table 3
Percent Occurrence of Dolphin Activities in 18 Subareas of the Sanibel Study Area

	Travel	Feed	Socialize	Travel/feed	Social/travel
Sanibel Pass	53	12	2	16	17
Causeway	31	22	8	23	16
Drawbridge	13	63	11	2	12
ICW	41	18	13	9	19
Sanibel Bay	43	18	10	24	5
Small Islands	21	49	5	5	21
Tarpon Bay Channel	40	33	5	20	2
Ft. Myers Beach	75	7	0	11	7
Matanzas Pass	43	29	5	17	5
Matlacha Pass	52	11	2	26	9
Pine Island Bay	55	7	6	18	14
Mainland	25	13	11	33	18
Tarpon Bay	26	22	3	45	3
Blind Pass	33	0	0	67	0
Captiva Bay	48	11	0	34	7
Captiva Gulf	91	4	1	3	1
Undeveloped Sanibel Gulf	47	8	8	20	18
Developed Sanibel Gulf	56	8	7	17	12
Overall %[a]	46	17	7	21	10

[a] The overall % given for each activity represents a summary for the whole study area. Note: Not all subareas are labeled on Figure 1.

relief and next to kelp-covered reefs (Heimlich-Boran, 1987). Spinner dolphins, *Stenella longirostris*, in Hawaii rested during the day in sheltered bays and fed offshore at night in the DSL (Norris and Dohl, 1980). Common dolphins fed over areas of high relief off southern California (Evans, 1971). The Sanibel data show that dolphins discriminated between ecologically different patches in their habitat and used these patches in various ways.

Activity Patterns: Biological Environmental Factors

Presence of Fish

My inability to determine fish abundance and distribution put a frustrating limitation on understanding what seems to be the key factor influencing bottlenose dolphin activities and movements. The use of a fish finder to determine fish presence did not prove as useful as I had expected. The fact that I could check for fish only in a narrow, cone-shaped area beneath my boat was one limitation of the method. Also, the presence of fish near dolphins does not necessarily indicate that the dolphins are feeding.

Direct visual observation of fish was the best indicator of fish presence and the best confirmation that dolphins were feeding. Of the fish I sighted visually, 55% (N = 33) were with feeding dolphins and 35% (N = 21) were with travel/feeding dolphins.

Influence of My Boat

The issue of how an observer's boat affects dolphins is of key importance, particularly in a behavioral study. In order to measure the degree of disturbance imposed by my boat and others early in the study, I recorded every instance where any boat was less than 10 m from dolphins during the period 11 May–14 July 1985 (N = 136 for my boat, N = 41 for other boats, N = 4 for my boat and another boat). I also noted dolphins bowriding my boat nine times and other boats four times during this period. I considered bowriding to be a positive response to a boat. To be as rigorous as possible in assessing the effect of boats on dolphins, I recorded any behavioral change as a negative response: changing direction, changing activity, peeking, tail slapping, diving, coughing, vanishing, fast traveling, lying at the surface, making "porpoise puddles" (surface boils produced by rapid subsurface fluke movements), and changing pod geometry. Dolphins responded negatively to my boat (26% of the encounters), even during the first part of the study, less than to other boats (37% of the encounters). Dolphins were less disturbed by my boat than by other boats, presumably because I moved slowly and cautiously around them. Of course, my definitions of "negative response" and "disturbance" may not include all such occurrences; some responses may be too subtle to detect but may still be of significance to the dolphins.

My boat was less than 10 m from dolphins during 143 instantaneous samples throughout the year of study. I was able to get close to dolphins that were socializing or social traveling significantly more often than I would have expected based on the frequency of occurrence of those activities ($\chi^2 = 56.6$, $df = 4$, $p < .001$). Perhaps the rate of disturbance by my boat was lowest when dolphins were socializing because they were focused on their own interactions more than on outside elements. My subjective impression was that most of the dolphins in the Sanibel area became habituated to my boat (i.e., recognized it and were not visibly disturbed by it) within 6 months.

Dolphin Associates

I recorded each instance when dolphins were within 100 m of feeding birds (at least 10 species), sport fishing boats, commercial fishing boats, West Indian manatees, *Trichechus manatus*, and loggerhead turtles, *Caretta caretta* (N = 861). In many instances such associations appeared to be incidental and of no biological significance to either associate. The most common associations were with brown pelicans, *Pelecanus occidentalis* (24%), terns, *Sterna* spp. (18%), sport fishing boats (17%), and mixed flocks of birds (15%). Sixty percent of the associations between feeding birds and dolphins occurred when dolphins were feeding or travel/feeding. When feeding birds were associated with dolphins, 40% of the time only a single bird was involved and 89% of the time there were five of fewer birds in the group. Terns, pelicans, gulls, and magnificent frigate birds, *Fregata magnificens*, often hovered, looking downward, above dolphin pods, seemingly evaluating the dolphins' activity. If the dolphins were feeding, the birds stayed and attempted to feed on fish chased to the surface by the mammals. If the dolphins were not feeding, the birds were usually able to determine this within several seconds and then depart.

Bird–dolphin associations at Sanibel were always initiated by the birds. In contrast, Würsig and Würsig (1980) proposed that dusky dolphins used sightings of feeding bird flocks to lead them to sites where other dolphins were already feeding. Tyack (1976) found that birds in Argentina associated with bottlenose dolphins only when the latter were feeding. The close association between feeding or travel/feeding dolphins and sport fishing boats at Sanibel suggests that there may be certain areas recognized by dolphins and humans as being more productive than others. Fishermen would not purposely associate with dolphins, which have the reputation of disrupting fishing, and there is no indication that dolphins are attracted directly to sport fishing boats.

Surface Behavior

Discrete surface behaviors may serve aggressive, social, sexual, alimentary, exploratory, play, or assisted-locomotion functions (see Appendix). The two most common surface behaviors were social splashes and feeding circles (13% each). I used a subset of the 11 most common surface behaviors to investigate the role played by surface behavior in the lives of the dolphins.

I analyzed the surface data to look for seasonal, time of day, activity, and tidal state effects on occurrence of surface behaviors, and the results, although statistically significant, were not biologically meaningful. Surface behaviors may have various functions depending on the context in which they occur. Therefore, the statistical associations between surface behaviors and environmental factors are complicated and impossible to disentangle in a meaningful way. The one environmental variable that did show a strong association with particular surface behaviors was water depth ($\chi^2 = 287.01$, $df = 30$, $p < .001$, Cramer's $V = 0.33$ indicating that depth explains 1/3 of the variation in surface behavior; $N = 869$). Forward leaps and social splashes were concentrated in the depth range where most socializing and social traveling occurred (see Fig. 7). Forward slaps occurred most frequently in the medium range (4–8 feet, 1.2–2.4 m). Feeding splashes and feeding circles were most common in the depth range where feeding and travel/feeding were concentrated (see Fig. 7).

Other observers have provided correlative data for bottlenose dolphin surface behavior elsewhere in the world. Saayman *et al.* (1973) associated leaping, chasing, and rubbing with mating behavior and leaping, splashing, and slapping with feeding (see also Tayler and Saayman, 1972). Würsig and Würsig (1979) associated "clean leaps" with possible feeding and "noisy leaps," "noseouts," and kelp tossing with social behavior. Goodwin (1985) associated 20 surface behaviors with sexual behavior, feeding, traveling, and milling; specifically, leaps (body out, enter head first) were linked with feeding and breaches (body out, enter side, belly, or back first) with sexual behavior. Shane (1977) linked leaping with socializing (mating). Clearly, it would be helpful to observe bottlenose dolphin surface behavior simultaneously from the surface and underwater (observers in South Africa and Argentina achieved nearly as good a viewpoint from cliffs overlooking clear water) to understand its functions better.

CONCLUSIONS

A complex array of temporal, environmental, and social factors is associated with the activity patterns of bottlenose dolphins at Sanibel Island and elsewhere. The frequency of occurrence of specific dolphin activities varied significantly from expected (using the χ^2 test) in the case of almost every factor considered in my analyses. However, in all cases where it was calculated, the Cramer's V statistic measuring the strength of association between the two variables (Zar, 1984) was between 0.09 and 0.33, indicating a rather weak level of association. This finding fuels the impression that dolphin behavior is dictated by a complicated web of interacting elements, and no single factor is very useful in explaining it.

ACKNOWLEDGMENTS

Financial support and donations of equipment and supplies were received from John and Meredith Alden, Bill and Ann Alquist, Mili Backus, Bill Boswell, Dolphin Biology Research Associates, Clyde Duchrow, Mary Dufner, John Francis, Kipp Frohlich, Barry and Nancy Gordon, Ron Hight, Dan McSweeney, Emily and Presson Shane, Jack and Star Thomas, Randy Wells, U.S. Fish and Wildlife Service–Denver Wildlife Research Center, University of California at Santa Cruz–Biology Board, Sanibel–Captiva Rotary Club, and Sanibel–Captiva Shell Club. I was assisted in data gathering and data analysis by Becky Hardenbergh, Lauri Holts, Jim Mulherin, Dave Garrison, and George Hall. Graeme Mellor provided support and encouragement throughout my year in the field. Marshall Sylvan guided me in deciphering the differences between my two Florida data sets, and he alerted me to ways to identify biases inherent in my data collection methods. Ken Norris was particularly helpful in reviewing drafts of this chapter, because he taught me that I should consider my reader as I write. Anonymous reviewers provided a great deal of guidance in making this paper more concise. Insightful comments on early drafts also were made by Jim Estes, Bernd Würsig, Carolyn Heath, Beth Mathews, and Greg Silber.

APPENDIX

LIST OF ACTIVITIES AND SURFACE BEHAVIORS PERFORMED BY BOTTLENOSE DOLPHINS AT SANIBEL ISLAND, FLORIDA

Activities. Five of the following activity patterns (marked by asterisks) were used in the data analyses.

Travel*: Moving steadily in one direction
Feed*: Any of a variety of behaviors distinguished by such things as repeated dives in varying directions in one location, feeding circles, feeding splashes, fish kicks, feeding rushes, and fish tosses (see section on Feeding for more detailed descriptions)
Socialize*: Some or all pod members in almost constant physical contact with one another, oriented toward one another, and often displaying surface behaviors; no forward movement; a calf engaged in repeated surface behavior on its own was also considered to be socializing
Rest: Moving very slowly or drifting in one direction
Travel/feed*: Moving steadily in one direction while presumably hunting for food and feeding intermittently
Social travel*: Moving steadily in one direction while socializing intermittently
Mill: Moving in varying directions in one location but showing no surface behaviors and no apparent physical contact between individuals; usually staying close to the surface
Mixed behaviors: Social/mill, social/feed, travel/rest, mill/feed, social/mill/feed, travel/mill, social/travel/rest (these refer to activities performed by different pod members simultaneously, e.g., one dolphin feeds while two others socialize)

Surface Behaviors. The 37 discrete surface behaviors (N = 1219 observations) are presented in categories indicating my interpretation of their functional significance. Eleven of the behaviors (marked by asterisks) were used in the data analyses.

Aggression
Tail slap: Flukes raised above the surface and ventral side slapped downward, usually making a loud, percussive sound
Upside-down tail slap: Flukes raised above the surface and dorsal side slapped downward
Flipper slap: Pectoral flipper slapping the surface
Cough*: A loud, forceful exhalation
Big underwater bubble: One large (diameter 10+ cm) bubble released underwater

Play or Assisted Locomotion
Surf: Riding a wave or boat wake

Exploratory
Peek*: Head raised out of the water far enough to expose the eyes; rostrum pointed forward; occurs during forward motion
Vertical spy hop: Head raised out of the water far enough to expose the eyes; rostrum pointed skyward; dolphin stationary

Social and/or Feeding
Leap: Entire body clears the water (any height)
  Forward leap*: Exit and enter head first with venter facing down
  Backward leap: Exit and enter with dorsum facing down
  Side leap: Exit and enter with either side facing down

Roll: Most of body excluding flukes and part of tail stock clears water
Forward roll*: Venter down
Backward roll: Dorsum down
Side roll: Side down
Tail sideswipe: Tail stock and flukes sweeping laterally at the surface
Undetermined splash: Splash occurring when dolphin activity is unknown

Social

Slap: Anterior half or less of body clears water and falls or is slapped downward against the surface
Forward slap*: Venter down
Backward slap: Dorsum down
Side slap: Side down
Head out*: Entire head exposed at surface; rostrum pointed at an angle; dolphin stationary
Nose out: Rostrum exposed at surface
Flipper wave: Pectoral flipper waving in the air
Tail wave*: Tail flukes waving in air for less than 2 seconds
Belly up: Venter exposed at surface
Tail upkick: Flukes rapidly lifted upward with dorsum up
Buck: Body bent forward and back hunched and exposed at surface while dolphin bounces vigorously
Social splash*: Splash made by two or more dolphins in contact with one another; splash obscures the details of their interaction
Beak/genital propulsion: One dolphin uses its rostrum to nudge the genital area of another dolphin who is lying stationary on its side
Erection: Erect penis visible at surface

Feeding

Head stand: Tail flukes and, often, tail stock exposed vertically in air for 2 or more seconds
Lie at surface*: Lie stationary with part of body exposed at surface for 5 or more seconds
Fish toss: Fish thrown into air by dolphin using its melon, rostrum, or teeth
Feeding splash*: Splash made as a dolphin makes a final grab for a fish it has pursued
Feeding circle*: Defined in text
Feeding rush: Defined in text
Fish kick: Defined in text
Surface tail slap: Defined in text

REFERENCES

Allyn, R. (1982). "Saltwater Florida Fishes." Great Outdoors Publishing Co., St. Petersburg, Florida. 80 pp.

Altmann, J. (1974). Observational study of behavior: Sampling methods. *Behaviour* **49**(3–4), 227–267.

Baker, A. N. (1983). "Whales and Dolphins of New Zealand and Australia: An Identification Guide." Victoria Univ. Press, Wellington, New Zealand. 103 pp.

Ballance, L. T. (1987). Ecology and behavior of the bottlenose dolphin, *Tursiops truncatus,* in the Gulf of California, Mexico. M.S. thesis, Moss Landing Marine Labs and San Jose State Univ., San Jose, California. 103 pp.

Barros, N. B. (1987). Food habits of bottlenose dolphins (*Tursiops truncatus*) in the southeastern United States, with special reference to Florida waters. M.S. thesis, Univ. of Miami, Miami, Florida. 139 pp.

Brower, K., and Curtsinger, W. R. (1979). "Wake of the Whale." Friends of the Earth, New York. 160 pp.

Busnel, R. G. (1973). Symbiotic relationship between man and dolphins. *Trans. N.Y. Acad. Sci.* **35,** 112–131.

Caldwell, D. K., and Caldwell, M. C. (1972). "The World of the Bottlenose Dolphin." Lippincott, Philadelphia, Pennsylvania. 157 pp.

Clark, J. (1976). "The Sanibel Report: Formulation of a Comprehensive Plan Based on Natural Systems." The Conservation Foundation, Washington, D.C. 305 pp.

Collins, M. R. (1985). Species profiles: Life histories and environmental requirements of coastal fishes and invertebrates (South Florida): Striped mullet. Biological Report 82 (11.34). U.S. Fish and Wildlife Service, 1010 Gause Blvd., Slidell, Louisiana 70458. 11 pp.

Connor, R. C., and Smolker, R. S. (1985). Habituated dolphins (*Tursiops* sp.) in western Australia. *J. Mammal.* **66**(2), 398–400.

Defran, R. H., and Pryor, K. (1980). The behavior and training of cetaceans in captivity. "Cetacean Behavior: Mechanisms and Functions" (L. M. Herman, ed.), pp. 319–362. Wiley, New York. 463 pp.

dos Santos, M. E., and Lacerda, M. (1987). Preliminary observations of the bottlenose dolphin (*Tursiops truncatus*) in the Sado estuary (Portugal). *Aquatic Mamm.* **13**(2), 65–80.

Evans, W. E. (1971). Orientation behavior of delphinids: Radio telemetric studies. *Ann. N.Y. Acad. Sci.* **188,** 142–160.

Gilbert, C. (1986). Species profiles: Life histories and environmental requirements of coastal fishes and invertebrates (South Florida): Florida pompano. Biological Report 82 (11.42). U.S. Fish and Wildlife Service, 1010 Gause Blvd., Slidell, Louisiana 70458.

Goodwin, D. E. (1985). Diurnal behavior patterns of *Tursiops truncatus* off Mobile Point, Alabama. M.S. thesis, San Francisco State Univ., San Francisco, California. 60 pp.

Gruber, J. A. (1981). Ecology of the Atlantic bottlenose dolphin (*Tursiops truncatus*) in the Pass Cavallo area of Matagorda Bay, Texas. M.S. thesis, Texas A & M University, College Station, Texas. 182 pp.

Gunter, G. (1942). Contributions to the natural history of the bottlenose dolphin, *Tursiops truncatus* (Montague), on the Texas coast, with particular reference to food habits. *J. Mammal.* **23**(3), 267–276.

Gunter, G. (1954). Mammals of the Gulf of Mexico. *In* "Gulf of Mexico: Its Origin, Waters, and Marine Life" (P. S. Galtsoff, ed.), pp. 543–551. *Fish. Bull.* **89.** 604 pp.

Harris, B. A., Haddad, K. D., Steidinger, K. A., and Huff, J. A. (1983). Assessment of fisheries habitat: Charlotte Harbor and Lake Worth, Florida. *Fin. Rep. Fla. Dept. Nat. Res., Bureau Mar. Res., Mar. Res. Labs*, St. Petersburg, Florida. 211 pp.

Harrison, R. J., and Ridgway, S. H. (1971). Gonadal activity in some bottlenose dolphins (*Tursiops truncatus*). *J. Zool. London* **165,** 355–366.

Heimlich-Boran, J. R. (1987). Habitat use patterns and behavioral ecology of killer whales (*Orcinus orca*) in the Pacific Northwest. M.S. thesis, San Jose State University and Moss Landing Marine Labs, Moss Landing, California. 30 pp.

Hoese, H. D. (1971). Dolphin feeding out of water in a salt marsh. *J. Mammal.* **52,** 222–223.

Hogan, T. (1975). Untitled draft of M.S. thesis, Univ. of Rhode Island, Narragansett, Rhode Island. 42 pp.

Hohn, A. A. (1980). Age determination and age related factors in the teeth of western north Atlantic bottlenose dolphins. *Sci. Rep. Whales Res. Inst., Tokyo* **32,** 39–66.

Hussenot, E. (1980). Le grand dauphin *Tursiops truncatus* en Bretagne: Types de fréquentation. *Penn ar Bed* **12,** 355–380.

Irvine, A. B., Scott, M. D., Wells, R. S., and Kaufmann, J. H. (1981). Movements and activities of the Atlantic bottlenose dolphin, *Tursiops truncatus,* near Sarasota, Florida. *Fish. Bull.* **79**(4), 671–688.

Lassuy, D. R. (1983a). Species profiles: Life histories and environmental requirements (Gulf of Mexico): Atlantic croaker. FWS/OBS 82 (11.3). U.S. Fish and Wildlife Service, 1010 gause Blvd., Slidell, Louisiana 70458. 12 pp.

Lassuy, D. R. (1983b). Species profiles: Life histories and environmental requirements (Gulf of Mexico): Spotted seatrout. FWS/OBS 82 (11.4). U.S. Fish and Wildlife Service, 1010 Gause Blvd., Slidell, Louisiana 70458. 14 pp.

Layne, J. N. (1965). Observations on marine mammals in Florida waters. *Bull. Fla. State Mus.* (*Biol. Sci.*) **9**(4), 131–181.

Leatherwood, S. (1975). Some observations of feeding behavior of bottle-nosed dolphins (*Tursiops truncatus*) in the Northern Gulf of Mexico and (*Tursiops* cf. *T. gilli*) off southern California, Baja California, and Nayarit, Mexico. *Mar. Fish. Rev.* **37**(9), 10–16.

Lockyer, C., and Morris, R. (1987). Observations on diving behaviour and swimming speeds in a wild juvenile *Tursiops truncatus. Aquatic Mamm.* **13**(1), 31–35.

McBride, A. F., and Kritzler, H. (1951). Observations on pregnancy, parturition, and post-natal behavior in the bottlenose dolphin. *J. Mammal.* **32**(3), 251–266.

Muncy, R. J., and Wingo, W. M. (1983). Species profiles: Life histories and environmental requirements of coastal fishes and invertebrates (Gulf of Mexico): Sea catfish and gafftopsail catfish. FWS/OBS 82 (11.5). U.S. Fish and Wildlife Service, 1010 Gause Blvd., Slidell, Louisiana 70458. 17 pp.

Norris, K. S., and Dohl, T. P. (1980). The behavior of the Hawaiian spinner porpoise, *Stenella longirostris. Fish. Bull.* **77,** 821–847.

Norris, K. S., and Prescott, J. H. (1961). Observations on Pacific cetaceans of Californian and Mexican waters. *Univ. Calif. Publ. Zool.* **63**(4), 291–402.

Ridgway, S. H. (1972). Homeostasis in the aquatic environment. *In* "Mammals of the Sea: Biology and Medicine" (S. H. Ridgway, ed.), pp. 590–747. Charles C. Thomas, Publ., Springfield, Illinois.

Rigley, L. (1983). Dolphins feeding in a South Carolina salt marsh. *Whalewatcher* **17**(2), 3–5.

Roughgarden, J. (1972). Evolution of niche width. *Am. Nat.* **106,** 683–718.

Saayman, G. S., Tayler, C. K., and Bower, D. (1973).

Diurnal activity cycles in captive and free-ranging Indian Ocean bottlenose dolphins (*Tursiops aduncus* Ehrenburg). *Behaviour* **44,** 212–233.

Schlais, J. F. (1984). Thieving dolphins: A growing problem in Hawaii's fisheries. *Sea Front.* **30**(5), 293–298.

Seaman, W., and Collins, M. (1983). Species profiles: Life histories and environmental requirements of coastal fishes and invertebrates (South Florida): Snook. FWS/OBS 82 (11.16). U.S. Fish and Wildlife Service, 1010 Gause Blvd., Slidell, Louisiana 70458. 16 pp.

Shane, S. H. (1977). The population biology of the Atlantic bottlenose dolphin, *Tursiops truncatus,* in the Aransas Pass area of Texas. M.S. Thesis, Texas A & M Univ., College Station, Texas. 239 pp.

Shane, S. H. (1980). Occurrence, movements, and distribution of bottlenose dolphin, *Tursiops truncatus,* in southern Texas. *Fish. Bull.* **78**(3), 593–601.

Shane, S. H., Wells, R. S., and Würsig, B. (1986). Ecology, behavior and social organization of the bottlenose dolphin: A review. *Mar. Mamm. Sci.* **2**(1), 34–63.

Stewart, V. N. (1985). Snook. Sea Stats No. 6. Fla. Dept. Nat. Res., Bureau of Education and Information, 3900 Commonwealth Blvd., Tallahassee, Florida 32303. 7 pp.

Tavolga, M. C., and Essapian, F. S. (1957). The behavior of the bottle-nosed dolphin (*Tursiops truncatus*): Mating, pregnancy, parturition and mother–infant behavior. *Zoologica* **42**(1), 11–31.

Tayler, C. K., and Saayman, G. S. (1972). The social organisation and behaviour of dolphins (*Tursiops aduncus*) and baboons (*Papio ursinus*): Some comparisons and assessments. *Ann. Cape Prov. Mus.* (*Nat. Hist.*) **9**(2), 11–49.

Tyack, P. (1976). Patterns of vocalization in wild *Tursiops truncatus.* Senior thesis, Harvard University, Cambridge, Massachusetts. 50 pp.

Wells, R. S., Irvine, A. B., and Scott, M. D. (1980). The social ecology of inshore odontocetes. *In* "Cetacean Behavior: Mechanisms and Functions" (L. M. Herman, ed.), pp. 263–317. Wiley, New York. 463 pp.

Wells, R. S., Scott, M. D., and Irvine, A. B. (1987). The social structure of free-ranging bottlenose dolphins. *In* "Current Mammalogy" (H. H. Genoways, ed.), Vol. 1, pp. 247–305. Plenum, New York and London.

Würsig, B. (1978). Occurrence and group organization of Atlantic bottlenose porpoises (*Tursiops truncatus*) in an Argentine Bay. *Biol. Bull.* **154**(2), 349–359.

Würsig, B. (1986). Delphinid foraging strategies. *In* "Dolphin Cognition and Behavior: A Comparative Approach" (R. J. Schusterman, J. A. Thomas, and F. G. Wood, eds.), pp. 347–359. Lawrence Erlbaum Assoc., Hillsdale, New Jersey. 393 pp.

Würsig, B., and Würsig, M. (1977). The photographic determination of group size, composition, and stability of coastal porpoises (*Tursiops truncatus*). *Science* **198,** 755–756.

Würsig, B., and Würsig, M. (1979). Behavior and ecology of the bottlenose dolphin, *Tursiops truncatus,* in the South Atlantic. *Fish. Bull.* **77**(2), 399–412.

Würsig, B., and Würsig, M. (1980). Behavior and ecology of the dusky dolphin, *Lagenorhynchus obscurus,* in the South Atlantic. *Fish. Bull.* **77**(4), 871–890.

Zar, J. H. (1984). "Biostatistical Analysis," 2nd Ed. Prentice-Hall, Englewood Cliffs, New Jersey. 718 pp.

Residence Patterns, Group Organization, and Surfacing Associations of Bottlenose Dolphins in Kino Bay, Gulf of California, Mexico

13

Lisa T. Ballance
Moss Landing Marine Laboratories
Moss Landing, California 95039
Present address: Department of Biology,
University of California, Los Angeles, California 92004

INTRODUCTION

Numerous studies spanning a year or more have examined the behavior, social systems, ranges, and other aspects of the life history of bottlenose dolphins, *Tursiops truncatus* (Irvine and Wells, 1972; Shane, 1977, 1980; Wells, 1978, 1986; Würsig, 1978; Würsig and Würsig, 1979; Wells *et al.*, 1980; Gruber, 1981; Irvine *et al.*, 1981; Goodwin, 1985; Hansen, Chapter 23, this volume). As a result, it is now possible to begin comparing aspects of dolphin biology and behavior in different areas.

The purpose of this chapter is twofold. First, I describe the social structure of a free-ranging community of bottlenose dolphins in a small bay in the Gulf of California. I analyze social structure in terms of three parameters: (1) residence patterns of individuals in this bay, (2) group composition and stability of this composition over time, and (3) associations between individuals at the surface. Second, I compare the social structure of this community to that of dolphins studied elsewhere. By comparing similarities and differences in social structure and corresponding similarities and differences in the habitats of each study, I attempt to illustrate which aspects of social structure may be modified with environmental variation and which are less subject to such modification.

METHODS

Research was conducted during May through July and October through December of 1984 in Kino Bay, Gulf of California, Mexico (Fig. 1). A preliminary study was conducted during the months of July and August of 1983 to refine data collection

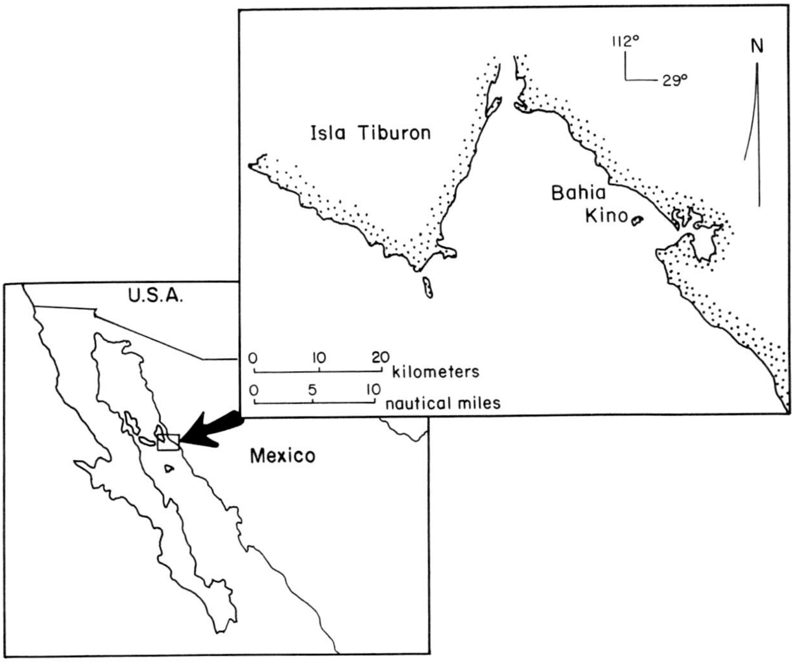

Figure 1 Map of the Kino Bay study area, Gulf of California, Mexico.

techniques; however, preliminary data are not included in the present analysis.

The study site is a small bay, extending approximately 5.5 km east to west and 13 km north to south, located along the west coast of mainland Mexico. Depths in the bay range from less than 1 m along the shore to 12 m along the western side where the bay meets the gulf. The bottom is uniformly composed of sand, and topography is characterized by a gradual slope from the shallow eastern side to the deeper western side. Water is relatively turbid; secchi disk depths were measured at 7 m or less.

A total of 398.5 hours was spent in boats on Kino Bay or at shore stations searching for dolphins. Animals were directly observed for 186.0 hours from one of two boats: a 17-foot, 165-hp inboard–outboard Sterncraft or a 30-foot, twin 155-hp Skipjack. A group of dolphins was followed from a distance of 50 m or more for periods of between 4 and 8 hours.

Bottlenose dolphins have dorsal fins that are frequently nicked and scarred. Using these marks, it is possible to identify individuals and to recognize them over long periods of time (Würsig and Würsig, 1977; Wells, 1986; Ballance, 1987). I used photographs of dorsal fins to record an individual's presence at a given time and location. All photographs were taken with an Olympus camera with a motor drive and data back, a 200-mm lens, and Kodachrome (K-64) color slide film. For these photo identifications, I took more than 7000 photographs.

Although the bulk of the data for this study consists of dorsal fin photographs, I also monitored group size, behavior, movements, and ecological and environmental parameters. These latter data are analyzed elsewhere (Ballance, 1987).

The following terms are used throughout this paper. "Group" refers to those dolphins sighted in Kino Bay at any one time. "Subgroup" refers to those individuals which were sighted in the same group on repeated occasions. "Complete sighting group" refers to groups in which all members were photographed and identified. These definitions make no distinctions with respect to spatial associations between individuals. While groups of dolphins were always recognizable as units, spatial associations between individuals in a group were not consistent. At some times, group members were scattered within an area of a few kilometers; at others all group members were located within a few body lengths of each other. The distinction between these three terms lies with respect to time or to the proportion of members which were identified.

The terms "resident" and "nonresident" are also used frequently. Nonresident dolphins were sighted in Kino Bay only one time. Resident dolphins were sighted more than once, and the term is used in a relative sense, that is, one animal can be said to exhibit a higher or lower degree of residence than another.

In analyses of group composition and stability, only complete sighting groups were included (Würsig and Würsig, 1977; Würsig, 1978). A group was considered to be a complete sighting group if every identifiable individual was represented by a minimum of four identifiable photographs (Fig. 2).

Figure 2 was constructed using photographs of individuals from groups of known size. All of the photographs of a particular group were randomized, and the identity of the individuals in each photograph was determined. Sequential photographs represented either individuals not yet identified or individuals who had been identified in previous photographs. At some point during the analysis, each individual so far identified was represented by two photographs. The number of identified individuals divided by the total known group size represented the proportion of all group members that were identified. Analysis continued until each individual was represented by three photographs, then four photographs, then five photographs. A larger number of photographs for a given individual in a group corresponded to a higher probability that all members in the group were photographed and identified. At four photographs of each member, there was a 95% chance that all members of a group had been photographed and identified. This value, a minimum of four identifiable photographs per individual, was

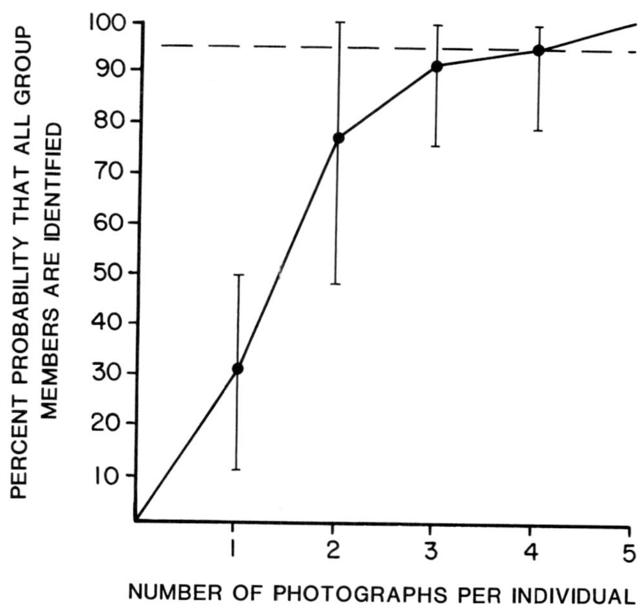

Figure 2 Probability that all members of a group are identified as a function of the number of photographs per individual in each group. Points represent mean values, vertical lines standard deviations. The dashed line illustrates 95% probability that all group members are identified (n = 10).

independently derived by Würsig (1978) and also was used to define a group in which all individual members had been identified.

To analyze associations between individuals during surfacing, 875 photographs of dolphins from the complete sighting groups were analyzed. In each of these photographs, from one to as many as six individuals were visible, and all were identifiable. Photographs of mother/calf pairs were not used.

A surfacing was defined as any portion of an animal's body breaking the surface of the water, and an association was scored between a pair or group of individuals when they surfaced at the same time. This definition of associations at surfacing implies no distance relationship. However, because photographs were taken with a 200-mm lens at an *f*-stop of 8 or 5.6 (shutter speed 1/500 to 1/1000 second), animals separated laterally by as little as 2 m were not both in focus; both were not identifiable, and such photographs were not used in the analysis. Therefore, surfacing associations involved animals within 2 m of each other.

A cluster analysis was used to quantify group membership stability patterns and surfacing associations (Anderberg, 1973; DeGhett, 1978). I used the Jaccard's association coefficient to compute a similarity matrix quantifying the number of times each individual was sighted in the same group or the number of times each individual surfaced together. The "unweighted pair group arithmetic average clustering algorithm" was used to construct clusters from data in the similarity matrix (Sokal and Michener, 1958). These clusters illustrate which animals were sighted in the same group or which individuals surfaced together most frequently and how frequently both of these types of associations occurred.

The ultimate results of a cluster analysis can be used as visual tools, but not to indicate significance of associations. In order to determine the possible significance of surfacing associations between pairs of animals the following procedure was used. Assuming that individuals surface together in a random pattern, the expected number of photographs containing both animal A and animal B (x) will be the probability of photographing animal A with animal B (p) multiplied by the total number of photographs of the group:

$$x = (1/\text{group size})(1/\text{group size})(\text{total no. photos of group})$$

$$x = p(\text{total no. photos of group})$$

The binomial distribution states that, assuming a random surfacing association between individuals, the probability of a sample containing x photographs of animal A with animal B, P, is:

$$P = \frac{(\text{no. photos of group})!\,(p)^x(1-p)^{(\text{no. photos of group}\,-\,x)}}{x!\,(\text{no. photos of group}-x)!}$$

When P is less than .05, the surfacing association between animal A and animal B is significantly different from random.

RESULTS

Occurrence Patterns and Residence

Most of the dolphins in Kino Bay were identified by the end of the study (Fig. 3). A large number of individuals were identified over a short period of time during the first few days of the study, but by mid-June most of the animals sighted in the bay had been previously identified. There was a second increase in the number of previously unidentified individuals during October, but by November most of the photographed dolphins in Kino Bay were reidentifications. One hundred fifty-five individuals were identified in Kino Bay of which 61 (39%) were sighted only once and 94 (61%) were resighted one or more times (Fig. 4). One animal was sighted on 11 separate occasions.

The time interval between first sighting and last resighting of any one dolphin ranged from 2 to 195 days (Fig. 5). Excluding resightings made within the same day, the average time between first and last sighting was 80 days (s = 75.3 days). The distribution of days between resightings is a multimodal one, with the span of 50–90 days not represented by a single animal. The average number of days between adjacent sightings for an individual was 30 days (s = 63.1 days, range 2–81 days; Fig. 6). There is a peak in the number of animals seen once every 1–4 days, and the distribution of values beyond 4 is slightly clustered toward low time spans.

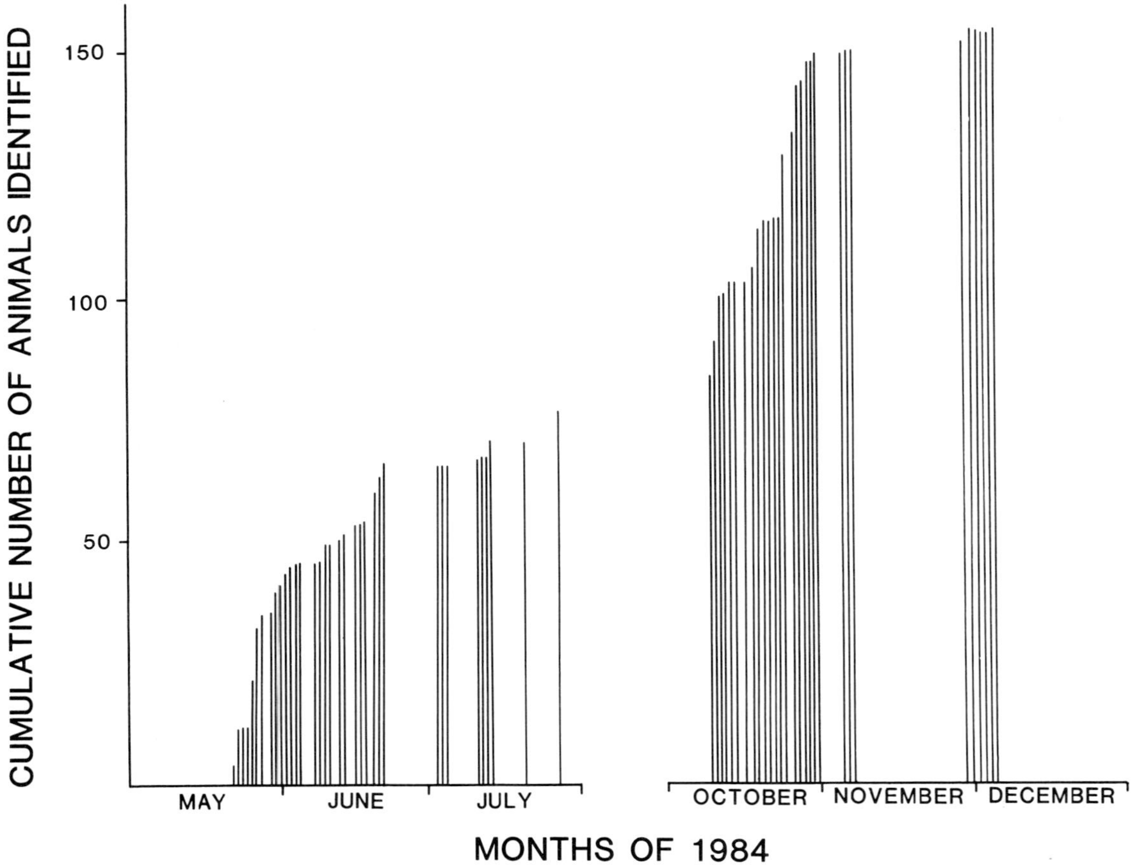

Figure 3 Cumulative number of animals in Kino Bay identified over time.

In summary, the variation in number of resightings of each dolphin, the multimodal distribution of time spans between the first and last sightings of each dolphin, and the broad distribution of average number of days between adjacent sightings indicate that dolphins exhibited varying degrees of residence to Kino Bay.

Group Membership and Stability Over Time

Sixty different groups were observed in Kino Bay. These groups contained an average of 15 individual dolphins (s = 23.6 animals), and sizes ranged from a single individual to groups of 125 animals. Almost half of the animals sighted only once visited Kino Bay in large groups containing more than 40 individuals (Fig. 4). These large groups also contained some of the 94 animals which were sighted in Kino Bay at a later date. Therefore, large groups appear to be composed of animals with varying degrees of residence to this bay.

Of the 60 groups observed during the study, 10 were complete sighting groups (i.e., all individual members were photographed and identified). These contained an average of 10 individuals (s = 6.2) and were comprised of 52 of the 155 individuals identified in Kino Bay (Fig. 7A).

Individuals could be classified into one of three patterns based on group membership and stability

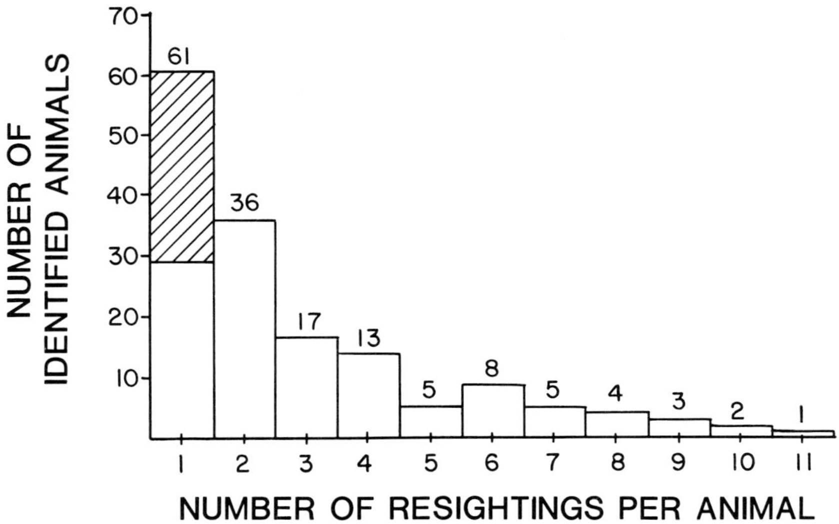

Figure 4 Number of resightings of dolphins in Kino Bay ($\bar{x}$ = 14 resightings per animal, s = 18.5, n = 155 animals). Hatching indicates percentage of animals sighted only once which occurred in groups larger than 40 individuals.

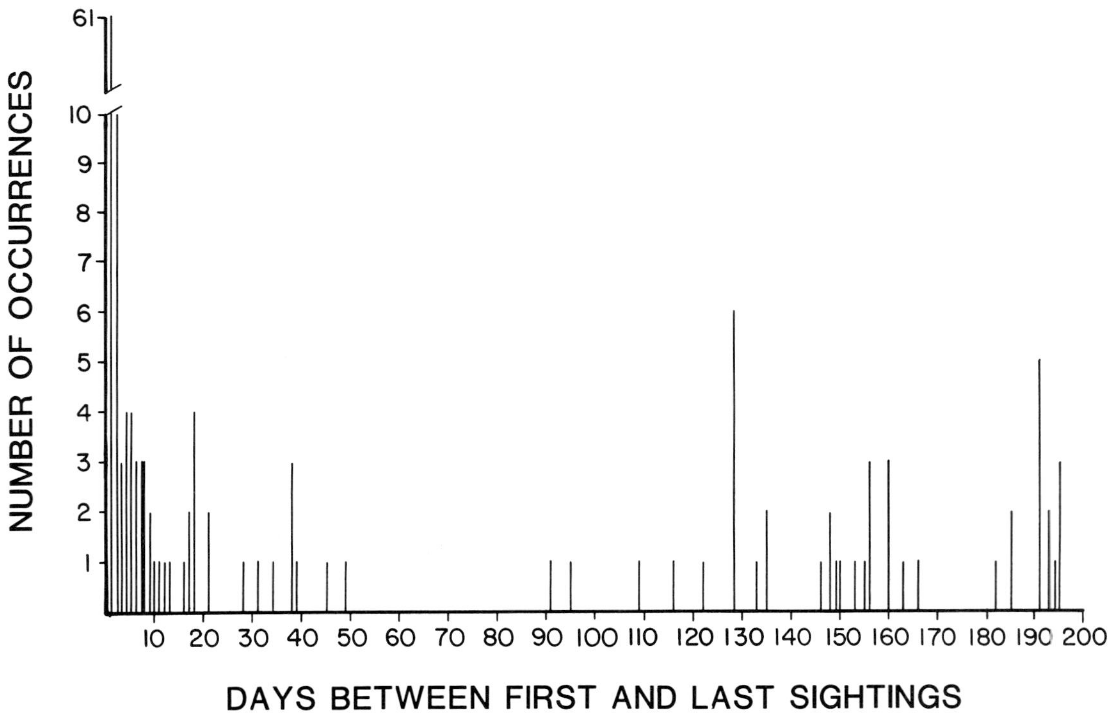

Figure 5 Frequency distribution of the time span between first and last sightings of each dolphin identified in Kino Bay ($\bar{x}$ = 79 days, s = 75.3, n = 155 animals).

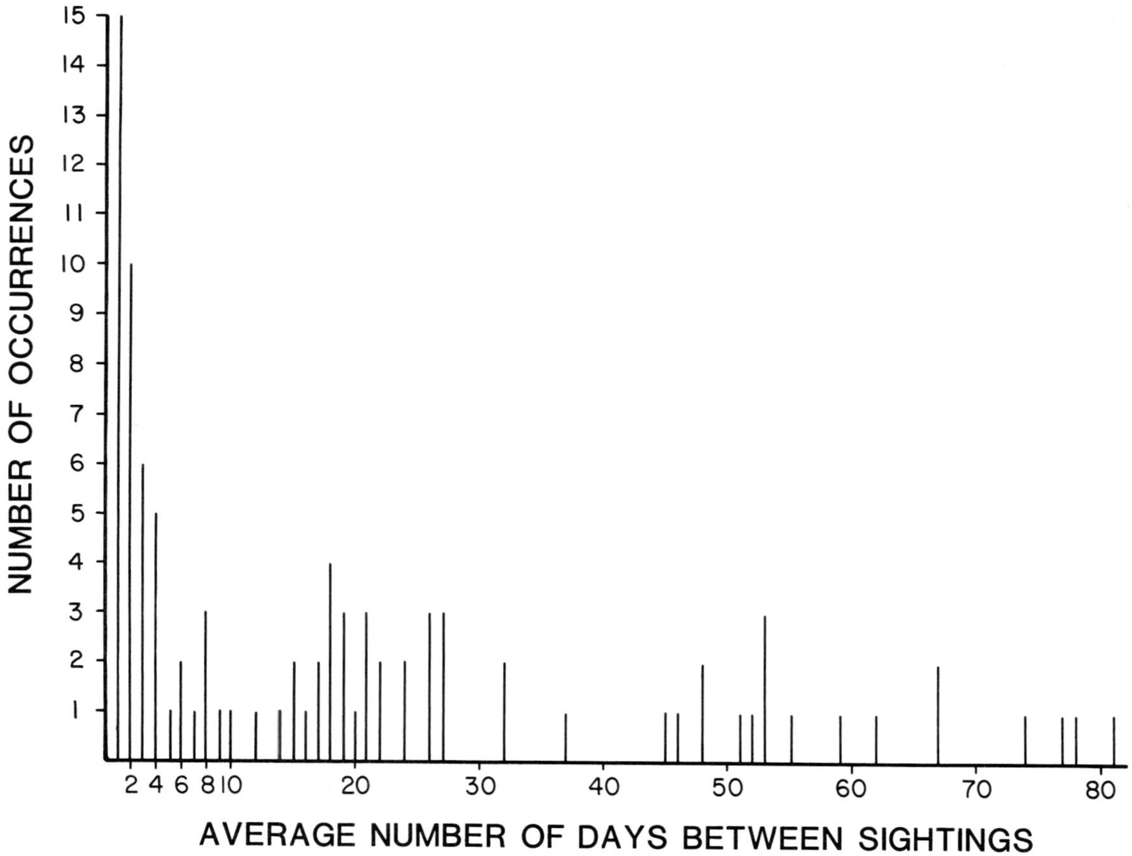

Figure 6 Frequency distribution of the average number of days between sightings for each dolphin identified in Kino Bay ($\bar{x}$ = 29 days, s = 63.1, n = 94 animals). Time between first and last sightings, was divided by the total number of sightings of each individual.

of associations. Some animals were members of groups that remained together over time. Some animals were present on only one day, although on this day they associated with individuals that were sighted on many different days. Some animals were present throughout the study period but could not be identified as members of a stable subgroup.

Eighteen of the 52 individuals were members of five stable subgroups that remained together over time (Fig. 7B). These subgroups were only a part of a larger group on any given day. However, all members of these subgroups were always seen together; when one individual was present, each of the members of a particular subgroup was present.

Many of these 18 individuals were also sighted in groups that were not complete sighting groups (Fig. 8). In these groups, some individuals of each subgroup appeared on various days without the other subgroup members. However, the general trend was for all subgroup members to appear together. For example, animals 73 and 99 were seen together on all but 3 days, 9, 19, and 23 October. Animals 7 and 92 were seen together on all but two days, 11 October and 30 November. Groups sighted on these days were not complete sighting groups (i.e., all members of the group were not photographed). This means that the other member of the pair may have been present even though a photograph of that individual was not obtained on that particular day.

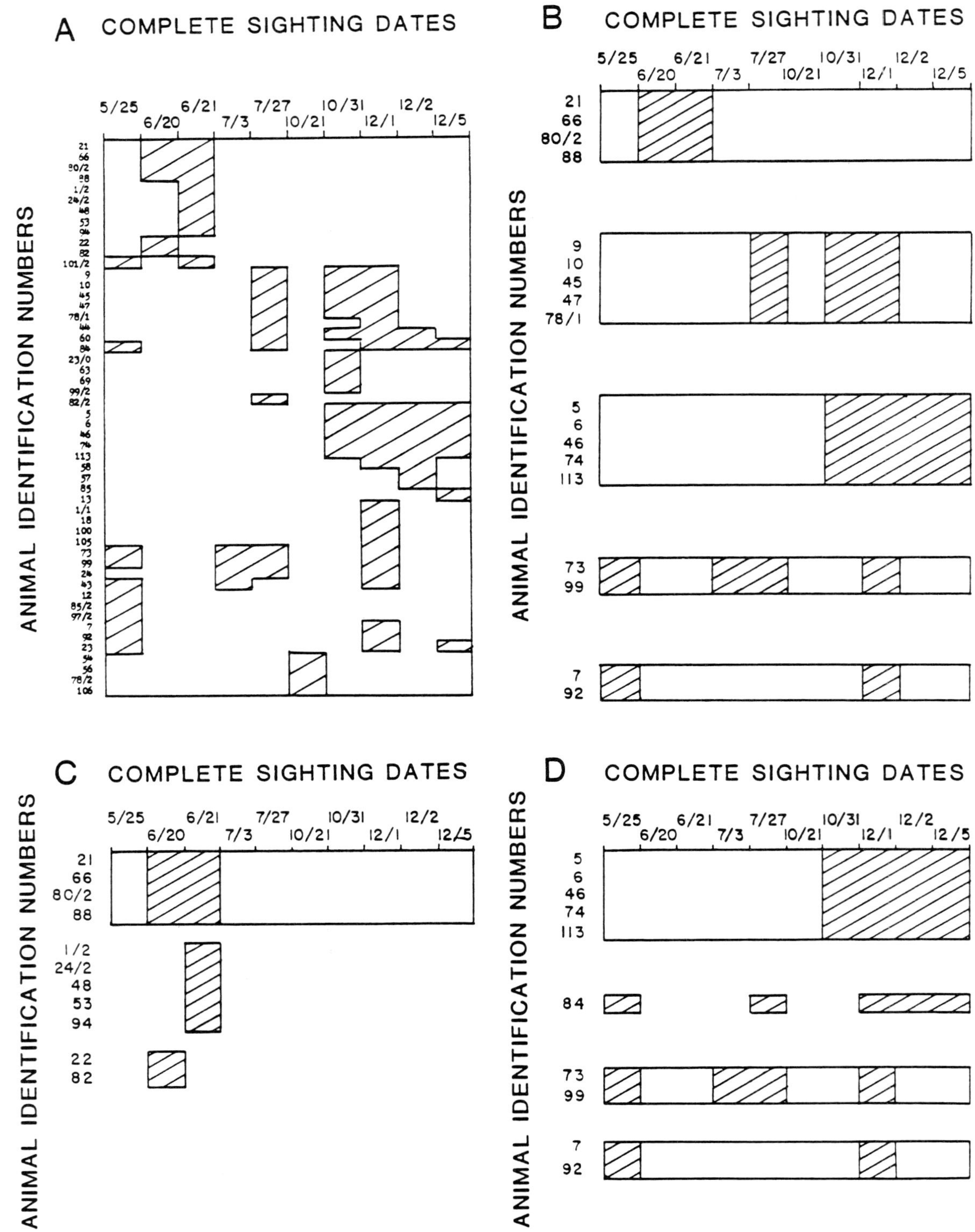

Figure 7 Group membership patterns of dolphins sighted in 10 complete sighting groups in Kino Bay. Hatching indicates presence of a dolphin on a particular day. (A) Occurrence of each dolphin sighted on the 10 days. (B) The first membership pattern is illustrated by 18 dolphins that were members of five different subgroups. Members of each subgroup remained together through time. (C) The second group membership pattern is illustrated by the bottom 7

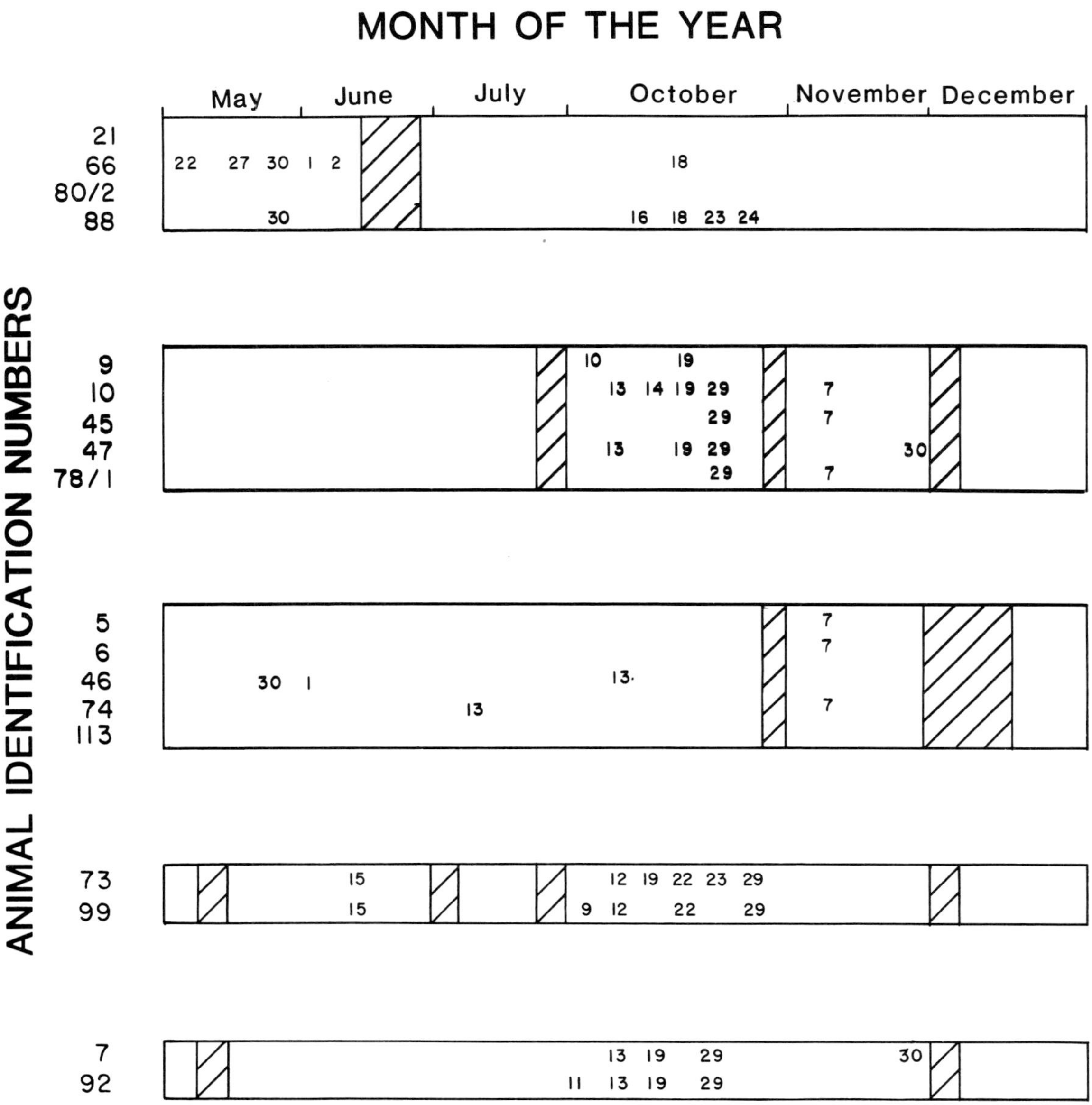

Figure 8 Dates of sightings of the 18 dolphins who were members of subgroups that were stable over time. Numbers indicate presence during that date. Hatching indicates presence during complete sighting days.

---

animals who were present on one complete sighting day only, in association with a particular stable subgroup. (D) The third membership pattern is illustrated by dolphin 84, who was present on many of the 10 days. This dolphin was not a member of a stable subgroup, but it associated with several of these subgroups over time.

A second pattern of group membership was exhibited by those individuals who were present on only a single complete sighting day, in association with only one stable subgroup (Fig. 7C). For example, two individuals, 22 and 82, were present with the stable subgroup shown in Fig. 7C, on 20 June. But these individuals were not present 21 June. On this day, five other individuals, not present on the 20th, were seen associating with this stable subgroup. Almost half of the 52 individuals seen in the 10 complete sighting groups exhibited this pattern of group membership.

A third pattern of group membership and association stability was exhibited by those individuals present throughout the study period but not members of a particular stable subgroup themselves. Instead, they associated with several different stable subgroups on different days (Fig. 7D). For example, dolphin 84 was sighted throughout the study in association with three different stable subgroups on different dates, although this animal itself was not a member of a stable subgroup.

Associations at Surfacing

Analysis of surfacing associations among individuals reveals three major trends (Fig. 9). First, the 52 individuals identified in the 10 complete sighting groups can be described in terms of nine clusters (22 and 82, 24 through 73, 48 through 94, 9 through 1/2, 101/2 through 84, 10 through 63, 57 through 58, 7 through 44, and 56 through 54; Fig. 9). This simply means that the members of each of these clusters associate when surfacing. In addition, there are 6 individuals, shown at the bottom of Fig. 9, that do not associate with any other individuals; they were always photographed alone.

Second, the surfacing associations between any pair or group of individuals are described by extremely low levels of similarity (Fig. 9). The highest level at which two individuals are joined is a value of 50 (22 and 82, 24 and 100, 85/2 and 97/2; Fig. 9); however, most pairs do not associate above a level of 20. This means that surfacing associations between individuals are rather weak. Finally, surfacing associations among groups comprised of more than three were described by similarity values of five or lower (Fig. 9). This indicates that the meaningful units, if there are any with respect to surfacing associations, are pairs or trios but not larger groups.

Of the 16 pairs of animals which surfaced together with the greatest frequency, only 7 pairs had surfacing associations that were significantly different from random surfacing patterns among all individuals within a group (Table 1). Most interestingly, none of these pairs with significant surfacing relationships were members of the stable subgroups described in the previous section. The significance of surfacing associations between members of these stable subgroups is mixed (Table 1). Only 5 of the 16 pairs analyzed were members of the same stable subgroup. One pair had no significant surfacing association. Four pairs had significant relationships on some days and no significant relationships on other days.

In summary, there are associations that occur between pairs and trios of animals when surfacing. But these relationships are weak and rarely significant. When significant associations do occur, they are found between individuals who come together in a group on one particular day and then separate, rather than between individuals who remain together in a group over an extended period of time.

DISCUSSION

The cumulative number of identified individuals leveled with time. Therefore, most of the dolphins using Kino Bay during the study period were identified, and the interactions recorded were representative of the social structure of the dolphins in this bay during this time. This is important because it applies to conclusions about occurrence patterns and residence, group membership, and surfacing associations.

Occurrence Patterns and Residence

Degree of residence was described in terms of three parameters: the number of times an animal was sighted in Kino Bay, the time between the first sighting and last resighting of an animal, and the

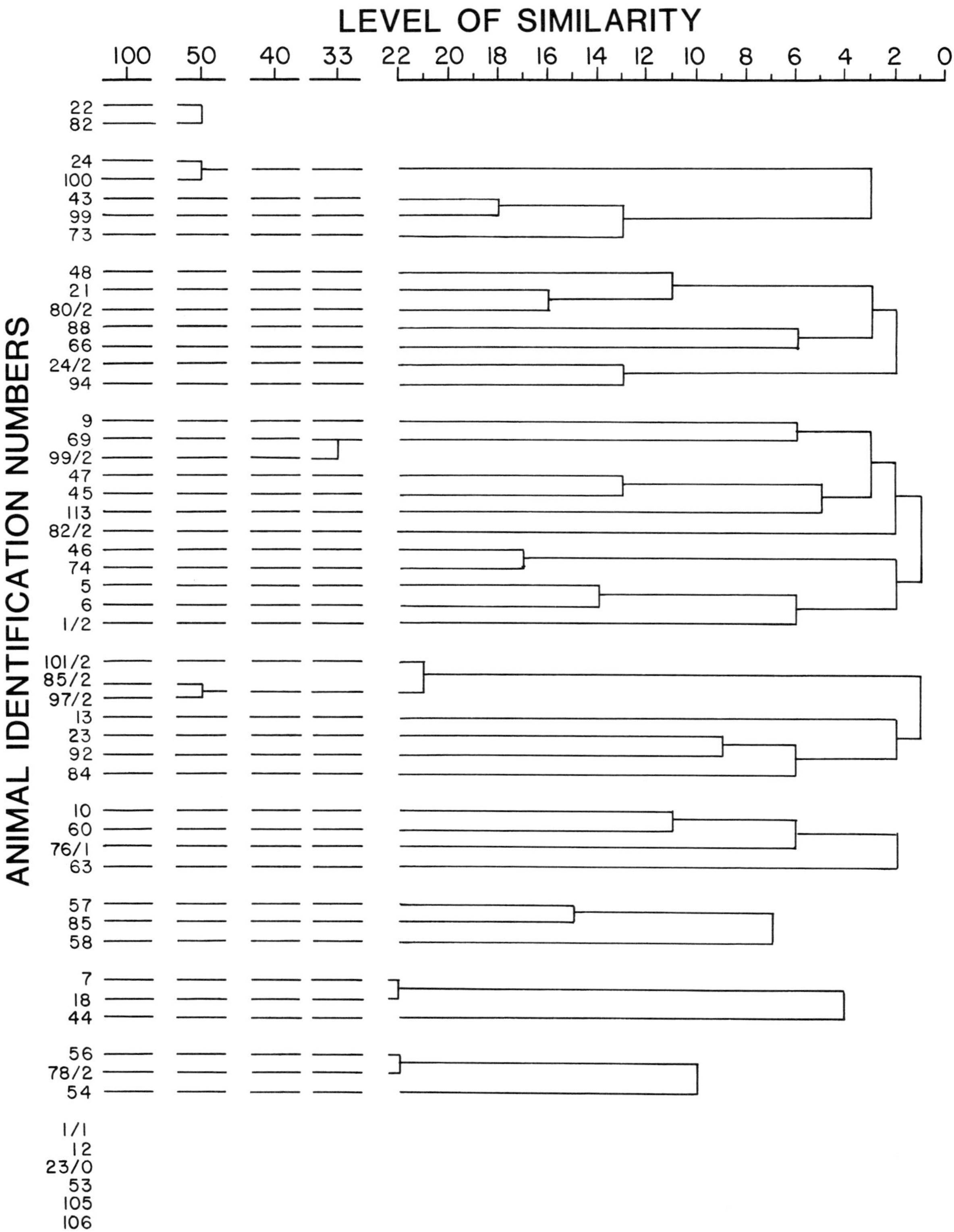

Figure 9 Associations at surfacing among 52 dolphins sighted on 10 complete sighting days.

Table 1
Significance of Associations at Surfacing between 16 Pairs of Dolphins in Kino Bay

Pairs of animals	Date sighted (1984)	P^a
22–82	6/20	.02*
24–100	12/1	<.001**
56–78/2	10/21	<.001**
24/2–94	6/21	.04*
69–99/2	10/31	<.001**
85/2–97/2	5/25	.005*
7–18	12/1	<.001**
88–66[b]	6/20	.37
	6/21	.24
57–85	12/2	.11
43–99	5/25	.24
	7/3	.21
	12/1	<.001**
21–80/2[b]	6/20	.08
	6/21	.04*
47–45[b]	7/27	.72
	10/31	.28
	12/1	.004*
46–74[b]	10/31	<.001**
	12/1	.004*
	12/2	.51
	12/5	.37
5–6[b]	10/31	.06
	12/1	.02*
	12/2	<.001**
	12/5	.37
23–92	5/25	<.001**
	12/1	.72
10–60	7/27	.72
	10/31	.72
	12/1	<.001**

[a] P is the probability that the observed surfacing association can be accounted for by random surfacing associations between individuals within the group (* indicates a probability less than .05; ** indicates a probability less than .001).
[b] Pair who are members of a stable subgroup.

average number of days between sightings for a particular animal. A high number of resightings, a long period of time between the first sighting and the last resighting, and a short time interval between adjacent sightings describes an animal with a high degree of residence.

Useful as they are, each measure of residence must be intrepreted with caution. Dolphins were frequently sighted on adjacent days. An animal which visited Kino Bay only a few times on adjacent days would have a high number of resightings yet a relatively low degree of residence. Alternatively, an animal may have been sighted over an extended period, up to 150 days, but may have been sighted only rarely during that period. This would indicate a high degree of residence for an animal which had visited Kino Bay only a few times. The average number of days between sightings takes into account both the number of times an animal was sighted and the time between first and last sightings. Yet this value might obscure residence patterns if an animal visited Kino Bay on only a few occasions spaced widely apart in time.

In summary, each parameter has biases. When results from each type of analysis are combined, however, one conclusion is certain: dolphins sighted in Kino Bay exhibited all degrees of residence. Most were sighted only once, yet some were sighted many more times. Many dolphins were sighted over a period of only a few days, yet many were sighted over periods spanning 150 days and longer. Some animals visited Kino Bay infrequently, only once each 2 months or so, yet some were sighted much more frequently. Therefore, this society appears to be relatively open, at least with respect to residence in this bay.

Group Membership and Stability Over Time

Dolphins in Kino Bay exhibited three patterns of group membership. Some were members of stable subgroups, some were seen only once in association with a single stable subgroup, and some were present throughout the study period but were not members of stable subgroups. Each group of animals was composed of individuals from two or all three of the categories; therefore, groups were not homogeneous with respect to membership patterns.

Associations at Surfacing

Mothers and calves have strong social bonds, and this is exemplified, in part, by their associations when surfacing (Würsig, 1978). Results from this study indicate, however, that bonds between individuals other than mothers and calves are rarely expressed as surfacing associations.

Würsig (1978) analyzed surfacing relationships between individual bottlenose dolphins on two separate days in Golfo San Jose, Argentina. Results were remarkably similar to those of the present study. Surfacing associations did occur between individual dolphins, but, aside from mother/calf pairs, these associations were never of greater significance than the $p < .25$ level. Further, these surfacing associations did not occur between animals who were members of the same stable subgroup.

While significant surfacing associations are rare, they do occur between some individuals at certain times and may represent temporary interactions that are quite strong for these short periods of time. However, the fact that these relationships are not consistent, even between these individuals, indicates that surfacing associations may not be reliable measures of social bonds. In any case, these are most likely not the same kind of relationships that hold individuals together over long periods of time, as the associations do not appear to exist between members of stable subgroups.

Comparison of Social Structure of Bottlenose Dolphins between Locations

The group organization of free-ranging bottlenose dolphins has been studied in two additional areas in sufficient detail to provide comparisons to the Kino Bay study. Würsig (1978) and Würsig and Würsig (1977, 1979) conducted a study in Golfo San Jose, Argentina. A total of 53 individuals were identified over a 21-month period, although the total population size was not known. These animals moved in groups averaging 14.9 individuals. Groups were described as dynamic units, continually changing in size and membership. However, there were some individuals that were repeatedly sighted together throughout the study. One of these subgroups was present in the study area the entire 21 months, while the others appeared and moved out of the area at different times of the year. In addition to these animals, some individuals were not members of a stable subgroup, but were present during a large part of the 21-month study. Finally, some individuals were seen infrequently, only once or a few times throughout the entire 21 months.

Irvine, Wells, Scott, and colleagues (Irvine and Wells, 1972; Wells, 1978, 1986; Wells *et al.*, 1980; Irvine *et al.*, 1981; Scott *et al.*, Chapter 11, this volume) conducted a study along the central west coast of Florida. The community in this area consisted of approximately 100 dolphins inhabiting a range of approximately 85 km^2. Group size averaged 4.8 animals during 1975–1976, and 7.0 animals during 1980–1984. As in Würsig's study, groups were described as dynamic with composition changing frequently, although there were some long-term associations between certain individuals. In addition, schools were segregated according to age, sex, reproductive condition, and familial relationships, and these groups utilized different areas of the home range preferentially. Notably, there was little immigration or emigration in the study area. Within the community home range, sightings of previously unidentified animals were rare, and most of the animals in the community were year-round residents. A number of individuals have been observed in the same waters throughout the study, during 1970–1987 (Wells, pers. commun.). The average number of days between sightings varied with age and sex of an individual but ranged between 3 and 6 days only.

The above summaries illustrate some interesting differences between the bottlenose dolphins of Kino Bay, Golfo San Jose, and Sarasota Bay and surrounding waters. Specifically, the number of animals using the area and the number of individuals per group were quite different. The number of animals identified in Kino Bay was one and one-half times larger than the Sarasota community. Group sizes were similar for Kino Bay and Golfo San Jose, but smaller for the Sarasota community.

Data from several other studies of bottlenose dolphins indicate that estimated numbers of animals in a particular area vary both between and within locations. Gruber (1981) found the number of dolphins in a 75-km^2 area off the coast of Texas to vary between 30 in the spring and 98 in the winter. Shane (1977) reported a low of 48.0 animals in the fall and a high of 164.2 animals in the winter for a much smaller 11.3-km^2 area 95 km to the southwest of Gruber's study area. Hansen (Chapter 23, this volume) estimated the population size for bottlenose dolphins off the San Diego coast as 173–240 individuals. These estimates are naturally dependent on the size of the area being considered, the method of estimation, and the time of year in which the estimations are made. Therefore, comparisons must be made with caution. Nevertheless, it appears that the number of dolphins using an area at a particular time varies with location.

Group sizes of bottlenose dolphins vary considerably between locations as well (see reviews by Wells *et al.*, 1980; Leatherwood and Reeves, 1982; Shane *et al.*, 1986). The number of individuals per group ranges from 1 to over 100 but usually averages between 2 and 15. However, off the southeastern coast of Africa groups of bottlenose dolphins average 140.3 animals, and groups of as many as 1000 individuals have been recorded (Saayman and Tayler, 1973). In addition, group size within a single location varies with distance from shore and water depth (Würsig, 1979).

Residence patterns were distinctly different between bottlenose dolphins in Kino Bay, Golfo San Jose, and Sarasota and surrounding waters. Dolphins in Kino Bay exhibited many different degrees of residence. Some were present often, others infrequently, and a large percentage of the total animals was sighted only once. Such nonresident animals were occasionally sighted in Golfo San Jose, but previously unidentified individuals were almost never recorded within the Sarasota community. (Previously unidentified animals were recorded, however, around the periphery of the community home range.)

The Sarasota area is much larger than Kino Bay or the study area within Golfo San Jose, and changes in dolphin distribution were recorded during different parts of the year (Irvine *et al.*, 1981). Had observations been made from a single location within the Sarasota area, the changes in distribution would have been recorded as different patterns of residence; this difference in scale of study area may contribute to the apparent differences between the systems. However, there were groups of dolphins that could reliably be sighted in quite localized areas within the Sarasota community (Wells, 1986). Additionally, it is possible that some of the dolphins sighted only once or a few times in Kino Bay would have been observed additional times had the study continued for a few more months. Nevertheless, even considering these caveats, it appears that residence patterns are different between dolphins of Kino Bay and Golfo San Jose and those of the Sarasota area.

Nonresident dolphins also appear to be a major component of animals sighted off the coast of southern California and northern Baja California (Defran *et al.*, 1985). Of 245 individuals identified over a period of 19 months, only 63 were sighted more than one time.

Habitat characteristics of Sarasota and vicinity, Golfo San Jose, and Kino Bay are different. The west coast of Florida is lined with a series of barrier islands that protect shallow bays and channels from open waters of the Gulf of Mexico. The Sarasota community home range opens by narrow channels to a larger bay to the north and to the gulf to the west. Boundaries are defined at the surface by geographical features. Depths range from less than 2 to 5 m except in the channels where water can be up to 11 m deep (Wells, 1978, 1986). Golfo San Jose is a large gulf which covers 750 km^2. The coastline is open and generally unprotected from winds and storms. Depths range from less than 1 to 35 m at increasing distances from shore (Würsig and Würsig, 1979). Kino Bay is a broad bay located along an open coastline with little protection from gulf winds and storms (Fig. 1). Depths within the bay are shallow, gradually increasing to 12 m where the bay meets the gulf. In summary, these habitats can be classified into two broad categories: closed systems with protected waters and surface boundaries in the form of geographical features, and open systems, unprotected from winds and currents and defined at the surface by few geographical features.

When habitat characteristics of Sarasota, Golfo San Jose, and Kino Bay are considered with the population parameters of the bottlenose dolphin communities, a pattern emerges. Smaller numbers of individuals, each exhibiting a fairly high degree of residence, are characteristic of the Sarasota community, a closed habitat. Larger numbers of individuals, many of whom are not resident to a localized area, are characteristic of the dolphins in Kino Bay and Golfo San Jose, both open habitats. This pattern was also suggested by Defran *et al.* (1985) and seems to apply to the system off southern California, an open habitat with few surface geographical boundaries.

Why do we see these patterns? Habitats with different physical characteristics undoubtedly vary with respect to biological characteristics as well; predation pressure and prey availability are two that have been previously suggested (Würsig, 1978; Norris and Dohl, 1980a; Wells *et al.*, 1980). Norris and Dohl (1980a) have proposed that variation in predation pressure may result in variation in group sizes. Several researchers (Saayman and Tayler, 1973, 1979; Würsig, 1978) have suggested that abundance and distribution of food sources may be related to the density or number of consumers using an area. And group size, home range, and foraging strategies of ungulates (Jarman, 1974) and primates (Clutton-Brock and Harvey, 1977) are related to characteristics of the habitat. Unfortunately, I have no data on predation pressure or distribution and abundance of food resources in Kino Bay. Thus, I can only speculate how, or if, these factors influence population parameters of Kino Bay dolphins.

In addition to the differences among communities of bottlenose dolphins inhabiting different areas, there are also some striking similarities. Group composition in the Sarasota area, Golfo San Jose, and Kino Bay can be described as dynamic, with individuals changing groups and sizes varying daily. However, in all three systems there are stable subgroups of individuals who remain together over long periods of time. In all three areas, there are also resident individuals who are not members of such stable subgroups but who associate with many different subgroups over time in a fluid manner.

Subgroups remain stable not only within a single location, but apparently between locations as well. Two animals identified off the coast of Sonora in locations separated by 25 km were seen together in both of these locations on dates 3 months apart. Two other animals sighted together on 25 July were sighted 3 months later 65 km to the north, one dolphin on 26 October, the other on 29 October (Ballance, 1987).

Evidence indicates that the social structure of bottlenose dolphins is similar in other locations as well. Shane (1977) reported associations between 20 identifiable individuals in Port Aransas, Texas, to be "extensive and quite flexible," with group composition varying frequently. She also noted that relationships between certain individuals were of a relatively permanent nature. Gruber (1981) concluded that group composition of bottlenose dolphins in the Pass Cavallo, Texas, area was "versatile and unpredictable, ranging from temporary, changeable bonds and encounters to long term, and enduring relationships among certain individuals."

It is interesting to note the similarities in group organization between the bottlenose dolphin and other small cetaceans. Hawaiian spinner dolphins, *Stenella longirostris,* associate in large schools that exhibit great fluidity in membership, but small groups of 4–10 animals may have some longer-term continuity (Norris and Dohl, 1980b). Humpbacked dolphins, *Sousa* sp., travel and interact with many different companions in groups of unstable and variable composition (Saayman and Tayler, 1979). Dusky dolphins, *Lagenorhynchus obscurus,* feed in large aggregations, but small groups remain together for several days at least (Würsig and Würsig, 1980; Würsig, 1982).

Even more intriguing are the parallels in group organization between dolphins and certain species of primates (Tayler and Saayman, 1972; Würsig, 1978; Saayman and Tayler, 1979; Würsig and Würsig, 1980). Societies of chimpanzees, *Pan troglodytes,* in particular, have been described as loosely organized, with groups constantly changing membership in an apparently random fashion (Goodall, 1965; Reynolds and Reynolds, 1965; Nishida, 1968). This organization seems to be related to food availability (Reynolds and Reynolds, 1965; Nishida, 1968), and Würsig (1978) and Saayman and Tayler (1979) have suggested this correlation for bottlenose and humpbacked dolphins, respectively.

In summary, the number of dolphins in a particular area, the average group size, and the residence patterns of these dolphins with respect to a particular site appear to be flexible parameters that vary across several geographic locations. Group membership and stability appear to be somewhat less flexible, as they have been described by similar patterns in several studies across a wide range.

ACKNOWLEDGMENTS

I would like to express my deepest gratitude to Harry and Vilma Ballance, Don Croll, and Bernd Würsig for their field assistance, intellectual stimulation, and financial and logistical support. Without their help, this research would not have been possible.

My sincere thanks also go to Lloyd Findley and Omar Vidal for support and assistance with permits, Bernd Würsig, Randy Wells, Greg Cailliet, and two anonymous reviewers for comments on the manuscript, Steven Young for final manuscript preparation, Don Thomson and the Department of Ecology and Evolutionary Biology at the University of Arizona for figure preparation, members of the Club Deportivo, especially Eldon and Joanie Heaston, Bob and Frankie Jarrett, and Roy and Mary Francis Morwood, and to Gregorio Camacho Molina, for friendship, knowledge, and constant logistical and emotional support.

I received funding from the American Museum of Natural History, the Packard Foundation, and the American Cetacean Society. I am indebted to the Secretaria de Pesca of Mexico for allowing me to work in the Gulf of California under Permit Number 3717.

REFERENCES

Anderberg, M. R. (1973). "Cluster Analysis for Applications." Academic Press, New York. 359 pp.

Ballance, L. T. (1987). Ecology and behavior of the bottlenose dolphin, *Tursiops truncatus,* in the Gulf of California, Mexico. M.S. thesis, Moss Landing Marine Laboratories, Moss Landing, California.

Clutton-Brock, T. H., and Harvey, P. H. (1977). Primate ecology and social organization. *J. Zool. London* **183**(1), 1–39.

Defran, R. H., Shultz, G. M., Espinosa Ley, M. A., and Martinez, A. Ch. (1985). The population biology of *Tursiops truncatus* along the southern California and northern Baja coastline. *Sixth Biennial Conf. Biol. Marine Mamm.*, Vancouver, British Columbia, Abstr.

DeGhett, V. J. (1978). Hierarchical cluster analysis. *In* "Quantitative Ethology" (P. Colgan, ed.), pp. 115–144. Wiley, New York. 364 pp.

Goodall, J. (1965). Chimpanzees of the Gombe stream reserve. *In* "Primate Behavior" (I. DeVore, ed.), pp. 425–473. Holt, Rinehart, and Winston, New York. 654 pp.

Goodwin, D. E. (1985). Diurnal behavior patterns of *Tursiops truncatus* off Mobile Point, Alabama. M.A. thesis, San Francisco State University, San Francisco, California.

Gruber, J. A. (1981). Ecology of the Atlantic bottlenosed dolphin (*Tursiops truncatus*) in the Pass Cavallo area of Matagorda Bay, Texas. M.S. thesis, Texas A & M University, College Station, Texas.

Irvine, B., and Wells, R. S. (1972). Results of attempts to tag Atlantic bottlenosed dolphins (*Tursiops truncatus*). *Cetology* **13,** 1–5.

Irvine, A. B., Scott, M. D., and Wells, R. S. (1981). Movements and activities of the Atlantic bottlenose dolphin, *Tursiops truncatus,* near Sarasota, Florida. *Fish. Bull.* **79**(4), 671–688.

Jarman, P. J. (1974). The social organization of antelope in relation to their ecology. *Behaviour* **48**(3), 215–267.

Leatherwood, S., and Reeves, R. R. (1982). Bottlenose dolphin (*Tursiops truncatus*) and other toothed cetaceans. *In* "Wild Mammals of North America" (J. A. Chapman and G. A. Feldhamer, eds.), pp. 369–414. Johns Hopkins University Press, Baltimore, Maryland. 1147 pp.

Nishida, T. (1968). The social group of wild chimpanzees in the Mahali mountains. *Primates* **9**(3), 167–224.

Norris, K. S., and Dohl, T. P. (1980a). The structure and function of cetacean schools. *In* "Cetacean Behavior: Mechanisms and Functions" (L. M. Herman, ed.), pp. 211–261. Wiley, New York. 463 pp.

Norris, K. S., and Dohl, T. P. (1980b). Behavior of the Hawaiian spinner dolphin, *Stenella longirostris. Fish. Bull.* **77**(4), 821–849.

Reynolds, V., and Reynolds, F. (1965). Chimpanzees of the Bundongo forest. *In* "Primate Behavior" (I. DeVore, ed.), pp. 368–424. Holt, Rinehart, and Winston, New York. 654 pp.

Saayman, G. S., and Tayler, C. K. (1973). Social organization of inshore dolphins (*Tursiops aduncus* and *Sousa*) in the Indian Ocean. *J. Mammal.* **54**(4), 993–996.

Saayman, G. S., and Tayler, C. K. (1979). The socioecology of humpback dolphins (*Sousa* sp.). *In* "Behavior of Marine Animals" (H. E. Winn and B. L. Olla, ed.) Vol. 3: Cetaceans, pp. 165–226. Plenum, New York. 438 pp.

Shane, S. H. (1977). The population biology of the Atlantic bottlenose dolphin, *Tursiops truncatus,* in the Aransas Pass area of Texas. M.S. thesis, Texas A & M University, College Station, Texas.

Shane, S. H. (1980). Occurrence, movements, and distribution of bottlenose dolphins, *Tursiops truncatus,* in southern Texas. *Fish. Bull.* **78**(3), 593–601.

Shane, S. H., Wells, R.S., and Würsig, B. (1986). Ecology, behavior and social organization of the bottlenose dolphin: A review. *Mar. Mamm. Sci.* **2**(1), 34–63.

Sokal, R. R., and Michener, C. D. (1958). A statistical method for evaluating systematic relationships. *Univ. Kansas Sci. Bull.* **38**(22), 1409–1438.

Tayler, C. K., and Saayman, G. S. (1972). The social organization and behaviour of dolphins (*Tursiops aduncus*) and baboons (*Papio ursinus*): Some comparisons and assessments. *Ann. Cape Prov. Mus. Nat. Hist.* **9,** 11–49.

Wells, R. S. (1978). Home range characteristics and group composition of Atlantic bottlenosed dolphins, *Tursiops truncatus*, on the west coast of Florida. M.S. thesis, University of Florida, Gainesville, Florida.

Wells, R. S. (1986). Structural aspects of dolphin societies. Ph.D. thesis, University of California, Santa Cruz, California.

Wells, R. S., Irvine, A. B., and Scott, M. D. (1980). The social ecology of inshore odontocetes. *In* "Cetacean Behavior: Mechanisms and Functions" (L. M. Herman, ed.), pp. 263–317. Wiley, New York. 463 pp.

Würsig, B. (1978). Occurrence and group organization of Atlantic bottlenose porpoises (*Tursiops truncatus*) in an Argentine bay. *Biol. Bull.* **154**(2), 348–359.

Würsig, B. (1979). Dolphins. *Sci. Am.* **240**(3), 136–148.

Würsig, B. (1982). Radio tracking dusky porpoises in the South Atlantic. *In* "Mammals in the Seas," FAO Fisheries Series 5, Vol. IV, pp. 145–160.

Würsig, B., and Würsig, M. (1977). The photographic determination of group size, composition, and stability of coastal porpoises (*Tursiops truncatus*). *Science* **198**, 755–756.

Würsig, B., and Würsig, M. (1979). Behavior and ecology of the bottlenose dolphin, *Tursiops truncatus*, in the South Atlantic. *Fish. Bull.* **77**(2), 399–412.

Würsig, B., and Würsig, M. (1980). Behavior and ecology of the dusky dolphin, *Lagenorhynchus obscurus*, in the South Atlantic. *Fish. Bull.* **77**(4), 871–890.

Aspects of the Behavioral Ecology of Inshore Dolphins *Tursiops truncatus* and *Sousa chinensis* in Moreton Bay, Australia

14

Peter J. Corkeron
Department of Anatomy
University of Queensland
St. Lucia, Queensland, Australia 4067

INTRODUCTION

There have been a number of studies of inshore or nearshore populations of bottlenose dolphins, *Tursiops truncatus,* since about 1970. In only a few instances, however, has it been possible to compare their behavioral ecology with that of dolphins in the same area belonging to a different species.

Würsig and Würsig (1980) reported that bottlenose dolphins were found in shallower water than were dusky dolphins, *Lagenorhynchus obscurus,* off the coast of Argentina and that the two species did not mix regularly (although both interacted with other marine mammals). It was unclear whether bottlenose and dusky dolphins competed for food. Although both species engaged in feeding activity in the same area, they did so at different times of day.

Saayman and Tayler (1973) found that Indo-Pacific humpbacked dolphins, *Sousa chinensis,* off the southeastern coast of Africa occurred in smaller groups than did sympatric bottlenose dolphins. Humpbacked dolphins were found close to shore, while bottlenose dolphins inhabited both nearshore and offshore waters (Saayman *et al.*, 1972). Food preferences of the two species varied: humpbacked dolphins apparently took mostly reef fish while bottlenose dolphins fed on both reef-dwelling and pelagic fishes (Saayman and Tayler, 1979). The humpbacked and bottlenose dolphins interacted occasionally.

Bottlenose dolphins have been observed following trawlers and feeding on their by-catch in Moreton Bay and elsewhere in the world (Corkeron *et al.*, Chapter 17, this volume). However, the effect of this method of feeding on aspects of the dolphins' behavioral ecology has not been investigated. Humpbacked dolphins have not been reported engaging in this type of feeding.

Bottlenose dolphins and Indo-Pacific humpbacked dolphins are sympatric in Moreton Bay, Queensland, off the northeastern coast of Australia (Bryden, 1978). From May 1984 to February 1987 I studied the behavioral ecology of dolphins in Moreton Bay. In this chapter I compare both species' grouping patterns and relate them to habitat variables in Moreton Bay.

MATERIALS AND METHODS

Study Area

Moreton Bay is bounded by the Queensland coast on the west and by Moreton, North Stradbroke, and South Stradbroke islands on the east. Waters of the northern bay are hydrographically distinct from those in the southern section (Newell, 1971). This study was conducted in the portion of the bay delineated by the southern end of Peel Island to the south and to the north by a line running through Comboyuro Point at 260°T (27° 04′ S, 153° 02′ E to 27° 31′ S, 153° 26′ E) (Fig. 1). This area is approximately 51 km long, has a maximum width of approximately 33 km, and encompasses approximately 1315 km^2. It is a subset of the northern section of the bay as defined by Newell (1971), but extending the study to the north and to the south was considered impractical.

The study area included the portion of the bay used by commercial trawlers. Commercial otter trawling for penaeid prawns (shrimp) commenced

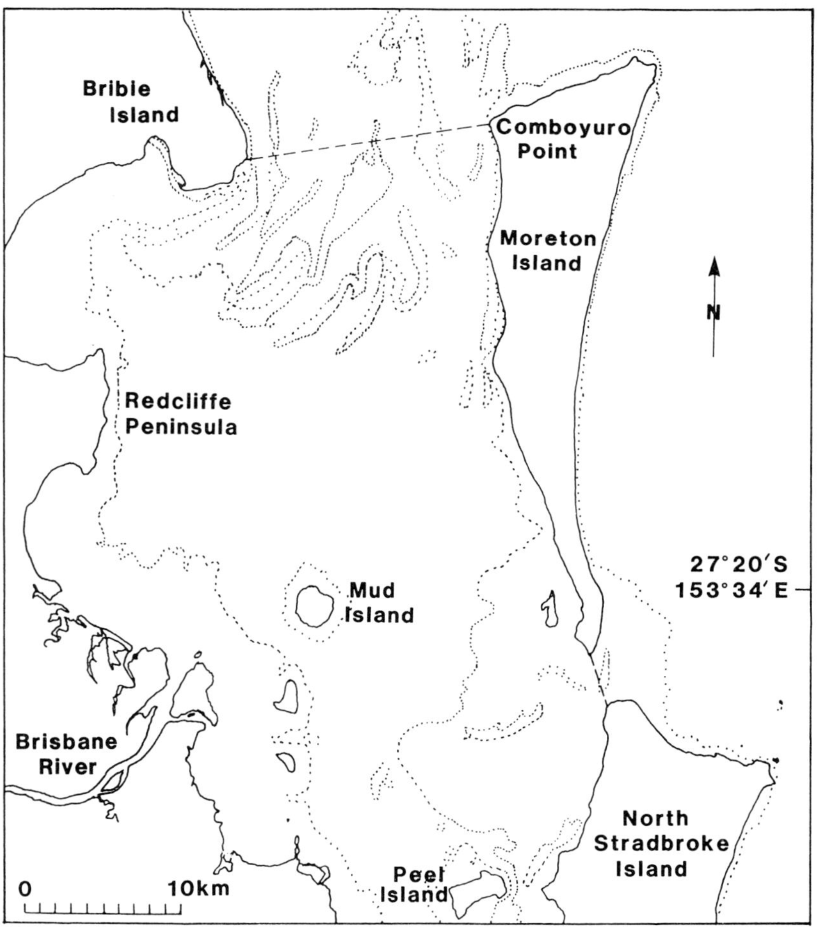

Figure 1 The Moreton Bay study area, Queensland, Australia. Dashed lines show 10-m isobath.

in Moreton Bay in 1950; in recent years the fishery has taken approximately 500,000 kg of prawns per year (Haysom, 1985). Trawling grounds extend throughout the bay north of Peel Island, in waters generally deeper than 5 m. Shallow sand banks prohibit fishing in the northeastern section.

Water temperatures in the bay range from 16° to 29°C at the surface and 17° to 26°C at 21 m, with maximum temperatures in January and minimum temperatures in August (Newell, 1971).

Dolphin Observations

Dolphins were observed from three different types of vessels, from May 1984 until February 1987. An average of four trips were made per month, with no seasonal variations. Most of the sightings of dolphins ($n = 238$) were made from the University of Queensland research trawler, a 16.2-m diesel-engined vessel. Observations were conducted following otter trawls of approximately 1 hour's duration. A few sightings ($n = 16$) were made from a 7.7-m sailboat. Whenever circumstances permitted (e.g., weather, crew availability), searches were made throughout the bay for dolphins and, on finding them, observations took place. The rest of the sightings were made from three speedboats, approximately 5 m long. Procedures for making observations were the same as those from the yacht.

Most sightings of bottlenose dolphins took place in a confined area west of Moreton Island (Fig. 2). Most of the work on the University trawler was carried out in this part of the bay, as large catches of fish could be made there regularly, and the water clarity was the best to be found in Moreton Bay. Owing to water clarity, this area was the best and safest place in the bay to conduct underwater observations of dolphins, in order to ascertain their sex and to carry out feeding studies.

Analyses of data were made by stratifying sightings of dolphin groups according to five classes of vessel and activity. These categories were (1) around the University trawler, (2) from the yacht following working trawlers (yacht–trawler), (3) from speedboats following working trawlers (speedboat–trawler), (4) from the yacht when not around working trawlers (yacht–other), and (5) from the speedboat not around working trawlers (speedboat–other). Groups of dolphins seen near trawlers (categories 1–3) were assumed to be feeding, and observations of dolphins ingesting fish reinforced this assumption (see Corkeron *et al.*, Chapter 17, this volume, for a discussion of bottlenose dolphins feeding around trawlers). Groups of dolphins observed in the latter two categories were not feeding on the large, easily located prey patches created by trawling. While these dolphins may have been feeding, they usually appeared to be engaged in other activities (e.g., travel, rest). There were insufficient sightings of dolphins not around trawlers to subdivide these categories according to activity.

Observations consisted of attempts to photograph animals, notation of the presence of known individuals, and recording of time and position. Observations were continued until I considered that good photographs were taken of all individual dolphins. The photographic equipment used throughout this project was a motor-driven Nikon FM2 with 80–200 Nikkor, 300-mm Nikkor, or Sigma 600-mm mirror lenses. Film used for most of the project was Kodachrome 64 and Ektachrome 200, although some Ektachrome 400 and 800 film was used also. Transparencies were viewed with a dissecting microscope at ×10 to ×25 power. Individual dolphins were identified by distinguishing natural marks (Würsig and Würsig, 1977) and were then catalogued.

Boat positions were determined using a hand compass (yacht, speedboat, and trawler) or by radar (trawler). Positions were plotted onto a hydrographic chart (Queensland Dept. of Harbours and Marine, Coochiemudlo to Mooloolaba, 1:100,000) that was divided into 1-km^2 grids, as grids of this size were within the limits of accuracy of the position-fixing methods used. From this chart, estimates of water depths (all at Mean Low Water Spring) and distance from shore were derived. Maximum and minimum depths (in meters) were taken from soundings or depth contours recorded in the grid, or from the sounding(s) or contour(s) nearest the grid. Depths were not recorded in the field as the yacht and speedboats used were not equipped with depth sounders. Distance from shore was recorded as that distance from the nearest point on shore (either mainland or island) to the nearest edge of the grid, and rounded to the nearest kilometer. Therefore, sightings in a grid that included parts of the shoreline have a distance value of zero.

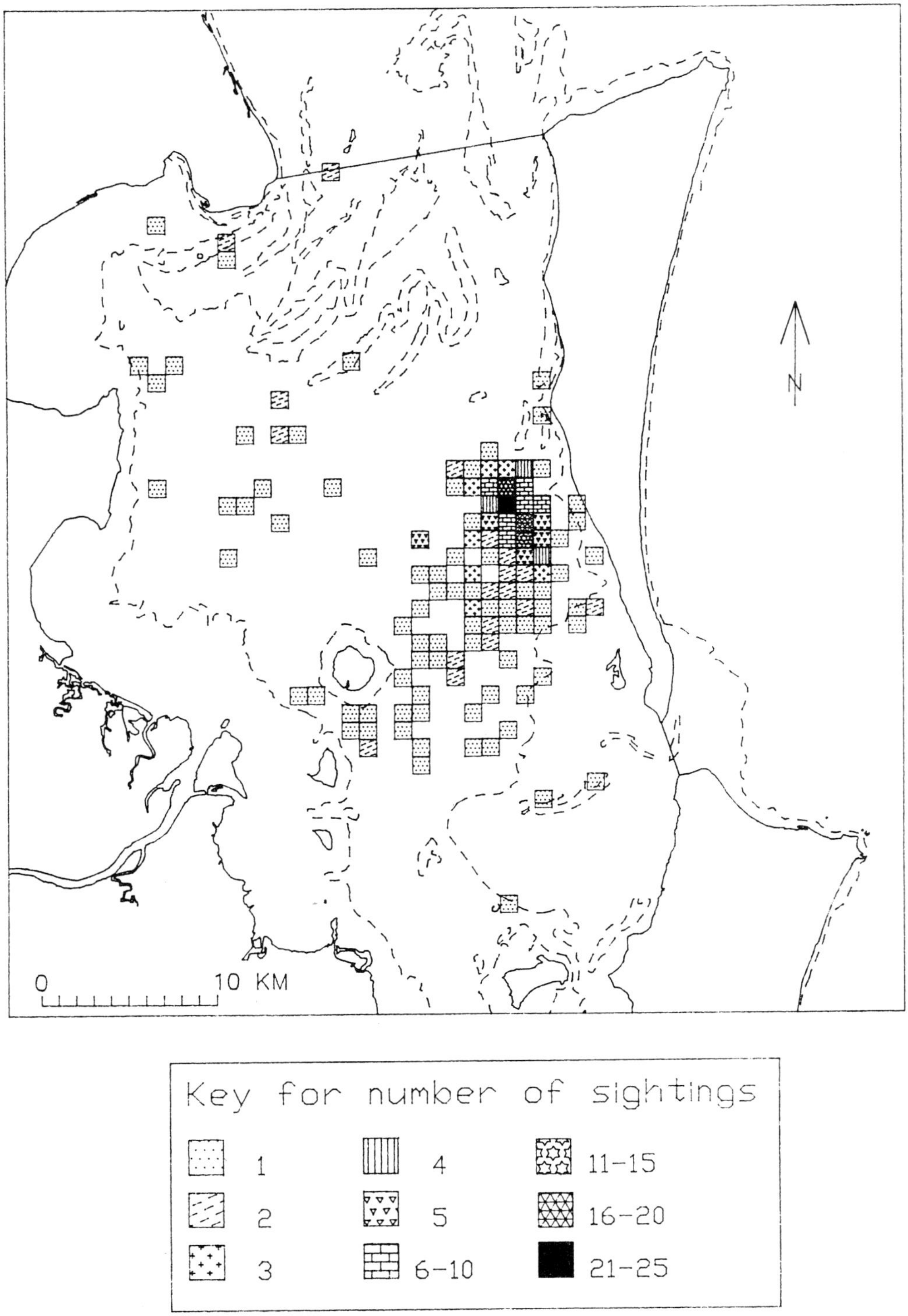

Figure 2 Bottlenose dolphin sightings in Moreton Bay.

In general, a "group" was defined as all animals seen within a radius of approximately 100 m (Wells *et al.*, 1980). For dolphins observed from the University trawler, a group was defined as all dolphins that came within this radius of the boat while a set of observations was underway (i.e., between hauling up the trawling gear and recommencing trawling). Dolphins could be sighted in "groups" several times in the same day.

During data collection, no attempt was made to count the number of dolphins in each group. Therefore, throughout this chapter, group size refers only to counts of those animals identified at a sighting. Hence, group size does not include counts of most calves and probably some juveniles that were present at sightings but were considered unidentifiable from natural marks. This means that group sizes reported here are not directly comparable with group sizes reported for other studies. Assessment of factors affecting group size within the study area can be attempted with greater accuracy than would be possible if estimates of numbers of unidentified animals were included in counts of each group. It could be suggested that the proportion of nonidentifiable dolphins in groups could vary with depth or distance from shore. Analysis of the ranging patterns of identified individual dolphins that would be expected to associate with such animals (e.g., adult females, identifiable juveniles) is being completed at present and demonstrates no such patterning.

The proportions of dolphins of each species that showed evidence of shark attack (see Corkeron *et al.*, 1987, for details) were compared to test the hypothesis that the two dolphin species suffered different rates of shark attack. All statistical tests are two-tailed, unless stated otherwise, and were performed using GLIM (Numerical Algorithms Group, 1985) and OZSTAT (Kennedy, 1985).

RESULTS

Three hundred thirty-four different bottlenose dolphins were identified, and some were resighted up to 91 times (median 3). Fifty humpbacked dolphins were identified, with as many as nine resightings (median 1). Overall, mean size of groups containing bottlenose dolphins only was 10.4 (S.D. = 7.00, n = 251). The overall mean size of groups of humpbacked dolphins only was 2.4 (S.D. = 1.13, n = 9). Sightings of humpbacked dolphins were classified according to their association with bottlenose dolphins. Of 40 sightings of humpbacked dolphins, 9 were of this species only and 31 were of groups containing both species (hereafter referred to as "mixed groups").

I tested the following hypotheses comparing bottlenose and humpbacked dolphins:

1. Mixed groups were associated with the vessel type from which observations were carried out. There was a significant difference between the number of sightings of humpbacked dolphin groups and mixed groups according to the vessel from which the observations took place (chi-square test, $p <$ .01). Mixed groups were seen only around trawlers (i.e., from the University trawler, speedboat–trawler, and yacht–trawler classes).
2. Group size varied according to species composition. A significant difference existed between the size of mixed and single-species group. Bottlenose dolphin groups were significantly larger than humpbacked dolphin groups (Mann–Whitney U tests, $p < .001$) and mixed groups ($p < .05$), and mixed groups were significantly larger than humpbacked dolphin groups ($p < .001$). There was a significant difference between the numbers of humpbacked dolphins in single-species groups and in mixed groups (t-test, $p < .001$).
3. Mixed groups were comprised of mostly one species. In mixed groups, the number of humpbacked dolphins was generally less than the number of bottlenose dolphins (t-test, $p < .001$) There was no significant difference between the number of bottlenose dolphins in, and the overall size of, mixed groups (one-tailed t-test, $p > .05$).
4. Bottlenose and humpbacked dolphins were found in waters of different depths and distances from shore. Kruskal–Wallis one-way ANOVAs were carried out on maximum depths within grids, changes of depth within grids, and distances of grids from shore, all stratified by the species composition of groups (Table 1). Mixed groups were found in

Table 1
Mean Depth Values and Distances Offshore for Different Dolphin Group Types

Group type	Number of sightings	Mean maximum depth (m)	Mean distance offshore (km)
Bottlenose only	251	18.4	4.0
Humpbacked only	9	9.0	6.0
Mixed	31	17.1	5.0
		$p < .001$	$p < .01$

waters of similar depths to those in which groups of bottlenose dolphins only were found. Humpbacked dolphin groups were found in waters that were shallower yet further offshore than groups of only bottlenose dolphins. The distributions of sightings of bottlenose and humpbacked dolphins in Moreton Bay are shown in Figs. 2 and 3, respectively.

5. Bottlenose and humpbacked dolphins showed evidence of differing rates of shark attack. Eighteen of 50 humpbacked dolphins and 122 of 334 bottlenose dolphins showed evidence of shark attack (see Corkeron *et al.*, 1987, for details of determining what constituted evidence of shark attack). There was no significant difference between the two species (chi-square test, $p > .05$).

DISCUSSION

Despite differences in the methods used to estimate dolphin group sizes, dolphin groups of both species recorded in the present study fall well within published ranges for both bottlenose dolphins (1 to over 1000, with groups of 2–15 animals most common; Shane *et al.*, 1986) and humpbacked dolphins (1–25; Saayman and Tayler, 1979).

In other areas where bottlenose dolphins have been observed in bays, they have been seen in shallow water close inshore. Würsig and Würsig (1979) found that, in their study area, bottlenose dolphins spent 92% of their time in depths of less than 39 m and within 1 km of shore. Shane (1980) observed bottlenose dolphins in waters up to 14 m deep. Irvine *et al.* (1981) showed that dolphins could be found in waters less than 2 m deep, and all their sightings were within 1 km of land. Saayman and Tayler (1979), studying coastal animals along exposed coasts, found bottlenose dolphins up to 5 km offshore, and humpbacked dolphins in the same study area were always within 1 km of shore.

In this study, humpbacked dolphins were found in smaller groups than bottlenose dolphins, and in shallower waters farther offshore, the latter because humpbacked dolphins were more likely to be seen on the western side of the bay, where waters remain fairly shallow for many kilometers offshore. Most research from the University trawler took place, and therefore most observations of dolphins were made, in the deepest water of the bay but relatively close to the western shore of Moreton Island.

The spatial distribution of mixed groups was similar to that of groups containing bottlenose dolphins only. Mixed groups were composed of relatively small groups of both species. These groups were seen only in association with trawlers (Fig. 4). Therefore, in Moreton Bay, small groups of both bottlenose and humpbacked dolphins come together to feed around trawlers. Saayman and Tayler (1979) observed mixed groups engaged in both traveling and play. They also observed humpbacked dolphins apparently avoiding bottlenose dolphins and groups of bottlenose dolphins aggressively chasing lone humpbacked dolphins. However, these authors were studying dolphins on an exposed coast, and the two species appeared to be using different food sources. No such interactions were noted in this study, although similar patterns of interaction have been observed off Point Lookout, on an exposed coast a few kilometers from the study area (P. J. Corkeron, pers. observ.). Bottlenose dolphins in the study area generally appeared to be dominant over humpbacked dolphins when around trawlers. Humpbacked dolphins would generally remain further from the stern of the University trawler and so not gain access to preferred food items (see Corkeron *et al.*, Chapter 17, this volume). This dominance may be because mixed groups were comprised mostly of bottlenose dolphins.

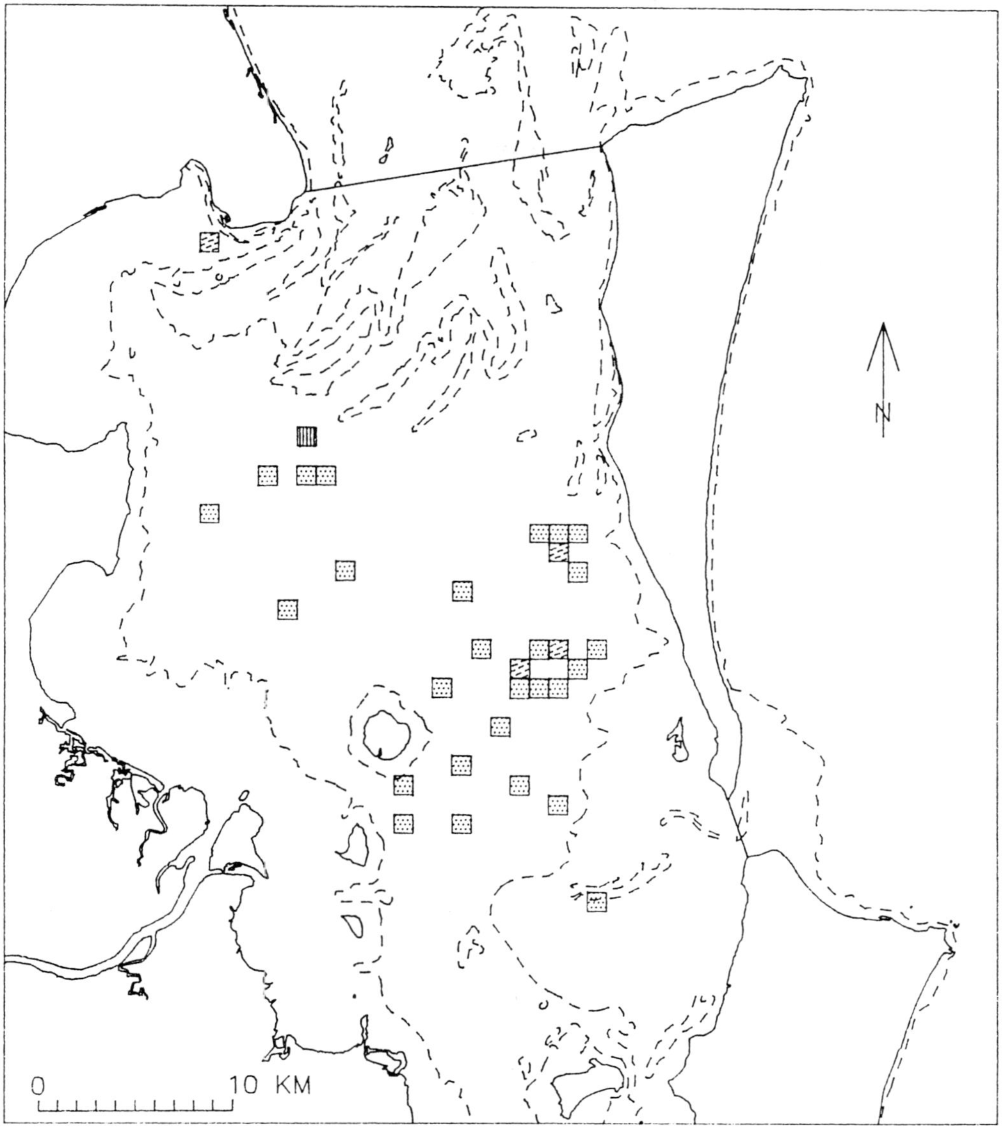

Figure 3 Humpbacked dolphin sightings in Moreton Bay. See Fig. 2 for key for number of sightings.

Sharks are known to attack dolphins in Moreton Bay (Corkeron *et al.*, 1987), and they are the only reported dolphin predators in the bay. Large sharks have been observed in the shallowest waters of the bay. I have seen a tiger shark, *Galeocerdo cuveri,* approximately 4 m long in waters classified in the shallowest depth stratum. Grant (1982) reported a 2.9-m great white shark, *Carcharodon carcharias,* from the southwestern part of the study area, in a depth probably less than 5 m. While it has been suggested that movement into shallow inshore waters is a strategy employed by dolphins for protection against predators, and that by moving inshore they do not have to guard

Figure 4 Mixed groups contained more bottlenose dolphins (A–top, B–bottom) than humpbacked dolphins (A–bottom, B–top) and were found in association with trawlers.

against attack from below (Norris and Dohl, 1980), dolphins could come into contact with predators even in the shallowest waters of Moreton Bay.

Until more is known about the distribution and abundance of different shark species in Moreton Bay, it is impossible to speculate on the role that predation plays in determining group size of either dolphin species in the bay. However, there is some evidence of the role that food patches created by trawlers plays in regulating dolphin group sizes. Social dominance appears to affect bottlenose dolphins' access to food from trawlers (Corkeron *et al.*, Chapter 17, this volume). Mixed groups are smaller than bottlenose-dolphin groups, and the number of humpbacked dolphins in mixed groups is less than that in humpbacked-dolphin groups. It may be that the size of food patches created by trawling and the interactions among dolphins in seeking access to the food patches are the factors limiting dolphin group sizes in Moreton Bay.

ACKNOWLEDGMENTS

J. Kelly, W. Hayes, L. Wale, and M. Ready assisted with the fieldwork, R. Carylon and R. Eggleton provided assistance with computing. M. M. Bryden, my Ph.D. supervisor, reviewed an earlier version of the manuscript and provided advice and encouragement throughout the project. Two anonymous referees commented on a previous draft also. The research was supported by a grant from the Australian Marine Sciences and Technologies Grants Scheme to Dr. Bryden.

REFERENCES

Bryden, M. M. (1978). Whales and whaling in Queensland waters. *Proc. R. Soc. Queensland* **89,** v–xii.

Corkeron, P. J., Morris, R. J., and Bryden, M. M. (1987). Interactions between bottlenose dolphins and sharks in Moreton Bay. *Aquatic Mamm.* **13,** 109–113.

Grant, E. M. (1982). "Guide to Fishes." Dept. of Harbours and Marine, Brisbane, Queensland. 896 pp.

Haysom, N. M. (1985). Review of the penaeid prawn fisheries of Australia. "Second Australian National Prawn Seminar" (P. C. Rothlisberg, B. J. Hill, and D. J. Staples, eds), pp. 195–203. NPS2, Brisbane, Queensland. 368 pp.

Irvine, A. B., Scott, M. D., Wells, R. S., and Kaufman, J. H. (1981). Movements and activities of the Atlantic bottlenose dolphin, *Tursiops truncatus,* near Sarasota, Florida. *Fish. Bull. (U.S.)* **79,** 671–688.

Kennedy, R. (1985). OZSTAT. Unpublished manuscript, Dept. of Zoology, University of Queensland. 39 pp.

Newell, B. S. (1971). The hydrological environment of Moreton Bay, Queensland 1967–1968. *CSIRO Div. Fish. Oceanogr. Tech. Pap.* **30,** 1–35.

Norris, K. S., and Dohl, T. P. (1980). The structure and functions of cetacean schools. *In* "Cetacean Behavior: Mechanisms and Functions" (L. M. Herman, ed.), pp. 211–261. Wiley, New York. 463 pp.

Numerical Algorithms Group. (1985). "The GLIM System Release 3.77 Manual." The Numerical Algorithms Group of the Royal Statistical Society. Oxford. 80 pp.

Saayman, G. S., and Tayler, C. K. (1973). Social organization of inshore dolphins (*Tursiops aduncus* and *Sousa*) in the Indian Ocean. *J. Mammal.* **54,** 993–996.

Saayman, G. S., and Tayler, C. K. (1979). The socioecology of humpback dolphins (*Sousa* sp). *In* "Behavior of Marine Animals" (H. E. Winn and B. L. Olla, eds.), Vol. 3:Cetaceans, pp. 165–226. Plenum, New York. 438 pp.

Saayman, G. S., Bower, D., and Tayler, C. K. (1972). Observations on inshore and pelagic dolphins on the south-eastern Cape coast of South Africa. *Koedoe* **15,** 1–24.

Shane, S. H. (1980). Occurrence, movements, and distribution of bottlenose dolphins, *Tursiops truncatus,* in southern Texas. *Fish. Bull. (U.S.)* **78,** 593–601.

Shane, S. H., Wells, R. S., and Würsig, B. (1986). Ecology, behavior and social organization of the bottlenose dolphin: A review. *Mar. Mamm. Sci.* **2,** 34–63.

Wells, R. S., Irvine, A. B., and Scott, M. D. (1980). The social ecology of inshore odontocetes. *In* "Cetacean Behavior: Mechanisms and Functions" (L. M. Herman, ed.), pp. 263–317. Wiley, New York. 463 pp.

Würsig, B., and Würsig, M. (1977). The photographic determination of group size, composition and stability of coastal porpoises (*Tursiops truncatus*). *Science* **198,** 755–756.

Würsig, B., and Würsig, M. (1979). Behavior and ecology of the bottlenose dolphin, *Tursiops truncatus,* in the South Atlantic. *Fish. Bull. (U.S.)* **77,** 399–412.

Würsig, B., and Würsig, M. (1980). Behavior and ecology of dusky porpoises, *Lagenorhynchus obscurus,* in the South Atlantic. *Fish. Bull. (U.S.)* **77,** 871–890.

Food and Feeding of the Indian Ocean Bottlenose Dolphin off Southern Natal, South Africa

15

V. G. Cockcroft
G. J. B. Ross
Port Elizabeth Museum
Humewood, Republic of South Africa

INTRODUCTION

Small cetaceans living in marine waters can be divided, crudely, into two broad groups based on their feeding strategies. Pelagic dolphins, by and large, traverse large areas of ocean in search of food, their movements probably related to the availability of food (Norris and Dohl, 1980; Wells *et al.*, 1980). In contrast, some offshore dolphins and particularly most coastal dolphins appear to range far less widely. Although some of the movements may be related to the movements of their prey, they probably exert considerably more pressure on local prey populations than do highly mobile pelagic forms (Irvine *et al.*, 1979, 1981). The authors cited above suggested that this dichotomy in feeding strategies would profoundly affect our understanding of cetacean ecology, indicating the necessity for comparative studies. The bottlenose dolphin (*Tursiops truncatus*) is widely distributed in coastal waters of the Indian Ocean (Ross, 1984). On the east coast of southern Africa it inhabits shallow, inshore water, rarely venturing beyond the 30-m isobath (Ross *et al.*, 1986). Off the Natal coast this species, the Indo-Pacific humpbacked dolphin, *Sousa chinensis*, and the common dolphin, *Delphinus delphis*, are killed sporadically in gill nets. These nets are set to catch and reduce the numbers of sharks potentially dangerous to bathers at 44 prime bathing beaches.

Stomach contents from bottlenose dolphins caught in the shark nets form the basis of this study, allowing quantitative analysis of the species' diet. The relationship of diet to movements and social structure in these bottlenose dolphins can provide a better understanding of the ecology of coastal species, particularly those exploited directly by man or caught incidentally during coastal fishing.

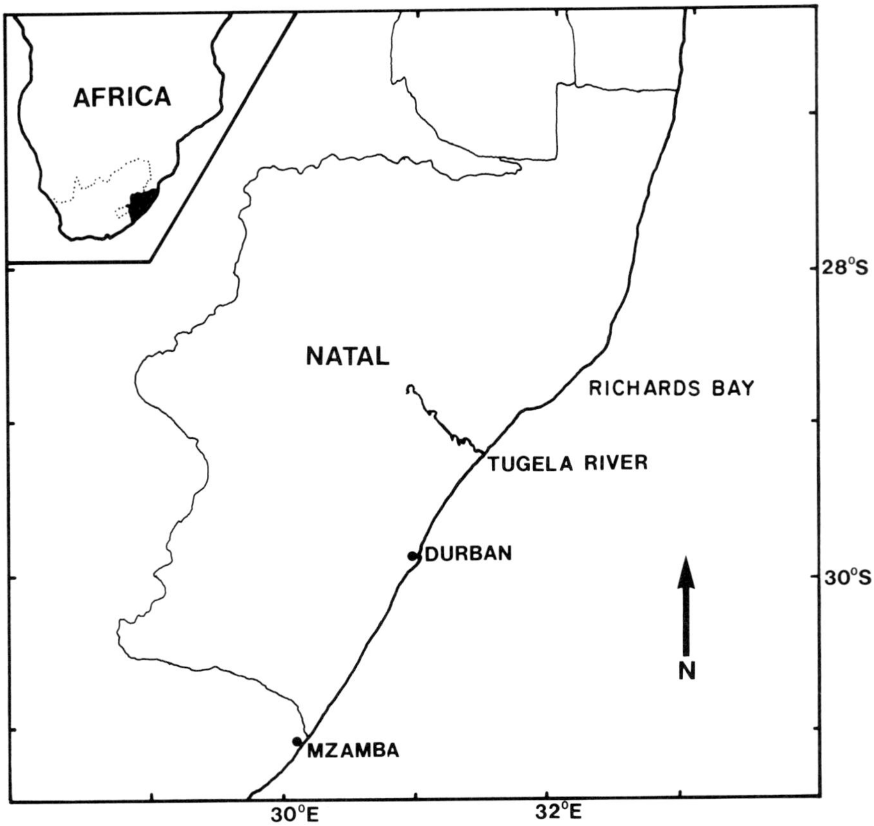

Figure 1 Natal, on the east coast of southern Africa. Shark net installations are positioned at 44 bathing beaches between Mzamba and the Tugela River. The area north of Durban is known as the north coast. The area between Durban and Mzamba is equally divided between the upper and lower south coasts.

STUDY AREA

The Natal coast of southern Africa is oriented north–south between about 27° and 31° S (Fig. 1). The climate is mild. Monthly air temperatures range between 15° and 27°C, and sea temperatures, warmed by the Agulhas Current, range between approximately 17° and 27°C. The southern half of this coast between the Tugela River (29° 13′ S) and Mzamba (31° 05′ S) is heavily populated and supports an extensive tourist industry. Almost all of this coastline slopes steeply offshore, reaching depths of 30 m within 1 km of the beach. The surf zone is highly energetic, extending 100–200 m offshore.

MATERIALS AND METHODS

Animals caught in shark nets were frozen as soon as possible after death. After thawing, the carcasses were weighed, measured, and dissected. Stomachs were removed whole, after tying and cutting the esophagus anterior to the forestomach and the intestine posterior to the pylorus. On occasion the esophagus was cut directly posterior to the larynx. On removal, stomachs were frozen until processing.

Thawed stomachs were weighed whole to the nearest gram and the contents washed into a plastic dish. Loose tissue was removed by agitating the contents of the dish and allowing water and loose tissue to flow gently over the lip. This was repeated until all light, loose tissue was removed. Loose cephalopod beaks and shells were removed and stored in 10% buffered formalin. Remaining material, such as bones, otoliths, and larger pieces of tissue, were sorted by hand. Beaks still encased in the buccal mass were removed and added to the formalin. All otoliths were removed from whole fish and skulls. Otoliths and any bones deemed important were stored dry. Whole fish were identified, measured, and weighed where possible. Empty stomachs were then weighed, for calculation of content mass, and discarded.

Where the stomach and entire esophagus were excised, both were weighed while containing food. With the esophagus held upright, both were filled with water until the esophagus overflowed. On all occasions the aspect of the stomach was kept the same, and it was not allowed to become overdistended. The esophagus was tied off and the stomach, plus contents and water, were weighed. After removal of the contents the empty stomach was again weighed. This gave an estimate of the maximum volume of the stomach.

Squid beaks and otoliths were identified using the prey-identification collection established for this purpose at the Port Elizabeth Museum. This collection includes over 15,000 specimens from which regressions for fish and cephalopods relating otolith or beak dimensions, respectively, to prey length and mass have been determined. Nomenclature for fish species follows Smith and Heemstra (1986).

Otoliths were measured to the nearest 0.1 mm across their greatest dimension. Some species with fragile points prone to breakage (e.g., *Scomber japonicus*) were measured across the shoulder. Beak lower rostral length was measured for all cephalopods, except octopods and sepiids, for which beak lower crest length was taken (Clarke, 1986a). The length and mass of prey species represented in each stomach were calculated from the otolith or beak measurements, using the appropriate regression. Otoliths that were visibly badly eroded were excluded from these calculations. The calculated weights of all fish and cephalopod remains found in each stomach were summed to give a reconstituted mass of prey for that stomach.

For species for which the collection contained insufficient numbers of specimens to calculate a regression, length and mass were indirectly estimated either from similar species in the same genus or from the few mass or length data in the prey reference collection. In those few species for which no information was available, even from closely related species, the mean mass values of all species (either fish or cephalopod) for which regressions were available were used to estimate reconstituted mass.

An index of relative importance (IRI) for each prey species was calculated according to Pinkas *et al.* (1971), where

$$\text{IRI} = (\%\ \text{number} + \%\ \text{reconstituted mass}) \times \%\ \text{frequency of occurrence}$$

Eight stomachs collected in 1986 were processed using the modified volume technique (Bigg and Perez, 1985) of assessing the importance of prey remains in stomachs, for comparison with estimates made using the reconstituted mass method.

RESULTS

The contents of 165 stomachs, representing bottlenose dolphins caught between 1975 and 1986, were examined. Unfortunately, date and locality of capture, length, mass, and sex information were not available for all animals caught. Thirty-four of the stomachs examined were from suckling calves, between 129 and 170 cm in length, and contained only a milky substance or this substance in addition to a few crustaceans, squid, and/or fish. Stomachs containing only the milky substance were excluded from calculations. Four of the remaining 131 stomachs were completely empty.

More than 72 species of prey representing 5610 prey items were found in the stomachs containing solid remains (Table 1). Of the fish prey categories, 56 were identifiable to species level. A further 4 and 6 were identifiable only to genus and family, respectively. All unidentified fish were pooled into a single category. Five cephalopod categories were identified, including three decapods and two octopods. Almost all decapod beaks were from the cuttlefish, *Sepia officinalis*, and an inshore squid, *Loligo reynaudi*. However, the latter may have

Table 1
Prey of Bottlenose Dolphins off Natal[a]

Prey species	Number	Freq. of occurrence	Total mass	% number	% freq. of occurrence	% mass	IRI
Elasmobranchs							
Rhinobatos spp.*	1	1	118	0.0	0.8	0.0	<1
Unidentified elasmobranch*	9	7	5,490	0.2	5.5	1.3	8
Fish							
Anguilla mossambica	1	1	1,660	0.0	0.8	0.4	<1
*Conger cinereus**	24	12	24,190	0.4	9.4	5.8	59
Conger wilsoni	23	6	1,097	0.4	4.7	0.3	3
Unidentified congrid sp. 1*	94	18	6,307	1.7	14.2	1.5	45
Etrumeus whiteheadi	30	4	594	0.5	3.1	0.1	2
*Hilsa kelee**	12	1	805	0.2	0.8	0.2	<1
Sardinops ocellatus	80	7	4,416	1.4	5.5	1.1	14
*Thryssa vitrirostris**	21	5	215	0.4	3.9	0.1	2
Unidentified clupeid*	1	1	67	0.0	0.8	0.0	<1
Galeichthys feliceps	4	2	1,528	0.1	1.6	0.4	1
*Saurida undosquamis**	22	8	1,615	0.4	6.3	0.4	5
*Synodus indicus**	1	1	67	0.0	0.8	0.0	<1
*Trachinocephalus myops**	2	2	134	0.0	1.6	0.0	<1
Merluccius capensis	2	2	134	0.0	1.6	0.0	<1
*Petalichthys capensis**	1	1	67	0.0	0.8	0.0	<1
*Cheilopogon furcatus**	29	9	5,014	0.5	7.1	1.2	12
*Scorpaenodes variepinnis**	1	1	67	0.0	0.8	0.0	<1
*Sorsogona prionata**	1	1	67	0.0	0.8	0.0	<1
Cheilodonichthys capensis	10	4	56	0.2	3.1	0.0	1
Cheilodonichthys queketti	1	1	67	0.0	0.8	0.0	<1
*Ambassis natalensis**	1	1	67	0.0	0.8	0.0	<1
Kuhlia mugil	6	1	402	0.1	0.8	0.1	<1
Epinephalus andersoni	1	1	1,500	0.0	0.8	0.4	<1
Epinephalus sp.*	1	1	1,500	0.0	0.8	0.4	<1
Priacanthus sp.*	54	3	11,234	1.0	2.4	2.7	9
Unidentified apogonid sp. 1*	7	3	469	0.1	2.4	0.1	1
Pomatomus saltatrix	60	21	8,516	1.1	16.5	2.0	51
Pomadasys commersonni	10	6	7,617	0.2	4.7	1.8	9
Pomadasys olivaceum	2,330	86	46,773	41.5	67.7	11.2	3,569
Pomadasys striatus	154	21	3,039	2.7	16.5	0.7	57
*Lutjanus argentimaculatus**	5	1	336	0.1	0.8	0.1	<1
Dinoperca petersi	2	2	201	0.0	1.6	0.0	<1
Cheimerius nufar	4	1	400	0.1	0.8	0.1	<1
Chrysoblephus puniceus	5	2	924	0.1	1.6	0.2	<1
*Crenidens crenidens**	2	1	134	0.0	0.8	0.0	<1
Diplodus cervinus	1	1	991	0.0	0.8	0.2	<1
Diplodus sargus	135	26	11,444	2.4	20.5	2.7	105
Lithognathus mormyrus	1	1	67	0.0	0.8	0.0	<1
Pagellus bellotti	493	44	10,274	8.8	34.6	2.5	389
*Rhabdosargus sarba**	13	2	1,216	0.2	1.6	0.3	1
Rhabdosargus thorpei	32	14	1,618	0.6	11.0	0.4	11
Sarpa salpa	105	21	10,606	1.9	16.5	2.5	73
Monodactylus falciformis	17	6	745	0.3	4.7	0.2	2
Argyrosomus hololepidotus	4	4	268	0.1	3.1	0.1	<1
*Upeneus vittatus**	18	1	583	0.3	0.8	0.1	<1

Table 1 (*continued*)

Prey species	Number	Freq. of occurrence	Total mass	% number	% freq. of occurrence	% mass	IRI
Argyrosomus thorpei	45	4	2,661	0.8	3.1	0.6	5
Otolithes ruber	8	3	2,032	0.1	2.4	0.5	1
Umbrina canariensis	28	17	771	0.5	13.4	0.2	9
*Gazza minuta**	1	1	67	0.0	0.8	0.0	<1
Secutor insidiator	2	1	151	0.0	0.8	0.0	<1
Trachurus delagoae	322	34	20,555	5.7	26.8	4.9	285
Caranx williamsi	1	1	67	0.0	0.8	0.0	<1
Liza dumerili	12	6	2,005	0.2	4.7	0.5	3
*Liza macrolepis**	7	2	1,134	0.1	1.6	0.3	1
Liza tricuspidens	1	1	162	0.0	0.8	0.0	<1
Mugil cephalus	49	6	10,977	0.9	4.7	2.6	17
Unidentified mugilid*	6	5	972	0.1	3.9	0.2	1
Sphyraena acutipinnis	33	17	980	0.6	13.4	0.2	11
Parapercis sp.	3	2	201	0.1	1.6	0.0	<1
Unidentified gobiid*	1	1	1	0.0	0.8	0.0	<1
Trichiurus lepturus	83	6	3,016	1.5	4.7	0.7	10
Scomber japonicus	184	66	21,648	3.3	52.0	5.2	439
Unidentified bothid*	52	5	3,487	0.9	3.9	0.8	7
Cynoglossus zanzibarensis	3	3	201	0.1	2.4	0.0	<1
Unidentified *cynoglossid**	2	2	134	0.0	1.6	0.0	<1
Unidentified fish	228	52	15,299	4.1	40.9	3.7	316
Cephalopods							
Sepia officinalis	415	66	136,837	7.4	52.0	32.7	2,082
Loligo spp.	241	43	14,573	4.3	33.9	3.5	263
Octopus sp. 1	38	13	4,184	0.7	10.2	1.0	17
Octopus sp. 2*	10	4	1,101	0.2	3.1	0.3	1
Unidentified ommastrephid*	4	1	895	0.1	0.8	0.2	<1
Totals	5,610		418,810				

[a] Prey data are expressed as number, frequency of occurrence, and reconstituted mass per stomach as well as percentage number, frequency of occurrence, and reconstituted weight over the whole sample. Derivation of the index of relative importance (IRI) is given in the text.

* No regressions available, length and mass estimated.

included a few specimens of a similar, smaller species, *Loligo dauvercelli*. Of a number of elasmobranchs found, only one was identified to genus. In addition to the above, three near-intact penaeid prawns and a few euphausiid exoskeletons were found in three stomachs. These have been ignored in all calculations because of their scarcity.

Six prey species had IRI values that were considerably higher than the others (Table 1). These six species contributed approximately 60% by mass of all prey taken. The estimated IRI of unidentified fish was also relatively high (316), and their estimated mass contribution is 3.7%. This group is made up of a number of unidentified species which individually would probably not be important.

Between 1982 and 1985, the 4 years for which sufficient data were available, the proportions of the six species were similar by number, frequency of occurrence, and mass, but variations in their IRI values were evident (Table 2). The greatest species variety was in 1982 (40) with 29, 35, and 28 taken in 1983, 1984, and 1985, respectively. The number of unidentified fish species was similar in all years (10–15).

Table 2
Index of Relative Importance (IRI) of the Six Major Prey of Bottlenose Dolphins off Natal[a]

	1982	1983	1984	1985
Trachurus delagoae	98	361	8	1873
Pomadasys olivaceum	3995	1515	4383	4270
Pagellus bellotti	51	830	140	637
Scomber japonicus	121	547	87	319
Sepia officinalis	1115	2467	1927	924
Loligo sp.	267	740	147	54
Number of stomachs	18	22	23	23

[a] Data were based only on stomachs where information was available.

Some seasonal patterns were evident when all data were stratified into quarters (Table 3), though inferences are limited due to the small numbers. Between October and December the IRI of both *Loligo* sp. and *T. delagoae* increased while that of *S. officinalis* decreased. An increased IRI of *P. olivaceum* between April and June was not concurrent with a decreased IRI in any other species.

The same six prey were dominant throughout the study area, although some variations between the north coast, upper south coast, and lower south coast were evident (Table 4). On the lower south coast only *P. olivaceum* and *S. officinalis* appeared of prime importance (Table 4). Conversely, the north coast showed a greater parity of IRI values, while on the upper south coast the IRI values of these prey were intermediate between those of the north and lower south coasts, except for an apparently greater importance of *Loligo* sp.

Both area and season were poorly correlated with the mean length of prey in stomachs, the variety of prey consumed, and the number of prey items in stomachs (all r values <0.31). A poor correlation ($r < 0.3$) was found between dolphin length and mean length of prey consumed. However, the maximum length of prey consumed increased with dolphin length (Fig. 2). Comparison of the mean prey length for each of five length, sex, and maturity classes showed a significant increase with predator size (Table 5). The mean lengths of prey found in calves was significantly less than those in all other groups ($p < .01$, student's t-test). The mean prey length of adolescents was significantly less than that found in all adult groups ($p < .01$, student's t-test). There were no significant differences between the mean prey lengths taken by any of the adults.

A plot of dolphin mass against maximum stomach volume is presented in Fig. 3. A power curve of the form

$$\log Y = 0.2143 + 1.596 \log X \qquad (r^2 = 0.72)$$

was fitted to these data.

Stomach content weights ranged from practically zero to a maximum of almost 5 kg. A poor

Table 3
Index of Relative Importance (IRI) of the Six Major Prey of Bottlenose Dolphins in Quarterly Periods during the Year[a]

	Jan–Mar	Apr–June	July–Sept	Oct–Dec
Trachurus delagoae	243	149	107	883
Pomadasys olivaceum	3538	8629	2718	2528
Pagellus bellotti	1114	335	65	1181
Scomber japonicus	49	227	85	249
Sepia officinalis	2213	1720	3147	184
Loligo sp.	197	456	250	2076
Number of stomachs	17	12	50	15

[a] Data were based only on stomachs where information was available.

Table 4
Index of Relative Importance (IRI) of the Six Major Prey of Bottlenose Dolphins in Three Regions of Natal[a]

	North coast	Upper south coast	Lower south coast
Trachurus delagoae	590	42	44
Pomadasys olivaceum	1198	2471	6092
Pagellus bellotti	1428	185	63
Scomber japonicus	592	109	<1
Sepia officinalis	653	1741	2017
Loligo sp.	317	826	81
Number of stomachs	33	28	43

[a] Data were based only on stomachs where information was available.

correlation ($r = 0.37$) was found between dolphin mass and stomach content mass, suggesting that dolphins were caught in all stages of repletion. Additionally, a low correlation was found between dolphin mass and the reconstituted mass of items contained in the stomach ($r = 0.26$) (Fig. 4). This suggests that many dolphins either voided the stomach of otoliths and/or beaks prior to or during capture or, in some instances, retained these remains from an earlier feed. The former would tend to underestimate the meal size and the latter to overestimate it.

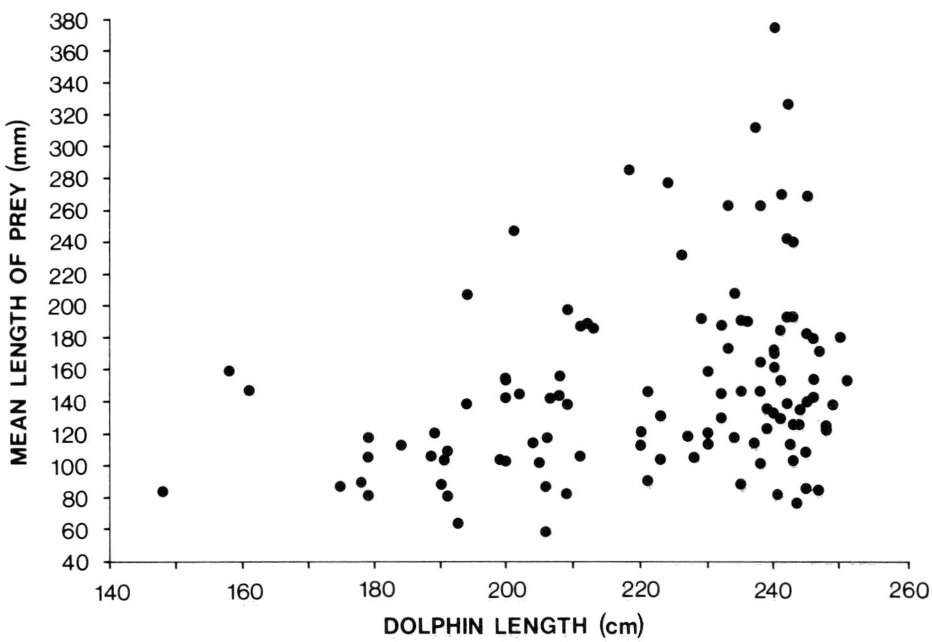

Figure 2 Relationship between the length of bottlenose dolphins and mean length of their prey.

Table 5
Index of Relative Importance, Rank of Importance, and Mean Length of the Eight Major Prey from Each of Five Maturity, Sex, and Size Classes of Bottlenose Dolphins off Natal[a]

	Calves (<200 cm)			Adolescents (>200, <230 cm)			Lactating females			Nonlactating mature females			Mature males (>230 cm)		
Species	IRI	IRI rank	Mean length	IRI	IRI rank	Mean length	IRI	IRI rank	Mean length	IRI	IRI rank	Mean length	IRI	IR rank	Mean length
Pomadasys olivaceum	7109	1	85	3990	2	106	2428	2	107	3885	1	103	3630	2	113
Sepia officinalis	196	3	79[a]	9769	1	129[a]	3155	1	141[a]	2371	2	136[a]	5957	1	142[a]
Pagellus bellotti	467	2	109	509	3	120							393	4	109
Sarpa salpa	181	4	162	135	8	158				91	7	165			
Conger cinereus	142+	5					570[b]	3							
Sardinops ocellatus	120	6	207				110	8	175	196	5				
Loligo sp.	108	7	109[a]	155	7	123[a]	498	4	128[a]				319	5	125[a]
Diplodus sargus	104	8	108				176	7	162	1417	3	172			
Pomadasys striatus							194	5	195						
Trachurus delagoae				232	5	163	187	6	141	546	4	154	442	3	171
Mugil cephalus				272	4	306									
Scomber japonicus				174	6	215							234	6	259
Priacanthus sp. 1													90	7	208
Octopus sp. 1													68	8	66[a]
Sphyraena acutipinnis										107	6	198			
Monodactylus falciformis										87[b]	8				
Total % mass		70.4			76.3			67.3			70.5			70.3	
Total number of species recorded		15			39			47			24			48	
Number of stomachs		16			32			18			7			32	

[a] Mantle length.
[b] Estimated, no regressions available.

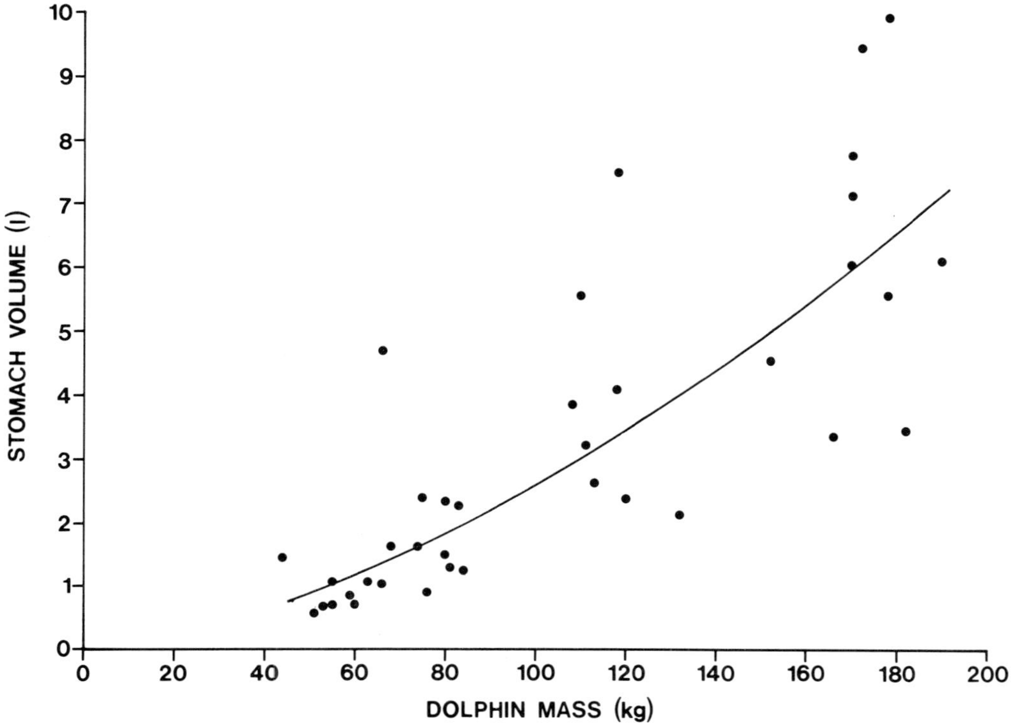

Figure 3 Relationship between bottlenose dolphin mass and estimated maximum volume of the stomach. A power curve ($\log Y = a + b \log x$, where $a = 0.2143$, $b = 1.596$) is fitted to these data ($r^2 = 0.72$).

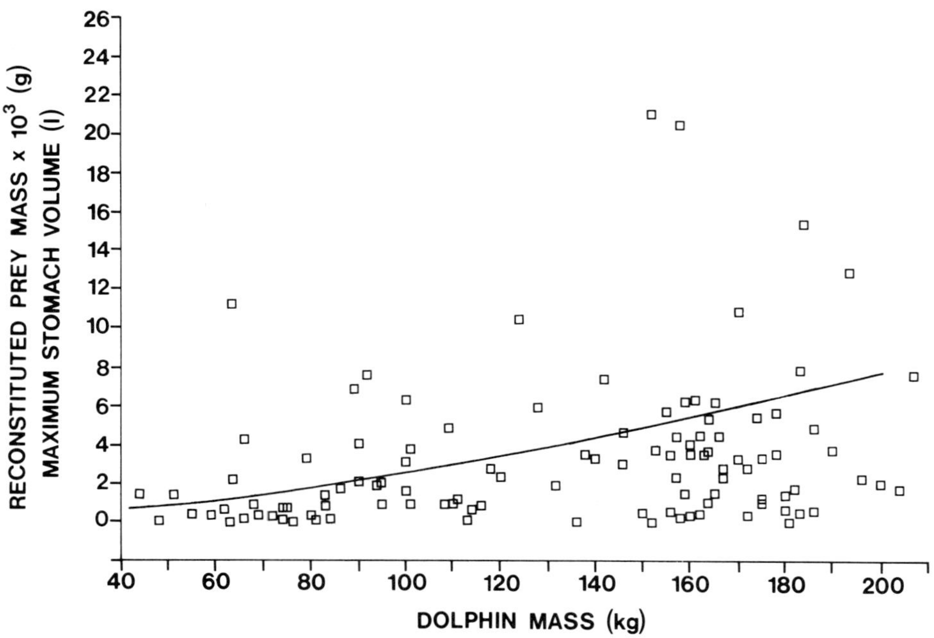

Figure 4 Relationship between bottlenose dolphin mass and reconstituted mass of stomach contents. The power curve from Fig. 3 is superimposed on these data.

Approximately 75% of the reconstituted mass estimates lie below the plot of the relationship between maximum stomach volume and dolphin mass (Fig. 4), assuming that prey volume and prey mass are equivalent. This suggests that 25% of stomachs may have contained otoliths and/or beaks from a previous meal. Four of the seven stomachs showing the highest reconstituted mass estimates contained large numbers of *S. officinalis* beaks. The indigestibility of cephalopod beaks and their subsequent retention are well documented (Clarke, 1986b). A further two stomachs contained otoliths of the eel, *Conger cinereus,* and a third contained a pair of otoliths of a species of *Epinephalus*. The mass estimates of both *C. cinereus* and *Epinephalus* sp. were made from smaller species of the same genus and may well be overestimates.

The mean number of prey items per stomach, deduced from the number of otoliths and/or beaks remaining in the stomach, was 48.5 (range 1–340). There was little correlation between the number of prey items found in stomachs and the length of dolphin ($r < 0.1$) (Fig. 5). However, the maximum number of prey items consumed increased with dolphin length and stomach volume. In general, the number of species present in any stomach was low (mean 6, range 1–15) but tended to increase with dolphin length (Fig. 6). There was nevertheless little correlation ($r = 0.03$) between the two.

Calves (mean 2.9) contained significantly fewer prey species than adolescents (mean 6), mature males (mean 6.5), and lactating females (mean 8.4) (Kruskal–Wallis test, $H = 22.325$, $p < .01$; Dunn's multiple contrast method, Zar, 1984). Numbers of prey species in nonlactating females (mean 5.5) did not differ from those in any other group. Supporting evidence that mothers' stomachs contained more species than their respective calves is shown by the proportions of species in four known mother and calf pairs (15:1, 6:1, 3:1, and 4:1). In all four cases all species in the calf's stomach were also present in the mother's stomach. No differences in mean prey length and numbers of prey taken were found between calves caught with or without their mothers or between mothers caught with or without calves.

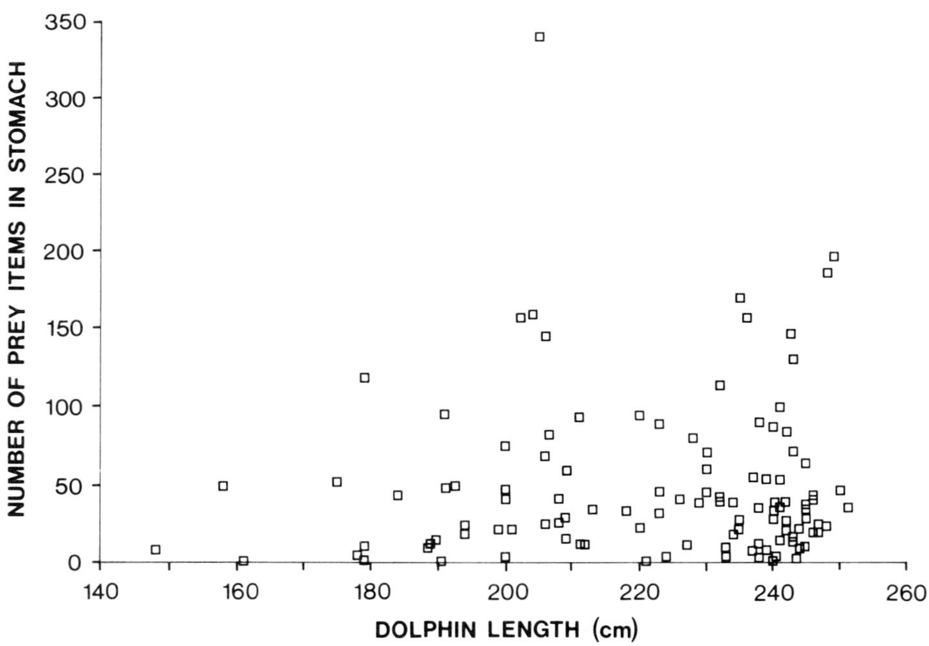

Figure 5 Relationship between bottlenose dolphin length and number of prey items contained in the stomach.

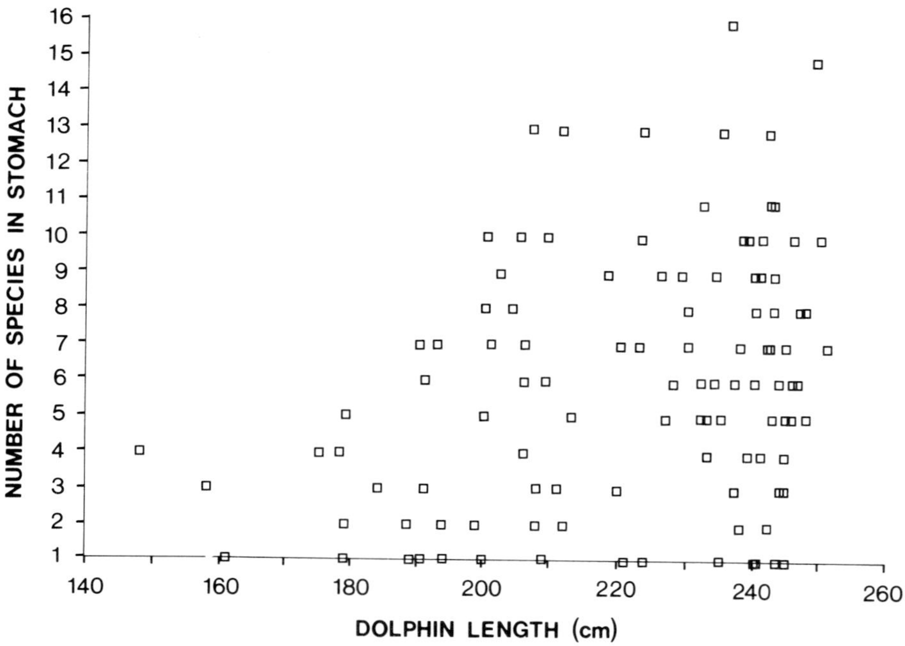

Figure 6 Relationship between bottlenose dolphin length and number of prey species consumed.

In each of the five length, sex, and maturity classes of bottlenose dolphins, eight prey species contributed approximately 70% by mass of all prey consumed (Table 5). *Pomadasys olivaceum* and *S. officinalis* were important to all classes, although their individual importance to each class varied. The importance of the remaining prey species varied between classes.

The sizes of *P. olivaceum, S. officinalis, Loligo* sp., and *D. sargus* taken by calves were significantly smaller ($p < .01$, student's t-test) than those taken by any other class. No significant differences were found between the mean sizes of individual prey species taken by adolescents and adults. However, there was a significant difference between the mean length of all prey taken by adolescents and that of the adults ($p < .01$, student's t-test).

No significant differences were found between the seasonal or overall geographic distribution of bottlenose dolphin size–maturity classes throughout the year. This suggests that the dietary differences observed are not due to sample variation.

The overall contributions of fish and squid to the diet of bottlenose dolphins were calculated as 69.9 and 30.1%, respectively, using the technique of Bigg and Perez, (1985). Calculation of fish and squid contributions in this study yielded figures of 74.8 and 25.2%, respectively.

DISCUSSION

The number of stomachs used to assess the similarity of methods employed in this study and that of Bigg and Perez (1985) was small. Despite this, there was a close agreement between the results of the two methods, suggesting that the approach used in the present study was valid.

The four fish and two cephalopod species that constitute the major portion of prey of bottlenose dolphins off Natal are all common inshore. Of the four fish species, two can be considered as benthic. *Pomadasys olivaceum* inhabits inshore reef areas, and *P. bellotti* prefers inshore sandy-bottom areas (van der Elst, 1981). The two remaining fish species, *T. delagoae* and *S. japonicus,* are both pelagic shoaling fish occurring in inshore waters in this area, particularly as juveniles (van der Elst, 1981). The two cephalopods *Loligo* sp. and *S. officinalis* are also common in the Natal inshore environment and are regularly seen around the shark nets (Natal Sharks Board personnel, pers. commun.). Of the 16 most important prey species (Table 5), all but *Priacanthus* sp. are known to be common inshore (van der Elst, 1981; Smith and Heemstra, 1986). *Priacanthus* inhabit deeper coastal waters (>50 m depth) (Smith and Heemstra, 1986) and were found in stomachs of mature males only. *Sardinops ocellatus* is a pelagic shoaling species occurring in Natal inshore regions only from July through September. *Conger cinereus* and *Sphyraena acutipinnis* are both predators of inshore reef fish and occur either singly or in small groups close inshore, while *Mugil cephalus* and *P. bellotti* occur predominantly above sandy bottoms (van der Elst, 1981).

This information suggests that bottlenose dolphins off Natal use a variety of food resources. These include benthic reef and sandy-bottom prey and their associated predators, pelagic shoaling fish and cephalopods, and deeper-water fish. The variety of prey consumed by bottlenose dolphins off Natal is one of the most extensive yet reported for any cetacean, superficially supporting the usual interpretation that the dolphins are opportunistic feeders. Such conclusions have been drawn about other bottlenose dolphin populations. Gunter (1942) reported that the diet of bottlenose dolphins off the coast of Texas was catholic but dominated by mullet (*M. cephalus*), one of the four most abundant benthic fish in the area. In British waters dolphins often feed on inshore herring and salmon shoals but were still considered as benthic feeders (Evans, 1980). Tomilin (1957) described bottlenose dolphins from the Black Sea as benthic feeders, only feeding on pelagic schooling fish when they formed dense aggregations. Walker (1981) noted that the prey of inshore bottlenose dolphins off southern California inhabited littoral and sublittoral zones.

Considering that only six species accounted for 60% by mass of the diet of bottlenose dolphins off Natal, it seems inappropriate to consider this species an opportunistic predator. Though there is some indication of annual, seasonal, and geographical variation in the proportions of the six major prey species, their continued importance throughout this study, for all age and sex classes, is particularly significant. The causes of these variations are uncertain, though they may reflect changes in the relative availability of each species.

The daily food requirement (as a percentage of body mass) of a captive, weaned bottlenose dolphin calf has been estimated at between 5 and 6.7%; the requirements for a lactating female and two nonlactating females were 8.3%, and 3.9 and 5%, respectively (Ross, 1984; Cockcroft and Ross, Chapter 27, this volume). The daily requirement of mature males is probably similar to that of nonlactating females and comparable to the 4.2% given for captive animals by Sergeant (1969).

If maximum stomach volume in equated to the amount of food in a full stomach, a number of assumptions can be made (Figs. 3 and 4). Mature males and mature quiescent (neither pregnant nor lactating) females would obtain their daily food requirements from about two full stomachs. However, a lactating female of 130 kg must consume the equivalent of between three and four stomachs full per day. Similarly, a weaned calf of about 75 kg must fill its stomach four or five times per day. These assumptions are based on biomass estimates and make no allowance for the nutritional value of the different prey species consumed by size/sex classes. Although the use of energy values is preferable in comparisons (Innes *et al.*, 1987), energy values for most of the prey species are not available. Where available, the energy values of prey species are similar (Ross and Cockcroft, Chapter 5, this volume), and although there are apparent differences in the diets of size/sex classes, they share many common species, particularly *P. olivaceum* and *S. officinalis*. This suggests that the use of energy values rather than biomass would not greatly alter the above pattern.

Saayman *et al.* (1973) observed two feeding peaks in free-ranging bottlenose dolphins, early morning and late evening, which is in keeping with the projected needs of adult males, quiescent females, and adolescents. In contrast, the present results suggest that calves and lactating females have to feed more often, perhaps almost continuously. The low numbers of mothers and calves would make such activities inconspicuous within the overall activity of mixed schools.

The energetic cost of lactation is an obvious and important component of the lactating female's increased nutritional requirements, particularly during the calf's early suckling. A second component is maternal care, which increases in importance during the weaning and postweaning periods (Cockcroft and Ross, Chapter 27, this volume). These increased costs may be reflected in the trend toward greater numbers of prey species observed in stomachs of lactating females. Time spent in maternal care may restrict the mother's foraging time and make it beneficial for her to cooperate with other female and calf pairs (subgroups) and to take any prey presenting itself. In contrast, the prey-species diversity in the stomachs of calves is low. Calves clearly feed with their mothers. In each of the four known mother and calf pairs the prey species consumed by the calf were also present in the mother. This implies that mothers assist the calves in foraging, possibly by herding prey, and that therefore the energetic cost of maternal care may be substantial.

The evidence that each class of dolphins prefers different prey sizes and, to some degree, different species (Table 5) suggests that subgroups within schools are likely to employ different feeding strategies. The partitioning of bottlenose dolphin populations into distinct size or sex subgroups and the postulated importance of these subgroups have been well documented (Mead, 1975; Irvine and Wells, 1972; Irvine *et al.*, 1981; Wells *et al.*, 1980) and have been reviewed for cetaceans in general by Norris and Dohl (1980). Nursery groups of mothers and calves have been reported by Leatherwood (1977), Würsig (1978), Irvine *et al.*, (1981), and Wells *et al.* (1980).

The implication that mothers and calves may need to feed more than other dolphins is of interest in view of the findings of Wells *et al.* (1980), who reported that this subgroup has the most extensive range of any within the school. The present study also indicates that calves used both reef- and sand-dwelling prey, perhaps maximizing their usage of the inshore area. Taken together these results suggest that foraging by nursery groups is an important determinant of the extent of their movements and may be a major factor in the delimination of apparent home range.

Shallow inshore reefs constitute the nursery areas for several fish species off Natal, including *P. olivaceum* (Joubert, 1981) which predominate in the diet of calves. Juveniles of this species are present inshore for at least 9 months of the year, indicating that calves are assured of a consistent food source by feeding in shallow water throughout the year. The seasonally occurring *S. ocellatus* was the only pelagic fish found in the stomachs of calves. In contrast, the stomachs of other age/sex classes contained other pelagic shoaling fish. The reasons for this difference are unclear. Calves may be too inexperienced to participate in cooperative feeding on pelagic fish, except in shallow water where *S. ocellatus* is at times packed tightly close to the surf zone by sea conditions and predators.

Though there was no significant difference between the mean lengths of their prey, adult male bottlenose dolphins consumed a greater proportion of larger fish than did adult females (Table 5). Additionally, males took species either not consumed at all (e.g., *Priacanthus* sp.) or consumed only in low numbers by other dolphins. Exclusively male subgroups within schools (Irvine *et al.*, 1981; Wells *et al.*, 1980) may facilitate this. The diet of females not under the stress of lactation or pregnancy differed from that of adult males only in the spectrum of species taken, suggesting that female subgroups may exist, though these have not been identified elsewhere.

Bottlenose dolphin schools off the Natal coast have been observed more frequently along particular regions, termed preferred areas (Ross *et al.*, 1987). There is some evidence that dolphins reside in preferred areas for relatively long periods. Consequently, they probably exert continuous pressure on available food resources in these areas. Under such circumstances inshore dolphins may have adapted socially and behaviorally to alleviate intraspecific competition for food. These behavioral adaptations may include partitioning of the school into subgroups and the use of different foraging ranges or different prey sizes and species by these subgroups. The present results support the conclusions of Irvine *et al.* (1979), who suggested that relatively mobile, wide-ranging pelagic dolphins may not exert the same pressure on their food resources, and that management and conservation strategies for each population need to be defined separately, particularly for those which are part of the more complex coastal ecosystems.

ACKNOWLEDGMENTS

The authors thank the Director and staff of the Natal Sharks Board for cooperation in obtaining the material for this study. We also thank Mzikayse Mahola for his valued work behind the scenes.

REFERENCES

Bigg, M. A., and Perez, M. A. (1985). Modified volume: A frequency volume method to assess marine mammal food habits. *In* "Marine Mammals and Fisheries" (J. R. Beddington, R. H. Beverton, and D. M. Lavigne, eds.), pp. 277–283. George Allen and Unwin, London.

Clarke, M. R., ed. (1986a). "A Handbook for the Identification of Cephalopod Beaks." Clarendon Press, Oxford.

Clarke, M. R. (1986b). Cephalopods in the diet of odontocetes. *In* "Research on Dolphins" (M. M. Bryden and R. Harrison, eds.), pp. 281–321. Clarendon Press, Oxford.

van der Elst, R. (1981). "A Guide to the Common Sea Fishes of Southern Africa." C. Struik, Cape Town, South Africa. 367 pp.

Evans, P. G. H. (1980). Cetaceans in British waters. *Mamm. Rev.* **10**(1), 1–52.

Gunter, G. (1942). Contributions to the natural history of the bottlenose dolphin, *Tursiops truncatus* (Montague), on the Texas coast, with particular reference to food habits. *J. Mammal.* **23**(3), 267–276.

Innes, S., Lavigne, D. M., Earle, W. M., and Kovacs, K. M. (1987). Feeding rates of seals and whales. *J. Anim. Ecol.* **56,** 115–130.

Irvine, B. A., and Wells, R. S. (1972). Results of attempts to tag Atlantic bottlenosed dolphins (*Tursiops truncatus*). *Cetology* **13,** 1–15.

Irvine, B. A., Scott, M. D., Wells, R. S., Kaufmann, J. H., and Evans, W. E. (1979). A study of the activities and movements of the Atlantic bottlenosed dolphin, *Tursiops truncatus,* including an evaluation of tagging techniques. *Nat. Tech. Inform Serv.* PB-298 042. 54 pp.

Irvine, B. A., Scott, M. D., Wells, R. S., and Kaufmann, J. H. (1981). Movements and activities of the Atlantic bottlenosed dolphin, *Tursiops truncatus,* near Sarasota, Florida. *Fish. Bull.* **79**(4), 671–688.

Joubert, C. S. W. (1981). Aspects of the biology of five species of inshore reef fishes on the Natal coast, South Africa. *Oceanogr. Res. Inst. Invest. Rep.* **51,** 1–16.

Leatherwood, S. (1977). Some preliminary impressions on the numbers of social behavior of free-swimming, bottlenosed dolphin calves (*Tursiops truncatus*) in the northern Gulf of Mexico. *In* "Breeding Dolphins: Present Status, Suggestions for the Future" (S. H. Ridgway and K. W. Benirschke, eds.), pp. 143–167. Marine Mammal Commission, Report MMC-76/07, Washington, D.C.

Mead, J. G. (1975). Preliminary report on the former net fisheries for *Tursiops truncatus* in the western north Atlantic. *J. Fish. Res. Board Can.* **32**(7), 1155–1162.

Norris, K. S., and Dohl, T. P. (1980). The structure and function of cetacean schools. *In* "Cetacean Behavior: Mechanisms and Functions" (L. M. Herman, ed.), pp. 211–262. Wiley, New York.

Pinkas, L., Oliphant, M. S., and Iverson, I. L. K. (1971). Food habits of albacore, bluefin tuna, and bonito in California waters. *Calif. Depart. Fish Game, Fish Bull.* No. 152. 105 pp.

Ross, G. J. B. (1984). The smaller cetaceans of the south east coast of southern Africa. *Ann. Cape Prov. Mus. (Nat. Hist.)* **11**(14), 259–327.

Ross, G. J. B., Cockcroft, V. G., and Butterworth, D. S. (1986). Offshore distribution of bottlenosed dolphins in Natal coastal waters and Algoa Bay, eastern Cape. *S. Afr. J. Zool.* **22**(1), 50–57.

Ross, G. J. B., Cockcroft, V. G., Cliff, G., and Richards, H. (1987). The longshore distribution of the bottlenosed dolphin *Tursiops aduncus* on the Natal coast. Paper presented at IWC meeting on Indian Ocean Sanctuary, Mahe, February 1987. Doc. SC/F87/5/12.

Ross, G. J. B., Cockcroft, V. G., Melton, D. A., and Butterworth, D. S. (1989). Population estimates for bottlenose dolphins *Tursiops truncatus* in Natal and Transkei waters. *S. Afr. J. Mar. Sci.* (in press).

Saayman, G. S., Tayler, C. K., and Bower, D. (1973). Diurnal activity cycles in captive and free-ranging Indian Ocean bottlenosed dolphins (*Tursiops aduncus* Ehrenberg). *Behaviour* **44,** 212–233.

Sergeant, D. E. (1969). Feeding rates of cetacea. *Fisk Dir. Skr. Ser. HavUnders.* **15,** 246–258.

Smith, M. M., and Heemstra, P. C., eds. (1986). "Smiths' Sea Fishes." Macmillan, Johannesburg, South Africa. 1047 pp.

Tomilin, A. G. (1957). "Mammals of the U.S.S.R. and Adjacent Countries. Volume IX. Cetacea." Israel Program for Scientific Translations, Jerusalem, 1967. 717 pp.

Walker, W. A. (1981). Geographical variation in morphology and biology of bottlenosed dolphins (*Tursiops*) in the eastern north Pacific. *NOAA Admin. Techn. Rep.,* No. LJ-81-23C. 19 pp.

Wells, R. S., Irvine, A. B., and Scott. M. D. (1980). The social ecology of inshore odontocetes. *In* "Cetacean Behavior: Mechanisms and Functions" (L. M. Herman, ed.), pp. 211–262. Wiley, New York.

Würsig, B. (1978). Occurrence and group organization of Atlantic bottlenose porpoises (*Tursiops truncatus*) in an Argentine bay. *Biol Bull.* **154,** 348–359.

Zar, J. H. (1984). "Biostatistical Analysis," 2nd Ed. Prentice Hall, Englewood Cliffs, New Jersey.

Food Habits of Bottlenose Dolphins in the Southeastern United States

16

Nélio B. Barros
Division of Biology and Living Resources
Rosenstiel School of Marine and Atmospheric Science
University of Miami
Miami, Florida 33149

Daniel K. Odell
Division of Biology and Living Resources
Rosenstiel School of Marine and Atmospheric Science
University of Miami
Miami, Florida 33149
Present address:
Sea World of Florida, Orlando, Florida 32821

INTRODUCTION

Bottlenose dolphins, *Tursiops truncatus,* are common in bays, gulfs, lagoons, and coastal waters of the Gulf of Mexico and along the southeastern United States. They are the cetaceans most frequently observed alive in Florida's coastal waters (Odell, 1976) and the most commonly stranded marine mammals in the southeastern United States (Odell, 1983).

Bottlenose dolphins have been implicated in interactions with fishery activities through predation, gear damage, and net entanglement (Caldwell and Caldwell, 1972; Leatherwood, 1975; Cato and Prochaska, 1976; Reynolds, 1985; Shane *et al.*, 1986). Their feeding behavior and diet have been described (Gunter, 1942; Kemp, 1949; Hoglund, 1965; Leatherwood, 1975; Shane, 1977; Leatherwood *et al.*, 1978; Gruber, 1981), but few studies have provided quantitative information. The objectives of our study were to identify and quantify the prey items in the diet of bottlenose dolphins in the southeastern United States and to investigate geographical, seasonal, and sex differences in the diet. The findings reported here derive solely from the examination of stomach contents of stranded animals and may involve biases with respect to wild and healthy populations.

MATERIALS AND METHODS

Stomach contents were collected from stranded bottlenose dolphins through the Southeastern U.S. Marine Mammal Stranding Network, which encompasses the states of Texas, Louisiana, Mississippi, Alabama, Florida, Georgia, North Carolina, and South Carolina (Odell, 1979). Of a total of 1506 bottlenose dolphins reported to have stranded in this region between 1973 and 1987, 234 were examined for stomach contents. Of these, 126 had empty stomachs and 108 had stomachs that contained food matter. Of the 108 containing food matter, 76 were available for further analysis and thus comprise the basis of the present study. Of the animals eligible for study, 57 were from Florida, 22 from Texas, 1 from Louisiana, 1 from Mississippi, and 1 from Alabama. Thirty-five were males, 36 females, and 5 of unknown sex. Animals in this sample ranged from 165 to 285 cm in total length.

Collection of stomach contents was opportunistic. Data collection depended primarily on the availability of volunteers involved in the stranding network and was not evenly distributed through the seasons. Twenty-three animals (30.3%) were collected during spring (March–May), 15 (19.7%) during summer (June–August), 16 (21.1%) during fall (September–November), and 22 (28.9%) during winter (December–February). This distribution does not reflect the seasonal stranding patterns of bottlenose dolphins in the study area, which appear to be uniform year-round, except in unusually cold winters (Hersh *et al.*, Chapter 8, this volume).

Stomach contents were usually preserved in a 70% ethanol solution. At least three samples were preserved in buffered and/or unbuffered formalin; this technique is not ideal because it damages structures like fish otoliths (Fitch and Brownell, 1968), dissolving distinguishing features even after short periods of fixation (McMahon and Tash, 1979). Thirteen stomachs were frozen on collection. Stomach contents were washed through a 1-mm-mesh sieve and weighed with a digital scale to the nearest gram. Food remains were hand-sorted into fish, cephalopod, crustacean, and other components (e.g., mollusc shells, plant material). Loose otoliths were stored dry in glass vials, and cephalopod breaks were preserved in 70% ethanol.

Fish otoliths from the stomach samples were identified using a preliminary reference collection and by comparison with published keys (Eziuzo, 1963; Schmidt, 1968; Chao, 1978; Hecht, 1978; Hecht and Hecht, 1978; Brouder, 1979; Morrow, 1979; Härkönen, 1986). Different species of seatrouts and weakfish (genus *Cynoscion*) were identified with the help of specific literature (Mohsin, 1973, 1981; Gonzáles, 1977; Chao, 1978). Cephalopod beaks were identified by S. Candela (University of Miami) using a reference collection (from the Invertebrate Museum of RSMAS) and with the aid of identification keys and pictorial guides (Clarke, 1962, 1986a; Iverson and Pinkas, 1971; Wolff, 1984). Undigested fish were identified to the lowest possible taxon by C. R. Robins (University of Miami) and measured to the nearest cm (total length).

Total number of prey individuals per alimentary tract was recorded. Sagittal otoliths outnumbered all other fish bones in frequency of occurrence and were used in the estimation of total fish eaten. Otoliths were identified as left or right, and the total number of fish was estimated by the largest number of otoliths of either side. In cases where separation of otoliths by side was not possible, the total number of fish was estimated by halving the number of otoliths. In a few cases (e.g., *Trichiurus lepturus, Callechelys* sp., *Bascanichthys* sp., scarids), other bones (jaws, pharyngeal teeth) were used to quantify fish occurrence. Cephalopod beaks were sorted into upper and lower, and the largest count of either provided the estimate of specimens ingested. Cephalopod eye lenses or gladii were used when beaks were not present. Shrimps were quantified by the number of rostra or telsons found.

Most dolphin carcasses from which stomach contents were collected were too decomposed, or necropsy data were insufficient, to determine cause of death. This was true of 45 (59.2%) of the dolphins. Six specimens exhibited shark bites (probably postmortem), mainly in the ventral region, and it was impossible to determine the sex in 4 of these dolphins. Eighteen (23.7%) probably died of various diseases, the most common (n = 10 or 55.6%) being pulmonary problems. Other sicknesses were related to kidney and pancreas malfunctions, and, in two cases, lobomycosis (a fungal disease) was present. Possible fishery-related mortalities were noted in 9 (11.8%) cases. Evidence suggesting such interactions included gunshot wounds, tail flukes cut off, and lacerations of various kinds (Hare and Mead, 1987). Eleven of the 18 aforementioned sick dolphins were chosen based on season and location of stranding for comparison with the 9 dolphins bearing indications of fishery-related mortality. We have assumed that these 9 dolphins were healthy and had a normal diet when they died or were killed.

The study area was divided into four subareas (Fig. 1): Area I, the states of Texas, Louisiana, Mississippi, and Alabama; Area II, from Tampa Bay southward to Cape Sable; Area III, the Florida Keys and the southeastern coast of Florida north to Palm Beach; Area IV, Palm Beach north to Cape Canaveral. The study area was divided primarily on the basis of the distribution patterns of the fish fauna. Cape Sable and Palm Beach have been considered the boundaries of fish fauna assemblies (Robins *et al.*, 1986), the former being the usual southern limit of temperate fishes coming from the northern Gulf of Mexico during winter, the latter the usual northern limit for many tropical fishes.

RESULTS

Seventy-five of the 76 dolphins from this study had preyed on fish (Table 1). In 28 (36.8%) of these, cephalopods were also found. Shrimp were found in 11 (14.5%) stomachs. A total of 17,782 free otoliths were retrieved from all stomachs. These were sagitta (n = 12,687) and lapilli and/or asterisci (n = 5095). The silver perch (*Bairdiella chrysoura*) possesses fairly large and distinctive lapilli and asterisci (Chao, 1978), and it was possible to assign 4590 lapilli and 97 asterisci to this species. Two hundred fifty-eight lapilli belonged to marine catfish (*Arius felis* or *Bagre marinus*). A total of 7454 prey items from the stomachs of dolphins in the study area are listed in Table 2. Of these, 7027 (94.3%) were fish, 387 (5.2%) cephalopods, 39 crustaceans (0.5%), and 1 a horseshoe crab. Figure 2 shows the frequency of occurrence of different prey taxa in the study area. The fish-only category dominated in Areas II–IV (62–80%), but only 32% of the stomachs from Area I were assigned to this category. Fish/cephalopod and fish/cephalopod/shrimp were present in 36 and 32%, respectively, of the stomachs from Area I. Cephalopods occurred in 68% of the samples in Area I, but were less frequent in Area II (31.8%), Area III (18.2%), and Area IV (11.2%). Prey groups involving cephalopods and shrimp occurred more frequently in Area I than in any of the other three areas (Fig. 2).

The fish found in the dolphin stomachs represented at least 43 species in 39 genera, distributed within 25 families. From these, at least 24 taxa (identified to lowest taxonomic level possible) represent new records for the known diet of bottlenose dolphins in the western North Atlantic (Table 2). With respect to cephalopods, 467 free beaks were found, belonging to 5 species, in 3 families.

Most important prey, by frequency of occurrence (i.e., presence or absence in the stomachs), were *Bairdiella chrysoura, Micropogonias undulatus, Cynoscion arenarius, Mugil* sp., *Lolliguncula brevis,* and *Leiostomus xanthurus,* all occurring in 20% or more of the stomachs (Table 2). Numerically important prey (by total number of individual remains in all stomachs) were *Bairdiella chrysoura, Micropogonias undulatus, Cynoscion arenarius, C. nothus, Lolliguncula brevis,* and *Trichiurus lepturus.* Fishes of the family Scianidae occurred in 46 (60.5%) stomachs and represented 78.4% (n = 4755) of all identified fish. Other frequent fish families were Mugilidae, Batrachoididae, and Haemulidae. Most fish prey are bottom dwellers (e.g., drums, croakers, seatrouts, toadfishes, midshipmen), but surface dwellers (e.g., mullets, clupeids) and pelagic fish (jacks, bluefish, cutlassfish) are also represented. Most of these fish occur in bays and sheltered lagoons, and many are estuarine-dependent (e.g., most sciaenids) (Darovec, 1983). The brief squid (*Lolliguncula brevis*) is associated with estuaries throughout its distribution and can live in salinities as low as 17% (Voss and Brakoniecki, 1985).

The wet weights of stomach contents ranged from 1 to 10,736 g (Table 1) with a large degree of variation ($\bar{x}$ = 682.6, S.D. = 1645.2, n = 67), and these did not differ significantly among areas (ANOVA, $p > .05$). These weights should not be taken as indicators of meal sizes, for most stomachs contained prey in varying stages of digestion. In 17 cases stomach contents weighed 10 g or less, falling into the "trace" category described by Perrin *et al.* (1973). The average weights of stomach

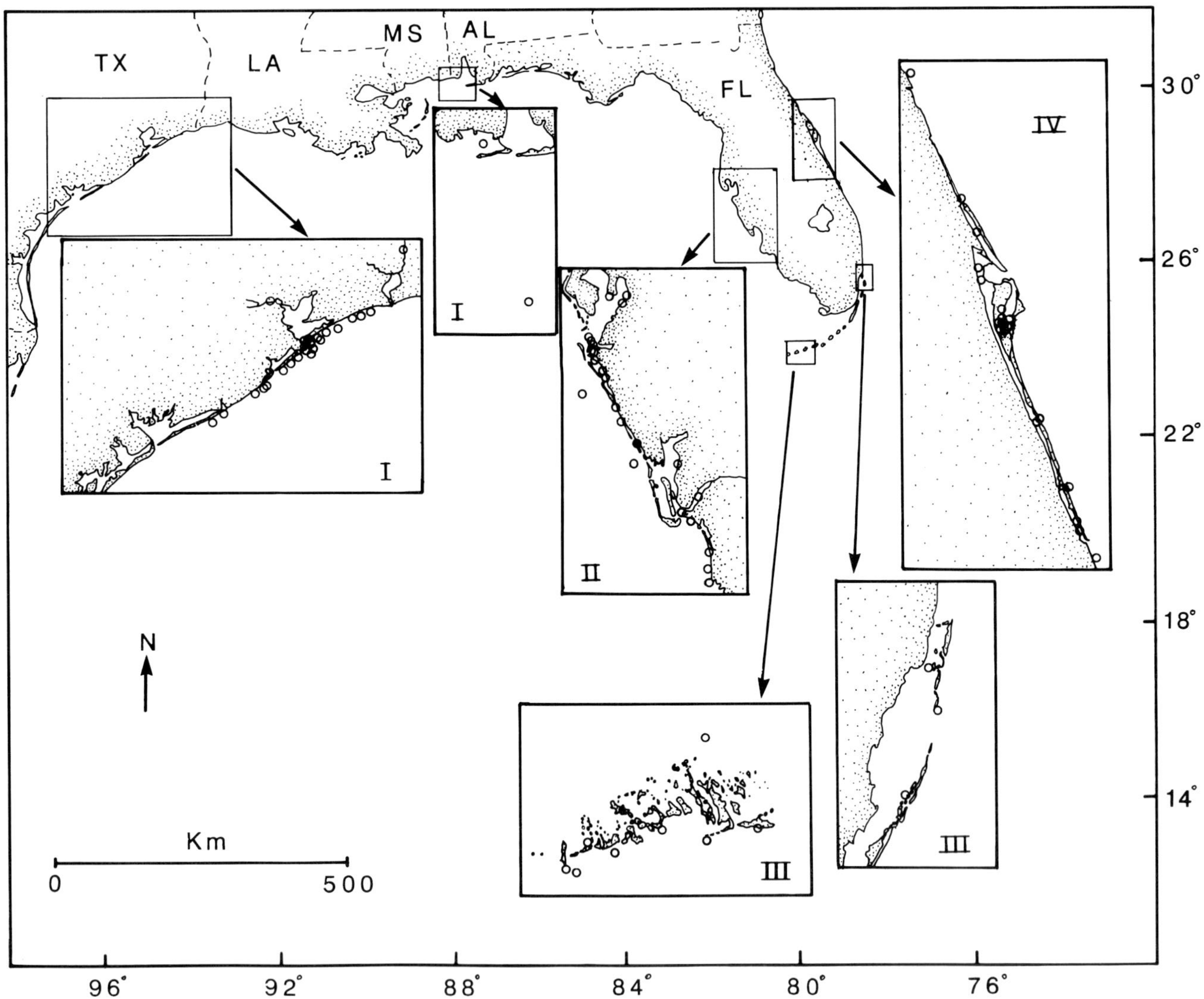

Figure 1 Collection sites of stomach contents from bottlenose dolphins stranded in the southeastern United States. Area I encompasses Texas, Louisiana, Mississippi, and Alabama; Area II, southwestern Florida; Area III, the Florida Keys; and Area IV, southeastern Florida.

Table 1
Stomach Contents of Bottlenose Dolphins from the Southeastern United States[a]

SEUS	Field No.	Area	Sex	Total Length (cm)	Season	Wet Weight (g)	Prey Type	Prey Items	Prey taxa
—	C-73-01	III	M	176.5	W	1	F, C	17	2
—	SWF 674	IV	F	266	W	660	F	2	2
—	SWF 702	IV	M	178	W	185	F	146	8
SE0095	C-78-08	III	M	226	Su	50	F	1	1
SE0412	C-80-18	III	M	226	Fa	1,094	F	11	1
SE0413	C-80-19	II	F	176	Fa	336	F, C	12	2
SE0424	SWF8028	IV	M	226	Fa	—	K, X	2	2
SE0442	C-80-23	II	F	257	W	798	F	23	8
SE0443	—	I	F	229	Fa	—	F	19	2
SE0464	C-81-03	II	M	236	W	1	F	8	2
SE0475	C-81-07	III	M	236	W	75	F	4	3
SE0518	C-81-09	II	M	(260)	W	252	F, C	67	6
SE0586	C-81-14	II	F	253.5	Su	10,736	F	20	6
SE0590	C-81-16	IV	U	252	Su	50+[b]	F	10	4
SE0599	C-81-18	IV	M	195	Su	920	F	18	6
SE0624[c]	—	I	F	187	Sp	—	F	2	1
SE0656	C-81-27	II	M	260	Fa	2	F	1	1
SE0731	SWF8204	IV	U	256	W	190	F	1	1
SE0767	SWF8222	IV	F	266	Sp	368	F	1	1
SE0863	C-82-10	III	M	199	Sp	556	F	5	1
SE1117	SWF8312	IV	F	225	Sp	244	F	1	1
SE1183	C-83-21	III	F	259	Su	2	F	1	1
SE1228	C-83-23	III	F	248	Su	496	F	4	4
SE1268	C-83-29	II	F	214	Su	56	F, C	125	11
SE1722	MML8402	II	M	252	Fa	—	F	1	1
SE1752	C-85-01	III	M	242	W	32	F	50	3
SE1788	SWF8501	III	M	244	W	3,629	F	4	3
SE1831	SWF8510	IV	M	195	Sp	118	F	18	2
SE1867	MML8511	II	M	205	W	—	F	4	2
SE1975	C-85-08	III	M	285	Su	1	F, C	10	3
SE2304	C-280	I	F	245	Sp	203+[d]	F, C	394	16
SE2400	SWF8682	IV	M	173	Fa	96	F	11	2
SE2437	C-351	I	U	(212)	Su	850	F, C, K	89	8
SE2441	C-355	I	U	U	Su	344	F, C	20	9
SE2459	MML8604	II	M	241	Su	—	F	4	1
SE2588	C-365	I	M	198	Fa	25	F	15	7
SE2663	MML8609	II	F	258	W	310	F	7	4
SE2664	MML8701	II	F	215	W	1	F, C	2	2
SE2665	MML8702	II	M	241	W	276	F	5	3
SE2702	MML8704	II	F	248	W	4,762	F	8	3
SE2712	SWF8714	IV	F	244	Sp	3	F	12	7
SE2717	SWF8719	IV	F	175	Sp	30	F, C, K	26	10
SE2731	MML8708	II	F	253	Sp	506	F, C	12	5
SE2732	MML8709	II	M	206	Sp	1,336	F, C, K	45	8
SE2746	C-87-02	IV	F	201	W	1	F	1	1
SE2759	C-374	I	F	236	W	9	F, C	70	8
SE2760	C-383	I	M	251	W	<1	F, C	5	4
SE2761	C-389	I	F	235	W	122	F	5	4
SE2762	C-391	I	M	237	Sp	6,550	F	82	4
SE2763	C-392	I	F	230	W	1	F	9	3
SE2764	C-415	I	F	235	Sp	434	F, C, K	203	14
SE2895	C-410	I	F	241	Sp	206	F, C	1,029	18
SE2896	C-420	I	F	225	Sp	4	F	8	2
SE2897	C-427	I	M	243	Sp	105	F, C, K	359	10
SE2898	C-428	I	F	243.5	Sp	35	F, C, K	340	11
SE2908	C-444	I	F	251	Sp	107	F, C, K	259	14
SE2917	C-453	I	F	243	Sp	514	F, C, K	308	13
SE2929	C-465	I	F	215	Sp	146	F, C	696	11
SE2941	C-477	I	M	255	Sp	559	F, C, K	929	16
SE2942	C-478	I	F	238	Sp	567	F, C	1,073	14

(Continued)

Table 1 *(continued)*

SEUS	Field No.	Area	Sex	Total Length (cm)	Season	Wet Weight (g)	Prey Type	Prey Items	Prey taxa
SE2945	C-481	I	F	240	Sp	1,204	F, C, K	100	12
SE2958	C-494	I	M	223	Sp	1,591	F	4	1
SE3036	MML8714	II	M	263	Fa	<1	F	1	1
SE3037	SWF8726	IV	M	(200)	Su	1+[b]	F	1	1
SE3038	SWF8727	IV	M	202	Su	522	F, K	231	9
SE3051	SWF8740	IV	U	(250)	Su	2	F	2	2
SE3054	SWF8743	IV	M	231	Fa	831	F, C	45	8
SE3085	C-87-06	III	F	206	Fa	<1	F	22	5
SE3134	MML8717	II	F	245	Fa	182	F	13	3
SE3135	MML8718	II	F	251	Fa	392	F, C	28	15
SE3155	SWF8747	IV	M	(215)	Fa	214	F	8	2
SE3171	C-510	I	F	249.5	Su	612	F, C	78	6
SE3178	C-517	I	F	219.3	Fa	1,377	F, C	200	13
SE3264	JW-002-87	II	M	199	W	8	F	10	3
SE3336	MML8802	II	F	262	W	86	F	84	4
SE3448	MML8715	II	M	164.6	Fa	8	F	13	4

[a] Key to abbreviations: SEUS, stranding network identification number; U, unknown; Sp, spring (March–May); Su, summer (June–August); Fa, fall (September–November); W, winter (December–February); F, fish; C, cephalopod; K, crustacean; X, xiphosuran. Values in parentheses are estimates. Areas I–IV refer to Fig. 1.
[b] Partial gut contents (stomach ripped open by sharks).
[c] L. Ogren (pers. commun., 12 February 1982).
[d] Weight of dried fish bones (material retrieved after being buried for 11 months).

contents weighing more than 10 g was 910.1 g (S.D. = 1851.9, n = 50).

Sergeant (1969) defined feeding rates of captive small cetaceans as percent body weight of food consumed per day. For two bottlenose dolphins he found an average feeding rate of 4.2%. Shapunov (1971) reported that Black Sea bottlenose dolphins (n = 50) consumed on average 5.6%/body weight/day of fish. Similarly, values of 5.3–6.5% for daily food consumption of a captive bottlenose dolphin were found by Spotte and Babus (1980). Bottlenose dolphins in captivity may receive anywhere from 5 to 32 kg of fish per day (Shapunov, 1971). At the Miami Seaquarium, bottlenose dolphins weighing an average of 180 kg consume from 12 to 24 ($\bar{x}$ = 20) kg of fish (Spanish mackerel, capelin, smelt, and herring) per day (G. Bossart, Miami Seaquarium, 1987, pers. commun.), which indicates a feeding rate of 3.0–6.0% ($\bar{x}$ = 5.0%).

In this study stomach contents averaged about 683 g (although much variation was found). Since in most cases contents were in different stages of digestion, only specimens in which recent ingestion of food was observed were considered for the following calculations. A 199-cm male (SE0863) had only remains of five undigested white grunts (*Haemulon plumieri*), contents weighing 556 g. Using a length–weight relationship (Ridgway and Fenner, 1982) for wild bottlenose dolphins, this animal was estimated to weight 150 kg. In this case, the ingested grunts would represent about 0.37% of the body weight, which would imply the need for 10 more meals to reach a feeding rate of 4%. By contrast, a 253-cm female (SE0586) had over 10,700 g of stomach contents, of which 2 snooks (*Centropomus undecimalis*) were responsible for 89% of the weight. This dolphin's weight was estimated to be 170 kg, giving a feeding rate of 6.32%, a value that better fits the range reported for bottlenose dolphins in the literature. Most stomach contents (62%), however, weighed less than 500g (see Table 1), and only in eight cases did values exceed 1 kg. Using Shapunov's (1971) equation expressing weight of mean daily ration as a function of the dolphin's weight, most values in this study fall below the expected weight of a daily food requirement.

The number of prey items found in the stomachs of bottlenose dolphins ranged from 1 to 1073 (Table 1) and was significantly different among areas (Table 1, F = 10.30, df = 3, 72, p < .001).

Table 2
Frequency of Occurrence (FO) and Number of Prey Taken by Bottlenose Dolphins in the Southeastern United States

Prey Species	Common Name[a]	Area I				Area II				Area III				Area IV			
		FO (n = 25)	%	No. Prey (N = 6296)	%	FO (n = 22)	%	No. Prey (N = 493)	%	FO (n = 11)	%	No. Prey (N = 129)	%	FO (n = 18)	%	No. Prey (N = 536)	%
Teleosts																	
Micropogonias undulatus	Atlantic croaker	22	88.0	828	13.2	1	4.5	3	0.6	—	—	—	—	3	16.7	146	27.2
Cynoscion arenarius	Sand seatrout	17	68.0	598	9.5	8	36.4	33	6.7	—	—	—	—	—	—	—	—
Bairdiella chrysoura	Silver perch	15	60.0	1999	31.8	8	36.4	66	13.4	—	—	—	—	4	22.2	126	23.5
Leiostomus xanthurus	Spot	12	48.0	67	1.1	2	9.1	5	1.0	—	—	—	—	4	22.2	19	3.5
Cynoscion nothus[b]	Silver seatrout	11	44.0	507	8.1	—	—	—	—	—	—	—	—	2	11.1	5	0.9
Mugil sp.	Mullet	11	44.0	25	0.4	8	36.4	25	5.1	1	9.1	1	0.8	3	16.7	22	4.1
Menticirrhus sp.[b]	Kingfish	9	36.0	70	1.1	—	—	—	—	—	—	—	—	1	5.6	1	0.2
Urophycis sp.	Hake	9	36.0	50	0.8	1	4.5	1	0.2	—	—	—	—	—	—	—	—
Trichiurus lepturus	Atlantic cutlassfish	8	32.0	173	2.7	—	—	—	—	—	—	—	—	1	5.6	1	0.2
Larimus fasciatus	Banded drum	8	32.0	120	1.9	—	—	—	—	—	—	—	—	—	—	—	—
Anchoa sp.	Anchovy	8	32.0	56	0.9	1	4.5	2	0.4	—	—	—	—	1	5.6	1	0.2
Porichthys plectrodon[b]	Atlantic midshipman	8	32.0	27	0.4	1	4.5	64	13.0	—	—	—	—	—	—	—	—
Cynoscion sp.	Seatrout	7	28.0	52	0.8	1	4.5	1	0.2	—	—	—	—	—	—	—	—
Arius felis/*Bagre marinus*	Sea catfish	5	20.0	123	2.0	1	4.5	1	0.2	—	—	—	—	1	5.6	1	0.2
Cynoscion nebulosus	Spotted seatrout	5	20.0	15	0.2	3	13.6	3	0.6	—	—	—	—	2	11.1	19	3.5
Atherinid[b]	Silverside	3	12.0	50	0.8	—	—	—	—	—	—	—	—	2	11.1	3	0.6
Sciaenops ocellatus[b]	Red drum	3	12.0	13	0.2	1	4.5	1	0.2	—	—	—	—	1	5.6	1	0.2
Pomatomus saltatrix[b]	Bluefish	3	12.0	3	—	1	4.5	1	0.2	1	9.1	1	0.8	1	5.6	27	5.0
Symphurus sp.	Tonguefish	2	8.0	80	1.3	—	—	—	—	—	—	—	—	—	—	—	—
Clupeid	Herring	2	8.0	22	0.3	2	9.1	5	1.0	—	—	—	—	1	5.6	1	0.2
Lagodon rhomboides	Pinfish	2	8.0	4	0.1	5	22.7	17	3.4	—	—	—	—	3	16.7	54	10.1
Anchoa hepsetus[b]	Striped anchovy	1	4.0	80	1.3	—	—	—	—	—	—	—	—	—	—	—	—
Pogonias cromis	Black drum	1	4.0	41	0.7	3	13.6	8	1.6	—	—	—	—	1	5.6	1	0.2
Callechelys sp./ *Bascanichthys* sp[b]	Snake eel	1	4.0	36	0.6	—	—	—	—	—	—	—	—	—	—	—	—
Pomoxis nigromaculatus[b]	Black crappie	1	4.0	25	0.4	—	—	—	—	—	—	—	—	—	—	—	—
Anguilliform	Eel	1	4.0	8	0.1	—	—	—	—	—	—	—	—	—	—	—	—
Archosargus probatocephalus	Sheepshead	1	4.0	4	0.1	—	—	—	—	—	—	—	—	—	—	—	—
Orthopristis chrysoptera	Pigfish	1	4.0	2	—	6	27.3	90	18.3	—	—	—	—	3	16.7	10	1.9
Synodus foetens	Inshore lizardfish	1	4.0	1	—	—	—	—	—	—	—	—	—	—	—	—	—
Opsanus beta[b]	Gulf toadfish	1	4.0	1	—	3	13.6	12	2.4	2	18.2	24	18.6	4	22.2	20	3.7
Haemulon sp.	Grunt	—	—	—	—	2	9.1	11	2.2	5	45.5	28	21.7	1	5.6	1	0.2
Sardinella aurita[b]	Spanish sardine	—	—	—	—	4	18.2	19	3.9	—	—	—	—	—	—	—	—
Brevoortia sp.	Menhaden	—	—	—	—	3	13.6	5	1.0	—	—	—	—	1	5.6	1	0.2
Cynoscion regalis[b]	Weakfish	—	—	—	—	—	—	—	—	—	—	—	—	4	22.2	5	0.9
Decapterus punctatus	Round scad	—	—	—	—	1	4.5	1	0.2	—	—	—	—	1	5.6	14	2.6

(Continued)

Table 2 *(continued)*

Prey Species	Common Name[a]	Area I				Area II				Area III				Area IV			
		FO (n = 25)	%	No. Prey (N = 6296)	%	FO (n = 22)	%	No. Prey (N = 493)	%	FO (n = 11)	%	No. Prey (N = 129)	%	FO (n = 18)	%	No. Prey (N = 536)	%
Haemulid	Grunt	—	—	—	—	—	—	—	—	1	9.1	4	3.1	1	5.6	10	1.9
Diapterus plumieri[b]	Striped mojarra	—	—	—	—	—	—	—	—	—	—	—	—	2	11.1	8	1.5
Sciaenid	Croaker	—	—	—	—	—	—	—	—	—	—	—	—	2	11.1	2	0.4
Scarid[b]	Parrotfish	—	—	—	—	—	—	—	—	1	9.1	1	0.8	1	5.6	1	0.2
Diapterus sp.	Mojarra	—	—	—	—	1	4.5	7	1.4	—	—	—	—	—	—	—	—
Haemulon sciurus[b]	Bluestriped grunt	—	—	—	—	—	—	—	—	1	9.1	6	4.7	—	—	—	—
Haemulon plumieri[b]	White grunt	—	—	—	—	—	—	—	—	1	9.1	5	3.9	—	—	—	—
Brevoortia patronus[b]	Gulf menhaden	—	—	—	—	1	4.5	3	0.6	—	—	—	—	—	—	—	—
Opisthonema oglinum[b]	Atlantic thread herring	—	—	—	—	—	—	—	—	—	—	—	—	1	5.6	2	0.4
Centropomus undecimalis	Snook	—	—	—	—	1	4.5	2	0.4	—	—	—	—	—	—	—	—
Caranx hippos	Crevalle jack	—	—	—	—	1	4.5	1	0.2	—	—	—	—	—	—	—	—
Caranx sp.	Jack	—	—	—	—	—	—	—	—	1	9.1	1	0.8	—	—	—	—
Gerres cinereus[b]	Yellowfin mojarra	—	—	—	—	—	—	—	—	1	9.1	1	0.8	—	—	—	—
Chaetodipterus faber[b]	Spadefish	—	—	—	—	1	4.5	1	0.2	—	—	—	—	—	—	—	—
Tilapia sp.[b]	Tilapia	—	—	—	—	—	—	—	—	—	—	—	—	1	5.6	1	0.2
Bodianus sp.[b]	Hogfish	—	—	—	—	—	—	—	—	1	9.1	1	0.8	—	—	—	—
Acanthurus sp.[b]	Surgeonfish	—	—	—	—	—	—	—	—	1	9.1	1	0.8	—	—	—	—
Lactophrys trigonus[b]	Trunkfish	—	—	—	—	1	4.5	1	0.2	—	—	—	—	—	—	—	—
Unidentified fish	—	16	64.0	844	13.4	12	54.5	66	13.4	10	90.9	44	34.1	9	50.0	20	3.7
Cephalopods																	
Lolliguncula brevis[b]	Brief squid	14	56.0	328	5.2	4	18.2	28	5.7	—	—	—	—	2	11.1	2	0.4
Doryteuthis sp.[b,c]	Longfin/arrow squid	2	8.0	4	0.1	3	13.6	7	1.4	—	—	—	—	—	—	—	—
Loliginid	Squid	3	12.0	5	0.1	—	—	—	—	—	—	—	—	—	—	—	—
Illex sp.	Shortfin squid	—	—	—	—	—	—	—	—	1	9.1	6	4.7	—	—	—	—
Ornithoteuthis antillarum[b]	Atlantic bird squid	—	—	—	—	—	—	—	—	1	9.1	1	0.8	—	—	—	—
Octopodid	Octopus	—	—	—	—	1	4.5	1	0.2	—	—	—	—	—	—	—	—
Unidentified cephalopod	—	—	—	—	—	1	4.5	1	0.2	2	18.2	4	3.1	—	—	—	—
Crustaceans																	
Penaeus sp.	Shrimp	7	28.0	34	0.5	1	4.5	1	0.2	—	—	—	—	2	11.1	2	0.4
Squilla empusa[b]	Mantis shrimp	1	4.0	1	—	—	—	—	—	—	—	—	—	—	—	—	—
Unidentified shrimp	Shrimp	—	—	—	—	—	—	—	—	—	—	—	—	1	5.6	1	0.2
Xiphosuran																	
Limulus polyphemus[b]	Horseshoe crab	—	—	—	—	—	—	—	—	—	—	—	—	2	5.6	1	0.2

[a] Common names of fishes follow those of Robbins *et al.* (1980, 1986).

[b] Not previously reported as prey of bottlenose dolphins in the Western North Atlantic.

[c] *Sensu* Brakoniecki (1986).

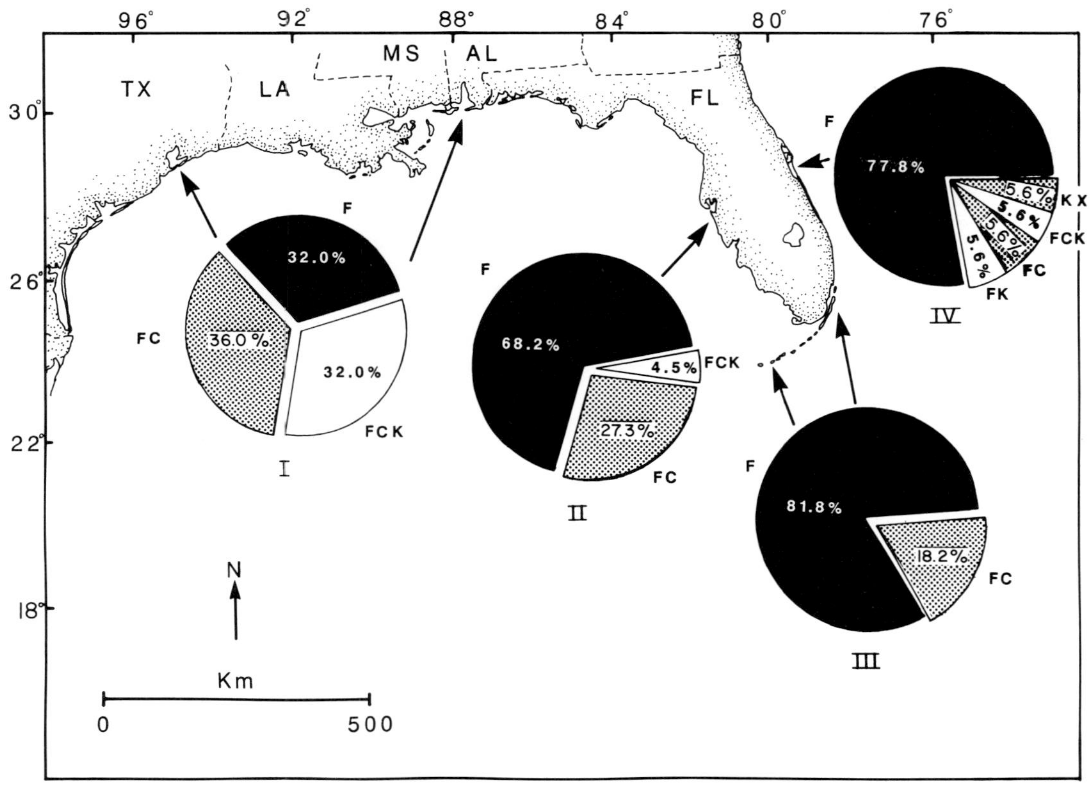

Figure 2 Geographical variation in the diet of bottlenose dolphins, expressed as frequency of occurrence of different prey groups. F, Fish; C, Cephalopod; K, Crustacean; X, Xiphosuran.

Likewise, the number of different prey taxa taken by the dolphins differed among areas (Table 1, $F = 7.54$, $df = 3, 72$, $p < .001$). These data were subjected to a natural logarithm transformation to comply with normality and homoscedasticity of variances (Sokal and Rolf, 1981). A Student–Newman–Keuls multiple range test showed Area I animals to have significantly greater values ($p < .001$) for both parameters. In eight cases (Table 1), more than 300 prey specimens were retrieved from single stomachs, all in Area I. Prey specimens in this area were often very small fish (judging from otolith size, mostly < 5 mm long), undigested squid, and shrimp. In other areas, prey specimens typically numbered less than 200 (Table 1), and squid and shrimp occurred less frequently (Fig. 2).

Two hundred eleven undigested fish belonging to at least 25 different taxa were measured (Table 3). Over 90% of all fish measured ranged from 5 to 30 cm in total length. Fifteen different taxa were recorded in the 20–25 cm category alone. A total of 109 undigested squid (mostly *Lolliguncula brevis*) was found in three stomachs. They ranged in size (mantle length) from 2.5 to 7.5 cm ($\bar{x} = 3.63$, S.D. $= 0.78$, $n = 99$).

One hundred twenty-six dolphins had empty stomachs. Of these, 28 (22.2%) were 97–167 cm long. In three instances when we did not personally examine the carcasses, a 132-cm female reportedly had remains of mangrove, seagrass, and seaweed leaves in the stomach, and a 169-cm female and a 170-cm male were reported to have stomachs "full of fish." These specimens were not included in the sample used for this study. The 5 smallest dolphins observed to have foot matter in their stomachs were a 165-cm male (SE3448), a

Table 3
Number of Undigested Fish, Categorized by Total Length, Found in the Stomachs of Bottlenose Dolphins from the Southeastern United States

Fish Species	N[a]	Total Length (cm)												Total
		0–5	5–10	10–15	15–20	20–25	25–30	30–35	40–45	45–50	50–55	75–80	80–85	
Callechelys sp./ *Bascanichthys* sp.	1					1			1	1	1			4
Brevoortia patronus	1					3								3
Brevoortia sp.	2						2							2
Anchoa hepsetus	1		80											80
Arius felis	1					1								1
Opsanus beta	3	1	7	5	4	1								18
Centropomus undecimalis	1											1	1	2
Pomoxis nigromaculatus	1				6	6	2							14
Diapterus plumieri	1					1								1
Diapterus sp.	1				2	1								3
Haemulon sciurus	1				2	4								6
H. plumieri	1					3	2							5
Haemulon sp.	2				2									2
Orthopristis chrysoptera	1				1									1
A. probatocephalus	1							3						3
Lagodon rhomboides	1			1										1
Bairdiella chrysoura	2			5	2									7
Cynoscion arenarius	1					1								1
C. nebulosus	2					1		1						2
C. nothus	2		1	4										5
Cynoscion sp.	1				1									1
Micropogonias undulatus	2			1		2								3
Pogonias cromis	1						1							1
Chaetodipterus faber	1				1									1
Bodianus sp.	1					1								1
Mugil sp.	4				3	10	7	2						22
Acanthurus sp.	1					1								1
Symphurus sp.	2	8		11										19
Lactophrys trigonus	1					1								1
Total		9	88	27	24	38	14	6	1	1	1	1	1	211
%		4.3	41.7	12.8	11.4	18.0	6.6	2.8	0.5	0.5	0.5	0.5	0.5	100.1

[a] *N* refers to the number of stomachs containing each species of fish. Nomenclature and phyletic listing are after Robins *et al.* (1980, 1986).

Table 4
Prey Composition of Sick and Presumably Healthy Bottlenose Dolphins[a]

Location	Presumably Healthy		Sick	
Area I	SE0443 *Micropogonias undulatus* *Opsanus beta*	SE2441 *Micropogonias undulatus* *Cynoscion arenarius* *C. nebulosus* *Menticirrhus* sp. *Porichthys plectrodon* *Mugil* sp. *Leiostomus xanthurus* *Lolliguncula brevis*	SE0624 *Micropogonias undulatus* SE2588 *Micropogonias undulatus* *Cynoscion nothus* *C. arenarius* *Leiostomus xanthurus* *Porichthys plectrodon* SE2760 *Porichthys plectrodon* *Cynoscion* sp. *Lolliguncula brevis*	SE2764 *Bairdiella chrysoura* *Trichiurus lepturus* *Leiostomus xanthurus* *Cynoscion arenarius* *Micropogonias undulatus* *Menticirrhus* sp. *Larimus fasciatus* *Arius felis/Bagre marinus* *Cynoscion nothus* *Pomatomus saltatrix* *Anchoa* sp.
Area II	SE0518 *Bairdiella chrysoura* *Cynoscion arenarius* *C. nebulosus* *Haemulon* sp. *Doryteuthis* sp. *Lolliguncula brevis* SE1867 *Brevoortia patronus* *Cynoscion* sp.	SE2663 *Orthopristis chrysoptera* *Cynoscion arenarius* *Brevoortia* sp. *Urophycis* sp. SE2731 *Mugil* sp. *Cynoscion arenarius* *Brevoortia* sp. *Sardinella aurita* *Lolliguncula brevis*	SE0464 *Cynoscion arenarius* *Bairdiella chrysoura* SE2459 Clupeid SE3134 *Opsanus beta* *Bairdiella chrysoura*	SE3135 *Mugil* sp. *Micropogonias undulatus* *Orthopristis chrysoptera* *Pogonias cromis* *Opsanus beta* *Lagodon rhomboides* *Bairdiella chrysoura* *Pomatomus saltatrix* *Lolliguncula brevis*
Area III	SE0412 *Haemulon sciurus* *Haemulon* sp.		SE0475 *Haemulon* sp.	
Area IV	SE1831 *Decapterus punctatus* *Cynoscion nothus*	SE2746 *Cynoscion regalis*	SE2712[b] *Bairdiella chrysoura* *Leiostomus xanthurus* *Cynoscion regalis* *Cynoscion nothus* Atherinid	SE2717[b] *Opsanus beta* *Arius felis/Bagre marinus* *Bairdiella chrysoura* *Micropogonias undulatus* Sciaenid Clupeid *Penaeus* sp. *Lolliguncula brevis*

[a] See text for definition of sick and healthy dolphin groups. Areas I–IV refer to Fig. 1.
[b] Inconclusive necropsy.

173-cm male (SE2400), a 176-cm female (SE0413), a 177-cm male (C-73-01), and a 178-cm male (SWF702). However, while this chapter was being written, food matter was found in the stomachs of a 131-cm male, a 137-cm female, and a 147-cm female; data from these specimens are not reported here.

Comparison of the prey composition of presumably healthy and sick dolphins (Table 4) indicates that important prey species (e.g., *Bairdiella chrysoura*, *Cynoscion* spp., *Micropogonias undulatus*, *Haemulon* ssp., *Mugil* sp., and *Lolliguncula brevis*) occurred in both groups. Of the five major prey taxa in the four geographical areas, sciaenids (*Micropogonias undulatus*, *Cynoscion arenarius*/*C. regalis*, *Bairdiella chrysoura*) were important, both by frequency of occurrence and by number (Table 2). In Area I, the five most important prey were present in half or more of the samples, indicating a clear dominance of those taxa. However, dolphins from Areas II and IV do not seem to prey disproportionately on any group, and frequency-of-occurrence values are much lower. The squid *Lolliguncula brevis* was found in large numbers in Area I, and only in this area were undigested specimens retrieved from the stomachs of dolphins.

Although sample sizes are small, male and female bottlenose dolphins appear to prey on the same species (Table 5). However, females seem to prey more frequently on *Bairdiella chrysoura* in Area I ($\chi^2 = 5.9$, $p < .05$) and on *Mugil* sp. in Area II ($\chi^2 = 7.1$, $p < .01$). Detailed seasonal considerations are hindered by the small sample size, but in all four areas of this study, the five major prey species occur in at least 3 seasons in proportional values. No clear seasonal preferences could be seen.

In most areas where bottlenose dolphins have been studied, there appear to be two ecotypes, a coastal form and an offshore form (Ross, 1977, 1984; Walker, 1981; Leatherwood and Reeves, 1982; Duffield *et al.*, 1983; Van Waerebeek *et al.*, Chapter 7, this volume). One of the dolphins from this study, a 285-cm male (SE1975), was diagnosed by morphometry (Hersh, 1987) as belonging to the offshore variety. Unlike the coastal dolphins, this animal had been feeding on ommastrephid squids (*Illex* sp. and *Ornithoteuthis antillarum*), and its stomach also contained a pair of unidentified fish otoliths.

DISCUSSION

Geographical Variation in Feeding

Sciaenids (drums, croakers, seatrouts) are the most diverse family of fish in the northwestern Gulf of Mexico and the dominant fish group in terms of biomass (Hoese and Moore, 1977). Off Texas, sciaenids are the dominant groundfish family, by weight, and comprise the majority of the by-catch in shrimping operations (Pellegrin, 1982). Sciaenids were the dominant group of organisms (by number and frequency of occurrence) found in stomachs of bottlenose dolphins in Areas I, II, and IV but were replaced by haemulids (grunts) in Area III. Sciaenids are the dominant fish family in temperate waters of the Gulf of Mexico; in the tropical and sub-tropical gulf, they are replaced in abundance by haemulids and gerreids (mojarras) (Hoese and Moore, 1977). This latitudinal change in fish fauna is also reflected in our results, suggesting that bottlenose dolphins change their diet according to local food availability.

Associations of bottlenose dolphins with shrimp boats in the Gulf of Mexico have been widely reported (Leatherwood, 1975; Shane, 1977; Gruber, 1981). Direct evidence of feeding on shrimp in Texas was presented by Gunter (1942, 1951) and Shane (1977), and in Mississippi by Leatherwood (1975). In this study, dolphins from Area I appeared to have a mixed diet of fish, shrimp, and cephalopods, all taken in large numbers, with individual prey items often being of small size. The similarity of dolphin prey composition wth that of the by-catch of shrimp boats (Bryan, 1980; Pellegrin, 1982; Lamkin, 1984), the presence of shrimp in the stomachs, and the greater diversity of dolphins' diet in Area I suggest that at least some dolphins may be feeding on organisms ("trash fish") discarded from shrimp boats.

Shane (Chapter 31, this volume) described several feeding techniques of the bottlenose dolphins she observed near Sanibel Island, Florida, and

Table 5
Frequency of Occurrence (FO) and Number of Prey Taken by Male and Female Bottlenose Dolphins for the Five Major Prey Species in Different Geographical Areas

		Males				Females			
Location[a]	Prey Species	FO	%	No. Prey	%	FO	%	No. Prey	%
Area I	*Micropogonias undulatus*	5	71.4	174	12.5	16	100.0	635	13.3
(7 males,	*Cynoscion arenarius*	3	42.9	251	18.0	13	81.3	345	7.2
16 females)	*Bairdiella chrysoura*	2	28.6	121	8.7	13	81.3	1878	39.2
	Lolliguncula brevis	3	42.9	81	5.8	10	62.5	273	5.7
	Leiostomus xanthurus	3	42.9	12	0.9	8	50.0	54	1.1
Area II	*Bairdiella chrysoura*	3	27.3	49	21.0	5	45.5	17	6.5
(11 males,	*Cynoscion arenarius*	4	36.4	25	10.7	4	36.4	8	3.1
11 females)	*Mugil* sp.	1	9.1	12	5.2	7	63.6	13	5.0
	Orthopristis chrysoptera	3	27.3	84	36.1	3	27.3	6	2.3
	Lagodon rhomboides	2	18.2	10	4.3	3	27.3	7	2.7
Area III	*Haemulon* sp.	3	37.5	26	25.2	2	66.7	2	7.4
(8 males,	*Opsanus beta*	1	12.5	21	20.6	1	33.3	3	11.1
3 females)	*Haemulon plumieri*	1	12.5	6	5.9	—	—	—	—
	Illex sp.	1	12.5	6	5.9	—	—	—	—
	Haemulon sciurus	1	12.5	5	4.9	—	—	—	—
Area IV	*Bairdiella chrysoura*	2	22.2	121	25.2	2	33.3	5	11.6
(9 males,	*Opsanus beta*	3	33.3	12	2.5	1	16.7	8	18.6
6 females)	*Leiostomus xanthurus*	2	22.2	12	2.5	1	16.7	3	7.0
	Cynoscion regalis	2	22.2	3	0.6	2	33.3	2	4.7
	Micropogonias undulatus	2	22.2	145	30.2	1	16.7	1	2.3

[a] Areas I–IV refer to Fig. 1.

compared them with a previous study (Shane, 1977, 1980). She concluded that the dolphins spent a significantly higher proportion of their time feeding in Florida than in Texas. This was possibly related to local fish abundance and distribution and the complexity of the habitats in the two study areas. Dolphins from Texas seem to have developed energy-saving behaviors such as gathering food from shrimp trawlers (Shane, 1977; Gruber, 1981).

Mullet Preference

Mullet often has been cited as the preferred prey of wild bottlenose dolphins (Harris, 1938; Mc Bride, 1940; Caldwell and Caldwell, 1972; Hamilton and Nishimoto, 1977; Irvine *et al.*, 1977, cited in Leatherwood, 1979; Irvine *et al.*, 1979, cited in Wells *et al.*, 1980; Lear and Bryden, 1980; Rein, 1982; Rigley *et al.*, 1985, 1987; Shane, Chapter 31, this volume; Weigle, 1987). In this study, although mullet occurred quite frequently in the stomachs (30.3%), they usually occurred in low numbers in each stomach. In fact, 13 of the 23 occurrences (56%) were represented by a single mullet specimen. This is in contrast with past studies on the Texas coast, where mullets comprised as much as 83% of all fish taken (Gunter, 1942) or were the most frequent (40%) fish in the stomach samples (Kemp, 1949). Bottlenose dolphins have been reported to pursue mullet and catch them leaping in midair (e.g., Harris, 1938; Caldwell and Caldwell, 1972; Wells *et al.*, 1980). This aerial behavior might cause observers to overestimate the importance of mullet in the dolphins' diet. For obvious reasons, studies of dolphin behavior at or above the surface provide little information on subsurface feeding behavior (Shane, Chapter 31, this volume).

Feeding of Sick and Healthy Dolphins

Some researchers have pointed out that stomach samples from stranded marine mammals may not be representative of feeding habits of active, healthy organisms (Ridgway and Dailey, 1972; Hall, 1977; Selzer *et al.*, 1986). Leatherwood *et al.* (1978) compared stomach contents of three netted and two stranded dolphins from North Carolina and New Jersey and found equal predominance of spot and croaker (the most abundant fish species in all stomachs) in both groups of dolphins. In contrast, Ross (1984) reported differences in the percentage of fish and squid in the stomachs of nine stranded and nine netted bottlenose dolphins off southern Africa. He suggested that cephalopod beaks may be retained for a longer period in the stomach of a stranded dolphin than are fish remains such as otoliths. This has been demonstrated experimentally to be true of northern fur seals, *Callorhinus ursinus* (Bigg and Perez, 1985), where passage rates for squid beaks and fish otoliths were shown to differ. This could result in overestimating the proportion of cephalopods in stranded animals (Finley and Gibb, 1982; Ross, 1984; Bigg and Perez, 1985). Finley and Gibb (1982) retrieved about 39,000 squid beaks from a "sick" narwhal, *Monodon monoceros,* and stated that healthy animals had a maximum of 1392 squid beaks in their stomachs. A stranded bottlenose dolphin from South Africa had 209 beaks of the squid *Loligo reynaldi* in the stomach, whereas seven netted dolphins had 1–3 beaks each (Ross, 1984).

Selzer *et al.* (1986) examined alimentary tracts of more than 500 harbor seals, *Phoca vitulina*, stranded along the Massachusetts coast. Most of the stomachs and intestines were found to be empty, and only 63 (13%) contained food matter. About 65% of the seals had died of acute pneumonia associated with influenza virus. The authors questioned the validity of using stomach contents of stranded seals to make inferences about prey preference or selection of all seals of that area, based on the fact that seals in a debilitated condition would probably not be feeding in regular feeding sites or would not be able to secure usual prey.

Empty stomachs are commonplace in cetacean mass strandings (e.g., Geraci *et al.*, 1978; Mead *et al.*, 1980), suggestive of little or no feeding some time prior to the stranding event. Cowan *et al.* (1986) described abnormalities found in small cetaceans stranded on the southern California coast. Although all the animals they examined had some kind of problem (e.g., parasites, cardiovascular diseases, damage to the middle ear apparatus, injuries) the authors observed that the dolphins were in full body weight and stated that "the complex neuromuscular functions involved in identifying, tracking, and capturing prey remained adequate." It has been pointed out, however, that when unusual prey items cause lesions in the esophagus lining, dolphins may suddenly stop eating or will not swallow the food (Dailey, 1985; Cowan *et al.*, 1986). Although information on the food of healthy and sick dolphins is limited, there is evidence suggesting that both groups prey on similar food species.

Weaning and Prey Size

Wild bottlenose dolphins up to 170 cm long have been reported to have milk in their stomachs (Gunter, 1942; Cockcroft and Ross, Chapter 15, this volume). The smallest bottlenose dolphins off the coast of southern Africa containing food matter in their stomachs were 172 and 179 cm long (Ross, 1977, 1984). Animals as small as 144 cm have been found to have solid food in their stomachs (Mead and Potter, Chapter 9, this volume). Although some solid food may be taken by dolphins 130–150 cm long, weaning is probably not complete in most cases until a body length of 170–180 cm is reached. In captivity, bottlenose dolphins have been observed to continue sucking until 18 months of age (McBride and Kritzler, 1951; Caldwell and Caldwell, 1972; Ross, 1977; Perrin and Reilly, 1984).

Most of the fish consumed by bottlenose dolphins from this study were 5–30 cm long. Although a larger sample would be necessary to make inferences about size selectivity, this range agrees fairly well with other studies (Norris and Prescott, 1961; Tomilin, 1967; Hamilton and Nishimoto, 1977; Mermoz, 1977; Pinedo, 1982; Connor and Smolker, 1985; Kasuya, 1985; Natarajan and Rajaguru, 1985; Shane, Chapter 12, this volume).

Seasonal Considerations

Cockcroft and Ross (Chapter 15, this volume) found the dietary differences between maturity classes not to be related to seasonal differences in their samples. Shane (Chapter 12, this volume) observed that dolphin feeding activities declined with decreasing water temperatures. In Florida, dolphins were seen most often in deep-water canals during the winter, apparently feeding on fish that sought these areas for refuge from cold water (Shane, Chapter 12, this volume). Weigle (1987) reported a summer increase in dolphin density in Tampa Bay, Florida, and attributed this, partly, to the inshore movement of animals, possibly from offshore, to forage on mullet (*Mugil cephalus*). Wells *et al.* (1980) also observed seasonal variation in the use of particular habitats by dolphins near Sarasota, Florida. Dolphin concentrations increased in the passes, along the beaches, and in the deeper waters of the Gulf of Mexico, and dolphin distribution was said to follow prey (i.e., mullet) distribution. In the present study bottlenose dolphins appear to have fed on mullet throughout the year.

Feeding of Males and Females

Mead and Potter (Chapter 9, this volume) observed no significant differences between the sexes of bottlenose dolphins in their major prey species (*Cynoscion, Micropogonias, Leiostomus*). They also found no differences in the sizes of prey taken by male and female dolphins. In contrast, Cockcroft and Ross (Chapter 15, this volume) observed significant differences in prey size for calves, adolescents, lactating females, nonlactating females, and mature males. Mean length of all prey taken by adolescents was significantly smaller than that of prey taken by adults. In our study, females were found to prey more frequently than males on mullet in Area II (see Fig. 2) and on silver perch in Area I. Wells *et al.* (1980) and Irvine *et al.* (1981) reported different age–sex distributions in Sarasota Bay (our Area II), and perhaps the food habits of at least some of the dolphins from our study reflect these observations. Clearly, more information is needed to evaluate this question further.

Inshore and Offshore Feeding

Reports of Atlantic bottlenose dolphins feeding in offshore waters are scarce. Irvine *et al.* (1981) predicted that the food habits and feeding strategies of bottlenose dolphins in Florida coastal waters would differ from those of dolphins in pelagic habitats (also see Kenney *et al.*, 1983). In our study an offshore (see Hersh and Duffield, Chapter 6, this volume) dolphin (SE1975) was found to have preyed primarily on ommastrephid squids (*Illex* sp. and *Ornithoteuthis antillarum*). Ommastrephid squids are active swimmers of the epipelagic zone (Voss, 1967), undergo diel vertical migrations (Arkhipkin and Fedulov, 1986), and can move onto the continental shelf at certain times of the year (Clarke, 1986b). They have been observed on the surface at night (Major, 1986) and can be captured from depths of 400–500 m during daytime (Roper and Young, 1975). Specimens of *Illex coindetti* collected from the Straits of Florida were captured between 183 and 531 m during the day and between 89 and 139 m at night, in waters 367–585 m deep (Cairns, 1976). *Illex illecebrosus* and two species of *Ommastrephes* were collected in slope waters in the Sargasso Sea and in cyclonic rings south of the Gulf Stream with open tows from a depth of 250 m (Lea, 1978). Mercer (1973) captured a 251-cm female bottlenose dolphin in the shelf-break area off North Carolina in waters 274 m deep; its stomach contained 150 beaks from the shortfin squid (*Illex illecebrosus*), 3 octopodid squids, and an unidentified fish otolith. *Illex illecebrosus* and other ommastrephid squids are typical prey of more pelagic predators in the Northwest Atlantic, such as swordfish, *Xiphias gladius* (Toll and Hess, 1981), long-finned pilot whales, *Globicephala melaena* (Squires, 1957), Atlantic white-sided dolphins, *Lagenorhyncus acutus* (Sergeant and Fisher, 1957), pygmy and dwarf sperm whales, *Kogia* spp. (Raun *et al.*, 1970; Candela, 1987), and common dolphins, *Delphinus delphis* (Major, 1986). This suggests that dolphin SE1975 could have been feeding offshore and in waters of considerable depth.

Mead and Potter (Chapter 9, this volume) found deep-water fish families to be important in the diet of six offshore bottlenose dolphins from the Northwest Atlantic. Rancurel (1964) retrieved mesopelagic fish, pasiphaeid crustaceans, and cephalopods (*Abralia veranyi* and *Ornithoteuthis antillarum*) from the stomachs of two bottlenose dolphins captured off Ivory Coast in waters 500–600 m deep. In the eastern North Pacific, the diet of offshore bottlenose dolphins consists primarily of squids, being less varied than that of inshore dolphins (Leatherwood and Reeves, 1978). Food habits and parasite load have also been used to differentiate inshore and offshore populations of bottlenose dolphins off California (Walker, 1981).

Prey Detection by Passive Listening

Although bottlenose dolphins seem opportunistic and nonselective in their feeding habits, as has often been postulated (Norris and Prescott, 1961; Hotta *et al.*, 1969; Leatherwood, 1975; Ross, 1977; Gaskin, 1982; Shane *et al.*, 1986), the dominant fish prey in their diet are known to produce relatively loud sound. These may include "grunts," "whistles," or "growls" (Tavolga, 1958; Gray and Winn, 1961; Breder, 1986) and can be made by stridulatory mechanisms, hydrodynamic phenomena, and muscular contraction involving the swimbladder (Tavolga, 1964). Myrberg (1983) pointed out that sound production in fishes may make them more susceptible to predation. Barros and Myrberg (1987) speculated that bottlenose dolphins may detect and perhaps orient to fish prey by listening for the sounds they produce.

ACKNOWLEDGMENTS

We thank the following persons and institutions for their cooperation in sending material through the U.S. Southeastern Marine Mammal Stranding Network: The Sea World Animal Care staff of Sea World of Florida, Orlando, Florida; Susan Candela, Michael Bergey, Douglas Burn, and Sandra Hersh, Marine Mammal Laboratory, University of Miami, Miami, Florida; Stephanie Fernandez and Chiara Jones, Texas A & M University, Galveston, Texas; Raymond Tarpley, Texas A & M University, College Station, Texas; Jeff Patton and staff of Mote Marine Laboratory, Sarasota, Florida. We thank Susan Candela for the identification of cephalopod beaks and C. Richard Robins for the identification of undigested fish. Vicki Credle assisted in the statistical analyses. Sharon Kelly-Fraga prepared the illustrations. We thank Sandra Hersh and anonymous reviewers for suggestions on the manuscript. All specimens from this study were collected under Letters of Authorization issued by the U.S. National Marine Fisheries Service, Southeast Region. This study was supported in part by a scholarship from the Brazilian National Research Council (CNPq Proc. 20.298/84-OC) to N.B.B. This is Sea World of Florida Technical Contribution Number 8807-F.

REFERENCES

Arkhipkin, A. I., and Fedulov, P. P. (1986). Diel movements of juvenile *Illex illecebrosus* and other cephalopods in the shelf-slope water frontal zone off the Scotian shelf in spring. *J. Northw. Atl. Fish. Sci.* **7**(1), 15–24.

Barros, N. B., and Myrberg, A. A., Jr. (1987). Prey detection by means of passive listening in bottlenose dolphins (*Tursiops truncatus*). *J. Acoust. Soc. Am.* **62,** Suppl. 1. S97 pp.

Bigg, M. A., and Perez, M. A. (1985). Modified volume: A frequency–volume method to assess marine mammal food habits. *In* "Marine Mammal and Fisheries" (J. R. Beddington, R. J. H. Beverton, and D. M. Lavigne, eds.), pp. 277–283. George Allen and Unwin, London. 354 pp.

Brakoniecki, T. F. (1986). A generic revision of the family Loliginidae (Cephalopoda; Myopsida) based primarily on the comparative morphology of the hectocotylus. Ph.D. dissertation, University of Miami, Coral Gables, Florida. 163 pp.

Breder, C. M., Jr. (1986). Seasonal and diurnal occurrences of fish sounds in a small Florida bay. *Bull. Am. Mus. Nat. Hist.* **138**(6), 325–378.

Brouder, R. D. (1979). "Guide to Otoliths of Some Northwest Atlantic Fishes." NMFS/NEFC Lab. Ref. 79–36. 70 pp.

Bryan, C. E. (1980). Organisms captured by the commercial shrimp fleet on the Texas brown shrimp (*Penaeus aztecus* Ives) grounds. M. S. thesis, Corpus Christi State University, Corpus Christi, Texas. 44 pp.

Cairns, S. D. (1976). Biological results of the University of Miami deep-sea expeditions. 118: Cephalopods collected in the straits of Florida by the R/V *Gerda. Bull. Mar. Sci.* **26**(2), 233–272.

Caldwell, D. K., and Caldwell, M. C. (1972). "The World of the Bottlenosed Dolphin." Lippincott, New York. 157 pp.

Candela, S. M. (1987). Cephalopod prey of pygmy and dwarf sperm whales (*Kogia breviceps* and *K. simus*) stranded in Florida and Georgia. *Abstr. 7th Biennial Conf. Biol. Mar. Mamm.*, Miami, Florida.

Cato, J. C., and Prochaska, F. J. (1976). Porpoise attacking hooked fish irk and injure Florida fishermen. *Natl. Fisherman* **56**(9), 1–4.

Chao, L. N. (1978). A basis for classifying Western Atlantic Sciaenidae (Teleostei, Perciformes). *NOAA Tech. Rep., NMFS Circ.*, No. 415. 64 pp.

Clarke, M. R. (1962). The identification of cephalopod "beaks" and the relationship between beak size and total body weight. *Bull. Br. Mus. (Nat. Hist.) Zool.* **8**(2), 419–480.

Clarke, M. R., ed. (1986a). "A Handbook for the Identification of Cephalopod Beaks." Clarendon Press, Oxford. 273 pp.

Clarke, M. R. (1986b). Cephalopods in the diets of odontocetes. *In* "Research on Dolphins" (M. M. Bryden and R. Harrison, eds.), pp. 281–321. Clarendon Press, Oxford. 478 pp.

Connor, R. C., and Smolker, R. S. (1985). Habituated dolphins (*Tursiops* sp.) in Western Australia. *J. Mammal.* **66**(2), 398–400.

Cowan, D. F., Walker, W. A., and Brownell R. L., Jr. (1986). Pathology of small cetaceans stranded along southern California beaches. *In* "Research on Dolphins" (M. M. Bryden and R. Harrison, eds.), pp. 323–367. Clarendon Press, Oxford. 478 pp.

Dailey, M. D. (1985). Diseases of Mammalia: Cetacea. *In* "Diseases of Marine Animals" (O. Kinne, ed.), Vol. 4, Part 2: Reptilia, Aves, Mammalia, pp. 805–847. Biologische Anstalt Helgoland, Hamburg. 884 pp.

Darovec, J. E., Jr. (1983). Sciaenid fishes (Osteichthyes: Perciformes) of western peninsular Florida. *Mem. Hourglass Cruises* **6**(3), 74 pp.

Duffield, D. A., Ridgeway, S. H. and Cornell, L. H. (1983). Hematology distinguishes coastal and offshore forms of dolphins (*Tursiops*). *Can. J. Zool.* **61**(4), 930–933.

Eziuzo, E. N. (1963). The identification of otoliths from West African demersal fish. *Inst. Fond. d'Afrique Noire. Bull. Ser. A* **25,** 488–512.

Finley, K. J., and Gibb, E. J. (1982). Summer diet of the narwhal (*Monodon monoceros*) in Pond Inlet, northern Baffin Island. *Can. J. Zool.* **60**(12), 3353–3363.

Fitch, J. E., and Brownell, R. L., Jr. (1968). Fish otoliths in cetacean stomachs and their importance in interpreting feeding habits. *J. Fish. Res. Board Can.* **25**(12), 2561–2574.

Gaskin, D. E. (1982). Diet and feeding behavior in Cetacea. *In* "The Ecology of Whales and Dolphins" (D. E. Gaskin, ed.), pp 30–78. Heinemann, London. 459 pp.

Geraci, J. R., Testaverde, S. A, St. Aubin, D. J., Loop, T. H. (1978). A mass stranding of the Atlantic white-sided dolphin, *Lagenorhyncus acutus:* A study into pathobiology and life history. *Mar. Mamm. Comm.: Fin. Rep.*, No. MMC-75/12. 155 pp.

Gonzáles, L. W. (1977). Aspectos técnicos de preparación de otolitos para estudios de edad en algunas especies del genero *Cynoscion* (Pisces: Sciaenidae). *Lagena* **39–40,** 43–48.

Gray, G. A., and Winn, H. E. (1961). Reproductive ecology and sound production of the toadfish, *Opsanus tau. Ecology* **42**(2), 274–282.

Gruber, J. A. (1981). Ecology of the Atlantic bottlenosed dolphin (*Tursiops truncatus*) in the Pass Cavallo area of Matagorda Bay, Texas. M. S. thesis, Texas A & M University, Austin, Texas. 182 pp.

Gunter, G. (1942). Contributions to the natural history of

the bottlenose dolphin, *Tursiops truncatus* (Montagu), on the Texas coast, with particular reference to food habits. *J. Mammal.* **23,** 267–276.

Gunter, G. (1951). Consumption of shrimp by the bottlenosed dolphin. *J. Mammal.* **32**(4), 465–466.

Hall, J. D. (1977). A nonlethal lavage device for sampling stomach contents of small marine mammals. *Fish. Bull.* **75**(3), 653–656.

Hamilton, P. V., and Nishimoto, R. T. (1977). Dolphin predation on mullet. *Fl. Sci.* **40**(3), 251–252.

Hare, M. P., and Mead, J. G. (1987). Handbook for determination of adverse human–marine mammal interactions from necropsies. *NWAFCA Proc. Rep.*, No. 87–06, NMFS Contract No. OABNR5 3224. 35 pp.

Härkönen, T. (1986). "Guide to the Otoliths of the Bony Fishes of the Northwest Atlantic." Danbiu ApS., Hellerup. 256 pp.

Harris, J. C. (1938). Porpoises feeding. *Life* **5**(12), 67.

Hecht, T. (1978). A descriptive systematic study of the otoliths of the neopterygean marine fishes of South Africa. Part I. Introduction. *Trans. R. Soc. S. Afr.* **43**(2), 191–197.

Hecht, T., and Hecht, A. (1978). A descriptive systematic study of the otoliths of the neopterygean marine fishes of South Africa. Part II. The delimitation of teleost orders, some systematic notes and a provisional new phyletic order sequence. *Trans. R. Soc. S. Afr.* **42**(2), 199–218.

Hersh, S. L. (1987). Characterization and differentiation of bottlenose dolphin populations (genus *Tursiops*) in the southeastern U.S. based on mortality patterns and morphometrics. Ph.D. dissertation. University of Miami, Coral Gables, Florida. 213 pp.

Hoese, H. D., and Moore, R. H. (1977). "Fishes of the Gulf of Mexico: Texas, Louisiana and Adjacent Waters." Texas A & M University Press, College Station, Texas. 327 pp.

Hoglund, R. (1965). Porpoises feeding. *Underwater Naturalist* **3**(2), 50–51.

Hotta, H., Mako, H., Okada, K., and Yamada, U. (1969). On the stomach contents of dolphins and porpoises off Kyushu. *Bull. Seikai Reg. Fish. Res. Lab.* **37,** 71–85.

Irvine, A. B., Scott, M. D., Wells, R. S., and Kaufmann, J. H. (1981). Movements and activities of the Atlantic bottlenose dolphin, *Tursiops truncatus,* near Sarasota, Florida. *Fish. Bull.* **79**(4), 671–688.

Iverson, I. L. K., and Pinkas, L. (1971). A pictorial guide to beaks of certain eastern Pacific cephalopods. *Calif. Dept. Fish Game, Fish. Bull.* **152,** 83–105.

Kasuya, T. (1985). Fishery–dolphin conflict in the Iki Island area of Japan. *In* "Marine Mammal and Fisheries" (J. R. Beddington, R. J. H. Beverton, and D. M. Lavigne, eds.), pp. 253–272. George Allen and Unwin, London. 354 pp.

Kemp, R. J. (1949). Report on stomach analysis from June 1, 1949 through August 31, 1949. *Ann. Rep. Mar. Lab. Texas Fish Game Oyster Comm.*, Fiscal Year 1948–1949, pp. 111–112, 126–127.

Kenney, R. D., Hyman, M. A. M., and Winn, H. E. (1983). Calculation of standing stocks and energetic requirements of the cetaceans of the United States outer continental shelf. *NMFS Final Rep.*, Contract No. NA-FA-C-00009. 154 pp.

Lamkin, J. T. (1984). A study of the Galveston Bay bait-shrimp fishery. M. S. thesis, Texas A & M University, College Station, Texas. 78 pp.

Lea, C. E. (1978). Pelagic cephalopods of Gulf Stream cyclonic rings. M. S. thesis, Texas A & M University, College Station, Texas. 59 pp.

Lear, R. J., and Bryden, M. M. (1980). a study of the bottlenose dolphin *Tursiops truncatus* in eastern Australian waters. *Aust. Nat. Parks Wildl. Serv. Occas. Pap.* **4,** 1–25.

Leatherwood, S. (1975). Some observations of feeding behavior of bottlenosed dolphins (*Tursiops truncatus*) in the northern Gulf of Mexico and (*Tursiops* cf. *T. gilli*) off southern California, Baja California, and Nayarit, Mexico. *Mar. Fish. Rev.* **37**(9), 10–16.

Leatherwood, S. (1979). Aerial survey of the bottlenose dolphin, *Tursiops truncatus,* and the West Indian manatee, *Trichechus manatus,* in the Indian and Banana rivers. *Fish. Bull.* **77**(1), 47–59.

Leatherwood, S., and Reeves, R. R. (1978). Porpoises and dolphins. *In* "Marine Mammals of Eastern North Pacific and Arctic Waters" (D. Haley, ed.), pp. 97–111. Pacific Search Press, Seattle.

Leatherwood, S., and Reeves, R. R. (1982). Bottlenose dolphin *Tursiops truncatus* and other toothed cetaceans. *In* "Wild Mammals of North America: Biology, Management and Economics" (J. A. Chapman and G. A. Feldhamer, eds.), pp. 369–414. Johns Hopkins University Press, Baltimore and London.

Leatherwood, S., Deerman, M. W., and Potter, C. W. (1978). Food and reproductive status of nine *Tursiops truncatus* from the northeastern United States coast. *Cetology*, No. 28, 6 pp.

McBride, A. F. (1940). Meet Mr. Porpoise. *Nat. Hist.* **45,** 16–29.

McBride, A. F., and Kritzler, H. (1951). Observations of pregnancy, parturition, and post-natal behavior in the bottle nose dolphin. *J. Mammal.* **32**(2), 251–266.

McMahon, T. E., and Tash, J. C. (1979). Effects of formalin (buffered and unbuffered) and hydrochloric acid on fish otoliths. *Copeia* **1979**(1), 155–156.

Major, P. F. (1986). Notes on a predator–prey interaction between common dolphins (*Delphinus delphis*) and short-finned squid (*Illex illecebrosus*) in Lydonia submarine canyon, Western North Atlantic ocean. *J. Mammal.* **67**(4), 769–770.

Mead, J. G., Odell, D. K., Wells, R. S., Scott, M. D.

(1980). Observations on a mass stranding of spinner dolphin, *Stenella longirostris,* from the west coast of Florida. *Fish. Bull.* **78**(2), 353–360.

Mercer, M. C. (1973). Observations on distribution and intraspecific variation in pigmentation patterns of odotocete cetacea in the western North Atlantic. *J. Fish. Res. Board Can.* **30**(8), 1111–1130.

Mermoz, J. F. (1977). Sobre el varamiento de un delfin nariz de botella, *Tursiops truncatus,* en la desembocadura del Rio de La Plata (Buenos Aires, Argentina). *Physis, Sec. C* **37**(93), 227–235.

Mohsin, A. K. M. (1973). Comparative osteology of the weakfishes (*Cynoscion*) of the Atlantic and Gulf coasts of the United States (Pisces, Sciaenidae). Ph.D. dissertation, Texas A & M University, College Station, Texas. 147 pp.

Mohsin, A. K. M. (1981). Comparative account of the otoliths of the weakfishes (*Cynoscion*) of the Atlantic and Gulf coasts of the United States. *Pertanika* **4**(2), 109–111.

Morrow, J. E. (1979). Preliminary keys to otoliths of some adult fishes of the Gulf of Alaska, Bering Sea, and Beaufort Sea. *NOAA Tech. Rep., NMFS Circ.* No. 420. 32 pp.

Myrberg, A. A., Jr. (1983). Sound communication and interception in fishes. *In* "Hearing and Sound Communication in Fishes" (W. N. Tavolga, A. N. Popper, and R. R. Fay, eds.), pp. 395–426. Springer-Verlag, New York. 608 pp.

Natarajan, R., and Rajaguru, A. (1985). Small cetaceans. *Mar. Mamm. Sci.* **1**(1), 89.

Norris, K. S., and Prescott, J. H. (1961). Observations of Pacific cetaceans of Pacific and Mexican waters. *Univ. Calif. Publ. Zool.* **63,** 291–400.

Odell, D. K. (1976). Distribution and abundance of marine mammals in south Florida: Preliminary results. *Univ. Miami Sea Grant Spec. Rep.* **5,** 203–212.

Odell, D. K. (1979). Proposal for the establishment of a southeastern United States regional marine mammal stranding network. *In* "Biology of Marine Mammals: Insights Through Strandings" (J. R. Gerarci and D. J. St. Aubin, eds.), pp. 311–314. Mar. Mamm. Comm.: *Final Rep.* No. MMC-77/13. 343 pp.

Odell, D. K. (1983). Southeastern United States marine mammal and sea turtle stranding and salvage network: 1 October 1981 to 31 December 1982. *NMFS Final Rep.* Contract No. NA81-GA-C-00026.

Pellegrin, G., Jr. (1982). Fish discards from the southeastern United States shrimp fishery. *In* "International Development Research Centre. FAO. Fish By-Catch—Bonus from the Sea" pp. 51–54. Rep. Tech. Consult. on Shrimp By-Catch Utilization, Georgetown, Guyana. 163 pp.

Perrin, W. F., and Reilly, S. B. (1984). Reproductive parameters of dolphins and small whales of the family Delphinidae. *Rep. int. Whal. Commn.* (Special Issue 6), 97–133.

Perrin, W. F., Warner, R. R., Fiscus, C. H. and Holts, D. B. (1973). Stomach contents of porpoise, *Stenella* spp., and yellowfin tuna, *Thunnus albacares,* in mixed-species aggregations. *Fish. Bull.* **71**(4), 1077–1092.

Pinedo, M. C. (1982). Análise dos conteúdos estomacais de *Pontoporia blainvillei* (Gervais & D'Orbigny, 1844) e *Tursiops gephyreus* (Lahille, 1908) (Cetacea, Platanistidae e Delphinidae) na zona estuarial e costeira de Rio Grande, RS, Brasil. M.S. thesis, Fundação Universidade do Rio Grande, Rio Grande, RS. Brazil. 95 pp.

Rancurel, P. (1964). Note sur la plongée profonde de *Tursiops truncatus. Cahiers ORSTOM (Océanogr.)* **2**(4), 135–141.

Raun, G. G., Hoese, H. D., Moseley, F. (1970). Pygmy sperm whales, genus *Kogia,* on the Texas coast. *Texas J. Sci.* **21**(3), 269–274.

Rein, B. B. (1982). "Dolphins Love our Florida Keys Home!" Litoky Publishing, Big Pine Key, Florida. 101 pp.

Reynolds, J. E., III. (1985). Evaluation of the nature and magnitude of interactions between bottlenose dolphins, *Tursiops truncatus,* and fisheries and other human activities in coastal areas of the southern United States. *Mar. Mamm. Comm. Final Rep.,* No. MMC-84/07, NTIS, PB86-162203. 38 pp.

Ridgway, S. H., and Dailey, M. D. (1972). Cerebral and cerebellar involvement of trematode parasites in dolphins and their possible role in stranding. *J. Wildl. Dis.* **8**(1), 33–43.

Ridgway, S. H., and Fenner, C. A. (1982). Weight–length relationships of wild-caught and captive Atlantic bottlenose dolphins. *J. Am. Vet. Med. Assoc.* **181**(11), 1310–1315.

Rigley, L., Hays, H., and Rigley, I. (1985). Terrestrial feeding by bottlenose dolphins. *Abstr. 6th Biennial Conf. Biol. Mar. Mamm.,* Vancouver, Canada.

Rigley, L., Hays, H., and Rigley, I. (1987). Terrestrial feeding by bottlenose dolphins. *Abstr. 7th Biennial Conf. Biol. Mar. Mamm.,* Miami, Florida.

Robins, C. R., Bailey, R. M., Bond, C. E. Brookes, J. R., Lachner, E. A., Lea, R. N., and Scott, W. B. (1980). A list of common and scientific names of fishes from the United States and Canada, 4th ed. *Am. Fish. Soc. Spec. Publ.,* No. 12. 174 pp.

Robins, C.R., Ray, G.C., Douglass, J., and Freund, R. (1986). "A Field Guide to Atlantic Coast Fishes of North America." Houghton Mifflin, Boston. 354 pp.

Roper, C. F. E., and Young, R. E. (1975). Vertical distribution of pelagic cephalopods. *Smithsonian Contrib. Zool.* **209,** 51 pp.

Ross, G. J. B. (1977). The taxonomy of bottlenose dolphins *Tursiops* species in South African waters, with notes on their biology. *Ann. Cape Prov. Mus. (Nat. Hist.)* **11**(9), 135–194.

Ross, G. J. B. (1984). The smaller cetaceans of the south east coast of southern Africa. *Ann. Cape Prov. Mus. (Nat. Hist.)* **15**(2), 173–410.

Schmidt, W. (1968). Vergleichend morphologische Studie über die otolithen mariner Knochenfishe. *Arch. Fischereiwiss.* **1,** 1–96.

Selzer, L. A., Early G., Fiorelli, P. M., Payne, P. M., and Prescott, R. (1986). Stranded animals as indicators of prey untilization by harbor seals, *Phoca vitulina concolor,* in southern New England. *Fish. Bull.* **84**(1), 217–220.

Sergeant, D. E. (1969). Feeding rates of Cetacea. *Fisk Dir. Skr. Ser. HavUnders.* **15,** 246–258.

Sergeant, D. E., and Fisher, H. D. (1957). The small cetaceans of eastern Canadian waters. *J. Fish. Res. Board Can.* **14**(1), 83–115.

Shane, S. H. (1977). The population biology of the Atlantic bottlenose dolphin, *Tursiops truncatus,* in the Aransas Pass area of Texas. M.S. thesis, Texas A & M University, College Station, Texas. 239 pp.

Shane, S. H. (1980). Occurrence, movements, and distribution of bottlenose dolphins, *Tursiops truncatus,* in southern Texas. *Fish. Bull.* **78**(3), 593–601.

Shane, S. H., Wells, R. S., and Würsig, B. (1986). Ecology, behavior and social orginazation of the bottlenose dolphin: A review. *Mar. Mamm. Sci.* **2**(1), 34–63.

Shapunov, V. M. (1971). Food requirements and energy balance in the Black Sea bottlenose dolphin (*Tursiops truncatus ponticus* Barabasch). *In* "Morphology and Ecology of Marine Mammals." (K. K. Chapskii, and E. E. Sokolov, eds.,) pp. 207–212. Israel Program for Scientific Translations, Jerusalem.

Sokal, R. R., and Rolf, F. J. (1981). "Biometry," 2nd Ed. Freeman, New York. 859 pp.

Spotte, S., and Babus, B. (1980). Does a pregnant dolphin (*Tursiops truncatus*) eat more? *Cetology* **39,** 7 pp.

Squires, H. J. (1957). Squid, *Illex illecebrosus* (Leseur), in the Newfoundland fishing area. *J. Fish. Res. Board Can.* **14**(5), 693–728.

Tavolga, W. N. (1958). Underwater sounds produced by two species of toadfish, *Opsanus tau* and *Opsanus beta. Bull. Mar. Sci. Gulf Carib.* **8**(3), 278–284.

Tavolga, W. N. (1964). Sonic characteristics and mechanisms in marine fishes. *In* "Marine Bio-Acoustics" (W. N. Tavolga, ed.), pp 195–211. MacMillan, New York. 413 pp.

Toll, R. B., and Hess, S. C. (1981). Cephalopods in the diet of the swordfish, *Xiphias gladius,* from the Florida straits. *Fish. Bull.* **79**(4), 765–774.

Tomilin, A. G. (1967). "Mammals of the USSR and Adjacent Countries. Volume IX: Cetacea," pp. 523–549. Israel Program for Scientific Translations, Jerusalem.

Voss, G. L. (1967). The biology and bathymetric distribution of deep-sea cephalopods. *Stud. Trop. Oceanogr. Miami* **5,** 511–535.

Voss, G. L., and Brakoniecki, T. (1985). The distribution and numbers of the commercial or potential squid of the Gulf of Mexico and southeastern Atlantic coast. *Gulf South Atlantic Fish. Dev. Found. Rep.* 61 pp.

Walker, W. A. (1981). Geographical variation in morphology and biology of bottlenose dolphins (*Tursiops*) in the eastern North Pacific. *NMFS/SWFC Admin. Rep.,* No. LJ-81-03C. 52 pp.

Weigle, B. L. (1987). Abundance, distribution, and movements of bottlenose dolphins, *Tursiops truncatus,* in lower Tampa Bay, Florida. M.S. thesis, University of South Florida, St. Petersburg, Florida. 47 pp.

Wells, R. S., Irvine, A. B., and Scott, M. D. (1980). The social ecology of inshore odontocetes. *In* "Cetacean Behavior: Mechanisms and Function" (L. M. Herman, ed.), pp. 263–317. Wiley, New York. 463 pp.

Wolff, G. A. (1984). Identification and estimation of size from the beaks of 18 species of cephalopods from the Pacific Ocean. *NOAA Tech. Rep. NMFS,* No. 17. 50 pp.

Feeding by Bottlenose Dolphins in Association with Trawling Operations in Moreton Bay, Australia

17

P. J. Corkeron
Department of Anatomy, University of Queensland, St. Lucia, Queensland, Australia

M. M. Bryden
Department of Anatomy, University of Queensland, St. Lucia, Queensland, Australia
Present address: Department of Veterinary Anatomy, University of Sydney, Sydney, New South Wales 2006, Australia

K. E. Hedstrom
Department of Anatomy, University of Queensland, St. Lucia, Queensland, Australia
Present Address: School of Australian Environmental Studies, Griffith University, Nathan, Queensland 4111, Australia

INTRODUCTION

Many interactions between bottlenose dolphins, *Tursiops truncatus*, and fishery operations have been reported. Examples of man–dolphin commensal fishing have been recorded for centuries, a well-known account being that of Pliny the Elder (see St. Denis, 1955). Huge schools of mullet, *Mugil* sp., would run from a pond called Latera in Gaul, through a narrow channel, and out into the open sea. The people of Narbonne (Gaul) would summon the dolphins from the deep and wait with their nets until the dolphins had cut off the fishes' escape and herded them toward shore.

This and other early accounts of dolphins aiding man in fishing operations have been described by Busnel (1973), who also outlined, from first-hand observations, methods of mullet fishing by the Imragen of Mauritania, in which dolphins play a part (also see Pelletier, 1975). Man–dolphin cooperation in fishing in southeastern Queensland, with many features similar to those instances cited above, was reported initially by Fairholme (1856). Australian Aborigines on Moreton and North Stradbroke islands, on seeing a school of mullet (Mugilidae: *Mugil cephalus*) or tailor (Pomatomidae: *Pomatomus saltatrix*), would slap the water with their spears, whereupon dolphins would come toward the shore and drive the mullet before them. As the dolphins approached the water's edge, the Aborigines would dash into the water with spears and handnets to capture some of the fish. Hall (1984) evaluated Fairholme's (1856) description, along with other evidence, and concluded that the report is probably accurate.

A common current man–dolphin interaction sometimes occurs during trawling operations. Leatherwood (1975) observed bottlenose dolphins following working shrimp boats, apparently feeding on organisms stirred up by the otter trawl, and reviewed previous reports of similar interactions. He outlined three types of feeding in association with shrimp trawlers: (1) behind actively trawling boats, (2) on "trash" fish discarded after a trawl, and (3) around nonworking, anchored shrimp boats, eating fish attracted by debris in the nets and discharges from the boats' bilges.

We have received verbal reports that dolphins feed behind prawn trawlers (shrimp boats) along many parts of the Australian coast. We have observed bottlenose dolphins engaged in this activity in Moreton Bay, southeastern Queensland. Dolphins were seen regularly engaged in apparent feeding behind the trawler, diving to take organisms disturbed by the boards and nets and possibly small organisms escaping through the nets, as well as feeding on the by-catch. This situation has provided us with the opportunity to evaluate the food preferences of bottlenose dolphins and to test the widely held opinion that these animals are catholic, opportunistic feeders (Gunter, 1954; Caldwell and Caldwell, 1972; Leatherwood, 1975; Ross, 1977; Lear and Bryden, 1980). In this chapter we describe investigations of feeding patterns of bottlenose dolphins around a trawler in Moreton Bay.

MATERIALS AND METHODS

Data were collected from the University of Queensland's research trawler as part of a larger study on the behavioral ecology of inshore dolphins in Moreton Bay (see Corkeron, Chapter 14, this volume, for details). At the completion of each trawl, the catch was sorted and returned selectively to the sea while the trawler was stationary. Components of the catch were identified according to Grant (1982). Some dolphins would aggregate within 10 m of the stern during this process. Known individual dolphins, previously identified from natural marks, were observed, and whenever possible the number and species of the fish and invertebrates eaten by an individual dolphin were recorded. Data were recorded only for animals within approximately 10 m of the trawler which were feeding at or very near (within 2 m of) the surface. A "feeding session" was considered to occur each time a recognizable individual dolphin was observed feeding. Feeding data were collected on 10 days between 21 April and 5 November 1986.

Data on the composition of a sample of the trawled catch were collected on 5 days from 5 August to 5 November 1986. Approximately one-quarter of the catch was sampled, and fish and invertebrates were sorted into the lowest taxonomic unit (generally species). These sorted groups were then counted and weighed on the boat. Stomach contents were collected from two dolphins found dead in trawlers' nets, on 10 February and 23 May 1986. Data analyses were performed using GLIM (Numerical Algorithms Group, 1985) and OZSTAT (Kennedy, 1985).

RESULTS

Data matrices were classified using Euclidean distance dissimilarity measures and UPGMA sorting strategies. In both cases clusters are numbered from left to right.

Figure 1 shows the classification of fish and invertebrates according to the frequency of their occurrence in trawls. This classification can be divided at the six-group level. The first cluster (Fig. 1A) consists of *Paramonacanthus oblongus*, *Caranx* spp., and *Saurida undosquamis*. The second, and main, cluster (Fig. 1B) consists of *Squilla granti*, *Priacanthus macracanthus*, *Pomatomus saltatrix*, *Sphoeroides hamiltoni*, *Apogon fasciata*, octopus, *Achylopa nigra*, *Platycephalus* sp., *Parupenius cyclostomus*, small crabs, *Parapercis nebulosus*, *Polydactylus plebejus*, *Portunus pelagicus*, and *Centropogon marmoratus*. The other four clusters (Fig. 1C–1F) contain single species: *Loligo* spp., *Sillago maculata*, *Callionymus limiceps*, and *Leiognathus moretoniensis*, respectively.

Figure 2 shows the classification of prey species according to the numbers of them that dolphins ate. This classification can be divided at the five-group level. Most prey species were taken occasionally and therefore constitute a single group (Fig. 2A), while *S. maculata*, *S. undosquamis*, *Loligo* spp., and *P. plebejus* form single clusters (Fig. 2B–2E, respectively).

Differences between males and females in the relationship between the number of times a dolphin was observed and the number of feeding sessions recorded on it were tested. As feeding-

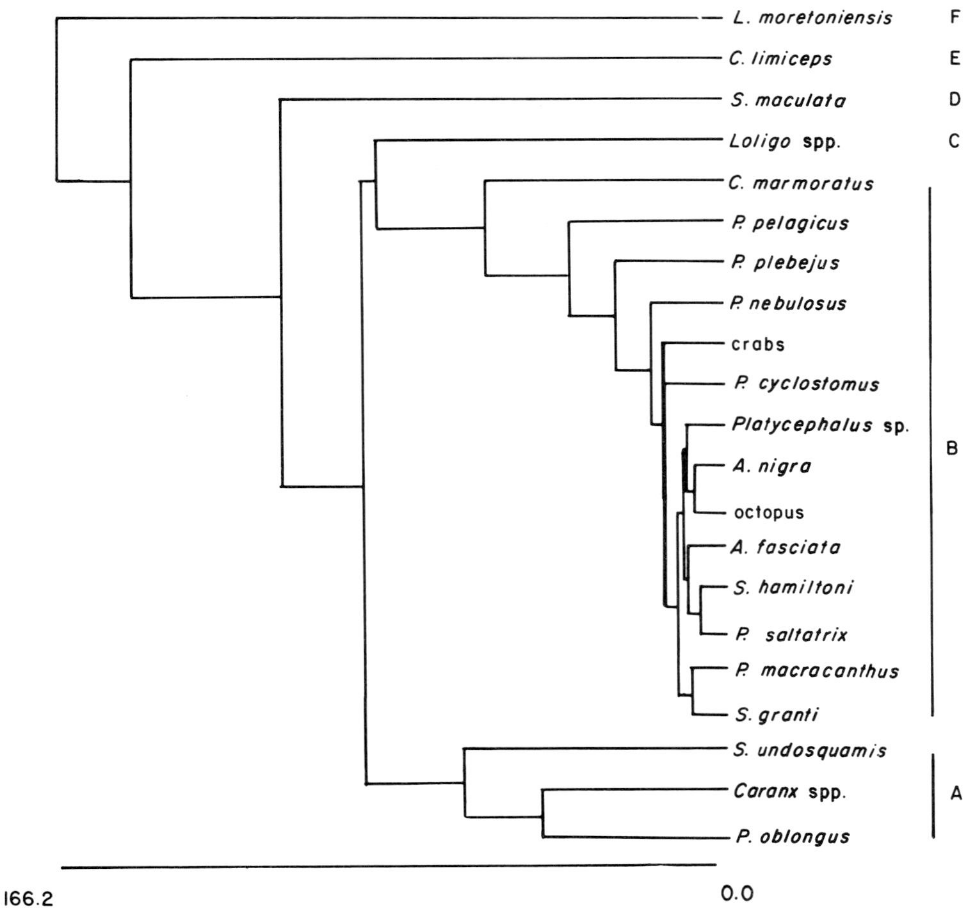

Figure 1 Fish and invertebrates caught in trawls, clustered into six groups on the basis of frequency of occurrence.

session data could be obtained only for animals very close to the trawler, these differences should reflect differing opportunities for access to the "first pick" of the prey. Dolphins recognized on only one occasion were not included, and only animals whose sex had been established positively were considered. This did not bias data collection, as most of the dolphins observed for feeding studies were well known to the observers. The sex of virtually all animals that came sufficiently close to the trawler to have feeding sessions recorded was known. Regressions of the number of feeding sessions recorded on the number of times that dolphins were observed were performed separately for males and females. The regression for males was significant ($F_{16} = 14.02$, $p < .001$, $r^2 = 0.47$), whereas that for females was nonsignificant ($F_{15,1} = 0.07$, $p > .1$, $r^2 = 0.01$).

The stomach contents of two juvenile dolphins are shown in Table 1. Maturity was determined by ovarian state.

DISCUSSION

Food Preferences

These simple observations have shown that although bottlenose dolphins are opportunistic feeders, if given a choice, they display distinct

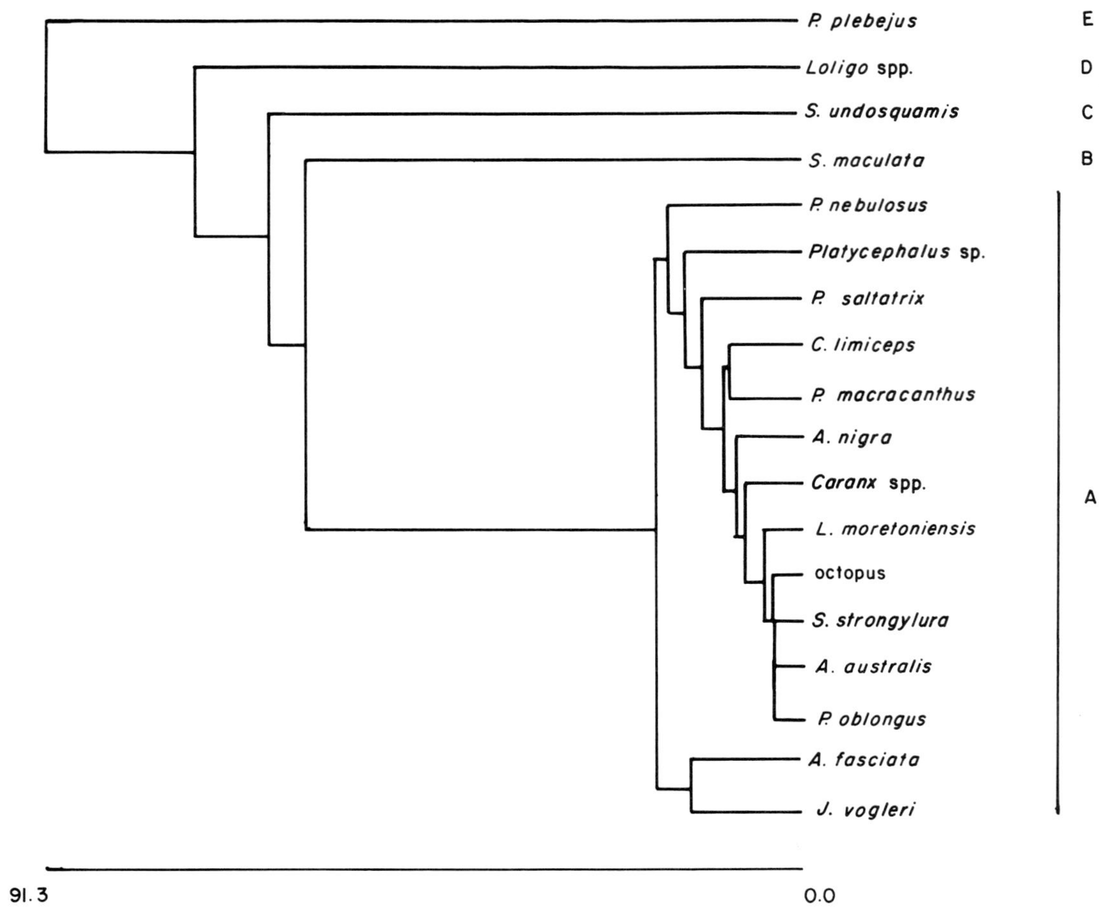

Figure 2 Fish and invertebrates eaten by dolphins, clustered into five groups on the basis of number eaten.

food preferences. Those individuals, notably large males, that had most ready access to a variety of food items were selective in their choice of food.

The two dendrograms are not directly comparable, as the catch composition was not sampled at every feeding session. However, some gross patterns are apparent. Whiting, *S. maculata*, was very common in the catch, and formed a single-species group (Fig. 1D). It was in a single-species cluster when its pattern of acceptance by dolphins was analyzed, but this was due to one feeding session of one adult female dolphin, during which she consumed approximately 50 whiting. On one other feeding session only was a dolphin observed to eat whiting. Many fish and invertebrate species clustered together in one group when their occurrence in trawls was analyzed (Fig. 1B). The acceptance of these species by dolphins varied greatly. Crustaceans, namely, squilla (*S. granti*), sand crabs (*P. pelagicus*), and small crabs, and some fish species, for example, toadfish (*S. hamiltoni*) and bullrouts (*C. marmoratus*), were never eaten. One species, the putty-nosed perch (*P. plebejus*), was a preferred prey item and formed a single-species group when its acceptance by dolphins was analyzed. Other species were eaten occasionally.

Two species were extremely common in trawl catches, and both formed single-species groups in the analysis of catch composition. These species, dragonets (*C. limiceps*) and sixpences (*L. moretoniensis*), are both very small fishes, weighing a few grams only. Despite their common occurrence in

Table 1
Stomach Contents of Dolphins Recovered from Trawl Nets

Dolphin	Prey Item	Number in Stomach
Female, juvenile, found 10 February 1986	*Loligo* spp.	30
	P. oblongus	14
	P. macracanthus	7
	S. strongylura	14
	E. moretoniensis	50
	S. undosquamis	6
	P. plebejus	1
	S. maculata	1
	Unknown	~15
Female, juvenile, found 23 May 1986	*Loligo* spp.	1
	E. moretoniensis	28
	P. plebejus	1
	Unknown	~15

trawls, both were taken infrequently by dolphins. The acceptance of squid (*Loligo* spp.) roughly equated to its presence in trawls, while grinner (*S. undosquamis*) was more frequently taken than the other two fish species with which it clustered, namely, golden trevalley (*Caranx* spp.) and leatherjackets (*P. oblongus*).

When offered items of food, dolphins would either seize and swallow them, approach but not mouth them, mouth and release them, or ignore them altogether (Fig. 3). It was not possible to discern what criteria the dolphins used as a basis for accepting or rejecting items of food. However, the fact that certain items were accepted readily and others ignored suggests that experience plays a part. Around the trawler dolphins ate dead prey more readily than live, and most dolphins refused fish that floated on the water surface.

Taste or "feel" may play a part in food selection, since on many occasions prey was seen to be mouthed and released. Bottlenose dolphins possess small taste buds in the root of the tongue, and some workers have suggested that they have a well-developed chemosensory system (for consideration of the chemosensory and gustatory faculties of dolphins, see Herman and Tavolga, 1980). Dolphins also possess acute vision both in the water and in air (Dawson, 1980), so vision may well be important in those instances where dolphins accept or reject items immediately. A further possibility is that dolphins are able to assess the composition of their prey by using sonar. Many click vocalizations could be heard in the region around the trawler during feeding sessions.

The species not eaten by dolphins were either crustaceans or venomous or poisonous fish. In an attempt to determine what made some species more acceptable, analyses of the fat, protein, moisture, and ash content of samples from the trawler's catch were undertaken. Insufficient samples were obtained to analyze examples of all species caught while trawling, and no clear association between chemical composition and prey acceptability could be discerned from the samples tested. However, one fish species selected preferentially by dolphins feeding close to the trawler (and therefore with ready access to all items offered) was *P. plebejus,* a fish with a high fat (and therefore a high energy) content.

Social Factors

When trash fish were discarded at the end of a trawl, most dolphins close to the boat and therefore with first access to food offered were males. Most lone females, females with calves, and juvenile dolphins remained some distance away (20 m or more) and apparently fed on species ignored or rejected by the closer animals.

There have been difficulties associated with the concept of social dominance (see Bernstein, 1981, and Peer Commentary that follows Bernstein's paper). However, de Waal (1986) has shown that the concept is useful in providing a context for the analysis of aspects of social interactions between animals. Aggression between individuals, and between groups of different age and sex, has been suggested as a mechanism for the maintenance of social spacing patterns in delphinids (Irvine *et al.*, 1981). However, Irvine *et al.* (1981) suggested that dominance hierarchies, observed in captive situations, may not influence the social structure of free-ranging dolphins.

Around the trawler, adult males generally had priority of access to the first choice of food. This subjective impression was borne out by the regression analysis conducted. The more often adult males were seen, the more likely they were to

Figure 3 Various views of bottlenose dolphins feeding in association with trawlers. In three instances, the individual dolphin and prey are identifiable: (B) BO19, a male, approaching *S. undosquamis;* (C) BO51, a male, investigating *P. nebulosis;* and (D) BO81, a male, eating *E. moretoniensis.*

have feeding data recorded for them. This was not so for females. Most females tended to avoid close contact with males near the boat, and, when the water was clear, females, calves, and juveniles could be seen swimming at greater depths (5 m down), eating fish not taken by dolphins feeding at the surface. Very occasionally, overt aggressive behavior between dolphins was observed directly. This behavior took the form of one dolphin's charging another, resulting in contact.

The stomach contents of two juvenile females that drowned accidentally in nets during trawling (Table 1) support the contention that juvenile animals fed largely on species rejected or at least not favored by adult males. While the stomach of one of the dolphons contained prey species eaten commonly by dolphins around the trawler, the relative abundance of *L. moretoniensis* and *P. oblongus* indicates a feeding pattern different from that of animals feeding close to the trawler. The total contents of the stomach were estimated to represent approximately 3 kg of food. The total stomach contents of the other dolphin were estimated to represent approximately 370 g of food, of which 300 g was *L. moretoniensis*. This further indicates a feeding pattern different from that of adults.

The dolphins of Moreton Bay must receive large quantities of food during the trawling season (September to May). We observed one adult female to eat an estimated 6.6 kg of food on one occasion. Judging from her behavior, this fed her to satiation. At Sea World, Southport (Queensland), dolphins that are not pregnant or lactating are fed 7–10 kg of food per day, in two feeding sessions.

We could not estimate the total quantity of food consumed by any dolphin over an entire day, as the animals also followed the trawler while it was trawling and presumably ate fish disturbed by the trawling operation and fish they could extract from the nets. The amounts of food obtained in this manner could not be estimated. Observations of this behavior would require a scuba-equipped observer stationed on the otter boards while trawling took place.

Costs and Benefits

Although dolphins benefit from feeding around trawlers, there are also some costs involved. They are captured in nets occasionally. Both animals examined by us that drowned accidentally in nets were juveniles, suggesting that feeding around nets during otter trawling is a learned behavior. Interrogation of fishermen suggests that the capture of dolphins in trawl nets is a relatively rare event in Moreton Bay, as in Matagorda Bay, Texas (Gruber, 1981, as cited in Shane *et al.*, 1986).

During our study, sharks were seen around the trawler on many occasions, and many dolphins bore shark-bite wounds (Corkerson *et al.*, 1987). At least one fresh wound was known to have been inflicted while the dolphin was following a trawler. Thus, it is interesting that dolphins are so selective of the organisms they take around trawlers. By spending seemingly "unnecessary" time taking preferred items, they appear to place themselves at a greater risk of shark attack.

It is difficult to assess the benefits of feeding around trawlers. The fish biomass in Moreton Bay not sampled by trawling may be insufficient to maintain the population of dolphins in the bay. Unfortunately, data to answer such questions are unavailable, so the extent to which dolphins' association with trawlers is obligate rather than facultative cannot be assessed.

CONCLUSIONS

Although this study was preliminary, some conclusions can be drawn. Bottlenose dolphons are catholic feeders, but they demonstrate clear preferences when given a choice of food items. Social factors that vary in effect between sex and age classes alter the degree of choice available to animals, and consequently the composition of their diet. The benefit of easy access to food is at least partially offset by the cost of risk of capture in the nets and of predation by sharks.

This method of observation of dolphins can provide insights into their feeding patterns that cannot be obtained from stomach-content analysis. Further work should concentrate on the following: (1) collecting data from more feeding sessions, with concurrent analysis of catch composition; (2) investigating the behavior of dolphins at the nets while fishing is taking place; (3) observing dominance interactions; and (4) recording detailed and accurate data on food consumption by individual dolphins throughout the year, to provide valuable information on the energetics of free-ranging animals.

ACKNOWLEDGMENTS

We thank J. Kelly, W. Hayes, T. Palagyi, and M. Connell for assistance during the field work, which was financed by a grant from the Marine Sciences and Technologies Grants Scheme, Australia, to M.M.B. We are grateful to H. Marsh and two anonymous referees for their comments on a previous draft of this paper.

REFERENCES

Bernstein, I. S. (1981). Dominance: The baby and the bathwater. *Behav. Brain Sci.* **4,** 419–457.

Busnel, R. G. (1973). Symbiotic relationship between man and dolphins. *Trans. N.Y. Acad. Sci., Ser. 2* **35,** 112–131.

Caldwell, D. K., and Caldwell, M. C. (1972). "The World of the Bottle-Nosed Dolphin." Lippincott, Philidelphia and New York. 157 pp.

Corkeron, P. J., Morris, R. J., and Bryden, M. M. (1987). Interactions between bottlenose dolphins and sharks in Moreton Bay. *Aquatic Mamm.* **13,** 109–114.

Dawson, W. W. (1980). The cetacean eye. *In* "Cetacean Behavior: Mechanisms and Functions." (L. M. Herman, ed.), pp. 53–100. Wiley, New York. 463 pp.

de Waal, F. B. M. (1986). The integration of dominance and social bonding in primates. *Q. Rev. Biol.* **61,** 459–479.

Fairholme, J. K. E. (1856). The blacks of Moreton Bay and the porpoises. *Proc. Zool. Soc. London* **24,** 353–354.

Grant, E. M. (1982). "Guide to Fishes." Dept. of Harbours and Marine, Brisbane, Queensland. 896 pp.

Gruber, J. A. (1981). Ecology of the Atlantic bottlenosed dolphin (*Tursiops truncatus*) in the Pass Cavallo area of Matagorda Bay, Texas. M.S. thesis, Texas A & M University, College Station, Texas 77843.

Gunter, G. (1954). Mammals of the Gulf of Mexico. *In* "Gulf of Mexico, its Origin, Waters and Marine Life." (P. S. Galtsoff. ed.), pp 543–551. *Fish. Bull. (U.S.)* **55.**

Hall, H. J. (1984). Fishing with dolphins? Affirming a traditional story in Moreton Bay, S.E. Queensland. *In* "Focus on Stradborke." (R. J. Coleman, J. Covacevich, and P. Davie, eds.,) pp. 16–22. Boolarong Publications, Brisbane, Queensland. 420 pp.

Herman L. M., and Tavolga, W. W. (1980). The communication systems of cetaceans. *In* "Cetacean Behavior: Mechanisms and Functions" (L. M. Herman, ed.,) pp. 144–209. Wiley, New York. 463 pp.

Irvine, A. B., Scott, M. D., Wells R. S., and Kaufman, J. H. (1981). Movements and activities of the Atlantic bottlenose dolphin, *Tursiops truncatus,* near Sarasota, Florida. *Fish. Bull. (U.S.)* **79,** 671–688.

Kennedy, R. (1985). OZSTAT Unpublished manuscript, Dept. of Zoology, University of Queesland, 39 pp.

Lear, R. J., and Bryden, M. M. (1980). A study of the bottlenose dolphin *Tursiops truncatus* in eastern Australian waters *ANPWS Occas. Pap.* **4,** 25 pp.

Leatherwood, S. (1975). Some observations of feeding behavior of bottlenose dolphins *Tursiops truncatus* in the northern Gulf of Mexico and (*Tursiops* cf. *T. gilli*) off southern california. *Mar. Fish. Rev.* **37**(9); 10–16.

Numerical Algorithms Group. (1985). "The GLIM System Release 3.77 Manual." The Numerical Algorithms Group of the Royal Statistical Society. Oxford. 80 pp.

Pelletier, F. X. (1975). Symboise entre l'Amrgi et le deuphin sur la cote Mauritanienne. *L'Homme et l'Animal, Ler Colloq. Ethnozool., Juin 1975, Inst. Int. Ethnosci., Paris.*

Ross, G. J. B. (1977). The taxonomy of the bottlenosed dolphins, *Tursiops* spp., in South African waters with notes on their biology. *Ann. Cape Prov. Mus. (Nat. Hist.)* **11,** 135–194.

Shane, Sh. H., Wells, R. S., and Würsig, B. (1986). Ecology, behavior and social organization of the bottlenose dolphin: A review. *Mar. Mamm. Sci.* **2**(1), 34–63.

St. Denis, E. de, transl. (1955). "Pline l'Ancien, Histoire Naturelle," Livre IX. Societe d'Edition less Belles Lettres, Collections des Universities de France, Paris.

Review of Incidents Involving Wild, Sociable Dolphins, Worldwide

18

Christina Lockyer
Sea Mammal Research Unit
c/o British Antarctic Survey
High Cross, Cambridge, England

INTRODUCTION

Dolphins have featured in the lives of humans and in mythology for thousands of years, and they form an important part of the culture of peoples like the Maori of New Zealand, who believe them to be messengers of the gods (Alpers, 1963). The pre-Hellenic Cretans honored dolphins as gods, and the ancient Greeks had a sanctuary in honor of the dolphin–god. The ancient Greeks, Romans, and Minoans regarded the dolphin as having special powers, and even as having a direct genealogical link with mankind through direct intervention by the gods. The Greek god of wine, Dionysus, turned men into dolphins when the crew of the ship on which he was traveling, in the guise of a mortal, decided to sell him into slavery on reaching land. The crew threw themselves overboard in terror when their plot was uncovered, and they were punished by the turning of the oars into serpents and the ship's becoming overgrown with vines. They were saved from drowning by Dionysus, who transformed them into dolphins after they had repented. Many ancient myths throughout the world incorporate the theme of transmutation from dolphin into man and vice versa; the theme is not peculiar to the Mediterranean. Humans clearly had a high regard for dolphins even in ancient times, and there is a history of intimate relationships with these mammals of the sea, which have subsequently been woven as stories into the native folklore.

This preoccupation with cetaceans is evident in the frescoes, mosaics, and sculptures of ancient Greece, Minoa (Crete), and Lydia (Etruria), where the ornamentation on buildings and domestic utensils has been eloquent and specific. The dolphin theme thus appears not to have been one of mere artistic license. Many of these items, such as the frescoes on the queen's bathroom wall in the temple at Knossus, Crete, date to 1600 B.C. There is a mosaic floor on the site of the Basilica at Olous, Crete, which depicts dolphins; the Dionysus cup which dates from 540 B.C. shows the Greek wine god (Bacchus in Roman culture) in a vessel loaded with grapes and surrounded by dolphins. The Euphronius cup, in the Louvre, Paris, dating from 500 B.C., is embellished with the scene of Theseus diving to the seabed surrounded by dolphins. Dolphins are often depicted accompanying ships, such as on the funeral fresco of a Tarquinian tomb of 600 B.C.

There are many scenes illustrating the famous story, of which there are several versions (Alpers, 1963), of the boy riding on a dolphin. The Roman philosopher Pliny the Younger related this story which featured a friendship between a local peasant lad and a dolphin called "Simo." Pliny's version was set in the Mediterranean Sea at Hippo, off the north African coast of present-day Tunisia, during the reign of Augustus. The boy befriended the dolphin after it saved him from drowning. Pliny the Elder also had a tale of a boy living by Lake Lucrino, Naples, who befriended a dolphin which carried him across the water each day to school. The boy died while young, and the dolphin supposedly pined to death for his company. These tales have a fanciful tone to them but do seem plausible, as is discussed in this chapter.

WHAT CONSTITUTES SOCIABLE?

The term "sociable" implies friendliness and seeking company; however, it also means a range of interactive behavior with other individuals including aggressive and affiliative behavior. Dolphins are generally regarded as social animals (Bradbury, 1986; Johnson and Norris, 1986), but many have been reported in the wild as actively seeking human contact, either as solitary individuals or in groups, sometimes with no obvious immediate reward or advantage. Connor and Norris (1982) concluded that dolphins are reciprocal altruists, investing energy and experience in each other in the hope of an eventual return favor which will be to the individual's benefit; furthermore, they reported instances of intergeneric altruism. Bradbury (1986) reviewed examples of reciprocal altruism in dolphins and examined them in a comparative context with other animals. Bradbury classified the types of reciprocal altruism as "threshold" and "scaled" cooperations, and "independent currency" and "barter." Such reciprocal altruism may form the basis of certain interactions between man and dolphin, for example, in symbiotic cooperative fishing, and be important to the solitary dolphin where a surrogate social "family" is sought. This chapter reviews incidents of dolphins specifically seeking and interacting with humans.

The contact between dolphin and man may be as tenuous as the dolphin's following in the wake of ships and bowriding at the stempost of fast boats (Gruber, 1981; Norris and Prescott, 1961; Shane and Schmidly, 1978; Shane *et al.*, 1986; Walker, 1975), or it can be as intimate as sustained hand-to-body contact between swimmers and individual dolphins. The reasons for dolphins choosing such contact may vary. Bowriding and chasing vessels are often merely ways of increasing the speed of travel, so the element of human contact is insignificant. Dolphins frequently position themselves in the bow-wave of large cetaceans (Leatherwood, 1974; Würsig and Würsig, 1979). The intimate body contact, however, may have an exclusive aspect, with the dolphin discriminating between familiar persons and strangers (Lockyer, 1978), perhaps using voice, appearance, and mannerisms as cues. It may select or shun human company accordingly. The behavioral interaction with familiar individual persons may be very specific. Dolphins have demonstrated an ability to identify individuals in different guises, for example, underwater clad in a wetsuit and scuba gear or clad in dry clothes aboard a small boat. For this kind of sociability to evolve, human encouragement is essential, although nearly all reported cases of "sociability" in dolphins started as expressions of extreme natural inquisitiveness on the part of the dolphin.

The sociability factor that arises from habituation in captive cetaceans in dolphinaria is one which is continually reinforced by carefully planned training, feeding, and body-contact schedules by the same people. Such habituation has also been reported in wild bottlenose dolphins (Connor and Smolker, 1985) and in wild populations of land mammals such as chimpanzees, *Pan troglo-*

dytes (Goodall, 1983). Although the behavioral interactions of captive dolphins with people appear similar to those of the wild sociable dolphins, the significant difference is that the latter group of dolphins, habituated or not, have the freedom to swim away completely and either choose or reject human company. The intimate relationships reported between wild dolphins and humans are remarkable, not least because they are between mammals each living entirely in a different environment: in water and on land. These relationships form the basis of this chapter.

OCCURRENCE OF SOCIABLE DOLPHINS

How Sociability Starts

Circumstances have to be such that dolphins and man frequent the same areas regularly, making peripheral contact more or less inevitable. Many contacts commence with dolphins following or swimming close to boats which are regularly in the area, such as ferries and fishing boats, or those which have a set course and routine (Stone, 1964; Norris, 1974; Leatherwood, 1975; Shane, 1977; Lockyer, 1978; Shane and Schmidly, 1978; Wells, 1978; Lockyer and Morris, 1985a, 1986). Dolphin encounters with surfers while riding in the waves alongside have been reported (Shane *et al.*, 1986). Direct approach to swimmers and divers is also frequently the beginning of a prolonged and regular social contact (Stone, 1964; Cousteau and Diolé, 1975; Dobbs, 1977, 1981; Lockyer, 1978; Doak, 1981; Lockyer and Morris, 1985a, 1986).

The sociability evolves as familiarity increases, especially with encouragement on the human side. The dolphin may regularly approach boats more closely or follow particular ones, as did "Percy" (see Appendix), a 4.1-m male bottlenose dolphin, *Tursiops truncatus,* off Portreath, Cornwall, England, who would follow the local lobster and crab fishing boat, stopping at each buoyed pot trap, and occasionally swim ahead to the next in anticipation of the routine (Lockyer and Morris, 1986).

Encounters with divers and swimmers, which have initially been wary, become closer, more prolonged, and even intimate. Lockyer and Morris (1986) reported that about 2 years elapsed between Percy's initial appearance in January 1981, when he was aloof, and the first time he permitted any body contact with people. Before then, but after becoming familiar with people, he would approach within a few centimeters, but always backed off out of range if a hand was extended to him. A 3.1-m male bottlenose dolphin, "Donald," also known as "Beaky" (Lockyer, 1978), off the west coast of England and Wales, approached people closely at a much earlier stage of contact, although actual handling and touching did not occur for several months. Doak (1981) reported several encounters with dolphin groups (both *T. truncatus* and common dolphins, *Delphinus delphis*) when dolphins spontaneously accompanied his Polynesian catamaran and would play with swimmers from on board. Hussenot (1980) documented the presence of "Jean-Louis," a 3.1-m female bottlenose dolphin off Pointe du Raz, Brittany, France, from September 1979. This dolphin, whose whereabouts are currently uncertain (April 1988), remained present in the same area where she had become well habituated to people, and she was still solitary after more than 7 years. However, despite the attention received from human visitors, she would still not permit direct tactile contact as recently as 1985 (Pelletier, 1985). In 1972, "Nina," a 2.44-m female bottlenose dolphin off La Coruna, Spain, first approached a clam diver, then sought body contact within only a few weeks (Cousteau and Diolé, 1975).

The famous "Opo," a female bottlenose dolphin at Opononi, New Zealand, chronicled by Alpers (1963), which would allow children to mount her and ride astride her back, was initially an inquisitive stray, patrolling Hokianga Harbour and chasing boats in 1955. An excursion further up to the beach area was greeted enthusiastically by local children and holidaymakers attempting to make physical contact. From that time on, the dolphin quickly became a highly sociable animal. One of the most emotionally significant points of this occurrence for the local people was that Hokianga was the place where the Maori explorer, Kupe, first landed in 925 A.D. Opo finally disappeared under mysterious circumstances. There are many rumors concerning her fate, none of which can be readily substantiated.

There is one reliable record of conflict between man and dolphin. An act of vengeance befell Donald off the Isle of Man (Dobbs, 1977; Lockyer, 1978), when he was shot in the side of the head, for allegedly interfering with fishing procedures, and he bore the scar for the rest of his life.

Form of Sociability

Solitary Animals

The most commonly recorded form of dolphin sociability has centered around apparently solitary individuals which seemingly chose, even if temporarily, human company, either in the absence of or in addition to the company of conspecifics. Most of the animals of this type have almost certainly been solitary, but Lockyer and Morris (1985b) systematically recorded, by means of photographs, the accumulation of fresh conspecific and other-species tooth-rake scars on the body of Percy over an extended period. This suggests that social contact with other dolphins had occurred periodically when he temporarily went missing from his usual home range for several days. Also, Donald (alias Beaky) had conspecific dolphin company temporarily off the Isle of Man (Dobbs, 1977; Lockyer, 1978). Nina had a brief encounter with members of a passing dolphin school, but elected not to join them during her 5-month stay in a bay (Cousteau and Diolé, 1975), The reason these dolphins became celebrities is that people were able to identify them easily, and their home range, the coastal area of regular daily usage, was usually confined to a small bay or an area of only a few square kilometers. The possibilities of close encounter were thus very great for people who perhaps had never seen a live dolphin before. Examples of dolphins living in small home ranges include Nina (Cousteau and Diolé, 1975) in a bay, Jean-Louis (Hussenot, 1980; A. Collet, pers. commun.) in a small area of bays, "Simo," a 2.71-m juvenile male bottlenose dolphin off Solva, Pembrokeshire, Wales, with a 25-km^2 range (Lockyer and Morris, 1987a), and Percy with a larger area of approximately 77 km^2 (Lockyer and Morris, 1986). These relatively small home ranges are not unusual for bottlenose dolphins (Wells *et al.*, 1980; Shane *et al.*, 1986).

Some of the recent well-documented solitary dolphins of a sociable disposition are introduced and discussed below. Each had a distinctive personality and made a deep impression on the people who "met" him or her.

"Pelorus Jack," a Risso's dolphin, *Grampus griseus*, was a well-known companion to ships which plied the Cook Strait off New Zealand from about 1888. The cetacean was a feature of the area for about 24 years (Cowan, 1911; Alpers, 1963; Baker, 1974) and regularly followed ships, although the animal never actually became more sociable than this.

"Dolly" was an ex-U.S. Navy trainee from Key West, Florida, who was released into the wild. In this respect she was not a truly wild animal, but she persisted in seeking human company after her liberation, the result of habituation rather than spontaneous sociability. Dolly was a bottlenose dolphin, female, and about 2.13 m in length and 193 kg in weight. Her age when liberated was estimated to be 4–5 years. Her home range became the channels of the Florida Keys where she befriended a number of families living close to the water, in particular the Ashbury family and their children, whom she would permit to pet and play with her (Cousteau and Diolé, 1975). "Carolina Snowball," a rare albino female bottlenose dolphin, frequented the U.S. Atlantic coast between South Carolina and Georgia (Caldwell, 1955; Caldwell and Caldwell, 1972). Another report from Florida (Burgess, 1982) concerns a subadult male bottlenose dolphin, "Nudgy," accidentally trapped for several months in a lake, Phillip's Inlet, after a storm. His affiliative behavior that subsequently developed with man was thought to be linked with the loss of the dolphin seen with Nudgy during the storm, and which may have been a companion. Nudgy displayed many behavioral similarities with Donald and Percy (see later) in his fascination for boats and engines. "Georgy Girl," a female bottlenose dolphin, befriended a family in Florida. The dolphin would not permit body contact for nearly 2 months after the initial human contact. Eventually, Georgy Girl would carry swimmers on her back by surfacing between their legs (Lund, 1970).

Opo, the female bottlenose dolphin off New Zealand in 1955 (Alpers, 1963; Dobbs, 1981), was so friendly with people, particularly children, that she would allow them to mount her, and she would then carry them through the water. She was believed to be only 1–2 years old when she first appeared in Hokianga Harbour. Opo appeared able to discriminate particular people who were her favorite companions, and she sought them out if they were swimming among other persons. The dolphin came into extremely shallow water to be with people, so that even small children could wade out to her.

Jean-Louis, the Brittany dolphin, has played for prolonged periods with divers and swimmers, chasing them and playing "tricks" on them, although she will still not permit body contact (Dobbs, 1984; Pelletier, 1985). This animal would now be either mature or close to maturity because she has been seen in the area for at least 7 years and was at least 1 year old when first sighted. Nina, the Spanish dolphin, was assumed to be a mature dolphin (Cousteau and Diolé, 1975), but, unlike Jean-Louis, she permitted body contact with swimmers and divers in only a few weeks.

A female bottlenose dolphin known as "Charlie" was a regular inhabitant of the Elie, Fife, coastal area of Scotland for 5–6 years (Gilchrist, 1967). She was 3.1 m in length and thus clearly mature. She followed ships and small boats and played enthusiastically with divers and swimmers alike, but she disappeared suddenly only to be reidentified further south off Northumberland, about 100 km (60 miles) distant (Mundey, 1967). Her identification was confirmed from a nick in the trailing edge of the dorsal fin as well as her behavioral characteristics.

Another bottlenose dolphin reidentified in different parts of the west coast of England and Wales was Donald (Beaky), who had so many distinguishing marks and scars (a healed bullet wound on the right side of the head and a bright white scar adjacent to the blowhole from an old and undocumented injury, but see Lockyer, 1978, for a complete list of the marks and scars and their history) that his movements south along the western British coast could be tracked in detail. Over a period of about 6 years from 1972, he covered a distance of more than 500 km (300 miles) southward from the Isle of Man to the Pembrokeshire coast, possibly via Ireland (Dobbs, 1981), to Cornwall. In each area he had a definable home range within which he could always be located, where he remained for many months or even years before suddenly moving away. When he moved from the Isle of Man the reason seemed to be the onset of harbor works in Port St Mary, which involved the use of explosives. Movements away from Pembrokeshire and later from Cornwall seemed less explicable, although the dolphin may have followed fishing boats or ferries. During his stay in Cornwall, his movements often seemed to be related to those of the local mackerel fishing fleet.

Donald became extremely friendly and would allow swimmers to hold on to his dorsal fin (but not mount him) and take them for long rides. The author once had the experience of being taken for a ride by this dolphin which returned to retrieve his passenger and resume the game after she fell off. Donald became very exuberant at times and frequently carried people out to sea (Webb, 1978a) rather than inshore as proponents of the lifesaving theory like to relate; he also became an immense nuisance, towing boats (Webb, 1978b) and overturning small moored craft, moving anchors and thus setting craft adrift (C. Lockyer and N. G. Webb, unpublished data), and disrupting small-yacht races by turning craft around at the wrong moment by surfacing under them and maneuvering them (Lockyer, 1978; Lockyer *et al.*, 1978). This behavior was far from popular, and appeared mischievous, a characteristic often apparent in this species. Another unpopular behavioral trait of Donald was pinning divers to the seabed for short periods (Dobbs, 1977; Lockyer, 1978).

Percy, the large elderly male dolphin off Cornwall between 1981 and 1985 (Lockyer and Morris, 1986), became sociable to the extent of permitting body contact and allowing people to hang on to the dorsal fin and go for a ride. His reaction to overzealous attention from visitors was not always friendly, however, and Lockyer and Morris (1986) and Dobbs (1987) reported a number of incidents of actual bodily harm befalling the people who were the object of sudden aggression. This dolphin, like Donald, developed a habit of pushing swimmers and even surfers out to sea, and actively preventing them from swimming ashore, to the extent of genuine distress on the part of the swimmer (Lockyer and Morris, 1986). Percy seemed aloof for about 1.5–2 years before allowing body contact, but then became almost

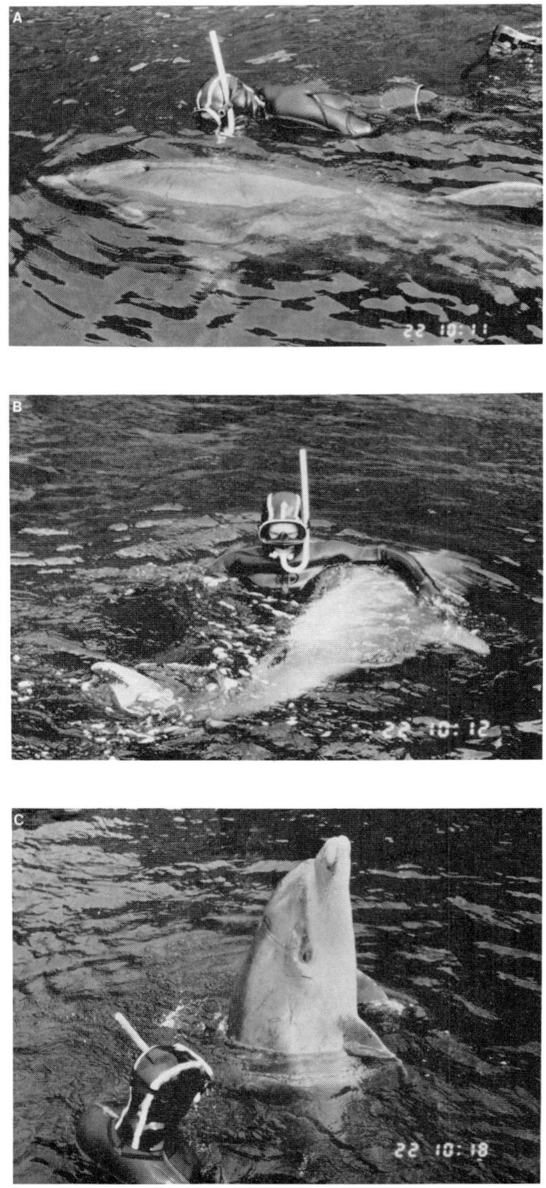

Figure 1 Simo, a wild, sociable juvenile bottlenose dolphin, off the coast of Wales, with a companion.

overexuberant in his behavior (Lockyer and Morris, 1986; Dobbs, 1987). The affiliative behavior became very unpredictable, with episodes of gentleness, sexual overtures, aggression, and also extreme violence when he would butt swimmers in the chest and push them out to sea, and smash up surfboards as people rode into shore aboard them. The tactics used for gaining attention became not only unpopular but dangerous, so that people became aware that wild animals, including dolphins, must be approached with caution and are never totally predictable. Fortunately, Percy disappeared shortly after these episodes, so that he was left alone to go his own way, rather than action being taken against him by local people. Accounts of this dolphin are given by Dobbs (1981) and Lockyer and Morris (1985a,b, 1986).

Three other male dolphins described by Dobbs (1981) were highly sociable. Of these, two were bottlenose dolphins: "Horace" off New Zealand and "Dobbie" off Eilat in the Red Sea. The third was "Sandy," a spinner dolphin, *Stenella longirostris,* which inhabited a small coastal area off San Salvador, Bahamas, between 1976 and 1978 (Doak, 1981). Dawson (1985) reported the presence of a solitary sociable male dusky dolphin, *Lagenorhynchus obscurus,* in the Tamaki estuary, New Zealand, for 3 months during 1984. He was called "Tammy" by local people and was very young. Another report of a solitary sociable bottlenose dolphin was of a 2.44-m individual off Chira, Costa Rica, which befriended and played with local swimmers and became the local "pet." Unfortunately, it was later killed (Shorn, 1983).

One of the most recent occurrences of a sociable dolphin is that of Simo off Solva, Pembrokeshire, from 1984. This animal disappeared sometime between fall 1985 and spring 1986. Unlike the other two male bottlenose dolphins, Donald and Percy, on the west coast of England and Wales, this one was young; his growth rate to 2.71 m was monitored by Lockyer and Morris (1987a), and some experimental studies of his diving abilities and swimming speeds were conducted (Lockyer and Morris, 1987b). Simo was extremely friendly, although he appeared to have favorite human companions both in the water and aboard boats. Like some of the other sociable bottlenose dolphins, he allowed swimmers to pet him (Fig. 1) and take hold of his dorsal fin. He would then pull them along at great speed for distances of several hundred meters. The author was carried in this manner on several occasions, each time the dolphin initiating the play by coming alongside and pressing the dorsal fin against the body of the swimmer. The most recent report of a human–dolphin association comes from the west coast of Ireland. Here, a solitary bottlenose male named Dorad, about 3.0 m in length, has become resident and will play and frolic with swimmers and divers (Holmes, 1987).

Group Encounter

Sometimes groups of dolphins have temporarily associated with man in the wild. Such a group encounter can be a casual one, such as described by Doak (1981) with passing schools of both bottlenose and common dolphins following his catamaran and playing with swimmers; a regular meeting and play with the same group of dolphins (Stone, 1964); or a regular meeting but less intimate interaction, such as the bottlenose dolphins associating with shrimp-fishing boats off the Texas coast (Leatherwood, 1975), following them, and consuming either the discarded catch and/or the fish consuming it and the symbiotic relationship of "Old Tom" and his pack with the whale boats. Old Tom, a killer whale, *Orcinus orca*, patrolled an area frequented by whaling vessels off Eden, Twofold Bay, eastern Australia, with his pack (Stead, 1933; Dakin, 1934). His fame originated from his habit of assisting the whalers chase their quarry in a kind of symbiotic manner. This cetacean was believed to have attained an age of 50–90 years, according to local tales. Historic records of sightings that are well documented cover a period of about 35 years; a tooth extracted from the skeleton hanging in the Eden Museum was used for age determination (Mitchell and Baker, 1980), and dentinal plus cemental growth layers indicated an age of about 35 years. Therefore, although the evidence is not in favor of Old Tom's having been nearly a century old, the animal was certainly of mature years.

Doak (1981) experimented to the extent of dressing his team in common dolphin-styled and patterned wetsuits in order to gain greater acceptability to the cetaceans they encountered. Certain body postures and the displaying of the white belly have a behavioral significance (Caldwell and Caldwell, 1972), and it was thought that communication by mimicry between man and dolphin might be easier if such costumes were worn. Many divers, including the author, have noted that bottlenose dolphins seem to be attracted to wetsuit-clad people in the water. It may be that the texture, color, or physical properties of the neoprene foam rubber are especially interesting to dolphins, which can probably "see" the texture by echolocation as well as visually.

In Shark Bay, Western Australia, at the site known as Monkey Mia, there is a group of habituated friendly bottlenose dolphins that regularly mingle with a larger group of unhabituated dolphins in the area (Doak, 1981; Gawain, 1982; Connor and Smolker, 1985). Currently, at least seven dolphins from the local population come inshore to water less than 1 m deep, to get close to the holiday visitors who hand-feed them fish. Dolphins have been reported doing this at Monkey Mia since the late 1960s. The fish are accepted, but not always eaten, often being used as play items and then dropped (Doak, 1981; Gawain, 1982). Lockyer (1978) and Lockyer *et al.* (1978) unsuccessfully attempted experimental feeding of Donald and also reported that although fish frequently offered to him on a buoy by local fishermen were taken, they were not eaten but dropped to the seafloor, which was littered with dead and rotting fish. Lund (1970) also reported failed attempts to feed Georgy Girl with fish. The hand-feeding of the Monkey Mia dolphins is thus interesting in that the dolphins have come to anticipate "handouts," as in chimpanzees (Wrangham, 1974; Goodall, 1983), and have also acclimated to eating dead and partially frozen fish.

The Monkey Mia dolphins are identifiable as individuals with names from dorsal fin shape and other visual cues, and the family relationships and the calves in the group are known (Connor and Smolker, 1985). Doak (1981) reported that the dolphins, however, will not always tolerate body contact with people, and they avoid persons wearing scuba gear. Gawain (1982) also reported "aloofness" and even avoidance of swimmers and snorkelers. During 1966–1972, one of these dolphins, known as "Old Charlie," became so friendly that children could sit astride its back (Doak, 1981). The sex of this dolphin was not known. This particular dolphin also reportedly drove the herring in toward the fishermen on shore. It was reported to have been eventually shot, although other dolphins have continued to approach people. Recent legislation has now provided some protection for the dolphins.

There have been numerous accounts of dolphins guiding troubled ships in storms, fog, and when drifting, and also of dolphins rescuing drowning men (Doak, 1981). However, there have also been authentic accounts of shipwreck caused by cetacean interference, such as killer whales breaking up and holing yachts by surfacing alongside and from beneath (Robertson, 1984). This may be construed as a mistake on the part of the animals, although the author feels that the primary interest of cetaceans in boats is a selfish one, emanating from curiosity and a love of speed associated with bowriding and not pleasure from associating with man. Also, cetaceans have a natural habit of pushing any unusual object to the surface (Pryor, 1975; Defran and Pryor, 1980).

There are a number of examples of associations described as symbiotic between man and dolphins, in addition to that between the whalermen and the Eden killer whales. The Mauritanian Imragen, an ancient people unrelated to the Berbers, Moors, or Yemenites, and numbering only a few hundred today, use a coastal netting technique that is centuries old (Anthonioz, 1967, 1968; Busnel, 1973; Cousteau and Diolé, 1975; Maigret, 1984; Maigret and Abdallahi, 1976). They fish for mullet in the Atlantic waters between Cape Blanc and Cape Timiris, a stretch of coast 130 km (80 miles) long. The fishermen beat the water with sticks and then the local bottlenose dolphins drive the mullet into the fishermen's nets as the fish migrate along the coast. The people are nomadic and very poor, and move from site to site to await the arrival of the dolphins which assist the efficiency of the fishing. Cousteau and Diolé (1975) reported that at Nauamghar, as the dolphins drive the mullet into dense shoals inshore, the Imragen run out from the beach in orderly teams and cast nets approximately 20–30 m in length, over the fish, while drawing them in toward the beach. About 100–130 kg of fish are trapped in each cast. The caught fish are subsequently wind- and sun-dried in huts.

Hall (1984) reviewed current and historic evidence for aboriginal fishing practices in Moreton Bay where bottlenose dolphins were guided to assist in herding mullet into nets. Hall described the association as commensal. Busnel (1973) recorded other examples of reportedly symbiotic activity between man and dolphins including the Irrawaddy dolphin, *Orcaella brevirostris*, in Burma, and the Amazon River dolphin, *Inia geoffrensis*, in the Tapajos River in South America. Thein (1977) reported cooperative fishing between man and Irrawaddy dolphins in Burma, where nets are weighted in the river and then the dolphin is "summoned" by oar-tapping on the sides of the fishing boat. The dolphin circles over the net, concentrating the fish which are then netted. Part of the catch is given to the dolphin. Lamb (1954) reported sociable interactions between Amazon River dolphins and fishermen, which included cooperative fishing when whistling and paddle-tapping summoned the dolphins. Herded fish were then harpooned.

Lasse Lofgren (pers. commun.) reported a cooperative fishery between local Brazilian fishermen and bottlenose dolphins, some of which were individually identifiable to the fishermen and had supposedly lived for ages in excess of 50 years. These fishermen used nets thrown from the shoreline on an estuary and trapped and caught the fish which the dolphins drove ashore into the nets. Other reports of cooperative fishing between man and river dolphins include the Ganges susu, *Platanista gangetica*, in India and the baiji, *Lipotes vexillifer*, in the Yangtse River in China (Lockley, 1979; Hall, 1984).

The killer whales at Eden, New South Wales, apparently entered the area before the northward migration of the right and humpback whales from southerly waters, and then remained there for the June–November whaling season (Dakin, 1934; Mead, 1962). The killers cooperated with the whalers (Lucas and Le Souef, 1909) by chasing and attacking right whales, and Hawkins and Cook (1908) reported that the whalermen would disentangle killers fouled in their lines. The Twofold Bay whaling commenced in 1828, and killers were recorded in 1843 by Oswald Brierly (Mitchell and Baker, 1980). In the 1870s, newly appearing killer whales were named after dead crewmen. During 1878 the killers were in three mobs called "Hookey"'s, "Stranger"'s, and "Cooper"'s, totalling 27 whales (Wellings, 1944). Confusion arose in the possibility of at least two whales with the same name: "Tom" and "Old Tom" in 1879, as well as "Ben" and "Old Ben." In 1904, Old Tom was cited as the leader of a pack which was cooperating with the whalers and included 5 whales identified by name (Mead, 1962). Whaling finally ceased in 1932, but by 1928, only 4–5 killers still frequented the area. In 1930, Old Tom died.

"Captive" Sociable Dolphins

Cetaceans, particularly bottlenose dolphins, have generally adapted well to a captive existence as far as social behavior is concerned, and they have proved to be friendly and cooperative animals with training. The author does not intend to cover dolphins maintained in dolphinaria for public display but wishes to mention briefly the current use of dolphins in naval exercises and training procedures as well as in scientific investigations.

Dolly, an ex-U.S. naval trainee dolphin, was liberated because of unpredictable behavior. Tasks assigned in naval undersea exercises involved carrying tools, acting as messengers and detectors, and retrieving certain items. Many of the training sessions were carried out in Hawaii in tanks open to the sea. The fact that the dolphins returned of their own free will seems to show that they were willing participants in the exercise (Cousteau and Diolé, 1975). "Tuffy," a famous male bottlenose dolphin U.S. naval recruit off California (Ridgway, 1972; Cousteau and Diolé, 1975), participated in many experimental trials and also served as a messenger and guide between seabed and surface in a series of undersea living experiments of Project Sealab including Sealab II and Sealab III.

Of the scientific investigations that have taken place, many involved voluntary participation both in captivity and in the wild. For example, dolphins can be trained to exhale underwater into collecting funnels for respiratory gas analysis or to dive and trigger cameras on the seabed to order to investigate diving performance and the effects of pressure on the body (Ridgway, 1972). Captive dolphins tolerate eye cups, which temporarily blind them, for tests of echolocation abilities.

The main point is that the dolphins voluntarily participate in these trials: the result of reinforcement training (Pryor, 1986), reflecting their trusting attitude to man. The close relationships that have developed between wild dolphins and man seem to affirm this trust, or rather represent the dolphin's experience of a reciprocal altruistic system where there is no human cheating.

Species and Distribution

The species recorded as "sociable" in the wild include *Delphinus delphis* (Doak, 1981), possible *Stenella coeruleoalba* (D. A. McBrearty, pers. commun.; see the detailed patterns in the dolphin frescoes of ancient Greece and Crete), *Stenella longirostris* (Doak, 1981; Dobbs, 1981), *Grampus griseus* (Cowan, 1911; Alpers, 1963), *Orcinus orca* (Mitchell and Baker, 1980), *Lagenorhynchus obscurus* (Dawson, 1985), *Inia geoffrensis* (Lamb, 1954; Busnel, 1973; Cousteau and Diolé, 1975), *Orcaella brevirostris* (Busnel, 1973; Cousteau and Diolé, 1975; Thein, 1977), *Platanista gangetica*, and *Lipotes vexillifer* (Lockley, 1979; Hall, 1984), as well as many incidents of *Tursiops truncatus* (see earlier examples). Bottlenose dolphins have been involved in many more reported instances of intimate sociability than have any other dolphins, and this may be partly because they inhabit inshore coastal waters where encounters with man are frequent.

The interesting fact is that sociable bottlenose dolphins have turned up all over the world: in the United States (Stenuit, 1969; Caldwell and Caldwell, 1972; Dobbs, 1981), New Zealand (Alpers, 1963; Dawson, 1985), Australia (Doak, 1981; Gawain, 1982; Connor and Smolker, 1985), France (Hussenot, 1980), Spain (Cousteau and Diolé, 1975), and the British Isles (Gilchrist, 1967; Dobbs, 1977, 1981; Lockyer, 1978; Webb, 1978a, b; Lockyer and Morris, 1986, 1987a). Symbiotic associations with man occur off Mauritania (Cousteau and Diolé, 1975; Maigret, 1984) and Brazil (L. E. Lofgren, pers. commun.). In conclusion, it seems that the sociable-dolphin phenomenon is worldwide and, while predominantly involving bottlenose dolphins, is possible with a variety of odontocetes.

Behavioral Traits

Is the Behavior of Wild Sociable Dolphins Similar to That of Captive Dolphins?

There is an interesting standard for comparison in Dolly, the ex-U.S. naval dolphin. She appeared to return satisfactorily to an independent existence in the Florida Keys, although she continued to seek human company. The types of behavior she exhibited have all been reported for captive dolphins, and the behavioral repertoire she displayed, once free, was similar to that of many truly wild sociable dolphins. Of particular interest to the author are Opo, Nina, Donald, Percy, and Simo, all of whose behavior has been well documented. None of these would actually eat fish proffered by humans, unlike the habituated Monkey Mia group, often accepting but then playing with the fish. In contrast, Percy has caught fish and then repeatedly "proffered" them to humans (Lockyer and Morris, 1986). The habit of taking swimmers for rides either astride or alongside appears not unusual, and this apparently natural habit is often reinforced in dolphinaria for shows. The habits of boat-towing and moving of objects are also traits exploited in captive dolphins on show. Donald enjoyed removing and picking up divers' tools (Lockyer, 1978), and Percy actively participated in fishing operations (Lockyer and Morris, 1986).

Much of the chest-butting and aggressive behavior displayed by the British adult male bottlenose dolphins is like that reported by McBride and Hebb (1947), Norris (1967), and Pryor (1975) for captive dolphins. The pinning down of divers to the seabed by Donald is characteristic of captive bottlenose dolphins as well (Defran and Pryor, 1980). Pushing of divers to the surface is also a frequently observed trait of captive dolphins.

The behavior of the adult male bottlenose dolphins Donald (Lockyer, 1978; Lockyer *et al.*, 1978) and Percy (Lockyer and Morris, 1986) was frequently overtly sexual, with strong aggressive tendencies. Cousteau and Diolé (1975) also reported sexual overtures by belly presentation and rubbing from the adult female Nina. In captivity the sexual component of male behavior can be very disruptive and is often apparently indiscriminate (McBride and Hebb, 1947; Caldwell and Caldwell, 1972). Donald and Percy apparently attempted copulation with the undersides of rubber inflatable boats and even divers. The behavior of wild sociable dolphins thus appears very similar to that of captives, given the environmental restrictions of the latter.

Is the Solitary Sociable Dolphin a Misfit?

The discussion in the previous section indicated that behavioral patterns in the solitary sociable dolphins were similar to those of captives, so that any abnormal traits, if they exist, in the wild animal must be subtle and perhaps undetectable to the human observer. This logic of course assumes that captive animals are not inherently abnormal! Of the dolphins that are well documented, Opo, Jean-Louis, Nina, Donald, Percy, and Simo all appear to fall into one of two categories, either very young or well mature when they made their debut. Although many were "resident" for several years, for example, Jean-Louis (and current fears are that she, too, may have disappeared, Anne Collet, pers. commun.), many either disappeared completely or died, so that we know their habits only for part of their lives which are potentially long, up to 27 years (as recorded from teeth in wild North Atlantic bottlenose dolphins (Hohn, 1980). The author knows of at least one captive bottlenose dolphin greater than 27 years old (from teeth layers) which was in captivity for at least 17 years.

Additionally, Lockyer and Morris (1985b) demonstrated from the regular accumulation of fresh tooth-rake scars that Percy clearly continued to have social contact with other dolphins during his apparently "solitary" phase. When he disappeared after a few years, he was never traced, and no carcass was found. Perhaps he joined a dolphin school and emigrated. There is evidence that the solitary state may be temporary. Georgy Girl was eventually joined by a male dolphin during her sociable phase, at which point she ceased close contact with her human "family" and stopped playing with them (Lund, 1970).

The author concludes from personal experience and the documented reports that such solitary animals are not necessarily misfits but are more likely to be the victims of circumstance. The young animals may either be orphans or temporary strays, and the older animals may simply have found interesting distractions while passing through an area. Dolphins are naturally inquisitive, and their sudden appearance and disappearance may coincide with a sudden increase or wane in local human activity or the reappearance of a passing school of dolphins. The presence of the dolphin may even be linked to human activities in a direct and practical way, for example, Percy liked to frequent the area near the sewage outfall of Portreath probably because this attracted the mullet in the area (Lockyer and Morris, 1986).

Representatives of both sexes, young and old,

Figure 2 Monument of a girl and a dolphin at Tower Bridge, London, England.

have been reported, so that the hypothesis of schools rejecting the aged animals is not supported. However, the social hierarchy within dolphin schools may result in excessive stresses on certain individuals through repeated aggressive encounters from other school members, to the extent that these individuals have no effective status and either are driven out or leave of their own volition. Such individuals may be of either sex and of any age. In the captive situation the author is aware of a particular instance where a juvenile female had to be removed from groups because of persistent aggression by other group members. Characteristically, this animal was covered in tooth-rake scars. In the wild, such emigrants or expelled individuals may, through curiosity and need, turn their attentions to man as a surrogate. We cannot be certain, however, why dolphins should spontaneously seek and apparently enjoy human company without immediate or any reward.

Fact and Fable

The opening to this chapter related some of the ancient myths and tales concerning the empathy between man and dolphins: tales of boys riding on dolphins and ships being guided by dolphins through stormy seas. The documented recent reports of similar incidents lend credence to these tales which, even if not precisely true, were not just "tall stories." They seem to have been based on characteristic behavioral traits of these mammals. The only tales that cannot be confirmed are those of the transformation of men into dolphins by the intervention of the gods.

CONCLUSIONS

All the sociable individuals described in this chapter are well documented (see Appendix); many other occurrences may have gone unreported. The wild animals that have become famous have often been honored in some way, first by legislation so that they may be protected by law from harassment (e.g., Opo: Alpers, 1963; Dobbs, 1981; Nina: Cousteau and Diolé, 1975; dolphins at Monkey Mia: Cousteau and Diolé, 1975; Doak, 1981; Dobbs, 1981) and second by a monument to their memory (e.g., Nina and Opo). Old Tom's skeleton was saved and erected in Eden Museum to his memory. These are specific monuments. However, one of the finest monuments the author has seen is that of a sculpture of a girl and a dolphin (Fig. 2) at the Thames riverside at Tower Bridge, London, England. This monument does not commemorate any particular dolphin, but symbolizes the enduring friendship between human and cetacean.

The phenomenon of the sociable dolphin is a fascinating one without complete explanation. Even as new incidents of such animals are reported, it is unlikely that there will be any real explanation. Certain species of animals appear to have a greater rapport with man, and it is also true that many of these have a vested interest in that there is a practical benefit to the animal from human contact. However, this is not obvious in many of the cases reported here for sociable dolphins, so that the real basis for the behavior is an enigma.

Appendix
Details of Specific Records of Sociability in Dolphins

Species	Locality	Number of Individuals	Given Names	Sex	Age Class	Period of Documented Sociability	Nature of Evidence	Reference
Unknown	Hippo, Mediterranean	Solitary	Simo	?	?	? Reign of Roman Emperor Augustus	Dolphin saved boy from drowning by permitting boy to ride astride its back	Alpers (1963)
Unknown	Lake Lucrino, Naples, Italy	Solitary	?	?	?	Ancient Roman history	Social interaction with a young boy	Alpers (1963)
Tursiops sp.	Portreath, S.W. England	Solitary	Percy	Male	Adult	January 1981–November 1984	Accompanying boats, swimmers, divers; permitting body contact	Lockyer and Morris (1985a,b; 1986), Dobbs (1981, 1987)
Tursiops sp.	Isle of Man, Wales, S.W. England	Solitary	Donald; Beaky	Male	Adult	April 1972–August 1978	Accompanying boats, swimmers, divers; permitting body contact	Lockyer (1978), Lockyer *et al.* (1978), Dobbs, (1977, 1981, 1987), Webb (1978a,b)
Tursiops sp.	La Coruna, Spain	Solitary	Nina	Female	Adult?	Summer 1972	Approaching boats, people; permitting body contact	Cousteau and Diolé (1975)
Tursiops sp.	Solva, Wales	Solitary	Simo	Male	Juvenile	1984–winter 1985/1986	Accompanying boats, swimmers, divers; permitting body contact	Lockyer and Morris (1987a,b)
Tursiops sp.	S.W. Ireland	Solitary	Dorad	Male?	Subadult?	Summer 1987	Accompanying boats, swimmers, divers; permitting body contact	Holmes (1987)
Tursiops sp.	Brittany, France	Solitary	Jean-Louis	Female	Adult	September 1979–1987	Accompanying swimmers, divers; wary of body contact	Hussenot (1980), Pelletier (1985), Dobbs (1984)
Tursiops sp.	Fife, Scotland, to Northumberland, England	Solitary	Charlie	Female	Adult	~1962–1967	Following boats; accompanying swimmers, divers	Gilchrist (1967), Mundey (1967)

(continued)

Appendix *(Continued)*

Species	Locality	Number of Individuals	Given Names	Sex	Age Class	Period of Documented Sociability	Nature of Evidence	Reference
Tursiops sp.	Opononi, New Zealand	Solitary	Opo	Female	Juvenile	1955, few months only	Chasing boats; seeking body contact with swimmers in shallows	Alpers (1963), Dobbs (1981)
Tursiops sp.	Key West, Florida	Solitary	Dolly	Female	Subadult	1960s	Ex-U.S. naval trainee liberated; seeking close human contact	Cousteau and Diolé (1975)
Turiops sp.	U.S. coastal, S. Carolina to Georgia	Solitary	Carolina Snowball	Female	?	1950s	Seeking human company	Caldwell (1955), Caldwell and Caldwell (1972)
Tursiops sp.	Phillip's Inlet, Florida	Solitary	Nudgy	Male	Subadult	Few months, 1981	Seeking boats and human company	Burgess (1982)
Tursiops sp.	Florida	Solitary; later joined by a male *Tursiops*	Georgy Girl	Female	Adult?	1960s	Close body contact, allowing swimmers to mount astride	Lund (1970)
Tursiops sp.	New Zealand	Solitary	Horace	Male	?	?	Accompanying swimmers and divers	Dobbs (1981)
Tursiops sp.	Eilat, Red Sea	Solitary	Dobbie	Male	?	?	Accompanying swimmers and divers	Dobbs (1981)
Tursiops sp.	Shark Bay, W. Australia	Group of >7 dolphins	All known individuals	Both sexes	Various	>20 years to present time	Body contact with humans in shallows; will accept fish from hand	Doak (1981), Gawain (1982), Connor and Smolker (1985)
Tursiops sp.	Chira, Costa Rica	Solitary	?	?	?	1983	Body contact with swimmers; following boats	Shorn (1983)
Lagenorhynchus obscurus	Tamaki, New Zealand	Solitary	Tammy	Male	Juvenile	3 months in 1984	Body contact with swimmers	Dawson (1985)
Stenella longirostris	San Salvador, Bahamas	Solitary	Sandy	Male	?	1976–1978	Accompanying swimmers, divers	Doak (1981), Dobbs (1981)
Grampus griseus	Cook Strait, New Zealand	Solitary	Pelorus Jack	Male	Adult	~24 years from 1888	Accompanying ships	Alpers (1963), Baker (1974)
Orcinus orca	Twofold Bay, E. Australia	Group	Old Tom and others	Male; ?	Adult; ?	>35 years in 1800s	Interaction with whale boats	Dakin (1934), Stead (1933)

REFERENCES

Alpers, A. (1963). "Dolphins." John Murray, New Zealand and London. 251 pp.

Anthonioz, R. (1967). Les Imraguens pêcheurs nomades de Mauritanie (El Memghar). *Bull. IZFAN* **29B** (3–4), 695–738.

Anthonioz, R. (1968). Les Imraguens pêcheurs nomades de Mauritanie II. *Bull. IZFAN* **30B**(2), 751–768.

Baker, A. N. (1974). Risso's dolphin in New Zealand waters, and the identity of "Pelorus Jack." *Rec. Dominion Mus.* **8**(16), 267–276.

Bradbury, J. W. (1986). Social complexity and cooperative behavior in delphinids. *In* "Dolphin Cognition and Behavior: A Comparative Approach" (R. J. Schusterman, J. A. Thomas, and F. G. Wood, eds.), pp. 361–372. Lawrence Erlbaum Associates, Hillsdale, New Jersey.

Burgess, R. F. (1982). The dolphin that came in from the cold. *Oceans* **15**(5), 13–15.

Busnel, R. G. (1973). Symbiotic relationship between man and dolphins. *Trans. N.Y. Acad. Sci. Ser. 2* **35**(2), 112–131.

Caldwell, D. K. (1955). Evidence of home range of an Atlantic bottlenose dolphin. *J. Mammal.* **36**(2), 304–305.

Caldwell, D. K., and Caldwell, M. C. (1972). "The World of the Bottlenose Dolphin." Lippincott, Philadelphia and New York. 157 pp.

Connor, R. C., and Norris, K. S. (1982). Are dolphins reciprocal altruists? *Am. Nat.* **119**(3), 358–374.

Connor R. C., and Smolker, R. S. (1985). Habituated dolphins (*Tursiops* sp.) in western Australia. *J. Mammal.* **66**(2), 398–400.

Cousteau, J. Y., and Diolé, P. (1975). "Dolphins." Cassell, London. 304 pp.

Cowan, J. (1911). "Pelorus Jack: The White Dolphin of French Pass, New Zealand, with Maori Legends." Whitcombe and Tombs, Christchurch, New Zealand. 47 pp.

Dakin, W. J. (1934). "Whalemen Adventurers." Angus and Robertson Ltd., Sydney, Australia. 263 pp.

Dawson, S. (1985). "The New Zealand Whale and Dolphin Digest—The Official Project Jonah Guidebook." Brick Row Publishing Co., New Zealand.

Defran, R. H., and Pryor, K. (1980). The behavior and training of cetaceans in captivity. *In* "Cetacean Behavior: Mechanisms and Functions" (L. M. Herman, ed.), pp. 211–261. Wiley, New York. 463 pp.

Doak, W. (1981). "Dolphin, Dolphin." Hodder and Stoughton, Auckland, New Zealand. 245 pp.

Dobbs, H. E. (1977) "Follow a Wild Dolphin." Souvenir Press, London. 237 pp.

Dobbs, H. E. (1981) "Save the Dolphins." Souvenir Press, London. 128 pp.

Dobbs, H. E. (1984). "The Magic of Dolphins." Butterworth Press, Guildford, Surrey. 77 pp.

Dobbs, H. E. (1987). "Tale of Two Dolphins." Jonathan Cape, London. 187 pp

Gawain, E. (1982). More than half way. *Oceans* **15**(5), 10–12.

Gilchrist, I. G. (1967). Dolphin days. *Animals* **10,** 104–107.

Goodall, J. (1983). Population dynamics during a 15-year period in one community of free-living chimpanzees in the Gombe National Park, Tanzania. *Z. Tierpsychol.* **61,** 1–60.

Gruber, J. A. (1981). Ecology of the Atlantic bottlenose dolphin (*Tursiops truncatus*) in the Pass Cavallo area of Montagorda Bay, Texas. M.S. thesis, Texas A & M University, College Station, Texas 77843. 182 pp.

Hall, H. J. (1984) Fishing with dolphins? Affirming a traditional aboriginal fishing story in Moreton Bay, S. E. Queensland. *In* "Focus on Stradbroke" (R. J. Coleman, J. Covacevich, and P. Davie, eds.), pp. 16–22. Boolarong Publications, Brisbane.

Hawkins, H. S., and Cook, R. H. (1908). Whaling at Eden, with some "Killer" yarns. *The Lone Hand* **3,** 265–273.

Hohn, A. A. (1980). Age determination and age related factors in the teeth of western North Atlantic bottlenose dolphins. *Sci. Rep. Whales Res. Inst., Tokyo* **32,** 39–66.

Holmes, B. (1987). A diary of my dives with "Dorad." *Diver* **32**(4), 30–31.

Hussenot, E. (1980). Le grand dauphin *Tursiops truncatus* en Bretagne: Types de fréquentation. *Penn ar Bed* **12,** 355–380.

Johnson, C. M., and Norris, K. S. (1986). Delphinid social organization and social behavior. *In* "Dolphin Cognition and Behavior: A Comparative Approach" (R. J. Schusterman, J. A. Thomas, and F. G. Wood, eds.), pp. 335–346. Lawrence Erlbaum Associates, Hillsdale, New Jersey.

Lamb, F. B. (1954). The fishermen's porpoise. *Nat. Hist.* **63**(5), 231–232.

Leatherwood, J. S. (1974). A note on gray whale behavioral interactions with other marine mammals. *Mar. Fish. Rev.* **36**(4), 50–51.

Leatherwood, J. S. (1975). Some observations on feeding behavior of bottlenosed dolphins (*Tursiops truncatus*) in the northern Gulf of Mexico and (*Tursiops* cf. *T. gilli*) off southern California, Baja California, and Nayarit, Mexico. *Mar. Fish. Rev.* **37,** 10–16.

Lockley, R. M. (1979). "Whales, Dolphins and Porpoises." Methuen, Sydney.

Lockyer, C. (1978). The history and behavior of a solitary wild, but sociable, bottlenose dolphin (*Tursiops truncatus*) on the west coast of England and Wales. *J. Nat. Hist.* **12,** 513–528.

Lockyer, C., and Morris, R. J. (1985a). A wild but sociable dolphin off Portreath, north Cornwall. *J. Zool., London (A)* **207,** 605–607.

Lockyer, C., and Morris, R. J. (1985b). Body scars of a resident, wild bottlenosed dolphin (*Tursiops truncatus*): Information on certain aspects of his behaviour. *Aquatic Mamm.* **11**(2), 42–45.

Lockyer C. and Morris, R. J. (1986). The history and behaviour of a wild, sociable bottlenose dolphin (*Tursiops truncatus*) off the north coast of Cornwall. *Aquatic Mamm.* **12**(1), 3–16.

Lockyer, C., and Morris, R. (1987a). Observed growth rate in a wild juvenile *Tursiops truncatus. Aquatic Mamm.* **13**(1), 27–30.

Lockyer, C., and Morris, R. J. (1987b). Observations on diving behaviour and swimming speeds in a wild juvenile *Tursiops truncatus. Aquatic Mamm.* **13**(1), 31–35.

Lockyer, C., Flewellen, C., Madgwick, A., and Morris, R. (1978). Some field observations and experiments on a bottlenosed dolphin. *Prog. Underwater Sci., Rep. Underwater Assoc.* **3,** 177–190.

Lucas, A. H. S., and Le Souef, W. H. D. (1909). "The Animals of Australia. Mammals, Reptiles and Amphibians." Whitcombe and Tombs Ltd., Melbourne, Australia.

Lund, V. (1970). Georgy Girl. *Underwater Nat.* **6**(3), 12–15.

McBride, A. F., and Hebb, D. O. (1947). Behavior of the captive bottle-nose dolphin, *Tursiops truncatus. J. Comp. Physiol. Psychol.* **41,** 111–123.

Maigret, J. (1984). Les Imraguen, pêcheurs des côtes de Mauritanie: Une technique ancestrale. "L'Exploitation de la mer," V-èmes Rencontres Internationales d'Archeologie et d'Histoire, Antibes, October, 1984, pp. 205–214.

Maigret, J., and Abdallahi, A. O. (1976). La pêche des Imraguen sur le banc d'Arguin et au cap Timiris (Mauritanie); techniques et méthodes de pêche. *Notes Africaines* **149,** 1–8.

Mead, T. (1962) "Killers of Eden, the Story of the Killer Whales of Twofold Bay." Angus and Robertson, Sydney. 222 pp.

Mitchell, E., and Baker, A. N. (1980). Age of reputedly old killer whale, *Orcinus orca,* "Old Tom" from Eden, Twofold Bay, Australia. *Rep. int. Whal. Commn.* (Special Issue 3), pp. 143–154.

Mundey, G. R. (1967). Charlie revisited. *Animals* **10,** 354–6.

Norris, K. S. (1967). Aggressive behaviour in Cetacea. *In* "Aggression and Defense: Natural Mechanisms and Social Patterns" (C. D. Clemente and D. B. Lindsley, eds.), pp. 225–241. University of California Press, Berkeley.

Norris, K. S. (1974). "The Porpoise Watcher." Norton, New York. 250 pp.

Norris, K. S., and Prescott, J. H. (1961). Observations on Pacific cetaceans of Californian and Mexican waters. *Univ. Calif. Publ. Zool.* **63,** 291–402.

Pelletier, F. X. (1985). Jean-Louis, le gentil dauphin breton. *30 Millions d'Amis, la Vie des Betes* **83,** 22–27.

Pryor, K. (1975)."Lads before the Wind." Harper and Row, New York.

Pryor, K. (1986). Reinforcement training. *In* "Dolphin Cognition and Behavior: A Comparative Approach" (R. J. Schusterman, J. A. Thomas, and F. G. Wood, eds.), pp. 253–260. Lawrence Erlbaum Associates, Hillsdale, New Jersey.

Ridgway, S. H. (1972). Homeostasis in the aquatic environment. *In* "Mammals of Sea—Biology and Medicine" (S. H. Ridgway, ed.), pp. 590–747. C. C. Thomas, Springfield, Illinois. 812 pp.

Robertson, D. (1984). "Survive the Savage Sea." Granada Publications.

Shane, S. H. (1977). The population biology of the Atlantic bottlenose dolphin, *Tursiops truncatus,* in the Aransas Pass area of Texas. M.S. thesis, Texas A & M University, College Station, Texas 77843. 239 pp.

Shane, S. H., and Schmidly, D. J. (1978). The population biology of the Atlantic bottlenose dolphin, *Tursiops truncatus,* in the Aransas Pass area of Texas. National Technical Information Service, DB-283 393. U. S. Dept of Commerce, Springfield, Virginia 22161. 130 pp.

Shane, S. H., Wells, R. S., and Würsig, B. (1986). Ecology, behavior and social organization of the bottlenose dolphin: A review. *Mar. Mamm. Sci.* **2**(1), 34–63.

Shorn, H. (1983). Dolphin killer struck dead by lightning bolt. *Weekly World News,* Nov. 29, p. 4.

Stead, D. G. (1933). "Giants and Pigmies of the Deep. A Story of Australian Sea Denizens." Shakespeare Head Australian Nature Books, No. 4. The Shakespeare Head Press, Sydney. 108 pp.

Stenuit, R. (1969). "The Dolphin, Cousin to Man." J. M. Dent and Sons, Ltd. 176 pp.

Stone, S. (1964). Girl on a dolphin. *Underwater Nat.* **2**(3), 8–10.

Thein, U. T. (1977). The Burmese freshwater dolphin. *Mammalia* **41**(2), 233–234.

Walker, W. (1975). Review of the live capture fishery for smaller cetaceans taken in southern California waters for public display, 1966–73. *J. Fish. Res. Board Can.* **32**(7), 1197–1211.

Webb, N. G. (1978a). Women and children abducted by a wild, but sociable adult male bottlenosed dolphin. *Carnivore* **1**(2), 89–94.

Webb, N. G. (1978b). Boat towing by a bottlenose dolphin. *Carnivore* **1**(1), 122–129.

Wellings. C. E. (1944). The killer whales of Twofold Bay, N.S.W., Australia, *Grampus orca*. *Aust. Zool.* **10**(3), 291–293.

Wells, R. S. (1978). Home range characteristics and group composition of Atlantic bottlenosed dolphins, *Tursiops truncatus*, on the west coast of Florida. M.S. thesis, University of Florida, Gainesville, Florida 32611. 110 pp.

Wells, R. S., Irvine, A. B., and Scott, M. D. (1980). The social ecology of inshore odontocetes. *In* "Cetacean Behavior: Mechanisms and Functions" (L. M. Herman, ed.), pp. 263–317. Wiley, New York. 463 pp.

Wrangham, R. W. (1974). Artificial feeding of chimpanzees in their natural habitat. *An. Behav.* **22,** 83–93.

Würsig, B., and Würsig, M. (1979). Behavior and ecology of bottlenose porpoises, *Tursiops truncatus*, in the South Atlantic. *Fish. Bull.* **77,** 399–442.

Quantitative Description of a Rare Behavioral Event: A Bottlenose Dolphin's Behavior toward Her Deceased Offspring

19

Richard C. Connor
Department of Biological Sciences
University of Michigan
Ann Arbor, Michigan 48109, and
Dolphins of Shark Bay Research Foundation, Inc.
Albuquerque, New Mexico 87106

Rachel A. Smolker
Pyschology Department
University of Michigan
Ann Arbor, Michigan 48109, and
Dolphins of Shark Bay Research Foundation, Inc.
Albuquerque, New Mexico 87106

INTRODUCTION

Opportunities to observe the behavior of dolphins toward their dead offspring are rare in nature, and existing reports are largely anecdotal (e.g., Hubbs, 1953; Moore, 1953, 1955; Norris and Prescott, 1961). Here we describe the behavior of a female bottlenose dolphin (*Tursiops* sp.) during the last hours in which she was sighted in the vicinity of her dead offspring.

The female and her dead calf were encountered in the context of regular behavioral observations on a population of bottlenose dolphins in Shark Bay, Western Australia. Several dolphins in Shark Bay are "tame" and frequent a shallow-water area by a campground where they accept dead fish and "petting" from people (Connor and Smolker, 1985). They even engage in an apparent form of play with people that involves the exchange of bits of seaweed. The tame dolphins are part of a much larger social network in the surrounding bay, of which about 70 individuals of known sex are encountered on a regular basis.

The female in question, "Crooked-fin," is tame and will approach boats offshore, but she rarely visited the campground shallows in the two years preceding the observations described here. Her previous surviving offspring, a female named "Puck," was 10 years old in early 1987. By comparison, another of the tame females, "Holey-fin," has three known surviving offspring born over the period 1975–1983 at 4-year intervals. Crooked-fin was without a calf throughout our 1986 field season (May–August), and our first 1987 sighting of her was with the calf, a male, on March 9. The last time the calf was seen alive was during 1434–1440 hours on April 3 with Crooked-fin and five other females plus a female calf. No unusual behavior was noted during that brief encounter which ended when Crooked-fin and her calf left the group. Sixty-six hours later between 0830 and 0850 on April 6, we discovered the calf dead, approximately 200 m offshore in about 5.5 m of water in the vicinity of Crooked-fin.

METHODS

Dolphins in Shark Bay are identified on the basis of scars and tears on the dorsal fin (Würsig and Würsig, 1977) in combination with the *shape* of the dorsal fin, which may be more variable in the Shark Bay population than in some others (e.g., Würsig and Würsig, 1977). Individuals are sexed by observing their genital area as they swim in an inverted position while bowriding, socializing, or chasing fish. Also, dolphins that are accompanied by calves on a regular basis are assumed to be females.

All behavioral observations were made from aboard two small (3.5–4.0 m) aluminum boats (much of the research at Shark Bay involves following dolphins in small boats, an activity that does not appear to disturb them) and recorded into hand-held cassette recorders for later transcription. Continuous focal observations of Crooked-fin's behavior when in the vicinity of the calf were begun at 0907 and ended at 1712 after Crooked-fin had not been sighted for 2 hours. Additionally, Crooked-fin's phonations were monitored for 22 minutes and 30 seconds through a hydrophone system with a frequency response of 20 kHz and a Sony TCD 5m tape recorder. After observations had ceased, the calf's body was retrieved for examination.

From 0907 until 1044 the water surface was glassy calm, which enabled us to keep the motors turned off and move with the use of oars. Between 1045 and 1130 ripples appeared and gave way to light chop. We switched from the use of oars to a 5-hp motor at 1145 in a moderate chop which became heavy by 1210.

BEHAVIOR TOWARD THE DEAD CALF

General Description

Crooked-fin's behavior over the observation period can be rather cleanly divided into two distinct periods. From 0907 to 1135 she typically remained in the vicinity of the calf, often "snagging." Snagging is a posture a dolphin takes when still or slowly moving at the surface for longer than required to breath and submerge. While snagging, the peduncle and flukes are often lowered so far that the dorsal fin is submerged or has only the rostral edge exposed. If the peduncle is lowered, the head is slightly elevated so that the tip of the rostrum and sometimes the upper tip of the mandible are above the water line (we avoid the more obvious term "resting," as the posture frequently occurs in social contexts when its primary function is clearly not "to rest").

From 1135 until she was last sighted at 1513, Crooked-fin engaged in a series of "excursions" in which she left the vicinity of the calf. Excursions were defined post hoc as ventures of more than about 30 m from the calf; an excursion ended when Crooked-fin was next sighted within 20 m of the calf. Excursions were further divided into "long trips," which lasted at least 5 minutes, and "short trips," which lasted 2.5 minutes or less. Following an excursion, Crooked-fin typically returned to the vicinity of the calf for a few seconds or minutes before leaving on another trip. When in the vicinity of the calf Crooked-fin often "milled." Milling is a general activity in which an individual makes little (<1.5 km/hour) or no directional progress and changes orientation with (nearly) every surfacing. Solitary milling is most often associated with hunting in Shark Bay.

The Snagging and Excursion Periods

Between 0907 and 1135, Crooked-fin snagged 51 times, of which 46 (90%) were timed (range 2–220 seconds). The closest approach to the calf during a snag ranged from an estimated 35 m to nearly touching. Within the snag period we also can identify a period, 0939–1013, of especially intense activity directed toward the calf. The five longest snags occurred during this "intense" period. A comparison of the "intense" period with the rest of the snag period for several parameters is given in Table 1. Crooked-fin's orientation relative to the calf during the intense period was variable. Typically, she flanked the calf when close, as though visually inspecting him. During four snags she

Table 1
Snagging[a]

Period	N	Duration (Seconds)	Approaches to 1 m	Interval (Seconds)
Regular	31	31.4	4 (13%)	122.7 (n = 28)
Intense	15	70.5	13 (87%)	68.1 (n = 14)
Total	46	44.2	19 (37%)	105.0 (n = 42)

[a] There were significant differences between the intense and regular periods in snag duration ($p < .01$, Mann–Whitney U test), number of approaches to within 1 m ($p < .0001$, two-sample Z test for proportions, Remington and Schork, 1985), and interval between successive snags ($p < .005$, Mann–Whitney U Test).

partially or fully circled the calf. The best example of this was a 176-second snag when she slowly completed 1⅓ circles around the calf while less than 1 m away. She never actually touched the calf while snagging. We are confident of this assessment as the glassy calm conditions would have enabled us to detect any slight, sudden movement of the calf.

No unusual sounds were heard while listening through the hydrophone at about 1–20 m from the calf. Crooked-fin was mostly silent during the 22-minute and thirty-second listening period, except for her usual burst-pulse "squeaks" directed toward the boats (she often approaches boats seeking fish handouts) and echolocation click trains. However, twice during the intense snag period Crooked-fin produced a vocalization in air that was not heard through the hydrophone. The vocalization was a faint, high-pitched sound reminiscent of steam escaping from a teakettle, but higher pitched and wavering more. Presumably, these "teakettle" vocalizations were produced by compressed air escaping through a small opening in the blowhole. The two occurrences were about 15 minutes apart during long snags of 220 and 155 seconds. The sounds, although not timed, were judged by two observers to be of very long duration, around 10–20 seconds. We have not heard the "teakettle" vocalization in other contexts, but that may simply be due to the unusually conducive conditions of the day (calm water and motors off).

During the excursion period (1135–1513), Crooked-fin made 17 long trips, two short trips, and two other probable short trips in which she was sighted 20–25 m moving away and subsequently was not seen returning to within 20 m for at least 2 minutes. Another trip was too poorly documented to be placed in either category. All 17 long trips ranged from just over 5 to just over 13 minutes in duration, while the short trips were 1.5–2.5 minutes. As the mean interval between all trips was only 111 seconds, the dominant component of the excursion period was the trips, not the time spent in the vicinity of the calf. This stands in contrast to the snag period in which Crooked-fin spent only two 8-minute periods away from the calf (at 25–45 m). There was no indication, however, that trips later in the excursion period were longer than those occurring earlier on. We do not know how far Crooked-fin ranged during the trips; the farthest confirmed sighting was 80 m, but we suspect that she traveled much further. Sixteen of the trips, including 12 in succession, were to the southeast, the direction the calf was drifting. The wind was NNW so Crooked-fin often surfed down the swells as she was leaving. On returning from a trip Crooked-fin often approached the calf closely (63% <4 m), but heavy chop precluded our seeing whether actual contact was made.

We suspect the purpose of Crooked-fin's trips was to travel to preferred hunting areas. Indicators of hunting (e.g., tail-out dives, n = 14) or fish chasing (e.g., splashes, n = 3) in the vicinity of the calf were infrequent during the snag and excursion bouts.

The only two sightings of other dolphins in the vicinity of the calf occurred during the excursion period, and, interestingly, both sightings were accompanied by the only observations of Crooked-fin snagging during the excursion period. In the first case, an unidentified dolphin was sighted 7 m from the calf moments before Crooked-fin snagged. At 1345, three male dolphins were seen surfing in the vicinity of the calf and a minute later they were milling at 1–2 m around the calf and Crooked-fin, who was snagging right beside the calf (0.5 m). The males moved off after a minute as Crooked-fin snagged briefly three more times within 1 m of the calf. Shortly after this occurred the only lengthy period (9 minutes) in which we lost contact with the calf and Crooked-fin.

Crooked-fin returned from her last trip at 1512 and was last sighted 25 m NW of the calf at 1513. After not sighting Crooked-fin for two hours, we retrieved the calf at 1712 for examination.

Figure 1 The dead calf when first encountered at 0900 hours, April 3, 1987.

DISCUSSION

During the 8.5 hour observation period the calf drifted 3 km. During both the snag and excursion periods when Crooked-fin was showing an interest in the calf, the calf was within the area where Crooked-fin was most commonly sighted. Crooked-fin may thus have failed to return to the calf when it drifted out of her "core" area. We suspect, however, that Crooked-fin abandoned the calf in order to avoid the three tame males that visit the campground. During our observations, the calf was drifting toward the campground and that is where we retrieved the calf at 1712. Two of the males were present at the campground during our observations. From late 1986 until their disappearance in January 1989, the three tame males often chased, herded, and harassed females they encountered near the campground or farther offshore. Crooked-fin was aggressively herded at least 15 times by the tame males in 1987. Further, she began visiting the campground on a regular basis again in February 1989 (as she had in previous years) following the disappearance of the males.

When first encountered, the calf was slightly bloated midflank and floating on its side. Rigor mortis had set in, the calf's eyes were sunken, but his skin color appeared normal (Fig. 1). By the time the carcass was retrieved 8 hours later, the bloating had slightly increased, and some discoloration was evident in the midventral area that had been above the water line. Postmortem examination was incomplete and did not yield a conclusive cause of death (but the lungs were congested with blood). The calf was not emaciated; in fact, shortly after we encountered Crooked-fin she flexed her peduncle forward ventrally and expelled a large cloud of milk. At 123 cm and 20.5 kg, the calf was large compared to other new calves we have seen in the area, suggesting an age of a few months at the time of death (adults in the area do not appear to exceed 230 cm).

The only external marks were a set of tooth rakes on the right side of the peduncle behind the dorsal fin (Fig. 2). Apparently caused by a small shark (S. Ridgway, pers. commun.), the wounds are not significant and may well have been inflicted posthumously. In fact, one of the splashes during the snag period was immediately followed by a 1-m shark (species unknown) swimming away from the vicinity of the calf, suggesting that Crooked-fin may have chased the shark. There were no marks on the calf's flippers, flukes, or elsewhere that would evince of Crooked-fin's having carried the dead calf with her mouth. In fact, the lack of obvious physical contact between Crooked-fin and her deceased calf during our observations stands in contrast to previously published accounts. (e.g., Hubbs, 1953; Moore, 1953, 1955; Norris and Prescott, 1961). Obviously, Crooked-fin may have directed other kinds of behavior toward the calf in the potentially many hours between the time of its death and our discovery of the situation.

The attention Crooked-fin gave to her dead offspring certainly reflects a strong mother–infant bond. Similar events are often recounted for primates (e.g., chimpanzees, *Pan troglodytes*, Goodall, 1987; baboons, genus *Papio*, Tayler and Saay-

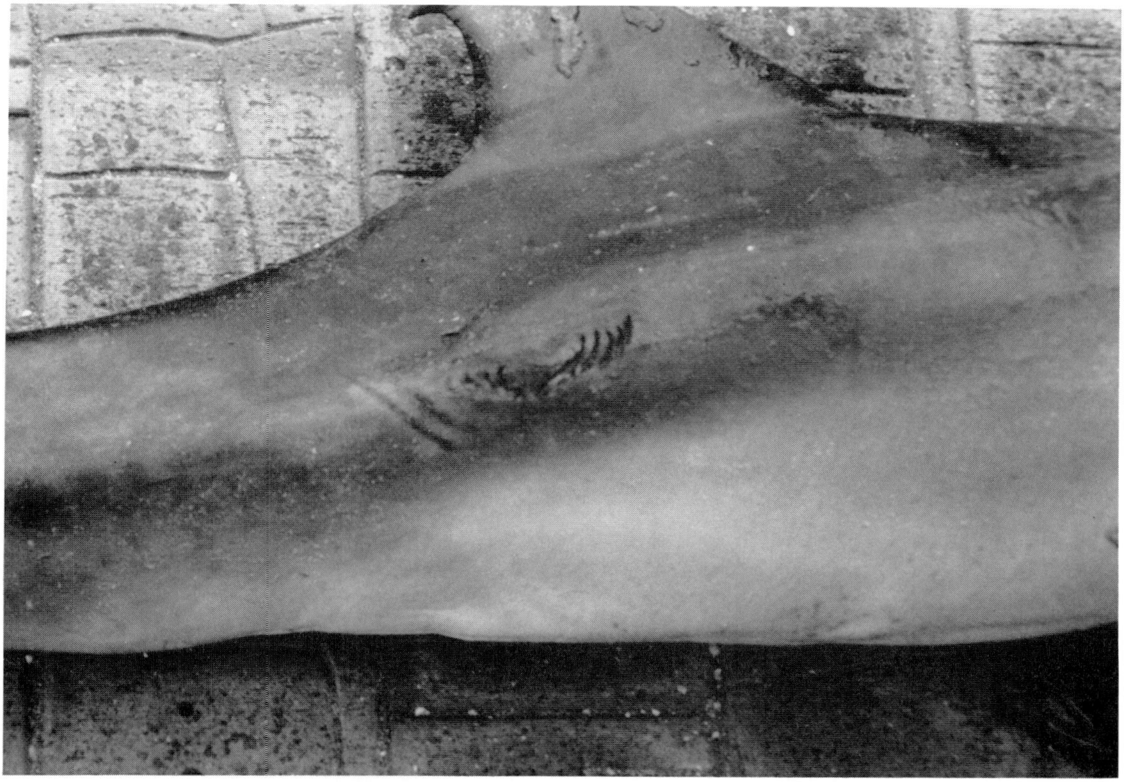

Figure 2 Tooth rakes on the right side of the peduncle of the dead calf. The marks appear to have been caused by a small shark rather than a dolphin.

man, 1972; Altman, 1980; Smuts 1987). A striking example for ungulates is an instance in which an impala, *Aepyceros melampus,* remained by her dead offspring for over 12 hours and watched as it was eaten by baboons (B. B. Smuts, pers. commun.). In the most extreme manifestation of this phenomenon, mothers have attended to dead offspring until they reached an advanced state of decomposition (e.g., dolphins, Tayler and Saayman, 1972; baboons, Altman, 1980; gorillas, *Gorilla gorilla,* D. Watts, pers. commun.) and sometimes until only a small part of the body remained (Moore, 1955; Altman, 1980). We would like to know if such extreme behavior has any adaptive value and if it is limited to large-brained, social mammals. Perhaps the phenomenon is simply underreported elsewhere. For cetaceans, most accounts are from the 1950s and 1960s in spite of a large increase in the number of active researchers since then. Only by consistent and detailed reporting of mothers' behavior toward deceased offspring can we begin to understand this puzzling and unusual phenomenon.

ACKNOWLEDGMENTS

We thank Joseph Hadchitti, Ruth Musgrave, and Bernhard Riegl for invaluable assistance in making the observations reported here. Barb Smuts, Andrew Richards, David Watts, Richard Wrangham, and an anonymous reviewer provided assistance and/or constructive criticism during manuscript preparation. Sam Ridgway provided critical eleventh-hour advice. R.C.C. was supported in 1987 by grants from the National Science Foundation and the National Geographic Society, and R.C.C. and R.A.S. were supported by a grant from the Ceebie Trust.

REFERENCES

Altman, J. (1980). "Baboon Mothers and Infants." Harvard University Press, Cambridge, Massachusetts. 242 pp.

Connor, R. C., and Smolker, R. A. (1985). Habituated dolphins (*Tursiops* sp.) in Western Australia. *J. Mammal.* **66,** 398–400.

Goodall, J. (1986). "The Chimpanzees of Gombe." Harvard University Press, Cambridge, Massachusetts, 672 pp.

Hubbs, C. L. (1953). Dolphin protecting dead young. *J. Mammal.* **34,** 498.

Moore, J. C. (1953). Distribution of marine mammals to Florida waters. *Am. Mid. Nat.* **49,** 117–158.

Moore, J. C. (1955). Bottle-nosed dolphins support remains of young. *J. Mammal.* **36,** 466–467.

Norris, K. S., and Prescott, J. H. (1961). Observations on Pacific cetaceans of Californian and Mexican waters. *Univ. Calif. Publ. Zool.* **63,** 291–402.

Remington, R. D., and Schork, M. A. (1985). "Statistics with Applications to the Biological and Health Sciences." Prentice-Hall, Inc., Englewood Cliffs, New Jersey. 415 pp.

Smuts, B. B. (1987). "Sex & Friendship in Baboons." Aldine, New York. 303 pp.

Tayler, C. K., and Saayman, G. S. (1972). The social organisation and behaviour of dolphins (*Tursiops aduncus*) and baboons (*Papio ursinus*): Some comparisons and assessments. *Ann. Cape Prov. Mus. (Nat. Hist.)* **9,** 11–49.

Würsig, B., and Würsig, M. (1977). The photographic determination of group size, composition, and stability of coastal porpoises (*Tursiops truncatus*). *Science* **198,** 755–756.

Site and Association Fidelity in Bottlenose Dolphins off Argentina

20

Bernd Würsig
Texas A&M University
Galveston, Texas 77550

Graham Harris
Puerto Madryn, Provincia Chubut, Argentina

INTRODUCTION

Individually recognizable bottlenose dolphins, *Tursiops truncatus*, have apparently frequented the same area for over 20 years in Western Australia (Gawain, 1981, 1984; Connor and Smolker, 1985). Wells (1986) has seen some individual dolphins of a study area in Western Florida since 1970, and it is generally believed that such site fidelity is a common feature of coastal bottlenose dolphins (reviews by Leatherwood and Reeves, 1982, and Shane *et al.*, 1986). Social relationships may also last for long periods, in many cases probably for life (Wells, 1986). Nevertheless, long-term studies are few, and we offer here further evidence that a certain amount of site and association fidelity are present in another community of coastal bottlenose dolphins. Our analysis comes from two resightings of a total of 10 members of a community of 53 animals studied in detail 8–12 years before.

STUDY ANIMALS

During a 21-month study from mid 1974 to early 1976, Würsig and Würsig (1977, 1979; Würsig, 1978) determined school sizes, affiliations, diurnality and seasonality of behavior, movement patterns, and calving seasonality of a community of bottlenose dolphins frequenting the southeast portion of Golfo San José, Chubut, Argentina, centered at 42° 23′ S, 64° 03′ W. They found that the animals exhibited a roughly 4-month cycle of occurrence. Schools numbering 8–22 individuals usually contained dolphins that were consistently found together and dolphins that appeared to have no particularly strong social affiliations with others. This created some group stability and a considerable amount of group "openness" as well, which Würsig (1978) compared to the openness of some societies of chimpanzees, *Pan troglodytes*.

During the 1974–1976 study period, bottlenose dolphins occurred within 1 km of the rocky Patagonian coastline on 191 of 433 sightable days (with winds less than 20 km/hour), or 44% of the time. Researchers of the New York Zoological Society who have been working at the study site from 1976 to the present, however, report that the frequency of occurrence dropped to less than 5% of sightable days in 1978 (information compiled from interviews by G.H.; since these researchers were not working directly with bottlenose dolphins, no actual numbers of sightings, nor dates, were recorded). This indicates that a drastic shift in areal usage took place, but we have no information on where these dolphins spent the majority of their time since the decline in use of the study area. We know from the main study in the mid 1970s that the study area was at the northern end of the dolphins' range, with animals never moving beyond a certain area about 8 km north of our main observation point. We also sighted some individuals as far as 300 km along the coastline to the south of the study area, and we can surmise that a shift of home range to the south may have been responsible for the decreased frequency of occurrence in the study area. Since this is an area with little human activity, it is likely that the shift in usage was due to natural ecological and not human factors. Wells *et al.* (Chapter 24, this volume) documented a dramatic temporary shift in the distribution of some bottlenose dolphins from southern to north-central California during a recent warm-water year, but the dolphins returned to their previous range in the south (Hansen, 1983, Chapter 23, this volume) when the water cooled. We have no information on general water mass or food availability changes for the Golfo San José area.

RESIGHTINGS

On 30 July 1984, G.H. photographed a part of a school of 10–15 bottlenose dolphins moving parallel to shore, within 500 m of land. Out of 12 "usable" dorsal fin photos (with enough size in the 35-mm image, and proper focus and clarity, for recognition capability), we recognized 7 individuals in 8 photos (one was photographed twice) as animals that were present in the study area in 1974–1976 (Fig. 1). The other four photos were of unmarked dorsal fins, probably, judging by general fin shape and size, of only two different animals.

On 21 August 1986, a school of about 15–20 bottlenose dolphins moved close to shore in similar manner as during the 1984 resighting. Out of 28 "usable" dorsal fin photos, we recognized 6 individuals (Fig. 1). Each of these was photographed two or three times. Five of the 28 photos were of unmarked dorsal fins, while one was of a marked animal not recognizable from the 1974–1976 study. It was either a new animal, or one whose dorsal fin changed beyond recognition in the intervening years. Three of the 6 dolphins recognized in 1986 were also present in 1984.

The 7 dolphins resighted on 30 July 1984 and 5 of the 6 dolphins resighted 21 August 1986 were first sighted toward the beginning of the 1974–1976 study, while one dolphin (DN) was first sighted October 1975 (Fig. 2). The dolphins depicted in Fig. 2 are grouped into four subgroups. Subgroup A was almost always present when dolphins occurred in the study area in 1974–1976. Subgroup B individuals disappeared September 1975, were resighted March 1976, over 300 km distant from camp, and reappeared in the study area December 1976. Subgroup C appeared when Subgroup B disappeared in late 1975, and Subgroup D includes dolphins that appeared and disappeared at various other times. The 1984 and 1986 resightings were of dolphins representative of each of the four original subgroups.

Dolphins Fuzzy, Small Nip, and LF were known mothers who had calves in early 1974, 1975, and 1976, respectively. Dolphins RN 2, OWS, and Small Flag had small calves beside them during the July 1984 resighting; dolphins B and LSN had small calves beside them during the August 1986 resighting. Würsig (1978) found that Argentine bottlenose dolphin calves do not appear to stray from their mothers during the first 3–5 months, but Wells (1986) has reported that calves off western Florida have spent minutes to hours with females not their mothers. It is likely, therefore, that even if not all animals which were in close proximity to calves were mothers, they were females. This suggests that six of the seven dolphins identified July 1984 and five of the six dolphins identified August 1984 were females. All of the calves photographed in 1984 and 1986 were

Figure 1 Ten dolphins recognized 30 July 1984 (LF, RN 2, B, OWS, WR, Sm Flag, and Sm Nip) and 21 August 1986 (RN 2, Fuzzy, B, DN, Sm Nip, and LSN), all animals which had been sighted previously during a long-term study from 1974 to 1976.

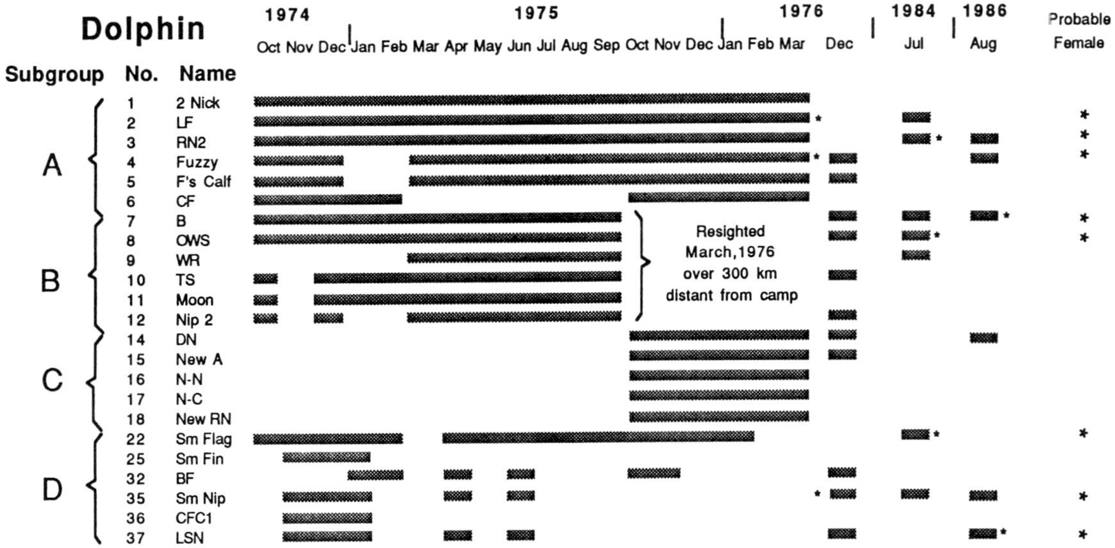

Figure 2 Month-by-month summary of the presence, marked by shading, of 23 bottlenose dolphins in Golfo San José, Argentina. October 1974 through March 1976 represent data from a long-term continuous study (which began June 1974, with a few data June–September 1974), while December 1976, July 1984, and August 1986 are represented by single sighting days. The animals shown here, part of a larger community of 53 dolphins recognized in 1974–1976, have been selected because they show relationships of occurrence within subgroups. Stars indicate dolphins accompanied by calves; and probable females, as deduced from presence of calves, are identified at right.

approximately 3–5 months old, and were therefore born in the first half of the year, or summer–fall. This hint at calving seasonality is consistent with three of four calves that were born in summer–fall during the 1974–1976 study.

CONCLUSIONS

It has been found in many studies that bottlenose dolphins tend to concentrate their activities in certain areas (e.g., Irvine and Wells, 1972; Wells *et al.*, 1980; Shane, 1980; Gruber, 1981; Irvine *et al.*, 1981; Ballance, 1986). The 1974–1976 Würsig study demonstrated this as well, with the more recent documentation of a decrease in use of a particular part of the community's range. It would be valuable to know enough about the ecology of the area to determine why this shift occurred.

The 10 recognizable dolphins of the 1984 and 1986 resightings appeared physically unchanged since they were seen in 1974–1976. A large wound below the dorsal fin of dolphin OWS in 1974 had healed to a pale white scar by 1976, and was still of the same paleness in 1984. A white spot on both sides of the dorsal fin of dolphin WR was still as prominent as when seen in 1974, and a small tatter of flesh on the dorsal fin of dolphin Small Nip had not changed between November 1974 and August 1986, almost 12 years. We do not imply that marks are unchanging for all animals; our photographic record of the 1984 and 1986 resightings was incomplete, with only 7 of 10–15 dolphins identified for the first resighting, and 6 of 15–20 dolphins identified for the second resighting. One individual had previously unseen marks on its dorsal fin, and the pattern of marks would, for example, have obscured the pattern of a previously recognized dolphin. Nevertheless, the fact that erosion of the marked area and changes in skin color did not take place to a noticeable degree on 10 dolphins bids good news for long-term recognition of bottlenose dolphins from other studies (e.g., Shane, 1980; Gruber, 1981; Hansen, Chapter 23, this volume; Connor and Smolker, 1985; Ballance, 1986).

Wells (1986) documented similar long-term marks as described here for the community studied in Florida.

It is not possible from discrete resightings such as these to make firm conclusions about long-term association patterns. However, two of the dolphins of Subgroup A and three of Subgroup B, both subgroups representing animals that were consistently together in 1974–1976, were found together again in 1984. This pattern of affiliation strongly hints that these individuals have been socially bound together over a long time, and it is likely that they were together at least part of the intervening years for which we do not have affiliation data.

The recurring affiliations between females found in this study indicate a social system similar to that found in more extensive work by Wells *et al.* (1980). In their study, associations were based to a large degree on the ages and sexes of animals; females (and their young) consistently traveled together, while adult males joined the females from time to time as the males moved between female groups (see also Shane *et al.*, 1986, and Wells, 1986). The scarcity (or possible absence) of males in our resightings indicates that the schools photographed on 30 July 1984 and 21 August 1986 may have been such female-and-young schools, with from none to, at most, only a few adult males present. Wells (1986) hypothesized from extensive work on sex-related affiliations that the social system of the community of dolphins of his study area is likely a form of promiscuous polygamy. In the Argentine bottlenose dolphin community, changes in affiliation hint at this as well, and the recent finding of a preponderance of females in some subgroups that are known to have traveled together for long periods of time (at least 21 months during the 1974–1976 study) further substantiates, but does not prove, this emerging picture of bottlenose dolphin social structure.

ACKNOWLEDGMENTS

Melany Würsig helped to gather data during the 1974–1976 study. Randy Wells made valuable comments on an early draft. Lynn McMaster drew Fig. 2, and Sandra Sanduski typed the manuscript. We thank them all.

REFERENCES

Ballance, L. T. (1986). Ecology and group organization of bottlenose dolphins, *Tursiops truncatus*, in the Gulf of California, Mexico. M.S. thesis, Moss Landing Marine Laboratories, Moss Landing, California. 60 pp.

Connor, R. C., and Smolker, R. S. (1985). Habituated dolphins (*Tursiops* sp.) in Western Australia. *J. Mammal.* **66**(2), 398–400.

Gawain, E. (1981). "The Dolphin's Gift." Whatever Publications, Mill Valley, California. 256 pp.

Gawain, E. (1984). Observations of a dolphin watcher. *Whalewatcher* **18**(3), 3–6.

Gruber, J. A. (1981). Ecology of the Atlantic bottlenosed dolphin (*Tursiops truncatus*) in the Pass Carvallo area of Matagorda Bay, Texas. M.S. thesis, Texas A & M University, College Station, Texas 77843. 182 pp.

Irvine, B., and Wells, R. S. (1972). Results of attempts to tag Atlantic bottlenose dolphins, *Tursiops truncatus*. *Cetology* **13**, 1–5.

Irvine, A. B., Scott, M. D., Wells, R. S., and Kaufman, J. H. (1981). Movements and activities of the Atlantic bottlenose dolphin, *Tursiops truncatus*, near Sarasota, Florida. *Fish. Bull. (U.S.)* **79**, 671–688.

Leatherwood, S., and Reeves, R. R. (1982). Bottlenose dolphin (*Tursiops truncatus*) and other toothed cetaceans. *In* "Wild Mammals of North America" (J. A. Chapman and G. A. Feldhamer, eds.), pp 369–414. Johns Hopkins University Press, Baltimore, Maryland. 1147 pp.

Shane, S. H. (1980). Occurrence, movements, and distribution of bottlenose dolphins, *Tursiops truncatus*, in southern Texas. *Fish. Bull. (U.S.)* **78**, 593–601.

Shane, S. H., Wells, R. S., and Würsig, B. (1986). Ecology, behavior, and social organization of the bottlenose dolphin: A review. *Mar. Mamm. Sci.* **2**, 34–63.

Wells, R. S. (1986). Structural aspects of dolphin societies. Ph.D. dissertation, University of California at Santa Cruz, California. 234 pp.

Wells, R. S., Irvine, A. B., and Scott, M. D. (1980). The social ecology of inshore odontocetes. *In* "Cetacean Behavior: Mechanisms and Functions" (L. M. Herman, ed.), pp. 263–317. Wiley, New York.

Würsig, B. (1978). Occurrence and group organization of Atlantic bottlenose porpoises (*Tursiops truncatus*) in an Argentine bay. *Biol. Bull.* **154**, 348–359.

Würsig, B., and Würsig, M. (1977). The photographic determination of group size, composition, and stability of coastal porpoises (*Tursiops truncatus*). *Science* **198**, 755–756.

Würsig, B., and Würsig, M. (1979). Behavior and ecology of bottlenose porpoises, *Tursiops truncatus*, in the South Atlantic. *Fish. Bull. (U.S.)* **77**,399–442.

PHOTO: Nearshore communities of bottlenose dolphins have been studied closely in such areas as Sarasota Bay, on the west coast of Florida (September 1980; Michael D. Scott).

Part VI

Distribution, Movements, and Abundance

Bottlenose Dolphins off the Northeastern United States

21

Robert D. Kenney
Graduate School of Oceanography
University of Rhode Island
Narragansett, Rhode Island 02882

INTRODUCTION

The Cetacean and Turtle Assessment Program (CETAP) was conducted from a base at the University of Rhode Island (URI) between 1978 and 1983. The primary goal was to provide data on the distribution and abundance of whales, dolphins, and sea turtles in the continental-shelf waters off the northeastern United States for use by federal agencies in making decisions about offshore oil and gas exploration and development (Edel *et al.*, 1981). Surveys between Cape Hatteras, North Carolina, and the northern Gulf of Maine (Fig. 1) covered over 460,000 km of trackline and resulted in more than 11,000 sightings involving a total of 26 species of cetaceans (CETAP, 1982). The CETAP data base also contains several thousand records from years before 1978. Previous distributional analyses from the CETAP data have focused on the entire cetacean community (Hain *et al.*, 1985; Kenney and Winn, 1986, 1987; Winn *et al.*, 1987) or on individual baleen whale species (e.g., humpback whale, *Megaptera novaeangliae:* Kenney, 1984; right whale, *Eubalaena glacialis:* Winn *et al.*, 1986). This chapter is the first detailed analysis of the distribution patterns of a small odontocete based on the CETAP sighting data.

The bottlenose dolphin, *Tursiops truncatus,* is distributed widely in temperate and tropical oceans. In the western Atlantic Ocean, it is common from at least the north coast of Argentina (Lichter and Hooper, 1984) to New England (Caldwell and Golley, 1965; Caldwell *et al.*, 1971a,b; CETAP, 1982; Erdman, 1970; Erdman *et al.*, 1973; Gunter, 1954; Katona *et al.*, 1978; Leatherwood *et al.*, 1976; Leatherwood and Reeves, 1982; Lowery, 1974; Moore, 1953; Schmidly, 1981; Shane, 1980; Taruski and Winn, 1976; Winn *et al.*, 1979). It is rare in the inshore waters of eastern Canada, with only two known records from the

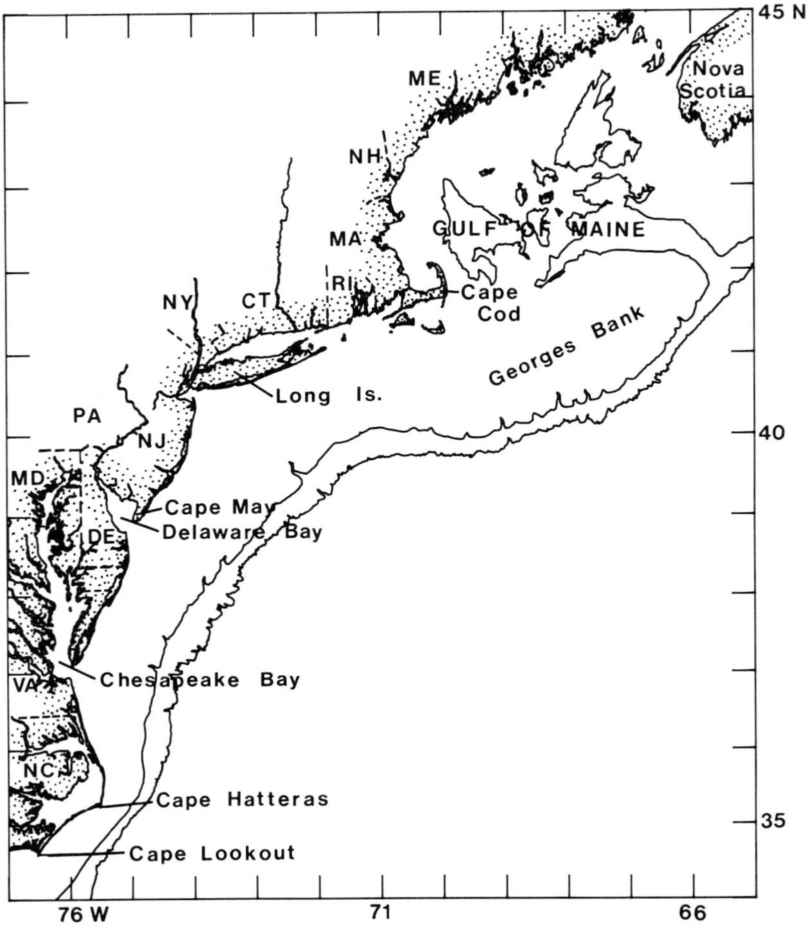

Figure 1 The CETAP study area, off the northeastern United States, showing important place names used in the text. Isobaths shown are 200 and 2000 m. This base map is used for all distribution plots.

upper Bay of Fundy (Sergeant and Fisher, 1957; Sergeant *et al.*, 1970). Its distribution extends further northward in the eastern North Atlantic, where it is considered the second most common stranded cetacean in the British Isles (Brown, 1975), and may occur occasionally as far north as northern Norway (Tomilin, 1957).

Bottlenose dolphins traditionally have been considered inhabitants of shallow coastal waters (e.g., Caldwell and Caldwell, 1972). Most authors now concur that there are coastal and offshore populations, which may be more or less isolated, with the larger, more robust animals generally farther offshore (Hersh, 1987; Hersh and Duffield, Chapter 6, this volume; Leatherwood *et al.*, 1976; Leatherwood and Reeves, 1982; Odell *et al.*, 1975; Schmidly, 1981; Walker, 1981). In some cases, inshore and offshore bottlenose dolphins have been described as separate nominal species, for example *T. truncatus* offshore and *T. aduncus* inshore off southern Africa (Ross, 1977, 1984; but see Ross and Cockcroft, Chapter 5, this volume). Mitchell (1975a) suggested that the larger form ranges farther into colder waters than the smaller.

The bottlenose dolphin was the small cetacean sighted most commonly during the CETAP study

(CETAP, 1982). It is a significant component of the marine ecosystem of the entire northeast U.S. continental shelf (Scott *et al.*, 1983), Georges Bank (Winn *et al.*, 1987), and the shelf-break region (Hain *et al.*, 1985). Based on estimated average seasonal standing stocks of bottlenose dolphins (Kenney *et al.*, 1985), and conservatively assuming an average consumption of 2% of body weight per day (Innes *et al.*, 1987; Lavigne *et al.*, 1986; Sergeant, 1969; Yasui and Gaskin, 1986), bottlenose dolphins off the northeastern United States consume over 8 million kg of prey annually. This chapter summarizes the CETAP data on the distribution and abundance of the bottlenose dolphin.

METHODS

Data Collection

The CETAP study area was defined as the waters of the U.S. continental shelf north of Cape Hatteras, from the shoreline to 5 nautical miles (9.3 km) seaward of the 1000-fathom (1829-m) isobath. Surveys were conducted from October 1978 through January 1982. Additional details of CETAP data collection procedures can be found in CETAP (1982), Edel *et al.* (1981), Hain *et al.* (1981), Kenney and Winn (1986), Scott and Gilbert (1982), and Scott *et al.* (1981). There were four phases of the CETAP data-collection effort and four resulting types of data.

Dedicated Aerial Surveys

Random-transect aerial surveys were conducted in defined blocks within the study area, including both regular surveys throughout the year and special-area surveys targeted at endangered species, particularly right whales. The primary objective of these surveys was to estimate the absolute abundance of each species in the study area, using line-transect sampling methods.

In the first sampling year (1979), there were nine unstratified blocks, designated A through I (blocks J and K, L and M, N and O, and P and Q were originally combined as blocks A, B, C, and D, respectively) (Fig. 2). In 1980, blocks were stratified by depth, with the 20- and 50-fathom isobaths comprising the stratum boundaries, and the Gulf of Maine/Georges Bank blocks partitioned as shown (Fig. 2). In 1981, the third and final year of sampling, regular aerial surveys were limited to blocks G, E, P, Q, and the newly added slope-water blocks R and S. Special-area surveys in all 3 years were flown mostly over the region east and southeast of Cape Cod, with some near Nova Scotia.

Surveys were flown in a Beechcraft AT-11 or a Cessna 337-G Skymaster, both twin-engine planes. All regular aerial surveys were flown with the AT-11; both aircraft were used for the special-area surveys. The AT-11 crew consisted of a pilot, a navigator, and four observers; two observers at a time were stationed in a clear acrylic observation bubble in the nose of the plane. The Skymaster carried two observers, who sat in the rear seats and watched out the side windows, in addition to

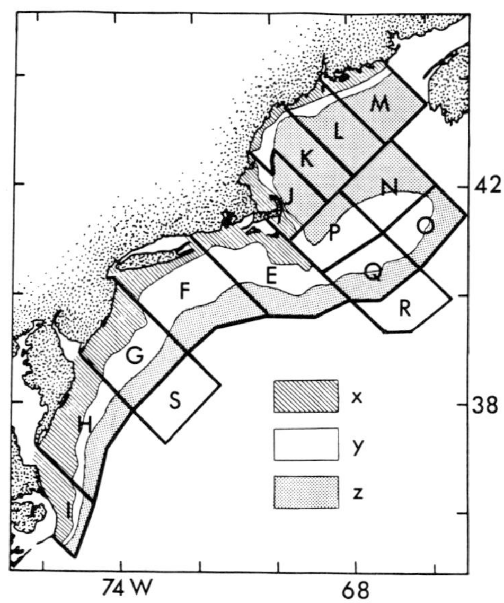

Figure 2 Blocks for CETAP dedicated aerial surveys. Depth stratification is as follows: *x*, < 20 fathoms; *y*, 20–50 fathoms; *z*, > 50 fathoms. Slope-water blocks R and S were not included in the stratification.

the pilot and navigator/data recorder. All surveys were conducted at an altitude of 750 feet (229 m) and a ground speed of approximately 120 knots (222 km/hour).

Platforms of Opportunity Program Surveys

For platforms of opportunity (POP) surveys, trained observers were placed aboard various ships and aircraft operating within the study area to collect distribution data to supplement the dedicated surveys. The platforms most often used were U.S. Coast Guard cutters, U.S. and foreign oceanographic- and fisheries-research vessels, and U.S. Coast Guard thermography and fisheries-patrol aircraft. The track of the ship or aircraft was determined by its primary mission. These data could not be used in abundance estimation because survey tracks were not random samples, the many platforms used were not comparable, and insufficient data were collected from each platform to develop the required sighting models. CETAP's shipboard POP effort was discontinued after 1980.

Observers on both dedicated aerial and POP surveys recorded a variety of data, including date, time, latitude and longitude, platform heading, beginning and end of periods when the observer(s) were actively on watch, and environmental information (air temperature, water temperature, depth, weather, visibility, sea state, wind direction, and cloud cover). The data were recorded at each sighting, as well as at periodic intervals (typically 5 minutes for aerial and 30 minutes for shipboard surveys) during all on-watch periods. This allowed for subsequent reconstruction of flight/cruise tracks. Additional data recorded at sightings included species, reliability of identification, number of animals, distance from the platform, animal heading, and behavior.

Opportunistic Data

Opportunistic sightings were contributed by mariners, aircraft pilots, other researchers working in the study area, and interested volunteers who, through their activities at sea, had the occasion to encounter cetaceans and turtles within the study area. These people were contacted, interviewed, and provided with a simple standardized checklist for recording information. Some unsolicited sighting reports also were received.

Historical Data

During the early portion of the study, an attempt was made to obtain sighting data previously collected by researchers or others working in the study area. These historical data go back to 1958.

Abundance Estimation

A major objective of CETAP was estimating the abundance of cetaceans in the study area using standard line-transect procedures. The methods are outlined generally in Burnham *et al.* (1980) and specifically in CETAP (1982), Scott and Gilbert (1982), and Scott *et al.* (1981, 1983). Density and abundance of each species seen were computed for each sampling block for each survey day. A variance was computed for each density estimate, and this was used to calculate the 95% confidence interval around the abundance (Scott *et al.*, 1983). In CETAP (1982), the individual density data were summarized as seasonal averages for the entire study area as well as for various smaller portions (Georges Bank, near-shore, shelf edge, etc.). Kenney *et al.* (1985), using somewhat different averaging methods, computed average seasonal species densities and abundances for each individual sampling block. For this chapter, the former estimates (CETAP, 1982) were simply summarized, while the latter (Kenney *et al.*, 1985) were summed across sampling blocks. These were then combined to produce a likely abundance range for each season. The "inshore stock" was assumed to be limited to waters less than 20 fathoms deep (stratum x in Fig. 2); the offshore stock, to waters deeper than 20 fathoms (strata y and z). No attempt has been made to incorporate any variance information into the summarized abundance estimates presented in this chapter; they are based only on the averaged point estimates. Detailed presentation, analysis, or discussion of the CETAP density estimates is beyond the scope of this chapter, and interested readers are directed to the above-cited literature for further details.

Distribution

All bottlenose dolphin sightings within the area shown on Fig. 1 were used for general analyses. Three sightings from a cruise aboard R/V *Westward* in May and June of 1986 were also included. Maps of geographic distribution were produced using a Mercator plotting program on the URI computer system. All other graphic products and statistical analyses were produced using SAS and SAS/GRAPH computer procedures.

Relative Density Distribution

Only the dedicated aerial data are useable in estimating absolute abundance, and the POP data are biased by uneven effort allocation. If effort could be measured, then both data sets could be combined and corrected for level of effort, resulting in an index of relative density which could be compared between discrete spatial or temporal units. In simplest terms, a relative density index allows one to confidently state that there are more animals in area A than in area B without knowing how many actually are in either. The dedicated aerial and POP data include associated trackline information, which can be used to quantify sighting effort. Those data can then be combined to examine density distribution without bias from uneven allocation of effort. Previous analyses have shown no significant differences in crude sighting rates between aerial and shipboard surveys (Hain *et al.*, 1981; Kenney and Winn, 1986).

All dedicated and POP data meeting defined criteria were included in the analysis. The criteria included the following: observer(s) formally on watch, clear visibility of at least 2 miles, and sea state of Beaufort 3 or lower. The study area was partitioned into quadrats measuring 10 minutes of latitude by 10 minutes of longitude, and the data were grouped by calendar seasons across all years of sampling. Because of extremely spotty coverage, all data from quadrats entirely seaward of the 2000-m isobath were deleted. Sighting effort was quantified as the length of trackline surveyed within the defined criteria in each quadrat and season (see Kenney and Winn, 1986, for details). The number of bottlenose dolphins sighted along those tracks was then divided by the track length to produce an index of dolphins per unit effort (DPUE), in units of animals sighted per 1000 km of trackline surveyed.

The DPUE distributions in each season and overall were partitioned into high-, medium-, and low-density quadrats. A high-density quadrat was defined as DPUE greater than the 95th percentile for that season; medium-density as DPUE between the 90th and 95th percentiles; and low-density as less than the 90th percentile but greater than 0. Those quadrats were then plotted on the same base map used for sighting plots, as in Kenney (1984), Kenney and Winn (1986), and Winn *et al.* (1986).

Distribution Relative to Environment

All depth and sea-surface temperature data recorded at bottlenose dolphin sightings were extracted from the data base for a preliminary look at the relationship of sightings to those parameters. The DPUE data, however, because they are corrected for uneven effort allocation, can be used for more rigorous statistical analyses. A digitized bathymetric data set was available, with depth values on a 2-minute latitude/longitude grid (25 depths within each 10-minute quadrat). Using these data, values were derived for each quadrat for mean, maximum, and minimum water depth (ZMEAN, ZMAX, ZMIN) and also for a bottom slope index (SI) modified from Hui's (1979) contour index:

$$SI = \frac{ZMAX - ZMIN}{2000} \times 100$$

In addition, the surface water temperature data in the CETAP data base were compiled, and values for mean, maximum, and minimum temperature (TMEAN, TMAX, TMIN), as well as the coefficient of variation of the temperature (TCV), were calculated for each quadrat and season. No whole-year temperature data were compiled.

Once the data set was completed, the environmental parameters associated with quadrats containing bottlenose dolphin sightings were examined first. Only the quadrats with DPUE values greater than 0 were used, and means were calculated for each parameter, both overall and by season. The data were tested for significant seasonal variability using the Kruskal–Wallis test. The final step in the analysis was a series of nonparametric statistical comparisons. The quadrats were partitioned into two classes: DPUE > 0 and DPUE = 0. Values of environmental parameters were then compared between the two partitions for each season and for the entire data set using the Wilcoxon rank sum test. The purpose of these comparisons was to test for any significant differences in the habitat in blocks inhabited by bottlenose dolphins compared to the remainder of the study area. Similar methods of comparison were previously used on the humpback (Kenney, 1984) and right whale (Winn *et al.*, 1986) data. The same test was used to compare the DPUE > 0 quadrats between inshore and offshore dolphin stocks, with ZMIN < 25 m used as the working definition of inshore.

RESULTS

General

There are 1155 sightings of bottlenose dolphins in the data base. Of these, 103 fall outside the area of the base map (Fig. 1) so were not included in the analysis, leaving 1052 sightings, representing 16,208 individual bottlenose dolphins. Group size ranged from 1 to 350, with a mean of 15.4 (S.E. 0.7). There was significant seasonal variation in group size (ANOVA, $F = 10.12$, $p < .001$), with seasonal means of 6.8 in winter, 14.7 in spring, 17.8 in summer, and 18.8 in fall. The winter mean group size was significantly smaller than the means at other seasons, which did not differ significantly from each other (Duncan's multiple range test, $\alpha = 0.05$).

Multispecies associations were commonly observed during CETAP surveys (CETAP, 1982). Bottlenose dolphins were observed with at least one other species in about 10% of all sightings. Nearly half of their associations were with pilot whales, *Globicephala* spp. They were also associated frequently with loggerhead sea turtles, *Caretta caretta*. Other infrequently observed associated species included fin whale, *Balaenoptera physalus*, minke whale, *B. acutorostrata*, sei whale, *B. borealis*, humpback whale, sperm whale, *Physeter catodon*, Risso's dolphin, *Grampus griseus*, Atlantic white-sided dolphin, *Lagenorhynchus acutus*, common dolphin, *Delphinus delphis*, and striped dolphin, *Stenella coeruleoalba*. The *Tursiops*/*Globicephala* association, observed on 84 occasions, was the third most common association seen during the program, surpassed only by fin whale/humpback whale and fin whale/white-sided dolphin associations (CETAP, 1982). The term "association" is used advisedly. Many of the observed multispecies sightings were simply of animals sighted in the same general vicinity. However, the *Tursiops*/*Globicephala* association seems actually to involve some interspecies interaction. On many occasions groups were observed with individuals of both species interspersed, often including calves and/or juveniles of one or both species (CETAP, 1982). The common association of bottlenose dolphins and pilot whales in mixed schools has been noted for both the North Atlantic and North Pacific (Kraus and Gihr, 1971; Leatherwood *et al.*, 1972, 1973, 1976, 1982; Norris and Dohl, 1980; Norris and Prescott, 1961; Walker, 1975).

Abundance

Table 1 summarizes the range of average seasonal abundance estimates obtained by combining the results from CETAP (1982) and Kenney *et al.* (1985) for both the inshore and offshore stocks. It should be emphasized, again, that these represent a range of average point abundance estimates summed across the study area, not a statistical confidence interval. [The 95% confidence interval on either side of an individual abundance estimate was typically about the same magnitude as the estimate itself (CETAP, 1982).] Bottlenose dolphins inhabited the study area throughout the year. Abundance was lowest in the winter, with no inshore sightings north of Cape Hatteras and a resulting abundance estimate of 0. Abundance was relatively high during the remainder of the year, with a maximum during the summer. The offshore stock was roughly 20 times more abundant than the inshore stock in all seasons. These estimates indicate that the northeast U.S. population of *Tursiops truncatus* contains approximately 10,000–13,000 individuals.

Table 1
Seasonal Average Estimates of Abundance of Bottlenose Dolphins for the Continental Shelf Waters off the Northeastern United States[a]

Stock	Season			
	Winter	Spring	Summer	Fall
Inshore	0	200–400	400–700	200–300
Offshore	1500–2300	7800–9200	9300–12,100	4600–8100
Total	1500–2300	8000–9600	9700–12,800	4800–8400

[a] Summarized from point estimates in CETAP (1982) and Kenney *et al.* (1985).

Distribution

The plot of geographic distribution of sightings of bottlenose dolphins off the northeastern United States (Fig. 3) clearly shows the inshore and offshore components. Inshore sightings were located in the extreme coastal waters from Cape Lookout to Delaware Bay. The gap in the inshore distribution near the Maryland–Virginia border is an artifact resulting from the restricted area (all civilian aircraft excluded) off the National Aeronautics and Space Administration (NASA) air base at Wallops Island, Virginia. Offshore sightings were generally distributed between the 200- and 2000-m isobaths throughout the entire study area, from Cape Hatteras to the eastern end of Georges Bank. There was an area of sparser sightings in the center of this distribution south of Long Island and New England. Sightings off the shelf were scattered widely throughout the deep-water regions, but this may simply reflect sparse sighting effort.

There was a cluster of sightings roughly 150 km beyond the shelf break east of the Chesapeake Bay region (approximately 38° N, 71° W). There is no bathymetric feature which might account for this phenomenon. The original data showed that the sightings were historical records from U.S. Coast Guard files, and most apparently were made from Ocean Weather Station Hotel. This concentration of sightings is therefore an artifact of increased effort from the level of activity associated with the weather station. (A large proportion of the historical data included in the CETAP data base resulted from a sighting program conducted by the U.S. Coast Guard. From the information provided on the sighting forms they used, the observers seemed to be well-trained and competent. Nevertheless, as a precaution in using data from unknown sources outside the program, the identification reliabilities in the historical records were downgraded one step when they were added to the data base. "Definite" identifications were changed to "probable," "probable" to "possible," and "possible" to "unidentified." Most of the off-shelf sightings in Fig. 3 are classified in the data base as probable bottlenose dolphins. The reader is cautioned that these identifications may be somewhat less reliable than those made by trained CETAP observers, but they should not be completely disregarded, either. In addition, one of

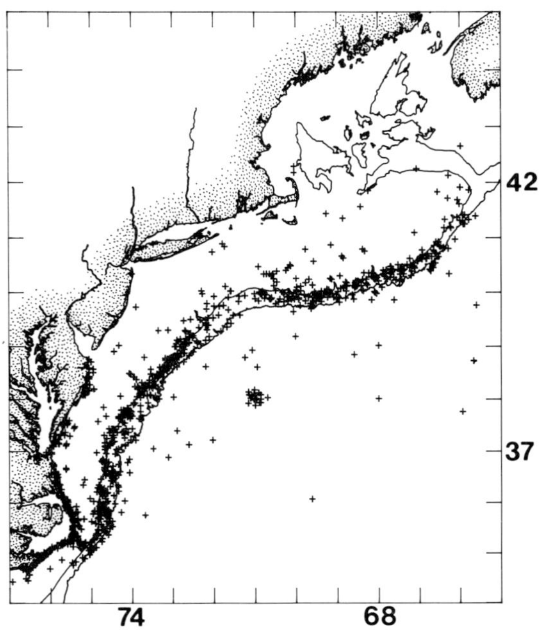

Figure 3 Distribution of all sightings of bottlenose dolphins contained in the CETAP data base. Each + represents a single sighting of one or more dolphins.

the sightings farthest offshore in Fig. 3 was made by the author from *Westward* in June 1986, confirming that bottlenose dolphins can be found in deep waters far beyond the shelf break.)

The temporal distribution of bottlenose dolphin occurrence is shown by month in Fig. 4. It reflects the same general pattern seen in the abundance data, with the caution that these are raw sighting data uncorrected for effort. Fewest dolphins were seen in January, February, and November; the numbers were generally high from May through October, with a peak in August.

Relative Density Distribution

Table 2 summarizes the valid sighting effort for CETAP dedicated aerial and POP surveys. Overall, 1467 10-minute quadrats were sampled in one or more seasons, with an average effort of about 253 km of trackline per quadrat. Nearly every quadrat in the study area received some level of sampling effort during the study. Bottlenose dolphins were sighted during valid survey tracks in a total of 296 quadrats, 20 in winter, 159 in spring, 144 in summer, and 84 in fall. The average number of dolphins per 1000 km of track (DPUE) was 60.8. Seasonal average DPUE values were 10.0 in winter, 39.2 in spring, 44.2 in summer, and 96.3 in fall.

The effort-corrected density distribution pattern is shown in Fig. 5. Most of the high- and medium-density blocks were located along the shelf break. There was another concentration of high-density blocks along the North Carolina coast and a few more along the coast between Chesapeake and Delaware bays. It should be reemphasized that the lack of any density distribution data beyond the shelf edge is due to lack of effort. The total of sighting data (Fig. 3) shows that bottlenose dolphins were present in deeper waters beyond the shelf, but the DPUE data are insufficient to compare relative densities in such areas to those of on-shelf areas.

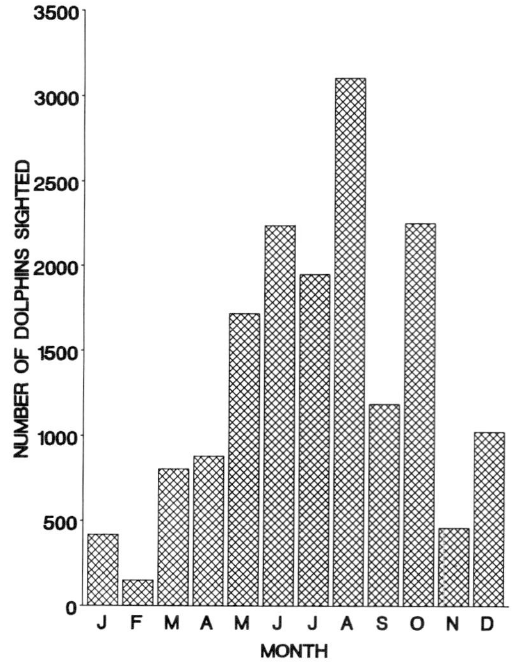

Figure 4 Monthly frequency distribution of bottlenose dolphins sighted off the northeastern United States.

The seasonal patterns of effort-corrected density distribution are shown in Fig. 6. In the winter, all quadrats with dolphin sightings fell in the high-density category because of the large number of quadrats with DPUE = 0. Inshore sightings

Table 2
Summary of Sighting Effort in 10-Minute Quadrats[a]

	Season				
	Winter	Spring	Summer	Fall	Overall
Quadrats sampled	1179	1344	1395	1169	1476
Mean effort per quadrat	40.3	108.0	80.7	58.2	252.9
Standard deviation	32.9	104.9	56.1	47.6	207.9
Maximum effort per quadrat	372	1137	596	546	2389
Total effort	47,506	145,204	112,576	67,994	373,280

[a] Effort data are expressed as kilometers of trackline surveyed within acceptable criteria, for CETAP dedicated and POP surveys.

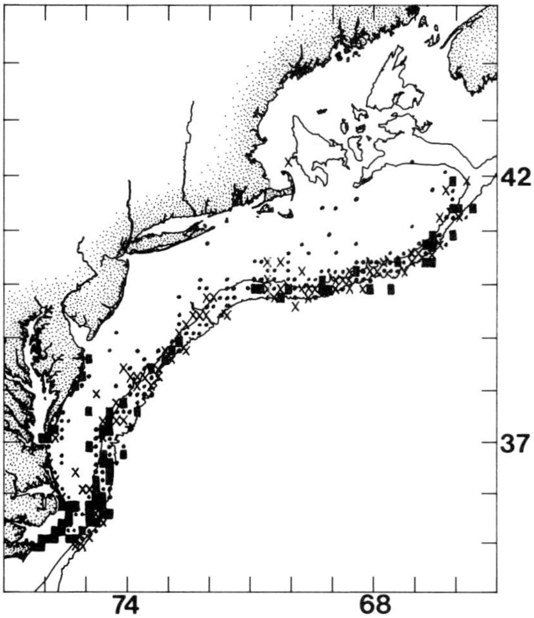

Figure 5 Effort-corrected relative density distribution of bottlenose dolphins in 10-minute quadrats. High-density quadrats are shown as filled boxes, medium-density as Xs, and low-density as small circles.

were absent north of Cape Hatteras but relatively common between Cape Hatteras and Cape Lookout. Offshore sightings tended to be infrequent and relatively scattered, with small concentrations off Cape Hatteras and on the southern edge of Georges Bank. The inshore distribution was most extensive in the spring, relatively reduced in the summer, except along the North Carolina Outer Banks, and even more reduced in the fall. The offshore distribution was most extensive in spring and summer, extending along the entire shelf break during both seasons. Again, there was a gap in the center in both seasons. In the fall, the offshore distribution was reduced and contracted southward, but there were still some high- and medium-density quadrats on Georges Bank.

Distribution Relative to Environment

There were 982 sightings of bottlenose dolphins for which there were associated water depths, and 607 with associated sea-surface temperatures. Water depth at bottlenose dolphin sightings ranged from 0 to 4,932 m ($\bar{x}$ = 587.9 m). Figure 7 shows the depth distribution of the sightings, uncorrected for effort. The inshore component can be seen easily; 31% of sightings occurred in waters less than 25 m deep. Defining the inshore/offshore separation at 25 m, the mean depth of the inshore sightings is then 10.3 m, that of the offshore sightings 845.6 m. There is no clear, narrow peak showing the offshore component; the apparent pattern in the figure is due primarily to the intervals used to show the detail at the lower depths. About 10% of the sightings occurred in 25–100 m, and 11% between 100 and 200 m. There was, then, a relatively even distribution from 200 to 2200 m, with 1–5% in any 100-m interval. Two things should be emphasized, however. One is that the pattern will be biased by any bias in effort allocation. The second is that the actual width of any 100-m depth zone decreases with greater depths, so the offshore sightings are actually concentrated in the geographically narrow band along the shelf edge seen in Fig. 3.

Surface temperature at sightings ranged from 1.1° to 31.1°C, with a mean of 19.7°C and a clear mode at 20°–22.5°C (Fig. 8). Again, using 25 m depth as the inshore/offshore boundary, there are 65 inshore and 542 offshore sightings with associated temperatures. The offshore temperature distribution looks essentially identical to the overall pattern, with the same range and a mean of 19.5°C. The inshore sightings, on the other hand, show a shift toward higher temperatures, with only eight (12%) below 17.5°C. Inshore sightings occurred in waters of 1.9°–29.4°C, with a mean of 21.0°C, significantly warmer than those of offshore sightings (t-test, p = .04). The narrower temperature distribution of the inshore sightings is also reflected in their coefficients of variation, 15.4% for inshore, 29.4% for offshore sightings. These data may be biased because of the greater decrease in inshore sightings during the winter, but comparisons within seasons (see below) show similar patterns.

Considering only those quadrats in which dolphins were observed (DPUE > 0), there was significant seasonal variation in DPUE, latitude, longitude, and all surface temperature parameters, but no significant variation in any of the four depth parameters (Table 3). The CETAP survey effort also showed significant seasonal variability (Table 3). High DPUEs in fall and winter resulted

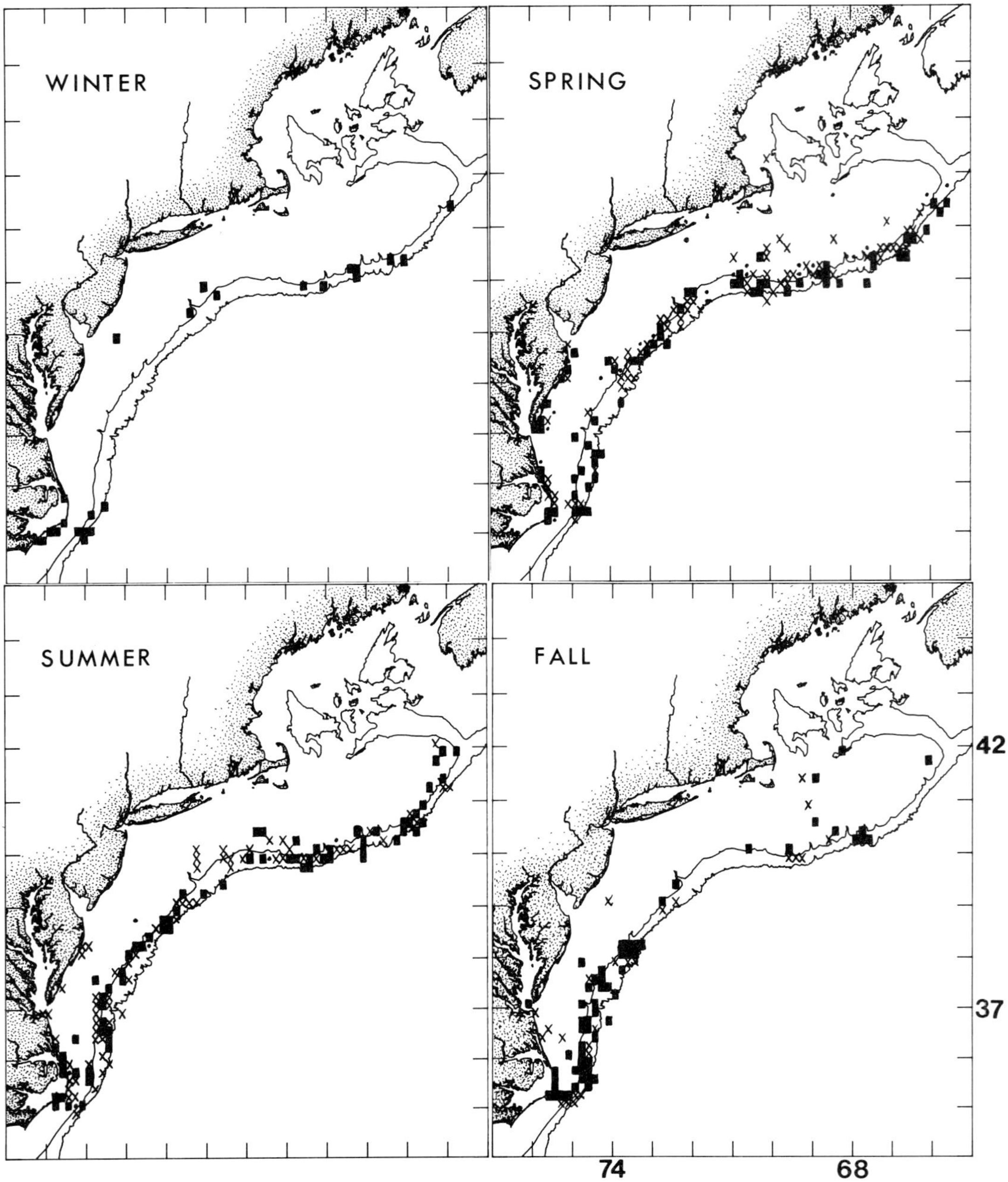

Figure 6 Seasonal effort-corrected relative density distributions of bottlenose dolphins in 10-minute quadrats. Quadrat density is identified as in Fig. 5

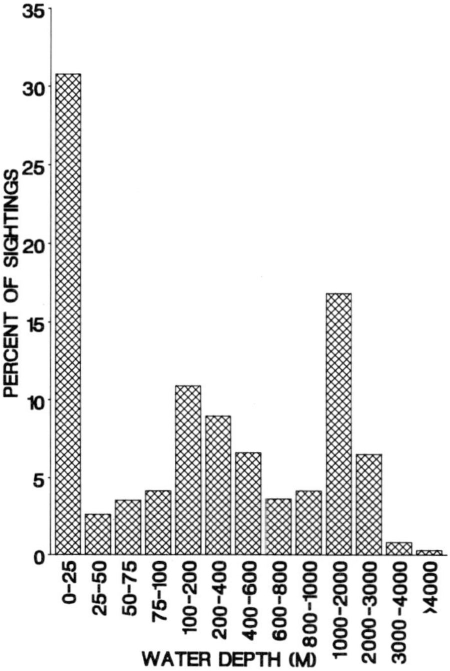

Figure 7 Frequency distribution of CETAP sightings of bottlenose dolphins relative to water depth at sighting, uncorrected for sighting effort.

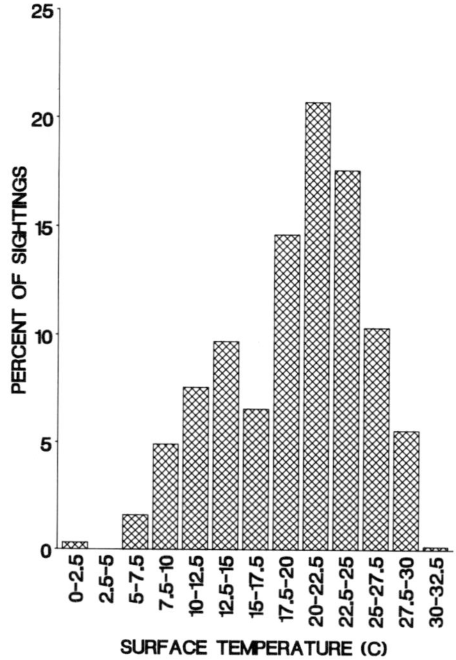

Figure 8 Frequency distribution of CETAP sightings of bottlenose dolphins relative to sea surface temperature at sighting, uncorrected for sighting effort.

Table 3
Summary of Relative Density, Effort, and Habitat-Related Data[a]

Variable	Overall Mean (n = 296)	Seasonal Means: Winter (n = 20)	Spring (n = 159)	Summer (n = 144)	Fall (n = 84)	p
DPUE (/1000 km)	300.77	589.52	331.74	424.53	1340.63	.001
Effort (km)	291.72	45.65	122.29	101.26	83.43	<.001
Latitude	38° 37′ N	38° 11′ N	38° 52′ N	38° 26′ N	37° 33′ N	<.001
Longitude	71° 55′ W	71° 47′ W	71° 36′ W	72° 06′ W	73° 19′ W	.003
ZMEAN (m)	708.79	706.24	710.16	680.60	760.80	.834
ZMAX (m)	975.87	1170.99	995.02	1003.75	1063.46	.774
ZMIN (m)	466.85	334.02	464.07	411.10	484.03	.990
SI (%)	25.45	41.85	26.55	29.63	28.97	.167
TMEAN (°c)	NA	13.55	14.50	23.60	19.75	<.001
TMAX (°c)	NA	16.13	18.56	25.84	23.21	<.001
TMIN (°c)	NA	9.26	9.61	21.42	15.82	<.001
TCV (%)	NA	18.60	22.86	7.03	15.57	<.001

[a] Data are from 10-minute quadrats with values greater than 0 for bottlenose dolphins sighted per unit effort and show the overall mean, seasonal means, and significance probability (p) for a Kruskal–Wallis test for differences among the seasonal data. NA, Not applicable.

from contraction of the bottlenose dolphin's distribution during those seasons. Seasonal variation in latitude/longitude without depth variation suggests that bottlenose dolphins migrate along the shelf, parallel to the depth contours, but without any significant inshore/offshore or on-shelf/off-shelf movements. Seasonal variation in surface temperature of quadrats with dolphins reflects normal annual changes.

Wilcoxon comparisons between 10-minute quadrats with and without bottlenose dolphins (Table 4) showed some interesting patterns. Quadrats with dolphins were significantly farther south and west and were in waters which were significantly deeper and had steeper bottoms and significantly warmer and more thermally stable surface waters. Repeating the tests using only those quadrats south of 41°N (to eliminate any bias which might have been added by including Georges Bank and the Gulf of Maine) did not change any results. Comparisons between the inshore and offshore quadrats with DPUE > 0 showed the expected results (i.e. inshore quadrats further south and west, shallower, and with flatter bottom slopes), but also showed that the inshore dolphins are found in significantly warmer waters during spring and summer.

DISCUSSION

Abundance

Of the total population of bottlenose dolphins off the northeastern United States of 10,000 to 13,000 individuals, the inshore stock comprised only 3–4%. Abundance was highest in the summer and lowest in the winter. Mitchell (1975b) estimated the size of the stock fished off Cape Hatteras as at least 13,748 in the 1880s. Since this was a near-shore net fishery (Mead, 1975), one would expect that it exploited the inshore stock, the present abundance of which is estimated as less than 1,000 (Table 1). There are at least four possible explanations: (1) the inshore stock is significantly smaller now than during the nineteenth century, (2) the CETAP results significantly underestimated abundance, (3) the CETAP surveys did not extend far enough south to census the majority of inshore dolphins in the Hatteras vicinity, and/or (4) the former fishery also exploited the offshore stock. It is not known what proportion of the entire western North Atlantic bottlenose dolphin population the CETAP estimates represent. No abundance estimates exist for the off-shelf regions. Estimates for the southeastern United States tend to be for relatively local, inshore areas only (e.g., Barham *et al.*, 1980; Leatherwood, 1979, 1982; Leatherwood and Reeves, 1983). Based simply on the bottlenose dolphin's apparent preference for warmer waters, one might expect populations of the southeastern United States to be substantially larger than those of the northeastern United States.

Despite being the most commonly sighted dolphin during the CETAP study, the bottlenose dolphin was not the most abundant small cetacean. The estimated populations of several species were similar or much higher [e.g., Atlantic white-sided dolphins (36,000–40,000), common dolphins (31,000–36,000), Risso's dolphins (11,000–14,000), and pilot whales (10,000–12,000) (CETAP, 1982; Kenney *et al.*, 1985)]. In addition, although the estimated abundance of striped dolphins was only about 4,000, most aerial sightings of dolphins of the genus *Stenella* were identified only to genus, and the estimated abundance of unidentified *Stenella* sp. was 40,000–43,000. Based on the proportion of striped dolphins in sightings of *Stenella* in which species was positively identified, roughly one-half to four-fifths of the unidentified *Stenella* sp. were likely striped dolphins, most of the remainder spotted dolphins (CETAP, 1982). The large number of bottlenose dolphin sightings was probably biased by two factors: the extreme near-shore distribution of the inshore stock, and the distribution of more sighting effort in the near-shore zone.

These abundance estimates must be considered as minimum values, since they are not adjusted for animals missed due to submergence below the surface during overflight of the survey aircraft. There are no quantitative data available on proportion of time spent submerged by bottlenose dolphins in the study area, but data from other species illustrate this point. CETAP data for fin, humpback, and right whales suggested a factor of 3–5 is needed to correct densities of those species for diving (CETAP, 1982). A juvenile male pilot whale fitted with a depth-encoding, satellite-tracked radio tag spent about 90% of its time submerged (Mate *et al.*, 1987).

Table 4
Results of Wilcoxon Rank Sum Comparisons of Environmental Parameters in 10-Minute Quadrats[a]

Variable	Season: Winter	Spring	Summer	Fall	Overall
LAT	−, ***	−, ***	−, ***	−, ***	−, ***
LONG	+, NS	+, ***	+, ***	+, ***	+, ***
ZMEAN	+, ***	+, ***	+, ***	+, ***	+, ***
ZMAX	+, ***	+, ***	+, ***	+, ***	+, ***
ZMIN	+, *	+, ***	+, ***	+, ***	+, ***
SI	+, ***	+, ***	+, ***	+, ***	+, ***
TMEAN	+, ***	+, ***	+, ***	+, ***	NA
TMAX	+, ***	+, ***	+, ***	+, ***	NA
TMIN	+, ***	+, ***	+, ***	+, ***	NA
TCV	−, *	−, **	−, *	−, NS	NA

[a] Comparisons are between data partitioned into two classes: those with values for bottlenose dolphins sighted per unit effort greater than 0 or equal to 0. Each entry shows whether the blocks with dolphins tended toward higher (+) or lower (−) values than those without. NA, Not applicable; NS, not significant; *, $p < .05$; **, $p < .01$; ***, $p < .001$.

Distribution

The distribution patterns of bottlenose dolphins off the northeastern United States are similar to those of other populations which have been studied. Geographically separate inshore and offshore stocks are present, although the term "stock" has been used here rather loosely, since there is as yet little evidence for reproductive isolation between the two populations in this region. Both stocks are seasonally migratory, with much lower abundance and a more southerly sighting distribution in the winter. Mead (1975) reported that the maximum catches in the Cape Hatteras fishery occurred during spring and fall, which would correspond to periods of migratory movement, when the largest numbers of dolphins would be expected to be moving along the beach. The migration apparently occurs primarily parallel to the isobaths, with no significant seasonal variation in depth distribution.

The inshore stock was distributed in the nearshore waters as far north as Delaware Bay. A bottlenose dolphin fishery existed at Cape May, New Jersey, in 1884 and 1885, and another fishery which possibly exploited bottlenose dolphins operated in eastern Long Island in the eighteenth century (Mead, 1975); CETAP surveys recorded no inshore sightings in either of those areas. Most of the inshore sightings were in very shallow water. Similar distributions have been noted for coastal bottlenose dolphins off southern Africa, with most sightings in depths of less than 15–30 m, depending on area (Ross and Cockcroft, Chapter 5, this volume), and off Argentina, where they showed a preference for water less than 10 m deep (Würsig and Würsig, 1979). The inshore stock showed a more temperature-limited distribution than the offshore. Inshore dolphins tended to be found in significantly warmer waters, had a distinct northern boundary to their distribution, and were absent from the study area during the winter.

The offshore stock was concentrated along the shelf break, with scattered sightings throughout the region beyond the continental shelf. It was much more widely distributed relative to geography and temperature than was the inshore stock. The sightings along the shelf break extended to the northern boundary of the study area. There is no reason to assume that the distribution stops there; however, no extensive survey program has been conducted off Nova Scotia to investigate the occurrence of small cetaceans there. In the absence of such data, it is also not possible to say whether the bottlenose dolphins seen in winter on Georges Bank are the same individuals which occur there in other seasons or are dolphins which have migrated south from Canadian waters.

The historical data include off-shelf sightings extending far beyond the boundaries of the map used in this chapter. The most northerly sighting occurred in September 1971 over the Mid-Atlantic Ridge about 1500 km south southwest of Iceland, where about 50 bottlenose dolphins were seen with 30–40 pilot whales by a U.S. Coast Guard observer aboard the cutter *Chase*. Here again, the observer was definite about the species identifications, but the sighting was classified as "probable" when included in the data base. The CETAP sighting distribution, as well as the statistical temperature comparisons reported here, confirm Mitchell's (1975) statement that the offshore form ranges farther into colder waters. The western North Atlantic distribution of the bottlenose dolphin, based on these data and previously published reports, can be surmised to extend from at least northern South America nearly to Iceland. The entire bottlenose dolphin distribution in the North Atlantic apparently includes most of the basin, and not just the continental-shelf waters as indicated by Marcuzzi and Pilleri (1971).

Separate inshore and offshore stocks of bottlenose dolphins are also known from the eastern North Pacific (Walker, 1981), the southeastern United States (Hersh, 1987; Hersh and Duffield, Chapter 6, this volume), southern Africa (Ross, 1977, 1984), Australia (Ross and Cockcroft, Chapter 5, this volume), and western South America (Van Waerebeek *et al.*, Chapter 7, this volume). There are too few reliable abundance estimates, however, to permit determination of the relative abundance of the inshore and offshore components of populations in these regions. In all areas, researchers must exercise caution in extrapolating biological information obtained from inshore stocks to offshore stocks. In addition to biological differences between the stocks and their habitats, the apparently greater abundance of inshore than offshore populations is probably extremely biased by the high visibility of the former and the increased logistical difficulties of studying the latter.

There appears to be a degree of habitat partitioning, by depth, among the shelf-edge odontocete species off the northeastern United States (CETAP, 1982, p. 407). The order of significant depth differences was as follows: bottlenose dolphin < common dolphin and pilot whale < Risso's dolphin < spotted dolphin < sperm whale < striped dolphin. From those results, bottlenose dolphins appear to be found in significantly shallower water than any other species. However, this pattern may well have been biased by the predominance of the inshore stock. Eliminating all bottlenose dolphin sightings in depths less than 25 m and repeating the same analysis, the mean depth for bottlenose dolphins is not significantly different from that for common dolphins and pilot whales (Duncan's multiple range test, $\alpha = 0.05$). The pattern for the remaining species is the same. There is still evidence for depth-related habitat partitioning, but inshore bottlenose dolphins are exploiting the extreme near-shore habitat, while offshore bottlenose dolphins share the innermost shelf-edge zone with common dolphins and pilot whales.

The CETAP data are extensive in their spatial and temporal coverage, and results presented herein demonstrate long-term average patterns of distribution of bottlenose dolphins off the northeastern U.S. These patterns, however, may not be static in space and time and may be poorly represented by plots of sightings data. Cetaceans are highly mobile and capable of moving relatively long distances. Würsig and Würsig (1977) sighted six photoidentified bottlenose dolphins over 300 km distant from their original location after 6 months. Wells *et al.* (Chapter 24, this volume) identified bottlenose dolphins known from photoidentification studies off southern California more than 600 km north of their previously documented range. A long-finned pilot whale satellite-tagged by Mate *et al.* (1987) and released off Cape Cod ranged from Virginia to Nova Scotia over the 3-month life of the tag. Daily movements ranged up to 130 km and averaged 74 km. (It should be noted that distances from such tag data are minimum straight-line distances between successive fixes, with the potential that actual travel distances are much longer.) One of 11 bottlenose dolphins satellite-tracked off Japan traveled 604 km in 18 days (Tanaka, 1987). Average travel speeds were 2.1–2.8 km/hour, with maximum speeds up to 7.4 km/hour.

The results presented herein demonstrate some environmental characteristics of the habitat of bottlenose dolphins off the northeastern United States. This is not to say, however, that these factors necessarily have any important direct influence on the distribution of these dolphins. Ross (1977) and Ross and Cockcroft (Chapter 5, this volume) relate larger body size in offshore ecotypes from southern Africa and Australia, respectively, to the colder water they encounter in offshore currents. Previous studies of other cetacean species in the western North Atlantic, based on the CETAP data, have concluded that the major factor influencing patterns of distribution directly is prey availability (Kenney, 1984; Kenney and Winn, 1986; Winn *et al.*, 1986), and environmental factors are more likely to influence prey distributions. Indeed, Gaskin (1982, p. 1) stated that ". . . study of the distributional ecology of these animals becomes largely a study of the distributional ecology of their prey species . . .". Most published reports indicate that the bottlenose dolphin is piscivorous (see Kenney *et al.*, 1985, for a review; Barros and Odell, Chapter 16, this volume; Cockcroft and Ross, Chapter 15, this volume); however, the vast majority of stomach-content data have come from inshore dolphins, and prey preferences of offshore dolphins are apparently very poorly known. Therefore, there is little basic biological information concerning the connection between the marine environment and the distributions of offshore bottlenose dolphins.

One might speculate about such connections, extrapolating from the distribution patterns shown here along with data from long-term, detailed studies of inshore populations. Such studies have shown a high degree of social organization into distinct local herds with long-term social relationships among the members (see Shane *et al.*, 1986, for a review). Winn *et al.* (1979) suggested that inshore bottlenose dolphins along the southeastern United States could be separated into "river" and "ocean" subgroups. It is also possible that offshore stocks (regardless of their actual taxonomic status relative to inshore stocks) are similarly separated into distinct groups, perhaps "shelf-break" and "pelagic" dolphins. The population could be even more finely subdivided. Perhaps the apparent gap in the center of the offshore distribution noted previously is evidence for separation of the "shelf-break" group into "mid-Atlantic" and "New England" subgroups. It seems likely that the same pattern seen in inshore stocks, with long-term, interrelated social groups and varying degrees of interchange between groups, would also be true of the offshore stocks. The density of animals within any specific zone would be a reflection of overall habitat suitability, primarily prey availability but also including temperature regime, presence of intra- and interspecific competitors, etc., resulting in a distribution pattern such as the one shown herein.

ACKNOWLEDGMENTS

The CETAP study was funded by the Outer Continental Shelf environmental studies office of the Bureau of Land Management, U.S. Department of the Interior, Contract No. AA551-CT8-48, and conducted under the scientific directorship of Dr. Howard E. Winn. The efforts of the many individuals who contributed to the success of all phases of the project are gratefully acknowledged. Special thanks go to Jerry Epstein, Bob Medved, Peter Sorensen, Saul Saila, Jerry Scott, and Howard Winn for their contributions to earlier work which made this analysis much easier. Thanks also to four anonymous referees for their suggestions for improving the manuscript.

REFERENCES

Barham, E. G., Sweeney, J. G., Leatherwood, S., Beggs, R. K., and Barham, C. L. (1980). Aerial census of the bottlenose dolphin, *Tursiops truncatus*, in a region of the Texas coast. *Fish. Bull.* **77**(3), 585–595.

Brown, S. G. (1975). Relation between stranding mortality and population abundance of smaller Cetacea in the Northeast Atlantic Ocean. *J. Fish. Res. Board Can.* **32**(7), 1055–1058.

Burnham, K. P., Anderson, D. R., and Laake, J. L. (1980). Estimation of density from line transect sampling of biological populations. *Wildl. Monogr.* **72,** 1–202.

Caldwell, D. K., and Caldwell, M. C. (1972). "The World of the Bottlenose Dolphin." Lippincott, New York. 157 pp

Caldwell, D. K., and Golley, F. B. (1965). Marine mammals from the coast of Georgia to Cape Hatteras. *J. Elisha Mitchell Sci. Soc.* **81**(1), 24–32.

Caldwell, D. K., Caldwell, M. C., Rathjen, W. F., and Sullivan, J. R. (1971a). Cetaceans from the Lesser Antillean island of St. Vincent. *Fish. Bull.* **69**(2), 303–312.

Caldwell, D. K., Neuhauser, H., Caldwell, M. C., and Coolidge, H. W. (1971b). Recent records of marine mammals from the coasts of Georgia and South Carolina. *Cetology* **5,** 1–12.

CETAP (Cetacean and Turtle Assessment Program). (1982). "A Characterization of Marine Mammals and Turtles in the Mid- and North-Atlantic Areas of the U.S. Outer Continental Shelf, Final Report," Contract AA551-CT8-48, U.S. NTIS PB83-215855. Bureau of Land Management, Washington, D.C. 576 pp.

Edel, R. K., Hyman, M. A. M., and Tyrrell, M. F. (1981). Introduction to 1979 annual report. *In* "A Characterization of Marine Mammals and Turtles in the Mid- and North-Atlantic Areas of the U.S. Outer Continental Shelf, Annual Report for 1979." (CETAP), Contract AA551-CT8-48, U.S. NTIS PB81-243289, pp. I.1–I.79. Bureau of Land Management, Washington, D.C. 850 pp.

Erdman, D. S. 1970. Marine mammals from Puerto Rico to Antigua. *J. Mammal* **51**(3), 636–639.

Erdman, D. S., Harms, J., and Flores, M. M. (1973). Cetacean records from the northeastern Caribbean region. *Cetology* **17,** 1–14.

Gaskin, D. E. (1982). "The Ecology of Whales and Dolphins." Heinemann, Portsmouth, New Hampshire. 471 pp.

Gunter, G. (1954). Mammals of the Gulf of Mexico. *Fish. Bull.* **55**(89), 543–551.

Hain, J. H. W., Edel, R. K., Hays, H. E., Katona, S. K., and Roanowicz, J. D. (1981). General distribution of cetaceans in the continental shelf waters of the northeastern United States. *In* "A Characterization of Marine Mammals and Turtles in the Mid- and North-Atlantic Areas of the U.S. Outer Continental Shelf, Annual Report for 1979" (CETAP), Contract AA551-CT8-48, U.S. NTIS PB81-243289, pp. II.1–II.345. Bureau of Land Management, Washington, D.C. 850 pp.

Hain, J. H. W., Hyman, M. A. M., Kenney R. D., and Winn, H. E. (1985). The role of cetaceans in the shelf edge region of the northeastern United States. *Mar. Fish. Rev.* **47**(1), 13–17.

Hersh, S. L. (1987). Characterization and differentiation of bottlenose dolphin populations (genus *Tursiops*) in the southeastern U.S. based on mortality patterns and morphometrics. Ph.D. dissertation, University of Miami, Coral Gables, Florida. 241 pp.

Hui, C. A. (1979). Undersea topography and distribution of dolphins of the genus *Delphinus* in the Southern California Bight. *J. Mammal.* **60**(3), 521–527.

Innes, S., Lavigne, D. M., Earle, W. M., and Kovacs, K. M. (1987). Feeding rates of seals and whales. *J. An. Ecol.* **56**(1), 115–130.

Katona, S. K., Steiner, W. W., and Winn, H. E. (1978). Marine mammals. *In* "A Summary and Analysis of Environmental Information on the Continental Shelf from Bay of Fundy to Cape Hatteras" (Center for Natural Areas), Vol. I, Book 2, Contract AA550-CT6-45, U.S. NTIS PB80-177470, pp. XIV.1–XIV.165 Bureau of Land Management, Washington, D.C. 783 pp.

Kenney, R. D. (1984). The distributional biology of the cetacean fauna of the northeast United States continental shelf. Ph.D. dissertation, University of Rhode Island, Kingston, Rhode Island. 158 pp.

Kenney, R. D., and Winn, H. E. (1986). Cetacean high-use habitats of the northeast United States continental shelf. *Fish. Bull.* **84**(2), 345–357.

Kenney, R. D., and Winn, H. E. (1987). Cetacean biomass densities near submarine canyons compared to adjacent continental shelf/slope areas. *Cont. Shelf Res.* **7**(2), 107–114.

Kenney, R. D., Hyman, M. A. M., and Winn, H. E. (1985). "Calculation of standing stocks and energetic requirements of cetaceans of the northeast United States outer continental shelf," NOAA Tech. Memo. NMFS-F/NEC-41. NOAA, Woods Hole, Massachusetts. 103 pp.

Kraus, C., and Gihr, M. (1971). On the presence of *Tursiops truncatus* in schools of *Globicephala melaena* off the Faeroe Islands. *Invest. Cetacea* **3**(1), 180–182.

Lavigne, D. M., Innes, S., Worthy, G. A. J., Kovacs, K. M., Schmitz, O. J., and Hickie, J. P. (1986). Metabolic rates of seals and whales. *Can. J. Zool.* **64**(2), 279–284.

Leatherwood, S. (1979). Aerial survey of the bottlenose dolphin. *Tursiops truncatus,* and the West Indian manatee, *Trichechus manatus,* in the Indian and Banana Rivers, Florida. *Fish. Bull.* **77**(1), 47–59.

Leatherwood, S. (1982). Size of bottlenose dolphin populations in the Indian River, Florida. *Rep. Int. Whal. Comm.* **32,** 567–568.

Leatherwood, S., and Reeves, R. R. (1982). Bottlenose dolphin (*Tursiops truncatus*) and other toothed cetaceans. *In* "Wild Mammals of North America" (J. A. Chapman and G. A. Feldhamer, eds.), pp. 369–414. Johns Hopkins University Press, Baltimore, Maryland. 1147 pp.

Leatherwood, S., and Reeves, R. R. (1983). Abundance of bottlenose dolphins in Corpus Christi Bay and coastal southern Texas. *Contrib. Mar. Sci.* **26,** 179–199.

Leatherwood, S., Evans, W. E., and Rice, D. W. (1972). The whales, dolphins and porpoises of the eastern North Pacific: A guide to their identification in the water. Naval Undersea Center (NUC), San Diego, Tech. Publ. 282. 175 pp.

Leatherwood, J. S., Evans, W. E., and Lingle, G. E. (1973). The Pacific pilot whale *Globicephala* c.f. *G. scammonii* - evaluation of its effects on high frequency sonar. Naval Undersea Center (NUC), San Diego, Tech. Note 1164. 46 pp.

Leatherwood, S., Caldwell, D. K., and Winn, H. E. (1976). "Whales, Dolphins, and Porpoises of the Western North Atlantic, A Guide to Their Identification," NOAA Tech. Rep. NMFS CIRC-369. NOAA, Seattle, Washington. 176 pp.

Leatherwood, S., Reeves, R. R., Perrin, W. F., and Evans, W. E. (1982). "Whales, Dolphins, and Porpoises of the Eastern North Pacific and Adjacent Arctic Waters, a Guide to Their Identification," NOAA Tech. Rep. NMFS CIRC-444. NOAA, Seattle, Washington. 250 pp.

Lichter, A., and Hooper, A. (1984). "Guia para el Reconocimiento de los Cetáceos del Mar Argentino." Fundación Vida, Silvestre Argentina, Buenos Aires. 96 pp.

Lowery, G. H. (1974). "The Mammals of Louisiana and Its Adjacent Waters." Louisiana State University Press, Baton Rouge, Louisiana. 588 pp.

Marcuzzi, G., and Pilleri, G. (1971). On the zoogeography of Cetacea. *Invest. Cetacea* **3**(1), 101–170.

Mate, B. R., Prescott, J. H., and Geraci, J. R. (1987). Free ranging movements of a pilot whale from a satellite-monitored radio. *7th Bienn. Conf. Biol. Mar. Mamm., 1987*, Abstr., p. 45.

Mead, J. G. (1975). Preliminary report on the former net fisheries for *Tursiops truncatus* in the western North Atlantic. *J. Fish. Res. Board Can.* **32**(7), 1155–1162.

Mitchell, E. D., ed. (1975a). Report of the meeting on smaller cetaceans, Montreal, April 1–11, 1974. *J. Fish. Res. Board Can.* **32**(7), 889–983.

Mitchell, E. D. (1975b). "Porpoise, Dolphin and Small Whale Fisheries of the World. Status and Problems." International Union for Conservation of Nature and Natural Resources, Morges, Switzerland. IUCN Monograph No. 3.

Moore, J. C. (1953). Distribution of marine mammals to Florida waters. *Am. Midl. Nat.* **49**(1), 117–158.

Norris, K. S., and Dohl, T. P. (1980). The structure and function of cetacean schools. *In* "Cetacean Behavior: Mechanisms and Functions" (L. M. Herman, ed.), pp. 211–261. Wiley, New York. 476 pp.

Norris, K. S., and Prescott, J. H. (1961). Observations on Pacific cetaceans of Californian and Mexican waters. *Univ. Calif. Publ. Zool.* **63**(4), 291–402.

Odell, D. K., Siniff, D. B., and Waring, G. H., eds. (1975). "*Tursiops truncatus* Assessment Workshop," U.S. Mar. Mamm. Comm. Rep. MMC-75/13, U.S. NTIS PB-291161. U.S. Marine Mammal Commission, Washington, D.C. 145 pp.

Ross, G. J. B. (1977). The taxonomy of bottlenosed dolphins *Tursiops* species in South African waters, with notes on their biology. *Ann. Cape Prov. Mus. (Nat. Hist.)* **11**(9), 135–194.

Ross, G. J. B. (1984). The smaller cetaceans of the south east coast of southern Africa. *Ann. Cape Prov. Mus. (Nat. Hist.)* **15**(2), 173–410.

Ross, G. J. B., Cockcroft, V. G., and Butterworth, D. S. (1987). Offshore distribution of bottlenosed dolphins in Natal coastal waters and Algoa Bay, eastern Cape. *S. Afr. Tydskr. Dierk.* **22**(1), 50–56.

Schmidly, D. J. (1981). "Marine Mammals of the Southeastern United States Coast and the Gulf of Mexico," U.S. Fish Wild. Serv. Rep. FWS/OBS-80/41. U. S. Fish and Wildlife Service, Washington, D. C. 167 pp.

Scott, G. P., and Gilbert, J. R. (1982). Problems and progress in the U.S. BLM-sponsored CETAP surveys. *Rep. int. Whal. Commn* **32,** 587–600.

Scott, G. P., Kenney, R. D., Gilbert, J. R., and Edel, R. K. (1981). Estimates of cetacean and turtle abundance in the CETAP study area with an analysis of factors affecting them. *In* "A Characterization of Marine Mammals and Turtles in the Mid- and North-Atlantic Areas of the U.S. Outer Continental Shelf, Annual Report for 1979" (CETAP), Contract AA551-CT8-48, U.S. NTIS PB81-243289, pp. IV.1–IV.72. Bureau of Land Management, Washington, D.C. 850 pp.

Scott, G. P., Kenney, R. D., Thompson, T. J., and Winn, H. E. (1983). "Functional Roles and Ecological Impacts of the Cetacean Community in the Waters of the Northeastern U.S. Continental Shelf," Int. Council Explor. Sea Document C.M.1983/N:12. International Council for the Exploration of the Sea, Copenhagen. 33 pp.

Sergeant, D. E. (1969). Feeding rates of Cetacea. *FiskDir. Skr. Ser. HavUnders.* **15**(3), 246–258.

Sergeant, D. E., and Fisher, H. D. (1957). The smaller Cetacea of eastern Canadian waters. *J. Fish. Res. Board Can.* **14**(1), 83–115.

Sergeant, D. E., Mansfield, A. W., and Beck, B. (1970). Inshore records of Cetacea for eastern Canada. *J. Fish. Res. Board Can.* **27**(11), 1903–1915.

Shane, S. H. (1980). Occurrence, movements, and distribution of bottlenose dolphins, *Tursiops truncatus,* in southern Texas. *Fish. Bull.* **78**(3), 593–601.

Shane, S. H., Wells, R. S., and Würsig, B. (1986). Ecology, behavior, and social organization of the bottlenose dolphin. *Mar. Mamm. Sci.* **2**(1), 34–63.

Tanaka, S. (1987). Satellite radio tracking of bottlenose dolphins. *Nippon Suisan Gakkaishi* **53**(8), 1327–1338.

Taruski, A. G., and Winn, H. E. (1976). Winter sightings of odontocetes in the West Indies. *Cetology* **22,** 1–12.

Tomilin, A. G. (1957). "Mammals of the U.S.S.R. and

Adjacent Countries," Vol. IX, Cetacea. Israel Program for Scientific Translation (No. 1124), Jerusalem. 738 pp.

Walker, W. (1975). Review of the live capture fishery for smaller cetaceans taken in southern California waters for public display, 1966–73. *J. Fish. Res. Board Can.* **32**(7), 1197–1211.

Walker, W. (1981). "Geographical Variation in Morphology and Biology of Bottlenose Dolphins (*Tursiops*) in the Eastern North Pacific." NMFS/SWFC Admin. Rep. No. LJ-81-03C. National Marine Fisheries Service/Southwest Fisheries Center, La Jolla, California.

Winn, H. E., Price, C. A., and Sorensen, P. W. (1986). Distributional biology of the right whale (*Eubalaena glacialis*) in the western North Atlantic. *Rep. int. Whal. Commn.* (Special Issue 10), pp. 129–138.

Winn, H. E., Hain, J. H. W., Hyman, M. A. M., and Scott, G. P. (1987). Whales, dolphins, and porpoises. *In* "Georges Bank" (R. H. Backus and D. W. Bourne, eds.), pp. 375–382. MIT Press, Cambridge, Massachusetts. 603 pp.

Winn, L. K., Winn, H. E., Caldwell, D. K., Caldwell, M. C., and Dunn, J. L. (1979). Marine mammals. *In* "A Summary and Analysis of Environmental Information on the Continental Shelf and Blake Plateau from Cape Hatteras to Cape Canaveral" (Center for Natural Areas), Vol. I, Book 2, Contract AA550-CT2-39, U.S. NTIS PB80-184104, pp. XII.1–XII.117. Bureau of Land Management, Washington, D.C. 637 pp.

Würsig, B., and Würsig, M. (1977). The photographic determination of group size, composition, and stability of coastal porpoises (*Tursiops truncatus*). *Science* **198,** 755–756.

Würsig, B., and Würsig, M. (1979). Behavior and ecology of the bottlenose dolphin, *Tursiops truncatus,* in the South Atlantic. *Fish. Bull.* **77**(2), 399–412.

Yasui, W. Y., and Gaskin, D. E. (1986). Energy budget of a small cetacean. *Ophelia* **25**(3), 183–197.

Distribution and Herd Structure of Bottlenose Dolphins in the Eastern Tropical Pacific Ocean

22

Michael D. Scott
Inter-American Tropical Tuna Commission
Scripps Institution of Oceanography
La Jolla, California 92038

Susan J. Chivers
Southwest Fisheries Center
National Marine Fisheries Service
La Jolla, California 92038

INTRODUCTION

Nearly all information currently available on bottlenose dolphins, *Tursiops truncatus,* has been gathered from coastal populations, even though the distribution of these animals extends far offshore. Little is known about populations that live in pelagic waters and around oceanic islands because of the many difficulties in studying dolphins in the open seas.

Because tuna purse-seine vessels in the eastern tropical Pacific Ocean actively search for herds of dolphins, the fishery can provide much useful information about pelagic dolphin populations. The U.S. National Marine Fisheries Service (NMFS) and the Inter-American Tropical Tuna Commission (IATTC) have collected data on pelagic dolphin species affected by the tuna purse-seine fishery. Scientific technicians from both organizations travel aboard tuna vessels of the international fleet to monitor the incidental mortality of dolphins during purse-seining operations, and they collect sightings and life-history information for population studies. The primary dolphin species affected by the fishery are the spotted dolphin, *Stenella attenuata,* and the spinner dolphin, *S. longirostris,* but sets are occasionally made on other species as well, including the bottlenose dolphin. The estimated incidental mortality of bottlenose dolphins ranges from 0 to almost 200 (Smith, 1979; Leatherwood and Reeves, 1982; NMFS, unpublished data) but makes up a small fraction (<0.5%) of the total dolphin mortality caused by the fishery.

From these data, we can determine some of the characteristics of pelagic bottlenose dolphin herds and compare them with those of better-known coastal populations. In this study, we were interested in patterns of distribution, herd size, and associations with other species—characteristics that may differ between open-ocean and coastal populations. In particular, we wished to test the hypothesis suggested by Norris and Dohl (1980) and Wells *et al.* (1980) that in the pelagic environment increased predation pressure and patchier food resources may combine to favor larger dolphin group sizes for protection and more efficient foraging. Groups may also become effectively larger by joining other cetacean species to form mixed-species herds. In the North Pacific, bottlenose dolphins are known to associate with other cetacean species, especially short-finned pilot whales, *Globicephala macrorhynchus* (Norris and Prescott, 1961; Leatherwood *et al.*, 1973, 1982; Walker, 1975). Also, the eastern tropical Pacific is an area where the formation of mixed-species herds, most notably containing spotted and spinner dophins, is common.

We used two sources of sightings data to examine these hypotheses. Most of the data derived from dolphin observations made during 1018 cruises by scientific technicians aboard tuna purse seiners from 1971 through mid 1986. Additional sightings were recorded by NMFS scientific technicians during 24 research cruises from 1974 through 1985. We have included sightings between 35° N and 20° S latitudes, and from the coast of the Americas west to 160° W longitude. This area covers the entire region known as the eastern tropical Pacific Ocean but extends northward of the Tropic of Cancer (23° 27′ N latitude) that marks, in a strict sense, its northern boundary. We refer to the entire study area as the eastern tropical Pacific, however, since the bulk of the sightings come from the tropical region. More detailed studies at the northern and southern edges of this study were reported by Walker (1981), Hansen (Chapter 23, this volume), and Van Waerebeek *et al.* (Chapter 7, this volume).

PATTERNS OF DISTRIBUTION AND HERD COMPOSITION

Distribution

Bottlenose dolphins are widely distributed in offshore waters of the eastern tropical Pacific Ocean (Fig. 1). Differences in morphology, feeding habits, and parasite loads have been reported between coastal and offshore bottlenose dolphins (Walker, 1981; Van Waerebeek *et al.*, Chapter 7, this volume; Mead and Potter, Chapter 9, this volume). In southern California, the coastal form generally has a limited range along the surfline (Hansen, Chapter 23, this volume), while the offshore form in the eastern tropical Pacific is suspected to range more widely (Leatherwood and Reeves, 1982). Owing to the pelagic nature of the tuna fishery, coastal sightings are probably underrepresented in Fig. 1. (The shallow waters along the shore favored by researchers studying coastal bottlenose dolphins are unappealing to large-draft vessels fishing for pelagic tuna.) The cumulative encounter rate with herds of bottlenose dolphins by monitored vessels in the eastern tropical Pacific for the years 1979 through 1985 is presented in Fig. 2 (see also Polachek, 1987). The area inhabited by bottlenose dolphins is shown to be surrounded by an area with at least moderate search effort in which there have been few sightings. This indicates that sightings of pelagic bottlenose dolphins outside the area searched are likely to be rare. The outlying sightings far offshore in Fig. 1 are associated with islands such as Hawaii to the west and the Marquesas and the Tuamotus to the southwest.

Au and Perryman (1985) compared dolphin distributions with oceanographic features in the eastern tropical Pacific. They suggested that two communities of dolphins occur, defined by the water masses in which they live. Au and Perryman found that spotted and spinner dolphins live predominantly in Tropical Surface Water (Wyrtki, 1966, 1967) in an area beginning off the coast of Mexico and extending westward along the 10° N parallel (see Figures 4 and 6 in Perrin *et al.*, 1983, for distribution maps of these species). This water is characterized by a sharp, shallow thermocline within 50 m of the surface, warm sea-surface temperatures (>25°C), and relatively low salinity (<34‰). Productivity is high owing to the divergence zone created by the westward-flowing North Equatorial Current and the eastward-flowing Equatorial Counter Current. A second oceanic regime, which Au and Perryman termed Upwelling-Modified Water, is more variable in character, with generally cooler sea-surface temperatures, weaker and deeper thermoclines, higher salinity (>35‰), and high productivity

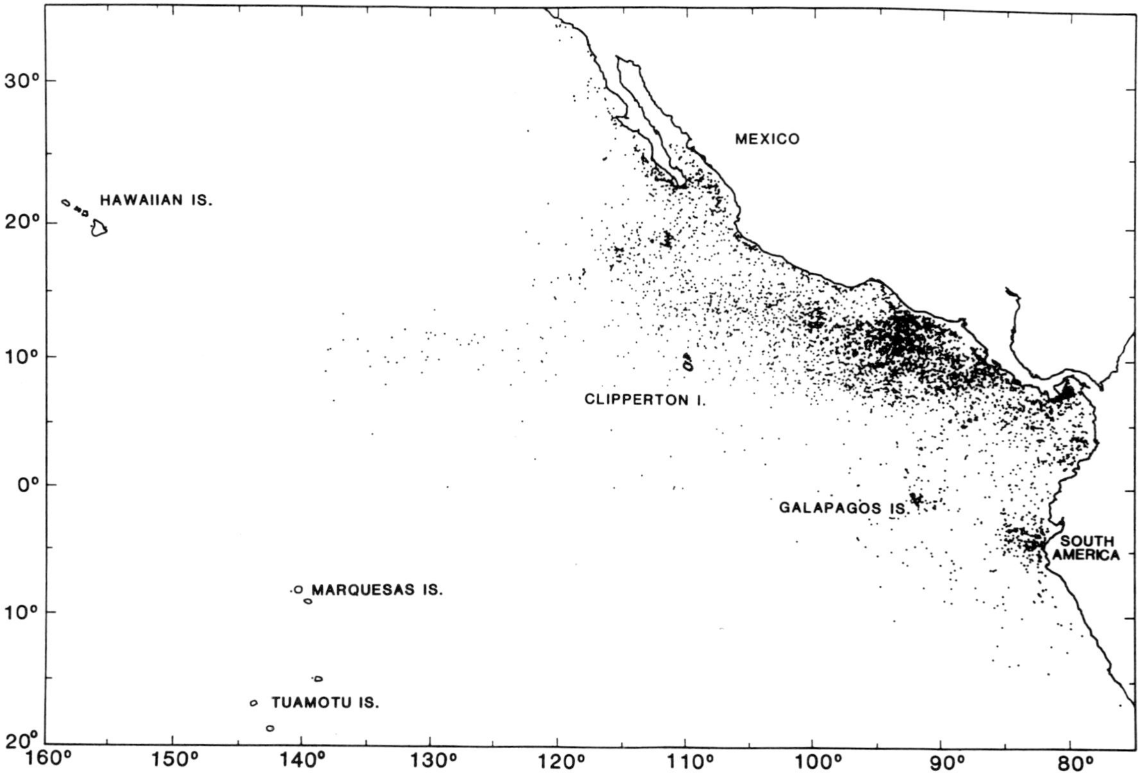

Figure 1 Distribution of all sightings of bottlenose dolphins in the eastern tropical Pacific Ocean ($n = 5461$).

induced primarily by upwelling. Such areas are found along the Equator and off the coasts of Baja California, Central America, and Peru, where common dolphins, *Delphinus delphis,* striped dolphins, *Stenella coeruleoalba,* and short-finned pilot whales are the predominant cetaceans. Bottlenose dolphins and Risso's dolphins, *Grampus griseus,* occur in both communities, and the bottlenose dolphin distribution (Fig. 1) appears to include the collective ranges of spotted dolphins and common dolphins (Perrin *et al.*, 1983). For example, they are found in both the Tropical Surface Water offshore of central Mexico and along the 10° N parallel (as are spotted dolphins) as well as in the Upwelling-Modified Water which extends westward along the Equator and along the coasts of Baja California and Peru (as are common dolphins).

Herd Size

The median herd size of bottlenose dolphins in the eastern tropical Pacific Ocean is 10 animals (mean 57), although some herds have been estimated to contain as many as 10,000[1] bottlenose dolphins (Fig. 3). There are potential biases, however, in the

[1] Large estimates should be viewed cautiously. Such herds are often described as being composed of small groups scattered from the ship out to the horizon. Given the difficulties in estimating large numbers of anything, let alone trying to estimate the sizes of moving groups scattered in all directions, and given the additional difficulties in properly identifying all the animals to species, the herd estimates of several thousands of bottlenose dolphins may be more of an indication of an "observer on overload" than an accurate representation of herd size. Suffice it to say, however, that the herds are undoubtedly very large.

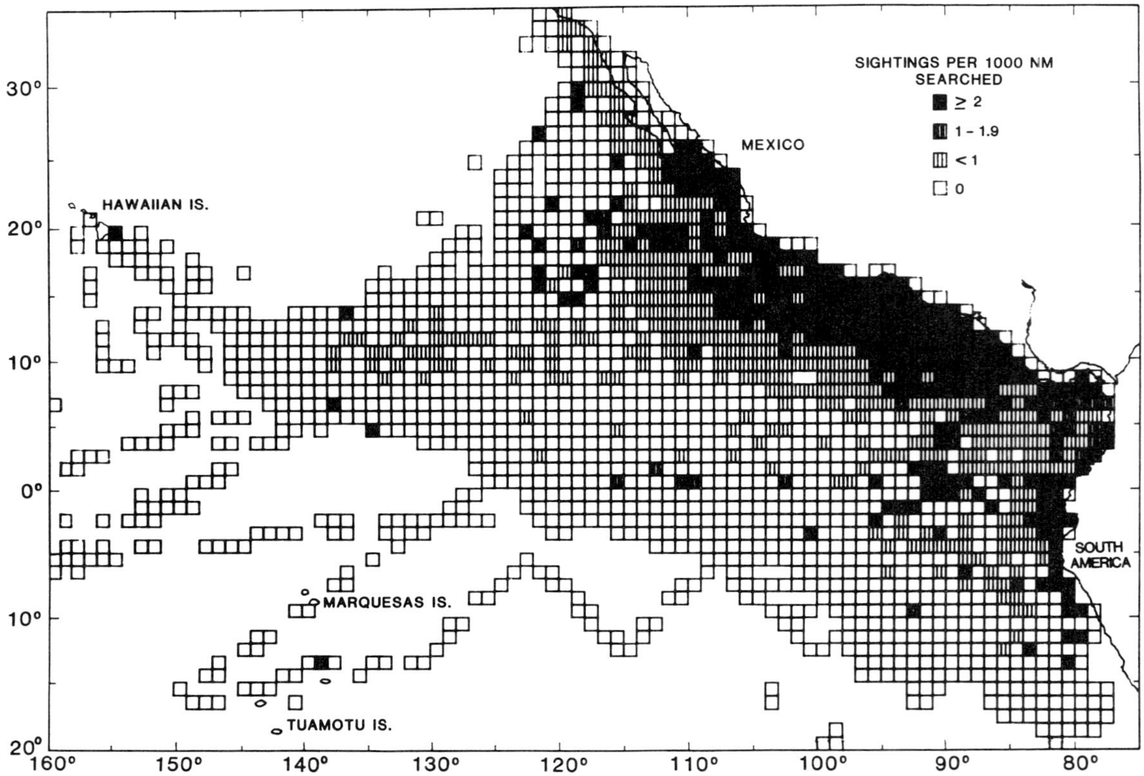

Figure 2 Density of bottlenose dolphins in the eastern tropical Pacific Ocean sighted in the years 1979–1985 from tuna purse seiners. Density is measured by the number of bottlenose dolphin herds sighted per 1000 nautical miles searched in a 1-degree quadrat.

estimates of herd size. The number of small herds may be underestimated, particularly for sightings from tuna boats, because such groups are typically more difficult to spot than large groups and because the fishermen search preferentially for large dolphin herds since they are thought to be associated with the largest groups of yellowfin tuna, *Thunnus albacares*. This bias may be counterbalanced by the tendency of bottlenose dolphins to approach boats to ride the bow wave or stern wake. The relative influence of these factors has not been assessed.

Estimates of herd size made from tuna boats can differ from estimates made from research vessels. Holt and Powers (1982) found that average herd size estimates made from tuna boats ($\bar{x}$ = 583, S.E. = 54, n = 167) were about three times those made from two research vessels ($\bar{x}$ = 166, 188; S.E. = 23, 27; n = 62, 49) for herds of *Stenella* or *Delphinus* greater than 14 animals. This factor probably does not significantly affect the median estimates for a species that occurs most often in groups small enough that the number of individuals can be counted, rather than estimated. The median herd size reported by research-vessel technicians is 8.5 animals (mean 20, n = 200), but the low sample size from the research vessels and the different amount of effort expended by the two types of vessels in different areas make comparison problematic.

To examine the effect of distance from shore on herd size, we stratified the sightings into five areas (Fig. 4):

Area 1. Coastal areas: 1-degree quadrats which contain portions of the coastline

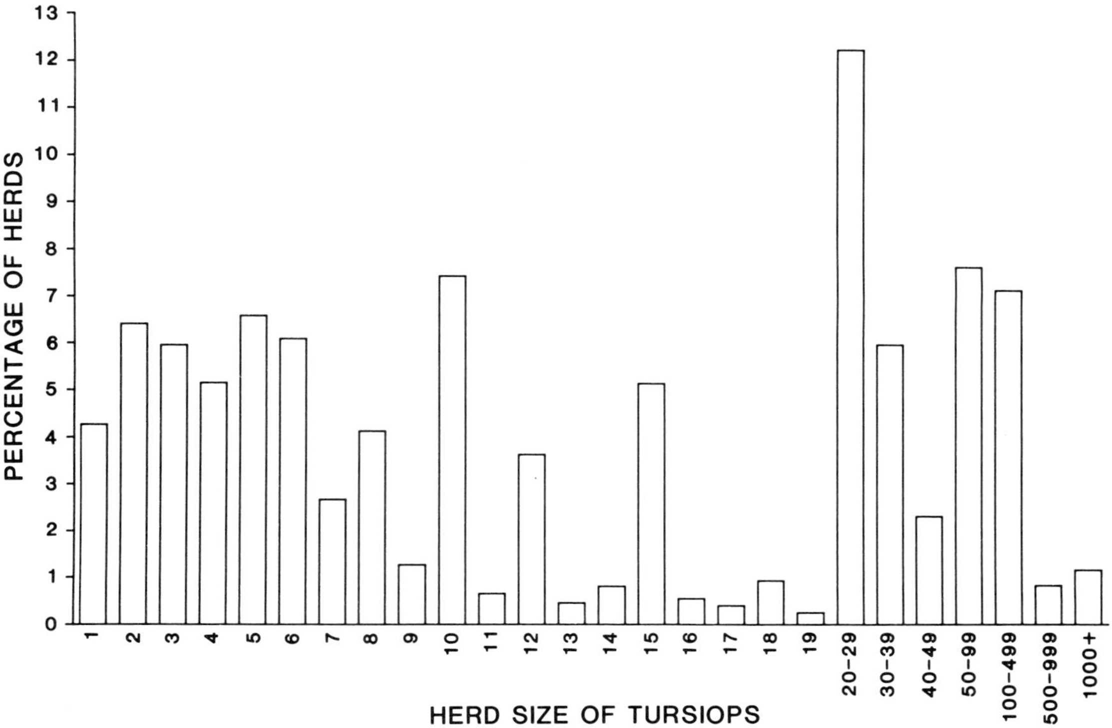

Figure 3 Frequency distribution of herd sizes of bottlenose dolphins in the eastern tropical Pacific Ocean ($n = 5461$). For mixed-species herds, the number of bottlenose dolphins is used, not the total herd size. Some of the spikes in the data are due to the tendency of technicians to round off herd size estimates to multiples of five.

Area 2. Island areas: 1-degree quadrats containing offshore islands
Area 3. Near-coastal areas: 1-degree quadrats adjacent to the coastal quadrants
Area 4. Far-western areas: the area west of 120° W longitude and south of 18° N latitude
Area 5. Offshore areas: the remaining areas

This somewhat arbitrary stratification scheme attempts to partition the areas by increasing distance offshore while incorporating features of the ocean's physiography and the dolphins' distribution. Thus, the coastal area roughly corresponds to waters inside the 2000-m contour, whereas the near-coastal area covers the band of pelagic waters generally just offshore of this contour. We have attempted to isolate the potential influences of island areas on the dolphin herds by separating them from the surrounding offshore area. The far-western area is the farthest from shore and is partitioned from the offshore areas by an apparent hiatus in the distribution of bottlenose dolphins. Such a hiatus occurs in other dolphin species in the eastern tropical Pacific as well (Perrin *et al.*, 1983).

There are significant differences among the above areas in sizes of herds containing only bottlenose dolphins (Kruskal–Wallis test, $p < .001$). There appears to be a decreasing onshore–offshore trend in mean herd sizes, although this trend is not reflected in the median herd sizes (Table 1). We favor the use of median over mean herd size in this study because of the heavily skewed distribution of the pelagic herd sizes; the presence of a relatively few, very large herds in offshore waters can greatly inflate the mean herd size. Thus, while the mean more accurately reflects the greater percentages of large herds in the coastal and island areas, the median offers a better representation of what a "typical"

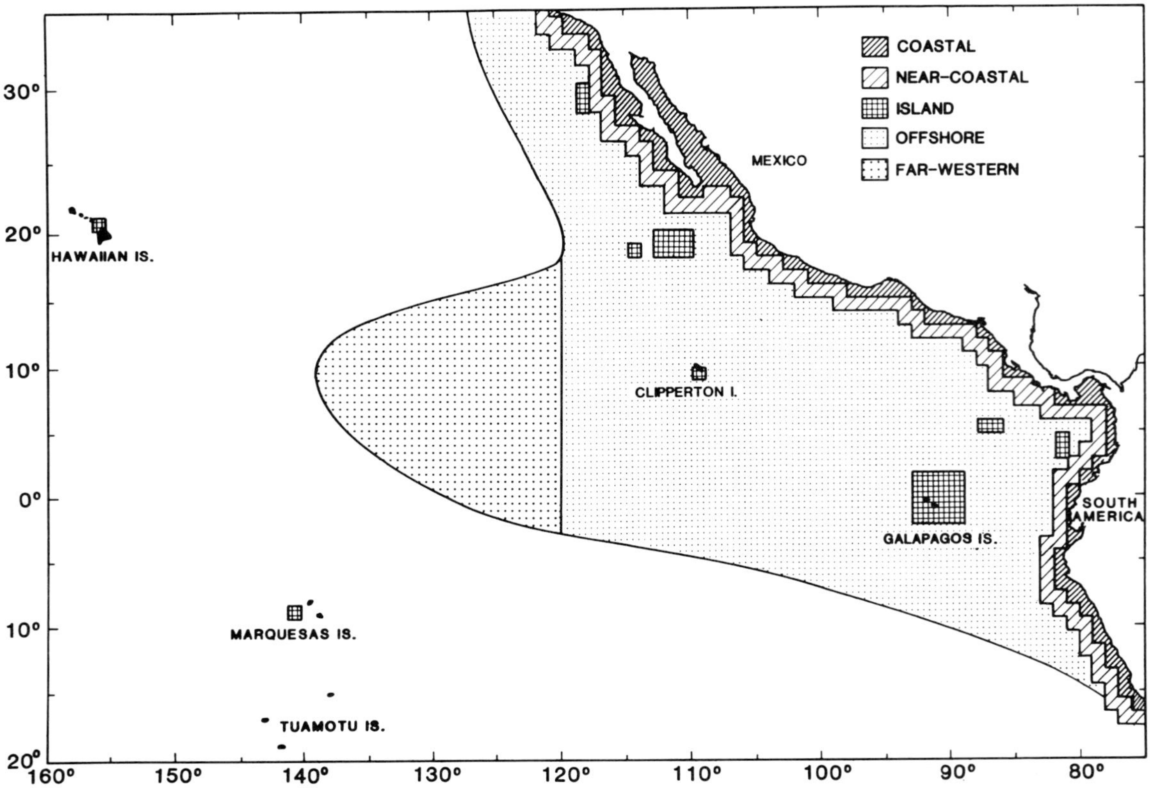

Figure 4 Areas used to stratify sightings of bottlenose dolphins.

Table 1
Herd Characteristics of Bottlenose Dolphins Stratified by Area in the Eastern Pacific Ocean

	Area[a]				
	Coastal (1)	Island (2)	Near-coastal (3)	Far-western (4)	Offshore (5)
Percentage of mixed-species herds	11	6	19	32	24
Percentage of herd sizes of					
1–10 dolphins	49	39	46	53	52
11–49 dolphins	34	41	39	34	38
50+ dolphins	17	20	15	13	10
Median herd size	12	20	12	9	10
Mean herd size	94	93	72	44	40
Sample size	867	228	1149	47	3170

[a] See text for definition of areas.

herd is like. Even the large herds of hundreds or thousands of bottlenose dolphins are usually described by observers as aggregations composed of numerous small groups scattered over an area of several miles, and the influence of such large, and perhaps ephemeral, herds on the mean herd size may be misleading.

Herd sizes were stratified by season and by year for the coastal (Area 1) and offshore (Area 5) regions. The median herd size did not change significantly with season in coastal and offshore areas north of the equator, although the medians in both areas were at their highest in April–June (Fig. 5). A similar comparison for areas south of the equator was not done because of inadequate sample sizes.

Variation in median herd sizes, however, was found between years (Fig. 6). Most strikingly, the median herd size was significantly lower in 1983 than in other years. It is possible that changes in median herd size were correlated with oceanographic conditions. Three El Niño events occurred during the period 1971–1986. Severe El Niño conditions, typified by higher sea-surface temperatures and decreased upwelling and productivity, occurred in 1972–1973 and 1982–1983, while a

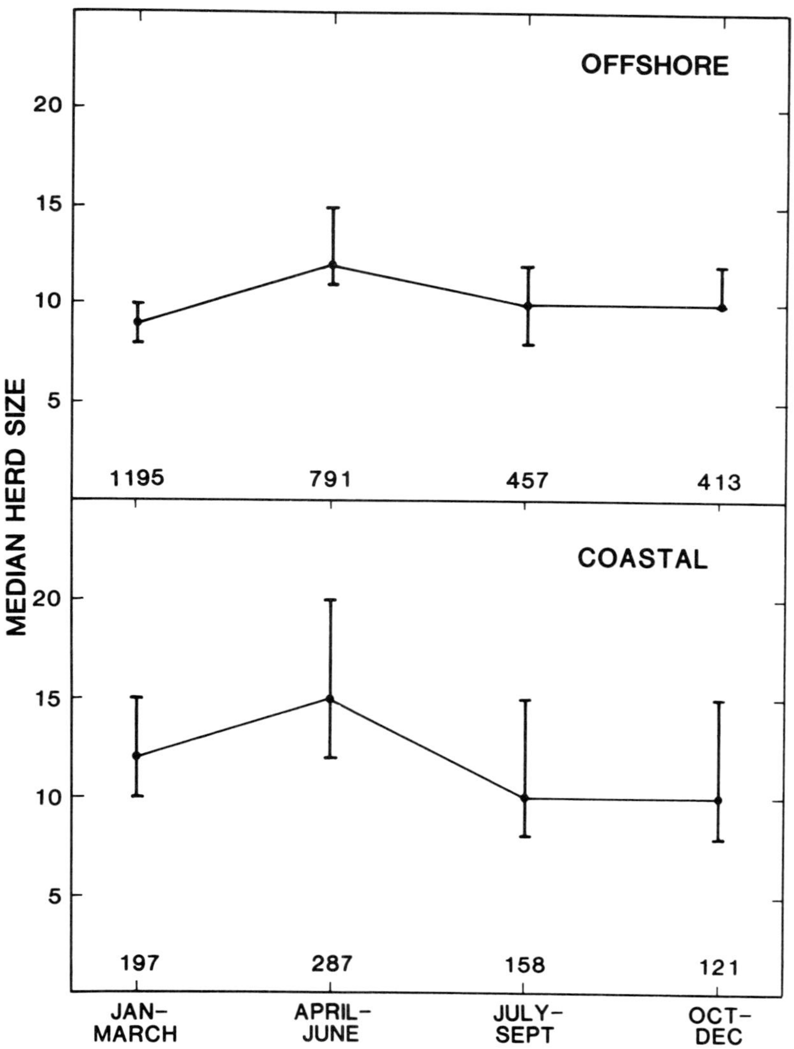

Figure 5 Median herd sizes of bottlenose dolphins plotted by quarter of the year. Included are 95% confidence intervals and the number of sightings per season. The coastal (Area 1) and offshore (Area 5) regions are as described in the text, except that the areas south of the equator have been excluded.

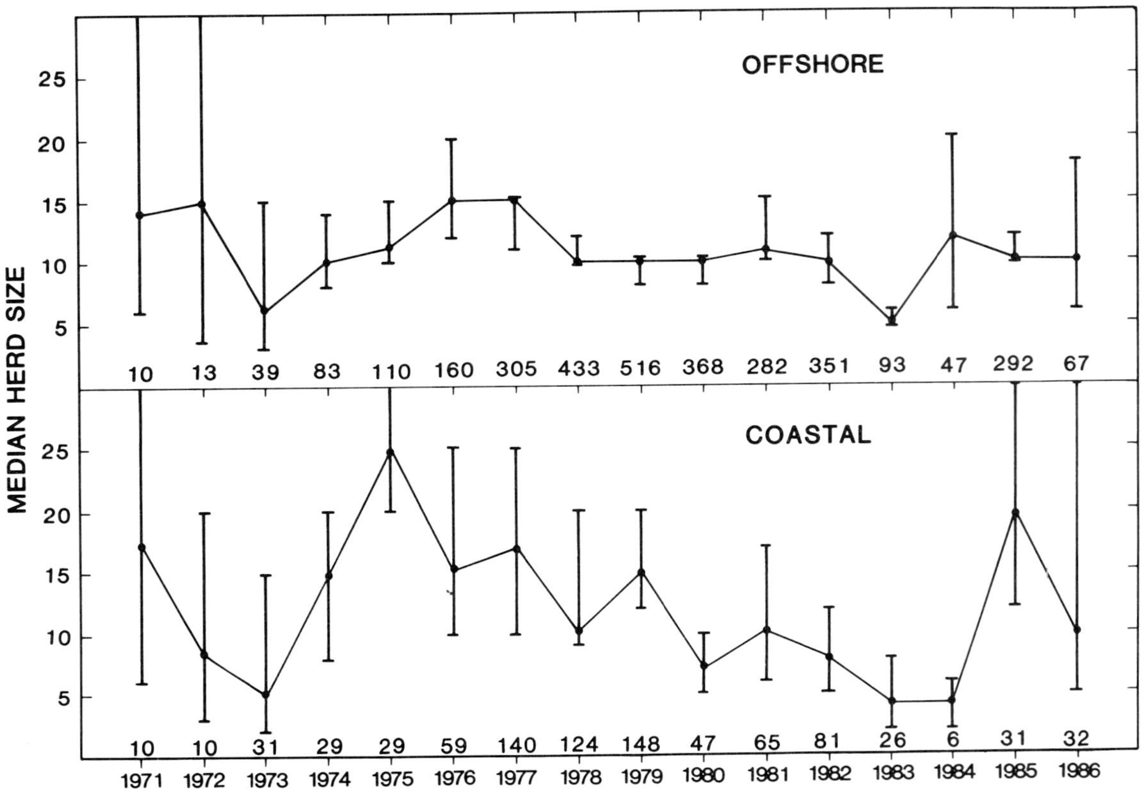

Figure 6 Median herd sizes of bottlenose dolphins plotted by year. Included are 95% confidence intervals and the number of sightings per year. The offshore and coastal regions are described in the text.

relatively mild event occurred in 1976 (see Fig. 45 in IATTC, 1984; IATTC, 1986). Declines in median herd size were observed at about the same periods as severe El Niño events (different lag times may have occurred in the different areas). Intense anti-El Niño events, typified by lower sea-surface temperatures, increased upwelling, and increased food production, occurred in 1970–1971, 1975, and 1984–1985; higher median herd sizes occurred at approximately the same times as these events. Another factor, however, that may influence these results is that average herd-size estimates by scientific technicians aboard tuna boats have declined for spotted and spinner dolphins during 1977–1983 (Hammond and Laake, 1983; Holt, 1985), perhaps influenced by increased training on herd-size estimation given the technicians during that time (Hammond and Laake, 1983).

Mixed-Species Herds

Many sightings of mixed-species herds containing bottlenose dolphins were recorded in the eastern tropical Pacific Ocean (Fig. 7). While the median and mean total sizes of mixed-species herds (median 50, mean 134) are significantly greater than the sizes of herds containing only bottlenose dolphins (median 10, mean 58, $p < .001$), the median and mean numbers of bottlenose dolphins in mixed-species herds (median 11, mean 56) are not significantly different from the numbers in herds containing only bottlenose dolphins (Wilcoxon rank sum tests, $p = .857$; ANOVA, $p = .829$). Thus, it appears that the presence of other species in the herd does not affect the number of bottlenose dolphins present. Eighty percent of the herds contained only bottlenose dolphins, while

16% contained one other species and 4% two or more other species.

The proportion of mixed-species herds is significantly different among areas (contingency test, $p < .001$); as illustrated in Table 1, there is a significant increasing trend in the percentage of mixed-species herds among areas that are progressively distant from shore, particularly when coastal and island areas are combined ($p < .001$; a test for linear trend in proportions was used with arbitrary scores being associated with areas to detect a linear component of the trend; see Snedecor and Cochran, 1973). Mixed-species herds are typically found well offshore, particularly along the 10° N parallel (Fig. 7).

Bottlenose dolphins were sighted with at least 13 other species of cetaceans (Fig. 8), in a total of 1102 recorded mixed-species herds. The short-finned pilot whale was the species most commonly associated with bottlenose dolphins, being present in 40% of all mixed-species herds. The next most commonly associated species was the spotted dolphin (36%). Bottlenose and spinner dolphins also associated (19%) but typically only when spotted dolphins were present as well. Only 7% of the mixed-species herds that contained spinner and bottlenose dolphins did not also contain spotted dolphins.

Sightings data from the tuna vessels are likely to be biased, however, because the tuna fishermen primarily search for, and set on, large herds of spotted dolphins, which are most often associated with yellowfin tuna. Thus, compared to other cetaceans, spotted dolphin herds are more likely to be approached by the vessel and are therefore more likely to be identified by technicians. The relatively high percentage of herds containing both bottlenose and spotted dolphins might be a direct result of this characteristic of the fishery. Data from the NMFS research cruises, however,

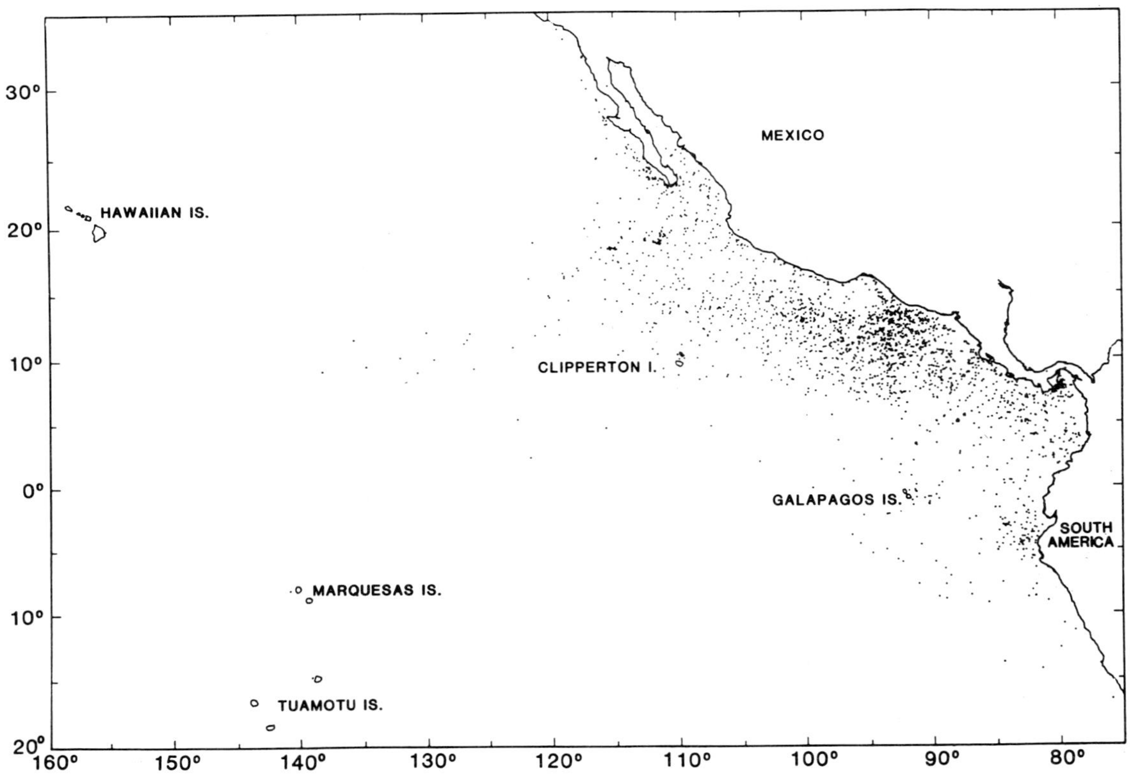

Figure 7 Distribution of all sightings of mixed-species herds containing bottlenose dolphins in the eastern tropical Pacific Ocean (n = 1102).

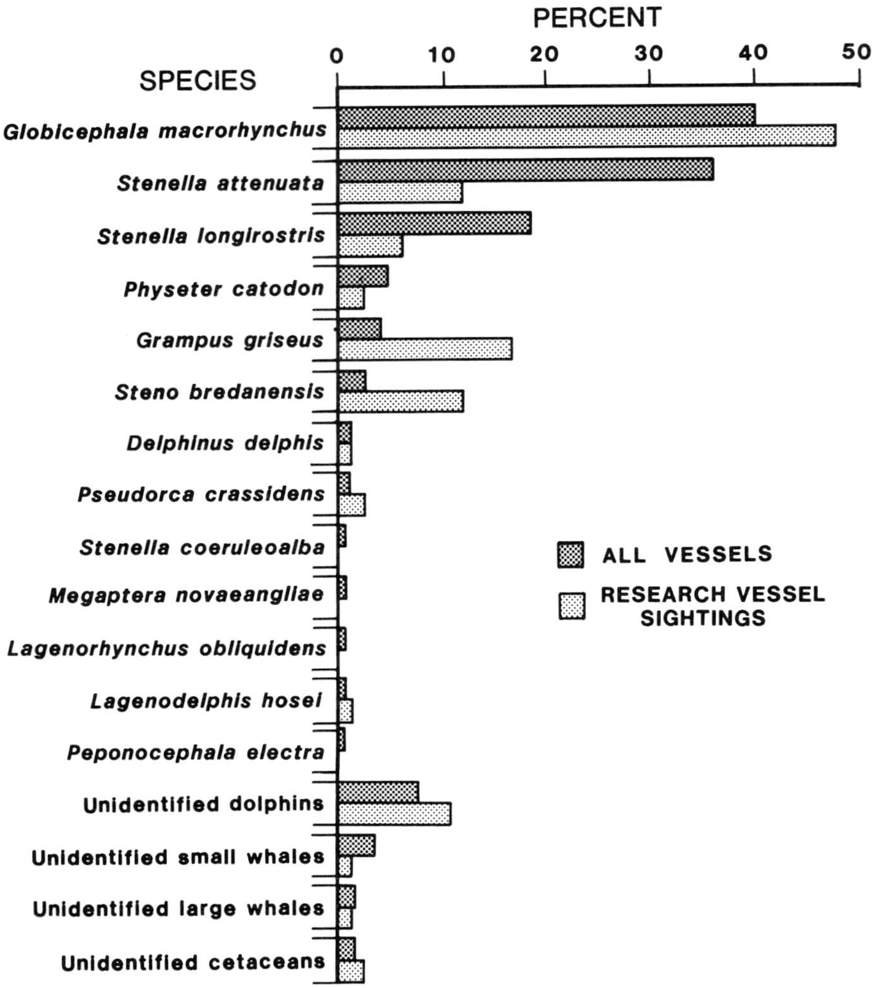

Figure 8 Associations of bottlenose dolphins with other cetacean species in mixed-species herds. Data are from both tuna and research vessels (n = 1100) and from research vessels only (n = 84).

are not biased in this manner because dolphin herds are generally approached regardless of species or herd size. When the relative frequencies of species associated with bottlenose dolphins from the research vessel data were compared with data from the tuna vessels (Fig. 8), the research cruise data contained a smaller percentage of mixed-species herds composed of bottlenose and spotted dolphins, with concurrent increases in associations with Risso's dolphins, rough-toothed dolphins, *Steno bredanensis,* and pilot whales.

The distribution of herds containing both bottlenose dolphins and pilot whales is limited primarily to three areas (Fig. 9): (1) along the coast of Baja California, (2) along the 10° N parallel, and (3) along the equator. In the westward-extending lobe along the 10° N parallel, however, there is a hiatus in sightings between 104° and 119° W longitude.

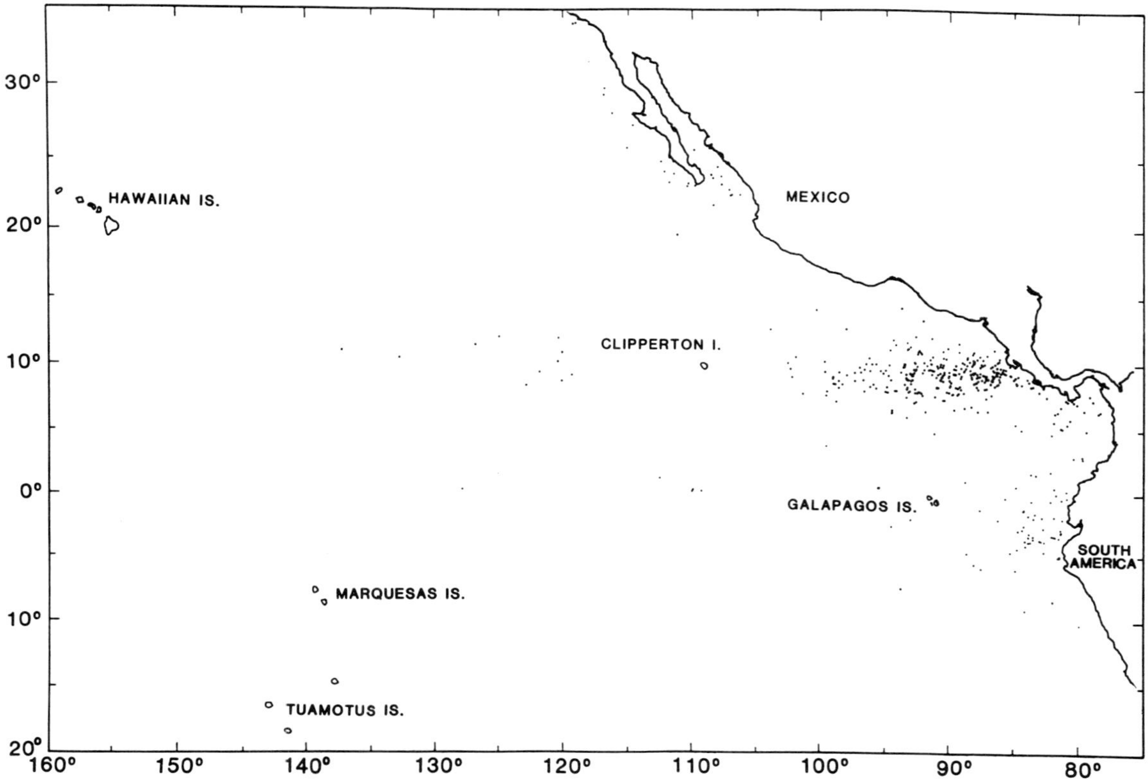

Figure 9 Distribution of sightings of mixed-species herds containing bottlenose dolphins and short-finned pilot whales in the eastern tropical Pacific Ocean (n = 434).

Within this gap lies Clipperton Island, which represents a "coastal" habitat in which pure herds of bottlenose dolphins are often sighted (Fig. 1). There is also a less-distinct gap in sightings of mixed herds of bottlenose dolphins and pilot whales near the Galapagos Islands, where small groups of bottlenose dolphins are frequently found (Lévêque, 1963; Hickin, 1979). Mixed-species herds containing bottlenose and spotted dolphins are found principally along the 10° N parallel with a hiatus near Clipperton Island as noted above (Fig. 10).

In mixed-species herds, bottlenose dolphins typically form only a small proportion of the total number of animals in the herd (Fig. 11), but this proportion appears to depend mainly on the other species in the herd. When in combination with species that typically occur in large herds of hundreds of animals, such as the spotted dolphin, bottlenose dolphins were a small minority, comprising usually 10% or less of the total animals in the herds (Fig. 11). When in combination with species, such as pilot whales, that typically occur in smaller herd sizes similar to those of bottlenose dolphins (Leatherwood *et al.*, 1982), the bottlenose dolphins can be either the majority or minority species in the herd.

DISCUSSION

How do pelagic and coastal populations of bottlenose dolphins differ? The mean herd sizes of pelagic bottlenose dolphins in the eastern tropical Pacific Ocean are considerably larger than those

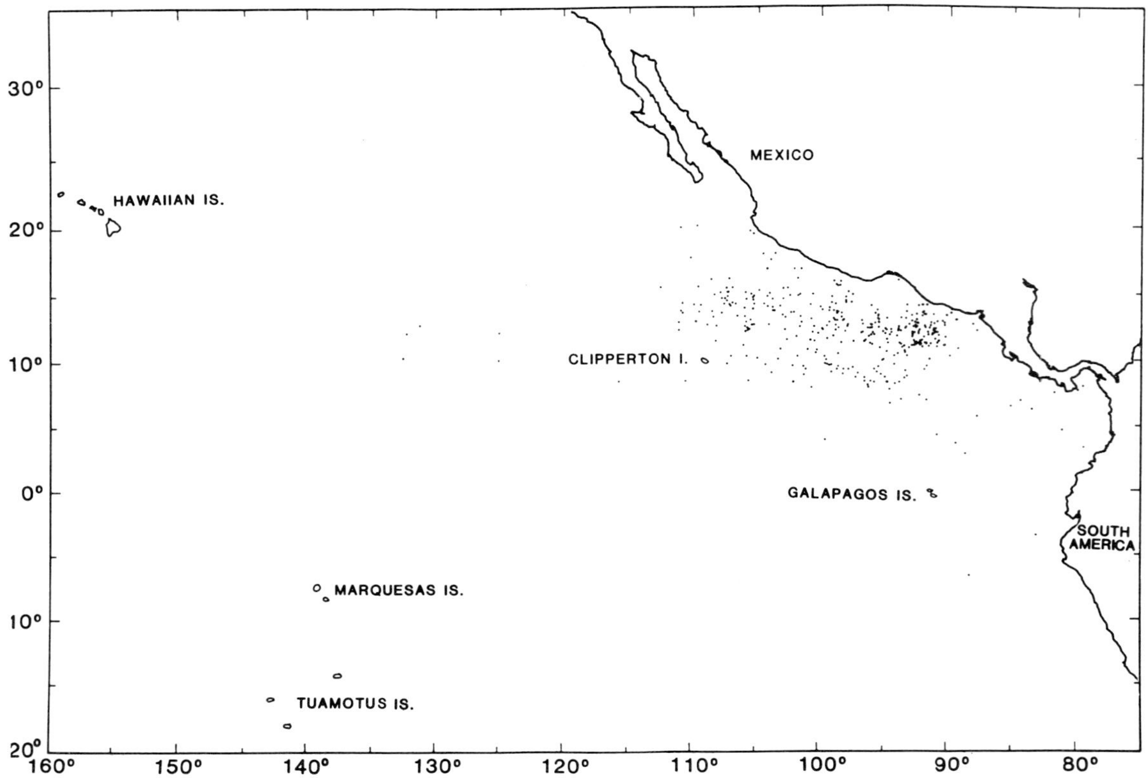

Figure 10 Distribution of sightings of mixed-species herds containing bottlenose dolphins and spotted dolphins in the eastern tropical Pacific Ocean (n = 387).

reported for coastal populations (Leatherwood and Reeves, 1982, table 18.4). The large majority of pelagic herds, however, are no larger than coastal herds, and the median herd size of offshore herds (10) is within the range of herd sizes typical of coastal populations (Leatherwood and Reeves, 1982; Shane *et al.*, 1986). Yet, the range of estimated sizes of bottlenose dolphin herds in the eastern tropical Pacific (from 1 to over 10,000) is much broader than that of any coastal population studied. Although Saayman and Tayler (1973) reported coastal and near-coastal herds that contained up to 1000 dolphins off southern Africa, herds in coastal populations typically contain fewer than 100 dolphins (Leatherwood and Reeves, 1982). Thus, it appears that in the pelagic waters of the eastern Pacific small herds are still the norm; it is the occasional occurrence of very large herds, sometimes containing thousands of animals, that distinguishes the pelagic from the coastal populations.

Do bottlenose dolphins associate with other cetaceans to a greater extent in the open ocean than in coastal areas? Coastal populations typically do not associate with other cetaceans, while mixed-species herds are more commonly seen in pelagic waters. The increasing trend observed in the proportion of mixed-species herds with distance offshore suggests that the formation of such congregations may be advantageous for pelagic life. Other pelagic cetaceans, such as Risso's dolphins, northern right whale dolphins, *Lissodelphis borealis,* Pacific white-sided dolphins, *Lagenorhynchus obliquidens,* spinner dolphins, and spotted dolphins, are also known to form mixed-species herds (Leatherwood *et al.*, 1982). Mixed-species

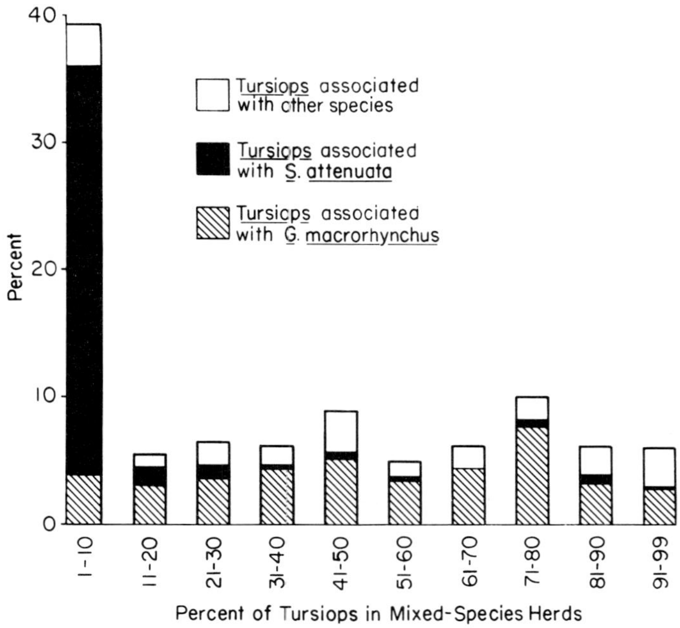

Figure 11 Relative frequencies of the proportion of bottlenose dolphins within mixed-species herds ($n = 1102$). For example, 39% of all mixed-species herds were ones in which bottlenose dolphins formed 10% or less of the herd; the spotted dolphin was the majority species in most of these herds.

herds could provide the same advantage of "safety in numbers" that large single-species herds do, while additionally providing more options in feeding strategies. Bernard and Thompson (1985) suggested that the same advantages exist for mixed-species bird flocks, observing that a species with opportunistic feeding habits may aggregate with species having more specialized feeding habits, thus allowing the generalist to mimic or capitalize on the more efficient strategies of the specialist.

Yet, even in the farthest offshore areas, 70% of the bottlenose dolphin herds observed were not associated with other cetacean species. Indeed, Fig. 12 demonstrates that small herds containing only bottlenose dolphins are distributed well offshore. While there appears to be some onshore–offshore effect on the size of bottlenose dolphin herds and their tendency to associate with other species, pelagic bottlenose dolphins obviously are not absolutely restricted to congregating in large or mixed-species herds to survive.

Clearly then, our initial hypothesis that pelagic herds are typically larger than coastal herds is not borne out by the data. This apparently contradicts observations by researchers studying coastal populations (Norris and Dohl, 1980; Wells *et al.*, 1980). However, these researchers typically have compared herd sizes of populations that live along shore to those of adjacent offshore populations, while we have included these adjacent populations in our definition of "coastal" dolphins because of the large area covered by this study. While an increasing trend in herd size may be apparent from the point of view of a coastal researcher, this trend does not extend to the pelagic populations of bottlenose dolphins when viewed on a broader scale. A more compelling notion, perhaps, is that the size of herds becomes more variable in the pelagic environment. To place this question in a clearer context, we need to reexamine our initial assumptions about characteristics of the pelagic environment that may influence the size and structure of bottlenose dolphin herds: increased predation pressure and the presence of rich, but patchy, prey resources.

Predation pressure may not exert a significant influence on bottlenose dolphin herd size in the pelagic environment. Sharks and killer whales, *Orcinus orca*, are potential predators of most dolphins. Shark attacks, and scars from previous attacks, have been observed on coastal bottlenose dolphins (Wood *et al.*, 1970; Wells *et al.*, 1980), but

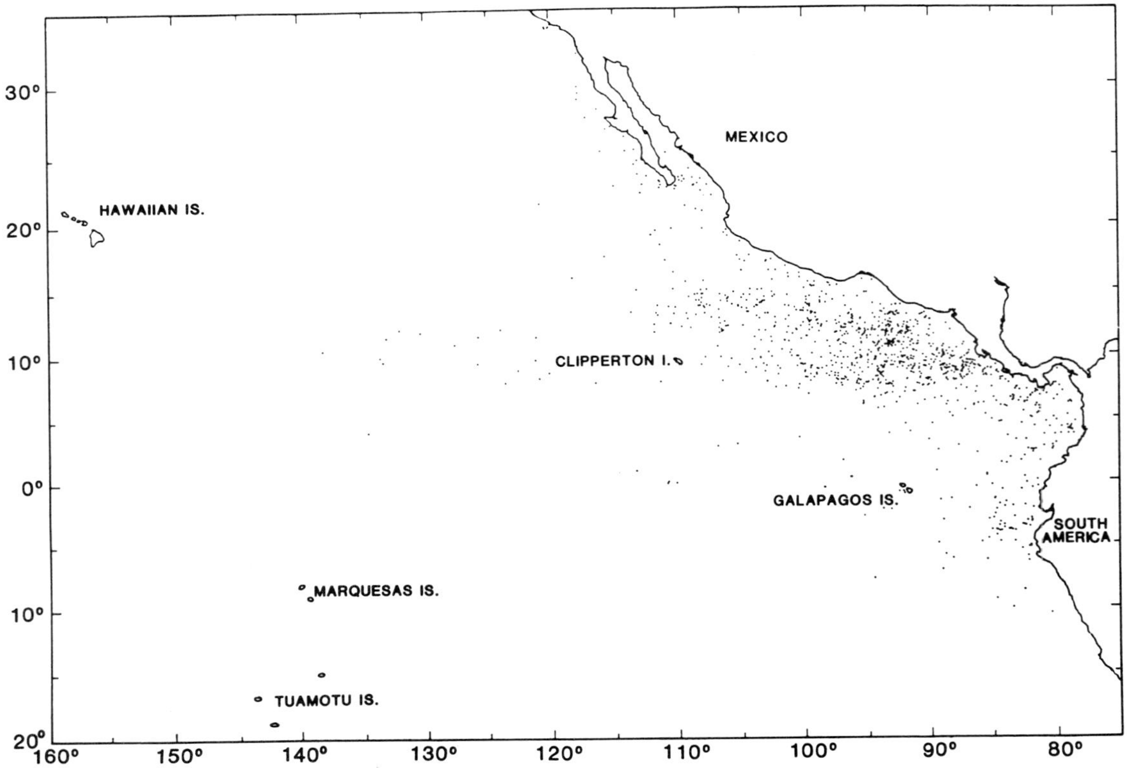

Figure 12 Distribution of sightings of herds containing only bottlenose dolphins with 10 or fewer animals in the eastern tropical Pacific Ocean (n = 2516).

shark-bite scars are less commonly observed on pelagic cetaceans (Wood *et al.*, 1970; Leatherwood *et al.*, 1971). Leatherwood *et al.* (1971) observed shark predation on dolphins (primarily dead or injured individuals) during tuna purse-seining operations, but they suggested that the rarity of shark-bite scars on *Stenella* spp. and *Delphinus* indicates either that shark attacks are typically fatal, or, more likely, that predation on healthy adults by sharks is low due to the speed and maneuverability of the dolphins which, in combination with the clarity and unobstructed visibility of pelagic waters, enables them to avoid attack. Possibly the pelagic populations of bottlenose dolphins face no greater, or perhaps even less, predation pressure than do coastal populations, with the risk of predation in both environments falling mainly on the young and the weak.

The relatively large size of bottlenose dolphins probably reduces the risk of attack by some predators. False killer whales, *Pseudorca crassidens*, for example, are known to prey on smaller dolphins such as *Stenella* spp. and *Delphinus* (Perryman and Foster, 1980; IATTC, unpublished data). Not surprisingly, false killer whales are not found in the same herds as these smaller dolphins, but they do associate occasionally with larger delphinids such as pilot whales and other small whales as well as with bottlenose dolphins (Fig. 8). Although circumstantial, this evidence suggests that bottlenose dolphins may be less vulnerable to predation than other species of dolphins in the eastern tropical Pacific.

The schooling behavior of prey species may influence the variability in dolphin herd size. Bottlenose dolphins are known to be opportunistic feeders with flexible feeding strategies (Caldwell and Caldwell, 1972; Leatherwood and Reeves, 1982; Shane *et al.*, 1986; Shane, Chapter 12, this volume). Walker (1981) recorded a variety of epipelagic prey in the stomachs of bottlenose dolphins in the eastern tropical Pacific, with frigate tuna, *Auxis thazard*, flying fish (Exocoetidae), and squid (Ommastrephidae and Enoplotreuthidae) being the most common. Little is known about the densities and schooling characteristics of these prey items, but observations on frigate tuna (see review by Uchida, 1981), flying fish (Beklemishev and Pasternak, 1960, cited in Kort, 1967), and squid (Wormuth, 1976) suggest that these prey do not always school in large numbers but instead show wide variations, similar to the variations in herd size exhibited by bottlenose dolphins. More detailed study of what constitutes a "typical" prey school is required to determine whether a correlation exists between the group sizes of bottlenose dolphins and their prey.

Bottlenose dolphins have exploited a wide range of habitats owing to their extreme plasticity in behavior and ecology. They have adapted to the different environmental regimes in the eastern tropical Pacific Ocean and are members in good standing of the two cetacean communities of the region. While the variability they display in herd size and association with other cetaceans likely plays a role in their ability to survive in such diverse environments, the variability they display in other factors such as their morphology and feeding strategies may prove to be even more important.

ACKNOWLEDGMENTS

The authors thank Karen Smith for help in the computer analyses, the many scientific technicians who have collected sightings data over the years, and our colleagues who reviewed the manuscript: Isadore Barrett, William Bayliff, Douglas DeMaster, Andrew Dizon, Eric Forsbergh, Martin Hall, Aleta Hohn, James Joseph, and William Walker.

REFERENCES

Au, D. W. K., and Perryman, W. L. (1985). Dolphin habitats in the eastern tropical Pacific. *Fish. Bull.* **83**(4), 623–643.

Bernard, C. J., and Thompson, D. B. A. (1985). "Gulls and Plovers: The Ecology and Behaviour of Mixed, Species Feeding Groups." Columbia Univ. Press, New York. 293 pp.

Caldwell, D. K., and Caldwell, M. C. (1972). "The World of the Bottlenosed Dolphin." Lippincott, Philadelphia, Pennsylvania. 157 pp.

Hammond, P. S., and Laake, J. L. (1983). Trends in estimates of abundance of dolphins (*Stenella* spp. and *Delphinus delphis*) involved in the purse-seine fishery for tunas in the eastern tropical Pacific, 1977–81. *Rep. int. Whal. Commn* **33,** 565–588.

Hickin, N. E. (1979). "Animal Life of the Galapagos." Springwood Books, London. 236 pp.

Holt, R. S. (1985). Estimates of abundance of dolphin stocks taken incidentally in the eastern tropical Pacific yellowfin tuna fishery. Natl. Mar. Fish. Serv., Southwest Fish. Center Admin. Rep., No. LJ-85-20. 32 pp.

Holt, R. S., and Powers, J. E. (1982). Abundance estimation of dolphin stocks involved in the eastern tropical Pacific yellowfin tuna fishery determined from aerial and ship surveys to 1979. NOAA Tech. Memo., NOAA-TM-NMFS-SWFC-23. 95 pp.

IATTC. (1984). Annual report of the Inter-American Tropical Tuna Commission, 1983. 272 pp. [In English and Spanish.]

IATTC. (1986). Annual report of the Inter-American Tropical Tuna Commission, 1985. 248 pp. [In English and Spanish.]

Kort, V. G. (1967). The Pacific Ocean: Biology of the Pacific Ocean. Book III. Fishes of the open waters. *Natl. Tech. Info. Serv. Trans.* **528,** No. AD-730633 [1971]. 273 pp.

Leatherwood, S., and Reeves, R. R. (1982). Bottlenose dolphin *Tursiops truncatus* and other toothed cetaceans. *In* "Wild Mammals of North America: Biology, Management, and Economics," (J. A. Chapman and G. A. Feldhamer, eds.), pp. 369–414. Johns Hopkins University Press, Baltimore, Maryland. 1147 pp.

Leatherwood, J. S., Perrin, W. F., Garvie, R. L., and LaGrange, J. C. (1971). Observations of sharks attacking porpoises (*Stenella* spp. and *Delphinus* cf. *D. delphis*). Naval Undersea Center, Tech. Note No. 908. San Diego, California. 7 pp.

Leatherwood, J. S., Evans, W. E., and Lingle, G. E. (1973). The pilot whale *Globicephapa* [sic] c.f. *G. scammonii*. Naval Undersea Center Report, NUC TN 1164. 45 pp.

Leatherwood, S., Reeves, R. R., Perrin, W. F., and Evans, W. E. (1982). "Whales, Dolphins, and Porpoises of the Eastern North Pacific and Adjacent Arctic Waters." NOAA Tech. Rep. NMFS Circ., No. 444. 245 pp.

Lévêque, R. (1963). Notes sur quatre cétacés de l'Océan Pacifique (Ecuador et Galapagos). *Mammalia* **27,** 608–609.

Norris, K. S., and Dohl, T. P. (1980). The structure and function of cetacean schools. *In* "Cetacean Behavior: Mechanisms and Functions" (L. M. Herman, ed.), pp. 211–261. Wiley, New York. 463 pp.

Norris, K. S., and Prescott, J. H. (1961). Observations on Pacific cetaceans of Californian and Mexican waters. *Univ. Calif. Publ. Zool.* **63,** 291–402.

Perrin, W. F., Scott, M. D., Walker, G. J., Ralston, F. M., and Au, D. W. K. (1983). Distribution of four dolphins (*Stenella* spp. and *Delphinus delphis*) in the eastern tropical Pacific, with an annotated catalogue of data sources. NOAA Tech. Memo., NOAA-NMFS-SWFC-38. 65 pp.

Perryman, W. L., and Foster, T. C. (1980). Preliminary report on predation by small whales, mainly the false killer whale, *Pseudorca crassidens,* on dolphins (*Stenella* spp. and *Delphinus delphis*) in the eastern tropical Pacific. Natl. Mar. Fish. Serv., Southwest Fish. Center Admin. Rep., No. LJ-80-05. 9 pp.

Polachek, T. (1987). Relative abundance, distribution and inter-specific relationship of cetacean schools in the eastern tropical Pacific. *Mar. Mamm. Sci.* **3**(1), 54–77.

Saayman, G. S., and Tayler, C. K. (1973). Social organisation in inshore dolphins (*Tursiops aduncus* and *Sousa*) in the Indian Ocean. *J. Mammal.* **54**(4), 993–996.

Shane, S. H., Wells, R. S., and Würsig, B. (1986). Ecology, behavior, and social organization of the bottlenose dolphin: A review. *Mar. Mamm. Sci.* **2**(1), 34–63.

Smith, T., ed. (1979). Report of the Status of Porpoise Stocks Workshop (August 27–31, La Jolla, California). Natl. Mar. Fish. Serv., Southwest Fish. Center Admin. Rep., No. LJ-79-41. 120 pp.

Snedecor, G. W., and Cochran, W. G. (1973). "Statistical Methods." Iowa State Univ. press, Ames. 593 pp.

Uchida, R. N. (1981). Synopsis of biological data on frigate tuna, *Auxis thazard,* and bullet tuna, *A. rochei.* NOAA Tech. Rep. NMFS Circ., No. 436. 63 pp.

Walker, W. A. (1975). Review of the live-capture fishery for smaller cetaceans taken in southern California waters for public display, 1966–1973. *J. Fish. Res. Board Can.* **32**(7), 1197–1211.

Walker, W. A. (1981). Geographical variation in morphology and biology of bottlenose dolphins (*Tursiops*) in the eastern North Pacific. Natl. Mar. Fish. Serv., Southwest Fish. Center Admin. Rep., No. LJ-81-03C. 54 pp.

Wells, R. S., Irvine, A. B., and Scott, M. D. (1980). The social ecology of inshore odontocetes. *In* "Cetacean Behavior: Mechanisms and Functions" (L. M. Herman, ed.), pp. 263–317. Wiley, New York. 463 pp.

Wood, F. G., Jr., Caldwell, D. K., and Caldwell, M. C. (1970). Behavioral interaction between porpoises and sharks. *Invest. Cetacea* **2,** 264–277.

Wormuth, J. H. (1976). The biogeography and numerical taxonomy of the oegopsid squid family Ommastrephidae in the Pacific Ocean. *Bull. Scripps Inst. Oceanogr.* **23.** 90 pp.

Wyrtki, K. (1966). Oceanography of the eastern equatorial Pacific Ocean. *Oceanogr. Mar. Biol. Annu. Rev.* **4,** 36–68.

Wyrtki, K. (1967). Circulation and water masses in the eastern equatorial Pacific Ocean. *Int. J. Oceanol. Limnol.* **1,** 117–147.

California Coastal Bottlenose Dolphins 23

Larry J. Hansen
Southeast Fisheries Center
Miami Laboratory
Coastal Resources Division
Miami, Florida 33149

INTRODUCTION

There are three forms of the bottlenose dolphin in the eastern North Pacific. The coastal form, occurring off southern California and Mexico, is distinguishable from the two offshore forms by differences in parasite faunas and cranial measurements (Walker, 1981). Ross (1977) has differentiated two types of *Tursiops* in southern African waters based mainly on cranial morphology, and Duffield *et al.* (1983) and Hersh and Duffield (Chapter 6, this volume) have been able to differentiate coastal and offshore varieties of bottlenose dolphins by hematology and morphology. In the Southern California Bight, the coastal form and one offshore form are parapatric. The systematics of the genus *Tursiops* are not resolved, but the coastal form off California does correspond to the nominal species *T. gilli* Dall, 1873 (Walker, 1981), which is considered by most workers (e.g., Mitchell, 1975) to be synonymous with *T. truncatus*. There are indications that some coastal populations of bottlenose dolphins may be small, genetically distinguishable stocks (Duffield *et al.*, 1987). With this possibility in mind, I began a study of the bottlenose dolphins found within a few kilometers of shore off southern California.

The historically recognized, "normal" range of the eastern Pacific coastal bottlenose dolphin is from about the northern end of Orange County, California (33° 30′ N), south along the coast of Mexico, and in the Gulf of California (Norris and Prescott, 1961; Walker, 1981; Leatherwood and Reeves, 1982). In addition, these dolphins probably occur much further south along the Pacific coast of Central America (Walker, 1981), and there is evidence that they may have occurred in the past in San Francisco Bay (Orr, 1963) and further north (Walker, 1981). They have been observed off Monterey (Wells *et al.*, Chapter 24, this volume) and between Point Mugu and Santa Barbara, California (Leatherwood *et al.*, 1987).

The presence of an inshore population of bottlenose dolphins along the coast of southern California was first noted in the scientific literature by Norris and Prescott (1961). Records of opportunistic sightings near La Jolla, California, have been kept by staff of the Southwest Fisheries Center (SWFC) since November 1971. The dolphins are seen there throughout the year, primarily just outside the surf zone. A project (sponsored by the SWFC) that used opportunistic sighting records collected by lifeguards in San Diego County was conducted during summer 1977, but my study was not initiated until September 1981.

Dohl *et al.* (1978) reported that during 1947–1973 the greatest number of sightings of California coastal and offshore bottlenose dolphins occurred from July through September, but the total number of animals was greatest from October through December. A similar pattern was noted by Dohl *et al.* (1978) for 32 sightings with herd sizes of 3–80 during 1975–1976 aerial surveys. These numbers are probably low because the survey transects were perpendicular to the shoreline and the distribution patterns of the coastal bottlenose dolphins. Norris and Prescott (1961) indicated that bottlenose dolphins were more common in summer along the coast north of San Diego Bay.

Prior to the passage of the Marine Mammal Protection Act in 1972, at least 25 bottlenose dolphins were taken from southern California waters. Seven taken from in and around San Diego Bay in the late 1950s (Norris and Prescott, 1961) were probably all coastal animals; 18 taken for public display in 1966–1972 by Marineland of the Pacific were all believed to have been offshore animals (Walker, 1975). Since 1972 only four southern California bottlenose dolphins have been taken legally, but these were all offshore animals (D. J. Seagers, pers. commun., 1987). During 1966–1986, at least 42 bottlenose dolphins stranded along the California coast from Santa Barbara south to the U.S.–Mexican border, and 8 of these may have been human-related mortalities (Bernard *et al.*, 1987; Seagers, *op. cit.*[1]).

The objectives of this study were (1) to make a population estimate, (2) to describe a minimum area of geographic distribution for some individuals, (3) to evaluate the home range of some individuals, (4) to estimate the population's reproductive rate, (5) to determine seasonal or other trends in group size and numbers, (6) to analyze the effects of the tidal state on group size and habitat use, and (7) to describe habitat-use patterns in the study area. Because the southern limit of the study area (the U.S.–Mexican border) was determined by geopolitical rather than zoogeographic factors, it could not be assumed that an entire, genetically isolated population was being studied. There probably was some movement of animals across the Mexican border, in either direction (north or south).

MATERIALS AND METHODS

Definitions

References to "dolphins" in this chapter are always to bottlenose dolphins. The "mark–recapture study area" is the area from the beach to about 400 m offshore between La Jolla and Oceanside Harbor (Fig. 1). The "extended study area" is from the Mexican border to Long Beach harbor from the beach out to 2900 m offshore, including San Diego Bay (Fig. 1). A "recognizable" dolphin is one whose dorsal fin is marked in such a manner that the animal can be distinguished from all other individuals. "Group" refers to any aggregation of one or more dolphins. "Minimum area of geographic distribution," or "range," refers to that area in which a dolphin or population of dolphins is known to occur. "Home range" is defined as a subunit of the "range," and the mark–recapture study area was considered part of a recognizable dolphin's home range if the dolphin was observed on five or more of the photographic surveys.

[1] Data from California Marine Mammal Stranding Network, NMFS, Southwest Region, 300 South Ferry Street, Terminal Island, California 90731.

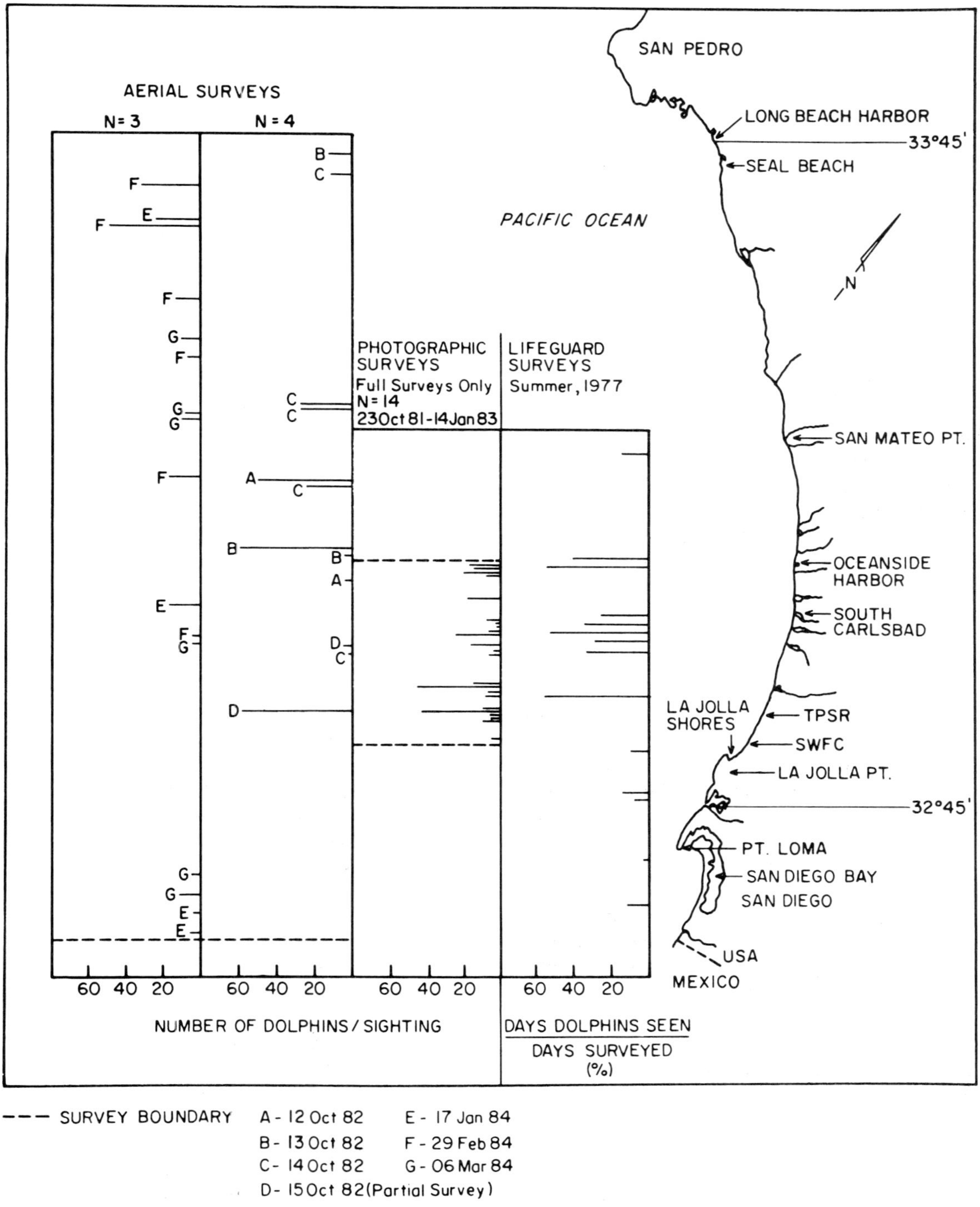

Figure 1 Map of the southern California coast from the Mexican border to San Pedro, which includes both the mark–recapture and extended study areas. Important geographic locations and main tributaries and lagoons are shown. The locations and sizes of dolphin groups are indicated for the aerial surveys and the photographic surveys. The frequency and location of bottlenose dolphin sightings by San Diego County lifeguards during 1977 are also shown.

Coastal bottlenose dolphins are thought to reside in discrete units in overlapping territories (Leatherwood and Reeves, 1982), and, for convenience, "population" is used to describe the dolphins found in the study area and does not refer to a condition of reproductive isolation. Based on Sergeant *et al.* (1973) and Ridgway and Benirschke (1977) two age classes were defined: calf, less than 1.5 m long; noncalf, greater than 1.5 m.

Types of Surveys and Sources of Data

Photographic Surveys

Photographic surveys allowed recognizable dolphins to be reidentified, in order to define the home range and number of dolphins in the mark–recapture study area. A total of 22 photographic surveys was conducted, 14 of the entire mark–recapture study area and 8 of portions thereof. Surveys were conducted from September 1981 through January 1983.

A 4.9-m outboard-powered Boston Whaler was used for the surveys. It was launched from the pier at the Scripps Institution of Oceanography in La Jolla, and then maneuvered north to Oceanside Harbor at about 22 km/hour, approximately 90–180 m offshore depending on the swell height and location of surf breaks. The southbound leg was usually about 1600–3600 m offshore at approximately 30 km/hour. The date, time at start, time at end, weather, and sea state were recorded for each survey.

When a group was sighted, photographs were taken until it was judged a good photograph of each dolphin had been obtained. The photographs were taken with a 35-mm Canon A-1 with a Canon 400-mm *f*4.5 lens or a Canon 80–300 mm *f*4.5 zoom lens; a motor drive was used at all times. Nearly all the photographs were taken with Kodachrome 64 film at a distance of 3–14 m. A few photographs were taken with Ektachrome 200 film. The time, location, numbers, age classes, behavior, and identity of any recognized individuals were recorded.

The photographs of recognizable individuals were sorted on a group by group basis. The best photographs from each sighting were catalogued by individual and dates sighted. Fins with few notches were difficult and sometimes impossible to use for recognizing individuals. Dolphins with such fins were considered nonrecognizable. Thus, there was a bias against resightings of poorly marked animals.

After all the photographs had been sorted, four methods were used to estimate the population size. All employed mark–recapture estimators for closed populations. Animals were considered captured when photographed, and marked if recognizable. The number of unrecognizable animals captured per survey was estimated by dividing the number of photographs of unrecognizable animals by the mean number of photographs per recognizable animal for that survey.

Two of the four estimation methods, those by Schnabel (1938) and Schumacher and Eschmeyer (1943), were developed from tag–recapture models. Both tend to provide low estimates because neither takes into account that not all of the animals in the population are subject to capture (Overton, 1971). The other methods, namely, those of Edwards and Eberhardt (1967) and Tanaka (1951), are based on a linear regression of frequency of capture; at least the former of these does assume that not all of the animals are subject to capture (Overton, 1971).

Schnabel's method assumes all animals are subject to capture, and that all of the animals have an equal probability of capture (Schnabel, 1938; Overton, 1971). Schumacher and Eschmeyer's (1943) method is similar to Schnabel's except it assumes animals have an unequal probability of capture (Seber, 1973). The population estimation formula for Schnabel's method was provided by Overton (1971, p. 442); the formula for Schumacher and Eschmeyer's method came from Seber (1973, p. 142). The confidence limits for Schnabel's method were calculated as shown by Overton (1971, p. 442); limits for Schumacher and Eschmeyer's method were calculated using a formula from Seber (1973, p. 142).

Edwards and Eberhardt's method assumes not all animals are subject to capture and provides an estimate of the number not captured by a linear regression of the number of animals captured x times against the number of times captured (1, 2, . . ., i times) (Edwards and Eberhardt, 1967). The confidence limits for the number of animals not captured were calculated by a standard formula (Sokal and Rohlf, 1973, p. 247). The other regression method used to estimate the population was developed by Tanaka (1951). It assumes a higher probability of capture for marked than for unmarked animals and requires that a plot of the log of proportion marked in the sample versus the log of the number marked before the sample was taken be fairly linear (Seber, 1973). The formulas use for estimating the population and calculating the confidence limits for Tanaka's method were provided by Seber (1973, p. 148).

Aerial Surveys

The aerial surveys were conducted to provide a minimum estimate of population size, information on offshore movements of the dolphins, and some information on their habitat preferences. They were flown in a Cessna Cardinal RG and usually covered the extended survey area. The survey team consisted of a pilot and two observers. The plane was flown at about 170 km/hour and approximately 150 m (for part of the first survey) and 215 m altitude (the rest of the surveys). Each survey consisted of two flights: one inshore (generally following the contour of the coast about 400 m out) and one approximately 1600 m off the coast. During the inshore flight the area from the surf zone to about 400 m offshore was surveyed; during the offshore flight the area from the outer edge of the inshore area to about 2900 m offshore was surveyed. On the inshore flight, both observers watched from the same side of the aircraft; on the offshore flights, they watched from opposite sides.

During the surveys, records were kept on survey effort, speed, altitude, positions, times, and environmental conditions. When a sighting was made, the aircraft circled the group several times while the dolphins were counted. The highest count was used as the estimate for each sighting.

Historical Sighting Records

Records of dolphin sightings were kept by knowledgeable staff at the SWFC and at the Torrey Pines State Reserve (TPSR), both near the south end of the mark–recapture study area (Fig. 1). Both locations are on bluffs about 90 m above sea level and provide excellent views of several miles of shoreline and adjacent sea. Sightings at both locations were made on a strictly opportunistic basis. No special effort was made to search actively for dolphins. It was assumed the records represented a random sample of dolphin sightings except possibly for number of sightings by time of day, as timing of lunch, morning, and afternoon break periods could have influenced the frequency of sightings. The sighting rate by time of day might have been positively biased during these periods, as staff would more likely be leaving offices and thus be in a better position to see dolphins. It is possible that the frequency of sightings in the winter underrepresented the frequency of dolphin occurrence owing to the poor sighting conditions at that time of year. Records of dolphin sightings were kept at the SWFC during November 1970–February 1983 and at TPSR during April 1978–March 1982.

Records from the SWFC and TPSR were reviewed for the following data: date, time, group size, age classes, direction of movement, and associated animals (such as sea lions or whales). The records were compared for six features of the sightings: (1) mean number of groups by month, (2) number of dolphins observed by month, (3) mean group size by month, (4) number of sightings by time of day, (5) mean group size by time of day, and (6) number of observations by group size. A one-way analysis of variance (ANOVA) was used to test the significance of the variation of number of dolphins observed by month at each location; a goodness-of-fit test was used for the other features, and the variation between locations was tested with a two-way ANOVA.

A tidal predictor program[2] that calculates the predicted tide for the south end of the study area was used to calculate the tidal state [tide height, above (plus) or below (minus) mean tide height, ebb or flood tide] for the SWFC sightings from January 1975 to February 1983. Tides were also calculated for 45 randomly chosen days on an hourly basis from 0500 to 2000 hours for the years

[2] Available from Scripps Institution of Oceanography, Shore Processes Laboratory, La Jolla, California 92038.

1975–1982. A variety of comparisons were made of various features of the SWFC sightings:

1. Mean group size by tide height
2. Number of sightings by month for ebb tides and flood tides
3. Mean group size by time of day for ebb tides and flood tides
4. Number of sightings by time of day for plus tides versus the normal distribution of plus tides by time of day
5. Number of sightings by time of day for minus tides versus the normal distribution of minus tides by time of day
6. Number of sightings for ebb tides and flood tides versus the expected number of ebb tides and flood tides
7. Mean number of calves by group size
8. Number of sightings of x number of calves per group
9. Number of sighting of groups with and without calves by month

The significance of the variation in features 1, 7, and 8 and in each component of the other comparisons was determined with a goodness-of-fit test. Two-way ANOVA was used to test the significance of the variation between the components of features 2–6 and 9.

Lifeguard Survey Data

During the summer of 1977, lifeguards along the coast of San Diego County participated in a survey of dolphin sightings. The lifeguards recorded all dolphin sightings, and the records were reviewed for frequency of sightings. A comparison by location was made of the frequency of days on which dolphins were seen (number of days seen/number of days surveyed).

Study Area

The study area has a coastline exposed to the full force of the ocean, with the exception of a few bays and artificial harbors (Fig. 1). Beaches are generally sandy and gently sloping. There are numerous rocky reefs and seagrass flats scattered throughout the area. There are kelp beds offshore, particularly between La Jolla and South Carlsbad and off Point Loma. There is a relatively low level of vessel traffic, mostly recreational, which increases on weekends. The highest density of vessel traffic usually occurs near harbor entrances. Average water temperature ranges from about 14° to 20°C, with the highest occurring from July through September and the lowest from December through May (Squire, 1978, 1982). Gray whales, *Eschrichtius robustus*, are fairly common during their southward winter migration. The California sea lion, *Zalophus californianus*, and the harbor seal, *Phoca vitulina*, are frequently found near the coast. Other marine mammals, including the common dolphin, *Delphinus delphis*, Pacific white-sided dolphin, *Lagenorhynchus obliquidens*, and pilot whale, *Globicephala* sp., are seen occasionally, but they normally occur further offshore. Concentrations of the northern anchovy, *Engraulis mordax*, are occasionally seen in or near the surf zone; feeding aggregations of fishes, pelicans, gulls, and other marine birds occasionally occur within 500 m of the coast.

RESULTS

Photographic Surveys

Dolphins were seen and photographed on all 22 photographic surveys except 2 covering the entire mark–recapture study area. Thirty-seven groups were observed, and about 2000 photographs were taken. In these photographs, 123 individual dolphins were distinguishable, and 197 recaptures were made. Of the 123 recognizable dolphins, 71 were seen more than once, and 21 were seen five or more times (Table 1). The distribution of the dolphin sightings was concentrated in the area between TPSR and South Carlsbad (Fig. 1). Thirty-six of the recognizable animals have been seen outside of the mark–recapture study area, and 20 of these, outside of the extended study area (8 of the 20 were also seen out of the mark–recapture study area but still within the extended study area). Of those outside both study areas, 11 were seen near Monterey, about 750 km north of the mark–recapture study area (Wells *et al.*, Chapter 24, this volume), and 3 near Ensenada, Mexico,

Table 1
Resightings of Bottlenose Dolphins along Southern California[a]

Fin	28 Sep 81	23 Oct 81	1 Nov 81	20 Nov 81	10 Dec 82	8 Jan 82	26 Jan 82	23 Apr 82	30 Apr 82	8 Jun 82	15 Jun 82	29 Jun 82	1 Jul 82	8 Jul 82	20 Jul 82	22 Jul 82	27 Jul 82	2 Aug 82	28 Oct 82	4 Nov 82	4 Jan 83	14 Jan 83	No. of times seen
	Survey Date																						
N		+	+			+			+			+	+					+		+		+	9
F				+		+						+	+		+		+	+			+	+	9
B		+				+	+		+		+	+			+			+			+		9
A		+		+		+	+		+			+						+			+		8
K		+					+		+						+			+			+	+	7
D			+				+				+					+	+	+				+	7
E		+		+		+	+		+									+			+		7
C															+	+	+	+			+	+	6
L												+			+		+	+		+		+	6
P		+					+		+								+	+				+	6
G									+			+						+			+	+	5
H		+	+										+					+			+		5
I									+		+						+	+		+			5
J			+												+			+		+	+		5
O				+									+			+		+				+	5
M												+			+	+		+			+		5
Q		+											+		+			+				+	5
U			+												+	+	+	+					5
R							+		+						+					+	+		5
S		+		+															+		+	+	5
T		+		+															+		+	+	5
Total	0	10	5	6	0	5	7	0	9	0	3	7	5	0	10	5	7	18	2	5	13	12	
Type[b]	P	F	F	F	F	F	F	P	P	P	P	P	F	F	P	F	F	P	F	F	F	F	

[a] Dolphins seen five or more times (on the dates indicated) are listed by fin code. +, Animal seen.
[b] Type of survey: P, partial; F, full.

Table 2
Summary of Data Used for Mark–Recapture Analyses

Survey No.	No. Captured (n_i)	No. of Resights in n_i (m_i)	No. Marked after n_{i-1}	No. of New Marks	No. Marked in Survey	Percent Marked in Survey
1[a]	46	0	0	37	37	80.4
2	23	6	37	8	14	60.9
3[b]	26	5	45	11	16	61.5
4	6	5	56	1	6	100.0
5[c]	20	10	57	4	14	70.0
6	43	14	61	19	33	76.7
7	13	6	80	3	9	69.2
8	16	8	83	5	13	81.3
9	20	8	88	4	12	60.0
10	15	10	92	1	11	73.3
11	12	8	93	1	9	75.0
12	33	19	94	9	28	84.9
13[d]	53	39	103	7	46	86.8
14	22	12	110	6	18	81.8
15	41	21	116	2	23	56.1
16	46	26	118	5	31	67.4
Total	435	197	123	123	320	

[a] Surveys for 28 Sep 81 and 23 Oct 81 were combined to form Survey 1.
[b] Surveys for 20 Nov 81 and 10 Dec 81 were combined to form Survey 3.
[c] Surveys for 26 Jan 82 and 23 Apr 82 were combined to form Survey 5.
[d] Surveys for 2 Aug 82 and 28 Oct 82 were combined to form Survey 13.

about 85 km south of La Jolla (Defran *et al.*).[3] The data used for the mark–recapture analyses are presented in Table 2, and the results are shown in Table 3.

Aerial Surveys

Four surveys were completed on consecutive days during October 1982. Three were of the extended study area and the last was of the area from the Mexican border to San Mateo Point (Fig. 1). During January–March 1984 three surveys were completed: two of the extended study area and one of only the inshore area from the Mexican border to Point Conception, California. Bottlenose dolphins were seen on all surveys but only during the inshore flights. A total of 29 dolphin groups (554 animals) was sighted, with the daily total of dolphins varying from 55 to 128, and a mean group size of 19.1. Locations of sightings and sizes of groups are shown on Fig. 1.

[3] Information concerning Defran *et al.* is taken from Southwest Fisheries Center Administrative Report LJ-86-36C (available from SWFC, P.O. Box 271, La Jolla, California 92038) and other presentations.

Table 3
Results of Mark–Recapture Estimators

Method	Population Estimate	Confidence Interval
Schnabel	173	150–199
Schumacher and Eschmeyer[a]	177	167–189
Edwards and Eberhardt	194	159–261
Tanaka	240	120–477

[a] Estimates number not marked at 71, confidence interval 36–138.

Historical Sighting Records

A total of 525 sightings of 9024 dolphins was recorded at the SWFC. Group size ranged from 1 to 139, with a mean of 19.5. Approximately 59% of the groups were of 1–15, 60% of 1–20, and 14% of 41 or more dolphins. The variation in the number of animals seen by month is not significant at $p < .05$. The variation in the mean number of groups observed by month and the mean group size by month is not significant at $p < .05$. The mean group size was low in the early morning, peaked from 1000–1100 hours, then dropped and gradually increased, but this variation was insignificant at $p < .05$. The number of sightings peaked in mid morning, dropped around noon, and then peaked again during the afternoon (Fig. 2); these differences were significant at $p < .05$.

The presence or absence of calves was noted in about 190 of the SWFC sightings; 7.15% of the dolphins in these sightings were calves. The number of calves per group ranged from 1 to 9, with most (43%) having one calf. The mean number of calves per group increased with group size, but not proportionally (number of calves increased somewhat geometrically). The number of sightings of groups with and without calves by month followed the same general pattern except for a peak in October of groups with calves; however, the variation within and between the number of groups with and without calves was not significant at $p < .05$.

The variability in the various features of the SWFC sightings at particular tidal states was not significant at $p < .05$ except for the mean group size by tide height (Fig. 3) and the number of sightings at ebb tides (247 or 57%) versus the number at flood tides (189 or 43%) compared to the normal distribution of ebb tides (46%) and flood tides (54%).

Sighting records at the TPSR during April 1978–March 1982 included 344 sightings of 6454 dolphins. The mean group size was 18, range 1–82. Fifty-six percent were of 1–15, 70% of 1–20, and only 6% of 41 or more dolphins. The variability in the mean group size by month and in the mean number of groups seen by month was not significant at $p < .05$. The variability in the number of dolphins seen by month was not significant at $p < .05$. The mean group size by time of day followed a pattern similar to SWFC sightings, but the variation was significant at $p < .05$ (Fig. 4).

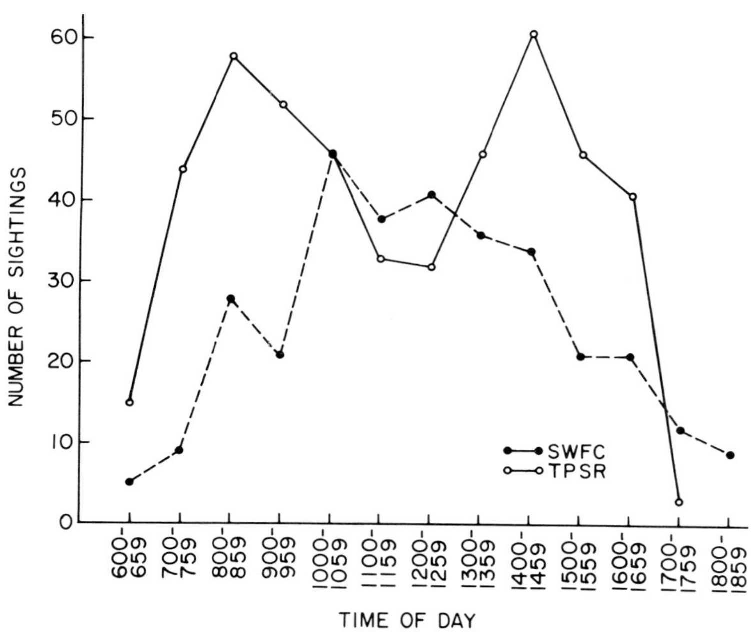

Figure 2 Number of sightings of bottlenose dolphin groups by time of day, from the Southwest Fisheries Center (SWFC) and the Torrey Pines State Reserve (TPSR).

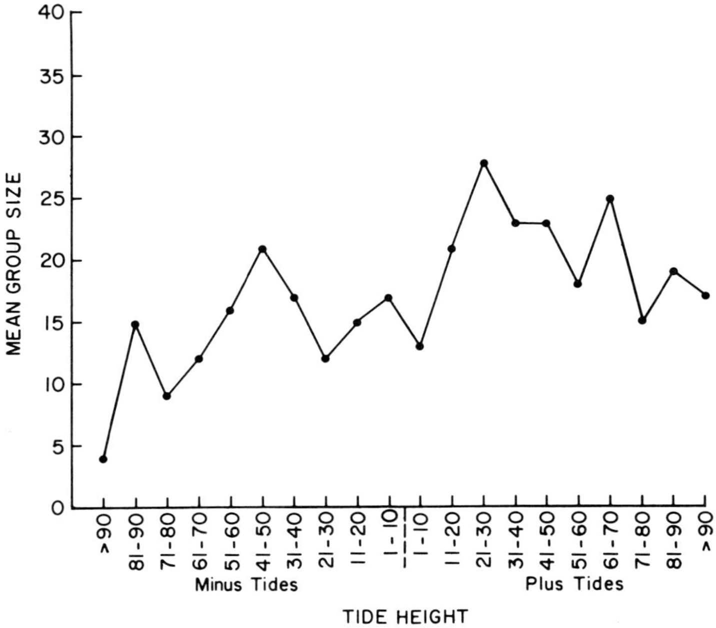

Figure 3 Mean group size of bottlenose dolphins by tide height (cm) observed from the Southwest Fisheries Center.

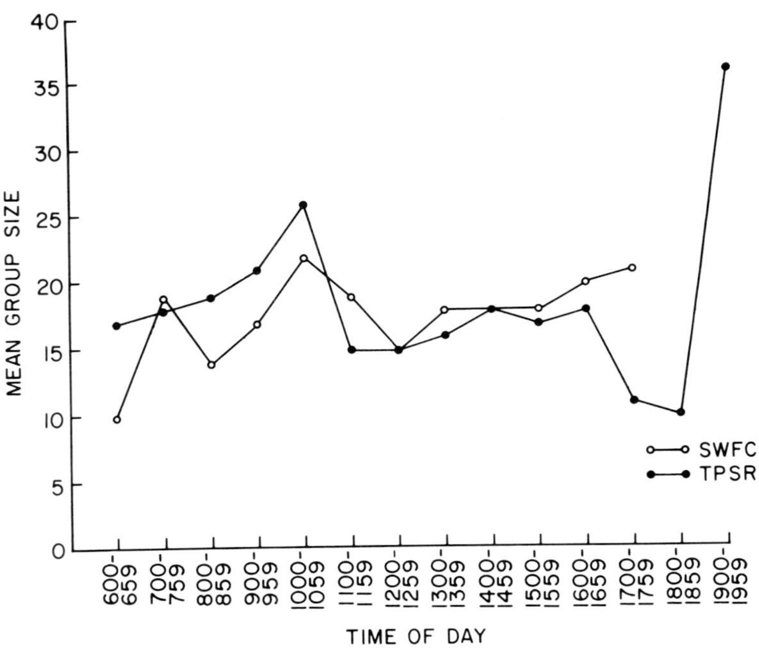

Figure 4 Mean group size of bottlenose dolphins by time of day observed from the Southwest Fishers Center (SWFC) and the Torrey Pines State Reserve (TPSR).

The number of sightings was low in the morning and afternoon, peaking just before noon, and the variation was significant at $p < .05$. Except for the number of animals seen by month, the variability in the various features of the sightings between the SWFC and the TPSR was not significant at $p < .05$. The SWFC and TPSR sightings are summarized in Table 4.

Lifeguard Survey Data

The lifeguards at fixed stations along the San Diego County coast recorded dolphin sightings on 245 of the 941 days surveyed. Most of the sightings were made in the northern part of the county between Oceanside and La Jolla (Fig. 1). The number of stations was greater in this area; however, dolphins were seen there on 41% of days surveyed versus 9% for the rest of the county.

Table 4
Summary of Sightings of Bottlenose Dolphins

	Total No. of Sightings		Total No. of Animals		Mean Group size	
Month	SWFC[a]	TPSR[b]	SWFC	TPSR	SWFC	TPSR
Jan	49	34	669	728	13	21
Feb	29	27	713	528	24	19
Mar	39	26	628	384	16	14
Apr	38	43	868	918	22	21
May	42	36	757	648	18	18
Jun	53	19	721	343	13	18
Jul	56	28	1060	507	18	18
Aug	39	27	501	419	12	15
Sep	48	17	956	260	19	15
Oct	49	36	966	722	19	20
Nov	39	33	785	676	20	20
Dec	44	18	400	321	9	17

[a] Southwest Fisheries Center.
[b] Torrey Pines State Reserve.

DISCUSSION

Population Estimate

The population of dolphins in the mark–recapture study area was estimated at 173–240 animals (Table 3), depending on the estimator used. Low estimates were given by the methods of Schnabel and Schumacher and Eschmeyer. According to Overton (1971), these estimators give low estimates because neither takes into account the fact that not all the animals are subject to capture. The method of Edwards and Eberhardt, which gave an estimate of 194, includes those animals not subject to capture (Edwards and Eberhardt, 1967; Overton, 1971). The highest estimate, 240, was given by the method of Tanaka. Tanaka's formula accounts for some, if not all, of the animals assumed by Edwards and Eberhardt's method not captured.

Tanaka's method takes into account animals not marked since it uses the number of animals captured during each capture event. Although the method of Edwards and Eberhardt does estimate the number not captured, the estimate it provided is probably low since it uses the number of animals marked per capture and not the total number captured. If all the animals captured had been marked, the estimate for Edwards and Eberhardt's method may have been larger. Therefore, the estimate given by Tanaka's method is probably the most reliable.

All of the mark–recapture methods used assume all the animals captured will be marked. Since this study relied on natural features for identification, that assumption was not met. However, Seber (1973) indicated that even though marking all anmals is the preferred situation, it is not necessary in order to use the mark–recapture estimators.

Population Range

The normal range of the population was considered to be the nearshore area from La Jolla to San Pedro (Fig. 1), thus an area along 155 km of coastline. This range was defined on the basis of several factors. The southern limit was based on historical-sighting, aerial-survey, and lifeguard-survey data. The SWFC records showed that whenever a group was observed traveling south and continuously watched until out of sight, it always turned north before reaching La Jolla Point and usually before it reached La Jolla Shores. Few dolphins were seen south of La Jolla on the aerial surveys, and the lifeguards south of La Jolla saw few dolphins (Fig. 1). Dolphins were apparently common in the San Diego Bay from 1950 to the mid 1960s and then absent for about 10 years, possibly because of increasing pollution levels in the bay (Food and Agriculture Organization, 1978, p. 112). Dolphins are currently seen in and near San Diego Bay, but the aerial survey data and my observations during other activities there indicate that dolphins are very rarely in the bay. Dolphin trainers working in San Diego Bay on a frequent basis report seeing dolphins (usually near the mouth of the bay) about four times a year; in 1985, one dolphin stayed in the bay for several months (S. Ridgway, pers. commun., 1986).

The topography of the sea bottom near the La Jolla area may explain why the dolphins seen from the SWFC usually did not travel south of this area. There are two deep canyons just offshore, and La Jolla Point is the only such northward-jutting point in the extended study area (Fig. 1). It has been theorized that delphinids may use the ambient acoustic characteristics of their environment to identify bottom topography (e.g., Norris, 1967; Evans, 1971). Würsig and Würsig (1979) found that dolphins in their Argentine study area used underwater geographical features as turning points, and Shane *et al.* (1986) reported that dolphins probably use underwater topography for navigation. Point Loma, a southward-jutting point south of the study area (Fig. 1), may serve as a marker to dolphins traveling northward. These two points may serve as boundary markers between populations of coastal dolphins. Since a few of the identified dolphins have been seen near Ensenada (Defran *et al.*[4]), it is clear that these boundaries are not inviolate and in fact may just coincidentally mark preferred areas.

The northern limit of the population was determined from the literature, aerial surveys, and observations by Walker (1981). Norris and Prescott (1961) described the northern end of the range as the northern border of Orange County, and Dohl *et al.* (1978) did not see any bottlenose dolphins along the coast north of San Pedro. Walker (1981) recorded few strandings of dolphins north of Seal Beach. All except one of the 29 sightings made during the aerial surveys occurred south of Seal Beach (Fig. 1). There is a large, jutting point at San Pedro (Fig. 1), which is just north of Seal Beach, that may serve as a northern boundary marker for the coastal dolphins. However, the northern end of the range may be subject to periodic extensions owing to local warm-water events.

Water temperature probably has more influence than topography in determining the range of this dolphin population. The areas at each end of the proposed range tend to be cooler (Squire, 1971, 1978). Cooler temperatures at the southern end are due to upwelling (Squire, 1978), and current eddies may cause cooler temperatures at the northern end (Owen, 1980). While bottlenose dolphins are animals of temperate to tropical waters (Leatherwood and Reeves, 1982), there is no information on specific temperature preferences. From November 1982 through January 1984, coastal bottlenose dolphins were seen as far north as Santa Cruz, some 450 km north of the proposed northern range limit (Wells *et al.*, Chapter 24, this volume). At least 11 of these animals have been seen in the study area, some of them numerous times. The waters were unusually warm during this period because of El Niño (Halpern *et al.*, 1983), and, therefore, the occurrence of dolphins in this area may have been anomalous. In addition, there were other, earlier northward penetrations of bottlenose dolphins along the U.S. Pacific coast that may have been related to local warm-water events (W. A. Walker, pers. commun., 1988). Bottlenose dolphins have been reported in North Atlantic waters off southern Greenland and Nova Scotia, "presumably in or near intrusions of temperate water under influence of the Gulf Stream" (Leatherwood and Reeves, 1982). Ross (1977) postulated that the distribution of bottlenose dolphins around the southeastern coast of Africa was related to warm-water conditions associated with a current.

The highest number of animals seen during a single aerial survey (128) was slightly more than half of the estimated total population (240). Considering the aerial survey covered the entire proposed range, this result requires some discussion. The proposed range may be part of a larger range which is separated into well-defined subranges by boundaries, such as La Jolla Point, La Jolla Canyon, and Point Loma. Thus, all members of the population may not be in the same subrange at a given time. The counts in the aerial surveys may have been low because of missed animals. For instance, dolphins could have been outside of the inshore strip during the inshore flight and then not seen during the offshore flight, or they may have been underwater and thus not visible. The most plausible explanation is that the actual range of the population is larger, and, although dolphins are present throughout the year, only a portion of the population is present in the proposed range at any given time. No statistically significant seasonal pattern has been found in the movements of these dolphins, and other researchers have noted that bottlenose dolphins will frequent an area for varying periods of time and then alter their pattern and relocate (Hogan, 1975; Shane, 1977; Würsig, 1978; Irvine *et al.*, 1981). In 1985, three of the identified dolphins were photographed off Ensenada, Mexico (Defran *et al.*[4]). Bottlenose dolphins are common near Ensenada, and the sighting of these animals indicates that the proposed range is definitely only part of a larger range and that the southern California population is not genetically isolated.

Caldwell (1955) provided the first evidence of a home range for bottlenose dolphins. Home ranges for bottlenose dolphins in the Sarasota Bay, Florida, area varied based on age and sex class (Wells, 1978, 1986; Wells *et al.*, 1980; Irvine *et al.*, 1981), and individual bottlenose dolphins have been resighted in this area for at least 15 years (Wells, 1986, p. 112; Scott *et al.*, Chapter 11, this volume). Shane (1980) found that bottlenose dolphins on the Texas coast had seasonal and year-round home ranges. Caldwell and Caldwell (1972) postulated that bottlenose dolphins may have seasonal home ranges joined by a traveling range.

During my study, it appeared the mark–recapture study area may have been part of a permanent home range for some dolphins and part of a seasonal home range for others. Of dolphins seen five or more times, some were seen generally throughout the study period while others were seen on a periodic basis (Table 1). The results of aerial surveys done on four consecutive days in October 1982 seem to indicate groups of dolphins did stay in or near the mark–recapture study area (Fig. 1). A report on dolphin photo-identification work in the mark–recapture study area from January 1984 through May 1986 (Defran *et al.*[4]) suggested that these coastal bottlenose dolphins exhibited no evidence of site fidelity for the mark–recapture study area. I believe that the differences observed in site fidelity are real and that the El Niño may have caused changes in the distribution and abundance of dolphins that used the mark–recapture study area. The El Niño began in the central and western Pacific in July 1982 and spread to the eastern Pacific where major effects occurred during late 1982 and 1983 (Halpren *et al.*, 1983). Considering that 86% of my photographic surveys were completed before November 1982, it appears that my study was nearly finished before the El Niño affected the study area. Perhaps what had been a stable community during my study was disrupted by the 1982–1983 El Niño and was still disrupted during the Defran *et al.* 1984–1986 study. The 1982–1983 El Niño caused changes in other marine mammal populations in southern California, including possible relocation of individuals, during 1983 and 1984 (DeLong and Antonelis, 1985; Ono *et al.*, 1985).

Behavior

The mean group size of dolphins in the study area was 18, with a range of 1–139; most groups were of 15 or fewer dolphins. Reported bottlenose dolphin group sizes are usually less than 100, with most in groups of 2–15 (Wells *et al.*, 1980; Leatherwood and Reeves, 1982; Shane *et al.*, 1986).

Seasonal and daily variation in group size does not appear to be significant. There was a pattern in the mean group size by month at SWFC and TPSR, but it was not significant at $p < .05$. The patterns in the number of animals seen by month were not significant at $p < .05$ at the SWFC and TPSR, but the variation between the numbers seen at the SWFC and TPSR was. This may be an artifact of some general difference in estimating numbers of dolphins between observers at the SWFC and TPSR. The mean group size by time of day showed the same general pattern at SWFC and TPSR, and it was significant only at TPSR and only if the largest mean group size by time, which is based on two sightings, is used. Shane *et al.* (1986) reported that "habitat structure and activity patterns" have the most influence on bottlenose dolphin group size and that "size tends to increase with increased water depth or openness of the habitat." In some coastal bottlenose dolphin populations in the Gulf of Mexico, the group composition and, thus, size changes often during the day (Wells *et al.*, 1980). The 1982 aerial survey data show the apparent stability and change in location and size of the same dolphin groups over a period of 4 days (Fig. 1). The group sizes were about the same for the first 2 days, changed on the third day, and then returned to that of the first 2 days. The locations of most animals changed daily.

Only two components of the tide seem to have affected the dolphins seen at SWFC. The variation in the mean group size by tide height was significant at $p < .05$, and there were significantly more sightings at ebb tides than at flood tides. The number of sightings at ebb tides and flood tides by month showed different patterns, with a peak of sightings during ebb tides in the winter months. However, the variation between the sightings at ebb tides and flood tides by month was not significant at $p < .05$. The mean group size at ebb tides and flood tides by time of day showed the same general pattern, but it was not statistically significant. The number of sightings at minus tides and at plus tides by time of day both showed definite patterns; however, the patterns were very similar to the normal distribution of minus and plus tides during the day. This similarity tends to validate the assumption that the records were collected at random, that is, the timing of breaks during the workday did not influence the sighting rates.

The reasons for the pattern in the mean group size by tide height are not readily evident. The mean group size is lowest (4) at the lowest tide height, and increases to 28 at a 30-cm plus tide, but drops to 17 at the highest tide (Fig. 3). Group size of bottlenose dolphins has been reported generally to increase as the water depth increases (Shane, 1977; Wells, 1978; Shane *et al.*, 1986). Larger group size is reported to be a more productive strategy for feeding on substantial but scattered food resources, such as schooling fishes in open waters (Scott and Chivers, Chapter 22, this volume); however, shallow inshore waters (such as are present throughout the study area) usually provide predictable food resources at reefs and other areas (Wells *et al.*, 1980). Bottlenose dolphins are reported to be opportunistic feeders with catholic tastes (Leatherwood and Reeves, 1982; Shane *et al.*, 1986), and Walker (1981) reported that a variety of inshore fishes and invertebrates were taken by California coastal bottlenose dolphins. Therefore, it is doubtful that feeding strategy is responsible for the variation in group size by tide height.

The reasons for more sightings of dolphins occurring during ebb tides than during flood tides is also unclear. Again, it appears that feeding strategy is not responsible. Of the SWFC sightings with tidal information where the presence or absence of feeding behavior was recorded (94 sightings), proportionally more feeding behavior occurred during flood tides than at ebb tides (feeding: 54% flood, 46% ebb; not feeding: 25% flood, 75% ebb). In addition, while feeding during ebb tides was split nearly equally between minus and plus tide heights, over 68% of the feeding during flood tides occurred during plus tide heights. There is a rocky reef below the SWFC, and dolphins have been observed chasing prey in and around the reef. It may be that during flood tides with above mean tide heights, prey associated with the reef is more available to the dolphins.

Bottlenose dolphins in the Aransas Pass, Texas, area stationed themselves against tidal flows for resting and feeding purposes (Shane, 1977, 1980; but see Shane, Chapter 31, this volume); however, this was in an area of protected bays and channels. The bottlenose dolphins in an Argentine bay tended to remain in shallower waters as the tide receded (Würsig and Würsig, 1979). Shane *et al.* (1986), who summarized tidal influence on coastal bottlenose dolphins, concluded that the "dolphins move with concentrations of food, move into shallow safe areas, move with or against the tide and show some . . . diurnal movement patterns," but that the "overriding theme is variability." They also believe it is probable bottlenose dolphins remember "particular areas very well and . . . when and where the best chances for finding prey" will be.

The distribution of observations of the dolphins in the study area was concentrated in the area between Torrey Pines and South Carlsbad (Fig. 1). Two features distinguish this area from the rest of the study area: offshore kelp beds and lagoon mouths. The prey species of eastern North Pacific bottlenose dolphins are generally available throughout the study area (e.g., North and Hubbs, 1968; Turner *et al.*, 1968; Feder *et al.*, 1974; Fitch and Lavenberg, 1975). However, the lagoons are probably relatively nutrient rich (e.g., Reid and Wood, 1976) and thus provide a greater abundance of prey. Additionally, the kelp beds provide habitat for a variety of fishes (North and Hubbs, 1968; Feder *et al.*, 1974). Therefore, the greater occurrence of dolphins near these lagoons and kelp beds may be related to the greater variety and abundance of prey species.

Reproduction

The estimated reproductive rate (percent calves in the population), 7.15%, is slightly below the mean of percentages given for other dolphin populations. The estimated percentages of calves in U.S. coastal bottlenose dolphin populations varied from 3.65% to 15.6%, with a mean of about 8% (Leatherwood and Reeves, 1982). Bottlenose dolphins off southern California have been shown to have the highest concentrations of DDT (dichlorodiphenyltrichloroethane) residues of any wild population of marine mammals (O'Shea *et al.*, 1980; Schafer *et al.*, 1984), and this may be having negative effects on their reproductive rate. DDT contamination may have been a factor in causing premature pupping in another marine mammal, the California sea lion (DeLong *et al.*, 1973; Gilmartin *et al.*, 1976; Reijnders, 1984), and it has affected reproductive rates in other animals (Peakall, 1970; Deichman and MacDonald, 1971).

In the study area, the majority of groups with calves included only one calf. The mean number of calves per group tended to increase geometrically, rather than proportionally, to the group size.

The calving interval of bottlenose dolphins is reported to be about 2 years (McBride and Kritzler, 1951; Mitchell, 1975). Spring and fall calving seasons have been postulated for Gulf coast waters of Mississippi and Florida (Harrison and Ridgway, 1971; Odell, 1975). Breeding and calving seasons are generally considered to be year-round, with peaks varying by area (Leatherwood and Reeves, 1982). Variation in the number of sightings of calves by month in my study area was not significant at $p < .05$, but there was a definite peak in October. Strandings of dolphin calves and females in the study area peak just before October (Henderson and Hansen, 1983). Thus, it appears calving occurs year-round in the study area, with a peak in the fall.

SUMMARY

The population of bottlenose dolphins in the study area at any time was less than the total number (240) estimated to use the area. Some individuals ranged from La Jolla north to at least Monterey, but Seal Beach was the normal northern limit of the range. A few individuals have been observed in Mexican waters, and interbreeding probably occurs with populations south of the study area. The area from La Jolla to Oceanside Harbor was probably part of a temporary or seasonal home range for some dolphins. The number and occurrence of dolphins did not appear to follow a seasonal pattern. Mean group size by month showed similar patterns at TPSR and SWFC, but the monthly variation was not statistically significant. The tide height influenced group size (generally, smaller at lower tides). It may have been advantageous for the dolphins to be in certain areas depending on the tidal state (ebb or flood, plus or minus tide). The dolphins preferred a particular portion of the study area (between TPSR and South Carlsbad), and this preference may be related to food resources. Breeding and calving occurred year-round, with a peak in the fall. Calves comprised slightly more than 7% of the population.

ACKNOWLEDGMENTS

The National Marine Fisheries Service (NMFS) is required by federal law to provide information on the population status of marine mammals to ensure proper management. Because the objectives of this study met several of the management needs of NMFS, all funding, equipment, and personnel for the field portion of this study were provided by NMFS through the Southwest

Fisheries Center, La Jolla, California. Many people contributed to the success of this study, and I thank them all, particularly Drs. W. F. Perrin, D. P. DeMaster, G. R. Trapp, and P. E. Hansen. The use of trade names does not indicate endorsement by NMFS. NOAA/NMFS/SEFC/Miami Laboratory Coastal Resources Division Contribution CRD-87/88-08.

REFERENCES

Bernard, H. J., Gilpatrick, J. W., Hansen, L. J., and Henderson, J. R. (1987). Cetacean strandings along the San Diego County, California, coast from 1966–86. Presented at Marine Mammal Stranding Workshop, 3–5 December 1987, University of Miami, Florida.

Caldwell, D. K. (1955). Evidence of home range of an Atlantic bottlenose dolphin. *J. Mammal.* **36,** 305–306.

Caldwell, D. K., and Caldwell, M. C. (1972). "The World of the Bottlenose Dolphin." Lippincott, New York. 157 pp.

Deichman, W. B., and MacDonald, W. E. (1971). Organochlorine pesticides and human health. *Food Cosmet. Toxicol.* **9**(1), 91–103.

DeLong, R. L., and Antonelis, G. A., Jr. (1985). Impact of the 1982–83 El Niño on the northern fur seal population at San Miguel Island. *Sixth Bienn. Conf. Biol. Mar. Mamm., 22–26 November 1985,* Vancouver, B.C., Abstr.

DeLong, R. L., Gilmartin, W. G., and Simpson, J. G. (1973). Premature pupping in California sea lions: Association with high organochlorine pollutant residue levels. *Science* **181,** 1168–1170.

Dohl, T. P., Norris, K. S., Guess, R. C., Bryant, J. D., and Honig, M. W. (1978). "Pinnipedia, Cetacea, and Parasitology of the Southern California Bight Area" Vol. III, No. PB-295-932. Natl. Tech. Information Service, Springfield, Virginia 22161. 472 pp.

Duffield, D. A., Ridgway, S. H., and Cornell, L. H. (1983). Hematology distinguishes coastal and offshore forms of dolphins (*Tursiops*). *Can. J. Zool.* **61,** 930–933.

Duffield, D. A., Wells, R. S., Odell, D. K., and Solangi, M. A. (1987). Biochemical genetics of wild bottlenose dolphin populations—Implications for population structure and use. *Seventh Bienn. Conf. Biol. Mar. Mamm., 5–9 December, 1987,* Miami, Florida, Abstr.

Edwards, W. R., and Eberhardt, L. L. (1967). Estimating cottontail abundance from livetrapping data. *J. Wildl. Manage.* **31**(1), 87–96.

Evans, W. E. (1971). Orientation behavior of delphinids: Radio telemetric studies. *Ann. N.Y. Acad. Sci.* **188,** 142–160.

Feder, H. M., Turner, C. H., and Limbaugh, C. (1974). Observations on fishes associated with kelp beds in southern California. *Calif. Dept. Fish Game, Fish Bull.* **160,** 1–144.

Fitch, J. E., and Lavenberg, R. J. (1975). "Tidepool and Nearshore Fishes of California." Univ. Calif. Press, Berkeley. 156 pp.

Food and Agriculture Organization. (1978). "Mammals in the Seas," Vol. 1, FAO Fish Ser. No. 5. Food and Agriculture Organization of United Nations, Rome. 264 pp.

Gilmartin, W. G., DeLong, R. L., Smith, A. W., Sweeney, J. C., de Lappe, B. W., Risebrough, R. W., Griner, L. A., Dailey, M. D., and Peakall D. B. (1976). Premature parturition in the California sea lion. *J. Wildl. Dis.* **12,** 104–115.

Halpren, D., Hayes, S. P., Leetmaa, A., Hansen, D. V., Philander, S.G. H. (1983). Oceanographic observations of the 1982 warming of the tropical eastern Pacific. *Science* **221,** 1173–1175.

Harrison, R. J., and Ridgway, S. H. (1971). Gonadal activity in some bottlenose dolphins (*Tursiops truncatus*). *J. Zool.* **165,** 355–366.

Henderson, J. R., and Hansen, L. J. (1983). Stranded marine mammals recovered by the Southwest Fisheries Center, 1966–1980. Southwest Fisheries Center, Admin. Rep. No. LJ-83-07, La Jolla, California 92038. 31 pp.

Hogan, T. (1975). Movements and behavior of the bottlenose dolphin in the Savannah River mouth area. M.S. thesis, Univ. of Rhode Island, Kingston, Rhode Island. 42 pp.

Irvine, A. B., Scott, M. D., Wells, R. S., and Kaufmann, J. H. (1981). Movements and activities of the Atlantic bottlenose dolphin, *Tursiops truncatus,* near Saratosa, Florida. *Fish Bull.* (*U.S.*) **79,** 671–688.

Leatherwood, S., and Reeves, R. R. (1982). Bottlenose dolphin *Tursiops truncatus* and other toothed cetaceans. *In* "Wild Mammals of North America: Biology, Management, Economics" (J. A. Chapman, and G. A. Feldhammer, eds.), pp. 369–414. Johns Hopkins Univ. Press, Baltimore, Maryland. 1147 pp.

Leatherwood, S., Stewart, B. S., and Folkens, P. A. (1987). Cetaceans of the Channel Islands National Marine Sanctuary. Prepared under Purchase Order 83-ABA-0271, NOAA, Channel Islands National Marine Sanctuary, Santa Barbara, California 93101. 66 pp.

McBride, A. F., and Kritzler, H. (1951). Observations on pregnancy, parturition, and postnatal behavior in the bottlenose dolphin. *J. Mammal.* **32,** 251–266.

Mitchell, E., ed. (1975). Review of biology and fisheries for smaller cetaceans. *J. Fish. Res. Board Can.* **32,** 899–983.

Norris, K. S. (1967). Some observations on the migration and orientation of marine mammals. *In* "Animal Orientation and Navigation" (R. H. Storm, ed.), pp. 101–125. Oregon State Univ. Press, Corvallis, Oregon.

Norris, K. S., and Prescott, J. H. (1961). Observations on Pacific cetaceans of Californian and Mexican waters. *Univ. Calif. Publ. Zool.* **63,** 291–402.

North, W. J., and Hubbs, C. L. (1968). Utilization of kelp-bed resources in southern California. *Calif. Dept. Fish Game, Fish Bull.* **139,** 1–264.

Odell, D. K. (1975). Status and aspects of the life history of the bottlenose dolphin, *Tursiops truncatus,* in Florida. *J. Fish. Res. Board Can.* **32,** 1055–1058.

Ono, K. A., Oftedal, O. T., Boness, D. J., and Iverson, S. J. (1985). Effects of the 1983 El Niño on the mother–pup relationship in the California sea lion. *Sixth Bienn. Conf. Biol. Mar. Mamm., 22–26 November 1985,* Vancouver, B.C., Abstr.

Orr, R. T. (1963). A Northern record for the Pacific bottlenose dolphin. *J. Mammal.* **44**(3), 424.

O'Shea, T. J., Brownell, R. L., Clark, D. R., Walker, W. A., Gay, M. L., and Lamont, T. G. (1980). Organochlorine pollutants in small cetaceans from the Pacific and South Atlantic oceans, November 1968–June 1976. *Pesticides Monitoring J.* **14**(2), 35–46.

Overton, W. S. (1971). Estimating the numbers of animals in wildlife populations. *In* "Wildlife Management Techniques" (R. H. Giles, ed.), pp. 403–456. The Wildlife Society, Washington, D.C. 633 pp.

Owen, R. W. (1980). Eddies of the California current system: Physical and ecological characteristics. *In* "The California Islands: Proceedings of a Multidisciplinary Symposium" (D. Power, ed.), pp. 237–262. Santa Barbara Mus. Nat. Hist., Santa Barbara, California. 787 pp.

Peakall, D. B. (1970). Pesticides and the reproduction of birds. *Sci. Am.* **222,** 73–78.

Reid, G. K., and Wood, R. D. (1976). "Ecology of Inland Waters and Estuaries." Van Nostrand, New York. 485 pp.

Reijnders, P. J. H. (1984). Man-induced environmental factors in relation to fertility changes in pinnipeds. *Environ. Conserv.* **11**(1), 61–65.

Ridgway, S. H., and Benirschke, K. W., eds. (1977). Breeding dolphins: Present status, suggestions for the future. NTIS No. PB-273 673, U.S. Dept. of Commerce, Natl. Tech. Information Service, Arlington, Virginia. 308 pp.

Ross, G. J. B. (1977). The taxonomy of bottlenosed dolphins *Tursiops* sp. in South African waters, with notes on their biology. *Ann. Cape. Prov. Mus. (Nat. Hist.)* **11,** 135–194.

Schafer, H.A., Gossett, R. W., Ward, C. F., and Westcott, A. M. (1984). Chlorinated hydrocarbons in marine mammals. *In* "Southern California Coastal Water Research Project Biennial Report, 1983–1984" (W. Bascom, ed.), pp. 109–114. Long Beach, California 90806.

Schnabel, Z. E. (1938). Estimation of the total fish population of a lake. *Am. Math. Monthly* **45,** 348–352.

Schumacher, F. X., and Eschmeyer, R. W. (1943). The estimation of fish populations in lakes or ponds. *J. Tenn. Acad. Sci.* **18,** 228–249.

Seber, G. A. F. (1973). "The Estimation of Animal Abundance." Hager Press, New York. 506 pp.

Sergeant, D. E., Caldwell, D. K., and Caldwell, M. C. (1973). Age, growth, and maturity of bottlenosed dolphin (*Tursiops truncatus*) from northeast Florida. *J. Fish. Res. Board Can.* **30,** 1009–1011.

Shane, S. H. (1977). The population biology of the Atlantic bottlenose dolphin, *Tursiops truncatus,* in the Aransas Pass area of Texas. M.S. thesis, Texas A & M University, College Station, Texas 77843. 239 pp.

Shane, S. H. (1980). Occurrence, movements, and distribution of bottlenose dolphins, *Tursiops truncatus,* in southern Texas. *Fish. Bull. (U.S.)* **78,** 593–601.

Shane, S. H., Wells, R. S., and Würsig, B. (1986). Ecology, behavior and social organization of the bottlenose dolphin: A review. *Mar. Mamm. Sci.* **2**(1), 34–63.

Sokal, R. R., and Rohlf, F. J. (1973). "Introduction to Bio-Statistics." Freeman, San Francisco. 368 pp.

Squire, J. L., Jr. (1971). Measurements of sea surface temperature on the eastern Pacific continental shelf using airborne infrared radiometry August 1963–July 1968. Oceanographic Report No. 47. U.S. Coast Guard Oceanographic Unit, Washington, D.C. 229 pp.

Squire, J. L., Jr. (1978). Sea surface temperature distributions obtained off San Diego, California, using an airborne infrared radiometer. NOAA Tech. Rep. NMFS SSRF-720. U.S. Govt. Printing Office, Washington, D.C. 30 pp.

Squire, J.L., Jr. (1982). Catch temperatures for some important marine species off California. NOAA Tech. Rep. NMFS SSRF-759. U.S. Govt. Printing Office, Washington, D.C. 19 pp.

Tanaka, R. (1951). Estimation of vole and mouse populations on Mt. Ishiauchi and on the uplands of southern Shikoku. *J. Mammal.* **32**(4), 450–458.

Turner, C. H., Ebert, E. E., and Given, R. R. (1968). The marine environment offshore from Point Loma, San Diego County. *Calif. Dept. Fish Game, Fish Bull.* **140,** 1–85.

Walker, W. A. (1975). Review of the live-capture fishery for smaller cetaceans taken in southern California waters for public display, 1966–73. *J. Fish. Res. Board Can.* **32**(7), 1197–1211.

Walker, W. A. (1981). Geographical variation in morphology and biology of bottlenose dolphins (*Tursiops*) in the eastern North Pacific. NMFS/SWFC Admin. Rep., No. LJ-81-03C. 52 pp.

Wells, R. S. (1978). Home range characteristics and group composition of Atlantic bottlenose dolphins, *Tursiops truncatus*, on the west coast of Florida. M.S. thesis, University of Florida, Gainesville, Florida 32611.

Wells, R. S. (1986). Structural aspects of dolphin societies. Ph.D. Dissertation, University of California, Santa Cruz, California. 234 pp.

Wells, R. S., Irvine, A. B., and Scott, M. D. (1980). The social ecology of inshore odontocetes. *In* "Cetacean Behavior: Mechanisms and Functions" (L. M. Herman, ed.), pp. 263–317. Wiley, New York. 463 pp.

Würsig, B. (1978). Occurrence and group organization of Atlantic bottlenose porpoise (*Tursiops truncatus*) in an Argentine bay. *Biol. Bull.* **154,** 348–359.

Würsig, B., and Würsig, M. (1979). Behavior and ecology of bottlenose porpoises, *Tursiops truncatus*, in the South Atlantic. *Fish. Bull. (U.S.)* **77,** 871–890.

Northward Extension of the Range of Bottlenose Dolphins along the California Coast

24

Randall S. Wells
Long Marine Laboratory
University of California
Santa Cruz, California 95060

Larry J. Hansen
Southeast Fisheries Center
National Marine Fisheries Service
Miami, Florida 33149

Alan Baldridge
Hopkins Marine Station
Stanford University
Pacific Grove, California 95060

Thomas P. Dohl
Long Marine Laboratory
University of California
Santa Cruz, California 95060

Dennis L. Kelly
Orange Coast College
Costa Mesa, California 92628

R. H. Defran
Department of Psychology
San Diego State University
San Diego, California 92182

INTRODUCTION

The movement patterns of bottlenose dolphins range from year-round residency in a well-defined home range to seasonal migrations (Shane *et al.*, 1986). These movement patterns appear to be linked, either directly or indirectly, to variations in water temperature. Along the east coast of the United States bottlenose dolphins inhabiting the northern extremes of the species' range migrate seasonally (True, 1891), whereas permanently resident communities can be found in at least some locations in the warmer waters of the Gulf of Mexico (Wells, 1986).

Along the west coast of the United States, records from the nineteenth century suggest that bottlenose dolphins may have occurred as far north as the Monterey and San Francisco Bay areas of California. The type specimen for *Tursiops gilli* was collected in Monterey Bay in 1871 (Dall, 1873). True (1889) reported another specimen from the same locality, but the collection date is unknown. A partial cranium dredged from San Francisco Bay and believed to be 50–100 years old was reported by Orr (1963). Walker (1981) reported the dredging in 1980 of another cranium from San Francisco Bay which he believed to be of much more recent origin than the skull reported by Orr.

In the recent past, bottlenose dolphins did not appear to range normally as far north as indicated by the historical accounts. Norris and Prescott (1961) reported that they were absent from the area where Dall's 1871 specimen was taken and uncommon north of Orange County, California. Based on review of earlier records, Leatherwood and Reeves (1982) considered southern Los Angeles County the northern limit of regular occurrence. Bottlenose dolphins apparently were absent from Monterey Bay during 1950–1955 (Barham, 1982). They were judged to be year-round inhabitants of the Southern California Bight during 1975–1978 (Dohl *et al.*, 1980), and observations were made as far north as Point Dume, California (Dohl *et al.*, 1980). More recently, the normal range of the nearshore bottlenose dolphin population along the California coast has been considered to extend approximately 155 km from La Jolla northward to San Pedro (Hansen, Chapter 23, this volume). Repeated sightings suggested that at least some members of this population were resident in this range over a minimum period of several years.

This chapter examines a shift in the range of bottlenose dolphins in apparent response to changes in water temperature. During the El Niño warm-water incursion of 1982–1983 (Linder and Breaker, 1983; Fiedler, 1984), bottlenose dolphins were identified in and near Monterey Bay, more than 600 km north of their normal range. Following the 1982–1983 El Niño, they continued to use waters to the north of their previous range. The distance traveled by these dolphins represents the longest documented movement for the species. The unusual shift in range (1) underscores the behavioral flexibility of the species, (2) suggests a mechanism for range expansion for the species, (3) provides new information on the stability of social associations between dolphins, and (4) offers a possible explanation for the historical records of specimens far north of the species' recent "normal" distribution.

DATA SOURCES

The normal pattern of bottlenose dolphin distribution along the southern California coast was determined from 71 marine mammal and seabird aerial surveys conducted by Dohl and colleagues during 1975–1978 (Dohl *et al.*, 1980). Transect lines were established and maintained throughout the 3-year study. These lines were 27.8 km apart and covered waters from the coast to 222.2 km offshore, from Pt. Conception to the U.S.–Mexican border. An aircraft with high-wing configuration was flown at 167 km/hour, at altitudes of 300 m (mammals) and 60 m (birds).

Similarly, 72 systematic aerial surveys for marine mammals and seabirds were conducted by Dohl and colleagues off central and northern California during 1980–1983 (Dohl *et al.*, 1983). Transect lines were established between Pt. Conception and the California–Oregon border. These lines were 9.3 km apart from the coast to 166.7 km offshore. Additional aerial surveys were conducted by Dohl over portions of these areas from 1984 through 1988.

A network for collecting reports of unusual sightings of marine mammals and seabirds along the central California coast has been coordinated by Baldridge since 1966. Many of the sightings listed in Table 1 as well as the identification photographs from Cannery Row were obtained through this network.

Regular vessel surveys using photographic identification techniques (e.g., Würsig and Würsig, 1977) have been conducted in southern California waters since 1981. Three research teams using small boats and 35-mm cameras with 200–400-mm telephoto lenses have worked along the coast between Santa Barbara and San Diego. Hansen (Chapter 23, this volume) conducted 22 surveys from September 1981 through January 1984 in the San Diego area. Defran *et al.* (1987) have continued the survey program initiated by Hansen. From January 1984 through December 1987 they conducted 97 surveys in the San Diego area and 10 around Ensenada, Mexico. In addition, they initiated a vessel survey program near Santa Barbara (Wolf *et al.*, 1987). During May–December 1987, 12 surveys were conducted in the Santa Barbara area. Kelly conducted 35 surveys in the Orange County area between Anaheim Bay and San Onofre Beach during the period October 1982–February 1986 (Defran *et al.*, 1987).

Individual dolphins were also identified in Monterey Bay from photographs. Wells obtained identification photographs from the shore in Santa Cruz on 2 days using a 35-mm camera with a 200-mm telephoto lens.

MOVEMENTS BY BOTTLENOSE DOLPHINS NORTH OF THE HISTORICAL RANGE

Prior to 1983, bottlenose dolphins were rarely seen north of the greater Los Angeles area. From May 1983 through April 1988, however, they were seen

64 times north of Point Conception (Table 1, Fig. 1), and on 3 occasions between Ventura and Santa Barbara, south of Point Conception (Wolf *et al.*, 1987).

The first sightings were during May–June 1983. They involved small numbers of dolphins seen three times near San Simeon. These sightings involved at least 6 dolphins, if all three sightings were of the same individuals, or 14 dolphins, if the sightings were of different animals. No photographs are available for comparing the compositions of the groups sighted.

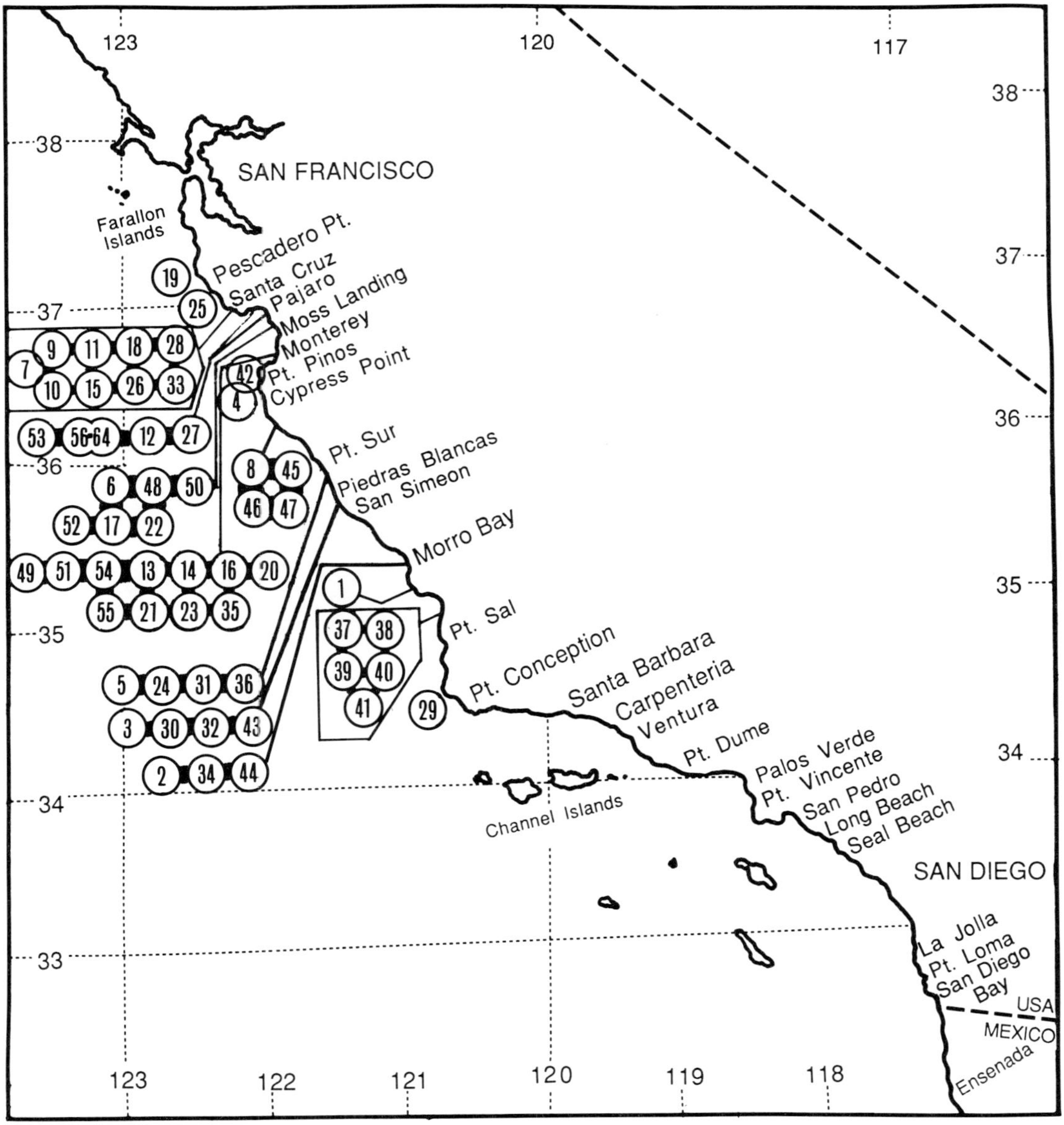

Figure 1 Map of the southern and central California coastline. Encircled numbers refer to the dolphin sightings listed in Table 1.

Table 1
Sightings of Bottlenose Dolphins along the Central California Coast, 1983–1988

#	Date	Time	Location	Latitude/Longitude	Number, Total/Calf	Dir.	Notes	Observers
1983								
1	May 18	854	Shell Beach	35° 09′ N 120° 40′ W	3	—	Milling near surfline	T. Dohl
2	Jun 8	953	Cayucos Pier	35° 25′ N 120° 52′ W	6	NW	Within surfline	T. Dohl
3	Jun 18	1022	San Simeon Cove	35° 38′ N 121° 12′ W	5	NNW	0.5 km offshore	T. Dohl
4	Sep 18	—	27.8 km W of Cypress Pt.	36° 35′ N 122° 14′W	30	—	—	K. Glover, P. Steele
5	Oct 25	750	7.4 km N of Pt. Sur	36° 22′ N 121° 55′ W	8	N	25 m offshore, slow	T. Dohl
6	Oct 27	1600	Moss Landing	36° 48′ N 121° 48′ W	6	N	In surf zone	J. Oakden, E. Baldridge
7	Oct 29	1700	West Cliff, Santa Cruz	36° 47′ N 122° 02′ W	9/1	W	Milling in surf zone	R. Wells, C. Schilt
8	Oct 31	1030	Carmel Beach	36° 33′ N 121° 56′ W	3/1	—	Milling just offshore	A. Baldridge
9	Nov 3	1030	Terrace Pt., Santa Cruz	36° 57′ N 122° 03′ W	6/1	W	Milling just offshore	R. Wells, K. Norris
10	Nov 4	1125	West Cliff, Santa Cruz	36° 57′ N 122° 02′ W	8–9/1	E	Close to shore	R. Wells, K. Fischer
11	Nov 7	1425	Cow Beach, Santa Cruz	36° 58′N 122° 07′ W	3	S	Close to shore	T. Dohl
12	Nov 9	900	Pajaro Dunes	36° 52′ N 121° 49′ W	9–10	S	Surfing	J. Warriner, R. Warriner
13	Nov 9	—	Monterey State Beach	36° 37′ N 121° 52′ W	5	—	—	Holiday Inn staff
14	Nov 11	—	Monterey State Beach	36° 37′ N 121° 52′ W	5	—	—	Holiday Inn staff
15	Nov 15	1635	Terrace Pt., Santa Cruz	36° 57′ N 122° 03′ W	2	—	Inside surfline	K. Miller
16	Nov 16	1100	Monterey State Beach	36° 37′ N 121°52′ W	2	S	—	A. Baldridge
17	Nov 18	945	Moss Landing	36° 48′ N 121° 48′ W	3–5	S	—	B. Würsig
18	Nov 20	1000	West Cliff, Santa Cruz	36° 57′ N 122° 02′ W	7–9/1	W	Milling	R. Wells, K. Miller
19	Nov 30	955	Pescadero Pt.	37° 14′ N 122° 24′ W	5	NW	0.5 km offshore	T. Dohl
20	Dec 23	—	Monterey State Beach	36° 37′ N 121° 52′ W	2	—	Seen over 3 days	R. Royer, M. Radakovich
21	Dec 30	—	Monterey State Beach	36° 37′ N 121° 52′ W	2	—	—	D. Shearwater, D. Serdeheley
1984								
22	Jan 5	1300	Moss Landing	36° 48′ N 121° 48′ W	7	S	Inside surf zone	Marine Lab. staff
23	Jan 7	700	Cannery Row, Monterey	36° 37′ N 121° 52′ W	25–50	—	Milling near kelp	P. Steele, J. Foott
24	Jan 12	1330	Pacific Valley	35° 36′ N 121° 29′ W	25–40	S	Followed 23.2 km	R. Jameson
25	May 24	1112	Waddell Creek	37° 05′ N 122° 18′ W	5	W	Moving slowly	T. Dohl
26	Jun 9	715	Wilder Beach, Santa Cruz	36° 55′ N 122° 05′ W	8	SE	Moving rapidly	T. Dohl
27	Aug 22	840	Pajaro Dunes	36° 56′ N 121° 52′ W	6	N	0.5 km offshore	T. Dohl
28	Sep 19	1020	Sand Hill, Santa Cruz	36° 59′ N 122° 08′ W	9	—	Milling 50 m offshore	T. Dohl
29	Nov 22	745	Pt. Conception	34° 27′ N 120° 28′ W	3	N	Moving rapidly	R. Hubbs
30	Dec 24	1152	16.7 km S of Piedras Blancas	—	10/2	S	Milling near surfline	J. Brodkin

1985								
31	Jan 1	—	Big Sur River mouth	—	10–15	—	Milling near surfline	J. Stein
32	Jan 2	930	Pt. Piedras Blancas	35° 38′ N 121° 12′ W	8	S	Moving slowly near surfline	J. Brodkin
33	Jan 3	1700	Terrace Pt., Santa Cruz	36° 57′ N 122° 04′ W	3–5/0	ENE	Moving slowing, 50 m offshore	K. Marten, K. Englund
34	Sep 27	1300	Morro Bay	35° 24′ N 120° 48′ W	4	N	Moving rapidly	B. Blackston
1986								
35	Dec 6	630	Monterey State Beach	36° 37′ N 121° 52′ W	3–5	N	Moving slowly near surfline	D. VenTresca
1987								
36	Jan 1	—	0.9 km S of Big Sur River	—	12 15	—	Milling in breakers	N. Fraley
37	Mar 27	—	Pt. Sal	34° 58′ N 120° 40′ W	11	—	Seen from aircraft	T. Dohl
38	Mar 27	—	Pt. Sal	35° 02′ N 120° 38′ W	8	—	Seen from aircraft	T. Dohl
39	Mar 28	—	Pt. Sal	35° 05′ N 120° 39′ W	12	SSE	Seen from aircraft	T. Dohl
40	Mar 29	—	Pt. Sal	35° 10′ N 120° 44′ W	7	W	Seen from aircraft	T. Dohl
41	May 21	—	Pt. Sal	34° 58′ N 120° 39′ W	9	SSW	Seen from aircraft	T. Dohl
42	Oct 30	—	14.8 km W of Pt. Pinos	36° 38′ N 122° 06′ W	1	W	With 1500 *L. obliquidens*	D. Goley, N. Black
43	Nov 8	—	Pt. Piedras Blancas	35° 38′ N 121° 12′ W	10	—	Just off beach	R. Jameson, G. Jameson
44	Nov 17	—	W of Cayucos	35° 25′ N 120° 52′ W	15	S	Near shore	R. Jameson
45	Dec 29	—	Carmel River State Beach	36° 33′ N 121° 56′ W	12/3	—	Milling near shore	A. Baldridge
1988								
46	Jan 2	1400	N of Carmel Pt.	36° 33′ N 121° 56′ W	12/3	—	Milling near shore	A. Baldridge
47	Jan 29	1000	Pt. Lobos Reserve	36° 32′ N 121° 57′ W	11–12/0	—	Milling in Whaler's Cove	J. Loomis, A. Baldridge
48	Jan 29	—	Moss Landing	36° 48′ N 121° 48′ W	8/1	S	In breakers	E. Baldridge
49	Feb 3	900	Monterey State Beach	36° 37′ N 121° 52′ W	8	N	In breakers	E. Haderlie
50	Feb 3	1345	Moss Landing	36° 48′ N 121° 48′ W	7–10/0	S	In breakers	E. Baldridge, B. Würsig
51	Feb 11	1600	Marina State Beach	36° 43′ N 120° 42′ W	5/0	S	Surfing	M. Radakovich
52	Feb 17	1430	Moss Landing	36° 48′ N 121° 48′ W	4/0	N	With 28 California sea lions	E. Baldridge, B. Würsig
53	Feb 20	1400	N of Pajaro River mouth	36° 48′ N 121° 49′ W	9/0	N	Just outside breakers	T. Dohl
54	Feb 24	—	Monterey State Beach	36° 37′ N 121° 52′ W	13/3	N	Just outside breakers	A. Baldridge, T. Jefferson
55	Feb 24	1433	Ft. Ord	36° 42′ N 121° 49′ W	8/0	—	Seen from aircraft	T. Dohl
56	Apr 2	—	Seacliff Beach, Aptos	36° 57′ N 121° 55′ W	10/1	—	Close to beach	N. Black
57	Apr 7	—	Seacliff Beach, Aptos	36° 57′ N 121° 55′ W	18/3	—	Close to beach	D. Shearwater, N. Black
58	Apr 9	—	Seacliff Beach, Aptos	36° 57′ N 121° 55′ W	24–26/2	—	Seen from aircraft	T. Dohl
59	Apr 10	—	Seacliff Beach, Aptos	36° 57′ N 121° 55′ W	24–26/2	—	Seen from aircraft	T. Dohl
60	Apr 11	—	Seacliff Beach, Aptos	36° 57′ N 121° 55′ W	24–26/2	—	Seen from aircraft	T. Dohl
61	Apr 11	1600	Seacliff Beach, Aptos	36° 57′ N 121° 55′ W	15/2	E	Near surfline	R. Wells
62	Apr 15	843	Seacliff Beach, Aptos	36° 57′ N 121° 55′ W	8–9/2	E	Near surfline	R. Wells
63	Apr 17	1830	Seacliff Beach, Aptos	36° 57′ N 121° 55′ W	10+	—	Close to beach	N. Black
64	Apr 23	900	Pajaro Dunes	36° 52′ N 121° 49′ W	8–10/1	N	Just outside breakers	T. Hahn

The second set of sightings occurred during September 1983–January 1984, when 21 sightings were made near Monterey Bay, probably consisting of more than 25 different individuals. Photographs of 3 dolphins seen off Santa Cruz in a school of 8–9 individuals (including a calf) on 4 November 1983 matched photographs of a school of 7–9 dolphins (including a calf) seen in the same area on 20 November 1983 (Table 1, sightings #10 and #18). However, none of these animals was identified from a school of 25–50 dolphins photographed on 7 January 1984 off Monterey (Table 1, #23).

No sightings were reported over the next 4.5 months, but dolphins were again seen near Monterey Bay on four occasions between May and September 1984. This third set of sightings apparently consisted of at least 9 dolphins, if all four sightings were of the same animals. If it is assumed that each sighting was of different dolphins, then as many as 28 individuals could have been involved.

A 3-month hiatus in northern sightings was followed by another set of sightings during late December 1984–early January 1985. Dolphins were seen four times between approximately San Simeon and Monterey Bay, involving 10–38 different individuals, depending on how many of the four sightings were repeated sightings. Only one northern sighting was reported between February 1985 and November 1986, involving a small school off Morro Bay during September 1985. Bottlenose dolphins were seen five times in the general vicinity of Point Sal during March–May 1987 (Table 1, #37–41). Sightings of two schools during aerial surveys on 27 March 1987 suggest that at least 19 individuals were in the area.

During May–June 1987 bottlenose dolphins from southern California were identified from photographs taken between Santa Barbara and Ventura (Wolf *et al.*, 1987). Three sightings were reported, totaling 80 individuals. Later surveys did not encounter dolphins in the same area, but occasional sighting reports continued to be received through 1987. Records from the Santa Barbara County Museum of Natural History indicate that the frequency of sightings of bottlenose dolphins in local waters increased greatly from 1983 to 1987 as compared to earlier years (C. Woodhouse, pers. commun. to R.H.D.).

A single sighting (Table 1, #42) 14.8 km west of Monterey Bay on 30 October 1987 did not follow the pattern of the previous 41 sightings. It involved a single large bottlenose dolphin swimming offshore in a large school of Pacific white-sided dolphins, *Lagenorhynchus obliquidens*. All previous sightings had involved single-species schools of bottlenose dolphins, primarily in coastal waters. This sighting is similar to one reported on 21 November 1982 (M. Webber, pers. commun. to A.B.), in which a single bottlenose dolphin was observed in a mixed school of Pacific white-sided dolphins, common dolphins, *Delphinus delphis*, and northern right whale dolphins, *Lissodelphis borealis*, near the Farallon Islands (37° 40′ N, 123° 00′ W).

Twenty-two sightings (Table 1, #43–64) were made during November 1987–April 1988. Most of these were in northeastern Monterey Bay during January–April. The reported range of group sizes and compositions suggests that at least 26 dolphins, including 2–3 calves, were in the bay during this period, but we do not know how many of the sightings involved the same dolphins. Preliminary examination of photographs did not result in any matches with dolphins photographed in Monterey Bay during 1983–1984.

It is difficult to assess the meaning of these sets of sightings. Do these sightings represent northward "pulses" in the movements of dolphins? How many animals were involved in the northward movements? Does a lack of sightings over a period of time demonstrate the absence of animals from northern waters, or could it be an artifact of uneven distribution of observers and effort along the coast? How many sightings might have been repeated observations of the same animals? Did more than one group of dolphins move into and out of the northern waters during a given set of sightings? Two lines of evidence provide answers to some of these questions. Comparison of school sizes gives a rough indication of how many animals were involved in a set of sightings. However, the most illuminating information comes from identifications of known individuals from photographs taken at several points along the California coast.

MOVEMENTS OF KNOWN INDIVIDUALS

The origins of some of the dolphins seen in Monterey Bay are known. Eleven distinctively marked individuals photographed in Monterey Bay during November 1983–January 1984 were originally photographed off southern California during 1979–1982 as part of local photographic-identification studies (Fig. 2). Other individuals were tentatively identified in Monterey Bay, but poor photographic quality precluded positive identifications. Three individuals, last seen in San Diego County during January 1983, were photographed off Santa Cruz during November 1983. Eight other dolphins were last identified in southern California between April 1982 and October 1983 before being photographed off Monterey during January 1984. Thus, at least 5% of the estimated southern California population of 240 bottlenose dolphins (Hansen, Chapter 23, this volume) was in Montery Bay during the winter of 1983–1984. If it is assumed that the unidentified associates of the positively identified individuals were also from the same southern California population, then approximately 12% of the population was in Monterey Bay. Seven of the southern California dolphins identified in Monterey Bay were subsequently reidentified in southern California waters between January 1984 and February 1985.

Because of the opportunistic nature of the sighting efforts, it is difficult to determine the timing of the northward movement made by the identifiable dolphins. However, based on the smallest time interval between sightings in the southern and northern ranges, we know that one dolphin (No. 02111) made the approximately 670-km transit to Monterey Bay in 74 days or less.

Once the dolphins had approached the northernmost extent of their movements, at least some of the animals remained in the Monterey Bay area for several weeks. The precise time of arrival of the identifiable animals to Monterey Bay cannot be confirmed, but a school of about eight dolphins was observed off Point Sur, heading north, on 25 October 1983. Similarly, a school of nine dolphins including one calf was seen off Santa Cruz on 29 October 1983. The school of eight to nine

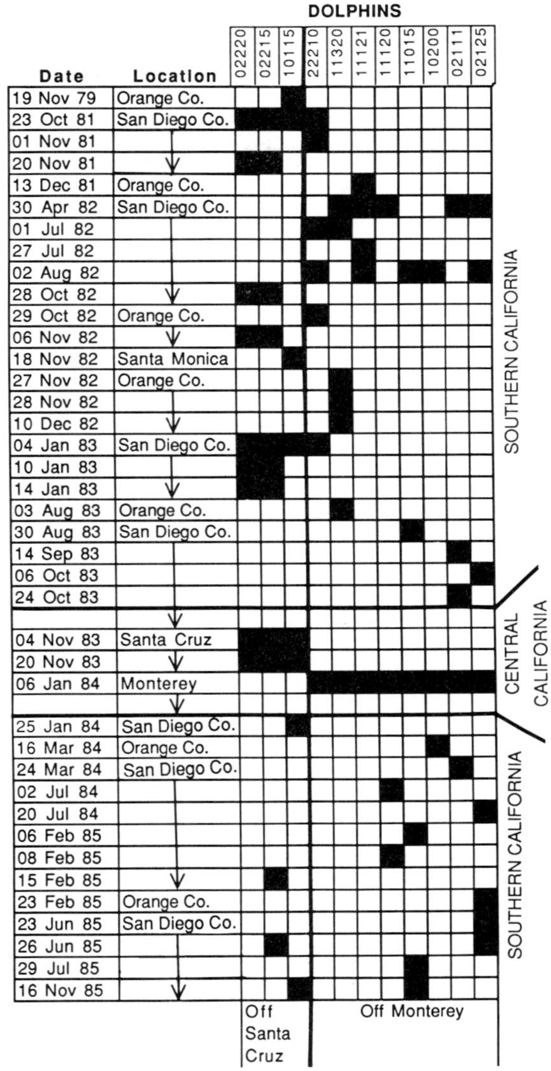

Figure 2 Sighting dates, locations, and associations of 11 known southern California bottlenose dolphins during 1979–1985. Filled boxes along a row indicate sightings of identifiable dolphins seen together. The three dolphins in the left-most columns were identified off Santa Cruz. All of the dolphins to the right were identified off Cannery Row, Monterey.

dolphins, including one calf, seen on 4 November 1983 was photographically identified again in the Santa Cruz area 16 days later. If these sightings were all of the same school, then it apparently remained in the Monterey Bay area for at least 27 days before returning to the southern range. The return trip of 670 km was made by one of these dolphins (No. 10115) in 66 days or less.

The school of more than 25 dolphins seen off Monterey on 7 January 1984 was apparently in the area for at least 5 days before returning south. A school of similar size was followed for 23.2 km on 12 January as it moved steadily southward along the coast south of Monterey Bay. One of the dolphins identified in Monterey Bay on 7 January (No. 10200) was photographed later in southern California on 16 March 1984. Thus, this animal, at least, completed the southward leg of the "migration" in 70 days or less.

The estimates of 66–74 days for completion of each 670-km leg of the trip indicate that the animals did not travel continuously. The calculated average speed, assuming continuous swimming, was approximately 0.4 km/hour, which is less than the traveling speeds of other bottlenose dolphins (6.1 km/hour, Würsig and Würsig, 1979; 2.1–2.7 km/hour, Tanaka, 1987). This suggests either that the California animals were at each terminus for quite some time before they were identified or that much time was spent in other areas along the way. A lack of continuous travel is further indicated by the reports of milling behavior for many of the sightings.

Wolf *et al.* (1987) found that 69% of the 36 dolphins photographically identified between Santa Barbara and Ventura in 1987 had originally been identified off San Diego. Of these, one had previously been identified off Ensenada, 483 km to the south. One individual was seen in both San Diego and Ventura over a 15-day period. Twelve dolphins were seen more than once off Santa Barbara during the 2 months, suggesting some degree of at least temporary residency in waters north of their previous range.

ENVIRONMENTAL CORRELATES

The initial northward movement of bottlenose dolphins from southern California to central California coincided with an El Niño warm-water incursion. Elevated sea-surface temperatures of 3.5°–5.0°C above the long-term mean persisted in coastal California during much of 1983 (Linder and Breaker, 1983; Fiedler, 1984). Based on comparisons of mean monthly sea-surface temperatures, as provided by Scripps Institution of Oceanography, the warm-water event resulted in an average increase in 1983 of about 0.9°C over 1982 temperatures in the San Diego area. In 1984, the temperatures were on average about 0.5°C higher than during 1982. In Santa Cruz, 1983 temperatures were, on average, 1.6°C higher than during 1982. In 1984, the Santa Cruz temperatures were 1.1°C higher than during 1982.

During 1983 and 1984, the temperatures in the Santa Cruz area, near the northward extent of the dolphins' movements, more closely approximated the 1982 temperatures in the San Diego area than Santa Cruz temperatures during most years. During 1982 the same dolphins that were seen in the Santa Cruz area during 1983 and 1984 were residing in the San Diego area. In 1982, the average difference in water temperature between San Diego and Santa Cruz was 3.6°C. The difference between the 1983 Santa Cruz water temperatures and the 1982 San Diego temperatures was on average 2.1°C. During 1984, the Santa Cruz temperatures were on average 3.1°C colder than the 1982 San Diego temperatures. Though the movements of the dolphins followed a northerly warming trend in water temperatures, the animals ventured into waters that were colder than those that they had experienced during the previous year in southern California. Thus, the animals do not appear to have moved strictly in response to the northward movement of isothermal water masses.

The northward-moving dolphins may have been responding more to the secondary effects of the warm-water incursion than to the movements of the warm water directly. Perhaps the dolphins were following changes in the distribution of prey. Hubbs (1948) pointed out that northern range extensions of fishes with normally southern distribution occurred during periods of ocean warming. He later noted that bottlenose dolphins and warm-water fishes and invertebrates were reported coincidentally during such ocean warming in northern California (Hubbs, 1960).

During the 1982–1983 El Niño event, fish species normally found in southern California and Mexican waters, such as the half moon, *Medialuna californiensis*, opaleye, *Girella nigricans*, Pacific bonito, *Sarda chilensis*, Pacific mackerel, *Scomber japonicus*, and California sheephead, *Pimelopetopon pulchrum*, were numerous in central and northern California (R. N. Lea, pers. commun. to A.B.; Lea, 1984; Hill and Haight, 1985). The red swimming crab, *Pleuroncodes planipes*, appeared in abundance in Monterey Bay at intervals between 26 December 1983 and 6 December 1984. The last published occurrence in January 1960 was the first since it was originally described from Monterey in 1859 (Glynn, 1961). Other odontocete cetaceans with warm-water affinities observed in the Monterey area during this period included false killer whales, *Pseudorca crassidens*, and common dolphins, the latter in numerous large schools.

Dohl noted that during March 1986–June 1988 bottlenose dolphins appeared to have replaced a population of 24–30 harbor porpoises, *Phocoena phocoena*, which had been observed regularly, year-round, between Point Sal and Morro Bay during aerial surveys in 1980–mid 1984 (Dohl, 1987; Dohl *et al.*, 1983). These harbor porpoises had been at the southern limit of their species' range along the west coast of North America. The nearest concentration of harbor porpoises was in Monterey Bay, about 180 km north.

SOCIAL STRUCTURE

The sightings of previously identified dolphins from southern California in the Monterey Bay area and then again in the San Diego area provide insights into the social structure of these animals. The dolphins observed in Santa Cruz (Table 1, #10 and #18) were not identified in the large school photographed off Monterey (#23). Photographic records from 1979 through 1985 (Fig. 2) indicate that members of these two groups of animals interacted infrequently off southern California. The school sizes in which the dolphins were seen were similar in southern and northern waters. While in southern California waters, the dolphins identified off Santa Cruz tended to be found in smaller schools (range 4–23 individuals) than were the animals identified off Monterey (range 7–50 individuals).

Association patterns observed in southern California waters were maintained while the identifiable animals were in Monterey Bay. There are no records of all 11 of the dolphins photographed together off Monterey swimming together in southern California waters, though occasional interactions between individual members of the Monterey school were recorded during 1979–1985 (Fig. 2). In contrast, two of the three animals identified together off Santa Cruz (Nos. 02220 and 02215) were always seen together between October 1981 and November 1983. The third identifiable dolphin from Santa Cruz (No. 10115) was observed with the pair in 67% of its sightings during 1979–1983. A similar pattern of close association between bottlenose dolphins while traveling a long distance was reported by Würsig (1978). Six dolphins that were seen together frequently over a 1-year period in Würsig's Argentina study area were resighted together approximately 300 km away after a 6-month absence. Nine months later, four of these six had returned and were together in the study area.

What was the age and sex composition of the schools of dolphins that moved into the northerly waters? Strong bonds between pairs or trios of bottlenose dolphins, apparently comparable to those noted for the dolphins off Santa Cruz, have been reported elsewhere among males (Wells *et al.*, 1980, 1987; Asper and Odell, 1980; Scott *et al.*, Chapter 11, this volume). In these cases it has also been noted that males occasionally travel relatively long distances. Lockyer (1978) described the movements of an apparently adult male bottlenose dolphin as covering over 500 km in 18 months.

The repeated associations, large body size, and absence of calves in close association suggest that the identified Santa Cruz trio consisted of adult males, but direct confirmation was not possible. The northerly movements were clearly not limited to adult males. Small calves were present in a number of the northern sightings, including those off Santa Cruz. Bottlenose dolphin calves typically remain with their mothers for at least 3–6 years along the central Gulf coast of Florida (Wells *et al.*, 1987); thus, adult females were probably present in the northern waters off California.

DISCUSSION

The variation in movement patterns of bottlenose dolphins from different areas is well documented. At one extreme, dolphins along the east coast of the United States migrate seasonally. At the other extreme, dolphins such as those in the resident community off Sarasota, Florida, show a high degree of site fidelity; some have been observed in the same area year-round for over 19 years (Wells, 1986, unpublished; Scott *et al.*, Chapter 11, this volume). One exceptional dolphin ("Ace") has been observed in a small deepwater hole (approximately 200 m by 75 m by 7 m deep) every time the survey vessels have passed through that area between June 1984 and January 1988 (n = 99) and has never been seen outside of the area.

The present case is unusual in that dolphins were observed over a period of several years both to maintain a home range and to undertake a lengthy migration. In thise case, 5–12% of a population that was previously considered to inhabit a well-established coastal home range, and with no previous migratory history, undertook a 670-km migration to the north of the previous extreme of the species' range. This raises the question of the relative importance of maintaining a home range of familiar waters versus abandoning the home range, at least temporarily, in response to changes in environmental conditions. Some of the southern California dolphins moved with the changing environment, shifting their range to include waters to the north of the range used prior to the warm-water event.

In contrast, resident bottlenose dolphins off Sarasota, Florida, remained in their established home range through fluctuations in water temperature greater than those experienced by the southern California dolphins, even considering El Niño. The Sarasota dolphins also remained in the home range during tremendous fish kills resulting from blooms of toxic dinoflagellates.

El Niño events are not uncommon along the California coast. If fringe-area bottlenose dolphins take advantage of such opportunities to explore new regions, then past warm-water events could explain the occurrence of specimens well north of the animals' present range.

ACKNOWLEDGMENTS

Jeff Foott provided photographs of excellent quality that were used to identify dolphins off Cannery Row, Monterey. Elizabeth Mathews prepared the figures, through the support of Dolphin Biology Research Associates, Inc. We thank Michael Scott for review of the manuscript.

REFERENCES

Asper, E. D., and Odell, D. K. (1980). Bottlenose dolphin local herd monitoring: Capture, marking, collection of biological data, and follow-up observations of marked animals. Final Rep. Natl. Mar. Fish. Serv., Contract No. NA79-6A-C-00027.

Barham, E. G. (1982). Marine mammals in Monterey Bay, California, during the years 1950–1955. *Calif. Fish Game* **68,** 213–223.

Dall, W. H. (1873). Preliminary description of three new species of Cetacea from the coast of California. *Proc. Calif. Acad. Sci.* **5,** 12–13.

Defran, R. H., Kelly, D. L., Schultz, G. M., Weaver, A. C., and Espinoza, M. A. (1987). Occurrence and movements of the bottlenose dolphin in the Southern California Bight. *Proc. 7th Conf. Biol. Mar. Mamm., Miami, Florida, December 5–9, 1987,* p. 16 (Abstr.).

Dohl, T. P. (1987). Atypical distribution, movement, and seasonality of five cetacean species in California waters. *Proc. 7th Conf. Biol. Mar. Mamm., Miami, Florida, December 5–9, 1987,* p. 17. (Abstr.).

Dohl, T. P., Norris, K. S., Guess, R. C., Bryant, J. D., and Honig, M. W. (1980). Cetacea of the Southern California Bight. *In* "Investigator's Reports, Summary of Marine Mammal and Seabird Survey of the Southern California Bight Area, 1975–1978." Vol. 3, Part 2, NTIS Publ. No. PB81-248-189. Natl. Tech. Information Serv., Springfield, Virginia. 414 pp.

Dohl, T. P., Guess, R. C., Duman, N. L., and Helm, R. C. (1983). Cetaceans of central and northern California, 1980–1983: Status, abundance, and distribution. *In* "Investigator's Final Report, Marine Mammal and Seabird Study of Central and Northern California," Vol. 3, Part 1. Pacific OCS Region, Minerals Management Service, U.S. Dept. of Interior, Contract No. 14-12-0001-29090. 270 pp.

Fiedler, P. C. (1984). Satellite observations of the 1982–1983 El Niño along the U.S. Pacific coast. *Science* **224,** 1251–1254.

Glynn, P. W. (1961). The first recorded mass stranding of pelagic red crabs *Pleuroncodes planipes,* at Monterey Bay, California since 1859, with notes on their biology. *Calif. Fish Game* **47,** 97–101.

Hill, K. T., and Haight, D. R. (1985). Northward range extension for the striped marlin. *Calif. Fish Game* **71,** 185–187.

Hubbs, C. L. (1948). Changes in the fish fauna of western North America correlated with changes in ocean temperature. *J. Mar. Res.* **7,** 459–482.

Hubbs, C. L. (1960). Marine vertebrates of the outer coast. *In* "Biogeography of Baja California and Adjacent Seas," Part. 2. *System. Zool.* **9,** 134–147.

Lea, R. N. (1984). Biological evidence of the 1982–83 El Niño off Caliornia. *CalCOFI Annu. Conf., Idyllwild, California, 30 October–1 November, 1984.* (Abstr.).

Leatherwood, J. S., and Reeves, R. R. (1982). Bottlenose dolphin (*Tursiops truncatus*) and other toothed cetaceans. *In* "Wild Mammals of North America: Biology, Management, Economics" (J. A. Chapman and G. A. Feldhammer, eds.), pp. 369–414. Johns Hopkins Univ. Press, Baltimore, Maryland. 1147 pp.

Linder, D., and Breaker, L. (1983). Warm sea surface temperatures along the California coast. *Coastal Oceanogr. Climatol. News* **5,** 31–32.

Lockyer, C. (1978). The history and behaviour of a solitary wild, but sociable, bottlenose dolphin (*Tursiops truncatus*) on the west coast of England and Wales. *J. Nat. Hist.* **12,** 513–528.

Norris, K. S., and Prescott, J. H. (1961). Observations on Pacific cetaceans of Californian and Mexican waters. *Univ. Calif. Publ. Zool.* **63,** 291–402.

Orr, R. T. (1963). A northern record of the Pacific bottlenose dolphin. *J. Mammal.* **44,** 424.

Shane, S. H., Wells, R. S., and Würsig, B. (1986). Ecology, behavior, and social organization of the bottlenose dolphin: A review. *Mar. Mamm. Sci.* **2,** 34–63.

Tanaka, S. (1987). Satellite radio tracking of bottlenose dolphins *Tursiops truncatus. Nippon Suishan Gakkaishi* **53,** 1327–1338.

True, F. W. (1889). A review of the family Delphinidae. *Bull. U.S. Nat. Mus.* **36,** 1–191.

True, F. W. (1891). Observations on the life history of the bottlenose porpoise. *U.S. Nat. Mus. Proc.* **13** (1890), 197–203.

Walker, W. A. (1981). Geographical variation in morphology and biology of bottlenose dolphins (*Tursiops*) in the eastern North Pacific. NMFS/SWFC Admin. Rep., No. LJ-81-03C.

Wells, R. S. (1986). Structural aspects of dolphin societies. Ph.D. dissertation, University of California, Santa Cruz. 233 pp.

Wells, R. S., Irvine, A. B., and Scott, M. D. (1980). The social ecology of inshore odontocetes. *In* "Cetacean Behavior: Mechanisms and Functions" (L. M. Herman, ed.), pp. 263–317. Wiley, New York. 463 pp.

Wells, R. S., Scott, M. D., and Irvine, A. B. (1987). The social structure of free-ranging bottlenose dolphins. *In* "Current Mammalogy" (H. Genoways, ed.), Vol. 1, pp. 247–305. Plenum, New York. 519 pp.

Wolf, J. M., Defran, R. H., and Shultz, G. M. (1987). Photographic surveys of the bottlenose dolphin (*Tursiops truncatus*) along the Ventura–Santa Barbara coastline of southern California. *Proc. 7th Conf. Biol. Mar. Mamm., Miami, Florida, December 5–9, 1987,* p. 75 (Abstr.).

Würsig, B. (1978). Occurrence and group organization of Atlantic bottlenose porpoises (*Tursiops truncatus*) in an Argentine Bay. *Biol. Bull.* **154,** 348–359.

Würsig, B., and Würsig, M. (1977). The photographic determination of group size, composition, and stability of coastal porpoises (*Tursiops truncatus*). *Science* **198,** 755–756.

Würsig, B., and Würsig, M. (1979). Behavior and ecology of bottlenose porpoises, *Tursiops truncatus,* in the South Atlantic. *Fish Bull.* (*U.S.*) **77,** 871–890.

PHOTO: In recent years, increasing effort has been applied to enhancing reproduction of bottlenose dolphins in captivity (Sea World of Florida, Orlando; Stephen Leatherwood).

Part VII

Husbandry and Captive Breeding

Breeding Bottlenose Dolphins in Captivity

25

J. Pete Schroeder
Naval Ocean Systems Center
Hawaii Laboratory
Kailua, Hawaii 96734

INTRODUCTION

Atlantic bottlenose dolphins, *Tursiops truncatus,* have been bred successfully in captivity since 1939 (McBride and Kritzler, 1951; Essapian, 1953; Tavolga and Essapian, 1957; Tavolga, 1966; Caldwell and Caldwell, 1972; Wood, 1977). The birth of a dolphin was formerly a newsworthy event; all too often, it came as a total surprise. The frequency of stillbirths was high, and the health and survival of captive-born calves often were compromised by the holding institution's lack of preparation for birth. In 1975, participants in a workshop reviewed what was known about the reproductive biology, physiology, and behavior of bottlenose dolphins (Ridgway and Benirschke, 1977). Among the goals of the workshop were encouragement of further research and the eventual establishment of a self-sustaining captive stock of bottlenose dolphins. The workshop stimulated research leading to major advances in our knowledge about dolphin reproductive physiology and in captive breeding. For example, Sawyer-Steffan *et al.* (1983) and Schroeder (1985) developed techniques to manipulate the estrous cycle of captive dolphins, and successful births at commercial oceanaria increased dramatically (Cornell *et al.*, 1987).

In 1977, the Naval Ocean Systems Center (NOSC) began developing a breeding population with Atlantic bottlenose dolphins from the Indian River, Florida. The principal goals of work with this population were to define further the reproductive cycles of males and females and to apply this knowledge both to the management of the captive population, in order to maximize its breeding potential and efficiency (Schroeder, 1983), and to the development of an artificial insemination program. This chapter deals primarily with the reproductive physiology research and breeding programs at the NOSC Hawaii Laboratory.

MATERIALS AND METHODS

Building on available knowledge (Harrison and Ridgway, 1971; Ridgway, 1972; Ridgway and Benirschke, 1977), we initiated studies of bottlenose dolphins at our facility in Kaneohe Bay, Oahu, Hawaii, in April 1981. One reproductively mature male and six reproductively mature females were housed in five seawater pens, each 2.5–3.5 m deep and 7–9 m square, enclosed by 16-cm square wire mesh. A cluster of five pens in the bay shared common walls, allowing auditory, pheromonal (if existing), and visual contact between the animals at all times. The male dolphin had one pen and was kept separate during the study, with no tactile access to the females. The seven animals ate a mixture of thawed Pacific mackerel, *Scomber japonicus,* Columbia River smelt, *Thaleicthys pacificus,* and Pacific herring, *Clupea harengus pallasi,* with a vitamin and mineral supplement. They also had access to any of the great variety of small marine organisms that swam, crawled, or floated through their pens. Five of the six females had previously delivered calves in captivity, and the male had sired three calves in captivity.

For three years, blood was sampled from the female dolphins at least once weekly, to monitor total serum estrogen and serum progesterone, and from the male twice monthly, to monitor serum testosterone levels by radioimmunoassay (RIA) (Kirby and Ridgway, 1984). During the same period, five of the females were subjects in exogenous ovulation studies. The protocol in the latter studies called for two intramuscular injections of pregnant mare's serum gonadotropin (PMSG, gestyl), a substance high in follicle-stimulating hormone (FSH). The purpose of these injections was to stimulate maturation of ovarian follicles. After an appropriate interval of 120 hours, human chorionic gonadotropin [HCG, Pregnyl (Organon Inc., West Orange, New Jersey)] was then injected. This gonadotropin preparation is high in luteinizing hormone (LH), which induces the ovarian follicle to rupture and release its egg. We determined the appropriate intervals and dosages of PMSG and HCG resulting in exogenous induction of ovulation (Schroeder, 1984, 1985).

To study reproductive physiology of the male, we trained the adult male dolphin for semen collection (Keller, 1987). By comparing his semen production and serum testosterone levels over 28 months, we developed a breeding profile (Schroeder *et al.*, 1983). In addition to sampling this single adult male in our breeding colony, we measured serum testosterone levels in 241 samples from 21 known-age dolphins at various times of the year.

RESULTS AND DISCUSSION

Anatomy of Reproduction

Male Anatomy

The penis of the male bottlenose dolphin is enclosed within the genital groove. The testes are retained in the abdominal cavity throughout life and vary in size seasonally in the species of small odontocetes examined to date (Ridgway and Green, 1967; Collet, 1981). When a male dolphin is viewed from below, his genital groove can be seen approximately 4–5 cm anterior to another small slit, the anal opening (Ridgway, 1972). This separation between the genital and anal slits is the most obvious external difference between male and female bottlenose dolphins. (The female genital slit is continuous with the anal slit.) Two small pores are present on the male, between the two slits, one on each side of an imaginary line joining the back of the genital slit and the front of the anal slit. If these pores are vestigial scent glands, such as those in the dog or the skunk, they could serve a pheromonal function. However, in my opinion they are most likely vestigial openings to vestigial mammary glands.

The penis of the dolphin (and other cetaceans) is of the fibroelastic type found in ruminants (Slijper, 1962). The cetacean penis has its origin by crura from the median surfaces of the pelvic bones. These crura fuse into a long, firm, ropelike body, round to oval in cross-section and forming a distinct sigmoid flexure lying in a horizontal plane

(Green, 1972). The distal portion of the dolphin's penis diminishes to a very small, sharply tapered tip, which seems to be well adapated to couple with the uniquely configured cervix of the female.

Female Anatomy

The female dolphin has one continuous anogenital slit, with one mammary gland, a short slit 1–2 cm lateral to the anogenital slit, on each side (see Harrison, 1969, and Green, 1972, 1977, for descriptions of internal anatomy). The female dolphin has a bicornate uterus, very similar in appearance to those of the horse, dog, and cow, but the dolphin has a unique, 6–7 cm, cervix/pseudocervix in the upper vagina. Some authors consider this cervix/pseudocervix as a true folding of the vaginal wall within a spermathecal recess (Harrison, 1969). My dissections of reproductive tracts of female bottlenose dolphins, Pacific white-sided dolphins, *Lagenorhynchus obliquidens,* common dolphins, *Delphinus delphis,* harbor porpoises, *Phocoena phocoena,* striped dolphins, *Stenella coeruleoalba,* spinner dolphins, *Stenella longirostris,* and white whales, *Delphinapterus leucas,* have demonstrated the presence of a pseudocervix in all. The upper section of the vagina in a platanistoid dolphin, the baiji (*Lipotes vexillifer*), is covered with thin leaflike folds, in addition to several rings of transverse folds facing the lower section of the vagina (Chen *et al.*, 1984).

Whichever terminology one uses—folds, rings, or pseudocervix—it is clear that the vaginas of these small odontocetes contain a remarkable anatomical adaptation for breeding in the marine environment. This "compartment," or spermathecal recess/space between the true cervix and the pseudocervix, is not seen in mammals that breed on land. The sow has a corkscrewlike (10–23 cm) cervix with two to three annular rings, and the cow an 8–10 cm cervix with three to four annular rings (McDonald, 1977), similarities to dolphins that may reflect the phylogenetic relationship of cetaceans and ungulates. The structure of the cervix in the cow and sow perhaps resemble anatomical conditions that were preadaptive for the successful transition of cetacean forebearers from terrestrial to wholly aquatic environments.

The folds of the pseudocervix contain some muscle bundles and appear capable of movement (Harrison, 1969). Copulatory behavior of bottlenose dolphins often has been observed in captivity. However, without being able to interrogate the female dolphin about her sexual sensitivities and responses, one cannot determine the role(s) of the pseudocervix. My own speculation is based on observations of equine breeding. When a mare is in estrus, the appearance of her cervix, as viewed through a vaginal speculum, is that of a hyperemic, relaxed structure. Evolutionary adaptation and selection for the pseudocervix–spermathecal recess–cervix combination in cetaceans would ensure breeding efficiency if, during estrus, the pseudocervix of the dolphin were relaxed (as is the cervix in the mare). It would become so by anatomically accommodating the small tapered tip of the dolphin's penis. After deposition of semen in the spermathecal space, the penis's withdrawal could trigger muscular contractions of the pseudocervix, effectively shutting out lethal (to sperm) seawater and allowing capacitation of the recently deposited sperm prior to penetration of the cervical mucus.

Reproductive Cycle Patterns

Female Cycle

The development and application of radioimmunoassay (RIA) since the 1960s (Lasley, 1980) makes it possible to measure hormone levels in biological fluids directly; these levels reflect specific physiological states. Using RIA, we determined that puberty in female bottlenose dolphins starts at 5–7 years of age. This means the female should be receptive to a male and could conceive if breeding were to occur at this age. It is important to distinguish between sexual maturity (puberty) and reproductive maturity. The latter occurs when the female is cycling on a regular basis and has attained her adult size. Reproductive maturity may be more important for successful births in terrestrial animals than in cetaceans, which lack a pelvic girdle. However, a female dolphin bred before she has attained her mature size may experience a difficult birth, stillbirth, or other problems associated with pregnancy in a system that is not fully developed. Additionally, the young mother's inexperience in birthing and nursing adds to the neonate's problems.

Reproductive maturity in bottlenose dolphins occurs between 7 and 10 years of age. Managed breeding of dolphins in this age range should result in fewer stillbirths, better maternal care, and healthier calves. Prior to 1975, approximately 40% of dolphin births in captivity were stillbirths (Ridgway and Benirschke, 1977). Since then, the rate of stillbirths has decreased to approximately 16% (Couver and Schroeder, 1984). This decrease is due to at least three factors: collectors avoid capturing pregnant females, which often experience stillbirths of that calf; captive dolphin colonies are being managed more carefully to prevent young (sexually mature but not reproductively mature) female dolphins from becoming pregnant; and the nutrition provided to pregnant dolphins has become better as husbandry practices and knowledge of nutrition have improved.

The bottlenose dolphin was classified as a reflex ovulator by some investigators (Harrison and Ridgway, 1971; Harrison, 1977). Recent evidence from histological examinations of ovaries, considered in conjunction with specified uterine sections from spinner and striped dolphins, supports the theory that dolphins of the genus *Stenella* may sometimes ovulate spontaneously (Benirschke *et al.*, 1980). Also, levels of progesterone and total immunoreactive estrogens have provided hormonal evidence of spontaneous ovulation in captive dolphins of the genera *Tursiops* and *Delphinus* (Kirby and Ridgway, 1984) and the species *Stenella longirostrus* (Wells, 1984). Findings at NOSC confirmed those results in bottlenose dolphins, in which spontaneous ovulation could be detected up to seven times in a 13-month period (Fig. 1). A "normal" reproductive year for NOSC captive dolphins in Hawaii includes two to three ovulations, usually occurring about August to November. It should be noted that captive dolphin births have occurred in all months of the year (Cornell and Asper, 1978). In all wild populations of delphinids studied to date, animals have been found to mate and calve seasonally (Perrin and Reilly, 1984).

At NOSC, we found that by measuring changes in serum progesterone levels from weekly blood samples it was possible to determine the estrous and anestrous periods for three of six female dolphins throughout the 3-year study period and for the other three dolphins for 1.5 of the years. Serum progesterone levels change daily, so it was necessary to sample blood from the dolphins at least weekly. We also determined the success of attempts to induce ovulation by injection of exogenous hormones (Schroeder, 1985). The latter capability provided a means of manipulating one aspect of the female dolphin's reproductive cycle to enhance breeding success (Fig. 2).

ESTRUS

The ability to monitor the onset and duration of estrus is essential to a successful breeding program, whether that program is based on natural or artificial insemination. However, estrus is difficult, sometimes impossible, to detect in many animals. The phenomenon of "silent heat" is well known in domestic animals and "estrous" behavior is not necessarily accompanied by a normal "heat" period which includes ovulation (Britt *et al.*, 1981). Changes in appearance of external genitalia, increased secretions, and/or behavioral changes are the most obvious, but not always the most reliable, signs indicating estrus in most domestic animals. It is difficult to detect estrus in captive animals, not only because there are generally few individuals in any single captive colony, but also because the normal cycling of physiological events can be disrupted by the conditions of captivity. Given that the reproductive efficiency of domestic animals is affected by such factors as season, temperature, confinement, group size, and pen structure, it is reasonable to expect that breeding groups of captive dolphins are affected similarly.

The female dolphin's genital slit is generally underwater, making the detection of estrus more complicated for the investigator than is the case for terrestrial animals. Behavioral signs, including increased acoustic activity, have been used in some cases to infer that captive dolphins are in estrus (Wood, 1953; Tavolga and Essapian, 1957). In our studies and observations, we found no consistent, objective criterion of behavior by which to judge whether or not a female dolphin is in estrus. Conditions traditionally taken to indicate dolphin estrus (e.g., the belly's becoming pinker than normal, the female's positioning close to or

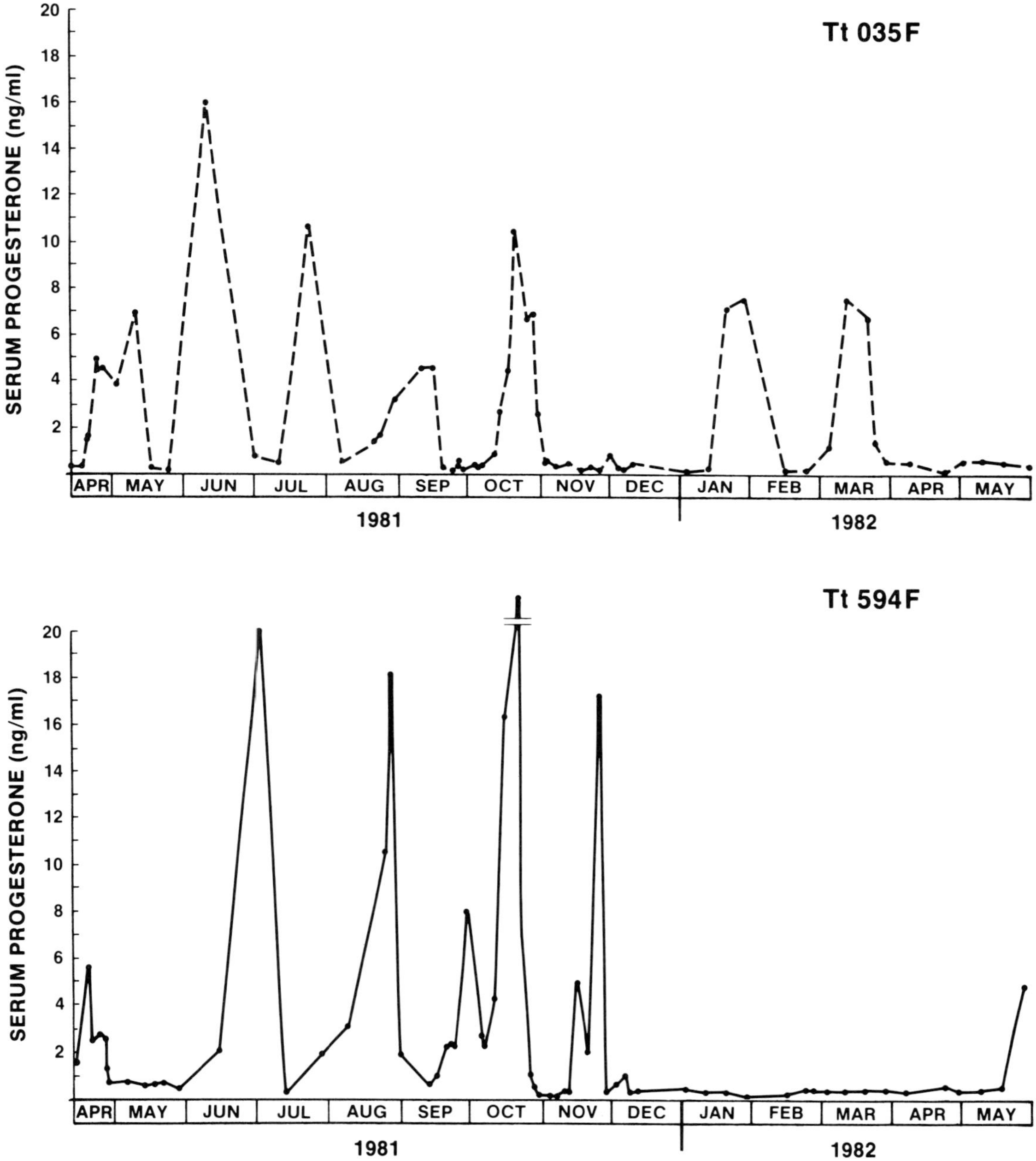

Figure 1 Serum progesterone levels of two captive adult female bottlenose dolphins. Each peak represents a spontaneous ovulation. More often, there are two to three ovulations each breeding season.

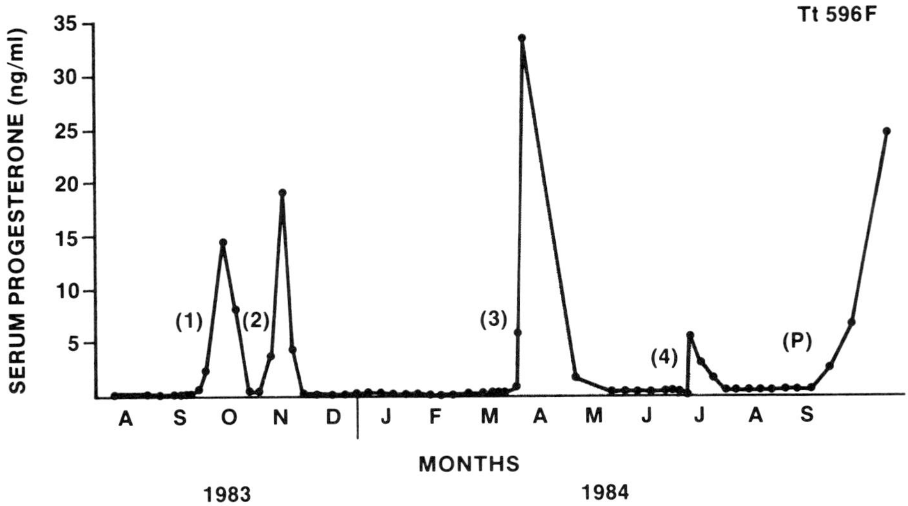

Figure 2 Serum progesterone levels from a captive adult female bottlenose dolphin showing progesterone peaks at two spontaneous ovulations, (1) and (2), and two exogenously induced ovulations, (3) and (4). This dolphin delivered a healthy calf 20 September 1985, proving pregnancy (P) occurred within 3 months of her second induced ovulation.

swimming closer than usual to the male in the area, two females, in the absence of a male, becoming more tactile than usual with each other, enlarged or puffy-appearing external genitalia, periodic anorexia) did not in this study correlate with induced or spontaneous ovulations. In no case did we observe a strong correlation between any of these conditions or behaviors and the detection of estrus through RIA of serum progesterone levels.

The most promising technique for detecting the onset of estrus and, even more important, ovulation is analysis of urinary estrone conjugates (Luskutoff *et al.*, 1983; Czekala *et al.*, 1986). This technique requires that the dolphin be trained for routine urine collection. Other methods of detecting estrus include the following: analysis of vaginal or cervical epithelium; laparotomy/laparoscopy (surgical entry into the abdomen and direct observation of the ovary through a laparoscope); palpation of the ovaries (impossible in small cetaceans); or ultrasound imaging of the developing follicle (at present a difficult procedure). At NOSC, we have used RIA analysis of progesterone and/or total estrogens to classify reproductive conditions of female dolphins. Baseline levels below 1 ng/ml of progesterone indicate the animal is anestrous or nonpregnant. Ovulations are indicated by episodic total estrogen levels above 120 pg/ml and/or episodic "spikes" of serum progesterone levels greater than 1 ng/ml. Progesterone levels during gestation are greater than 2 ng/ml, sometimes as high as 56 ng/ml (Schroeder and Keller, Chapter 26, this volume).

We developed the abilty to regulate estrus by the injection of exogenous hormones and to monitor each individual female's serum progesterone levels so that she could be placed with a male for breeding at an opportune time. In our experience, three pairs of reproductively mature males and females have been housed together continuously for 4 to over 7 years without producing young. On the other hand, placement of a reproductively mature male with a reproductively mature female, when the pair has not been housed together previously, usually results in conception at the first ovulation.

PREGNANCY DETERMINATION AND GESTATION

The bottlenose dolphin's gestation period is 12 months, plus or minus 10–14 days. Determination of pregnancy after breeding is an important component of management. At NOSC, as soon as an animal is diagnosed as pregnant, she is allowed to eat *ad libitum,* and her diet is supplemented to insure a calcium : phosphorus ratio of 1.25 : 1. The mother's nutritional state affects the quality of the environment in which the fetus grows and develops.

Probable pregnancy may be detected as soon as 2 weeks after breeding if the serum progesterone level reaches 3 ng/ml or higher. We have determined through monitoring of serum progesterone levels that the reproductive cycle of the female dolphin extends from 29 to 42 days. This cycle includes an increase of total immunoreactive estrogens for 5–7 days. Once the levels have peaked, ovulation and the onset of estrus occur, causing a rise in progesterone levels during the 24- to 36-day luteal phase. After conception, progesterone levels range between 3 and 52 ng/ml. We have recorded 115 levels in this range from 24 bottlenose dolphin pregnancies resulting in births. A similar range was found for six female bottlenose dolphins followed on a monthly basis from shortly after conception to delivery (Schroeder and Keller, Chapter 26, this volume).

If serum progesterone levels remain above 3 ng/ml for three consecutive blood samples over a period of 6 weeks, the dolphin is very likely pregnant. Confusion can arise (a false positive) if the second or third sample is taken during the following estrous period, but any doubt can be removed by further sampling after that animal's breeding season. The use of an obstetrical doppler confirms pregnancy by measuring the fetal heartbeat from 4 to 6 months after conception (Gray and Chapman, 1974; Cornell *et al.*, 1987). The fetal hearbeat is generally 120–130 beats per minute, or about double the mother's. Imaging of the fetus may be possible using ultrasound somewhat sooner than 4 months (Cornell *et al.*, 1987).

With proper conditioning, it is practical (and desirable) to monitor serum progesterone levels and weights of pregnant dolphins once monthly throughout the gestation period. In horses, it takes a serum progesterone level of about 4 ng/ml or higher to sustain pregnancy. The level may fall below 4 ng/ml when other corpora lutea develop between 40 and 140 days. This is the critical period for the equine fetus, since the progesterone level may drop too low to sustain pregnancy during the transfer of function from the primary corpus luteum to the placenta. There is some suggestion that this is the case in dolphins as well (Yoshioka *et al.*, 1986). In horses, after about 100–140 days, the placenta produces enough progesterone to sustain pregnancy. It is not known whether this also occurs in dolphins. If the serum progesterone level of a pregnant dolphin drops below 3 ng/ml, injection of progesterone might be indicated, particularly for animals which have a history of failing to reach term or which exhibit very low serum progesterone levels in the fourth or fifth month of pregnancy. I must emphasize that this is an area of great uncertainty, and data from dolphins are needed to test the validity of the analogy with horses. In cattle, the corpus luteum is the major source of progesterone and its removal before day 200 terminates pregnancy (Britt *et al.*, 1981).

BIRTH AND LACTATION

As the female dolphin approaches term, her abdomen continues to enlarge; her mammary glands develop rapidly within 2–3 days of delivery. In our experience, birth of the 95–115 cm calf usually takes place during the night, nearer dawn than dusk. We have had seven births at night and one at 3:35 in the afternoon. In general, labor lasts for as little as 20 minutes, but it can take up to 2 hours. During difficult births, it is sometimes helpful to apply gentle traction on the calf's tail (caudal presentation is normal). (A colleague, Dr. David Mackay, told me he attended a dolphin birth in which a head-first presentation of the calf stopped labor. He was able, with some difficulty, to insert an obstetrical instrument in the blowhole of the calf, enabling him to extract the dead fetus slowly from the birth canal. This procedure saved the mother.)

Following a normal delivery, the calf usually nurses within the first hour and as often as four times per hour for the first 4–8 days. The calf usually begins eating some fish at 6–7 months of age. Lactation lasts for at least 18 months if the calf is left with the mother, less time if it is separated. We wean the calves at 18–24 months, at which time the mother's mammary glands are still somewhat swollen. The separation of the mother from the calf produces no evident concern in either parent or offspring. However, they may be left together longer than 24 months unless there is a need to separate them for such specific purposes as decreasing the intercalf interval.

CARE AND HUSBANDRY OF THE NEWBORN

The first 2 weeks after birth are especially critical; nearly 40% of 89 calves born between 1977 and 1982 in oceanaria around the world died during their first month (Fig. 3) (Couver and Schroeder, 1984). The situation has improved since 1982. There is no substitute for good mothering, but the quality varies substantially among individuals. We have found that it helps to have first-time pregnant females housed with older, experienced mothers when the latter give birth. In two such cases, inexperienced females we housed with older females that gave birth, successfully delivered, and nursed their own first calves. It is not clear whether the first-time pregnant females learn by observation or profit in some other way from the exposure to a successful birth.

It also has proved useful to have an experienced mother (the proverbial "auntie") present in the birthing area with the pregnant female. We exclude males from the birthing area entirely. Our birthing enclosure is a fenced-off portion of a natural lagoon, measuring 35 × 35 m. The first week postpartum, the mother and calf swim in a 7- to 10-m circle, using little of their available space. As the calf grows, more of the area is explored. We have no data to recommend the best size for a birthing area.

Female dolphins control movements of their calves and can be quite rough with them. For example, one captive bottlenose dolphin flipped her calf out of the pool onto the asphalt surrounding the tank. The calf was rapidly returned to the water and suffered no obvious ill effects. It remains healthy at the time of this writing, more than 3 years after the traumatic event.

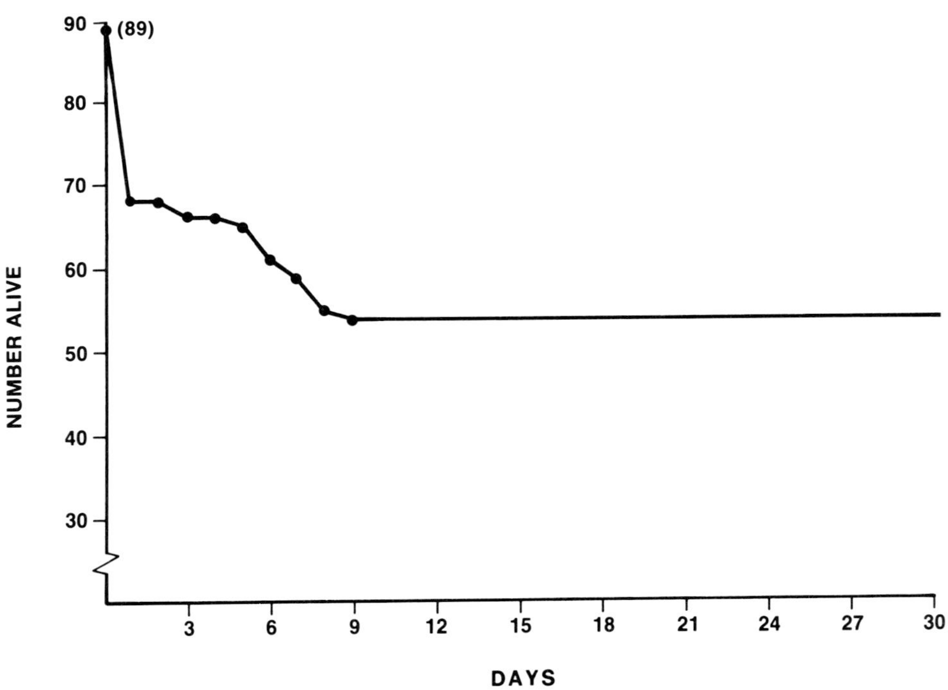

Figure 3 Calf mortality during the first month.

Male Cycle

Male bottlenose dolphins are sexually precocious. Reflex penile erection occurs within 48 hours of birth (Caldwell and Caldwell, 1972). Within several months, the young male will have approached and examined closely every female in its tank.

Serum testosterone levels fluctuate greatly throughout the year, as does semen production. Peak testosterone levels precede peak sperm production levels by 2–6 weeks (Schroeder, 1985; Schroeder and Keller, 1989, Chapter 26, this volume). The reproductively mature male bottlenose dolphin produces 800–1300 × 10^6 viable, healthy sperm per milliliter of ejaculate. This condition generally is reached at 10–13 years of age. We consider serum testosterone levels of 0–3 mg/ml to indicate an immature or prepubescent animal, 3–5 ng/ml a pubescent maturing animal, and 5–54 ng/ml a reproductively mature male dolphin. However, we recorded serum testosterone levels below 3 ng/ml during the breeding season from reproductively mature males that had sired calves. Because peak serum testosterone levels and peak sperm levels are out of phase, it is risky to use testosterone levels from a single sample to assess the degree of maturity.

Male dolphins can be trained for ejaculation and semen collection (Schroeder and Keller, Chapter 26, this volume). Semen samples collected by manual stimulation of an animal so trained have ranged from 0.1 to 39.5 ml ($\bar{x}$ = 4.3, S.E.M. = 0.14, n = 1332). Total sperm counts per ejaculate have ranged from near zero during the nonbreeding season to 54,596 × 10^9/ml ($\bar{x}$ = 1.04 × 10^9, S.E.M. = 0.102 × 10^9, n = 1332). The greatest sperm density in a high-quality semen sample is important for reproductive success. Motility percentages of all samples were from 90 to 100; motility status was from 4 to 5. Bottlenose dolphin sperm production is clearly seasonal, so reproductive efficiency would be enhanced by scheduling breeding or semen collection at times of the year (September–October in Hawaii) when the maximum number of sperm per cubic centimeter of ejaculate (sperm density) is present.

Spermatogenesis was found to occur throughout the year, not surprising since captive bottlenose dolphin births have occurred in every month of the year. Dolphin semen can be frozen after collection and later thawed to a 85% viable sperm. We stored dolphin semen for up to 955 days. Once thawed, it contained 60% live sperm with 90% motility and no abnormal sperm (Schroeder and Keller, Chapter 26, this volume). Portions of the same sample were thawed more recently, after 2 additional years in storage, with the same results.

Techniques for Improving Reproductive Efficiency

Bryden and Harrison (1986) stated that one of the major obstacles to the detailed study of reproduction in dolphins is the difficulty of breeding them in captivity. However, in our experience at the NOSC Hawaii Laboratory, the bottlenose dolphin has proved to be a remarkably efficient subject for a captive breeding program. Other institutions, such as the Miami Seaquarium, Sea Life Park in Hawaii, and the Sea World Parks in the United States (Cornell *et al.*, 1987), also have experienced consistently successful reproduction with those animals designated as breeding stock. In the past, lack of funds and lack of an explicit commitment to captive breeding have been major obstacles to the study of dolphin reproductive physiology.

Among the factors that I consider essential to an efficient dolphin breeding program are the following: accurate record keeping, knowledge of the reproductive cycles of male and female dolphins, a consistent supply of high-quality food of at least three fish species, careful monitoring of water quality, and water temperature of at least 15°C for newborns. Because mammalian reproduction relies on a finely tuned series of physiological and behavioral processes or events, a good preventive medicine program and adequate nutrition are absolute necessities.

We were concerned that the injection of exogenous hormones and the artificial manipulation of reproductive cycles of female dolphins could impair their reproductive capacities and efficiency. At the NOSC Hawaii Laboratory, there were six deliveries of live calves to dolphins that had been monitored for spontaneous ovulations and had undergone the exogenously induced ovulation protocol two to six times (Fig. 2). No ill effects have been detected.

SUMMARY AND CONCLUSIONS

When the dolphin breeding and physiology research program began at NOSC, we knew that the bottlenose dolphin had a gestation period of about 12 months, that this animal could be bred in captivity, and that conception could occur in any month of the year. Females were suspected of being seasonally polyestrous and sexually mature at 5–8 years of age. The view that dolphins were spontaneous ovulators was somewhat controversial. There was some evidence that ovulation could be induced and some data that plasma progesterone indicated pregnancy and/or ovulation had occurred (Kirby and Ridgway, 1984). No information was available on the length of the estrus cycle, whether the estrus cycle could be manipulated to induce "heat," the techniques of artificial insemination, or early and accurate means of diagnosing pregnancy.

Female bottlenose dolphins tend to have peak breeding seasons. NOSC dolphin breeding and births concentrate in the spring in San Diego and in the fall in Hawaii, though there was one birth in Hawaii in July 1985 and one in March 1988. Spontaneous ovulation occurs seasonally, and it is possible to induce ovulation by the injection of exogenous hormones. The interval between doses of PMSG and HCG is critical. Serum progesterone levels above 1 ng/ml indicate a reproductive event other than anestrus. Levels as high as 89 ng/ml have been recorded. Serum progesterone levels during pregnancy range from 3 to 52 ng/ml but may drop to 3 ng/ml without interruption of the pregnancy. The length of the estrus cycle, that is, from anestrus to anestrus, during the breeding season is 24–36 days. Female dolphins start reproductive cycles at 5 years of age and can deliver healthy calves up to age 30. Bottlenose dolphins can be inseminated artificially with appropriate conditioning and minimal restraint, using fiberoptic equipment (Schroeder and Keller, Chapter 26, this volume). It is possible to diagnose a nonpregnant dolphin within 2 weeks of breeding and a pregnant dolphin within 7–8 weeks. Analysis of serum progesterone levels has provided accurate pregnancy diagnosis, confirmed by 31 separate births. The repeated injection of exogenous hormones to induce ovulation is a safe procedure. It has allowed us to confirm previous findings concerning the reproductive biology of the bottlenose dolphin (Sawyer-Steffan *et al.*, 1983; Schroeder, 1984; Kirby and Ridgway, 1984).

When our program began, it was known that male bottlenose dolphins exhibit seasonal variation in the size of gonads and testosterone levels (Ridgway and Green, 1967; Harrison and Ridgway, 1971). They were known to be sexually mature at 10–12 years of age.

We found seasonal fluctuations in sperm production and serum testosterone levels. The highest sustained levels of sperm production of the dolphin in our program were in March, April, May, August, September, and October; peak testosterone levels precede peak sperm levels by 2–6 weeks. Different stages of sexual maturity can be classified according to serum testosterone levels: 0–3 ng/ml for immature, 3–5 ng/ml for pubescent, and over 5 ng/ml for mature males. Testosterone levels of an individual dolphin ranged in 1 year from 1 to 54.4 ng/ml; so, multiple sampling is necessary to determine accurately the maturity state of males.

Trained behavior to facilitate semen collection can be sustained over a wide range of conditions and for a long period of time. The male dolphin in this study has maintained the behavior for 6 years with some breaks in collection of 6–8 months. His performance of trained behavior was so reliable during the study that we were able to prepare a female dolphin for artificial insemination and collect fresh semen for immediate insemination. Frozen dolphin semen can be stored and/or shipped to facilities with immature or no male dolphins.

Regarding the management of captive dolphins for breeding and physiological research, we have found it possible to allow access of male and female dolphins to one another without extinguishing important behaviors, such as those facilitating semen collection and blood sampling from the flukes. Our dolphins have been handled and their blood sampled up to 5 times per week without any change in their trained behavior. Housing of reproductively mature males with reproductively mature females in the same pen for many years is not appropriate for managing captive dolphins for breeding. However, it can be an effective way to stop breeding, as we have experienced on three occasions.

There is still much to learn about reproductive

biology of bottlenose dolphins and about techniques for improving breeding success in captivity. Analyses of pituitary reproductive hormones, FSH and LH, will increase knowledge of reproductive processes. The best time for artificial insemination relative to ovulation is unknown. With the ability to freeze semen successfully, *in vitro* fertilization and embryo transfer have become interesting possibilities.

ACKNOWLEDGMENTS

For their direction and support I am grateful to Sam Ridgway and Bill Lasley. I also thank Vicky Kirby for RIA of reproductive hormones in dolphin serum, conducted at the research department of the San Diego Zoological Society.

REFERENCES

Benirschke, K., Johnson, M. L., and Benirschke, R. V. (1980). Is ovulation in dolphins, *Stenella longrostris* and *Stenella attenuata*, always copulation induced? *Fish. Bull.* **78,** 507–528.

Britt, J. H., Cox, N. M., and Stevenson, J. S. (1981). "Advances in Reproduction in Dairy Cattle." Paper No. 6657 of the Journal Series of the North Carolina Agricultural Research Service, Raleigh, North Carolina. 25 pp.

Bryden, M. M., and Harrison, W. (1986). Gonads and reproduction. *In* "Research on Dolphins" (M. M. Bryden and R. V. Harrison, eds.), pp. 149–159. Clarendon Press, Oxford.

Caldwell, D. K., and Caldwell, M. C. (1972). Behavior of marine mammals. *In* "Mammals of the Sea, Biology and Medicine" (S. H. Ridgway, ed.), pp. 419–465. Charles Thomas Publ., Springfield, Illinois, 812 pp.

Chen, P. X., Liu, R. V., and Lin, K. V. (1984). Reproduction and the reproductive system in the beiji, *Lipotes vexillifer. In* "Reproduction in Whales, Dolphins and Porpoises" (W. F. Perrin, R. L. Brownell, and D. P. DeMaster, eds.), pp. 445–450. International Whaling Commission, Cambridge, Special Issue 6.

Collet, A. (1981). Biology of the common dolphin, *Delphinus delphis,* in the N. E. Atlantic. Ph.D. thesis, L'Université de Poitiers, France. 155 pp.

Cornell, L. H., and Asper, E. D. (1978). A census of captive marine mammals in North America. *Int. Zool. Yearb.* **18,** 220–224.

Cornell, L. H., Asper, E. D., Antrim, J. E., Searles, S. S., Young, W. G., and Goff, T. (1987). Progress report: Results of a long range captive breeding program, *Tursiops truncatus* and *Tursiops truncatus gilli. Zool. Biol.* **6,** 41–53.

Couver, R. L., and Schroeder, J. P. (1984). A status report on a survey of *Tursiops* breeding programs. *Progr. Summ., Annu. Symp. Eur. Assoc. Aquatic Mamm.,* p. 34.

Czekala, N. M., Galluser, S., Meier, J. E., and Lasley, B. L. (1986). The development and application of an enzyme immunoassay for urinary estrone conjugates. *Zool. Biol.* **5,** 1–61.

Essapian, F. S. (1953). The birth and growth of a porpoise, *Nat. Hist.* **62**(9), 392–399.

Gray, K. N., and Chapman, C. G. (1974). Pregnancy diagnosis in bottle-nosed dolphins. *Annu. Proc., Am. Assoc. Zoo Vet.,* pp. 157–160.

Green, R. F. (1972). Observations on the anatomy of some cetaceans and pinnipeds. *In* "Mammals of the Sea, Biology and Medicine" (S. H. Ridgway, ed.), pp. 247–297. Charles Thomas Publ., Springfield, Illinois. 812 pp.

Green, R. F. (1977). Anatomy of the reproductive organs in dolphins. *In* "Breeding Dolphins: Present Status, Suggestions for the Future" (S. H. Ridgway and K. Benirschke, eds.), pp. 185–194. U.S. Dept. of Commerce, NTIS PB-273-673, Washington, D.C. 308 pp.

Harrison, R. J. (1969). Reproduction and reproductive organs. *In* "Biology of Marine Mammals" (H. R. Anderson, ed.), pp. 253–348. Academic Press, New York.

Harrison, R. J. (1977). Ovarian appearances and histology in *Tursiops truncatus. In* "Breeding Dolphins, Present Status, Suggestions for the Future" (S. H. Ridgway and K. Benirschke, eds.). U.S. Dept. of Commerce, NTIS PB-273-673, Washington, D.C. 308 pp. 195–204.

Harrison, R. J., and Ridgway, S. H. (1971). Gonadal activity in some bottlenose dolphins, *Tursiops truncatus. J. Zool., London* **165,** 355–366.

Keller, K. V. (1987). Training of Atlantic bottlenose dolphins, *Tursiops truncatus,* for artificial insemination. *Proc., 14th Annu. IMATA Conf.,* pp. 22–24.

Kirby, V. L., and Ridgway, S. H. (1984). Hormonal evidence of spontaneous ovulation in captive dolphins, *Tursiops truncatus* and *Delphinus delphi. In* "Reproduction in Whales, Dolphins and Porpoises" (W. F. Perrin, R. L. Brownell, Jr., and D. P. DeMartin, ed.), pp. 459–464. International Whaling Commission, Cambridge, Special Issue 6.

Lasley, B. L. (1980). Endocrine research advances in breeding endangered species. *Int. Zoo Yearb.* **20,** 521.

Luskutoff, N. M., Ott, J. E., and Lasley, B. L. (1983).

Strategies for assessing ovarian function in exotic species. *J. Zoo Anim Med.* **14,** 3–12.

McBride, A. F., and Kritzler, H. (1951). Observations on pregnancy, parturition and post-natal behavior in the bottlenose dolphin. *J. Mammal.* **32**(3), 251–266.

McDonald, L. E. (1977). "Veterinary Endocrinology and Reproduction." Lea & Febiger, Philadelphia, Pennsylvania. 493 pp.

Perrin, W. F., and Reilly, S. B. (1984). Reproductive parameters of dolphins and small whales of the family Delphinidae. *In* "Reproduction in Whales, Dolphins and Porpoises" (W. F. Perrin, R. L. Brownell, Jr., and D. P. DeMartin, eds.), pp. 97–133. International Whaling Commission, Cambridge, Special Issue 6. 195 pp.

Ridgway, S. H. (1972). Homeostasis in the aquatic environment. *In* "Mammals of the Sea, Biology and Medicine" (S. H. Ridgway, ed.), pp. 590–747. Charles Thomas, Publ., Springfield, Illinois. 812 pp.

Ridgway, S. H., and Green, R. F. (1967). Evidence for a sexual rhythm in male porpoises, *Lagenorhynchus obliquidens* and *Delphinus delphis bairdi. Norsk. Hvalfangst-Tidende,* No. 1, pp. 1–8.

Ridgway, S. H., and Benirschke, K. eds. (1977). "Breeding Dolphins: Present Status, Suggestions for the Future." U.S. Dept. of Commerce, NTIS PB-273-673, Washington, D.C. 308 pp.

Sawyer-Steffan, J. E., Kirby, V. L., and Gilmartin, W. G. (1983). Progesterone and estrogens in the pregnant and non-pregnant dolphin, *Tursiops truncatus,* and the effect of induced ovulation. *Biol. Reprod.* **28,** 897–901.

Schroeder, J. P. (1983). Techniques for improving reproductive efficiency in *Tursiops. 14th Int. Assoc. Aquatic Anim. Med.,* p. 10 (Abstr).

Schroeder, J. P. (1984). Induced reproductive cycle events in *Tursiops truncatus. In* "Reproduction in Whales, Dolphins, and Porpoises" (W. F. Perrin, R. L. Brownell, Jr., and D. P. DeMaster, eds.), p. 483. (Abstr). International Whaling Commission, Cambridge, Special Issue 6. 495 pp.

Schroeder, J. P. (1985). Artificial insemination of the bottlenose dolphin *Tursiops truncatus. In* "American Association of Zoo Veterinarians Annual Meeting" (M. S. Silberman and S. Silberman, eds.), abstracts and papers, pp. 122–124.

Schroder, J. P., and Keller, K. V. (1989). Seasonality of testosterone and sperm density in *Tursiops truncatus. J. Exp. Zool.* **249,** 316–326.

Schroeder, J. P., Keller, K. V., and Kirby, V. L. (1983). Testosterone levels and sperm production in *Tursiops truncatus. Fifth Bienn. Conf. Biol. Mar. Mamm., Boston, Massachusetts,* p. 91 (Abstr).

Slijper, E. J. (1962). "Whales." New York, Basic Books.

Tavolga, M. C. (1966). Behavior of the bottlenose dolphin (*Tursiops truncatus*): Social interactions in a captive colony. *In* "Whales, Dolphins and Porpoises" (K. S. Norris, ed.), pp. 718–730. Univ. of California Press, Berkeley.

Tavolga, M. C., and Essapian, F. S. (1957). The behavior of the bottlenose dolphin *Tursiops truncatus*: Mating, pregnancy, parturition and mother–infant behavior. *Zoologica* **42**(1), 11–31.

Wells, R. S. (1984). Reproductive behavior and hormonal correlates in Hawaiian spinner dolphins, *Stenell longirostris. In* "Reproduction in Whales, Dolphins and Porpoises" (W. F. Perrin, R. L. Brownell, Jr., and D. P. DeMartin, eds.), pp. 465–472. International Whaling Commission, Cambridge, Special Issue 6.

Wood, F. G. (1953). Underwater sound production and concurrent behavior of captive porpoises, *Tursiops truncatus* and *Stenella plagiodon. Bull. Mar. Sci., Gulf Caribbean* **32,** 120–133.

Wood, F. G. (1977). Birth of porpoises at Marineland, Florida, 1939 to 1969, and comments on problems involved in breeding small cetaceans. *In* "Breeding Dolphins: Present Status, Suggestions for the Future" (S. H. Ridgway and K. Benirschke, eds.), pp. 46–70. U.S. Dept. of Commerce, NTIS PB-273-673. Washington, D.C. 308 pp.

Yoshioka, M., Mohri, E., Tobayama, T., Aida, K., and Hanyu, I. (1986). Annual changes in serum reproductive hormone levels in the captive female bottle-nose dolphins. *Bull. Jpn. Soc. Fish.* **52**(11), 1939–1946.

Artificial Insemination of Bottlenose Dolphins

26

J. Pete Schroeder
Karl V. Keller
Naval Ocean Systems Center
Hawaii Laboratory
Kailua, Hawaii 96734

INTRODUCTION

Techniques for artificial insemination (AI) are well established for breeding of domestic animals. Those techniques have been extended in recent years to the breeding of endangered species of wild terrestrial animals in captivity. Based on the suggestions and recommendations of a 1975 workshop report (Ridgway and Bernirschke, 1977), the Naval Oceans Systems Center (NOSC), Hawaii Laboratory, started research on the reproductive biology of bottlenose dolphins, *Tursiops truncatus*, with the goals of establishing a managed breeding program and developing techniques for artificially inseminating these and other small odontocetes.

Kirby and Ridgway (1984) presented hormonal evidence of spontaneous ovulation in captive dolphins. Sawyer-Steffan *et al.* (1983) indicated that ovulation can be induced by injections of exogenous hormones. Schroeder (1985) presented a method of artificial insemination of bottlenose dolphins. This chapter describes a study of a group of bottlenose dolphins to validate initial reports of hormone activity, refine drug dosages for exogenous induction of ovulation, and adapt AI techniques traditionally used with terrestrial animals for use with cetaceans.

METHODS AND RESULTS

Animal Subjects

Seven reproductively mature bottlenose dolphins, one male and six females, were housed in a cluster of five seawater pens in Kaneohe Bay, Hawaii. The male had previously sired three calves born in captivity. Five of the six females had previously calved in captivity. Each pen was 8 by 8 by 3.5 m deep; all were enclosed by 16 cm square wire mesh so that auditory, chemical, and visual contact was possible among all animals. The male was kept separate in one pen. The females shared the other four. Gates allowed control of access among pens for social interaction and research manipulation. The dolphins were fed thawed mackerel, *Scomber japonicus,* Columbia River smelt, *Thaleichtys pacificus,* and herring *Clupea harengus,* plus multivitamin and mineral supplements.

Blood Sampling

The dolphins were conditioned to swim onto a stranding device and lie in a stretcher for blood sampling with minimal restraint. Three of the seven dolphins, the male and two females, were trained to present their flukes while remaining in the water, allowing blood sampling without restraint.

Blood was collected weekly from dorsal or ventral vessels of the flukes. Care was taken to draw samples at the same time of day, before feeding, using consistent techniques. Blood was placed in serum separation tubes (SST, blood serum separation tubes for determinations requiring serum; Becton Dickinson, Franklin Lakes, NJ) for radioimmunoassay (RIA) determinations of each female's total estrogen levels and progesterone levels, and the male's testosterone levels. Serum collected from centrifuged samples was stored at −4°C within 1 hour of collection. Blood for complete blood counts was collected, in 2-ml tubes containing potassium EDTA, for routine health monitoring. Initially, reproductive hormone RIAs were performed at the Research Department of the San Diego Zoo (Kirby and Ridgway, 1984). By analyzing matching sample data sets, however, we found we could obtain comparable results of progesterone and testosterone RIA from a commercial laboratory (Accupath, a SmithKline Bio-Science Laboratory) (Wells, 1984) which provided clinically accurate results within 72 hours.

Determination of the male dolphin's reproductive cycle required taking blood samples twice monthly on the same day as a semen sample, and semen samples an average of twice per week for 28 months, through three breeding seasons. The dolphin's serum testosterone levels fluctuated throughout the year, each year, as is observed for other bottlenose dolphins (Schroeder and Keller, 1989). Of 21 dolphins sampled during a 3-year period, serum testosterone levels ranged from less than 1.0 to 54.4 ng/ml ($\bar{x}$ = 9.5, S.E.M. = 0.81, n = 241), with peaks occurring in June and July. Serum testosterone levels of the male dolphin used in this study ranged from 2.0 to 54.4 ng/ml ($\bar{x}$ = 20.4, S.E.M. = 1.6, n = 79) over 28 months, showing peaks similar to those of the larger sample (Fig. 1).

Semen Collection

The male dolphin was trained to extend his penis on command for manual collection of semen (Fig. 2). The operant conditioning steps included hand stationing; positioning of body near the water surface, ventral side up; penis erection; presentation of the penis; and semen collection (Keller, 1986). The behavior allowing semen collection was operational after 55 training days, each consisting of one 20- to 30-minute session or up to six 3- to 5-minute sessions. Once trained, the behavior remained consistent and reliable throughout the study.

Semen was obtained by manual manipulation an average of twice weekly. Each sample was collected in a 15-ml sterile glass test tube with a 40-ml plastic sleeve attached. The sleeve fit the tip of the dolphin's penis and was held in place by the operator's hand. Within 30 minutes of collection, each sample was analyzed for appearance, sperm count, sperm density, motility (0–99%), quality of motility (1–5), and percentage of clumping (Seager and Platz, 1977). The best quality ejaculates were then placed in a cryoprotectant/extender solution (Table 1). The mixture of semen and cryoprotectant agent was gradually cooled, then frozen by pipetting onto a block of dry ice (early in the study) or on a polyethylene/polystyrene platform in liquid nitrogen surrounded by liquid nitrogen vapor. The number of sperm per pellet was recorded, and the pellets were stored in liquid nitrogen.

In the period 28 August 1981 through 2 December 1983 a total of 1332 ejaculates was collected, at a rate of 1 to 12 per session during twice-weekly

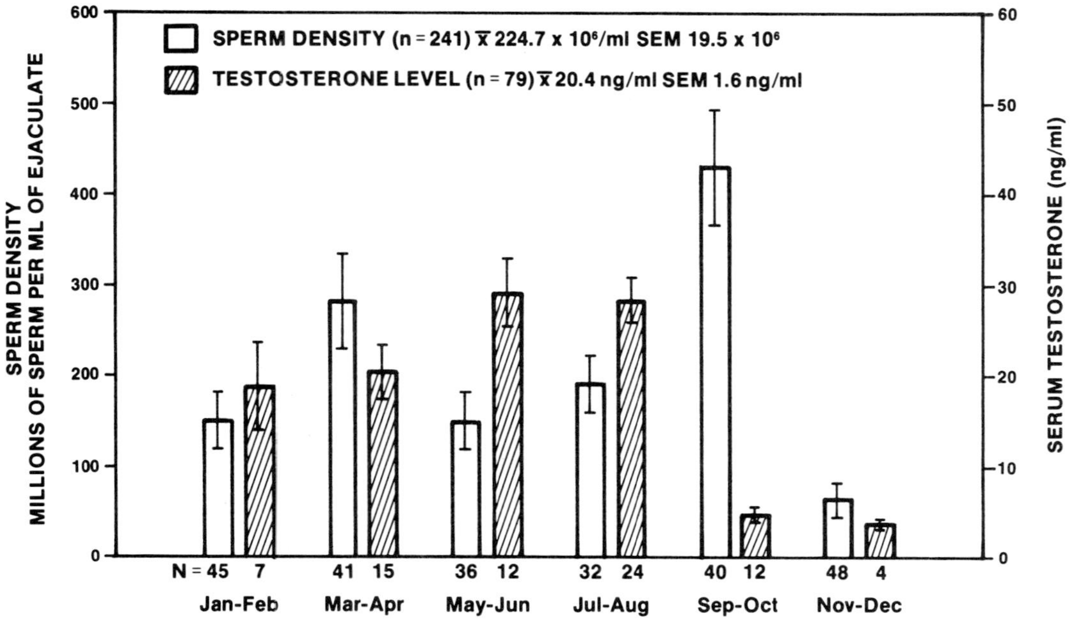

Figure 1 Sperm density of 241 daily combined semen samples, collected twice weekly from 28 August 1981 to 2 December 1983, and serum testosterone levels from 79 serum samples, collected between June 1981 and November 1983 twice monthly. Levels represent mean values for all samples collected during that 28-month period; standard errors of the means are also indicated.

sessions, averaging four ejaculates per 20- to 30-minute session. The volume of individual ejaculates ranged from 0.1 to 39.5 ml ($\bar{x}$ = 4.3, S.E.M. = 0.14, n = 1332), with sperm counts ranging from near zero to 54,569 × 10^9 sperm per ejaculate ($\bar{x}$ = 1.104 × 10^9; S.E.M. = 0.102 × 10^9, n = 1332) Data from each day's samples were combined to calculate the daily sperm output. Sperm density calculated from each session/day's output (expressed as the number of sperm times

Figure 2 A male bottlenose dolphin presents its penis (A) for semen collection (B). Photo: J. Pete Schroeder.

Table 1
Cryoprotectant/Extender Agents Used to Protect Bottlenose Dolphin Semen during Freezing[a]

	In Fructose	In Lactose
% Glycerol (v/v)[b]	6	6
% Egg yolk (v/v)	20	20
% Fructose (w/v)[c]	11	0
% Lactose (w/v)	0	11
Antibiotics		
Penicillin G (units/ml)	1000	1000
Streptomycin sulfate (mg/ml)	1.25	1.25

[a] Solutions are made up to 100 ml with deionized water.
[b] Percent volume in 100 ml of solution.
[c] Percent weight in 100 ml of solution.

10^6 per milliliter) (Linde *et al.*, 1981) ranged from less than 1 to 1100×10^6/ml ($\bar{x} = 224.7 \times 10^6$; S.E.M. $= 19.5 \times 10^6$; $n = 241$). The highest daily sperm output and sperm densities were observed in spring (March and April) and fall (September and October) (Fig. 1). Peak sperm density levels followed peak levels of serum testosterone by 2–6 weeks (Schroeder *et al.*, 1983; Schroeder and Keller, 1989).

We found the "hanging drop" in liquid nitrogen vapor to be the most successful method of freezing semen for storage. Hanging drop refers to the method of controlled dripping of semen from a glass pipette to a polyethylene film stretched on a styrofoam platform. When thawed and analyzed, one sample frozen for 955 days had 60% progressively motile sperm, with grade 5 motility (Table 2). Very few abnormal sperm were noted based on criteria of Flemming *et al.* (1981).

Three pellets were thawed periodically to determine the best technique of freezing (i.e., that in which the highest percentage of sperm remained viable) as well as the difference between lactose and fructose as constituents of the cryoprotective solution (Table 2). The last semen sample selected for this study was frozen with lactose in the cryoprotectant in May 1985. When it was thawed 1 month later, it had 85% motile sperm, grade 5 motility, and no visible abnormal sperm. A more recent thawing, February 1988, revealed 65% motile sperm, grade 5 motility, and still no visible abnormal sperm.

Spontaneous Ovulation

In analyses of blood samples collected once weekly (or more frequently) for 42 months from three female bottlenose dolphins, 24 months from three others, and 18 months from one dolphin, the females were found to ovulate spontaneously, one to four times within a limited season. Ovulation was evidenced by sudden pulses in serum progesterone levels from a baseline of less than 1 ng/ml to levels ranging from 4.2 to 22.2 ng/ml ($\bar{x} = 11.7$, S.E.M. $= 0.92$, $n = 37$) (Table 3). Pulses were followed by significant declines to baseline. The time interval between serum progesterone elevation from baseline to return to baseline represents the luteal phase of the estrous cycle.

Table 2
Evaluation of Cryoprotectant/Extender Agents in Dolphin Semen

Sample Date	Sperm/ml of Ejaculate (10^6)	Days Frozen at Thawing[a]	% Motile Sperm at Thaw[b]	
			In Lactose	In Fructose
5/3/82	812	955	60	65
5/5/82	702	700	60	30
5/29/84	1225	199	65	60
6/11/84	378	186	50	10
10/12/84	950	24	45	5
5/9/85	1253	30	85	45

[a] Pellets were frozen in liquid nitrogen vapor and stored in liquid nitrogen.
[b] All samples had 95% motile sperm before freezing.

When consecutive spontaneous ovulations occurred it was impossible (because of sampling intervals) to separate the follicular phase from the luteal phase of the estrous cycle accurately. In cows and sows the follicular phase, from the regression of the corpus luteum to ovulation, is 3–6 days. This short follicular phase does not reflect the true duration of graafian follicle growth. Thus, the length of the estrous cycle is closely related to the duration of the luteal phase (Levasseur and Thibault, 1980). The time interval between successive ovulations in the dolphins was therefore considered to represent both follicular and luteal phases of estrous cycles, ranging from 24 to 35 days ($\bar{x}$ = 27.25 days, S.E.M. = 3.4, n = 37).

Large numbers of spontaneous ovulations were recorded during the first 9 months of the study as compared to the following years. There is obviously individual variability in each female's reproductive cycle, but the trend in dolphins toward seasonality is confirmed by 20 of 37 ovulations during August, September, and October and 8 of 37 during the adjacent months July and November (Fig. 3). The two pregnancies shown in Fig. 3, plus Tt-580's becoming pregnant in October 1985, support the concept of a fall breeding season for this group of dolphins in Hawaii.

Inducing Ovulation

Several protocols have been developed for exogenous induction of estrus and ovulation in the female dog using pregnant mare's serum gonadotropin (PMSG) and human chorionic gonadotropin (HCG) (Thun *et al.*, 1977; Wright, 1980). In our female dolphins, two intramuscular injections of PMSG (Gestyl Gonadotrophinum Sericum; Organon, N.V. Oss, Holland) 48 hours apart, followed 5 days later by one intramuscular injection of HCG (Pregnyl Human Chorionic Gonadotropin for injection, U.S.P.; Organon, Inc., West Orange, NJ), induced ovulation. Injection of less than 800 international units (IU) of PMSG did not stimulate physiological processes leading to ovulation. In one female dolphin (Tt-009) an initial dose of 2000 IU of PMSG produced signs of a persistent corpus luteum.

The optimum dose of PMSG was 1600 IU, injected during anestrus, to initiate folliculogenesis. It was necessary to follow the initial injection 48 hours later with an injection of 800 or 1000 IU of PMSG. The second PMSG injection ensured that the follicle would continue developing to maturation. HCG (1000 or 3000 IU) was injected 7 days after the first PMSG injection to cause the follicle to rupture and shed an egg. Ovulations were

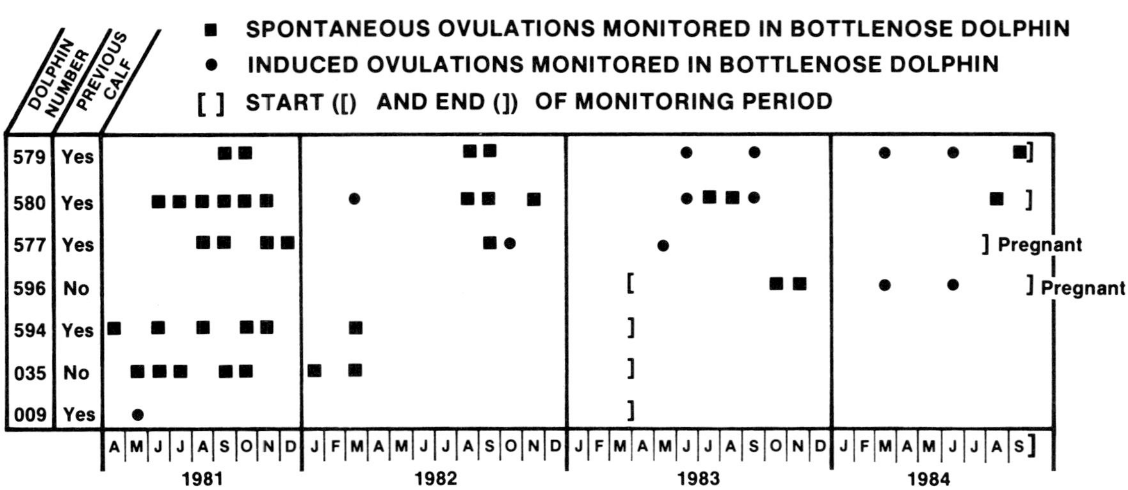

Figure 3 The large number of ovulations the first year of the study point out the variability of each individual's reproductive cycle. The trend toward seasonality is confirmed by 20 of 37 ovulations during August, September, and October and 8 of 37 during the adjacent months July and November. The two pregnancies shown support a fall breeding season for this group of dolphins in Hawaii. Additionally, Tt-580 became pregnant in October 1985.

Table 3
Serum Progesterone Levels in Bottlenose Dolphins in Various Reproductive Circumstances

		Serum Progesterone (ng/ml)			
Dolphin	Circumstance	Sample Size	Range	Mean	Standard Error
—	All bottlenose dolphins in NOSC monitoring program impregnated naturally and delivering live calves*	115	2.0–52.0	19.6	9.0
Tt-577	Naturally impregnated; monitored from conception to birth	18	11.5–26.2	16.9	1.5
Tt-596	Naturally impregnated, monitored from conception to birth	20	9.5–28.4	18.9	2.1
Tt-580	Naturally impregnated, monitored from conception to birth	15	3.0–48.0	18.6	2.9
Tt-653	Naturally impregnated, monitored from conception to birth	17	3.0–31.0	17.4	2.2
Tt-580	AI, monitored for 14 weeks until spontaneous abortion	16	12.2–55.3	36.0	2.5
Tt-650	AI, monitored for 10 weeks until spontaneous abortion	15	2.7–56.0	19.7	7.5
Tt-009	Persistent corpus luteum from excessive dose of PMSG	39	1.1–87.0	38.0	3.5
—	All bottlenose dolphins in NOSC Hawaii program monitored for spontaneous ovulations	37	4.2–22.2	11.7	0.9
—	All bottlenose dolphins in NOSC Hawaii program monitored for luteal phase of exogenously induced ovulations[a]	11	1.3–48.0	15.0	1.2

[a] Persistent corpus luteum and 2 AI pregnancies are not included. *Includes data on animals at cooperating institutions.

induced in 14 instances, 8 with 1000 IU and 6 with 3000 IU (Table 4). Following these injection protocols and comparing serum total estrogen and progesterone levels to previously established criteria, it appeared that, in this study, ovulation was induced in 14 of 20 subjects by injection of exogenous hormones (Table 3 and Table 4). Figure 4 presents typical results of ovulation induction in dolphin Tt-580, including data on total estrogen increases during folliculogenesis.

Artificial Insemination

To inseminate a female dolphin artificially, approximately 100 frozen semen pellets were placed in a sterile Whirl-pak, thawed as rapidly and gently as possible in a 37°C water bath and placed in the reproductive tract of the dolphin 8 and 24 hours following induced ovulation. In Trials 1 and 4, freshly collected semen was used (Table 4). Following procedures used in AI of domestic animals, the intervals between injection of HCG and introduction of sperm were arbitrary. We analyzed aliquots of the semen before and after AI.

We placed thawed semen into the spermathecal recess (Green, 1977) (between the external opening of the pseudocervix and the external opening of the true cervix), using a flexible fiber-optic laryngoscope (AO Scientific Instruments, Southbridge, Massachusetts 01550), 50 cm long and 6.5 mm in diameter, with remote end-tip control. The laryngoscope was introduced aseptically through a modified 16-mm tracheal tube which had been placed in the dolphin's vagina up to the pseudocervix. After the opening of the pseudocervix was clearly seen, the laryngoscope tip was manipulated through the pseudocervix into the spermathecal space. The semen was injected from a 3-ml syringe through the 2-mm aspiration channel of the laryngoscope into the spermathecal space.

After injection of the semen, the laryngoscope was withdrawn to clean the optics and obtain a swab for microbiological culture. Confirmation of correct placement of the semen was possible by reinserting the scope into the vaginal speculum and noting the presence or absence of semen in the speculum. On confirmation of proper placement of the semen, the dolphin was inclined about 30°, head down, for 5 minutes to facilitate capacitation (Viriyapanich and Bedford, 1981) and hasten entry of sperm into the uterus. There are no data to support the effectiveness of this technique.

Diagnosing Pregnancy

Radioimmunoassay of serum progesterone was used to determine pregnancy (Sawyer-Steffan *et al.*, 1983). Our criterion for a positive diagnosis of pregnancy was serum progesterone levels of 3–55 ng/ml (Fig. 5) sustained over 8 weeks (three separate serum samples). The criterion for a negative diagnosis of pregnancy was the presence of less than 1 ng/ml of progesterone in the serum any time post-AI. Serum progesterone levels of four dolphins impregnated by natural breeding and then monitored monthly from the month of conception to birth were as follows:

Dolphin Tt-577, $\bar{x}$ = 16.9 ng/ml, range 11.5–26.2, S.E.M. = 1.5 (n = 18)
Dolphin Tt-596, $\bar{x}$ = 18.9 ng/ml, range 9.5–28.4, S.E.M. = 2.1 (n = 20)
Dolphin Tt-580, $\bar{x}$ = 18.6 ng/ml, range 3.0–48.0, S.E.M. = 2.9 (n = 15)
Dolphin Tt-653, $\bar{x}$ = 17.4 ng/ml, range 3.0–31.0, S.E.M. = 2.2 (n = 17)

We detected the fetal heartbeat (roughly two times faster than that of the female) with an obstetrical doppler when the mother was 4–6 months pregnant, and the fetus was visualized with ultrasound at various times during gestation (Stone *et al.*, 1985; Cornell *et al.*, 1987).

Evaluating Artificial Insemination

The AI procedure was technically successful and caused no evident distress to any of the three female dolphins. Five AI procedures, each consisting of inseminations 8 and 24 hours post-HCG injection, resulted in indications of two pregnancies (Table 4).

The first AI procedure was with freshly collected semen in dolphin Tt-580, in March 1982 using a straight equine laryngoscope. The impossibility of entering the curved vagina to the opening of the pseudocervix with this instrument con-

Table 4
Induction of Ovulation in Bottlenose Dolphins Using PMSG and HCG

		PMSG (IU[a])						
Dolphin	Date	First Injection	Second Injection	HCG (IU)	Ovulation	AI	Trial no.	Comments
Tt-009	March 81	2000	None	1000	Yes	No		Persistent corpus luteum, 32 weeks
Tt-580	March 82	800	400	1000	Yes	Yes	1	No pregnancy
Tt-035	June 82	1000	None	1000	No	No		—
Tt-577	June 82	1000	None	1000	No	No		—
Tt-035	October 82	800	400	1000	No	No		—
Tt-577	October 82	800	400	1000	Yes	No		Placed with male, no pregnancy
Tt-577	March 83	1200	1000	1000	Yes	No		—
Tt-580	June 83	1600	800	1000	Yes	No		—
Tt-579	June 83	1600	800	1000	Yes	No		—
Tt-580	September 83	1600	1000	3000	Yes	Yes	2	Pregnant 14 weeks
Tt-579	September 83	1600	1000	3000	Yes	No		Placed with male, no pregnancy
Tt-580	March 84	1600	800	3000	No	No		—
Tt-596	March 84	1600	800	3000	Yes	No		No previous calf
Tt-579	March 84	1600	800	3000	Yes	No		Placed with male, no pregnancy
Tt-580	June 84	1600	800	1000	No	No		—
Tt-579	June 84	1600	800	1000	Yes	No		—
Tt-596	June 84	1600	800	1000	Yes	No		Placed with male, no pregnancy
SLP	February 85	1600	1000	3000	No	Yes	3	No pregnancy
Tt-580	April 85	1600	800	3000	Yes	Yes	4	No pregnancy
Tt-650	July 85	1600	1000	3000	Yes	Yes	5	Pregnant 10 weeks

[a] IU, International units.

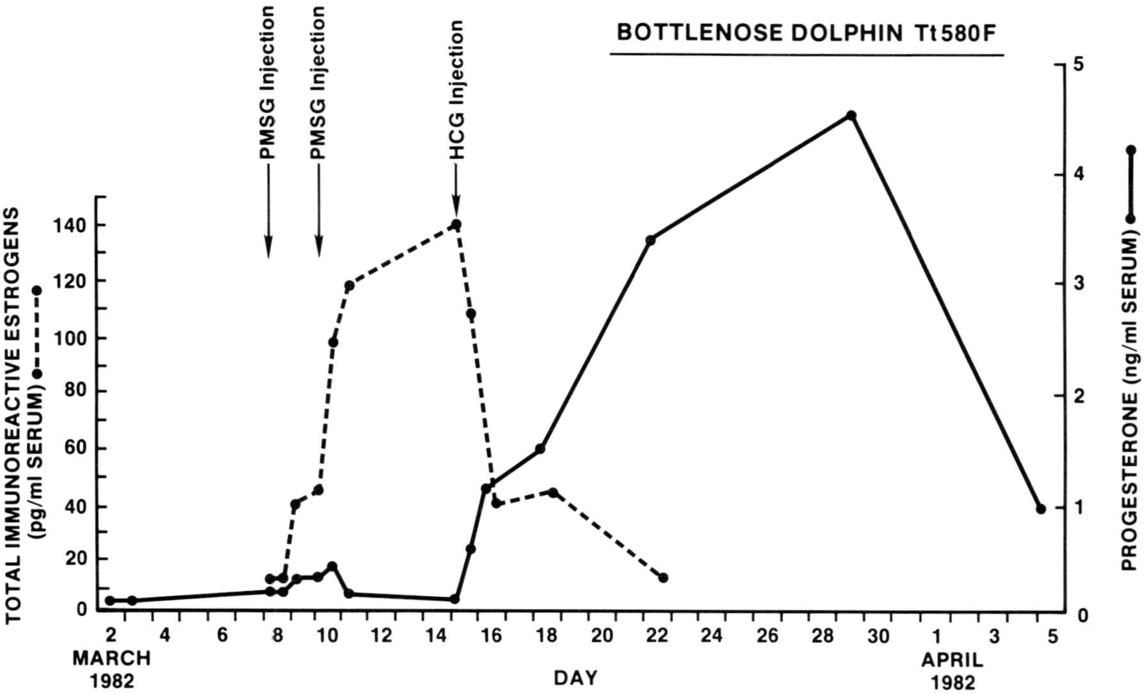

Figure 4 Hormone levels related to induction of ovulation of bottlenose dolphin Tt-580, March–April 1982. With injection of HCG, serum estrogen drops precipitously and serum progesterone begins increasing, indicating that ovulation has occurred.

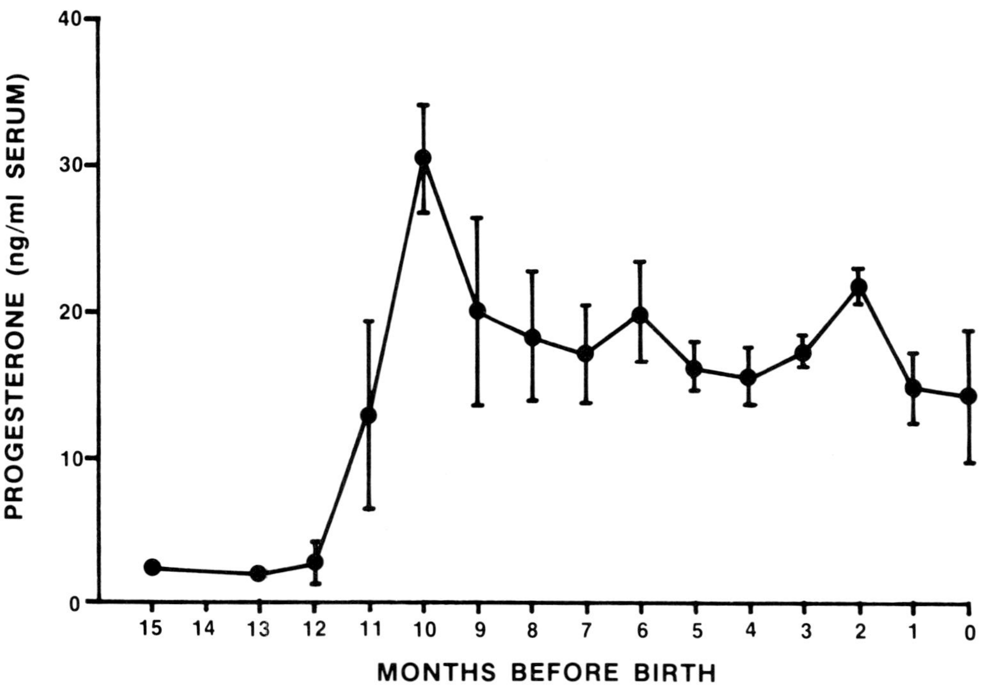

Figure 5 Mean serum progesterone levels of four pregnant bottlenose dolphins, sampled monthly (n = 68). Each experienced a successful delivery. Bars represent the standard errors of the mean.

tributed to the failure of the procedure. This led to replacement of the instrument with the flexible fiber-optic laryngoscope, allowing entry to the spermathecal space. The second AI trial using dolphin Tt-580 and the fiber-optic system resulted in indications of pregnancy. The dolphin's mean serum progesterone level over 14 weeks after insemination was 36 ng/ml (S.E.M. = 2.56, range 12.2–55.3, n = 16). A later attempt (Trial 4, April 1985) to artificially inseminate dolphin Tt-580 was unsuccessful.

In July 1985, we induced ovulation in dolphin Tt-650 and artificially inseminated her (Trial 5). Her mean progesterone level in the 10 weeks after insemination was 19.7 ng/ml (S.E.M. = 7.5, range 2.7–56, n = 15). This was her first pregnancy. Both the above pregnancies, however, ended spontaneously (Tt-580 after 14 weeks, Tt-650 after 10 weeks) as judged by rapid drops in progesterone to anestrous levels.

One of the five AI attempts (Trial 3) was performed at Sea Life Park (SLP), Oahu, Hawaii. We followed the protocol outlined in the Appendix to induce ovulation in a mature, 220-kg, 18-year-old female dolphin that had never calved. AI was successfully completed. Analysis of serum progesterone levels, however, indicated ovulation had not occurred. Dolphins Tt-580 and Tt-650, and all other dolphins in which we have attempted to induce ovulation, weighed 133–178 kg. Further studies are necessary to determine if larger doses of PMSG and/or HCG are necessary to induce ovulation in dolphins over 200 kg.

DISCUSSION

Artificial Insemination Trials

An orderly progression was made from determining normal reproductive patterns in our bottlenose dolphins, to implementing techniques for AI, to determining pregnancy by AI. In spite of the small numbers of subjects and procedures, the results were promising. The two diagnosed pregnancies from AI, though spontaneously terminated after 14 and 10 weeks, demonstrated progesterone level patterns similar to those recorded from 27 dolphin pregnancies from natural insemination that resulted in live births (Table 3). Of the three AI trials producing negative signs of pregnancy, the wrong equipment was used in Trial 1, a large female with a history of no breeding success was involved in Trial 3, and fresh sperm with suboptimal density, collected in April (the off-breeding season for the male), was used in Trial 4 (Table 4). The progesterone levels from the two AI pregnancies differed from those of dolphin Tt-009, who was given an excessive dose of PMSG, causing a persistent corpus luteum and resulting in progesterone levels higher than any previously reported. In that animal, levels ranged from 1.1 to 87 ng/ml ($\bar{x}$ = 38, S.E.M. = 3.52, n = 39) (Table 3). Levels peaked 3 weeks after injection, then gradually declined to baseline level (1 ng/ml) over the next 29 weeks.

Progesterone levels of pregnant dolphins in our monitoring program that delivered live calves ranged from 2 to 52 ng/ml ($\bar{x}$ = 19.6, S.E.M. = 9, n = 115) (Table 3). After sustaining levels of progesterone normal for pregnant bottlenose dolphins for 10–14 weeks, the two artificially inseminated dolphins experienced sharp drops in progesterone levels to less than 1 ng/ml, indicating spontaneous abortion similar to that seen in domestic animals. We recorded progesterone levels of 30 ng/ml from dolphin Tt-580 5 days before delivery of her normal calf and less than 1 ng/ml within 12 hours after birth (J. P. Schroeder, unpublished data). This indicates that dolphin progesterone has a short half-life and decreases sharply with expulsion of the fetus. The physiological half-life of progesterone is only 22–36 minutes in the cow. This is a surprisingly short time when one considers that this hormone is essential to maintaining pregnancy (McDonald, 1977). A constant secretion therefore is essential to maintaining the circulating levels.

False Pregnancy

Increased progesterone levels also may result from a persistent corpus luteum causing prolonged luteal activity. Such increased progesterone levels have been recorded from mares with prolonged luteal activity (Henry *et al.*, 1981). This "pseudopregnant" condition requires a progesterone source (Ginther, 1979). In horses with a persistent corpus luteum, progesterone levels tend to pulse to a peak and then decline gradually, similar to the pattern of dolphin Tt-009. One female bottlenose dolphin showed prolonged luteal phases over a 5-month period, supposedly indicating false pregnancy (Yoshioka *et al.*, 1986). However, in this animal, as in the instances noted above, progesterone levels returned to normal gradually. Harrison (1977) reported that corpora lutea of pseudopregnancy have not been found in the ovaries of bottlenose dolphins.

Effects of Ovulation-Induction Procedures on Breeding Success

By monitoring reproductive cycles of female dolphins, we were able to program natural breeding of three dolphins in which we had previously attempted to induce ovulation (dolphin Tt-580, six times; dolphin Tt-577, three times; dolphin Tt-596, two times) (Table 4). Natural pregnancies in these animals were diagnosed within 8 weeks, allowing for changes in nutrition and housing. Pregnancies were confirmed by doppler at 6 months and, in one case, by ultrasound at 11 months. All births were normal, occurring on 20 July 1985, 20 September 1985, and 21 October 1986, respectively. These births are significant because the females had undergone repeated ovulation-induction procedures without harm to their normal reproductive processes. The safety of the procedure has been established.

Aspects of a Successful AI Program

Biological and environmental factors influence reproductive success. Physiological stress provoked by disruption of parental and social bonds affects pregnant women, terrestrial mammals, and, according to Geraci (1984), undoubtedly whales and dolphins. There have been nine births from natural breeding at NOSC Hawaii from 1978 to 1988, so we can assume the environment provided by our facilities is suitable for successful dolphin reproduction. However, considerably more research is needed to determine the optimal timing of AI and to refine various aspects of post-AI husbandry.

Time of ovulation and the optimal interval between ovulation and insemination have yet to be determined. To date, no dolphin has been impregnated as a result of induced ovulation coupled with natural breeding, although sexual behavior has been observed following successful induction of ovulation. Analysis of urinary estrogen conjugates may prove to be a more accurate and faster method for diagnosing ovulation, whether spontaneous or induced (Czekala, 1986). Dolphins can be trained to cooperate with daily urine-collection procedures.

At a workshop on the artificial insemination of cetaceans, held at Victoria, British Columbia, Canada, in 1984, it was agreed that a research protocol such as that outlined in the Appendix would be of great benefit to oceanaria. AI would give them the abilities to control breeding, ship frozen sperm (instead of the animal), make genetic selections, and control disease. Thorough and consistent record-keeping and the use of a standardized protocol to induce ovulation (Appendix) are essential for any successful AI program. When freezing semen, each individual sperm must be considered a living entity, and the entire sample must be handled gently at each step of the freezing and thawing processes.

New Possibilities for the Future

Embryo transplantation is well established for use in laboratory and farm animals. The collection and transfer of embryos is generally achieved by surgical methods, but nonsurgical techniques apply to some species. The fiber-optic system used in artificial insemination of our dolphins could be useful for collecting eggs nonsurgically from a dolphin induced to ovulate, eggs which could be fertilized *in vitro* and then implanted in the uterus of another female dolphin. In this study, we demonstrated that banking of frozen semen is possible; the next step is to apply similar techniques to storage of frozen embryos.

We have described the development of techniques necessary for successful artificial insemination of the bottlenose dolphin. With additional knowledge of this animal's reproductive physiology, it should become possible to enhance captive breeding programs with bottlenose dolphins and other small odontocetes as well. It is advisable when developing reproductive technologies for rare and endangered species to begin by studying reproductive biology of a related common species that can serve as a model (Wildt *et al.*, 1986). The model for bottlenose dolphins reported herein may well be applicable to the propagation of endangered species such as the baiji, *Lipotes vexillifer* (Ralls, 1989). With further research on additional subjects and species, AI techniques may be applicable to other toothed cetaceans.

ACKNOWLEDGMENTS

We thank Steve Seager for starting us on the road to successful cryopreservation of dolphin semen. Thanks are also due to Mike Cates, Gary Purdy, and Randy Cates for their technical assistance. Ingrid Kang Shallenberger and her staff at Sea Life Park generously cooperated in our attempt to inseminate their dolphin artificially. We also thank Vicky Kirby for analysis of dolphin reproductive hormones at the research department of the San Diego Zoological Society.

APPENDIX

NOSC PROTOCOL FOR INDUCING OVULATION IN BOTTLENOSE DOLPHINS

1. Collect one blood sample in the morning before feeding. Have a reliable laboratory radioimmunoassay (RIA) performed on the serum as soon as possible. A serum progesterone level of less than 1 ng/ml indicates a nonpregnant, anestrous dolphin.

2. If the serum progesterone level is below 1 ng/ml, take a blood sample and administer the first of two PMSG injections at 900 hours on Day 1. (Day 1 should occur as soon as possible after the positive results from the serum sample have been obtained). On subsequent days, also beginning at 900 hours, unless otherwise noted, proceed as follows:

Day		Action
Day 2		Observe female dolphin's behavior
3		Sample blood, inject second dose of PMSG
4		Observe
5		Observe
6		Observe
7		Observe
8		Sample blood, inject HCG
8,	1600 hours	Sample blood
8,	after 1600 hours	Inseminate or introduce male*
9		Sample blood and inseminate or reintroduce to male
15		Sample blood

3. Submit all blood samples from Days 1–15 to a laboratory for RIA.

4. Dolphins used in the study should be conditioned to the blood-sampling procedure before the start of a study. The preferred technique is to sample from the flukes presented to the operator, while the dolphin remains in the water.

5. Blood samples should total 40 ml; the collected blood should be placed in tubes as follows:
   a. Place 1 ml in a 2-ml lavender-top heparinized test tube (a complete blood count should be done as soon as possible).
   b. Divide the remainder among three serum separation tubes.

All tubes of blood must be placed in ice immediately and centrifuged within 1 hour. The serum should be separated, labeled, and frozen for storage. This should provide at least five serum samples of 3 oml each. One sample from each day should go to the laboratory for RIA, the others to a serum bank for additional analysis if necessary.

6. If a male dolphin is to be introduced at the time of ovulation (Day 8 or 9),* his blood should have been sampled once every 2 weeks for 1 month before the study; his blood should be sampled again 2 weeks after Day 9.

* In the present study males were left with females only during daylight hours, so that reproductive activity could be observed. Success may well be improved by leaving females with males overnight or longer.

REFERENCES

Cornell, L. H., Asper, E. D., Antrim, J. E., Searles, S. S., Young, W. G., and Goff, T. (1987). Progress report: Results of a long term captive breeding program, *Tursiops truncatus* and *Tursiops truncatus gilli*. *Zoo Biol.* **6,** 41–53.

Czekala, N. M., Gallusser, S., Meier, J. E., and Lasley, B. L. (1986). The development and application of an enzyme immunoassay for urinary estrone conjugates. *Zoo Biol.* **5,** 1–61.

Flemming, A. D., Yanagimachi, R., and Yanagamachi, H. (1981). Spermatozoa of the Atlantic bottlenose dolphin *Tursiops truncatus*. *J. Reprod. Fert.* **63,** 509.

Geraci, J. R. (1984). Influence of stress and disease on reproductive success in cetaceans. *Rep. Int. Whal. Comm.* (Special Issue 6), p. 477.

Ginther, O. J., ed. (1979). "Reproductive Biology of the Mare, Basic and Applied Aspects." McNaughton and Gunn, Inc., Cross Plaines, Wisconsin. 413 pp.

Green, R. F. (1977). Anatomy of the reproductive organs in dolphins. *In* "Breeding Dolphins: Present Status, Suggestions for the Future" (S. H. Ridgway and K. Benirschke, eds.), pp. 185–194. U.S. Dept. of Commerce, NTIS PB-273-673, Washington, D.C. 308 pp.

Harrison, R. J. (1977). Ovarian appearances in histology in *Tursiops truncatus*. *In* "Breeding Dolphins: Present Status, Suggestions for the Future" (S. H. Ridgway and K. Benirschke, eds.), pp. 195–204. U.S. Dept. of Commerce, NTIS PB-273-673, Washington, D.C. 308 pp.

Henry, M., Coryn, M., Spincemaille, J., and Vandeplassche, M. (1981). Prolonged luteal activity in the mare. *Zuchthyg.* **16,** 145–153.

Keller, K. V. (1986). Training of Atlantic bottlenose dolphins (*Tursiops truncatus*) for artificial insemination. *14th Ann. IMATA Conf. Proc.*, pp. 22–24.

Kirby, V. L., and Ridgway, S. H. (1984). Hormonal evidence of spontaneous ovulation in captive dolphins (*Tursiops truncatus* and *Delphinus delphis*). *Rep. int. Whal. Commn* (Special Issue 6), pp. 459–464.

Levasseur, M. G., and Thibault, C. (1980). Reproductive life cycles. *In* "Reproduction in Farm Animals" (E. S. E. Hafly, ed.), pp. 130–149. Lea & Febiger, Philadelphia, Pennsylvania. 626 pp.

Linde, R., Doelle, G. C., Alexander, N., Kirchner, F., Vale, W., Rivier, J., and Rabin, D. (1981). Reversible inhibition of testicular steroidogenesis and spermatogenesis by a potent gonadotropin-releasing hormone agonist in normal men: An approach toward the development of a male contraceptive. *New Engl. J. Med.* **305,** 663–667.

McDonald, L. E. (1977). Female reproductive system. *In* "Veterinary Endocrinology and Reproduction" (L. E. McDonald, ed.), pp. 247–303. Lea & Febiger, Philadelphia, Pennsylvania. 493 pp.

Ralls, K. (1989). A semi-captive breeding program for the beiji, *Lipotes vexillifer*: Genetic and demographic considerations. *In* "Biology and Conservation of the River Dolphins," IUCN/SSC Occasional Papers No. 3.

Ridgway, S. H., and Benirschke, K. eds. (1977). "Breeding Dolphins: Present Status, Suggestions for the Future." U.S. Dept. of Commerce, NITS PB-273-673, Washington, D.C. 308 pp.

Sawyer-Steffan, J. E., Kirby, V. L., and Gilmartin, W. G. (1983). Progesterone and estrogens in the pregnant and non-pregnant dolphin, *Tursiops truncatus*, and the effects of induced ovulation. *Biol. Reprod.* **28,** 897–901.

Schroeder, J. P. (1985). Artificial insemination of the bottlenose dolphin *Tursiops truncatus*. *In* "American Association of Zoo Veterinarians Annual Meeting" (M. S. Silberman and S. Silberman, eds.), abstracts and papers, pp. 122–124.

Schroeder, J. P., and Keller, K. V. (1989). Seasonality of testosterone and sperm density in *Tursiops truncatus*. *J. Exp. Zool.* **249,** 316–321.

Schroeder, J. P., Keller, K. V., and Kirby, V. L. (1983). Testosterone levels of sperm production in *Tursiops truncatus*. *Fifth Bienn. Conf. Biol. Mar. Mamm., Boston*, p. 91 (Abstr.).

Seager, S. W. J., and Platz, C. C. (1977). Collection and evaluation of canine semen. *Vet. Clin. North Am. Small Anim. Pract.* **4,** 765–773.

Stone, L. R., Phillips, B., and Sweeney, J. C. (1985). Diagnostic ultrasonography of the bottlenose dolphin. *In* "American Association of Zoo Veterinarians Annual Meeting" (M. S. Silberman and S. Silberman, eds.), abstracts and papers. 94 pp.

Thun, R., Watson, P., and Jackson, G. L. (1977). Induction of estrus and ovulation in the bitch using exogenous gonadotropins. *Am. J. Vet. Res.* **138,** 483–486.

Viriyapanich, P., and Bedford, J. M. (1981). Sperm capacitation in the fallopian tube of the hamster and its suppression by endocrine factors. *J. Exp. Zool.* **217,** 403–407.

Wells, R. S. (1984). Reproductive behavior and hormonal correlates in Hawaiian spinner dolphins, *Stenella longirostris*. *Rep. int. Whal. Commn* (Special Issue 6), pp. 465–472.

Wildt, D. F., Schiewe, M. C., Schmidt, P. M., Goodrowe, K. I., Howard, J. G., Phillips, I. G., O'Brian, S. U., and Buch, M. (1986). Developing animal model systems for embryo technologies in rare and endangered wildlife. *Theriogenology* **25,** 33–51.

Wright, P. J. (1980). The induction of oestrus and ovulation in the bitch using pregnant mare serum gonadotrophin and human chorionic gonadotrophin. *Aust. Vet. J.* **56,** 137–140.

Yoshioka, M., Mohri, E., Tobayama, T., Aida, K., and Hanyu, I. (1986). Annual changes in serum reproductive hormone levels in the captive female bottle-nose dolphins. *Bull. Jpn. Soc. Fish.* **52**(11), 1939–1946.

Observations on the Early Development of a Captive Bottlenose Dolphin Calf

27

V. G. Cockcroft
G. J. B. Ross
Port Elizabeth Museum
Port Elizabeth
Republic of South Africa

INTRODUCTION

The relatively easy access to lactating female pinnipeds and their pups has facilitated studies of mother and calf interactions, calf energetics, suckling behavior, and early growth (Tedman *et al.*, 1981; Fedak and Anderson, 1982; Lavigne *et al.*, 1982). Bonner (1984) reviewed pinniped lactation strategies. Apart from the constituent analysis of cetacean milk, similar energetic and early developmental studies of cetaceans are almost completely lacking.

This is partly due to the difficulties inherent in observing free-ranging cetaceans. In the wild, access to mother and calf pairs is rarely practical, and such pairs are often secretive and elusive. Despite these difficulties a few short-term studies of wild births and subsequent behavioral development have been undertaken. For example, Norris *et al.* (1977) studied the movements and respiration patterns of mother and calf pairs of gray whales, *Eschrichtius robustus*, in Magdalena Bay, Mexico, and Taber and Thomas (1982) observed the first four months of calf development, mother–calf spatial relationships, and yearling behavior in southern right whales, *Eubalaena australis*, off Argentina.

Captive births of small cetaceans provide an ideal opportunity for studies of mother–calf interactions and the development of the calf. As a result, many behavioral and developmental studies of captive small cetaceans have been undertaken (McBride and Hebb, 1948; McBride and Kritzler, 1951; Tavolga and Essapian, 1957; Essapian, 1953; Tayler and Saayman, 1972; Saayman and Tayler, 1977; Drinnan and Sadlier, 1981). However, even in captivity there is a natural reluctance to allow investigators physical contact at the risk of imposing additional stress on neonates and their mothers. Consequently, the majority of studies on captive mother and calf pairs have been brief or cursory. This chapter presents the results of a study of the relationships among suckling, respiration, growth, and behavioral development of an Indian Ocean bottlenose dolphin, *Tursiops truncatus*, observed from birth to 2 years of age.

MATERIALS AND METHODS

Observations were made in the L-shaped dolphin lake (4.72×10^6 liters, maximum depth 5 m) at the Port Elizabeth Oceanarium. An underwater viewing facility provides a view of over 80% of the pool area.

"Simo" was born at 1636 hours on 3 February 1982. His mother, "Thandi," had been captured 36 days prior and was still largely untrained. Observations were initiated at 1715 hours and maintained until last light at approximately 1915 hours. The following day observations began at 800 hours and were intermittent throughout the day for a total of about 5 hours. Subsequently, a 12-hour continuous watch was kept once per week and a 24-hour watch once per month. This sequence was altered occasionally to allow more or less observation when needed. Over the 2 years, we used three additional observers trained by one or both of us. Night observations were aided by a 1500-W tungsten light which illuminated most of the pool area sufficiently. The light was illuminated before sunset and extinguished after dawn to prevent any sudden changes in light intensity. A total of 829 hours was spent observing Simo from birth to 16 December 1983. Data were recorded in diary fashion, specifying time and the relevant behavior.

Initially, almost any action or behavior by mother or calf was noted. Subsequently, only distinctive features of mother and calf activity were recorded. Of the calf's behaviors these included all feeding or attempted feeding, the side (left or right) from which he fed, the length of the feed, and the calf's pre- and postsuckling behavior. Suckling entailed the insertion of the rostral and mandibular tips into either of the mammary slits and the ensuing intake of milk (Tavolga and Essapian, 1957). Suckling time was measured from the moment of contact with the nipple to the moment of release. Other relevant behaviors, not directly associated with feeding, were also recorded. One of these, termed bumping, consisted of the calf's swimming beneath the mother and bumping her mammary area with the top of his head. It occurred both before and after suckling.

Three types of presentation by the mother during suckling were recognized and recorded. In side presentations Thandi turned to one side, arching her back and presenting the calf with her mammary area. In vertical presentations she arched her back while in the normal swimming position (without turning to either side). The mammary area was thus below her, forcing the calf to feed while on his side. Combined presentations were combinations of these two forms. If no presentation occurred prior to a feed this was also recorded.

Respiration times of the calf were recorded for a maximum of 15 minutes in each hour of observation. Respiration was considered to have occurred if Simo broke the surface. The time of each respiration was recorded as was the identity of any associated adult. Only if a minimum of five consecutive breaths with any one partner were recorded were these included in calculations. This excluded very short periods of partner association which were not representative of longer-term patterns. The general activity of the calf, resting or active, was also noted. If an activity precluded the monitoring of respirations, recording was terminated and only resumed when respirations could again be adequately monitored.

Handling of Simo was prohibited during his first year, and estimates of his body length were made from a series of photographs of him with his mother. To avoid errors of parallax, these were usually taken when he was directly below the mother. Thandi measured 241 cm at capture and 242 cm 19 months later. Photographs were taken frequently, at irregular intervals, for the first 9 months and less often thereafter. On a few occasions after the first year, it was possible to estimate Simo's length from the number of floor tiles between the tip of his rostrum and fluke notch. As the tiles were square and of known dimensions, length could be estimated, a suitable account being taken of intertile spacing. The first direct measurement of Simo was made 14 months after birth. Simo's length and age were used to estimate mass, judging from data obtained from 74 shark-net-caught bottlenose dolphin calves less than 210 cm in length.

To estimate mammary gland volume, a mammary gland was excised from PEM N1182, a heavily lactating female caught in a shark net. The gland, measuring 360 × 100 × 38 mm, was washed, filled with X-ray-opaque fluid, and radiographed. Subsequently, a resin cast was made by filling the gland with slow-acting resin (7 days to mature). This dolphin's uterus was slightly distended, and a large corpus was evident in her right ovary. Unfortunately, no information about her calf is available.

RESULTS

Suckling and Feeding

Eighty-four (5.2%) of 1622 recorded suckling bouts were either untimed or uncertain and have been excluded from these results. There was no significant difference in Simo's use of left and right mammary glands ($p > .05$, Student's t-test; 829 left, 793 right; mean suckling time on both sides 4.7 seconds), and therefore no distinction has been made between suckling from the left or right.

No diurnal trends were evident in Simo's mean hourly suckling times during the first 16 weeks, suggesting no regular peak suckling time and no daily suckling pattern (Spearman's rank order correlation coefficient, all p values >0.25). Suckling almost always took place while mother and calf circuited the deepest part of the pool, directly in front of the viewing window. Circuiting was characterized by a repetitious circling of the same route and a slow, regular breathing pattern.

Simo's hourly suckling frequency and total number of suckles per hour decreased with age, particularly over the first 4 months (Fig. 1). Conversely, the mean suckling time per feed was highly variable but increased with age, reached a plateau between 5 and 16 months, and then decreased rapidly (Fig. 1). The date of Simo's last suckle is unknown. During the last observation, on day 710, he suckled once in the 24-hour period.

Simo's first known solid-food intake was on day

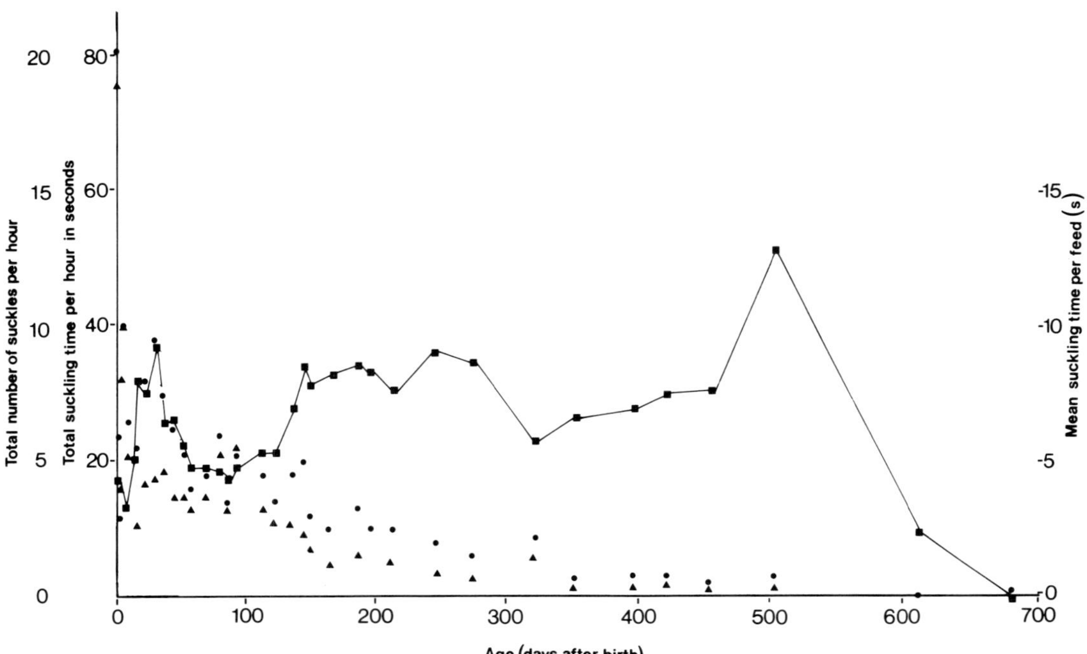

Figure 1 Simo's total number of suckling bouts per hour (▲), total time spent suckling per hour (●), and mean time of each suckle (■) plotted against his age.

321 (600 g of maasbanker, *Trachurus trachurus*). Regular solid-food intake did not begin until day 331. Simo's mean daily fish intake increased with his mass, although the mass of this food as a percentage of his body mass remained fairly stable at around 4% (Fig. 2).

Less than 1% of suckles were followed, within 3 minutes, by bumping. The percentage of suckles preceded by bumping (within 3 minutes prior to suckling) ranged between 58% to 96% throughout, except during initial observations when bumping was not specifically recorded (Fig. 3). Bumping frequency outside these times remained constant with age, but its proportion to the total number of bumps increased as suckling and presuckling bumping frequency decreased.

All Simo's suckles over the first 4 weeks were associated with side presentations by the mother (Fig. 4A). At about 6 weeks, Thandi began combining side presentations with vertical presentations. Subsequently, the frequency of side presentations decreased, while that of vertical and combined presentations showed high variability and no obvious pattern (Fig. 4). By 1 year only half of all suckles were accompanied by some change in the mother's posture.

The mammograph of PEM N1182 and a resin cast of the mammary gland are shown in Figs. 5 and 6, respectively. One hundred forty milliliters of X-ray-opaque fluid was used to fill the mammary gland for radiography, and 147 ml of resin was used in the drip to make the cast. The similarity of these two volumes suggests that they are reasonable estimates of the total volume of this gland. The volume of the cistern and major ducts was estimated from the resin cast only, by displacement, at 40 ml.

Respiration

During the study 13,908 respirations were recorded. Simo rarely respired with females in the pool other than his mother, and respirations with these animals were either single or double breaths. Therefore, no age-related respiration patterns were evident from these data. At day 9, when respiration partners were first recorded, 98% of Simo's respirations were with his mother (Fig. 7). Despite marked variations over the following weeks, this proportion had declined to 20% by day

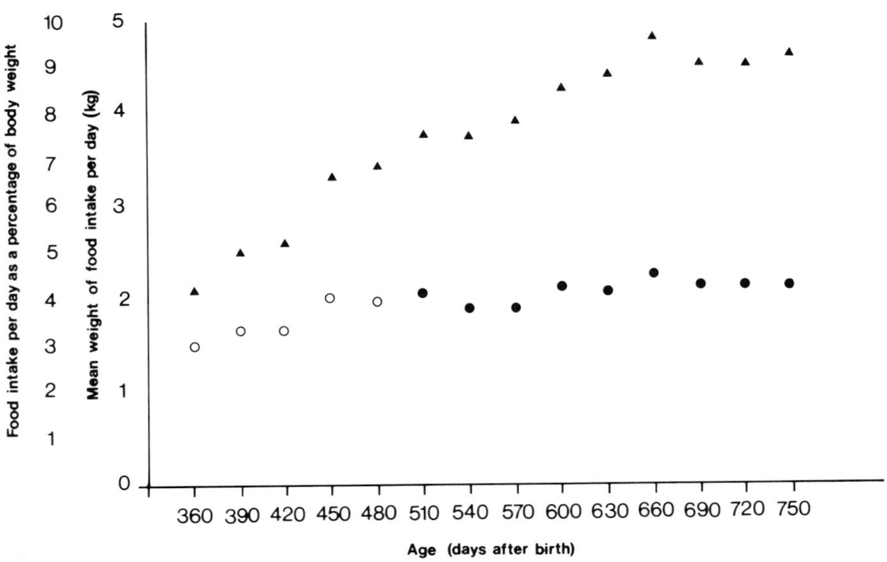

Figure 2 Simo's mean daily weight of ingested food after his first solid-food intake at day 321 (▲). The food intake as a percentage of body weight is also shown (○, calculated from estimated body weights; ●, calculated from actual weights).

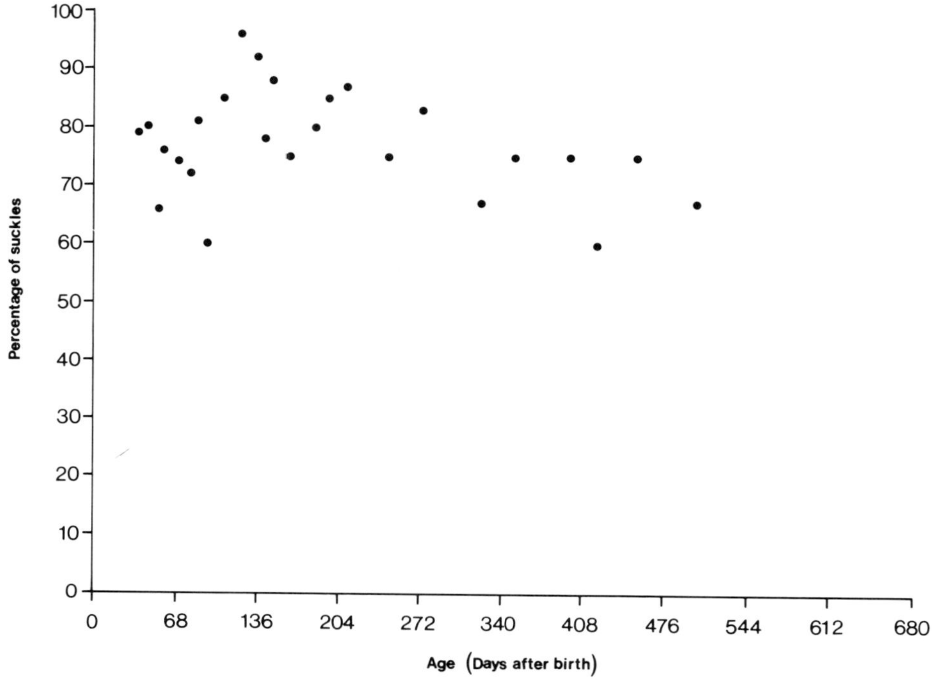

Figure 3 Influence of age on the percentage of Simo's suckles preceded by his bumping his mother's mammary area.

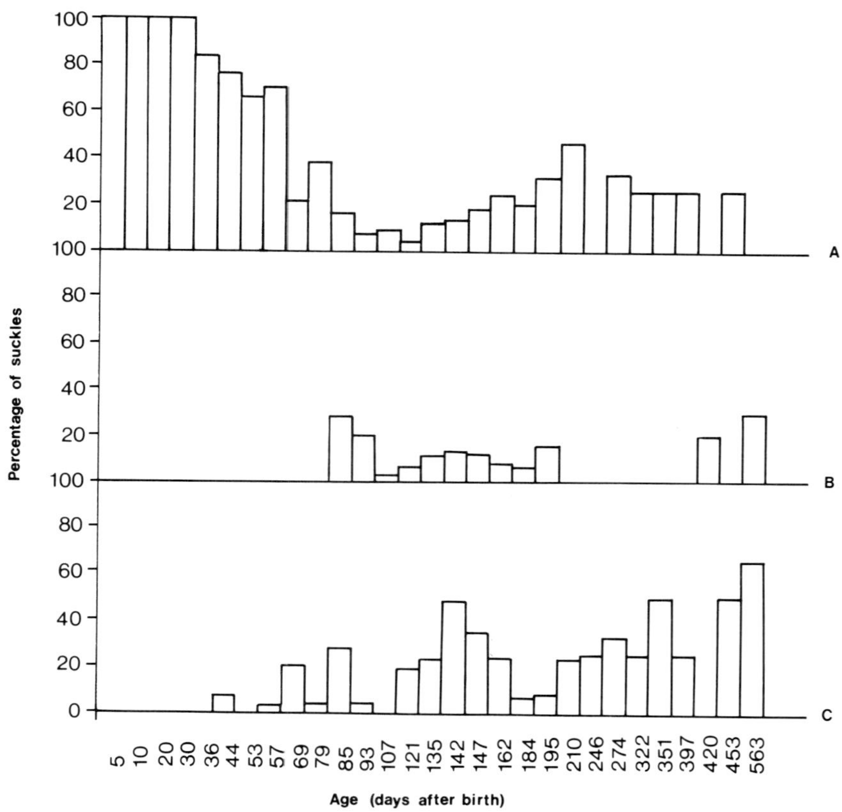

Figure 4 Change with age in the percentage of Simo's suckles preceded by Thandi's presenting sideways (A), vertically (B), or a combination of the two (C).

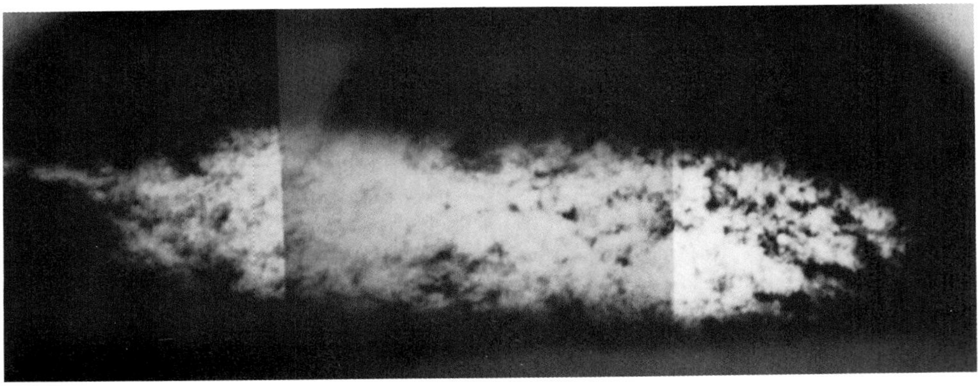

Figure 5 X-Ray (mammograph) of the mammary gland of a lactating bottlenose dolphin (PEM N1182).

140. Conversely, the proportion of respirations taken alone rose from 0.5% at day 9 to approximately 70% by day 140 and remained at this level for the next 200 days (Fig. 7). Subsequently, the proportion of breaths taken with the mother appears to have increased and those taken alone decreased. This is because most respiration times were recorded when Simo was with Thandi.

When with his mother Simo's mean interrespiration times remained between 23 and 43 seconds, with a mean of 31.6 seconds (Fig. 8). In general, Simo's solo respirations up to day 30 were restricted to single, irregular sojourns to the surface. After day 30 he began to respire alone regularly (Fig. 8), and the mean time between breaths (21 seconds) varied quite markedly between observations but was consistently less than when breathing with his mother (29 seconds). Between days 30 and 200 his respiration times remained at approximately the same level (19 seconds), in general 8–10 seconds less than when breathing with his mother. Subsequently, his time between respirations increased steadily. From day 400 there was very little difference between his mean respiration time when alone (33.5 seconds) and when with his mother (35.1 seconds).

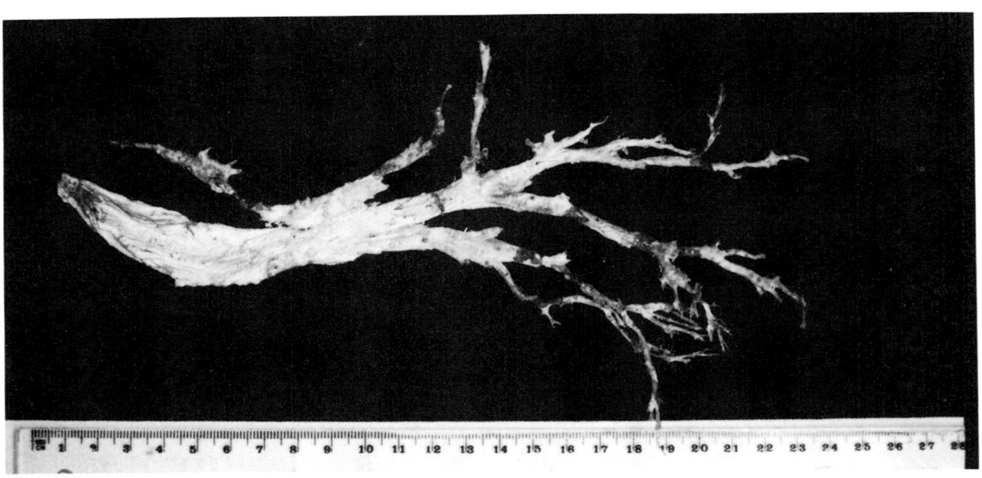

Figure 6 Resin cast of the mammary gland in Fig. 5. The main ducts and mammary cistern are shown.

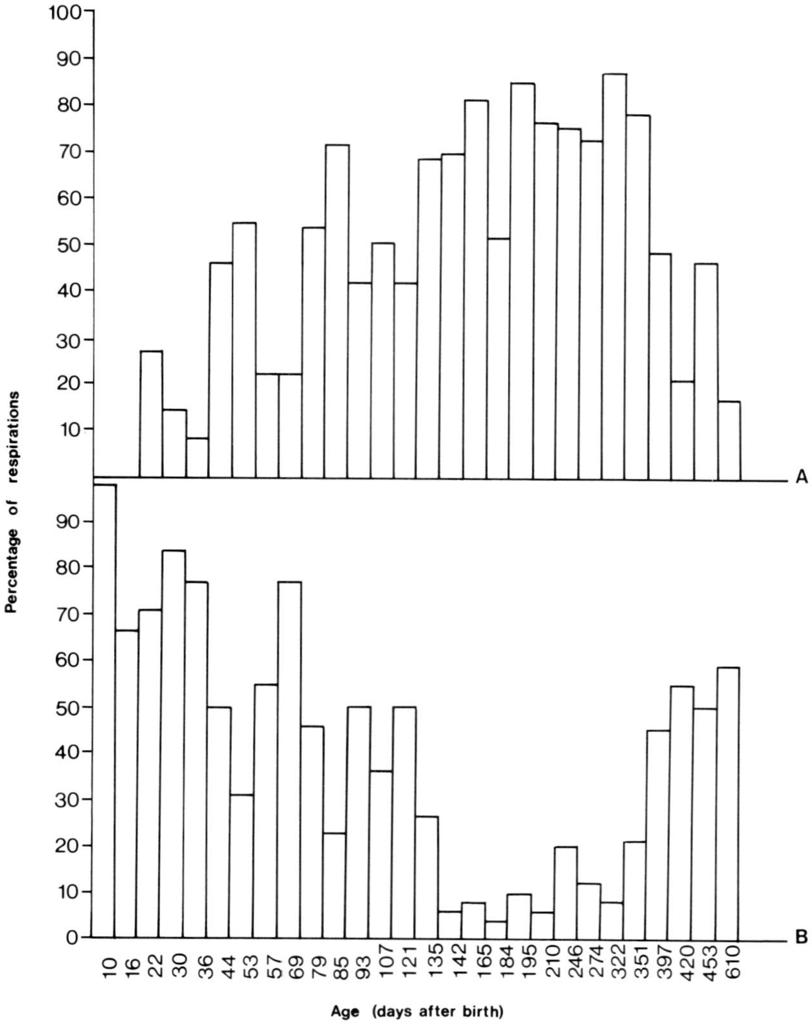

Figure 7 Change with age in proportional amount of time that Simo respired alone (A) and with his mother (B).

The progression in Simo's ability to remain submerged is shown in three notes made during observations. On day 2 he breathed after each suckle, but by day 22 he was able to stay submerged for two successive suckles. At 5 months he stayed submerged for five successive suckles over a period of 1 minute 53 seconds.

Growth and Development

Simo's growth pattern over 2 years was based on 61 length estimates from photographs and other measurements, and his estimated body mass at age was based on the length/mass relationship of 71 calves caught in shark nets (V. G. Cockcroft,

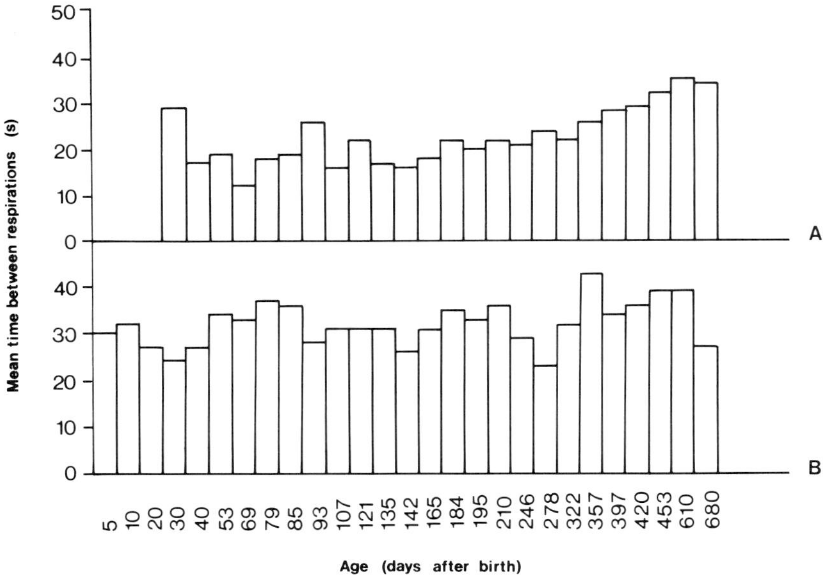

Figure 8 Influence of age on Simo's mean time between respirations while swimming alone (A) and with his mother (B).

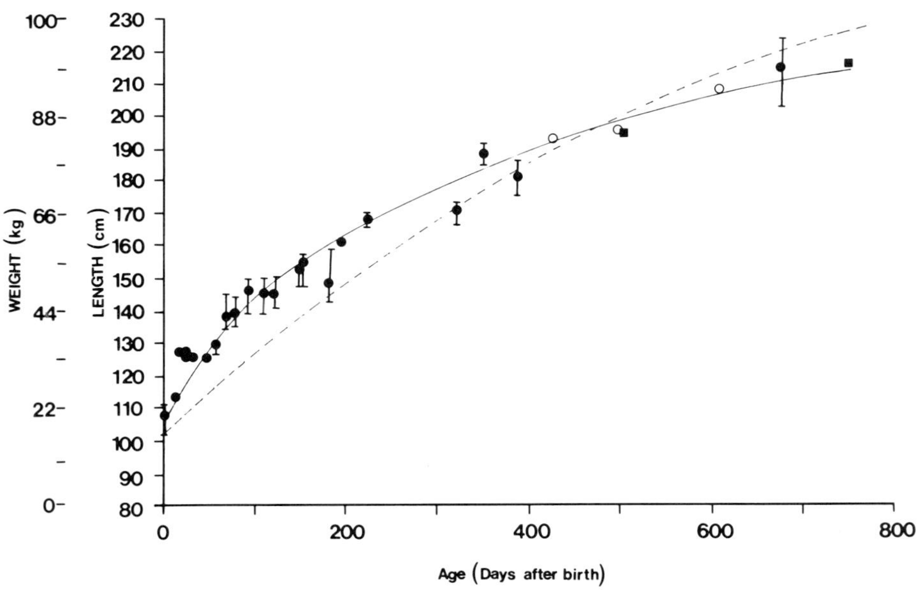

Figure 9 Simo's growth from birth to 2 years, determined as the mean of photographic length estimates (●), showing ranges where more than one estimate was made, direct measurements (○), and tile count estimates (■). Simo's calculated weight increase over this period (---) is also shown.

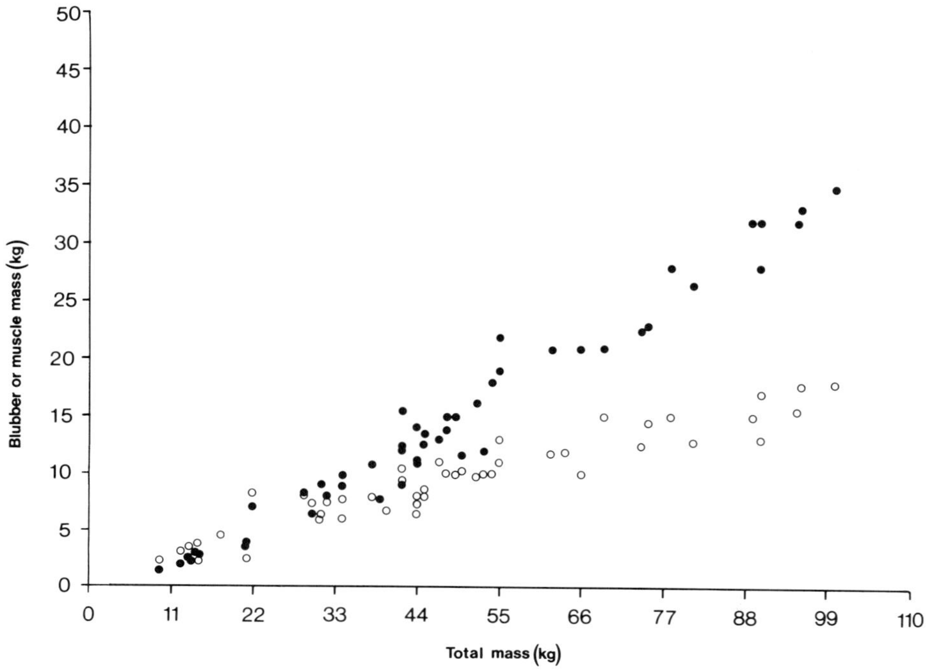

Figure 10 Plot of muscle mass (●) and blubber mass (○) against total mass for bottlenose dolphins under 2.1 m total length.

unpublished data) (Fig. 9). Simo's estimated length at birth was 108–111 cm; the best photographic estimate was 109 cm.

At birth the proportion of blubber to total mass is greater than that of muscle (Fig. 10). This trend is reversed at a mass of approximately 22 kg or an age of between 4 and 5 weeks. Simo displayed this phenomenon quite graphically. At birth he was noticeably thin, with lean flanks and a large head in relation to his body. This leanness appeared to be primarily due to underdevelopment of the epaxial musculature. Only after day 30 did he fill out and develop the sleek proportions of adults.

At birth Simo's dorsal surface was darkly pigmented, and the ventral "anchor" mark, anterior to the flipper insertions, was very evident. Most of this coloration had changed by day 10 and given way to the typical gray cape pattern (Ross, 1984). The umbilical wound had healed by day 22 and was reduced to a small scar by day 44. By day 22 the rostral whiskers had also disappeared, and the fetal creases were apparent only as lightly pigmented lines but were still visible, under certain light conditions, up to day 62. The first maxillary teeth erupted by day 91, and most teeth were visible in both jaws by day 163. Associated with the acquisition of teeth, Simo began mouthing the other dolphins in the pool. Fresh rake marks were often evident on the other animals, particularly on their flippers and dorsal caudal areas. Simo was first heard whistling on day 2, but sonar clicks were not heard until day 42. Directly after birth Simo swam high up on his mother's flank, within a few centimeters of her, maintaining this position most of the day by infrequent fluke strokes. Whenever Thandi accelerated, as in times of fright or emergency, Simo moved along next to her. At such times she appeared to adjust her course to maintain the calf's proximity. Simo's relative passivity in this position was often shown when the females in the pool competed for him. Any female was able to take him from another by swimming past very closely at high speed.

The spatial relationship of the mother and calf

changed with time. By day 9 Simo's positioning was no longer as close to the mother's flank as previously. He still used irregular fluke beats to maintain his position but began to swim underneath his mother, between her flippers, and sometimes rose alone to breathe. By day 16 he was further removed from the mother and breathed alone more often. Subsequent to day 22 he swam approximately 70–100 cm from the mother's flank and constantly moved his flukes much as the other animals did. By day 36 he swam beneath Thandi as much as at her side, appearing to maintain contact beneath her with his dorsal fin. Only some 50% of his time was, however, spent swimming with his mother during circuits of the deepest part of the pool. Circuiting occurred irregularly, lasting from a few minutes to half an hour. Simo swam beneath his mother increasingly, and by 3 months very little swimming was done at her side. After day 79 Simo swam at his mother's side only when alarmed or frightened, at which time Thandi would accelerate and swim away rapidly with him at her flank. This behavior continued until Simo was well over 1 year old. Between about 3 and 15 months Simo progressively spent more of his time independently of his mother (Fig. 7). When together, he swam beside her increasingly as the other animals did. By 2 years he swam beneath Thandi only when distressed or when bumping and suckling during short periods of slow circuiting.

Simo's development of swimming control was also progressive and linked to his physical development. At birth he was unable to maintain an even keel and was constantly making minor adjustments. He was positively buoyant and tended to drift upward, and he was unable to neutralize his buoyancy, by releasing air, until 6 months of age. This is consistent with the proportionally high blubber mass of early neonates (Fig. 10).

On a number of occasions Thandi exhibited maternal care when Simo wandered from her side. For example, between days 2 and 23 she sometimes pushed him above the surface (Fig. 11) or held him on the pool bottom (Fig. 12). On other occasions Thandi whistled and Simo immediately returned to her. These whistles were not recorded, and no audible distinction was possible between these and other whistles made by Thandi.

Figure 11 Thandi chastises Simo by lifting him clear of the surface.

Figure 12 Thandi chastises Simo by rolling on him and forcing him to remain on the floor of the pool.

DISCUSSION

The development of young dolphins is marked by a progressive independence from the mother through physiological changes and the acquisition of behavioral skills. This is especially marked in the first 4 weeks after birth. Tavolga and Essapian (1957) described how calves become more independent of their mothers at about 2 weeks (see also Tavolga, 1966; Gurevich 1977). Taber and Thomas (1982) outlined a similar process in right whale calves 1–3 months old.

Positioning of the Calf in Relation to Its Mother

The newborn calf appears unable to swim efficiently, probably because it lacks both muscular development and motor coordination. Its proportionally high blubber mass acts as an insulator and probably increases its buoyancy, making submergence difficult. Under these circumstances the neonate's ability to maintain a position at the mother's side is of prime importance (Essapian, 1953).

The positioning of newborn dolphins, high on and probably touching the mother's flank, has been noted frequently (McBride and Hebb, 1948; McBride and Kritzler, 1951; Tavolga and Essapian, 1957; Prescott, 1977; Dudock van Heel, 1977; Herman and Tavolga, 1980; Norris and Dohl, 1980). In this position newborn dolphins are probably riding in the pressure wave caused by the mother's body (Norris and Prescott, 1961), thus conserving energy (Prescott, 1977). Though Norris and Dohl (1980) claimed that calves lack movement in this position, Simo actively maintained his station by infrequent fluke strokes.

The first few weeks are also a learning period for the calf, during which the mother constantly monitors its position, behavior, and safety (Tavolga and Essapian, 1957; Tavolga, 1966; Taber and Thomas, 1982) and punishes it for misbehavior (Tavolga, 1966; Tayler and Saayman, 1972). Simo spent his early days apparently resting and sleeping, as Tavolga and Essapian (1957) observed for other neonates. Right whale calves spend the majority of their first month in close proximity to and almost in continuous contact with the mother (Taber and Thomas, 1982). It is during this early, close association with its mother that a calf learns how to recognize her and find the nipples (Herman and Tavolga, 1980). Only after learning this essential behavior can the young dolphin acquire a measure of independence.

At about 3 weeks Simo began to gain independence from Thandi and the assistance of her pressure wave. This coincided with the development of axial musculature and a reduction in proportional blubber mass. The loss of rostral vibrissae at this time is also of interest, for the calf may be aware of the mother's pressure wave through its sensory rostral pits (Ling, 1974, 1977). The sensitivity of these pits may be enhanced by vibrissae during the critical first weeks of life. McBride and Kritzler (1951) recorded the loss of vibrissae in bottlenose dolphin calves at approximately 1 month, but Herman and Tavolga (1980) noted that in *Tursiops* and *Stenella* spp. the vibrissae may remain until weaning.

Protective Coloration

The color patterns of mother and calf may be important for the protection of the dependent calf. During early infancy the calf is protected by its ability to remain in its mother's pressure wave and be accelerated away from danger. The blending of the calf's dark dorsal cape with that of the mother and a similar blending of their lighter flank areas may provide additional calf protection (Fig. 13). When the calf begins to swim beneath the mother, it is no longer assisted by the pressure wave but is still very dependent. In this position the calf is beneath the posterior, ventrally sweeping flank pattern of the mother such that its flank color and light ventral area appear as extensions of hers (Fig. 14). This makes the pair appear as one large animal, particularly at low light intensities such as would be found in the natural environment. It is important that, in later life, Simo returned immediately to this position whenever alarmed.

Respiratory Development

Simo's respiration frequency decreased with age, as did that of the calves observed by McBride and Kritzler (1951) and Gurevich (1977). When with Thandi, Simo's behavior was characterized by rhythmic breathing (resting), and his respiration frequency probably was determined by hers. For about 3 weeks after birth, when he was dependent on her, Simo's respiration frequency more or less matched his mother's. For the next 4 weeks, his dependence decreased and his respiration frequency increased markedly (Fig. 8). In the subsequent 3 weeks, Simo's respiration frequency declined gradually, though the reasons are unclear. Perhaps the calf's rate is determined by the mother when he is completely dependent on her. However, as the calf's swimming proficiency increases and his dependence decreases, he begins establishing his own respiration rate.

Gurevich (1977) found that the respiration rate of a young bottlenose dolphin varied depending on time of day and activity. During his first year Simo displayed no time preference for rest or play, and no diurnal pattern was observed in his respiration rate, the rate varying considerably with activity. Thereafter, his daylight activities were similar to those of the other adults. At night, however, Simo's activity pattern remained similar to his daylight activity. The other adults appeared to rest, with a reduced respiration rate, in the early morning hours, either lying on the pool floor or hanging upside down, rostrum on the floor, for about 3.5 minutes.

Suckling Behavior

All feeds in the first 4 weeks of Simo's life were preceded by side presentations. This posture may make suckling easier for the calf when its respiratory and swimming efficiency are low. It is possible that in early infancy the calf is stimulated to suckle by the mother's presentation, the mother initiating suckling in response to pressure in her mammary glands. The change in the mother's presentation posture coincided with an increase in the number of suckling bouts being preceded by bumping.

Although bumping was not specifically recognized and monitored for the first 4 weeks, its high frequency of occurrence, once identified, suggests it probably did not occur early in the calf's life. Once identified and monitored, the high proportion of feeds preceded by bumping remained fairly constant (Fig. 3). This suggests some rela-

Figure 13 Successive frames of Simo's "riding" at Thandi's flank, showing the matching coloration. Note the effect of the light in the bottom frame. The tips of the calf's dorsal fin and flippers are barely visible.

tionship between bumping and suckling. It is possible that bumping marks the onset of the calf's initiation of suckling bouts, indicating to the mother its need to feed and allowing the mother time to ready herself and present. It is also possible that bumping provides the calf with information on the relative fullness of the mammary gland.

On a number of occasions just prior to suckling, it was obvious that Thandi's mammary glands were full. When she twisted or turned, milk was expelled. This is common in lactating delphinids (Tavolga and Essapian, 1957; Arvy, 1974; Drinnan and Sadlier, 1981), and it seems probable, therefore, that the milk reservoir is normally full prior to suckling, despite the calf's relatively high suckling frequency. Bumping may hasten or stimulate the "let down" of milk from the alveoli into an

Figure 14 Simo's swimming below Thandi in typical fashion. The close match in pigmentation of mother and calf provides disruptive coloration.

empty gland, allowing the calf to maintain its high suckling frequency. Ely and Peterson (1941, cited in Folley, 1952) postulated that let down is normally governed by circulating adrenalin and oxytocin released when the enervated teat is tactually stimulated. This reflex is conditionable, apparently, and bumping may be the operant stimulus.

Bumping behavior appears either not to have occurred in other mother–calf pairs observed or was not noticed. Gurevich (1977) noted that a bottlenose dolphin calf initiated nursing by nudging its mother in the mammary region. This nudging behavior was observed in practically all of Simo's sucklings, but it is distinct from bumping.

Drinnan and Sadlier (1981) showed that, during the first month after its birth, a beluga, *Delphinapterus leucas,* displayed no circadian suckling pattern. Simo's suckling behavior similarly showed no diurnal pattern. His first day was characterized by a high number of suckles (19/hour). The reasons for this high number of suckles may include the mother's inability to provide milk and/or the calf's inexperience. Over the following weeks, Simo's suckling frequency, although highly variable, was similar to that observed for another calf of the same age (Gurevich, 1977). The latter nursed four times per hour, each feed lasting 7–10 seconds. Simo's nursings, at this age, ranged between 3 and 9 seconds, with a mean of 5.6 seconds (Fig. 1). Gurevich (1977) noted a decrease in the suckling time per hour and an increase in the number of bouts per hour at 3 months. At 6 months the calf increased the time of each suckling bout while the number of suckles per hour decreased. He attributed these variations to changes in the mother's posture during bouts (Gurevich, 1977).

Simo showed none of these variations as Thandi's presentation posture changed. His suckling rate and the duration of each event decreased steadily. Similar reductions in nursing frequency have been observed for bottlenose dolphins (McBride and Kritzler, 1951) and belugas (Drinnan and Sadlier, 1981). This apparent decrease in milk, and perhaps energy, consumption during the course of the calf's development is interesting.

Energetics of Suckling

There are two methods of estimating the amount of milk and energy Simo consumed. The amount and energy value of milk consumed can be estimated from the volume of the mammary gland. The amount of energy consumed can also be estimated from the difference in Thandi's food consumption during and after pregnancy and lactation. Our estimates from both methods are based on single animals and should only be regarded as first approximations of milk and energy consumption during suckling.

The mammary-gland volume of the heavily lactating female was estimated at approximately 145 ml. Given an hourly suckling frequency of 4.5 (less than 6 weeks old) and presuming the entire mammary is emptied by each suckle, Simo's daily milk intake would have been 15.7 liters. However, the mammary alveoli (Fig. 5) are minute, and their flow resistance is probably too great for the milk to be available to the calf during the brief suckling period. It is more probable that only the cistern and larger ducts of the mammary gland are emptied. The volume of the ducts and cistern were estimated at 40 ml (Fig. 6). On this basis, Simo's daily intake would have been approximately 4.3 liters, based on a suckling frequency of 4.5/hour.

Information on the fat, lactose, and protein composition of bottlenose dolphin milk is scanty, and reported values vary considerably. Eichelberger *et al.* (1940, cited in Arvy, 1974) reported a range of values for fat, lactose, and protein, with means of 14.1, 0.6, and 10.4%, respectively. Ackman *et al.* (1971) found a fat content of 19%. If the fat, protein, and lactose proportions of Thandi's milk are assumed to be 14.5, 10, and 1, respectively, the estimated energy value of the milk would be approximately 827 kJ/100 g of milk, and Simo's daily intake would have been approximately 35,600 kJ. (Energy conversions of fat, protein, and lactose were calculated using the figures of Brody, 1945, cited in Crisp, 1971.)

Thandi's mean food intake during lactation was 13.5 kg of fish per day (6.5 kg *Sardinops ocellatus*, 4.5 kg *Merluccius capensis*, and 2.5 kg *Scomber japonicus*) but had dropped to 8.5 kg per day (4.5 kg *M. capensis* and 4 kg *S. ocellatus*) 3 years after Simo's birth. Thandi's mean mass during this period was 163 kg. During the stress of lactation she required 8.3% of her body mass in food per day compared to the 5.2% subsequent to birth and lactation. The additional energy required for lactation was contained in 5 kg of fish per day. The energy value of this fish was estimated at about 37,000 kJ from energy values for food species (Ross and Cockcroft, Chapter 5, this volume). The estimated daily milk energy consumed by Simo represents 96% of the additional food energy ingested by Thandi. The close agreement of these derived figures and consideration of Thandi's assimilation efficiency suggest that great energetic demands are placed on the mother during lactation. Unless she has sufficient fat reserves to last the nearly 5 months of high suckling frequency, she must be capable of either reducing her own energy requirements or increasing her food consumption. In a captive situation, both alternatives may be possible. However, the ability of free-ranging lactating females to conserve energy is unknown. The existence of nursery groups within free-ranging schools may facilitate energy reduction and the ability of lactating females to obtain additional food, as discussed by Cockcroft and Ross (Chapter 15, this volume).

The pattern of Simo's estimated daily energy intake, from birth, shows an apparently gradual decrease in the energy consumed until weaning (Fig. 15), a pattern clearly related to the decreased suckling rate observed in this and other studies (see above). Immediately after solids were consumed, energy intake increased dramatically. Thereafter, it remained approximately constant and at about the same level as initial milk-energy consumption. It is improbable that the calf's energy requirements would decrease prior to weaning. Energy requirements to overcome thermoregulatory stress may decrease with growth, but more energy is required for growth and activity. It is unlikely that milk production or milk energy value increases during lactation. Arvy (1974) suggested that milk production at the beginning of lactation is probably greater than at the end. Further, although nutritional changes in milk during lactation are well known (Arvy, 1974; Oftedal, 1984), this has not been recorded in cetaceans.

Simo's estimated mean daily mass gain be-

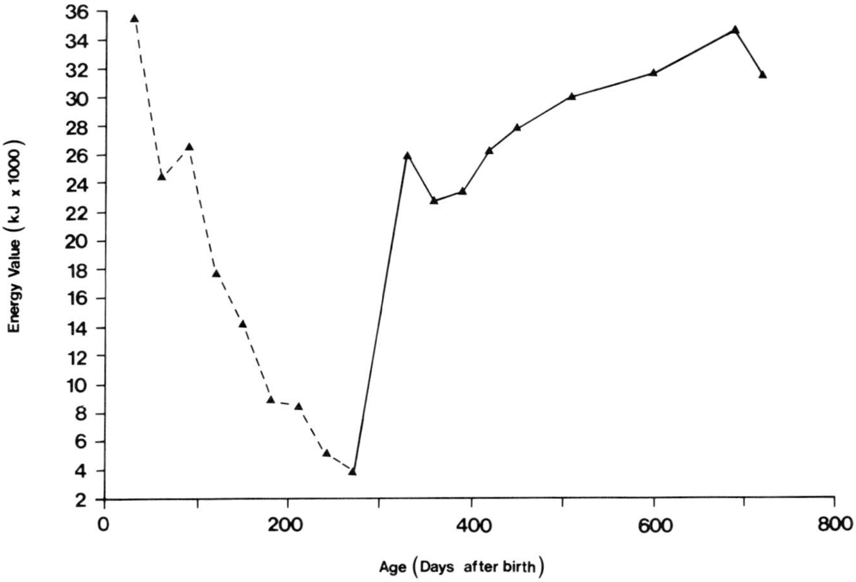

Figure 15 Estimated energy content of Simo's daily food intake: milk only (---), fish plus milk (——).

tween days 100 and 150 was 240 g. The daily mass gain of blubber and muscle as a proportion of this increase can be estimated as 49.2 g of muscle and 53.8 g of blubber (Fig. 10). The energy value of this mass of muscle and blubber is 310 kJ and 2127 kJ, respectively (Lockyer, 1978). At this age (day 100–150) it appeared that the calf's energy intake was equal to or less than that needed to maintain this level of growth (Fig. 11). One logical interpretation of these phenomena is that weaning is a response to the increase in energy requirement which the calf is unable to fulfill through nursing. This interpretation is supported by the increased growth which occurred concurrent with the commencement of solid-food intake (Fig. 9).

Although Simo continued suckling past his first year, and probably beyond the second, it appears that he received a progressively decreasing energetic contribution from his mother's milk. Suckling probably remains important in mother–calf interactions after the calf begins taking solid food. During its second year, the calf continues acquiring the necessary behavioral, respiratory, and feeding skills for complete independence, which may not be attained until a number of years later. Progressive independence may be reflected in free-ranging bottlenose dolphin schools by the presence within them of subgroups of larger calves (Leatherwood, 1977). These subgroups may themselves be important in the continuing learning process.

All reported values for bottlenose dolphin milk-fat are markedly less than values reported for other delphinids. Milk-fat values for *Stenella attenuata*, *S. longirostris*, *S. coeruleoalba*, and *Delphinus delphis* were all greater than 29% (Pilson and Waller, 1970; Kawai and Fukushima, 1980; Ross, 1984). It is impossible to evaluate why these differences between dolphin species occur without additional information on growth, metabolic, and suckling rates. However, if suckling frequency of calves of other species is similar to Simo's, it suggests that females of these species would be subject to enormous stress during lactation. This may lead to earlier weaning and a reduction in the period of mother–calf interaction, apparently so important in the calf's development. It is possibly a reflection of the smaller school size and inshore habitat of some bottlenose dolphins, which may require greater maternal care and a longer learning period for the young to attain total independence. These possibilities are as yet unexplored in the literature, but they suggest that delphinid species may have evolved diverse strategies for the nurturing and care of their young.

ACKNOWLEDGMENTS

We gratefully acknowledge James Jackleman, Michele Simmons, and Wayne Brown for many hours of observation, especially those in the early morning. We appreciate also the unstinting cooperation and assistance of the staff of the Port Elizabeth Oceanarium.

REFERENCES

Ackman, R. G., Eaton, C. A., and Mitchell, E. D. (1971). The bottlenosed dolphin *Tursiops truncatus*: Fatty acid composition of milk triglycerides. *Can. J. Biochem.* **49**(10), 1172–1174.

Arvy, L. (1974). Mammary glands, milk and lactation in cetaceans. *In* "Investigations on Cetacea" (G. Pilleri, ed.), Vol. 5, pp. 157–202. Institute of Brain Anatomy, University of Berne, Berne, Switzerland.

Bonner, W. N. (1984). Lactation strategies in pinnipeds: Problems for a marine mammalian group. *Symp. Zool. Soc. London* **51,** 253–272.

Crisp, D. J. (1971). Energy flow measurements. *In* "Methods for the Study of Marine Benthos" (N. A. Holme and A. D. McIntyre, eds.), pp. 197–279. IBP Handbook No. 16, Blackwell, London.

Drinnan, R. L., and Sadlier, R. M. F. S. (1981). The suckling behavior of a captive beluga (*Delphinapterus leucas*) calf. *Appl. Anim. Ethol.* **7,** 179–185.

Dudock van Heel, W. H. (1977). Dolphin reproduction in western Europe *In* "Breeding Dolphins: Present Status, Suggestions for the Future" (S. H. Ridgway and K. Benirschke, eds.), pp. 71–76. U.S. Marine Mammal Commission Report MMC-76/07, Washington, D.C.

Essapian, F. S. (1953). The birth and growth of a porpoise. *Nat. Hist.* **62**(9), 392–399.

Fedak, M. A., and Anderson, S. S. (1982). The energetics of lactation: Measurements from a large wild mammal, the grey seal. *J. Zool., London* **198,** 473–479.

Folley, S. J. (1952). Lactation. *In* "Marshall's Physiology of Reproduction" (A. S. Parkes, ed.), Vol. 2, pp. 525–647. Longmans, London.

Gurevich, V. S. (1977). Post-natal behavior of an Atlantic bottlenosed dolphin calf (*Tursiops truncatus,* Montagu) born at Sea World. *In* "Breeding Dolphins: Present Status, Suggestions for the Future" (S. H. Ridgway and K. Benirschke, eds.), pp. 168–184. U.S. Marine Mammal Commission Report MMC-76/07, Washington, D.C.

Herman, L. M., and Tavolga, W. N. (1980). The communciation systems of cetaceans. *In* "Cetacean Behavior" (L. M. Herman, ed.), pp. 149–209. Wiley, New York.

Kwai, S., and Fukushima, M. (1980). Relation between the lipid composition of organochlorine compounds in the various organs of striped dolphins (*Stenella coeruleoalba*). *In* "Studies on the Levels of Oragnochlorine Compounds and Heavy Metals in Marine Organisms" (T. Fujiyama, ed.), pp. 85–96. Univ. of Ryukus, Okinawa, Japan.

Lavigne, D. M., Stewart, R. E., and Fletcher, F. (1982). Changes in composition and energy content of Harp seal milk during lactation. *Physiol. Zool.* **55,** 1–9.

Leatherwood, S. (1977). Some preliminary impressions on the numbers and social behavior of free-swimming bottlenosed dolphin calves (*Tursiops truncatus*) in the northern gulf of Mexico. *In* "Breeding Dolphins: Present Status, Suggestions for the Future" (S. H. Ridgway and K. Benirschke, eds.), pp. 143–147. U.S. Marine Mammal Commission Report MMC-76/07, Washington, D.C.

Ling, J. K. (1974). The integument of marine mammals. *In* "Functional Anatomy of Marine Mammals" (R. J. Harrison, ed.), Vol. 1, pp. 1–44. Academic Press, London and New York.

Ling, J. K. (1977). Vibrissae of marine mammals. *In* "Functional Anatomy of Marine Mammals" (R. J. Harrison, ed.), Vol. 3, pp. 387–415. Academic Press, London and New York.

Lockyer, C. (1978). A theoretical approach to the balance between growth and food consumption in fin and sei whales, with special reference to the female reproductive cycle. *Rep. int. Whal. Commn* **28,** 243–249.

McBride, A. F., and Hebb, D. O. (1948). Behavior of the captive bottle-nose dolphin (*Tursiops truncatus*). *J. Comp. Physiol. Psychol.* **41**(2), 111–123.

McBride, A. F., and Kritzler, H. (1951). Observations on pregnancy, parturition, and post-natal behavior in the bottlenose dolphin. *J. Mammal.* **32**(3), 251–266.

Norris, K. S., and Dohl, T. P. (1980). The structure and functions of cetacean schools. *In* "Cetacean Behavior" (L. M. Herman, ed.), pp. 149–209. Wiley, New York.

Norris, K. S., and Prescott, J. H. (1961). Observations on Pacific cetaceans of Californian and Mexican waters. *Univ. Calif. Publ. Zool.* **63,** 291–402.

Norris, K. S., Goodman, R. M., Villa-Ramirez, B., and Hobbs, L. (1977). Behavior of California gray whale, *Eschrichtius robustus,* in southern Baja California, Mexico. *Fish. Bull.* **75,** 159–172.

Oftedal, O. T. (1984). Milk composition, milk yield and energy output at peak lactation: A comprative review. *Symp. Zool. Soc. London* **51,** 33–85.

Pilson, M. E., and Waller, D. W. (1970). Composition of milk from spotted and spinner porpoises. *J. Mammal.* **51,** 74–79.

Prescott, J. H. (1977). Comments on captive births of *Tursiops truncatus* at Marineland of the Pacific. *In* "Breeding Dolphins: Present Status, Suggestions for the Future" (S. H. Ridgway and K. Benirschke, eds.), pp. 71–76. Marine Mammal Commission Report MMC-76/07, Washington, D.C.

Ross, G. J. B. (1984). The smaller cetaceans of the south east coast of southern Africa. *Ann. Cape Prov. Mus.* (*Nat. Hist.*), **15**(2), 173–410.

Saayman, G. S., and Tayler, C. K. (1977). Observations on the sexual behavior of Indian Ocean bottlenosed dolphins (*Tursiops aduncus*). *In* "Breeding Dolphins: Present Status, Suggestions for the Future" (S. H. Ridgway and K. Benirschke, eds.), pp. 113–129. U.S. Marine Mammal Commission Report MMC-76/07, Washington, D.C.

Taber, S., and Thomas, P. (1982). Calf development and mother–calf spatial relationships in southern right whales. *Anim. Behav.* **30**(4), 1072–1083.

Tavolga, M. C. (1966). Behavior of the bottlenose dolphin (*Tursiops truncatus*): Social interactions in a captive colony. *In* "Whales, Dolphins, and Porpoises" (K. S. Norris, ed.), pp. 718–730. Univ. of California Press, Berkeley and Los Angeles.

Tavolga, M. C., and Essapian, F. S. (1957). The behavior of the bottle-nosed dolphin (*Tursiops truncatus*): Mating, pregnancy, parturition and mother–infant behavior. *Zoologica* **42**(1), 11–31.

Tayler, C. K., and Saayman, G. S. (1972). The social organisation and behavior of dolphins (*Tursiops aduncus*) and baboons (*Papio ursinus*): some comparisons and assessments. *Ann. Cape Prov. Mus.* (*Nat. Hist.*) **9**(2), 11–49.

Tedman, R. A., Bryden, M. M., and Green, B. (1981). Milk intake of Weddell seal young estimated by tritiated water and ^{22}sodium turnover. *Antarct. J. U.S.* **16**(5), 152–153.

Hematology and Serum Chemistry Values in Bottlenose Dolphins

28

Edward D. Asper
Sea World, Inc.
Orlando, Florida 32821

Lanny H. Cornell
Pacific Research Laboratory, Inc.
El Cajon, California 92022

Deborah A. Duffield
Department of Biology
Portland State University
Portland, Oregon 97201

Daniel K. Odell
University of Miami
Rosenstiel School of Marine and Atmospheric Science
Miami, Florida 33149
Present address: Sea World of Florida
Orlando, Florida 32821

Brian E. Joseph
Sea World of California
San Diego, California 92109
Present address: Chicago Zoological Society, Brookfield Zoo, Brookfield, Illinois 60513

Bernice I. Stark
Sea World of California
San Diego, California 92109

Cathy A. Perry
Sea World of Florida
Orlando, Florida 32821

INTRODUCTION

In 1983, there were 304 bottlenose dolphins, *Tursiops truncatus*, in zoological parks and aquaria in the United States and Canada (Asper *et al.*, 1988). From the late 1930s through 1983, over 180 bottlenose dolphin births were recorded at North American facilities. The years of routine clinical tests associated with the care and handling of bottlenose dolphins in captivity have permitted workers to accumulate considerable data on hematology and serum chemistry of this species. Statistics for some of these parameters have been reported (Medway and Geraci, 1964, 1965, 1986; Lenfant, 1969; Ridgway *et al.*, 1970; Ridgway, 1972). In this chapter we summarize hundreds of clinical evaluations of hematology and serum chemistry tests made during routine health examinations of bottlenose dolphins at Sea World of California and Sea World of Florida. The animals at these facilities are sampled regularly for information on "normal" clinical values which can be used to monitor changes in health status. These long-term records provide an unprecedented opportunity for examining age- and sex-related differences in normal hematology and serum chemistry values. Values for creatinine, triglycerides, and amylase have not been previously reported for the bottlenose dolphin.

For wild dolphins, hematology and serum chemistry values determined from stranded animals cannot be assumed to be normal, since such animals generally have at least one pathological condition. Heretofore, normal values of wild dolphins have been estimated from studies of captive dolphins. Some investigators have been concerned that these values of captive dolphins reflect adaptations to a specialized environment and, thus, are not normal for healthy, free-ranging dolphins.

Field studies in the Indian–Banana River complex of eastern Florida, 1979–1986, have involved the capture, marking, and release of 109 bottlenose dolphins (Odell and Asper, 1982, Chapter 30, this volume). Although these studies were aimed principally at learning about dolphin movements in the area, workers also obtained hematology and serum chemistry data from the animals captured and released. Blood samples from the wild animals were tested for the same hematology and serum chemistry panel routinely examined for captive animals at the Sea World facilities. This permits comparison of "normal" values from long-term studies of captive dolphins with a large sample from wild dolphins.

MATERIALS AND METHODS

"Normal" Hematology and Serum Chemistry Values for Captive Dolphins

Hematology and serum chemistry data were compiled from the routine health records of 35 bottlenose dolphins (13 males, 22 females) maintained at Sea World of California and Sea World of Florida. Only dolphins for which at least 3 years of data were available were incorporated into this study. Clinical records for a 9-year period (1977–1986) were used as the basis for defining and comparing "normal" values and ranges. Samples from periods when a consistent change in behavior, appearance, or food consumption indicated a change in an animal's health status were excluded as they were considered not representative of normal values. This sample included dolphins originating from coastal waters of the Atlantic and Gulf of Mexico and coastal and offshore waters of the North Pacific, as well as dolphins born in the Sea World parks. In 1986, at the end of the study period, the 35 dolphins included in this analysis had been in captivity from 5 to 18 years and ranged in age class from juveniles (6–8 years of age) to older adults (25 years and older).

Blood samples were collected at monthly or bimonthly intervals as part of a health management program. Pregnant and postpartum females were excepted from this regimen in the latter third of their pregnancy and while they were nursing young calves, respectively. Blood was collected from the fluke by venipuncture, using a 4-cm (18–20 gauge) needle, and placed in anticoagulant (EDTA) and clot (SST) tubes. All clinical analyses were performed by Sea World laboratories in California or Florida. Within 1 hour of collection, the blood was centrifuged to separate the serum. Hematology and serum chemistry values were determined within 24 hours. Test procedures have been previously reported (Cornell, 1983). Complete blood counts were performed using Coulter instrumentation (Coulter ZBI with a Coulter hemoglobinometer or a Coulter M430; Coulter Electronics, Hialeah, Florida 33014) and routine microscopic procedures. Serum chemistry analyses were performed using a Gilford semiautomated chemistry system and Gilford reagents (Gilford Instrument Laboratories Inc., Oberlin, Ohio 54074). All tests were standardized daily and calibrated for use with marine mammals.

From 15 to 80 routine sample results were available per animal for each clinical test. The values for each test were examined by box plot analysis (Chambers *et al.*, 1977). This analysis calculates and plots individual animal medians, 25–75% quartiles (the distribution of 50% of the values around this median), and the distribution of 90% of the values about the median. Individual animal medians, quartiles, and 90% limits were averaged for all animals and for age and sex classes to obtain "normal" values, a narrow range (quartiles), and a broad range (90% limits) for each clinical test.

Statistical analysis of age and sex differences was performed by analysis of variance (Zar, 1984). In this sample, dolphins were scored as "adults" if they were 12 years of age or older and as "juveniles" if they were 2–5 years of age when they entered the study. The juveniles were 9 years of age or younger at the end of the study. Dolphins less than 2 years old are not routinely handled at the parks. Age was roughly estimated from the length of the animal when brought into captivity or from birth date, if born in the park.

Clinical Values for Wild Dolphins

Odell and Asper (1982, Chapter 30, this volume) captured, marked, sampled, and released 109 dolphins (69 males, 40 females) over a 3-year period in the Indian–Banana River complex on the east coast of Florida. Blood samples were taken for clinical evaluation of health status, analysis of reproductive hormones, and genetic analyses. Hematology and serum chemistry tests were performed by the Sea World of Florida laboratory, using the same procedures and reagents described above.

Teeth were extracted and relative ages estimated by counting growth layer groups (GLGs) (see Perrin and Myrick, 1980; Myrick and Cornell, Chapter 34, this volume). Thus, it was possible to test clinical blood values from the field studies for age- and sex-related differences, as was done for the long-term park study. For comparison with the park animals, dolphins with 2–9 GLGs were considered "juveniles" and animals with 10 or more GLGs "adults." Statistical comparison of the values for captive and wild bottlenose dolphins was made by the two-tailed Student's *t*-test (Zar, 1984).

RESULTS AND DISCUSSION

"Normal" hematology and serum chemistry values and ranges are given in Tables 1 and 2, respectively. Box plot analysis of the routine samplings from captive animals is a useful way of representing the clinical values for individual animals. In this analysis, medians are calculated as central tendency because they not only represent the middle value of a long-term data set for each clinical test, but also because they are less sensitive than means to occasional extreme values or outliers. The 25–75% quartiles from the box plot analyses (50% of the values around the median) are taken to represent a "narrow" range of normal values for each animal, the 90% values a "broader" range of normal values. Values for different animals vary within the overall species range. Therefore, the individual box plots have been used to define an animal's normal profile and to place it in the overall range of values. A significant deviation from median, narrow, and broad range values is usually seen during illness. A consistent shift in an individual's values toward the limits of its narrow range, and especially of its broad range, can be taken to indicate a change in health status. The observation of such a shift facilitates early diagnosis and treatment.

The normal values and ranges reported here for hematology and serum chemistry parameters are in general agreement with values previously given for bottlenose dolphins (Medway and Geraci, 1964, 1965, 1986; Ridgway *et al.*, 1970). The only major differences are in bilirubin values (higher in Ridgway *et al.*, 1970), ALP values (lower by half in Ridgway *et al.*, 1970; overlapping range in Medway and Geraci, 1986), and TIBC values (higher in both Ridgway *et al.*, 1970, and Medway and Geraci, 1986). The tendency of results of these tests to vary with technique may account for the observed differences. This report adds normal values for creatinine, triglycerides, and amylase.

In the data on captive animals, age-related differences were seen in white blood cell (WBC) count (ANOVA, $p < .01$), in the percentage of neutrophils ($p < .01$) and lymphocytes ($p < .05$), in glucose levels ($p < .01$), and in alkaline phosphatase (ALP or AP, $p < .01$), alanine aminotransferase (ALT or SGPT, $p < .01$), aspartate aminotransferase (AST or SGOT, $p < .05$), creatine kinase (CK, $p < .01$), and lactate dehydrogenase (LDH, $p < .01$) activities. Juveniles had higher values for WBC, percentage of lymphocytes, glucose, ALP, ALT, AST, CK, and LDH. The percentage of neutrophils was higher in adults. Sex-related differences were seen for iron ($p < .01$) and total iron-binding capacity (TIBC, $p < .01$): values for both tests were higher in females. Age- and/or sex-related differences for many of these clinical blood tests have been reported for various domestic and wild animals (Seal and Erickson, 1969; Schalm *et al.*, 1975; Kaneko,

Table 1
Hematology Values for Bottlenose Dolphins Maintained at Sea World of California and Sea World of Florida[a]

Test	Sea World, California and Florida			Indian River, Mean (S.D.)
	Mean (S.D.)	25–75% Narrow Range	5–95% Broad Range	
Erythrocytes				
Hemoglobin (g/dl)				
Coastal[b]	14.6 (0.6)	14.4–14.9	14.0–15.2	14.3 (0.9)
Intermediate	16.1 (0.2)	15.9–16.3	15.5–16.7	16.4 (0.5)
Offshore	18.6 (0.9)	17.9–19.2	16.8–19.8	
Hematocrit (%)				
Coastal	42 (2)	42–43	41–44	41 (3)
Intermediate	45 (1)	44–47	43–47	47 (3)
Offshore	52 (1)	51–53	47–55	
RBC count (10/mm^3)				
Coastal	3.52 (0.2)	3.40–3.64	3.18–3.89	3.56 (0.2)
Intermediate	3.73 (0.2)	3.61–3.89	3.43–4.12	3.96 (0.1)
Offshore	4.48 (0.2)	4.40–4.69	4.36–4.60	
MCV (fl)	120 (6)	116–125	109–130	114 (5)
MCH (pg)	41 (2)	40–42	40–43	41 (2)
MCHC (g/dl)	35 (1)	34–35	33–36	35 (1)
White blood cells				
Total WBC count (/mm^3)				
Adults	6400 (1100)	5800–6900	5600–7300	11000 (3000)
Juveniles	**8000 (1500)	7200–8800	6700–9500	**9660 (3000)
Differential				
Neutrophils (%)				
Adults	67 (6)	64–69	62–72	(all) 47 (11)
Juveniles	**60 (8)	56–64	54–67	
Lymphocytes (%)				
Adults	18 (5)	15–20	13–22	17 (7)
Juveniles	*24 (6)	19–28	19–29	*** 26 (9)
Monocytes (%)	2 (1)	1–4	1–4	1 (1)
Eosinophils (%)	12 (5)	11–14	9–16	30 (11)

[a] Mean values, standard deviations, narrow ranges (an average of 50% of the values about individual medians), and broad ranges (an average of 90% of individual values) have been calculated for the group as a whole or for male versus female or adult versus juvenile if the differences in values were statistically significant (ANOVA). These data summarize over 1600 separate measurements per clinical test. Coastal, intermediate, and offshore hematologic ecotype values are indicated for hemoglobin, hematocrit, and red blood cell count (Duffield *et al.*, 1983). There were no differences in other hematologic or serum chemistry tests based on these ecotypes. The values for captive dolphins are compared with mean values from 109 wild dolphins from the Indian–Banana River complex in Florida. Levels of significance: *, .05; **, .01; ***, .001 for comparisons within groups. See text for comparisons between groups.
[b] Coastal ecotype, $N = 29$ for park animals, $N = 97$ for the Indian River animals; intermediate ecotype, $N = 4$ for park animals, $N = 12$ for the Indian River animals; offshore ecotype, $N = 2$ for park animals, $N = 0$ for the Indian River animals. For all other tests, $N = 35$ for the park animals, $N = 109$ for the Indian River animals.

Table 2
Serum Chemistry Values for Bottlenose Dolphins at Sea World of California and Sea World of Florida[a]

	Sea World, California and Florida			
Test	Mean (S.D.)	25–75% Narrow Range	5–95% Broad Range	Indian River, Mean (S.D.)
Glucose (mg/dl)				
Adults	121 (8)	117–125	106–134	125 (28)
Juveniles**	132 (13)	125–139	117–147	*** 149 (34)
BUN (mg/dl)	50 (4)	47–53	44–56	61 (11)
Creatinine (mg/dl)	1.5 (0.2)	1.3–1.5	1.2–1.6	1.4 (0.7)
Bilirubin, total (mg/dl)	0.23 (0.94)	0.21–0.24	0.19–0.31	0.20 (0.1)
Cholesterol (mg/dl)	247 (40)	227–267	208–286	190 (53)
Triglycerides (mg/dl)	110 (30)	84–139	77–157	74 (23)
Total protein (g/dl)	6.9 (0.2)	6.7–6.8	6.5–7.2	7.6 (0.8)
Albumin (g/dl)	4.6 (0.3)	4.3–4.7	4.2–4.9	3.9 (0.8)
Globulin (g/dl)	2.2 (0.4)	2.0–2.5	1.8–2.6	3.7 (0.9)
Amylase (U/liter)	14 (3)	12–15	9–20	22 (17)
ALP, AP (U/liter)				
Adults	259 (127)	204–323	196–341	203 (96)
Juveniles	**401 (113)	323–491	307–515	*** 404 (175)
ALT, SGPT (U/liter)				
Adults	25 (5)	23–27	19–34	(All) 31 (18)
Juveniles	**32 (6)	27–37	24–41	
AST, SGOT (U/liter)				
Adults	115 (18)	107–123	102–133	(All) 139 (76)
Juveniles	140 (23)	125–155	118–163	
CK (U/liter)				
Adults	60 (13)	54–65	47–74	60 (20)
Juveniles	**80 (14)	68–91	64–96	*** 81 (27)
LDH (U/liter)				
Adults	194 (26)	181–208	177–216	(M) 250 (62)
Juveniles	**244 (34)	225–264	217–281	*** (F) 214 (62)
Calcium (mg/dl)	9.2 (0.3)	8.8–9.6	7.9–10.4	9.6 (0.7)
Phosphorus (mg/dl)	4.9 (0.4)	4.4–5.4	3.6–6.5	5.6 (0.7)
Chloride (mEq/liter)	110 (3)	109–111	105–114	113 (13)
Iron (μg/dl)				
Males	132 (27)	110–154	110–175	(All) 130 (40)
Females	**178 (38)	160–195	143–220	
TIBC (μg/dl)				
Males	242 (37)	210–288	196–294	(A) 298 (90)
Females	**301 (31)	284–319	252–342	** (J) 254 (90)

[a] Means, standard deviations, narrow, and broad ranges are given (as in Table 1). Values for captive dolphins are compared with mean values from free-ranging dolphins from the Indian–Banana River complex. Levels of significance: *, .5; **, .01; ***, .001.

1980), including other cetaceans (Cornell, 1983), and for humans (Byrne *et al.*, 1981; Tietz, 1983).

Hematology and serum chemistry values from the captive and wild dophins were similar for the majority of the tests. However, statistically significant differences were seen in white blood cell (WBC) count, in the percentages of neutrophils and eosinophils, in total protein, albumin, and globulin, and in serum urea nitrogen (BUN), cholesterol,and triglyceride values.

Wild dolphins, both adults and juveniles, had significantly higher WBC counts (two-tailed *t*-test, $p < .002$), a lower percentage of neutrophils ($p < .05$), a higher percentage of eosinophils ($p < .002$), and higher total protein, albumin, and globulin values ($p < .002$). Because these tests are indicative of immune system activity, the observed differences suggest that wild cetaceans are exposed to greater antigenic challenge. Elevated globulins in wild versus captive animals have been reported for terrestrial mammals (Weber and Bliss, 1972; Schalm *et al.*, 1975). Pronounced eosinophilia (increased eosinophil values) is seen in domestic animals in response to parasites that migrate and invade tissues (Soulsby, 1975; Saror *et al.*, 1979). Free-ranging dolphins are known to have the following tissue-invasive parasites: *Anisakis* sp. (stomach), *Halocerus* sp. (lungs), *Nasitrema* sp. (air sinuses), *Braunina* sp. (stomach and intestines), and *Campula* sp. (liver) (Dailey and Brownell, 1972; Dailey, 1986; D. K. Odell, unpublished data). A major contribution to increased antigenic stimulation in the free-ranging dolphins may come from parasitic infections. Parasite shedding often occurs after an animal has been brought into captivity (L. H. Cornell, 1983, unpublished data). Because there is little opportunity for reinfection in a properly controlled environment, antigenic stimulation decreases with time in captive animals. This could account for the lower WBC, total protein, albumin, and globulin values and the differences in the relative percentages of neutrophils and eosinophils seen in the captive dolphins.

BUN levels were lower in the captive dolphins ($p < .002$) than in the wild dolphins, while cholesterol ($p < .002$) and triglyceride ($p < .002$) levels were higher. The levels of all three of these tests are influenced by diet (Weber and Bliss, 1972; Kaneko, 1980; Byrne *et al.*, 1981). Fish vary considerably in fat and protein composition (Geraci, 1986), and it is therefore likely that these differences are in part related to the types and quantities of food eaten by the dolphins. Degree of activity may also be a major contributing factor (Kaneko, 1980; Byrne *et al.*, 1981).

CONCLUSIONS

The values given in this chapter represent hematology and serum chemistry "normals" for captive and wild dolphins. The clinical values developed from long-term health records of animals maintained in stable park environments have provided a useful comparison with values from free-ranging dolphins. Most of the clinical tests revealed similar values in the two groups. Differences of potential biological significance between captive and wild dolphins lie in hematology and serum chemistry values which suggest a greater antigenic stimulation in wild dolphins (thought to be related, among other things, to increased parasitism) and in serum chemistry values influenced by diet and exercise.

The normal values given here provide a baseline for diagnosing pathology in captive and wild dolphins. They also may prove helpful for assessing the effects of environmental stresses (e.g., pollutants) on the health of wild dolphin populations.

ACKNOWLEDGMENTS

We thank Dr. Deane Clarkson and Darey Shell for development of the clinical statistics package used to evaluate the long-term park data and the Sea World of California and Florida laboratory personnel and Animal Care staff for their many hours of dedicated work. This study was supported in part by a grant from Sea World Enterprises, Inc., contracts from the National Marine Fisheries Service (NA79-GA-C-00027 and NA80-GA-C-00063), and NIH Biomedical Research Support Grant 2 S07 RR07128 to Portland State University. This is Sea World Contribution No. 8812-F.

REFERENCES

Asper, E. D., Cornell, L. H., Duffield, D. A., and Dimeo-Ediger, N. (1988). Marine mammals in zoos, aquaria and marine zoological parks in the United States and Canada; 1983 census report. *Int. Zoo Yearbk.* **27,** 287–294.

Byrne, C. J., Saxton, D. F., Pelikan, P. K., and Nugent, P. M. (1981). "Laboratory Tests: Implications for Nurses and Allied Health Professionals." Addison-Wesley, Menlo Park, California. 384 pp.

Chambers, J. M., Cleveland, W. S., Kleiner, B., and Tukey, P. A. (1977). "Graphical Methods for Data Analysis." Wadsworth Int. Gp., Belmont, California. 210 pp.

Cornell, L. H. 1983. Hematology and clinical chemistry values in the killer whale, *Orcinus orca* L. *J. Wildl. Dis.* **19**(3), 259–264.

Dailey, M. (1986). Parasitology—Basic considerations. *In* "Zoo and Wild Animal Medicine" (M. E. Fowler, ed.), pp. 781–784. Saunders, Philadelphia, Pennsylvania.

Dailey, M., and Brownell, R. L., Jr. (1972). A checklist of marine mammal parasites. *In* "Mammals of the Sea: Biology and Medicine" (S. H. Ridgway, ed.), pp. 528–589. Charles C. Thomas, Springfield, Illinois.

Duffield, D. A., Ridgway, S. H., and Cornell, L. H. (1983). Hematology distinguishes coastal and offshore forms of dolphins (*Tursiops*). *Can. J. Zool.* **61**(4), 930–933.

Geraci, J. R. (1986). Nutrition and nutritional disorders. *In* "Zoo and Wild Animal Medicine" (M. E. Fowler, ed.), pp. 760–764. Saunders, Philadelphia, Pennsylvania.

Kaneko, J. J., ed. (1980). "Clinical Biochemistry of Domestic Animals." Academic Press, San Francisco, California. 832 pp.

Lenfant, C. (1969). Physiological properties of blood of marine mammals. *In* "The Biology of Marine Mammals" (H. T. Andersen, ed.), pp. 95–116. Academic Press, New York.

Medway, W., and Geraci, J. R. (1964). Hematology of the bottlenose dolphin (*Tursiops truncatus*). *Am. J. Physiol.* **207**(6), 1367–1370.

Medway, W., and Geraci, J. R. (1965). Blood chemistry of the bottlenose dolphin. *Am. J. Physiol.* **209**(1), 169–172.

Medway, W., and Geraci, J. R. (1986). Clinical pathology of marine mammals. *In* "Zoo and Wild Animal Medicine" (M. E. Fowler, ed.), pp. 791–797. Saunders, Philadelphia, Pennsylvania.

Odell, D. K., and Asper, E. D. (1982). Live capture, marking and resighting of bottlenose dolphins, *Tursiops truncatus.* NMFS Contract NA80-GA-C-00063. 195 pp.

Perrin, W. F., and Myrick, A. C., Jr, eds. (1980). "Age Determination of Toothed Whales and Sirenians." International Whaling Commission, Cambridge, England. Special issue 3, 229 pp.

Ridgway, S. H. (1972). Homeostasis in the aquatic environment. *In* "Mammals of the Sea: Biology and Medicine" (S. H. Ridgway, ed.), pp. 590–747. Charles C. Thomas, Springfield, Illinois.

Ridgway, S. H., Simpson, J. G., Patton, G. S., and Gilmartin, W. G. (1970). Hematological findings in certain small cetaceans. *J. Am. Vet. Assoc.* **157,** 566–575.

Saror, D. I., Schillhorn Van Veen, T. W., and Adevabju, J. B. (1979). The hemogram of dogs with intestinal parasites in Zaria, Nigeria. *J. Small Anim. Pract.* **20,** 243–250.

Schalm, O. W., Jain, N. C., and Carroll, E. J. (1975). "Veterinary Hematology." Lea & Febiger, Philadelphia, Pennsylvania. 256 pp.

Seal, U. S., and Erickson, A. W. (1969). Hematology, blood chemistry and protein polymorphisms in the white-tailed deer (*Odocoileus virginianus*). *Comp. Biochem. Physiol.* **30,** 695–713.

Soulsby, E. J. (1975). "Textbook of Veterinary Clinical Parasitology," Vol. 1: Helminths, F. A. Davis, Philadelphia, Pennsylvania. 450 pp.

Tietz, N. W., ed. (1983). "Clinical Guide to Laboratory Tests." Saunders, Philadelphia, Pennsylvania. 695 pp.

Weber, Y. B., and Bliss, M. L. (1972). Blood chemistry of roosevelt elk (*Cervus canadensis roosevelti*). *Comp. Biochem. Physiol.* **43A,** 649–653.

Zar, J. H. (1984). "Biostatistical Analysis." Prentice-Hall, Englewood Cliffs, New Jersey. 680 pp.

PHOTO: Cryogenic marking, or freeze-branding can be a useful way of supplementing natural marks used in photoidentification studies (Sea World of Florida, Orlando, August 1978; Daniel K. Odell).

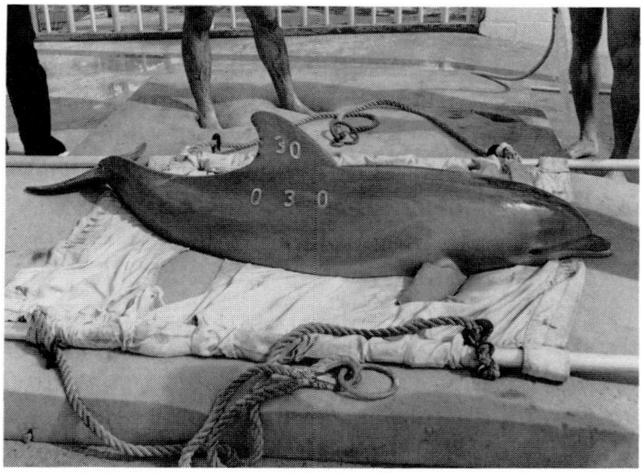

Part VIII

Research Techniques

Tagging and Marking Studies on Small Cetaceans

29

Michael D. Scott
Inter-American Tropical Tuna Commission
Scripps Institution of Oceanography
La Jolla, California 92038

Randall S. Wells
Dolphin Biology Research Associates
Sarasota, Florida, 34242, and
Long Marine Laboratory
University of California at Santa Cruz
Santa Cruz, California 95060

A. Blair Irvine
Dolphin Biology Research Associates
Sarasota, Florida 34242

Bruce R. Mate
Hatfield Marine Science Center
Oregon State University
Newport, Oregon 92110

INTRODUCTION

One of the first things that a field biologist usually attempts while observing animals is to differentiate individuals. If they can be reliably differentiated, the biologist can monitor recognizable individuals and gather information about their daily movements and activities, home ranges, reproductive histories, growth rates, and social behavior. When the life-history data of such individuals are combined, much information can be inferred about the size, vital rates, and social structure of the population.

During the 1970s, several investigators began field studies using techniques for identifying individuals in local populations of odontocetes (Irvine and Wells, 1972; Würsig and Würsig, 1977; Shane, 1980; Irvine *et al.*, 1981; Balcomb *et al.*, 1982; Bigg, 1982). Bottlenose dolphins, *Tursiops truncatus*, and killer whales, *Orcinus orca*, were particularly amenable for such studies because they commonly occur in coastal waters where they can be predictably found and easily observed. Since that time, a wide range of species of small cetaceans has been studied using natural marks, tags, and radiotelemetry.

In this chapter we review the various methods that have been used for identifying small cetaceans. We first examine the methods for identifying animals from tags and marks. We also discuss methods for attaching radio transmitters because these techniques, like tagging methods, must deal with the question of how to keep a package safely on an animal. The advantages and disadvantages of each technique and the use of these techniques in the field are discussed.

METHODS FOR IDENTIFYING INDIVIDUALS

Tags are often considered to be physical devices attached to an animal, while marks are natural or artificial alterations to the animal that are distinctive enough for identification. Evans *et al.* (1972) discussed the use of several types of tags (freeze-brands, spaghetti tags, and button tags) and radio transmitters. Irvine *et al.* (1982) conducted extensive field tests of several of these methods on a localized population of dolphins in 1974–1976. During 1978–1980, the Porpoise Tagging Project of the National Marine Fisheries Service (directed by J. Jennings) conducted an extensive series of laboratory and captive-dolphin tests of tags prior to their use on pelagic dolphins; they evaluated the effectiveness of tattoos, freeze brands, spaghetti tags, flag tags, and button tags (White *et al.*, 1981; see also Hobbs, 1982, for photographs of the various types of tags in field use).

When evaluating the usefulness of a tagging or marking technique, a number of criteria need to be considered. The ease of application influences the logistics and the potential number of animals that can be identified. The visibility of a tag or mark (how far away can it be seen?) and its legibility (how far away can it be identified?) indicate how reliable and effective the method of identification will be. One must also consider the potential harm and disturbance to the animals that may offset the potential results that can be attained from each technique.

The goals of the study need to be served by the identification method. Mark–recapture studies require that a large percentage of the population be marked, and for sizable populations this puts a premium on the ease and cost of application and resighting. Some studies do not require that individuals be recognized, only that they be categorized as marked or unmarked (e.g., the Peterson method for population size estimates); legibility of individual marks in such cases is less important. The longevity required of the tag or mark will certainly depend largely on the time scale of the study. In reality, most researchers desire a method which will simultaneously maximize the ease of application, visibility, legibility, longevity, safety, and recovery of data (all at minimal cost)—a phenomenon known as biogreed (S. Tomkiewicz, Telonics, Inc., pers. commun.).

Application Methods That Do Not Require Capture

Natural Marks

The most commonly used method for recognizing individual dolphins is photoidentification of natural marks. Reports of repeated sightings in Florida of a bottlenose dolphin with a distinctive dorsal fin (Caldwell, 1955) and of an albino bottlenose dolphin in Georgia (Essapian, 1962; Caldwell and Golley, 1965; Caldwell and Caldwell, 1972) gave one of the earliest indications that dolphins inhabit discrete home ranges. The first investigations in which naturally marked individuals were used in cetacean population studies were of killer whales (Bigg *et al.*, 1976; Balcomb *et al.*, 1982; Bigg, 1982) and bottlenose dolphins (Irvine and Wells, 1972; Würsig and Würsig, 1977; Shane 1977, 1980; Würsig, 1978). Most subsequent field studies of free-ranging small cetaceans have involved at least some use of natural marks for identification of individuals. For many dolphin species, the pattern of notches along the easily tattered rear edge of the dorsal fin allows identification of an individual; unique coloration, scars, or shapes of the flukes and flippers have been used to a lesser extent. Table 1 lists studies of small cetaceans employing natural-mark identification and describes the types of data obtained.

The extensive use of photography greatly increases the number of dolphins that can be identified. Relatively few individuals can be positively identified in the field by eye compared to the number of animals that can be identified from photographs; subtle notches or differences in fin shapes that would be missed in the field are often visible in photographs (Fig. 1). For example, during a field study of Hawaiian spinner dolphins, *Stenella longirostris*, almost 10 times as many dolphins were recognized from photographs as were recognized by eye in the field (Norris *et al.*, 1985).

Because recognition of natural marks does not require any handling of the animals, it is a relatively inexpensive method for identification and it

Table 1
Results from Field Studies Using Natural Markings

Species and Location	Description of Results	Number of Marked Animals	References
Tursiops truncatus			
W. Florida	Home range	1	Caldwell (1955)
Georgia, South Carolina	Home range	1	Essapian (1962), Caldwell and Golley (1965), Caldwell and Caldwell (1972)
W. Florida	Supplemented tagging studies to determine home ranges, long-term residency, association patterns	12	Irvine and Wells (1972), Wells *et al.* (1980, 1987), Irvine *et al.* (1981)
Argentina	Home range, long-range movements, association patterns, diving behavior	53	Würsig and Würsig (1977, 1979), Würsig (1978)
Texas	Core areas, movements	20	Shane (1977, 1980), Shane and Schmidly (1978)
S.W. England	Foraging, social interactions	1	Lockyer (1978), Lockyer and Morris (1985)
E. Florida	Supplemented tagging studies to determine home ranges	13	Odell and Asper (Chapter 30, this volume)
Texas	Core areas, group integrity	20	Gruber (1981)
S. California	Population size, group integrity, long-range movements	123	Hansen (1983, Chapter 23, this volume), Wells *et al.* (Chapter 24, this volume)
W. Florida	Population discreteness, home range	246	Weigle (1987)
N.W. Mexico	Home range	6	de la Parra and Galvan (1985)
Australia	Behavior	7	Connor and Smolker (1985)
S. California	Population size, group integrity	207	Defran *et al.* (1986)
W. Florida	Population size, home range	126	Shane (Chapter 12, this volume)
N.W. Mexico	Group integrity, site fidelity, behavior	155	Ballance (Chapter 13, this volume)
Stenella longirostris			
Hawaii	Group integrity, movements, population size	224	Norris and Dohl (1980), Norris *et al.* (1985)
Sousa chinensis			
South Africa	Home range	"Several"	Saayman and Tayler (1973, 1979)

(*continued*)

Table 1 (*continued*)

Species and Location	Description of Results	Number of Marked Animals	References
Orcinus orca			
British Columbia–Washington	Range, group integrity and structure, reproductive and mortality rates, relative growth rates, acoustics, behavior	332	Bigg *et al.* (1976, 1987), Erickson (1978), Balcomb *et al.* (1982), Bigg (1982), Ford and Fisher (1983), Balcomb and Bigg (1986), Haenel (1986), J. Heimlich-Boran (1986), S. Heimlich-Boran (1986), Hoelzel and Osborne (1986), Jacobsen (1986), Osborne (1986)
S. Alaska	Group structure, behavior	298	Ellis (1984), Hall *et al.*(1986)
Iceland	Association patterns	59	Lyrholm *et al.* (1987)
Grampus griseus			
S. California	Movements	1	Dohl *et al.* (1978)
Globicephala macrorhynchus			
S. California	Group structure, seasonal residency	8	Shane (1984, 1986)
Phocoena phocoena			
Bay of Fundy	Behavior and respiration rate	—	Watson and Gaskin (1983)

does not entail any risk of injuring the animal. The marks appear to be long-lasting and change little over periods of several years. Notches experimentally cut out of the dorsal fin of a killer whale remained visible for at least 7 years (Bigg, 1982). Slight and gradual changes did occur, however, with the notches becoming more elongate and shallow with time. Norris *et al.* (1985) also documented slight changes in natural marks over periods of 1–2 years. Figure 2 illustrates the changes that occurred in the dorsal-fin notches of a bottlenose dolphin over a 4-year period. Small scratches or bite marks may be used for identification over periods of 6 months to 1 year (Würsig and Würsig, 1977), but Bigg (1982) found that if a scar remained visible for 2 years, the mark could be considered permanent. Even short-term scars can tell a story: Lockyer and Morris (1985) inferred information about interactions of an individual bottlenose dolphin with its prey, conspecifics, and other species from the scars on that individual's body.

Depending on the degree to which the dorsal fin has been tattered, the visibility of the marks varies and much effort may be required to match photographs taken at different times of marked dolphins. Recent studies have attempted to cope with the labor-intensive task of matching photographs of individual dolphins by computerizing much of the process. Hansen (1983) and Defran *et al.* (1986) have developed methods for categorizing and matching photographs of bottlenose dolphins in southern California.

Mark-recapture techniques have been used to estimate population size from sightings of natural marks (Hansen, 1983, Chapter 23, this volume; Defran *et al.*, 1986). Hammond (1986) examined the problems of treating data from natural-mark studies so that they can be used in conventional mark–recapture analyses. For example, because

Figure 1 The notching patterns on the dorsal fins of dolphins may be subtle and can often be overlooked by eye in the field. Photographs, such as this one of a bottlenose dolphin from Florida, allow more frequent and reliable identifications. In this case, the small notches on the trailing edge of the fin and the fin shape are distinctive.

natural marks vary more in their visibility than tags (which are usually presumed to be all equally identifiable), care must be taken that more easily identifiable individuals are not overrepresented in the sample. Selection of the photographs must be based only on photographic quality and not on the ability to identify the animal; low-quality photographs which may identify only the most obvious natural marks could bias the population estimates. Hammond recommended that only those naturally marked individuals that can consistently be identifed in good quality photographs taken at another time be included in the number of marked animals, as inclusion of individuals that are so subtly marked that they may not be recognized in subsequent photographs can lead to biased population estimates.

Two potential problems should be kept in mind when analyzing photographs of naturally marked individuals. One concern, suggested by Pennycuick (1978), is that two or more animals may have identical markings. Payne *et al.* (1983) argued that the chances of finding identical patterns in the callosity patterns of southern right whales, *Eubalaena australis,* are exceedingly slim. We have found some individual bottlenose dolphins in Florida that appeared to have very similar marking patterns. Although identifications of these individuals were initially confused in the field, examination of several photographs taken during multiple sightings of each individual uncovered the subtle differences that distinguished them (Fig. 3). The second consideration is the potential for change in the marking pattern. While most notches appear to be permanent, the researcher must monitor the appearance of additional marks or gradual changes in existing marks that may occur during the study. To test the assumption about the permanency of natural marks, two or more independent marks on the same individual should be used to monitor changes in the marks. For example, Payne *et al.* (1983) used dorsal markings to validate the use of callosity patterns for individual identification of southern right whales,

Figure 2 Photographs of a single dolphin taken 4 years apart. The first photograph, taken in July 1976, shows a hole near the rear edge of the fin where a tag bolt had been and a high notch caused by shedding of a rototag.

and Irvine *et al.* (1982) used freeze brands to confirm identification of bottlenose dolphins with distinctive fin notches. Both of these problems (i.e., the presence in the population of two or more animals with the same markings and the changeability of natural marks) point up the need, first, to compile an extensive catalog of good-quality photographs and, second, to challenge regularly the validity of the identifications in the catalog.

Unlike tagging techniques that may require capturing the animal, reliance on natural marks makes it difficult to determine the sex, size, and age of a small cetacean. When an individual dolphin is closely and consistently accompanied by a calf, then it may be reasonable to assume that the animal is a female. Also, in some species (e.g., killer whales and the eastern stock of spinner dolphins), adult males can be distinguished by the size or shape of the dorsal fin. The majority of animals from most species of small cetaceans, however, cannot be sexed unless the reproductive histories of individuals have been studied for many years or their genital areas clearly seen and photographed. Aerial photogrammetry offers the possibility of obtaining length estimates without capture of the animal, but because of the logistics involved in getting a group of identifiable dolphins, an aircraft from which to photograph them, and a boat from which to identify them, all together under the near-ideal conditions needed for high-resolution photography, it is not a simple matter.

The second photograph, taken in September 1980, shows the addition of a notch at the tip of the dorsal fin and widening of the original notch. The hole in the fin remained distinctive.

Tattoo and Paint Marks

Remote tattooing was attempted in 1949–1950 on belugas, *Delphinapterus leucas*, by modifying a harpoon head so that a marking plate with ink-filled pins could be attached long enough to leave a tattoo mark on the animal (Doan and Douglas, 1953). The device was tested on a carcass, and then applied to 11–15 live belugas, but no recaptures were reported. White *et al.* (1981) tested several tattooing techniques, but none was applied to live dolphins.

Watkins and Schevill (1976) used paint marks for short-term identification. They first tested two commercially available paints on a captive bottlenose dolphin. The paints, applied with pressurized cans from above water, adhered to the skin underwater. The dolphin did not react to the application process, but much of the mark was gone after 24 hours because of skin sloughing. Paint was applied to a free-ranging Atlantic white-sided dolphin, *Lagenorhynchus acutus*, in a similar manner. The animal did not react to the marking and was subsequently followed for 30 minutes, but it was not resighted after the herd left the area. Shane (1977), who also used the technique on a free-ranging bottlenose dolphin in Texas, stated that the paint "appeared to disturb the animal excessively," and no further marking was done.

Paint marks are relatively easy and inexpensive

A
B

to apply to bow-riding dolphins. They appear to be visible at a distance, and could be designed so that each is unique. The very short-term nature of the marks, however, severely limits their usefulness.

Discovery Tags

Although Discovery tags are normally applied to large whales, Brown (1975) reported the tagging of 1 dusky dolphin, *Lagenorhynchus obscurus*, and 75 small odontocete whales (killer whales, ziphiids, and pilot whales, *Globicephala* sp.). Standard Discovery tags are shot into the blubber layer of large whales with a 12-gauge shotgun, but smaller tags used with a .410-caliber gun were developed for small whales (Brown, 1975, 1978). White streamers can be attached to facilitate resightings, but usually the tags are recovered only when the animals are killed in a fishery. Only 2 of 75 tags were recovered in a small-whale fishery, both within 1 day after tagging. The dependence on a fishery for recovery of tags limits their usefulness since fishery catches are almost always biased samples of a population. Brown also did not recommend the use of Discovery tags for dolphins or small whales less than 4.6 m long because of the potential for serious injury. Christensen (1978) tagged 2–3 killer whales (presumably with Discovery tags, although the tag type was not identified) and reported that the whales became elusive when reapproached by the tagging vessel. Lockyer (1979), however, reported that killer whales did not alter their behavior after being tagged in this manner.

Spaghetti Tags

Spaghetti tags are vinyl-covered strands of wire cable connected to steel or plastic dart tips. These tags were initially developed for tagging fish (Mather, 1963; Beckett, 1968). Spaghetti tags have been applied to free-ranging or captured dolphins with a jab stick or crossbow, usually while the animal was riding the bow of a boat. The spaghetti tag was first tested on four captive dolphins (bottlenose dolphins and rough-toothed dolphins, *Steno bredanensis*) by Nishiwaki *et al.* (1966). They found that abscessing and shedding occurred within 2 months and concluded that the method was not effective. Various modified spaghetti tags were tested on captive common dolphins, *Delphinus delphis* (White *et al.*, 1981), and used in field studies on several dolphin species (Table 2; see also Evans *et al.*, 1972; Perrin *et al.*, 1979). In a study of free-ranging, regularly monitored bottlenose dolphins, Irvine *et al.* (1982) demonstrated that the rates of abscessing and subsequent shedding were high. Common dolphins in captivity also have been observed to pull on the tags of tankmates, probably hastening the shedding process (White *et al.*, 1981). Irvine *et al.* (1982) also observed during a recapture of one of their tagged dolphins that it had a recent, superficial shark bite that appeared to have been directed toward a dangling spaghetti tag.

The main advantage of spaghetti tags is that they can easily be applied to free-swimming animals, and relatively large numbers of animals can be tagged. The tags are difficult to spot, however, and the symbols are usually too small to be read from a free-swimming animal. Braided tags with unique color and length combinations have been used to facilitate behavioral observations of dolphins captured in tuna purse seines (Perrin *et al.*, 1979), and color-coded tags also have been used to identify free-ranging dusky dolphins (Würsig and Bastida, 1986). The maximum longevity of the tags is several years, reaching almost 8 years in dusky dolphins (Würsig and Bastida, 1986). The high shedding rate and the low resighting rate observed in the field, however, indicate that the longevity of the tags is typically low, of the order of a few weeks or months. Irvine *et al.* (1982), in their comparison of tagging techniques, recommended against the use of spaghetti tags on bottlenose dolphins.

Figure 3 Two bottlenose dolphins from Florida (A, dolphin at right, and B) were initially thought to be a single individual until several photographs taken at different times could be compared for subtle differences in fin shape. The large notch on the fin of the dolphin in B was caused by shedding of a rototag; all other notches on the two animals were natural.

Table 2
Field Studies Involving the Use of Spaghetti Tags on Small Cetaceans[a]

Species and Location	Number of Tagged Animals	Number of Resightings	Maximum Observed Longevity	References
Tursiops truncatus				
W. Florida	14	43	1 year	Irvine and Wells (1972), Irvine *et al.* (1982)
S. and Baja California	8	1	3 months	Evans *et al.* (1972)
E. Japan	1	0	—	Kasuya and Oguro (1972)
S. California	39	2	10 days	Dohl *et al.* (1978), Norris *et al.* (1978)
E. Pacific	125	0	—	Perrin *et al.* (1979), Hedgepeth (1985)
Stenella attenuata				
E. Pacific	2905	91	1 year	Evans *et al.* (1972), Perrin (1975), Perrin *et al.* (1979, 1985), Hedgepeth (1985)
Stenella longirostris				
E. Pacific	332	7	2.1 years	Perrin (1975), Perrin *et al.* (1979), Hedgepeth (1985)
Stenella coeruleoalba				
W. Pacific	9	0	—	Kasuya and Oguro (1972)
E. Pacific	4	0	—	Perrin *et al.* (1979), Hedgepeth (1985)
Delphinus delphis				
S. and Baja California	240	3	"Several months"	Evans *et al.* (1972)
S. California	525	7	1.5–2.5 months	Dohl *et al.* (1978), Norris *et al.* (1978)
E. Pacific	193	0	—	Perrin *et al.* (1979) Hedgepeth (1985)
Lagenorhynchus obliquidens				
S. and Baja California	10	0	—	Evans *et al.* (1972)
S. California	38	0	—	Dohl *et al.* (1978), Norris *et al.* (1978)
S. and Baja California	15	0	—	Perrin *et al.* (1979), Hedgepeth (1985)
Lagenorhynchus obscurus				
Argentina	30	10	7.9 years	Würsig and Würsig (1980), Würsig (1982), Würsig and Bastida (1986)
Lissodelphis borealis				
S. California	14	0	—	Dohl *et al.* (1978), Norris *et al.* (1978)
Globicephala macrorhynchus				
S. California	41	0	—	Dohl *et al.* (1978), Norris *et al.* (1978)
E. Pacific	2	0	—	Perrin *et al.* (1979), Hedgepeth (1985)
N.E. Florida	3	1	6 days	Irvine *et al.* (1979)
Delphinapterus leucas				
Hudson Bay	694	3	1 year	Sergeant and Brodie (1969)
Phocoenoides dalli				
E. Japan	3	0	—	Kasuya and Oguro (1972)
S. California	24	0	—	Dohl *et al.* (1978), Norris *et al.* (1978)
Alaska	23	1	3 months	Hall (1981)

[a] Tagged dolphins that were not identified to species are not included (Perrin *et al.*, 1979, 1985).

Laser Brands

Dohl (1981) has proposed using a laser for marking dolphins at a distance. He tested several types of lasers on two captive Pacific white-sided dolphins, *Lagenorhynchus obliquidens,* and on dolphin carcasses. Some of the tests resulted in the desired depigmented marks, and no reaction by the dolphins to the marking process was observed. Most of the marks faded within a short time, however, with the longest lasting mark remaining visible until the dolphin died of natural causes 5 months later. Dohl concluded that more testing was needed to develop the technique.

Application Methods That Require Capture

Fin Notching

Fin notching has been attempted on bottlenose dolphins (n = 2, Asper and Odell, 1980), killer whales (n = 1, Bigg, 1982), spotted dolphins, *Stenella attenuata* (n = 651, Powers *et al.*, 1979; Hedgepeth, 1985), and spinner dolphins (n = 15, Powers *et al.*, 1979; Hedgepeth, 1985; n = 1, Norris *et al.*, 1985). Notching the dorsal fins of dolphins that are captured and released provides the same advantage of longevity that natural marks have, while at the same time providing the opportunity to collect information on sex and length as well as other biological data from the animals. To date, fin notching has been used primarily to supplement other tagging methods; Powers *et al.* (1979) used the marks to identify the date and location where the dolphins were captured and not for identification of individuals. The degree to which the marks can be sighted and identified varies with the shape and the depth of the notch that is made, but, potentially, a variety of distinctively shaped marks can be created that are uncommon in the wild. Just as for natural marks, an intensive photo-identification effort is required for individual recognition. Asper and Odell (1980) mentioned that they stopped the notching of fins in their study because of minor but persistent bleeding of the fins. Other authors have not reported such problems.

Rototags

Rototags, normally used to tag the ears of cattle, were first used on rough-toothed dolphins and spotted dolphins by Evans (1967) and Norris and Pryor (1970). These tags clip along the thin rear edge of the dorsal fin with a pair of specially designed pliers. Application may be facilitated and tissue trauma reduced by creating a small hole with a sterile awl first, before the tag is clipped through the hole. While the numbers on the tag are too small to be read at a distance in the field, the color, number, and positions of the tags on the fin can provide a means of identification.

Rototags have been used in a variety of ways (Table 3). These tags have often been used as backup identification to other types of tags (Norris *et al.*, 1978, 1985; Asper and Odell, 1980; Irvine *et al.*, 1981, 1982) or as the primary means of identification. Rototags were used for studies of the acoustics (Evans, 1967), movements (Perrin *et al.*, 1979), association patterns and population structure (Wells, 1986; Wells *et al.*, 1987) of free-ranging dolphins, and for studying the subsequent fate of stranded cetaceans after they were returned to the wild (Irvine *et al.*, 1979; Mead *et al.*, 1980).

Because the rototags cannot be read from a distance, identification is limited by the combinations of colors and different fin locations available. This limitation makes the technique more useful for small-scale tagging programs, or for animals that may be taken in a fishery and examined closely. The retention time of the tags has proved to be short on coastal animals, although some tags have remained on the animals for several years (Norris and Pryor, 1970; Perrin *et al.*, 1979). Apparent tooth marks on the fin of a bottlenose dolphin rototagged by Wells *et al.* (1987) suggest that other dolphins may pull on the tags and contribute to the tag-shedding problem.

When rototags are shed, typically a small hole or notch along the rear edge of the dorsal fin remains. This allows subsequent identification of the animal based on the position of the small hole or notch just as if it were a natural mark. Thus, a rototag can identify an animal long after the tag itself has dropped off. Wells *et al.* (1987) have used the well-healed notches and holes left by rototags

Table 3
Field Studies Involving the Use of Rototags on Small Cetaceans

Species and Location	Number of Tagged Animals	Number of Resightings	Maximum Observed Longevity	References
Tursiops truncatus				
W. Florida	38	342	2 years	Irvine *et al.* (1981, 1982), Wells *et al.* (1980, 1987)
Japan	31	—	—	Tanaka and Takao (1986)
E. Florida	8	—	—	Asper and Odell (1980)
Stenella attenuata				
Hawaii	1	"Several"	3.5 years	Evans (1967), Norris and Pryor (1970)
E. Pacific	98	4	4 years	Perrin *et al.* (1979)
Stenella longirostris				
W. Florida	1	1	1 day	Mead *et al.* (1980)
Hawaii	3	—	—	Norris *et al.* (1985)
Delphinus delphis				
S. California	3	—	—	Norris *et al.* (1978)
Steno bredanensis				
Hawaii	2	0	—	Evans (1967); Norris and Pryor (1970)
Lagenorhynchus obliquidens				
S. California	2	0	—	Norris *et al.* (1978)
N. California	2	0	—	Norris and Hall (1979)
Japan	5	—	—	Tanaka and Takao (1986)
Globicephala macrorhynchus				
N.E. Florida	16	1	6 days	Irvine *et al.* (1979)
Phocoena phocoena				
Bay of Fundy	11	2	>2 years	Gaskin *et al.* (1975, 1985), A. Read (pers. commun.)

to identify bottlenose dolphins for over 11 years (Figs. 2–4).

Rototags have other advantages: they are relatively easy to apply, and the risk of injury is low. Norris and Pryor (1970) reported no problems when these tags were tested on captive dolphins. Occasional problems with pinching when the tag was too tight (Asper and Odell, 1980) and with fouling by barnacles and algae (Irvine *et al.*, 1982) have been reported, but, overall, the use of rototags has proved to be a successful method for dolphin identification.

Freeze Brands

Freeze-branding was first used for identification of terrestrial animals (Farrell *et al.*, 1966, 1969). The technique usually involves cooling a metal brand with liquid nitrogen or with a mixture of dry ice and either alcohol or acetone. The brand is usually applied to the dorsal fin or the body just below the dorsal fin, causing depigmentation of the skin. In bottlenose dolphins, a readable symbol usually appears within 10 days after marking. Tests on captive dolphins (bottlenose, common, spinner, and Pacific white-sided dolphins, and false killer whales, *Pseudorca crassidens*) indicated that successful marking times ranged from 8 to over 24 seconds applied at pressures of 1.0–1.4 kg/cm^2 (Cornell *et al.*, 1979; White *et al.*, 1981; Odell and Asper, Chapter 30, this volume). Both liquid nitrogen and a dry ice–alcohol mixture have produced adequate marks, although dry ice is usually more economical and easier to handle. White *et al.* (1981) also tested equipment with a view to improving the technique's use in the field, such as application of Freon spray. The use of Freon spray and stencils has produced mixed success on captive dolphins (Tobayama *et al.*, 1982; Tanaka *et al.*, 1987).

Freeze-branding has been successfully used in field studies on dolphins (Table 4), with marks remaining visible for up to 11 years (Fig. 4). Freeze brands are not as visible in the field as rototags or button tags, but a freeze-branded dolphin can be readily identified at distances less than 30 m (Irvine *et al.*, 1982). With the exception of dorsal fin notches, freeze brands are the most long-lived of all the marks or tags tested. Freeze brands do fade with time, but even when they cannot be read easily in the field, the marks can be discerned from good photographs. Fading, however, may occur more rapidly and more completely on calves (Irvine *et al.*, 1982). Dolphins display little or no reaction to the freeze-branding process, although skin tissue may be damaged if the brand is kept on the skin for too long. Price (cited in Gaskin, 1982) suggested that freeze-branding might cause damage to the underlying muscle, but no evidence of this has been observed in field or captive studies of dolphins.

Button Tags

Because the dorsal fin is so prominent, various types of tags have been designed to be bolted through the dorsal fin for maximal visibility in the field (Table 5). Tags of this type, collectively termed button tags in this chapter, are often disks or rectangular plates made of fiberglass or plastic. They are usually attached with stainless steel bolts, although pins made of other materials have also been used. Sergeant and Brodie (1969) first attempted to attach a Peterson disk tag, normally used on fish, with a bolt through the dorsal ridge of belugas. Several types of button tags were also field tested on bottlenose dolphins (Irvine and Wells, 1972; Irvine *et al.*, 1982).

Disk tags developed by the U.S. National Marine Fisheries Service for use on dolphin species in the eastern tropical Pacific Ocean underwent hydrodynamic evaluation (National Fisheries Engineering Laboratory, 1978), and then several designs were tested on captive dolphins (White *et al.*, 1981). The final design was a convex disk made of a biocompatible plastic (delrin) that was perforated to relieve buildup of water pres-

Table 4
Field Studies Involving the Use of Freeze Brands on Small Cetaceans

Species and Location	Number of Marked Animals	Maximum Observed Longevity	References
Tursiops truncatus			
W. Florida	95	11 years	Evans *et al.* (1972), Irvine and Wells (1972), Wells *et al.* (1980, 1987, unpublished data), Irvine *et al.* (1981, 1982)
Mississippi Sound	53	21 days	Solangi and Dukes (1983)
Florida	132	8 years	Odell and Asper (Chapter 30, this volume)
Stenella longirostris			
Hawaii	1	37 days	Norris *et al.* (1985)
Delphinus delphis			
S. California	3	—	Norris *et al.* (1978)
Lagenorhynchus obliquidens			
S. California	2	—	Norris *et al.* (1978)
N. California	3	—	Norris and Hall (1979)
Globicephala macrorhynchus			
S. California	1	—	Norris *et al.* (1978)

Figure 4 Photograph of a bottlenose dolphin that had been rototagged and freeze-branded (No. 45) 11 years previously. A hole at the top of the fin caused by shedding of the rototag and the faded, but visible, freeze brands still allow identification of the dolphin. The natural marks at the base of the fin also assist in identification.

sure underneath the tag. These tags were attached with stainless steel bolts to 650 spotted and spinner dolphins; 4 of these tagged dolphins were recovered in the tuna fishery, and 115 were resighted, but not identified (Hedgepeth, 1985). A similar tag design was tested on captive and free-ranging bottlenose dolphins in Japan (Tanaka *et al.*, 1987). Only 3 resightings of 34 tagged dolphins were made; additionally, 2 dolphins showed evidence that tag migration through the fin had occurred. Tanaka *et al.* attempted to stabilize the tag by adding a fiberglass belt around the leading edge of the dorsal fin to prevent the tag from migrating posteriorly, but the design was not effective in a test on a captive dolphin.

Button tags can be the most easily sighted and legible of all the tags, particularly for small-scale tagging efforts. For large-scale efforts, the size of the symbols on the tag has to be reduced as the number of dolphins to be tagged and the number of symbols on the tags increase. Thus, in the eastern tropical Pacific tagging study where hundreds of animals were tagged (Powers *et al.*, 1979; Hedgepeth, 1985), the tag symbols were too small to be read unless the dolphin was caught in a tuna purse-seine net and examined on the deck. An additional problem reported by Irvine *et al.* (1982) and Tanaka *et al.* (1987) was that tags attached to coastal bottlenose dolphins were subject to fouling by algae and barnacles, which

Table 5
Field Studies Involving the Use of Button Tags on Small Cetaceans

Species and Location	Type of Tag	Number of Tagged Animals	Maximum Observed Longevity	References
Tursiops truncatus				
W. Florida	Plastic disk tags, nylon bolts; fiberglass rectangular tags, teflon pins or stainless steel bolts	23	3 months	Irvine and Wells (1972), Irvine *et al.* (1982)
Japan	Plastic disk tags, stainless steel bolts	34	11 months	Tanaka and Takao (1986), Tanaka *et al.* (1987)
Stenella attenuata				
E. Pacific	Plastic disk tags, stainless steel bolts	641	4.5 years	Powers *et al.* (1979), Hedgepeth (1985)
Stenella longirostris				
E. Pacific	Plastic disk tags, stainless steel bolts	15	—	Powers *et al.* (1979), Hedgepeth (1985)
Delphinus delphis				
S. and Baja California	Plastic disk tags, nylon bolts	46	1.8 years	Evans *et al.* (1972)
Lagenorhynchus obliquidens				
S. and Baja California	Plastic disk tags, nylon bolts	6	"Almost 1 year"	Evans *et al.* (1972)
Japan	Plastic disk tags, stainless steel bolts	5	—	Tanaka and Takao (1986), Tanaka *et al.* (1987)
Delphinapterus leucas				
Hudson Bay	Peterson disk tags, Monel metal pin	118	—	Sergeant and Brodie (1969)
Phocoena phocoena				
Bay of Fundy	Plastic disk tag, stainless steel bolt	1	—	Gaskin *et al.* (1975), A. Read (pers. commun.)

obscured the symbols on the tag, hindered identification, and sometimes caused skin abrasions on the fin.

Button tags have lasted several years on pelagic dolphins, but Irvine *et al.* (1982) found that bottlenose dolphins gradually broke the tags (probably by battering the tags against the shallow bottom), usually within a few weeks or months. The potential for injury varies with tag design; Irvine *et al.* found that rectangular plates secured by a single bolt could migrate posteriorly or twist and damage the dorsal fin, but that adding a second bolt stabilized the tag and decreased the risk of injury. Nevertheless, Irvine *et al.* removed virtually all of these tags from the dolphins at the conclusion of their study, partly because of the potential for injury owing to fouling by barnacles. The hydrodynamic design of the disk tags used in the eastern tropical Pacific may have alleviated some of the above problems, but the low number of tag recoveries prevented an assessment of the potential usefulness.

Other Tagging and Marking Methods

Several other types of tags have been designed and tested on captive dolphins. Clips designed to attach to the forward or rear edge of the dorsal fin were evaluated by White *et al.* (1981), but were not tested further because of potential tissue damage. Tomilin (1983) produced readable marks by clamping a stencil onto the dorsal fins of captive dolphins for up to 19 hours. Four days after the stencil was removed, a light mark appeared, arising from depigmentation of the skin. The mark remained visible until the death of the animals 2 years later. While it was believed that the method could produce a mark which would be visible for several years, a technique that requires holding a dolphin in captivity for at least a day has limited usefulness for field studies.

The relative narrowness of the caudal peduncle of the dolphin makes it an attractive area for attachment. Nishiwaki *et al.* (1966) wrapped vinyl-covered copper wires around the peduncles of 65 captive dolphins of several species. They concluded that the method was ineffective due to the high shedding rate of the tag and due to the injuries to the flukes caused by the wire. White *et al.* (1981) came to a similar conclusion about peduncle belts made of softer materials (soft rubber, surgical tubing, or polypropylene) because of the skin abrasions caused by the tags. Even if such problems could be resolved, visibility of the tags would be limited unless they were attached to the caudal peduncle with a tether. A tether, however, poses a potential danger of entanglement around the animal or other objects and could impair the animal's swimming. "Breakaway" tethers have been safely attached to the peduncles of manatees, *Trichechus manatus*, in Florida, but the attachment is feasible because a manatee's hide is tough and thick and its movements typically slow (Rathbun *et al.*, 1989).

RADIOTAGS AND SATELLITE TAGS

Unlike tags that can be identified only by sight, radiotags permit remote tracking of the animals. Radiotracking studies conducted on small cetaceans have been reviewed by Norris *et al.* (1974), Leatherwood and Evans (1979), Dietz (1986), and Würsig *et al.* (1990). Dietz (1986) compiled a list of researchers and companies currently involved in radio- and satellite tracking studies. The increased availability of radiotracking methods has been a major impetus for the proliferation of field studies on cetaceans (Table 6). These studies have produced information on movements, activities, and diving behavior. Telemetry offers the potential for gathering data on physiological and environmental conditions as well.

This section focuses primarily on radiotag attachment designs used on small cetaceans because it is the lack of an adequate method that continues to limit the longevity of these devices to just a few days or weeks for most small cetaceans. The most common method of attachment for dolphins has been to mount the radio on a plastic or fiberglass saddle bolted to the dorsal fin or ridge. Some of these saddles wrap around the front edge or surround the base of the dorsal fin. Other radios are mounted on plates bolted to the fin. The nuts securing the package are often made of magnesium or zinc, which will dissolve in seawater, so that the package will eventually be shed. Many of these packages have been found to migrate through the fin, resulting in loss of the package (Irvine *et al.*, 1982). A package that oriented the transmitter along the leading edge with a cross-pinning arrangement through the dorsal fin of killer whales, however, remained attached for at least 4.5 months (Erickson, 1978). Whether the greater longevity in the Erickson study was due to this attachment procedure or to the greater thickness and size of the killer whale's dorsal fin is unclear.

Mate (1989) attached transmitters to pilot whales, *Globicephala macrorhynchus*, using delrin plastic pins and high-grade stainless steel nuts. A heat-molded plastic saddle carrying both a satellite and a VHF radio transmitter was wrapped around the upper, anterior edge of the dorsal fin and secured with six pins. To a second whale, they attached a VHF radio transmitter on a plastic plate with two delrin pins. The satellite package lasted for at least the 97-day life of the transmitter, while the radio package lasted for at least 24 days. Mate *et al.* also demonstrated that the delrin-pin design had a breakaway feature. When one package on a third whale entangled in a net, the pins broke free, allowing the whale to escape.

Table 6
Radio- and Satellite Tracking of Small Cetaceans

Species and Location	Transmitter Type	Number of Animals Tracked	Maximum Observed Longevity (days)	References
Tursiops truncatus				
W. Florida	OAR,[a] 27 MHz	10	21	Irvine *et al.* (1981, 1982)
S.E. Japan	Toyocom,[b] ARGOS, 401 MHz	6	18	Nakamura *et al.* (1985), Soma and Tsutsumi (1986), Tanaka (1987), Tanaka *et al.* (1987)
Stenella longirostris				
E. Pacific	OAR, 27 MHz	1	1	Leatherwood and Ljungblad (1979), Perrin *et al.* (1979)
Hawaii	OAR, 27 MHz	1	3	Norris and Dohl (1980)
Hawaii	Telonics,[c] 148 MHz	3	6	Norris *et al.*(1985)
Stenella attenuata				
E. Pacific	OAR, 27 MHz	12	19	Leatherwood and Evans (1979), Leatherwood and Ljungblad (1979), Perrin *et al.* (1979)
E. Pacific	Telonics, 148 MHz; OAR, 27 MHz	6	2	Powers *et al.* (1979), Butler and Jennings (1980)
Hawaii	Handar,[d] NIMBUS-6, 401 MHz	2	7	Southwest Fisheries Center (1981), Jennings (1982)
Hawaii	Telonics, 148 MHz	1	5	Scott and Wussow (1983)
Delphinus delphis				
S. California	OAR, 27 MHz	14	> 30	Evans (1971, 1974, 1975), Martin *et al.* (1971), Evans *et al.* (1972), Leatherwood and Evans (1979)
S. California	OAR, 27 MHz	7	34	Dohl *et al.* (1978), Norris *et al.* (1978)
Lagenorhynchus obliquidens				
S. California	OAR, 27 MHz	2	<1	Dohl *et al.* (1978), Norris *et al.* (1978)
S. California	OAR, 27 MHz	1	>50	Leatherwood and Evans (1979)
Lagenorhynchus obscurus				
Argentina	OAR, 27 MHz	10	15	Würsig (1982)
New Zealand	Telonics, 148 MHz	3	>7	Cipriano (1985), Würsig *et al.* (1990)
Globicephala macrorhynchus				
S. California	OAR, 27 MHz	1	—	Evans (1974)
S. California	OAR, 27 MHz	1	7	Norris *et al.*(1978)
N.W. Atlantic	Telonics, ARGOS, 401 MHz	1	95	Mate (1989)

(continued)

Table 6 (*continued*

Species and Location	Transmitter Type	Number of Animals Tracked	Maximum Observed Longevity (days)	References
N.W. Atlantic	Telonics, 148 MHz	2	24	Mate (1989)
Orcinus orca				
Puget Sound	ATS,[e] 164 MHz	2	144	Erickson (1978)
Delphinapterus leucas				
Alaska	OAR, 164 MHz	2	14	Frost *et al.* (1985)
Phocoena phocoena				
Bay of Fundy	OAR, 27 MHz	18	22	Gaskin *et al.* (1975, 1985), Read and Gaskin (1983, 1985), Watson and Gaskin (1983), A. Read (pers. commun.)

[a] Ocean Applied Research Corp., 10447 Roselle St., San Diego, CA 92121.
[b] Toyo Communication Equipment Co., 753 Koyato, Samukawa-Machi, Koza-Gun, Kanagawa-Pref., Code 253-01, Japan.
[c] Telonics, Inc., 923 E. Impala Ave., Mesa, AZ 85204.
[d] Handar Co., 1380 Borragas Ave., Sunnyvale,CA 94089.
[e] Advanced Telemetry Systems, Inc., Box 398, 470 First Ave., N., Isanti, MN 55040 (formerly Cedar Creek Laboratory, University of Minnesota).

Radio transmitters attached to suction cups were tested on captive dolphins (Baldwin, 1965) and shot from a crossbow to track free-ranging whales (Goodyear, 1981), but they have not remained on an animal for more than a few days. Harnesses, typically used in some form for most terrestrial animals, have also been tested on dolphins (e.g., Starks, 1966; Bowers and Henderson, 1972), but Irvine (1970) found that harnesses seriously impaired the dolphins' locomotion. More recently, harnesses were used to attach radio packages to captive and free-ranging bottlenose dolphins in Japan (Kato, 1986; Tanaka *et al.*, 1987). The package consisted of a plastic saddle wrapped around the base of the dorsal fin secured by a band around the body; sometimes the package was bolted through the dorsal fin for additional stability. Diving and other behavioral changes occurred initially, and the bands left white traces on the skin after 1 month. The combination of the belly band and bolt attachment also caused injury to the dorsal fin. Attachment with the belly band alone did not cause serious damage, but the saddles were susceptible to being lifted off the dorsal fin by the water flow and eventually were shed. Tanaka *et al.* did obtain signals for 35 days from a satellite transmitter placed on a free-ranging dolphin, but they could not determine whether the package was actually on the animal for the entire time or whether the transmitter had been shed at some point and was just drifting at the surface.

Compromises must be made among various characteristics, such as the weight and strength of the radio package, hydrodynamic streamlining, battery life, range of the transmitter, and potential for injury to the animal. Thus, for example, efforts to increase the range of the transmitter are usually counterbalanced by decreases in longevity or increases in weight. Tradeoffs must also be considered in selecting among the frequencies available for wildlife radiotracking (e.g., see Kenward, 1987). Most dolphin radiotracking studies have used either HF (high frequency, 27 MHz) or VHF (very high frequency, 148–172 MHz) bands. The HF transmitters are larger and, at 27 MHz, more prone to interference from citizen-band radios, but the transmission ranges can be greater because the signals can bend over the horizon. VHF radios, although smaller, have a line-of-sight transmission which is limited and prone to signal bounce which can produce false directional headings (see Würsig *et al.*, 1990, for a more detailed discussion). In a comparison of HF and VHF systems used to track spotted dolphins, Butler and Jennings (1980) found their particular VHF system more reliable.

Many of the tradeoffs in attachment design are greatly influenced by the species and habitat involved. Bottlenose dophins, for example, are large and powerful enough to destroy a tag by battering it against the bottom in the coastal shallow waters in which they are typically studied. Thus, the radio packages used on these animals have to be "armored" if they are to endure more than a few days. Pelagic dolphins, on the other hand, may retain radio packages and tags more readily because the seafloor is not as accessible to them; the packages used on them need not be reinforced so heavily. The pelagic environment poses other logistical problems, however. For example, the fates of the individual dolphins and their tags are much more difficult to monitor.

Radiotracking is labor-intensive and expensive, and it can be logistically difficult in remote areas. Tracking animals from satellites can offset some of these difficulties. Satellite transmitters designed to handle the physical stresses imposed by diving marine mammals, however, are currently relatively expensive and bulky, although the cost and size of the transmitters have decreased with time. After extensive development work (Gandy *et al.*, 1977; Jennings and Gandy, 1980; Stogner, 1981), NIMBUS-6 satellite transmitters located two free-ranging spotted dolphins off Hawaii for 1–7 days (Southwest Fisheries Center, 1981; Jennings, 1982). One dolphin, recaptured 8 days after tagging, showed signs of weight loss, presumably due to the effort required to carry the bulky radio package. Also, the package bore tooth marks, indicating that other dolphins had been tugging on it.

Owing to miniaturization and more efficient designs of current transmitters, the tracking of free-ranging cetaceans has become possible with the ARGOS satellite system. ARGOS transmitters (termed platform transmitter terminals or PTTs) transmit on a UHF band (ultra high frequency, 401 MHz). The signals are picked up by receivers aboard two satellites, which are kept active at all times in separate orbits that cross the poles. The receivers pick up PTT messages within 2500 km on either side of the satellite's ground path. Positions are determined from the doppler shift in the transmitter's frequency as the satellite passes overhead. Additional environmental or behavioral data can be encoded in the signal.

Fourteen ARGOS satellite transmitters were placed on bottlenose dolphins off Japan during 1983–1986, and signals were received from one of these transmitters for up to 35 days (Nakamura *et al.*, 1985; Soma and Tsutsumi, 1986; Tanaka, 1987). One dolphin was tracked a distance of at least 600 km over an 18-day period.

In 1987, Mate (1989) tracked a pilot whale for 95 days in the western North Atlantic. The PTT transmitted every time the whale surfaced during two 8-hour periods each day, sending data on dive duration and ambient temperature. The whale traveled at least 7588 km during the track, and data on 194,000 dives were received. Satellite monitoring showed that deep dives began just before sunset and occurred almost exclusively at night, coinciding with the nocturnal rise of the deep scattering layer and of squid, the primary prey of pilot whales. Multiple sightings of the tagged whale in a herd of other pilot whales and the consistency of the whale's diving habits suggested that the animal was in good health throughout the track. This is the most detailed example of satellite telemetry of cetaceans to date and shows some of the potential of the technique.

These results suggest that satellite monitoring can be a cost-effective means of collecting movement data and associated behavioral and oceanographic data for relatively long periods. Although satellite tracking can be expensive, it becomes more cost-effective relative to conventional radiotracking as the duration of the track increases (Harrington *et al.*, 1987). This would be particularly true for offshore populations for which the cost of close-range monitoring is high. Attachment techniques need to be perfected for the smaller cetaceans, however, to ensure that the satellite package will stay attached long enough to make it worth the cost.

As field biologists, we see a somewhat intangible benefit of radiotracking which may make it preferable over satellite tracking. Nothing can give one a "feel" for how an animal lives better than by dogging its flukeprints continuously for periods of days or weeks. On-the-spot interpretations of behavior displayed by the tagged animal and its associates at a particular time and location allow the field biologist to develop a sense for what is important in the day-to-day life of the animal. Regular observations of the tagged animal allow the researcher to judge whether the data being received are accurate and whether the behavior of the animal is normal.

DISCUSSION

Of the many types of tags proposed for use on dolphins, few have proven useful in the field. During a 17-year tagging study of bottlenose dolphins, we have had the greatest success with a combination of freeze brands and rototags for individual identification (Wells *et al.*, 1987). Initially, the freeze-brand provides the primary means of identification, while the rototag draws attention to the animals from a distance. After the rototag is shed, a notch along the rear edge of the dorsal fin usually results. The notched fin can then be photographically catalogued, and the combination of the notching pattern and the freeze brand permits identification of the animal. Even after the freeze brand fades, the notching pattern is adequate for identification.

Photoidentification of natural marks, like a parliamentary motion to adjourn, is always in order. Either as the sole study method or in conjunction with other tagging methods, natural-mark identification can provide data that are relatively inexpensive and easy to obtain. Analytical techniques, such as the testing of assumptions about the uniqueness of natural marks (Payne *et al.*, 1983) and their suitability for estimating population size with mark–recapture methods (Hammond, 1986), have been stimulated by the increasing use of photoidentification of free-ranging dolphins.

The problem of long-term attachment of radio transmitters to dolphins remains to be solved. Studies on wound-healing and the reaction of skin tissue to implants have identified some of the problems with current attachment methods (Bruce-Allen and Geraci, 1985; Geraci *et al.*, 1985). Tags that break the skin can wick bacteria into the wound and prevent it from healing. A combination of movement by the tag and a series of cellular mechanisms involved in the healing process encourages extrusion of implanted portions of the tags. Long-term attachment should be designed with the following features: a composite material which encourages growth of tissue into the implanted portion of the tag to immobilize the tag and prevent rejection, an epidermal seal to prevent wicking of pathogens, and an elasticity of the implanted portion which matches that of the tissue yet is strong enough to resist mechanical stress (Bruce-Allen and Geraci, 1985; Geraci *et al.*, 1985). The success of the delrin-pin design employed by Mate (1989), along with the decreasing trend in transmitter size, shows promise for giving us more than a glimpse into the lives of small cetaceans, but more research into the biocompatibility and design of future attachment devices is needed if long-term monitoring is to become commonplace.

What has been learned from tagging and marking studies of dolphins? Bottlenose dolphins have been the most intensively studied of all the small cetaceans, and radiotracking and individual identification have played a major role in field studies of these animals. Daily movements and home ranges have been described using radiotelemetry and tags (Irvine *et al.*, 1981) and natural marks (Caldwell and Caldwell, 1972; Würsig and Würsig, 1977; Shane, 1980). Long-range movements have been documented for naturally marked (Würsig and Würsig, 1977; Würsig, 1978; Wells *et al.*, Chapter 24, this volume) and satellite-tracked dolphins (Soma and Tsutsumi, 1986; Tanaka, 1987). Individual associations and population structure have been studied using tags and photographs of natural markings (Würsig, 1978; Wells *et al.*, 1980, 1987; Wells, 1986). Observation of marked animals has been the key to studying behavior in several studies (Würsig, 1978; Würsig and Würsig, 1979; Shane, 1980; Wells *et al.*, 1980; Irvine *et al.* 1981; Ballance, Chapter 13, this volume). Both tags and natural marks have been used to estimate population sizes (Irvine *et al.*, 1981; Hansen, 1983, Chapter 23, this volume; Defran *et al.*, 1986; Wells, 1986).

It is axiomatic that the disturbance to an animal under study should be minimized, both for scientific and ethical reasons. Tags that cause serious injury are not only harmful to the animals but can be counterproductive to the research as well, since aberrant behavior by a tagged animal may yield misleading information. One question that should be considered, then, is whether tagging is indeed harmful or disruptive of the animal's behavior. Short-term behavioral changes can occur in some radiotagged dolphins, but experience has shown that after an initial adjustment of 1 or 2 days, the movements and activities of most radio-tagged animals do not differ from those of other dolphins (Leatherwood and Evans, 1979; Irvine *et al.*, 1982; Würsig, 1982). Irvine *et al.* (1982) reported that the large radio packages used in the mid 1970s could damage the dorsal fin, but the trend in more recent studies toward the use of smaller radio packages and identification tags appears to have reduced the severity of this problem.

Capturing free-ranging dolphins also poses risks to both dolphins and researchers, and the potential risks must be weighed against the potential benefits. Are the risks worth the additional information that handling the animal can provide? We believe that many times they are. By capturing, tagging, and releasing the dolphins, a suite of information on the sex, size, age, reproductive condition, weight, body condition, and genetic relationships to other members of the herd becomes available—information that cannot be gathered in any other way. Under admittedly ideal conditions in the coastal bays of Florida, we have found the risks to bottlenose dolphins to be minimal and the diversity and volume of data to be gained great. Although capturing dolphins is not always possible or affordable, we believe that capture-and-release methods should be used when safe and feasible. A decision to forego collecting some information based on the concern for individual animals carries its own risks, since the fate of populations and species may very well rest on how much is known about their biology.

ACKNOWLEDGMENTS

We wish to thank William Perrin for letting us pillage his reprint collection. We also acknowledge helpful reviews by William Bayliff, Martin Hall, Andrew Read, and William Watkins.

REFERENCES

Asper, E. D., and Odell, D. K. (1980). Bottlenose dolphin local herd monitoring: Capture, marking, collection of biological data, and followup observations of marked animals. Hubbs–Sea World Res. Inst. Tech. Rep. No. 80-122. 119 pp.

Balcomb, K. C., III, and Bigg, M. A. (1986). Population biology of the three resident killer whale pods in Puget Sound and off southern Vancouver Island. *In* "Behavioral Biology of Killer Whales" (B. C. Kirkevold and J. S. Lockard, eds.), Zoo Biology Monographs, Vol. 1, pp. 85–95. Alan R. Liss, New York. 457 pp.

Balcomb, K. C., III, Boran, J. R., and Heimlich, S. L. (1982). Killer whales in greater Puget Sound. *Rep. int. Whal. Commn* **32,** 681–685.

Baldwin, H. A. (1965). Marine biotelemetry. *BioScience* **15**(2), 95–97.

Beckett, J. S. (1968). A harpoon adapter for tagging large free-swimming fish at the surface. *J. Fish. Res. Board Can.* **25**(1), 177–179.

Bigg, M. (1982). An assessment of killer whale (*Orcinus orca*) stocks off Vancouver Island, British Columbia. *Rep. int. Whal. Commn* **32,** 655–666.

Bigg, M. A., MacAskie, I. B., and Ellis, G. (1976). Abundance and movements of killer whales off eastern and southern Vancouver Island, with comments on management. Unpublished report, Arctic Biol. Stn., Ste. Anne de Bellevue, Quebec. 21 pp.

Bigg, M. A., Ellis, G. M., Ford, J. K. B., and Balcomb, K. C. (1987). "Killer Whales: A Study of Their Identification, Genealogy and Natural History in British Columbia and Washington State." Phantom Press and Publishers, Inc., Nanaimo, British Columbia. 79 pp.

Bowers, C. A., and Henderson, R. S. (1972). Project Deep Ops: Deep object recovery with pilot whales and killer whales. Naval Undersea Center Tech. Pub. No. 306. 86 pp.

Brown S. G. (1975). Marking of small cetaceans using "Discovery" type whale marks. *J. Fish. Res. Board Can.* **32**(7), 1237–1240.

Brown, S. G. (1978). Whale marking techniques. *In* "Animal Marking: Recognition Marking of Animals in Research" (B. Stonehouse, ed.), University Park Press, Baltimore, Maryland. 257 pp.

Bruce-Allen, L. J., and Geraci, J. R. (1985). Wound healing in the bottlenose dolphin (*Tursiops truncatus*). *Can. J. Fish. Aquatic Sci.* **42,** 21–44.

Butler, R. W., and Jennings, J. G. (1980). Radio tracking of dolphins in the eastern tropical Pacific using VHF and HF equipment. *In* "A Handbook on Biotelemetry and Radio Tracking" (C. J. Amlaner, Jr., and D. W.

MacDonald, eds.), pp. 757–759. Pergamon, New York. 804 pp.

Caldwell, D. K. (1955). Evidence of home range in the bottlenosed dolphin. *J. Mammal.* **36**(2), 304–305.

Caldwell, D. K., and Caldwell, M. C. (1972). "The World of the Bottlenosed Dolphin." Lippincott, Philadelphia, Pennsylvania. 157 pp.

Caldwell, D. K., and Golley, F. B. (1965). Marine mammals from the coast of Georgia to Cape Hatteras. *J. Elisha Mitchell Sci. Soc.* **81**(1), 24–32.

Christensen, I. (1978). Spekkhoggeren (*Orcinus orca*) i det nordostlige Atlanterhav. [The killer whale (*Orcinus orca*) in the northeast Atlantic.] *Fisken Hav., 1978* **1,** 23–31.

Cipriano, F. (1985). Dusky dolphin research at Kaikoura, New Zealand: A progress report. *Mauri Ora* **12,** 151–158.

Connor, R. C., and Smolker, R. S. (1985). Habituated dolphins (*Tursiops* sp.) in Western Australia. *J. Mammal.* **66**(2), 398–400.

Cornell, L. H., Asper, E. D., Osborn, K., and White, M. J. (1979). Investigations on cryogenic marking procedures for marine mammals. Natl. Tech. Inform. Serv. Rep. No. PB-291570. 24 pp.

Defran, R. H., Kelly, D. L., Schultz, G. M., Weaver, A. C., and Espinoza, M. A. (1986). The occurrence and movements of the bottlenose dolphin in the Southern California Bight. Natl. Mar. Fish. Serv., Southwest Fish. Center Admin. Rep. No. LJ-86-36C. 26 pp.

Dietz, R. (1986). A feasibility study on satellite and VHF tracking of marine mammals. Report to The Commission for Scientific Research in Greenland and Greenland Fisheries and Environment Research Institute. 93 pp.

Doan, K. H., and Douglas, C. W. (1953). Beluga of the Churchill Region of Hudson Bay. *Fish. Res. Board Can. Bull.*, No. 98, 1–27.

Dohl, T. P. (1981). Remote laser branding of marine mammals. Natl. Tech. Inform. Serv. Rep. No. PB-213449. 34 pp.

Dohl, T. P., Norris, K. S., Guess, R. C., Bryant, J. D., and Honig, M. W. (1978). Cetacea of the Southern California Bight. *In* "Investigator's Reports, Summary of Marine Mammal and Seabird Surveys of the Southern California Bight Area, 1975–1978," Part II, Vol. III. Bureau of Land Management Contract Report No. AA550-CT7-36. 414 pp.

Ellis, G. (1984). Killer whales of southern Alaska: A catalogue of individuals photo-identified in 1984. Hubbs–Sea World Res. Inst. Tech. Rep. No. 84-176. 73 pp.

Erickson, A. W. (1978). Population studies of killer whales (*Orcinus orca*) in the Pacific Northwest: A radio-marking and tracking study of killer whales. Natl. Tech. Inform. Serv. Rep. No. PB-285615. 34 pp.

Essapian, F. S. (1962). An albino bottlenosed dolphin, *Tursiops truncatus,* captured in the U.S. *Norsk Hvalfangst-tidende* **9,** 341–344.

Evans, W. E. (1967). Vocalization among marine mammals. *In* "Marine Bio-Acoustics" (W. B. Tavolga, ed.), Vol. 2, pp. 159–186. Pergamon, New York. 353 pp.

Evans, W. E. (1971). Orientation behavior of delphinids: Radio telemetric studies. *Ann. N.Y. Acad. Sci.* **188,** 142–160.

Evans, W. E. (1974). Radio-telemetric studies of two species of small odontocete cetaceans. *In* "The Whale Problem: A Status Report" (W. E. Schevill, ed.), pp. 385–394. Harvard Univ. Press, Cambridge, Massachusetts. 419 pp.

Evans, W. E. (1975). Distribution, differentiation of populations, and other aspects of the natural history of *Delphinus delphis* Linnaeus in the northeastern Pacific. Ph.D. dissertation, Univ. of California, Los Angeles. 145 pp.

Evans, W. E., Hall, J. D., Irvine, A. B., and Leatherwood, J. S. (1972). Methods for tagging small cetaceans. *Fish. Bull.* **70**(1), 61–65.

Farrell, R. K., Koger, L. M., and Winward, L. D. (1966). Freeze-branding of cattle, dogs and cats for identification. *J. Am. Vet. Med. Assoc.* **149,** 745–752.

Farrell, R. K., Laisner, G. A., and Russell, T. S. (1969). An international freeze-mark animal identification. *J. Am. Vet. Med. Assoc.* **154,** 1561–1572.

Ford, J. K. B., and Fisher, H. D. (1983). Group-specific dialects of killer whales (*Orcinus orca*) in British Columbia. *In* "Communication and Behavior of Whales" (R. Payne, ed.), pp. 129–161. Westview Press, Boulder, Colorado. 643 pp.

Frost, K. J., Lowry, L. F., and Nelson, R. R. (1985). Radiotagging studies of belukha whales (*Delphinapterus leucas*) in Bristol Bay, Alaska. *Mar. Mamm. Sci.* **1**(3), 191–202.

Gandy, W. F., Vanselous, T. M., and Jennings, J. G. (1977). Tracking marine animals by satellite. Natl. Mar. Fish. Serv., Southeast Fish. Center Contrib. No. 77-30F. 9 pp.

Gaskin, D. E. (1982). "The Ecology of Whales and Dolphins." Heinemann, London. 459 pp.

Gaskin, D. E., Smith, G. J. D., and Watson, A. P. (1975). Preliminary study of movements of harbor porpoises (*Phocoena phocoena*) in the Bay of Fundy using radiotelemetry. *Can. J. Zool.* **53**(10), 1466–1471.

Gaskin, D. E., Read, A. J., Watts, P. F., and Smith, G. J. D. (1985). Population dispersal, size, and interactions of harbour porpoises in the Bay of Fundy and Gulf of Maine. Can. Tech. Rep. Fish. Aquatic Sci. No. 1291. 28 pp.

Geraci, J. R., St. Aubin, D. J., Smith, G. J. D., and Hicks, B. D. (1985). Cutaneous response to plastic and metallic implants of potential use for marking cetaceans. Final Report, Part II, Contract No. 14-12-

0001-29169, U.S. Dept. of Interior, Minerals Management Service. 120 pp.

Goodyear, J. (1981). A new radio tag; the Remora, and behavior of a humpback whale (*Megaptera novaeangliae*). *J. Ceta-Res.* **2,** 1–2.

Gruber, J. A. (1981). Ecology of the Atlantic bottlenosed dolphin (*Tursiops truncatus*) in the Pass Cavallo area of Matagorda Bay, Texas. M.S. thesis, Texas A & M Univ., College Station, Texas. 182 pp.

Haenel, N. J. (1986). General notes on the behavioral ontogeny of Puget Sound killer whales and the occurrence of allomaternal behavior. *In* "Behavioral Biology of Killer Whales" (B. C. Kirkevold and J. S. Lockard, eds.), Zoo Biology Monographs, Vol. 1, pp. 285–300. Alan R. Liss, New York. 457 pp.

Hall, J. D. (1981). Aspects of the natural history of cetaceans of Prince William Sound, Alaska. Ph.D. dissertation, Univ. of California, Santa Cruz. 148 pp.

Hall, J. D., Cornell, L. H., and Ellis, G. (1986). Killer whales of Prince William Sound, Alaska: Results of 1985 field research. Sea World Tech. Contrib. No. 8611C. 15 pp.

Hammond, P. (1986). Estimating the size of naturally marked whale populations using capture–recapture techniques. *In* "Behaviour of Whales in Relation to Management" (G. P. Donovan, ed.), pp. 253–282. Int. Whaling Commission, Cambridge, England. Special Issue 8, 282 pp.

Hansen, L. J. (1983). Population biology of the coastal bottlenose dolphin (*Tursiops truncatus*) of Southern California. M.A. thesis. California State Univ., Sacramento. 104 pp.

Harrington, F. H., Veitch, A. M., and Luttich, S. N. (1987). Tracking barren-ground and woodland caribou by satellite: The more the need for PTTs, the better they work. *In* "Proceedings of the Service ARGOS, Inc. International Users Conference and Exhibit, Greenbelt, Maryland" pp. 221–242. September, 15–17, 1987. 300 pp.

Hedgepeth, J. B. (1985). Database for dolphin tagging operations in the eastern tropical Pacific, 1969–1978, with discussion of 1978 tagging results. Natl. Mar. Fish. Serv., Southwest Fish. Center Admin. Rep. No. LJ-85-03. 40 pp.

Heimlich-Boran, J. R. (1986). Photogrammetric analysis of growth in Puget Sound *Orcinus orca*. *In* "Behavioral Biology of Killer Whales" (B. C. Kirkevold and J. S. Lockard, eds.), Zoo Biology Monographs, Vol. 1, pp. 97–111. Alan R. Liss, New York. 457 pp.

Heimlich-Boran, S. L. (1986). Cohesive relationships among Puget Sound killer whales. *In* "Behavioral Biology of Killer Whales" (B. C. Kirkevold and J. S. Lockard, eds.), Zoo Biology Monographs, Vol. 1, pp. 251–284. Alan R. Liss, Inc., New York. 457 pp.

Hobbs, L. (1982). Tags on whales, dolphins, and porpoises. Appendix A. *In* "Whales, Dolphins, and Porpoises of the Eastern North Pacific and Adjacent Arctic Waters" (S. Leatherwood, R. R. Reeves, W. F. Perrin, and W. E. Evans, eds.), pp. 218–230. NOAA Tech. Rep. NMFS Circ. 444. 245 pp.

Hoelzel, A. R., and Osborne, R. W. (1986). Killer whale call characteristics: Implications for cooperative foraging strategies. *In* "Behavioral Biology of Killer Whales" (B. C. Kirkevold and J. S. Lockard, eds.), Zoo Biology Monographs, Vol. 1, pp. 373–403. Alan R. Liss, New York. 457 pp.

Irvine, B. (1970). Conditioning marine mammals to work in the sea. *Mar. Tech. Soc. J.* **4**(3), 47–52.

Irvine, A. B., and Wells, R. S. (1972). Results of attempts to tag Atlantic bottlenosed dolphins (*Tursiops truncatus*). *Cetology* **13,** 1–5.

Irvine, A. B., Scott, M. D., Wells, R. S., and Mead, J. G. (1979). Stranding of the pilot whale, *Globicephala macrorhynchus*, in Florida and South Carolina. *Fish. Bull.* **77(2),** 511–513.

Irvine, A. B., Scott, M. D., Wells, R. S., and Kaufmann, J. H. (1981). Movements and activities of the Atlantic bottlenose dolphin, *Tursiops truncatus*, near Sarasota, Florida. *Fish. Bull* **79**(4), 671–688.

Irvine, A. B., Wells, R. S., and Scott, M. D. (1982). An evaluation of techniques for tagging small odontocete cetaceans. *Fish. Bull.* **80**(1), 135–143.

Jacobsen, J. K. (1986). The behavior of *Orcinus orca* in the Johnstone Strait, British Columbia. *In* "Behavioral Biology of Killer Whales" (B. C. Kirkevold and J. S. Lockard, eds.), Zoo Biology Monographs, Vol. 1, pp. 135–185. Alan R. Liss, New York. 457 pp.

Jennings, J. G. (1982). Tracking marine mammals by satellite—Status and technical needs. *In* "Oceans '82 Conference Record" pp. 751–754. Marine Tech. Soc. Washington, D.C., September 20–22, 1982.

Jennings, J. G., and Gandy, W. F. (1980). Tracking pelagic dolphins by satellite. *In* "A Handbook on Biotelemetry and Radio Tracking" (C. J. Amlaner, Jr., and D. W. MacDonald, eds.), pp. 753–755. Pergamon, New York. 804 pp.

Kasuya, T., and Oguro, N. (1972). A new tagging method of dolphins. *Sci. Rep. Whales Res. Inst., Tokyo* **24,** 81–85.

Kato, N. (1986). [Development of dolphin tagging techniques: Satellite biotelemetry: Harness development]. *In* "Gyogyou kougai (yūgai seibutsu kujo) taisaku chousa itaku jigyou" [Report of a research project for countermeasures to fishery damage (caused by small cetaceans), 1981–1985] (S. Ohsumi, ed.), Chap. IV, Sect. 2B, pp. 106–114. Committee Report of the Fisheries Agency. 285 pp.

Kenward, R. (1987). "Wildlife Radio Tagging." Academic Press, Orlando, Florida. 222 pp.

Leatherwood, S., and Evans, W. E. (1979). Some recent uses and potentials of radiotelemetry in field studies of cetaceans. *In* "Behavior of Marine Animals" (H. E.

Winn and B. L. Olla, eds.), Vol. 3: Cetaceans, pp. 1–31. Plenum, New York, 438 pp.

Leatherwood, S., and Ljungblad, D. K. (1979). Nighttime swimming and diving behavior of a radio-tagged spotted dolphin, *Stenella attenuata*. *Cetology* **34,** 1–6.

Lockyer, C. (1978). The history and behaviour of a solitary wild, but sociable, bottlenose dolphin (*Tursiops truncatus*) on the west coast of England and Wales. *J. Nat. Hist.* **12,** 513–528.

Lockyer, C. (1979). Response of orcas to tagging. *Carnivore* **2**(3), 19–21.

Lockyer, C., and Morris, R. J. (1985). Body scars of a resident, wild bottlenosed dolphin (*Tursiops truncatus*): Information on certain aspects of his behaviour. *Aquatic Mamm.* **11**(2), 42–45.

Lyrholm, T., Leatherwood, S., and Sigurjonsson, J. (1987). Photoidentification of killer whales (*Orcinus orca*) off Iceland, October 1985. *Cetology* **52,** 1–14.

Martin, H., Evans, W. E., and Bowers, C. A. (1971). Methods for radio tracking marine mammals in the open sea. *In* "Transactions of the IEEE Conference on Engineering in the Ocean Environment" pp. 44–49. September, 1971, San Diego, California. 389 pp.

Mate, B. R. (1989). Watching habits and habitats from Earth satellites. *Oceanus* **32**(2), 14–18.

Mather, F. J., Jr. (1963). Tags and tagging techniques for large pelagic fishes. North Atlantic Fish Marking Symposium, Int. Comm. Northwest Atlantic Fish. Spec. Publ. 4, pp. 288–293.

Mead, J. G., Odell, D. K., Wells, R. S., and Scott, M. D. (1980). Observations on a mass stranding of spinner dolphin, *Stenella longirostris,* from the west coast of Florida. *Fish. Bull.* **78**(2), 353–360.

Nakamura, A., Soma, M., and Tsutsumi, M. (1985). Jînkō eisei wo riyō shita iruka no kaiyū chōsayō sōshin-ki to sono Ōyō. [A satellite-linked transmitter to study migration of dolphin and its application.] *J. Fac. Mar. Sci. Tech. Tokai Univ.* **21,** 65–77. [English abstr.]

National Fisheries Engineering Laboratory. (1978). Porpoise tagging progress report. Natl. Mar. Fish. Serv., Southeast Fish. Center Rep. No. 78-46F. 86 pp.

Nishiwaki, M., Nakajima, M., and Tobayama, T. (1966). Preliminary experiments for dolphin marking. *Sci. Rep. Whales Res. Inst., Tokyo* **20,** 101–107.

Norris, K. S., and Dohl, T. P. (1980). Behavior of the Hawaiian spinner dolphin, *Stenella longirostris. Fish. Bull.* **77**(4), 821–849.

Norris, K. S., and Hall, J. D. (1979). Development of techniques for estimating trophic impact of marine mammals. Natl. Tech. Inform. Serv. Rep. No. PB-290399. 16 pp.

Norris, K. S., and Pryor, K. W. (1970). A tagging method for small cetaceans. *J. Mammal.* **51**(3), 609–610.

Norris, K. S., Evans, W. E., and Ray, G. C. (1974). New tagging and tracking methods for the study of marine mammal biology and migration. *In* "The Whale Problem: A Status Report" (W. E. Schevill, ed.), pp. 395–408. Harvard Univ. Press, Cambridge, Massachusetts. 419 pp.

Norris, K. S., Dohl, T. P., Guess, R. C., Hobbs, L. J., and Honig, M. W. (1978). Cetacea: Numbers, distribution, and movements in the Southern California Bight. *In* "Marine Mammal and Seabird Survey of the Southern California Bight Area," Vol. III: Principal investigator's reports, Book I: Pinnipedia, Cetacea and Parasitology, Sect. II, pp. 270–441. Nat. Tech. Inform. Serv. No. PB-295932. 472 pp.

Norris, K. S., Würsig, B., Wells, R. S., Würsig, M., Brownlee, S. M., Johnson, C., and Solow, J. (1985). The behavior of the Hawaiian spinner dolphins, *Stenella longirostris.* Natl. Mar. Fish. Serv., Southwest Fish. Center Admin. Rep. No. LJ-85-06C. 213 pp.

Osborne, R. W. (1986). A behavioral budget of Puget Sound killer whales. *In* "Behavioral Biology of Killer Whales" (B. C. Kirkevold and J. S. Lockard, eds.), Zoo Biology Monographs, Vol. 1, pp. 211–249. Alan R. Liss, New York. 457 pp.

de la Parra, R., and Galvan, B. E. (1985). Observaciones del tursión costero Pacífico en el sistema Topolobampo-Ohuira, Sinaloa (con notas acerca de comportamiento, ritmo respiratorio e identificación individual). Publication of the Universidad de Occidente, Sección de Hidrobiologia. 32 pp. [English abstr.]

Payne, R., Brazier, O., Dorsey, E. M., Perkins, J. S., Rowntree, V. J., and Titus, A. (1983). External features in southern right whales (*Eubalaena australis*) and their use in identifying individuals. *In* "Communication and Behavior of Whales" (R. Payne, ed.), pp. 371–445. Westview Press, Boulder, Colorado. 643 pp.

Pennycuick, C. J. (1978). Identification using natural markings. *In* "Animal Marking: Recognition Marking of Animals in Research" (B. Stonehouse, ed.), pp. 147–159. University Park Press, Baltimore, Maryland. 257 pp.

Perrin, W. F. (1975). Distribution and differentiation of populations of dolphins of the genus *Stenella* in the eastern tropical Pacific.. *J. Fish. Res. Board Can.* **32**(7), 1059–1067.

Perrin, W. F., Evans, W. E., and Holts, D. B. (1979). Movements of pelagic dolphins (*Stenella* spp.) in the eastern tropical Pacific as indicated by the results of tagging, with summary of tagging operations, 1969–76. NOAA Tech. Rep. NMFS SSRF-737. 14 pp.

Perrin, W. F., Scott, M. D., Walker, G. J., and Cass, V. L. (1985). Review of geographical stocks of tropical dolphins (*Stenella* spp. and *Delphinus delphis*) in the eastern Pacific. NOAA Tech. Rep. NMFS 28. 28 pp.

Powers, J. E., Butler, R. W., Jennings, J. G., McLain, R., Peters, C. B., and DeBeer, J. (1979). Summary of research results from the fourth cruise of the dedicated vessel, 12 September to 31 October 1978. Natl.

Mar. Fish. Serv., Southwest Fish. Center Admin. Rep. No. LJ-79-14. 45 pp.

Rathbun, G. B., Reid, J. P., and Carowan, G. (1989). Manatee (*Trichechus manatus*) distribution and movement patterns in northwestern peninsular Florida. Florida Marine Research Publications. St. Petersburg, Florida.

Read, A. J., and Gaskin, D. E. (1983). The application of radio tracking techniques to the study of harbour porpoises (*Phocoena phocoena*) in the Bay of Fundy. *In* "Proceedings of the Fourth International Wildlife Biotelemetry Conference" (D. G. Pincock, ed.), Halifax, Nova Scotia. 364 pp.

Read, A. J., and Gaskin, D. E. (1985). Radio tracking the movements and activities of harbor porpoises, *Phocoena phocoena* (L.), in the Bay of Fundy, Canada. *Fish. Bull.* **83**(4), 543–552.

Saayman, G. S., and Tayler, C. K. (1973). Social organisation of inshore dolphins (*Tursiops aduncus* and *Sousa*) in the Indian Ocean. *J. Mammal.* **54**(4), 993–996.

Saayman, G. S., and Tayler, C. K. (1979). The socioecology of humpback dolphins (*Sousa* spp.). *In* "Behavior of Marine Animals" (H. E. Winn and B. L. Olla, eds.), Vol. 3: Cetaceans, pp. 165–226. Plenum, New York. 438 pp.

Scott, M. D., and Wussow, P. C. (1983). Movements of a Hawaiian spotted dolphin. *In* "Proceedings of the Fourth International Wildlife Biotelemetry Conference" (D. G. Pincock, ed.), pp. 353–364. Halifax, Nova Scotia. 364 pp.

Sergeant, D. E., and Brodie, P. F. (1969). Tagging white whales in the Canadian Arctic. *J. Fish. Res. Board Can.* **26**(8), 2201–2205.

Shane, S. H. (1977). The population biology of the Atlantic bottlenose dolphin, *Tursiops truncatus*, in the Aransas Pass area of Texas. M.S. Thesis, Texas A & M Univ., College Station, Texas. 239 pp.

Shane, S. H. (1980). Occurrence, movements, and distribution of bottlenose dolphins, *Tursiops truncatus*, in southern Texas. *Fish. Bull.* **78**(3), 593–601.

Shane, S. H. (1984). Pilot whales and other marine mammals at Santa Catalina Island, California. Natl. Mar. Fish. Serv., Southwest Fish. Center Admin. Rep. No. LJ-84-28C. 29 pp.

Shane, S. H. (1986). Detailed observations of a single pod of pilot whales and sightings of other marine mammals at Catalina Island, California. Natl. Mar. Fish. Serv., Southwest Fish. Center Contract Rep. No. 40-ABNF-5-702. 49 pp.

Shane, S. H., and Schmidly, D. J. (1978). The population biology of the Atlantic bottlenose dolphins, *Tursiops truncatus*, in the Aransas Pass area of Texas. Natl. Tech. Inform. Serv. Rep. No. PB-283393. 130 pp.

Solangi, M. A., and Dukes, G. E. (1983). Atlantic bottlenose dolphin, *Tursiops truncatus*, herd studies in the Mississippi Sound, USA: Capture, freeze marking and biological sampling. Final contract to the Natl. Mar. Fish. Serv., Rep. No. NA 82-GA-C-00023. 178 pp.

Soma, M., and Tsutsumi, M. (1986). Biological telemetry utilizing the Argos system. *Mesogee* **46**(2), 87–92.

Southwest Fisheries Center. (1981). Dolphin satellite tracking experiment successfully completed. *In* "Southwest Fisheries Center Monthly Report, June," pp. 19–23. 44 pp.

Starks, W. H. (1966). Instrumentation for ecological studies. *In* "Whales, Dolphins, and Porpoises" (K. S. Norris, ed.), pp. 497–500. Univ. of California Press, Berkeley. 789 pp.

Stogner, L. (1981). Evaluation of porpoise surfacing data for design changes to the NIMBUS-6 marine mammal transmitter. Natl. Mar. Fish. Serv., Natl. Fish. Engineering Program. 33 pp.

Tanaka, S. (1987). Satellite radio tracking of bottlenose dolphins *Tursiops truncatus*. *Nippon Suisan Gakkaishi* **53**(8), 1327–1338.

Tanaka, S., and Takao, K. (1986). [Development of dolphin tagging techniques; Description of tags.] *In* "Gyogyou kougai (yūgai seibutsu kujo) taisaku chousa itaku jigyou" [Report of a research project for countermeasures to fishery damage (caused by small cetaceans), 1981–1985] (S. Ohsumi, ed.), Chapter IV, Sect. 3, pp. 149–160. Committee Report of the Fisheries Agency. 285 pp.

Tanaka, S, Takao, K., and Kato, N. (1987). Tagging techniques for bottlenose dolphins *Tursiops truncatus*. *Nippon Suisan Gakkaishi* **53**(8), 1317–1325.

Tobayama, T., Arai, K., Shimizu, H., and Mouri, E. (1982). Kaijyu-rui eno reitō mākingu no Ōyō. [The application of freezebranding to marine mammals.] *J. Jpn. Assoc. Zool. Gard. Aquar.* **24,** 52. [Abstr. in Japanese.]

Tomilin, A. G. (1983). A new method of marking small cetaceans. *Rep. Int. Whal. Comm.* **33,** 643–645.

Watkins, W. A., and Schevill, W. E. (1976). Underwater paint marking of porpoises. *Fish. Bull.* **74**(3), 687–689.

Watson, A. P., and Gaskin, D. E. (1983). Observations on the ventilation cycle of the harbour porpoise *Phocoena phocoena* (L.) in coastal waters of the Bay of Fundy. *Can. J. Zool.* **61**(1), 126–132.

Weigle, B. L. (1987). The abundance, distribution, and movements of bottlenose dolphins, *Tursiops truncatus*, in lower Tampa Bay, Florida, M.S. thesis, Univer. of South Florida, Tampa, Florida. 47 pp.

Wells, R. S. (1986). Structural aspects of dolphin societies. Ph.D. dissertation, Univ. of California, Santa Cruz. 243 pp.

Wells, R. S., Irvine, A. B., and Scott, M. D. (1980). The social ecology of inshore odontocetes. *In* "Cetacean Behavior: Mechanisms and Functions" (L. M. Herman, ed.), pp. 263–317. Wiley, New York. 463 pp.

Wells, R. S., Scott, M. D., and Irvine, A. B. (1987). The

social structure of free-ranging bottlenose dolphins. *In* "Current Mammalogy" (H. Genoways, ed.), Vol. 1, pp. 247–305. 519 pp.

White, M. J., Jr., Jennings, J. G., Gandy, W. F., and Cornell, L. H. (1981). An evaluation of tagging, marking, and tattooing techniques for small delphinids. NOAA Tech. Memo. NOAA-TM-NMFS-SWFC-16. 142 pp.

Würsig, B. (1978). Occurrence and group organization of Atlantic bottlenose porpoises (*Tursiops truncatus*) in an Argentine bay. *Biol. Bull.* **154**(2), 348–359.

Würsig, B. (1982). Radio tracking dusky porpoises in the South Atlantic. *In* "Mammals in the Seas" Vol. 4, pp. 145–160. FAO Fish Ser., No. 5. 531 pp.

Würsig, B., and Bastida, R. (1986). Long-range movement and individual associations of two dusky dolphins (*Lagenorhynchus obscurus*) off Argentina. *J. Mammal.* **67**(4), 773–774.

Würsig, B., and Würsig, M. (1977). The photographic determination of group size, composition, and stability of coastal porpoises (*Tursiops truncatus*). *Science* **198,** 755–756.

Würsig, B., and Würsig, M. (1979). Behavior and ecology of bottlenose porpoises, *Tursiops truncatus,* in the South Atlantic. *Fish. Bull.* **77**(2), 399–442.

Würsig, B., and Würsig, M. (1980). Behavior and ecology of dusky porpoises, *Lagenorhynchus obscurus,* in the South Atlantic. *Fish. Bull.* **77**(4), 871–890.

Würsig, B., Cipriano, F., and Würsig, M. (1990). Dolphin movement patterns: Information from radio and theodolite tracking studies. *In* "Dolphin Societies: Discoveries and Puzzles" (K. Pryor and K. S. Norris, eds.).

Distribution and Movements of Freeze-Branded Bottlenose Dolphins in the Indian and Banana Rivers, Florida

30

Daniel K. Odell
Division of Biology and Living Resources
Rosenstiel School of Marine and Atmospheric Science
University of Miami
Miami, Florida 33149
Present address: Sea World of Florida
Orlando, Florida 32821

Edward D. Asper
Sea World of Florida
Orlando, Florida 32821

INTRODUCTION

More than 1500 bottlenose dolphins were brought from the wild into captivity from 1938 to 1980 (Leatherwood and Reeves, 1982); an undetermined but high proportion of these came from Florida. At least 88 were captured and permanently removed from the Indian–Banana River complex, on the east central coast of Florida (Fig. 1), from 1970 through March 1988 (Reeves and Leatherwood, 1984; Scott, Chapter 36, this volume).

Capture permits are issued for various provisionally identified regional "stocks" or "populations," including one in the Indian River, based on quotas representing 2% or less of the best available population estimate (Scott, Chapter 36, this volume). While it seems unlikely, under this conservative management scheme, that any bottlenose dolphin population will be adversely affected by live-capture activities, the U.S. National Marine Fisheries Service (NMFS), the display industry, and the research community all have a stake in the continued close monitoring of selected stocks. Accordingly, in-depth, multidisciplinary studies of bottlenose dolphins have been, and continue to be, conducted in many different regions of the western and southeastern United States, including the Indian River (e.g., Shane, 1980, 1987; Wells *et al.*, 1980, 1987; Irvine *et al.*, 1981; Gruber, 1981; Scott *et al.*, Chapter 11, this volume; Hansen, 1983, Chapter 23, this volume; Weigle, 1987).

This chapter reviews the history of research on bottlenose dolphins in the Indian and Banana rivers, Florida, 1974–1984, and presents the results of a freeze-branding program. Other aspects of the program have been summarized by Asper and Odell (1980), Odell and Asper (1982), and Asper *et al.* (Chapter 28, this volume).

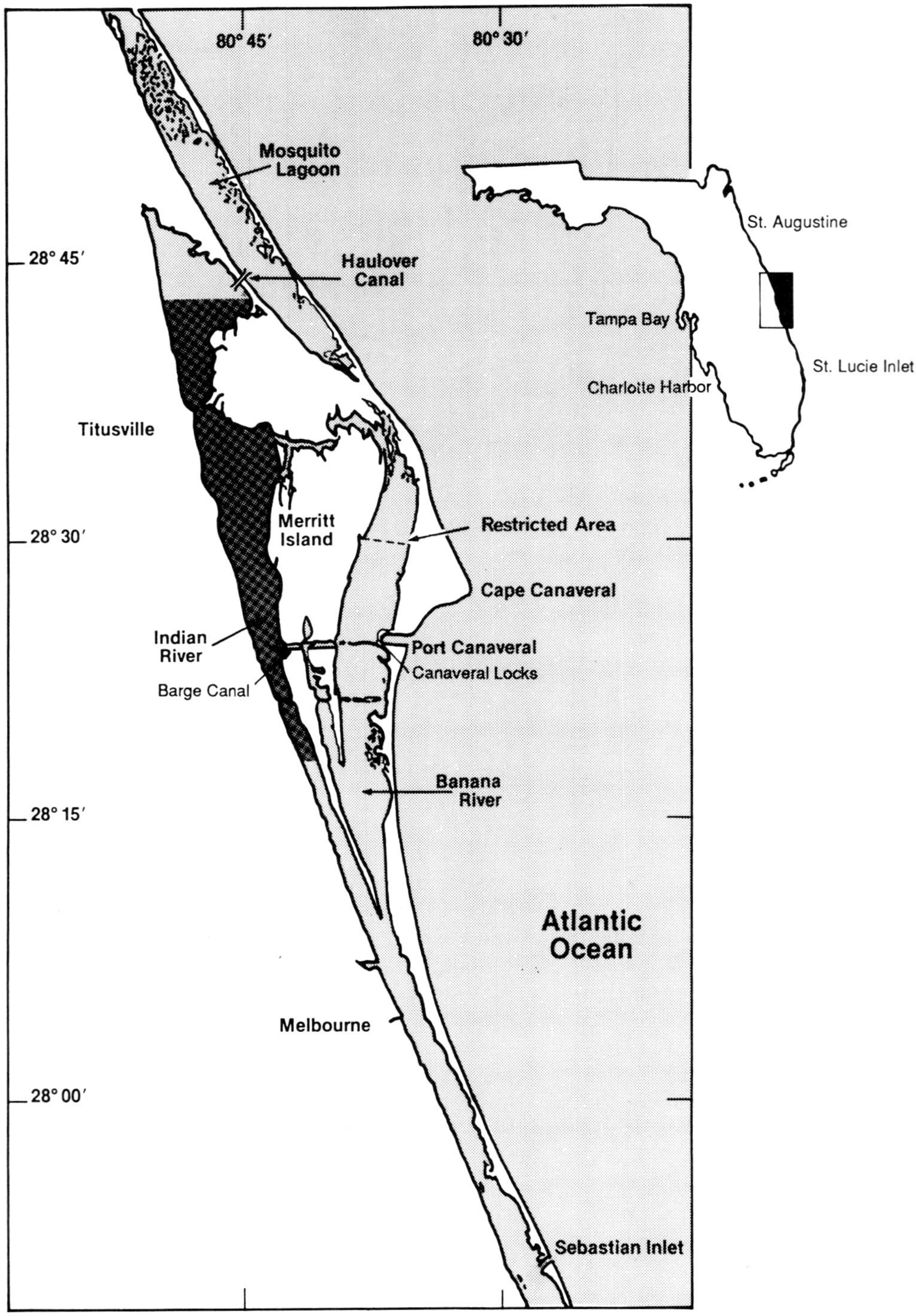

Figure 1 The Florida study area, encompassing the Indian and Banana rivers from Mosquito Lagoon to Sebastian Inlet, showing the subregion (cross-hatched) in which dolphins were captured between August 1979 and October 1981 (see Appendix).

STUDY AREA

The Indian–Banana river complex is a sea-flushed lagoon network on the east central coast of Florida, 27° 10′ to 28° 48′ N (Fig. 1). A dredged ship channel (the Intracoastal Waterway, ICW) lies along the longitudinal center of the river and continues through similar waterways both north and south of this river complex. The Banana River is connected to the Atlantic Ocean by an inlet at Cape Canaveral which is controlled by a set of locks and is only intermittently navigable. A barge canal across Merritt Island connects the Indian and Banana rivers. The two rivers converge at the southern tip of Merritt Island. South of this convergence, the Indian River is connected to the Atlantic Ocean by passes at Sebastian, Ft. Pierce (just north of St. Lucie Inlet), and St. Lucie inlets.

The Indian and Banana rivers together cover a length of just over 185 km. Widths vary from less than 0.93 to over 9.3 km. The surface area is approximately 800 km^2. Much of the area is shallow (<1 m at high tide), but the vast majority of it is nevertheless potential habitat for bottlenose dolphins, as they are seen frequently in shallow areas. Depths, excluding those in dredged channels, are generally 3 m or less, and there are numerous low spoil islands and sandbars, especially inside the passes, along the ICW, and near bridges. During our studies, water temperatures ranged from about 8° to 35°C, salinities from 16 to 41 ppt. Tides are negligible, varying less than 0.5 m even at spring extremes. Currents are mostly local and dominated by tidal flow through the inlets, fresh water flowing out of the major streams, and protracted winds. Such tides and currents are thought to have little direct effect on bottlenose dolphin distribution and movements in this area (Leatherwood, 1979; Leatherwood and Show, 1980).

HISTORY OF THE RESEARCH PROGRAM

Between March 1977 and March 1979, 36 bottlenose dolphins were caught, processed, and released immediately in the Indian River (Appendix). Ten more captured between 23 August 1978 and 15 February 1979 were held at Sea World of Florida for up to 90 days, used in experiments to monitor the development of freeze-brands after initial application, and then released. Two animals were processed similarly in 1980 and 8 in 1981 (Appendix).

Phase I: 1979 Program

In 1979, a program was initiated to (1) capture, brand, collect biological data from, and release 25 bottlenose dolphins; (2) observe branded animals to determine effectiveness of the branding technique and movements of branded individuals; and (3) prepare a photographic catalog of flukes and dorsal fins of branded animals handled and of distinctive features on the bodies of all naturally marked animals encountered during the study. Between 23 August and 2 September 1979, 26 bottlenose dolphins were captured, handled, branded, and released (Appendix). NMFS-sponsored boat searches for branded dolphins were attempted 5 days per week between 17 September and 14 December 1979 (Asper and Odell, 1980) and 3 consecutive days every 2 weeks between 18 August and 10 December 1980 (Patton *et al.*, 1981). In this and subsequent resighting efforts, dolphin herds were approached closely to observe and photograph brand numbers and evaluate brand condition.

Phase II: 1980–1981 Program

A second phase of the branding program, with similar objectives, was initiated in November 1980. It was designed to capture and brand 50 dolphins and to recapture at least 25 dolphins branded during Phase I or in the first Phase II field activities in November 1980. Between 3 and 22 November 1980, 49 dolphins were captured, branded, sampled, and released, and 13 previously branded animals were recaptured, sampled, and released (Appendix). Between 5 and 17 October 1981, 21 previously branded animals were captured, sampled, and released. Of those, 15 had been first branded in November 1980 and 6 prior to 1980.

MATERIALS AND METHODS

Capture and Handling

Captures were made using techniques similar to those described by Asper (1975). Three boats were involved: an 8.5-m inboard-powered boat modified to carry and deploy a seine net 300 m long and 4 m deep with a stretched mesh size of 20 cm; a 7.3-m outboard-powered boat which served as the principal diver-transport vessel and assisted with net-hauling and stacking; and a 5.5-cm outdrive-powered boat which served as an auxiliary scouting vessel and was later used for resighting branded animals.

During capture expeditions, the boats were in continuous radio contact. When a group of dolphins was sighted where a set could be made (i.e., in an area of adequate depth and away from navigation channels), the net boat approached and began to circle the animals. The set was aborted if small calves were present or if, during attempts at recapture, it became apparent that no branded animals were present. If the herd was judged suitable, the net was deployed around it at high speed. The diver-transport vessel followed the net boat, and divers untangled the net when necessary and supported any dolphins that became entangled. Boarded animals were placed on thick foam pads, covered with a wet sheet, and kept cool with an intermittent spray of salt water.

Freeze-Branding

The branding "irons" have wooden handles and 5-cm brass heads on a 35-cm brass shaft and come in a variety of configurations. Dolphins in numbered series (see Appendix) were branded with two numbers on each side of the dorsal fin and three numbers on each side of the base of the dorsal fin. Those in alphanumeric series were branded with a number and a letter on both sides of the dorsal fin and the body. The area to be branded was towel-dried to avoid formation of an ice layer between the brand head and the skin during the branding procedure. The brand was thoroughly cooled in liquid nitrogen held in a stainless steel container. The iron was considered cooled when bubbles were no longer forming on the brand head. The brand was applied with "firm" pressure (Fig. 2) to the cleaned surface for about 24 seconds (dorsal fin) or 16 seconds (body). The branding process was subjectively fine-tuned by the brander. In particular, the brander adjusted the time of application slightly to account for the shape and surface area of the brand head and the location and shape of the surface being branded. Once branding was complete, the iron was removed and the area rinsed with warm salt water.

Tooth Extraction and Age Determination

One tooth was extracted from near the center of the left or right ramus, following the technique described by Ridgway *et al.* (1975). Each animal was given 22 mg oxytetracycline/kg body weight, injected at one or more sites in the dorsal musculature. The tetracycline served as a time marker in the layers deposited in growing teeth (see Myrick and Cornell, Chapter 34, this volume; Hohn, Chapter 33, this volume). Each tooth was sectioned longitudinally, etched in 5% formic acid for 3 hours, and its growth layer groups counted (see Hersh, 1987) to estimate the animal's age (also see Hohn, Chapter 33, this volume, for a discussion of age determination in *Tursiops*).

Resighting

The most important sources of follow-up information on quality of brands and movements of branded animals were directed searches by project members. Searches covered primarily the area between Mosquito Lagoon and Sebastian Inlet (Fig. 3), with occasional forays as far north as Ponce Inlet (just south of St. Augustine) and as far south as St. Lucie Inlet.

In 1979, resighting searches began in September and concluded in mid December, for a total of 390 hours. Follow-up searches from boats were resumed in December 1980. These were undertaken daily for the first 2 weeks, 3 days per week for the next 4 weeks, and 3 days every other week from then until 26 March 1982, for a total of 830 hours.

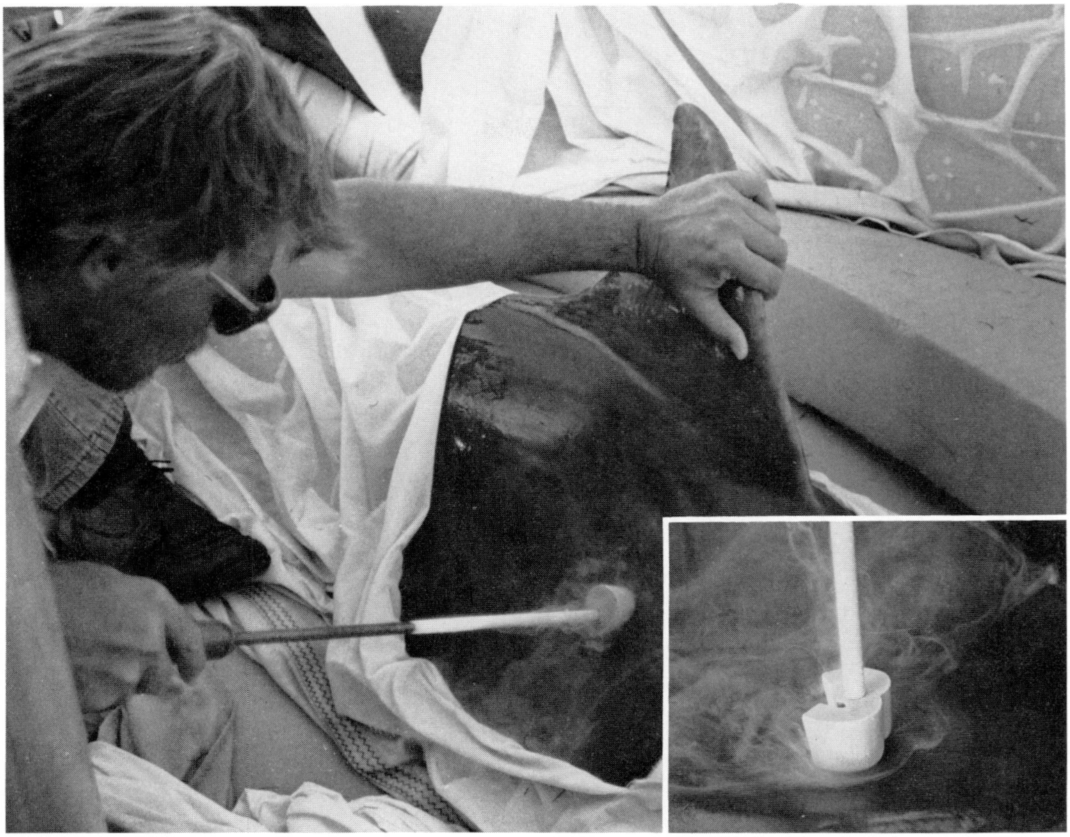

Figure 2 Application of a freeze brand to the body of a bottlenose dolphin. Note the metal shaft and the wooden handle on the branding iron. (Inset) Close-up of a brand head on the skin of a dolphin.

The entire study area could not be surveyed by boat in a single day. Therefore, crews rotated among subsections of the river, following preplanned search patterns which assured complete coverage of the area under investigation that day. Two or three observers aboard the 5.5-m boat searched all navigable waters (≥1.5 m) in the day's area. Whenever dolphins were seen, they were approached closely to search for branded animals and assess the condition of any brands present. To obtain adequate information on all brands in a group, it often was necessary to remain with the dolphin(s) for an hour or more.

To broaden coverage, a poster was distributed to local newspapers, marine patrol offices, marinas, boat landings, bait shops, dive shops, etc. Sighting reports contributed by the public were not used in our analysis of movements within the "home ranges" defined by our own observations. However, reports indicating the presence of branded animals outside our principal study area were investigated carefully and, if deemed reliable (e.g., included photodocumentation), were used to support overall conclusions about the population's movements.

Branded animals were photographed on Kodak Plus-X film (ASA 125, black and white) using a Canon AT-1 with autowinder and a Vivitar 100–300-mm zoom lens or a motor-driven Nikon F2S with a 200-mm fixed or 100–300-mm zoom lens.

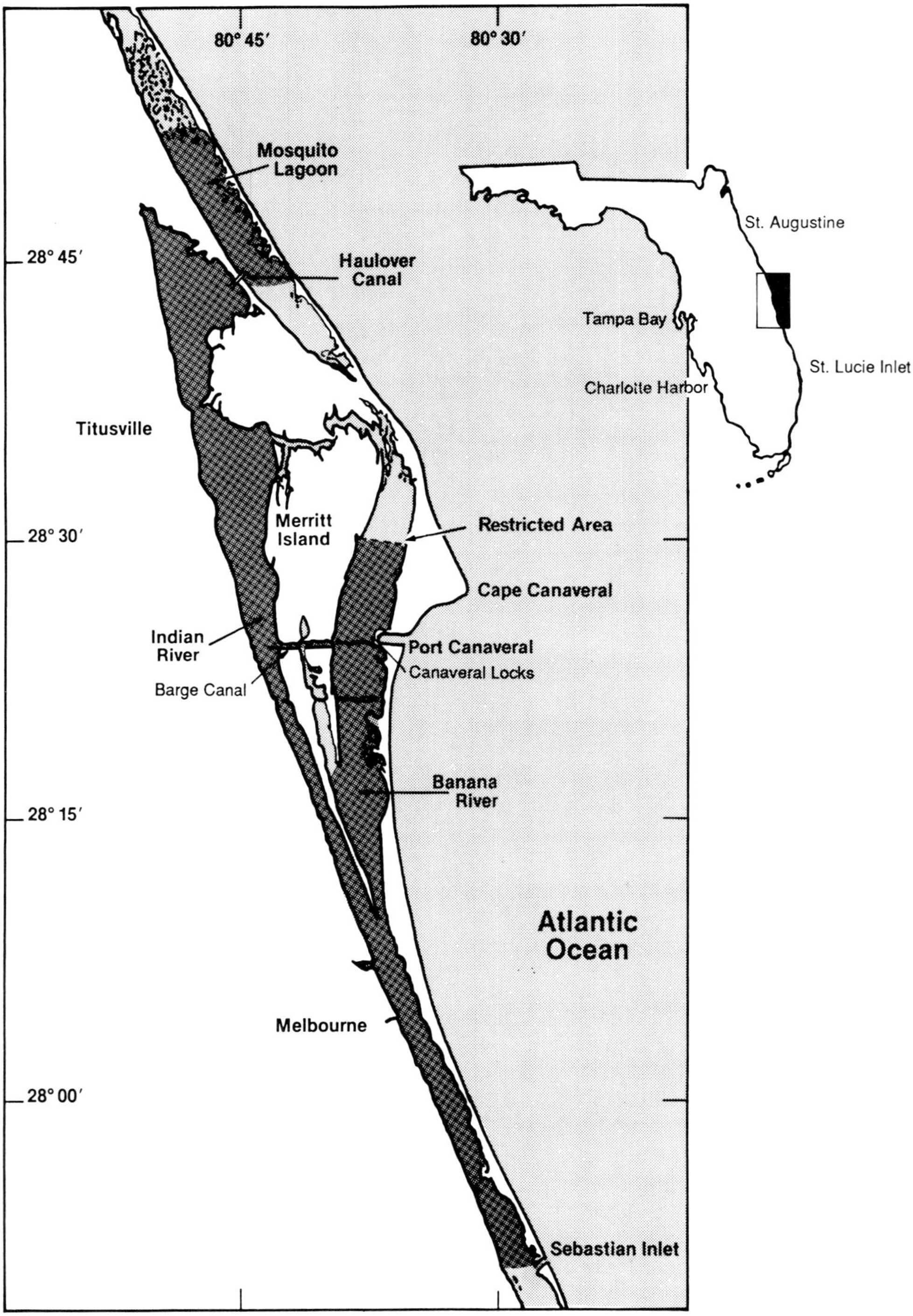

Figure 3 Study area, showing the main areas searched (cross-hatched) for marked dolphins during boat-based surveys.

RESULTS AND DISCUSSION

Capture, Handling, and Marking

Between March 1977 and October 1981, we captured, handled, and released a total of 133 bottlenose dolphins in the Indian River (Appendix). This included recaptures of 29 animals once and 2 animals twice (Appendix, Tables 1 and 2). One hundred thirty-four animals were freeze-branded. Of those, 37 had brands retouched during second or third captures. Captures concentrated in the Titusville area but occurred as far north as the northern end of the Indian River and as far south as the end of Merritt Island (Fig. 1).

Capture, handling, marking, and release activities went smoothly, even though handling time was protracted by a requirement of the contract that animals captured as a group be released as a group. During work conducted prior to August 1979, individual dolphins were processed and released in less than 30 minutes. In this study, mean handling time ranged from 1.1 (for recaptures) to 1.3 hours (for unbranded animals) and increased with group size. Despite these longer-than-usual handling times, only one animal died during the entire study. Animal 60, a 193-cm, 86-kg male, collided with the net immediately after release and died. Necropsy revealed an advanced stage of intestinal enteritis, raising the possibility that stress of prolonged handling exacerbated his illness and was a factor in his disorientation and the resultant accident. There is no other evidence that capture, handling, or marking significantly affected any of the dolphins involved.

Resighting

In 398 boat hours of searching (Fig. 3) in 1979, we located 443 groups of dolphins containing 2263 individuals. In 860 hours of searching between 1980 and 1982, we encountered 742 groups containing 4266 individuals. From 18 August to 10 December 1980, Patton *et al.* (1981) searched for 187 hours and located 188 herds containing 928 dolphins. In addition to our own observations and those of Patton *et al.* (1981), we received from volunteer participants 57 and 113 reports of marked dolphins for the periods 1979 and 1980–1982, respectively. Many dolphins appeared to remain within very small areas, some near our base at Titusville, and were therefore more likely to be seen than other farther-ranging individuals. Some individuals approached boats and were, therefore, seen and reported many times. Also, once observers located a marked individual, they remained with it until all marked animals in its group had been identified; so, associated marked individuals probably were seen more frequently than solitary marked individuals (assuming stability of the associations and the solitary status).

Resightings were plotted and examined for geographical patterns. Branded animals were seen throughout the entire area searched (Fig. 4). Eight of the freeze-branded dolphins (37, 90, C2, C5, C8, D1, D8, and D9) were never reidentified positively following their release. Others were seen from one to 42 times (animal 56). Information on resightings of all marked dolphins is contained in Tables 1 and 2 and in the Appendix. Locations of resightings of two of the more frequently seen animals are shown in Fig. 5. Although some individuals moved freely between the Indian and Banana rivers through the barge canal west of Port Canaveral or around the southern tip of Merritt Island, no marked animals were seen or reported in the ocean outside the river system. Most animals remained in the general vicinity of the area where they were captured and marked.

Using the northern and southern limits of resighted marked dolphins' distributions as the measure, the average home range for individuals resighted only in the Indian River was 12 km in length. The average for animals seen in both Indian and Banana rivers was 30 or 45 km in length, depending on the assumed travel route (i.e., transit through the barge canal versus around the south end of Merritt Island, see Fig. 1). Sightings of marked individuals outside our main survey area were reported to us from the north (animal 70 on 24 January 1980 and animal D4 on 22 April 1981) in the ICW between Marineland and St. Augustine and from the south (animal 66 on 3 January 1980) just north of Vero Beach (north of St. Lucie Inlet) in the ICW.

Evaluation of Brand Longevity

Whenever freeze-branded animals were handled, the brands were evaluated and their legibility

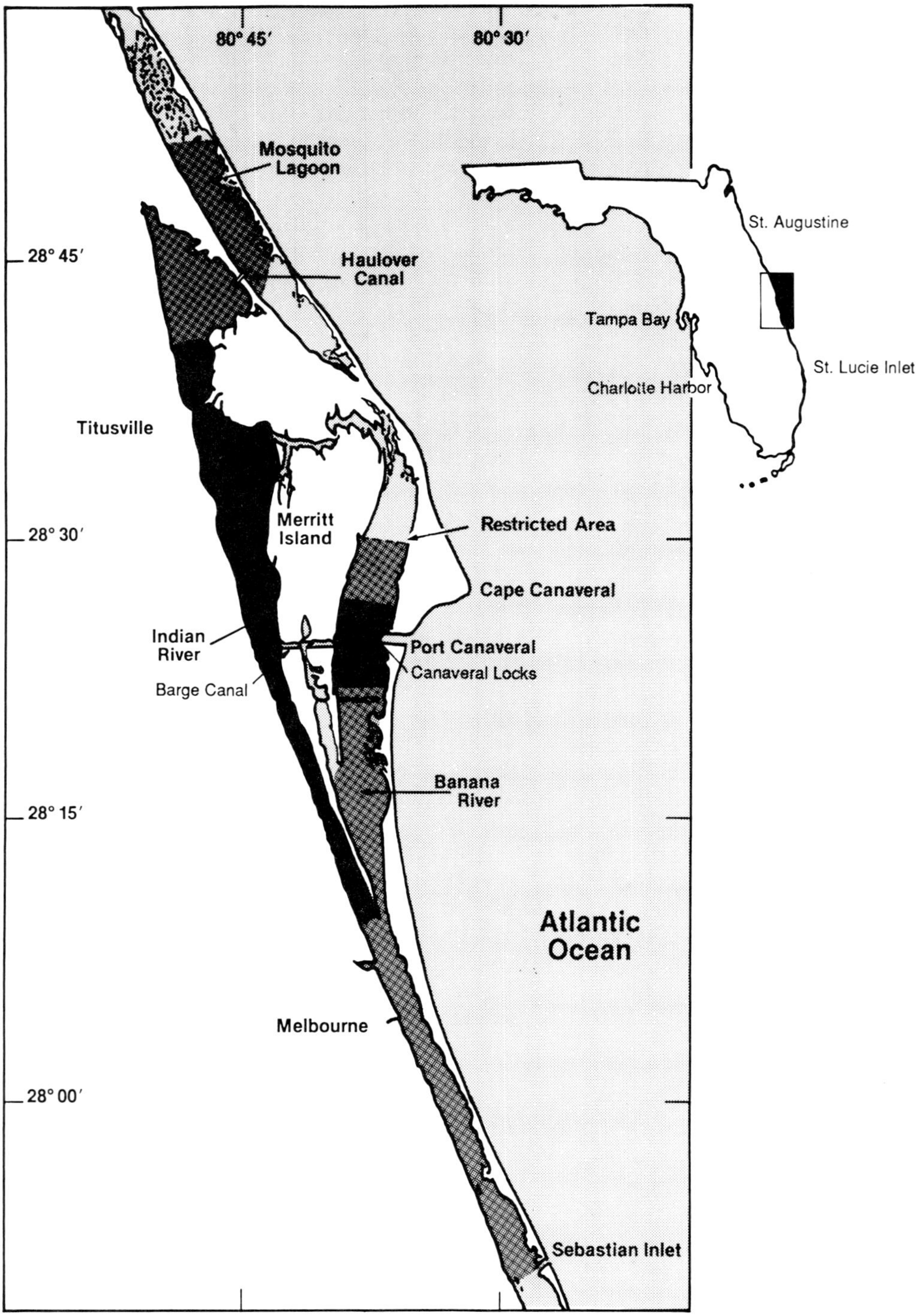

Figure 4 Study area, showing the overall area (cross-hatched) in which marked animals were resighted during boat-based surveys. The darker cross-hatched areas show where most of the sightings were made.

rated as 1–4, after Cornell *et al.* (1979): 1, poor: little of the mark remained and it could not be read; 2, fair: extensive repigmentation had reduced the mark's legibility, but it could be read; 3, good: the mark displayed little or no repigmentation and could be read easily; or 4, excellent: the mark was crisp, bright, and highly legible. Because brands on the base of the dorsal fin were expected to repigment within about 12 months, they were rated excellent if the mark could be read on close inspection. Legibility could not be rated in the above manner from observations or photographs of brands on free-ranging animals. However, ability of observers to determine a brand number was taken as evidence of continued legibility.

Brand Development in Captive Dolphins

Results of marking and observation of the Indian River dolphins held in captivity between 15 April 1977 and 11 May 1980 are comparable to those reported by Cornell *et al.* (1979). These 11 animals were branded 1 to 90 days prior to their release in the river (Table 1 and Appendix). Condition of each brand was evaluated subjectively by close examination and/or photography throughout the day following branding, daily for the next 1–2 weeks, then weekly until release.

The progression of brand development, by area of branding, was as follows. On the dorsal fin, the brand was not visible immediately after application. The branded area became edematous within 1 hour of branding. Tissue necrosis, resulting in skin sloughing off the branded area, was evident within 4–5 days, and tissue regeneration was extensive within 20 days of marking (Fig. 6). The brands usually began to be visible at other than very close range by about 10 days. On the side at the base of the dorsal fin, the brand remained as a distinct depression for about 30 minutes following branding, after which the depression disappeared. Development of edema, necrosis, and regeneration at this site were as described above for the dorsal fin. Little or no repigmentation, and therefore little or no degeneration in brand legibility, had occurred on any of the experimental brands by the time of release; thus, all marks older than 21 days were rated as 3 or 4 at the time of the animal's release (Table 1).

Brand Development in Free-Ranging Dolphins

In our experience, the intensity, and thus the legibility, of brands has varied by individual, reflecting numerous differences in individual morphology and marking procedures. Irrespective of their initial legibility, however, legibility of all brands has declined with time (e.g., Irvine and Wells, 1972; Cornell *et al.*, 1979; Hobbs, 1982). In this study, too, the legibility and usefulness of brands varied and declined with time. Brands on the dorsal fin deteriorated more quickly than those on the side of the body, although those on both areas decreased in legibility with time (Tables 1 and 2). These changes in brand usefulness with time can best be assessed by examining the progressive condition of marks on animals recaptured once or twice and by noting the persistence of legible brands on free-ranging dolphins (Tables 1 and 2; Fig. 7).

In addition to the quality of the brand itself, an observer's ability to read a brand is affected by such factors as observer experience; distance from the animal; animal behavior, especially swimming speed, time at the surface, and predictability of surfacing patterns; platform stability; available light; angle of sun glare; and viewing angle. When a brand must be read exclusively from photographs, the above factors are compounded by differences in camera and lens quality, film grain, degree of enlargement, etc.

From observations and/or photographs, 8 of the experimentally marked animals described above were reidentified in the river (Table 1). Five of these were not recaptured, so the condition of their brands could not be evaluated at close range. However, their brands were sufficiently legible to be read from observation vessels or shore after 404–1106 days. Three of these animals were reportedly seen within 30 days after our resighting effort concluded on 26 March 1982, and the brands undoubtedly remained legible for some time thereafter. For example, animal 42 was seen and identified seven times between 11 August 1981 and 26 March 1982; from photographs taken this last day the legibility of the brands appeared to be at least 3/3. The dorsal-fin-base brands on two of the three animals recaptured in November 1980 were still excellent (4) (animal 13) or good (3) (animal 33). Although neither of these dolphins

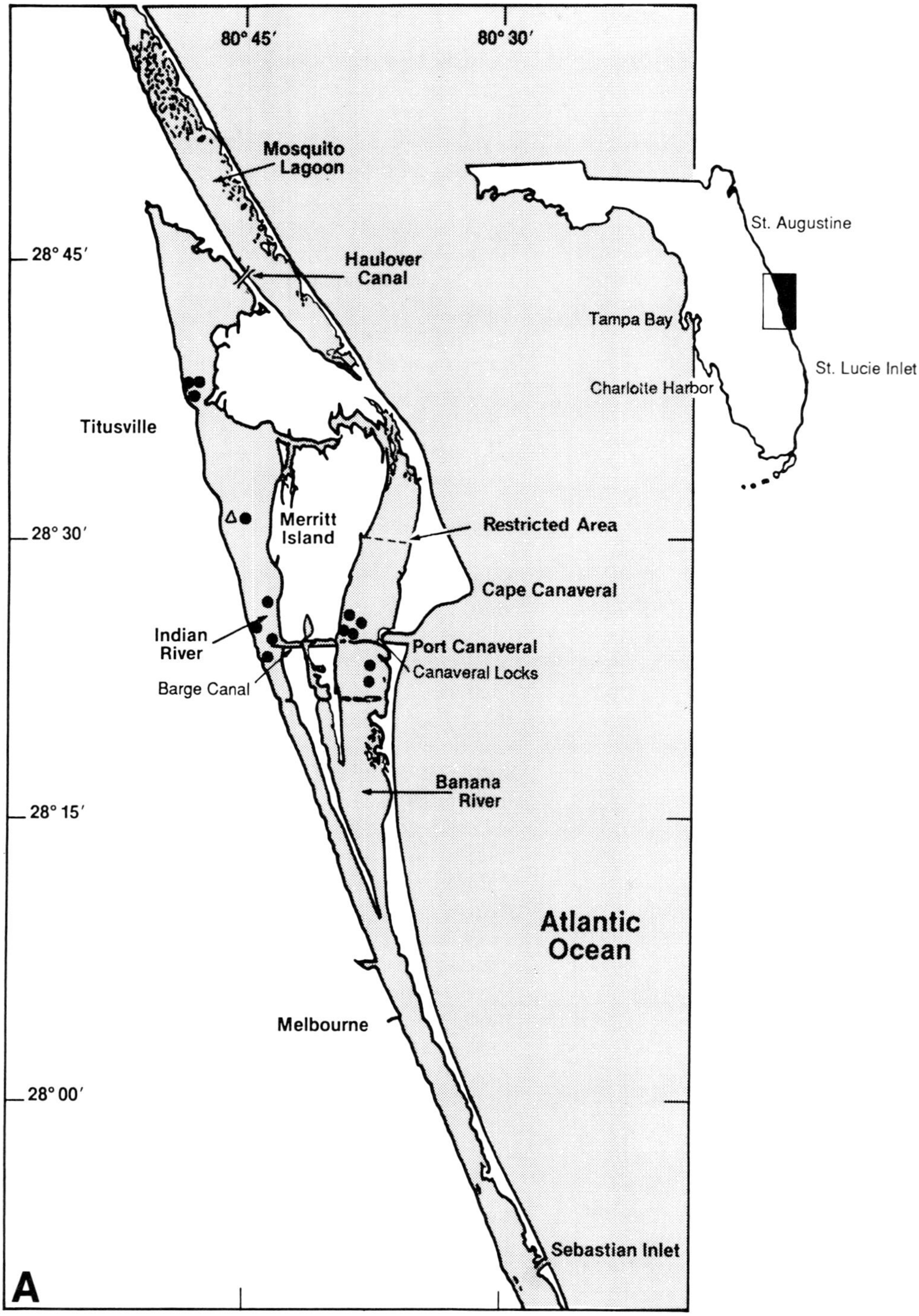

Figure 5 Study area, showing the distribution of sightings of dolphins 26 (A) and 56 (B). The triangle indicates the initial capture site.

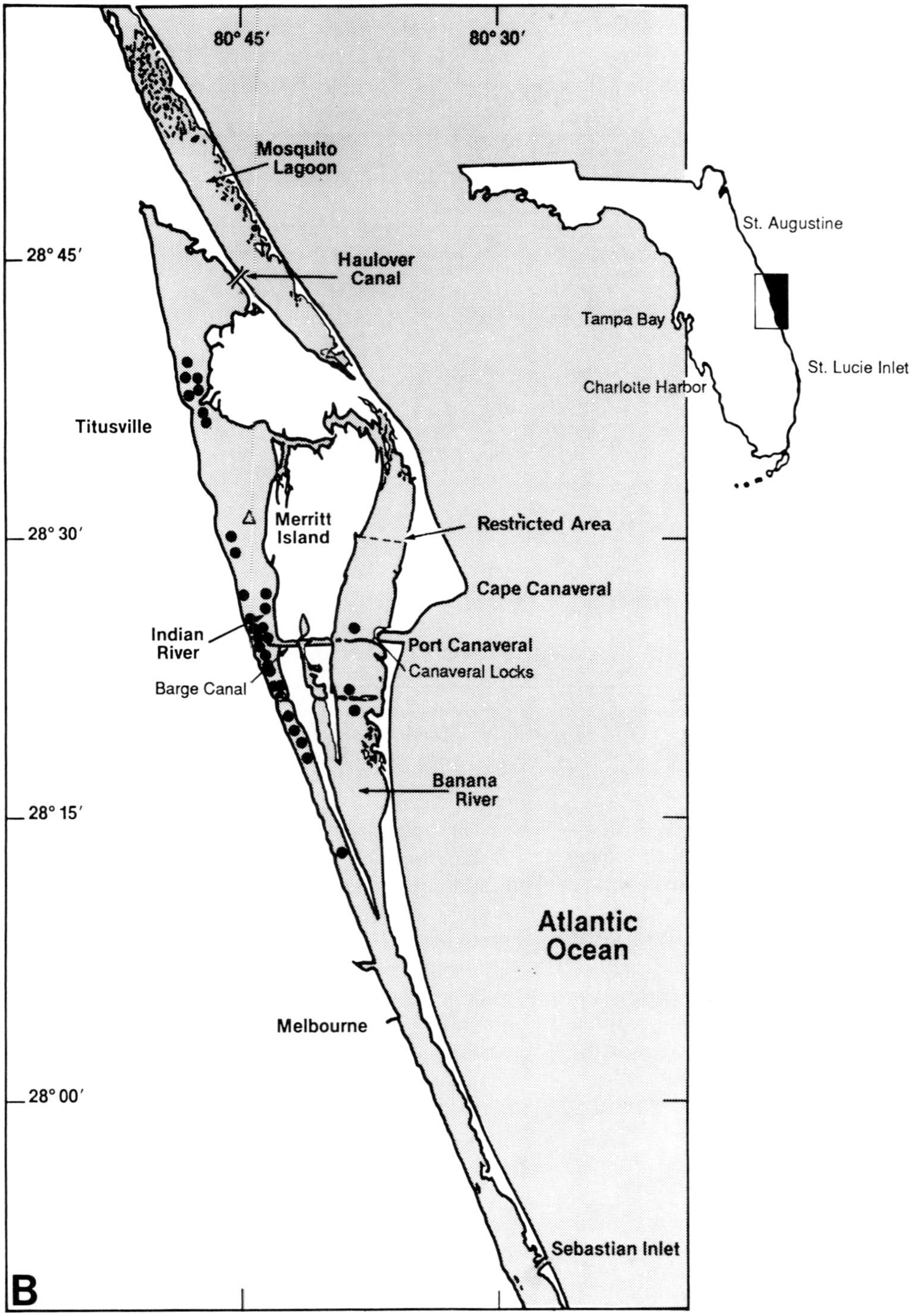

80° 45′
80° 30′
Mosquito Lagoon
28° 45′
Haulover Canal
St. Augustine
Tampa Bay
St. Lucie Inlet
Charlotte Harbor
Titusville
Merritt Island
28° 30′
Restricted Area
Cape Canaveral
Indian River
Port Canaveral
Canaveral Locks
Barge Canal
Banana River
28° 15′
Atlantic Ocean
Melbourne
28° 00′
Sebastian Inlet
B

Table 1
History and Longevity of Freeze Brands for Nine Animals Observed in Captivity Prior to Release and in the Wild at Various Intervals after Release[a]

	Animal/Brand Number								
Event(s)/Date(s)	13	15	30	32	33	42	50	81	82
Date captured	15 Apr 77	5 May 77	23 Aug 78	24 Aug 78	24 Aug 78	14 Feb 79	15 Feb 79	7 Feb 80	5 May 80
Date branded	Unknown	Unknown	Unknown	Unknown	Unknown	Unknown	Unknown	2 May 80	10 May 80
Date released	5 May 77	30 Jun 77	31 Oct 78	31 Oct 78	28 Aug 78	16 Mar 79	16 Mar 79	5 May 80	11 May 80
Most recent resight before recapture of 22 Nov 80	23 Oct 79	14 Dec 79	None	27 Nov 79	None	2 Oct 80	None	None	None
Recaptures 3–22 Nov 80	14 Nov 80	—	—	—	16 Nov 80	—	—	—	—
Brand legibility at recapture	1/4(T)	—	—	—	2/3(T)	—	—	—	—
Most recent resight before recapture of 17 Oct 81	7 Oct 81	14 Dec 79	8 Dec 80	25 Nov 80	18 Mar 81	11 Aug 81	7 Oct 81	None	None
Recaptures 5–17 Oct 81	—	—	—	—	—	—	9 Oct 81	—	—
Brand legibility at recapture	—	—	—	—	—	—	1/2(T)	—	—
Most recent resighting	22 Oct 81	14 Dec 79	8 Dec 80	25 Nov 80	18 Mar 81	26 Mar 82	26 Feb 82	None	27 Feb 82
Minimum days brand legible									
To last resight before retouch	**901**	—	—	—	None	—	**965**	—	—
To retouch (for code 3/4)	1289	—	—	—	**811**	—	967	—	—
To most recent resighting overall	1631	**897**	**404**	**756**	933	**1106**	1107	—	**658**

[a] Brand legibility codes are for dorsal fin/dorsal fin base: 1, poor; 2, fair; 3, good; 4, excellent. The notation "T" following legibility code indicates that the brand was retouched after legibility was scored. The boldface numbers in the last three rows represent the minimum period (in days) the brand was legible to shore- or boat-based observers without retouching.

had been observed and reidentified for a considerable time before recapture (animal 13, 387 days before recapture), we do not consider this to mean that their brands were illegible.

Animals branded on the boat and released immediately could not be observed as closely as the experimentally marked dolphins during the 3-week period of healing; therefore, the chronology of edema, necrosis, and regeneration for the brands on the immediately released dolphins is not well documented. However, we do know that at least some brands were legible after about 6 days. Eighteen of the animals branded in November 1980 were first identified by field observers an average of 29 days later (range 6–39 days). All the brands on captive dolphins have been visible by 21 days after branding (Cornell *et al.*, 1979; D. K. Odell and E. D. Asper, unpublished data).

The animals recaptured once (29) or twice (2) (Table 2) provide additional data for (a) assessing brand longevity from close-up examination, (b) rating the brands, and (c) utility in field observations from the number of days after initial branding (or touch-up) that each brand was still legible to boat- or shore-based observers. We interpreted all such data conservatively. For example, for brands retouched at recapture, we took as the minimum length of brand utility the number of days from initial branding to the most recent resighting before the brand was retouched. In some cases, even though there was a long hiatus between the most recent sighting and the recapture, the brands were rated on examination to be good (3) to excellent (4). Therefore, a lack of resightings does not necessarily indicate that the brands were illegible during the periods the

Figure 6 Development of a freeze brand on a captive dolphin over a 20-day period. A–D: Dorsal fin base 5, 11, 15, and 20 days after branding. E–H: Dorsal fin immediately and 5, 11, and 20 days after branding.

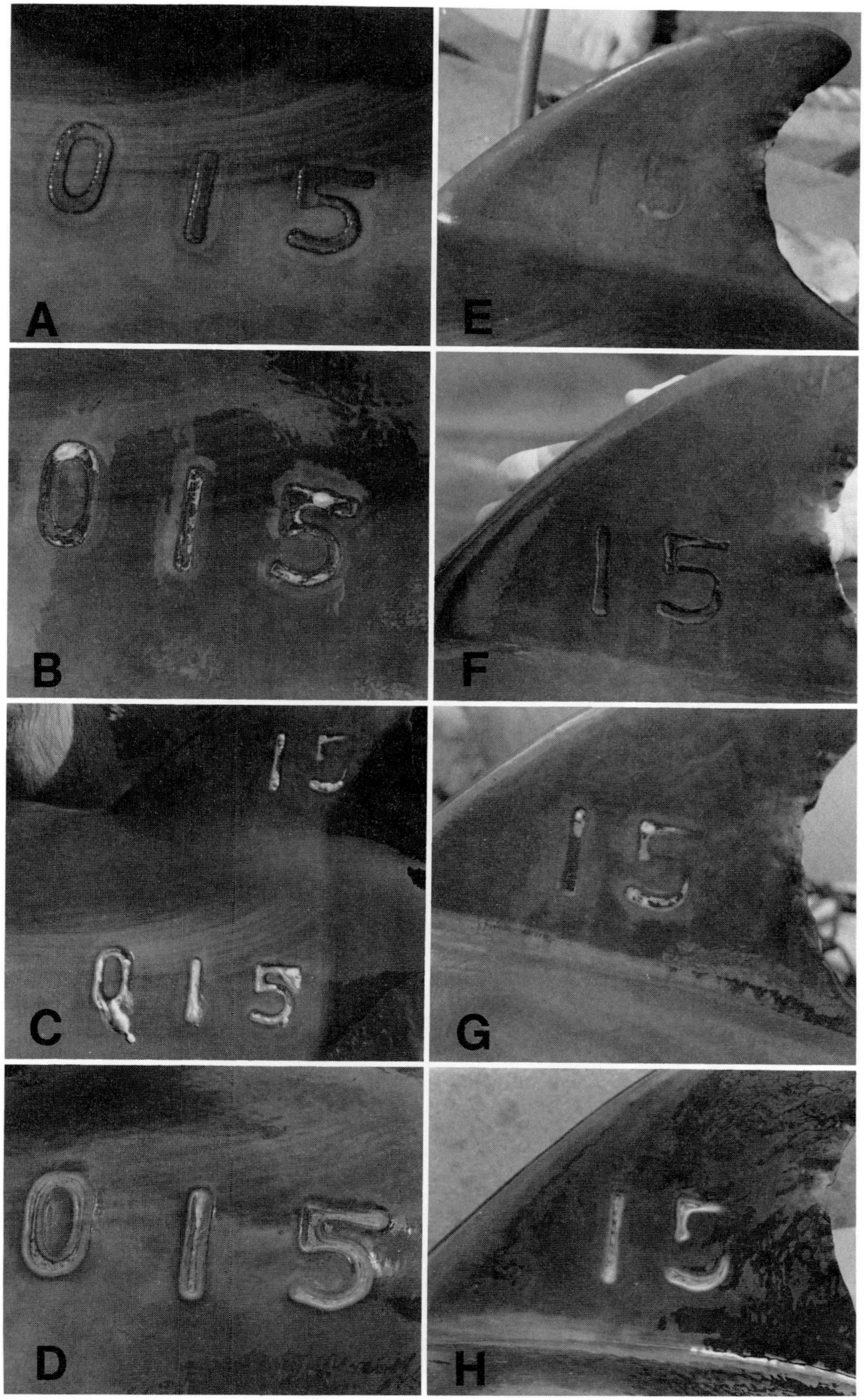
A
B
C
D
E
F
G
H

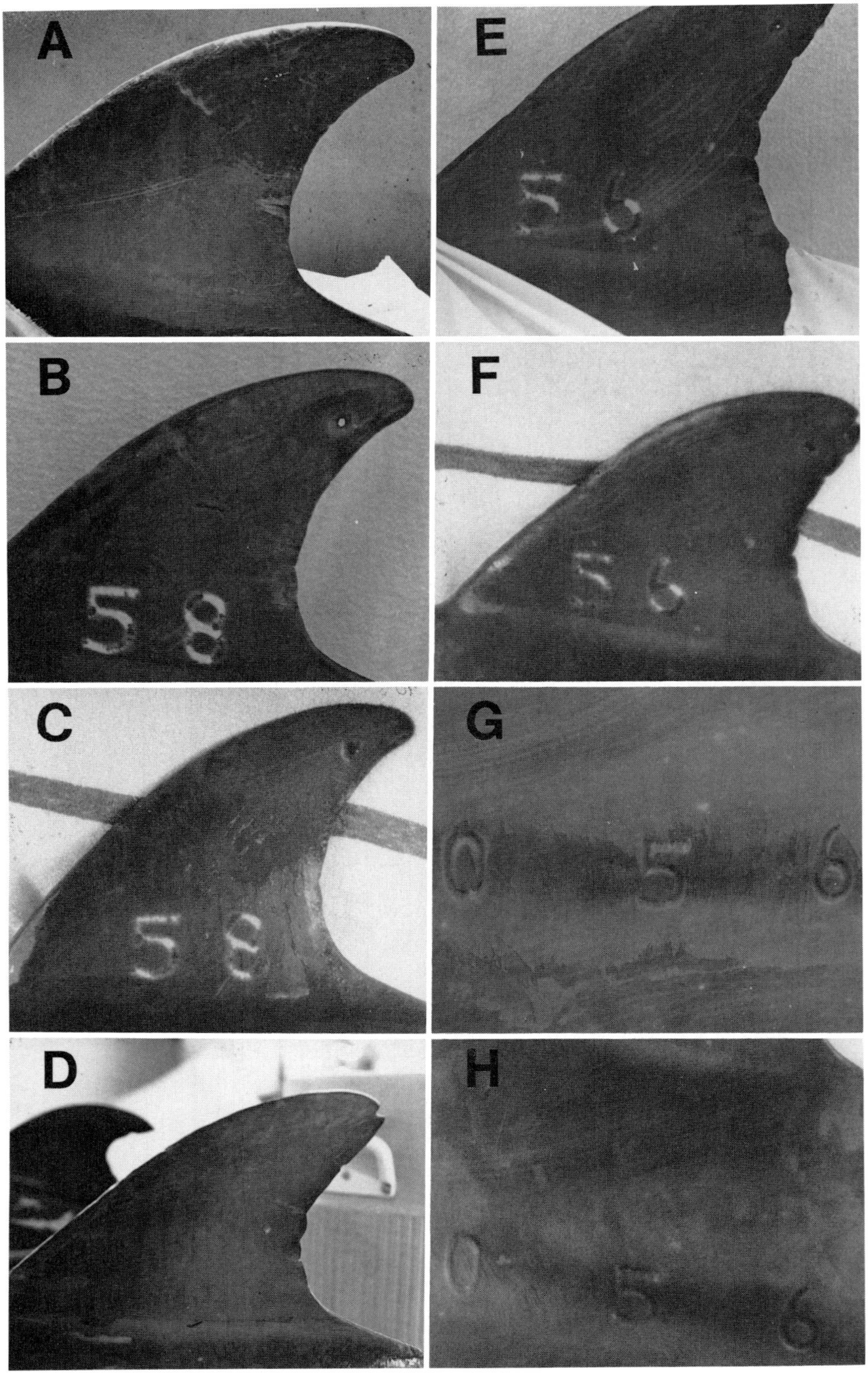

A
B
58
C
58
D
E
56
F
56
G
0 5 6
H
0 5 6

animals were not positively identified. In this sample, brands were legible in the field after at least 39 to 1046 days ($\bar{x}$ = 458.4, S.D. = 280.2, n = 26). The brands of the 57 dolphins which were never recaptured but which were positively identified in field encounters were legible for a minimum of 35–1742 days ($\bar{x}$ = 576.7, S.D. = 372.3).

Summary of Evaluation

If properly applied, freeze brands of the kind used in the present study can be adequate for identifying individuals in the field for 4.8 years or longer. Their usefulness in a program of regular resighting effort is thus demonstrated. Some of the brands were still being reported at or near the end of the study period. For example, animal 14, whose brand had not been retouched, was seen 11 February 1982, 1742 days after initial branding. Animal 56 was photographed in the Indian River on 15 October 1987, 2990 days after its rebranding on 4 November 1980. The dorsal-fin brand was clearly legible on close inspection of the photos.

Mortality of Branded Dolphins

Nine freeze-branded dolphins were found dead during the period of this study (8, 18, 38, 60, 62, 64, 98, D0, and D1). Proportions of branded animals found dead cannot be used reliably to calculate mortality rates. While brands obviously can be read on fresh or moderately decayed specimens, they are not likely to be evident on badly decomposed carcasses. Also, some (unknown) proportion of stranded carcasses, including those of branded dolphins, are never detected or examined; thus, the record of mortality is incomplete. Mortality rates can be determined more reliably from much larger samples (Hersh *et al.*, Chapter 8, this volume).

"Home Range"

For each freeze-branded dolphin seen more than once after its release, the "home range" was determined as follows. For those seen only within Indian River (including 3 seen in Mosquito Lagoon), the linear distance from the northernmost to the southernmost sightings were measured. For those moving between the Indian and Banana rivers, "home range" was determined (1) by measuring as above for sightings in both rivers and then adding 7.4 km, the length of the barge canal, to represent the shortest distance and (2) by measuring from the northernmost sighting in each river to the southern tip of Merritt Island, to represent the longest direct-line distance the animals might have traveled between those two points (i.e., around the southern tip of Merritt Island rather than through the barge canal).

Of 81 freeze-branded dolphins for which "home range" data were available, 60 were resighted exclusively within the Indian River (although 3 moved short distances into Mosquito Lagoon). These 60 dolphins had linear ranges of 1.8–80.6 km (mean = 32.8, S.D. = 18.0 km). Twenty-one dolphins were seen in both the Indian and Banana rivers; they had linear ranges of 14.8–90.8 km (mean 55.6, S.D. = 20.5 km) if they used the barge canal exclusively or 55.6–100.0 km (mean 84.1, S.D. = 12.1 km) if they went around the tip of Merritt Island to travel between the two bodies of water. Because both rivers are essentially slender North–South-oriented bodies of water, we take the travel distances to be rough approximations of the dolphins' minimum "home ranges." We use the term advisedly for several reasons. We did not search extensively outside the core areas (see Fig. 3), so it is possible that some animals ranged farther than our sample indicates. We note, for example, that volunteer observers reported a few freeze-branded animals in the ICW as far north as Marineland and as far south as St. Lucie Inlet. Also, we did not identify all animals in

Figure 7 Freeze brands on selected dolphins 1 and 2 years after branding. A–C: Dorsal fin of animal 58 at capture in 1979, at recapture in October 1980, showing the rototag post hole near the tip of the fin, and at recapture in November 1981. D–H: Dorsal fin of animal 56 at capture in August 1979 and at recapture in November 1981, and dorsal fin base at recapture in October 1980 and at recapture in November 1981.

Table 2
Chronology of Branding, Resighting, and Rebranding of 29 Animals Recaptured Once or Twice[a]

	Animal/Brand Number								
Event(s)/Date(s)	10	11	18	25	26	39	56	58	59
Date captured	17 Apr 77	19 Apr 77	30 Jun 77	10 Mar 78	10 Mar 78	13 Feb 79	28 Aug 79	28 Aug 79	28 Aug 79
Most recent resight before recapture of 22 Nov 80	27 Feb 80	—	—	13 Dec 79	28 Oct 80[b]	20 Oct 80[b]	29 Oct 80[b]	29 Oct 80[b]	29 Oct 80[b]
Recaptures 3–22 Nov 80	—	7 Nov 80	18 Nov 80	19 Nov 80	19 Nov 80	6 Nov 80	4 Nov 80	4 Nov 80	11 Nov 80
Brand legibility at recapture	—	1/4(T)	1/4(T)	2/3(T)	2/4(T)	1/3(T)	3/4(T)	3/4(T)	2/4(T)
Most recent resight before recapture of 17 Oct 81	—	25 May 81	18 Mar 81	25 Feb 81	11 Sep 81	18 Feb 81	5 Oct 81	5 Oct 81	7 Oct 81
Recaptures 5–17 Oct 81	16 Oct 81	—	—	—	—	—	7 Oct 81	7 Oct 81	—
Brand legibility at recapture	1/4	—	—	—	—	—	3/4	3/4	—
Most recent resighting	28 Dec 81	28 May 81	18 Mar 81	28 Jan 82	25 Feb 82	25 Feb 82	25 Mar 82	27 Feb 82	28 Jan 82
Minimum days brand legible									
To last resight before retouch	**1046**	—	—	**643**	**963**	**554**	**428**	**428**	**428**
To retouch (for code 3/4)	1643	1298	1237	985	985	632	434/771[c]	434/771[c]	441
To most recent resighting overall	1716	1500	1357	1420	1448	1108	940	914	884

the population of which the marked (freeze-branded or naturally tagged) dolphins were a sample. The Indian River is estimated to contain 200–300 dolphins (Scott, Chapter 36, this volume), and we are not certain how well the sample of around 130 marked animals represents the population overall.

Casual observations of 13 naturally marked animals defined the "home range" of this population almost as exactly as did the more organized, labor-intensive observations of freeze-branded animals. However, the handling associated with freeze-branding allowed us to sex the animals immediately and to obtain other useful biological data (see Asper and Odell 1980; Odell and Asper 1982; Asper *et al.*, Chapter 28, this volume; Duffield and Chamberlin-Lea, Chapter 35, this volume). While freeze-branding may appear from these results to be redundant for studying inshore populations of bottlenose dolphins, it may well be an effective way of tagging species that are less abundant, more wide-ranging, and less accessible for study using natural marks alone.

ACKNOWLEDGMENTS

The authors gratefully acknowledge the assistance of J. Provancha, B. Schroeder, and S. Hersh, who gathered and analyzed field data. K. Goodrich, R. Ford, and W. Hoggard also assisted in the field. C. Perry, G. Case, and S. Gaby assisted with laboratory work. This study would not have been possible without the collective expertise of the Sea World of Florida Animal Care Department. The work was funded by U.S. National Marine Fisheries Service Contracts NA79-GA-C-00027 and NA80-GA-C-00063. This is Sea World of Florida Technical Contribution No. 8813-F.

Animal/Brand Number									
62	65	67	75	80	92	93	95	A4	A7
29 Aug 79	30 Aug 79	30 Aug 79	1 Sep 79	2 Sep 79	17 Nov 80	17 Nov 80	19 Nov 80	9 Nov 80	9 Nov 80
20 Aug 80	15 Sep 80	14 Nov 80	12 Nov 79?	14 Nov 79	—	—	—	—	—
7 Nov 80	—	—	11 Nov 80	14 Nov 80	—	—	—	—	—
2/4(T)	—	—	3/4	3/4	—	—	—	—	—
12 May 81	7 Oct 81	5 Feb 81	14 Jul 81	25 Sep 81	7 Oct 81	—	5 Feb 81	31 Mar 81	9 Oct 81
—	7 Oct 81	7 Oct 81	—	—	9 Oct 81	8 Oct 81	5 Oct 81	7 Oct 81	17 Oct 81
—	2/3(T)	3/4	—	—	3/4(T)	3/4(T)	3/4(T)	2/4(T)	?/4(T)
12 May 81	7 Oct 81	28 Feb 82	27 Feb 82	9 Oct 81	11 Mar 82	1 Dec 81	25 Feb 82	10 Mar 82	10 Mar 82
357	**769**	525	437	439	**324**	None	**78**	**142**	**324**
436	769	769	437	439	326	325	320	332	342
622	769	**911**	**910**	**768**	479	379	463	486	486

(continued)

[a] Brand legibility codes are for dorsal fin/dorsal fin base: 1, poor; 2, fair; 3, good; 4, excellent. The notion "T" following legibility code indicates the brand was retouched after legibility was scored. The boldface numbers in the last three rows represent the minimum period (in days) the brand was legible to shore- or boat-based observers without retouching.
[b] Reported by Patton *et al.*, 1981.
[c] First and second recaptures, respectively.

Table 2 (*continued*)

Event(s)/Date(s)	Animal/Brand Number A8	A9	C0	C1	C6	C7	D2	D4	V2	V5
Date captured	9 Nov 80	10 Nov 80	7 Nov 80	7 Nov 80	8 Nov 80	8 Nov 80	10 Nov 80	14 Nov 80	4 Nov 80	4 Nov 80
Most recent resight before recapture of 22 Nov 80	—	—	—	—	—	—	—	—	—	—
Recaptures 3–22 Nov 80	—	—	—	—	—	—	—	—	—	—
Brand legibility at recapture	—	—	—	—	—	—	—	—	—	—
Most recent resight before recapture of 17 Oct 81	4 Jun 81	19 Dec 80	5 Oct 81	7 Oct 81	20 May 81	9 Sep 81	25 Sep 81	26 May 81	25 Sep 81	12 Aug 81
Recaptures 5–17 Oct 81	10 Oct 81	10 Oct 81	9 Oct 81	9 Oct 81	8 Oct 81	10 Oct 81	7 Oct 81	10 Oct 81	9 Oct 81	6 Oct 81
Brand legibility at recapture	3/3(T)	3/3(T	3/4(T)	3/4(T)	4/4(T)?	3/4(T)	4/4(T)	4/4(T)	3/4	2/4(T)
Most recent resighting	28 Jan 82	30 Dec 81	5 Oct 81	7 Oct 81	16 Nov 81	26 Feb 82	25 Mar 82	26 Feb 82	25 Sep 81	12 Aug 81
Minimum days brand legible										
To last resight before retouch	**207**	**39**	**363**	**334**	**193**	**305**	**319**	193	**325**	**281**
To retouch (for code 3/4)	335	334	367	336	334	336	331	330	339	336
To most recent resighting overall	445	415	367	336	373	475	500	**469**	339	336

Appendix

Chronology of Collection, Marking, Biological Sampling, and Resighting of Bottlenose Dolphins in Charlotte Harbor and in the Indian and Banana Rivers, Florida

		Animals by Brand Number (Sex)																	
Dates	Activities	01 (F)	02 (F)	03 (M)	04 (M)	05 (F)	06 (M)	07 (F)	08[i] (M)	09 (F)	10 (F)	11 (F)	12[a] (F)	13 (F)	14 (M)	15 (F)	16 (M)	17 (F)	18 (M)
September 1976–January 1977	Capture, sample, mark, release (Charlotte Harbor)	I 246	I 244	I 218	I 213	I 246													
April 1977–March 1979	Capture, sample, mark, release						I 208	I 236	I 229	I 210	I 253	I 208	I 175		I 179		I 254	I 231	I 259
August 1978–February 1979, May 1980, March–August 1981	Capture, sample, mark, hold 90 days, release													I 229		I 204			
August 1979–September 1979	Capture, sample, mark, release																		
September 1979–December 1979	Resighting						0	3	0	0	3	0	0	1	1	4	2	1	0
August 1980–December 1980	Resighting[e] (Mote Marine Lab)						0	0	0	0	0	0	0	0	1	0	0	0	0
November 1980	Capture, sample, mark, release of 50 new dolphins																		
	Recapture, sample, release of 13 previously marked dolphins											R		R					R
October 1981	Recapture, sample, release of 21 previously marked dolphins										R								
November 1980–March 1982	Resighting						0	1	0	1	1	2	0	5	1	0	10	0	2

Note: I, Initial capture. Length (cm) and estimated age (years) (if known) are listed immediately below I codes (e.g., I/250/10 indicates an animal that was 250 cm long and estimated to be 10 years old at capture). R, Recapture. Numbers in resighting rows indicate the number of times an individual was observed during that period.

[a] Animal handled but not freeze-branded.

[b] Number not assigned to an animal.

[c] Animal found dead 26 October 1978.

[d] Numbers 83–89 not assigned to animals.

[e] After Patton *et al.* (1981).

[f] Animal 18 found dead 12 February 1988, 260 cm, in the Indian River.

[g] Animal 60 died at release (see text).

[h] Animal 81 released 5 May 1980; found dead 10 May 1980.

[i] Animals found dead: 08, 28 June 1977; 62, 26 August 1981; 64, 02 May 1982; 98, 14 January 1981; D0, 7 July 1982; D1, June 1981.

		Animals by Brand Number (Sex)															
Dates	Activities	19[a] (F)	20[a] (M)	21[a] (F)	22[a] (F)	23[b] (—)	24[b] (—)	25 (M)	26 (M)	27 (M)	28 (F)	29 (F)	30 (F)	31 (F)	32 (F)	33 (F)	34 (F)
September 1976–January 1977	Capture, sample, mark, release (Charlotte Harbor)																
April 1977–March 1979	Capture, sample, mark, release			I[a] 201	I[a] 239	—	—	I 225	I 219	I 222	I 237			I 244			I 252
August 1978–February 1979, May 1980, March–August 1981	Capture, sample, mark, hold 90 days, release	I 241	I 257									I 239	I 185		I 183	I 229	
August 1979–September 1979	Capture, sample, mark, release																
September 1979–December 1979	Resighting	—	—	—	—	—	—	3	5	0	0	0	0	0	1	0	0
August 1980–December 1980	Resighting[c] (Mote Marine Lab)	—	—	—	—	—	—	0	4	0	0	0	1	0	2	0	0
November 1980	Capture, sample, mark, release of 50 new dolphins																
	Recapture, sample, release of 13 previously marked dolphins							R	R							R	
October 1981	Recapture, sample, release of 21 previously marked dolphins																
November 1980–March 1982	Resighting	—	—	—	—	—	—	8	16	0	0	0	0	0	0	1	0

Animals by Brand Number (Sex)																								
35 (F)	36 (M)	37 (F)	38[c] (F)	39 (F)	40 (M)	41 (F)	42 (M)	43 (F)	44 (M)	45 (M)	46 (M)	47 (M)	48 (F)	49 (F)	50 (F)	51 (F)	52 (F)	53 (F)	54[b] (—)	55 (M)	56 (M)	57 (F)	58 (M)	59 (M)
I	I	I	I	I	I	I		I	I	I	I	I	I	I		I	I	I	—					
255	205	255	197	233	258	196		230	251	202	189	162	160	240		203	238	244	—					
							I								I									
							212								218									
																				I	I	I	I	I
																				223	238	216	233	234
0	1	0	—	1	3	0	3	3												7	12	6	13	8
									2	5	1	0	0	6	11	0	0	0	—	17	18	8	12	18
0	4	0	—	2	0	0	7	0																
									0	4	0	0	0	1	4	1	0	0	—	6	6	1	6	5
				R																				
																					R		R	R
															R						R		R	
0	2	0	—	6	0	0	11	0	10	0	0	0	0	1	8	0	0	0	—	14	24	7	7	7

(continued)

Dates	Activities	Animals by Brand Number (Sex)															
		60[g] (M)	61 (F)	62[i] (M)	63 (F)	64[i] (M)	65 (M)	66 (M)	67 (M)	68 (M)	69 (M)	70 (F)	71 (M)	72 (M)	73 (M)	74 (M)	75 (M)
September 1976–January 1977	Capture, sample, mark, release (Charlotte Harbor)																
April 1977–March 1979	Capture, sample, mark, release																
August 1978–February 1979, May 1980, March–August 1981	Capture, sample, mark, hold 90 days, release																
August 1979–September 1979	Capture, sample, mark, release	I 193	I 246	I 260	I 260	I 205	I 225	I 218	I 255	I 255	I 258	I 228	I 250	I 246	I 237	I 245	I 220
September 1979–December 1979	Resighting	4 0	14 2	17 3	14 2	3 2	12 14	5 2	12 13	15 9	— 11	8 6	14 2	10 4	11 6	15 10	9 13
August 1980–December 1980	Resighting[e] (Mote Marine Lab)										1	0	0	2	0	5	2
November 1980	Capture, sample, mark, release of 50 new dolphins	0	0	1	6	6	2	5	3	2							
	Recapture, sample, release of 13 previously marked dolphins			R													R
October 1981	Recapture, sample, release of 21 previously marked dolphins						R		R								
November 1980–March 1982	Resighting	—	0	4	7	6	3	4	13	6	3	4	6	6	1	11	12

Animals by Brand Number (Sex)																								
76 (F)	77 (F)	78 (M)	79 (F)	80 (F)	81[h] (—)	82 —)	90[d] (F)	91 (F)	92 (M)	93 (F)	94 (M)	95 (M)	96 (M)	97 (F)	98 (M)	99 (M)	V0 (M)	V1 (M)	V2 (M)	V3 (M)	V4 (M)	V5 (M)	V6 (M)	V7 (M)
					I	I																		
					202	226																		
I	I	I	I	I																				
210	230	198	240	251																				
8			12	8																				
23	3	0	5	4	—	—																		
7	1	1	1	0													0	1	0	1	1	1	0	0
							I	I	I	I	I	I	I	I	I		I	I	I	I	I	I	I	I
							263	243	194	249	230	219	233	204	199		205	212	216	215	224	247	185	244
							12	12	3	10	9	7	5	10	3		7	6	6	6	6	11	3	11
				R																				
									R	R		R				I								
																251			R			R		
11	1	3	0	6	—	7	0	2	3	1	10	3	11	1	0	12	9	9	7	16	8	11	3	2

(*continued*)

Dates	Activities	Animals by Brand Number (Sex)															
		V8 (F)	V9 (M)	C0 (M)	C1 (F)	C2 (F)	C3 (F)	C4 (F)	C5 (F)	C6 (M)	C7 (M)	C8 (M)	C9 (F)	A0 (M)	A1 (M)	A2 (M)	A3 (F)
September 1976–January 1977	Capture, sample, mark, release (Charlotte Harbor)																
April 1977–March 1979	Capture, sample, mark, release																
August 1978–February 1979, May 1980, March–August 1981	Capture, sample, mark, hold 90 days, release																
August 1979–September 1979	Capture, sample, mark, release																
September 1979–December 1979	Resighting																
August 1980–December 1980	Resighting[e] (Mote Marine Lab)	0	0	0	1	0	0	0	0	1	0	0	1	0	0	0	0
November 1980	Capture, sample, mark, release of 50 new dolphins	I 251	I 241	I 190	I 251	I 251	I 231	I 220	I 258	I 212	I 231	I 171	I 241	I 250	I 25[illegible]	I 269	I 209
	Recapture, sample, release of 13 previously marked dolphins	13	11	2	12	11	8	2	12	4	5	—	13	15	12	10	3
October 1981	Recapture, sample, release of 21 previously marked dolphins			R	R					R	R						
November 1980–March 1982	Resighting	11	8	6	13	0	13	1	0	3	11	0	0	9	4	4	12

	Animals by Brand Number (Sex)																							
	A4 (M)	A5 (M)	A6 (F)	A7 (M)	A8 (M)	A9 (F)	D0[i] (M)	D1[i] (F)	D2 (M)	D3 (M)	D4 (F)	D5 (F)	D6 (M)	D7 (F)	D8 (F)	D9 (M)	10A[a] (F)	X0 (M)	X1 (M)	X2 (M)	X3 (M)	X4 (F)	X5 (F)	X6 (—)	X (F)
																		I	I	I	I	I	I	I	I
																		240	230	228	229	221	215	200	191
	1	1	0	2	0	0	0	0	0	0	0	0													
	I	I	I	I	I	I	I	I	I	I	I	I	I	I	I	I	I								
	276	230	254	230	253	234	223	226	236	186	246	242	195	234	160	205	198								
	15	8	12	13	10	13	12	12	9	2	—	14	3	14	—	2	—								
	R			R	R	R					R														
	8	16	11	20	4	6	13	4	12	2	6	3	6	0	13	0	—	3	3	0	3	3	2	1	

REFERENCES

Asper, E. D. (1975). Techniques of live capture of smaller Cetacea. *J. Fish. Res. Board Can.* **32,** 1191–1196.

Asper, E. D., and Odell, D. K. (1980). Bottlenose dolphin local herd monitoring: Capture, marking, collection of biological data, and follow-up observations. Final report, U.S. National Marine Fisheries Service Contract NA79-GA-C-00027. 127 pp.

Cornell, L. H., Asper, E. D., Osborn, K., and White, M. J., Jr. (1979). Investigations on cryogenic marking procedures for marine mammals. Final report, U.S. Marine Mammal Commission Contract MM6AC003. NTIS, Springfield, Virginia, PB-291 570. 24 pp.

Gruber, J. A. (1981). Ecology of the Atlantic bottlenosed dolphin (*Tursiops truncatus*) in the Pass Cavallo area of Matagorda Bay, Texas. M.S. thesis, Texas A & M University, College Station, Texas. 182 pp.

Hansen, L. J. (1983). Population biology of the coastal bottlenose dolphin (*Tursiops truncatus*) of southern California. M.S. thesis, California State Univ., Sacramento, California. 104 pp.

Hersh, S. L. (1987). Characterization and differentiation of bottlenose dolphin populations (genus *Tursiops*) in the southeastern U.S. based on mortality patterns and morphometrics. Ph.D. dissertation. Univ. of Miami, Coral Gables, Florida. 213 pp.

Hobbs, L. (1982). Tagging and marking. Appendix III. *In* "Whales, Dolphins, and Porpoises of the Eastern North Pacific and Adjacent Arctic Waters: a Guide to Their Identification" (S. Leatherwood, R. R. Reeves, W. E. Perrin, and W. E. Evans). NOAA Tech. Rep. NMFS Circ. 444:245 pp.

Irvine, A. B., and Wells, R. S. (1972). Results of attempts to tag Atlantic bottlenosed dolphins, *Tursiops truncatus. Cetology* **13,** 1–5.

Irvine, A. B., Scott, M. D., Wells, R. S., and Kaufman, J. H. (1981). Movements and activities of the Atlantic bottlenose dolphin, *Tursiops truncatus,* near Sarasota, Florida. *Fish. Bull. (U.S.)* 79, 671–688.

Leatherwood, S. (1979). Aerial survey of the bottlenose dolphin, *Tursiops truncatus,* and the West Indian manatee, *Trichechus manatus,* in the Indian and Banana Rivers, Florida. *Fish. Bull. (U.S.)* **77,** 47–59.

Leatherwood, S., and Reeves, R. R. (1982). Bottlenose dolphin *Tursiops truncatus* and other toothed cetaceans. *In* "Wild Mammals of North America: Biology, Management, and Economics" (J. A. Chapman and G. A. Feldhamer, eds.), pp. 369–414. Johns Hopkins Univ. Press, Baltimore, Maryland. 1147 pp.

Leatherwood, S., and Show, I. T. (1980). Development of systematic procedures for estimating sizes of "populations" of bottlenose dolphins and estimates of sizes of "populations" of bottlenose dolphins in three geographical areas, with incidental observations on densities of West Indian manatees and marine turtles. Final report, U.S. National Marine Fisheries Service Contract NA79-GA-C-0038. 98 pp.

Odell, D. K., and Asper, E. D. (1982). Live capture, marking, and resighting of bottlenose dolphins. Final report, U.S. National Marine Fisheries Service Contract NA80-GA-C-00063. 325 pp.

Patton, G. W., Gilliland, S. D., and Gross, D. T. (1981). Marine resighting and monitoring of 25 *Tursiops truncatus* (dolphins) in the coastal area of mid-eastern Florida. Final report, U.S. National Marine Fisheries Service Contract NA80-GA-C-0041W. 82 pp.

Reeves, R. R., and Leatherwood, S. (1984). Live-capture fisheries for cetaceans in USA and Canadian waters, 1973–1982. *Rep. int. Whal. Commn* **34,** 497–508.

Ridgway, S. H., Green, R. F., and Sweeney, J. C. (1975). Mandibular anesthesia and tooth extraction in the bottlenosed dolphin. *J. Wildl. Dis.* **11,** 415–418.

Shane, S. H. (1980). Occurrence, movements, and distribution of bottlenose dolphins, *Tursiops truncatus,* in southern Texas. *Fish. Bull. (U.S.)* **78,** 593–601.

Shane, S. H. (1987). The behavioral ecology of the bottlenose dolphin. Ph.D. dissertation, University of California, Santa Cruz. vi + 97 pp.

Weigle, B. W. (1987). Abundance, distribution, and movements of bottlenose dolphins, *Tursiops truncatus,* in lower Tampa Bay, Florida. M.S. thesis, University of South Florida, Tampa, Florida. 47 pp.

Wells, R. S., Irvine, A. B., and Scott, M. D. (1980). The social ecology of inshore odontocetes. "Cetacean Behavior: Mechanisms and Functions" (L. M. Herman, ed.), pp. 263–317. Wiley, New York. 463 pp.

Wells, R. S., Scott, M. D., and Irvine, A. B. (1987). The social structure of free-ranging bottlenose dolphins. *In* "Current Mammalogy" (H. H. Genoways, ed.), Vol. 1, pp. 247–305. Plenum, New York. 519 pp.

Comparison of Bottlenose Dolphin Behavior in Texas and Florida, with a Critique of Methods for Studying Dolphin Behavior

31

Susan H. Shane
West Coast Whale Research Foundation
Santa Cruz, California 95060

INTRODUCTION

The bottlenose dolphin, *Tursiops truncatus*, inhabits a remarkably wide range of habitats. It has a cosmopolitan distribution in the temperate and tropical zones and occupies open coasts, lagoons, estuaries, oceanic island coasts, and pelagic waters in the eastern tropical Pacific as far as 2000 km offshore (Ross, 1977; Leatherwood and Reeves, 1982; Scott and Chivers, Chapter 22, this volume). Bottlenose dolphins survive well in captivity and often adapt to human activities such as fishing and boating in densely populated coastal zones (Norris and Prescott, 1961; Caldwell and Caldwell, 1972; Busnel, 1973; Leatherwood, 1975; Shane, 1977; Hussenot, 1980; Gruber, 1981; Schlais, 1984).

In this chapter I try to show that behavioral flexibility has contributed to the dolphin's success in diverse habitats. Behavioral flexibility in a species can be studied from at least three viewpoints: (1) individual differences in behavior within one population ("between-phenotype variation" according to Roughgarden, 1972); (2) diverse behaviors serving a common function (e.g., food gathering) used by all individuals in one population (Roughgarden's "within-phenotype variation"); and (3) different behaviors with a single function developed by separate populations in distinct habitats. Here, I address the third perspective on behavioral flexibility by comparing the behavior (especially feeding) and population characteristics of dolphins in Texas and Florida.

The best example of the bottlenose dolphin's flexibility is the wide variety of recorded feeding behaviors, each of which can be associated with specific ecological conditions (Table 1). Norris and Prescott (1961) first remarked on the bottlenose dolphin's flexibility in feeding behavior, and Caldwell and Caldwell (1972, p. 74) suggested that ". . .the many types of fishes eaten [by the bot-

Table 1
Feeding Behaviors Described for Bottlenose Dolphins

Feeding Behavior	Locations	References
Feeding behind trawling shrimp boats	Texas, N. Gulf of Mexico, Florida's Atlantic coast, Georgia, S. Carolina, Gulf of California	Gunter (1942, 1954), Norris and Prescott (1961), Caldwell and Caldwell (1972), Hogan (1975), Leatherwood (1975), Shane (1977), Gruber (1981)
Feeding on fish discarded by shrimpers	Texas, N. Gulf of Mexico, Florida's Atlantic Coast, Georgia, S. Carolina, Gulf of California	Norris and Prescott (1961), Caldwell and Caldwell (1972), Hogan (1975), Leatherwood (1975), Shane (1977), Gruber (1981)
Midair capture of mullet	Florida, Mauritania	Harris (1938), Caldwell and Caldwell (1972), Busnel (1973)
Diving to feed on bottom or midwater fishes	Texas, S. California, Florida, Portugal	Gunter (1954), Norris and Prescott (1961), Shane (1977, Chapter 12, this volume), dos Santos and Lacerda (1987)
Partially beaching while feeding on fish chased onto shore	Georgia, S. Carolina, Mauritania	Hoese (1971), Busnel (1973), Rigley (1983)
Throwing mullet onto beach and partially stranding during fish retrieval	Portugal	dos Santos and Lacerda (1987)
Shallow-water pursuit of fish	Texas, Florida, Georgia, S. Carolina	Gunter (1954), Hogan (1975), Hamilton and Nishimoto (1977), Shane (1977, Chapter 12, this volume), Irvine *et al.* (1981)
Feeding where garbage was dumped daily	S. California	Norris and Prescott (1961)
Following ferry to feed on organisms stirred up by prop wash	S. California, Texas	Norris and Prescott (1961), Shane (1977)
Hanging vertically in the water column and sweeping head around in a circle, pursuing small fish fleeing from the tip of the rostrum	Florida	Shane (Chapter 12, this volume)
Shaking, rubbing, or tossing fish to break into smaller pieces	Florida, Gulf of California, Argentina, Texas	Norris and Prescott (1961), Caldwell and Caldwell (1972), Shane (1977), Würsig and Würsig (1979)
Head slapping with cuttlefish in mouth; decapitating prey and consuming the head	Portugal	dos Santos and Lacerda (1987)
Kicking fish into air with flukes and swimming over to ingest dead or injured fish when it falls in water	Florida	Wells *et al.* (1987), Shane (Chapter 12, this volume)
Facing into a strong current and capturing fish carried with the current	Texas, Florida	Shane (1977, Chapter 12, this volume)
Chasing and pinwheeling to capture fish (often on side or upside down)	Numerous	Leatherwood (1975), Shane (1977, Chapter 12, this volume), Irvine *et al.* (1981), dos Santos and Lacerda (1987)
Feeding on fish attracted to anchored boats	Numerous	Leatherwood (1975)

Table 1 (*continued*)

Feeding Behavior	Locations	References
Cooperative herding of fish against shore	Georgia, S. Carolina, S. Africa, Black Sea, Florida	Caldwell and Caldwell (1972), Tayler and Saayman (1972), Hogan (1975), Bel'kovich *et al.* (1978)
Cooperative driving and trapping of fish between dolphin groups	S. Africa, Black Sea	Saayman *et al.* (1973), Bel'kovich *et al.* (1978)
Cooperative encircling of fish with individual dolphins charging through fish school to feed	Texas, N. Gulf of Mexico, Black Sea, Georgia, S. Carolina	Caldwell and Caldwell (1972), Tayler and Saayman (1972), Hogan (1975), Leatherwood (1975), Shane (1977), Bel'kovich *et al.* (1978)
Cooperative capture of moray eel by using a scorpion fish to coax eel out of hole	Captivity	Brown and Norris (1956)
Taking bait from hands of people on boats or on beach	Georgia, Florida, Black Sea, W. Australia	Dean (1979), Connor and Smolker (1985), Shane *et al.* (1986), Anonymous (1987)
Taking bait and catch from hook and line or nets	Hawaii, Florida	Cato and Prochaska (1976), Leatherwood (1979), Schlais (1984)
Individual herding of fish against shore or other barrier	Black Sea, Mauritania, Georgia, S. Carolina, Florida, S. Africa	Tayler and Saayman (1972), Busnel (1973), Baker (1983), Shane (Chapter 12, this volume)
Percussive leaps for driving fish	Black Sea, S. Africa	Tayler and Saayman (1972), Bel'kovich *et al.* (1978)
Intense echolocation to separate individual fish from school, followed by striking fish underwater with ventral side of flukes	Captivity	Hult (1982)
Encircling a mullet school and slapping with posterior half of body against the water's surface followed by swimming through the encircled area to ingest fish	Florida	Hamilton and Nishimoto (1977)
Diving amidst feeding pilot whales presumably feeding on prey initially detected by the whales	S. California	Brown and Norris (1956), Shane (1986)

tlenose dolphin] must require a number of kinds of behavior on the part of dolphins in order to make successful captures." Bel'kovich *et al.* (1978) emphasized the flexibility of the bottlenose dolphin's feeding behavior in the Black Sea. Leatherwood (1975) contended that small home ranges and short-term movements in the bottlenose dolphin would favor flexible food habits. Würsig (1986) noted that bottlenose dolphins are "highly adaptive animals" in terms of their ability to use different feeding techniques in different circumstances.

Acquisition of new behaviors, often in response to human activities, is one way that animals display behavioral flexibility (e.g., Nilgiri langurs, *Presbytis johnii,* learned to feed on cultivated crops: Poirier, 1969; Japanese macaques, *Macaca fuscata,* learned to wash sweet potatoes in salt water: Itani, 1958). Bottlenose dolphins display this type of flexibility. Shane *et al.* (1986) suggested that bowriding, exhibited by bottlenose dolphins worldwide, could be energy-saving locomotory behavior. Norris and Prescott (1961) described bottlenose dolphins' attraction to a site where the

U.S. Navy dumped garbage each day and the dolphins' habit of following a ferry, presumably eating food stirred up in its wake. A few bottlenose dolphins have been fed and petted by people at a beach in Shark Bay, Western Australia, for over 20 years (Connor and Smolker, 1985). Dolphins follow shrimp boats and feed on prey disturbed by or caught in the trawl, as well as on fish discarded by the fishermen, in many locations around the world (Norris and Prescott, 1961; Leatherwood, 1975; Hogan, 1975; Shane, 1977; Gruber, 1981; Goodwin, 1985; Corkeron *et al.*, Chapter 17, this volume). Symbiotic associations between bottlenose dolphins and fishermen catching mullet in nets along shore were reported in Mauritania and Australia (Busnel, 1973; Baker, 1983). Schlais (1984) described dolphins which habitually stole bait and catch from line fishermen in Hawaii. An individual bottlenose dolphin apparently started this behavior 35 years ago, and it then spread throughout the main island chain.

Bottlenose dolphins in captivity sometimes combine cooperative techniques with ingenuity in capturing prey. One example of such flexibility is the case of two dolphins that stationed themselves on each side of a crevice in which a moray eel, *Gymnothorax funebris*, was hiding. One dolphin swam away, killed a scorpion fish, *Scorpaena guttata*, returned to its station at the crevice, and poked the eel with the fish's spine until the eel left its hiding place and was captured by the other waiting dolphin (Brown and Norris, 1956).

Bottlenose dolphins exhibit flexibility, not only in locomotory and feeding behavior, but also in a number of population characteristics. Ballance (1987, Chapter 13, this volume) identified differences in density, group size, and residence patterns of bottlenose dolphins in Mexico, Argentina, and Florida and attributed these differences to habitat variation. She generalized that one finds small populations of highly resident dolphins in geographically closed habitats and larger populations of both resident and transient animals in open habitats.

My first objective in this chapter is to test the hypothesis that bottlenose dolphin behavior, particularly feeding, and population characteristics vary in different habitats by comparing dolphins in Texas and Florida. The comparative approach has been used effectively and often in studies of mammalian social structure (e. g., for primates: Crook and Gartlan, 1966; Eisenberg *et al.*, 1972; Jolly, 1972; for canids and felids: Kleiman and Eisenberg, 1973; for antelopes: Jarman, 1974) but less frequently in studies of the behavioral repertoires of related species or of different populations of one species (e.g., for sea otters, *Enhydra lutris:* Estes *et al.*, 1981; for coyotes, *Canis latrans:* Bekoff, 1978; Kleiman and Brady, 1978). My second objective is to review and critique methods for studying dolphin behavior. To date, methodology has not been critically addressed by researchers studying bottlenose dolphin behavior, and there has been no attempt to standardize methods. I hope that, by bringing this issue into focus in this chapter, I can encourage the use of uniform methods for studying cetacean behavior.

MATERIALS AND METHODS

I studied bottlenose dolphin behavior and population biology at Port Aransas, Texas, from June 1976 through May 1977 (Shane, 1977, 1980; Shane *et al.*, 1986) and dolphin behavior and ecology at Sanibel Island, Florida, from April 1985 through May 1986 (Shane, 1987, Chapter 12, this volume) (Fig. 1). In both studies I photographically identified naturally marked dolphins and determined minimal home range sizes, seasonal movements, and residence patterns from sightings of these animals. I recorded pod size and estimated population size in each study area. I compared the seasonal and daily activity patterns of dolphins in Texas and Florida and assessed the relationship between various environmental factors and movement patterns and activities.

Figure 1 Two areas in the Gulf of Mexico where bottlenose dolphin behavior and ecology were studied. The Florida study site (top) was centered around Sanibel Island, and the Texas study site (bottom) was in the vicinity of Port Aransas.

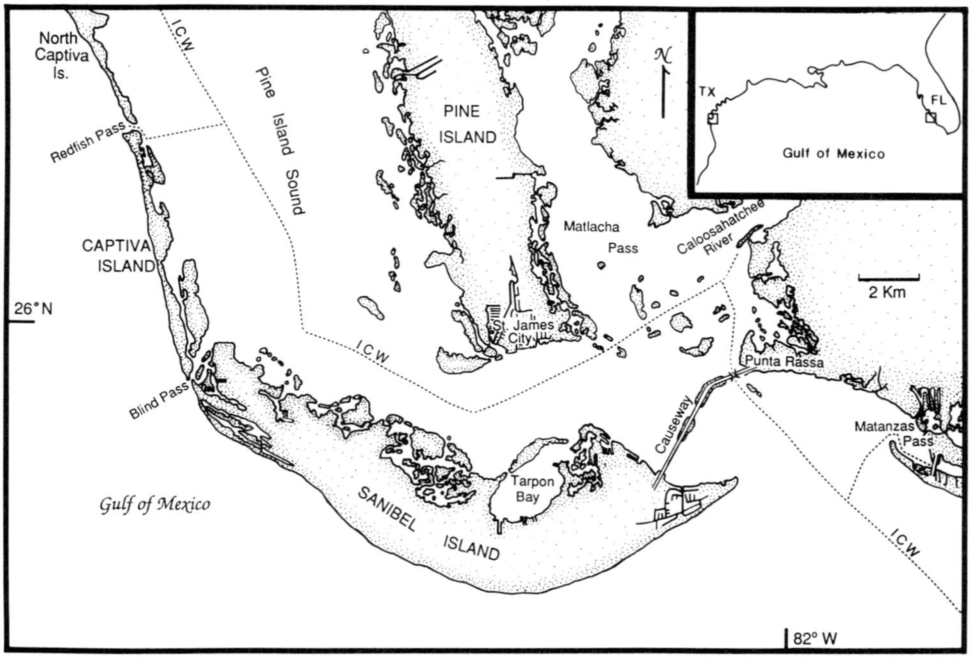
North Captiva Is.
I.C.W.
Pine Island Sound
PINE ISLAND
Redfish Pass
CAPTIVA ISLAND
Matlacha Pass
Caloosahatchee River
2 Km
26° N
St. James City
I.C.W.
Punta Rassa
Blind Pass
Causeway
Matanzas Pass
Tarpon Bay
Gulf of Mexico
SANIBEL ISLAND
I.C.W.
82° W
N
TX
FL
Gulf of Mexico

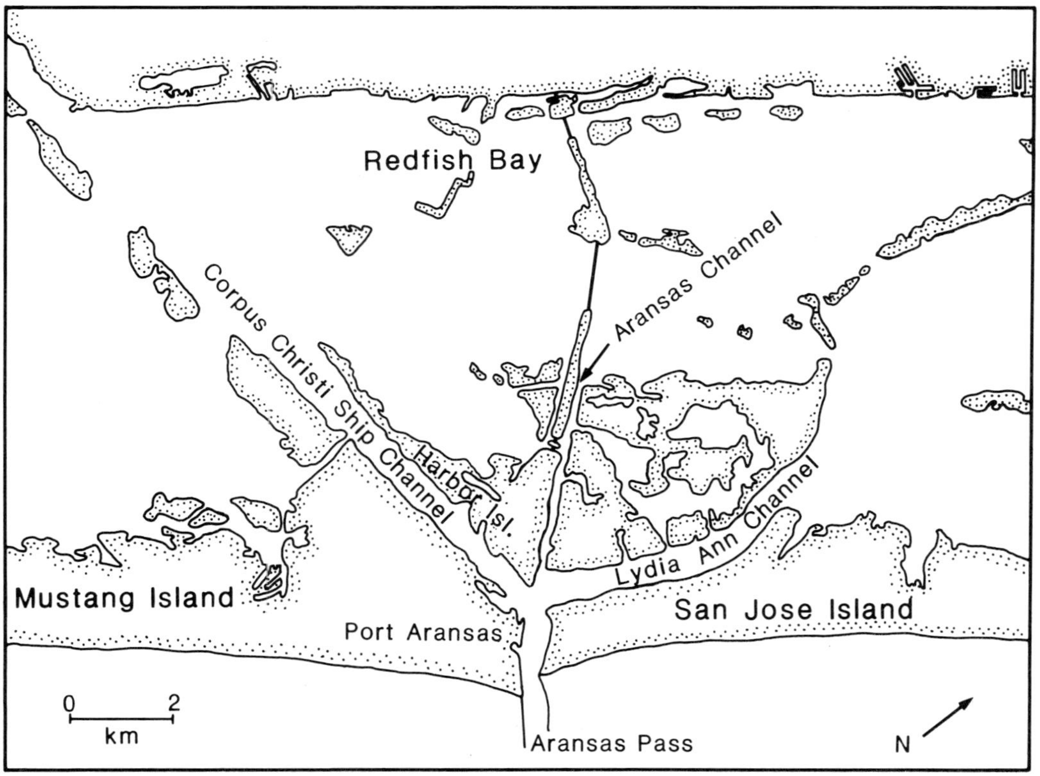
Redfish Bay
Aransas Channel
Corpus Christi Ship Channel
Harbor Isl.
Lydia Ann Channel
Mustang Island
San Jose Island
Port Aransas
0
2
km
Aransas Pass
N

I spent 1064.5 hours in the field in Texas, the majority of which was spent directly observing dolphins. A very conservative estimate of direct observation time would be 585 hours (55% of the total). The Texas data set (N = 2325 records of dolphin activity) consists of observations of dolphins made primarily from a boat and occasionally from land (N = 1954), as well as observations made from fixed points on land from sunrise to sunset on chosen days thoughout the study (N = 371). The Florida data set (N = 6196 records of dolphin activity) covers 881 hours in the field, 450 hours of which were spent observing dolphins. Practically all of the Florida observations were made from a boat.

To make a fair comparison of dolphin activity patterns in Texas and Florida it was necessary to readjust some of the data in each set to create comparable categories of behavior. The major behavior patterns recorded in Texas were traveling, feeding, mating, playing, and resting (see Shane, 1977, and Shane *et al.*, 1986, for definitions). In Florida, activity patterns included traveling, feeding, socializing, travel/feeding, and social traveling (see Shane, Chapter 12, this volume, for definitions). I combined mating and playing in the Texas data set to establish an activity category directly equivalent to socializing in Florida. I believe that I misinterpreted the function of the behavior I called "resting" in Texas (Shane, 1977; Shane *et al.*, 1986); a review of my Texas field notes and Florida observations led me to conclude that the behavior observed was, in fact, against-current feeding (Shane, Chapter 12, this volume). Accordingly, resting was combined with feeding in the Texas data. A total of 666 records of dolphin behavior in Texas could not be placed into activity categories unambiguously equal to the activities recorded in Florida. Therefore, they were discarded from the original data set reported in Shane (1977). Included in the discarded records were 247 instances of bowriding, 230 leaps and slaps (specifically noted by Shane, 1977, to be unrelated to her definition of mating), 104 racing dives, and 85 miscellaneous surface behaviors.

I reorganized the Florida data set by lumping all travel/feeding into feeding, based on the assumptions that dolphins were likely to be searching for food as they traveled and that searching is a legitimate component of feeding behavior when calculating an animal's time budget (MacArthur and Pianka, 1966). For the same reason, I lumped all social traveling with socializing. Since I did not distinguish travel/feeding from feeding or social traveling from socializing in Texas, this restructuring of the Florida data made it more consistent with the Texas data.

Study Areas

Although both are coastal regions in the Gulf of Mexico, the Texas and Florida study sites differed in a number of respects, many of which may be relevant to interpreting dolphin behavior.

Mean water temperatures in Texas were colder during every season than in Florida, and the minimum water temperature in Texas was much colder than that in Florida (Table 2). The water depth and bottom topography in the two areas differed. Generally, the Texas study site was dominated by narrow, deeply dredged channels; the Florida study area, by open, shallow bays which included abundant seagrass flats. Water depth in the Texas study area ranged from less than 1 to 14 m. Water depth in the Florida study area ranged from less than 1 to 8 m, and most dolphin pods (80%) were sighted in water 4 m or less in depth.

Tidal currents within certain parts of each study area were very strong. Maximum diurnal currents in Aransas Pass were 1.9 knots during ebb tide and 2.0 knots during flood tide (Brown *et al.*, 1976). Although current rates were unavailable for the Florida study area, Clark (1976, p. 276) stated that construction of the Sanibel Causeway "narrowed the effective width of the inlet" by 50% and caused an increase in tidal velocities there.

The amount of habitat available to fishes consumed by dolphins differed between the two study areas. Mangroves and seagrasses are key components of productive subtropical estuarine systems in that they provide food and shelter for fish and invertebrates. It is likely that most fishes preyed on by coastal bottlenose dolphins are estuary dependent. For example, approximately 85% of the weight and value of commercial fish and shellfish (including striped mullet, *Mugil cephalus*, and spotted seatrout, *Cynoscion nebulosus*, key bottlenose dolphin prey items) caught in south Flor-

Table 2
Characteristics of the Two Study Areas

Characteristic	Texas[a]	Florida[b]
Study area size (km^2)	34	140
Latitude, longitude	27° 50′ N, 97° 02′ W	26° 27′ N, 82° 05′ W
Annual rainfall (cm)	89	107
Water temperature (°C)		
Minimum	7.2	13.3
Maximum	30.0	30.6
Mean ± S.D.		
Summer	28.3 ± 1.7	28.9 ± 0.5
Fall	13.3 ± 13.9	25.6 ± 1.6
Winter	8.9 ± 7.8	18.9 ± 3.3
Spring	19.4 ± 10.6	23.3 ± 3.4
Prevailing winds	SE (Mar–Sept) NE (Oct–Feb)	SE (spring–fall) NE (winter)
Tides	Diurnal and semidiurnal	Diurnal and semidiurnal
Mean tidal range (m)	0.52	0.55
Maximum storm tides (m)	7.6	4.6

[a] Port Aransas, Texas.
[b] Sanibel Island, Florida.

ida are estuary dependent (Lindall, 1973). Both study sites had approximately 50% coverage by seagrasses, but the Florida area had a higher proportion of mangroves (39%) than did the Texas area (14%) (Table 3). The Florida vegetation data were collected in 1982 and represent a loss of 2576 hectares of seagrasses and an increase by 206 hectares of mangroves since 1945 (Harris *et al.*, 1983). The Texas vegetation data were collected in 1979 and represent an expansion of seagrass flats (not quantified) in the area as compared with a 1958 study; the change in mangrove coverage could not be determined (White *et al.*, 1983).

Human impacts on the habitat in the Texas study area were more visible than in the Florida study area, but human activities around Sanibel [specifically, dredging of the Intracoastal Waterway (ICW) and construction of the Sanibel Causeway] caused a 29% decrease in seagrass beds between 1945 and 1982 (Harris *et al.*, 1983). Much of the Port Aransas area was developed for oil-related activities. Channels were dredged to depths of 14 m and lined with the resulting spoil. The lower portion of Harbor Island was filled to accommodate oil storage tanks and a facility for the construction of offshore oil rigs. Oil tankers, container ships, shrimp vessels, and recreational boats all contributed to the heavy use of Aransas Pass and adjacent dredged channels. Some shrimp fishing occurred within the Texas study area bays, but most shrimping took place in the Gulf of Mexico from less than 1 km to many kilometers offshore. The most obvious human impacts in the Sanibel study area were the Causeway and its spoil islands, the ICW and its spoil islands, seawalls along most developed bay shorelines, and abundant recreational boat traffic. Commercial mullet and crab fishing as well as some gill netting and bait shrimping occurred in the Sanibel area.

Both commercial and sport fisheries in Texas and Florida were significant and may have affected the abundance of several dolphin prey species directly (Table 4). Although commercial fishery landings, subject to the vagaries of the marketplace, are a poor indicator of dolphin prey abundance, they do show the significant role of fisheries in the two study areas. French and Parsons (1983) noted that the catch of some fish (especially spotted seatrout) by sport fishermen in Florida often equals or exceeds the commercial

Table 3
Coverage of Mangroves and Seagrass Beds (as Indicators of Habitat Available to Dolphin Prey Species) in the Two Study Areas

	Texas[a]	Florida[b]
Bay waters accessible to dolphins (hectares)	3400	9500
Bay waters including those probably too shallow for dolphins (hectares)	5000	9500
Mangroves (hectares)	700	3665
Percentage of mangrove acreage within study area	14%	39%
Mangrove-lined shoreline (km)	~13	~48
Seagrasses (hectares)	2500	4888
Percentage of all bay waters in study area covered by seagrasses	50%	50%

[a] Data from White *et al.* (1983).
[b] Data from Harris *et al.* (1983).

catch. In Texas, from September 1975 to August 1976, sport fishermen in Corpus Christi Bay caught 84,000 kg of spotted seatrout (versus an average annual commercial catch of 41,000 kg).

RESULTS

Behavior

Activity Patterns

There was a significant difference between the seasonal activity patterns of dolphins in Texas and Florida, (Fig. 2; $x^2 = 274.53$, $df = 6$, $p < .001$; N=8702). Dolphins spent the most time traveling in Florida in spring and summer and in Texas in winter. Feeding peaked in both areas in fall. Socializing exhibited the most distinctly different pattern in the two areas, peaking in spring in Texas and in fall and winter in Florida.

The difference in activity patterns at the two sites according to time of day also was significant (Fig. 3; $\chi^2 = 609.52$, $df = 6$, $p < .001$; N=8689). In this case, the proportion of time spent feeding and traveling in Texas and Florida was the source of the main difference (also see Table 5). Dolphins spent more time feeding in Florida than in Texas at all times of the day except evening. On the other hand, Texas dolphins spent a greater proportion of time traveling throughout the day.

Table 4
Fisheries Statistics for Some Bottlenose Dolphin Prey Items in the Two Study Areas

Species	Commercial catch (kg/year)
Texas[a]	
Red drum (*Sciaenops ocellatus*)	210,000 44,000
Spotted seatrout (*Cynoscion nebulosus*)	90,000 41,000
Black drum (*Pogonias cromis*)	48,000 57,000
Flounder (all species)	19,000 9,400
Sheepshead (*Archosargus probatocephalus*)	16,000 20,000
Florida[b]	
Striped mullet (*Mugil cephalus*)	2,200,000[c]
Spotted seatrout	180,000[d]

[a] Data are for 1968–1977 (Shew *et al.*, 1981). The first value for Texas landings is from Copano/Aransas bays which include the Port Aransas study area but also cover a much larger area. The second value for Texas landings is from Corpus Christi Bay which is contiguous with the Port Aransas study area.
[b] Data are for 1985 (Florida Dept. of Natural Resources, 1986). Florida landings were taken from Lee County. Lee County includes the entire Sanibel study area as well as other areas outside the study area.
[c] 13,573 trips.
[d] 12,099 trips.

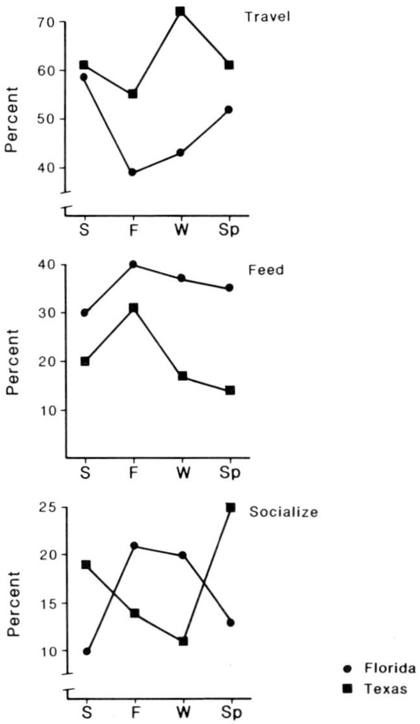

Figure 2 Percent occurrence of three bottlenose dolphin activities in Florida and Texas according to season. S, Summer (June–August); F, fall (September–November); W, winter (December–February); Sp, spring (March–May).

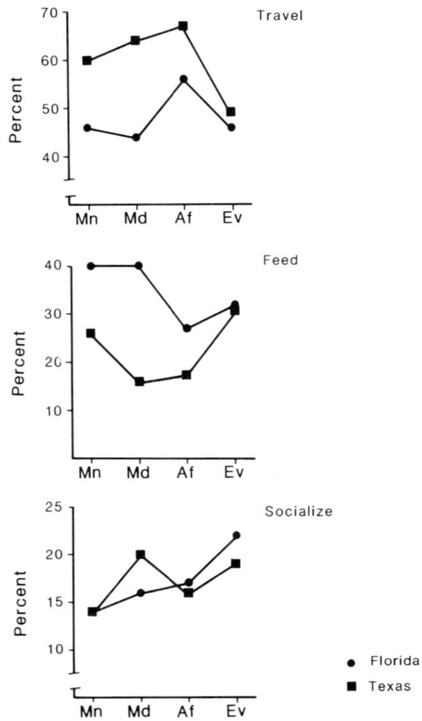

Figure 3 Percent occurrence of three bottlenose dolphin activities in Florida and Texas during four time periods. Mn, Morning (0700–0959); Md, midday (1000–1259); Af, afternoon (1300–1559); Ev, evening (1600–2059).

Types of Feeding

Seven different types of feeding behavior were seen in both Texas (Shane, 1977) and Florida (Shane, Chapter 12, this volume). Definitions of all feeding types are in Shane (Chapter 12, this volume). Two types dominated the dolphins' feeding activity in Texas: subsurface feeding ("random feeding" in Shane, 1977) was most common, followed by against-current feeding ("resting" in Shane, 1977). Horizontal circle feeding ("circle feeding" in Shane, 1977), feeding rushes ("shallow water feeding" in Shane, 1977), cooperative feeding, feeding behind trawling shrimpers, and feeding on fish discarded by shrimpers were all seen relatively infrequently in Texas (Shane, 1977). In Florida dolphins devoted approximately equal amounts of time to each of the seven feeding behaviors: subsurface, against-current, horizontal circle, vertical circle, and edge feeding, feeding rushes, and fish kicking (Shane, Chapter 12, this volume).

Population Characteristics

The Texas and Florida dolphin populations differed in the number of identified animals, the number of sightings per individual, and overall size (Table 5). The estimated size of the population was much larger in Florida. Also, in Florida the mean number of sightings per photographically identified individual was 3.8 (S.D. 3.17), whereas in Texas it was 36.9 (S.D. 44.49). Residence patterns at the two sites were similar; some individuals were sighted year-round, while others were

Table 5
Results of Comparisons of Bottlenose Dolphin Behavior and Population Characteristics in Texas and Florida

Comparison	Texas	Florida
Proportion of time spent at each activity	Travel 63% Feed 21% Socialize 17% $N = 2325$	48% 36% 16% $N = 6378$
$\chi^2 = 199.88$ $df = 2$ $p < .001$		
Movement in relation to tidal flow	Against 76% With 24% $N = 662$	65% 35% $N = 845$
Water depth and feeding	Mostly in 15-m-deep holes	Mostly in 0.3–0.9 m
No. of identified individuals	20	126
No. of sightings per individual	1–156	1–19
Estimated population size	Monthly minimum (Oct) = 48.0 Monthly maximum (Jan) = 164.2	Maximum in area during year = 286 (95% CI = 224–397) (Shane, 1987)
Residence patterns	Year-round and seasonal	Year-round and seasonal
Mean pod size	3.8 (± 2.85) (against-current feed and travel) to 6.9 (± 5.84) (socialize) $N = 1954$	2.4 (± 1.51) (feed and travel/feed) to 7.4 (± 3.96) (socialize) $N = 6378$
	Mean pod size during other activities intermediate between these values	

seen only during one or a few seasons (Shane, 1977, Chapter 12, this volume). Mean pod size at the two sites also was similar (Table 5).

DISCUSSION AND CONCLUSIONS

Methods of Studying Dolphin Behavior: A Critique

In animal behavior studies different research methods can produce widely varying results. For example, the reported amount of time spent feeding by gorillas, *Gorilla gorilla,* in one population changed from 30% of the day to 45% of the day when sampling methods were altered (Harcourt and Stewart, 1984). Ideally, when making any comparison, the same methodology should be used at the compared study sites. It is vital to ask three questions about a behavioral study in order to assess the results produced: (1) How were the data collected? (2) How were the data analyzed? (3) How much time (number of hours and seasonal coverage) was invested in the study?

Descriptive accounts of bottlenose dolphin behavior have proliferated in the literature. They represent a first step in understanding behavior in that an animal's behavior must be described and defined before it can be quantified and before precise questions can be addressed. Shane *et al.* (1986) reviewed bottlenose dolphin behavior and described four major activities (feeding, traveling, socializing, and idling) that have been used by most researchers who have quantified behavior. Despite the use of different terms for behavior, each author usually defines functional categories, so that one can determine whether they are comparable from one study to another. In many cases, behaviors that are of unclear function to the observer but involve animals in a pod moving in different directions are categorized as "milling."

Methods for collecting and analyzing data on the behavior of bottlenose dolphins have developed from mostly opportunistic descriptions (Gunter, 1942; Norris and Prescott, 1961) to quantitative studies (Saayman *et al.*, 1973; Würsig and Würsig, 1979; Goodwin, 1985). In the studies where sampling methods were not explicitly stated, it is likely that the nonsystematic *ad libitum* method disparaged by Altmann (1974) was used. The *ad libitum* method of data collection involves the irregular notation of various aspects of behavior.

I used *ad libitum* sampling in Texas. Such data may be biased, because highly visible behaviors (especially surface behaviors, which are most closely associated with socializing) may be overrepresented in the data set, and less spectacular behaviors like traveling or subsurface feeding may be underrepresented. As Taber and Thomas (1982) pointed out, no studies of cetacean behavior prior to their own had employed any of the six standard behavioral sampling techniques described in Altmann's classic 1974 paper: sociometric matrix completion, focal animal sampling, sampling all occurrences of some behaviors, sequence sampling, one-zero sampling, and instantaneous sampling. In Florida I used Altmann's sampling techniques (Shane, Chapter 12, this volume). For activity data I used both instantaneous sampling every 3 minutes and the duration of each activity during an observation session (a technique discussed by Altmann within "focal animal sampling"). After comparing the results of each sampling method, I concluded that instantaneous sampling best represented the dolphins' activity patterns (for details, see Shane, Chapter 12, this volume).

Knowledge of the completeness of the data set in terms of observation time and seasonal coverage is the final element needed for assessing the results of a behavioral study. Certainly, opportunistic sightings have provided insight into cetacean behavior and represent a critical first step in any behavioral study. Hoese's (1971) report of bottlenose dolphins sliding out of the water onto mud banks to capture fish is one example of such a useful account. However, if one is interested in documenting typical behavior for a chosen population of dolphins, the study should be conducted over at least one annual cycle and under as many different temporal and environmental conditions as possible. The time spent directly observing bottlenose dolphins during quantified, long-term studies has ranged from 186 hours (partial year: Ballance 1987, Chapter 13, this volume) to 260 hours (full year: Würsig and Würsig, 1979). Gruber (1981) spent the longest time in the field (1121 hours, full year) and probably spent the most time directly observing dolphins, but this variable is not quantified separately.

In conclusion, the most useful study of dolphin behavior is one that is conducted nearly full-time for at least 1 year and that employs one or more of Altmann's (1974) standard behavioral sampling methods. Instantaneous sampling seems to be a better method than recording activity durations for determining the activity patterns of bottlenose dolphins. The use of the same terms and definitions as in previous studies allows the unambiguous comparison of results.

Behavior

Activity Patterns

To assess the significance of differences between dolphin activity patterns in Texas and Florida, one must first try to determine the possible function of each activity. First, why do dolphins travel? They probably travel primarily to locate food and conspecifics and, possibly, to avoid predators. Thermoregulatory needs also may induce dolphins to travel. Seasonal or daily changes in these factors could account for changes in the amount of traveling done by dolphins. The increase in time spent traveling in spring and summer in Florida could be due to the fact that many fishes that are restricted to warmer, deeper waters in winter disperse widely as shallow water temperatures increase in spring and summer (Shane, Chapter 12, this volume). I cannot explain the increased amount of traveling in winter in Texas.

Daily food consumption by adult bottlenose dolphins has been calculated to be 4–6% of body weight (Sergeant, 1969). In agreement with this estimate, Spotte and Babus (1980) found that an adult female dolphin, whose weight ranged from 148 to 205 kg (when pregnant) and who was allowed to regulate her own feeding rate, consumed an average of about 8–9 kg of food each day. Ross (1984) reported that three captive bottlenose dolphins consumed 4.7–6.7% of their body weight per day. At this rate a wild bottlenose dolphin weighing about 200 kg would consume about twenty 0.5 kg, 32-cm-long mullet daily (Collins, 1985). However, energy requirements in the wild could be even higher, since wild coyotes have been found to have energy requirements as much as three times those of captives (Bekoff and Wells, 1986).

The significantly higher percentage of time devoted to feeding by Florida dolphins as compared with Texas dolphins (Figs. 2 and 3, Table 5) may be a function of local fish abundance and distribution. Passes such as Aransas Pass act as funnels that concentrate fish and invertebrates migrating between the bays and the Gulf, daily with the tides as well as seasonally (Copeland, 1965). Other studies showed that dolphins themselves were not evenly distributed in inland waters in Texas; dolphins concentrated in passes and were found in unusually low densities within the open bays (Barham *et al.*, 1980; Leatherwood and Reeves, 1983a). Therefore, the dolphins in my Texas study area, centered around Aransas Pass, presumably had access to a concentrated and abundant source of food. In contrast, the Florida study area was large, open, and composed of numerous, diverse patches (e.g., seagrass flats, channels, open bays, mangrove-lined shores, canals bordered by concrete seawalls, oyster bars, the Gulf). Prey were presumably dispersed among these patches in such a way that dolphins had to cover more distance and devote more time to hunting and feeding than in Texas.

A reduction in prey abundance caused by human activities in Florida could account for increased feeding time for dolphins there. Whether the productivity of Florida waters has been more reduced by human activities than that of Texas waters could not be determined. However, the decrease in seagrass beds in Florida as compared with the increase in seagrass beds in Texas at the times of the respective studies suggests that such a difference in productivity may exist. The human pressure on the Florida environment is extreme: there are 1/2 million registered boats in the state, and 36 million tourists visit the state annually (French and Parsons, 1983); approximately 75% of the state's population lives along the coast (Seaman, 1985). Pollution, dredging, and filling (e.g., for the ICW and the Sanibel Causeway) and reduced freshwater runoff into Florida's estuaries may have caused declines in abundance of estuarine fishes (Lindall, 1973; Harris *et al.*, 1983). According to French and Parsons (1983), a decline in spotted seatrout catches in Florida from 1951 to 1976 was possibly due to habitat destruction. Sport and commercial fishing also may have reduced fish abundance. Reduced abundance, especially of large fish, would mean increased search time for dolphins looking for prey.

The peaks in feeding frequency seen in Texas and Florida in the fall may relate to dolphins building up fat stores in preparation for winter. The fall peak in feeding was most pronounced in Texas, where water temperatures dropped significantly lower than in Florida.

In Texas, as in southern Africa (Tayler and Saayman, 1972; Saayman *et al.*, 1973), feeding peaked in the early morning and evening. Crepuscular periods are typically times for increased piscivore activity and success (Helfman, 1986; Estes *et al.*, 1982). Both diurnal and nocturnal fishes are more susceptible to capture at dawn and dusk because of low light levels (Munz and McFarland, 1973). There was no clear-cut diurnal trend in feeding patterns in Florida, perhaps because dolphin behavior there was influenced by a "complicated web of interacting elements," and time alone was not a powerful explanatory factor (Shane, Chapter 12, this volume).

Dolphins socialize for reproductive purposes as well as for protection, efficient food gathering, learning, and possibly other undiscovered reasons (Norris and Dohl, 1980a). Subtle forms of socializing probably occur continuously in dolphin pods, but only the most overt socializing can be observed. Socializing peaked dramatically in the spring in Texas, corresponding with a peak in calf abundance (Shane, 1977) and suggesting a discrete breeding season (with another, less dramatic increase in the summer). Socializing in Florida occurred most frequently in fall and winter, possibly suggesting a protracted breeding season.

Types of Feeding

The different feeding behaviors described for the two study areas seem to fit the respective ecological conditions. Although the number of different behaviors seen at the two study sites was the same, each of the Florida feeding techniques was seen frequently, whereas feeding in Texas was dominated by subsurface feeding, with against-current feeding playing an important role as well (Shane, 1977). Each type of feeding in Florida was specifically adapted to function in a particular part of that variable, patchy environment. For example, edge feeding allowed dolphins to exploit fish which aggregated along the edges of oyster bars, sand bars, channels, and mangrove shores. Alternatively, dolphins trapped small fish against the water's surface using vertical circle feeding and captured larger fish near the surface by kicking them out of the water with their flukes and then retrieving the dead or injured prey (fish kicking). Deep channels and Aransas Pass were the predominant features of the Texas study area. Presumably, subsurface and against-current feeding were the most efficient means of procuring prey in such a relatively uniform habitat.

My current interpretation of flexibility in bottlenose dolphin feeding behavior assumes that each individual dolphin at a given study site performs the entire spectrum of feeding behaviors described (within-phenotype variation). However, it is possible that, instead, individual dolphins specialize in a few particular behaviors (between-phenotype variation). Individual humpback whales, *Megaptera novaeangliae,* showed their own unique feeding styles when bubble-net feeding off Massachusetts (Weinrich, 1982). Individual specialization has been shown for black oystercatchers, *Haematopus ostralegus,* which are either "stabbers" or "hammerers" when feeding on mussels (Sutherland, 1987). Lyons and Estes (1985) demonstrated individual differences in sea otter prey choice and foraging patterns. Another step in the analysis of bottlenose dolphin behavioral flexibility will be to test the hypothesis that flexibility in feeding behaviors is based on differences between individuals.

Bottlenose dolphins probably evolved in variable, complex near-shore environments where behavioral flexibility would have been selected for if dolphins were to exploit the diverse prey items available (Roughgarden, 1972). The variability of near-shore habitats and the patchy distribution of prey within them (e.g., changing with tides, temperature, salinity, depth, vegetational cover) produced selective pressures on the bottlenose dolphin to become a generalist, capable of taking advantage of a variety of prey items (MacArthur and Pianka, 1966; Magurran, 1986). Several families of fishes (Mugilidae, Sciaenidae, Atherinidae, Carangidae, Clupeidae, and Engraulididae, to name a few) share a very similar worldwide distribution with the bottlenose dolphin (as exemplified by striped mullet, Fig. 4) (Nelson, 1984, E. Anderson, California Academy of Sciences, pers. commun.). Dolphins prey on a number of species within these groups, perhaps most notably mullet (*Mugil* spp.), which are taken in the Gulf of Mexico (Gunter, 1942; Kemp, 1949; Leatherwood, 1975; Shane, 1977; Hamilton and Nishimoto, 1977; Irvine *et al.*, 1981; Barros, 1987), off the east coast of the United States (Harris, 1938; Caldwell and Caldwell, 1972; Hogan, 1975; Rigley, 1983; Barros, 1987; Barros and Odell, Chapter 16, this volume), off Australia (Lear and Bryden, 1980), off Mauritania in West Africa (Busnel, 1973), off Portugal (dos Santos and Lacerda, 1987), and in the Black Sea (Bel'kovich *et al.*, 1978). The selective pressures that encouraged flexible food habits in the bottlenose dolphin also worked in favor of the dolphin's becoming "anatomically the most generalized of modern delphinids" (Gaskin, 1982, p. 179). For instance, bottlenose dolphins have a generalist dentition, having 72–100 medium-sized teeth as compared with the approximately 200 small teeth of *Stenella* spp. and the 28–40 large teeth of *Globicephala* spp. (Nishiwaki, 1972).

Although other small odontocetes are fairly flexible in terms of their food habits (Gaskin, 1982), I am not aware of any data demonstrating flexibility equivalent to that of the bottlenose dolphin in feeding behaviors. A cetacean that appears to be a specialist in its food habits and relatively inflexible in its behavior is the pilot whale, *Globicephala* spp. Pilot whales feed almost exclusively on squid (Sergeant, 1962; Seagars and Henderson, 1985). In the Northwest Atlantic long-finned pilot whales, *G. melaena,* follow their primary prey, the short-finned squid, *Illex illecebrosus,* very closely, and they seem reluctant to switch to other prey in that they usually move with the squid rather than altering their diet (Sergeant, 1962). At Catalina Island in southern California I have found the behavior of short-finned pilot whales, *G. macro-*

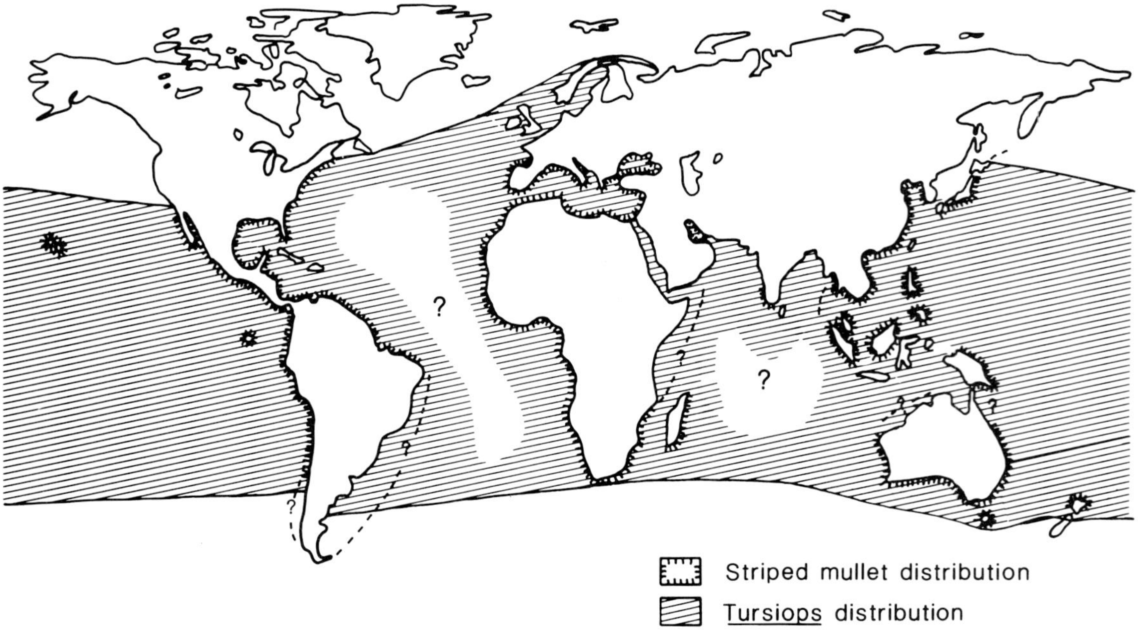

Figure 4 Worldwide distribution of the striped or black mullet (*Mugil cephalus*) (Futch 1966) and bottlenose dolphins (Tomilin, 1967; Baker, 1983; Leatherwood and Reeves, 1983b; Leatherwood, 1986). Bottlenose dolphins are also found in Milford Sound and at Dunedin on the South Island of New Zealand (S. Leatherwood, pers. commun.).

rhynchus, to be relatively inflexible and regimented: the whales perform particular activities in predictable geometric formations in certain depths at prescribed distances from shore (Shane, 1986). For instance, they usually feed in a widely dispersed group 0.4–0.8 km offshore at depths of 27–44 m.

Another cetacean with quite specialized food habits is the Hawaiian spinner dolphin, *Stenella longirostris*, which feeds primarily on lantern fishes (family Myctophidae) (Perrin *et al.*, 1973; Norris *et al.*, 1985). In Hawaii spinners feed almost exclusively on prey less than 20 cm long (Norris *et al.*, 1985). These dolphins exhibit relatively rigid, invariable activity patterns, feeding offshore during the night and then moving into shallow, protected bays to rest for several hours during the day (Norris and Dohl, 1980b; Norris *et al.*, 1985). Spinners make a predictable transition from resting to an active state in the afternoon; aerial behavior increases until the entire school is active and ready to head offshore for another night of feeding.

Population Characteristics

Far more individual dolphins were identified in Florida (126) than in Texas (20). There were four factors that contributed to this difference: (1) The Florida study area was four times as large as the Texas site. (2) I took more photographs of dolphins in Florida (5375) than in Texas (3110). (3) I was more conservative about considering a dolphin "identifiable" in Texas than in Florida. In Texas I included only dolphins that could be recognized visually in the field, whereas in Florida I included such animals as well as those identified only from photographs on 2 or more days. (4) Dolphins in Florida appeared to be more transient in the study area than were dolphins in Texas.

Supportive evidence for a greater degree of transience in Florida dolphins is found in the huge discrepancy between the number of sightings per identified individual in Florida (3.8) and Texas (36.9). Even though I surveyed each study area thoroughly each week, part, but certainly not all, of this difference could be due to a sampling artifact: the large size of the Florida study area decreased my chances of encountering specific individuals very frequently and, thus, is a contributing factor to the disparity in frequency of sightings.

Dolphins in both Texas and Florida had a variety of residence patterns: some individuals were present year-round while others were seen only during one to three seasons (Shane, 1977, 1980, 1987). Ballance (1987, Chapter 13, this volume) found a similar wide range of residence patterns for bottlenose dolphins in the Gulf of California. In contrast, the community of dolphins studied by Irvine *et al.* (1981) and Wells (1986) at Sarasota, Florida, was resident year-round (see Scott *et al.*, Chapter 11, this volume). Such variable residence patterns at two Florida study sites within 110 km of each other provide another example of flexibility in the bottlenose dolphin.

Bottlenose dolphin group size, though nearly identical in Texas and Florida, is another variable that shows flexibility in different habitats. Saayman and Tayler (1973) found group size of bottlenose dolphins in inshore and offshore waters off southern Africa to range from 3 to 1000 with a mean of 140.3 ± 21.4 (S.E.). In the pelagic waters of the eastern tropical Pacific bottlenose dolphin groups ranged in size from 1 to over 10,000, although the median size was 10 (Scott and Chivers, Chapter 22, this volume). The rule appears to be that in inshore waters bottlenose dolphins form consistently small groups, while in offshore waters the range in group size is much greater.

Adaptation to Human Activities

One of the most striking aspects of bottlenose dolphin behavioral flexibility is the species' adaptability to human activities in the environment (e.g., fisheries for food and ships for bowriding, as described in the Introduction). A terrestrial species that shows the same type of adaptability to human incursions into its habitat is the coyote (Kleiman and Brady, 1978). Like the bottlenose dolphin, the coyote is a medium-sized member of its family with catholic tastes (Kleiman and Brady, 1978). Also like bottlenose dolphins, coyotes use varied techniques to gather food; for instance, some coyotes live and hunt in packs while others are solitary (Bekoff and Wells, 1986). Coyotes living under these different social conditions show differences in their time budgets (Bekoff and Wells, 1986), as do dolphins under different conditions. The coyote, like the bottlenose dolphin, inhabits a wide range of habitats, including deserts, forests, and even cities (such as Los Angeles, California). Generalized anatomy, catholic food habits, variable time budgets, intelligence, flexible social structures, and wide distributions all contribute to the behavioral flexibility exhibited by bottlenose dolphins and coyotes. One could test whether these two species are the most flexible members of their taxonomic groups by quantifying the degree of variability in the preceding list of characteristics in other delphinids and canids.

ACKNOWLEDGMENTS

Ken Norris, Bernd Würsig, Jim Estes, Beth Mathews, and several anonymous reviewers contributed perceptive, constructive suggestions for improving this chapter. Cindy Zenker produced the graphics. I am grateful to Jody Gruber, Jim Mead, Greg Cailliet, and Eric Anderson for providing information and reference materials.

REFERENCES

Altmann, J. (1974). Observational study of behavior: Sampling methods. *Behaviour* **49**(3–4), 227–267.

Anonymous. (1987). Something fishy (photo of boater feeding dolphin). *San Jose Mercury News*, 8 April 1987.

Baker, A. N. (1983). "Whales and Dolphins of New Zealand and Australia: An Identification Guide." Victoria Univ. Press, Wellington, New Zealand. 103 pp.

Ballance, L. T. (1987). Ecology and behavior of the bottlenose dolphin, *Tursiops truncatus*, in the Gulf of California, Mexico. M.S. thesis, Moss Landing Marine Labs and San Jose State Univ., Moss Landing, California. 103 pp.

Barham, E. G., Sweeney, J. C., Leatherwood, S., Beggs, R. K., and Barham, C. L. (1980). Aerial census of the bottlenose dolphin, *Tursiops truncatus*, in a region of the Texas coast. *Fish. Bull.* **77**(3), 585–595.

Barros, N. B. (1987). Food habits of bottlenose dolphins (*Tursiops truncatus*) in the southeastern United States, with special reference to Florida waters. M.S. thesis, Univ. of Miami, Miami, Florida. 139 pp.

Bekoff, M. (1978). Behavioral development in coyotes and eastern coyotes. *In* "Coyotes: Biology, Behavior and Management" (M. Bekoff, ed.), pp. 97–126. Academic Press, New York. 384 pp.

Bekoff, M., and Wells, M. C. (1986). Social ecology and behavior of coyotes. *Adv. Study Behav.* **16,** 251–338.

Bel'kovich, V. M., Ivanova, Y., Yefremenkova, O. V., Kozarovitskiy, L. B., and Kharitonov, S. P. (1978). Description of searching–hunting behavior of dolphins. *In* "Behavior and Bioacoustics of Dolphins" (V. M. Bel'kovich, ed.), pp. 24–56. FBIS-JPRS L/8786, 28 November 1979. Alexandria, Virginia.

Brown, D. H., and Norris, K. S. (1956). Observations of captive and wild cetaceans. *J. Mammal.* **37**(3), 311–326.

Brown, L. F., Brewton, J. L., McGowen, J. H., Evans, T. J., Fisher, W. L., and Groat, C. G. (1976). Environmental geologic atlas of the Texas coastal zone—Corpus Christi area. Bureau of Economic Geology, Univ. of Texas, Austin, Texas 78712. 123 pp.

Busnel, R. G. (1973). Symbiotic relationship between man and dolphins. *Trans. N.Y. Acad. Sci.* **35,** 112–131.

Caldwell, D. K., and Caldwell, M. C. (1972). "The World of the Bottlenose Dolphin." Lippincott, Philadelphia, Pennsylvania. 157 pp.

Cato, J. C., and Prochaska, F. J. (1976). Porpoise attacking hooked fish irk and injure Florida fishermen. *Natl. Fisherman* **53**(9).

Clark, J. (1976). "The Sanibel Report: Formulation of a Comprehensive Plan Based on Natural Systems." The Conservation Foundation, Washington, D.C. 305 pp.

Collins, M. R. (1985). Species profiles: Life histories and environmental requirements of coastal fishes and invertebrates (South Florida): Striped mullet. Biological Report 82 (11.34). U.S. Fish and Wildlife Service, 1010 Gause Blvd., Slidell, Louisiana 70458. 11 pp.

Connor, R. C., and Smolker, R. S. (1985). Habituated dolphins (*Tursiops* sp.) in western Australia. *J. Mammal.* **66**(2), 398–400.

Copeland, B. J. (1965). Fauna of the Aransas Pass Inlet, Texas. I. Emigration as shown by tide trap collections. *Publ. Inst. Mar. Sci. Univ. Texas* **10,** 9–21.

Crook, J. H., and Gartlan, J. S. (1966). Evolution of primate societies. *Nature (London)* **210,** 1200–1203.

Dean, N. (1979). Friends at sea (photo of crab fisherman feeding dolphin). *Sea Secrets* **23**(4), 13.

dos Santos, M. E., and Lacerda, M. (1987). Preliminary observations of the bottlenose dolphin (*Tursiops truncatus*) in the Sado estuary (Portugal). *Aquatic Mamm.* **13**(2), 65–80.

Eisenberg, J. F., Muckenhirn, N. A., and Rudran, R. (1972). The relation between ecology and social structure in primates. *Science* **176,** 863–874.

Estes, J. A., Jameson, R. J., and Johnson, A. M. (1981). Food selection and some foraging tactics of sea otters. *In* "Worldwide Furbearer Conference Proceedings" (J. A. Chapman and D. Pursley, eds.), Vol. 1, pp. 606–641. August 3–11, 1980, Frostburg, Maryland. 652 pp.

Estes, J. A., Jameson, R. J., and Rhode, E. B. (1982). Activity and prey selection in the sea otter: Influence of population status on community structure. *Am. Nat.* **120**(2), 242–258.

Florida Dept. of Natural Resources. (1986). Marine Fisheries Info. System, 1986 Annual Landings Summary. Florida DNR, St. Petersburg, Florida.

French, C. O., and Parsons, J. W., eds. (1983). Coastal ecological characterization: A socioeconomic study of the southwest region, Vol. I. U.S. Fish and Wildlife Service, Div. of Biol. Serv., Washington, D.C. FWS/OBS-83/14.

Futch, C. R. (1966). Lisa, the Florida black mullet. Salt Water Fisheries, Leaflet 6.

Gaskin, D. E. (1982). "The Ecology of Whales and Dolphins." Heinemann, London and Exeter. 459 pp.

Goodwin, D. E. (1985). Diurnal behavior patterns of *Tursiops truncatus* off Mobile Point, Alabama. M.S. thesis, San Francisco State Univ., San Francisco, California. 60 pp.

Gruber, J. A. (1981). Ecology of the Atlantic bottlenosed dolphin (*Tursiops truncatus*) in the Pass Cavallo area of Matagorda Bay, Texas. M.S. thesis, Texas A & M University, College Station, Texas. 182 pp.

Gunter, G. (1942). Contributions to the natural history of the bottlenose dolphin, *Tursiops truncatus* (Montague), on the Texas coast, with particular reference to food habits. *J. Mammal.* **23**(3), 267–276.

Gunter, C. (1954). Mammals of the Gulf of Mexico. *In* "Gulf of Mexico: Its Origin, Waters, and Marine Life" (P. S. Galtsoff, ed.), pp. 543–551. *Fish. Bull.* **89,** 604 pp.

Hamilton, P. V., and Nishimoto, R. T. (1977). Dolphin predation on mullet. *Fla. Sci.* **40**(3), 251–252.

Harcourt, A. H., and Stewart, K. J. (1984). Gorillas' time feeding: Aspects of methodology, body size, competition and diet. *Afr. J. Ecol.* **22,** 207–215.

Harris, J. C. (1938). Porpoises feeding. *Life* **5**(12), 67.

Harris, B. A., Haddad, K. D., Steidinger, K. A., and Huff, J. A. (1983). Assessment of fisheries habitat: Charlotte Harbor and Lake Worth, Florida. Final report, Fla. Dept. of Nat. Res., Bureau of Marine Research, Marine Research Labs, St. Petersburg, Florida. 211 pp.

Helfman, G. S. (1986). Fish behavior by day, night and twilight. *In* "The Behaviour of Teleost Fishes" (T. J. Pitcher, ed.), pp. 367–387. Croom Helm, London.

Hoese, H. D. (1971). Dolphin feeding out of water in a salt marsh. *J. Mammal.* **52,** 222–223.

Hogan, T. (1975). Untitled draft of M.S. thesis, Univ. of Rhode Island, Narragansett, Rhode Island. 42 pp.

Hult, R. W. (1982). Another function of echolocation for bottlenosed dolphins (*Tursiops truncatus*). *Cetology* **47,** 1–7.

Hussenot, E. (1980). Le grand dauphin *Tursiops truncatus* en Bretagne: Types de fréquentation. *Penn ar Bed* **12,** 355–380.

Irvine, A. B., Scott, M. D., Wells, R. S., and Kaufmann, J. H. (1981). Movements and activities of the Atlantic bottlenose dolphin, *Tursiops truncatus,* near Sarasota, Florida. *Fish Bull.* **79**(4), 671–688.

Itani, J. (1958). On the acquisition and propagation of a new habit in the natural group of the Japanese monkeys at Takasakiyama. *Primate* **2,** 84–98.

Jarman, P. J. (1974). The social organisation of antelope in relation to their ecology. *Behaviour* **58**(3–4), 215–267.

Jolly, A. (1972). "The Evolution of Primate Behaviour." Macmillan, New York. 397 pp.

Kemp, R. J. (1949). Report on stomach analysis. *Annu. Rept. Mar. Lab. Texas Game, Fish Oyster Comm.,* Fiscal Year 1948–1949.

Kleiman, D. G., and Brady, C. A. (1978). Coyote behavior in the context of recent canid research: Problems and perspectives. *In* "Coyotes: Biology, Behavior and Management" (M. Bekoff, ed.), pp. 163–188. Academic Press, New York. 384 pp.

Kleiman, D. G., and Eisenberg, J. F. (1973). Comparisons of canid and felid social systems from an evolutionary perspective. *An. Behav.* **21,** 637–659.

Lear, R. J., and Bryden, M. M. (1980). A study of the bottlenose dolphin *Tursiops truncatus* in eastern Australian waters. *Aust. Natl. Parks Wildl. Serv. Occas. Pap.* No. 4, 1–25.

Leatherwood, S. (1975). Some observations of feeding behavior of bottle-nosed dolphins (*Tursiops truncatus*) in the Northern Gulf of Mexico and (*Tursiops* cf. *T. gilli*) off southern California, Baja California, and Nayarit, Mexico. *Mar. Fish. Rev.* **37**(9), 10–16.

Leatherwood, S. (1979). Aerial survey of the bottlenosed dolphin, *Tursiops truncatus,* and the West Indian manatee, *Trichechus manatus,* in the Indian and Banana Rivers, Florida. *Fish Bull.* **77**(1), 47–59.

Leatherwood, S. (1986). Whales, dolphins, and porpoises of the Indian Ocean Sanctuary; a catalogue of available information. *Hubbs Mar. Res. Center Tech. Rep.,* No. 87-197.

Leatherwood, S., and Reeves, R. (1982). Bottlenose dolphin, *Tursiops truncatus,* and other toothed cetaceans. *In* "Wild Mammals of North America" (J. A. Chapman and G. A. Feldhamer, eds.), pp. 369–414. Johns Hopkins Press, Baltimore, Maryland, 1147 pp.

Leatherwood, S., and Reeves, R. R. (1983a). Abundance of bottlenose dolphins in Corpus Christi Bay and coastal southern Texas. *Contrib. Mar. Sci.* **26,** 179–199.

Leatherwood, S., and Reeves, R. R. (1983b). "The Sierra Club Handbook of Whales and Dolphins." Sierra Club Books, San Francisco, California. 302 pp.

Lindall, W. N. (1973). Alterations of estuaries of South Florida: A threat to its fish resources. *Mar. Fish. Rev.* **35**(10), 26–33.

Lyons, K. J., and Estes, J. A. (1985). Individual variation in diet and foraging strategy in the female California sea otter, *Enhydra lutris. Sixth Biennial Conf. Biol. Mar. Mamm.,* Vancouver, B.C., Abstr.

MacArthur, R. H., and Pianka, E. R. (1966). On optimal use of a patchy environment. *Am. Nat.* **100,** 603–609.

Magurran, A. E. (1986). Individual differences in fish behaviour. *In* "The Behaviour of Teleost Fishes" (T. J. Pitcher, ed.), pp. 338–365. Croom Helm, London. 553 pp.

Munz, F. W., and McFarland, W. N. (1973). The significance of spectral position in the rhodopsins of tropical marine fishes. *Vision Res.* **13**(10), 1829–1874.

Nelson, J. S. (1984). "Fishes of the World," 2nd Ed. Wiley, New York.

Nishiwaki, M. (1972). General biology. *In* "Mammals of the Sea, Biology and Medicine" (S. H. Ridgway, ed.), pp. 3–204. Charles C. Thomas, Publ., Springfield, Illinois. 812 pp.

Norris, K. S., and Dohl, T. P. (1980a). The structure and function of cetacean schools. *In* "Cetacean Behavior: Mechanisms and Functions" (L. M. Herman, ed.), pp. 211–261. Wiley, New York, 463 pp.

Norris, K. S., and Dohl, T. P. (1980b). Behavior of the Hawaiian spinner dolphin, *Stenella longirostris. Fish. Bull. (U.S.)* **77,** 821–849.

Norris, K. S., and Prescott, J. H. (1961). Observations on Pacific cetaceans of Californian and Mexican waters. *Univ. Calif. Publ. Zool.* **63**(4), 291–402.

Norris, K. S., Würsig, B., Wells, R. S., Würsig, M., Brownlee, S. M., Johnson, C., and Solow, J. (1985). The behavior of the Hawaiian spinner dolphin, *Stenella longirostris. Natl. Mar. Fish. Serv., S.W. Fish. Center Admin. Rep.,* No. LJ-85-06C.

Perrin, W. F., Warner, R. R., Fiscus, C. H., and Holts, D. B. (1973). Stomach contents of porpoise, *Stenella* spp., and yellowfin tuna, *Thunnus albacares,* in mixed-species aggregations. *Fish. Bull.* **71**(4), 1077–1092.

Poirier, F. E. (1969). Behavioral flexibility and intertroop variation among Nilgiri langurs (*Presbytis johnii*) of south India. *Folia Primat.* **11,** 119–133.

Rigley, L. (1983). Dolphins feeding in a South Carolina salt marsh. *Whalewatcher* **17**(2), 3–5.

Roughgarden, J. (1972). Evolution of niche width. *Am. Nat.* **106,** 683–718.

Ross, G. J. B. (1977). The taxonomy of bottlenosed dolphins *Tursiops* species in South African waters, with notes on their biology. *Ann. Cape Prov. Mus. (Nat. Hist.)* **11,** 135–194.

Ross, G. J. B. (1984). The smaller cetaceans of the south east coast of southern Africa. *Ann. Cape Prov. Mus. (Nat. Hist.)* **15**(2), 173–410.

Saayman, G. S., and Tayler, C. K. (1973). Social organisation of inshore dolphins (*Tursiops aduncus* and *Sousa*) in the Indian Ocean. *J. Mammal.* **54,** 993–996.

Saayman, G. S., Tayler, C. K., and Bower, D. (1973). Diurnal activity cycles in captive and free-ranging Indian Ocean bottlenose dolphins (*Tursiops aduncus* Ehrenburg). *Behaviour* **44,** 212–233.

Schlais, J. F. (1984). Thieving dolphins: A growing problem in Hawaii's fisheries. *Sea Front.* **30**(5), 293–298.

Seagars, D. J., and Henderson, J. R. (1985). Cephalopod remains from the stomach of a short-finned pilot whale collected near Santa Catalina Island, California. *J. Mammal.* **66**(4), 777–779.

Seaman, W., ed. (1985). Florida aquatic habitat and fishery resources. Florida American Fisheries Soc., 207 W. Carroll St., Kissimmee, Florida 32741.

Sergeant, D. E. (1962). The biology of the pilot or pothead whale *Globicephala melaena* (Traill) in Newfoundland waters. *Bull. Fish. Res. Board Can.* **132,** 1–84.

Sergeant, D. E. (1969). Feeding rates of cetacea. *Fisk. Dir. Skr. Ser. Havunders., Norway* **15**(3), 246–258.

Shane, S. H. (1977). The population biology of the Atlantic bottlenose dolphin, *Tursiops truncatus,* in the Aransas Pass area of Texas. M.S. thesis, Texas A & M Univ., College Station, Texas. 239 pp.

Shane, S. H. (1980). Occurrence, movements, and distribution of bottlenose dolphin, *Tursiops truncatus,* in southern Texas. *Fish. Bull.* **78**(3), 593–601.

Shane, S. H. (1986). Detailed observations of a single pod of pilot whales and sightings of other marine mammals at Catalina Island, California in winter 1984–5. Contract Report to U.S. Marine Mammal Commission and NMFS–Southwest Fisheries Center. 49 pp.

Shane, S. H. (1987). The behavioral ecology of the bottlenose dolphin. Ph.D. dissertation, Univ. of California, Santa Cruz, California.

Shane, S. H., Wells, R. S., and Würsig, B. (1986). Ecology, behavior and social organization of the bottlenose dolphin: A review. *Mar. Mamm. Sci.* **2**(1), 34–63.

Shew, D. M., Baumann, R. H., Fritts, T. H., and Dunn, L. S. (1981). Texas barrier islands region ecological characterization: Environmental synthesis papers. U.S. Fish and Wildlife Service, Biological Serv. Program, Washington, D.C. FWS/OBS-81/32. 413 pp.

Spotte, S., and Babus, B. (1980). Does a pregnant dolphin (*Tursiops truncatus*) eat more? *Cetology* **39,** 1–7.

Sutherland, W. J. (1987). Why do animals specialize? *Nature (London)* **325,** 483–484.

Taber, S., and Thomas, P. (1982). Calf development and mother–calf spatial relationships in southern right whales. *An. Behav.* **30,** 1072–1083.

Tayler, C. K., and Saayman, G. S. (1972). The social organisation and behaviour of dolphins (*Tursiops aduncus*) and baboons (*Papio ursinus*): Some comparisons and assessments. *Ann. Cape Prov. Mus. (Nat. Hist.)* **9**(2), 11–49.

Tomilin, A. G. (1967). "Mammals of the USSR and Adjacent Countries," Vol. IX: Cetacea. [Transl. from Russian.] Israel Program for Scientific Translation, Jerusalem. 717 pp.

Weinrich, M. (1982). The humpback whales of Stellwagen Bank. *Whalewatcher* **16**(4), 12–15.

Wells, R. S. (1986). The population structure of bottlenose dolphins, Chap. 1, pp. 1–95. Ph.D. dissertation, Univ. of California, Santa Cruz, California.

Wells, R. S., Scott, M. D., and Irvine, A. B. (1987). The social structure of free-ranging bottlenose dolphins. *In* "Current Mammalogy" (H. H. Genoways, ed.), Vol. 1, pp. 247–305. Plenum, New York and London.

White, W. A., Calnan, T. R., Morton, R. A., Kimble, R. S., Littleton, T. G., McGowen, J. H., Nance, H. S., and Schmedes, K. E. (1983). Submerged lands of Texas, Corpus Christi area: sediments, geochemistry, benthic macroinvertebrates, and associated wetlands. Bureau of Economic Geology, Univ. of Texas, Austin, Texas. 154 pp.

Würsig, B. (1986). Delphinid foraging strategies. *In* "Dolphin Cognition and Behavior: A Comparative Approach" (R. J. Schusterman, J. A. Thomas, and F. G. Wood, eds.), pp. 347–359. Lawrence Erlbaum Assoc., Hillsdale, New Jersey. 393 pp.

Würsig, B., and Würsig, M. (1979). Behavior and ecology of the bottlenose dolphin, *Tursiops truncatus,* in the south Atlantic. *Fish Bull.* **77**(2), 399–412.

Ergometry (Exercise Testing) of the Bottlenose Dolphin

32

Harold W. Goforth, Jr.
Naval Health Research Center
San Diego, California 92138

INTRODUCTION

The metabolic characteristics of marine mammals are generally thought to differ significantly from those of terrestrial mammals, considering the obvious differences between the environments inhabited by the two groups. For example, some marine mammals (e.g., the Weddell seal, *Leptonychotes weddelli*, northern elephant seal, *Mirounga angustirostris*, and sperm whale, *Physeter catodon*) can dive to depths of 600–2,000 m and remain submerged for 35–90 minutes (Norris and Harvey, 1972; Kooyman *et al.*, 1980; LeBoeuf *et al.*, 1986). This capability could derive from highly effective mechanisms for storing oxygen in blood and muscles, preferential circulation, an improved overall metabolic efficiency, and a greater potential to derive adenosine triphosphate (ATP) from metabolic pathways operating independently of oxygen (i.e., anaerobic glycolysis); it most likely derives from a combination of these adaptations (Hochachka and Storey, 1975; Zapol *et al.*, 1979; Kooyman *et al.*, 1981; Castellini, 1986).

Attempts to characterize the metabolism of marine mammals have taken several approaches, including measuring oxygen uptake during swimming (Karandeeva *et al.*, 1973; Ridgway, 1973; Worthy, 1985), enzyme activities in muscle samples (DuBois *et al.*, 1948; Simon *et al.*, 1974; Storey and Hochachka, 1974; Shoubridge *et al.*, 1976; Ponganis and Pierce, 1978; Castellini *et al.*, 1981; Suzuki *et al.*, 1983), and metabolite concentrations in blood and muscle after such stresses as swimming, diving, and apnea (Irving *et al.*, 1941; Hochachka *et al.*, 1975; Kooyman *et al.*, 1980; Davis, 1983; Castellini *et al.*, 1985). To date, inferences about metabolic values in marine mammals have often been questioned because thay have been based on studies of tissues from dead animals, live animals under stress, or animals under other than natural or free-swimming conditions.

The adult bottlenose dolphin, with its large body mass, great muscular power, and aquatic existence, presents unique problems for the study of energy metabolism during exercise. To study the effect of work intensity on the energy metabolism of any animal, one must be able to control such critical aspects of the exercise task as the type, rate, and relative intensity of the muscular contractions, exercise duration, exercise history, health, and diet. The main reason is that these factors strongly influence fiber recruitment patterns and the animal's hormonal status, which in turn affect energy metabolism. Unfortunately, in most metabolic studies of marine mammals these factors have not been controlled to an acceptable degree. The extent to which they are controlled significantly affects the reliability of the results and their value for comparison with those of other studies. Furthermore, when subjects have been conditioned (behaviorally and physically) to perform exercise tests, they exhibit reduced stress, and the consistency of results is greater (Gleser and Vogel, 1973; Stromme *et al.*, 1977; Freund *et al.*, 1986). This chapter describes the development of a technique for exercise testing of captive dolphins that permits control of many important exercise parameters.

BACKGROUND

Ergometry

There is a considerable body of literature on metabolic characteristics of large terrestrial mammals (wild and domestic) derived from controlled exercise testing (ergometry). The treadmill, the most common exercise device, has been used to determine maximum oxygen uptake (aerobic capacity) (Seeherman *et al.*, 1981; Persson, 1983) and rates of oxygen consumption, glycogen depletion (glycogenolysis), and lactate accumulation (glycolysis) at different exercise intensities (Table 1). Marine mammal ergometry has been limited to kinematic and metabolic studies of pinnipeds swimming in "watermills" or flumes (Elsner and Ashwell-Erickson, 1982; Castellini *et al.*, 1985; Davis *et al.*, 1985; Feldkamp, 1985; Williams and Kooyman, 1985). There have been no quantitative exercise

Table 1
Ergometric Studies of Terrestrial Mammals

Species	Weight (kg)	Max VO_2 (ml/kg/minute)	Peak Lactate: Blood (m*M*)	Peak Lactate: Muscle (μmol/g d.w.)	Reference
Zebu cattle (*Bos indicus*)	160–254	27–28	20–21	—	Taylor *et al.* (1980)
Wildebeest (*Connochaetes taurinus*)	114	46	18	—	Taylor *et al.* (1980)
Horse (*Equus caballus*)					
Standardbred	105	97	23	54	Seeherman *et al.* (1981)
Thoroughbred	475–525	Max	10–35	96–204	Snow *et al.* (1985)
Thoroughbred	500	130–140	35–40	—	Bayly *et al.* (1987)
Red kangaroo (*Megaleia rufa*)	18–38	67		—	Dawson and Taylor (1973)
Tammar wallaby (*Macropies eugenii*)	5	70	4	—	Hochachka *et al.* (1985)
Man (*Homo sapiens*)					
Average	70–80	38–57	10–20	90–150	Astrand (1960), Sahlin *et al.* (1976), Pollock (1977)
Elite athlete	57–72	71–84	10–20	90–150	

tests and relatively few measurements of oxygen uptake of dolphins, either at "rest" or during exercise (Irving *et al.*, 1941; Kleinenberg, 1956; Pierce, 1970; Hampton *et al.*, 1971; Kolchinskaya *et al.*, 1971; Ridgway and Patton, 1971; Karandeeva *et al.*, 1973; Ridgway, 1973; Hampton and Whittow, 1976; Worthy, 1985; Feldkamp *et al.*, 1987; Table 2).

Metabolic Effects of Stress from Exercise, Diving, and Apnea

Dolphins

Inferences about dolphin muscle metabolism based on enzyme and metabolite values have often been suspect because they have been based on studies of tissues from dead or dying animals (DuBois *et al.*, 1948; Simon *et al.*, 1974; Storey and Hochachka, 1974; Tulsi, 1975; Ponganis and Pierce, 1978; Castellini *et al.*, 1981; Castellini and Somero, 1981; Suzuki *et al.*, 1983; Bello *et al.*, 1985). Only two studies have reported lactate levels in the blood and/or muscles of healthy dolphins following periods of apnea (Irving *et al.*, 1941; Hochachka *et al.*, 1975). Both studies indicated a low to moderate level of glycolytic activity (i.e., low lactate levels).

Pinnipeds

Exercise studies with seals conditioned (or forced) to swim in flumes or dive in chambers have indicated only small (2–4 m*M*) increases in blood lactate (Castellini *et al.*, 1985; Davis *et al.*, 1985). Unfortunately, swimming intensities achieved in these studies may have been too low to elicit more than moderate glycolytic activity. Free-diving Weddell seals exhibit peak postdive (0–5 minutes) blood lactate values of 1.1 m*M* after 90% of their dives; baseline levels are 0.5 m*M* (Kooyman *et al.*, 1980). The blood lactate data from other studies (Murphy *et al.*, 1980; Guppy *et al.*, 1986) also suggest a modest involvement of anaerobic glycolysis in seals under routine swimming and diving conditions. Muscle metabolites were not measured.

Limitations of Previous Exercise Protocols

Measurements of oxygen consumption in dolphins (and pinnipeds) during "rest" or exercise (Table 2) have often been made under loosely defined or other than natural conditions (i.e., forced dives, forced swimming, forced apnea, and free diving), which preclude valid comparisons

Table 2
Oxygen Consumption Studies of Delphinids

Species	Weight (kg)	Test Condition	VO_2 (ml/kg/minute)	Reference
Atlantic bottlenose dolphin (*Tursiops truncatus*)	170	Resting	5.9	Irving *et al.* (1941)
	170	Estimated maximum	21.1	Irving *et al.* (1941)
	156	Resting	6.2	Hampton *et al.* (1971)
	156	Swimming, 1 m/second	16.7	Hampton *et al.* (1971)
	128	Resting	7.0	Ridgway and Patton (1971)
	138	Resting	7.6	Ridgway (1973)
	138	"Leisure" swimming	10.4	Ridgway (1973
	140	Resting	4.0–4.5	Feldkamp *et al.* (1987)
Black Sea bottlenose dolphin (*Tursiops* sp.)	?	Resting	3.0	Kleinenberg (1956)
	120–220	Resting	4.1–4.4	Kolchinskaya *et al.* (1971)
Harbor porpoise (*Phocoena phocoena*)	29/42	Swimming, 2 m/second	12.0/15.5	Worthy (1985)
Hawaiian spinner dolphin (*Stenella longirostris*)	68	Resting	3.0–4.4	Hampton and Whittow (1976)

with oxygen consumptions in man or other terrestrial mammals in which measurements were made under more natural and controlled conditions (Lavigne *et al.*, 1986). Unfortunately, conditions in each set of experiments vary and affect muscle energy metabolism somewhat differently. For example, oxygen availability and an animal's autonomic and hormonal responses differ significantly between forced and voluntary apnea or swimming. Even if conditions under which swimming studies were conducted had been more natural, measurements of oxygen uptake by the whole body would have provided only limited insight into the activity of specific metabolic pathways (e.g., glycolysis and glycogenolysis) operating at the muscle level during the exercise.

Purpose of the Study

The methods presented below were developed to permit researchers to quantify the level of muscular work performed by a captive dolphin (with unrestricted access to atmospheric oxygen), given positive reinforcement (fish) and real-time performance feedback while exercising at a preestablished intensity. They were developed to support experiments to determine effects of different exercise intensities on lactate accumulation and glycogenolytic activity in the locomotor muscles. This also permitted measurements (reported in this chapter) of tailbeat forces during burst and sustained swimming. Other results, on metabolic properties of locomotor muscles at rest and during various exercise intensities, are reported elsewhere (Goforth, 1986; Goforth *et al.*, 1987). These studies are part of ongoing research at the Naval Ocean Systems Center (NOSC) and California universities to develop an understanding of the musculotendinous structure (Bello *et al.*, 1985), kinematics, and energetics (Feldkamp *et al.*, 1987) of the bottlenose dolphin.

MATERIALS AND METHODS

Specimen

The bottlenose dolphin used for exercise testing was a 3-m-long, 309-kg, 21-year-old adult male captured off the Pacific coast of Baja California. He was maintained at NOSC and fed a standard mixed ration of freshly thawed fish (3.5% body weight/day). The average daily ration of 10.9 kg consisted of 1.8 kg smelt *(Thaleichtys vetulus)*, 2.3 kg herring *(Clupea harengus)*, and 6.8 kg Pacific mackerel *(Scomber japonicus)* and contained 13, 380 kcal or 43.3 kcal/kg body weight. The dolphin was housed in a floating pen moored alongside a pier in San Diego Bay. The pen measured 7.6 × 7.6 × 3 m deep and was made of 10-cm square mesh netting. A 1-m wide wooden walkway surrounded the pen.

Equipment

A portable microcomputer (Compaq) equipped with 640K RAM, DOS 2.02 operating system, and an analog-to-digital (A/D) conversion board (Tecmar Inc., PC-Mate Lab Master) was used in training (see Fig. 1). Software was developed using compiler language (TURBO PASCAL) to collect analog data on force and determine if the physical work performed during a trial satisfied a predetermined set of criteria (e.g., tailbeat force, frequency, and work duration).

The force produced by the dolphin was measured as he pushed his rostrum against a well-padded load cell with a capacity of 15,000 N (3,000 lb) (Lebow, Model 3169, force transducer) mounted on the corner of his pen (see Fig. 4C, below). Forces were measured in compression along a single axis. The load cell was used in conjunction with a Daytronic, Model 3170, signal conditioner with the following specifications: 150% overload capacity, 0.02% nonlinearity, 0.03% hysteresis, ringing frequency of 2.44 kHz, and a usable temperature range of −65 to +200° F. The load cell was calibrated in compression with an INSTRON, Model 1125, tensile/compression unit on three occasions. A 10% drift in the calibration developed after several months of exposure to the marine environment, necessitating an update of calibration values.

A differential amplifier (Princeton Applied Research, Model 113) transmitted the signal from the load cell to an oscilloscope (Tektronix, Model SC 503) and a microcomputer. The microcomputer collected real-time force data at a rate of 144 Hz, evaluated the forces, and sent commands to a Tektronix, Model FG 501, voltage-control generator (VCG). In turn, the VCG sent signals to an LC-10 underwater transducer that projected one of four tones (1, 3, 5, or 10 kHz) to the dolphin. The trainer monitored these tones throughout the trials with the aid of a hydrophone receiver and headset. This integrated system (Fig. 1) was capable of receiving, analyzing, and storing force data while simultaneously controlling the tones projected to the exercising dolphin. In effect, the microcomputer system made the critical quantitative, real-time decisions regarding the work performance and controlled the reinforcement stimuli (acoustic signals) necessary to condition the dolphin's behavior (Fig. 2).

During training trials, force data were stored temporarily in the 640 K RAM microcomputer memory. At the conclusion of a trial, the data were analyzed by the microcomputer and transferred to a $5\frac{1}{2}$-inch flexible disk. The relative muscular effort involved in maintaining a mean force/tailbeat while exercising against the load cell was quantified and characterized using the following parameters (Fig. 3):

Primary Parameters
- Relative work load (% of maximum mean force/stroke)
- Absolute work load (mean force/stroke) (N/stroke)
- Percentage of maximum voluntary effort (% MVE)
- Work duration (total seconds rhythmic force production)
- Total number of tailbeat cycles (*n*)
- Percentage of irregular (double-peaked) tailbeats
- Tailbeat frequency (*n*/second or Hz)

Secondary Parameters
- Mean duration of force application/stroke (seconds)
- Mean duration of positive and negative phases of force application (seconds)
- Percent difference in mean duration of positive and negative phases of force application
- Rate of positive and negative force change (*df*/*dt*)
- Mean integral of force/stroke (impulse) (N-second)
- Total integral of force (mean integral × total number of tailbeats) (N-seconds)
- Mean integral of force/time (total impulse/exercise duration) (N-seconds/second)

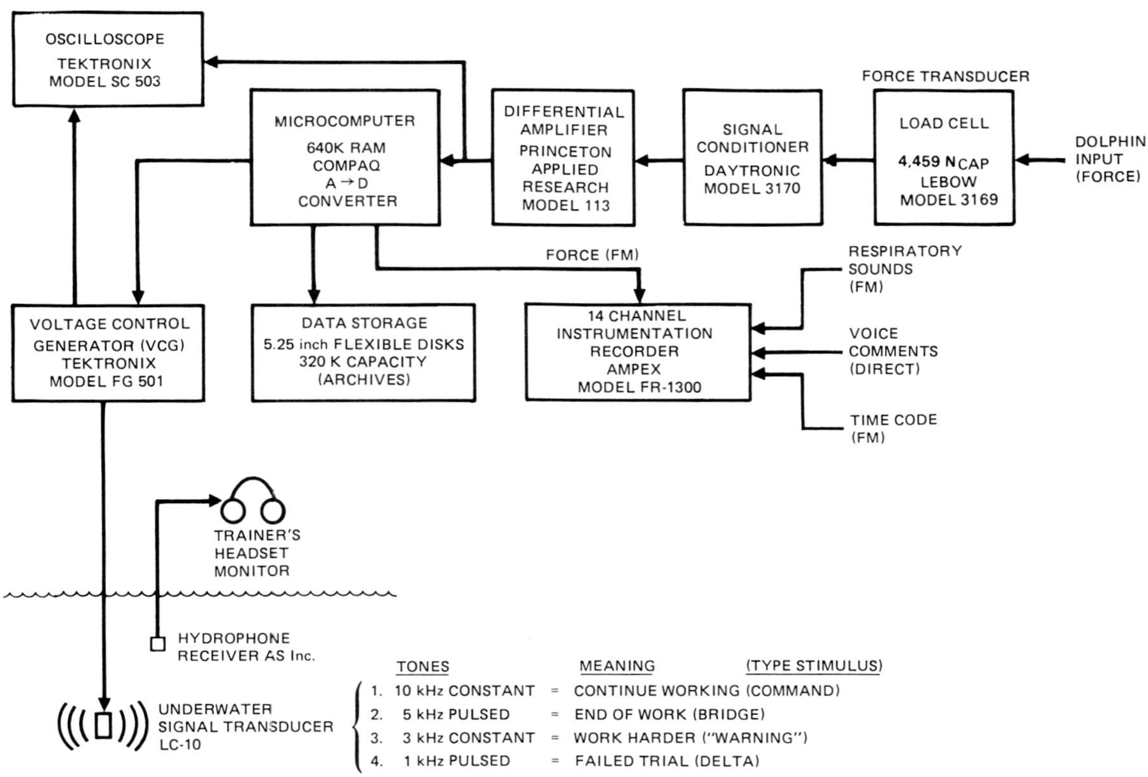

Figure 1 Hardware interfacing for behavioral conditioning and data collection.

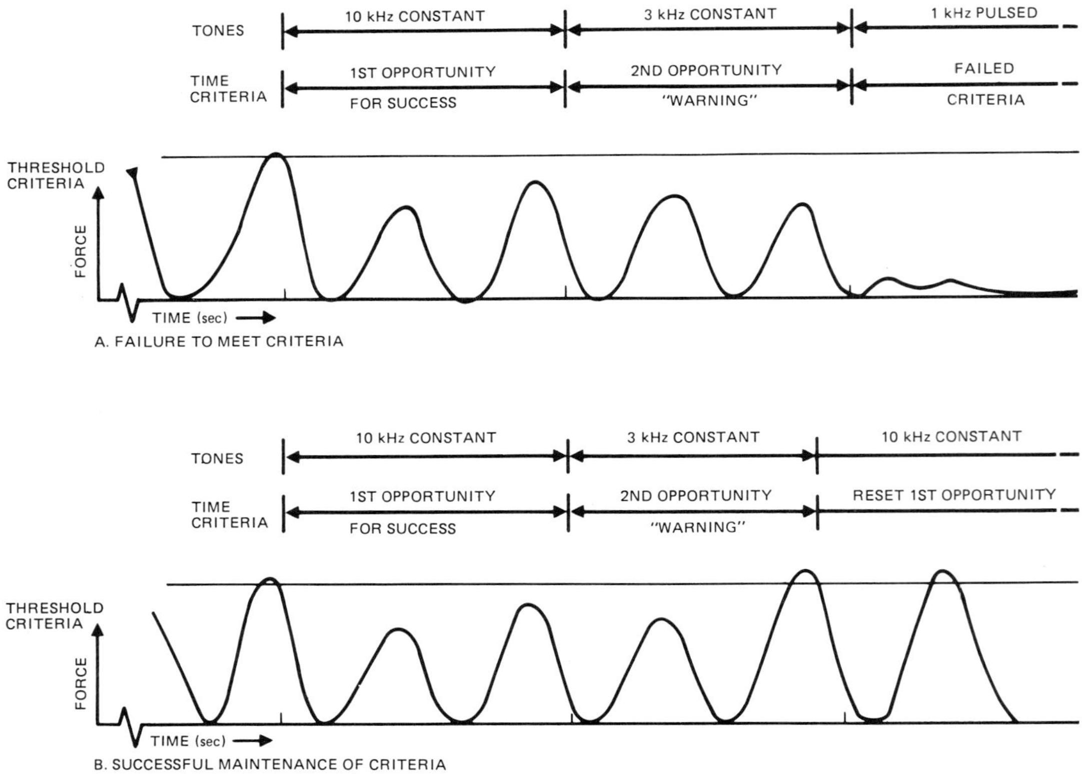

Figure 2 Microcomputer control of behavioral conditioning during dolphin ergometry.

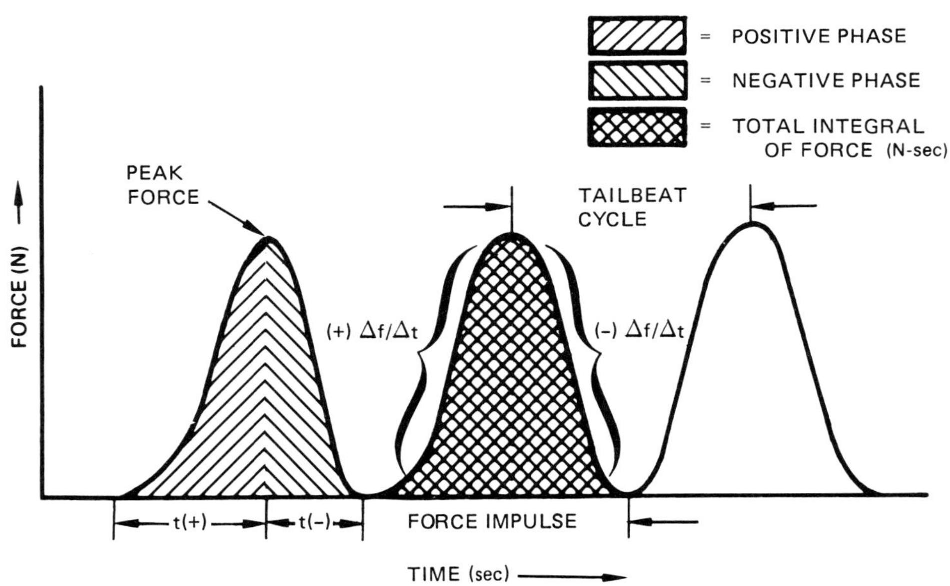

Figure 3 Typical force profile showing the basic and additional parameters characterizing the force production.

Mean integral of force during positive and negative phases (N-seconds)
Percent difference of integral of force between positive and negative phases
Frequency distribution of peak force (% occurrence)
Frequency distributions of tailbeat frequencies (% occurrence)

The method used to measure muscular force was deemed appropriate in the present study, given limitations imposed by equipment, facilities, and the dolphin's repertoire of behavior. To increase the accuracy of measuring force exerted against the load cell, the dolphin was trained to respond to the touch of a padded pole or "target" (Fig. 4B) by aligning himself along the axis measured by the load cell. The load cell mounting bracket (Fig. 4C) was adjusted until it was located to allow (1) the dolphin's force vector to coincide with that of the load cell, (2) the dolphin to perform full flexion and extension of its tail and flukes without breaking the surface, and (3) the dolphin to have unrestricted access to atmospheric air. The absolute force values recorded by the load cell were comprised to an undetermined extent by the indirect method used and the measurement of a single force vector.

During exercise trials, a 14-channel analog tape recorder (Ampex, Model FR-1300) operating at a speed of 19.05 cm/second (7.5 inches/second) collected data on force, respiratory sounds, and time code as FM signals with a response range of DC to 2.5 kHz. Voice comments and marks to permit synchronizing data with tailbeats were recorded directly with a signal response range of 0.100–150 kHz.

The load cell was mounted on a 1.3-cm aluminum plate bolted diagonally across the corner of the pen and was secured to a stainless-steel, adjustable mounting bracket, set at an 8-degree angle from the horizontal (Fig. 4C). A wind-protected microphone (Electro-Voice, Model 660) was suspended directly above the dolphin's blowhole to record respiratory sounds. During exercise tests, a black and white video camera (Video Science Inc.) was used to record underwater movement of the dolphin's tail and flukes from the right side. A 5.1-cm-thick foam pad on the walkway edge to the dolphin's left protected his tailstock when he voluntarily presented his flukes to permit collection of resting or postexercise blood samples.

Behavioral and Physical Conditioning

The dolphin was trained, using operant conditioning techniques, to exercise continuously for 3 minutes against the load cell at a predetermined (i.e., computer-programmed threshold value) mean force/tailbeat (Fig. 2 and 5). This was accomplished by conditioning him to "swim" at a specified mean peak force/stroke for increasingly longer periods until he could maintain it for slightly more than 3 minutes. The microcomputer allowed immediate evaluation of the forces produced by the exercising dolphin and provided real-time discriminative and reinforcement stimuli to him as he worked to satisfy the programmed force and duration criteria (Fig.2). To achieve the desired criteria, training was conducted for 5–6 days/week and in two daily sessions of 1–1.5 hours each separated by a minimum of 3 hours. Individual exercise trials within a session lasted for 20–190 seconds and were conducted on a randomized schedule with a 1:4 work to rest ratio. Daily exercise time averaged 8.8 minutes (S.D. 3.5, range 5–16 minutes), with the longest sessions occurring in the weeks immediately prior to testing as the animal's condition peaked. During exercise trials a "warning" tone (constant 3 kHz) was used to alert the dolphin when his thrust dropped below the preset level (Fig. 2). This reduced the number of failures and the associated stress and frustration to both trainer and dolphin (especially during the high-intensity 3-minute tests).

After 6 months of conditioning, the dolphin was considered to have reached the desired level (i.e., ability to sustain the highest mean force/tailbeat he had reached during training trials for a minimum of 3 minutes). I assumed the dolphin either could not or would not increase his force beyond this level or his endurance time at this level. Therefore, this value (1,324 N) was considered a reasonable approximation of the maximum sustainable force of which this dolphin was capable under these conditions.

Additional physical conditioning in open water probably would have improved this dolphin's exercise endurance somewhat (but probably not his maximum performance for 8 seconds). However, as this was not possible, the 2 months of load-cell training immediately prior to testing [consisting of numerous high-intensity, short-

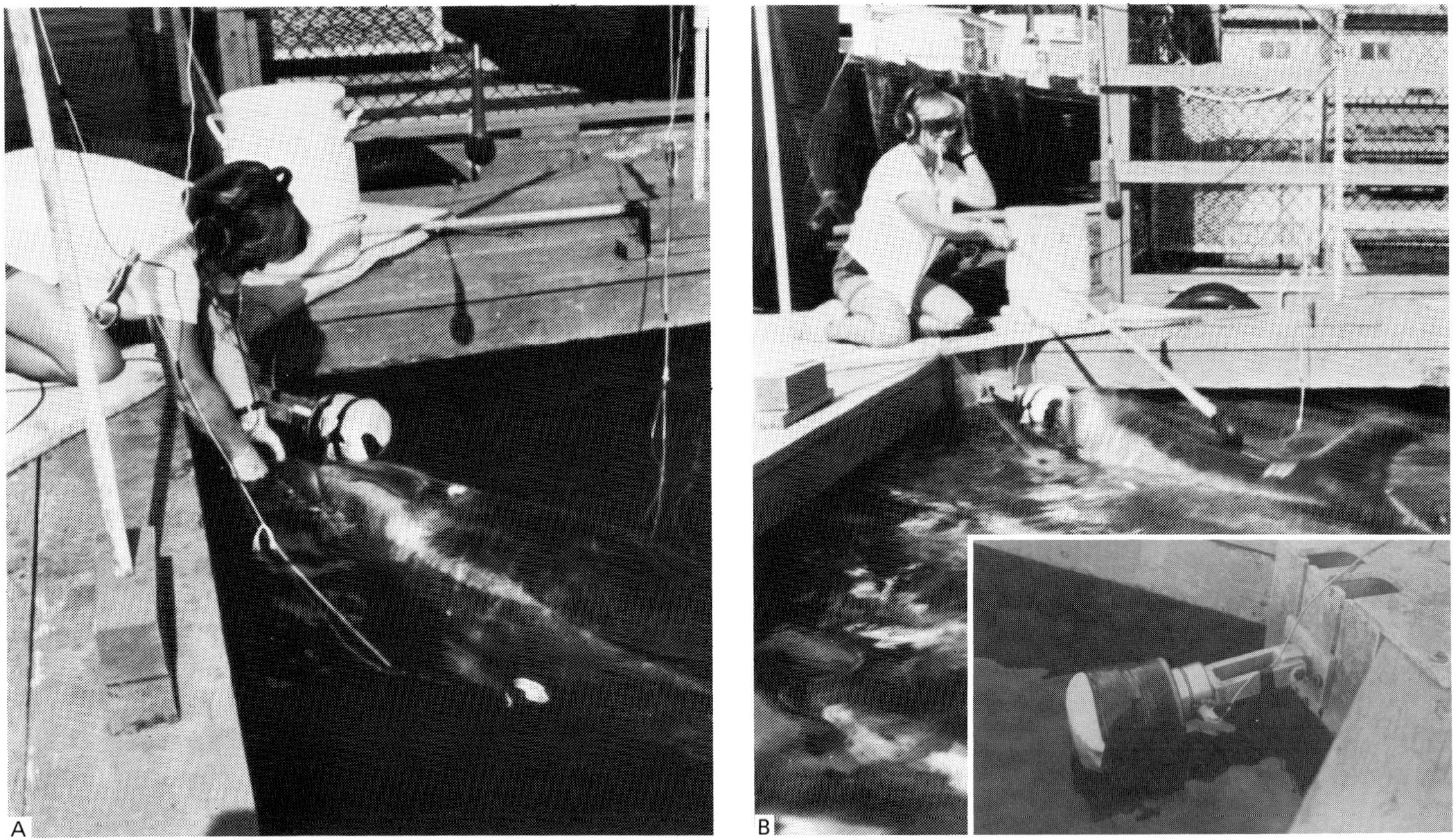

Figure 4 (A) Preparing the dolphin for exercise testing. (B) Dolphin performing an exercise test. (Inset) Padded, fix-mounted load cell (force transducer).

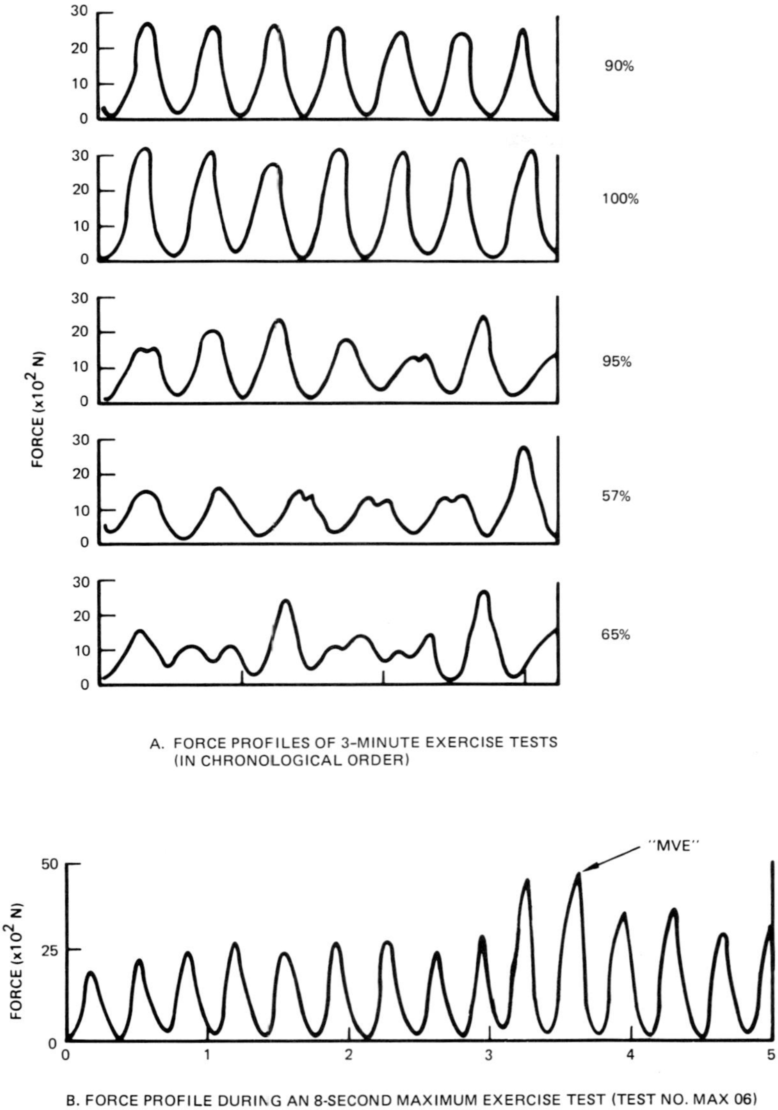

Figure 5 Representative force profiles during exercise tests.

duration (1–3 minutes) exercise bouts totaling 5–16 minutes/day] was accepted as adequate under these conditions.

After the dolphin had achieved consistent performance at a specified level of effort (i.e., two to four training sessions at the desired mean force/tailbeat), exercise tests were conducted with the full complement of data collection. Since all test levels were preset based on the 100% level achieved during the first test, this test was replicated to confirm that the dolphin would not, in fact, increase his output. The average values for these tests were 1,324 and 1,319 N/tailbeat. Following the procedures described above, the dolphin was retrained, for 4 consecutive weeks, to exercise for 3 minutes at the same tailbeat frequency (2.0 Hz) but at decreasing levels of mean force/stroke (1,246, 1,207, 863, and 755 N/tailbeat, equivalent to 95, 90, 65, and 57% of the highest mean thrust sustained for 3 minutes.

After the 3-minute tests were completed, an additional 2 weeks of training was devoted to increasing gradually the force criterion for an 8-second, maximum "exercise burst." This was designed to determine the maximum force of a single tailbeat and was considered to represent the dolphin's maximum voluntary muscular effort (MVE). This MVE value was also used to calculate the % MVE for the mean sustained tailbeat forces of each 3-minute exercise test.

To facilitate collection of postexercise muscle and blood samples, the dolphin was also trained to "beach" voluntarily on a specially designed, padded restraint platform. This was located at the end of the pen opposite the force transducer. The beaching behavior was practiced before and after many of the daily training trials so that it would be performed reliably when an actual exercise test was being conducted. Maintenance of the beaching behavior required routine training and reinforcement with food. The behavior was also necessary for periodic medical check-ups and body weight determinations.

RESULTS AND DISCUSSION

Selected Force Production Characteristics

General

A consistent rhythmic pattern of tailbeat forces with one peak/tailbeat cycle was produced by the dolphin during all (except two) exercise tests (Fig. 5). After analyzing the metabolic data (i.e., muscle lactate, glucose 6-phosphate, and glycogen levels; see Goforth, 1986), the mean peak force/stroke was selected as the best indicator of muscular effort. Other tailbeat force measures varied inconsistently or insufficiently to be of obvious value in quantifying relative effort. For instance, neither the total-time–tension integral nor the force integral/second correlated with metabolite levels or the mean peak force. On the other hand, mean tailbeat frequency and mean integral of force (impulse) were generally constant and unrelated to the mean peak force. For example, during the maximum 8-second exercise test the mean peak force/stroke was 2.5 times larger than that of the 100% level 3-minute exercise test (3,263 versus 1,324 N), yet the mean impulse/stroke for this test (9,414 N-seconds) was well within the range (10,228–7,100 N-seconds) observed across all trials.

Maximum Voluntary Effort (MVE)

The maximum effort (force) for a single tailbeat was 4,730 N (Fig. 5). This value was taken to be the MVE for this dolphin, under the test conditions of the study. This value was 66% of the theoretical maximum flexion force potential (7,144 N) calculated from the architectural and anatomical data obtained from ventral muscles of dead dolphins (Bello *et al.*, 1985; Goforth, 1986). During the 8-second test in which the dolphin achieved the MVE, he produced 24 tailbeats at a frequency of 2.9 Hz with a mean force/stroke of 3,123 N (66% of MVE) and a mean integral of force/stroke of 9,385 N-seconds.

Three-Minute Exercise Tests

Tailbeat frequency during the 3-minute exercise tests (1.6–2.4 Hz) was always lower than during the high-intensity exercise tests (2.7–2.9 Hz). The mean peak force/stroke during 3-minute tests varied from 755 N (S.D. 156) to 1,324 N (S.D. 11.7), representing forces equivalent to 16–28% of the MVE. During these tests the respiratory rate increased from 3/minute at rest to 9–12/minute, but it was unrelated to exercise intensity (mean peak force/stroke).

Eight-Second Exercise Tests

Forty 8-second tests were performed during the last 14 days of the study to obtain the MVE and maximum burst values for this dolphin, under the test conditions. These exercise efforts were characterized by very regular, high-frequency (2.7–2.9 Hz) tailbeats, with mean peak forces/stroke of 2,648–3,263 N, equivalent to 56–69% of the MVE. During all 8-second tests the dolphin took only an initial breath.

The force production pattern during the maximum 8-second burst tests differed from that during the maximum 3-minute exercise tests in the following ways: (1) the mean peak force/stroke was 60% greater, (2) the mean tailbeat frequency was 30% greater, (3) the rate of positive force development (N/second) was 73% greater, and (4) the mean duration of the positive phase almost equalled that of the negative phase. Despite the major differences between these two type of exercise tests, however, the mean integral of force/stroke did not differ (9,154 versus 9,133 N-seconds). This occurred because the lower, broader, profiles of the lower force strokes contained the same area as the narrow, sharp-peaked profiles of the high-force strokes. A constant total integral of force across all speeds has also been observed in other animals (Cavagna *et al.*, 1977; Taylor *et al.*, 1980; Taylor and Heglund, 1982; Kuehl *et al.*, 1985). During all 8-second tests thrusts were characterized by a single peak/tailbeat cycle. This phenomenon was also observed during all but a few of the 3-minute tests (Fig. 5). Based on these observations it appears that the dolphin is capable of producing thrusts in a variety of patterns depending on its needs and test conditions.

Relationship of Stationary Exercise Test to Free Swimming Velocity

Previous studies designed to determine the swimming speed of dolphins have experienced limitations of some type which require caution when interpreting the results. So it is with the present study. The following is an attempt to relate this study's findings to data obtained from delphinids under various test conditions which might be extrapolated to free-swimming dolphins.

During the 3-minute exercise tests the dolphin typically employed a regular 2-Hz tailbeat frequency (range 1.9–2.4). This rate is almost double the 1.1-Hz rate measured for a captive bottlenose dolphin while cruising slowly at 2.35 m/second or 0.94 body lengths (*L*)/second (Videler and Kamermans, 1985) but below the 3–4 Hz observed during rapid swimming at 4.0–6.5 m/second or 1.6–2.6 *L*/second (Kayan and Pyatetskiy, 1977). Pershin (1971) reported the preferred tailbeat frequencies for wild and captive dolphins (*Delphinus delphis* and *Phocoena phocoena*) to be 1.9 and 1.5 Hz, respectively. The tailbeat frequencies during 3-minute exercise tests (1.9–2.4 Hz) coincide with frequencies recorded from film for dolphins swimming at velocities of 4–5 m/second (1.6–2.0 *L*/second) (Smith *et al.*, 1976; Kayan and Pyatetskiy, 1977). During the maximum 8-second tests the dolphin generated maximum "power" from high mean peak forces/stroke at tailbeat frequencies in the range 2.7–2.9 Hz.

Although short burst and sustained swimming speeds of dolphins have been the subject of many popular articles, rarely have they been accurately measured. The most reliable such measurements are presented in Table 3. Given that the human speed records shown in Table 3 are from exceptional individuals, it is intriguing that the speed records of presumably less exceptional dolphin individuals are similar in magnitude of velocity, duration, and contraction (step/tailbeat) frequency (i.e., 11 m/second at 5 Hz and 3–5 m/second at 2–3 Hz).

The burst and sustained swimming speeds in Table 3 can be compared to the level of effort observed during the exercise tests in the present study. If one assumes that the test dolphin was exercising maximally during the 8-second burst tests, then this may be considered equivalent to the muscular effort required to swim at 8–10 m/second. On the other hand, the levels of effort sustained by the dolphin during the 3-minute exercise tests may be considered equivalent to sustainable speeds in the range of 2–5 m/second. One may further speculate that the 6.09 m/second speed sustained for only 50 seconds by the 89-kg bottlenose dolphin of Lang and Norris (1966) may have been sustainable for 3 minutes by the larger (309 kg) dolphin in the present study.

Limitations in Stationary Exercise Tests

For force transducer measurements, the forces (thrusts) recorded during all exercise tests were produced by the dolphin's complex propulsion system. This system involves the transduction of forces, generated from the locomotor muscles at their tendinous attachments, to the propulsive surfaces (flukes), which in turn produce thrust by oscillating with varying angles of attack in a fluid. One must consider the possibility that the exercise

Table 3
Selected Measurements or Estimates of Maximum Swimming Speeds of Delphinids and a Fish Species and Maximum Running Speeds of Humans[a]

Species	Test Condition	Sustained (m/second)	Duration (minutes)	Burst (m/second)	Duration (seconds)	References
Bottlenose dolphin (*Tursiops truncatus*)	Lagoon (conditioned to chase lure)	11.9 knots (6.09)	~1	16.2 knots (8.3)	7.5	Lang and Norris (1966)
Common dolphin (*Delphinus delphis*)	Free swimming (radio-tracked)	2.3 knots (1.2)	Days	—	—	Perrin *et al.* (1979)
	Circular tank (photography)	3.1 knots (1.6)	Days	—	—	Hui (1987)
Northern right whale dolphin (*Lissodelphis borealis*)	Flight reaction (sonabuoy positioning)	15 knots (7.8)	26	—	—	Leatherwood and Walker (1979)
Spotted dolphin (*Stenella* spp.)	Lagoon (conditioned to chase lure)	—	—	21.4 knots (11)	2–3	Lang and Pryor (1966)
	Free swimming (shipboard observations)	5.1–8.8 knots (4.5)	16–40	—	—	Au and Perryman (1982)
Unidentified dolphin (with white sides)	Free swimming (shipboard observations)	14–18 knots (7.2–9.2)	8–25	—	—	Johannessen and Harder (1960)
Skipjack tuna (*Katsuwonus pelamis*)	Circular tank (hook and line)	7.8–9.8 knots (4–5)	60+	18 knots (8–10)	7–10 (minutes)	Guppy *et al.* (1979), Stevens and Dizon (1982)
Man (*Homo sapiens*)	Competetive events	10.7 knots (5.5)	128	21.4 knots (11)	9	Riegel (1981)

[a] Figures in parentheses are converted from reported values.

testing conditions of the study caused the dolphin to produce thrust in an aberrant manner which may differ from that of a free-swimming dolphin. To resolve this question properly, one needs kinematic studies of a free-swimming dolphin to compare with results of a dolphin exercising against a load cell. Nevertheless, the forces recorded during these exercise tests represent carefully measured levels of muscular effort documented under controlled conditions.

Advantages of Load Cell Exercise Tests

The method described in this chapter allows for (1) control, characterization, and quantification of the relative intensity of muscular effort, (2) collection of postexercise muscle and blood samples, (3) measurement of respiration rate, and (4) adaptations to obtain heart rate, electromyographic signals, and data on oxygen uptake and substrate turnover rates from an exercising bottlenose dolphin. Properly controlled exercise studies will expand our understanding of the metabolic and force production characteristics of this familiar and adaptable delphinid.

ACKNOWLEDGMENTS

A multidisciplinary project of this nature would have been impossible without the following professional and creative contributions: hardware design and development, Don Miller, Morris "Mo" Wintermantle, and "Pete" A. Peters; electronic engineering and computer programming, Eric Cartagena, Donald A. Carder, James Kadane, Franklin Borkat, and Ralph R. Johnson; behavioral conditioning, Tricia Kamolnick, Debbie Skaar, Joy Norton, Billy L. Scronce, and Clark A. Bowers; veterinary support, Sam H. Ridgway, Alexandra Bakarich, Major (USAF), and HM1 Erik Huber; photography and underwater video, Bob White and Dave Nelson; biochemistry, Paul A. Ponganis; animal ergometry, C. Richard Taylor and Ivan "the not-so-terrible"; administrative support, Jeffrey E. Haun and Lester W. Bivens; and my dissertation chairman, V. Reggie Edgerton, and committee, Kenneth M. Baldwin, Earl Eldred, Robert J. Gregor, Gerald L. Kooyman, Roland Roy, and Arthur C. Vailas.

REFERENCES

Astrand, I. (1960). Aerobic work capacity in men and women, with special reference to age. *Acta Physiol. Scand.* **49** (Suppl. 169), 83.

Au, D., and Perryman, W. (1982). Movement and speed of dolphin schools responding to an approaching ship. *Fish. Bull.* **80,** 371–379.

Bayly, W. M., Schultz, D. A., Hodgson, D. R., and Gollnick, P. D. (1987). Ventilatory responses of the horse to exercise: effect of gas collection systems. *J. Appl. Physiol.* **63,** 1210–1217.

Bello, M. A., Roy, R. R., Martin, T. P., Goforth, H. W., and Edgerton, V. R. (1985). Axial musculature in the dolphin (*Tursiops truncatus*): Some architectural and histochemical characteristics. *Mar. Mamm. Sci.* **1**(4), 324–336.

Castellini, M. A. (1986). Metabolic depression in tissues and organs of marine mammals during diving: Living longer with less oxygen. *Mol. Physiol.* **8,** 427–437.

Castellini, M. A., and Somero, G. N. (1981). Buffer capacity of vertebrate muscle: Correlations with potentials for anaerobic function. *J. Comp. Physiol.* **143,** 191–198.

Castellini, M. A., Somero, G. N., and Kooyman, G. L. (1981). Glycolytic enzyme activities in tissues of marine and terrestrial mammals. *Physiol. Zool.* **54**(2), 242–252.

Castellini, M. A., Murphy, B. J., Fedak, M., Ronald, K., Gofton, N., and Hochachka, P. W., (1985). Potentially conflicting metabolic demands of diving and exercise in seals. *J. Appl. Physiol.* **58**(2), 392–399.

Cavagna, G. A., Heglund, N. C., and Taylor, C. R. (1977). Mechanical work in terrestrial locomotion: Two basic mechanisms for minimizing energy expenditure. *Am. J. Physiol.* **233,** R243–R261.

Davis, R. W. (1983). Lactate and glucose metabolism in the resting and diving harbor seal (*Phoca vitulina*). *J. Comp. Physiol.* **153,** 275–288.

Davis, R. W., Williams, T. M., and Kooyman, G.L. (1985). Swimming metabolism of yearling and adult harbor seals *Phoca vitulina*. *Physiol. Zool.* **58**(5), 590–596.

Dawson, T. J., and Taylor, C. R. (1973). Energetic cost of locomotion in kangaroos. *Nature (London)* **246,** 313–314.

DuBois, K. P., Geiling, E. M. K., McBride, A. F., and Thompson, J. F. (1948). Studies on the intermediary carbohydrate metabolism of aquatic animals. *J. Biol. Chem.* **359,** 347–359.

Elsner, R. E., and Ashwell-Erickson, S. (1982). Maximum oxygen consumption of exercising harbor seals. *Physiologist* **25**(4), 279. (abstr.).

Feldkamp, S. D. (1985). Swimming and diving in the California sea lion (*Zalophus californianus*). Ph.D. dissertation, Scripps Inst. of Oceanography, La Jolla, California, Univ. of California, San Diego. 176 pp.

Feldkamp, S. D., Costa, D. P., Schroeder, J. P., Friedl, W., and Haun, J. E., (1987). Oxygen consumption and thermoregulation of the bottlenose dolphin *(Tursiops truncatus). Abstr., Second Workshop on Pinniped Energetics, Miami, Florida, Dec. 5.*

Freund, B. J., Allen, D., and Wilmore, J. H. (1986). Interaction of test protocol and inclined run training on maximal oxygen uptake. *Med. Sci. Sports Exerc.* **18**(6), 588–592.

Gleser, M. A., and Vogel, J. A. (1973). Endurance capacity for prolonged exercise on the bicycle ergometer. *J. Appl. Physiol.* **34**(4), 438–442.

Goforth, H. W., Jr. (1986). Glycogenolytic responses and force production characteristics of a bottlenose dolphin (Tursiops truncatus) while exercising against a force transducer. Ph. D. Dissertation, University of California, Los Angeles. 144 pp.

Goforth, H. W., Jr. Ponganis, P. J., and Edgerton, V. R. (1987). Glycogen levels and depletion rates in the locomotor muscles of the bottlenose dolphin (*Tursiops truncatus*). *Abstr., Second Workshop on Pinniped Energetics, Miami, Florida, Dec. 5.*

Guppy, M., Hulbert, W. C., and Hochachka, P. W. (1979). Metabolic sources of heat and power in tuna muscles, II. Enzyme and metabolite profiles. *J. Exp. Biol.* **82,** 303–320.

Guppy, M., Hill, R. D., Schneider, R. C., Qvist, J., Liggins, G. C., Zapol. M. W., and Hochachka, P. W. (1986). Microcomputer-assisted metabolic studies of voluntary diving of Weddell seals. *Am. J. Physiol.* **250,** R175–R187.

Hampton, I. F. G., and Whittow, G. C. (1976). Body temperature and heat exchange in the Hawaiian spinner dolphin (*Stenella longirostris*). Comp. Biochem. Physiol. **55A,** 195–197.

Hampton, I. F. G., Whittow, G. C., Szerkerczes, J., and Rutherford, S. (1971). Heat transfer and body temperature in the Atlantic bottlenose dolphin (*Tursiops truncatus*) *Int. J. Biometeor.* **15**(2–4), 247–253.

Hochachka, P. W., and Storey, K. B. (1975). Metabolic consequences of diving in animals and man. *Science* **187,** 613–621.

Hochachka, P. W., Owen, T. G., Allen, J. F., and Whittow, G. C. (1975). Multiple end products of anaerobiosis in diving vertebrates. *Comp. Biochem. Physiol.* **50,** 17–22.

Hochachka, P. W., Runciman, W. B., and Baudinette, R. V. (1985). Why exercising tammar wallabies turn over lactate rapidly. Implications for models of mammalian exercise metabolism. *Mol. Physiol.* **7**(1), 17–27.

Hui, C. A. (1987). Power and speed of swimming dolphins. *J. Mammal.* **68,** 126–132.

Irving, L., Scholander, P. F., and Grinnel, S. W. (1941). The respiration of the porpoise, *Tursiops truncatus. J. Cell. Comp. Physiol.* **17,** 145–168.

Johannessen, C. L., and Harder, J. A. (1960). Sustained swimming speed of dolphins. *Science* **132,** 1549–1550.

Karandeeva, O. G., Matisheva, S. K., and Shapunova, V. M. (1973). Features of external respiration in the Delphinidae. *In* "Morphology and Ecology of Marine Mammals (Seals, Dolphins, and Porpoises)" (K. K. Chapskii and V. E. Sokolov, eds.), pp. 196–205. Wiley New York. [Transl. from Russian (1973) by H. Mills, Israel Program for Scientific Translations, Jerusalem.]

Kayan, V. P., and Pyatetskiy, V. Ye. (1977). Kinematics of bottlenosed dolphin swimming as related to acceleration mode. *Bionika* **11,** 36–41. [Transl. from Russian by Israel Program for Scientific Translations, Jerusalem.]

Kleinenberg, S. Ye. (1956). Features of cetacean respiration. *Uspekhi Sovrem Biol.* **41**(3), 366–380. [Transl. from Russian by Israel Program for Scientific Translations, Jerusalem.]

Kolchinskaya, A. Z., Karandeyeva, O. G. Mishchenko, V. S., Shapunov, V. M., Matisheva, S. K., and Stepanov, Ye. V. (1971). Respiration of the bottlenose dolphin. *Bionika* **5,** 9–16. [Transl. from Russian by Israel Program for Scientific Translations, Jerusalem.]

Kooyman, G. L., Wahrenbrock, E. A., Castellini, M. A., Davis, R. W., and Sinnett, E. E. (1980). Aerobic and anaerobic metabolism during voluntary diving in Weddell seals: Evidence of preferred pathways from blood chemistry and behavior. *J. Comp. Physiol.* **138,** 335–346.

Kooyman, G. L., Castellini, M. A., and Davis, R. W. (1981). Physiology of diving in marine mammals. *Annu. Rev. Physiol.* **43,** 343–346.

Kuehl, M. Roy, R. R., Gregor, R. J., and Edgerton, V. R. (1985). Electromyographic activity of selected cat hindlimb muscles during unrestrained locomotion at varying speeds and grades. *Soc. Neurosci. 15th Annual Meeting, Dallas, Texas, Oct. 20–15,* Abstr. No. 302.7.

Lang, T. G., and Norris, K. S. (1966). Swimming speed of the Pacific bottlenose porpoise. *Science* **151,** 588.

Lang, T. G., and Pryor, K. (1966). Hydrodynamic performance of porpoises (*Stenella attenuata*). *Science* **152,** 531–533.

Lavigne, D. M., Innes, S., Worthy, G. A. J., Kovacs, K. M., Schmitz, O. J., and Hickie, J. P. (1986). Metabolic rates of seals and whales. *Can J. Zool.* **64,** 279–284.

Leatherwood, S., and Walker, W. A. (1979). The northern right whale dolphin *Lissodelphis borealis* in the eastern North Pacific. *In* "Behavior of Marine Animals, Vol. 3, Cetaceans," (H. E. Winn and B. L. Olla, Eds.), pp. 85–141. Plenum, New York .

LeBoeuf, B. J., Costa, D., Huntley, A. C., Kooyman, G. L., and Davis, R. W. (1986). Patterns and depth of dives in two northern elephant seals (*Mirounga angustirostris*). *J. Zool. Ser. A* **2087**(1), 1–8.

Murphy, B., Zapol, W. M., and Hochachka, P. W. (1980). Metabolic activities of heart, lung and brain during diving and recovery in the Weddell seal. *J. Appl. Physiol.* **48**(4), 596–605.

Norris, K. S., and Harvey, G. W. (1972). A theory of the function of the spermaceti organ of the sperm whale (*Physeter catadon* L.). *In* "Animal Orientation and Navigation" (S. R. Galler, K. Schmidt-Koenig, G. J. Jacobs, and R. E. Belleville, eds.), pp. 397–417. NASA, Washington, D. C.

Perrin, W. F., Evans, W. E., and Holts, D. B. (1979). Movements of pelagic dolphins (*Stenella* spp.) in the eastern tropical Pacific as indicated by results of tagging with summary of tagging operations, 1969. *NOAA Tech. Rep., NMFS*, SSRF-737. 14 pp.

Pershin, S. V. (1971). Frequency characteristics of aquatic animals. *Bionika* **4,** 37–43. [Transl. from Russian by Israel Program for Scientific Translations, Jerusalem.]

Persson, S. G. B. (1983). Evaluation of exercise intolerance and fitness in the performance horse. *In* "Equine Exercise Physiology" (D. H. Snow, S. G. B. Persson, and R. J. Rose, eds.), pp. 441–457. Granta Editions, Cambridge.

Pierce, R. W. (1970). The design and operation of a metabolic chamber for marine mammals. Ph.D. Dissertation, Univ. of California, Berkeley. 83 pp.

Pollock, M. L., (1977). Submaximal and maximal working capacity of elite distance runners. Part I: Cardiorespiratory aspects. *In* "The Marathon: Physiological, Medical, Epidemiological and Psychological Studies," (P. Milvey, ed.), pp. 310–322. *Ann. N. Y. Acad. Sci.* **301.**

Ponganis, P. J., and Pierce, R. W. (1978). Muscle metabolism profiles and fiber type composition in some marine mammals. *Comp. Biochem. Physiol.* **59B,** 99–102.

Ridgway, S. H. (1973). Control mechanisms in diving dolphins and seals. Ph.D. dissertation, Univ. of Cambridge, Cambridge. 73 pp.

Ridgway, S. H., and Patton, G. S. (1971). Dolphin thyroid: Some anatomical and physiological findings. *Z. Vergl. Physiol.* **71,** 129.

Riegel, P. S. (1981). Athletic records and human endurance. *Am. Sci.* **69,** 285–290.

Sahlin, K., Harris, R. C., Nylind, B., and Hultman, E. (1976). Lactate content and pH in muscle samples obtained after dynamic exercise. *Pflugers Arch.* **367,** 143–149.

Seeherman, H. J., Taylor, C. R., Maloiy, G. M. O., and Armstrong, R. B. (1981). Design of the mammalian respiratory system. II. Measuring maximum aerobic capacity. *Resp. Physiol.* **44,** 11–23.

Shoubridge, E. A., Carscadden, J. E., and Leggett, W. C. (1976). LDH isozyme patterns in cetaceans: Evidence for a biochemical adaptation to diving. *Comp. Biochem. Physiol.* **53B,** 357–359.

Simon, L. M., Robin, E. D., Elsner, R., Van Kessel, A. L. G. J., and Theodore, J. (1974). A biochemical basis for differences in maximum diving time in aquatic mammals. *Comp. Biochem. Physiol.* **47B,** 209–215.

Smith, G. J. D., Brown, K. W., and Gaskin, D. E. (1976). Functional myology of the harbour porpoise *Phocoena phocoena* (L.). *Can. J. Zool.* **54,** 716–729.

Snow, D. H., Harris, R. C., and Gash, S. P. (1985). Metabolic responses of equine muscle to intermittent maximal exercise. *J. Appl. Physiol.* **58**(5), 1689–1697.

Stevens, E. D., and Dizon, A. E. (1982). Energetics of locomotion in warm-bodied fish. *Annu. Rev. Physiol.* **44,** 121–131.

Storey, K. B., and Hochachka, P. W. (1974). Glycolytic enzymes in muscles of the Pacific dolphin: Role of pyruvate kinase in aerobic/anaerobic transition during diving. *Comp. Biochem. Physiol.* **49B,** 119–128.

Stromme, S. B., Ingjer, F., and Meen, H. D. (1977). Assesment of maximal aerobic power in specifically trained athletes. *J. Appl. Physiol.* **42**(6), 833–837.

Suzuki, A., Tsuchiya, T., Takahashi, Y., and Tamate, H. (1983). Histochemical properties of myofibers in longissimus muscle of common dolphins (*Delphinus delphis*). *Acta Histochem. Cytochem.* **3,** 223–231.

Taylor, C. R., and Heglund, N. C. (1982). Energetics and mechanics of terrestrial locomotion. *Annu. Rev. Physiol.* **44,** 97–107.

Taylor, C. R., Heglund, N. C., McMahon, T. A., and Looney, T. R. (1980). Energetic cost of generating muscular force during running. *J. Exp. Biol.* **86,** 9–18.

Tulsi, R. S. (1975). Observations on the structure of the dorsal muscle in the bottle-nose dolphin (*Tursiops truncatus*). *J. Anat.* **119**(1), 39–48.

Videler, J., and Kamermans, P. (1985). Dolphin swimming performance: Difference between upstroke and downstroke. *Aquatic Mamm.* **11**(2), 46–52.

Williams, T. M., and Kooyman, G. L. (1985). Swimming performance and hydrodynamic characteristics of harbor seals *Phoca vitulina*. *Physiol. Zool.* **58**(5), 576–589.

Worthy, G. A. J. (1985). "Bioenergetics of the Harbour Porpoise *Phocoena phocoena* in the Bay of Fundy." Freshwater Institute, Dept. of Fisheries and Oceans, Winnipeg, Manitoba, Canada. Contract No. 1sf-85-00053. 46 pp.

Zapol, W. M., Liggins, G. C., Schneider, R. C., Qvist, J., Snider, M. T., Creasy, R. K., and Hochachka, P. W. (1979). Regional blood flow during simulated diving in the conscious Weddell seal. *J. Appl. Physiol.* **47,** 968–973.

Reading between the Lines: Analysis of Age Estimation in Dolphins

33

Aleta A. Hohn
National Marine Fisheries Service
Southwest Fisheries Center
La Jolla, California 92038

INTRODUCTION

The discovery of a method to estimate the age of dolphins from teeth was a major breakthrough in the study of dolphin life history and population biology. The pioneering work of Scheffer (1950), describing annual growth layers in teeth of fur seals and demonstrating their value for population studies, led to a search for similar structures in teeth of dolphins. Early work by Nishiwaki and Yagi (1953) to calibrate growth layers in teeth of striped dolphins, *Stenella coeruleoalba*, proved unsuccessful. A significant breakthrough was made in Sergeant's (1959) study of growth layers in teeth of captive bottlenose dolphins, *Tursiops truncatus*, which included two animals born in captivity and another that had been in captivity for over 9 years. The number of dentinal growth layers equaled the known (for captive born) or approximately known age (number of years in captivity plus the estimated age at capture) of the animals. These results established the usefulness of such layers for at least bottlenose dolphin age estimation.

In the years since Sergeant's discovery, the technique of age estimation from teeth has been applied to many species of odontocetes. Problems remain, however, in the use of tooth layering for age estimation. For example, (1) age estimates from teeth of known-age animals have rarely been made without the reader knowing in advance the ages of the specimens or other age-related information about them, (2) growth layers have been calibrated in very few species, and (3) growth-layer calibrations have been based on captive animals.

The purposes of this chapter are to discuss the current status of age estimation in dolphins and to address some of the criticisms of the use of teeth for dolphin age estimation. I use examples from the literature to demonstrate the difficulties inherent in age estimation from teeth. I also discuss how age estimates may be biased and how such biased estimates may affect estimates of life history parameters.

Growth layer group (GLG) is the term currently used for the structures defined and counted in cetacean and sirenian teeth for age estimation (Perrin and Myrick, 1980, pp. 48–49). The term was developed because the structures "read" for age estimation consist of a series of incremental growth layers. A GLG, by definition, is not necessarily an annual layer, but rather a repeating pattern that can be counted. For example, in a decalcified and stained thin section, a GLG can be considered to consist of two pairs of alternating but adjacent stainable and nonstainable layers, a typical pattern in mammal teeth (Klevezal', 1980) representing 1 year's growth. On the other hand, the same pattern can be considered as two GLGs, each consisting of one stainable and one nonstainable layer, so that two GLGs represent 1 year's growth. In practice, because it is the annual layering pattern that is of interest for dolphin age estimation, GLGs are generally identified and described as patterns thought to represent 1 year's growth. The situation may be different in some nondelphinid odontocetes, for example, the beluga, *Delphinapterus leucas* (see Goren *et al.*, 1987). To prevent confusion between GLGs and annual layers, I refer to layering patterns that correspond to one year's growth as annual layers rather than as GLGs, even if they are equivalent.

CURRENT STATUS OF AGE ESTIMATION

Direct evidence of annual growth layers in dolphin teeth is available for only a few species, and most information comes from studies of bottlenose dolphins. For bottlenose dolphins, we know that annual layers are present and can be identified and counted repeatedly (Fig. 1). Studies to calibrate the annual layering patterns and to show that such patterns are consistent between animals have involved three approaches: (1) estimating age from teeth of animals of known age; (2) estimating age from teeth marked by tetracycline; and (3) estimating age from teeth that have been removed at known intervals during the life of a dolphin. Even when known-age specimens were unavailable, valuable information has been obtained from animals for which an approximate or minimum age was known.

Known-Age Specimens

Teeth from known-age animals have been used most commonly to define and validate annual growth layers. Most known-age dolphins formerly available for study have been animals that spent most or all of their lives in captivity. From such individuals annual growth layers have been identified, at least to the extent that age estimates have been equal or close to the known ages of the specimens when teeth were read without previous knowledge of the data (see, e.g., Table 19, p. 30, in Perrin and Myrick, 1980). Even when the investigator knew the ages of the specimens, the correct number of layers could be found to match the known age of a specimen.

Sergeant's (1959) important study was based on two known-age and one minimum-known-age bottlenose dolphin (as well as one individual that survived only 4 months in captivity). Further evidence for an annual deposition rate of growth layers came from 13 captive bottlenose dolphins from Florida (Sergeant *et al.*, 1973): 3 were of known age, ranging from 0.2 to 21 years old, and 10 were of minimum known age. The minimum-known-age dolphins had spent at least half of their lives, based on estimated age, in captivity, so the resulting age estimates supported the unlikelihood that two layers had been deposited per year. The authors did not state whether the teeth from any of these specimens were read without previous knowledge of the ages.

Hui (1978) described growth layers in teeth from two known-age captive-born bottlenose dolphins, both of which died during their fourth year. Knowing the ages of the animals, he was able to identify annual layers (Hui, 1978, his Figs. 1–2).

Teeth from 26 known-age, or approximately known-age, and 19 minimum-known-age, free-ranging bottlenose dolphins were examined (Hohn *et al.*, 1989a). When age estimates were made from these teeth without prior knowledge of the available data, the estimated age was the same

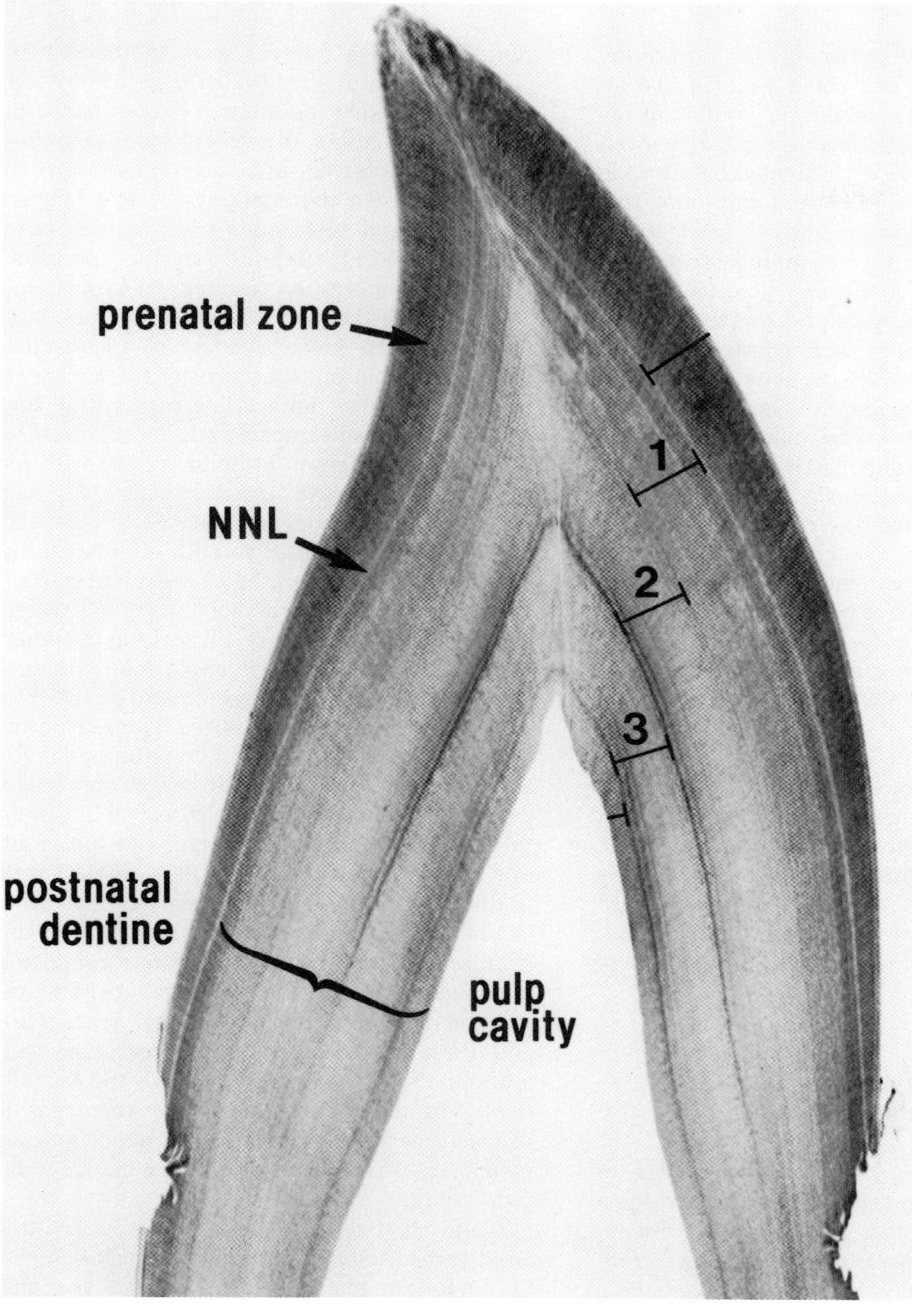

Figure 1 Decalcified and stained section from a tooth of a known-age bottlenose dolphin from Sarasota, Florida. The neonatal line is well-defined, picking up little, if any, stain. Accessory layers are apparent but not disruptively conspicuous. The boundary layer between the first and second GLGs is also not conspicuous, a condition common in tooth sections from bottlenose dolphins as well as other delphinids. The subsequent boundary layers between GLGs are well-defined. The tooth is from a male known to be 3.2–3.8 years old when the tooth was extracted. The age estimated from the tooth section was 3.4 years. (NNL, Neonatal line; numbers, GLG numbers.) The bars in the postnatal dentine demark likely GLGs. The unusual indentations on the lower outer edges of the section resulted during tooth extraction. (From Hohn *et al.*, 1989a.)

as or very close to the known age for most of the specimens. Discrepancies could generally be attributed to poor quality of the preparation of the tooth section or unusual layering patterns. Even so, ages could be correctly estimated over a wide range of known ages (2–16 years), indicating that annual growth layers in dentine can be identified and read correctly. These results are important for two reasons. First, for the first time teeth from known-age, free-ranging dolphins were available. Hence, there were no possible effects on growth layer deposition caused by captivity. Second, the teeth were read independently. The reader had no knowledge of any data pertaining to the specimens, and the known-age teeth were mixed randomly with teeth from animals whose age was not known. This prevented any reading bias associated with the estimates.

Besides those for bottlenose dolphins, studies on dentinal layering have been published for only one other known-age specimen, a spinner dolphin, *Stenella longirostris,* from Hawaii. The specimen was born in captivity and died at about 3.7 years of age (Myrick *et al.*, 1984). Subsequently, the teeth were found to contain the expected number of annual layers, although they were not read without knowledge of the age. This specimen was one of seven captive spinner dolphins in the Myrick *et al.* study. Three had spent most of their lives in captivity (see Wells, 1984, for data on these animals). For these specimens, an annual layering pattern was identified to match the approximately known ages of the animals.

Tetracycline Marking and Multiple Extractions

In all of the studies based on known-age or approximately known-age dolphins, the investigators have concluded that growth layers were annual. Nevertheless, because relatively few specimens of known age have been available, and then, generally, annual layers were identified only after the ages of the specimens were known, additional evidence from tetracycline marks has been important in confirming the existence of annual growth layers by allowing direct calibration of layering patterns.

Application of a form of the antibiotic tetracycline produces a discrete mark in the growing parts of teeth and bone (Milch *et al.*, 1957; Milch, 1958). By marking the tooth at known times, the amount of dentine or cement deposited in a known time interval can be determined using the interval between two markings or that between a marking and extraction. Because tetracycline marking does not require that the specimen's absolute age or even approximate age be known, it is a technique that can be applied to more specimens and to a greater variety of species than techniques requiring specimens of known age. In practice, however, tetracycline marks have been administered over comparatively short time spans relative to the lifespan of dolphins and with few marks per individual. The marks generally have provided data about deposition rates and patterns of layers between marks, and this information has been applied to previous and subsequent layers to determine indirectly whether the correct pattern and number of annual layers have been identified. The marks themselves, as applied in studies to date, do not validate age estimates directly.

Tetracycline marking has been reported for four dolphin species. [The first successful marking of an odontocete for age estimation was on a harbor porpoise, *Phocoena phocoena* (Nielsen, 1972).] In the first use of tetracycline in dolphins to specifically calibrate growth layers, Best (1979) marked teeth of three dusky dolphins, *Lagenorhynchus obscurus.* In one animal, no mark was found, presumably because the pulp cavity was almost completely occluded, indicating little, if any, dentine deposition. In another, the interval between marking and death was 1.93 years; two annual layers were found in the teeth between the mark and the pulp cavity. In the third animal, 3.86 years passed between marking and death, and about four annual layers were found between the mark and the pulp cavity.

Gurevich *et al.* (1980) tetracycline-marked four adult female common dolphins, *Delphinus delphis.* Of the four animals, two died almost a year after marking and two had teeth extracted after 1.2 years and were then remarked. No mark was found in the extracted teeth of one of the living animals, apparently because of an occluded pulp cavity. The teeth of the other three animals had tetracycline marks. The dentine between the mark and the pulp cavity in these animals was used to identify annual layering patterns and to measure dentinal growth rates. Using these data as a guide, the previous growth layers were determined to be annual.

More elaborate studies were conducted on spinner (Myrick *et al.*, 1984) and bottlenose dolphins (Myrick and Cornell, Chapter 34, this volume). The spinner dolphin study included the one captive-born animal (mentioned above) that had received tetracycline clinically and six wild animals, three which had been treated with tetracycline clinically and had subsequently died and three which were marked every 3 months for 1 year as well as at other times for clinical treatments. In the bottlenose dolphin study, the animals were treated at regular intervals, and their teeth were extracted periodically, different animals having different treatment and extraction regimes. The results from both studies were used to determine the annual deposition rate of layers and to define and describe the annual layers.

A similar approach to that of tetracycline marking is the periodic extraction of teeth from the same individual. Hui (1978) extracted a tooth from a captive bottlenose dolphin estimated to be 9 years of age and then, 2.5 years later, extracted another tooth. He identified 2.5 annual layers between the time of the extractions (Hui, 1978, his Fig. 3). In a study of age estimation in free-ranging bottlenose dolphins (Hohn *et al.*, 1989a), six animals had two teeth extracted. The time between extractions ranged from 4 months to 3 years. When the age estimates were conducted, the reader did not know that a second tooth had been taken from these animals. As with the tetracycline studies, the results in both of these studies were used to verify that annual growth layers were correctly identified, to assist in describing annual growth layer patterns, and to determine the deposition rate of those layers.

In all of the studies using tetracycline or involving extraction of more than one tooth from an individual, the authors concluded that annual growth layers were deposited in identifiable, repeating, and predictable patterns. These results, combined with the studies of known-age specimens, provide conclusive evidence that annual growth layers exist, at least in some species. Obvious questions remain, however. For instance, what are the effects of captivity and can these calibration studies be applied to other odontocete species? These questions are less easily answered.

Effects of Captivity

Little information is available on possible differences in growth layer deposition between captive and free-ranging animals. Although additional data for free-ranging animals will be available (e.g., more than 75 bottlenose dolphins have been tetracycline-marked in the Indian River in Florida; see Odell and Asper, Chapter 30, this volume, for a description of the study), except for the Hohn *et al.* (1989a) study, evidence documenting annual growth layers in dolphin teeth has come from captive animals. If captivity does alter the deposition rate or layering pattern of annual layers, models developed from captive animals cannot be applied to free-ranging animals for purposes of age estimation.

Currently, it is unknown whether the annual periodicity of tooth layering is due to seasonally varying factors, such as temperature or diet, or to endogenous factors. An endogenous pattern has been suggested by many authors (e.g., Kasuya, 1977; Klevezal', 1980; Myrick *et al.*, 1984). If the layering were dependent on environmental factors or on seasonal changes in the wild animal's diet, however, results of captive-animal studies would differ from results of studies of free-ranging animals. In captivity, environmental conditions and diet are controlled and would be expected to remain relatively constant throughout the year. The conditions experienced by an animal in captivity, therefore, may differ appreciably from those experienced in the wild.

Results from two studies indicate that captivity does not affect the overall annual growth layer pattern in bottlenose dolphins. Hohn *et al.* (1989a) compared the width of annual layers in the free-ranging dolphins to the width of layers in the captive-born dolphins whose parents were from Florida waters (as reported by Sergeant, 1959). The widths were virtually identical. In another study (Hohn *et al.*, 1989b), a model of annual layer deposition in captive animals, developed by Myrick and Cornell (Chapter 34, this volume), was applied to estimate the ages of the free-ranging animals. The estimates were the same as or close to the known ages of the free-ranging dolphins. We concluded that, because the model developed for teeth from captive bottlenose dolphins could be applied to the free-ranging animals for accurate age estimation, captivity did not alter the overall pattern or widths of the annual layers.

Other lines of evidence also support the notion that captivity does not significantly alter the annual growth layer patterns. There are no reported changes in the general growth layer patterns in teeth from captive animals occurring after the animals were brought into captivity. Some captive individuals show indistinct layers (e.g., see Nielsen, 1972), as do some free-ranging animals. There are also no reports of significant differences in growth layer patterns between captive-born dolphins and their wild parents. Given the available evidence, it appears that calibration of annual growth layer patterns from captive animals is a valid approach to estimating age of their free-ranging counterparts.

ANALYSIS OF CRITICISMS OF AGE ESTIMATION AND CALIBRATION

With the information outlined above, one can evaluate the criticisms directed toward age-estimation techniques and indicate how errors most likely occurred in the past. The criticism that captivity may affect the layering pattern was addressed in the previous section. Below, I address two others: (1) bias introduced when estimates are made with prior knowledge of age and other data pertaining to specimens and (2) application of age-estimation techniques to species for which layers have not been calibrated.

Reading without Prior Knowledge of Age: Accessory Layer Influence

An important and valid criticism of the verification of annual layers based on known-age or tetracycline-marked specimens is that often the estimates have been made by someone who had prior knowledge of the specimen's age (or other related data). There is one anecdote that illustrates the point well. A researcher obtained the tooth of a bottlenose dolphin that had been born in captivity and lived for a few years. When given the tooth, the researcher was also told the age of the animal so that the growth layer pattern could be mapped. After the researcher mapped the layers according to the "known age," it was discovered that an incorrect date of death had accompanied the tooth and that the real "known age" was actually almost 4 years, rather than 3.3, the age that had been used to map the growth layer pattern (for details, see Myrick *et al.*, 1984). Had the tooth been read before knowing the age, the reader would not have been constrained to a layering pattern which matched the "known age." Rather, the reader may have been able to determine that the wrong age had been assigned to the tooth. The consequence of not reading known-age specimens "in the blind" is the "choosing" of a layering pattern that fits the age rather than adhering to a defined pattern and determining, after reading, whether that pattern is correct.

As anyone who has examined odontocete tooth sections knows, the identification of an incorrect, presumed annual layering pattern is easily explained. Annual growth layers consist of many finer growth layers. A variety of terms has been applied to these finer layers, including accessory lines or layers, lunar layers, or other terms depending on the point an author wishes to make (e.g., Hohn, 1980a,b; Myrick, 1980). It was because of these incremental, but less than annual, growth layers that the term "growth layer group" (GLG) was coined.

The subannual layers vary in their contrast within annual layers, but some can be quite distinct. In many delphinid species, there is a prominent subannual layer (accessory line) at about the midpoint of the annual layers (Hohn, 1980a; Myrick *et al.*, 1983). These subannual layers are responsible for the description of annual deposition rates of two layers per year in some species (Klevezal', 1980).

Often the accessory layer, in the first one to two GLGs, is more prominent than the boundary layer between those annual layers, partly because the boundary layer is frequently relatively indistinct. When this condition occurs, confusion in interpreting annual layering patterns may occur. Kimura (1980) described an experiment in which six researchers were asked to estimate the age of a bottlenose dolphin tooth independently and to map on a photograph the presumed annual layering pattern. The tooth was from a 3.3-year-old known-age animal. Four of the readers did not know the age prior to reading. The age estimates of those four readers were 3.3, 4.7, 5.6, and 6.2 years. In the cases where the age estimate was incorrect, the prominent, midpoint accessory layers were counted as the end of an assumed annual layer (Kimura, 1980, his Fig. 2).

In a study of food habits of bottlenose dolphins, Leatherwood *et al.* (1978) reported estimates made by Hui from tooth sections of seven specimens. Hohn (1980a) included those seven specimens in a larger study on age estimation in the same population and found the published estimates to be twice her estimates for three animals less than 3 years old. It is likely that earlier estimates had included accessory layers as annual layers.

An experiment illustrating between-reader differences in age estimates is presented in the report of a workshop on age determination in odontocetes and sirenians (Perrin and Myrick, 1980, see their tables throughout pp. 4–39). Estimates were made independently by different readers from the teeth of 16 species of odontocetes, and the individual counts illustrate wide variation for some specimens and some species. Reader experience varied and the quality of preparation of the tooth sections was not standardized; so, some of the variation can be explained by these two factors. Most of the differences between readers, however, were the result of confusion caused by misinterpretation of the accessory layers.

Some authors have provided, or at least discussed, results from multiple and independent readings of a sample (e.g., Kasuya, 1976; Ross, 1977; Asper and Odell, 1980; Hohn, 1980b). Within- and between-reader differences in multiple age estimates for a large sample of spotted dolphins, *Stenella attenuata*, were reported by Reilly *et al.* (1983).

Subannual layers are responsible for most of the difficulty in distinguishing annual patterns and for allowing an investigator to find exactly the number of "annual" layers "expected" when the age of the animal is known beforehand. The prominent accessory layers produced during the first and second years of life are especially troublesome for correct identification of annual layers. However, incorrect identification of the first two annual layers is not trivial. Once the midpoint subannual layer has been mistakenly identified as the end of the first year, the error is compounded, biasing age estimates upward. The end of the first annual layer will most likely be mistaken to be the end of the second year, and the end of the second annual layer will be mistaken as the end of the fourth year. In general, after the second annual layer, the accessory layer becomes less prominent, the annual layers are thinner, and the boundary layers between annual layers are more distinct; all combine to minimize the chances of further error. For animals younger than 2 years, the estimated age may be twice the actual age. For animals much older than 2 years, the estimated age will be biased upward but will not be twice the actual age.

A "model" of GLG deposition can help investigators avoid counting accessory layers as annual layers. Such a model defines absolute and relative widths of GLGs, lists the characteristics of boundary layers, and provides information about the relative location and intensity of predictable accessory layers. Such models have been developed from known-age or tetracycline-marked teeth from spinner dolphins (Myrick *et al.*, 1984) and bottlenose dolphins (Hohn *et al.*, 1989a). There are, unfortunately, few other species for which known-age or marked specimens are available for confirming the existence of annual growth layers and for developing a model describing those layers.

Application to Other Species

Although growth layers have been calibrated for very few species, age estimation from dental growth layers has been applied to most odontocetes. To calibrate growth layers in a species, it is necessary to obtain known-age or tetracycline-marked specimens for that species. Until such samples become available, however, can age estimates be made reliably for the "uncalibrated" species? I believe it may be valid to apply models developed for a few species to others.

Fortunately, described dentinal growth-layer patterns in delphinids are remarkably similar across species (e.g., Nishiwaki and Yagi, 1953; Sergeant, 1959, 1962; Hohn, 1980a; Myrick *et al.*,

1983). This similarity is illustrated well in Perrin and Myrick (1980), where descriptions of layering patterns are provided for many species of odontocetes and for sirenians. In addition, Klevezal' (1980) described some patterns of annual layer formation in hard tissues "which seem to be the same for mammals of different systematic and ecological groups" (see also Klevezal' and Kleinenberg, 1967).

The most parsimonious explanation for similarities in growth layer patterns across species is that layer deposition is regulated by mechanisms common among (at least) related groups of animals and that layers that look the same represent the same amount of time. Thus, until the patterns for some species can be calibrated, it should be possible to apply to them the general models of annual layering patterns developed from known-age and tetracycline-marked specimens from other, similar species. The application of a general model of the calibrated annual layering pattern in bottlenose dolphin teeth to layering patterns in other delphinid teeth would ensure more consistency in age estimation among various readers. The inconsistencies or differences in age-estimation techniques often account for the apparent differences between species and populations in age structure and corresponding life-history parameters.

EFFECTS OF COMMON AGE-ESTIMATION ERRORS ON ESTIMATES OF LIFE-HISTORY PARAMETERS

Any systematic deviation, or bias, in age estimation that is caused by misinterpretation of annual layers will affect estimates of life-history parameters. Yet how does one judge whether a particular investigator's age estimates are biased? Direct comparisons of reading techniques can be done only if high-quality photographs illustrating the "annual" pattern are included in publications. Many papers do not include photographs. When photographs are included, often (1) the structures identified as annual layers are not marked, (2) the quality of the photograph or tooth preparation is too poor to allow the reader to differentiate annual layers, (3) the magnification at which a tooth section is photographed is too great, even if a series of high-magnification photos are taped together, to allow the reader to identify the annual layers, or (4) any combination of the above features is present. The result is that it is difficult or impossible to compare various authors' techniques for age estimation even though, in many cases, they differ.

When acceptable photographs are provided, they can show inconsistencies in age-estimation techniques within, as well as between, studies. Photographs of two individuals from a sample of 15 captive and 46 free-ranging bottlenose dolphins from Florida (Sergeant *et al.*, 1973) indicate that accessory layers may have been counted as annual layers in the actual first two annual layers, at least in the free-ranging specimens. Sergeant *et al.* provided photos of tooth sections from four of their specimens, but they did not mark the "annual" layers. In their photographs from the two older animals, the quality is not adequate to see growth layers. For the two younger animals, however, growth layers are visible. One of the young animals was a 1.25-year-old known-age female, and one and one-quarter annual growth layers can be seen. The other young animal was not of known age, and its age was estimated to be 5 years. If the first annual layer is taken to be the same thickness in this animal as in the 1.25-year-old animal, however, the second animal would more likely be 2.5 years old (see Sergeant *et al.*, 1973, their Fig. 1b). In obtaining an estimated age of 5 years, they probably counted midpoint accessory layers as annual layers in the first two annual layers.

When photographs have not been provided or are unsuitable for comparison of annual-layer patterns, indirect methods can sometimes be used to compare age-estimation techniques. For example, differences in age-estimation techniques can be determined by comparing independent readings from the same samples by different investigators. In two studies (Leatherwood *et al.*, 1978; Hohn, 1980a), the same sample of teeth was read. As part of their study, Leatherwood *et al.* (1978) were interested in the age at weaning of bottlenose dolphin calves. In the stomach of one specimen (504218) estimated from teeth to have been 2 years old, but estimated from length to have been from just less than 1 to just less than 2 years old, Leatherwood *et al.* found fish, but no evidence of milk. They concluded that the animal had been at least partially self-sufficient. The animal was estimated by Hohn (1980a) to be close to 1 year of age. Such inconsistencies in estimated age affect estimates of age at weaning.

Dolphin growth rates can be used to verify age-estimation techniques indirectly if the rates are similar among related species. Age estimation was probably accurate if estimated growth rates from a sample are similar to expected growth rates. For example, on the basis of growth and age-at-length data for bottlenose dolphins from Florida and the east coast of the United States (Harrison *et al.*, 1972; Hohn, 1980a; Hohn *et al.*, 1989a), animals estimated to be 2 years old in Sergeant *et al.'s* (1973) sample are the length of 1-year-old, known-age and known-length animals from the same region, again supporting the observation that accessory layers were counted as annual boundaries. The effect of estimating twice the actual number of annual layers in young animals is to underestimate growth rates and overestimate age-at-length.

Conversely, and to a limited extent, length data can be used to estimate the age of dolphins. Although not useful for older animals, estimated age from length is relatively precise during the first 2 years (Hohn and Hammond, 1985). This relationship can have implications for management. For example, minimum-size limits are applied to live-capture fisheries for bottlenose dolphins in the United States (Scott, Chapter 36, this volume). If the length-at-1-year data of Sergeant *et al.* (1973) were used, then 1-year-old animals could be collected under the mistaken belief that they were 2 years old. This estimated age, from length, may also affect later estimates of "known" age when the specimen dies in captivity and its teeth are used to map annual growth layers. The errors are compounded.

Obviously, any bias in age estimation will affect all estimates of life history parameters that depend on age. The parameters most sensitive to biases in age estimates are those that use data from relatively young animals, such as early postnatal growth and age at maturation, and those that involve life-table parameters, such as age-specific mortality and fecundity. Estimates of other parameters, such as longevity, will be less sensitive. Estimates of population growth rates are, in turn, sensitive to two of these parameters, mortality and reproductive rates (Reilly and Barlow, 1986). Population growth rates are relatively insensitive to small changes in age at maturation in late-maturing species, such as dolphins, but age at maturation has been shown to be one of the most useful parameters for measuring density-dependent changes in populations (Fowler, 1987). Thus, the parameters which are most important in estimating population growth rates are the ones most sensitive to biased age estimates.

The differences in definitions and interpretation of annual layer patterns (e.g., one versus two GLGs per year), the effects of method on readability of tooth sections (e.g., Asper and Odell, 1980), and the effects of reader experience or the availability of a good model to use as a guide will affect the accuracy of age estimates from teeth. Comparative studies of populations or species that depend on these estimates will not be valid unless it is known that the same age-estimation methods were used.

CONCLUSIONS AND RECOMMENDATIONS

Annual growth layers are present in the teeth of dolphins. Nonetheless, age estimates are often systematically incorrect. Most of the confusion in age estimation appears to be the result of misidentification of the annual structures rather than the lack of such structures. Accessory layers found in the first two annual layers probably cause the most difficulty. Further confusion stems from the lack of a consistent definition of what constitutes annual layers. Until species-specific annual layering patterns can be calibrated, the pattern calibrated for one species may be used as a model for others.

I have two recommendations that should help improve studies of age estimation. First, in the absence of species-specific models, a model developed from species where known-age or tetracycline-marked specimens have been available should be used on species with similar dentinal patterns (e.g., within the Delphinidae). This would prevent some of the common errors in age estimation and would at least increase the consistency of readings from study to study. Second, a photograph mapping the structures assumed to be annual layers must be included in papers in which age is estimated. This would allow other investigators to determine whether those readings are comparable to those in their own or others' studies. It would also facilitate comparative studies, which are presently plagued by differences in age-estimation techniques.

ACKNOWLEDGMENTS

I am grateful to Susan Chivers, Douglas DeMaster, Andrew Dizon, Christina Lockyer, Michael Scott, and an anonymous reviewer for critical reviews of the manuscript.

REFERENCES

Asper, E. D., and Odell, D. K. (1980). Bottlenose dolphin local herd monitoring: Capture, marking, collection of biological data and follow-up observations of marked animals. Final report to National Marine Fisheries Service, Contract No. NA79-GA-C-00027. Unpaginated.

Best, P. B., (1979). Tetracycline marking and the rate of growth layer formation in the teeth of a dolphin *(Lagenorhynchus obscurus). S. Afr. J. Sci.* **72,** 216–218.

Fowler, C. W. (1987). A review of density dependence in populations of large mammals. *In* "Current Mammalogy" (H. H. Genoways, ed.), Vol. 1, pp. 401–441. Plenum Publishing Corp., New York.

Goren, A. D., Brodie, P. F., Spotte, S., Ray, G. C., Kaufman, H. W., Gwinnett, A. J., Sciubba, J. J., and Buck, J. D., (1987). Growth layer groups (GLGs) in the teeth of an adult belukha whale (*Delphinapterus leucas*) of known age: Evidence for two annual layers. *Mar. Mamm. Sci.* **3**(1), 14–21.

Gurevich, V. S., Stewart, B. S., and Cornell, L. H. (1980). The use of tetracycline in age determination of common dolphins, *Delphinus delphis. Rep. int. Whal. Commn* (Special Issue 3), pp. 165–170.

Harrison, R. J., and Brownell, R. L., Jr., and Boice, R. C. (1972). Reproduction and gonadal appearances in some odontocetes. *In* "Functional Anatomy of Marine Mammals" (R. J. Harrison, ed.), Vol. 1, pp. 362–429. Academic Press, New York. 451 pp.

Hohn, A. A. (1980a). Age determination and age related factors in the teeth of western North Atlantic bottlenose dolphins. *Sci. Rep. Whales Res. Inst., Tokyo* **32,** 39–66.

Hohn, A. A. (1980b). Analysis of growth layers in the teeth of *Tursiops turncatus* using light microscopy, microradiography, and SEM. *Rep. int. Whal. Commn* (Special Issue 3), pp. 155–160.

Hohn, A. A., and Hammond, P. S. (1985). Early postnatal growth of the spotted dolphin, *Stenella attenuata,* in the offshore eastern tropical Pacific. *Fish. Bull.* **83**(4) 553–566.

Hohn, A. A., Scott, M. D., Wells, R. S., Sweeney, J. S., and Irvine, A. B. (1989a). Growth layers in teeth from known-age, free-ranging bottlenose dolphins. *Mar. Mamm. Sci.* **5**(4).

Hohn, A. A., Myrick, A. C., Jr., Wells, R. S., Scott, M. D., and Irvine, A. B. (1989b). Dentinal growth layer similarities in captive and free-ranging known-age bottlenose dolphins. In preparation.

Hui, C. A. (1978). Reliability of using dentin layers for age determination in *Tursiops turncatus.* U. S. Dept. of Commerce, Natl. Tech. Info. Ser. Publ. No. PB-288 444. 25 pp.

Kasuya, T. (1976). Reconsideration of life history parameters of the spotted and striped dolphins based on cemental layers. *Sci. Rep. Whales Res. Inst., Tokyo* **28,** 73–106.

Kasuya, T. (1977). Age determination and growth of Baird's beaked whale with a comment on the fetal growth rate. *Sci. Rep. Whales Res. Inst., Tokyo* **29,** 1–20.

Kimura, M. (1980). Variability in techniques of counting dentinal growth layer groups in a tooth of a known-age dolphin, *Tursiops truncatus. Rep. int. Whal. Commn* (Special Issue 3), pp. 161–164.

Klevezal', G. A. (1980). Layers in the hard tissues of mammals as a record of growth rhythms of individuals. *Rep. int. Whal. Commn* (Special Issue 3), pp. 89–94.

Klevezal', G. A., and Kleinenberg, S. E. (1967). "Opredelenie Vosrosta Mlekopitayushchikh po Sloistym Struckturam Zubov i Kosti." Izdatel'stvo Nauka, Moscow. 144 pp. [Transl.: 1969. "Age Determination of Mammals from Annual Layers in Teeth and Bones." Israeli Prog. for Scientific Translation, Jerusalem. 126 pp.]

Leatherwood, S., Deerman, M. W., and Potter, C. W. (1978). Food and reproductive status of nine *Tursiops truncatus* from the northeastern United States coast. *Cetology* **28,** 1–6.

Milch, R. A. (1958). Fluorescence of tetracycline antibiotics in bone. *J. Bone Joint Surg.* **40A,** 897–910.

Milch, R. A., Rall, D. P., and Tobie, J. E. (1957). Bone localization of the tetracyclines. *J. Natl. Cancer Inst.* **19,** 87–91.

Myrick, A. C., Jr. (1980). Examination of layered tissues of odontocetes for age determination using polarized light microscopy. *Rep. int. Whal. Commn* (Special Issue 3), pp. 105–112.

Myrick, A. C., Jr., Hohn, A. A., Sloan, P. A., Kimura, M., and Stanley, D. D. (1983). Estimating age of spotted and spinner dolphins (*Stenella attenuata* and *Stenella longirostris*) from teeth. NOAA Technical Memorandum NMFS-SWFC-30. 17 pp.

Myrick, A. C., Jr., Shallenberger, E. W., Kang, I., and Mackay, D. B. (1984). Calibration of dental layers in seven captive Hawaiian spinner dolphins, *Stenella longirostris*, based on tetracycline labeling. *Fish. Bull.* **82**(1), 207–225.

Nielsen, H. G. (1972). Age determination of the harbour porpoise *Phocoena phocoena* (L.) (Cetacea). *Vidensk. Meddr kansk Naturh. Foren.* **135,** 61–84.

Nishiwaki, M., and Yagi, T. (1953). On the age and growth of the teeth in a dolphin (*Prodelphinus caeruleoalbus*). *Sci. Rep. Whales Res. Instit., Tokyo* **8,** 133–146.

Perrin, W. F., and Myrick, A. C., Jr., eds. (1980). Report of the workshop. *Rep. int. Whal. Commn* (Special Issue 3), pp. 1–50.

Reilly, S. B., and Barlow, J. (1986). Rates of increase in dolphin population size. *Fish. Bull.* **84**(3), 527–534.

Reilly, S. B., Hohn, A. A., and Myrick, A. C., Jr. (1983). Precision of age determination of northern offshore spotted dolphins. NOAA Technical Memorandum NMFS-SWFC-35. 27 pp.

Ross, G. J. B. (1977). The taxonomy of bottlenosed dolphins *Tursiops* species in South African waters, with notes on their biology. *Ann. Cape Prov. Mus.* **11**(9), 135–194.

Scheffer, V. B. (1950). Growth layers in the Pinnipedia as an indication of age. *Science* **112,** 309–311.

Sergeant, D. E. (1959). Age determination of odontocete whales from dentinal growth layers. *Norwegian Whaling Gaz. 1959,* No. 6, pp. 273–288.

Sergeant, D. E. (1962). The biology of the pilot or pothead whale *Globicephala melaena* (Traill) in Newfoundland waters. *Fish. Res. Board Can. Bull.* **132,** 1–84.

Sergeant, D. E., Caldwell, D., and Caldwell, M. (1973). Age, growth, and maturity of bottlenosed dolphins (*Tursiops truncatus*) from northeast Florida. *J. Fish. Res. Board Can.* **30,** 1009–1011.

Wells, R. S. (1984). Reproductive behavior and hormonal correlates in Hawaiian spinner dolphins, *Stenella longirostris. Rep. int. Whal. Commn* (Special Issue 6), pp. 465–474.

Calibrating Dental Layers in Captive Bottlenose Dolphins from Serial Tetracycline Labels and Tooth Extractions

34

Albert C. Myrick, Jr.
Southwest Fisheries Center
National Marine Fisheries Service
La Jolla, California 92038

Lanny H. Cornell
Pacific Research Laboratories, Inc.
El Cajon, California 92022

INTRODUCTION

Scientists studying marine mammal populations have become increasingly dependent on age estimates from tooth layer counts to assess the population dynamics (Scheffer and Myrick, 1981). Such dependence has underscored the importance of understanding the patterns and rates of tooth layer deposition.

Cetacean teeth have complex patterns of layers, or growth layer groups (GLGs, terminology of Perrin and Myrick, 1981), which are similar to each other in detail. The GLG has been assumed to represent deposition during a 1-year period or some other constant unit of time (Sergeant, 1959). However, neither GLGs nor the variably prominent finer layers that they contain can be interpreted with certainty until GLG patterns are calibrated with units of absolute time.

Rough calibrations of GLGs with absolute time can be achieved by using layers known to have been deposited between two dates (Myrick, 1981a; Myrick *et al.*, 1984; Hohn, Chapter 33, this volume). However, calibration based only on two dates (usually birth and death) gives only a mean annual rate that provides no information on whether the GLG pattern and formation period are dependent on endogenous or exogenous factors. If rates are factor-dependent, then GLGs cannot be used with confidence to estimate ages unless the nature and timing of the factors are known.

Three techniques have been used to calibrate GLGs with two dates. In one, layer patterns were examined in single teeth from a few captive-born (known-age) bottlenose dolphins, *Tursiops truncatus* (Sergeant, 1959; Sergeant *et al.*, 1973; Hui, 1978). In another, a single lead acetate or tetracycline label was introduced into the layering pattern, and tooth layers deposited after the label were examined (*Stenella coeruleoalba*, Nishiwaki and Yagi, 1953; *Phocoena phocoena*, Nielsen, 1972; *Lagenorhynchus obscurus*, Best, 1976; and *Delphinus delphis*, Gurevich *et al.*, 1981). In the third, two teeth from a bottlenose dolphin were extracted a few years apart, and the difference in the number of GLGs was noted (Hui, 1978).

The successful uses of these techniques have led to recommendations calling for more sophisticated studies that include combinations of the following: (1) monitoring captive-born animals, whose ages are exactly known, together with wild captives of the same species to compare GLG patterns; (2) introducing tetracycline labels weeks or months apart over several years to mark segments of time in tissue that could be studied alone or with other segments; (3) studying the timing of unique or repeated physical events to determine any association with changes in GLG patterns during a monitored experimental period; and (4) using serial extractions to compare tissue deposition between teeth from one animal as well as between teeth of different animals to detect similarities and differences produced over the same period (Myrick, 1981a,b; Myrick *et al.*, 1984).

In addition, after it was realized that multiple tetracycline labels may be inadvertently produced in teeth of captive dolphins through repeated therapeutic treatments of the animals, Myrick (1981a) suggested studying layering rates retrospectively by matching labels to treatment dates from the medical records maintained by oceanaria. Myrick *et al.* (1984) conducted such a study of seven Hawaiian spinner dolphins (*Stenella longirostris*) (four dead, three living), held at Sea Life Park in Hawaii, that had been treated intermittently with tetracycline for up to 8 years. To ensure that at least some labels could be documented if therapeutic records and labels could not be matched, experimental injections of tetracycline were given and three tooth extractions were made during a 1-year period for each of the three live animals.

The Hawaiian spinner dolphin study showed annual dentinal GLGs to be of age-specific thickness, which varied only slighty from tooth to tooth in each individual and between animals of the same age. Annually deposited dentine revealed that a GLG consists of two light layers, each followed by a dark layer. In addition to the coarse GLG pattern, about 13 pairs of finer layers were visible in annual GLGs, each pair consisting of a dark layer and a light layer. Where fine layers were especially distinct and tetracycline labels closely spaced, label dates showed that a pair of fine layers was formed each month. No differences were noted between the GLG patterns formed while the animals lived in the wild and the patterns formed in captivity. This led to the conclusion that neither the captive environment nor the natural environment influenced dental layering patterns in the specimens studied (Myrick *et al.*, 1984).

In 1979, as part of a continuing effort to examine GLG patterns and deposition rates in delphinids, we began a 3.5-year project to calibrate tooth layers in bottlenose dolphins. Our two objectives were to use multiple tetracycline injections and tooth extractions in captive bottlenose dolphins to monitor rates of tooth tissue deposition, in relation to measurable factors that might affect deposition, and to calibrate GLGs with absolute time. Here we describe the protocol and results of the project.

METHODS AND MATERIALS

The Sample

We used data from 12 bottlenose dolphins (three females and nine males), hereafter called "animals" or "dolphins," maintained for public display at the three parks operated by Sea World, Inc., in San Diego, California (SWSD), Orlando, Florida (SWO), and Aurora (near Cleveland), Ohio (SWA). The dolphins ranged from captive-born yearlings to wild animals that had been held in captivity for up to 9 years (Table 1). All but one of the wild animals were collected from coastal waters off Florida; the other was captured in the Pacific Ocean off California. One of the three females (animal M) died 1 year and 3 months after the project began.

Tetracycline Labeling and Tooth Extractions

Treatments, begun in the spring of 1979, were of four types: tooth extraction and tetracycline labeling, tetracycline labeling only, extraction only, and

Table 1
Background Data and Age Estimates of 12 Captive Bottlenose Dolphins Used in Dental Layer Calibration Project

Animal	Sex	Origin[a]	Date of Birth or Capture	Capture Length/Age or Age By GLG Inspection (Years)[b]	Preproject Time in Captivity (Years)	Age at End of Project (Years)
A	M	cb:P/F	5-78		1.1	4.5
B	M	cb:F/F	5-78		1.1	4.5
C	M	cb:F/P	4-78		1.2	4.6
D	F	wc:F/F	9-77	195 cm/2	1.8	7.2
E	M	wc:F/F	1-77	201 cm/2	2.3	8.0
F	M	wc:F/F	10-75	2	3.7	8.5
I	F	wc:F/F	6-70	1–2*	9.0	14.0
J	M	wc:F/F	12-72	3–4*	6.4	13.0
M	F	wc:F/F	2-79	196 cm/2	0.3	3.0[c]
P	M	wc:P/P	2-78	3*	1.3	7.0
Controls						
K	M	wc:F/F	12-72	1–2	6.4	11.0
L	M	wc:F/F	11-72	4	6.5	14.0

[a] cb, Captive born; wc, wild captured; P/F or F/P, hybrid; origin of mother/sire; P/P, Pacific purebred; F/F, Florida coastal purebred.
[b] An asterisk indicates that age was estimated based on GLG counts relative to label introduced at capture.
[c] Animal M died in the second year of the project.

handling (i.e., sham procedures) only. Biomycin, a tetracycline, was the labeling agent used; it was injected into the dorsal musculature slightly off the midline of the body and anterior to the dorsal fin, at a dosage of 20–30 mg/kg body weight. Teeth were extracted from the middle of the row of either lower jaw using anesthetization and extraction methods similar to those described by Ridgway *et al.* (1975) and used successfully on Hawaiian spinner dolphins by Myrick *et al.* (1984).

Each animal had one tooth extracted and received one injection at the beginning of the study. All except the two animals (K and L) that were used as controls received tetracycline injections approximately every 3 months and had one or (occasionally) two teeth extracted each year for the next 3.5 years. All intact teeth were permanently labeled each time tetracycline was injected. A label introduced after the first tooth was extracted appeared in the second and third teeth and so on, but not in the first tooth (Table 2). In sham procedures, dolphins were lifted out of the water, weighed, measured, and returned to the pool. Sham procedures were done at random in lieu of real treatment or extraction to determine whether merely removing an animal from the water and weighing and measuring it affected its tooth-layering patterns.

The two control animals were not experimentally treated after the initial treatment and extraction at the start of the project. A second tooth was extracted from them only at the termination of the experiment (fall 1982) when a final extraction was carried out on all animals involved in the project.

Dolphin Health and Environment

We recorded average weekly changes in water temperature and salinity, and the average daily amount and species of food consumed by each animal, to identify any effects of these factors on dental-layering pattern or deposition rate (water

Table 2
Extraction and Labeling Schedule for Animal F as an Example of How Serial Data Were Generated for Analysis

Date or Period	Tooth Extracted	Labels Present	Labels Introduced
6-5-79	F1	None	A
9-7-79			B
9-11-79 to 10-9-79 daily			C (therapeutic)
11-1-79 to 11-4-79 daily			D (therapeutic)
12-3-79			E
3-10-80			F
5-28-80	F2	A through F	G
9-5-80			H
12-4-80	F3	A through H	I
12-10-80			J (therapeutic)
3-5-81			K
6-9-81			L
9-8-81			M
12-14-81	F4	A through M	N
1-15-82			O (therapeutic)
3-5-82			P (therapeutic)
6-4-82			Q
9-7-82			R
1-11-82	F5[a]	A through R	None

[a] Last tooth extracted; end of project.

temperatures and salinities were artificially controlled at Ohio and Florida parks). In addition, daily records were made of behavior, including type of activity (e.g., free feeding, training, show), periods of illness and treatment, and within-park transfers from one tank or pool to another. Finally, because SWA is 11° latitude farther north than SWSD and 14° latitude farther north than SWO, we tracked transfer of dolphins between parks (interpark transport) to try to detect possible effects of long-distance transport or changes in latitude on dental-layering patterns.

Tooth Preparation and Examination

Preparation and examination of teeth followed the methods described by Myrick *et al.* (1983). Each tooth was prepared in untreated mid-longitudinal thin section between approximately 100 and 150 μm thick. In a few cases, decalcified and hematoxylin stained thin sections, about 40 μm thick, were prepared from the remaining fragments (leftovers) to enhance resolution of layering patterns. Microscopic examination was in plain and polarized transmitted light to determine layering structure and in UV reflected light to locate tetracycline labels, visible as fluorescent lines (Fig. 1A,B).

The best sections of the series of teeth extracted from each animal were photographed in plain and UV light, and their interstructural (e.g., between the neonatal line and the pulp-cavity margin) and interlabel distances were measured. Distance measurements (in micrometers) were taken in a step-down fashion toward the pulp cavity from the neck region of the tooth near the base of and perpendicular to the neonatal line (Fig. 1B; see

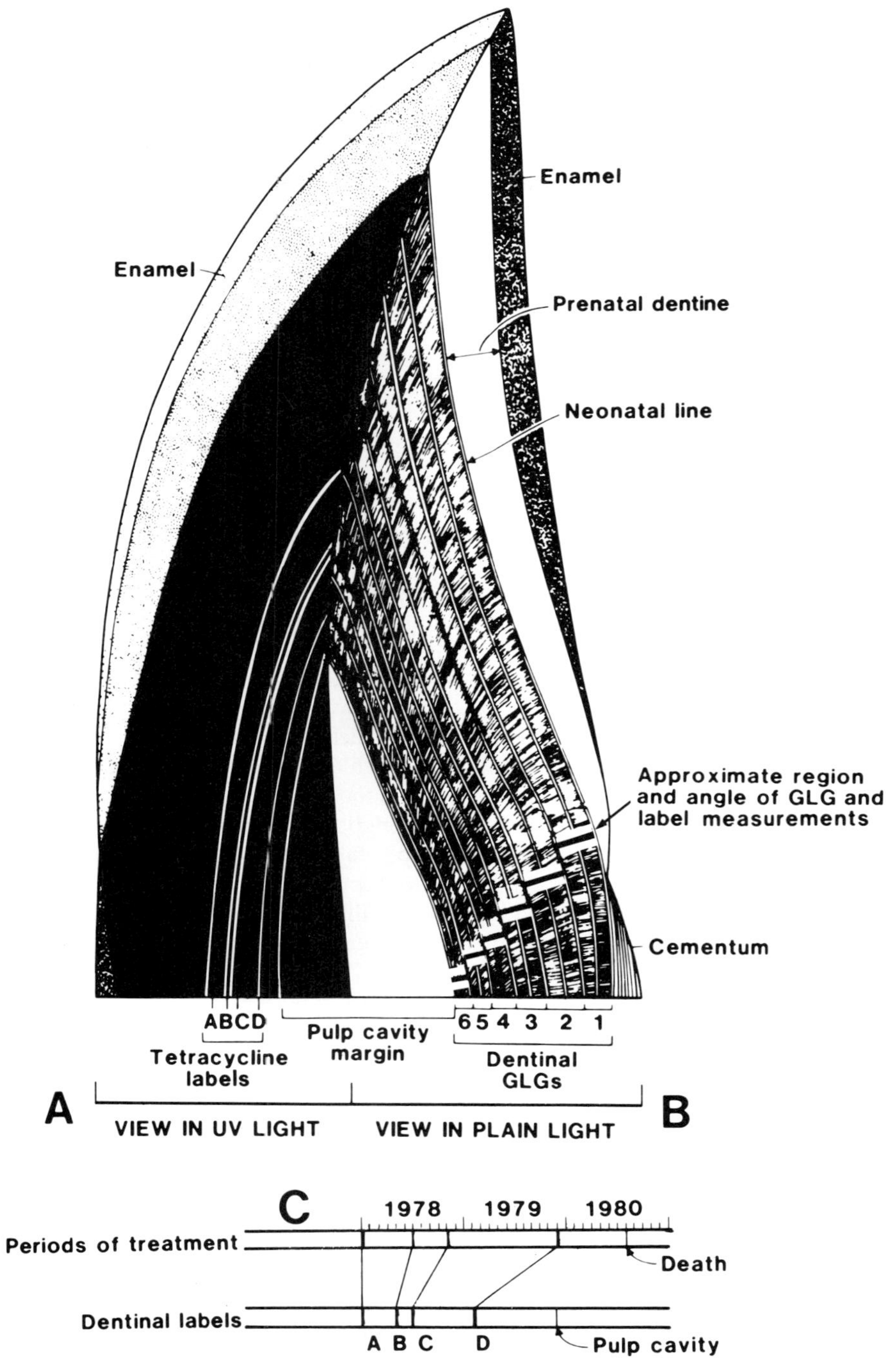

Figure 1 Diagram of tooth thin section from hypothetical project dolphin showing tetracycline labels, A–D under ultraviolet light (A, left-hand side) and dentinal growth layer group (GLG) layering patterns under plain light and standard positions in tooth where labels and GLGs are measured (B, right-hand side). (C) Method of identifying labels introduced into tooth and determining elapsed time between labels by comparing relative spacing of labels. (Modified from Myrick *et al.*, 1984.)

also Myrick *et al.*, 1984). (In this region the layers are usually the least distorted, neither highly expanded as in the apex, nor strongly narrowed as in the root base.)

The mean of three to five series of measurements was used for final values of GLG thickness (Table 3). Each series of thickness measurements was taken at a different time to minimize measurer bias. At least one full series of measurements was taken from the last tooth pulled from an animal because it contained the complete continuous record of the project period. Because the full 3.5-year monitored record had not been completed in teeth extracted before the end of the project, measurements taken from such teeth were augmented by measurements from teeth that were extracted later in the project, including the last tooth extracted. This helped to ensure, as well as possible, that GLG and interlabel distances in most of the teeth extracted from each animal were represented in calculating final mean values.

Data Analysis

Label locations on the UV micrographs were marked on clear plastic overlays, which were then aligned on the corresponding plain light micrographs to determine the location of the labels among the layers. Label positions for each animal were transcribed onto a calibration chart as a row of vertical lines, each separated from others and tooth landmarks (such as the neonatal line) on a scale representing 10-μm intervals. Above this row of label distances, the treatment and extraction data were transcribed as a row of lines scaled in months, representing time of treatment. The two data rows completed for each animal were used to identify corresponding project treatments and labels, designated alphabetically in the order that they were introduced (Fig. 1C).

Time-Calibrated Dentinal Patterns

To define annual or subannual depositional segments within the dentinal patterns, distance measurements between labels spanning a year or half-years were added together. Because the animals were often treated on different days or months, spaces between labels in the teeth represented various lengths of time. These different time-calibrated segments provided many opportunities to examine layering patterns in the continuum within 3.5 years of monitored deposition. Repeating patterns of yearly deposition could be identified, annual GLGs could be defined, and time represented by GLG component (fine) layers could be investigated.

Monthly Depositional Rates

There were enough data points from most of the dolphins to calculate average monthly depositional rates by dividing segment distance (in micrometers) by elapsed time of segment deposition. Although all twelve project animals were useful in determining annual GLGs, only eight of the twelve animals were useful in studying monthly rates. This was because the two controls did not have enough labels, and in one other animal (J) distances between labels were so short that measurements lacked precision. For the fourth animal (M, which died after 1 year and 3 months), the period was too short to yield enough data. For each of the eight measurable animals, average monthly depositional rates were compared with season, water temperature, and salinity data, and with periods of illness, behavior and feeding, and interpark transport.

Calibration of Preproject Layers

Because nine of the animals were captured at about 2 years old or older, the earliest formed layers were accumulated before capture and it was not possible to calibrate those GLGs directly. We characterized the patterns of initial GLGs exhibited in the three captive-born animals (A, B, and C), which, by the end of the project, had four and a half completely documented GLGs. However, we were reluctant to use these patterns before interpreting initial annual GLGs in the remaining animals for three reasons: (1) two of the captive-born dolphins were hybrids (mother and sire were from different populations), and it was not known whether hybridization affected dentinal-layer patterns; (2) all three had always been in captivity,

Table 3
Number and Mean Thickness (μm) of Completed Annual Growth Layer Groups in Dentine of 12 Captive Bottlenose Dolphins[a]

	GLG Number														
Animal	1	2	3	4	5	6	7	8	9	10	11	12	13	14	15
M	900*	500*	343												
	(41)	(44)	(69)												
	4	4	5												
A	858	536	458	461	( )										
	(38)	(56)	(19)	(11)											
	4	3	3	3											
B	881	500	348	375	( )										
	(22)	(56)	(38)	(21)											
	4	4	4	4											
C	471	327	298	281	( )										
	(13)	(15)	(5)	(13)											
	5	5	5	5											
D	825*	500*	370*	340*	234	239	190	( )							
	(35)	(0)	(28)	(14)	(11)	(18)	(10)								
	3	3	3	3	5	5	5								
P	505*	375*	350*	339*	352	350	337	( )							
	(10)	(50)	(0)	(14)	(22)	(35)	(32)								
	4	4	4	4	4	3	3								
E	875*	500*	425*	347*	300*	303	215	211	( )						
	(29)	(0)	(25)	(42)	(35)	(16)	(11)	(14)							
	4	3	5	4	5	5	5	4							
F	925?	500?	450?	375?	300?	195	291	191	( )						
	(29)	(0)	(0)	(29)	(0)	(31)	(41)	(53)							
	4	3	3	4	3	5	5	5							
I	813*	465*	440*	360*	300*	280*	225*	195*	180*	185*	170*	185	245	237	( )
	(35)	(21)	(10)	(10)	(0)	(0)	(15)	(15)	(0)	(25)	(65)	(39)	(66)	(32)	
	3	3	3	3	3	3	3	3	3	3	3	3	3	3	
J	890*	463*	450*	407*	353*	293*	230*	220*	207*	187*	167	140	143		
	(17)	(12)	(0)	(23)	(25)	(12)	(10)	(20)	(12)	(12)	(29)	(10)	(12)		
	3	3	3	3	3	3	3	3	3	3	3	3	3		
Controls															
K	817?	520?	450?	373?	320?	240?	237?	182?	160	137	(117)	( )			
	(58)	(27)	(0)	(58)	(35)	(30)	(35)	(17)	(10)	(15)	(6)				
	3	3	3	3	3	3	3	3	3	3	3				
L	820?	483?	450?	417?	407?	367?	283?	263?	243?	170?	170?	150	130	125	
	(27)	(29)	(0)	(58)	(51)	(29)	(29)	(47)	(60)	(17)	(17)	(0)	(0)	(0)	
	3	3	3	3	3	3	3	3	3	3	3	3	3	3	
Florida coast averages (A, B, D, E, F, I, J, K, L, M) (up to 14 GLGs)															
	862	497	418	384	316	274	239	210	198	170	156	158	173	181	
	(40)	(23)	(46)	(39)	(54)	(55)	(36)	(30)	(36)	(23)	(26)	(24)	(63)	(80)	
Pacific coast averages (C, P) (7 GLGs in one of two)															
	488	351	324	310	352	350	337								
	(24)	(34)	(37)	(41)	(0)	(0)	(0)								

[a] ?, Estimated annual GLG by inspection plus time in captivity; *, estimated annual using some documentation; empty parentheses indicate additional partial GLG present; values are means calculated from three to five measurements of GLG thickness; parenthetical values are rounded standard deviations of mean values; third row of values for each specimen is number of measurements used to calculate the mean.

and the effects of captivity were not known; and (3) the dentinal thicknesses of the first three GLGs in one (animal C) of the two hybrids was obviously different from those of the other two captive-born animals (A and B).

To try to calibrate preproject and precapture GLGs of the wild animals, we used estimated age at capture based on capture body length, or on increasing body lengths during the project, plus time in captivity to get a total age in years. (We knew that each animal must have been at least 1 year old when captured and thus the preproject age would have had to be at least 1 year plus total time in captivity.) We then looked for the same number of repeating layer groups as our age estimate. If our GLG count was one or two different than the age in years estimated for a young animal based on body length, we concluded that the body-length age was incorrect and that the GLGs were annual units. For a hypothetical example, based on a capture body length of about 225 cm, an animal estimated to be 3 years old is held captive for 1 year before its first label is introduced. Its teeth should have approximately four repeating GLG units between the neonatal line and the first project label. If only three GLGs occur, we would have to consider the capture-length age estimate of 3 years at 225 cm too high.

In some cases, we tried to augment the calibration of preproject layers by also using certain undocumented labels to define annual preproject GLGs. Here, we assumed that the label closest to the neonatal line and formed before the project began had been introduced at about the time of capture. We hoped to determine whether the number of GLGs formed prior to and after the "capture label" corresponded to age (in years) based on length at capture and time in captivity, respectively (Table 1).

When we decided that the first preproject label in a tooth was introduced at or near the time of capture (as was done in three of five animals), we used the label with the other, better documented labels from project treatments to calibrate all dentine, by inspection, behind and in front of the label. (In a case where the undocumented label was close to the first project label and the animal was captive for several years, we regarded the undocumented label as not a capture label, and calibration by this method was not attempted.) After preproject GLGs were defined in wild specimens by this method, we compared their thicknesses and patterns with those of the three captive-born animals to determine what similarities and differences existed. Adding the three captive-born dolphins to the animals with useful presumed capture labels brought to six the number of animals for which the initial GLGs were defined. We defined initial GLGs in the remaining six animals by inspection, estimated age at length, and time in captivity.

RESULTS AND DISCUSSION

Tetracycline Labels

Preproject Labels

Examinations of sections of the extracted teeth from captive-born dolphin C revealed a fluorescent label in the postnatal dentine approximately 25 μm from the neonatal line in the direction of the pulp cavity (Fig. 2A, C). This male calf was about 3 weeks old and nursing when its mother became sick and was placed on a 10-day treatment of 5 g of tetracycline daily. The closeness of the label to the neonatal line in animal C's teeth is evidence that tetracycline was transferred from mother to calf through the milk during the 10 days of treatment. It is also compatible with the popular assumption that the neonatal line is formed near the time of birth (Nishiwaki and Yagi, 1953; Myrick, 1981b). The calf was never given tetracycline before the project and could not have ingested medicated fish intended for other dolphins sharing the pool because he could not yet eat solid food.

In addition to the label produced in C through nursing, labels introduced before the start of the project were found in the teeth of five of the wild project animals, dolphins I, J, K, L, and P. There were no medical records of tetracycline treatment to account for these labels. Nevertheless, the earliest formed label in each of the three other noncontrol animals (I, J, and P) was used as though it had been introduced at or about the time of the animal's capture (Table 1). This was done to see if estimates of age at capture plus time in captivity in

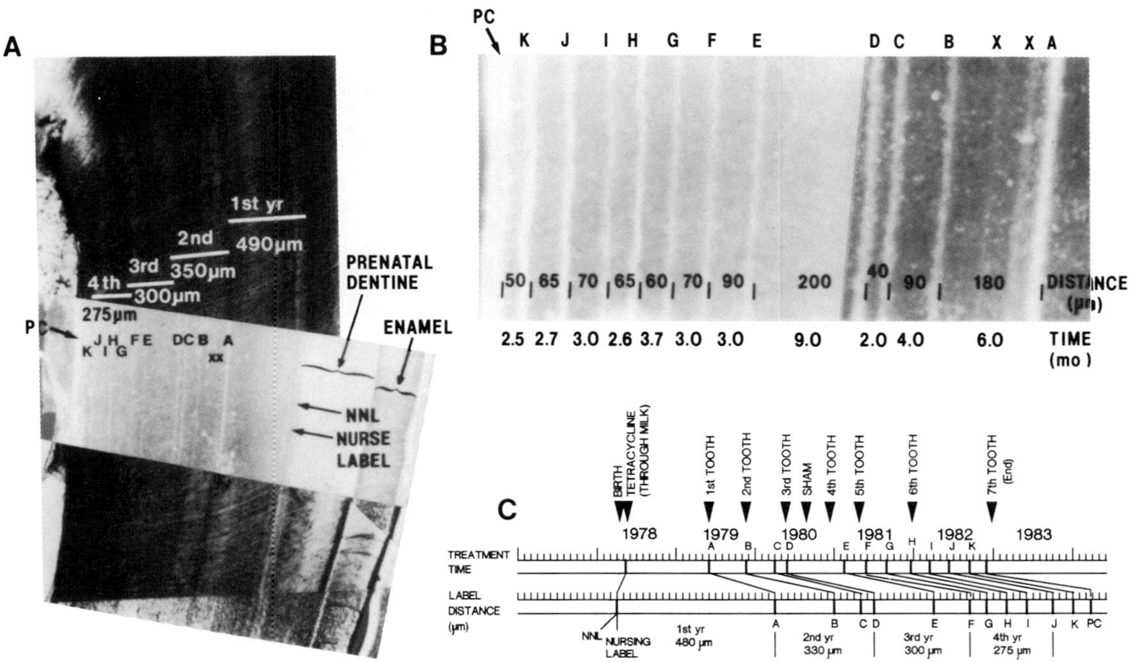

Figure 2 Dentine calibration of animal C. (A) Swatch of UV micrograph with tetracycline labels (lettered) superimposed on light micrograph of dentinal pattern (GLGs marked and measured); ×39. Nurse label internal to the NNL reflects introduction of tetracycline imparted by the mother through milk at 3 weeks after birth. Distance between the NNL and label A is the thickness of dentine deposited in the first 1.1 years. Faint labels between A and B and marked with X are without records. (B) Magnified (×150) UV micrograph of labels showing distances (micrometers) between labels (lettered) and time elapsed between labels (months). (C) Calibration chart showing labeling treatments and tooth extractions, label identifications, and annual dentinal thicknesses. (Abbreviations in this and other figures: PC, pulp-cavity margin; mo, months; yr, year; End, termination of project; 1st TOOTH, 2nd TOOTH, etc., tooth extractions; NNL, neonatal line formed at birth.)

years matched the number of GLGs defined in thickness and pattern that were counted in front of and behind the label.

Preproject labels were not used to estimate age at capture for the two control animals (K and L). This decision was made for animal K because its tooth layers lacked clear patterns in regions critical to interpretation of the labels. The two preproject labels in the teeth of L were too close to the project label (<300 μm), given the animal's 6.5 years in captivity, to have been introduced at capture.

For animal I, a pair of bright, closely spaced labels (X and Y, Fig. 3) occurred about two GLGs inward from the neonatal line (formed at birth) and nine GLGs external to the first project label introduced 20 April 1979. If the XY label is taken as having been introduced at or near capture, June 1970, the project would have begun approximately nine annual GLGs later, which is what the layering pattern showed. Accordingly, animal I would have been about 2 years old at capture, and birth would have been in 1968 (Fig. 3B, D).

For animal J, the presumed capture label was separated from the first project label (A, produced 11 April 1979) by six GLGs, which if formed annually would be compatible with our interpretation that the undocumented label was introduced at capture (December 1972) (Table 1). For animal P, we estimated that the preproject label (X, Fig. 4) was introduced at or about the time of capture for the following reasons. First, P was captured in February 1978 (Table 1), only slightly more than a year before the project began. Second, label X was slightly more than one GLG external to the first

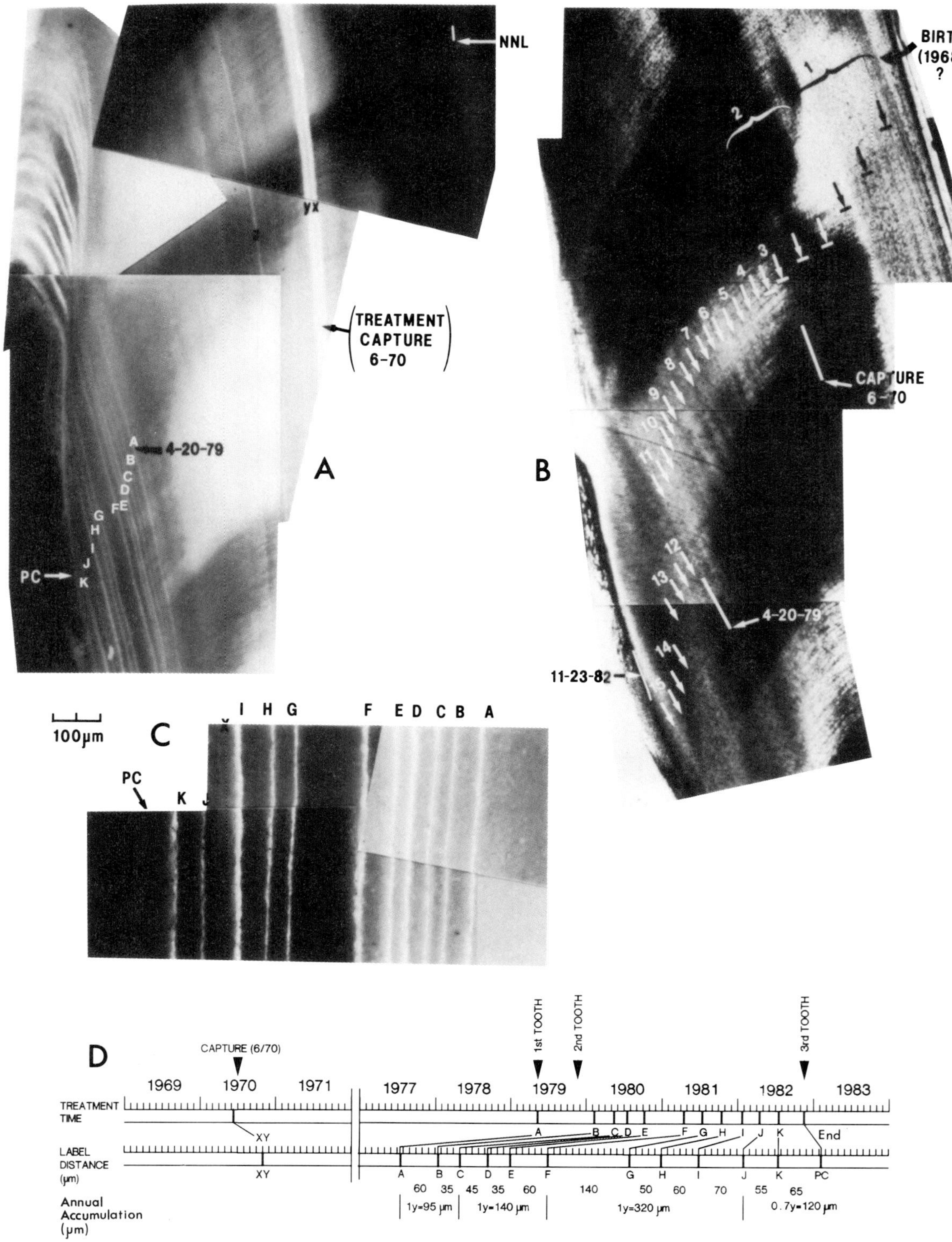

NNL
yx
TREATMENT CAPTURE 6-70
A
4-20-79
B
C
D
E
F
G
H
I
J
K
PC
BIRTH (1968) ?
1
2
3
4
5
6
7
8
9
10
11
12
13
14
15
CAPTURE 6-70
4-20-79
11-23-82
100μm
C
PC
K J
X I H G F E D C B A
D
CAPTURE (6/70)
1st TOOTH
2nd TOOTH
3rd TOOTH
1969 1970 1971 1977 1978 1979 1980 1981 1982 1983
TREATMENT TIME
XY
A B C D E F G H I J K End
LABEL DISTANCE (μm)
XY
A B C D E F G H I J K PC
60 35 45 35 60 140 50 60 70 55 65
Annual Accumulation (μm)
1y=95 μm
1y=140 μm
1y=320 μm
0.7y=120 μm

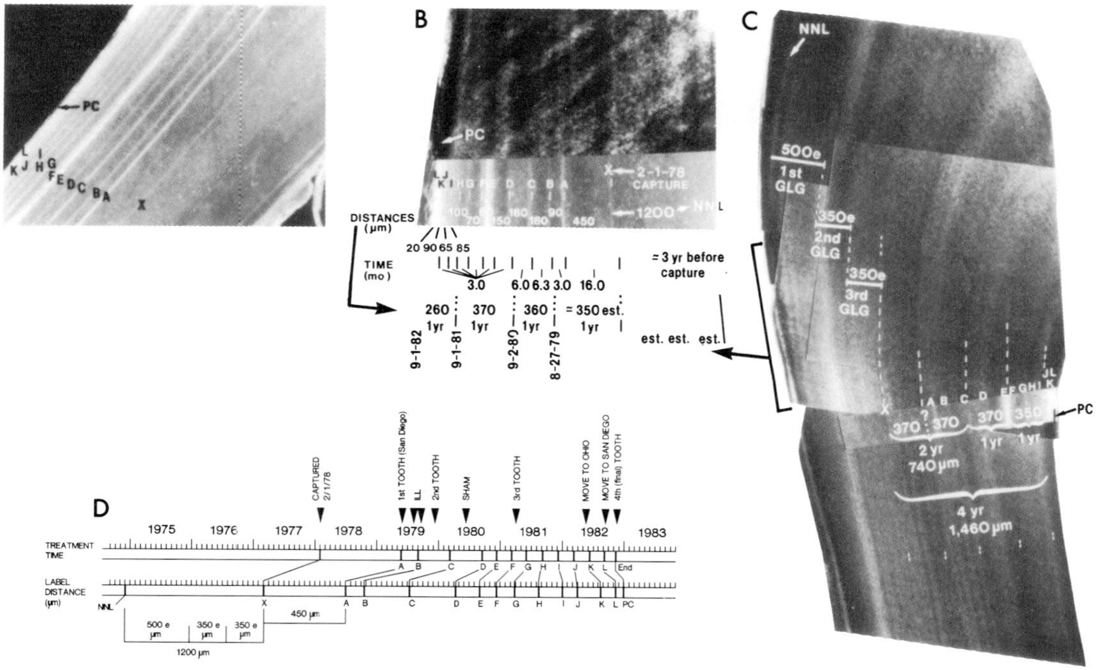

Figure 4 Dentine calibration of animal P augmented by a capture label. (A) UV micrograph showing all labels including one, X, introduced before the beginning of project, probably at about capture. (B) Light micrograph with swatch of UV micrograph showing positions, distances (in micrometers), and time (in months), and estimated annual depositional thicknesses. (C) Position of labeled portion of dentine and total presumed annual GLGs (marked by dashed lines) in last tooth extracted. Based on project body length increases, P was about 3 years old when captured. (D) Calibrated labels and estimated annual GLG thicknesses deposited before the project label (label A). (Original magnification of A and B, ×150; C, ×39. e, Limits of annual GLGs estimated by inspection and consideration of preproject labels and background data; see Fig. 2 for other abbreviations.)

project label (A). Third, the distance of label X from label A was about the same as or slightly greater than the distance between successive project labels representing almost the same amount of elapsed time. Finally, four GLGs occurred between the neonatal line and the first project label (A), and the body length of P, when label A was introduced, was 268 cm (length continued to increase through the project period to 284 cm). The length at the beginning of the project indicates that of a juvenile of about 4 years old (Table 1).

Thus, there is support, from the three specimens in which the earliest preproject label was assumed to have been introduced at capture, that GLGs defined in this study represent annual

Figure 3 Dentine calibration of animal I. (A and B) UV and light micrographs, respectively, of final extracted tooth (pulled 11-23-82). Three preproject labels (X, Y, Z) are contained in the dentine. Labels XY may have been introduced at or about the time of capture considering the number of apparent GLGs (i.e., double dark and light layer pairs) between the XY labels and the neonatal line and between the labels XY and A, introduced at the beginning of the project, 4-20-79, for this animal. This estimate would be compatible with the capture date of June 1970 given by park records. Numbered paired arrows indicate paired dark layers in the second GLG; triple arrows point out three dark layers in GLG number one. (C and D) The gap between labels F and G is unexpected and represents a break in the rather regular deposition rate. Label X is a faint label without a record. (See Fig. 2 for other abbreviations.)

units. Nonetheless, whether the labels we used actually represented treatments at or near capture is not directly proved.

Intentional Project Labels

In all specimens, all intentionally introduced tetracycline labels were identified in the dentine (e.g., Figs. 2, 3, 4, and 5) but not in the cementum. Apparently the tetracycline doses (20–30 mg/kg body weight) used to label dentine were too weak to produce distinct cemental labels. Because of this, we concentrated mainly on calibrating dentinal layers, and only when strong therapeutic concentrations of tetracycline had been used were labels in dentine and cementum compared.

Therapeutic Project Labels

Labels other than those scheduled in the project were inadvertently introduced when tetracycline was used medicinally during ill health of some of the animals. These therapeutic treatments occurred in the teeth as exceptionally bright, wide labels (e.g., Fig. 5B). We used them as additional data points and to demarcate tissue formed during sickness and recovery.

Unidentified Project Labels

Dolphins A, C, and I each had undocumented labels, introduced during the project, that were much fainter and more irregularly spaced than

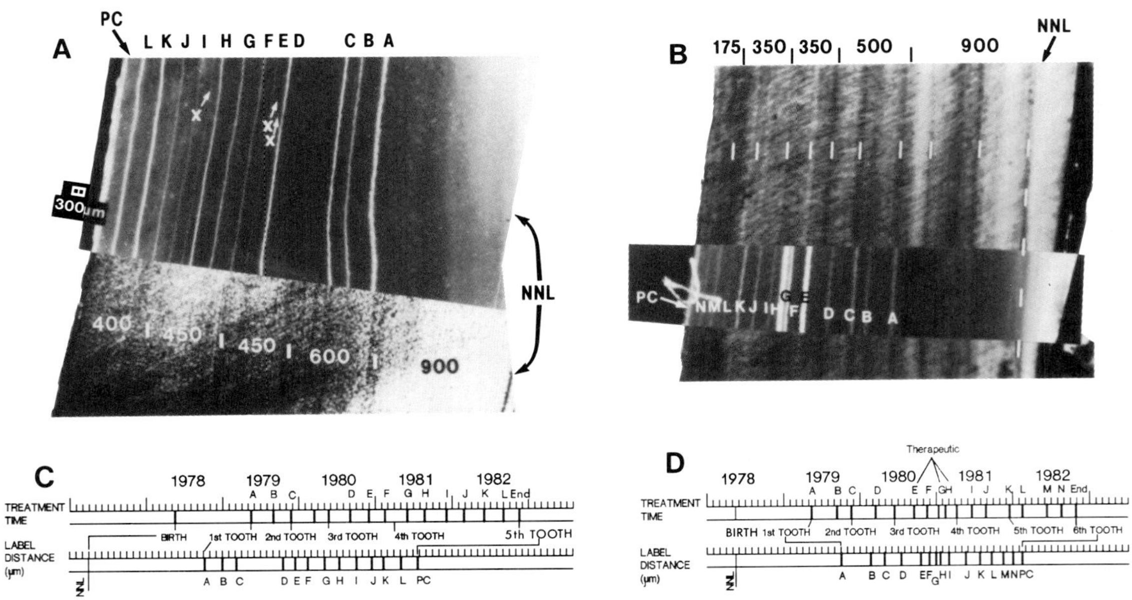

Figure 5 Dentinal calibration for animals A and B. (A and B) UV micrographs with tetracycline labels (lettered) superimposed on GLG patterns (measurements) of the final tooth extracted from animals A and B, respectively. (B) White vertical bars indicate the tripartite nature of first (900-m-thick) GLG and bipartite nature of subsequent GLGs. Faint labels indicated by X's in A are without dates. (C and D) Calibration charts for animals A and B with dated tetracycline labelings and tooth extractions connected with labels, lettered in order of introduction. The substantial gap between labels C and D in micrograph A reflects protracted period, shown in chart C, during which no tetracycline was administered to animal A. Chart D for animal B indicates therapeutic treatments of tetracycline as evidenced by thick, bright labels E, G, and H shown in micrograph B. (Original magnification of photos, ×39. See Fig. 2 for abbreviations.)

other labels (Figs. 2A, 3C, and 5A). Because all intentional labels were accounted for in these specimens, we ignored the faint labels. Presumably the labels were formed when the animals ingested fish medicated with tetracycline, intended for other dolphins sharing a common tank. A similar case was reported by Myrick *et al.* (1984) in their study of tetracycline-labeled Hawaiian spinner dolphins.

Dentinal Growth Layer Groups

Untreated Sections

In transmitted light, well-prepared untreated sections of teeth from all project animals had a repeating pattern of layers comprising a GLG similar to those described by Myrick *et al.* (1984) for Hawaiian spinner dolphins. Except for the first-formed GLG, which exhibited three pairs of layers when resolution was adequate, a GLG consisted of two light layers each followed by a darker layer. The more external light layer formed the outer boundary of the GLG, and the second light layer usually was found about midway through the GLG (i.e., "mid-GLG layer," Myrick *et al.*, 1984). Each new GLG began with a light boundary layer formed at the internal margin of the second dark layer of the previous GLG; the dark layer usually highlighted the light boundary layer (Figs. 3B and 5B).

Decalcified and Stained Sections

Because leftovers from the sides of a central untreated section were used to produce decalcified and stained thin sections, most were noticeably off-center, and few were of sufficiently high quality for detailed GLG thickness comparison with their untreated counterparts. The best specimens showed GLGs with thin, dark-stained boundary and mid-GLG layers and a more lightly stained background (Fig. 8A).

The first GLG consisted of three pairs, instead of two pairs, of thin, dark-stained layers and thick, light-stained layers. The two-pair pattern was common in the dentine formed after the first GLG. However, in some cases, the GLG boundary layer was indistinct because one or more thin, dark-stained accessory layers occurred nearby [e.g., note the boundary layer(s) between GLGs 4 and 5 in Fig. 8A].

Calibration of Growth Layer Groups

Age-Specific Growth Layer Group Thickness

Mean thicknesses of GLGs in the 12 animals are presented in Table 3 along with rounded standard deviations. In comparisons of thickness between GLGs, annual dentine deposition had decreased with increased age in all 12 of the project dolphins. Additionally, serial measurements showed that each annual dentinal GLG was deposited at a more or less predetermined thickness according to the age of the animal.

The age-specific GLG thickness pattern (especially for the first three GLGs) was similar in all wild animals from the same geographic area whether the layers were formed in captivity or in the wild. All wild dolphins had already formed at least two GLGs before capture, and the teeth of two captive-born dolphins exhibited GLG thicknesses similar to the Florida specimens. The other captive-born dolphin (animal C) had GLG thicknesses resembling those of the wild animal (P) from California.

Population-Specific Patterns

Comparisons of mean age-specific dentinal thicknesses for the 12 individuals shown in Table 3 indicate the possibility that two depositional patterns exist: one for Florida-coast animals and one for California-coast animals. All wild Florida-coast animals deposited more than 800 μm of dentine the first year, more than 450 μm the second year, and usually more than 350 μm (418 μm average) the third year. This was also true of the captive-born Florida purebred (animal B) and the captive-born hybrid of a Florida male and a Pacific-coast female (animal A) (refer to Table 1). The one wild animal from California (P) apparently deposited about 500, 375, and 350 μm of dentine in his first, second, and third years, respectively. The captive-born hybrid (C) sired by a purebred California male shows a dentinal-depositional pattern (i.e., 471, 327, and 298 μm) much nearer that of animal P than those from Florida animals (Table 3 and Fig. 6).

The Pacific-coast sample (n = 2, including animal C) was too small for conclusive results. Nonetheless, it is at least interesting that a thickness pattern so different from the other 10 project animals was found in the wild specimen from California and the one captive-born hybrid male sired by a wild California animal. Figure 6 includes mean values and standard deviations for GLGs one and two. In these GLGs, the two specimens have values more than three standard deviations below mean values of the Florida-coast animals. This indicates that the probability of these two patterns being part of the same population is very low; for the first GLG the probability that the two patterns are from the same population is less than 1.0%.

Monthly Layers

Labels were spaced closely enough and fine dentinal layering was clear enough in several specimens (especially in B, D, and F) to investigate the timing of what have been thought to be monthly layers (e.g., Laws, 1962; Kasuya, 1977; Myrick, 1979) or lunar monthly layers (Myrick, 1981b; Myrick *et al.*, 1984). In the second extracted tooth of D, labels A and B had been introduced 3 months apart, and 3 additional months were represented between label B and the pulp-cavity margin (Fig. 7). Almost four full layers occurred between labels A and B, and there were three layers between label B and the edge of the pulp cavity. In the third extracted tooth of animal D, 12 months were represented between labels A and D. Thirteen pairs of dark–light layers were visible within that segment. In the sixth (last) tooth taken from D, further deposition of monthly layers was identified by comparing layers with elapsed time between labels on the calibration chart (Figs. 7 and 8). In other animals, tooth sections with well-developed fine layering show similar results. This supports calibration observations made earlier by Myrick *et al.* (1984) that fine dentinal layers in dolphins are deposited with lunar (or at least monthly) regularity. Thus, monthly layers have

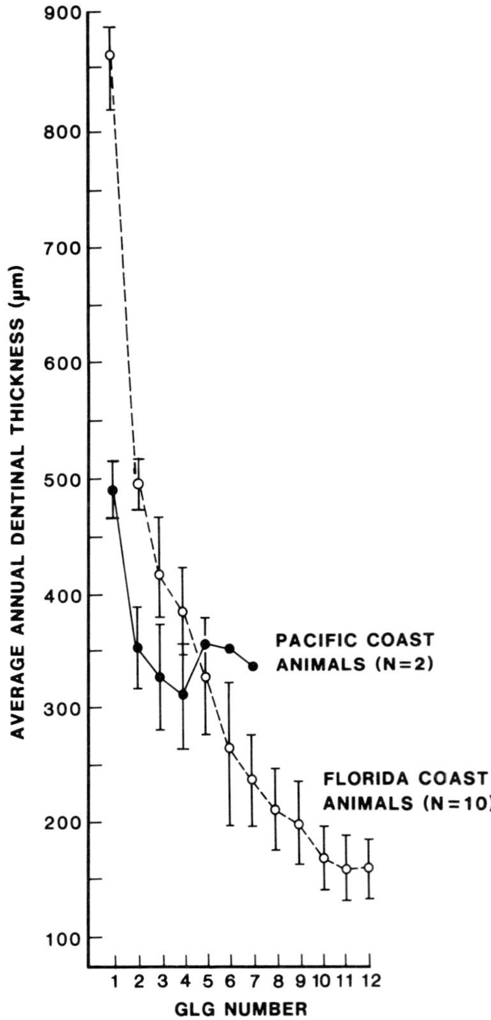

Figure 6 Average dentinal GLG thickness in Pacific-coast and Florida-coast bottlenose dolphins. Vertical lines represent one standard deviation from mean values.

now been verified for time in specimens of two species, Hawaiian spinner dolphins and bottlenose dolphins.

Monthly Rates of Deposition

Enough between-label measurements were available to plot mean monthly accumulated thicknesses (in micrometers) of dentine during the

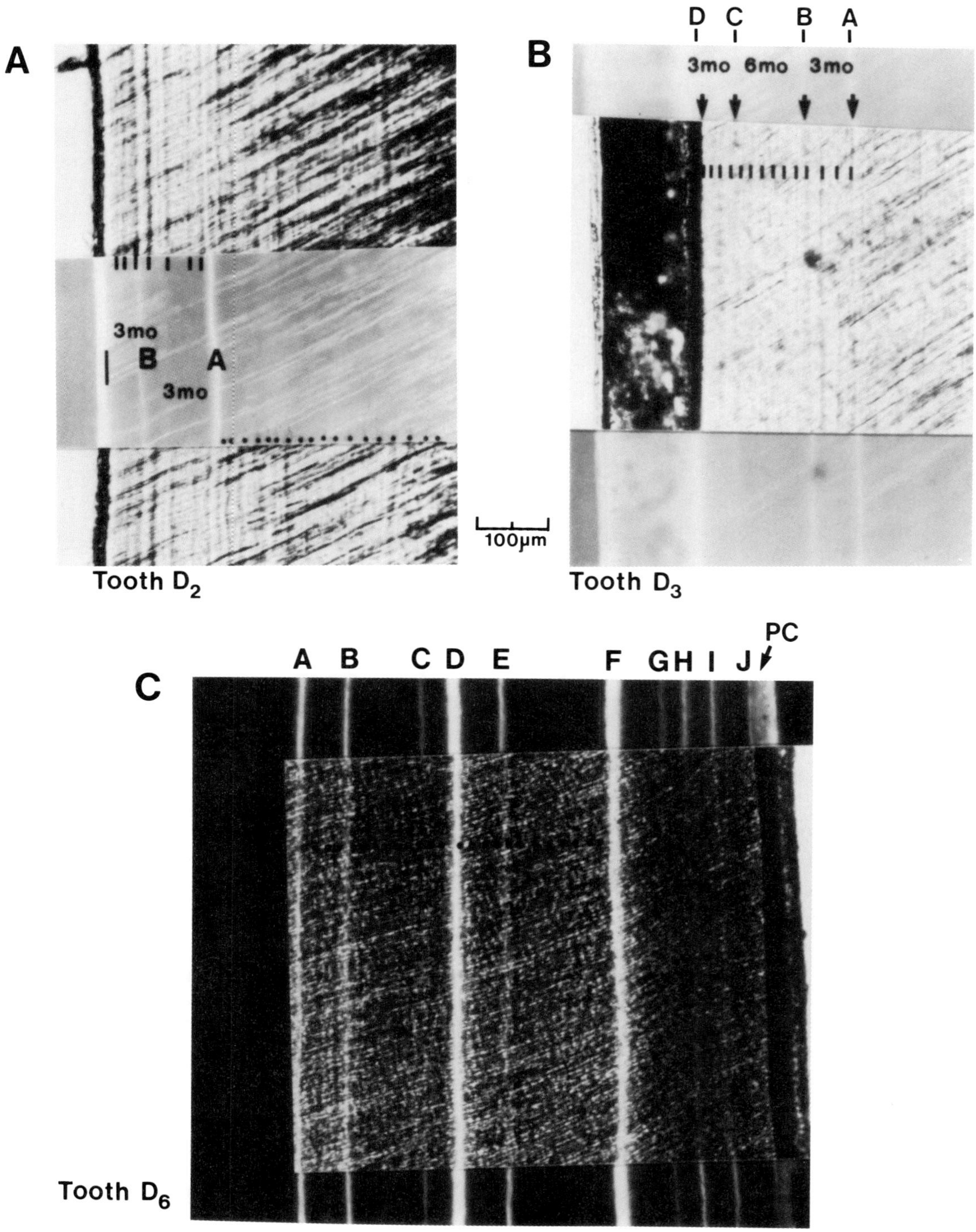

Figure 7 Accumulation of labeled dentine in animal D. (A and B) UV micrograph swatches on light micrographs of dentine in second and third tooth extracted, respectively. The accumulating labels, dentine, and elapsed time are indicated. Drawn marks suggest numbers of fine layers between labels. (C) Swatch of light micrograph on UV micrograph of dentine in the last tooth extracted from animal D. (Original magnification, ×150.)

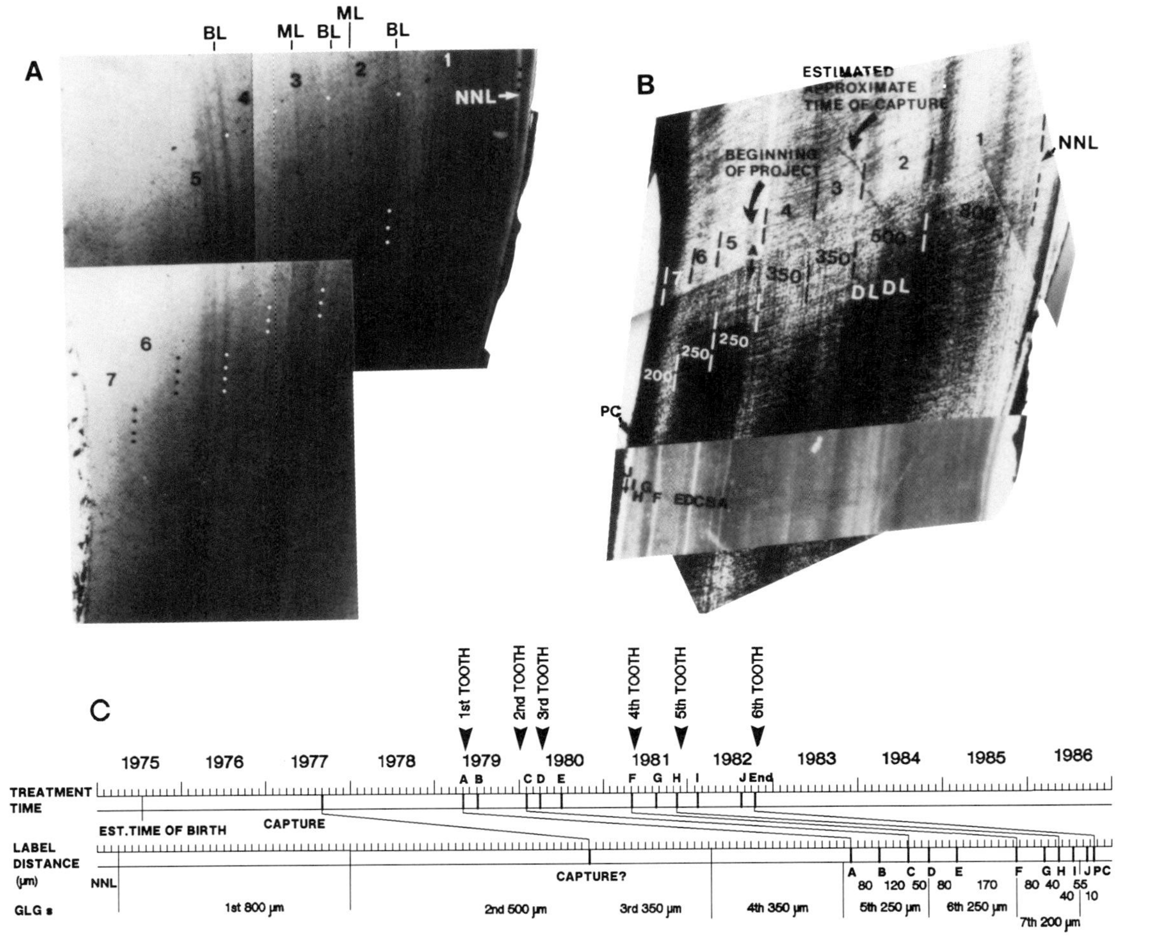

Figure 8 Calibration of GLGs in the last extracted tooth of animal D. (A) Decalcified and stained section showing GLG patterns featuring a single thin dark boundary layer (BL) (dotted lines) and a single thin dark mid-GLG (ML) layer. The boundary between the apparent fourth and apparent fifth GLG is indistinct due to the occurrence of several thin, dark-stained layers. (B) Untreated section with UV swatch showing labels (lettered). GLGs (bounded by dashed lines) are numbered and show thicknesses in micrometers. (C) Calibration chart showing birth and capture dates (estimated from capture body length and time in captivity) and annual GLG thicknesses based on estimated and monitored dentine deposition. (Original magnification of A and B, ×39. DLDL indicates dark–light–dark–light layer pattern in second GLG; GLG number one has a tripartite pattern; see Fig. 2 for other abbreviations.)

3.5-year project for eight of the twelve dolphins. Rates of deposition were computed for each animal by dividing thickness by the elapsed time in tenths of months. Because of age-specifc depositional patterns, younger individuals accumulated considerably more dentine per month than others. For this reason, and because of the possibility of population-specific thickness patterns, average monthly deposition by each animal during the monitored period was plotted on the same graph. The abscissa is scaled in 10-μm intervals, with absolute values omitted, and the period of the project, from spring 1979 to fall 1982, is represented on the ordinate with vertical lines set at successive January firsts (Fig. 9).

A common pattern in dentine depositional rates was discerned to some extent in all specimens: increased rates in spring and summer with de-

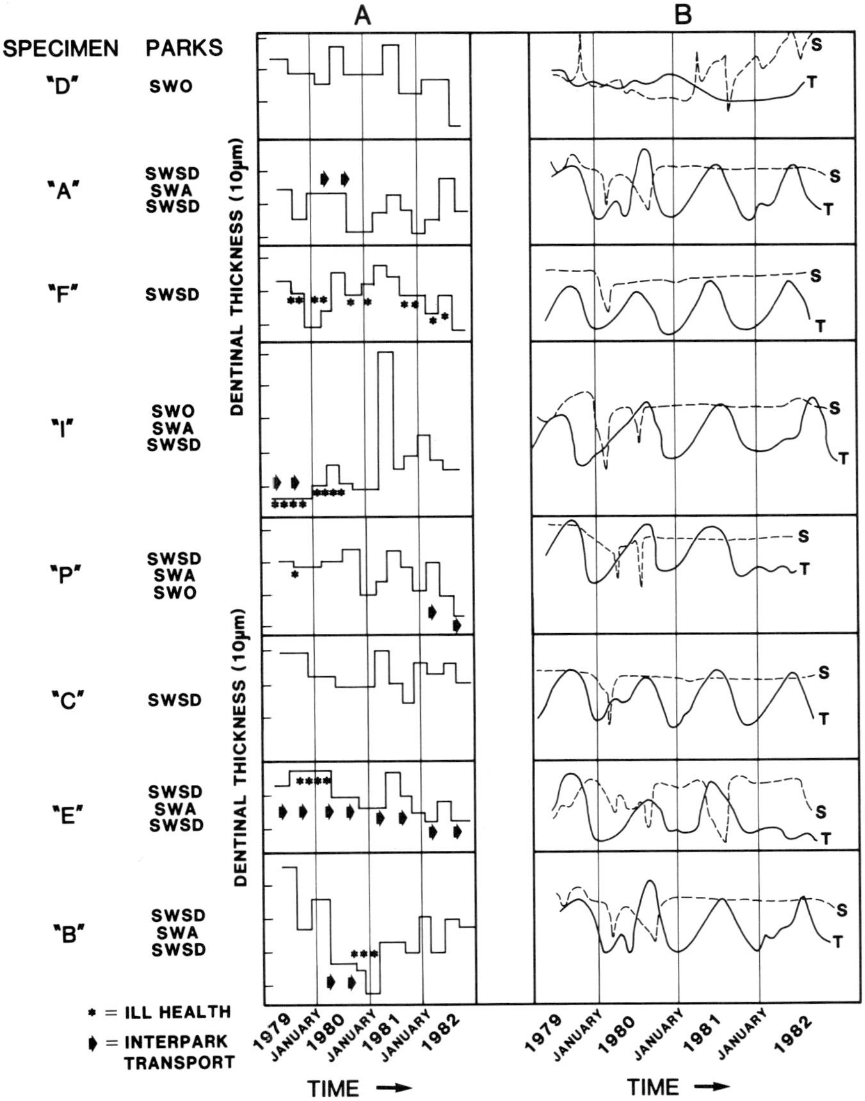

Figure 9 Monthly depositional rates and external environmental factors. (A) Relative average monthly rates of dentine deposition in eight bottlenose dolphins over a 3.5-year period, with plots of weekly measurements of water temperature (T) and salinity (S) (smoothed by eye) shown for each. (B) Vertical lines indicate January first of years shown. Rates tend to increase in individuals in spring–summer months and to drop in fall–winter months, although considerable wandering from this pattern is apparent. Depositional rate changes roughly correspond to but are not preceded by fluctuations in water temperature. They are not correlated with changes in salinity, nor with transport between Sea World parks (SWO, Florida; SWA, Ohio; SWSD, California). Arrows represent times of interpark transport, and asterisks indicate periods of ill health.

creased rates in fall and winter (Fig. 9A). This pattern, typified in plots for dolphins D and F, showed distinct and synchronized peaks and dips in mean rate over the entire 3.5-year period. The resemblance of relative changes in rates between these two animals was especially remarkable considering that D remained at SWO (Florida), where temperature and salinity are manipulated, and F remained at SWSD (California), where these two parameters are subject to ambient fluctuation.

Despite the striking similarities among many of the animals in depositional-rate cycles, however, there was noticeable variation in timing (e.g., B and E) and amplitude (e.g., I and P) within and between individuals. To test whether the spring–summer rate increase and fall–winter decrease amounted to a significant pattern, we compared synchronized drops and rises in rates among the eight animals for the 3.5-year period. For an animal, in each time segment (spring–summer or fall–winter) there can be only one of three possible outcomes: a peak matching a spring–summer segment, a drop matching a fall–winter segment, or something else that disrupts the ideal model of synchronized rate changes. If there is a common seasonal pattern that animals follow, there should be little difference in the frequency distribution of synchronized rate changes among animals.

A contingency table was set up with the rows representing animals and columns representing the three possible outcomes. Calculations gave a χ^2 value of 13.174 with $(8 - 1)(3 - 1) = 14$ df ($p = .5$). This indicates that the frequency distribution of the three outcomes is independent of the animals, and we conclude that the animals follow the synchronized cyclic model. Out of a potential 32 peaks and 32 dips in an ideal model of eight eight-season rate changes, the sample showed a 40% probability of peak (26 out of 64), a 40% probability of dip (again, 26 out of 64), and 12 out of 64 or a 20% probability of neither a synchronized dip nor peak.

Factors Possibly Affecting Dentinal Deposition

Exogenous Factors

The cyclic depositional rates of D illustrate how parameters were evaluated as possible influencing factors. Animal D had one of the most regular, cyclelike depositional patterns, yet experienced the most uniform environment. She was not transported between parks during the project, was never ill, and was not subjected to much or sudden fluctuation in water temperature (although water salinities temporarily fluctuated sharply several times). She usually ate no less than about 7.5–13 kg (20–35 pounds) of fish daily. She was not trained and did not participate in shows in the first year of the project. In the second year, training and show activity was frequent but did not affect her daily fish consumption; nor was it reflected by any noticeable difference in layering patterns or depositional rates in her teeth. In addition, D was kept at a lower latitude (Orlando, Florida) than any other animal in the project, yet she had one of the most distinctly cyclic changes in rate. This is contrary to what one might expect if day length were an important *zeitgeber* for changes in dentinal-deposition rates.

For all individuals, with the important exception of D, there were regular seasonal changes in water temperature which corresponded roughly with depositional rate changes in dentine, but none of the changes in water temperature or other physical parameters that we monitored consistently preceded changes in depositional rate (Figs. 9A, B). Thus, they do not seem to be causally related. In some cases, when water temperature had risen or had begun to rise, dentinal rate remained unchanged (e.g., P and C in 1980) or dropped (e.g., B and E in 1980). In other cases, when water temperature had started to decrease or had already reached a nadir, dentinal-deposition rate rose (e.g., E and B in winter 1979) or remained unchanged (e.g., C and E in winter 1980).

Perhaps more importantly, at various times depositional rates changed while water temperature (and other parameters) did not change. The rates of D in all years and of P and E from fall of 1981 to fall of 1982 followed a pattern of fall–winter decrease and spring–summer increase even though water temperature remained fairly constant. For D, mean water temperature was 16.4°C, with gradual changes between 13° and 20°C; for P and E water temperatures for 1982 fluctuated slowly between 14° and 18°C (Fig. 9B).

Changes in types of food given had no detectable effect on depositional rate. For example, A was fed fish of only one species in 1979, fish of mixed species in 1980, and fish and squid in 1981–1982. None of these changes were reflected in A's near-perfect cyclicity of winter decreases and summer increases in dentinal deposition.

Depositional rate changes and layering patterns typified by A, D, and F were unaffected by interpark transport. This is illustrated in A in 1980 and P in 1982. It is shown especially well in E. Dentinal accumulation for E was normal in 1981 and 1982 despite semiannual airplane transport of this dolphin (shown as arrows in Fig. 9A) between California and Ohio in all years during the project.

Endogenous Factors

A rare accident permitted us to examine possible effects of short-term stress in measurably altering layering patterns. A project animal leaped out of its tank and apparently was out of the water on the ground for hours before it was discovered. It was returned to its pool, presumably somewhat stressed. Using tetracycline labels to pinpoint the region of dentine formed at the time the dolphin was out of the tank, we were unable to detect any unusual lines or layers that might have reflected a physiological response to that incident.

Periods of sickness, however, did appear to affect depositional rate. The rate plot for F (Fig. 9A) is accompanied by several clusters of asterisks symbolizing bouts of sickness, usually indicated by inappetence. Distribution of these symbols corresponds with diminution of amplitude in changes of rate. For example, the drop in the rate in January 1980–1981, common in other dolphins plotted, is shallow in F, and the expected rise in spring–summer of 1982, common in most of the other animals, is equivocal in F. Conversely, the peaks in spring–summer 1980 and 1981, when F was not sick, are present in the plot. Similarly, P did not exhibit a decrease in depositional rate during fall–winter 1979, when he had a poor appetite and was receiving medication. However, his depositional pattern cycled normally during other years.

Animal E was not well during fall–winter of 1979. Beginning at about the same time, his dentinal-deposition rates departed from the presumed common pattern, which was not reestablished until winter 1980–1981. For I, symbols for interpark transport are concurrent with those of ill health in 1979, and the condition apparently was not normalized until mid 1980. During the period of sickness, there was very little change in the slow dentinal-deposition rate. The common pattern appeared in I in the spring of 1981 in what could be considered a compensatory acceleration of the rate. In fact, however, we have no data that provide a satisfactory explanation of this sudden escalation in rate. It is possible that this spurt in deposition represents variation in one tooth. It was documented in only the last tooth that was pulled.

For C, we have no data to account for either the lack of a peak in mid 1980 or an early increase in rate in winter of 1982. He remained in California during the entire period and was never ill or medicated. Salinity decreased sharply in early 1980 in C's tank, but similar sudden drops in salinity for other animals on various occasions (e.g., A in 1980 and E in 1981) did not result in disruptions of normal depositional cycles. If ill health was a factor in the apparent distortion of C's pattern, it was not detected by our observers.

Depositional rates of B differ markedly from those of A, although the two dolphins were subjected to almost identical conditions during the project (Fig. 9A, B). Part of this difference may be artifactual; the broad peak in the first half of 1980 for A is due in part to a mean rate computed using one measurement over 9 months during which no labels were introduced (Fig. 5A, C). However, comparisons of thicknesses of the second, third, and fourth GLGs for the two animals show that annual deposition was indeed less in B (Table 2). Animal B had a decreased appetite shortly before returning to San Diego from Ohio in 1980, and his record for the next 6 months is a picture of intermittently low food consumption accompanied by therapeutic treatments to help correct the condition (Figs. 5D and 9A). He was fully recovered by spring 1981, but his dentinal-deposition rate pattern remained largely unsynchronized relative to that exhibited in the other animals for the remaining 17 months of the study.

Cementum

Although cemental GLGs, each consisting of a light layer and a dark layer, were visible and countable, calibration of cementum was not possible in this study. The labels that we tried to introduce experimentally did not show up distinctly, apparently because our dosages of tetracycline were too weak for detection in cementum. We also were unable to match dentinal GLG counts with those of cementum using therapeutic labels. Often the cementum was too thin, or had fewer GLGs than the dentine, or showed the therapeutic labels in the wrong relative position. For example, in animal I the *XY* label, visible in about the third annual dentinal GLG (Fig. 3), was identified in about the eighth or ninth cemental GLG (Fig. 10). Cemental GLG counts, therefore, may be difficult to use as reliable indicators of age for this species.

CONCLUSIONS

Results of this long-term monitoring and marking study suggest that bottlenose dolphins deposit a specific annual thickness of dentine at a given age in a predetermined pattern which is largely under endogenous control. In healthy animals of our sample, the pattern of GLG deposition was generally unaffected by conditions of captivity, including changes in water temperature or salinity, physical activity, at least some kinds of short-term stress, type of food consumed, and possibly even differences in latitude.

Rates of dentinal deposition were fairly cyclic, tending to increase in the spring and summer and to decrease in the fall and winter. They seem to have altered somewhat when the animals were ill and for some period after the illnesses. Apparently, they were not strongly entrained by any external *zeitgeber* that we were able to detect. However, the following observations suggest that some entrainment cue or complex of cues exists: (1) there was a rough correspondence of depositional rate changes with regular changes in water temperature, with a number of important exceptions, (2) to some extent rate peak and depression varied in time and in amplitude in an animal and

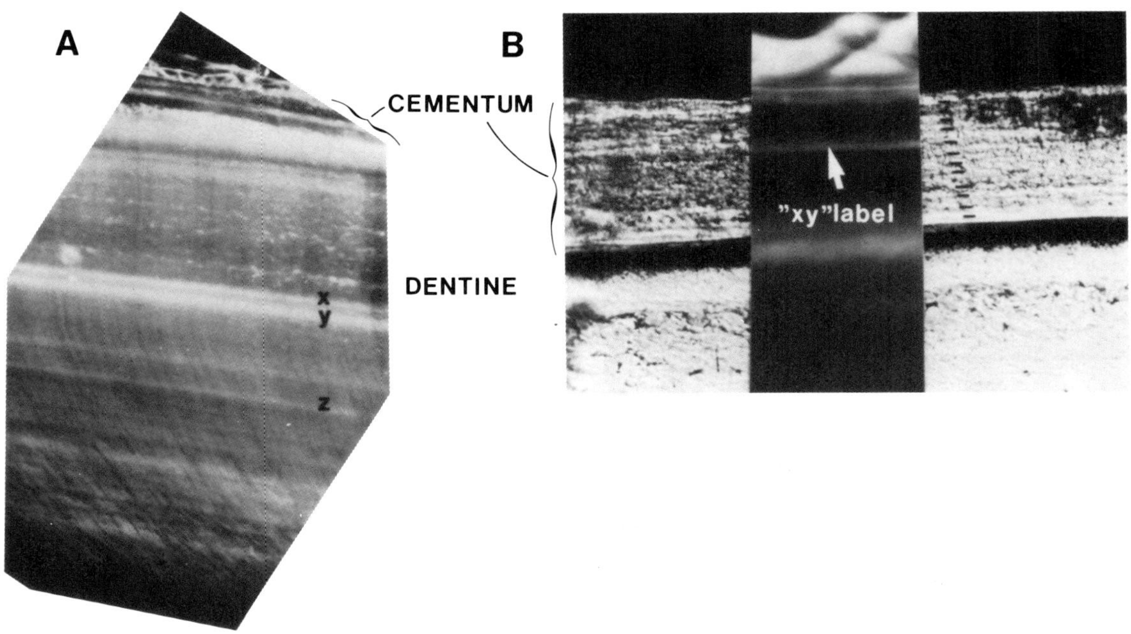

Figure 10 Unreliability of cemental layers for estimating age. (A) UV micrograph of area of tooth tissue including cementum from animal I (more detail and calibration are shown in Fig. 3). Preproject labels *X* and *Y* were presumably introduced at or near capture, when I was only a few years old. The labels occur in approximately the third dentinal GLG (×39). (B) Light micrograph of cementum from I with swatch of UV micrograph showing what is interpreted to be the *XY* label in approximately the eighth or ninth cemental GLG (×150). This example suggests that cemental GLG deposition is not necessarily correlated with that of dentine.

between animals, and (3) in some specimens, after being unsynchronized for a time, deposition cycles usually became resynchronized. The *zeitgeber* was not identified in this study.

Although the sample from the California coast consisted of only two animals including one captive-born hybrid, comparisons with the sample from Florida suggest that the age-specific annual thickness of dentine may be manifested as two population-specific patterns: one for Atlantic animals and one for Pacific animals. Studies using labels in additional captive animals from the Pacific should help to decide this question.

In untreated thin sections viewed in plain light, the annual GLG consists of two light layers each followed by a darker layer. In decalcified and stained thin sections, in plain transmitted light, each annual GLG tends to have one or more dark-stained layers near its boundary and one or more near its midpoint. The dark-stained mid-GLG layer(s) bisects the otherwise lightly stained GLG. All GLGs tend to be bipartite except the first GLG formed, which tends to be tripartite. Fine layers within the GLGs are probably formed monthly.

We were not able to calibrate cemental layers with dentinal layers in this study. Tetracycline labels could not be used to match GLGs in dentine and cementum. This suggests that cemental layers may have no simple relationship to annual GLGs deposited in dentine. In our opinion, cemental GLGs alone should not be relied on for estimating ages of bottlenose dolphins.

A tetracycline label immediately internal to the neonatal line in one captive-born animal was coincident with tetracycline treatment of its mother. This supports the suggestion by Myrick *et al.* (1984) that it may be possible to treat nursing calves in ill health with certain drugs that combine with constituents of milk by medicating the lactating mothers. The label's closeness to the neonatal line is also corroborating evidence that the neonatal line forms at or near the time of birth.

ACKNOWLEDGMENTS

We are grateful to V. Gurevich, J. Antrim, and W. E. Evans for seeing to the many needs of the project. We are especially indebted to P. Yochem for her tireless efforts in coordinating events, tracking the progress of the project, and collecting, organizing, and delivering the data. C. Dargan, D. Stanley, and L. Arnell prepared many of the specimens for study. M. Pedroarena and R. Smith ordered and tabled data for analysis. R. Allen and H. Orr were meticulous in their preparation of the figures. N. Lo gave statistical advice and help. We thank several anonymous reviewers as well as A. Dizon and D. DeMaster whose comments were helpful in preparing the final draft of the manuscript.

REFERENCES

Best, P. B. (1976). Tetracycline marking and the rate of growth layer formation in the teeth of a dolphin (*Lagenorhynchus obscurus*). *S. Afr. J. Sci.* **72,** 216–218.

Gurevich, V. S., Stewart, B. S., and Cornell, L. H. (1981). The use of tetracycline in age determination of common dolphins, *Delphinus delphis. In* "Age Determination of Toothed Whales and Sirenians"(W. F. Perrin and A. C. Myrick, Jr., eds.), pp. 165–169. *Rep. int. Whal. Commn,* Spec. Issue 3.

Hui, C. A. (1978). Reliability of using dentin layers for age determination in *Tursiops truncatus.* U.S. Marine Mammal Comm. Rep. No. MMC-77/09 (Natl. Tech. Info. Serv. PB288444). 25 pp.

Kasuya, T. (1977). Age determination and growth of Baird's beaked whale with a comment on the fetal growth rate. *Sci. Rep. Whales Res. Inst., Tokyo* **29,** 1–30.

Laws, R. M., (1962). Age determination of pinnipeds with special reference to growth layers in the teeth. *Saugetier. Mitt.* **27**(3), 129–146.

Myrick, A. C., Jr. (1979). Variation, taphonomy, and adaptation of the Rhabdosteidae (=Eurhinodelphidae) (Odontoceti, Mammalia) from the Calvert Formation of Maryland and Virginia. Ph.D. thesis, Univ. of California, Los Angeles. 411 pp. (Unpublished.)

Myrick, A. C., Jr. (1981a). Some approaches to calibration of age in odontocetes using layered hard tissues. *In* "Age Determination of Toothed Whales and Sirenians"(W. F. Perrin and A. C. Myrick, Jr., eds.), pp. 95–97. *Rep. int. Whal. Commn,* Spec. Issue 3.

Myrick, A. C., Jr., (1981b). Examination of layered tissues of odontocetes for age determination using polarized light microscopy. *In* "Age Determination of Toothed Whales and Sirenians" W. F. Perrin and A. C. Myrick, Jr., eds.), pp. 105–112. *Rep. int. Whal. Commn,* Spec. Issue 3.

Myrick, A. C., Jr., Hohn, A. A., Sloan, P. A., Kimura, M., and Stanley, D. D. (1983). Estimating age of spotted and spinner dolphins (*Stenella attenuata* and *Stenella longirostris*) from teeth. NOAA-TM-NMFS-SWFC-30. 17 pp.

Myrick, A. C., Jr., Shallenberger, E. W., Kang, I., and MacKay, D. B. (1984). Calibration of dental layers in seven captive Hawaiian spinner dolphins, *Stenella longirostris,* based on tetracycline labeling. *Fish. Bull. (U.S.)* **82**(1), 207–225.

Nielsen, H. G. (1972). Age determination of the harbour porpoise *Phocoena phocoena* (L.) (Cetacea). *Vidensk. Medd. Dan. Naturhist. Foren.* **135,** 61–84.

Nishiwaki, M., and Yagi, T. (1953). On the age and the growth of teeth in a dolphin, (*Prodelphinus caeruleo-albus*). (I). *Sci. Rep. Whales Res. Inst., Tokyo* **8,** 133–146.

Perrin, W. F., and Myrick, A. C., Jr., eds. (1981). "Age Determination of Toothed Whales and Sirenians." *Rep. int. Whal. Commn,* Spec. Issue 3. 229 pp.

Ridgway, S. H., Green, R. F., and Sweeny, J. C. (1975). Mandibular anesthesia and tooth extraction in the bottlenose dolphin. *J. Wildl. Dis.* **11,** 415–418.

Scheffer, V. B., and Myrick, A. C., Jr., (1981). A review of studies to 1970 of growth layers in the teeth of marine mammals. *In* "Age determination of Toothed Whales and Sirenians" (W. F. Perrin and A. C. Myrick, Jr., eds.), pp. 51–63. *Rep. int. Whal. Commn,* Spec. Issue 3.

Sergeant, D. E. (1959). Age determination in odontocete whales from dentinal growth layers. *Nor. Hvalfangst-tidende* **48,** 273–288.

Sergeant, D. E., Caldwell, D. K., and Caldwell, M. C. (1973). Age, growth, and maturity of bottlenosed dolphins *(Tursiops truncatus)* from northeast Florida. *J. Fish. Res. Board Can.* **30,** 1009-1011.

Use of Chromosome Heteromorphisms and Hemoglobins in Studies of Bottlenose Dolphin Populations and Paternities

35

Deborah A. Duffield
Jan Chamberlin-Lea
Department of Biology
Portland State University
Portland, Oregon 97201

INTRODUCTION

Chromosome analyses have often been applied in studies of animal species and populations. Similarities and differences in number and banding pattern of chromosomes have commonly been used to investigate evolutionary and phylogenetic relationships (e.g., de Grouchy *et al.*, 1978; Ryder *et al.*, 1978; Baker and Bickham, 1980; Dutrillaux and Couturier, 1983; Stock, 1983; Wayne *et al.*, 1987). In addition, however, the chromosomes in many species have regions which may vary in size and appearance. These normal chromosomal variations (chromosome heteromorphisms) have been used in the study of population and species differences, reproductive isolation, and the role of heterochromatin in karyotypic evolution (Patton and Sherwood, 1982; Sites, 1983; Barros and Patton, 1985; Turner *et al.*, 1985; Davis *et al.*, 1986; Matayoshi *et al.*, 1987). In human populations, chromosome heteromorphisms have been used in determining the parental origin of chromosomes (Chamberlin and Magenis, 1980), as genetic markers for family studies (Olson *et al.*, 1983), and as discriminators of paternity (Olson *et al.*, 1986). Heteromorphisms have no phenotypic effects. They are stable from tissue to tissue and are inherited.

In cetaceans, chromosomes have been used to address a variety of questions. For example, karyotypes have been used to investigate evolutionary and phylogenetic relationships (Duffield *et al.*, 1971; Arnason, 1972, 1974, 1980, 1981; Duffield Kulu, 1972; Duffield, 1974, 1977; Arnason *et al.*, 1977), to map genes (Arnason *et al.*, 1978; Duffield *et al.*, 1978), and to examine patterns of chromosomal variability within and between species (Arnason *et al.*, 1980; Stock, 1981; Duffield, 1986a; Lambertsen and Duffield, 1987). Standard giemsa, G-banded, C-banded, and R-banded karyotypes (see below) have been presented for the bottlenose dolphin, *Tursiops truncatus* (Walen and Madin, 1965; Duffield *et al.*, 1967, 1978; Prasad *et al.*, 1970; Arnason, 1974; Duffield, 1977, 1986b). This species has a diploid chromosome number (2*N*) of 44. Its general chromosome banding pattern is similar to those of other delphinids.

Chromosome banding techniques were first described by Caspersson *et al.* (1968), who used the fluorochrome stain quinacrine mustard to give a banded appearance to chromosomes exposed to ultraviolet light. The banding patterns (Q-bands) were unique to each chromosome pair, allowing, for the first time, certain identification of each homologue. Since then, a variety of banding techniques have been developed (e.g., Yunis, 1974; Schweizer, 1981; Sumner, 1982). The techniques used principally in animal and human studies are Q-banding (quinacrine mustard), G-banding (giemsa banding), R-banding (reverse banding), and C-banding (heterochromatin banding). G-banding produces a pattern similar to Q-banding but does not require fluorescence. Both G- and Q-banding differentially band the chromosomes, and in some species, such as humans, Q-banding also visualizes chromosome heteromorphisms. Fluorescent R-banding (Schweizer, 1980; Sahar and Latt, 1978) produces a pattern opposite to those of Q- and G-bands, and visualizes different chromosome heteromorphisms. C-banding also stains heteromorphisms. However, it does not differentially band the rest of the chromosome, and, therefore, with this stain, it is not always possible to distinguish chromosome pairs, i.e., homologues.

In cetaceans, fluorescent R-banding is a particularly useful chromosome banding technique for population studies because it visualizes a number of their chromosome heteromorphisms, while simultaneously providing for homologue identification (Duffield, 1986a; Lambertsen and Duffield, 1987). Q-banding does not visualize heteromorphisms in cetaceans (personal observation) and C-banding, although effectively demonstrating heteromorphisms in cetaceans (see for example, Duffield, 1972; Arnason *et al.*, 1980), does not allow homologue identification for all chromosome pairs. Homologue identification is essential for tracking the inheritance patterns of chromosomes and for establishing paternity.

In the genus *Tursiops*, North Pacific bottlenose dolphins have two electrophoretically distinct hemoglobins, designated fast and slow to indicate their relative electrophoretic mobility (Baluda *et al.*, 1972; Sharp, 1975). Approximately 30% of their total hemoglobin is fast, 70% slow. Coastal Northwest Atlantic and Gulf of Mexico bottlenose dolphins have a single hemoglobin which is electrophoretically equivalent to the fast hemoglobin of the North Pacific specimens (Baluda *et al.*, 1972; Sharp, 1975). Northwest Atlantic offshore dolphins have a two-hemoglobin electrophoretic profile like that of the North Pacific dolphins (Hersh and Duffield, Chapter 6, this volume).

In this chapter, we consider the range of R-band chromosome heteromorphisms and hemoglobin types found in a sample of bottlenose dolphins from the North Atlantic, Gulf of Mexico, North Pacific, and Gulf of California. The potential for using these chromosome heteromorphisms and hemoglobin types to help track social-unit relationships in wild dolphin populations was validated by establishing their consistency and heritability in four mother–calf pairs with known sires from breeding colonies of bottlenose dolphins at Sea World of California.

MATERIALS AND METHODS

Samples

Blood was drawn for chromosome and hemoglobin analyses from 66 captive bottlenose dolphins. Ten of the dolphins sampled were from the Florida east coast: Indian River and areas off Miami Beach and the Florida Keys. Forty-six dolphins came from various collecting areas throughout the Gulf of Mexico: the west coast of Florida (20), Mississippi Sound (21), and the Florida Panhandle (5). Four dolphins were from the Channel Islands area of Southern California and 2 from the Gulf of California; 4 had been born in captivity.

Chromosome Analysis

Blood samples taken by fluke venipuncture were placed in heparinized vacutainer tubes, refrigerated, and shipped cold to Portland State University. Chromosome preparations were made using a standard leukocyte culture technique (Hack and Lawce, 1980). Whole blood aliquots (~7 drops from a 1-ml disposable pipette) were put in 5 ml of RPMI-1640, supplemented with 25% fetal calf serum, penicillin–streptomycin (Gibco; 10,000 U penicillin, 10,000 μg streptomycin per 100 ml RPMI), and 1.5% phytohemagglutinin (Burroughs Wellcome). Cultures were incubated for 4 days at 36°C, synchronized by the addition of colcemid (Gibco, 50μl per 5 ml culture media) for 2 hours, spun down, and recultured with supplemented media for approximately 21 hours. Metaphase cells were collected by the addition of colcemid (same concentration) for 10 minutes and then harvested. The hypotonic solution was 75 m*M* KCl. The cells were fixed in 3:1 absolute ethanol:glacial acetic acid. A fluorescent R-banding technique employing chromomycin A-3 and distamycin A (Sahar and Latt, 1978; Schweizer, 1980) was used to band the chromosomes and to visualize heteromorphisms. Photographs were taken on a Zeiss epifluorescence microscope equipped with an ultraviolet light source, using Kodak Technical Pan 2415 film.

Chromosome Heteromorphism Analysis

Five to ten R-band karyotypes were prepared for each animal. Homologous chromosomes were paired, and composites of the heteromorphic pairs were made to ensure that all scored heteromorphisms were real variants and not artifacts of staining or of differential chromosome contraction. Heteromorphisms were scored for size, intensity (brightness), and the presence of satellites, following criteria used in the scoring of heteromorphisms in humans (Chamberlin and Magenis, 1980; Olson *et al.*, 1983, 1986) and in other species being studied in our laboratory (Duffield, 1986a). Each animal was independently scored by both authors. For 20 of the dolphins, additional karyotypes were made from repeat blood samples to check for the consistency of heteromorphisms within the same individual and from sample to sample. The range of variants was established by comparing the chromosome pairs of all animals studied. Inheritance of heteromorphisms was verified by comparing all heteromorphic pairs in four known mother–calf–sire combinations from the Sea World of California bottlenose dolphin breeding colony.

Hemoglobin Electrophoresis

Cellulose acetate electrophoresis (pH 7.6) was used to identify the hemoglobin profile of each dolphin (see Hersh and Duffield, Chapter 6, this volume). If two hemoglobins were present, their relative percentages were determined using a Beckman microzone densitometer (model R-110).

RESULTS

A fluorescent R-band karyotype of a male bottlenose dolphin is shown in Fig. 1. This dolphin had six chromosome pairs with heteromorphisms (Fig. 2). In the karyotypes of the 66 dolphins studied, a total of eleven chromosome pairs had heteromorphic regions (Fig. 3). These were located near the centromeres, interstitially in either the short (p) or long (q) arms, or terminally on the p arms of the chromosomes. Four of the chromosome pairs had more than one heteromorphic region. From two to seven heteromorphisms were found for each heteromorphic region (Fig. 4), making a total of 52 different heteromorphisms identified in this sample of bottlenose dolphins from around the southeastern and southwestern coasts of the United States and from the Gulf of California. The karyotypes of Atlantic and Pacific bottlenose dolphins were similar in basic banding pattern, with the same chromosomal location of heteromorphic regions and a similar range of heteromorphisms. Each animal had its own distinctive combination of heteromorphisms. This was consistent in all repeat samples.

In all four mother–calf pairs with known paternity examined to validate inheritance of the chromosomal heteromorphisms, heteromorphisms found in the R-band karyotypes of the calves could be clearly identified in the karyotypes of their respective parents (Fig. 5). Two of the captive matings involved parents of the same hemoglobin type, both having a single hemoglo-

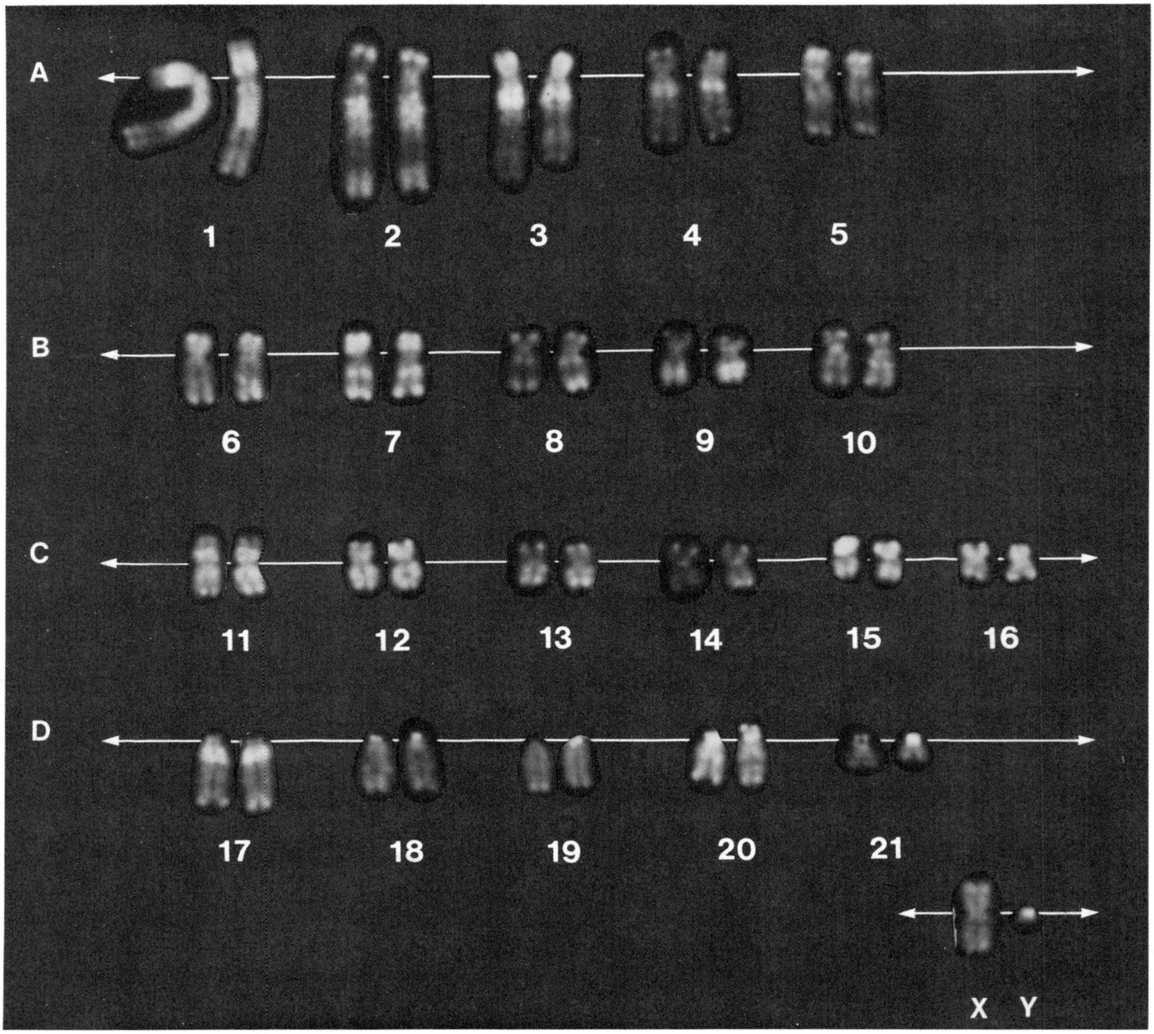

Figure 1 Fluorescent R-band karyotype of a male bottlenose dolphin. The chromosomes are arranged by centromere position and size into groups A–D and are numbered consecutively. The X and Y chromosomes are indicated. Chromosome pairs 1, 3, 4, 19, 20, and 21 have heteromorphic regions (see Fig. 2).

bin (100% fast). The offspring of these crosses had a single fast hemoglobin. The other two matings were between dolphins with a single hemoglobin (100% fast) and dolphins with two hemoglobins (30% fast, 70% slow). These matings resulted in offspring with two hemoglobins, but in different percentages than the parents (65% fast, 35% slow), reflecting the relative contributions from each parent (Fig. 6). Among the 62 wild dolphins in this study, all from the North Pacific had a two-hemoglobin profile (30% fast, 70% slow), like those previously reported (Baluda *et al.*, 1972; Sharp, 1975). Four Atlantic coastal and Gulf of Mexico animals had a two-hemoglobin pattern (65% fast, 35% slow) and the remaining dolphins had a single hemoglobin (100% fast).

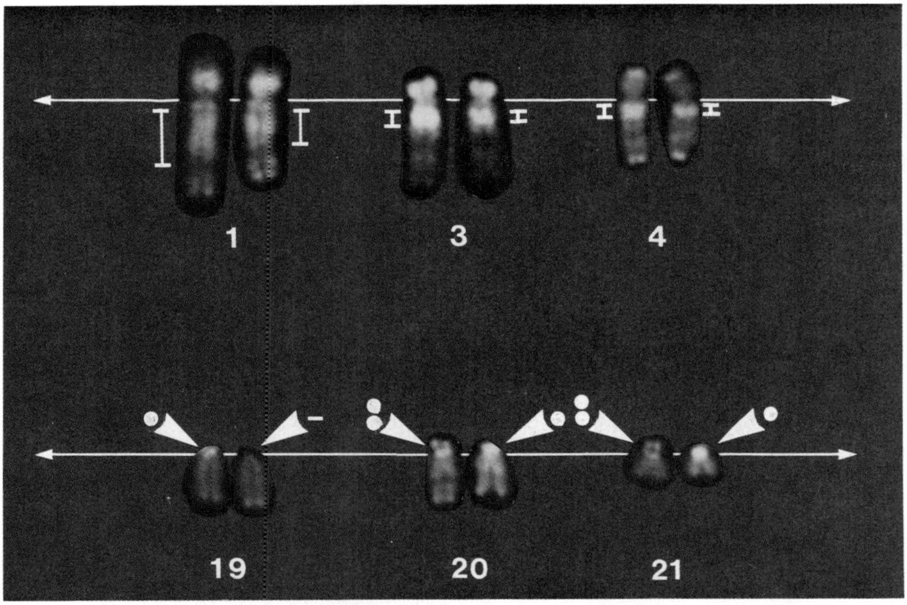

Figure 2 A composite showing in more detail the heteromorphic chromosome pairs in the male dolphin of Fig. 1. Brackets and symbols indicate the specific heteromorphic regions and differences in size, presence or absence of a bright band, and presence or absence of satellites. Each pair was taken from a single karyotype of this animal, but the different pairs were selected from karyotypes that best illustrated the heteromorphism. These heteromorphisms are identifiable in all karyotypes of this individual and are not due to staining or contraction differences in the chromosome preparations. Their presence was verified in repeat cultures and preparations.

DISCUSSION

A large number of chromosome heteromorphisms can be seen in the bottlenose dolphin karyotype by fluorescent R-banding. Because of the range of karyotypic combinations that this offers, R-band heteromorphisms provide an effective system for identifying individuals. The level of R-band heteromorphism in the dolphin approaches the level of Q-band heteromorphisms in the human (Olson *et al.*, 1986). In general, chromosome heteromorphisms can be used in studies of population differentiation and regional gene flow, as can other genetic variants, such as protein polymorphisms (e.g., Ayala, 1976). However, given the amount of heteromorphic variability in the dolphin karyotype, its greatest benefit is that it permits one to detect specific relationships between animals. For example, in situations where a female–calf relationship is known but paternity is uncertain, analysis of the chromosome heteromorphisms of female, calf, and all potential breeding males can be used to exclude males that are not the sire.

In population studies, chromosome heteromorphism analysis makes it possible to investigate relationships among animals of known associations and to establish patterns of reproductive exchange among dolphin groups. In order to realize the full potential of chromosome heteromorphism anlaysis (or any other type of genetic variant analysis) in the study of wild dolphin populations, it is important to couple biological sampling with long-term observational data (e.g., see Scott *et al.*, Chapter 11, this volume). Doing so provides important information on demography,

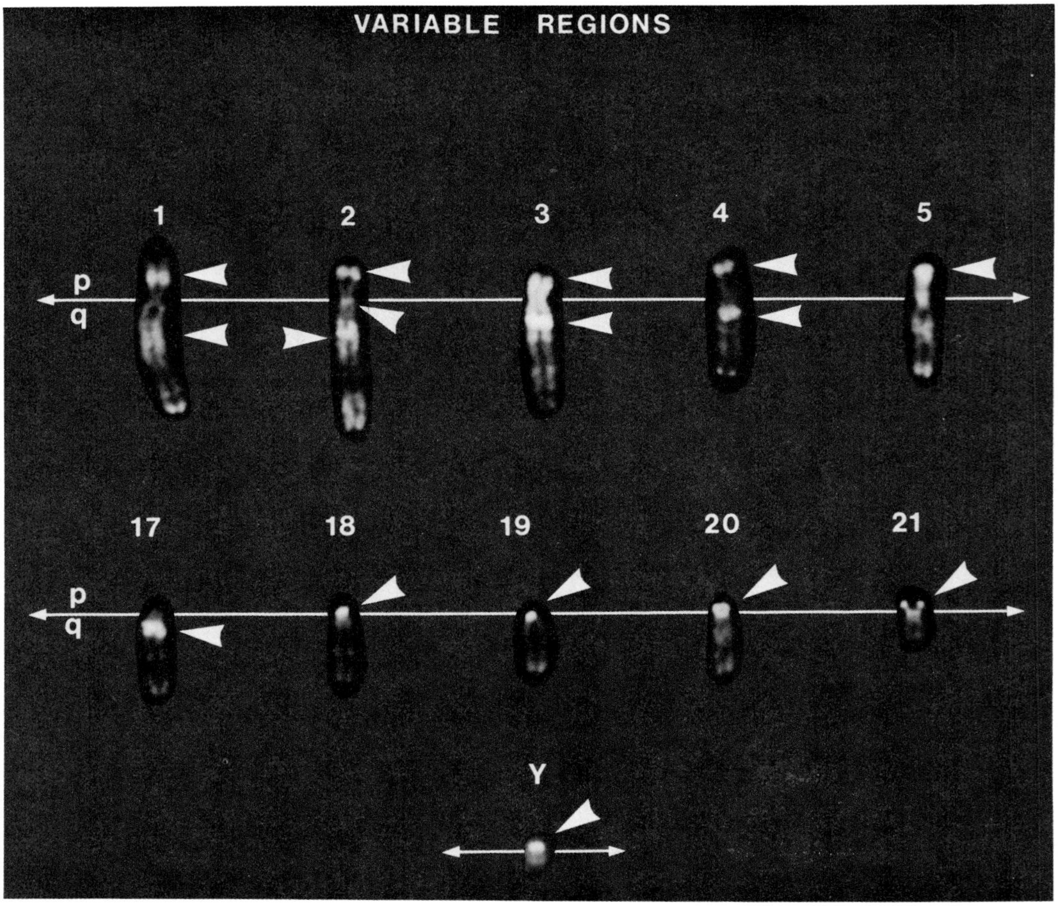

Figure 3 Chromosomal locations of heteromorphic regions found in a survey of 62 dolphins from the North Pacific, North Atlantic, and Gulf of Mexico are indicated for the dolphin karyotype. Eleven pairs had heteromorphisms in the short (p) arms, long (q) arms, or both. Arrows point to the general area of the heteromorphism on the chromosome.

Figure 4 (See pages 615–616) Range of variants for the eleven heteromorphic chromosome pairs found in a sample of 62 wild dolphins. The karyotypic number of the chromosome pair is given; arrows indicate the heteromorphic regions. The variants in these heteromorphic regions are consistent and heritable. They differ in size, intensity, whether they are present or absent, and, for chromosome pairs 20 and 21, whether they have satellites.

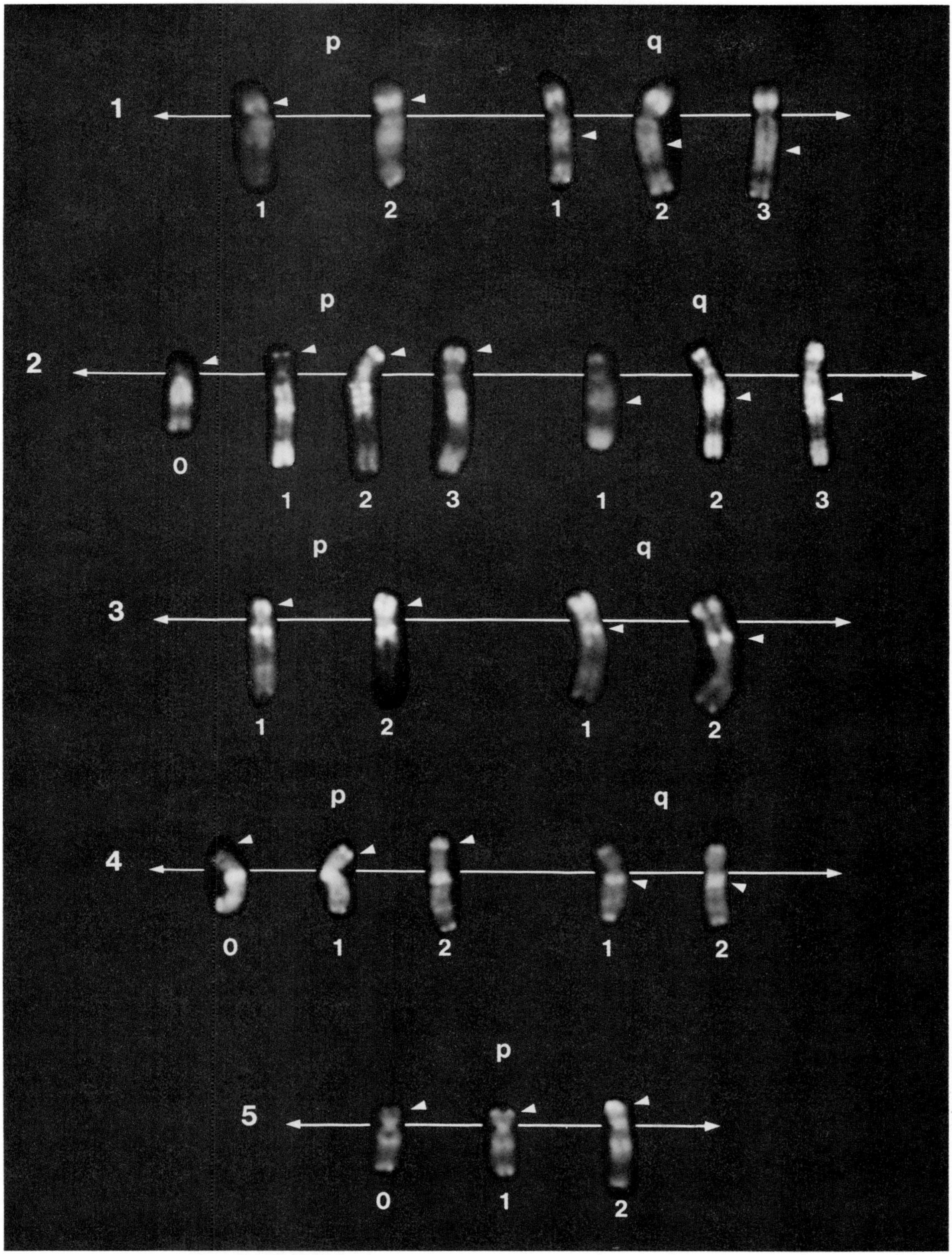
p
q
1
1
2
1
2
3
p
q
2
0
1
2
3
1
2
3
p
q
3
1
2
1
2
p
q
4
0
1
2
1
2
p
5
0
1
2

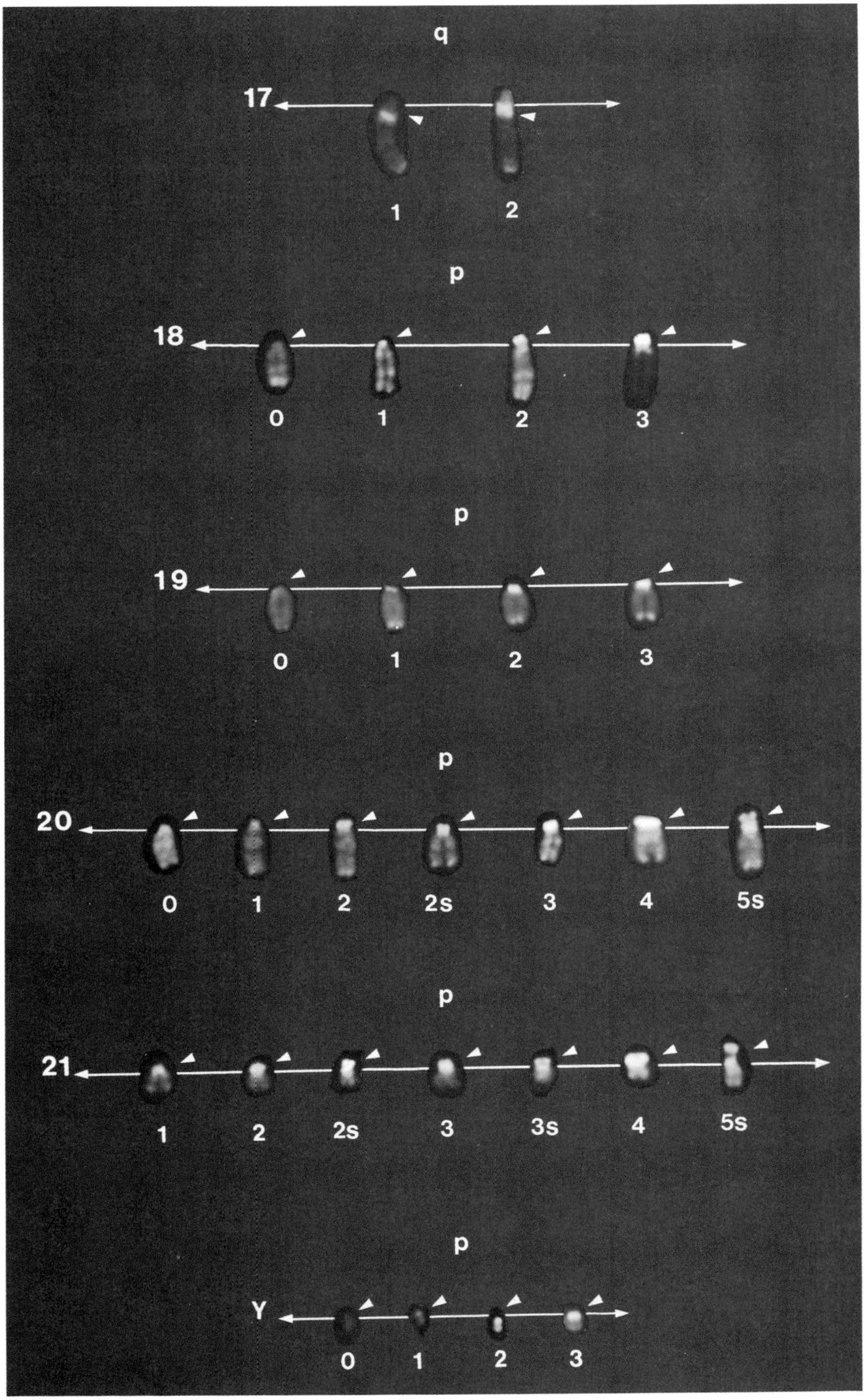

Figure 4 (*continued*)

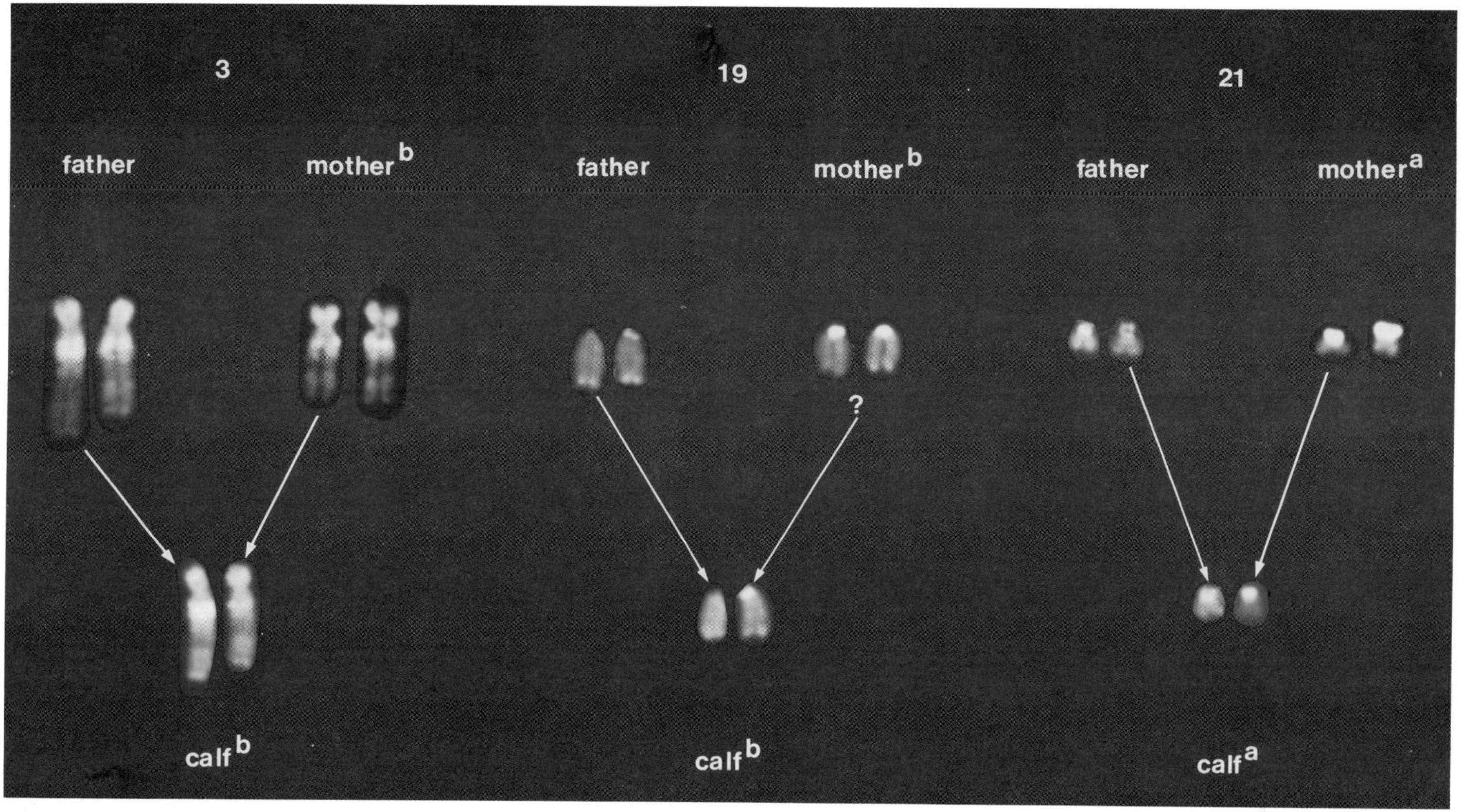

Figure 5 Three examples of the inheritance of chromosomal heteromorphisms in mother–calf pairs with known sires. In chromosome pair 3, the mother has a small, single bright band in the heteromorphic region of both chromosomes, and the father has a larger, double bright band on both of his chromosomes. The calf inherited one from each parent. In chromosome pair 19, the calf inherited either one of the mother's chromosomes with a large, bright p-arm band and the father's chromosome with no band at all. In chromosome pair 21, the calf inherited the smaller of the two bright p-arm chromosomes from the mother and the satellited one from the father. The inheritance of these heteromorphisms was evaluated by making a composite of the chromosome pairs in question from the karyotypes of the dam, sire, and calf.

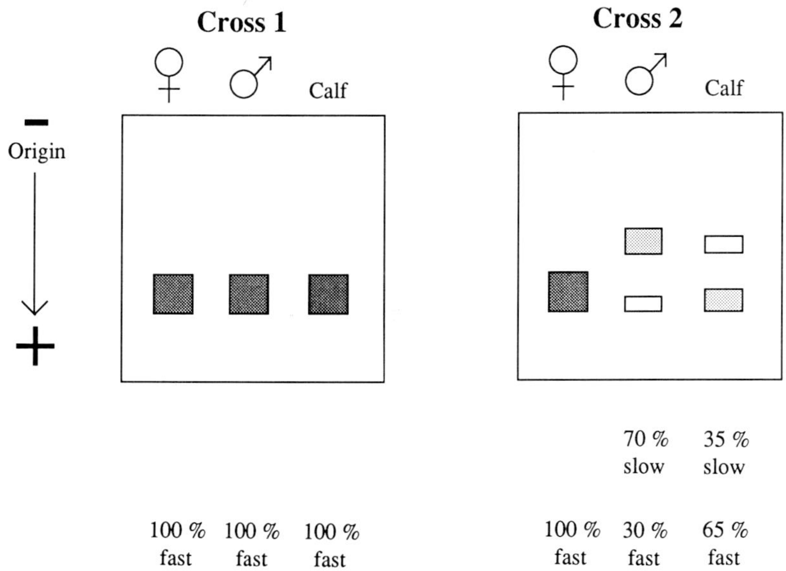

Figure 6 Inheritance of hemoglobins in four bottlenose dolphins with known parents. Hemoglobin types were separated by cellulose acetate electrophoresis (pH 7.6).

long-term associations, movements, and potential reproductive interactions. Such information, in turn, makes it possible to ask specific questions of chromosome heteromorphism analysis, such as, "How are two or more dolphins related?" or, "What is the paternity of a particular individual?"

Hemoglobin analysis offers an additional genetic system for the study of dolphin populations. In captive-bred animals a 65% fast, 35% slow hemoglobin electrophoretic pattern resulted from matings between dolphins with 100% fast hemoglobin and those with 30% fast, 70% slow. We conclude from this that wild dolphins that exhibit the 65% fast, 35% slow pattern also come from matings between these two parental types. As there are differences in hemoglobin profiles between offshore (30% fast, 70% slow) and coastal (100% fast) dolphins in the western North Atlantic (Hersh and Duffield, Chapter 6, this volume), reproductive exchange between the two forms can be inferred by the 4 out of 62 dolphins which exhibited the 65% fast, 35% slow pattern. This suggests infrequent but occasional gene flow between the coastal and offshore dolphin populations.

Together, the R-band chromosome heteromorphisms and hemoglobin variants offer ways of genetically examining breeding success. With the chromosome heteromorphisms, parentage in known social groups can be determined, and with the hemoglobin variants, it is possible to look at the broader aspects of reproductive interaction between dolphins of different geographic regions. These techniques are useful for gaining a greater understanding of population structure and dynamics in bottlenose dolphins.

ACKNOWLEDGMENTS

Blood samples were made available by Sea World, Marineland of the Pacific, the Institute for Delphinid Research, Miami Seaquarium, Marine World, Marine Animal Productions, Sea Life Park, the Naval Ocean Systems Center, and the Aquarium of Niagara Falls. This work was supported in part by grants from Sea World, Inc., NMFS Contract NA83-GA-C-00063, PSU Biomedical Research Grant 90-262-0301, and PSU Research and Publication Grant 90-050-5801.

REFERENCES

Arnason, U. (1972). The role of chromosomal rearrangement in mammalian speciation with special reference to Cetacea and Pinnipedia. *Hereditas* **70,** 113–118.

Arnason, U. (1974). Comparative chromosome studies in Cetacea. *Hereditas* **77,** 1–36.

Arnason, U. (1980). C- and G-banded karyotypes of three delphinids: *Stenella clymene, Lagenorhynchus albirostris* and *Phocoena phocoena. Hereditas* **92,** 179–187.

Arnason, U. (1981). Banding studies on the gray and sperm whale karyotypes. *Hereditas* **95,** 277–281.

Arnason, U., Benirschke, K., Mead, J. G., and Nichols, W. W. (1977). Banded karyotypes of three whales: *Mesoplodon europaeus, M. carlhubbsi* and *Balaenoptera acutorostrata. Hereditas* **87,** 189–200.

Arnason, U., Purdom, I. F., and Jones, K. W. (1978). Conservation and chromosomal localization of DNA satellites in balenopterid whales. *Chromosoma* **66,** 141–159.

Arnason, U., Lutley, R., and Sandholt, B. (1980). Banding studies on six killer whales: An account of C-band polymorphism and G-band patterns. *Cytogenet. Cell Genet.* **28,** 71–78.

Ayala, F. J. (1976). "Molecular Evolution." Sinauer, Boston. 277 pp.

Baker, R. J., and Bickham, J. W. (1980). Karyotypic evolution in bats: Evidence of extensive and conservative chromosomal evolution in closely related taxa. *Syst. Zool.* **29,** 235–253.

Baluda, M. C., Duffield Kulu, D. A., and Sparkes, R. S. (1972). Cetacean hemoglobins: Electrophoretic findings in nine species. *Comp. Biochem. Physiol.* **41B,** 647–653.

Barros, M. A., and Patton, J. L. (1985). Genome evolution in pocket gophers (genus *Thomomys*). III. Fluorochrome-revealed heterochromatin heterogeneity. *Chromosoma* **92,** 337–343.

Caspersson, T., Farber, S., Foley, G.E., Kudynowski, J., Modest, E. J., Simonsson, E., Wagh, U., and Zech, L. (1968). Chemical differentiation along metaphase chromosomes. *Exp. Cell Res.* **489,** 219–222.

Chamberlin, J., and Magenis, R. E. (1980). Parental origin of de novo chromosome rearrangements. *Hum. Genet.* **53,** 343–347.

Davis, K. M., Smith, S. A., and Greenbaum, I. F. (1986). Evolutionary implications of chromosomal polymorphisms in *Peromyscus boylii* from southwestern Mexico. *Evolution* **40**(3), 645–549.

Duffield Kulu, D. A. (1972). Evolution and cytogenetics. *In* "Mammals of the Sea: Biology and Medicine" (S. H. Ridgway, ed.), pp. 503–526. Charles C. Thomas, Springfield, Illinois.

Duffield, D. A. (1974). Fluorescent karyotype of the California gray whale. *Mar. Fish. Rev.* **36**(4), 25–28.

Duffield, D. A. (1977). Phylokaryotypic evaluation of the Cetacea. Ph.D. dissertation, Univ. of California, Los Angeles, California.

Duffield, D. A. (1986a). *Orcinus orca:* Taxonomy and evolution—Cytogenetics and population structure. *In* "Behavioral Biology of Killer Whales" (J. Lockhard and B. Kirkevold, eds.), pp. 19–33. Alan R. Liss, New York.

Duffield, D. A. (1986b). Investigation of genetic variability in stocks of the bottlenose dolphin (*Tursiops truncatus*) and the loggerhead sea turtle (*Caretta caretta*). NMFS Tech. Rep. NA83-GA-C-00036.

Duffield, D. A., Ridgway, S. H., and Sparkes, R. S. (1967). Cytogenetic studies of two species of porpoise. *Nature (London)* **213,** 189–190.

Duffield, D. A., Veomett, I., and Sparkes, R. S. (1971). Cytogenetic studies of four species of cetaceans. *J. Mammal.* **52,** 828–832.

Duffield, D. A., Spence, M., and Paquin, C. (1978). Evaluation of metaphase association patterns of acrocentric chromosomes in the order Cetacea. *Cytologia* **43,** 541–549.

Dutrillaux, B., and Couturier, J. (1983). The ancestral karyotype of Carnivora: Comparison with that of platyrrhine monkeys. *Cytogenet. Cell Genet.* **35,** 200–208.

Grouchy, J. de, Turleau, C., and Finaz, C. (1978). Chromosomal phylogeny of the primates. *Annu. Rev. Genet.* **12,** 289–328.

Hack, M. S., and Lawce, H. J., eds. (1980). "Association of Cytogenetic Technologists Cytogenetics Laboratory Manual." Assoc. of Cytogenet. Tech., 677-S, Univ. of California, San Francisco, California. 430 pp.

Lambertsen, R. H., and Duffield, D. A. (1987). Biopsy studies of the humpback whale, *Megaptera novaeangliae.* NMFS and Int. Whaling Comm. Tech. Rep. 50-EANF-5-00094. 72 pp.

Matayoshi, T., Seuanez, H. N., Nasazzi, N., Nagle, C., Armada, J. L., Freitas, L., Alves, G., Barroso, C. M., and Howlin, E. (1987). Heterochromatic variation in *Cebus apella* (Cebidae, Platyrrhini) of different geographic regions. *Cytogenet. Cell Genet.* **44,** 158–162.

Olson, S. B., Magenis, R. E., Rowe, S. I., and Lovrien, E. W. (1983). Chromosome heteromorphism analysis in cases of disputed paternity. *Am. J. Med. Genet.* **15,** 47–55.

Olson, S. B., Magenis, R. E., and Lovrien, E. W. (1986). Human chromosome variation: The discriminatory power of Q-band heteromorphism (variant) analysis in distinguishing between individuals with specific application to cases of questionable paternity. *Am. J. Hum. Genet.* **38,** 235–252.

Patton, J. L., and Sherwood, S. W. (1982). Genome evolution in pocket gophers (genus *Thomomys*). I. Heterochromatin variation and speciation potential. *Chromosoma* **85,** 149–162.

Prasad, N., Mumford, D. M., Barsales, P. B., Whithman, T., and Wilbur, J. R. (1970). Cytogenetic studies of dolphin (*Tursiops truncatus*) by an extended tissue culture technique. *Experientia* **26,** 1167–1168.

Ryder, O. A., Epel, N. C., and Benirschke, K. (1978). Chromosome banding studies of the Equidae. *Cytogenet. Cell Genet.* **20,** 323–350.

Sahar, E., and Latt, S. A. (1978). Enhancement of the banding patterns in human metaphase chromosomes by energy transfer. *Proc. Natl. Acad. Sci. U.S.A.* **75**(11), 5650–5654.

Schweizer, D. (1980). Simultaneous fluorescent staining of R-bands and specific heterochromatic regions (DA-DAPI) bands in human chromosomes. *Cytogenet. Cell Genet.* **27,** 190–193.

Schweizer, D. (1981). Counterstain-enhanced chromosome banding. *Hum. Genet.* **57,** 1–14.

Sharp, G. D. (1975). A study of the electrophoretic and O_2 dissociation properties of the hemoglobins of some delphinid cetaceans. *Comp. Biochem. Physiol.* **51A,** 673–681.

Sites, J. W., Jr. (1983). Chromosome evolution in the iguanid lizard *Sceloporus grammicus*. I. Chromosome polymorphisms. *Evolution* **37,** 38–53.

Stock, A. D. (1981). Chromosomal variation and constitutive heterochromatin in three porpoise species (genus *Stenella*). *Cytogenet. Cell Genet.* **31,** 91–100.

Stock, A. D. (1983). Chromosomal homologies and phylogenetic relationships of the vespertilionid bat genera *Euderma, Idionycteris,* and *Plecotus*. *Cytogenet. Cell Genet.* **35,** 126–140.

Sumner, A. T. (1982). The nature and mechanisms of chromosome banding. *Cancer Genet. Cytogenet.* **6,** 59–87.

Turner, B. J., Grudzien, T. A., Adkisson, K. P., and Worrell, R. A. (1985). Extensive chromosomal divergence within a single river basin in the goodeid fish, *Ilyodon furcidens*. *Evolution* **39**(1), 122–134.

Walen, K. H., and Madin, S. H. (1965). Comparative chromosome analyses of the bottlenosed dolphin (*Tursiops truncatus*) and the pilot whale (*Globicephala scammoni*). *Am. Nat.* **99,** 349–354.

Wayne, R. K., Nash, W. G., and O'Brien, S. J. (1987). Chromosomal evolution of Canidae. I. Species with high diploid numbers. *Cytogenet. Cell Genet.* **44,** 123–133.

Yunis, J. J., ed.(1974). "Human Chromosome Methodology," 2nd Ed. Academic Press, New York. 377 pp.

PHOTO: Deaths of hundreds of bottlenose dolphins off the United States east coast in 1987 and 1988, believed to have been caused in part by red-tide toxins, have prompted a reevaluation of the management program for this species (Corolla, North Carolina, September 1974; courtesy James G. Mead, USNM 504123).

Part IX

Management

Management-Oriented Research on Bottlenose Dolphins by the Southeast Fisheries Center

36

Gerald P. Scott
National Oceanic and Atmospheric Administration
National Marine Fisheries Service
Southeast Fisheries Center
Miami Laboratory
Miami, Florida 33149

INTRODUCTION

Since the Marine Mammal Protection Act (MMPA) of 1972, more than 530 Atlantic bottlenose dolphins, *Tursiops truncatus,* have been permanently removed from the waters of the southeastern United States (Table 1). The Southeast Fisheries Center (SEFC), the research arm of the National Marine Fisheries Service (NMFS) in the southeastern region, began its Marine Mammal Program (MMP) in October 1978 to satisfy the requirement that it offer management advice on the live-capture fishery for bottlenose dolphins to the management arms of NMFS: the Southeast Regional Office in St. Petersburg, Florida, and the Washington, D. C., Office. Marine mammal research at the SEFC has been shaped to its present form by recommendations developed during a continuing series of internal and external reviews, during the SEFC Stock Assessment Workshops in August 1982 and July 1984 (Powers, 1983), and by the U.S. Marine Mammal Commission (MMC) in February 1983.

The goals of MMP are (1) to provide estimates of abundance and recommendations of live-capture quotas for the stocks of bottlenose dolphins occurring in the southeast, (2) to determine the dynamics and discreteness of alongshore and inshore–offshore "populations" of bottlenose dolphins, and (3) to determine the validity of the live-capture quota guidelines developed by the MMC. These research goals are being pursued using a variety of research tools and methods. The purpose of this chapter is to describe the marine mammal research activities at the SEFC and their application to the management of the live-capture fishery for bottlenose dolphins.

Table 1
Summary of Permanent Removals of Bottlenose Dolphins from the Wild[a,b]

	1973	1974	1975	1976	1977	1978	1979	1980	1981	1982	1983	1984	1985	1986	1987	1988[c]	Total
Florida Indian–Banana River complex	—	—	10	6	15	5	7	9	7	3	2	—	—	—	—	4	68
Florida Keys	15	—	—	—	—	—	—	—	—	—	—	—	—	—	—	—	15
Florida West Coast: Charlotte Harbor to Crystal River	—	—	16	5	16	5	—	8	5	7	11	12	21	9	6	13	134
(Tampa Bay)	—	—	—	(1)	(2)	—	—	—	—	—	—	(1)	(6)	—	—	(2)	(12)
(Charlotte Harbor)	—	—	(8)	—	(5)	—	—	(3)	(4)	(5)	(5)	(2)	(4)	(4)	—	(3)	(43)
(North of Charlotte Harbor to Crystal River, not including Tampa Bay)	—	—	(8)	(4)	(5)	(5)	—	(5)	(1)	(2)	(6)	(9)	(11)	(5)	(6)	(8)	(75)
Florida Panhandle: Crystal River to Mobile Bay, Alabama	—	5	—	7	3	4	—	4	2	—	—	6	2	1	1	8	43
(Apalachicola–St. Joseph bays)	—	(5)	—	(7)	—	—	—	(2)	—	—	—	—	—	—	—	—	(14)
(Destin–Ft. Walton Beach)	—	—	—	—	—	(2)	—	(1)	(1)	—	—	(1)	(1)	—	(1)	(2)	(9)
Mississippi Sound	—	15	7	2	8	24	14	11	—	11	14	31	4	6	24	31	202
Texas Coast: North of Corpus Christi Bay up to and including Matagorda Bay	6	—	5	—	2	17	13	12	8	8	—	—	—	—	—	—	71
(Aransas–Copano–San Antonio Bay complex)	(5)	—	5	—	—	(2)	(6)	—	(2)	—	—	—	—	—	—	—	(20)
(Matagorda Bay)	(1)	(—)	(—)	(—)	(—)	(—)	(1)	(6)	(4)	(8)	(—)	(—)	(—)	(—)	—	—	(20)
Total	21	20	38	20	44	55	34	44	22	29	27	49	27	16	31	56	533

[a] Includes animals accidentally killed during research which did not authorize permanent removal.
[b] Information provided by P. Montanio and W. Cain, NMFS, Washington, D.C. and E. Roche, NMFS, St. Petersburg, Florida.
[c] 1988 data may not be complete. Data current as of January 24, 1989.

RESEARCH TO ADDRESS GOALS

The MMP focuses on three main research areas to address the stated program goals: abundance estimation, stock differentiation, and life-stage modeling. Within each theme, one or more research methods and tools have been or are being applied.

Abundance Estimation

The principal approach for estimating abundance has been to conduct aerial sampling surveys designed to serve two purposes. First, some surveys are intended to provide the data necessary to estimate abundance and recommend quotas in "priority" inshore areas, defined as those for which live-capture permits have been or are expected to be requested. Sampling surveys in priority areas have been called "local surveys." Second, based on recommendations of the first SEFC Stock Assessment Workshop (Powers, 1983), other surveys were designed to provide the data necessary to estimate abundance and determine distribution patterns of bottlenose dolphins and other cetacean species over larger geographical regions. The results of these "regional surveys" are expected to facilitate definition of stock boundaries by correlation with distribution patterns.

The localized surveys were conducted between 1979 and 1982 using techniques developed by Leatherwood and Show (1980) under contract to the SEFC, and by 1982 they included data for 11 specific geographical localities (Leatherwood and Show, 1980; Carew 1980, 1981a,b,c,d, 1982a,b,c; 1983) (see Fig. 1). Each survey involved sampling along randomly placed transects. During a survey period, up to six replicate samples were taken (six independent random draws). Surveys were conducted seasonally for at least 1 year in each area. Abundance estimates for those areas where quota recommendations were required have been made

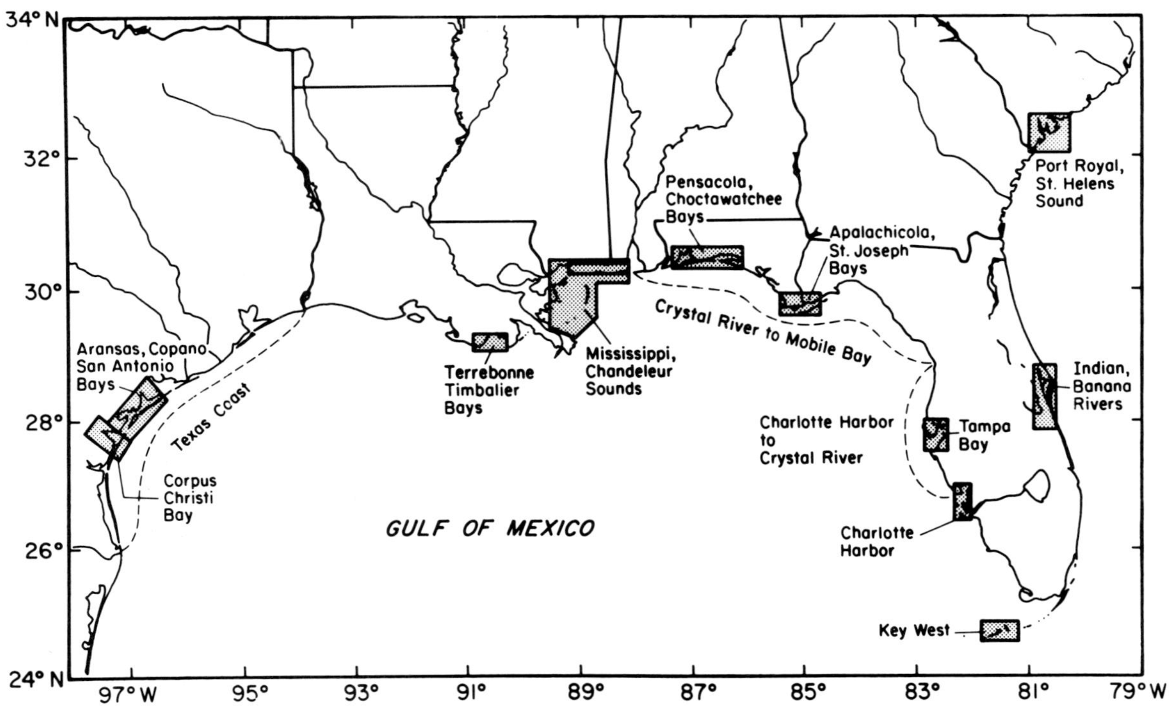

Figure 1 Map of the southeastern United States, indicating areas of SEFC-sponsored research on bottlenose dolphins. Shading indicates areas of localized and regional surveys completed to date. Management areas or subareas listed in text are indicated by dashed lines.

using line-transect analytical techniques (Burnham *et al.*, 1980). The estimates resulting from these and previous localized aerial surveys are given in Table 2.

With the completion of the last localized aerial survey in priority inshore areas, a regional effort was initiated to produce estimates of bottlenose dolphin density and abundance, as well as estimates of the density and abundance of other cetaceans, turtles, and other species of interest throughout the Gulf of Mexico. The first region sampled was "the Northwestern Gulf," defined as "all waters from the coast, including embayments and estuaries, to 5 n mi beyond the surface projection of the 100 fathom isobath and along the coast from the mouth of the Rio Grande River to the Mississippi River delta" (see Figs. 1 and 2). The second region was "the Northeastern Gulf of Mexico," defined as "all waters from the Mississippi River delta to the Florida Keys, including the estuarine, neritic, and pelagic waters, as above." The third region was "the South Atlantic," defined as "waters from the Florida Keys to Cape Hatteras, North Carolina and from the coast to the western wall of the Gulf Stream." Data sufficient for estimating the abundance of bottlenose dolphins and other cetaceans in the South Atlantic were collected by the SEFC Southeast Turtle Survey, which started in 1981 and finished in 1984 (Thompson, 1984). Sampling in the Northwestern Gulf started in September 1983 and was completed in July 1984 (Carter 1983, 1984; Carter and Derman, 1984a, b). Sampling in the Northeastern Gulf began in June 1985 and ended in March 1986 (Carter and Derman, 1985a,b, 1986).

The live-capture fishery for bottlenose dolphins is concentrated in the bay and inshore portions of the study area. For this reason, inshore sampling effort in the Gulf of Mexico regional survey was more intensive than that in the offshore areas. Higher sampling intensity is expected to result in greater precision of density and abundance estimates. Some areas of the region were sampled previously by other investigators (Leatherwood and Platter 1975; Leatherwood *et al.*, 1978; Barham *et al.*, 1979; Leatherwood and Show, 1980; Odell and Reynolds, 1980; Leatherwood, 1982; Thompson 1981a,b,c,d, 1982a,b; Fritts *et al.*, 1983; Leatherwood and Reeves, 1983). The bay and inshore blocks were duplicates of the sample areas of previous surveys, so that direct comparisons of density estimates could be made without poststratifying the data.

Stock Differentiation

To make responsible management decisions, it is necessary to define the stock units involved in the live-capture fishery. The SEFC has used two principal approaches to address this need: (1) tag and resighting surveys, and (2) biochemical genetics studies.

Tag and Resighting

Tagging studies involving fin-notching, rototagging, and freeze-branding have been used to examine the dynamics of herd associations and to help define the discreteness of local populations of bottlenose dolphins (e.g., Asper and Odell, 1980; Odell and Asper, Chapter 30, this volume; Scott *et al.*, Chapter 29, this volume). These studies have been modeled after research conducted on individually identifiable dolphins in portions of the Gulf of Mexico (e.g., Irvine and Wells, 1972; Shane and Schmidly, 1978; Shane, 1980; Wells *et al.*, 1980; Gruber, 1981; Irvine *et al.*, 1981, 1982). Herds of dolphins were captured using seine nets, marked (usually by freeze-branding), released, and subsequently resampled using systematic- and random-design boat surveys, land-based surveys, and information supplied by unofficial observers. SEFC-sponsored research using this approach has been conducted in the Indian and Banana rivers (Asper and Odell, 1980; Odell and Asper, 1982, Chapter 30, this volume), Sarasota Bay (Wells *et al.*, 1981), and the Mississippi Sound (Solangi and Dukes, 1983). For the Sarasota Bay work, however, SEFC sponsored only resighting efforts and a summary of observations made over 10 years by Irvine, Wells, Scott, and others (see Scott *et al.*, Chapter 11, this volume).

On initial capture, biological data were obtained from individual dolphins, including blood and other tissue biopsies, morphometrics, bacteriological samples, and teeth following methods developed by Asper and Odell (1980). These data are useful in assessing the health, age, growth, and

Table 2
Nature and Results of Local Surveys for Bottlenose Dolphins

Area Surveyed	Date	Density (Dolphins/km^2)	Abundance (95% CI)	References
Florida East Coast: Indian and Banana rivers	Nov 1979	0.274	222 (±34)	Leatherwood and Show (1980), Leatherwood (1982)
	Jan 1980	0.264	214 (±42)	
	May 1980	0.255	206 (±170)	Thompson (1981a,b)
	Aug 1980	0.539	435 (±172)	
	Nov 1980	0.251	202 (±106)	
Florida Keys[a]	—	—	—	
Florida West Coast: Charlotte Harbor to Crystal River				
Tampa area	Sep 1979	0.120	364 (±326)	Leatherwood and Show (1980)
	Nov 1979	0.210	634 (±180)	Thompson (1981c)
Charlotte Harbor	Jul 1980	0.204	189 (±186)	Thompson (1981d)
	Oct 1980	0.170	157 (±454)	
	Jan 1981	0.469	434 (±275)	
	Apr 1981	0.206	191 (±140)	
Charlotte Harbor to Crystal River	Jul 1975–Jun 1976	0.070	569 (—)	Odell and Reynolds (1980)
Florida Panhandle: Crystal River to Mobile Bay				
Apalachicola–St. Joseph bays	Jun 1980	0.093	58 (±242)	Thompson (1982a)
	Sep 1980	0.056	35 (±120)	
	Dec 1980	0.091	57 (±34)	
	Mar 1981	0.067	42 (±96)	
Crystal River to Pensacola	Jul 1975-Jun 1976	0.085	936 (—)	Odell and Reynolds (1980)
Mississippi Sound				
Mississippi Sound	Jun 1980	0.103	111 (±104)	Thompson (1982a)
	Sep 1980	0.130	140 (±172)	
	Dec 1980	0.087	93 (±44)	
	Mar 1981	0.105	113 (±72)	

(*continued*)

Table 2 (*continued*)

Area Surveyed	Date	Density (Dolphins/km^2)	Abundance (95% CI)	References
Mississippi, Chandeleur, and Breton sounds plus portion of the Gulf of Mexico	Jul 1974	0.148 (sounds)	1342 (847, S.D.)[b]	Leatherwood and Platter (1975)
		0.127 (marsh)	438 (294, S.D.)	
	Jul 1975	0.097 (sounds)	879 (368, S.D.)	Leatherwood *et al.* (1978)
Texas Coast: North of Corpus Christi Bay including Matagorda Bay				
Aransas–Copano–San Antonio bays	May 1980	0.134	131 (±32)	Thompson (1982a)
	Aug 1980	0.303	296 (±58)	
	Nov 1980	0.274	268 (±56)	
	Jan 1981	0.364	356 (±154)	
Matagorda–Aransas–Copano San Antonio bays	Apr 1978	0.752	1319 (±260)	Barham *et al.* (1979)
Corpus Christi	Sep 1979	1.134	115 ($81 < u < 421$)	Thompson (1982b)
		1.016	103 (±36)	Leatherwood and Show (1980), Leatherwood and Reeves (1983)
Coastal southern Texas	Sept 1979	0.314	300 (±74)	Leatherwood and Reeves (1983)
Laguna Madre and portion of the Gulf of Mexico	Sep 1979	0.421	100 (±11)	Leatherwood and Show (1980)
Other areas				
Atchafalaya Bay, Louisiana	Jul 1975	0.099	897 (461, S.D.)	Leatherwood *et al.* (1978)
Virgin Islands	Aug 1981	0.45 (St. Croix)	49 (±88)	Scott (1982)
		0.05 (St. Thomas)		

[a] Data for the Florida Keys area are not yet available.
[b] S.D., Standard deviation.

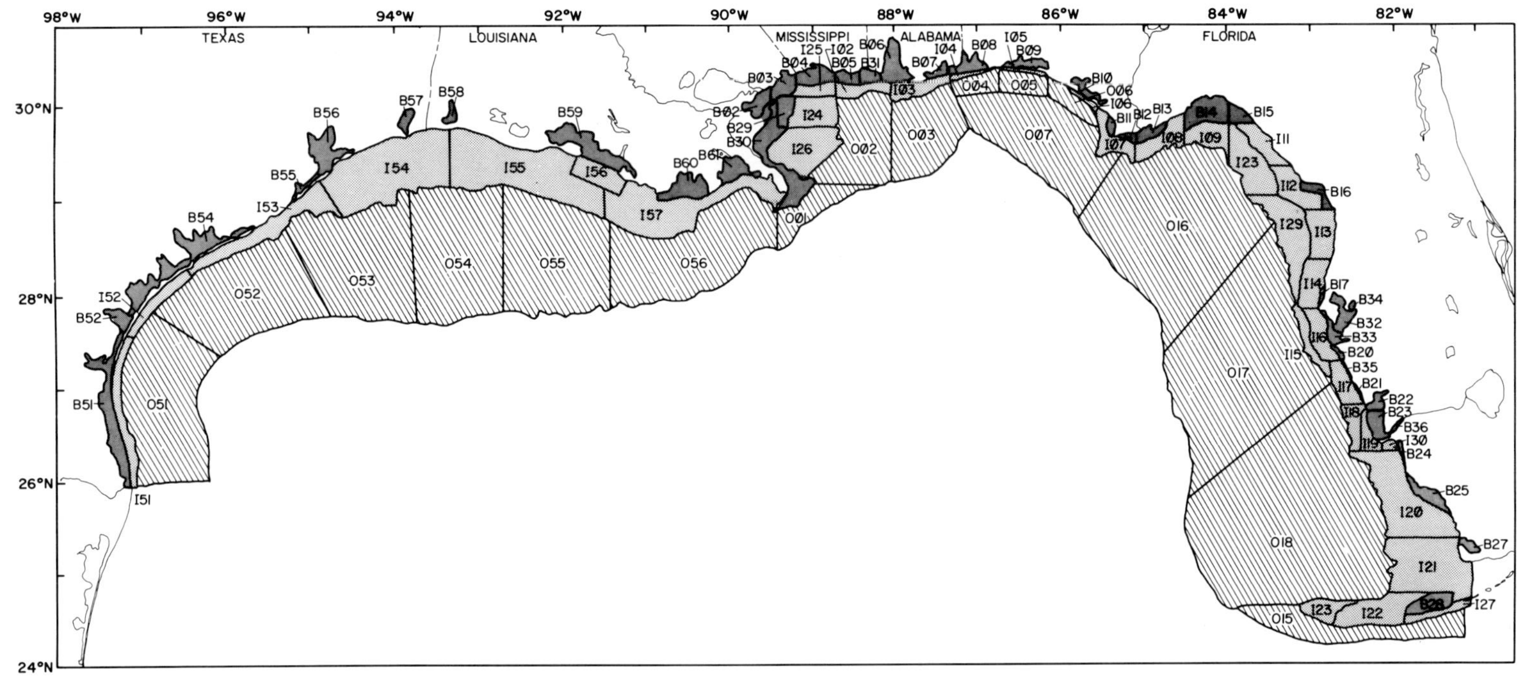

Figure 2 Gulf of Mexico study area. Sampling blocks O01–O56 represent the offshore strata, I01–I57, the inshore strata, and B01–B61 the bay strata.

genetic profile of the population being studied. In the Indian and Banana rivers 109 dolphins have been captured, tagged, and released (Odell and Asper, Chapter 30, this volume); 76 of which were captured under SEFC sponsorship (Asper and Odell, 1980). In Mississippi Sound, the total stands at 57 individuals (NMFS, unpublished data; Solangi and Dukes, 1983). In addition to the study of herd dynamics, mark–recapture estimators of abundance are being examined for application to the resighting data.

Biochemical Genetics

Although long-term residency within a coastal ecosystem may be interpreted as supporting the hypothesis that there are resident, geographically local stocks of bottlenose dolphins throughout southeastern U.S. waters, this condition is neither sufficient nor necessary to prove that these animals are genetically isolated from the balance of the wild gene pool. To support that assumption, the degree of isolation and the rate of genetic exchange among geographically local groups need to be evaluated. As a part of the tagging studies conducted in the southeast, the biochemical genetics of animals sampled in the Indian and Banana rivers (Duffield, 1982) and Mississippi Sound (Toom, 1983) have been studied.

In response to recommendations of the Stock Assessment Workshop (Powers, 1983) and the favorable review of the MMC, in October 1983 the SEFC let a contract for a study of genetic variability in samples obtained from captive bottlenose dolphins with known capture locations. The objective of this study was to compare the genetic profiles of animals collected from different locations and, by reference to hematological, enzyme, and chromosome banding analyses, to test the hypothesis that there are various resident stocks. The comparative analysis was conducted using data and samples from the Indian and Banana rivers and Mississippi Sound collected during tagging studies as well as those collected from captive animals with known capture locations (Duffield, 1986).

Duffield (1986) concluded that there was evidence of genetic differentiation that supported the hypothesis of localized "populations," but these "populations" are not genetically isolated but have some unmeasured rate of exchange among them. Wells (1986) concluded similarly that for Sarasota Bay, Florida, there was both behavioral and genetic evidence supporting the hypothesis of localized populations but that animals there were not genetically isolated. Dowling and Brown (1987), using the mitochondrial DNA (mtDNA) approach, with a relatively small sample of individuals, found a relative lack of genetic divergence within Gulf of Mexico dolphins sampled, but they did observe differences between animals from the Gulf of Mexico and those from the western North Atlantic.

Life-Stage Modeling

To assess the effects of the current management scheme on the stock(s) of bottlenose dolphins in the southeastern United States, it was considered necessary to study the population life- and death-rate processes so that the risk to the dolphin stock(s) of past and future exploitation could be assessed (see Breiwick *et al.*, 1984). Given sufficient age-structured data, an appropriate technique for assessing this risk is a Leslie matrix approach (Leslie, 1945, 1948). In the absence of sufficient age-structured data, life-stage data can be used (Lefkovitch, 1965). Estimates of the vital parameters can come from a variety of sources, including literature, tagging studies, and photogrammetric studies.

Tagging studies have provided data on age and growth of bottlenose dolphins in the southeast (Asper and Odell, 1980; Odell and Asper, 1982; Odell and Schneyer, 1983; Solangi and Dukes, 1983). Long-term observational studies and literature sources provide estimates of birth- and death-rate ranges (Scott *et al.*, Chapter 11, this volume; Wells, 1986; Reilly and Barlow, 1986). Photogrammetric studies produce large length-structured samples of the populations of interest. The photogrammetric method is currently being used to investigate whether or not length-structure data can be used in estimating vital rates and in population-projection models. Test samples were obtained off Cape Hatteras, North Carolina, in September 1983, and a sample of the imagery has been analyzed (Owen and Carter, 1987).

APPLICATION OF RESEARCH RESULTS

After a permit application has been reviewed by the MMC and its Committee of Scientific Advisors on Marine Mammals, the Secretary of Commerce is empowered to grant permits for the taking and importation of marine mammals for the purposes of scientific research and public display. In practice, the responsibility for the permitting process is held by NMFS and implemented through its Protected Species Office in Washington, D.C. Implementation of the MMPA in 1973 restricted the taking of bottlenose dolphins (and other marine mammals) for public display, and in 1977 a quota system was introduced for bottlenose dolphins in the southeastern United States. Under the MMPA, the management objective is to maintain the stocks of marine mammals in the range of optimum sustainable population (OSP). The MMPA as amended defines OSP to range from carrying capacity downward to the level resulting in maximum net production (MNP).

The management scheme originally devised for bottlenose dolphins in the southeast provided for seven management areas, essentially defined according to where bottlenose dolphins had been captured historically. The primary management regions for this fishery are as follows: (1) Florida East Coast (Indian and Banana rivers), (2) Florida Keys, (3) Florida West Coast from Charlotte Harbor to Crystal River, (4) Florida Panhandle from Crystal River to Mobile Bay, (5) Mississippi Sound, (6) Texas Coast north of Corpus Christi Bay to Matagorda Bay, and (7) other areas (see Fig. 1).

The management scheme first recommended by the MMC and adopted by the Secretary was based on a quota system limiting annual removals from wild populations to not more than 2% of the minimum estimated number of dolphins in any management area. The basic assumptions behind this management were (1) that the MNP for bottlenose dolphins is 2% and (2) that each management area or subarea represents a stock of dolphins. Given the lack of empirical data for testing these assumptions, this conservative management scheme was seen as a responsible way of allowing the fishery to continue while at the same time preventing overexploitation of dolphin stocks.

The quota system currently in place has evolved as additional scientific information has become available. Initial annual quotas were 7 animals from the Florida East Coast, 0 from the Florida Keys, 6 from the Florida West Coast, 10 from the Florida Panhandle, 35 from the Mississippi Sound, and 17 from the Texas Coast. These quotas were based on estimates of the abundance of bottlenose dolphins in each region. In 1979 the localized aerial sampling surveys of "priority" inshore locations for which improved management advice was required were started under contract to the SEFC as described previously. In June 1982 the quota scheme was modified in response to recommendations from the SEFC, based on analyses of the available aerial survey data collected under contract. These modifications and their bases are listed in Table 3.

Subsequent to these modifications, and after further review of the management scheme and recommendations for further research by the MMC in February 1983, the available data were analyzed for evidence of seasonal trends in abundance of dolphins in areas surveyed. The results of this analysis were used to test the hypothesis of the existence of resident stocks.

Resident Stocks Hypothesis and Current Management Recommendations

Seasonal trends in abundance of bottlenose dolphins in embayments along the southeastern United States could be analyzed only for those sampling areas where comparable seasonal abundance estimates were available: Florida East Coast (Indian and Banana rivers), Florida West Coast (Charlotte Harbor), Florida Panhandle (Apalachicola and St. Joseph bays), Mississippi Sound, and Texas Coast (Aransas and Copano bays) (Table 2 and Fig. 3). The uncertainty in the seasonal estimates is such that a few estimates can be determined as being different from any others using normal criteria. However, given the expressed concern of possible overexploitation of local, resident stocks, a conservative interpretation of the available data and correspondingly conservative management actions are believed necessary.

SEFC-sponsored mark and resighting studies

Table 3
Initial Bottlenose Dolphin Live-Capture Quotas and Modifications Implemented in 1982 by NMFS Management Arm after Recommendations from the SEFC

Management Area (Subarea)	Initial Quota	1982 Modifications[a]
Florida East Coast	7	Reduce quota to 6 animals per year based on estimates of Thompson (1981a,b)
Florida Keys	0	No change
Florida West Coast	6	Increase quota to 23 animals per year based on estimates of Odell and Reynolds (1980) and impose quotas for subareas surveyed
(Tampa Bay)	(—)	Limit take in subarea to 10 animals per year based on estimate of Thompson (1981c)
(Charlotte Harbor)	(—)	Limit take in subarea to 5 animals per year based on estimate of Thompson (1981d)
Florida Panhandle	10	Retain the annual quota of 10 animals based on Odell and Reynolds (1980), but impose quotas for subareas surveyed
(Apalachicola–St. Joseph bays)	(—)	Impose quota of 1 animal per year based on estimate of Thompson (1982a)
(Destin–Ft. Walton Beach)	(—)	Impose quota of 2 animals based on estimate of Thompson (1982a)
Mississippi Sound	35	No change
Texas Coast	17	No change in overall quota which was based on work by Barham *et al.* (1979), but impose quotas for subareas surveyed
(Aransas–Copano–San Antonio bays)	(5)	Limit take in subarea based on estimate of Thompson (1982a)
(Matagorda Bay)	(12)	Limit take in subarea to difference between Aransas–Copano–San Antonio bays and the total quota for the Texas Coast management area

[a] Modifications made in 1982 represent the current management scheme referenced in Table 4.

conducted in the Indian River demonstrated that individual dolphins either reside in or return to the river over a period of at least 10 years (Odell and Asper, Chapter 30, this volume). Similar studies along the Florida west coast suggest residency of up to 10 or more years by an individual dolphin in Sarasota Bay (Wells *et al.*, 1981; Scott *et al.*, Chapter 11, this volume). These observations support the hypothesis that dolphin populations in certain estuarine embayments in the southeast region are resident.

There is a high degree of genetic similarity among the dolphins sampled in the Indian and Banana rivers (Duffield, 1982; Hersh and Duffield, Chapter 6, this volume). Distinct genetic differences were found between a single individual from outside the system and those within the system (Rodriquez *et al.*, 1984). This observation supports the hypothesis that the dolphins sampled in the Indian River are from the same stock, but distinct from animals outside the system.

Assuming that the population of bottlenose dolphins along the southeastern coast of the United States is structured such that there are local, resident stocks in certain embayments and transient stocks which migrate into and out of these embayments seasonally, there is a risk of overexploitation of the resident stocks under the current quota system. This risk can be substantially lowered if quotas are based on the estimated abundance of the resident stock in each em-

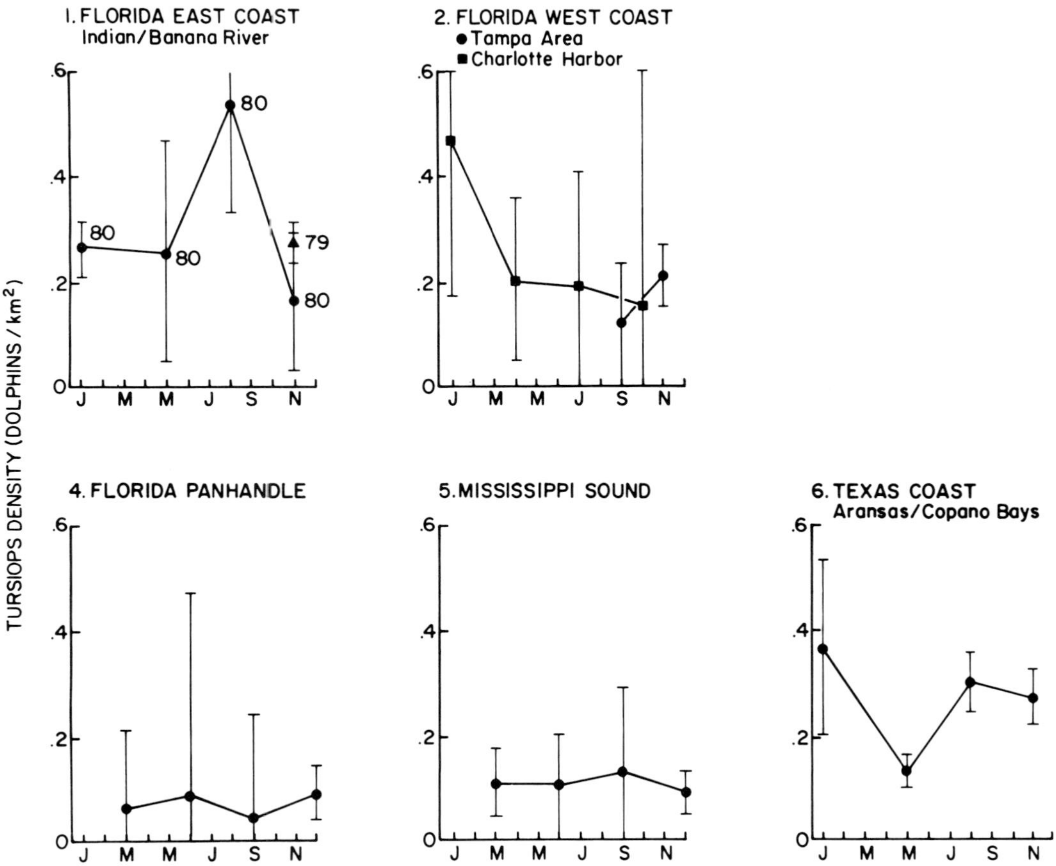

Figure 3 Seasonal trends in dolphin density for the areas listed. Point estimates and 95% confidence limits (bars) are shown. The year of the sample is depicted for the Florida East Coast area; all other years are as indicated in the text.

bayment, rather than on the average annual abundance of dolphins. Additional management options that would mitigate this risk to a lesser degree include restricting catch to certain levels during the different seasons. These options, however, run higher risks of overexploitation of the resident stock than does the current management scheme if the catch is not proportionally distributed between the resident and transient stocks and is principally from the resident stock. For these reasons, and following the recommendation of the MMC, in 1983 the SEFC proposed new quotas based on the estimated abundance of the assumed resident stock of dolphins in each management area or subarea (Table 4).

Florida East Coast Management Area

INDIAN AND BANANA RIVERS

The peak in estimated abundance in the Indian and Banana rivers was observed during August 1980. Based on the criterion of no overlap in 95% confidence intervals, the August abundance estimate is significantly higher than both the January and November estimates. Although there is considerable overlap in the intervals about the May and August estimates (see Table 2 and Fig. 3), the probability of the true means from the populations sampled in May and August jointly occurring in

Table 4
Recommended Changes to the Management Scheme Implemented in 1982 for the Live-Capture Fishing for *Tursiops truncatus* in the Southeastern United States Based on Available Data[a]

Management Area (Subarea)	Annual Allowable Take	Recommended Change
All areas		Limit catch of females to not more than 50% of the annual total allowable catch within any management area or subarea
Florida East Coast	6	Reduce allowable annual take to 4 animals in the Indian/Banana River
Florida Keys	0	No change recommended
Florida West Coast	23	1. Define Florida west coast to include the waters between the southern end of Pine Island Sound (26° 27′ N) and Crystal River
(Tampa Bay)	(10)	2. Establish an allowable annual quota of 10 animals from Pine Island Sound to Crystal River, exclusive of Charlotte Harbor, Tampa Bay, and Sarasota Bay
(Charlotte Harbor)	(5)	
(North of Charlotte Harbor exclusive of Tampa Bay)	(8)	3. Establish an allowable annual quota of 2 animals in the Sarasota Bay subarea and lower the allowable quota in the Charlotte Harbor subarea to 3 animals
		4. Remove Tampa Bay as an allowable subarea for capture
Florida Panhandle	10	1. Increase the annual allowable catch from 10 to 18 animals
(Apalachicola–St. Joseph bays)	(1)	2. Retain the current limitations of 2 animals from the Destin–Ft. Walton Beach subarea and 1 animal from the Apalachicola–St. Joseph bays subarea
(Destin–Ft. Walton Beach)	(2)	
Mississippi Sound	35	1. Define the Mississippi Sound management area to include the waters of Mississippi, Chandeleur, and Breton sounds and that portion of the Gulf of Mexico lying between the coast and 88° W longitude and 29° 07′ N latitude and maintain the current capture quota of 35 animals
		2. Define the Mississippi Sound subarea as the waters lying between the coast and the island chain ranging from Cat Island to Petit Bois Island and between 88° 30′ W and 89° 10′ W longitude
		3. Establish an annual allowable quota of 2 animals from the Mississippi Sound subarea
Texas Coast	17	1. Reduce the allowable take in the Aransas–Copano–San Antonio bays subarea to 2 animals
(Aransas–Copano–San Antonio bays)	(5)	2. Establish an allowable annual take of 2 animals in the Corpus Christi Bay subarea
(Matagorda Bay)	(12)	3. Reduce the annual allowable catch in the Matagorda Bay subarea to 6 animals
		4. Establish an allowable annual catch of 2 animals in the Laguna Madre and nearshore Gulf of Mexico waters from Laguna Madre south
Other areas	0	Establish an annual allowable catch of 17 animals in the area off the mouth of the Atchafalaya Bay, Louisiana; define the management area as the water lying between Marsh Island and Pt. Au Fer Island to 10 nautical miles offshore

[a] Changes were recommended by the SEFC to NMFS management arm in 1983.

the confidence-interval overlap is approximately 0.07. The sample estimates for January, May, and November of 1980 and November of 1979 appear not to differ significantly. These results suggest that the August 1980 abundance of bottlenose dolphins in the Indian and Banana rivers was higher than during November 1979, January 1980, May 1980, and November 1980 and provide some supporting evidence for the hypothesis of a summertime influx of dolphins from outside of the Indian and Banana River system. The alternate hypothesis that the significantly higher August abundance estimate is due to better sighting conditions during the August survey cannot be tested owing to insufficient data. The August peak in abundance is therefore assumed to have been due to the influx of a transient stock of animals during that sampling period. The estimated abundance of the resident stock of dolphins in the Indian and Banana rivers is taken as the average of the estimates other than that of August 1980: 211 individuals.

Florida Keys Management Area

Although the SEFC has collected seasonal sampling survey data for bottlenose dolphins in the Key West area, abundance estimates are not yet available.

Florida West Coast Management Area

The Florida West Coast management area is defined to include the Florida coast from Charlotte Harbor to Crystal River. This area extends westerly to the mid Gulf of Mexico, but in practice, live-capture removals are generally limited to very shallow coastal and estuarine waters. The available data from surveys conducted in 1975 and 1976 (Odell and Reynolds, 1980) support the concept of extending the management region to the southern end of Pine Island Sound, Florida.

TAMPA BAY

Abundance estimates previously reported from Tampa Bay are based on data collected in the nearshore region from the mouth of Tampa Bay to Hudson, Florida. There are no data presently available to support any level of take inside Tampa Bay, and it was therefore recommended that Tampa Bay be removed as an allowable capture location, until such time as the requisite data become available. The Tampa area sampled by SEFC-supported surveys overlaps a portion of the area sampled by Odell and Reynolds (1980) in 1975 and 1976. The estimated abundance of dolphins in the Tampa area sampled from November 1979 is significantly higher than the average annual abundance estimate for the same region sampled in 1975 and 1976 by Odell and Reynolds. Assuming (conservatively) that the abundance reported by Odell and Reynolds more closely represents the resident population level, then quota levels for this management area should be based on the estimate of Odell and Reynolds (505 individuals).

CHARLOTTE HARBOR

For Charlotte Harbor, the January 1981 estimate of abundance is more than double that of any other estimate. The variability about all of the estimates is quite high and does not allow discrimination between the January sample estimate and the remaining estimates (Fig. 3). However, maintaining a conservative interpretation of the available data, the high January abundance may be considered the result of a winter influx of transient dolphins into Charlotte Harbor. Therefore, the resident stock abundance is assumed to be the average of April, June, and October estimates for Charlotte Harbor, that is, 179 individuals.

SARASOTA BAY

The SEFC had not until 1985 sampled in the Sarasota Bay complex. Data from the surveys remain to be fully analyzed. However, long-term tagging and resighting studies in Sarasota Bay (Wells *et al.*, 1981; Irvine *et al.*, 1981) support an abundance estimate of 105 individuals within the system.

Florida Panhandle Management Area

The data from Odell and Reynolds (1980) support an abundance estimate of 936 bottlenose dolphins in the Florida Panhandle management area (Crystal River, Florida, to Mobile Bay, Alabama).

APALACHICOLA AND ST. JOSEPH BAYS

The seasonal abundance data collected from the Apalachicola and St. Joseph Bay system do not support the assumption of seasonal trends in bottlenose dolphin abundance there (see Fig. 3 and Table 2). The assumed resident stock in the system, taken as the average of the seasonal sampling estimates, numbers about 48 individuals.

DESTIN–FT. WALTON BEACH

Although the SEFC has collected seasonal sampling survey data for bottlenose dolphins in the Pensacola and Choctowhatchee bays, abundance estimates are not yet available.

Mississippi Sound Management Area

The estimate for the Mississippi Sound management area is based on surveys conducted in 1974 and 1975 by Leatherwood and others (Leatherwood and Platter, 1975; Leatherwood *et al.*, 1978). The study area that Leatherwood sampled included the Mississippi, Chandeleur, and Breton sounds proper and a portion of the Gulf of Mexico (see Fig. 1). The SEFC sampled a portion of the Mississippi Sound in 1980 and 1981 and, although dolphin densities were comparable to those observed by Leatherwood *et al.*, the estimated abundance was lower because of the smaller sampling area. The seasonal sampling estimates derived from SEFC local aerial survey data do not support the hypothesis of peaks in seasonal abundance in the region. The estimated average abundance of bottlenose dolphins in the Mississippi Sound based on the SEFC samples is 114 individuals. To prevent possible overexploitation of any resident population of bottlenose dolphins within the Mississippi Sound proper, and to remain consistent in management strategy, it was recommended that catch be limited within the Mississippi Sound to the level supported by the available data. Thus, the SEFC recommended that this would be best accomplished by defining a subarea within the Mississippi Sound management area and limiting catch from the smaller subarea while maintaining the current quota based on the estimates of Leatherwood *et al.* for the entire management area.

Texas Coast Management Area

ARANSAS–COPANO–SAN ANTONIO BAYS

In the Aransas–Copano–San Antonio bays sampling area of the Texas Coast management area, the May 1980 sample abundance estimate is significantly lower ($p < .003$) than the August 1980, November 1980, and January 1981 estimates (see Fig. 3 and Table 2). The assumed resident stock level in the Aransas–Copano–San Antonio Bay Complex, then, is taken as 131 individuals.

MATAGORDA BAY

The SEFC has not sampled Matagorda Bay on a seasonal basis. Reanalysis of the data from Barham *et al.* (1979) collected in 1978 and stratified to represent Matagorda Bay gives an estimate of 318 dolphins in the bay.

CORPUS CHRISTI BAY AND LAGUNA MADRE

The available SEFC data from 1979, 1980, and 1981 support abundance estimates of 100 individuals in Laguna Madre and the nearshore portion of the Gulf of Mexico from Laguna Madre south, and 109 individuals in Corpus Christi Bay.

Other Areas

ATCHAFALAYA BAY

Leatherwood *et al.* (1978) sampled an area outside Atchafalaya Bay, Louisiana, in July 1975 and found a high density of dolphins in the area at the mouth of Atchafalaya Bay lying between Marsh Island and Pt. Au Fer Island to 10 n mi (18 km) offshore. Presently, no quota is established for this region although available data support some take. The data support an abundance estimate of 897 dolphins in the region. Application of the 2% rule results in a recommended quota of 17 animals from this area.

TERREBONNE AND TIMBALIER BAYS AND PORT ROYAL SOUND

The SEFC has recently sampled two additional areas for which no quotas have been established. These are Terrebonne and Timbalier Bays, Louisiana, and Port Royal Sound, South Carolina. Abundance estimates for these areas are not yet available.

Management Recommendations

In addition, the MMC has recommended that the female take be no more than 50% of the allowable annual quota. This has not been the practice in the current management scheme, and 66% of all bottlenose dolphins removed from the wild have been females. Capture concentrated on the female portion of the stock significantly increases the risk of overexploitation by lowering the potential for future recruitment. As such, limiting catch of females in a quota area to no more than 50% of the annual allowable take is a recommended modification to the current management scheme.

Acting on (1) the MMC recommendation to base quotas on resident stock abundance, (2) the MMC policy of a 2% annual allowable take, and (3) the available supporting data on abundance estimates by season and area, in 1983 the SEFC recommended the changes in the 1982 management scheme to the management arm of NMFS. The recommendations, outlined in Table 4, are under review.

ACKNOWLEDGMENT

NOAA/NMFS/SEFC/Fishery Analysis Division Contribution Number ML1-85-25, September, 1985.

REFERENCES

Asper, E. D., and Odell, D. K. (1980). Bottlenose dolphin local herd monitoring, capture, marking, collection of biological data, and follow-up observations of marked animals. NMFS Final Contract Report NA-79-GA-C-00027.

Barham, E. G., Sweeney, J. C., Leatherwood, S., Beggs, R. K., and Barham, C. L. (1979). Aerial census of the bottlenose dolphin, *Tursiops truncatus,* in a region of the Texas coast. *Fish. Bull.* **77**(3), 585–595.

Breiwick, J. M., Eberhardt, L. L. and Braham, H. W. (1984). Population dynamics of western Arctic bowhead whales (*Balaena mysticetus*). *Can. J. Fish. Aq. Sci.* **41**(3), 484–496.

Burnham, K. P., Anderson, D. R., and Laake, J. L. (1980). Estimation of density from line transect sampling of biological populations. *Wildl. Monogr.* **72,** 1-202.

Carew, R. J. (1980). Aerial surveys to estimate herd size and density of populations of bottlenose dolphins. Item 1: Indian and Banana Rivers, Florida. NMFS Final Contract Report NA-80-GA-C-00029.

Carew, R. J. (1981a). Aerial surveys to estimate herd size and density of populations of bottlenose dolphins. Item 2: Aransas Bay, Copano Bay, and San Antonio Bay, Texas. NMFS Final Contract Report NA-80-GA-C-00029.

Carew, R. J. (1981b). Aerial surveys to estimate herd size and density of populations of bottlenose dolphins. Item 3: Apalachicola Bay, St. George Sound, and St. Joseph Bay, Florida. NMFS Final Contract Report NA-80-GA-C-00029.

Carew, R. J. (1981c). Aerial surveys to estimate herd size and density of populations of bottlenose dolphins. Item 4: Mississippi Sound, near Pascagoula, Mississippi. NMFS Final Contract Report NA-80-GA-C-00029.

Carew, R. J. (1981d). Aerial surveys to estimate herd size and density of populations of bottlenose dolphins. Item 5: Charlotte Harbor and areas of Florida. NMFS Final Contract Report NA-80-GA-C-00029.

Carew, R. J. (1982a). Aerial surveys to estimate herd size and density of populations of bottlenose dolphins in the area of Terrebonne and Timbalier Bay, Louisiana. NMFS Final Contract Report NA-81-GA-C-00017.

Carew, R. J. (1982b). Aerial surveys to estimate herd size and density of populations of bottlenose dolphins in the Key West, Florida area. NMFS Final Contract Report NA-81-GA-C-00016.

Carew, R. J. (1982c). Aerial surveys to estimate herd size and density of populations of bottlenose dolphins in the area of Savannah River, Port Royal, and St. Helena Sounds. NMFS Final Contract Report NA-81-GA-C-00008.

Carew, R. J. (1983). Aerial surveys to estimate herd size and density of populations of bottlenose dolphins in the area of Pensacola and Choctawhatchee Bays, Florida. NMFS Final Contract Report NA-82-GA-C-00015.

Carter, G. (1983). Bottlenose dolphin (*Tursiops truncatus*) Gulf of Mexico aerial survey. Aero Systems, Inc., Survey No. 1 Report, September–October 1983. NMFS Final Contract Report NA-83-GA-C-00031.

Carter, G. (1984). Bottlenose dolphin (Tursiops truncatus) Gulf of Mexico aerial survey. Aero Systems,

Inc., Survey No. 2 Report, January–February 1984. NMFS Final Contract Report NA-83-GA-C-00031.

Carter, G. R., and Derman, B. J. (1984a). Aircraft and data collection services in support of aerial surveys of *Tursiops truncatus* and other cetaceans in the Gulf of Mexico waters. April–May 1984. Biological Surveys Inc., Final Report. Survey 3. NMFS Final Contract Report NA-84-WC-C-06082.

Carter, G. R., and Derman, B. J. (1984b). Aircraft and data collection services in support of aerial surveys of *Tursiops truncatus* and other cetaceans in the Gulf of Mexico waters. July–August 1984. Biological Surveys, Inc., Final Report. Survey 4. NMFS Final Contract Report NA-84-WC-C-06082,

Carter, G. R. and Derman, B. J. (1985a). Aircraft and data collection services in support of aerial surveys of *Tursiops truncatus* and other cetaceans in the Gulf of Mexico waters. June 16–August 1, 1985. Biological Surveys Inc., Final Report. Survey 5. NMFS Final Contract Report NA-84-WC-C-06082.

Carter, G. R., and Derman, J. B. (1985b). Aircraft and data collection services in support of aerial surveys of *Tursiops truncatus* and other cetaceans in the Gulf of Mexico waters. September 15–October 29, 1985. Biological Surveys Inc., Final Report. Survey 6. NMFS Final Contract Report NA-84-WC-C-06082.

Carter, G. R., and Derman, J. B. (1986). Aircraft and data collection services in support of aerial surveys of *Tursiops truncatus* and other cetaceans in the Gulf of Mexico waters. January 10–March 2, 1986. Biological Surveys Inc., Final Report. Survey 7. NMFS Final Contract Report NA-84-WC-C-06082.

Dowling, T. E., and Brown, W. M. (1987). Preliminary analysis of dolphinid mtDNA, with special emphasis on population discreteness of *Tursiops truncatus* in the Gulf of Mexico. MMC Contract Report MM3309818-6. 22p.

Duffield, D. A. 1982. *Tursiops truncatus* genetics studies: Indian River 1980–1981. *In* "Indian River Herd Biodynamics," (D. K. Odell and E. D. Asper, eds.). NMFS/SEFC Final Contract Report NA80-GA-C-00063.

Duffield, D. A. (1986). Investigations of genetic variability in stocks of the bottlenosed dolphin (*Tursiops truncatus*) and the loggerhead sea turtle (*Caretta caretta*). Final Contract Report NMFS/SEFC/NA83-GA-C-00036.

Duffield, D. A., and Wells, R. S. (1986). Population structure of bottlenose dolphins: Genetic studies of bottlenose dolphins along the central west coast of Florida. Final Contract Report NMFS/SEFC/45-WCNF-5-00366.

Fritts, T. H., Irvine, A. B., Jennings, R. D., Collum, L. A., Hoffman, W., and McGekee, M. A. (1983). Turtles, birds, and mammals in the northern Gulf of Mexico and nearby Atlantic waters. An overview based on aerial surveys of OCS areas, with emphasis on oil and gas effects. Minerals Management Service Final Contract Report 14-16-0009-81-949.

Gruber, J. A. (1981). Ecology of the Atlantic bottlenose dolphin (*Tursiops turncatus*) in the Pass Cavallo area of Matagorda Bay, Texas. M.S. thesis, Texas A & M University, College Station, Texas. 182 pp.

Irvine, A. B., and Wells, R. S. (1972). Results of attempts to tag Atlantic bottlenosed dolphins, *Tursiops truncatus*. *Cetology* **13,** 1–5.

Irvine, A. B., Scott, M. D., Wells, R. S., and Kaufmann, J. H. (1981). Movements and activities of Atlantic bottlenose dolphins, *Tursiops truncatus,* near Sarasota, Florida. *Fish. Bull.* **79** 671–688.

Irvine, A. B., Wells, R. S., and Scott, M. D. (1982). An evaluation of techniques for tagging small odontocete cetaceans. *Fish Bull.* **80,** 135–143.

Leatherwood, S. (1982). Size of bottlenose dolphin population(s) in Indian River, Florida. *Rep. int. Whal. Commn* **32,** 567–568.

Leatherwood, S., and Platter, M. F. (1975). Aerial assessment of bottlenose dolphins off Alabama, Mississippi, and Louisiana. *In* "Final Report: *Tursiops truncatus* Assessment Workshop, Rosenstiel School of Marine and Atmospheric Science" (D. K. Odell, D. B. Siniff, and G. H. Waring, eds.), pp. 46–86. University of Miami, Miami, Florida.

Leatherwood, S., and Reeves, R. R. (1983). Abundance of bottlenose dolphins in Corpus Christi Bay and coastal southern Texas. *Contrib. Mar. Sci.* **26,** 179–199.

Leatherwood, S., and Show, I. T. (1980). Development of systematic procedures for estimating sizes of "population(s)" of bottlenose dolphins and estimates of sizes of "population(s)" of bottlenose dolphins in three geographical areas; with incidental observations on densities of West Indian manatees and marine turtles. Final Report to the U.S. National Marine Fisheries Service, Southeast Fisheries Center. Contract NA79-GA-C-0038.

Leatherwood, S., Gilbert, J. R., and Chapman, D. G. (1978). An evaluation of some techniques for aerial census of bottlenose dolphins. *J. Wildh. Manage.* **42,** 239–250.

Lefkovitch, L. P. (1965). The study of population growth in organisms grouped by stages. *Biometrica* **21,** 1–18.

Leslie, P. H. (1945). On the use of matricies in certain population mathematics. *Biometrica* **33,** 183–212.

Leslie, P. H. (1948). Some further notes on the use of matricies in population mathematics. *Biometrica* **35,** 213–245.

Odell, D. K. and Asper, E. D. (1982). Indian River herd biodynamics. NMFS/SEFC Final Contract Report NA80-GA-C-00063.

Odell, D. K., and Reynolds, J. E. (1980). Abundance of bottlenose dolphin, *Tursiops truncatus,* on the west coast of Florida. MMC Final Report MMC-75116.

Odell, D. K., and Schneyer, A. (1983). Age estimation and hormone analysis for bottlenose dolphins, *Tursiops truncatus,* from Mississippi. *In* "Atlantic Bottlenose Dolphins, *Tursiops truncatus,* Herd Studies in the Mississippi Sound, U.S.A." (M.A. Solangi and G. E. Dukes, eds.) NMFS/SEFC Final Contract Report NA82-GA-C-00023.

Owen, R. E., and Carter, G. R. (1987). Photogrammetric studies of bottlenose dolphins. NMFS/SEFC Report ML-CRD-87/88-2.

Powers, J. E., ed. (1983). Report of the Southeast Fisheries Center Stock Assessment Workshop. August 3–6, 1984. NOAA Tech. Mem. NMFS-SEFC-127. 229 pp.

Reilly, S. B., and Barlow, J. (1986). Rates of increase in dolphin population size. *Fish. Bull.* **84**(3), 527–534.

Rodriquez, J., Rodriquez, M., Garrigan, C., Asper, E. D., Duffield, D. A., and Hersh, S. (1984). Different forms of *Tursiops* in the Atlantic. Paper presented at 15th Annual Conference and Workshop of the International Association for Aquatic Animal Medicine. Tampa, Florida, 30 April–2 May 1984.

Scott, G. P. (1982). Estimates of the abundance of *Tursiops truncatus* in waters surrounding the Virgin Islands based on aerial sampling survey data. NMFS/SEFC/Miami Laboratory, Fishery Data Analysis Division Report.

Shane, S. H. (1980). Occurrence, movements, and distribution of bottlenose dolphin, *Tursiops truncatus,* in southern Texas. *Fish. Bull.* **78,** 593–601.

Shane, S. H. and Schmidly, D. J. (1978). The population biology of the Atlantic bottlenose dolphin, *Tursiops truncatus,* in the Aransas pass area of Texas. U.S. Marine Mammal Commission, Washington, D.C. NTIS PB-283-392. 130 pp.

Solangi, M. A., and Dukes, G. E. (1983). Atlantic bottlenose dolphin, *Tursiops truncatus,* herd studies in the Mississippi Sound, U.S.A. NMFS/SEFC Contract NA82-GA-C-00023.

Thompson, N. B. (1981a). Estimation of abundance of *Tursiops truncatus* in the Indian–Banana River complex, May, August, and November. NMFS/SEFC/Miami Laboratory, Fishery Data Analysis Division Report.

Thompson, N. B. (1981b). Estimation of the abundance of *Tursiops truncatus* in the Indian–Banana River complex, Florida, 1980. Fourth Biennial Conf. on Biol. and Conserv. of Marine Mammals, San Francisco, California, December 14–19, 1981 (Abstr.).

Thompson, N. B. (1981c). Estimates of abundance of *Tursiops truncatus* in Tampa Bay, Florida. NMFS/SEFC/Miami Laboratory, Fishery Data Analysis Division Report.

Thompson, N. B. (1981d). Estimates of abundance of *Tursiops truncatus* in Charlotte Harbor, Florida. NMFS/SEFC/Miami Laboratory, Fishery Data Analysis Division Report.

Thompson, N. B. (1982a). Estimates of abundance of *Tursiops truncatus,* bottlenose dolphin, St. Joseph–Apalachicola Bays, Florida; Mississippi Sound, Mississippi; and the Aransas–Copano–San Antonio Bay complex, Texas. NMFS/SEFC/Miami Laboratory, Fishery Data Analysis Division Report.

Thompson, N. B. (1982b). Estimate of abundance of *Tursiops truncatus* in Corpus Christi Bay, Texas, September, 1979. NMFS/SEFC/Miami Laboratory, Fishery Data Analysis Division Report.

Thompson, N. B. (1984). Progress report on estimating density and abundance of marine turtles: Results of first year pelagic surveys in the southeast U.S. SAW/84/MMT/FE.

Toom, P. M. (1983). Serum protein and hemoglobin electrophoretic profiles in *Tursiops* from the northern Gulf. *In* "Atlantic Bottlenose Dolphin, *Tursiops truncatus,* Herd Studies in the Mississippi Sound, U.S.A.", (M. A. Solangi and G. E. Dukes, eds.). NMFS/SEFC Contract NA82-GA-C-00023.

Wells, R. S. (1986). Population structure of bottlenose dolphins: Behavioral studies along the central west coast of Florida. Final Contract Report NMFS/SEFC/45-WCNF-5-00366.

Wells, R. S., Irvine, A. B., and Scott, M. D. (1980). The social ecology of inshore odontocetes. *In* "Cetacean Behavior: Mechanisms and Functions" L. H. Herman, ed)., pp. 263–317. Wiley, New York. 463 pp.

Wells, R. S., Scott, M. D., Irvine, A. B., and Page, P. T. (1981). Observations during 1980 of bottlenose dolphins, *Tursiops truncatus,* marked during 1970–1976, on the west coast of Florida. NMFS Final Report NA80-GA-A-195.

Index